# IMPORTANT NAMES, DATES AND ADDRESSES

This book belongs to

______________________

______________________

______________________

phone ______________________

examination date: ____________ hours: ________

examination location: ______________________

______________________

______________________

______________________

*tape your cancelled check here*

phone number of your registration board: ______________________

address of your registration board: ______________________

______________________

______________________

names of contacts at your registration board: ______________________

______________________

______________________

*tape your proof of mailing here*

date you sent your application: ______________________

registered/certified mail receipt number: ______________________

date confirmation was received: ______________________

names of examination proctors: ______________________

______________________

______________________

tape dime here

booklet number: ____________ (A.M.) ____________ (P.M.)

tape dime here

problems you disagreed with on the examination

| problem no. | reason |
|---|---|
| | |
| | |
| | |

# CIVIL ENGINEERING REFERENCE MANUAL

**Fifth Edition**

**Michael R. Lindeburg, P.E.**

*PROFESSIONAL PUBLICATIONS, INC.*
*Belmont, CA 94002*

**In the ENGINEERING REVIEW MANUAL SERIES**

Engineer-In-Training Review Manual
  Engineering Fundamentals Quick Reference Cards
  Mini-Exams for the E-I-T Exam
  1001 Solved Engineering Fundamentals Problems
  E-I-T Review: A Study Guide
Civil Engineering Reference Manual
  Civil Engineering Quick Reference Cards
  Civil Engineering Sample Examination
  Civil Engineering Review Course on Cassettes
  Seismic Design for the Civil P.E. Exam
  Timber Design for the Civil P.E. Exam
Structural Engineering Practice Problem Manual
Mechanical Engineering Review Manual
  Mechanical Engineering Quick Reference Cards
  Mechanical Engineering Sample Examination
  101 Solved Mechanical Engineering Problems
  Mechanical Engineering Review Course on Cassettes
  Consolidated Gas Dynamics Tables
Electrical Engineering Review Manual
Chemical Engineering Reference Manual
  Chemical Engineering Practice Exam Set
Land Surveyor Reference Manual
Metallurgical Engineering Practice Problem Manual
Petroleum Engineering Practice Problem Manual
Expanded Interest Tables
Engineering Law, Design Liability, and Professional Ethics
Engineering Unit Conversions

**In the ENGINEERING CAREER ADVANCEMENT SERIES**

How to Become a Professional Engineer
The Expert Witness Handbook—A Guide for Engineers
Getting Started as a Consulting Engineer
Intellectual Property Protection—A Guide for Engineers
E-I-T/P.E. Course Coordinator's Handbook
Becoming a Professional Engineer

Distributed by: Professional Publications, Inc.
1250 Fifth Avenue
Department 77
Belmont, CA 94002
(415) 593-9119

**CIVIL ENGINEERING REFERENCE MANUAL**
**Fifth Edition**



Printed in the United States of America

ISBN: 0-912045-16-7

Professional Publications, Inc.
1250 Fifth Avenue, Belmont, CA 94002

Current printing of this edition (last number): 6 5 4 3 2 1

# TABLE OF CONTENTS


**PREFACE** ........ xiii

**ACKNOWLEDGMENTS** ........ xv

**1 MATHEMATICS AND RELATED SUBJECTS**

1 Introduction ........ 1-1
2 Symbols Used in This Book ........ 1-1
3 The Greek Alphabet ........ 1-1
4 Mensuration ........ 1-2
5 Significant Digits ........ 1-4
6 Algebra ........ 1-4
- Polynomial Equations ........ 1-4
- Simultaneous Linear Equations ........ 1-6
- Simultaneous Quadratic Equations ........ 1-7
- Exponentiation ........ 1-7
- Logarithms ........ 1-7
- Logarithm Identities ........ 1-8
- Partial Fractions ........ 1-8
- Linear and Matrix Algebra ........ 1-9

7 Trigonometry ........ 1-11
- Degrees and Radians ........ 1-11
- Right Triangles ........ 1-11
- General Triangles ........ 1-12
- Hyperbolic Functions ........ 1-12

8 Straight Line Analytic Geometry ........ 1-12
- Equations of a Straight Line ........ 1-12
- Points, Lines, and Distances ........ 1-12
- Linear and Curvilinear Regression ........ 1-13
- Vector Operations ........ 1-15
- Direction Numbers, Direction Angles, and Direction Cosines ........ 1-16
- Curvilinear Interpolation ........ 1-17

9 Tensors ........ 1-17
10 Planes ........ 1-18
11 Conic Sections ........ 1-19
- Circle ........ 1-19
- Parabola ........ 1-19
- Ellipse ........ 1-20
- Hyperbola ........ 1-20

12 Spheres ........ 1-21
13 Permutations and Combinations ........ 1-21
14 Probability and Statistics ........ 1-22
- Probability Rules ........ 1-22
- Probability Density Functions ........ 1-22
- Statistical Analysis of Experimental Data ........ 1-23

15 Basic Hypothesis Testing ........ 1-27
16 Differential Calculus ........ 1-28
- Terminology ........ 1-28
- Basic Operations ........ 1-28
- Transcendental Functions ........ 1-28
- Variations on Differentiation ........ 1-29

17 Integral Calculus ........ 1-31
- Fundamental Theorem ........ 1-31
- Integration by Parts ........ 1-31
- Indefinite Integrals ........ 1-31
- Uses of Integrals ........ 1-32

18 Differential Equations ........ 1-32
- First Order Linear ........ 1-32
- Second Order Homogeneous With Constant Coefficients ........ 1-33

19 Laplace Transforms ........ 1-33
20 Applications of Differential Equations ........ 1-34
- Fluid Mixture Problems ........ 1-34
- Decay Problems ........ 1-35
- Surface Temperature ........ 1-35
- Surface Evaporation ........ 1-36

21 Fourier Analysis ........ 1-36
22 Critical Path Techniques ........ 1-37
- Introduction ........ 1-37
- Definitions ........ 1-37
- Solving a CPM Problem ........ 1-38
- Probabilistic Critical Path Models ........ 1-39

**A** Laplace Transforms ........ 1-40
**B** Conversion Factors ........ 1-41
**C** Computational Values of Fundamental Constants ........ 1-43
Practice Problems ........ 1-44

**2 ENGINEERING ECONOMIC ANALYSIS**

1 Equivalance ........ 2-1
2 Cash Flow Diagrams ........ 2-1
3 Typical Problem Format ........ 2-2
4 Calculating Equivalence ........ 2-2
5 The Meaning of "Present Worth" and "$i$" ........ 2-4
6 Choice Between Alternatives ........ 2-4
7 Treatment of Salvage Value in Replacement Studies ........ 2-6
8 Basic Income Tax Considerations ........ 2-7
9 Depreciation ........ 2-7
10 Advanced Income Tax Considerations ........ 2-10
- Investment Tax Credit ........ 2-10
- Gain on the Sale of a Depreciated Asset ........ 2-10
- Capital Gains and Losses ........ 2-10

11 Rate and Period Changes ........ 2-11
12 Probabilistic Problems ........ 2-11
13 Estimating Economic Life ........ 2-13
14 Basic Cost Accounting ........ 2-13
15 Break-Even Analysis ........ 2-14
16 Handling Inflation ........ 2-15
17 Learning Curves ........ 2-15
18 Economic Order Quantity ........ 2-16
19 Consumer Loans ........ 2-16
- Simple Interest ........ 2-16
- Loans with Constant Amount Paid Towards Principal ........ 2-17
- Direct Reduction Loans ........ 2-17
- Direct Reduction Loan with Balloon Payment ........ 2-17

20 Sensitivity Analysis ........ 2-18
- Standard Cash Flow Factors ........ 2-19

Practice Problems ........ 2-37

**3 FLUID STATICS AND DYNAMICS**

**Part 1: Fluid Properties** ........ 3-2
1 Fluid Density ........ 3-2
2 Specific Volume ........ 3-2
3 Specific Gravity ........ 3-2
4 Viscosity ........ 3-3

5 Vapor Pressure ... 3-4
6 Surface Tension ... 3-4
7 Capillarity ... 3-5
8 Compressibility ... 3-5
9 Bulk Modulus ... 3-5
10 Speed of Sound ... 3-5

**Part 2: Fluid Statics** ... 3-7

1 Measuring Pressures ... 3-7
2 Manometers ... 3-7
3 Hydrostatic Pressure Due to Incompressible Fluids ... 3-8
  Force on a Horizontal Plane Surface ... 3-8
  Vertical and Inclined Rectangular Plane Surfaces ... 3-8
  General Plane Surface ... 3-9
  Curved Surfaces ... 3-9
4 Hydrostatic Pressure Due to Compressible Fluids ... 3-10
5 Buoyancy ... 3-11
6 Stability of Floating Objects ... 3-12
7 Fluid Masses Under Acceleration ... 3-12
8 Modification for Other Gravities ... 3-13

**Part 3: Fluid Flow Parameters** ... 3-14

1 Introduction ... 3-14
2 Fluid Energy ... 3-14
3 Bernoulli's Equation ... 3-14
4 Impact Energy ... 3-15
5 Hydraulic Grade Line ... 3-16
6 Reynolds' Number ... 3-16
7 Equivalent Diameter ... 3-16
8 Hydraulic Radius ... 3-17

**Part 4: Fluid Dynamics** ... 3-19

1 Fluid Conservation Laws ... 3-19
2 Head Loss Due to Friction ... 3-19
3 Minor Losses ... 3-20
4 Head Additions/Extractions ... 3-23
5 Discharge From Tanks ... 3-23
6 Culverts and Siphons ... 3-24
7 Multiple Pipe Systems ... 3-25
  Series Pipe Systems ... 3-25
  Parallel Pipe Systems ... 3-25
  Reservoir Branching Systems ... 3-27
  Pipe Networks ... 3-28
8 Flow Measuring Devices ... 3-29
  Velocity Measurement ... 3-29
  Flow Measurement ... 3-31
9 The Impulse/Momentum Principle ... 3-33
  Jet Propulsion ... 3-34
  Open Jet on Vertical Flat Plate ... 3-34
  Open Jet on Horizontal Flat Plate ... 3-34
  Open Jet on Single Stationary Blade ... 3-34
  Open Jet on Single Moving Blade ... 3-34
  Confined Streams in Pipe Bends ... 3-35
  Water Hammer ... 3-35
  Open Jet on Inclined Plate ... 3-36
10 Lift and Drag ... 3-36
11 Similarity ... 3-38
  Viscous and Inertial Forces Dominate ... 3-38
  Inertial and Gravitational Forces Dominate ... 3-39
  Surface Tension Dominates ... 3-39
12 Effects of Non-Standard Gravity ... 3-39
13 Choice of Piping Materials ... 3-39
14 Flow of Compressible Fluids and Steam ... 3-39
A Properties of Water at Atmospheric Pressure ... 3-41
B Properties of Air at Atmospheric Pressure ... 3-41
C Viscocity of Water ... 3-42
D Important Fluid Conversions ... 3-42
E Area, Wetted Perimeter and Hydraulic Radius of Partially Filled Circular Pipes ... 3-43
F Dimensions of Welded and Seamless Steel Pipe ... 3-44
G Dimensions of Copper Water Tubing ... 3-48
H Dimension of Brass and Copper Tubing ... 3-49
I Typical Dimensions and Weights of Concrete Sewer Pipe ... 3-50
J Cast Iron Pipe Dimensions ... 3-51
K American Standard Piping Symbols ... 3-52
L Equivalent Length of Straight Pipe for Various Fittings (feet) ... 3-53
M Hazen-Williams Nomograph ... 3-54
N Manning Equation Nomograph ... 3-55
  Practice Problems ... 3-56


## 4 HYDRAULIC MACHINES


1 Introduction ... 4-1
2 Types of Pumps ... 4-1
3 Types of Centrifugal Pumps ... 4-2
4 Pump and Head Terminology ... 4-3
5 Net Positive Suction Head and Cavitation ... 4-5
6 Pumping Hydrocarbons and Other Liquids ... 4-6
7 Recirculation ... 4-7
8 Pumping Power and Efficiency ... 4-7
9 Specific Speed ... 4-10
10 Affinity Laws—Centrifugal Pumps ... 4-11
11 Pump Similarity ... 4-12
12 The Cavitation Number ... 4-12
13 Pump Performance Curves ... 4-12
14 System Curves ... 4-14
15 Pumps in Series and Parallel ... 4-15
16 Consideration for Wastewater Plants ... 4-15
17 Impulse Turbines ... 4-17
18 Reaction Turbines ... 4-17
19 Turbine Specific Speed ... 4-18
20 Hydroelectric Generating Plants ... 4-18
A Atmospheric Pressure versus Altitude ... 4-19
B Upper Limits of Specific Speeds (Single and Double Suction Pumps) ... 4-20
C Upper Limits of Specific Speeds (Mixed and Axial Flow Pumps) ... 4-21
D Volumetric Conversion Factors ... 4-22
E Pump Performance Correction Factor Chart ... 4-23
  Practice Problems ... 4-24


## 5 OPEN CHANNEL FLOW


1 Introduction ... 5-1
2 Definitions ... 5-2
3 Parameters Used in Open Channel Flow ... 5-3
4 Governing Equations for Uniform Flow ... 5-3
5 Variations in the Manning Roughness Constant, $n$ ... 5-4
6 Normal Depth ... 5-4
7 Energy and Friction Relationships ... 5-5
8 Most Efficient Cross Section ... 5-6
9 Analysis of Natural Watercourses ... 5-6
10 Flow Measurement with Weirs ... 5-7
11 Flow Measurement with Parshall Flumes ... 5-9
12 Steady Flow ... 5-9
13 Critical Flow in Rectangular Channels ... 5-11
14 Critical Flow in Non-rectangular Channels ... 5-12
15 The Froude Number ... 5-12
16 Predicting Open Channel Flow Behavior ... 5-12
17 Occurrences of Critical Flow ... 5-13
18 Controls on Flow ... 5-14
19 Flow Choking ... 5-14

20 Varied Flow Calculations ........ 5-14
21 Hydraulic Jump ........ 5-15
22 Length of Hydraulic Jump ........ 5-16
23 Spillways ........ 5-16
24 Sluiceways ........ 5-17
25 Erodible Canals ........ 5-17
26 Applications of Open Channel Flow to Culvert Design ........ 5-17
27 Determining Type of Culvert Flow ........ 5-18
A Design Use Values of Manning's $n$ ........ 5-23
B Manning Nomograph ........ 5-24
C Circular Channel Ratios ........ 5-25
D Critical Depths in Circular Channels ........ 5-26
Practice Problems ........ 5-27

## 6 HYDROLOGY

1 Important Conversions ........ 6-1
2 Definitions ........ 6-2
3 Precipitation ........ 6-3
4 Flood Considerations ........ 6-5
5 Subsurface Water ........ 6-6
6 Well Drawdown in Aquifers ........ 6-6
7 Seepage and Flow Nets ........ 6-7
8 Total Surface Runoff from Stream Hydrographs ........ 6-8
9 Peak Runoffs from the Unit Hydrograph ........ 6-10
10 The SCS Unit Hydrograph (1957) ........ 6-11
11 Hydrograph Synthesis ........ 6-12
Lagging Storm Method ........ 6-12
S-Curve Method ........ 6-12
12 Peak Runoff from the Rational Method ........ 6-13
13 Peak Runoff by SCS Methods ........ 6-15
The Cook Equation (1940) ........ 6-16
The SCS Drainage Coefficient Method (1939) ........ 6-17
Curve Number Method (1975) ........ 6-17
14 Reservoir Yield and Reservoir Sizing ........ 6-20
A Rational Method Runoff Coefficients ........ 6-24
B Random Numbers ........ 6-25
C Map of the U.S. Showing Average Annual Precipitation in Inches for the Period 1889-1938 ........ 6-26
Practice Problems ........ 6-27

## 7 WATER SUPPLY ENGINEERING

1 Conversions ........ 7-1
2 Definitions ........ 7-2
3 Chemistry Review ........ 7-4
Valence ........ 7-4
Chemical Reactions ........ 7-4
Stoichiometry ........ 7-5
Equivalent Weights ........ 7-5
Solutions of Solids in Liquids ........ 7-6
Solutions of Gases in Liquids ........ 7-7
Acids and Bases ........ 7-7
Reversible Reactions ........ 7-7
Solubility Product ........ 7-9
4 Qualities of Supply Water ........ 7-10
Acidity and Alkalinity ........ 7-10
Hardness ........ 7-11
Iron Content ........ 7-12
Manganese Content ........ 7-12
Fluoride Content ........ 7-12
Chloride Content ........ 7-13
Phosphorous Content ........ 7-13
Nitrogen Content ........ 7-13
Color ........ 7-13
Turbidity ........ 7-13
Suspended and Dissolved Solids ........ 7-14
Water-Borne Diseases ........ 7-14
5 Trihalomethanes ........ 7-14
6 Comparison of Alkalinity and Hardness ........ 7-15
7 Water Quality Standards ........ 7-16
8 Water Demand ........ 7-16
9 Methods of Water Distribution ........ 7-17
10 Storage of Water ........ 7-18
11 Pipe Materials ........ 7-18
12 Loads on Buried Pipes ........ 7-18
13 Water Supply Treatment Methods ........ 7-21
Aeration ........ 7-21
Plain Sedimentation (Clarification) ........ 7-22
Mixing and Flocculation ........ 7-24
Clarification with Flocculation ........ 7-25
Filtration ........ 7-25
Disinfection ........ 7-26
Fluoridation ........ 7-27
Iron and Manganese Removal ........ 7-27
Water Softening ........ 7-27
Turbidity Removal ........ 7-30
Taste and Odor Control ........ 7-31
Demineralization/Desalination ........ 7-32
14 Typical Municipal Systems ........ 7-32
A Conversions from mg/l as a Substance to mg/l as $CaCO_3$ ........ 7-35
B Atomic Weights of Elements Referred to Carbon (12) ........ 7-36
C Inorganic Chemicals Used in Water Treatment ........ 7-37
Practice Problems ........ 7-38

## 8 WASTE-WATER ENGINEERING

1 Conversions ........ 8-1
2 Definitions ........ 8-2
3 Wastewater Quality Characteristics ........ 8-3
Dissolved Oxygen ........ 8-3
Biochemical Oxygen Demand ........ 8-4
Relative Stability ........ 8-5
Chemical Oxygen Demand ........ 8-6
Chlorine Demand ........ 8-6
Grease ........ 8-7
Volatile Acids ........ 8-7
Suspended Solids ........ 8-7
4 Disinfection ........ 8-8
5 Typical Composition of Domestic Sewage ........ 8-8
6 Wastewater Quality Standards ........ 8-8
7 Design Flow Quantity ........ 8-8
8 Collection Systems ........ 8-9
Storm Drains and Inlets ........ 8-9
Manholes ........ 8-10
Pipes ........ 8-10
9 Pipe Flow Velocities ........ 8-10
10 Pumps Used in Wastewater Plants ........ 8-11
11 Dilution Purification ........ 8-12
12 Small Volume Disposal ........ 8-14
Cesspools ........ 8-14
Septic Tank ........ 8-14
Imhoff Tank ........ 8-14
13 Wastewater Plant Siting Considerations ........ 8-15
14 Pretreatment of Industrial Wastes ........ 8-15
15 Wastewater Processes ........ 8-15
Preliminary Treatment ........ 8-15
Primary Treatment ........ 8-17
Secondary Treatment ........ 8-17
Advanced Tertiary Treatment ........ 8-25
16 Sludge Disposal ........ 8-26

Sludge Quantities ............ 8-26
Sludge Thickening ............ 8-27
Sludge Dewatering ............ 8-28
Digestion and Stabilization ............ 8-28
Additional Sludge Processing Methods ............ 8-30
Residual Disposal ............ 8-30
17 Effluent Disposal ............ 8-30
18 Sanitary Landfills ............ 8-30
Introduction ............ 8-30
Design of Sanitary Landfills ............ 8-31
Gas Production ............ 8-31
19 Hazardous Waste Disposal ............ 8-31
Introduction ............ 8-31
Hazardous Waste Landfills ............ 8-32
Hydrogeological Site Characteristics ............ 8-32
Pretreatment Required ............ 8-32
Liner and Disposal Site Design ............ 8-32
Liners for Disposal Sites ............ 8-32
Covers for Disposal Sites ............ 8-33
Leachate Collection/Recovery Systems ............ 8-33
20 Injection Wells ............ 8-33
21 Slurry Trench Containment ............ 8-33
22 Steps in Cleaning Up Hazardous Waste Spills ............ 8-34
A Selected "10-States' Standards" ............ 8-35
B Saturated Oxygen Concentrations ............ 8-36
C Typical Sequences Used in Wastewater Plants ............ 8-37
Practice Problems ............ 8-39


## 9 SOILS


1 Conversions ............ 9-1
2 Definitions ............ 9-2
3 Soil Types ............ 9-2
4 Soil Indexing ............ 9-5
5 Aggregate Soil Properties ............ 9-6
6 Soil Testing and Mechanical Properties ............ 9-9
Penetration Resistance Test ............ 9-9
Moisture-Density Relationships ............ 9-9
Modified Proctor Test ............ 9-12
In-Place Density Test ............ 9-12
Unconfined Compressive Strength Test ............ 9-12
Sensitivity Tests ............ 9-12
Atterberg Limit Tests (Consistency Tests) ............ 9-12
Permeability Tests ............ 9-13
Consolidation Tests ............ 9-14
Triaxial Stress Tests ............ 9-15
California Bearing Ratio Test: Shearing Resistance ............ 9-17
Plate Bearing Value Test: The Subgrade Modulus ............ 9-18
Hveem's Resistance Value Test: The R-Value ............ 9-19
Practice Problems ............ 9-21


## 10 FOUNDATIONS AND RETAINING WALLS


1 Conversions ............ 10-1
2 Definitions ............ 10-2
3 Comparison of Sand and Clay as Foundation Materials ............ 10-2
4 General Considerations for Footings ............ 10-2
5 Allowable Soil Pressures ............ 10-3
6 General Footing Design Equation ............ 10-3
7 Footings on Clay and Plastic Silt ............ 10-4
8 Footings on Sand ............ 10-5
9 Footings on Rock ............ 10-6
10 Moments on Footings ............ 10-6
11 General Considerations for Rafts ............ 10-7
12 Rafts on Clay ............ 10-7
13 Rafts on Sand ............ 10-7
14 General Considerations for Piers ............ 10-7
15 Piers in Clay ............ 10-8
16 Piers in Sand ............ 10-8
17 General Considerations for Piles ............ 10-8
Capacity of Individual Piles ............ 10-8
Capacity of Pile Groups ............ 10-10
18 Soil Pressures Due to Applied Loads ............ 10-10
Boussinesq's Equation ............ 10-10
Influence Chart Method ............ 10-11
Stress Contour Charts ............ 10-12
19 Primary Settlement in Clay ............ 10-12
20 Time Rate of Primary Consolidation ............ 10-13
21 Secondary Consolidation ............ 10-13
22 Slope Stability ............ 10-14
Homogeneous, Soft Clay ($\phi = 0°$) ............ 10-14
Homogeneous, Cohesive Soil ($c > 0$, $\phi = 0°$) ............ 10-15
Cohesionless Sand ($c = 0$) ............ 10-16
23 Earth Pressure Theories ............ 10-16
The Rankine Theory ............ 10-16
Wedge Theories ............ 10-16
24 Sloped and Broken Slope Backfill ............ 10-16
25 Surcharge Loading ............ 10-17
Uniform Surcharge ............ 10-17
Point Load ............ 10-17
Line Load ............ 10-17
26 Horizontal Pressures from Saturated Sand ............ 10-18
27 Retaining Walls ............ 10-18
28 Flexible Anchored Bulkheads ............ 10-21
29 Braced Cuts ............ 10-21
Introduction ............ 10-21
Pressure Distributions in Braced Cuts ............ 10-22
Design of Braced Cuts ............ 10-22
A Active Components for Retaining Walls with Straight Slope Backfill ............ 10-23
B Active Components for Retaining Walls with Broken Slope Backfill ............ 10-24
C Boussinesq Stress Contour Charts—Infinitely Long and Square Footings ............ 10-25
D Boussinesq Stress Contour Chart—Uniformly Loaded Circular Footings ............ 10-26
Practice Problems ............ 10-27


## 11 STATICS


1 Concentrated Forces and Moments ............ 11-1
2 Distributed Loads ............ 11-2
3 Pressure Loads ............ 11-3
4 Resolution of Forces and Moments ............ 11-3
5 Conditions of Equilibrium ............ 11-3
6 Free-Body Diagrams ............ 11-4
7 Reactions ............ 11-4
8 Influence Lines for Reactions ............ 11-5
9 Axial Members ............ 11-5
10 Trusses ............ 11-6
Method of Joints ............ 11-6
Cut-and-Sum Method ............ 11-7
Method of Sections ............ 11-8
11 Superposition of Loadings ............ 11-8
12 Cables ............ 11-9
Cables under Concentrated Loads ............ 11-9
Cables under Distributed Loads ............ 11-9
Parabolic Cables ............ 11-10
The Catenary ............ 11-10
13 3-Dimensional Structures ............ 11-11

14 General Tripod Solution ........ 11-12
15 Properties of Areas ........ 11-13
Centroids ........ 11-13
Moment of Inertia ........ 11-14
Product of Inertia ........ 11-15
16 Rotation of Axes ........ 11-16
17 Properties of Masses ........ 11-16
Center of Gravity ........ 11-16
Mass Moment of Inertia ........ 11-16
18 Friction ........ 11-17
A Centroids and Area Moments of Inertia ........ 11-18
B Mass Moments of Inertia ........ 11-19
Practice Problems ........ 11-20

## 12 MECHANICS OF MATERIALS

**Part 1: Strength of Materials** ........ 12-1
1 Properties of Structural Materials ........ 12-2
The Tensile Test ........ 12-2
Fatigue Tests ........ 12-3
Estimates of Material Properties ........ 12-3
2 Deformation under Loading ........ 12-3
3 Thermal Deformation ........ 12-4
4 Shear and Moment Diagrams ........ 12-4
5 Stresses in Beams ........ 12-5
Normal Stress ........ 12-5
Shear Stress ........ 12-6
6 Stresses in Composite Structures ........ 12-7
7 Allowable Stresses ........ 12-9
8 Beam Deflections ........ 12-9
Double Integration Method ........ 12-9
Moment Area Method ........ 12-10
Strain Energy Method ........ 12-11
Conjugate Beam Method ........ 12-11
Table Look-up Method ........ 12-12
Method of Superposition ........ 12-12
9 Truss Deflections ........ 12-12
Strain-Energy Method ........ 12-12
Virtual Work Method (Hardy Cross Method) ........ 12-13
10 Combined Stresses ........ 12-14
11 Dynamic Loading ........ 12-15
12 Influence Diagrams ........ 12-15
Influence Diagrams for Beam Reactions ........ 12-15
Finding Reaction Influence Diagrams Graphically ........ 12-16
Influence Diagrams for Beam Shears ........ 12-16
Shear Influence Diagrams by Virtual Displacement ........ 12-17
Moment Influence Diagrams by Virtual Displacement ........ 12-18
Shear Influence Diagrams on Cross-Beam Decks ........ 12-19
Influence Diagrams on Cross-Beam Decks ........ 12-19
Influence Diagrams for Truss Members ........ 12-19
13 Moving Loads on Beams ........ 12-20
Global Maximum Moment Anywhere on Beam ........ 12-20
Placement of Load Group to Maximize Local Moment ........ 12-21
14 Columns ........ 12-21

**Part 2: Application to Design** ........ 12-23
1 Springs ........ 12-23
2 Thin-Walled Cylinders ........ 12-23
3 Rivet and Bolt Connections ........ 12-23
4 Fillet Welds ........ 12-24
5 Shaft Design ........ 12-24
6 Eccentric Connector Analysis ........ 12-25
7 Surveyor's Tape Corrections ........ 12-26
Temperature Correction ........ 12-26
Tension Correction ........ 12-26
8 Stress Concentration Factors ........ 12-26
9 Cables ........ 12-26
10 Thick-Walled Cylinders under External and Internal Pressure ........ 12-27
Stresses ........ 12-27
Strains ........ 12-28
Press Fits ........ 12-28
11 Flat Plates under Uniform Pressure Loading ........ 12-29
A Beam Formulas ........ 12-31
B Centroids and Area Moments of Inertia ........ 12-33
C Typical Properties of Structural Steel, Aluminum, and Magnesium (ksi) ........ 12-34
Practice Problems ........ 12-35

## 13 INDETERMINATE STRUCTURES

1 Introduction ........ 13-1
2 Consistent Deformation ........ 13-1
3 Using Superposition with Statically Indeterminate Beams ........ 13-3
4 Indeterminate Trusses: Dummy Unit Load Method ........ 13-4
5 Introduction to the Moment Distribution Method ........ 13-5
6 Signs of Fixed-End Moments ........ 13-6
7 Beam Stiffness ........ 13-6
8 Distribution Factors ........ 13-7
9 Moment Distribution Method Procedure ........ 13-8
10 Drawing Shear and Moment Diagrams from Fixed-End Moments ........ 13-10
11 Direct Moments and Free Ends ........ 13-11
12 Flexible and Yielding Supports ........ 13-12
13 Frames without Sidesway ........ 13-14
14 Frames with Sidesway ........ 13-15
15 Frames with Concentrated Joint Loads ........ 13-16
16 Approximate Methods ........ 13-17
Assumed Inflection Points ........ 13-17
Use of Design Moments (Moment Coefficients) ........ 13-18
Use of Design Shears (Shear Coefficients) ........ 13-19
A Moment Distribution Worksheet ........ 13-20
B Fixed-End Moments ........ 13-21
C Beam Formulas—Fixed-End Beams ........ 13-23
Practice Problems ........ 13-28

## 14 REINFORCED CONCRETE DESIGN

1 Conversions ........ 14-2
2 Definitions ........ 14-2
3 Concrete Mixing ........ 14-3
Types of Concrete ........ 14-3
Aggregates ........ 14-4
Admixtures ........ 14-4
Proportioning Concrete ........ 14-4
In-Place Volume ........ 14-4
4 Properties of Concrete ........ 14-5
Slump ........ 14-5
Compressive Strength ........ 14-6
Tensile Strength ........ 14-6
Shear Strength ........ 14-6
Density ........ 14-6
Modulus of Elasticity ........ 14-6
5 Steel Reinforcing ........ 14-6
6 Development Length ........ 14-7
7 Ultimate Strength Design ........ 14-8
8 Beams ........ 14-8
Introduction ........ 14-8

Factored Loads .......... 14-8
General Provisions of Beam Design .......... 14-9
Spacing of Reinforcement in Beams .......... 14-9
Locating the Neutral Axis .......... 14-9
Introduction to Strength Theory .......... 14-10
Strength Design of Beams .......... 14-11
Crack Checking .......... 14-13
Beam Deflections .......... 14-14
Shear Reinforcement in Beams .......... 14-15
Anchorage of Shear Reinforcement .......... 14-17
9 One-Way Floor Slabs .......... 14-18
10 T-Beams .......... 14-19
Introduction .......... 14-19
Flange Width Limitations .......... 14-19
T-Beam Design and Analysis .......... 14-19
11 Deep Beams .......... 14-20
12 Doubly-Reinforced Beams .......... 14-21
Introduction .......... 14-21
Capacity of Doubly-Reinforced Beams .......... 14-22
Design of Doubly-Reinforced Beams .......... 14-22
13 Prestressed and Post-Tensioned Beam Design .......... 14-22
Introduction .......... 14-22
ACI Code Provisions for Stress .......... 14-23
Analysis of Prestressed Construction .......... 14-23
14 Columns .......... 14-24
Introduction .......... 14-24
Tied Columns .......... 14-25
Spiral Columns .......... 14-26
Column Eccentricity .......... 14-26
Large Eccentricities .......... 14-27
Slenderness Effects .......... 14-28
15 Footing Design .......... 14-29
Failure Mechanisms .......... 14-29
Factored Load .......... 14-29
Allowable Shear .......... 14-29
Reinforcement in Footings .......... 14-30
ACI Code Provisions for Footings .......... 14-30
Bearing Pressure .......... 14-30
Dowel Bars .......... 14-30
Footing Analysis .......... 14-31
Footing Design .......... 14-31
Eccentrically-Loaded Footings .......... 14-31
16 Retaining Walls .......... 14-33
General Design Characteristics .......... 14-33
General Design Procedure .......... 14-33
A Miscellaneous ACI Detailing Requirements .......... 14-37
B The Alternate (Working Stress) Method of Design .. 14-39
Practice Problems .......... 14-41

## 15 STEEL DESIGN AND ANALYSIS

1 Review of Steel Nomenclature .......... 15-1
2 Conversions .......... 15-2
3 Types of Steels .......... 15-2
4 Steel Properties .......... 15-2
5 Structural Shapes .......... 15-2
6 Standard Combinations of Shapes .......... 15-3
7 Reinforcement of Mill Shapes .......... 15-3
8 Fatigue Loading .......... 15-4
9 Allowable Stresses for Impact, Wind, and Earthquake Loads .......... 15-4
10 The Most Economical Shape .......... 15-4
11 Compact Sections .......... 15-4
12 Beam Bending Planes .......... 15-4
13 Lateral Bracing .......... 15-5
14 Beam Deflections .......... 15-5
15 Bending Stress in Steel Beams .......... 15-5
16 Allowable Bending Stress .......... 15-6
W-Shapes Bending About Major Axis .......... 15-6
Weak-Axis Bending .......... 15-6
Intermediate Cases .......... 15-6
17 Shear Stress in Steel Beams .......... 15-6
18 Local Buckling .......... 15-6
19 Beam Design and Analysis .......... 15-7
20 Beam Design by Table and Chart .......... 15-7
Beam Table Use .......... 15-7
Allowable Moments Chart .......... 15-7
21 Load Factor Beam Design .......... 15-8
Basic Procedure .......... 15-8
Ultimate Moments .......... 15-10
Ultimate Shears .......... 15-10
Constructing Plastic Moment Diagrams .......... 15-11
22 Columns with Axial Loads .......... 15-12
Introduction .......... 15-12
Geometric Terminology .......... 15-12
Effective Length .......... 15-13
Slenderness Ratio .......... 15-14
Allowable Compressive Stress .......... 15-14
Column Analysis .......... 15-14
Column Design .......... 15-15
23 Stability of Plates in Compression .......... 15-16
24 Miscellaneous Combinations in Compression .......... 15-17
25 Members in Bearing .......... 15-18
Projected Areas in Connections .......... 15-18
Projected Areas in Pinned Connections .......... 15-18
Bearing Stiffeners .......... 15-18
Bearing on Masonry Supports at Beam Ends .......... 15-18
Column Base Plates .......... 15-19
26 Tension Members .......... 15-20
Introduction .......... 15-20
Allowable Tensile Stress .......... 15-20
Slenderness Ratios for Tension Members .......... 15-20
Threaded Members in Tension .......... 15-20
Plates and Members with Holes .......... 15-20
27 Beam-Column Analysis .......... 15-22
Introduction .......... 15-22
Small Axial Compressions .......... 15-22
Large Axial Compressions .......... 15-22
Web Stiffeners .......... 15-24
28 Beam-Column Design .......... 15-24
29 Plate Girders .......... 15-24
Dimensions and Nomenclature .......... 15-24
Depth-Thickness Ratios .......... 15-24
Shear Stress .......... 15-25
Design of Girder Flanges .......... 15-25
Width-Thickness Ratios .......... 15-26
Reduction in Flange Stress .......... 15-26
Final Check .......... 15-26
Location of First (Outboard) Stiffeners .......... 15-26
Location of Interior Stiffeners .......... 15-26
Maximum Bending Stress .......... 15-26
Design of Intermediate Stiffeners .......... 15-26
Design of Bearing Stiffeners .......... 15-27
Web Crippling at Points of Loading .......... 15-28
30 Bending with Axial Tension .......... 15-28
31 Bolts and Rivets .......... 15-28
Introduction .......... 15-28
Hole Spacing and Edge Distances .......... 15-28
Stress Area of Fasteners .......... 15-28
Allowable Connector Stresses .......... 15-28
Concentric Tension Connections .......... 15-29
Tension Effects Due to Eccentricity .......... 15-30
Bolt Preloading .......... 15-31
32 Framing Connections .......... 15-32

Introduction ........ 15-32
Simple (Type 2) Framing Connections ........ 15-32
Moment-Resisting (Type 1) Framing Connections ........ 15-33
33 Bolted and Riveted Eccentric Shear Connections ........ 15-33
Introduction ........ 15-33
Traditional Elastic Approach ........ 15-34
Reduced Eccentricity Models ........ 15-35
Ultimate Strength Analysis ........ 15-35
34 Welds ........ 15-35
35 Welded Connections ........ 15-36
Concentric Tension Connections ........ 15-36
Moment Resisting Connections ........ 15-36
Balancing Weld Groups ........ 15-37
Combining Shear and Bending Stresses ........ 15-37
Torsion Connections ........ 15-38
36 Composite Construction ........ 15-40
A Steel Used for Buildings and Bridges ........ 15-41
B Properties of Structural Steel at High Temperatures ........ 15-43
C Properties of Welds Treated as Lines ........ 15-44
D The Moment Gradient Multiplier ........ 15-45
Practice Problems ........ 15-46


## 16 TRAFFIC ANALYSIS, TRANSPORTATION, AND HIGHWAY DESIGN


1 Definitions ........ 16-2
2 Translational Dynamics ........ 16-3
3 Simple Roadway Banking ........ 16-4
4 Sight and Stopping Distances ........ 16-5
5 Length of Circular Horizontal Curve for Stopping Distance ........ 16-6
6 Length of Vertical Curves for Sight Distances ........ 16-6
7 Speed Parameters ........ 16-6
8 Design Speeds ........ 16-7
9 Volume Parameters ........ 16-7
10 Truck, Bus, and RV Equivalents ........ 16-7
11 Level of Service ........ 16-8
12 Calculation of Freeway Capacity ........ 16-8
13 Speed, Flow, and Density Relationships for Uninterrupted Flow ........ 16-9
14 Determining the Level of Service ........ 16-11
15 At-Grade Signalized Intersection Capacity ........ 16-12
16 Standard Truck Loadings ........ 16-12
17 Types of Pavement ........ 16-13
Rigid Pavement ........ 16-13
Prestressed Concrete Pavement ........ 16-14
Flexible Asphalt Concrete ........ 16-14
Full-Depth Asphalt Pavement ........ 16-14
Deep-Lift Asphalt Pavement ........ 16-14
18 Sulfur Extended Asphalt Binder ........ 16-14
19 Minimum Layer Thicknesses ........ 16-15
20 Pavement Design Parameters ........ 16-15
Layer Strengths ........ 16-15
Equivalent Axle Loadings ........ 16-15
Predicting Traffic Growth ........ 16-16
21 AASHTO Method of Flexible Pavement Design ........ 16-16
22 CALTRANS Method of Flexible Pavement Design ........ 16-19
23 Full-Depth Asphalt Pavements—Simplified Asphalt Institute Method ........ 16-22
24 Designing Portland Cement Concrete Pavement ........ 16-22
25 AASHTO Method of Rigid Pavement Design ........ 16-24
26 Fatigue Strength Method of Design ........ 16-25
27 Roadway Detailing ........ 16-28
28 Joints in Pavement ........ 16-28
29 Grooving Pavements ........ 16-29
30 Geotextiles ........ 16-29
31 Subgrade Drainage ........ 16-29
32 Controlling Frost Damage ........ 16-29
33 Parking Design ........ 16-30
34 Intersection Signaling ........ 16-30
Conditions Requiring Signaling ........ 16-30
Signal Controllers ........ 16-30
Determining Fixed-Time Cycle Lengths ........ 16-30
Determining On-Demand Timing ........ 16-31
Time-Space Diagrams ........ 16-32
35 Highway Interchange Design ........ 16-33
36 Pedestrian Levels of Service ........ 16-34
37 Economic Justification of Highway Safety Features ........ 16-34
38 Queuing Models ........ 16-35
General Relationships ........ 16-35
The M/M/1 System ........ 16-35
The M/M/s System ........ 16-36
A Approximate Correlation between California Bearing Ratio and Subgrade Modulus ........ 16-37
B Revised Soil Support Correlations ........ 16-38
C Approximate Correlation between Subgrade Modulus and Soil R-value ........ 16-39
D AASHTO Equivalence Factors—Flexible Pavement ........ 16-40
E AASHTO Equivalence Factors—Flexible Pavement ........ 16-41
Practice Problems ........ 16-42


## 17 SURVEYING


1 Error Analysis ........ 17-1
Measurements of Equal Weight ........ 17-1
Measurements of Unequal Weight ........ 17-2
Errors in Computed Quantities ........ 17-3
2 Orders of Accuracy ........ 17-4
3 Distance Measurement ........ 17-4
4 Elevation Measurement ........ 17-5
Curvature and Refraction ........ 17-5
Direct Leveling ........ 17-5
Indirect Leveling ........ 17-6
Differential Leveling ........ 17-6
5 Equipment and Methods Used to Measure Angles ........ 17-7
6 Angle Measurement ........ 17-7
7 Closed Traverses ........ 17-8
Adjusting Closed Traverse Angles ........ 17-8
Latitudes and Departures ........ 17-9
Adjusting Closed Traverse Latitudes and Departures ........ 17-10
Reconstructing Missing Sides and Angles ........ 17-11
Area of a Traverse ........ 17-11
Areas by Double Meridian Distances ........ 17-12
8 Areas with Irregular Boundaries ........ 17-12
9 Curves ........ 17-13
10 Horizontal Curves ........ 17-13
Introduction to Symbols ........ 17-13
Tangent Offsets for Circular Curves ........ 17-14
Deflection Angles ........ 17-14
11 Vertical Curves ........ 17-15
12 Vertical Curves with Obstructions ........ 17-16
13 Spiral Curves ........ 17-17
14 Photogrammetry ........ 17-17
15 Triangulation ........ 17-18
16 Trilateration ........ 17-18
17 Solar Observations ........ 17-18
18 Celestial Basis ........ 17-18
19 Sides (Angles) of the Astronomical Triangle ........ 17-19
20 Zenith Angle ........ 17-19
21 Line Azimuth ........ 17-20
22 Polaris Observations ........ 17-20

23 Accuracy of Polaris Azimuths ...................... 17-20
24 Celestial Basis ................................... 17-20
25 Sides (Angles) of the Astronomical Triangle......... 17-21
26 Zenith Angle for Polaris Observations............... 17-21
27 Line Azimuth from Polaris Observations ............ 17-22
A Oblique Triangle Equations......................... 17-23
B Circle and Circular Curve Geometry ................ 17-24
C Surveying Conversion Factors....................... 17-25
D Critical Constants.................................. 17-26
Practice Problems.................................. 17-27

## 18 MANAGEMENT THEORIES

1 Introduction......................................... 18-1
2 Behavioral Science Key Words........................ 18-1
3 History of Behavioral Science Studies and Theories... 18-1
Hawthorne Experiments.............................. 18-1
Bank Wiring Observation Room Experiments .... 18-2
Need Hierarchy Theory ............................. 18-2
Theory of Influence ................................ 18-2
Herzberg Motivation Studies ....................... 18-2
Theory X and Theory Y ............................ 18-3
4 Job Enrichment ...................................... 18-3
5 Quality Improvement Programs....................... 18-3
Zero Defects Program............................... 18-3
Quality Circles/Team Programs..................... 18-4

## 19 MISCELLANEOUS TOPICS

**Part 1: Accuracy and Precision Experiments** ........ 19-1
1 Accuracy............................................. 19-1
2 Precision............................................ 19-1
3 Stability ............................................ 19-1

**Part 2: Dimensional Analysis** ........................ 19-2

**Part 3: Reliability** ..................................... 19-5
1 Item Reliability ..................................... 19-5
2 System Reliability.................................... 19-5
Serial Systems ..................................... 19-5
Parallel Systems.................................... 19-6
k-out-of-n Systems.................................. 19-6
General System Reliability ......................... 19-6

**Part 4: Replacement** .................................. 19-8
1 Introduction.......................................... 19-8
2 Deterioration Models................................. 19-8
3 Failure Models ...................................... 19-8
4 Replacement Policy .................................. 19-9

**Part 5: FORTRAN Programming** .................. 19-10
1 Structural Elements.................................. 19-10
2 Data................................................. 19-10
3 Variables............................................ 19-10
4 Arithmetic Operations ............................... 19-11
5 Program Loops........................................ 19-11
6 Input/Output Statements ............................. 19-12
7 Control Statements................................... 19-13
8 Library Functions .................................... 19-13
9 User Functions....................................... 19-14
10 Subroutines ........................................ 19-14

**Part 6: Fire Safety Systems** ......................... 19-16
1 Introduction ........................................ 19-16
2 Detection Devices.................................... 19-16
3 Alarm Devices ...................................... 19-17
4 Signal Transmission.................................. 19-17
5 Auxiliary Control .................................... 19-17
6 Special Considerations............................... 19-18
7 Special Suppression Systems.......................... 19-18

**Part 7: Nondestructive Testing** ...................... 19-19
1 Magnetic Particle Testing ........................... 19-19
2 Eddy Current Testing................................ 19-19
3 Liquid Penetrant Testing............................ 19-19
4 Ultrasonic Testing................................... 19-19
5 Infrared Testing..................................... 19-19
6 Radiography ........................................ 19-19

**Part 8: Environmental Impact Assessment** ......... 19-20

**Part 9: Mathematical Programming**................. 19-22
1 Introduction ........................................ 19-22
2 Formulation of a Linear Programming Problem ..... 19-22
3 Solution to 2-Dimensional Problems ................ 19-22

## 20 SYSTEMS OF UNITS

1 Consistent Systems of Units ......................... 20-1
2 The Absolute English System......................... 20-1
3 The English Gravitational System ................... 20-1
4 The English Engineering System...................... 20-2
5 The cgs System ..................................... 20-2
6 The mks System ..................................... 20-3
7 The SI System ...................................... 20-3
A Selected Conversion Factors to SI Units............. 20-7

## 21 ENGINEERING LICENSING ............ 21-1

## 22 POSTSCRIPTS ............................ 22-1

## INDEX

## INDEX OF FIGURES AND TABLES

# PREFACE

In the 3rd century B.C., Aristotle set out to record all of the knowledge then known to man. The result was a vast body of writings which occupied his entire life.

The CIVIL ENGINEERING REFERENCE MANUAL is somewhat like Aristotle's works. In it, I have tried to organize and record the most frequently used aspects of civil engineering practice. Of course, the emphasis on what is important is my own, and this emphasis has been greatly colored by my interpretations of the subjects that concern civil engineers in their day-to-day activities.

In my teaching career, and during the long development period that this book has enjoyed, I have come to recognize the importance of proper background material for engineering subjects. For example, structural design depends on statics and mechanics of materials. Water supply engineering depends on chemistry. Open channel flow and hydrology depend on fluid mechanics. There are many such examples. Therefore, this CIVIL ENGINEERING REFERENCE MANUAL has become, in effect, a condensed four-year engineering degree in a single volume.

This edition differs from previous editions in several ways. New topics have been added, and extensive reorganization and revisions have been made to existing topics. Many new examples have been included to illustrate key concepts, and a significant number of new end-of-chapter problems have been added. Many new tables of data have been added to make this book one of the most valuable resources in your reference library. Also, this book has been typeset, as opposed to the typewriter type of earlier editions.

This book is based on the most current versions of the concrete and steel codes. The chapter on concrete design is based on the 1983 version of ACI 318. Similarly, the 8th edition of the AISC Manual was used to write the steel design chapter. I was also fortunate to have an early copy of the 1985 edition of the Transportation Research Board's Highway Capacity Manual to use in writing the chapter on traffic analysis. No changes in these documents are expected for several years.

This edition was written for a much more diverse audience than were previous editions of the CIVIL ENGINEERING REVIEW MANUAL, which were primarily intended for engineers preparing for the professional engineering licensing examinations. However, so many engineers told me that their REFERENCE MANUAL had become one of their most valuable sources for day-to-day referencing after the examination, that I set out to expand the depth and breadth of subjects covered.

The CIVIL ENGINEERING REFERENCE MANUAL is still the best resource available for you to prepare for the professional engineering licensing examination in civil engineering. However, to call this a 'review book' would unnecessarily restrict its usefulness. The subjects covered and the depth of subject development are based on my concept of a modern civil engineer's technical needs. You will be able to use this book long after you have received your engineering license.

Several examination subjects are conspicuously absent. Specifically, timber design, seismic design, and such professional subjects as law, liability, and ethics are not mentioned in this book. These subjects are covered in books from many publishers, including Professional Publications, Inc. I considered these examination subjects to be too specialized to include them in a general engineering reference book.

Nevertheless, you will be able to prepare effectively with this book for the engineering licensing examination in civil engineering. The approaches to and development of subjects, solved examples, and end-of-chapter problems have been tested in my classes, and they will prove to be highly valuable during your preparation. If you choose to use this book as one of your examination references, you should carefully read chapter 21, which describes the licensing procedure in detail.

I see this book as a place marker in the growing body of knowledge of civil engineering. It is the result of feedback from professors at universities in every state, students in my own courses, and purchasers of previous editions. I suspect that some years from now, there will be another edition. So, in the years, ahead, I urge and invite you to contribute to, as well as draw from, the reservoir of knowledge which the CIVIL ENGINEERING REFERENCE MANUAL represents.

Michael R. Lindeburg, P.E.
September 1989
Belmont, CA

# Concentrate Your Studies . . .

**Studying for the exam requires a significant amount of time and effort on your part. Make every minute count towards your success with these additional study materials in the Engineering Review Manual Series:**

## Solutions Manual for the Civil Engineering Reference Manual

136 pages, $8\frac{1}{2} \times 11$, paper 0-932276-64-4

Don't forget that there is a companion **Solutions Manual** that provides step-by-step solutions to the practice problems given at the end of each chapter in this reference manual. This important study aid will provide immediate feedback on your progress. Without the **Solutions Manual**, you may never know if your methods are correct.

## Civil Engineering Quick Reference Cards

48 pages, $8\frac{1}{2} \times 11$, spiral bound 0-932276-59-8

Because speed is important during the exam, you will welcome the advantage provided by the **Civil Engineering Quick Reference Cards**. This handy resource gives you quick access to equations, methods, and data needed during the exam. The cards are divided into specific exam subjects following the **Civil Engineering Reference Manual** in organization and nomenclature.

## Civil Engineering Sample Exam

24 pages, $8\frac{1}{2} \times 11$, paper 0-932276-68-7

Increase your speed and confidence in solving the types of questions that may appear on the exam by taking the **Civil Engineering Sample Examination**. This facsimile examination includes 20 typical problems (with complete, worked-out solutions) divided into two 10-problem sets to acquaint you with the exam format and teach you the correct—and most efficient—methods for solving Civil Engineering exam problems.

## Seismic Design for the Civil P.E. Examination

122 pages, $8\frac{1}{2} \times 11$, paper 0-932276-92-X

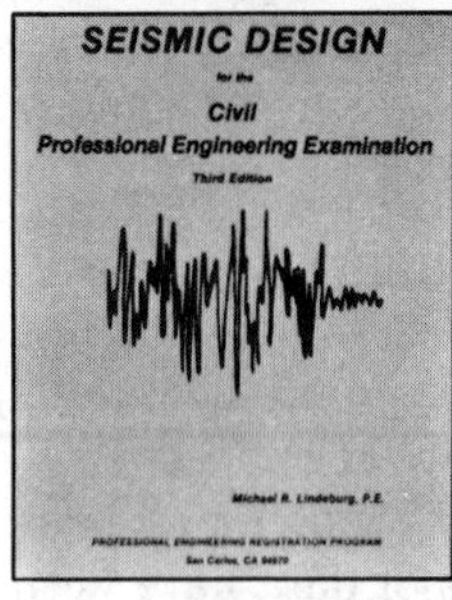


Specialized areas of the professional engineering exam require additional review materials for adequate preparation. **Seismic Design** is intended to help you prepare for problems relating to earthquake activity. This compilation of seismic problems (with solutions) is taken from actual previous exams. It presents and illustrates lateral force theory, from simple vibration analysis through the application of modern SEAOC codes. Each concept builds on previous material to gradually introduce you to new ideas.

## Civil Engineering Review Course on Cassettes

The discipline of a formalized class is provided through the **Civil Engineering Review Course on Cassettes** taken directly from Michael Lindeburg's outstanding lecture series. By listening to these tapes and following along with the accompanying written material, you will benefit from Mr. Lindeburg's insights into exam content and test-taking strategies. This cassette course was taped before a live audience of 100 practicing engineers. Included are case studies, examples, and questions raised during the classes. To save you the time and the inconvenience of taking notes, each lecture is accompanied by detailed notes duplicating material illustrated in class.

The cassette course has been divided into two sections so that you can select the topics you need most.

- Non-Structural Series (15 tapes)
  Approximate lecture time: 19 hours 0-932276-85-7
- Structural Series (14 tapes)
  Approximate lecture time: 19 hours 0-932276-84-9

**Professional Publications, Inc.**
**Department 77**
**1250 Fifth Avenue**
**Belmont, CA 94002**
**(415) 593-9119**

# ACKNOWLEDGMENTS

When a typical book is published, usually only one name appears on the cover. Although the illusion of self-sufficiency may appeal to the ego of the author, it is only that—an illusion. A book may be written by one person only, but it is produced only by great team effort. Nowhere is this more true than in the publication of a technical book.

This book, more so than any other book that I have written, depended on regular and significant contributions from a large number of individuals. It seems unfair that these individuals should receive only a brief acknowledgment, but it would be even more unfair not to acknowledge their essential contributions at all.

Thank you, Joanne Bergeson, for your growing project and company management skills. Without your hand at the helm of Professional Publications, Inc., I would not have been able to devote the majority of my time to writing this book.

Thank you, Rhonda Jones, for turning my sketches and visual concepts into finished illustrations, and for laying out the pages. It was you, after all, who composed the book and gave the readers a finished project. Thank you, too, for coordinating all aspects of the book typesetting, production, and printing.

Thank you, Susan Madden, David Goldstein, and Wendy Nelson for reading and rereading manuscripts, galleys, and page proofs, and for improving my work in ways I never could. I know it must have been rough for you, as non-engineers, to read all the equations and technical jargon. Your editing skills produced a polished work that we are all proud of.

Thank you, Yasuko Kitajima and Richard Weyhrauch of Aldine Press for typesetting another of Professional Publications' books. With your technical knowledge and familiarity with my writing style, you were not afraid to make corrections to the manuscript on the fly. You produced the book on time, according to an almost-impossible schedule.

And, of course, thank you to my family, which has learned that a writer's life can be as high-pressured and turbulent as a stock market trader's. Elizabeth, Jennifer, and Katherine, you have learned to capture bits of a normal family life whenever they become available. You have never complained about the isolated environment that I need in which to write.

Thanks to you all!

Michael R. Lindeburg

# 1 MATHEMATICS AND RELATED SUBJECTS

## 1 INTRODUCTION

Engineers working in design and analysis encounter mathematical problems on a daily basis. Although algebra and simple trigonometry are often sufficient for routine calculations, there are many instances when a quick review of certain subjects is needed. This chapter, in addition to supporting the calculations used in other chapters, consolidates the mathematical concepts most often needed by engineers in their daily activities.

## 2 SYMBOLS USED IN THIS BOOK

Many symbols, letters, and Greek characters are used to represent variables in the formulas used throughout this book. These symbols and characters are defined in the nomenclature section of each chapter. However, some of the symbols which are used as operators in this book are listed here.

**Table 1.1**
Symbols Used in This Book

| Symbol | Name | Use | Example |
|---|---|---|---|
| $\sum$ | sigma | series addition | $\sum_{i=1}^{3} x_i = x_1 + x_2 + x_3$ |
| $\prod$ | pi | series multiplication | $\prod_{i=1}^{3} x_i = x_1 x_2 x_3$ |
| $\Delta$ | delta | change in quantity | $\Delta h = h_2 - h_1$ |
| $-$ | over bar | average value | $\overline{x}$ |
| $\cdot$ | over dot | per unit time | $\dot{Q}$ = quantity flowing per second |
| ! | factorial | | $x! = x(x-1)(x-2)\cdots(2)(1)$ |
| $\mid\;\mid$ | absolute value | | $\lvert -3 \rvert = +3$ |
| $\approx$ | approximately equal to | | $x \approx 1.5$ |
| $\propto$ | proportional to | | $x \propto y$ |
| $\infty$ | infinity | | $x \rightarrow \infty$ |
| log | base 10 logarithm | | $\log(5.74)$ |
| ln | natural logarithm | | $\ln(5.74)$ |
| EE | scientific notation | | EE–4 |
| exp | exponential power | | $\exp(x) = e^x$ |

## 3 THE GREEK ALPHABET

**Table 1.2**
The Greek Alphabet

| | | | | | |
|---|---|---|---|---|---|
| $A$ | $\alpha$ | alpha | $N$ | $\nu$ | nu |
| $B$ | $\beta$ | beta | $\Xi$ | $\xi$ | xi |
| $\Gamma$ | $\gamma$ | gamma | $O$ | $o$ | omicron |
| $\Delta$ | $\delta$ | delta | $\Pi$ | $\pi$ | pi |
| $E$ | $\epsilon$ | epsilon | $P$ | $\rho$ | rho |
| $Z$ | $\varsigma$ | zeta | $\Sigma$ | $\sigma$ | sigma |
| $H$ | $\eta$ | eta | $T$ | $\tau$ | tau |
| $\Theta$ | $\theta$ | theta | $\Upsilon$ | $\upsilon$ | upsilon |
| $I$ | $\iota$ | iota | $\Phi$ | $\phi$ | phi |
| $K$ | $\kappa$ | kappa | $X$ | $\chi$ | chi |
| $\Lambda$ | $\lambda$ | lambda | $\Psi$ | $\psi$ | psi |
| $M$ | $\mu$ | mu | $\Omega$ | $\omega$ | omega |

## 4 MENSURATION

*Nomenclature*

- A total surface area
- d distance
- h height
- p perimeter
- r radius
- s side (edge) length, arc length
- V volume
- $\theta$ vertex angle, in radians
- $\phi$ central angle, in radians

*Circle*

$$p = 2\pi r \qquad 1.1$$

$$A = \pi r^2 = \frac{p^2}{4\pi} \qquad 1.2$$

*Circular Segment*

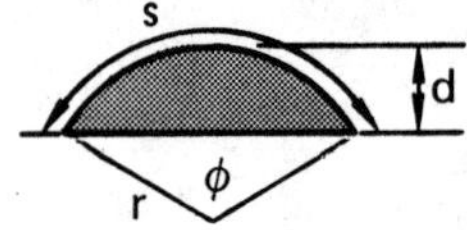


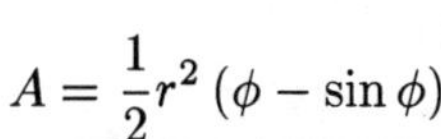

$$A = \frac{1}{2}r^2\,(\phi - \sin\phi) \qquad 1.3$$

$$\phi = \frac{s}{r} = 2\left(\arccos\frac{r-d}{r}\right) \qquad 1.4$$

*Triangle*

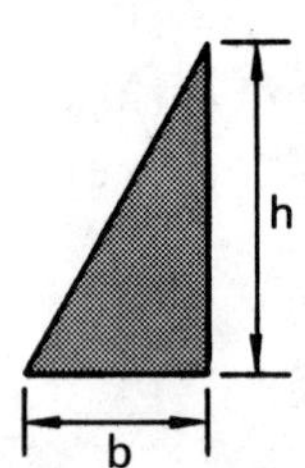


$$A = \frac{1}{2}bh \qquad 1.5$$

*Parabola*

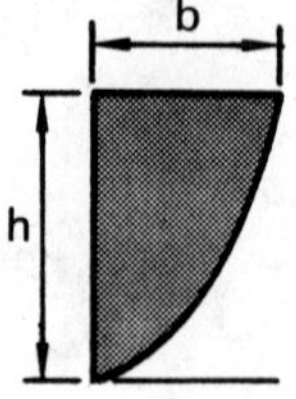


$$A = \frac{2bh}{3} \qquad 1.6$$

$$A = \frac{1}{3}bh \qquad 1.7$$

*Circular Sector*

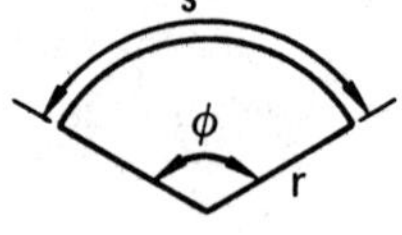


$$A = \frac{1}{2}\phi r^2 = \frac{1}{2}sr \qquad 1.8$$

$$\phi = \frac{s}{r} \qquad 1.9$$

*Ellipse*

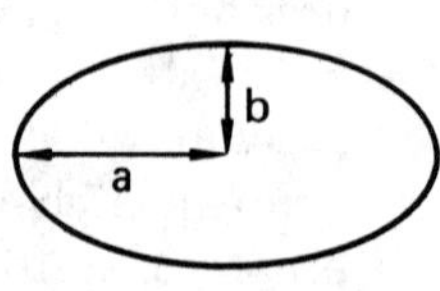


$$A = \pi ab \qquad 1.10$$

$$p = 2\pi\sqrt{\frac{1}{2}(a^2 + b^2)} \qquad 1.11$$

*Trapezoid*

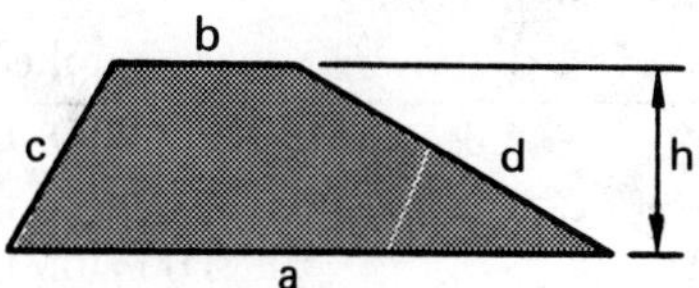


$$p = a + b + c + d \qquad 1.12$$

$$A = \frac{1}{2}h(a + b) \qquad 1.13$$

The trapezoid is isosceles if $c = d$.

*Parallelogram*

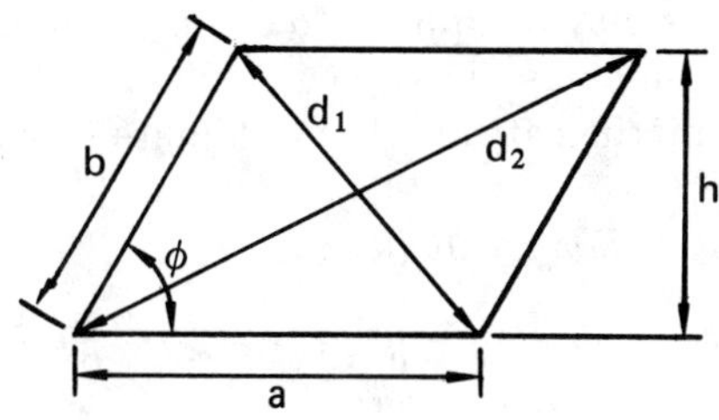


$$p = 2(a + b) \qquad 1.14$$

$$d_1 = \sqrt{a^2 + b^2 - 2ab(\cos\phi)} \qquad 1.15$$

$$d_2 = \sqrt{a^2 + b^2 + 2ab(\cos\phi)} \qquad 1.16$$

$$d_1^2 + d_2^2 = 2(a^2 + b^2) \qquad 1.17$$

$$A = ah = ab(\sin\phi) \qquad 1.18$$

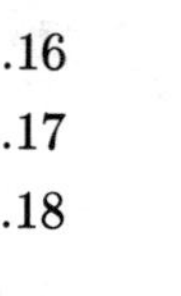

If $a = b$, the parallelogram is a rhombus.

*Regular Polygon*

($n$ equal sides)

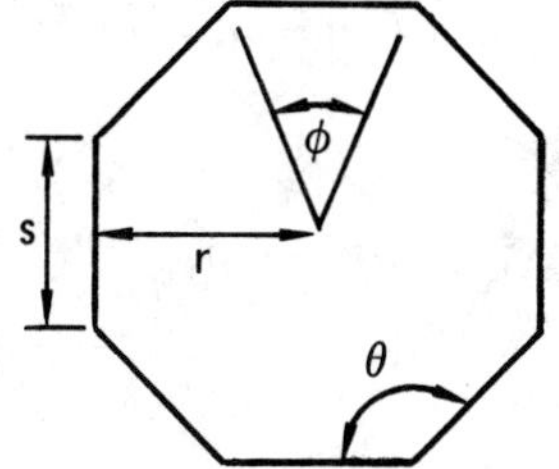


$$\phi = \frac{2\pi}{n} \qquad 1.19$$

$$\theta = \frac{\pi(n - 2)}{n} \qquad 1.20$$

$$p = ns \qquad 1.21$$

$$s = 2r\left(\tan\left(\frac{\phi}{2}\right)\right) \qquad 1.22$$

$$A = \frac{1}{2}nsr \qquad 1.23$$

**Table 1.3**
Polygons

| Number of Sides | Name of Polygon |
|---|---|
| 3 | triangle |
| 4 | rectangle |
| 5 | pentagon |
| 6 | hexagon |
| 7 | heptagon |
| 8 | octagon |
| 9 | nonagon |
| 10 | decagon |

*Sphere*

$$V = \frac{4\pi r^3}{3} \qquad 1.24$$

$$A = 4\pi r^2 \qquad 1.25$$

*Right Circular Cone*

$$V = \frac{\pi r^2 h}{3} \qquad 1.26$$

$$A = \pi r\sqrt{r^2 + h^2} \qquad 1.27$$

(does not include base area)

*Right Circular Cylinder*

$$V = \pi h r^2 \qquad 1.28$$

$$A = 2\pi r h \qquad 1.29$$

(does not include end area)

*Paraboloid of Revolution*

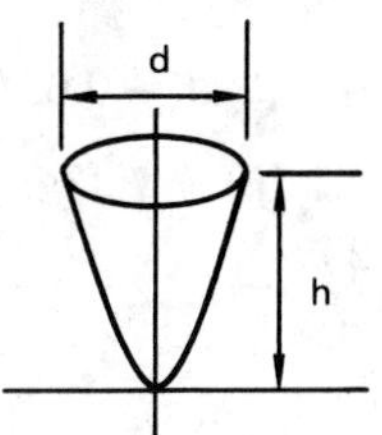


$$V = \frac{\pi h d^2}{8} \qquad 1.30$$

*Regular Polyhedron*

The radius of a sphere inscribed within a regular polyhedron is

$$r = \frac{3V}{A} \qquad 1.31$$

**Table 1.4**
Polyhedrons

| Number of Faces | Form of Faces | Total Surface Area | Volume |
|---|---|---|---|
| 4 | equilateral triangle | 1.7321 $s^2$ | 0.1179 $s^3$ |
| 6 | square | 6.0000 $s^2$ | 1.0000 $s^3$ |
| 8 | equilateral triangle | 3.4641 $s^2$ | 0.4714 $s^3$ |
| 12 | regular pentagon | 20.6457 $s^2$ | 7.6631 $s^3$ |
| 20 | equilateral triangle | 8.6603 $s^2$ | 2.1817 $s^3$ |

*Example 1.1*

What is the hydraulic radius of a 6″ pipe filled to a depth of 2″?

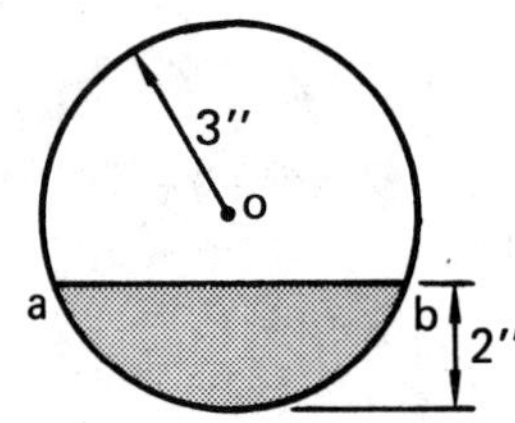


The hydraulic radius is defined as

$$r_h = \frac{\text{area in flow}}{\text{length of wetted perimeter}} = \frac{A}{s}$$

Points $o$, $a$, and $b$ may be used to find the central angle of the circular segment.

$$\frac{1}{2}(\text{angle}\,aob) = \arccos\left(\frac{1}{3}\right) = 70.53°$$
$$\phi = 141.06° = 2.46 \text{ radians}$$

Then,

$$A = \frac{1}{2}(3)^2(2.46 - 0.63) = 8.235 \text{ in}^2$$
$$s = (3)(2.46) = 7.38 \text{ in}$$
$$r_h = \frac{8.235}{7.38} = 1.12 \text{ in}$$

## 5 SIGNIFICANT DIGITS

The significant digits in a number include the left-most, non-zero digits to the right-most digit written. Final answers from computations should be rounded off to the number of decimal places justified by the data. The answer can be no more accurate than the least accurate number in the data. Of course, rounding should be done on final calculation results only. It should not be done on interim results.

| number as written | number of significant digits | implied range |
|---|---|---|
| 341 | 3 | 340.5 to 341.5 |
| 34.1 | 3 | 34.05 to 34.15 |
| 0.00341 | 3 | 0.003405 to 0.003415 |
| 3410. | 4 | 3409.5 to 3410.5 |
| 341 EE7 | 3 | 340.5 EE7 to 341.5 EE7 |
| 3.41 EE−2 | 3 | 3.405 EE−2 to 3.415 EE−2 |

## 6 ALGEBRA

Algebra provides the rules which allow complex mathematical relationships to be expanded or condensed. Algebraic laws may be applied to complex numbers, variables, and numbers. The general rules for changing the form of a mathematical relationship are given here:

*Commutative law for addition*:

$$a + b = b + a \qquad 1.32$$

*Commutative law for multiplication*:

$$ab = ba \qquad 1.33$$

*Associative law for addition*:

$$a + (b + c) = (a + b) + c \qquad 1.34$$

*Associative law for multiplication*:

$$a(bc) = (ab)c \qquad 1.35$$

*Distributive law*:

$$a(b + c) = ab + ac \qquad 1.36$$

### A. POLYNOMIAL EQUATIONS

*1. Standard Forms*

$$(a + b)(a - b) = a^2 - b^2 \qquad 1.37$$
$$(a \pm b)^2 = a^2 \pm 2ab + b^2 \qquad 1.38$$
$$(a \pm b)^3 = a^3 \pm 3a^2b + 3ab^2 \pm b^3 \qquad 1.39$$
$$(a^3 \pm b^3) = (a \pm b)(a^2 \mp ab + b^2) \qquad 1.40$$
$$(a^n + b^n) = (a + b)(a^{n-1} - a^{n-2}b + \cdots + b^{n-1})(\text{factorable only for } n \text{ odd}) \qquad 1.41$$
$$(a^n - b^n) = (a - b)(a^{n-1} + a^{n-2}b + \cdots + b^{n-1})(n \text{ odd or even}) \qquad 1.42$$

*2. Quadratic Equations*

Given a quadratic equation $ax^2 + bx + c = 0$, the roots $x_1^*$ and $x_2^*$ may be found from

$$x_1^*, x_2^* = \frac{-b \pm \sqrt{b^2 - 4ac}}{2a} \qquad 1.43$$

$$x_1^* + x_2^* = -\frac{b}{a} \qquad 1.44$$

$$x_1^* x_2^* = \frac{c}{a} \qquad 1.45$$

*3. Cubic Equations*

Cubic and higher order equations occur infrequently in most engineering problems. However, they usually are difficult to factor when they do occur. Trial and error solutions are usually unsatisfactory except for finding the general region in which a root occurs. Graphical means can be used to obtain only a fair approximation to the root.

Numerical analysis techniques must be used if extreme accuracy is needed. The more efficient numerical analysis techniques are too complicated to present here. However, the bisection method illustrated in example 1.2 usually can provide the required accuracy with only a few simple iterations.

The bisection method starts out with two values of the independent variable, $L_0$ and $R_0$, which straddle a root. Since the function has a value of zero at a root, $f(L_0)$ and $f(R_0)$ will have opposite signs. The following algorithm describes the remainder of the bisection method:

> Let $n$ be the iteration number. Then, for $n = 0, 1, 2, \ldots$ perform the following steps until sufficient accuracy is attained.
>
> Set $m = \frac{1}{2}(L_n + R_n)$
>
> Calculate $f(m)$
>
> If $f(L_n)f(m) \leq 0$, set $L_{n+1} = L_n$ and $R_{n+1} = m$
>
> Otherwise, set $L_{n+1} = m$ and $R_{n+1} = R_n$
>
> $f(x)$ has at least one root in the interval $(L_{n+1}, R_{n+1})$
>
> The estimated value of that root, $x^*$, is

$$x^* \approx \frac{1}{2}(L_{n+1} + R_{n+1}) \qquad 1.46$$

The maximum error is $1/2(R_{n+1} - L_{n+1})$. The iterations continue until the maximum error is reasonable for the accuracy of the problem.

*Example 1.2*

Use the bisection method to find the roots of

$$f(x) = x^3 - 2x - 7$$

The first step is to find $L_0$ and $R_0$, which are the values of $x$ which straddle a root and have opposite signs. A table can be made and values of $f(x)$ calculated for random values of $x$.

| $x$ | $-2$ | $-1$ | $0$ | $+1$ | $+2$ | $+3$ |
|---|---|---|---|---|---|---|
| $f(x)$ | $-11$ | $-6$ | $-7$ | $-8$ | $-3$ | $+14$ |

Since $f(x)$ changes sign between $x = 2$ and $x = 3$,

$$L_0 = 2 \text{ and } R_0 = 3$$

*Iteration 0:*

$$m = \frac{1}{2}(2 + 3) = 2.5$$

$$f(2.5) = (2.5)^3 - 2(2.5) - 7 = 3.625$$

Since $f(2.5)$ is positive, a root must exist in the interval (2, 2.5). Therefore,

$$L_1 = 2 \text{ and } R_1 = 2.5$$

At this point, the best estimate of the root is

$$x^* \approx \frac{1}{2}(2 + 2.5) = 2.25$$

The maximum error is $\frac{1}{2}(2.5 - 2) = 0.25$.

*Iteration 1:*

$$m = \frac{1}{2}(2 + 2.5) = 2.25$$

$$f(2.25) = -0.1094$$

Since $f(m)$ is negative, a root must exist in the interval (2.25, 2.5). Therefore,

$$L_2 = 2.25 \text{ and } R_2 = 2.5$$

The best estimate of the root is

$$x^* \approx \frac{1}{2}(2.25 + 2.5) = 2.375$$

The maximum error is $\frac{1}{2}(2.5 - 2.25) = 0.125$.

This procedure continues until the maximum error is acceptable. Of course, this method does not automatically find any other roots that may exist on the real number line.

*4. Finding Roots to General Expressions*

There is no specific technique that will work with all general expressions for which roots are needed. If graphical means are not used, some combination of factoring and algebraic simplification must be used. However, multiplying each side of an equation by a power of a variable may introduce extraneous roots. Such an extraneous root will not satisfy the original equation, even though it was derived correctly according to the rules of algebra.

Although it is always a good idea to check your work, this step is particularly necessary whenever you have squared an expression or multiplied it by a variable.

*Example 1.3*

Find the value of $x$ which will satisfy the following expression:

$$\sqrt{x-2} = \sqrt{x} + 2$$

First, square both sides.

$$x - 2 = x + 4\sqrt{x} + 4$$

Next, subtract $x$ from both sides and combine constants.

$$4\sqrt{x} = -6$$

Solving for $x$ yields $x^* = 9/4$. However $9/4$ does not satisfy the original expression since it is an extraneous root.

## B. SIMULTANEOUS LINEAR EQUATIONS

Given $n$ independent equations and $n$ unknowns, the $n$ values which simultaneously solve all $n$ equations can be found by the methods illustrated in example 1.4.

*1. By Substitution (shown by example)*

*Example 1.4*

Solve

$$2x + 3y = 12\,(a)$$
$$3x + 4y = 8\,(b)$$

*step 1*: From equation (a), solve for $x = 6 - 1.5y$

*step 2*: Substitute $(6-1.5y)$ into equation (b) wherever $x$ appears. $3(6 - 1.5y) + 4y = 8$ or $y^* = 20$

*step 3*: Solve for $x^*$ from either equation:

$$x^* = 6 - 1.5(20) = -24$$

*step 4*: Check that $(-24,\ 20)$ solves both original equations.

*2. By Reduction (same example)*

*step 1*: Multiply each equation by a number chosen to make the coefficient of one of the variables the same in each equation.

$$3\times \text{ equation (a)}: 6x + 9y = 36\text{(c)}$$
$$2\times \text{ equation (b)}: 6x + 8y = 16\text{(d)}$$

*step 2*: Subtract one equation from the other. Solve for one of the variables.

$$\text{(c)} - \text{(d)} : \ y^* = 20$$

*step 3*: Solve for the remaining variable.

*step 4*: Check that the calculated values of $(x^*, y^*)$ solve both original equations.

*3. By Cramer's Rule*

This method is best for 3 or more simultaneous equations. (The calculation of determinants is covered later in this chapter.)

To find $x^*$ and $y^*$ which satisfy

$$a_1x + b_1y = c_1$$
$$a_2x + b_2y = c_2$$

calculate the determinants

$$\mathbf{D}_1 = \begin{vmatrix} a_1 & b_1 \\ a_2 & b_2 \end{vmatrix} \qquad 1.47$$

$$\mathbf{D}_2 = \begin{vmatrix} c_1 & b_1 \\ c_2 & b_2 \end{vmatrix} \qquad 1.48$$

$$\mathbf{D}_3 = \begin{vmatrix} a_1 & c_1 \\ a_2 & c_2 \end{vmatrix} \qquad 1.49$$

Then, if $\mathbf{D}_1 \neq 0$, the unique numbers satisfying the two simultaneous equations are:

$$x^* = \frac{\mathbf{D}_2}{\mathbf{D}_1} \qquad 1.50$$

$$y^* = \frac{\mathbf{D}_3}{\mathbf{D}_1} \qquad 1.51$$

If $\mathbf{D}_1$ (the determinant of the coefficients matrix) is zero, the system of simultaneous equations may still have a solution. However, Cramer's rule cannot be used to find that solution. If the system is homogeneous (i.e., has the general form $\mathbf{A}x = \mathbf{0}$), then a non-zero solution exists if and only if $\mathbf{D}_1$ is zero.

*Example 1.5*

Solve the following system of simultaneous equations:

$$2x + 3y - 4z = 1$$
$$3x - y - 2z = 4$$
$$4x - 7y - 6z = -7$$

Calculate the determinants:

$$\mathbf{D}_1 = \begin{vmatrix} 2 & 3 & -4 \\ 3 & -1 & -2 \\ 4 & -7 & -6 \end{vmatrix} = 82$$

$$\mathbf{D}_2 = \begin{vmatrix} 1 & 3 & -4 \\ 4 & -1 & -2 \\ -7 & -7 & -6 \end{vmatrix} = 246$$

$$\mathbf{D}_3 = \begin{vmatrix} 2 & 1 & -4 \\ 3 & 4 & -2 \\ 4 & -7 & -6 \end{vmatrix} = 82$$

$$\mathbf{D}_4 = \begin{vmatrix} 2 & 3 & 1 \\ 3 & -1 & 4 \\ 4 & -7 & -7 \end{vmatrix} = 164$$

Then,

$$x^* = \frac{\mathbf{D}_2}{\mathbf{D}_1} = 3$$
$$y^* = \frac{\mathbf{D}_3}{\mathbf{D}_1} = 1$$
$$z^* = \frac{\mathbf{D}_4}{\mathbf{D}_1} = 2$$

## C. SIMULTANEOUS QUADRATIC EQUATIONS

Although simultaneous non-linear equations are best solved graphically, a specialized method exists for simultaneous quadratic equations. This method is known as *Eliminating the Constant Term.*

*step 1*: Isolate the constant terms of both equations on the right-hand side of the equalities.

*step 2*: Multiply both sides of one equation by a number chosen to make the constant terms of both equations the same.

*step 3*: Subtract one equation from the other to obtain a difference equation.

*step 4*: Factor the difference equation into terms.

*step 5*: Solve for one of the variables from one of the factor terms.

*step 6*: Substitute the formula for the variable into one of the original equations and complete the solution.

*step 7*: Check the solution.

*Example 1.6*

Solve for the simultaneous values of $x$ and $y$:

*step 1*:

$$2x^2 - 3xy + y^2 = 15$$
$$x^2 - 2xy + y^2 = 9$$

*steps 2 & 3*:

$$\begin{array}{r} 6x^2 - 9xy + 3y^2 = 45 \\ -(5x^2 - 10xy + 5y^2) = 45 \\ \hline x^2 + xy - 2y^2 = 0 \end{array}$$

*steps 4 & 5*: $x^2 + xy - 2y^2$ factors into $(x+2y)(x-y)$ from which we obtain $x = -2y$.

*step 6*: Substituting $x = -2y$ into $(2x^2 - 3xy + y^2 = 15)$ gives $y^* = \pm 1$, from which $x^* = \pm 2$ can be derived by further substitution.

## D. EXPONENTIATION

($x$ is any variable or constant)

$$x^m x^n = x^{(n+m)} \quad 1.52$$
$$\frac{x^m}{x^n} = x^{(m-n)} \quad 1.53$$
$$(x^n)^m = x^{(mn)} \quad 1.54$$
$$a^{m/n} = \sqrt[n]{a^m} \quad 1.55$$
$$\left(\frac{a}{b}\right)^n = \frac{a^n}{b^n} \quad 1.56$$
$$\sqrt[n]{x} = (x)^{1/n} \quad 1.57$$
$$x^{-n} = \frac{1}{x^n} \quad 1.58$$
$$x^0 = 1 \quad 1.59$$

## E. LOGARITHMS

Logarithms are exponents. That is, the exponent $x$ in the expression $b^x = n$ is the logarithm of $n$ to the base $b$. Therefore, $(\log_b n) = x$ is equivalent to $(b^x = n)$.

The base for common logs is 10. Usually, *log* will be written when common logs are desired, although $\log_{10}$ appears occasionally. The base for *natural (naperien) logs* is 2.718..., a number which is given the symbol $e$. When natural logs are desired, usually *ln* will be written, although $\log_e$ is also used.

Most logarithms will contain an integer part (the *characteristic*) and a fractional part (the *mantissa*). The logarithm of any number less than one is negative. If the number is greater than one, its logarithm is positive.

Although the logarithm may be negative, the mantissa is always positive.

For common logarithms of numbers greater than one, the characteristics will be positive and equal to one less than the number of digits in front of the decimal. If the number is less than one, the characteristic will be negative and equal to one more than the number of zeros immediately following the decimal point.

*Example 1.7*

What is $\log_{10}(.05)$?

Since the number is less than one and there is one leading zero, the characteristic is $-2$. From the logarithm tables, the mantissa of 5.0 is 0.699. Two ways of combining the mantissa and characteristic are possible:

*Method 1*: $\bar{2}.699$

*Method 2*: $8.699 - 10$

If the logarithm is to be used in a calculation, it must be converted to operational form: $-2+0.699 = -1.301$. Notice that $-1.301$ is not the same as $\bar{1}.301$.

## F. LOGARITHM IDENTITIES

$$x^a = \text{antilog}[a\log(x)] \qquad 1.60$$

$$\log(x^a) = a\log(x) \qquad 1.61$$

$$\log(xy) = \log(x) + \log(y) \qquad 1.62$$

$$\log\left(\frac{x}{y}\right) = \log(x) - \log(y) \qquad 1.63$$

$$ln(x) = \frac{\log_{10} x}{\log_{10} e} \approx 2.3(\log_{10} x) \qquad 1.64$$

$$\log_b(b) = 1 \qquad 1.65$$

$$\log(1) = 0 \qquad 1.66$$

$$\log_b(b^n) = n \qquad 1.67$$

*Example 1.8*

The surviving fraction, $x$, of a radioactive isotope is given by

$$x = e^{-0.005t}$$

For what value of $t$ will the surviving fraction be 7%?

$$.07 = e^{-0.005t}$$

Taking the natural log of both sides,

$$ln(0.07) = ln(e^{-0.005t})$$

$$-2.66 = -0.005t$$

$$t = 532$$

## G. PARTIAL FRACTIONS

Given some rational fraction $H(x) = P(x)/Q(x)$ where $P(x)$ and $Q(x)$ are polynomials, the polynomials and constants $A_i$ and $Y_i(x)$ are needed such that

$$H(x) = \sum_i \frac{A_i}{Y_i(x)} \qquad 1.68$$

*Case 1*: $Q(x)$ factors into $n$ different linear terms. That is,

$$Q(x) = (x-a_1)(x-a_2)\cdots(x-a_n) \qquad 1.69$$

Then,

$$H(x) = \sum_{i=1}^{n} \frac{A_i}{x-a_i} \qquad 1.70$$

*Case 2*: $Q(x)$ factors into $n$ identical linear terms. That is,

$$Q(x) = (x-a)(x-a)\cdots(x-a) \qquad 1.71$$

Then,

$$H(x) = \sum_{i=1}^{n} \frac{A_i}{(x-a)^i} \qquad 1.72$$

*Case 3*: $Q(x)$ factors into $n$ different quadratic terms, $(x^2 + p_i x + q_i)$. Then,

$$H(x) = \sum_{i=1}^{n} \frac{A_i x + B_i}{x^2 + p_i x + q_i} \qquad 1.73$$

*Case 4*: $Q(x)$ factors into $n$ identical quadratic terms, $(x^2 + px + q)$. Then,

$$H(x) = \sum_{i=1}^{n} \frac{A_i x + B_i}{(x^2 + px + q)^i} \qquad 1.74$$

*Case 5*: $Q(x)$ factors into any combination of the above. The solution is illustrated by example 1.9.

*Example 1.9*

Resolve

$$H(x) = \frac{x^2 + 2x + 3}{x^4 + x^3 + 2x^2}$$

into partial fractions.

Here, $Q(x) = x^4 + x^3 + 2x^2$ which factors into $x^2(x^2 + x + 2)$. This is a combination of cases 2 and 3. We set

$$H(x) = \frac{A_1}{x} + \frac{A_2}{x^2} + \frac{A_3 + A_4 x}{x^2 + x + 2}$$

Cross multiplying to obtain a common denominator yields

$$\frac{(A_1{+}A_4)x^3 + (A_1{+}A_2{+}A_3)x^2 + (2A_1{+}A_2)x + 2A_2}{x^4 + x^3 + 2x^2}$$

Since the original numerator is known, the following simultaneous equations result:

$$\begin{aligned} A_1 + A_4 &= 0 \\ A_1 + A_2 + A_3 &= 1 \\ 2A_1 + A_2 &= 2 \\ 2A_2 &= 3 \end{aligned}$$

The solutions are: $A_1^* = 0.25$; $A_2^* = 1.5$; $A_3^* = -0.75$; $A_4^* = -0.25$. So,

$$H(x) = \frac{1}{4x} + \frac{3}{2x^2} - \frac{x+3}{4(x^2+x+2)}$$

## H. LINEAR AND MATRIX ALGEBRA

A matrix is a rectangular collection of variables or scalars contained within a set of square or round brackets. In the discussion that follows, matrix **A** will be assumed to have $m$ rows and $n$ columns. There are several classifications of matrices:

If $n = m$, the matrix is *square.*

A *diagonal* matrix is a square matrix with all zero values except for the $a_{ij}$ values, for all $i = j$.

An *identity* matrix is a diagonal matrix with all non-zero entries equal to '1'. (This usually is designated as '**I**'.)

A *scalar* matrix is a square diagonal matrix with all non-zero entries equal to some constant.

A *triangular* matrix has zeros in all positions above or below the diagonal. This is not the same as an *echelon* matrix since the diagonal entries are non-zero.

Matrices are used to simplify the presentation and solution of sets of linear equations (hence the name 'linear algebra'). For example, the system of equations in example 1.4 can be written in matrix form as:

$$\begin{pmatrix} 2 & 3 \\ 3 & 4 \end{pmatrix} \begin{pmatrix} x \\ y \end{pmatrix} = \begin{pmatrix} 12 \\ 8 \end{pmatrix}$$

The above expression implies that there is a set of algebraic operations that can be performed with matrices. The important algebraic operations are listed here, along with their extensions to linear algebra.

(a) *Equality of Matrices*: For two matrices to be equal, they must have the same number of rows and columns. Corresponding entries must all be the same.

(b) *Inequality of Matrices*: There are no 'less-than' or 'greater than' relationships in linear algebra.

(c) *Addition and Subtraction of Matrices*: Addition (or subtraction) of two matrices can be accomplished by adding (or subtracting) the corresponding entries of two matrices which have the same shape.

(d) *Multiplication of Matrices*: Multiplication can be done only if the left-hand matrix has the same number of columns as the right-hand matrix has rows. Multiplication is accomplished by multiplying the elements in each right-hand matrix column, adding the products, and then placing the sum at the intersection point of the involved row and column. This is illustrated by example 1.10.

(e) *Division of Matrices*: Division can be accomplished only by multiplying by the inverse of the denominator matrix.

*Example 1.10*

$$\begin{pmatrix} 1 & 4 & 3 \\ \\ 5 & 2 & 6 \end{pmatrix} \begin{pmatrix} 7 & 12 \\ 11 & 8 \\ 9 & 10 \end{pmatrix} = \mathbf{C}$$

$$\begin{aligned} [(1)(7) + (4)(11) + (3)(9)] &= 78 \\ [(1)(12) + (4)(8) + (3)(10)] &= 74 \\ [(5)(7) + (2)(11) + (6)(9)] &= 111 \\ [(5)(12) + (2)(8) + (6)(10)] &= 136 \end{aligned}$$

$$\mathbf{C} = \begin{pmatrix} 78 & 74 \\ 111 & 136 \end{pmatrix}$$

Other operations which can be performed on a matrix are described and illustrated below.

1. The *transpose* is an $(n \times m)$ matrix formed from the original $(m \times n)$ matrix by taking the ith row and making it the ith column. The diagonal is unchanged in this operation. The transpose of a matrix **A** is indicated as $\mathbf{A}^t$.

*Example 1.11*

What is the transpose of

$$\mathbf{A} = \begin{pmatrix} 1 & 6 & 9 \\ 2 & 3 & 4 \\ 7 & 1 & 5 \end{pmatrix}?$$

$$\mathbf{A}^t = \begin{pmatrix} 1 & 2 & 7 \\ 6 & 3 & 1 \\ 9 & 4 & 5 \end{pmatrix}$$

2. The *determinant*, **D**, is a scalar calculated from a square matrix. The determinant of a matrix is indicated by enclosing the matrix in vertical lines.

For a (2 × 2) matrix,

$$\mathbf{A} = \begin{pmatrix} a & b \\ c & d \end{pmatrix}$$

$$\mathbf{D} = \begin{vmatrix} a & b \\ c & d \end{vmatrix} = ad - bc \qquad 1.75$$

For a (3 × 3) matrix,

$$\mathbf{A} = \begin{pmatrix} a & b & c \\ d & e & f \\ g & h & i \end{pmatrix}$$

$$D = a\begin{vmatrix} e & f \\ h & i \end{vmatrix} - d\begin{vmatrix} b & c \\ h & i \end{vmatrix} + g\begin{vmatrix} b & c \\ e & f \end{vmatrix} \qquad 1.76$$

There are several rules governing the calculation of determinants:

- If **A** has a row or column of zeros, the determinant is zero.
- If **A** has two identical rows or columns, the determinant is zero.
- If **A** is triangular, the determinant is equal to the product of the diagonal entries.
- If **B** is obtained from **A** by multiplying a row or column by a scalar $k$, then $\mathbf{D}_B = k(\mathbf{D}_A)$.
- If **B** is obtained from **A** by switching two rows or columns, then $\mathbf{D}_B = -\mathbf{D}_A$.
- If **B** is obtained from **A** by adding a multiple of a row or column to another, then $\mathbf{D}_B = \mathbf{D}_A$.

*Example 1.12*

What is the determinant of

$$\begin{pmatrix} 2 & 3 & -4 \\ 3 & -1 & -2 \\ 4 & -7 & -6 \end{pmatrix}?$$

$$\mathbf{D} = 2\begin{vmatrix} -1 & -2 \\ -7 & -6 \end{vmatrix} - 3\begin{vmatrix} 3 & -4 \\ -7 & -6 \end{vmatrix} + 4\begin{vmatrix} 3 & -4 \\ -1 & -2 \end{vmatrix}$$

$$= 2(6 - 14) - 3(-18 - 28) + 4(-6 - 4)$$

$$= 82$$

3. The *cofactor* of an entry in a matrix is the determinant of the matrix formed by omitting the entry's row and column in the original matrix. The sign of the cofactor is determined from the following positional matrices:

For a (2 × 2) matrix,

$$\begin{pmatrix} + & - \\ - & + \end{pmatrix}$$

For a (3 × 3) matrix,

$$\begin{pmatrix} + & - & + \\ - & + & - \\ + & - & + \end{pmatrix}$$

*Example 1.13*

What is the cofactor of the (−3) in the following matrix?

$$\begin{pmatrix} 2 & 9 & 1 \\ -3 & 4 & 0 \\ 7 & 5 & 9 \end{pmatrix}$$

The resulting matrix is

$$\begin{pmatrix} 9 & 1 \\ 5 & 9 \end{pmatrix}$$

with determinant 76. The cofactor is −76.

4. The *classical adjoint* is a matrix formed from the transposed cofactor matrix with the conventional sign arrangement. The resulting matrix is represented as $\mathbf{A}_{adj}$.

*Example 1.14*

What is the classical adjoint of

$$\begin{pmatrix} 2 & 3 & -4 \\ 0 & -4 & 2 \\ 1 & -1 & 5 \end{pmatrix}?$$

The matrix of cofactors (considering the sign convention) is

$$\begin{pmatrix} -18 & 2 & 4 \\ -11 & 14 & 5 \\ -10 & -4 & -8 \end{pmatrix}$$

The transposed cofactor matrix is

$$\mathbf{A}_{adj} = \begin{pmatrix} -18 & -11 & -10 \\ 2 & 14 & -4 \\ 4 & 5 & -8 \end{pmatrix}$$

5. The *inverse*, $\mathbf{A}^{-1}$, of **A** is a matrix such that $(\mathbf{A})(\mathbf{A}^{-1}) = \mathbf{I}$. (**I** is a square matrix with ones along the left-to-right diagonal and zeros elsewhere.)

For a (2 × 2) matrix

$$\begin{pmatrix} a & b \\ c & d \end{pmatrix}$$

the inverse is

$$\frac{1}{\mathbf{D}}\begin{pmatrix} d & -b \\ -c & a \end{pmatrix} \qquad 1.77$$

For larger matrices, the inverse is best calculated by dividing every entry in the classical adjoint by the determinant of the original matrix.

*Example 1.15*

What is the inverse of

$$\begin{pmatrix} 4 & 5 \\ 2 & 3 \end{pmatrix}?$$

The determinant is 2. The inverse is

$$\frac{1}{2}\begin{pmatrix} 3 & -5 \\ -2 & 4 \end{pmatrix} = \begin{pmatrix} \frac{3}{2} & -\frac{5}{2} \\ -1 & 2 \end{pmatrix}$$

## 7 TRIGONOMETRY

### A. DEGREES AND RADIANS

360 degrees = one complete circle = $2\pi$ radians

90 degrees = right angle = $\frac{1}{2}\pi$ radians

one radian = 57.3 degrees

one degree = 0.0175 radians

multiply degrees by $\left(\frac{\pi}{180}\right)$ to obtain radians

multiply radians by $\left(\frac{180}{\pi}\right)$ to obtain degrees

### B. RIGHT TRIANGLES

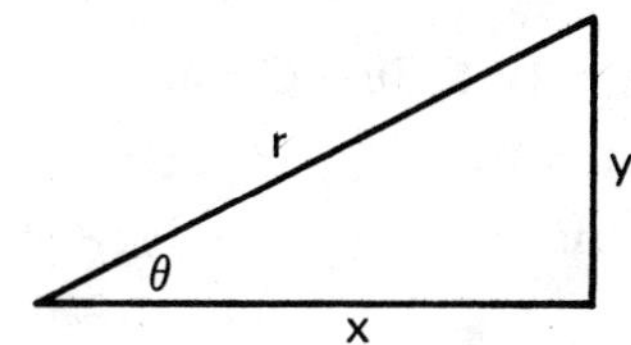


**Figure 1.1** A Right Triangle

*1. Pythagorean Theorem*

$$x^2 + y^2 = r^2 \qquad 1.78$$

*2. Trigonometric Functions*

$$\sin\theta = \frac{y}{r} \qquad 1.79$$

$$\cos\theta = \frac{x}{r} \qquad 1.80$$

$$\tan\theta = \frac{y}{x} \qquad 1.81$$

$$\cot\theta = \frac{x}{y} \qquad 1.82$$

$$\csc\theta = \frac{r}{y} \qquad 1.83$$

$$\sec\theta = \frac{r}{x} \qquad 1.84$$

*3. Relationship of the Trigonometric Functions to the Unit Circle*

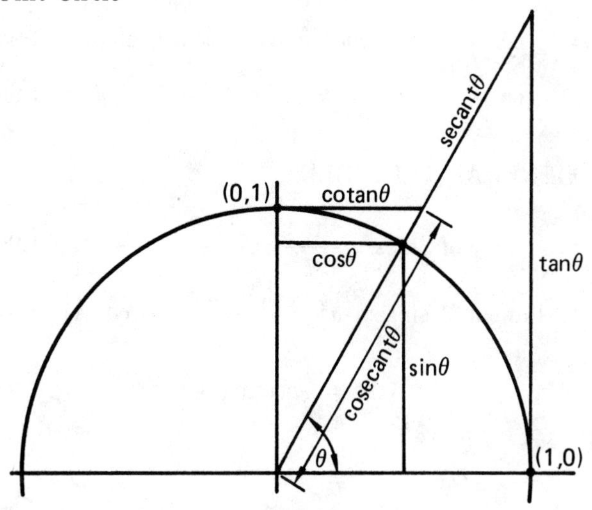


**Figure 1.2** The Unit Circle

*4. Signs of the Trigonometric Functions*

| quadrants | | quadrant | I | II | III | IV |
|---|---|---|---|---|---|---|
| II | I | sin | + | + | − | − |
| | | cos | + | − | − | + |
| III | IV | tan | + | − | + | − |

*5. Functions of the Related Angles*

| $f(\theta)$ | $-\theta$ | $90-\theta$ | $90+\theta$ | $180-\theta$ | $180+\theta$ |
|---|---|---|---|---|---|
| sin | $-\sin\theta$ | $\cos\theta$ | $\cos\theta$ | $\sin\theta$ | $-\sin\theta$ |
| cos | $\cos\theta$ | $\sin\theta$ | $-\sin\theta$ | $-\cos\theta$ | $-\cos\theta$ |
| tan | $-\tan\theta$ | $\cot\theta$ | $-\cot\theta$ | $-\tan\theta$ | $\tan\theta$ |

*6. Trigonometric Identities*

$$\sin^2\theta + \cos^2\theta = 1 \qquad 1.85$$

$$1 + \tan^2\theta = \sec^2\theta \qquad 1.86$$

$$1 + \cot^2\theta = \csc^2\theta \qquad 1.87$$

$$\sin 2\theta = 2(\sin\theta)(\cos\theta) \qquad 1.88$$

$$\cos 2\theta = \cos^2\theta - \sin^2\theta = 1 - 2\sin^2\theta \qquad 1.89$$

$$\sin\theta = 2\left[\sin\left(\frac{\theta}{2}\right)\cos\left(\frac{\theta}{2}\right)\right] \qquad 1.90$$

$$\sin\left(\frac{\theta}{2}\right) = \pm\sqrt{\frac{1}{2}(1-\cos\theta)} \qquad 1.91$$

*7. Two-Angle Formulas*

$$\sin(\theta+\phi) = [\sin\theta][\cos\phi] + [\cos\theta][\sin\phi] \qquad 1.92$$

$$\sin(\theta-\phi) = [\sin\theta][\cos\phi] - [\cos\theta][\sin\phi] \qquad 1.93$$

$$\cos(\theta+\phi) = [\cos\theta][\cos\phi] - [\sin\theta][\sin\phi] \qquad 1.94$$

$$\cos(\theta-\phi) = [\cos\theta][\cos\phi] + [\sin\theta][\sin\phi] \qquad 1.95$$

C. GENERAL TRIANGLES

Law of Sines: $\frac{\sin A}{a} = \frac{\sin B}{b} = \frac{\sin C}{c}$ 1.96

Law of Cosines: $a^2 = b^2 + c^2 - 2bc(\cos A)$ 1.97

$$\text{Area} = \frac{1}{2}ab(\sin C) \qquad 1.98$$

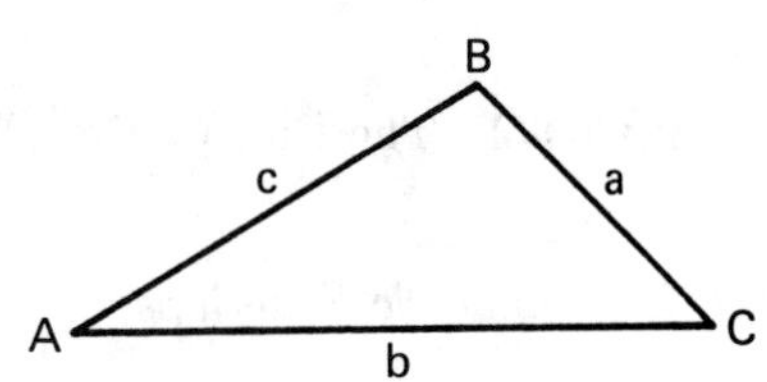


**Figure 1.3** A General Triangle

D. HYPERBOLIC FUNCTIONS

Hyperbolic functions are specific equations containing the terms $e^x$ and $e^{-x}$. These combinations of $e^x$ and $e^{-x}$ appear regularly in certain types of problems. In order to simplify the mathematical equations in which they appear, these hyperbolic functions are given special names and symbols.

$$\sinh x = \frac{e^x - e^{-x}}{2} \qquad 1.99$$

$$\cosh x = \frac{e^x + e^{-x}}{2} \qquad 1.100$$

$$\tanh x = \frac{e^x - e^{-x}}{e^x + e^{-x}} = \frac{\sinh x}{\cosh x} \qquad 1.101$$

$$\coth x = \frac{e^x + e^{-x}}{e^x - e^{-x}} = \frac{\cosh x}{\sinh x} \qquad 1.102$$

$$\operatorname{sech} x = \frac{2}{e^x + e^{-x}} = \frac{1}{\cosh x} \qquad 1.103$$

$$\operatorname{csch} x = \frac{2}{e^x - e^{-x}} = \frac{1}{\sinh x} \qquad 1.104$$

The hyperbolic identities are somewhat different from the standard trigonometric identities. Several of the most common identities are presented below.

$$\cosh^2 x - \sinh^2 x = 1 \qquad 1.105$$

$$1 - \tanh^2 x = \operatorname{sech}^2 x \qquad 1.106$$

$$1 - \coth^2 x = \operatorname{csch}^2 x \qquad 1.107$$

$$\cosh x + \sinh x = e^x \qquad 1.108$$

$$\cosh x - \sinh x = e^{-x} \qquad 1.109$$

$$\sinh(x+y) = [\sinh x][\cosh y] + [\cosh x][\sinh y] \qquad 1.110$$

$$\cosh(x+y) = [\cosh x][\cosh y] + [\sinh x][\sinh y] \qquad 1.111$$

## 8 STRAIGHT LINE ANALYTIC GEOMETRY

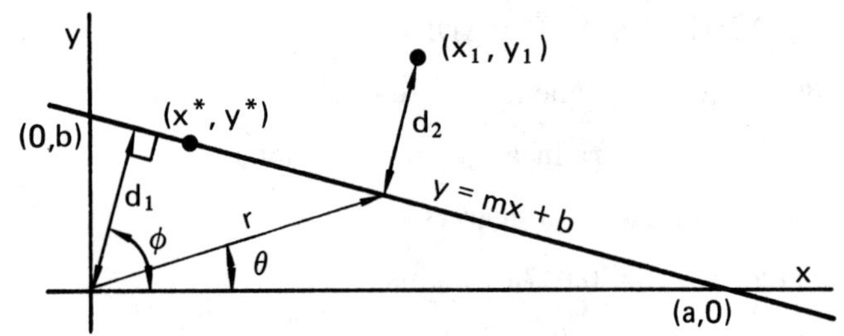


**Figure 1.4** A Straight Line

A. EQUATIONS OF A STRAIGHT LINE

General Form: $Ax + By + C = 0$ 1.112

Slope Form: $y = mx + b$ 1.113

Point-Slope Form: $(y - y^*) = m(x - x^*)$ 1.114

$(x^*, y^*)$ is any point on the line.

Intercept Form: $\frac{x}{a} + \frac{y}{b} = 1$ 1.115

Two-Point Form: $\frac{y - y_1^*}{x - x_1^*} = \frac{y_2^* - y_1^*}{x_2^* - x_1^*}$ 1.116

Normal Form: $x(\cos\phi) + y(\sin\phi) - d_1 = 0$ 1.117

Polar Form: $r = \frac{d_1}{\cos(\phi - \theta)}$ 1.118

B. POINTS, LINES, AND DISTANCES

The distance $d_2$ between a point and a line is:

$$d_2 = \frac{|Ax_1 + By_1 + C|}{\sqrt{A^2 + B^2}} \qquad 1.119$$

The distance between two points is:

$$d = \sqrt{(x_2 - x_1)^2 + (y_2 - y_1)^2} \quad 1.120$$

Parallel lines:

$$\frac{A_1}{A_2} = \frac{B_1}{B_2} \quad 1.121$$

$$m_1 = m_2 \quad 1.122$$

Perpendicular lines:

$$A_1A_2 = -B_1B_2 \quad 1.123$$

$$m_1 = \frac{-1}{m_2} \quad 1.124$$

Point of intersection of two lines:

$$x_1 = \frac{B_2C_1 - B_1C_2}{A_2B_1 - A_1B_2} \quad 1.125$$

$$y_1 = \frac{A_1C_2 - A_2C_1}{A_2B_1 - A_1B_2} \quad 1.126$$

Smaller angle between two intersecting lines:

$$\tan\phi = \frac{A_1B_2 - A_2B_1}{A_1A_2 + B_1B_2} = \frac{m_2 - m_1}{1 + m_1m_2} \quad 1.127$$

$$\phi = |\arctan(m_1) - \arctan(m_2)|\,. \quad 1.128$$

*Example 1.16*

What is the angle between the lines?

$$y_1 = -0.577x + 2$$
$$y_2 = +0.577x - 5$$

*method 1*:

$$\arctan\left[\frac{m_2 - m_1}{1 + m_1m_2}\right]$$
$$= \arctan\left[\frac{0.577 - (-0.577)}{1 + (0.577)(-0.577)}\right] = 60°$$

*method 2*: Write both equations in general form:

$$-0.577x - y_1 + 2 = 0$$

$$0.577x - y_2 - 5 = 0$$

$$\arctan\left[\frac{A_1B_2 - A_2B_1}{A_1A_2 + B_1B_2}\right]$$
$$= \arctan\left[\frac{(-0.577)(-1) - (0.577)(-1)}{(-0.577)(0.577) + (-1)(-1)}\right] = 60°$$

*method 3*:

$$\phi = |\arctan(-0.577) - \arctan(0.577)|$$
$$= |-30° - 30°| = 60°$$

C. LINEAR AND CURVILINEAR REGRESSION

If it is necessary to draw a straight line through $n$ data points $(x_1, y_1), (x_2, y_2), \ldots, (x_n, y_n)$, the following method based on the theory of least squares can be used:

*step 1*: Calculate the following quantities.

$$\sum x_i \quad \sum x_i^2 \quad (\sum x_i)^2 \quad \overline{x} = \left(\frac{\sum x_i}{n}\right) \quad \sum x_iy_i$$

$$\sum y_i \quad \sum y_i^2 \quad (\sum y_i)^2 \quad \overline{y} = \left(\frac{\sum y_i}{n}\right)$$

*step 2*: Calculate the slope of the line $y = mx + b$.

$$m = \frac{n\sum(x_iy_i) - (\sum x_i)(\sum y_i)}{n\sum x_i^2 - (\sum x_i)^2} \quad 1.129$$

*step 3*: Calculate the $y$ intercept.

$$b = \overline{y} - m\overline{x} \quad 1.130$$

*step 4*: To determine the goodness of fit, calculate the correlation coefficient.

$$r = \frac{n\sum(x_iy_i) - (\sum x_i)(\sum y_i)}{\sqrt{\left[n\sum x_i^2 - (\sum x_i)^2\right]\left[n\sum y_i^2 - (\sum y_i)^2\right]}} \quad 1.131$$

If $m$ is positive, $r$ will be positive. If $m$ is negative, $r$ will be negative. As a general rule, if the absolute value of $r$ exceeds .85, the fit is good. Otherwise, the fit is poor. $r$ equals 1.0 if the fit is a perfect straight line.

*Example 1.17*

An experiment is performed in which the dependent variable ($y$) is measured against the independent variable ($x$). The results are as follows:

| $x$ | $y$ |
|---|---|
| 1.2 | 0.602 |
| 4.7 | 5.107 |
| 8.3 | 6.984 |
| 20.9 | 10.031 |

What is the least squares straight line equation which represents this data?

*step 1:*

$$\sum x_i = 35.1$$
$$\sum y_i = 22.72$$
$$\sum x_i^2 = 529.23$$
$$\sum y_i^2 = 175.84$$
$$\left(\sum x_i\right)^2 = 1232.01$$
$$\left(\sum y_i\right)^2 = 516.38$$
$$\overline{x} = 8.775$$
$$\overline{y} = 5.681$$
$$\sum x_i y_i = 292.34$$
$$n = 4$$

*step 2:*

$$m = \frac{(4)(292.34) - (35.1)(22.72)}{(4)(529.23) - (35.1)^2} = 0.42$$

*step 3:*

$$b = 5.681 - (0.42)(8.775) = 2.0$$

*step 4:* From equation 1.131, $r = 0.91$.

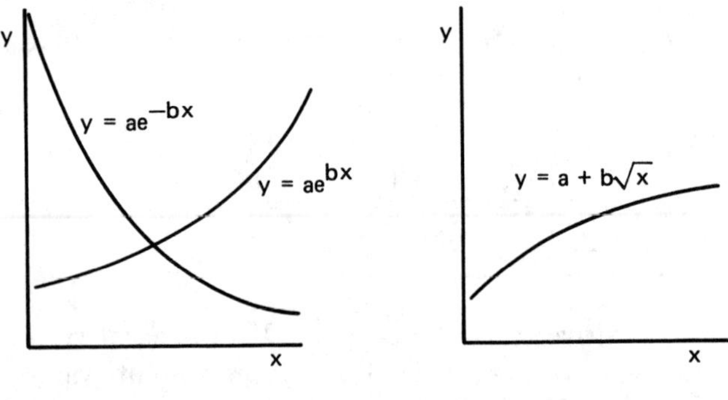


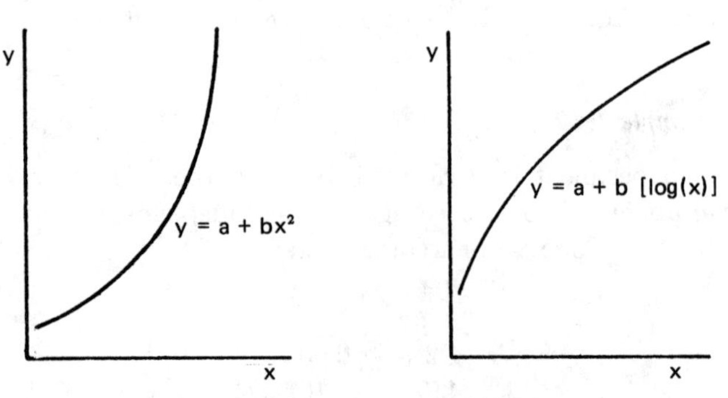


**Figure 1.5** Non-Linear Data Plots

A low value of $r$ does not eliminate the possibility of a non-linear relationship existing between $x$ and $y$. It is possible that the data describes a parabolic, logarithmic, or other non-linear relationship. (Usually this will be apparent if the data are graphed.) It may be necessary to convert one or both variables to new variables by taking squares, square roots, cubes, or logs, to name a few of the possibilities.

The apparent shape of the line through the data will give a clue to the type of variable transformation that is required. The following curves may be used as guides to some of the simpler variable transformations.

*Example 1.18*

Repeat example 1.17 assuming that the relationship between the variables is non-linear.

The first step is to graph the data. Since the graph has the appearance of the fourth case, it can be assumed that the relationship between the variables has the form of $y = a + b[\log(x)]$. Therefore, the variable change $z = \log(x)$ is made, resulting in the following set of data:

| z | $y$ |
|---|---|
| 0.0792 | 0.602 |
| 0.672 | 5.107 |
| 0.919 | 6.984 |
| 1.32 | 10.031 |

If the regression analysis is performed on this set of data, the resulting equation and correlation coefficient are:

$$y = -0.036 + 7.65z$$
$$r = 0.999$$

This is a very good fit. The relationship between the variable $x$ and $y$ is approximately

$$y = -0.036 + 7.65[\log(x)]$$

Figure 1.6 illustrates several common problems encountered in trying to fit and evaluate curves from experimental data. Figure 1.6(a) shows a graph of clustered data with several extreme points. There will be moderate correlation due to the weighting of the extreme points, although there is little actual correlation at low values of the variables. The extreme data should be excluded or the range should be extended by obtaining more data.

Figure 1.6(b) shows that good correlation exists in general, but extreme points are missed, and the overall correlation is moderate. If the results within the small linear range can be used, the extreme points should be excluded. Otherwise, additional data points are needed, and curvilinear relationships should be investigated.

Figure 1.6(c) illustrates the problem of drawing conclusions of cause and effect. There may be a predictable relationship between variables, but that does not imply a cause and effect relationship. In the case shown, both variables are functions of a third variable, the city population. But, there is no direct relationship between the plotted variables.

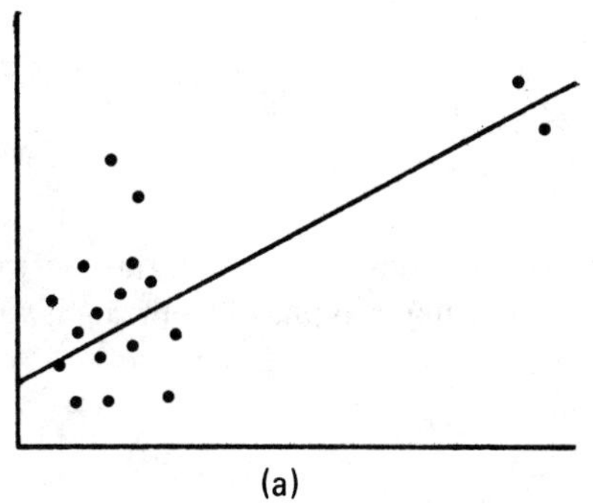
(a)

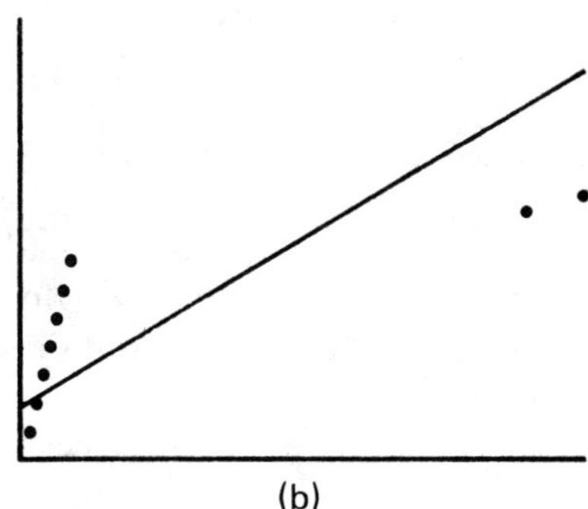
(b)

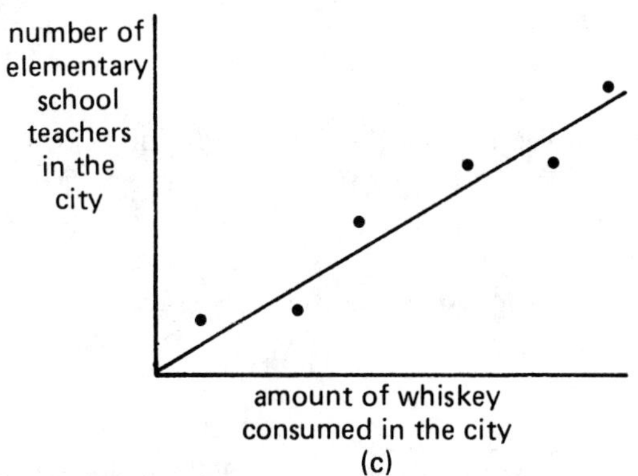

(c)

**Figure 1.6** Common Regression Difficulties

D. VECTOR OPERATIONS

A vector is a directed straight line of a given magnitude. Two directed straight lines with the same magnitudes and directions are said to be *equivalent.* Thus, the actual end points of a vector often are irrelevant as long as the direction and magnitude are known.

A vector defined by its end points and direction is designated as

$$\mathbf{V} = \overrightarrow{p_1}\overrightarrow{p_2}$$

Usually, $p_1$ will be the origin, in which case $\mathbf{V}$ will be designated by its end-point, $p_2 = (x, y)$. Such a zero-based vector is equivalent to all other vectors of the same magnitude and direction. Any vector $p_1p_2$ can be transformed into a zero-based vector by subtracting $(x_1, y_1)$ from all points along the vector line.

A vector also can be specified in the terms of the unit vectors $(\mathbf{i}, \mathbf{j}, \mathbf{k})$. Thus,

$$\mathbf{V} = (x, y) = (x\mathbf{i} + y\mathbf{j})$$

Or,

$$\mathbf{V} = (x, y, z) = (x\mathbf{i} + y\mathbf{j} + z\mathbf{k})$$

Important operations on vectors based at the origin are:

$$c\mathbf{V} = (cx, cy) \quad \text{(vector multiplication by a scalar)}$$

$$\mathbf{V}_1 + \mathbf{V}_2 = (x_1 + x_2, y_1 + y_2) \qquad 1.132$$

$$|\mathbf{V}| = \sqrt{x^2 + y^2} \quad \text{(vector magnitude)} \qquad 1.133$$

$$\alpha = \text{angle between vector } \mathbf{V} \text{ and x axis}$$

$$= \arccos\left(\frac{x}{|\mathbf{V}|}\right) = \arcsin\left(\frac{y}{|\mathbf{V}|}\right) \qquad 1.134$$

$$m = \text{slope of vector} = \frac{y}{x} \qquad 1.135$$

$$\theta = \left\{\begin{matrix}\text{angle between}\\ \text{two vectors}\end{matrix}\right\}$$

$$= \arccos\frac{(x_1x_2 + y_1y_2)}{|\mathbf{V}_1||\mathbf{V}_2|} \qquad 1.136$$

$$\mathbf{V}_1 \cdot \mathbf{V}_2 = \text{dot product}$$

$$= |\mathbf{V}_1||\mathbf{V}_2|\cos\theta = x_1x_2 + y_1y_2 \qquad 1.137$$

When equation 1.137 is solved for $\cos\theta$, it is known as the *Cauchy-Schwartz theorem.*

$$\cos\theta = \frac{x_1x_2 + y_1y_2}{|\mathbf{V}_1||\mathbf{V}_2|} \qquad 1.138$$

$$\mathbf{V}_1 \times \mathbf{V}_2 = \text{cross product} = \begin{vmatrix} \mathbf{i} & x_1 & x_2 \\ \mathbf{j} & y_1 & y_2 \\ \mathbf{k} & z_1 & z_2 \end{vmatrix} \quad 1.139$$

$$\mathbf{V}_1 \times \mathbf{V}_2 = -\mathbf{V}_2 \times \mathbf{V}_1 \quad 1.140$$

$$|\mathbf{V}_1 \times \mathbf{V}_2| = |\mathbf{V}_1||\mathbf{V}_2| \sin\theta \quad 1.141$$

*Example 1.19*

What is the angle between the vectors $\mathbf{V}_1 = (-\sqrt{3}, 1)$ and $\mathbf{V}_2 = (2\sqrt{3}, 2)$?

$$\cos\theta = \frac{\mathbf{V}_1 \cdot \mathbf{V}_2}{|\mathbf{V}_1||\mathbf{V}_2|} = \frac{(-\sqrt{3})(2\sqrt{3}) + (1)(2)}{\sqrt{3+1}\sqrt{12+4}} = -\frac{1}{2}$$

$$\theta = 120°$$

(Graph and compare this result to example 1.16, in which the lines were not directed.)

*Example 1.20*

Find a unit vector orthogonal to $\mathbf{V}_1 = \mathbf{i} - \mathbf{j} + 2\mathbf{k}$ and $\mathbf{V}_2 = 3\mathbf{j} - \mathbf{k}$.

The cross product is orthogonal to $\mathbf{V}_1$ and $\mathbf{V}_2$, although its length may not be equal to one.

$$\mathbf{V}_1 \times \mathbf{V}_2 = \begin{vmatrix} \mathbf{i} & 1 & 0 \\ \mathbf{j} & -1 & 3 \\ \mathbf{k} & 2 & -1 \end{vmatrix} = -5\mathbf{i} + \mathbf{j} + 3\mathbf{k}$$

Since the length of $|\mathbf{V}_1 \times \mathbf{V}_2|$ is $\sqrt{35}$, it is necessary to divide $|\mathbf{V}_1 \times \mathbf{V}_2|$ by this amount to obtain a unit vector. Thus,

$$\mathbf{V}_3 = \frac{-5\mathbf{i} + \mathbf{j} + 3\mathbf{k}}{\sqrt{35}}$$

The orthogonality can be proved from

$$\mathbf{V}_1 \cdot \mathbf{V}_3 = 0 \text{ and } \mathbf{V}_2 \cdot \mathbf{V}_3 = 0$$

That $\mathbf{V}_3$ is a unit vector can be proved from

$$\mathbf{V}_3 \cdot \mathbf{V}_3 = +1$$

## E. DIRECTION NUMBERS, DIRECTION ANGLES, AND DIRECTION COSINES

Given a directed line from $(x_1, y_1, z_1)$ to $(x_2, y_2, z_2)$, the direction numbers are:

$$L = x_2 - x_1 \quad 1.142$$

$$M = y_2 - y_1 \quad 1.143$$

$$N = z_2 - z_1 \quad 1.144$$

The distance between two points is:

$$d = \sqrt{L^2 + M^2 + N^2} \quad 1.145$$

The direction cosines are:

$$\cos\alpha = \frac{L}{d} \quad 1.146$$

$$\cos\beta = \frac{M}{d} \quad 1.147$$

$$\cos\gamma = \frac{N}{d} \quad 1.148$$

Note that

$$\cos^2\alpha + \cos^2\beta + \cos^2\gamma = 1 \quad 1.149$$

The direction angles are the angles between the axes and the lines. They are found from the inverse functions of the direction cosines. That is,

$$\alpha = \arccos\left(\frac{L}{d}\right) \quad 1.150$$

$$\beta = \arccos\left(\frac{M}{d}\right) \quad 1.151$$

$$\gamma = \arccos\left(\frac{N}{d}\right) \quad 1.152$$

Once the direction cosines have been found, they can be used to write the equation of the straight line in terms of the unit vectors. The line $\mathbf{R}$ would be defined as

$$\mathbf{R} = \mathbf{i}\cos\alpha + \mathbf{j}\cos\beta + \mathbf{k}\cos\gamma \quad 1.153$$

Similarly, the line may be written in terms of its direction numbers,

$$\mathbf{R} = L\mathbf{i} + M\mathbf{j} + N\mathbf{k} \quad 1.154$$

Given two directed lines, $\mathbf{R}_1$ and $\mathbf{R}_2$, the angle between $\mathbf{R}_1$ and $\mathbf{R}_2$ is defined as the angle between the two arrow heads.

$$\cos\phi = \cos\alpha_1\cos\alpha_2 + \cos\beta_1\cos\beta_2 + \cos\gamma_1\cos\gamma_2 = \frac{L_1L_2 + M_1M_2 + N_1N_2}{d_1d_2} \quad 1.155$$

If $\mathbf{R}_1$ and $\mathbf{R}_2$ are parallel and in the same direction, then

$$\alpha_1 = \alpha_2$$

$$\beta_1 = \beta_2$$

$$\gamma_1 = \gamma_2$$

If $\mathbf{R}_1$ and $\mathbf{R}_2$ are parallel but in opposite directions, then

$$\alpha_1 + \alpha_2 = 180 \text{ (etc.)}$$

If $\mathbf{R}_1$ and $\mathbf{R}_2$ are normal to each other, then

$$\phi = 90^\circ \text{ and } \cos\phi = 0$$

*Example 1.21*

A line passes through the points (4,7,9) and (0,1,6). Write the equation of the line in terms of its direction cosines and direction numbers.

$$L = 4 - 0 = 4$$
$$M = 7 - 1 = 6$$
$$N = 9 - 6 = 3$$

Now, the line may be written in terms of its direction numbers.

$$\mathbf{R} = 4\mathbf{i} + 6\mathbf{j} + 3\mathbf{k}$$

The distance between the two points is

$$d = \sqrt{(4)^2 + (6)^2 + (3)^2} = 7.81$$

The line now may be written in terms of its direction cosines.

$$\begin{aligned} \mathbf{R} &= \frac{4\mathbf{i} + 6\mathbf{j} + 3\mathbf{k}}{7.81} \\ &= 0.512\mathbf{i} + 0.768\mathbf{j} + 0.384\mathbf{k} \end{aligned}$$

## F. CURVILINEAR INTERPOLATION

A situation which occurs frequently is one in which a function value must be interpolated from other data along the curve. Straight-line interpolation typically is used because of its simplicity and speed. However, straight-line interpolation ignores all but two of the points on the curve and is, therefore, unable to include any effects of curvature.

A more powerful technique is the *Lagrangian Interpolating Polynomial.* It is assumed that $(n+1)$ values of $f(x)$ are known (for $x_0, x_1, x_2, \ldots, x_n$) and that $f(x)$ is a continuous, real-valued function on the interval $(x_0, x_n)$. The value of $f(x)$ at $x^*$ can be estimated from the following equations:

$$f(x^*) = \sum_{k=0}^{n} f(x_k) L_k(x^*) \qquad 1.156$$

where the Lagrangian Interpolating Polynomial is

$$L_k(x^*) = \prod_{\substack{i=0 \\ i \neq k}}^{n} \frac{x^* - x_i}{x_k - x_i} \qquad 1.157$$

Example 1.22 illustrates use of the Lagrangian Interpolating Polynomial.

## 9 TENSORS

A scalar has magnitude only. A vector has magnitude and a definite direction. A *tensor* has magnitude in a specific direction, but the direction is not unique. An example of a tensor is stress. From the combined stress equation, stress at a point in a solid depends on the direction of the plane passing through that point. Tensors frequently are associated with *anisotropic materials* which have different properties in different directions. Other examples are dielectric constant and magnetic susceptibility.

A vector in a three-dimensional space is defined completely by three quantities, $F_x$, $F_y$, and $F_z$. A tensor in three-dimensional space requires nine quantities for complete definition. These nine values are given in ma-

*Example 1.22*

A real-valued function has the following values:

$$f(1) = 1.5709 \quad f(4) = 1.5727 \quad f(6) = 1.5751$$

What is $f(3.5)$?

$$\begin{array}{llcccl} & & \underline{i=0} & \underline{i=1} & \underline{i=2} & \\ \underline{k=0}: & L_0(3.5) = & \cancel{\left(\frac{3.5-1}{1-1}\right)} & \left(\frac{3.5-4}{1-4}\right) & \left(\frac{3.5-6}{1-6}\right) & = 0.08333 \\ \underline{k=1}: & L_1(3.5) = & \left(\frac{3.5-1}{4-1}\right) & \cancel{\left(\frac{3.5-4}{4-4}\right)} & \left(\frac{3.5-6}{4-6}\right) & = 1.04167 \\ \underline{k=2}: & L_2(3.5) = & \left(\frac{3.5-1}{6-1}\right) & \left(\frac{3.5-4}{6-4}\right) & \cancel{\left(\frac{3.5-6}{6-6}\right)} & = -0.12500 \end{array}$$

$$\begin{aligned} f(x^*) &= (1.5709)(0.08333) + (1.5727)(1.04167) \\ &\quad + (1.5751)(-0.12500) \\ &= 1.57225 \end{aligned}$$

trix form. The tensor definition for stress at a point is

$$\begin{pmatrix} \sigma_{xx} & \sigma_{xy} & \sigma_{xz} \\ \sigma_{yx} & \sigma_{yy} & \sigma_{yz} \\ \sigma_{zx} & \sigma_{zy} & \sigma_{zz} \end{pmatrix} \quad 1.158$$

## 10 PLANES

A plane **P** is uniquely determined by one of three combinations of parameters:

1. three non-collinear points in space
2. two non-parallel vectors ($\mathbf{V}_1$and$\mathbf{V}_2$) and their intersection point $p_0$
3. a point $p_0$ and a normal vector **N**

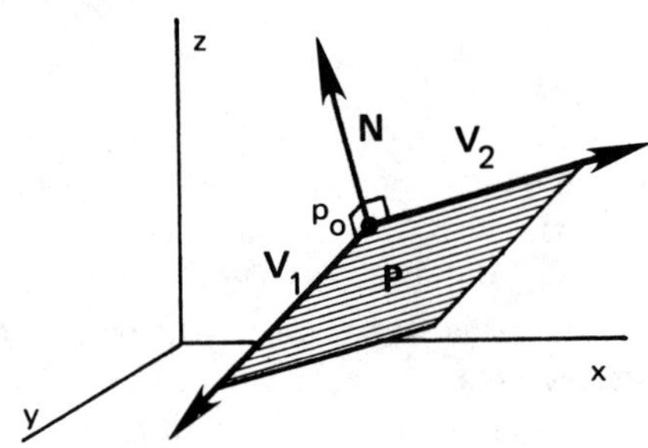


**Figure 1.7** A Plane in 3-Space

The plane consists of all points such that the coordinates can be written as a linear combination of $\mathbf{V}_1$ and $\mathbf{V}_2$. That is, points in the plane can be written as

$$(x, y, z) = s\mathbf{V}_1 + t\mathbf{V}_2 \quad 1.159$$

where $s$ and $t$ are constants and

$$\mathbf{V}_1 = a_1\mathbf{i} + b_1\mathbf{j} + c_1\mathbf{k} \quad 1.160$$

$$\mathbf{V}_2 = a_2\mathbf{i} + b_2\mathbf{j} + c_2\mathbf{k} \quad 1.161$$

If the intersection point $p_0 = (x_0, y_0, z_0)$ is known, then points in the plane can be represented by the parametric equations given below. Notice the similarity to the slope form of an equation for a straight line.

$$x = sa_1 + ta_2 + x_0 \quad 1.162$$

$$y = sb_1 + tb_2 + y_0 \quad 1.163$$

$$z = sc_1 + tc_2 + z_0 \quad 1.164$$

The plane also is defined by its rectangular equations:

$$A(x - x_0) + B(y - y_0) + C(z - z_0) = 0 \quad 1.165$$

or

$$Ax + By + Cz + D = 0 \quad 1.166$$

where

$$D = -(Ax_0 + By_0 + Cz_0) \quad 1.167$$

Constants A, B, and C are found from the cross product giving the normal vector **N**.

$$\mathbf{N} = \mathbf{V}_1 \times \mathbf{V}_2 = A\mathbf{i} + B\mathbf{j} + C\mathbf{k} \quad 1.168$$

*Example 1.23*

A plane is defined by a point $(2, 1, -4)$ and two vectors:

$$\mathbf{V}_1 = (2\mathbf{i} - 3\mathbf{j} + \mathbf{k}) \qquad \mathbf{V}_2 = (2\mathbf{j} - 4\mathbf{k})$$

Find the parametric and rectangular plane equations.

The parametric equations (for any values of $s$ and $t$) are:

$$x = 2 + 2s$$
$$y = 1 - 3s + 2t$$
$$z = -4 + s - 4t$$

The normal vector is found by evaluating the determinant

$$\mathbf{N} = \begin{vmatrix} \mathbf{i} & 2 & 0 \\ \mathbf{j} & -3 & 2 \\ \mathbf{k} & 1 & -4 \end{vmatrix}$$

$$= \mathbf{i}\,(12 - 2) - 2\,(-4\mathbf{j} - 2\mathbf{k}) = 10\mathbf{i} + 8\mathbf{j} + 4\mathbf{k}$$

One form of the rectangular equation is

$$10(x - 2) + 8(y - 1) + 4(z + 4) = 0$$

Another form can be derived from equations 1.166 and 1.167

$$D = -[(10)(2) + (8)(1) + (4)(-4)] = -12$$
$$\mathbf{P} = 10x + 8y + 4z - 12 = 0$$

Three noncollinear points can be used to describe a plane with the following procedure:

*step 1*: Form vectors $\mathbf{V}_1$ and $\mathbf{V}_2$ from two pairs of the points.

*step 2*: Find the normal vector $\mathbf{N} = \mathbf{V}_1 \times \mathbf{V}_2$.

*step 3*: Write the rectangular form of the plane using A, B, and C from the normal vector and any one of the three points.

If the rectangular form of the plane is known, it can be used to write parametric equations. In this case, two of the three variables $(x, y, z)$ replace the parameters $s$ and $t$.

*Example 1.24*

Find the rectangular and parametric equations of a plane containing the following points: $(2,1,-4)$; $(4,-2,-3)$; $(2,3,-8)$.

Use the first two points to find $\mathbf{V}_1$:

$$\begin{aligned}\mathbf{V}_1 &= (4-2)\mathbf{i} + (-2-1)\mathbf{j} + (-3-(-4))\mathbf{k}\\ &= 2\mathbf{i} - 3\mathbf{j} + \mathbf{k}\end{aligned}$$

Similarly,

$$\begin{aligned}\mathbf{V}_2 &= (2-2)\mathbf{i} + (3-1)\mathbf{j} + (-8-(-4))\mathbf{k}\\ &= 2\mathbf{j} - 4\mathbf{k}\end{aligned}$$

From the previous example,

$$\begin{aligned}\mathbf{N} &= 10\mathbf{i} + 8\mathbf{j} + 4\mathbf{k}\\ \mathbf{P} &= 10x + 8y + 4z - 12 = 0\end{aligned}$$

Dividing the rectangular form by 4 gives

$$2.5x + 2y + z - 3 = 0$$

or

$$z = 3 - 2y - 2.5x$$

Using $x$ and $y$ as the parameters, the parametric equations are

$$\begin{aligned}x &= x\\ y &= y\\ z &= 3 - 2y - 2.5x\end{aligned}$$

The angle between two planes is the same as the angle between their normal vectors, as calculated from the following equation:

$$\begin{aligned}\cos\phi &= \frac{|\mathbf{N}_1 \cdot \mathbf{N}_2|}{|\mathbf{N}_1||\mathbf{N}_2|}\\ &= \frac{|A_1A_2 + B_1B_2 + C_1C_2|}{\sqrt{A_1^2+B_1^2+C_1^2}\sqrt{A_2^2+B_2^2+C_2^2}}\end{aligned} \quad 1.169$$

A vector equation of the line formed by the intersection of two planes is given by the cross product $(\mathbf{N}_1 \times \mathbf{N}_2)$. The distance from a point $(x', y', z')$ to a plane is given by

$$d = \frac{Ax' + By' + Cz' + D}{\sqrt{A^2+B^2+C^2}} \quad 1.170$$

## 11 CONIC SECTIONS

### A. CIRCLE

The center-radius form of a circle with radius $r$ and center at $(h,k)$ is

$$(x-h)^2 + (y-k)^2 = r^2 \quad 1.171$$

The $x$-intercept is found by letting $y = 0$ and solving for $x$. The $y$-intercept is found similarly.

The general form is

$$x^2 + y^2 + Dx + Ey + F = 0 \quad 1.172$$

This can be converted to the center-radius form.

$$\left(x + \frac{D}{2}\right)^2 + \left(y + \frac{E}{2}\right)^2 = \frac{1}{4}(D^2 + E^2 - 4F) \quad 1.173$$

If the right-hand side is greater than zero, the equation is that of a circle with center at $(-\frac{1}{2}D, -\frac{1}{2}E)$ and radius given by the square root of the right-hand side. If the right-hand side is zero, the equation is that of a point. If the right-hand side is negative, the plot is imaginary.

### B. PARABOLA

A parabola is formed by a locus of points equidistant from point F and the *directrix*.

$$(y-k)^2 = 4p(x-h) \quad 1.174$$

Equation 1.174 represents a parabola with *vertex* at $(h,k)$, focus at $(p+h,k)$, and directrix equation $x = h-p$. The parabola points to the left if $p > 0$ and points to the right if $p < 0$.

$$(x-h)^2 = 4p(y-k) \quad 1.175$$

Equation 1.175 represents a parabola with vertex at $(h,k)$, focus at $(h, p+k)$, and directrix equation $y = k-p$. The parabola points down if $p > 0$ and points up if $p < 0$.

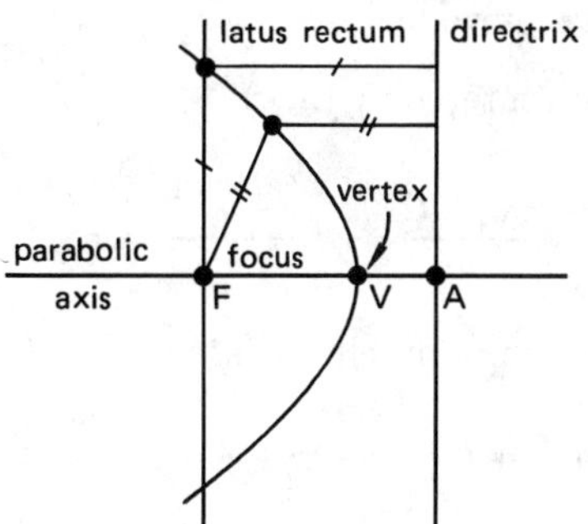


**Figure 1.8** A Parabola

An alternate form of the vertically-oriented parabola is

$$y = Ax^2 + Bx + C \quad 1.176$$

This parabola has a vertex at

$$\left(\frac{-B}{2A}, C - \frac{B^2}{4A}\right)$$

and points down if $A > 0$ and points up if $A < 0$.

C. ELLIPSE

An ellipse is formed from a locus of points such that the sum of distances from the two foci is constant. The distance between the two foci is $2c$. The sum of those distances is

$$F_1P + PF_2 = 2a \qquad 1.177$$

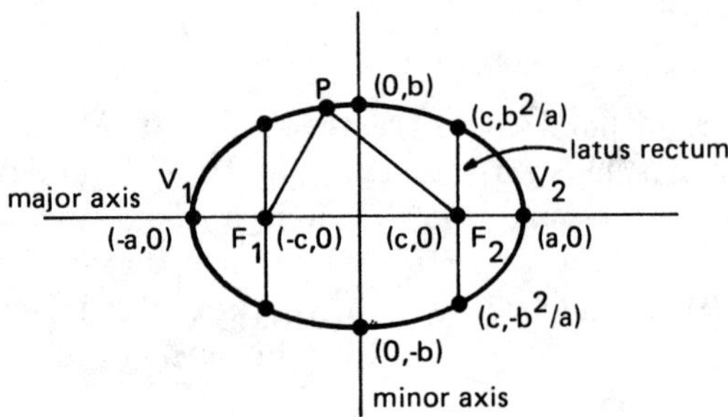


**Figure 1.9** An Ellipse

The eccentricity of an ellipse is less than 1, and is equal to

$$e = \frac{\sqrt{a^2 - b^2}}{a} \qquad 1.178$$

For an ellipse centered at the origin,

$$\left(\frac{x}{a}\right)^2 + \left(\frac{y}{b}\right)^2 = 1 \qquad 1.179$$

$$b^2 = a^2 - c^2 \qquad 1.180$$

If $a > b$, the ellipse is wider than it is tall. If $a < b$, it is taller than it is wide.

For an ellipse centered at $(h, k)$,

$$\frac{(x-h)^2}{a^2} + \frac{(y-k)^2}{b^2} = 1 \qquad 1.181$$

The general form of an ellipse is

$$Ax^2 + Cy^2 + Dx + Ey + F = 0 \qquad 1.182$$

If $A \neq C$ and both have the same sign, the general form can be written as

$$A\left(x + \frac{D}{2A}\right)^2 + C\left(y + \frac{E}{2C}\right)^2 = M \qquad 1.183$$

$$M = \frac{D^2}{4A} + \frac{E^2}{4C} - F \qquad 1.184$$

If $M = 0$, the graph is a single point at

$$\left(\frac{-D}{2A}, \frac{-E}{2C}\right).$$

If $M < 0$, the graph is the null set.

If $M > 0$, then the ellipse is centered at

$$\left(-\frac{D}{2A}, -\frac{E}{2C}\right)$$

and the equation can be rewritten

$$\frac{\left(x + \frac{D}{2A}\right)^2}{\frac{M}{A}} + \frac{\left(y + \frac{E}{2C}\right)^2}{\frac{M}{C}} = 1 \qquad 1.185$$

D. HYPERBOLA

A hyperbola is a locus of points such that $F_1P - PF_2 = 2a$. The distance between the foci is $2c$, and $a < c$.

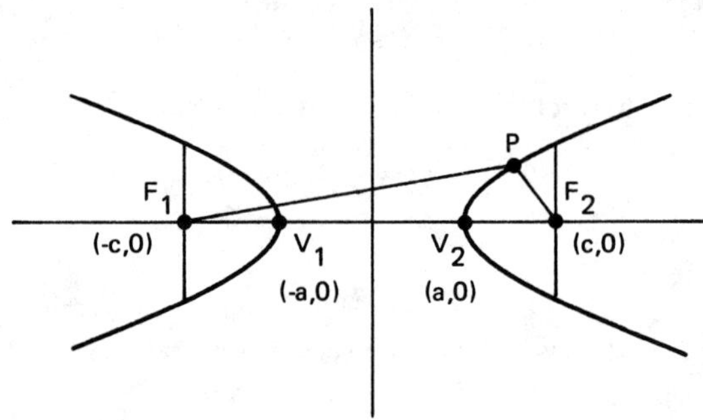


**Figure 1.10** A Hyperbola

For a hyperbola centered at the origin with foci on the x-axis,

$$\left(\frac{x}{a}\right)^2 - \left(\frac{y}{b}\right)^2 = 1 \quad \text{with} \quad b^2 = c^2 - a^2 \qquad 1.186$$

If the foci are on the y-axis,

$$\left(\frac{y}{a}\right)^2 - \left(\frac{x}{b}\right)^2 = 1 \qquad 1.187$$

The coordinates and length of the *latus recta* are the same as for the ellipse. The hyperbola is asymptotic to the lines

$$y = \pm\left(\frac{b}{a}\right)x \qquad 1.188$$

The asymptotes need not be perpendicular, but if they are, the hyperbola is known as a *rectangular hyperbola.* If the asymptotes are the $x$ and $y$ axes, the equation of the hyperbola is

$$xy = \pm a^2 \qquad 1.189$$

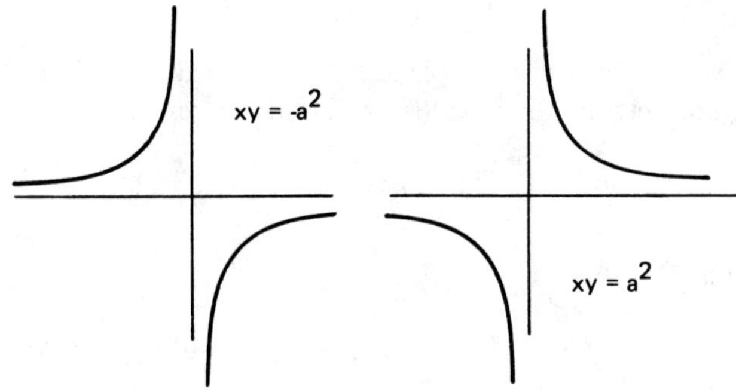


**Figure 1.11** Rectangular Hyperbolas

In general, for a hyperbola with transverse axis parallel to the x-axis and center at $(h, k)$,

$$\frac{(x-h)^2}{a^2} - \frac{(y-k)^2}{b^2} = 1 \qquad 1.190$$

The general form of the hyperbolic equation is

$$Ax^2 + Cy^2 + Dx + Ey + F = 0 \qquad 1.191$$

If $AC < 0$, the equation can be rewritten as

$$A\left(x + \frac{D}{2A}\right)^2 + C\left(y + \frac{E}{2C}\right)^2 = M \qquad 1.192$$

where

$$M = \frac{D^2}{4A} + \frac{E^2}{4C} - F \qquad 1.193$$

If $M = 0$, the graph is two intersecting lines.

If $M \neq 0$, the graph is a hyperbola with center at

$$\left(-\frac{D}{2A}, -\frac{E}{2C}\right)$$

The transverse axis is horizontal if $(M/A)$ is positive. It is vertical if $(M/C)$ is positive.

## 12 SPHERES

The equation of a sphere whose center is at the point $(h, k, l)$ and whose radius is $r$ is

$$(x-h)^2 + (y-k)^2 + (z-l)^2 = r^2 \qquad 1.194$$

If the sphere is centered at the origin, its equation is

$$x^2 + y^2 + z^2 = r^2 \qquad 1.195$$

## 13 PERMUTATIONS AND COMBINATIONS

Suppose you have $n$ objects, and you wish to work with a subset of $r$ of them. An order-conscious arrangement of $n$ objects taken $r$ at a time is known as *permutation*. The permutation is said to be order-conscious because the arrangement of two objects (say A and B) as AB is different from the arrangement BA. There are a number of ways of taking $n$ objects $r$ at a time. The total number of possible permutations is

$$P(n, r) = \frac{n!}{(n-r)!} \qquad 1.196$$

*Example 1.25*

A shelf has room for only three vases. If four different vases are available, how many ways can the shelf be arranged?

$$P(4, 3) = \frac{4!}{(4-3)!} = \frac{(4)(3)(2)(1)}{(1)} = 24$$

The special cases of $n$ objects taken $n$ at a time are illustrated by the following examples.

*Example 1.26*

How many ways can seven resistors be connected end-to-end into a single unit?

$$P(7, 7) = \frac{7!}{(7-7)!} = \frac{7!}{0!} = 7! = 5040$$

*Example 1.27*

Five people are to sit at a round table with five chairs. How many ways can these five people be arranged so that they all have different companions?

This is known as a *ring permutation.* Since the starting point of the arrangement around the circle does not affect the number of permutations, the answer is

$$(5-1)! = 4! = 24$$

An arrangement of $n$ objects taken $r$ at a time is known as a *combination* if the arrangement is not order-conscious. The total number of possible combinations is

$$C(n, r) = \frac{n!}{(n-r)!r!} \qquad 1.197$$

*Example 1.28*

How many possible ways can six people fit into a four-seat boat?

$$C(6, 4) = \frac{6!}{(6-4)!4!} = \frac{(6)(5)(4)(3)(2)(1)}{(2)(1)(4)(3)(2)(1)} = 15$$

## 14 PROBABILITY AND STATISTICS

### A. PROBABILITY RULES

The following rules are applied to sample spaces **A** and **B**:

$$\mathbf{A} = [A_1, A_2, A_3, \ldots, A_n] \text{ and } \mathbf{B} = [B_1, B_2, B_3, \ldots, B_n]$$

where the $A_i$ and $B_i$ are independent.

Rule 1:

$$p\{\emptyset\} = \text{probability of an impossible event} = 0 \quad 1.198$$

*Example 1.29*

An urn contains five white balls, two red balls, and three green balls. What is the probability of drawing a blue ball from the urn?

$$p\{\text{blue ball}\} = p\{\emptyset\} = 0$$

Rule 2:

$$p\{A_1 \text{ or } A_2 \text{ or } \ldots \text{ or } A_n\} = p\{A_1\} + p\{A_2\} + \cdots + p\{A_n\} \quad 1.199$$

*Example 1.30*

Returning to the urn described in example 1.29, what is the probability of getting either a white ball or a red ball in one draw from the urn?

$$p\{\text{red or white}\} = p\{\text{red}\} + p\{\text{white}\} = 0.2 + 0.5 = 0.7$$

Rule 3:

$$p\{A_i \text{ and } B_i \text{ and } \ldots Z_i\} = p\{A_i\}p\{B_i\}\ldots p\{Z_i\} \quad 1.200$$

*Example 1.31*

Given two identical urns (as described in example 1.29), what is the probability of getting a red ball from the first urn and a green ball from the second urn, given one draw from each urn?

$$p\{\text{red and green}\} = p\{\text{red}\}\,p\,\{\text{green}\} = (0.2)(0.3) = 0.06$$

Rule 4:

$$p\{\text{not A}\} = \text{probability of event A not occurring} = 1 - p\{A\} \quad 1.201$$

*Example 1.32*

Given the urn of example 1.29, what is the probability of not getting a red ball from the urn in one draw?

$$p\{\text{not red}\} = 1 - p\{\text{red}\} = 1 - 0.2 = 0.8$$

Rule 5:

$$p\{A_i \text{ or } B_i\} = p\{A_i\} + p\{B_i\} - p\{A_i\}p\{B_i\} \quad 1.202$$

*Example 1.33*

Given one urn as described in example 1.29 and a second urn containing eight red balls and two black balls, what is the probability of drawing either a white ball from the first urn or a red ball from the second urn, given one draw from each?

$$\begin{aligned} p\{\text{white or red}\} &= p\{\text{white}\} + p\{\text{red}\} \\ &\quad - p\{\text{white}\}p\{\text{red}\} \\ &= 0.5 + 0.8 - (0.5)(0.8) = 0.9 \end{aligned}$$

Rule 6:

$p\{A|B\}$ = probability that A will occur given that B has already occurred, where the two events are dependent.

$$= \frac{p\{A \text{ and } B\}}{p\{B\}} \quad 1.203$$

The above equation is known as *Bayes Theorem.*

### B. PROBABILITY DENSITY FUNCTIONS

Probability density functions are mathematical functions giving the probabilities of numerical events. A *numerical event* is any occurrence that can be described by an integer or real number. For example, obtaining heads in a coin toss is not a numerical event. However, a concrete sample having a compressive strength less than 5000 psi is a numerical event.

Discrete density functions give the probability that the event $x$ will occur. That is,

$$f\{x\} = \text{probability of a process having a value of } x$$

Important discrete functions are the binomial and Poisson distributions.

*1. Binomial*

$n$ is the number of trials

$x$ is the number of successes

$p$ is the probability of a success in a single trial

$q$ is the probability of failure, $1 - p$

$\binom{n}{x}$ is the binomial coefficient $= \frac{n!}{(n-x)!x!}$

$x! = x(x-1)(x-2)\cdots(2)(1)$

Then, the probability of obtaining $x$ successes in $n$ trials is

$$f\{x\} = \binom{n}{x} p^x q^{(n-x)} \qquad 1.204$$

The mean of the binomial distribution is $np$. The variance of the distribution is $npq$.

*Example 1.34*

In a large quantity of items, 5% are defective. If seven items are sampled, what is the probability that exactly three will be defective?

$$f\{3\} = \binom{7}{3}(0.05)^3(0.95)^4 = 0.0036$$

*2. Poisson*

Suppose an event occurs, on the average, $\lambda$ times per period. The probability that the event will occur $x$ times per period is

$$f\{x\} = \frac{e^{-\lambda}\lambda^x}{x!} \qquad 1.205$$

$\lambda$ is both the distribution mean and the variance. $\lambda$ must be a number greater than zero.

*Example 1.35*

The number of customers arriving in some period is distributed as Poisson with a mean of eight. What is the probability that six customers will arrive in any given period?

$$f\{6\} = \frac{e^{-8}8^6}{6!} = 0.122$$

Continuous probability density functions are used to find the cumulative distribution functions, $F\{x\}$. Cumulative distribution functions give the probability of event $x$ or less occurring.

$x =$ any value, not necessarily an integer

$$f\{x\} = \frac{dF\{x\}}{dx} \qquad 1.206$$

$F\{x\} =$ probability of $x$ or less occurring

*3. Exponential*

$$f\{x\} = u(e^{-ux}) \qquad 1.207$$

$$F\{x\} = 1 - e^{-ux} \qquad 1.208$$

The mean of the exponential distribution is $\frac{1}{u}$. The variance is $\left(\frac{1}{u}\right)^2$.

*Example 1.36*

The reliability of a unit is exponentially distributed with mean time to failure (MTBF) of 1000 hours. What is the probability that the unit will be operational at $t = 1200$ hours?

The reliability of an item is (1− probability of failing before time $t$). Therefore,

$$R\{t\} = 1 - F\{t\} = 1 - (1 - e^{-ux}) = e^{-ux}$$

$$u = \frac{1}{\textbf{MTBF}} = \frac{1}{1000} = 0.001$$

$$R\{1200\} = e^{-(0.001)(1200)} = 0.3$$

*4. Normal*

Although $f\{x\}$ can be expressed mathematically for the normal distribution, tables are used to evaluate $F\{x\}$ since $f\{x\}$ cannot be easily integrated. Since the $x$ axis of the normal distribution will seldom correspond to actual sample variables, the sample values are converted into standard values. Given the mean, $u$, and the standard deviation, $\sigma'$, the standard normal variable is

$$z = \frac{\text{sample value} - u}{\sigma'} \qquad 1.209$$

Then, the probability of a sample exceeding the given sample value is equal to the area in the tail past point $z$.

*Example 1.37*

Given a population that is normally distributed with mean of 66 and standard deviation of five, what percent of the population exceeds 72?

$$z = \frac{72 - 66}{5} = 1.2$$

Then, from table 1.5,

$$p\{\text{exceeding } 72\} = 0.5 - 0.3849 = 0.1151 \text{ or } 11.5\%$$

## C. STATISTICAL ANALYSIS OF EXPERIMENTAL DATA

Experiments can take on many forms. An experiment might consist of measuring the weight of one cubic foot of concrete. Or, an experiment might consist of measuring the speed of a car on a roadway. Generally, such experiments are performed more than once to increase the precision and accuracy of the results.

Of course, the intrinsic variability of the process being measured will cause the observations to vary, and we would not expect the experiment to yield the same result each time it was performed. Eventually, a collection of experimental outcomes (observations) will be available for analysis.

One fundamental technique for organizing random observations is the *frequency distribution.* The frequency distribution is a systematic method for ordering the observations from small to large, according to some convenient numerical characteristic.

*Example 1.38*

The number of cars that travels through an intersection between 12 noon and 1 p.m. is measured for 30 consecutive working days. The results of the 30 observations are:

79, 66, 72, 70, 68, 66, 68, 76, 73, 71, 74, 70, 71, 69, 67, 74, 70, 68, 69, 64, 75, 70, 68, 69, 64, 69, 62, 63, 63, 61

What is the frequency distribution using an interval of 2 cars per hour?

| cars per hour | frequency of occurrence |
|---|---|
| 60–61 | 1 |
| 62–63 | 3 |
| 64–65 | 2 |
| 66–67 | 3 |
| 68–69 | 8 |
| 70–71 | 6 |
| 72–73 | 2 |
| 74–75 | 3 |
| 76–77 | 1 |
| 78–79 | 1 |

In example 1.38, two cars per hour is known as the *step interval.* The step interval should be chosen so that the data is presented in a meaningful manner. If there are too many intervals, many of them will have zero frequencies. If there are too few intervals, the frequency distribution will have little value. Generally, 10 to 15 intervals are used.

Once the frequency distribution is complete, it can be represented graphically as a histogram. The procedure in drawing a histogram is to mark off the interval limits on a number line and then draw bars with lengths that are proportional to the frequencies in the intervals. If it is necessary to show the continuous nature of the data, a frequency polygon can be drawn.

*Example 1.39*

Draw the frequency histogram and frequency polygon for the data given in example 1.38.

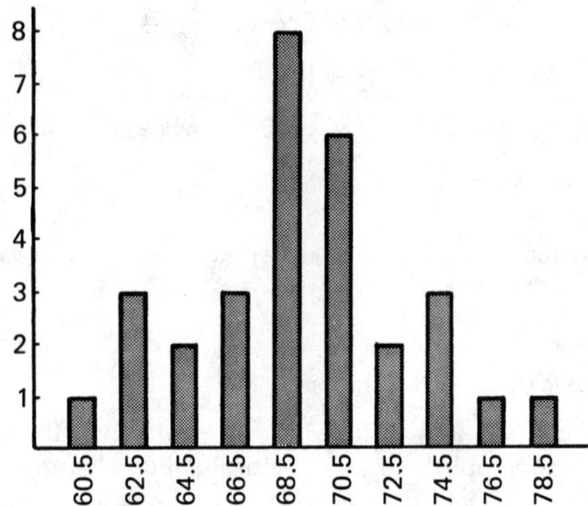


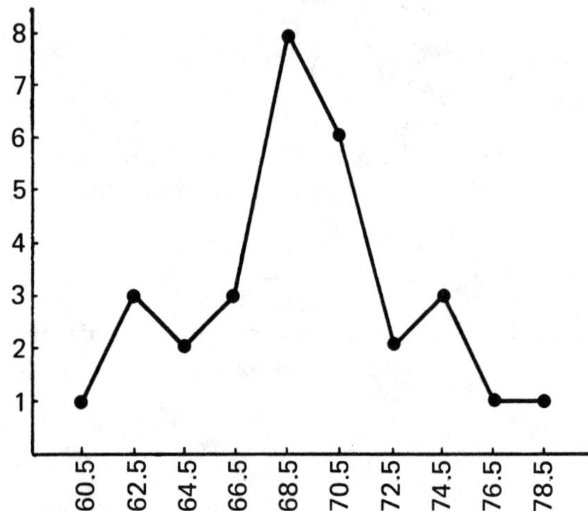


If it is necessary to know the number or percentage of observations that occur up to and including some value, the cumulative frequency table can be formed. This procedure is illustrated in the following example.

*Example 1.40*

Form the cumulative frequency distribution and graph for the data given in example 1.38.

| cars per hour | frequency | cumulative frequency | cumulative percent |
|---|---|---|---|
| 60–61 | 1 | 1 | 3 |
| 62–63 | 3 | 4 | 13 |
| 64–65 | 2 | 6 | 20 |
| 66–67 | 3 | 9 | 30 |
| 68–69 | 8 | 17 | 57 |
| 70–71 | 6 | 23 | 77 |
| 72–73 | 2 | 25 | 83 |
| 74–75 | 3 | 28 | 93 |
| 76–77 | 1 | 29 | 97 |
| 78–79 | 1 | 30 | 100 |

**Table 1.5**
Areas Under The Standard Normal Curve
(0 to z)

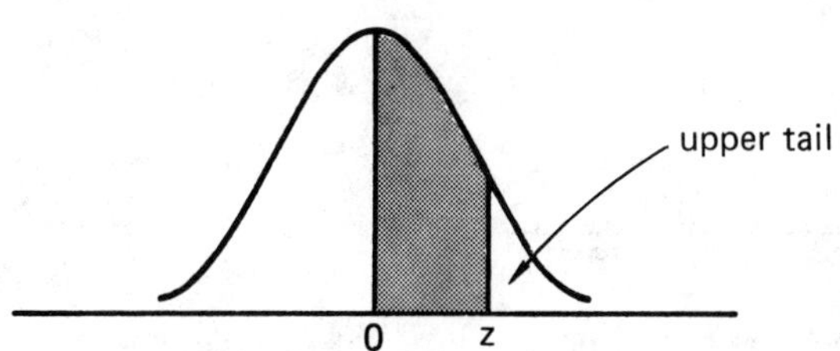


| z | 0 | 1 | 2 | 3 | 4 | 5 | 6 | 7 | 8 | 9 |
|---|---|---|---|---|---|---|---|---|---|---|
| 0.0 | .0000 | .0040 | .0080 | .0120 | .0160 | .0199 | .0239 | .0279 | .0319 | .0359 |
| 0.1 | .0398 | .0438 | .0478 | .0517 | .0557 | .0596 | .0636 | .0675 | .0714 | .0754 |
| 0.2 | .0793 | .0832 | .0871 | .0910 | .0948 | .0987 | .1026 | .1064 | .1103 | .1141 |
| 0.3 | .1179 | .1217 | .1255 | .1293 | .1331 | .1368 | .1406 | .1443 | .1480 | .1517 |
| 0.4 | .1554 | .1591 | .1628 | .1664 | .1700 | .1736 | .1772 | .1808 | .1844 | .1879 |
| 0.5 | .1915 | .1950 | .1985 | .2019 | .2054 | .2088 | .2123 | .2157 | .2190 | .2224 |
| 0.6 | .2258 | .2291 | .2324 | .2357 | .2389 | .2422 | .2454 | .2486 | .2518 | .2549 |
| 0.7 | .2580 | .2612 | .2642 | .2673 | .2704 | .2734 | .2764 | .2794 | .2823 | .2852 |
| 0.8 | .2881 | .2910 | .2939 | .2967 | .2996 | .3023 | .3051 | .3078 | .3106 | .3133 |
| 0.9 | .3159 | .3186 | .3212 | .3238 | .3264 | .3289 | .3315 | .3340 | .3365 | .3389 |
| 1.0 | .3413 | .3438 | .3461 | .3485 | .3508 | .3531 | .3554 | .3577 | .3599 | .3621 |
| 1.1 | .3643 | .3665 | .3686 | .3708 | .3729 | .3749 | .3770 | .3790 | .3810 | .3830 |
| 1.2 | .3849 | .3869 | .3888 | .3907 | .3925 | .3944 | .3962 | .3980 | .3997 | .4015 |
| 1.3 | .4032 | .4049 | .4066 | .4082 | .4099 | .4115 | .4131 | .4147 | .4162 | .4177 |
| 1.4 | .4192 | .4207 | .4222 | .4236 | .4251 | .4265 | .4279 | .4292 | .4306 | .4319 |
| 1.5 | .4332 | .4345 | .4357 | .4370 | .4382 | .4394 | .4406 | .4418 | .4429 | .4441 |
| 1.6 | .4452 | .4463 | .4474 | .4484 | .4495 | .4505 | .4515 | .4525 | .4535 | .4545 |
| 1.7 | .4554 | .4564 | .4573 | .4582 | .4591 | .4599 | .4608 | .4616 | .4625 | .4633 |
| 1.8 | .4641 | .4649 | .4656 | .4664 | .4671 | .4678 | .4686 | .4693 | .4699 | .4706 |
| 1.9 | .4713 | .4719 | .4726 | .4732 | .4738 | .4744 | .4750 | .4756 | .4761 | .4767 |
| 2.0 | .4772 | .4778 | .4783 | .4788 | .4793 | .4798 | .4803 | .4808 | .4812 | .4817 |
| 2.1 | .4821 | .4826 | .4830 | .4834 | .4838 | .4842 | .4846 | .4850 | .4854 | .4857 |
| 2.2 | .4861 | .4864 | .4868 | .4871 | .4875 | .4878 | .4881 | .4884 | .4887 | .4890 |
| 2.3 | .4893 | .4896 | .4898 | .4901 | .4904 | .4906 | .4909 | .4911 | .4913 | .4916 |
| 2.4 | .4918 | .4920 | .4922 | .4925 | .4927 | .4929 | .4931 | .4932 | .4934 | .4936 |
| 2.5 | .4938 | .4940 | .4941 | .4943 | .4945 | .4946 | .4948 | .4949 | .4951 | .4952 |
| 2.6 | .4953 | .4955 | .4956 | .4957 | .4959 | .4960 | .4961 | .4962 | .4963 | .4964 |
| 2.7 | .4965 | .4966 | .4967 | .4968 | .4969 | .4970 | .4971 | .4972 | .4973 | .4974 |
| 2.8 | .4974 | .4975 | .4976 | .4977 | .4977 | .4978 | .4979 | .4979 | .4980 | .4981 |
| 2.9 | .4981 | .4982 | .4982 | .4983 | .4984 | .4984 | .4985 | .4985 | .4986 | .4986 |
| 3.0 | .4987 | .4987 | .4987 | .4988 | .4988 | .4989 | .4989 | .4989 | .4990 | .4990 |
| 3.1 | .4990 | .4991 | .4991 | .4991 | .4992 | .4992 | .4992 | .4992 | .4993 | .4993 |
| 3.2 | .4993 | .4993 | .4994 | .4994 | .4994 | .4994 | .4994 | .4995 | .4995 | .4995 |
| 3.3 | .4995 | .4995 | .4995 | .4996 | .4996 | .4996 | .4996 | .4996 | .4996 | .4997 |
| 3.4 | .4997 | .4997 | .4997 | .4997 | .4997 | .4997 | .4997 | .4997 | .4997 | .4998 |
| 3.5 | .4998 | .4998 | .4998 | .4998 | .4998 | .4998 | .4998 | .4998 | .4998 | .4998 |
| 3.6 | .4998 | .4998 | .4999 | .4999 | .4999 | .4999 | .4999 | .4999 | .4999 | .4999 |
| 3.7 | .4999 | .4999 | .4999 | .4999 | .4999 | .4999 | .4999 | .4999 | .4999 | .4999 |
| 3.8 | .4999 | .4999 | .4999 | .4999 | .4999 | .4999 | .4999 | .4999 | .4999 | .4999 |
| 3.9 | .5000 | .5000 | .5000 | .5000 | .5000 | .5000 | .5000 | .5000 | .5000 | .5000 |

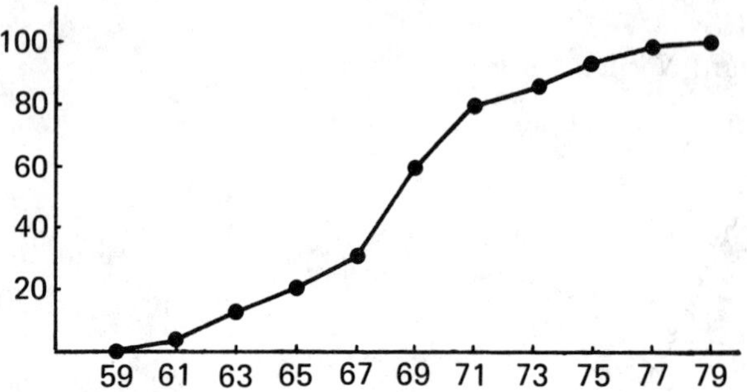


It is often unnecessary to present the experimental data in its entirety, either in tabular or graphical form. In such cases, the data and distribution can be represented by various parameters. One type of parameter is a measure of *central tendency.* Mode, median, and mean are measures of central tendency. The other type of parameter is a measure of dispersion. Standard deviation and variance are measures of dispersion.

The *mode* is the observed value which occurs most frequently. The mode may vary greatly between series of observations. Therefore, its main use is as a quick measure of the central value since no computation is required to find it. Beyond this, the usefulness of the mode is limited.

The *median* is the point in the distribution which divides the total observations into two parts containing equal numbers of observations. It is not influenced by the extremity of scores on either side of the distribution. The median is found by counting up (from either end of the frequency distribution) until half of the observations have been accounted for. The procedure is more difficult if the median falls within an interval, as illustrated in example 1.41.

Similar in concept to the median are *percentile ranks, quartiles,* and *deciles.* The median could also have been called the *50th percentile* observation. Similarly, the 80th percentile would be the number of cars per hour for which the cumulative frequency was 80%. The quartile and decile points on the distribution divide the observations or distribution into segments of 25% and 10%, respectively.

The *arithmetic mean* is the arithmetic average of the observations. The *mean* may be found without ordering the data (which was necessary to find the mode and median). The mean can be found from the following formula:

$$\overline{x} = \left(\frac{1}{n}\right)(x_1 + x_2 + \cdots + x_n) = \frac{\sum x_i}{n} \qquad 1.210$$

The *geometric mean* is used occasionally when it is necessary to average ratios. The geometric mean is calculated as

$$\text{geometric mean} = \sqrt[n]{x_1 x_2 x_3 \ldots x_n} \qquad 1.211$$

The *harmonic mean* is defined as

$$\text{harmonic mean} = \frac{n}{\frac{1}{x_1} + \frac{1}{x_2} + \cdots + \frac{1}{x_n}} \qquad 1.212$$

The *root-mean-squared (rms) value* of a series of observations is defined as

$$x_{rms} = \sqrt{\frac{\sum x_i^2}{n}} \qquad 1.213$$

*Example 1.41*

Find the mode, median, and arithmetic mean of the distribution represented by the data given in example 1.38.

The mode is the interval 68–69, since this interval has the highest frequency. If 68.5 is taken as the interval center, then 68.5 would be the mode.

Since there are 30 observations, the median is the value which separates the observations into two groups of 15. From example 1.40, the median occurs someplace within the 68–69 interval. Up through interval 66–67, there are nine observations, so six more are needed to make 15. Interval 68–69 has eight observations, so the median is found to be $\left(\frac{6}{8}\right)$ or $\left(\frac{3}{4}\right)$ of the way through the interval. Since the real limits of the interval are 67.5 and 69.5, the median is located at

$$67.5 + \frac{3}{4}(69.5 - 67.5) = 69$$

The mean can be found from the raw data or from the grouped data using the interval center as the assumed observation value. Using the raw data,

$$\overline{x} = \frac{\sum x}{n} = \frac{2069}{30} = 68.97$$

The simplest statistical parameter which describes the variation in observed data is the *range.* The range is found by subtracting the smallest value from the largest. Since the range is influenced by extreme (low probability) observations, its use as a measure of variability is limited.

The *standard deviation* is a better estimate of variability because it considers every observation. The standard deviation can be found from:

$$\sigma = \sqrt{\frac{\sum (x_i - \overline{x})^2}{n}} = \sqrt{\frac{\sum x_i^2}{n} - (\overline{x})^2} \qquad 1.214$$

The above formula assumes that $n$ is a large number, such as above 50. Theoretically, $n$ is the size of the entire population. If a small sample (less than 50) is used to calculate the standard deviation of the distribution, the formulas are changed. The *sample standard deviation* is

$$s = \sqrt{\frac{\sum (x_i - \overline{x})^2}{n-1}} = \sqrt{\frac{\sum x_i^2 - \frac{\left(\sum x_i\right)^2}{n}}{n-1}} \qquad 1.215$$

The difference is small when $n$ is large, but care must be taken in reading the problem. If the *standard deviation of the sample* is requested, calculate $\sigma$. If an estimate of the *population standard deviation* or *sample standard deviation* is requested, calculate $s$. (Note that the standard deviation of the sample is not the same as the sample standard deviation.)

The *relative dispersion* is defined as a measure of dispersion divided by a measure of central tendency. The *coefficient of variation* is a relative dispersion calculated from the standard deviation and the mean. That is,

$$\text{coefficient of variation} = \frac{s}{\overline{x}} \qquad 1.216$$

*Skewness* is a measure of a frequency distribution's lack of symmetry. It is calculated as

$$\text{skewness} = \frac{\overline{x} - \text{mode}}{s} \qquad 1.217$$

$$\approx \frac{3(\overline{x} - \text{median})}{s} \qquad 1.218$$

*Example 1.42*

Calculate the range, standard deviation of the sample, and population variance from the data given in example 1.38.

$$\sum x = 2069 \quad \left(\sum x\right)^2 = 4280761 \quad \sum x^2 = 143225$$

$$n = 30 \quad \overline{x} = 68.97$$

$$\sigma = \sqrt{\frac{143225}{30} - (68.97)^2} = 4.16$$

$$s = \sqrt{\frac{143225 - \frac{(4280761)}{30}}{29}} = 4.29$$

$$s^2 = 18.4 \quad \text{(sample variance)}$$

$$\sigma^2 = 17.3 \quad \text{(population variance)}$$

$$R = 79 - 61 = 18$$

Referring again to example 1.38, suppose that the hourly through-put for 15 similar intersections is measured over a 30 day period. At the end of the 30 day period, there will be 15 ranges, 15 medians, 15 means, 15 standard deviations, and so on. These parameters themselves constitute distributions.

The mean of the sample means is an excellent estimator of the average hourly through-put of an intersection, $\mu$.

$$\mu = \left(\frac{1}{15}\right) \sum \overline{x}$$

The standard deviation of the sample means is known as the *standard error of the mean* to distinguish it from the standard deviation of the raw data. The standard error is written as $\sigma_{\overline{x}}$.

The standard error is not a good estimator of the population standard deviation, $\sigma'$.

In general, if $k$ sets of $n$ observations each are used to estimate the population mean ($\mu$) and the population standard deviation ($\sigma'$), then

$$\mu \approx \left(\frac{1}{k}\right) \sum \overline{x} \qquad 1.219$$

$$\sigma' \approx \sqrt{k}\sigma_{\overline{x}} \qquad 1.220$$

## 15 BASIC HYPOTHESIS TESTING

Suppose a distribution is $\sim N(\mu, \sigma'^2)$.[1] If samples of size $n$ are taken $k$ times, the values of the sample means $\overline{x}$ will form a distribution themselves. These means also will be distributed normally with the form

$$\sim N\left(\mu, \frac{\sigma'^2}{n}\right)$$

That is, the mean of the sample means will be identical to the original population, but the variance and standard deviation will be much smaller. This is known as the *Central Limit Theorem.*

Thus, the probability that $\overline{x}$ exceeds some value, say $x^*$, is

$$P\left\{z > \frac{x^* - \mu}{\frac{\sigma'}{\sqrt{n}}}\right\} \qquad 1.221$$

This can be solved as an *exceedance problem* (see example 1.37), or a hypothesis test can be performed. A *hypothesis test* has the following characteristics:

- a sample is taken in an experiment
- a parameter (usually $\overline{x}$) is measured
- it is desired to know if the sample could have come from a population $\sim N(\mu, \sigma'^2)$

There are many types of hypothesis tests, depending on the type of population (i.e., whether or not normal), the parameter being tested (i.e., central tendency or dispersion), and the size of the sample.

[1] This is the standard method of saying the distribution is normally distributed with mean $\mu$ and variance $\sigma'^2$.

If the sample size is not much greater than 30, if the native population is assumed to be normal, and if $\mu$ and $\sigma'$ are known, the following procedure can be used.

*step 1*: Assume random sampling from a normal population.

*step 2*: Choose the desired confidence level, C. Usually, a 95% confidence level result is said to be *significant.* 99% test results are said to be *highly significant.*

*step 3*: Decide on a 1-tail or 2-tail test. If the question is worded as "Has the population mean changed?" or "Are the populations the same?", then a *2-tail test* is needed. If the question is "Has the mean increased?" or "... decreased?", then a *1-tail test* is needed.

*step 4*: From the normal table, find the value $z'$ for a table entry equal to

$$\frac{1-C}{\#\text{tails in the test}} \qquad 1.222$$

*step 5*: Calculate

$$z = \left| \frac{\overline{x} - \mu}{\frac{\sigma'}{\sqrt{n}}} \right| \qquad 1.223$$

If $z \geq z'$, then the distributions are not the same.

*Example 1.43*

When operating properly, a chemical plant has a product output which is normally distributed with mean 880 tons/day and standard deviation of 21 tons. The output is measured on 50 consecutive days, and the mean output is 871 tons/day. Is the plant operating correctly?

*step 1*: Assume random sampling from the normal distribution

*step 2*: Choose $C = 0.95$ for significant results.

*step 3*: Wanting to know if the plant is operating correctly is the same as asking, "Has anything changed?" There is no mention of *direction* (i.e., the question was not, "Has the output decreased?"). Therefore, choose a 2-tail test.

*step 4*: $\frac{1}{2}(1-C) = 0.025$. The 0.025 outside lower limit in table 1.5 is $z' = 1.96$. (This corresponds to an area under the curve of $0.5 - 0.025 = 0.475$.)

*step 5*:

$$z = \left| \frac{871 - 880}{\frac{21}{\sqrt{50}}} \right| = 3.03$$

Since $3.03 > 1.96$, the distributions are not the same. There is a 95% chance that the plant is not operating correctly.

## 16 DIFFERENTIAL CALCULUS

### A. TERMINOLOGY

Given $y$, a function of $x$, the first derivative with respect to $x$ may be written as

$$\mathbf{D}y,\ y',\ \text{or}\ \left(\frac{dy}{dx}\right)$$

The first derivative corresponds to the slope of the line described by the function $y$. The second derivative may be written as

$$\mathbf{D}^2y,\ \left(\frac{d^2y}{dx^2}\right),\ \text{or}\ y''$$

### B. BASIC OPERATIONS

In the formulas that follow, $f$ and $g$ are functions of $x$. $\mathbf{D}$ is the derivative operator. $a$ is a constant.

$$\mathbf{D}(a) = 0 \qquad 1.224$$

$$\mathbf{D}(af) = a\mathbf{D}(f) \qquad 1.225$$

$$\mathbf{D}(f+g) = \mathbf{D}(f) + \mathbf{D}(g) \qquad 1.226$$

$$\mathbf{D}(f-g) = \mathbf{D}(f) - \mathbf{D}(g) \qquad 1.227$$

$$\mathbf{D}(f \cdot g) = f\mathbf{D}(g) + g\mathbf{D}(f) \qquad 1.228$$

$$\mathbf{D}\left(\frac{f}{g}\right) = \frac{g\mathbf{D}(f) - f\mathbf{D}(g)}{g^2} \qquad 1.229$$

$$\mathbf{D}(x^n) = nx^{n-1} \qquad 1.230$$

$$\mathbf{D}(f^n) = nf^{n-1}\mathbf{D}(f) \qquad 1.231$$

$$\mathbf{D}(f(g)) = \frac{df(g)}{dg}\mathbf{D}(g) \qquad 1.232$$

$$\mathbf{D}(lnx) = \frac{1}{x} \qquad 1.233$$

$$\mathbf{D}(e^{ax}) = ae^{ax} \qquad 1.234$$

*Example 1.44*

A function is given as $f(x) = x^3 - 2x$. What is the slope of the line at $x = 3$?

$$y' = 3x^2 - 2$$

$$y'(3) = 27 - 2 = 25$$

### C. TRANSCENDENTAL FUNCTIONS

$$\mathbf{D}(\sin x) = \cos x \qquad 1.235$$

$$\mathbf{D}(\cos x) = -\sin x \qquad 1.236$$

$$\mathbf{D}(\tan x) = \sec^2 x \qquad 1.237$$

$$\mathbf{D}(\cot x) = -\csc^2 x \qquad 1.238$$

$$\mathbf{D}(\sec x) = (\sec x)(\tan x) \quad 1.239$$

$$\mathbf{D}(\csc x) = (-\csc x)(\cot x) \quad 1.240$$

$$\mathbf{D}(\arcsin x) = \frac{1}{\sqrt{1-x^2}} \quad 1.241$$

$$\mathbf{D}(\arctan x) = \frac{1}{(1+x^2)} \quad 1.242$$

$$\mathbf{D}(\mathrm{arcsec}\, x) = \frac{1}{x\sqrt{x^2-1}} \quad 1.243$$

$$\mathbf{D}(\arccos x) = -\mathbf{D}(\arcsin x) \quad 1.244$$

$$\mathbf{D}(\mathrm{arccot}\, x) = -\mathbf{D}(\arctan x) \quad 1.245$$

$$\mathbf{D}(\mathrm{arccsc}\, x) = -\mathbf{D}(\mathrm{arcsec}\, x) \quad 1.246$$

## D. VARIATIONS ON DIFFERENTIATION

### *1. Partial Differentiation*

If the function has two or more independent variables, a partial derivative is found by considering all extraneous variables as constants. The geometric interpretation of the partial derivative $(\partial z/\partial x)$ is the slope of a line tangent to the 3-dimensional surface in a plane of constant $y$ and parallel to the $x$ axis. Similarly, the interpretation of $(\partial z/\partial y)$ is the slope of a line tangent to the surface in a plane of constant $x$ and parallel to the $y$ axis.

*Example 1.45*

A surface has the equation $x^2+y^2+z^2=9$. What is the slope of a line tangent to (1,2,2) and parallel to the $x$ axis?

$$z = \sqrt{9-x^2-y^2}$$

$$\frac{\partial z}{\partial x} = \frac{-x}{\sqrt{9-x^2-y^2}}$$

At the point (1,2,2),

$$\frac{\partial z}{\partial x} = -\frac{1}{2}$$

### *2. Implicit Differentiation*

If a relationship between $n$ variables cannot be manipulated to yield an explicit function of $(n-1)$ independent variables, the relationship implicitly defines the $n$th remaining variable. The derivative of the implicit variable taken with respect to any other variable is found by a process known as implicit differentiation.

If $f(x,y)=0$ is a function, the implicit derivative is

$$\frac{dy}{dx} = -\frac{\partial f}{\partial x} \bigg/ \frac{\partial f}{\partial y} \quad 1.247$$

If $f(x,y,z)=0$ is a function, the implicit derivatives are

$$\frac{\partial z}{\partial x} = -\frac{\partial f}{\partial x} \bigg/ \frac{\partial f}{\partial z} \quad 1.248$$

$$\frac{\partial z}{\partial y} = -\frac{\partial f}{\partial y} \bigg/ \frac{\partial f}{\partial z} \quad 1.249$$

*Example 1.46*

If $f = x^2+xy+y^3$, what is $\frac{dy}{dx}$?

Since this function cannot be written as an explicit function of $x$, implicit differentiation is required.

$$\frac{\partial f}{\partial x} = 2x+y$$

$$\frac{\partial f}{\partial y} = x+3y^2$$

$$\frac{dy}{dx} = \frac{-(2x+y)}{(x+3y^2)}$$

*Example 1.47*

Solve example 1.45 using implicit differentiation.

$$f = x^2+y^2+z^2-9$$

$$\frac{\partial f}{\partial x} = 2x$$

$$\frac{\partial f}{\partial z} = 2z$$

$$\frac{\partial z}{\partial x} = \frac{-2x}{2z} = -\frac{x}{z}$$

and at (1,2,2),

$$\frac{\partial z}{\partial x} = -\frac{1}{2}$$

### *3. The Gradient Vector*

The slope of a function is defined as the change in one variable with respect to a distance in another direction. Usually, this direction is parallel to an axis. However, the maximum slope at a point on a 3-dimensional object may not be in a direction parallel to one of the coordinate axes.

The gradient vector function $\nabla f(x,y,z)$ (pronounced "del f") gives the maximum rate of change of the function f(x,y,z). The gradient vector function is defined as

$$\nabla f(x,y,z) = \frac{\partial f(x,y,z)}{\partial x}\mathbf{i} + \frac{\partial f(x,y,z)}{\partial y}\mathbf{j} + \frac{\partial f(x,y,z)}{\partial z}\mathbf{k} \quad 1.250$$

*Example 1.48*

Find the maximum slope of $f(x,y) = 2x^2 - y^2 + 3x - y$ at the point $(1,-2)$. What is the equation of the maximum-slope tangent?

This is a 2-dimensional problem.

$$\frac{\partial f(x,y)}{\partial x} = 4x + 3$$
$$\frac{\partial f(x,y)}{\partial y} = -2y - 1$$
$$\nabla f(x,y) = (4x+3)\mathbf{i} + (-2y-1)\mathbf{j}$$

The equation of the maximum-slope tangent is

$$\nabla f(1,-2) = 7\mathbf{i} + 3\mathbf{j}$$

The magnitude of the slope is

$$\sqrt{(7)^2 + (3)^2} = \sqrt{58}$$

*4. The Directional Derivative*

The rate of change of a function in the direction of some given vector **U** can be found from the directional derivative function, $\nabla_u f(x,y,z)$. This directional derivative function depends on the gradient vector and the direction cosines of the vector **U**.

$$\nabla_u f(x,y,z) = \frac{\partial f(x,y,z)}{\partial x}\cos\alpha + \frac{\partial f(x,y,z)}{\partial y}\cos\beta + \frac{\partial f(x,y,z)}{\partial z}\cos\gamma \qquad 1.251$$

*Example 1.49*

What is the rate of change of $f(x,y) = 3x^2 + xy - 2y^2$ at the point $(1,-2)$ in the direction $4\mathbf{i} + 3\mathbf{j}$?

$$\cos\alpha = \frac{4}{\sqrt{(4)^2 + (3)^2}} = \frac{4}{5}$$
$$\cos\beta = \frac{3}{5}$$
$$\frac{\partial f(x,y)}{\partial x} = 6x + y$$
$$\frac{\partial f(x,y)}{\partial y} = x - 4y$$
$$\nabla_u f(x,y) = \left(\frac{4}{5}\right)(6x+y) + \left(\frac{3}{5}\right)(x-4y)$$
$$\nabla_u f(1,-2) = \left(\frac{4}{5}\right)[(6)(1) - 2] + \left(\frac{3}{5}\right)[1 - (4)(-2)]$$
$$= 8.6$$

*5. Tangent Plane Function*

Partial derivatives can be used to find the tangent plane to a 3-dimensional surface at some point $p_o$. If the surface is defined by the function $f(x,y,z) = 0$, the equation of the tangent plane is

$$T(x_o,y_o,z_o) = (x - x_o)\frac{\partial f(x,y,z)}{\partial x}\Big|_{p_o} + (y - y_o)\frac{\partial f(x,y,z)}{\partial y}\Big|_{p_o} + (z - z_o)\frac{\partial f(x,y,z)}{\partial z}\Big|_{p_o} \qquad 1.252$$

*Example 1.50*

What is the equation of the plane tangent to $f(x,y,z) = 4x^2 + y^2 - 16z$ at the point (2,4,2)?

$$\frac{\partial f(x,y,z)}{\partial x}\Big|_{p_o} = 8x\,\Big|_{(2,4,2)} = (8)(2) = 16$$

$$\frac{\partial f(x,y,z)}{\partial y}\Big|_{p_o} = 2y\,\Big|_{(2,4,2)} = (2)(4) = 8$$

$$\frac{\partial f(x,y,z)}{\partial z}\Big|_{p_o} = -16\,\Big|_{(2,4,2)} = -16$$

Therefore,

$$\mathbf{T}(2,4,2) = 16(x-2) + 8(y-4) - 16(z-2)$$
$$= 2x + y - 2z - 4$$

*6. Normal Line Function*

Partial derivatives can be used to find the equation of a straight line normal to a 3-dimensional surface at some point $p_o$. If the surface is defined by the function $f(x,y,z) = 0$, the equation of the normal line is

$$\mathbf{N} = A\mathbf{i} + B\mathbf{j} + C\mathbf{k} \qquad 1.253$$

where

$$A = \frac{\partial f(x,y,z)}{\partial x}\Big|_{p_o} \qquad 1.254$$
$$B = \frac{\partial f(x,y,z)}{\partial y}\Big|_{p_o} \qquad 1.255$$
$$C = \frac{\partial f(x,y,z)}{\partial z}\Big|_{p_o} \qquad 1.256$$

*7. Extrema and Optimization*

Derivatives can be used to locate local *maxima*, *minima*, and *points of inflection*. No distinction is made between local and global extrema. The end points of the interval always should be checked against the local extrema located by the method below. The following rules define the extreme points.

$f'(x) = 0$ at any extrema
$f''(x) = 0$ at an inflection point
$f''(x)$ is negative at a maximum
$f''(x)$ is positive at a minimum

There is always an inflection point between a maximum and a minimum.

*Example 1.51*

Find the global extreme points of the function $f(x) = x^3 + x^2 - x + 1$ on the interval [−2, +2].

$$f'(x) = 3x^2 + 2x - 1$$

$$f'(x) = 0 \text{ at } x = \frac{1}{3} \text{ and } x = -1$$

$$f''(x) = 6x + 2$$

$$f(-1) = 2$$

$$f''(-1) = -4$$

So, $x = -1$ is a maximum.

$$f\left(\frac{1}{3}\right) = \frac{22}{27}$$

$$f''\left(\frac{1}{3}\right) = +4$$

So, $x = -\frac{1}{3}$ is a minimum.

Checking the end points,

$$f(-2) = -1$$

$$f(+2) = +11$$

Therefore, the absolute extreme points are the end points.

## 17 INTEGRAL CALCULUS

### A. FUNDAMENTAL THEOREM

The *Fundamental Theorem of Calculus* is

$$\int_{x_1}^{x_2} f'(x) = f(x_2) - f(x_1) \qquad 1.257$$

### B. INTEGRATION BY PARTS

If $f$ and $g$ are functions, then

$$\int f\,dg = fg - \int g\,df \qquad 1.258$$

*Example 1.52*

Evaluate the following integral: $\int xe^x dx$

Use integration by parts.

Let $f = x$. Then, $df = dx$.

Let $dg = e^x dx$. Then, $g = \int e^x dx = e^x$.

Therefore,

$$\int xe^x dx = xe^x - \int e^x dx + c$$
$$= xe^x - e^x + c$$

### C. INDEFINITE INTEGRALS ("$\cdots + C$" omitted)

$$\int dx = x \qquad 1.259$$

$$\int au\;dx = a\int u\;dx \qquad 1.260$$

$$\int (u+v)dx = \int u\;dx + \int v\;dx \qquad 1.261$$

$$\int x^m dx = \frac{x^{(m+1)}}{m+1} \qquad m \neq -1 \qquad 1.262$$

$$\int \frac{dx}{x} = ln|x| \qquad 1.263$$

$$\int e^{ax} dx = \frac{1}{a}e^{ax} \qquad 1.264$$

$$\int xe^{ax} dx = \frac{1}{a^2}e^{ax}(ax - 1) \qquad 1.265$$

$$\int \cosh x\;dx = \sinh x \qquad 1.266$$

$$\int \sinh x\;dx = \cosh x \qquad 1.267$$

$$\int \sin x\;dx = -\cos x \qquad 1.268$$

$$\int \cos x\;dx = \sin x \qquad 1.269$$

$$\int \tan x\;dx = ln|\sec x| \qquad 1.270$$

$$\int \cot x\;dx = ln|\sin x| \qquad 1.271$$

$$\int \sec x\;dx = ln|\sec x + \tan x| \qquad 1.272$$

$$\int \csc x\;dx = ln|\csc x - \cot x| \qquad 1.273$$

$$\frac{\int dx}{(1+x^2)} = \arctan x \qquad 1.274$$

$$\frac{\int dx}{\sqrt{1-x^2}} = \arcsin x \qquad 1.275$$

$$\frac{\int dx}{x\sqrt{x^2-1}} = \operatorname{arcsec} x \qquad 1.276$$

D. USES OF INTEGRALS

*1. Finding Areas*

The area bounded by $x = a$, $x = b$, $f_1(x)$ above, and $f_2(x)$ below is given by

$$A = \int_a^b [f_1(x) - f_2(x)]dx \qquad 1.277$$

*2. Surfaces of Revolution*

The surface area obtained by rotating $f(x)$ about the $x$ axis is

$$A_s = 2\pi \int_a^b f(x)\sqrt{1 + [f'(x)]^2}dx \qquad 1.278$$

*3. Rotation of a Function*

The volume of a function rotated about the $x$ axis is

$$V = \pi \int_a^b (f(x))^2 dx \qquad 1.279$$

The volume of a function rotated about the $y$ axis is

$$V = 2\pi \int_a^b x f(x) dx \qquad 1.280$$

*4. Length of a Curve*

The length of a curve given by f(x) is

$$L = \int_a^b \sqrt{1 + (f'(x))^2}dx \qquad 1.281$$

*Example 1.53*

For the shaded area shown, find (a) the area, and (b) the volume enclosed by the curve rotated about the $x$ axis.

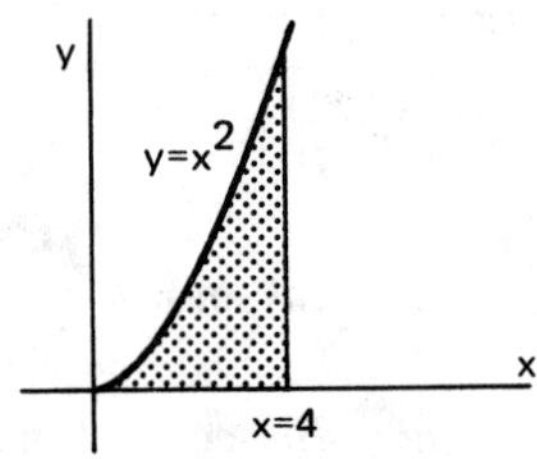


(a)

$$f_2(x) = 0 \qquad f_1(x) = x^2$$

$$A = \int_0^4 x^2 dx = \left[\frac{x^3}{3}\right]_0^4 = 21.33$$

(b)

$$V = \pi \int_0^4 (x^2)^2 dx = \pi \left[\frac{x^5}{5}\right]_0^4 = 204.8\pi$$

## 18 DIFFERENTIAL EQUATIONS

A differential equation is a mathematical expression containing a dependent variable and one or more of that variable's derivatives. First order differential equations contain only the first derivative of the dependent variable. Second order equations contain the second derivative.

The differential equation is said to be *linear* if all terms containing the dependent variable are multiplied only by real scalars. The equation is said to be *homogeneous* if there are no terms which do not contain the dependent variable or one of its derivatives.

Most differential equations are difficult to solve. However, there are several forms which are fairly simple. These are presented here.

A. FIRST ORDER LINEAR

The first order linear differential equation has the general form given by equation 1.282. $p(t)$ and $g(t)$ may be constants or any function of $t$.

$$x' + p(t)x = g(t) \qquad 1.282$$

The solution depends on an *integrating factor* defined as

$$u = \exp\left[\int p(t)\ dt\right] \qquad 1.283$$

The solution to the first order linear differential equation is

$$x = \frac{1}{u}\left[\int ug(t)\ dt + c\right] \qquad 1.284$$

*Example 1.54*

Find a solution to the differential equation

$$x' - x = 2te^{2t} \qquad x(0) = 1$$

This meets the definition of a first order linear equation with

$$p(t) = -1 \text{ and } g(t) = 2te^{2t}$$

The integrating constant is

$$u = \exp\left[\int -1\ dt\right] = e^{-t}$$

Then, $x$ is

$$x = \left(\frac{1}{e^{-t}}\right)$$

$$\left[\int e^{-t}2te^{2t}dt + c\right]$$

$$= e^t\left[\int 2te^t dt + c\right]$$

$$= e^t[2te^t - 2e^t + c]$$

But, $x(0) = 1$, so

$$c = +3, \text{ and}$$
$$x = e^t[2e^t(t-1)+3]$$

B. SECOND ORDER HOMOGENEOUS WITH CONSTANT COEFFICIENTS

This type of differential equation has the following general form:

$$c_1x'' + c_2x' + c_3x = 0 \qquad 1.285$$

The solution can be found by first solving the characteristic quadratic equation for its roots $k_1^*$ and $k_2^*$. This characteristic equation is derived directly from the differential equation:

$$c_1k^2 + c_2k + c_3 \qquad 1.286$$

The form of the solution depends on the values of $k_1^*$ and $k_2^*$. If $k_1^* \neq k_2^*$ and both are real, then

$$x = a_1\left(e^{k_1^*t}\right) + a_2\left(e^{k_2^*t}\right) \qquad 1.287$$

If $k_1^* = k_2^*$, then

$$x = a_1\left(e^{k_1^*t}\right) + a_2t\left(e^{k_2^*t}\right) \qquad 1.288$$

If $k^* = (r \pm iu)$, then

$$x = a_1(e^{rt})\cos(ut) + a_2(e^{rt})\sin(ut) \qquad 1.289$$

In all three cases, $a_1$ and $a_2$ must be found from the given initial conditions.

*Example 1.55*

Solve the following differential equation for $x$.

$$x'' + 6x' + 9x = 0 \quad x(0) = 0,\ x'(0) = 1$$

The characteristic equation is

$$k^2 + 6k + 9 = 0$$

This has roots of $k_1^* = k_2^* = -3$; therefore, the solution has the form

$$x(t) = a_1e^{-3t} + a_2te^{-3t}$$

But, $x(0) = 0$,

$$0 = a_1(e^0) + a_2(0)(e^0)$$
$$0 = a_1(1) + 0$$
$$0 = a_1$$

Also, $x'(0) = 1$. The derivative of x(t) is

$$x'(t) = -3a_2te^{-3t} + a_2e^{-3t}$$
$$1 = -3a_2(0)(e^0) + a_2e^0$$
$$1 = 0 + a_2$$
$$1 = a_2$$

The final solution is

$$x = te^{-3t}$$

## 19 LAPLACE TRANSFORMS

Traditional methods of solving non-homogeneous differential equations are very difficult. The Laplace transformation can be used to reduce the solution of many complex differential equations to simple algebra.

Every mathematical function can be converted into a Laplace function by use of the following transformation definition.

$$\mathcal{L}[f(t)] = \int_0^\infty e^{-st}f(t)dt \qquad 1.290$$

The variable $s$ is equivalent to the derivative operator. However, it may be thought of as a simple variable.

*Example 1.56*

Let f(t) be the unit step. That is, $f(t) = 0$ for $t < 0$ and $f(t) = 1$ for $t \geq 0$.

Then, the Laplace transform of $f(t) = 1$ is

$$\mathcal{L}[f(t)] = \int_0^\infty e^{-st}(1)dt = -\frac{e^{-st}}{s}\Big]_0^\infty$$
$$= 0 - \left(-\frac{1}{s}\right) = \frac{1}{s}$$

*Example 1.57*

What is the Laplace transformation of $f(t) = e^{at}$?

$$\mathcal{L}[e^{at}] = \int_0^\infty e^{-st}e^{at}dt$$
$$= \int_0^\infty e^{-(s-a)t}dt$$
$$= -\frac{e^{-(s-a)t}}{(s-a)}\Big]_0^\infty$$
$$= \frac{1}{s-a}$$

Generally it is unnecessary to actually obtain a function's Laplace transform by use of equation 1.290. Tables of these transforms are readily available. A small collection of the most frequently required transforms is given at the end of this chapter.

The Laplace transform method can be used with any linear differential equation with constant coefficients. Assuming the dependent variable is $x$, the basic procedure is as follows:

*step 1*: Put the differential equation in standard form.

*step 2*: Use superposition and take the Laplace transform of each term.

*step 3*: Use the following relationships to expand terms.

$$\mathcal{L}(x'') = s^2\mathcal{L}(x) - sx_0 - x_0' \quad 1.291$$

$$\mathcal{L}(x') = s\mathcal{L}(x) - x_0 \quad 1.292$$

*step 4*: Solve for $\mathcal{L}(x)$. Simplify the resulting expression using partial fractions.

*step 5*: Find $x$ by applying the inverse transform.

This method reduces the solutions of differential equations to simple algebra. However, a complete set of transforms is required.

Working with Laplace transforms is simplified by the following two theorems:

*Linearity Theorem*: If $c$ is constant, then

$$\mathcal{L}[cf(t)] = c\mathcal{L}[f(t)] \quad 1.293$$

*Superposition Theorem*: If $f(t)$ and $g(t)$ are different functions, then

$$\mathcal{L}[f(t) \pm g(t)] = \mathcal{L}[f(t)] \pm \mathcal{L}[g(t)] \quad 1.294$$

*Example 1.58*

Suppose the following differential equation results from the analysis of a mechanical system:

$$x'' + 2x' + 2x = \cos(t)$$

$$x_0 = 1,\ x_0' = 0$$

$x$ is the dependent variable. Start by taking the Laplace transform of both sides:

$$\mathcal{L}(x'') + 2\mathcal{L}(x') + 2\mathcal{L}(x) = \mathcal{L}(\cos(t))$$

$$s^2\mathcal{L}(x) - sx_0 - x_0' + 2s\mathcal{L}(x) - 2x_0 + 2\mathcal{L}(x) = \mathcal{L}\cos(t)$$

But, $x_0 = 1$ and $x_0' = 0$. Also, the Laplace transform of $\cos(t)$ can be found from the appendix of this chapter.

$$s^2\mathcal{L}(x) - s + 2s\mathcal{L}(x) - 2 + 2\mathcal{L}(x) = \frac{s}{s^2+1}$$

$$\mathcal{L}(x)[s^2 + 2s + 2] - s - 2 = \frac{s}{s^2+1}$$

$$\mathcal{L}(x) = \frac{s^3 + 2s^2 + 2s + 2}{(s^2+1)(s^2+2s+2)}$$

This is now expanded by partial fractions:

$$\begin{aligned}
&\frac{s^3 + 2s^2 + 2s + 2}{(s^2+1)(s^2+2s+2)} \\
&= \frac{A_1 s + B_1}{s^2+1} + \frac{A_2 s + B_2}{s^2+2s+2} \\
&= [s^3(A_1 + A_2) + s^2(2A_1 + B_1 + B_2) \\
&\quad + s(2A_1 + 2B_1 + A_2) + 2B_1 + B_2] \\
&\quad \div [(s^2+1)(s^2+2s+2)]
\end{aligned}$$

The following simultaneous equations result:

$$\begin{array}{lllllllll}
A_1 & + & A_2 & & & & & = & 1 \\
2A_1 & & & + & B_1 & + & B_2 & = & 2 \\
2A_1 & + & A_2 & + & 2B_1 & & & = & 2 \\
 & & & & 2B_1 & + & B_2 & = & 2
\end{array}$$

These equations have the solutions

$$A_1^* = \frac{1}{5} \quad A_2^* = \frac{4}{5}$$

$$B_1^* = \frac{2}{5} \quad B_2^* = \frac{6}{5}$$

Therefore, $x$ can be found by taking the following inverse transform:

$$x = \mathcal{L}^{-1}\left[\frac{\frac{s}{5} + \frac{2}{5}}{s^2+1} + \frac{\frac{4s}{5} + \frac{6}{5}}{s^2+2s+2}\right]$$

The solution is

$$x = \frac{1}{5}\cos(t) + \frac{2}{5}\sin(t) + \frac{4}{5}e^{-t}\cos(t) + \frac{2}{5}e^{-t}\sin(t)$$

## 20 APPLICATIONS OF DIFFERENTIAL EQUATIONS

### A. FLUID MIXTURE PROBLEMS

The typical fluid mixing problem involves a tank containing some liquid. There may be an initial solute in the liquid, or the liquid may be pure. Liquid and solute are added at known rates. A drain usually removes some of the liquid which is assumed to be thoroughly mixed. The problem is to find the weight or concentration of solute in the tank at some time $t$. The following symbols are used.

| | |
|---|---|
| $C(t)$ | concentration of solute in tank at time $t$ |
| $I(t)$ | liquid inflow rate from all sources at time $t$ |
| $k$ | a constant |
| $\Phi(t)$ | liquid outflow rate due to all drains at time $t$ |

$S_1(t)$ solute inflow rate at time $t$ (this may have to be calculated from the incoming concentration and $I(t)$)
$S_2(t)$ solute outflow rate at time $t$
$V_o$ original volume of tank at time $= 0$
$V(t)$ volume of tank at time $= t$ (equal to $V_o + \int I(t)\,dt - \int \Phi(t)\,dt$)
$W_o$ initial weight of solute in tank at $t = 0$
$W(t)$ weight of solute in tank at time $= t$

*Case 1*: Constant Volume

$$\begin{aligned} W'(t) &= S_1(t) - S_2(t) \\ &= S_1(t) - \frac{\Phi(t)W(t)}{V_o} \end{aligned} \qquad 1.295$$

The differential equation is

$$W'(t) + \frac{\Phi W(t)}{V_o} = S_1(t) \qquad 1.296$$

This is a first order linear equation because $\Phi$, $V_o$, and $S_1$ are constants.

*Case 2*: Changing Volume

$$W'(t) = S_1(t) - S_2(t) \qquad 1.297$$

$$= S_1(t) - \frac{\Phi(t)W(t)}{V(t)} \qquad 1.298$$

The differential equation is

$$W'(t) + \frac{\Phi(t)U_t)}{V(t)} = S_1(t) \qquad 1.299$$

*Example 1.59*

A tank contains 100 gallons of pure water at the beginning of an experiment. 1 gpm of pure water flows into the tank, as does 1 gpm of water containing $\frac{1}{4}$ pound of salt per gallon. A perfectly mixed solution drains from the tank at the rate of 2 gpm. How much salt is in the tank 8 minutes after the experiment has begun?

Choose $W(t)$ as the variable giving the weight of salt in the tank at time $t$. $\frac{1}{4}$ pound of salt enters the tank per minute. What goes out depends on the concentration in the tank. Specifically, the leaving salt is

$$\begin{aligned} \text{salt leaving} &= (2 \text{ gpm})(\#\text{ lbs salt per gallon}) \\ &= (2 \text{ gpm})\left(\frac{\#\text{ lbs salt total}}{100}\right) = 0.02W(t) \end{aligned}$$

The difference between the inflow and the outflow is given by equation 1.295.

$$W'(t) = \frac{1}{4} - 0.02W(t)$$

This is a first order linear differential equation. It can be solved using the integrating factor (equation 1.283), simple constant coefficient methods, or Laplace transforms. The solution is

$$W(t) = 12.5 - 12.5e^{-0.02t}$$

At $t = 8$, $W(t) = 1.85$ lbs.

B. DECAY PROBLEMS

A given quantity is known to decrease at a rate proportional to the amount present. The original amount is known, and the amount at some time $t$ is desired.

$k$ a negative proportionality constant
$Q_o$ original amount present
$Q(t)$ amount present at time t
$t$ time
$t_{1/2}$ half-life

The differential equation is

$$Q'(t) = kQ(t) \qquad 1.300$$

The solution is

$$Q(t) = Q_o e^{kt} \qquad 1.301$$

If $Q^*$ is known for some time $t^*$, $k$ can be found from

$$k = \left(\frac{1}{t^*}\right) ln \left(\frac{Q^*}{Q_o}\right) \qquad 1.302$$

k also can be found from the half-life:

$$k = \frac{-0.693}{t_{1/2}} \qquad 1.303$$

C. SURFACE TEMPERATURE

$k$ a constant
$t$ time
$T$ absolute temperature of the surface
$T_o$ ambient temperature

Assuming that the surface temperature changes at a rate proportional to the difference in surface and ambient temperatures, the differential equation is

$$\frac{dT}{dt} = k(T - T_o) \qquad 1.304$$

Equation 1.304 is known as *Newton's Law of Cooling.*

D. SURFACE EVAPORATION

| | |
|---|---|
| $A$ | exposed surface area |
| $k$ | proportionality constant |
| $r$ | radius |
| $s$ | side length |
| $t$ | time |
| $V$ | object volume |

The equation is

$$\frac{dV}{dt} = -kA \qquad 1.305$$

For a spherical drop, this reduces to

$$\frac{dr}{dt} = -k \qquad 1.306$$

For a cube, this reduces to

$$\frac{ds}{dt} = -2k \qquad 1.307$$

## 21 FOURIER ANALYSIS

Any periodic waveform can be written as the sum of an infinite number of sinusoidal terms. In practice, it is possible to obtain a close approximation to the original waveform with a limited number of sinusoidal terms since most series converge rapidly.

Fourier's theorem is given in equation 1.308. The object of Fourier analysis is to determine the coefficients $a_n$ and $b_n$.

$$\begin{aligned} f(t) &= a_o + a_1 \cos \omega_o t + a_2 \cos 2\omega_o t + \cdots \\ &+ b_1 \sin \omega_o t + b_2 \sin 2\omega_o t + \cdots \end{aligned} \qquad 1.308$$

$\omega_o$ is known as the *fundamental frequency* of the waveform. It depends on the actual waveform period.

$$\omega_o = \frac{2\pi}{T} \qquad 1.309$$

To simplify the analysis, the time domain can be normalized to the radian scale. The normalized scale is obtained by dividing all frequencies by $\omega_o$. Then, the Fourier series becomes

$$\begin{aligned} f(t) &= a_o + a_1 \cos t + a_2 \cos 2t + \cdots \\ &+ b_1 \sin t + b_2 \sin 2t + \cdots \end{aligned} \qquad 1.310$$

The coefficients $a_n$ and $b_n$ can be found from the following relationships:

$$a_o = \frac{1}{2\pi}\int_o^{2\pi} f(t)\, dt \qquad 1.311$$

$$a_n = \frac{1}{\pi}\int_o^{2\pi} f(t) \cos\ nt\ dt \qquad 1.312$$

$$b_n = \frac{1}{\pi}\int_o^{2\pi} f(t) \sin\ nt\ dt \qquad 1.313$$

Notice that $a_o$ is the average value of the function. Usually, this average value can be determined by observation without having to go through the integration process. The equation for $a_n$ cannot be used to find $a_o$.

*Example 1.60*

Find the Fourier series for

$$f(t) = \begin{Bmatrix} 1 & 0 < t < \pi \\ 0 & \pi < t < 2\pi \end{Bmatrix}$$

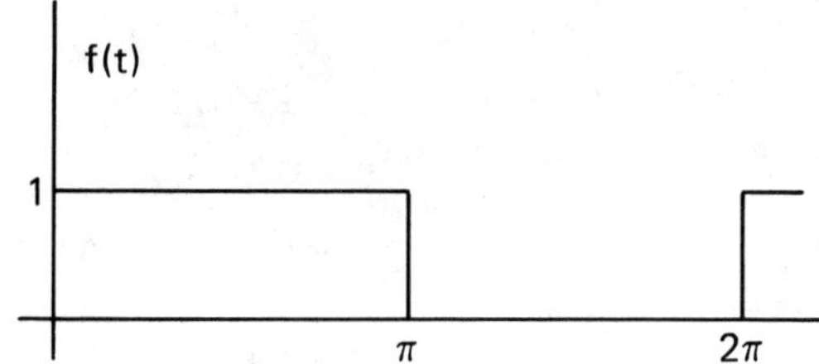


$$a_o = \frac{1}{2\pi}\int_o^{\pi} (1) dt + \frac{1}{2\pi}\int_{\pi}^{2\pi} (0) dt = \frac{1}{2}$$

This value of $\frac{1}{2}$ corresponds to the average value of $f(t)$. It could have been found by observation.

$$\begin{aligned} a_1 &= \frac{1}{\pi}\int_o^{\pi} (1) \cos t\ dt + \frac{1}{\pi}\int_{\pi}^{2\pi} (0) \cos t\ dt \\ &= \frac{1}{\pi}\left[\sin t\right]_o^{\pi} + 0 = 0 \end{aligned}$$

In general,

$$a_n = \frac{1}{\pi}\left[\frac{\sin\ nt}{n}\right]_o^{\pi} = 0$$

$$\begin{aligned} b_1 &= \frac{1}{\pi}\int_o^{\pi} (1) \sin t\ dt + \frac{1}{\pi}\int_{\pi}^{2\pi} (0) \sin t\ dt \\ &= \frac{1}{\pi}\left[-\cos t\right]_o^{\pi} = \frac{2}{\pi} \end{aligned}$$

In general,

$$b_n = \frac{1}{\pi}\left[\frac{-\cos\ nt}{n}\right]_o^{\pi} = \begin{Bmatrix} 0 \text{ for n even} \\ \frac{2}{\pi n} \text{ for n odd} \end{Bmatrix}$$

The series is

$$f(t) = \frac{1}{2} + \frac{2}{\pi}\left[\sin t + \frac{1}{3}\sin\ 3t + \frac{1}{5}\sin\ 5t + \cdots\right]$$

The sum of the first few terms is illustrated.

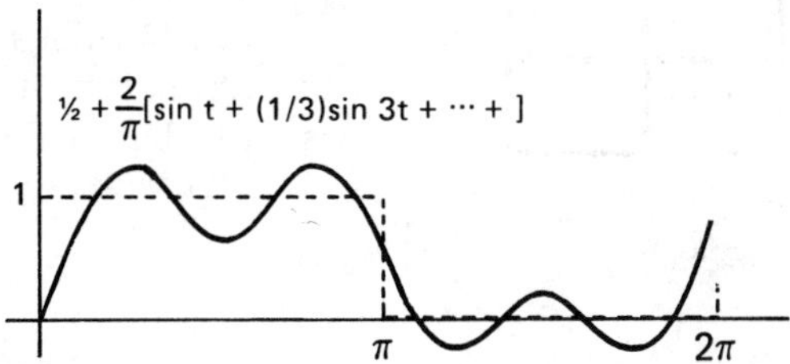


It may be possible to eliminate some of the $a_n$ or $b_n$ coefficients if the function $f(t)$ is symmetrical. There are four types of *symmetry*.

A function is said to have *even symmetry* if $f(t) = f(-t)$. The cosine is an example of this type of waveform. Even symmetry can be detected from the graph of the function. The function to the left of $x = 0$ is a reflection of the function to the right of $x = 0$. With even symmetry, all $b_n$ terms are zero.

A function is said to have *odd symmetry* if $f(t) = -f(-t)$. The sine is an example of this type of waveform. With odd symmetry, all $a_n$ terms are zero (but not necessarily $a_o$).

A function is said to have *rotational symmetry* or *half-wave symmetry* if $f(t) = -f(t+\pi)$. Functions of this type are identical on alternate $\frac{1}{2}$-cycles, except for a sign reversal. All $a_n$ and $b_n$ are zero for even values of $n$.

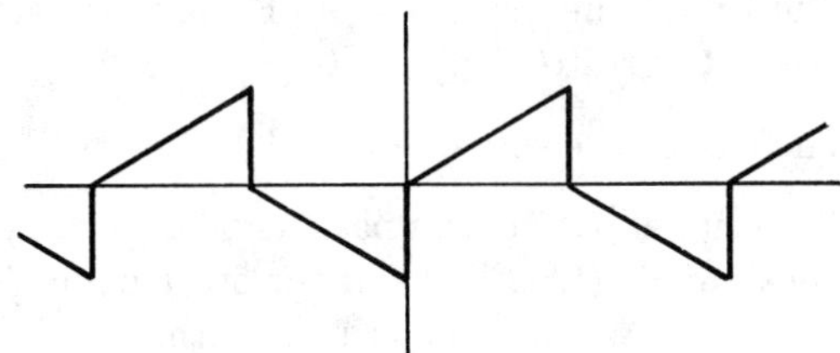

**Figure 1.12** Rotational Symmetry

These types of symmetry are not mutually exclusive. For example, it is possible for a function with rotational symmetry to have either odd or even symmetry also. Such a case is known as *quarter-wave symmetry*.

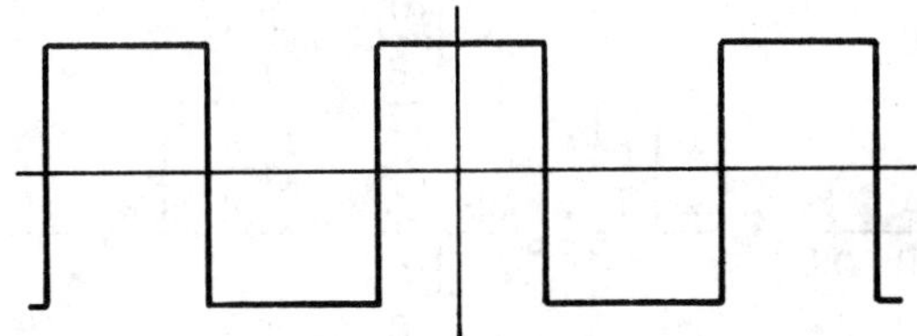

**Figure 1.13** Quarter-Wave Symmetry

## 22 CRITICAL PATH TECHNIQUES

### A. INTRODUCTION

Critical path techniques are used to represent graphically the multiple relationships between stages in complicated projects. The graphical networks show the dependencies or *precedence relationships* between the various activities and can be used to control and monitor the progress, cost, and resources of projects. Critical path techniques will also identify the most critical activities in projects.

### B. DEFINITIONS

Activity: Any subdivision of a project whose execution requires time and other resources.

Critical path: A path connecting all activities which have minimum or zero slack times. The critical path is the longest path through the network.

Duration: The time required to perform an activity. All durations are *normal durations* unless otherwise referred to as *crash durations*.

Event: The beginning or completion of an activity.

Event time: Actual time at which an event occurs.

Float: Same as slack time.

Slack time: The maximum time that an activity can be delayed without causing the project to fall behind schedule. Slack time is always minimum or zero along the critical path.

Critical path techniques use directed graphs to represent a project. These graphs are made up of arcs (arrows) and nodes (junctions). The placement of the arcs and nodes completely specifies the precedences of the project. Durations and precedences usually are given in a *precedence table* or matrix.

One specific technique is known as the *Critical Path Method, CPM*. This deterministic method is applicable when all activity durations are known in advance. CPM usually is represented as an *activity-on-node model*. Arcs are used to specify precedence, and the nodes actually represent the activities. Events are not present on the graph, other than as the heads and tails of the arcs. Two dummy nodes taking zero time can be used to specify the start and finish of the project.

*Example 1.61*

Given the project listed in the precedence table, construct the precedence matrix and draw an activity-on-node-network.

| Activity | Time (days) | Predecessors |
|---|---|---|
| A, start | 0 | – |
| B | 7 | A |
| C | 6 | A |
| D | 3 | B |
| E | 9 | B,C |
| F | 1 | D,E |
| G | 4 | C |
| H, finish | 0 | F,G |

The precedence matrix is given below.

| predecessor \ successor | A | B | C | D | E | F | G | H |
|---|---|---|---|---|---|---|---|---|
| A | | X | X | | | | | |
| B | | | | X | X | | | |
| C | | | | | X | | X | |
| D | | | | | | X | | |
| E | | | | | | X | | |
| F | | | | | | | | X |
| G | | | | | | | | X |
| H | | | | | | | | |

The activity-on-node-network is shown.

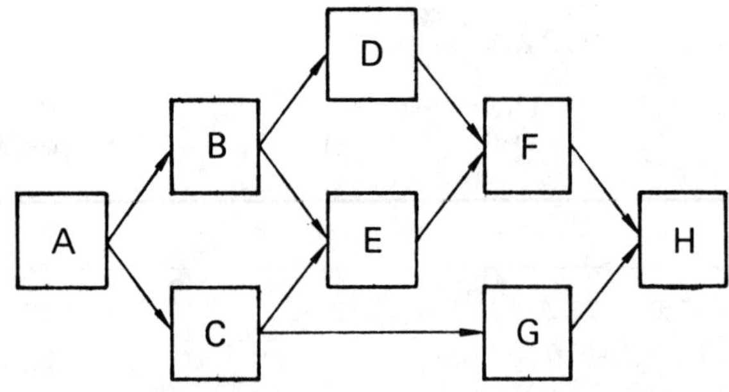


## C. SOLVING A CPM PROBLEM

The solution of a critical path problem results in a knowledge of the earliest and latest times that an activity can be started and finished. It also identifies the critical path and generates the slack time for each activity.

To facilitate the solution method, each node should be replaced by a square which has been quartered. The compartments have the meanings indicated by the following key.

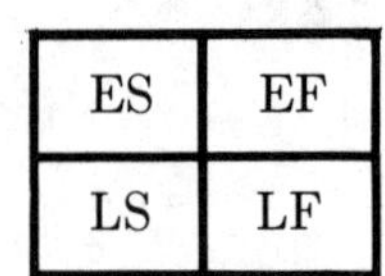


Key
ES: Earliest Start
EF: Earliest Finish
LS: Latest Start
LF: Latest Finish

The following procedure will find the earliest and latest starts and finishes of each node.

1. Place the project start time or date in the ES and EF positions of the start activity. The start time is zero for relative calculations.
2. Consider any unmarked activity whose predecessors have all been marked in the EF and ES positions. (Go to step 4 if there is none.) Mark in its ES position the largest number marked in the EF position of those predecessors.
3. Add the activity time to the ES time and write this in the EF box. Return to step 2.
4. Place the value of the latest finish date in the LS and LF boxes of the finish node.
5. Consider any unmarked activity whose successors have all been marked in the LS and LF positions. The LF is the smallest LS of the successors. Go to step 7 if there are no unmarked activities.
6. The LS for the new node is LF minus its activity time. Return to step 5.
7. The slack for each node is (LS-ES) or (LF-EF).
8. The critical path encompasses nodes for which the slack equals (LS-ES) from the start node. There may be more than one critical path.

*Example 1.62*

Complete the network for the previous example and find the critical path. Assume the desired completion date is in 19 days.

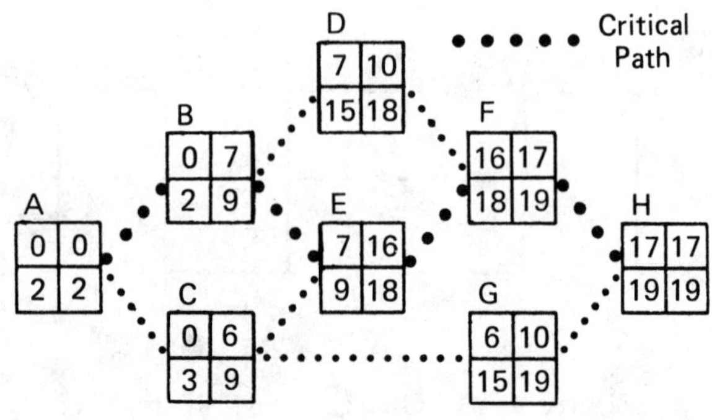

## D. PROBABILISTIC CRITICAL PATH MODELS

Probabilistic networks differ from deterministic networks only in the way in which the activity durations are found. Whereas durations are known explicitly for a deterministic network, the time for a probabilistic activity is distributed as a random variable.

This variable nature complicates the problem greatly since the actual distribution of times often is unknown. For this reason, such a problem usually is solved as a deterministic model using the mean of the duration distribution as the activity duration.

The most common probabilistic critical path model is *PERT*, which stands for *Program Evaluation and Review Technique*. In PERT, all duration variables are assumed to come from a beta distribution, with mean and standard deviation given as:

$$t_{mean} = \left(\frac{1}{6}\right)(t_{minimum} + 4t_{most\ likely} + t_{maximum}) \qquad 1.314$$

$$\sigma = \frac{1}{6}(t_{maximum} - t_{minimum}) \qquad 1.315$$

The project completion time for large projects is assumed to be distributed normally with mean ($\mu$) equal to the critical path length and overall variance ($\sigma^2$) equal to the sum of the variances along the critical path.

If necessary, the probability that a project duration will exceed some length ($D$) can be found from the normal table and the following relationship:

$$P\{\text{duration} > D\} = p\{x > z\} \qquad 1.316$$

where $z$ is the standard normal variable equal to

$$z = \frac{D - \mu}{\sigma} \qquad 1.317$$

*Example 1.63*

Determine the probability that the project illustrated will be completed on or before 19 days.

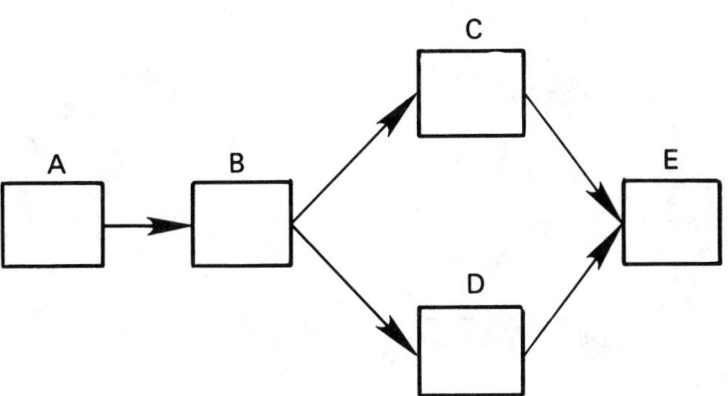


| event | $t_{min}$ | $t_{likely}$ | $t_{max}$ |
|---|---|---|---|
| A | 2 days | 3 days | 5 days |
| B | 4 | 6 | 7 |
| C | 1 | 3 | 5 |
| D | 2 | 3 | 7 |
| E | 3 | 5 | 10 |

Start by calculating the mean time of completion and the variance for each event. Using equations 1.314 and 1.315 for event A,

$$t_{A,mean} = \frac{2 + 4 \times 3 + 5}{6} = 3.2$$

$$\sigma_A^2 = \left[\frac{5-2}{6}\right]^2 = 0.25$$

The remaining times and variances are similarly calculated.

| event | $t_{mean}$ | $\sigma^2$ |
|---|---|---|
| A | 3.2 | 0.25 |
| B | 5.8 | 0.25 |
| C | 3.0 | 0.44 |
| D | 3.5 | 0.69 |
| E | 5.5 | 1.36 |

Since the mean time for event D is greater than the mean time for event C, the critical path is the sequence of events A-B-D-E. The sums of mean times and variances along the critical path are

$$\mu = 3.2 + 5.8 + 3.5 + 5.5 = 18$$

$$\sigma^2 = 0.25 + 0.25 + 0.69 + 1.36 = 2.55$$

The project has a 50% probability of being completed within 18 days, since 18 is the distribution mean. To calculate the probability of being completed in 19 days, the standard normal variable is calculated from equation 1.317.

$$z = \frac{19 - 18}{\sqrt{2.55}} = 0.63$$

The area under the standard normal curve corresponding to $z = 0.63$ is 0.2357. Therefore, the probability of finishing the project on or before 19 days is $0.5 + 0.2357 = 0.7357$ (73.6%).

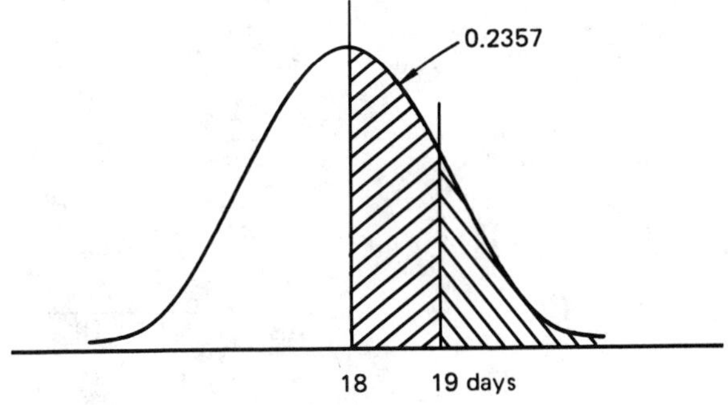

# Appendix A: Laplace Transforms

| f(t) | $\mathcal{L}[f(t)]$ |
|---|---|
| Unit impulse at t=0 | 1 |
| Unit impulse at t=c | $e^{-cs}$ |
| Unit step at t=0 | (1/s) |
| Unit step at t=c | $\frac{e^{-cs}}{s}$ |
| t | $\frac{1}{s^2}$ |
| $\frac{t^{n-1}}{(n-1)!}$ | $\frac{1}{s^n}$ |
| sinAt | $\frac{A}{s^2+A^2}$ |
| At − sinAt | $\frac{A^3}{s^2(s^2+A^2)}$ |
| sinh(At) | $\frac{A}{s^2-A^2}$ |
| t sinAt | $\frac{2As}{(s^2+A^2)^2}$ |
| cosAt | $\frac{s}{s^2+A^2}$ |
| 1 − cosAt | $\frac{A^2}{s(s^2+A^2)}$ |
| cosh(At) | $\frac{s}{s^2-A^2}$ |
| t cosAt | $\frac{s^2-A^2}{(s^2+A^2)^2}$ |
| $t^n$ (n is a positive integer) | $\frac{n!}{s^{(n+1)}}$ |
| $e^{At}$ | $\frac{1}{s-A}$ |
| $e^{At}\sin Bt$ | $\frac{B}{(s-A)^2+B^2}$ |
| $e^{At}\cos Bt$ | $\frac{s-A}{(s-A)^2+B^2}$ |
| $e^{At}t^n$ (n is positive integer) | $\frac{n!}{(s-A)^{n+1}}$ |
| $1 - e^{-At}$ | $\frac{A}{s(s+A)}$ |
| $e^{-At} + At - 1$ | $\frac{A^2}{s^2(s+A)}$ |
| $\frac{e^{-At} - e^{-Bt}}{B-A}$ | $\frac{1}{(s+A)(s+B)}$ |
| $\frac{(C-A)e^{-At} - (C-B)e^{-Bt}}{B-A}$ | $\frac{s + C}{(s+A)(s+B)}$ |
| $\frac{1}{AB} + \frac{Be^{-At} - Ae^{-Bt}}{AB(A-B)}$ | $\frac{1}{s(s+A)(s+B)}$ |

# Appendix B: Conversion Factors

| To Convert | Into | Multiply by |
|---|---|---|
| Acres | hectares | 0.4047 |
| Acres | square feet | 43,560.0 |
| Acres | square miles | 1.562 EE−3 |
| Ampere hours | coulombs | 3,600.0 |
| Angstrom units | inches | 3.937 EE−9 |
| Angstrom units | microns | 1 EE−4 |
| Astronomical units | kilometers | 1.495 EE8 |
| Atmospheres | cms of mercury | 76.0 |
| BTU's | horsepower-hrs | 3.931 EE−4 |
| BTU's | kilowatt-hrs | 2.928 EE−4 |
| BTU/hr | watts | 0.2931 |
| Bushels | cubic inches | 2,150.4 |
| Calories, gram (mean) | BTU (mean) | 3.9685 EE−3 |
| Centares | square meters | 1.0 |
| Centimeters | kilometers | 1 EE−5 |
| Centimeters | meters | 1 EE−2 |
| Centimeters | millimeters | 10.0 |
| Centimeters | feet | 3.281 EE−2 |
| Centimeters | inches | 0.3937 |
| Chains | inches | 792.0 |
| Coulombs | faradays | 1.036 EE−5 |
| Cubic centimeters | cubic inches | 0.06102 |
| Cubic centimeters | pints (U.S. liq.) | 2.113 EE−3 |
| Cubic feet | cubic meters | 0.02832 |
| Cubic feet/min. | pounds water/min. | 62.43 |
| Cubic feet/sec. | gallons/min. | 448.831 |
| Cubits | inches | 18.0 |
| Days | seconds | 86,400.0 |
| Degrees (angle) | radians | 1.745 EE−2 |
| Degrees/sec. | revolutions/min. | 0.1667 |
| Dynes | grams | 1.020 EE−3 |
| Dynes | joules/meter (newtons) | 1 EE−5 |
| Ells | inches | 45.0 |
| Ergs | BTU's | 9.480 EE−11 |
| Ergs | foot-pounds | 7.3670 EE−8 |
| Ergs | kilowatt-hours | 2.778 EE−14 |
| Faradays/sec. | amperes (absolute) | 96,500 |
| Fathoms | feet | 6.0 |
| Feet | centimeters | 30.48 |
| Feet | meters | 0.3048 |
| Feet | miles (nautical) | 1.645 EE−4 |
| Feet | miles (statute) | 1.894 EE−4 |
| Feet/min. | centimeters/sec. | 0.5080 |
| Feet/sec. | knots | 0.5921 |
| Feet/sec. | miles/hour | 0.6818 |
| Foot-pounds | BTU's | 1.286 EE−3 |
| Foot-pounds | kilowatt-hours | 3.766 EE−7 |
| Furlongs | miles (U.S.) | 0.125 |
| Furlongs | feet | 660.0 |
| Gallons | liters | 3.785 |
| Gallons of water | pounds of water | 8.3453 |
| Gallons/min. | cubic feet/hour | 8.0208 |
| Grams | ounces (avoirdupois) | 3.527 EE−2 |
| Grams | ounces (troy) | 3.215 EE−2 |
| Grams | pounds | 2.205 EE−3 |
| Hectares | acres | 2.471 |
| Hectares | square feet | 1.076 EE5 |

| To Convert | Into | Multiply by |
|---|---|---|
| Horsepower | BTU/min. | 42.42 |
| Horsepower | kilowatts | 0.7457 |
| Horsepower | watts | 745.7 |
| Hours | days | 4.167 EE−2 |
| Hours | weeks | 5.952 EE−3 |
| Inches | centimeters | 2.540 |
| Inches | miles | 1.578 EE−5 |
| Joules | BTU's | 9.480 EE−4 |
| Joules | ergs | 1 EE7 |
| Kilograms | pounds | 2.205 |
| Kilometers | feet | 3,281.0 |
| Kilometers | meters | 1,000.0 |
| Kilometers | miles | 0.6214 |
| Kilometers/hr. | knots | 0.5396 |
| Kilowatts | horsepower | 1.341 |
| Kilowatt-hours | BTU'S | 3,413.0 |
| Knots | feet/hour | 6,080.0 |
| Knots | nautical miles/hr. | 1.0 |
| Knots | statute miles/hr. | 1.151 |
| Light years | miles | 5.9 EE12 |
| Links (surveyor's) | inches | 7.92 |
| Liters | cubic centimeters | 1,000.0 |
| Liters | cubic inches | 61.02 |
| Liters | gallons (U.S. liq.) | 0.2642 |
| Liters | milliliters | 1,000.0 |
| Liters | pints (U.S. liq.) | 2.113 |
| Meters | centimeters | 100.0 |
| Meters | feet | 3.281 |
| Meters | kilometers | 1 EE−3 |
| Meters | miles (nautical) | 5.396 EE−4 |
| Meters | miles (statute) | 6.214 EE−4 |
| Meters | millimeters | 1,000.0 |
| Microns | meters | 1 EE−6 |
| Miles (nautical) | feet | 6,080.27 |
| Miles (statute) | feet | 5,280.0 |
| Miles (nautical) | kilometers | 1.853 |
| Miles (statute) | kilometers | 1.609 |
| Miles (nautical) | miles (statute) | 1.1516 |
| Miles (statute) | miles (nautical) | 0.8684 |
| Miles/hour | feet/min. | 88.0 |
| Milligrams/liter | parts/million | 1.0 |
| Milliliters | liters | 1 EE−3 |
| Millimeters | inches | 3.937 EE−2 |
| Newtons | dynes | 1 EE5 |
| Ohms (international) | ohms (absolute) | 1.0005 |
| Ounces | grams | 28.349527 |
| Ounces | pounds | 6.25 EE−2 |
| Ounces (troy) | ounces (avoirdupois) | 1.09714 |
| Parsecs | miles | 19 EE12 |
| Parsecs | kilometers | 3.084 EE13 |
| Pints (liq.) | cubic centimeters | 473.2 |
| Pints (liq.) | cubic inches | 28.87 |
| Pints (liq.) | gallons | 0.125 |
| Pints (liq.) | quarts (liq.) | 0.5 |
| Pounds | kilograms | 0.4536 |
| Pounds | ounces | 16.0 |
| Pounds | ounces (troy) | 14.5833 |
| Pounds | pounds (troy) | 1.21528 |
| Quarts (dry) | cubic inches | 67.20 |

| To Convert | Into | Multiply by |
|---|---|---|
| Quarts (liq.) | cubic inches | 57.75 |
| Quarts (liq.) | gallons | 0.25 |
| Quarts (liq.) | liters | 0.9463 |
| Radians | degrees | 57.30 |
| Radians | minutes | 3,438.0 |
| Revolutions | degrees | 360.0 |
| Revolutions/min. | degrees/sec. | 6.0 |
| Rods | meters | 5.029 |
| Rods | feet | 16.5 |
| Rods (surveyor's measure) | yards | 5.5 |
| Seconds | minutes | 1.667 EE−2 |
| Slugs | pounds | 32.17 |
| Tons (long) | kilograms | 1,016.0 |
| Tons (short) | kilograms | 907.1848 |
| Tons (long) | pounds | 2,240.0 |
| Tons (short) | pounds | 2,000.0 |
| Tons (long) | tons (short) | 1.120 |
| Tons (short) | tons (long) | 0.89287 |
| Volt (absolute) | statvolts | 3.336 EE−3 |
| Watts | BTU/hour | 3.4129 |
| Watts | horsepower | 1.341 EE−3 |
| Yards | meters | 0.9144 |
| Yards | miles (nautical) | 4.934 EE−4 |
| Yards | miles (statute) | 5.682 EE−4 |

# Appendix C: Computational Values of Fundamental Constants

| Constant | SI | English |
|---|---|---|
| charge on electron | −1.602 EE−19 C | |
| charge on proton | +1.602 EE−19 C | |
| atomic mass unit | 1.66 EE−27 kg | |
| electron rest mass | 9.11 EE−31 kg | |
| proton rest mass | 1.673 EE−27 kg | |
| neutron rest mass | 1.675 EE−27 kg | |
| earth weight | | 1.32 EE25 lb |
| earth mass | 6.00 EE24 kg | 4.11 EE23 slug |
| mean earth radius | 6.37 EE3 km | 2.09 EE7 ft |
| mean earth density | 5.52 EE3 kg/m³ | 3.45 lbm/ft³ |
| earth escape velocity | 1.12 EE4 m/s | 3.67 EE4 ft/sec |
| distance from sun | 1.49 EE11 m | 4.89 EE11 ft |
| Boltzmann constant | 1.381 EE−23 J/°K | 5.65 EE−24 $\frac{\text{ft-lbf}}{°\text{R}}$ |
| permeability of a vacuum | 1.257 EE−6 H/m | |
| permittivity of a vacuum | 8.854 EE−12 F/m | |
| Planck constant | 6.626 EE−34 J·s | |
| Avogadro's number | 6.022 EE23 molecules/gmole | 2.73 EE26 $\frac{\text{molecules}}{\text{pmole}}$ |
| Faraday's constant | 9.648 EE4 C/gmole | |
| Stefan-Boltzmann constant | 5.670 EE−8 W/m²−K⁴ | 1.71 EE−9 $\frac{\text{BTU}}{\text{ft}^2-\text{hr}-°\text{R}^4}$ |
| gravitational constant (G) | 6.672 EE−11 m³/s²−kg | 3.44 EE−8 $\frac{\text{ft}^4}{\text{lbf}-\text{sec}^4}$ |
| universal gas constant | 8.314 J/°K−gmole | 1545 $\frac{\text{ft-lbf}}{°\text{R}-\text{pmole}}$ |
| speed of light | 3.00 EE8 m/s | 9.84 EE8 ft/sec |
| speed of sound, air, STP | 3.31 EE2 m/s | 1.09 EE3 ft/sec |
| speed of sound, air, 70°F, one atmosphere | 3.44 EE2 m/s | 1.13 EE3 ft/sec |
| standard atmosphere | 1.013 EE5 N/m² | 14.7 psia |
| standard temperature | 0°C | 32°F |
| molar ideal gas volume (STP) | 22.4138 EE−3 m³/gmole | 359 ft³/pmole |
| standard water density | 1 EE3 kg/m³ | 62.4 lbm/ft³ |
| air density, STP | 1.29 kg/m³ | 8.05 EE−2 lbm/ft³ |
| air density, 70°F, 1 atm | 1.20 kg/m³ | 7.49 EE−2 lbm/ft³ |
| mercury density | 1.360 EE4 kg/m³ | 8.49 EE2 lbm/ft³ |
| gravity on moon | 1.67 m/s² | 5.47 ft/sec² |
| gravity on earth | 9.81 m/s² | 32.17 ft/sec² |

**Practice Problems: MATHEMATICS**

Untimed

1. A state law requires a statistical analysis of the average speed driven by motorists on a road prior to the use of radar speed control. The following speeds were observed in a random sample of 40 cars: 44, 48, 26, 25, 20, 43, 40, 42, 29, 39, 23, 26, 24, 47, 45, 28, 29, 41, 38, 36, 27, 44, 42, 43, 29, 37, 34, 31, 33, 30, 42, 43, 28, 41, 29, 36, 35, 30, 32, 31 (all in mph).

(a) Tabulate the frequency distribution of the above data. (b) Draw the frequency histogram. (c) Draw the frequency polygon. (d) Tabulate the cumulative frequency distribution. (e) Draw the cumulative frequency graph. (f) What is the upper quartile speed? (g) What are the mode, median, and mean speeds? (h) What is the standard deviation of the sample data? (i) What is the sample standard deviation? (j) What is the sample variance?

2. Activities constituting a project are given. The project starts at time zero.

| activity | predecessors | successors | duration |
|---|---|---|---|
| start | – | A | 0 |
| A | start | B,C,D | 7 |
| B | A | G | 6 |
| C | A | E,F | 5 |
| D | A | G | 2 |
| E | C | H | 13 |
| F | C | H,I | 4 |
| G | D,B | I | 18 |
| H | E,F | finish | 7 |
| I | F,G | finish | 5 |
| finish | H,I | – | 0 |

(a) Draw the CPM network. (b) Indicate the critical path. (c) What is the earliest finish? (d) What is the latest finish? (e) What is the slack along the critical path? (f) What is the float along the critical path?

3. Activities constituting a short project are given.

| activity | predecessors | successors | $t_{min}$ | $t_{likely}$ | $t_{max}$ |
|---|---|---|---|---|---|
| start | – | A | 0 | 0 | 0 |
| A | start | B,D | 1 | 2 | 5 |
| B | A | C | 7 | 9 | 20 |
| C | B | D | 5 | 12 | 18 |
| D | A,C | finish | 2 | 4 | 7 |
| finish | D | – | 0 | 0 | 0 |

If the project starts at $t=15$, what is the probability that the project will be completed by $t=42$ or sooner?

4. A pipe with an inside diameter of 18.812″ contains fluid to a depth of 15.7″. What is the hydraulic radius?

5. What is the determinant of the following matrix?

$$\begin{bmatrix} 8 & 2 & 0 & 0 \\ 2 & 8 & 2 & 0 \\ 0 & 2 & 8 & 2 \\ 0 & 0 & 2 & 4 \end{bmatrix}$$

6. The number of cars entering a toll plaza on a bridge during the hour following midnight is distributed as Poisson with a mean of 20. What is the probability that 17 cars will pass through the toll plaza during that hour on any given night? What is the probability that 3 or fewer cars will pass through the toll plaza at that hour on any given night?

7. The time taken by a toll taker to collect the toll from vehicles crossing a bridge is an exponential distribution with mean of 23 seconds when a line of vehicles exists waiting to enter the toll booth. What is the probability that a random vehicle will be processed in 25 seconds or more (i.e., will take longer than 25 seconds)?

8. The average number of vehicles lining up behind a flashing railroad crossing has been observed for five trains of different lengths. What is the mathematical formula which relates the two variables?

| # cars in train | # vehicles |
|---|---|
| 2 | 14.8 |
| 5 | 18.0 |
| 8 | 20.4 |
| 12 | 23.0 |
| 27 | 29.9 |

9. Holes drilled in structural steel parts for bolts are normally distributed with a mean of 0.502″ and standard deviation of .005″. Holes are defective if their diameters are less than 0.497″ or more than 0.507″. (a) What is the probability that a hole chosen at random will be defective? (b) What is the probability that 2 holes out of a sample of 15 will be defective?

10. The oscillation exhibited by a certain 1-story building in free motion is given by the following differential equation:

$$x'' + 2x' + 2x = 0 \qquad x(0) = 0; \quad x'(0) = 1$$

(a) What is $x$ as a function of time? (b) What is the building's natural frequency of vibration? (c) What is the amplitude of oscillation? (d) What is $x$ as a function of time if a lateral wind load is applied with form of sin(t)?

11. Using the bisection method, find the root of the equation

$$x^3 + 2x^2 + 8x - 2 = 0$$

12. Consider the yield data from five treatment plants. Develop a mathematical equation to correlate the data.

| treatment plant | average yield | average temperature |
|---|---|---|
| 1 | 92.30 | 207.1 |
| 2 | 92.58 | 210.3 |
| 3 | 91.56 | 200.4 |
| 4 | 91.63 | 201.1 |
| 5 | 91.83 | 203.4 |

13. The following data is given from a waterflow experiment. What is the mathematical formula which relates the two variables?

| x | y |
|---|---|
| −1 | 0 |
| 0 | 1 |
| 1 | 1.4 |
| 2 | 1.7 |
| 3 | 2 |
| 4 | 2.2 |
| 5 | 2.4 |
| 6 | 2.6 |
| 7 | 2.8 |
| 8 | 3 |

14. Four recruits whose shoe sizes are 7, 8, 9, and 10 report to the supply clerk to be issued shoes. The supply clerk selects one pair of shoes in each of the four required sizes and hands them at random to the men. What is the probability that (a) no man will receive the correct size? (b) exactly 3 men will receive the correct size?

15. In an ammeter, two resistances are connected in parallel. Most of the current passing through the meter goes through the shunt. In order to determine the accuracy of the resistance of shunts being made for ammeters, a manufacturer tested a sample of 100 shunts. The resistance of each, to the nearest hundredth of an ohm, is indicated in the following data (number of shunts followed by resistance in ohms).

1-.200, 2-.210, 2-.290, 3-.280, 1-.210, 2-.220, 3-.230, 2-.270, 4-.260, 28-.250, 3-.220, 3-.230, 7-.240, 5-.260, 12-.250, 4-.230, 5-.240, 4-.260, 4-.270, 5-.240.

(a) Draw the frequency distribution. (b) Find the arithmetic mean. (c) Find the sample standard deviation. (d) Draw a frequency polygon. (e) Find the median. (f) Find the sample variance.

16. Listed is a set of activities and sequence requirements to start a warehouse construction project. Prepare a CPM project diagram. Indicate the critical path.

| activity | letter code | code of immediate predecessor |
|---|---|---|
| move-in | A | |
| job layout | B | A |
| excavations | C | B |
| make-up forms | D | A |
| shop drawing, order rebar | E | A |
| erect forms | F | C,D |
| rough in plumbing | G | F |
| install rebars | H | E,F |
| pour, finish concrete | I | G,H |

17. Activities constituting a bridge construction project are listed. (a) Draw the CPM network showing the critical path. (b) Compute ES, EF, LS, and LF for each of the activities. Assume a target time for completing the project which, for the bridge is 3 days after the EF time.

| job number | immediate predecessors | time (days) |
|---|---|---|
| a Start | | 0 |
| b | a | 4 |
| c | b | 2 |
| d | c | 4 |
| e | d | 6 |
| f | c | 1 |
| g | f | 2 |
| h | f | 3 |
| i | d | 2 |
| j | d,g | 4 |
| k | i,j,h | 10 |
| l | k | 3 |
| m | l | 1 |
| n | l | 2 |
| o | l | 3 |
| p | e | 2 |
| q | p | 1 |
| r | c | 1 |
| s | o,t | 2 |
| t | m,n | 3 |
| u | t | 1 |
| v | q,r | 2 |
| w | v | 5 |
| x Finish | s,u,w | 0 |

18. Prepare a PERT diagram for the activities and sequence requirements given in problem 17.

19. Listed is a set of activities, sequence requirements, and estimated activity times required for the renewal of a pipeline. (a) Prepare a PERT project diagram.

| activity | letter code | code of immediate predecessor | activity time requirement (days) |
|---|---|---|---|
| assemble crew for job | A | | 10 |
| use old line to build inventory | B | D | 28 |
| measure and sketch old line | C | A | 2 |
| develop materials list | D | C | 1 |
| erect scaffold | E | D | 2 |
| procure pipe | F | D | 30 |
| procure valves | G | D | 45 |
| deactivate old line | H | B | 1 |
| remove old line | I | E,H | 6 |
| prefabricate new pipe | J | F | 5 |
| place valves | K | I,G | 1 |
| place new pipe | L | I,J | 6 |
| weld pipe | M | L | 2 |
| connect valves | N | K,M | 1 |
| insulate | O | N | 4 |
| pressure test | P | N | 1 |
| remove scaffold | Q | O,P | 1 |
| clean up and turn over to operating crew | R | O,Q | 1 |

(b) There is additional information in the form of optimistic, most likely, and pessimistic time estimates for the project. Compute the expected mean time, $t_m$, and the variance, $\sigma^2$, for the activities. Which activities have the greatest uncertainty in their completion schedules?

| activity code | optimistic time ($t_{min}$) | most likely time ($t_{ml}$) | pessimistic time ($t_{max}$) |
|---|---|---|---|
| A | 8 | 10 | 12 |
| B | 26 | 26.5 | 36 |
| C | 1 | 2 | 3 |
| D | 0.5 | 1 | 1.5 |
| E | 1.5 | 1.63 | 4 |
| F | 28 | 28 | 40 |
| G | 40 | 42.5 | 60 |
| H | 1 | 1 | 1 |
| I | 4 | 6 | 8 |
| J | 4 | 4.5 | 8 |
| K | 0.5 | 0.9 | 2 |
| L | 5 | 5.25 | 10 |
| M | 1 | 2 | 3 |
| N | 0.5 | 1 | 1.5 |
| O | 3 | 3.75 | 6 |
| P | 1 | 1 | 1 |
| Q | 1 | 1 | 1 |
| R | 1 | 1 | 1 |

(c) Suppose that due to penalties in the contract each day the pipeline renewal project can be shortened is worth $100. Which of the following possibilities would you follow and why?

- Shorten $t_m$ of activity B by 4 days at a cost of $100.
- Shorten $t_{ml}$ of activity G by 5 days at a cost of $50.
- Shorten $t_m$ of activity O by 2 days at a cost of $150.
- Shorten $t_m$ of activity O by 2 days by drawing resources from activity N, thereby lengthening its $t_e$ by 2 days.

20. An 8% solution, a 10% solution, and a 20% solution of nitric acid are to be mixed in order to get 100 ml of a 12% solution. If the volume of acid from the 8% solution equals half the volume of acid from the other two solutions, how much of each is needed?

21. A tank contains 100 gallons of brine made by dissolving 60 pounds of salt in water. Salt water containing 1 pound of salt per gallon runs in at the rate of 2 gallons per minute. A well-stirred mixture runs out at the rate of 3 gallons per minute. Find the amount of salt in the tank at the end of 1 hour.

Timed

1. A survey field crew measures one leg of a traverse four times. The following results are obtained:

| repetition | measurement | direction |
|---|---|---|
| 1 | 1249.529 | forward |
| 2 | 1249.494 | backward |
| 3 | 1249.384 | forward |
| 4 | 1249.348 | backward |

The crew chief is under orders to obtain readings with confidence limits of 90%. (a) Which readings are acceptable? (b) Which readings are not acceptable? (c) Explain how to determine which readings are not acceptable. (d) What is the most probable value of the distance? (e) What is the error in the most probable value (at 90% confidence)? (f) If the distance is one side of a square traverse whose sides are all equal, what is the most probable closure error? (g) What is the probable error of part (f) expressed as a fraction? (h) Is this error of closure within the second order of accuracy? (i) Define accuracy and distinguish it from precision. (j) Give an example of a systematic error.

2. A 90-pound bag of a chemical is accidentally dropped in an aerating lagoon. The chemical is water soluble and non-reacting. The lagoon is 120 feet in diameter and filled to a depth of 10 feet. The aerators circulate and distribute the chemical evenly throughout the lagoon.

Water enters the lagoon at the rate of 30 gallons per minute. Fully mixed water is pumped into a reservoir at the rate of 30 gpm.

The established safe concentration of this chemical is 1 ppb (part per billion). How long will it take (in days) for the concentration of the discharge water to reach this level?

# 2 ENGINEERING ECONOMIC ANALYSIS

*Nomenclature*

| | | |
|---|---|---|
| A | annual amount or annuity | \$ |
| B | present worth of all benefits | \$ |
| $BV_j$ | book value at the end of the *j*th year | \$ |
| C | cost, or present worth of all costs | \$ |
| d | declining balance depreciation rate | decimal |
| $D_j$ | depreciation in year $j$ | \$ |
| D.R. | present worth of after-tax depreciation recovery | \$ |
| e | natural logarithm base (2.718) | – |
| EAA | equivalent annual amount | \$ |
| EUAC | equivalent uniform annual cost | \$ |
| f | federal income tax rate | decimal |
| F | future amount or future worth | \$ |
| G | uniform gradient amount | \$ |
| i | effective rate per period (usually per year) | decimal |
| k | number of compounding periods per year | – |
| n | number of compounding periods, or life of asset | – |
| P | present worth or present value | \$ |
| $P_t$ | present worth after taxes | \$ |
| ROR | rate of return | decimal |
| ROI | return on investment | \$ |
| r | nominal rate per year (rate per annum) | decimal |
| s | state income tax rate | decimal |
| $S_n$ | expected salvage value in year $n$ | \$ |
| t | composite tax rate, or time | decimal,- |
| z | a factor equal to $\frac{(1+i)}{(1-d)}$ | decimal |
| $\phi$ | effective rate per period | decimal |

## 1 EQUIVALENCE

Industrial decision makers using engineering economics are concerned with the timing of a project's cash flows as well as with the total profitability of that project. In this situation, a method is required to compare projects involving receipts and disbursements occurring at different times.

By way of illustration, consider \$100 placed in a bank account which pays 5% effective annual interest at the end of each year. After the first year, the account will have grown to \$105. After the second year, the account will have grown to \$110.25.

Assume that you will have no need for money during the next two years and that any money received would immediately go into your 5% bank account. Then, which of the following options would be more desirable?

**option a:** \$100 now

**option b:** \$105 to be delivered in one year

**option c:** \$110.25 to be delivered in two years

In light of the previous illustration, none of the options is superior under the assumptions given. If the first option is chosen, you will immediately place \$100 into a 5% account, and in two years the account will have grown to \$110.25. In fact, the account will contain \$110.25 at the end of two years regardless of the option chosen. Therefore, these alternatives are said to be *equivalent.*

## 2 CASH FLOW DIAGRAMS

Although they are not always necessary in simple problems (and they are often unwieldy in very complex problems), *cash flow diagrams* may be drawn to help visualize and simplify problems having diverse receipts and disbursements.

The conventions below are used to standardize cash flow diagrams.

- The horizontal (time) axis is marked off in equal increments, one per period, up to the duration or horizon of the project.

- All disbursements and receipts (cash flows) are assumed to take place at the end of the year in which they occur. This is known as the *year-end convention.* The exception to the year-end convention is any initial cost (purchase cost) which occurs at $t = 0$.
- Two or more transfers in the same year are placed end-to-end, and these may be combined.
- Expenses incurred before $t = 0$ are called *sunk costs.* Sunk costs are not relevant to the problem.
- Receipts are represented by arrows directed upward. Disbursements are represented by arrows directed downward. The arrow length is proportional to the magnitude of the cash flow.

*Example 2.1*

A mechanical device will cost \$20,000 when purchased. Maintenance will cost \$1000 each year. The device will generate revenues of \$5000 each year for 5 years after which the salvage value is expected to be \$7000. Draw and simplify the cash flow diagram.

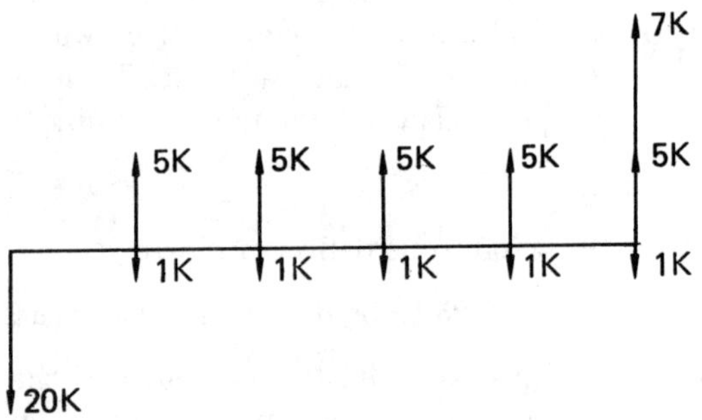


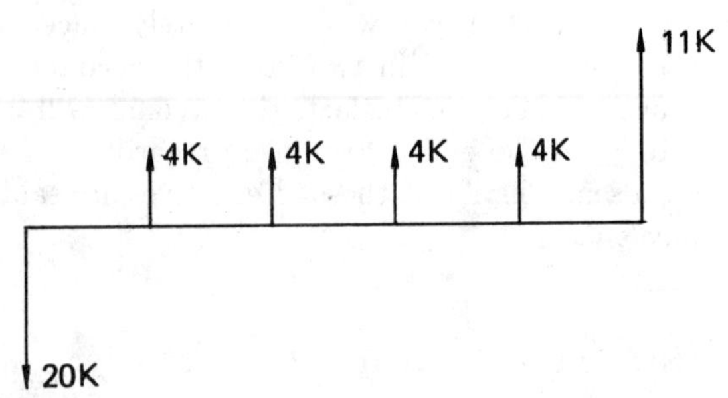


## 3 TYPICAL PROBLEM FORMAT

With the exception of some investment and rate of return problems, the typical problem involving engineering economics will have the following characteristics:

- An interest rate will be given.
- Two or more alternatives will be competing for funding.
- Each alternative will have its own cash flows.
- It is necessary to select the best alternative.

*Example 2.2*

Investment **A** costs \$10,000 today and pays back \$11,500 two years from now. Investment **B** costs \$8000 today and pays back \$4500 each year for two years. If an interest rate of 5% is used, which alternative is superior?

The solution to this example is not difficult, but it will be postponed until methods of calculating equivalence have been covered.

## 4 CALCULATING EQUIVALENCE

It was previously illustrated that \$100 now is equivalent at 5% to \$105 in one year. The equivalence of any present amount, $P$, at $t = 0$ to any future amount, $F$, at $t = n$ is called the *future worth* and can be calculated from equation 2.1.

$$F = P(1+i)^n \qquad 2.1$$

The factor $(1+i)^n$ is known as the *compound amount factor* and has been tabulated at the end of this chapter for various combinations of $i$ and $n$. Rather than actually writing the formula for the compound amount factor, the convention is to use the standard functional notation $(F/P, i\%, n)$. Thus,

$$F = P(F/P, i\%, n) \qquad 2.2$$

Similarly, the equivalence of any future amount to any present amount is called the *present worth* and can be calculated from

$$P = F(1+i)^{-n} = F(P/F, i\%, n) \qquad 2.3$$

The factor $(1+i)^{-n}$ is known as the *present worth factor*, with functional notation $(P/F, i\%, n)$. Tabulated values are also given for this factor at the end of this chapter.

*Example 2.3*

How much should you put into a 10% savings account in order to have \$10,000 in 5 years?

This problem could also be stated: What is the equivalent present worth of \$10,000 5 years from now if money is worth 10%?

$$P = F(1+i)^{-n} = 10{,}000(1+0.10)^{-5} = 6209$$

The factor 0.6209 would usually be obtained from the tables.

A cash flow which repeats regularly each year is known as an *annual amount.* When annual costs are incurred due to the functioning of a piece of equipment, they are often known as *operating and maintenance* (*O&M*) *costs.* The annual costs associated with operating a business in general are known as *general, selling, and administrative* (*GS&A*) *expenses.* Although the equivalent value for each of the $n$ annual amounts could be calculated and then summed, it is much easier to use one of the *uniform series factors*, as illustrated in example 2.4.

*Example 2.4*

Maintenance costs for a machine are \$250 each year. What is the present worth of these maintenance costs over a 12 year period if the interest rate is 8% ?

Notice that

$$\begin{aligned}(P/A, 8\%, 12) &= (P/F, 8\%, 1) + (P/F, 8\%, 2) \\ &\quad + \cdots + (P/F, 8\%, 12)\end{aligned}$$

Then,

$$\begin{aligned}P = A(P/A, i\%, n) &= -250(7.5361) \\ &= -1884\end{aligned}$$

A common complication involves a uniformly increasing cash flow. Such an increasing cash flow should be handled with the *uniform gradient factor*, $(P/G, i\%, n)$. The uniform gradient factor finds the present worth of a uniformly increasing cash flow which starts in year 2 (not year 1) as shown in example 2.5.

*Example 2.5*

Maintenance on an old machine is \$100 this year but is expected to increase by \$25 each year thereafter. What is the present worth of 5 years of maintenance? Use an interest rate of 10%.

In this problem, the cash flow must be broken down into parts. Notice that the 5-year gradient factor is used even though there are only 4 non-zero gradient cash flows.

$$\begin{aligned}P &= A(P/A, 10\%, 5) + G(P/G, 10\%, 5) \\ &= -100(3.7908) - 25(6.8618) = -551\end{aligned}$$

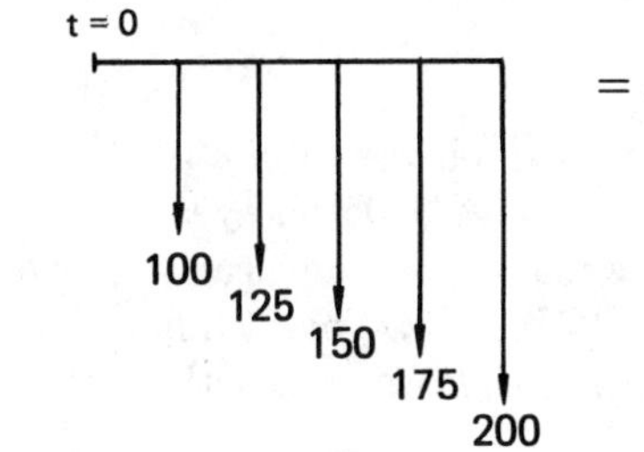


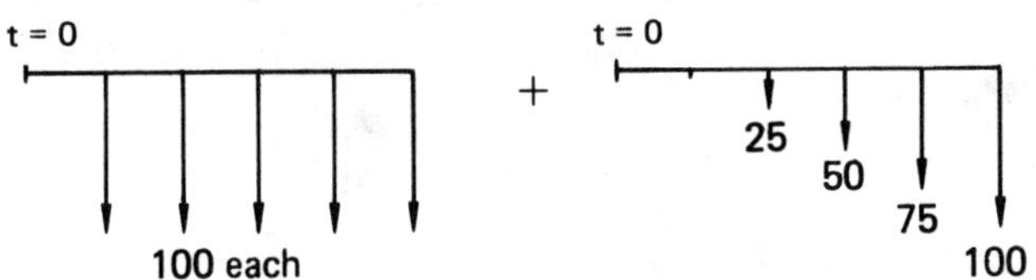


**Table 2.1**
Discount Factors for Discrete Compounding

| factor name | converts | symbol | formula |
|---|---|---|---|
| single payment compound amount | $P$ to $F$ | $(F/P, i\%, n)$ | $(1+i)^n$ |
| present worth | $F$ to $P$ | $(P/F, i\%, n)$ | $(1+i)^{-n}$ |
| uniform series Sinking Fund | $F$ to $A$ | $(A/F, i\%, n)$ | $\frac{i}{(1+i)^n-1}$ |
| capital recovery | $P$ to $A$ | $(A/P, i\%, n)$ | $\frac{i(1+i)^n}{(1+i)^n-1}$ |
| compound amount | $A$ to $F$ | $(F/A, i\%, n)$ | $\frac{(1+i)^n-1}{i}$ |
| equal series present worth | $A$ to $P$ | $(P/A, i\%, n)$ | $\frac{(1+i)^n-1}{i(1+i)^n}$ |
| uniform gradient | $G$ to $P$ | $(P/G, i\%, n)$ | $\frac{(1+i)^n-1}{i^2(1+i)^n} - \frac{n}{i(1+i)^n}$ |

Various combinations of the compounding and discounting factors are possible. For instance, the annual cash flow that would be equivalent to a uniform gradient may be found from

$$A = G(P/G, i\%, n)(A/P, i\%, n) \qquad 2.4$$

Formulas for all of the compounding and discounting factors are contained in table 2.1. Normally, it will not be necessary to calculate factors from the formulas. The tables at the end of this chapter are adequate for solving most problems.

## 5 THE MEANING OF "PRESENT WORTH" AND "$i$"

It is clear that \$100 invested in a 5% bank account will allow you to remove \$105 one year from now. If this investment is made, you will clearly receive a *return on investment* (ROI) of \$5. The cash flow diagram and the present worth of the two transactions are

$$\begin{aligned} P &= -100 + 105(P/F, 5\%, 1) \\ &= -100 + 105(0.9524) = 0 \end{aligned}$$

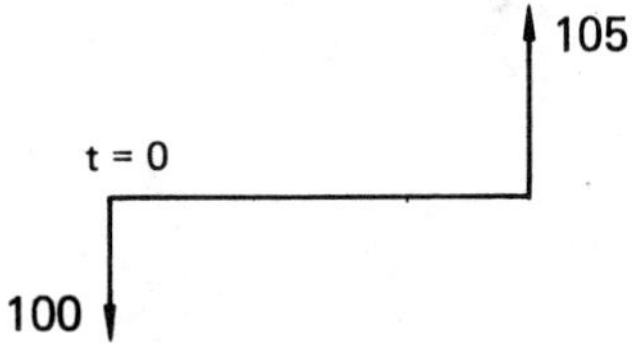


**Figure 2.1**

Notice that the present worth is zero even though you did receive a 5% return on your investment.

However, if you are offered \$120 for the use of \$100 over a one-year period, the cash flow diagram and present worth (at 5%) would be

$$\begin{aligned} P &= -100 + 120(P/F, 5\%, 1) \\ &= -100 + 120(0.9524) = 14.29 \end{aligned}$$

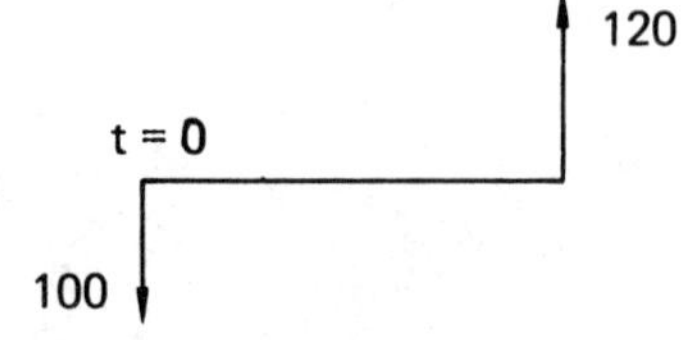


**Figure 2.2**

Therefore, it appears that the present worth of an alternative is equal to the equivalent value at $t = 0$ of the increase in return above that which you would be able to earn in an investment offering $i\%$ per period. In the above case, \$14.29 is the present worth of (\$20−\$5), the difference in the two *ROI's*.

Alternatively, the actual earned interest rate, called *rate of return* (*ROR*), can be defined as the rate which makes the present worth of the alternative zero.

The *present worth* is also the amount that you would have to be given to dissuade you from making an investment, since placing the initial investment amount along with the present worth into a bank account earning $i\%$ will yield the same eventual *ROI*. Relating this to the previous paragraphs, you could be dissuaded against investing \$100 in an alternative which would return \$120 in one year by a $t = 0$ payment of \$14.29. Clearly, (\$100 + \$14.29) invested at $t = 0$ will also yield \$120 in one year at 5%.

The selection of the interest rate is difficult in engineering economics problems. Usually it is taken as the average rate of return that an individual or business organization has realized in past investments. Fortunately, an interest rate is usually given. A company may not know what effective interest rate to use in an economic analysis. In such a case, the company can establish a minimum acceptable return on its investment. This *minimum attractive rate of return* (*MARR*) should be used as the effective interest rate $i$ in economic analyses.

It should be obvious that alternatives with negative present worths are undesirable, and that alternatives with positive present worths are desirable because they increase the average earning power of invested capital.

## 6 CHOICE BETWEEN ALTERNATIVES

A variety of methods exists for selecting a superior alternative from among a group of proposals. Each method has its own merits and applications.

### Present Worth Method

The *Present Worth Method* has already been implied. When two or more alternatives are capable of performing the same functions, the superior alternative will have the largest present worth. This method is suitable for ranking the desirability of alternatives. The present worth method is restricted to evaluating alternatives that are mutually exclusive and which have the same lives.

Returning to example 2.2, the present worth of each alternative should be found in order to determine which alternative is superior.

*Example 2.2, continued*

$$P(\mathbf{A}) = -10{,}000 + 11{,}500(P/F, 5\%, 2) = 431$$
$$P(\mathbf{B}) = -8000 + 4500(P/A, 5\%, 2) = 367$$

Alternative **A** is superior and should be chosen.

### Capitalized Cost Method

The present worth of a project with an infinite life is known as the *capitalized cost* or life cycle cost. Capitalized cost is the amount of money at $t = 0$ needed to perpetually support the project on the earned interest only. Capitalized cost is a positive number when expenses exceed income.

$$\text{Capitalized Cost} = \text{Initial Cost} + \frac{\text{Annual Costs}}{i} \qquad 2.5$$

Capitalized cost is the present worth of an infinitely-lived project. Normally, it would be difficult to work with an infinite stream of cash flows since most economics tables don't list factors for periods in excess of 100 years. However, the $(A/P)$ discounting factor approaches the interest rate as $n$ becomes large. Since the $(P/A)$ and $(A/P)$ factors are reciprocals of each other, we would expect to divide an infinite series of equal cash flows by the interest rate in order to calculate the present worth of the infinite series. This is the basis of equation 2.5.

Equation 2.5 can be used when the annual costs are equal in every year. The "Annual Cost" in that equation is assumed to be the same each year. If the operating and maintenance costs occur irregularly instead of annually, or if the costs vary from year to year, it will be necessary to somehow determine a cash flow of equal annual amounts (EAA) which is equivalent to the stream of original costs.

The equal annual amount may be calculated in the usual manner by first finding the present worth of all the actual costs, and then multiplying the present worth by the interest rate (the $(A/P)$ factor for an infinite series). However, it is not even necessary to convert the present worth to an equal annual amount, since equation 2.6 will convert the equal annual amount back to the present worth.

$$\text{Capitalized Cost} = \text{Initial Cost} + \frac{\text{EAA}}{i} \qquad 2.6$$

In comparing two alternatives, each of which is infinitely lived, the superior alternative will have the lowest capitalized cost.

### Annual Cost Method

Alternatives which accomplish the same purpose but which have unequal lives must be compared by the *Annual Cost Method.* The annual cost method assumes that each alternative will be replaced by an identical twin at the end of its useful life (infinite renewal). This method, which may also be used to rank alternatives according to their desirability, is also called the *Annual Return Method* and *Capital Recovery Method.*

Restrictions are that the alternatives must be mutually exclusive and infinitely renewed up to the duration of the longest-lived alternative. The calculated annual cost is known as the *Equivalent Uniform Annual Cost, EUAC.* Cost is a positive number when expenses exceed income.

*Example 2.6*

Which of the following alternatives is superior over a 30 year period if the interest rate is 7%?

| | A | B |
|---|---|---|
| type | brick | wood |
| life | 30 years | 10 years |
| cost | \$1800 | \$450 |
| maintenance | \$5/year | \$20/year |

$$\text{EUAC}(A) = 1800(A/P, 7\%, 30) + 5 = 150$$
$$\text{EUAC}(B) = 450(A/P, 7\%, 10) + 20 = 84$$

Alternative **B** is superior since its annual cost of operation is the lowest. It is assumed that three wood facilities, each with a life of 10 years and a cost of \$450, will be built to span the 30 year period.

### Benefit-Cost Ratio Method

The *Benefit-Cost Ratio Method* is often used in municipal project evaluations where benefits and costs accrue to different segments of the community. With this method, the present worth of all benefits (regardless of the beneficiary) is divided by the present worth of all costs. The project is considered acceptable if the ratio exceeds *one.*

When the benefit-cost ratio method is used, disbursements by the initiators or sponsors are *costs.* Disbursements by the users of the project are known as *disbenefits.* It is often difficult to determine whether a cash flow is a cost or a disbenefit (whether to place it in the numerator or denominator of the benefit-cost ratio calculation).

Regardless of where the cash flow is placed, an acceptable project will always have a benefit-cost ratio greater than one, although the actual numerical result will depend on the placement. For this reason, the benefit-cost ratio method should not be used to rank competing projects.

The benefit-cost ratio method may be used to rank alternative proposals only if an *incremental analysis* is used. First, determine that the ratio is greater than one for each alternative. Then, calculate the ratio of benefits to costs

$$\frac{B_2 - B_1}{C_2 - C_1}$$

for each possible pair of alternatives. If the ratio exceeds one, alternative 2 is superior to alternative 1. Otherwise, alternative 1 is superior.

### Rate of Return Method

Perhaps no method of analysis is less understood than the *Rate of Return* (ROR) *Method.* As was stated previously, the ROR is the interest rate that would yield identical profits if all money were invested at that rate. The present worth of any such investment is zero.

The ROR is defined as the interest rate that will discount all cash flows to a total present worth equal to the initial required investment. This definition is used to determine the ROR of an alternative. The advantage of the ROR method is that no knowledge of an interest rate is required.

To find the ROR of an alternative, proceed as follows:

*step 1*: Set up the problem as if to calculate the present worth.

*step 2*: Arbitrarily select a reasonable value for $i$. Calculate the present worth.

*step 3*: Choose another value of $i$ (not too close to the original value) and again solve for the present worth.

*step 4*: Interpolate or extrapolate the value of $i$ which gives a zero present worth.

*step 5*: For increased accuracy, repeat steps (2) and (3) with two more values that straddle the value found in step (4).

A common, although incorrect, method of calculating the ROR involves dividing the annual receipts or returns by the initial investment. However, this technique ignores such items as salvage, depreciation, taxes, and the time value of money. This technique also fails when the annual returns vary.

Once a rate of return is known for an investment alternative, it is typically compared to the *minimum attractive rate of return* (*MARR*) specified by a company. However, ROR should not be used to rank alternatives. When two alternatives have ROR's exceeding the MARR, it is not sufficient to select the alternative with the higher ROR.

An *incremental analysis*, also known as a *rate of return on added investment study*, should be performed if ROR is to be used to select between investments. In an incremental analysis, the cash flows for the investment with the lower initial cost are subtracted from the cash flows for the higher-priced alternative on a year-by-year basis. This produces, in effect, a third alternative representing the cost and benefits of the added investment. The added expense of the higher-priced investment is not warranted unless the ROR of this third alternative exceeds the MARR as well.

*Example 2.7*

What is the return on invested capital if \$1000 is invested now with \$500 being returned in year 4 and \$1000 being returned in year 8?

First, set up the problem as a present worth calculation.

$$P = -1000 + 500(P/F, i\%, 4) + 1000(P/F, i\%, 8)$$

Arbitrarily select $i = 5\%$. The present worth is then found to be \$88.15. Next take a higher value of $i$ to reduce the present worth. If $i = 10\%$, the present worth is $-\$192$. The ROR is found from simple interpolation to be approximately 6.6%.

## 7 TREATMENT OF SALVAGE VALUE IN REPLACEMENT STUDIES

An investigation into the retirement of an existing process or piece of equipment is known as a *replacement study.* Replacement studies are similar in most respects to other alternative comparison problems: an interest rate is given, two alternatives exist, and one of the previously mentioned methods of comparing alternatives is used to choose the superior alternative.

In replacement studies, the existing process or piece of equipment is known as the *defender.* The new process or piece of equipment being considered for purchase is known as the *challenger.*

Because most defenders still have some market value when they are retired, the problem of what to do with the salvage arises. It seems logical to use the salvage value of the defender to reduce the initial purchase cost of the challenger. This is consistent with what would actually happen if the defender were to be retired.

By convention, however, the salvage value is subtracted from the defender's present value. This does not seem logical, but it is done to keep all costs and benefits related to the defender with the defender. In this case, the salvage value is treated as an opportunity cost which would be incurred if the defender is not retired.

If the defender and the challenger have the same lives and a present worth study is used to choose the superior alternative, the placement of the salvage value will have no effect on the net difference between present worths for the challenger and defender. Although the values of the two present worths will be different depending on the placement, the difference in present worths will be the same.

If the defender and the challenger have different lives, an annual cost comparison must be made. Since the salvage value would be 'spread over' a different number of years depending on its placement, it is important to abide by the conventions listed in this section.

There are a number of ways to handle salvage value. The best way is to think of the EUAC of the defender as the cost of keeping the defender from now until next year. In addition to the usual operating and maintenance costs, that cost would include an opportunity interest cost incurred by not selling the defender and also a drop in the salvage value if the defender is kept for one additional year. Specifically,

$$\begin{aligned}\text{EUAC(defender)} &= \text{maintenance costs}\\ &\quad + i(\text{current salvage value})\\ &\quad + (\text{current salvage–next year's salvage})\end{aligned} \qquad 2.7$$

It is important in retirement studies not to double count the salvage value. That is, it would be incorrect to add the salvage value to the defender and at the same time subtract it from the challenger.

## 8 BASIC INCOME TAX CONSIDERATIONS

Assume that an organization pays $f\%$ of its profits to the federal government as income taxes. If the organization also pays a state income tax of $s\%$, and if state taxes paid are recognized by the federal government as expenses, then the composite tax rate is

$$t = s + f - sf \qquad 2.8$$

The basic principles used to incorporate taxation into economic analyses are listed below.

a. Initial purchase cost is unaffected by income taxes.

b. Salvage value is unaffected by income taxes.

c. Deductible expenses, such as operating costs, maintenance costs, and interest payments, are reduced by $t\%$ (e.g., multiplied by the quantity $(1-t)$).

d. Revenues are reduced by $t\%$ (e.g., multiplied by the quantity $(1-t)$).

e. Depreciation is multiplied by $t$ and added to the appropriate year's cash flow, increasing that year's present worth.

Income taxes and depreciation have no bearing on municipal or governmental projects since municipalities, states, and the U.S. Government pay no taxes.

*Example 2.8*

A corporation which pays 53% of its revenue in income taxes invests \$10,000 in a project which will result in \$3000 annual revenue for 8 years. If the annual expenses are \$700, salvage after 8 years is \$500, and 9% interest is used, what is the after-tax present worth? Disregard depreciation.

$$\begin{aligned}P_t &= -10{,}000 + 3000(P/A, 9\%, 8)(1-0.53)\\ &\quad - 700(P/A, 9\%, 8)(1-0.53)\\ &\quad + 500(P/F, 9\%, 8)\\ &= -3766\end{aligned}$$

It is interesting that the alternative evaluated in example 2.8 is undesirable if income taxes are considered but is desirable if income taxes are omitted.

## 9 DEPRECIATION

Although depreciation calculations may be considered independently in examination questions, it is important to recognize that depreciation has no effect on engineering economic calculations unless income taxes are also considered.

Generally, tax regulations do not allow the cost of equipment[1] to be treated as a deductible expense in the year of purchase. Rather, portions of the cost may be allocated to each of the years of the item's economic life (which may be different from the actual useful life). Each year, the book value (which is initially equal to the purchase price) is reduced by the depreciation in that year. Theoretically, the book value of an item will equal the market value at any time within the economic life of that item.

Since tax regulations allow the depreciation in any year to be handled as if it were an actual operating expense, and since operating expenses are deductible from the income base prior to taxation, the after-tax profits will be increased. If $D$ is the depreciation, the net result to the after-tax cash flow will be the addition of $tD$.

[1] The IRS tax regulations allow depreciation on almost all forms of *property* except land. The following types of property are distinguished: *real* (e.g., buildings used for business), *residential* (e.g., buildings used as rental property), and *personal* (e.g., equipment used for business). Personal property does *not* include items for personal use, despite its name. *Tangible* personal property is distinguished from *intangible property* (e.g., goodwill, copyrights, patents, trademarks, franchises, and agreements not to compete).

The present worth of all depreciation over the economic life of the item is called the *depreciation recovery.* Although originally established to do so, depreciation recovery can never fully replace an item at the end of its life.

Depreciation is often confused with amortization and depletion. While depreciation spreads the cost of a fixed asset over a number of years, *amortization* spreads the cost of an intangible asset (e.g., a patent) over some basis such as time or expected units of production.

*Depletion* is another artificial deductible operating expense designed to compensate mining organizations for decreasing mineral reserves. Since original and remaining quantities of minerals are seldom known accurately, the *depletion allowance* is calculated as a fixed percentage of the organization's gross income. These percentages are usually in the 10%–20% range and apply to such mineral deposits as oil, natural gas, coal, uranium, and most metal ores.

There are four common methods of calculating depreciation. The book value of an asset depreciated with the *Straight Line* (SL) *Method* (also known as the *Fixed Percentage Method*) decreases linearly from the initial purchase at $t = 0$ to the estimated salvage at $t = n$. The depreciated amount is the same each year. The quantity $(C - S_n)$ in equation 2.9 is known as the *depreciation base.*

$$D_j = \frac{C - S_n}{n} \qquad 2.9$$

*Double Declining Balance*[2] (DDB) depreciation is independent of salvage value. Furthermore, the book value never stops decreasing, although the depreciation decreases in magnitude. Usually, any remaining book value is written off in the last year of the asset's estimated life. Unlike any of the other depreciation methods, DDB depends on accumulated depreciation.

$$D_j = \frac{2(C - \sum_{i=1}^{j-1} D_i)}{n} \qquad 2.10$$

In *Sum-of-the-Years'-Digits* (SOYD) depreciation, the digits from 1 to $n$ inclusive, are summed. The total, $T$, can also be calculated from

$$T = \frac{1}{2}n(n+1) \qquad 2.11$$

[2] Double declining balance depreciation is a particular form of *declining balance depreciation,* as defined by the IRS tax regulations. Declining balance depreciation also includes 125% declining balance and 150% declining balance depreciations which can be calculated by substituting 1.25 and 1.50 respectively for the 2 in equation 2.10.

The depreciation can be found from

$$D_j = \frac{(C - S_n)(n - j + 1)}{T} \qquad 2.12$$

The *Sinking Fund Method* is seldom used in industry because the initial depreciation is low. The formula for sinking fund depreciation (which increases each year) is

$$D_j = (C - S_n)(A/F, i\%, n)(F/P, i\%, j-1) \qquad 2.13$$

The above discussion gives the impression that any form of depreciation may be chosen regardless of the nature and circumstances of the purchase. In reality, the IRS tax regulations place restrictions on the higher-rate ("accelerated") methods such as DDB and SOYD. Furthermore, the *Economic Recovery Act of 1981* substantially changed the laws relating to personal and corporate income taxes.

Property placed into service in 1981 or after must use the *Accelerated Cost Recovery System* (*ACRS*). Other methods (straight-line, declining balance, etc.) cannot be used except in special cases.

Property placed into service in 1980 or before must continue to be depreciated according to the method originally chosen (e.g., straight-line, declining balance, or sum-of-years-digits). *ACRS* cannot be used.

Under *ACRS*, the cost recovery amount in the $j$th year of an asset's cost recovery period is calculated by multiplying the initial cost by a factor.

$$D_j = (\text{initial cost})(\text{factor}) \qquad 2.14$$

The initial cost used is not reduced by the asset's salvage value for either the regular or alternate *ACRS* calculations. The factor used depends on the asset's cost recovery period. Such factors are subject to continuing legislation changing them. Current tax publications should be consulted before using the *ACRS* method.

Three other depreciation methods should be mentioned, not because they are currently accepted or in widespread use, but because they are occasionally called for by name.

The *sinking-fund plus interest on first cost* depreciation method, like the following two methods, is an attempt to include the *opportunity interest cost* on the purchase price with the depreciation. That is, the purchasing company not only incurs an annual loss due to the drop in book value, but it also loses the interest on the purchase price. The formula for this method is

$$D_j = (C - S_n)(A/F, i\%, n) + (C)(i) \qquad 2.15$$

The *straight-line plus interest on first cost* method is similar. Its formula is

$$D_j = \left(\frac{1}{n}\right)(C - S_n) + (C)(i) \qquad 2.16$$

The *straight-line plus average interest method* assumes that the opportunity interest cost should be based on the book value only, not on the full purchase price. Since the book value changes each year, an average value is used. The depreciation formula is

$$D_j = \left(\frac{1}{n}\right)(C - S_n) + \frac{1}{2}(i)(C - S_n)\left(\frac{n+1}{n}\right) + iS_n \qquad 2.17$$

These three depreciation methods are not to be used in the usual manner (e.g., in conjunction with the income tax rate). These methods are attempts to calculate a more accurate annual cost of an alternative. Sometimes they work, and sometimes they give misleading answers. Their use cannot be recommended. They are included in this chapter only for the sake of completeness.

*Example 2.9*

An asset is purchased for $9000. Its estimated economic life is 10 years, after which it will be sold for $1000. Find the depreciation in the first three years using SL, DDB, and SOYD.

**SL:** $D = \dfrac{(9000 - 1000)}{10} = 800$ each year

**DDB:** $D_1 = \dfrac{2(9000)}{10} = 1800$ in year 1

$D_2 = \dfrac{2(9000 - 1800)}{10} = 1440$ in year 2

$D_3 = \dfrac{2(9000 - 3240)}{10} = 1152$ in year 3

**SOYD:** $T = \frac{1}{2}(10)(11) = 55$

$D_1 = \left(\frac{10}{55}\right)(9000 - 1000) = 1455$ in year 1

$D_2 = \left(\frac{9}{55}\right)(8000) = 1309$ in year 2

$D_3 = \left(\frac{8}{55}\right)(8000) = 1164$ in year 3

*Example 2.10*

For the asset described in example 2.9, calculate the book value during the first three years if SOYD depreciation is used.

The book value at the beginning of year 1 is $9000. Then,

$$BV_1 = 9000 - 1455 = 7545$$
$$BV_2 = 7545 - 1309 = 6236$$
$$BV_3 = 6236 - 1164 = 5072$$

*Example 2.11*

For the asset described in example 2.9, calculate the after-tax depreciation recovery with SL and SOYD depreciation methods. Use 6% interest with 48% income taxes.

**SL:** $D.R. = 0.48(800)(P/A, 6\%, 10) = 2826$

**SOYD:** The depreciation series can be thought of as a constant 1,454 term with a negative 145 gradient.

$$\begin{aligned} D.R. &= 0.48(1454)(P/A, 6\%, 10) \\ &\quad - 0.48(145)(P/G, 6\%, 10) \\ &= 3076 \end{aligned}$$

Finding book values, depreciation, and depreciation recovery is particularly difficult with DDB depreciation, since all previous years' quantities seem to be required. It appears that the depreciation in the 6th year cannot be calculated unless the values of depreciation for the first five years are calculated first. Questions asking for depreciation or book value in the middle or at the end of an asset's economic life may be solved from the following equations:

$$d = \frac{2}{n} \qquad 2.18$$

$$z = \frac{1+i}{1-d} \qquad 2.19$$

$$(P/EG) = \frac{z^n - 1}{z^n(z-1)} \qquad 2.20$$

Then, assuming that the remaining book value at $t = n$ is written off in one lump sum, the present worth of the depreciation recovery is

$$D.R. = t\left[\frac{(d)(C)}{(1-d)}(P/EG) + (1-d)^n(C)(P/F, i\%, n)\right] \qquad 2.21$$

$$D_j = (d)(C)(1-d)^{j-1} \qquad 2.22$$

$$BV_j = C(1-d)^j \qquad 2.23$$

*Example 2.12*

What is the after-tax present worth of the asset described in example 2.8 if SL, SOYD, and DDB depreciation methods are used?

The after-tax present worth, neglecting depreciation, was previously found to be −3766.

Using SL, the depreciation recovery is

$$\begin{aligned} D.R. &= (0.53)\frac{(10{,}000-500)}{8}(P/A,9\%,8) \\ &= 3483 \end{aligned}$$

Using SOYD, the depreciation recovery is calculated as follows:

$$T = \frac{1}{2}(8)(9) = 36$$

Depreciation base = (10,000 − 500) = 9500

$$\begin{aligned} D_1 &= \frac{8}{36}(9500) = 2111 \\ G &= \text{gradient} = \frac{1}{36}(9500) \\ &= 264 \end{aligned}$$

$$\begin{aligned} D.R. &= (0.53)\,[2111(P/A,9\%,8) \\ &\quad - 264\,(P/G,9\%,8)] \\ &= 3829 \end{aligned}$$

Using DDB, the depreciation recovery is calculated as follows:

$$d = \frac{2}{8} = 0.25$$

$$z = \frac{1.09}{0.75} = 1.453$$

$$(P/EG) = \frac{(1.453)^8 - 1}{(1.453)^8(0.453)} = 2.096$$

$$\begin{aligned} D.R. &= 0.53\left[\frac{(0.25)(10{,}000)}{0.75}(2.096)\right. \\ &\quad \left. +(0.75)^8(10{,}000)(P/F,9\%,8)\right] \\ &= 3969 \end{aligned}$$

The after-tax present worths including depreciation recovery are:

**SL:** $P_t = -3766 + 3483 = -283$
**SOYD:** $P_t = -3766 + 3829 = 63$
**DDB:** $P_t = -3766 + 3969 = 203$

## 10 ADVANCED INCOME TAX CONSIDERATIONS

There are a number of specialized techniques that are needed infrequently. These techniques are related more to the accounting profession than to the engineering profession. Nevertheless, it is occasionally necessary to use these techniques.

### A. INVESTMENT TAX CREDIT

An investment tax credit (also known as a tax credit or an investment credit) is a one-time credit against income taxes. The investment tax credit is calculated as a fraction of the initial purchase price of certain types of equipment purchased for industrial, commercial, and manufacturing use.

$$\text{credit} = (\text{initial cost})(\text{fraction}) \qquad 2.24$$

The fraction is subject to continuing legislation changing its value and applicability. The fraction, which typically is taken as 10% for initial estimates, actually depends on the asset life, year of acquisition, and number of years the asset is held before being disposed of. The current tax laws should be studied before using the investment tax credit.

Since the investment tax credit reduces the buyer's tax liability, the credit should only be used in after-tax analyses.

### B. GAIN ON THE SALE OF A DEPRECIATED ASSET

If an asset is sold for more than its current book value, the difference between selling price and book value is taxable income. The gain is taxed at capital gains rates. Excluded from this preferential treatment is non-residential real property depreciated under regular *ACRS* provisions. However, non-residential real property depreciated under the straight-line alternate method qualifies for the capital gains rate.

### C. CAPITAL GAINS AND LOSSES

A *gain* is defined as the difference between selling and purchase prices of a capital asset. The gain is called a *regular gain* if the item sold has been kept less than one year. The gain is called a *capital gain* if the item sold has been kept for longer than one year. Capital gains are taxed at the taxpayer's usual rate, but 60% of the gain is excluded from taxation.

*Regular* (as defined above) *losses* are fully deductible in the year of their occurrence. The IRS tax regulations should be consulted to determine the treatment of *capital losses.*

## 11 RATE AND PERIOD CHANGES

All of the foregoing calculations were based on compounding once a year at an *effective interest rate*, $i$. However, some problems specify compounding more frequently than annually. In such cases, a *nominal interest rate*, $r$, will be given. The nominal rate does not include the effect of compounding and is not the same as the effective rate, $i$. A nominal rate may be used to calculate the effective rate by using equation 2.25 or 2.26.

$$i = \left(1 + \frac{r}{k}\right)^k - 1 \qquad 2.25$$

$$= (1 + \phi)^k - 1 \qquad 2.26$$

A problem may also specify an effective rate per period, $\phi$, (e.g., per month). However, that will be a simple problem since compounding for $n$ periods at an effective rate per period is not affected by the definition or length of the period.

The following rules may be used to determine which interest rate is given in a problem:

- Unless specifically qualified in the problem, the interest rate given is an annual rate.
- If the compounding is annually, the rate given is the effective rate. If compounding is other than annually, the rate given is the nominal rate.
- If the type of compounding is not specified, assume annual compounding.

In the case of continuous compounding, the appropriate discount factors may be calculated from the formulas in table 2.2.

**Table 2.2**
Discount Factors for
Continuous Compounding

| | |
|---|---|
| (F/P) | $e^{rn}$ |
| (P/F) | $e^{-rn}$ |
| (A/F) | $(e^r - 1)/(e^{rn} - 1)$ |
| (F/A) | $(e^{rn} - 1)/(e^r - 1)$ |
| (A/P) | $(e^r - 1)/(1 - e^{-rn})$ |
| (P/A) | $(1 - e^{-rn})/(e^r - 1)$ |

*Example 2.13*

A savings and loan offers $5\frac{1}{4}\%$ compounded daily. What is the annual effective rate?

method 1: $r = 0.0525$, $k = 365$

$$i = \left(1 + \frac{0.0525}{365}\right)^{365} - 1 = 0.0539$$

method 2: Assume daily compounding is the same as continuous compounding.

$$i = (F/P) - 1$$
$$= e^{0.0525} - 1 = 0.0539$$

## 12 PROBABILISTIC PROBLEMS

Thus far, all of the cash flows included in the examples have been known exactly. If the cash flows are not known exactly but are given by some implicit or explicit probability distribution, the problem is *probabilistic.*

Probabilistic problems typically possess the following characteristics:

- There is a chance of extreme loss that must be minimized.
- There are multiple alternatives that must be chosen from. Each alternative gives a different degree of protection against the loss or failure.
- The outcome is independent of the alternative chosen. Thus, as illustrated in example 2.15, the size of the dam that is chosen for construction will not alter the rainfall in successive years. However, it will alter the effects on the down-stream watershed areas.

Probabilistic problems are typically solved using annual costs and expected values. An *expected value* is similar to an 'average value' since it is calculated as the mean of the given probability distribution. If cost 1 has a probability of occurrence of $p_1$, cost 2 has a probability of occurrence of $p_2$, and so on, the expected value is

$$E(\text{cost}) = p_1(\text{cost 1}) + p_2(\text{cost 2}) + \cdots \qquad 2.27$$

*Example 2.14*

Flood damage in any year is given according to the table below. What is the present worth of flood damage for a 10-year period? Use 6%.

| Damage | Probability |
|---|---|
| 0 | 0.75 |
| $10,000 | 0.20 |
| $20,000 | 0.04 |
| $30,000 | 0.01 |

The expected value of flood damage is

$$E(\text{damage}) = (0)(0.75) + (10{,}000)(0.20) + (20{,}000)(0.04) + (30{,}000)(0.01)$$
$$= 3100$$

$$\text{present worth} = 3100(P/A, 6\%, 10)$$
$$= 22{,}816$$

Probabilities in probabilistic problems may be given to you in the problem (as in the example above) or you may have to obtain them from some named probability distribution. In either case, the probabilities are known explicitly and such problems are known as *explicit probability problems.*

*Example 2.15*

A dam is being considered on a river which periodically overflows and causes \$600,000 damage. The damage is essentially the same each time the river causes flooding. The project horizon is 40 years. A 10% interest rate is being used.

Three different designs are available, each with different costs and storage capacities.

| design alternative | cost | maximum capacity |
|---|---|---|
| **A** | 500,000 | 1 unit |
| **B** | 625,000 | 1.5 units |
| **C** | 900,000 | 2.0 units |

The U.S. Weather Service has provided a statistical analysis of annual rainfall in the area draining into the river.

| units annual rainfall | probability |
|---|---|
| 0 | 0.10 |
| 0.1–0.5 | 0.60 |
| 0.6–1.0 | 0.15 |
| 1.1–1.5 | 0.10 |
| 1.6–2.0 | 0.04 |
| 2.1 or more | 0.01 |

Which design alternative would you choose assuming the dam is essentially empty at the start of each rainfall season?

The sum of the construction cost and the expected damage needs to be minimized. If alternative **A** is chosen, it will have a capacity of 1 unit. Its capacity will be exceeded (causing \$600,000 damage) when the annual rainfall exceeds 1 unit. Therefore, the annual cost of **A** is

$$\begin{aligned}\text{EUAC}(\mathbf{A}) &= 500{,}000(A/P, 10\%, 40)\\ &\quad + 600{,}000(0.10 + 0.04 + 0.01)\\ &= 141{,}150\end{aligned}$$

Similarly,

$$\begin{aligned}\text{EUAC}(\mathbf{B}) &= 625{,}000(A/P, 10\%, 40)\\ &\quad + 600{,}000(0.04 + 0.01)\\ &= 93{,}940\end{aligned}$$

$$\begin{aligned}\text{EUAC}(\mathbf{C}) &= 900{,}000(A/P, 10\%, 40)\\ &\quad + 600{,}000(0.01)\\ &= 98{,}070\end{aligned}$$

Alternative **B** should be chosen.

In other problems, a probability distribution will not be given even though some parameter (such as the life of an alternative) is not known with certainty. Such problems are known as *implicit probability problems* since they require a reasonable assumption about the probability distribution.

Implicit probability problems typically involve items whose expected times to failure are known. The key to such problems is in recognizing that an expected time to failure is not the same as a fixed life.

Reasonable assumptions can be made about the form of probability distributions in implicit probability problems.

One such reasonable assumption is that of a *rectangular distribution.* A rectangular distribution is one which is assumed to give an equal probability of failure in each year. Such an assumption is illustrated in example 2.16.

*Example 2.16*

A bridge is needed for 20 years. Failure of the bridge at any time will require a 50% reinvestment. Assume that each alternative has an annual probability of failure that is inversely proportional to its expected time to failure. Evaluate the two design alternatives below using 6% interest.

| design alternative | initial cost | expected time to failure | annual costs | salvage at $t = 20$ |
|---|---|---|---|---|
| **A** | 15,000 | 9 years | 1200 | 0 |
| **B** | 22,000 | 43 years | 1000 | 0 |

For alternative **A**, the probability of failure in any year is $\left(\frac{1}{9}\right)$. Similarly, the annual failure probability for alternative **B** is $\left(\frac{1}{43}\right)$.

$$\begin{aligned}\text{EUAC}(\mathbf{A}) &= 15{,}000(A/P, 6\%, 20)\\ &\quad + 15{,}000(0.5)\left(\frac{1}{9}\right) + 1200\\ &= 3341\end{aligned}$$

$$\begin{aligned}\text{EUAC}(\mathbf{B}) &= 22{,}000(A/P, 6\%, 20)\\ &\quad + 22{,}000(0.5)\left(\frac{1}{43}\right) + 1000\\ &= 3174\end{aligned}$$

Alternative **B** should be chosen.

## 13 ESTIMATING ECONOMIC LIFE

As assets grow older, their operating and maintenance costs typcially increase each year. Eventually, the cost to keep an asset in operation becomes prohibitive, and the asset is retired or replaced. However, it is not always obvious when an asset should be retired or replaced.

As the asset's maintenance is increasing each year, the amortized cost of its initial purchase is decreasing. It is the sum of these two costs that should be evaluated to determine the point at which the asset should be retired or replaced. Since an asset's initial purchase price is likely to be high, the amortized cost will be the controlling factor in those years when the maintenance costs are low. Therefore, the EUAC of the asset will decrease in the initial part of its life.

However, as the asset grows older, the change in its amortized cost decreases while maintenance increases. Eventually the sum of the two costs reaches a minimum and then starts to increase. The age of the asset at the minimum cost point is known as the *economic life* of the asset. The economic life is, generally, less than the mission and technological lifetimes of the asset.

The determination of an asset's economic life is illustrated by example 2.17.

*Example 2.17*

A bus in a municipal transit system has the characteristics listed below. When should the city replace its buses if money can be borrowed at 8% ?

Initial cost: $120,000

| year | maintenance cost | salvage value |
|---|---|---|
| 1 | 35,000 | 60,000 |
| 2 | 38,000 | 55,000 |
| 3 | 43,000 | 45,000 |
| 4 | 50,000 | 25,000 |
| 5 | 65,000 | 15,000 |

If the bus is kept for 1 year and then sold, the annual cost will be

$$\begin{aligned} \text{EUAC}(1) &= 120{,}000(A/P, 8\%, 1) + 35{,}000(A/F, 8\%, 1) \\ &\quad - 60{,}000(A/F, 8\%, 1) \\ &= 104{,}600 \end{aligned}$$

If the bus is kept for 2 years and then sold, the annual cost will be

$$\begin{aligned} \text{EUAC}(2) &= [120{,}000 + 35{,}000(P/F, 8\%, 1)](A/P, 8\%, 2) \\ &\quad + (38{,}000 - 55{,}000)(A/F, 8\%, 2) \\ &= 77{,}300 \end{aligned}$$

If the bus is kept for 3 years and then sold, the annual cost will be

$$\begin{aligned} \text{EUAC}(3) &= [120{,}000 + 35{,}000(P/F, 8\%, 1) \\ &\quad + 38{,}000(P/F, 8\%, 2)](A/P, 8\%, 3) \\ &\quad + (43{,}000 - 45{,}000)(A/F, 8\%, 3) \\ &= 71{,}200 \end{aligned}$$

This process is continued until EUAC begins to increase. In this example, EUAC(4) is 71,700. Therefore, the bus should be retired after 3 years.

## 14 BASIC COST ACCOUNTING

*Cost accounting* is the system which determines the cost of manufactured products. Cost accounting is called *job cost accounting* if costs are accumulated by part number or contract. It is called *process cost accounting* if costs are accumulated by departments or manufacturing processes.

Three types of costs (direct material, direct labor, and all indirect costs) make up the total manufacturing cost of a product.

*Direct material costs* are the costs of all materials that go into the product, priced at the original purchase cost.

*Indirect material and labor costs* are generally limited to costs incurred in the factory, excluding costs incurred in the office area. Examples of indirect materials are cleaning fluids, assembly lubricants, and temporary routing tags. Examples of indirect labor are stock-picking, inspection, expediting, and supervision labor.

Here are some important points concerning basic cost accounting:

- The sum of direct material and direct labor costs is known as the *prime cost.*
- Indirect costs may be called *indirect manufacturing expenses* (IME).
- Indirect costs may also include the overhead sector of the company (e.g., secretaries, engineers, and corporate administration). In this case, the indirect cost is usually called *burden* or *overhead.* Burden may also include the EUAC of non-regular costs which must be spread evenly over several years.
- The cost of a product is usually known in advance from previous manufacturing runs or by estimation. Any deviation from this known cost is called a *variance.* Variance may be broken down into *labor variance* and *material variance.*

- Indirect cost per item is not easily measured. The method of allocating indirect costs to a product is as follows:

  **step 1**: Estimate the total expected indirect (and overhead) costs for the upcoming year.

  **step 2**: Decide on some convenient vehicle for allocating the overhead to production. Usually, this vehicle is either the number of units expected to be produced or the number of direct hours expected to be worked in the upcoming year.

  **step 3**: Estimate the quantity or size of the overhead vehicle.

  **step 4**: Divide expected overhead costs by the expected overhead vehicle to obtain the unit overhead.

  **step 5**: Regardless of the true size of the overhead vehicle during the upcoming year, one unit of overhead cost is allocated per product.

- Although estimates of production for the next year are always somewhat inaccurate, the cost of the product is assumed to be independent of forecasting errors. Any difference between true cost and calculated cost goes into a variance account.
- *Burden (overhead) variance* will be caused by errors in forecasting both the actual overhead for the upcoming year and the vehicle size. In the former case, the variance is called *burden budget variance*; in the latter, it is called *burden capacity variance.*

*Example 2.18*

A small company expects to produce 8000 items in the coming year. The current material cost is \$4.54 each. Sixteen minutes of direct labor are required per unit. Workers are paid \$7.50 per hour. 2133 direct labor hours are forecast for the product. Miscellaneous overhead costs are estimated at \$45,000.

Find the expected direct material cost, the direct labor cost, the prime cost, the burden as a function of production and direct labor, and the total cost.

- The direct material cost was given as \$4.54.
- The direct labor cost is $\left(\frac{16}{60}\right)(\$7.50) = \$2.00$.
- The prime cost is $\$4.54 + \$2.00 = \$6.54$.
- If the burden vehicle is production, the burden rate is $\$\frac{45{,}000}{8000} = \$5.63$ per item, making the total cost $\$4.54 + \$2.00 + \$5.63 = \$12.17$.
- If the burden vehicle is direct labor hours, the burden rate is $\left(\frac{45{,}000}{2133}\right) = \$21.10$ per hour, making the total cost $\$4.54 + \$2.00 + \frac{16}{60}(\$21.10) = \$12.17$.

*Example 2.19*

The actual performance of the company in example 2.18 is given by the following figures:

actual production: 7560
actual overhead costs: \$47,000

What are the burden budget variance and the burden capacity variance?

The burden capacity variance is

$$\$45{,}000 - 7560(\$5.63) = \$2437$$

The burden budget variance is

$$\$47{,}000 - \$45{,}000 = \$2000$$

The overall burden variance is

$$\$47{,}000 - 7560(\$5.63) = \$4437$$

## 15 BREAK-EVEN ANALYSIS

*Break-even analysis* is a method of determining when costs exactly equal revenue. If the manufactured quantity is less than the break-even quantity, a loss is incurred. If the manufactured quantity is greater than the break-even quantity, a profit is incurred.

Consider the following special variables:

$f$ a fixed cost which does not vary with production

$a$ an incremental cost which is the cost to produce one additional item. It may also be called the *marginal cost* or *differential cost.*

$Q$ the quantity sold

$p$ the incremental revenue

$R$ the total revenue

$C$ the total cost

Assuming no change in the inventory, the *break-even point* can be found from $C = R$, where

$$C = f + aQ \quad 2.28$$

$$R = pQ \quad 2.29$$

An alternate form of the break-even problem is to find the number of units per period for which two alternatives have the same total costs. Fixed costs are to be spread over a period longer than one year. One of the alternatives will have a lower cost if production is less than the break-even point. The other will have a lower cost for production greater than the break-even point.

The *cost per unit* problem is a variation of the breakeven problem. In the typical cost per unit problem, data will be available to determine the direct labor and material costs per unit, but some method is needed to additionally allocate part of the annual overhead (burden) and initial facility purchase/construction costs.

Annual overhead is allocated to the unit cost simply by dividing the overhead by the number of units produced each year. The initial purchase/construction cost is multiplied by the appropriate $(A/P)$ factor before similarly dividing by the production rate. The total unit cost is the sum of the direct labor, direct material, prorata share of overhead, and prorata share of the equivalent annual facility investment costs.

*Example 2.20*

Two plans are available for a company to obtain automobiles for its salesmen. How many miles must the cars be driven each year for the two plans to have the same costs? Use an interest rate of 10%.

*Plan A* Lease the cars and pay \$0.15 per mile

*Plan B* Purchase the cars for \$5000. Each car has an economic life of three years, after which it can be sold for \$1,200. Gas and oil cost \$0.04 per mile. Insurance is \$500 per year. (Assume the year-end convention applies to the insurance.)

Let $x$ be the number of miles driven per year. Then, the EUAC for both alternatives is:

$$\begin{aligned}\text{EUAC}(A) &= 0.15x\\ \text{EUAC}(B) &= 0.04x + 500 + 5000(A/P, 10\%, 3)\\ &\quad - 1200(A/F, 10\%, 3)\\ &= 0.04x + 2148\end{aligned}$$

Setting EUAC(A) and EUAC(B) equal and solving for $x$ yields 19,527 miles per year as the break-even point.

## 16 HANDLING INFLATION

It is important to perform economic studies in terms of *constant value dollars.* One method of converting all cash flows to constant value dollars is to divide the flows by some annual *economic indicator* or price index. Such indicators would normally be given to you as part of a problem.

If indicators are not available, this method can still be used by assuming that inflation is relatively constant at a decimal rate $e$ per year. Then, all cash flows can be converted to '$t = 0$' dollars by dividing by $(1+e)^n$ where $n$ is the year of the cash flow.

*Example 2.21*

What is the uninflated present worth of \$2000 in two years if the average inflation rate is 6% and $i$ is 10% ?

$$P = \frac{\$2000}{(1.10)^2(1.06)^2} = \$1471.07$$

An alternative is to replace $i$ with a value corrected for inflation. This corrected value, $i'$, is

$$i' = i + e + ie \qquad 2.30$$

This method has the advantage of simplifying the calculations. However, pre-calculated factors may not be available for the non-integer values of $i'$. Therefore, table 2.1 will have to be used to calculate the factors.

*Example 2.22*

Repeat example 2.21 using $i'$.

$$i' = 0.10 + 0.06 + (0.10)(0.06) = 0.166$$

$$P = \frac{\$2000}{(1.166)^2} = \$1471.07$$

## 17 LEARNING CURVES

The more products that are made, the more efficient the operation becomes due to experience gained. Therefore, direct labor costs decrease. Usually, a *learning curve* is specified by the decrease in cost each time the quantity produced doubles. If there is a 20% decrease per doubling, the curve is said to be an 80% learning curve.

Consider the following special variables:

$T_1$ time or cost for the first item
$T_n$ time or cost for the $n$th item
$n$ total number of items produced
$b$ learning curve constant

Then, the time to produce the $n$th item is given by

$$T_n = T_1(n)^{-b} \qquad 2.31$$

**Table 2.3**
Learning Curve Constants

| learning curve | b |
|---|---|
| 80% | 0.322 |
| 85% | 0.234 |
| 90% | 0.152 |
| 95% | 0.074 |

The total time to produce units from quantity $n_1$ to $n_2$ inclusive is

$$\int_{n_1}^{n_2} T_n dn \approx \frac{T_1}{(1-b)}\left[\left(n_2+\frac{1}{2}\right)^{1-b} - \left(n_1-\frac{1}{2}\right)^{1-b}\right] \tag{2.32}$$

The average time per unit over the production from $n_1$ to $n_2$ is the above total time from equation 2.32 divided by the quantity produced, $(n_2 - n_1 + 1)$.

It is important to remember that learning curve reductions apply only to direct labor costs. They are not applied to indirect labor or direct material costs.

*Example 2.23*

A 70% learning curve is used with an item whose first production time was 1.47 hours. How long will it take to produce the 11th item? How long will it take to produce the 11th through 27th items?

First, find $b$.

$$\frac{T_2}{T_1} = 0.7 = (2)^{-b} \quad \text{or } b = 0.515$$

Then, $T_{11} = 1.47(11)^{-0.515} = 0.428$ hours.

The time to produce the 11th item through 27th item is approximately

$$T = \frac{1.47}{1-0.515}\left[(27.5)^{1-0.515} - (10.5)^{1-0.515}\right]$$
$$= 5.643 \text{ hours}$$

## 18 ECONOMIC ORDER QUANTITY

The *economic order quantity* (EOQ) is the order quantity which minimizes the inventory costs per unit time. Although there are many different EOQ models, the simplest is based on the following assumptions:

- Reordering is instantaneous. The time between order placement and receipt is zero.
- Shortages are not allowed.
- Demand for the inventory item is deterministic (i.e., is not a random variable).
- Demand is constant with respect to time.
- An order is placed when the on-hand quantity is zero.

The following special variables are used:

| | |
|---|---|
| $a$ | the constant depletion rate $\left(\frac{\text{items}}{\text{unit time}}\right)$ |
| $h$ | the inventory storage cost $\left(\frac{\$}{\text{item-unit time}}\right)$ |
| $H$ | the total inventory storage cost between orders ($) |
| $K$ | the fixed cost of placing an order ($) |
| $Q_0$ | the order quantity |

If the original quantity on hand is $Q_0$, the stock will be depleted at

$$t^* = \frac{Q_0}{a}$$

The total inventory storage cost between $t_0$ and $t^*$ is

$$H = \frac{1}{2}h\frac{Q_o^2}{a} \tag{2.33}$$

The total inventory and ordering cost per unit time is

$$C_t = \frac{aK}{Q_0} + \frac{1}{2}hQ_0 \tag{2.34}$$

$C_t$ can be minimized with respect to $Q_0$. The EOQ and time between orders are:

$$Q_0^* = \sqrt{2\frac{aK}{h}} \tag{2.35}$$

$$t^* = \frac{Q_0^*}{a} \tag{2.36}$$

## 19 CONSUMER LOANS

Many consumer loans cannot be handled by the equivalence formulas presented up to this point. Many different arrangements can be made between lender and borrower. Four of the most common consumer loan arrangements are presented below. Refer to a real estate or investment analysis book for more complex loans.

### A. SIMPLE INTEREST

Interest due does not compound with a *simple interest* loan. The interest due is merely proportional to the length of time the principal is outstanding. Because of this, simple interest loans are seldom made for long periods (e.g., longer than one year).

*Example 2.24*

A $12,000 simple interest loan is taken out at 16% per annum. The loan matures in one year with no intermediate payments. How much will be due at the end of the year?

$$\text{Amount due} = (1+0.16)(\$12{,}000) = \$13{,}920$$

For loans less than one year, it is commonly assumed that a year consists of 12 months of 30 days each.

*Example 2.25*

$4000 is borrowed for 75 days at 16% per annum simple interest. How much will be due at the end of 75 days?

$$\text{Amount due} = \$4000 + (0.16)\left(\frac{75}{360}\right)(4000) = \$4133$$

B. LOANS WITH CONSTANT AMOUNT PAID TOWARDS PRINCIPAL

With this loan type, the payment is not the same each period. The amount paid towards the principal is constant, but the interest varies from period to period. The following special symbols are used.

| | |
|---|---|
| $BAL_j$ | principal balance after the $j$th payment |
| $LV$ | total value loaned (cost minus down payment) |
| $j$ | payment or period number |
| $N$ | total number of payments to pay off the loan |
| $PI_j$ | $j$th interest payment |
| $PP_j$ | $j$th principal payment |
| $PT_j$ | $j$th total payment |
| $\phi$ | effective rate per period $(r/k)$ |

The equations which govern this type of loan are

$$BAL_j = LV - (j)(PP) \qquad 2.37$$

$$PI_j = \phi(BAL_{j-1}) \qquad 2.38$$

$$PT_j = PP + PI_j \qquad 2.39$$

C. DIRECT REDUCTION LOANS

This is the typical 'interest paid on unpaid balance' loan. The amount of the periodic payment is constant, but the amounts paid towards the principal and interest both vary.

The same symbols are used with this type of loan as are listed above.

$$N = -\frac{ln\left[\frac{-\phi(LV)}{PT}+1\right]}{ln(1+\phi)} \qquad 2.40$$

$$BAL_{j-1} = PT\left[\frac{1-(1+\phi)^{j-1-N}}{\phi}\right] \qquad 2.41$$

$$PI_j = \phi(BAL_{j-1}) \qquad 2.42$$

$$PP_j = PT - PI_j \qquad 2.43$$

$$BAL_j = BAL_{j-1} - PP_j \qquad 2.44$$

*Example 2.26*

A $45,000 loan is financed at 9.25% per annum. The monthly payment is $385. What are the amounts paid toward interest and principal in the 14th period? What is the remaining principal balance after the 14th payment has been made?

The effective rate per month is

$$\phi = \frac{r}{k} = \frac{0.0925}{12}$$

$$= 0.007708$$

$$N = -\frac{ln\left[\frac{-(0.007708)(45{,}000)}{385}+1\right]}{ln(1+0.007708)} = 301$$

$$BAL_{13} = 385\left[\frac{1-(1+0.007708)^{14-1-301}}{0.007708}\right]$$

$$= \$44{,}476.39$$

$$PI_{14} = (0.007708)(\$44{,}476.39) = \$342.82$$

$$PP_{14} = \$385 - \$342.82 = \$42.18$$

$$BAL_{14} = \$44,476.39 - \$42.18 = \$44,434.21$$

Equation 2.40 calculates the number of payments necessary to pay off a loan. This equation can be solved with effort for the total periodic payment (PT) or the initial value of the loan (LV). It is easier, however, to use the $(A/P, i\%, n)$ factor to find the payment and loan value.

$$PT = (LV)(A/P, \phi\%, n)$$

If the loan is repaid in yearly installments, then $i$ is the effective annual rate. If the loan is paid off monthly, then $i$ should be replaced by the effective rate per month ($\phi$ from equation 2.26). For monthly payments, $n$ is the number of months in the payback period.

D. DIRECT REDUCTION LOAN WITH BALLOON PAYMENT

This type of loan has a constant periodic payment, but the duration of the loan is insufficient to completely pay back the principal. Therefore, all remaining unpaid principal must be paid back in a lump sum when the loan matures. This large payment is known as a *balloon payment.*

Equations 2.40 through 2.44 can also be used with this type of loan. The remaining balance after the last payment is the balloon payment. This balloon payment must be repaid along with the last regular payment calculated.

## 20 SENSITIVITY ANALYSIS

Data analysis and forecasts in economic studies represent judgment on costs which will occur in the future. There are always uncertainties about these costs. However, these uncertainties are insufficient reason not to make the best possible estimates of the costs. Nevertheless, a decision between alternatives often can be made more confidently if it is known whether or not the conclusion is sensitive to moderate changes in data forecasts. Sensitivity analysis provides this extra dimension to an economic analysis.

The sensitivity of a decision is determined by inserting a range of estimates for critical cash flows. If radical changes can be made to a cash flow without changing the decision, the decision is said to be insensitive to uncertainties regarding that cash flow. However, if a small change in the estimate of a cash flow will alter the decision, that decision is said to be very sensitive to changes in the estimate.

An established semantic tradition distinguishes between risk analysis and uncertainty analysis. Risk analysis addresses variables which have a known or estimated probability distribution. In this regard, statistics and probability theory can be used to determine the probability of a cash flow varying between given limits. On the other hand, uncertainty analysis is concerned with situations in which there is not enough information to determine the probability or frequency distribution for the variables involved.

As a first step, sensitivity analysis should be applied one at a time to the dominant cost factors. Dominant cost factors are those which have the most significant impact on the present value of the alternative. If warranted, additional investigation can be used to determine the sensitivity to several cash flows varying simultaneously. Significant judgment is needed, however, to successfully determine the proper combinations of cash flows to vary.

It is common to plot the dependency of the present value on the cash flow being varied in a two-dimensional graph. Simple linear interpolation is used (within reason) to determine the critical value of the cash flow being varied.

## STANDARD CASH FLOW FACTORS

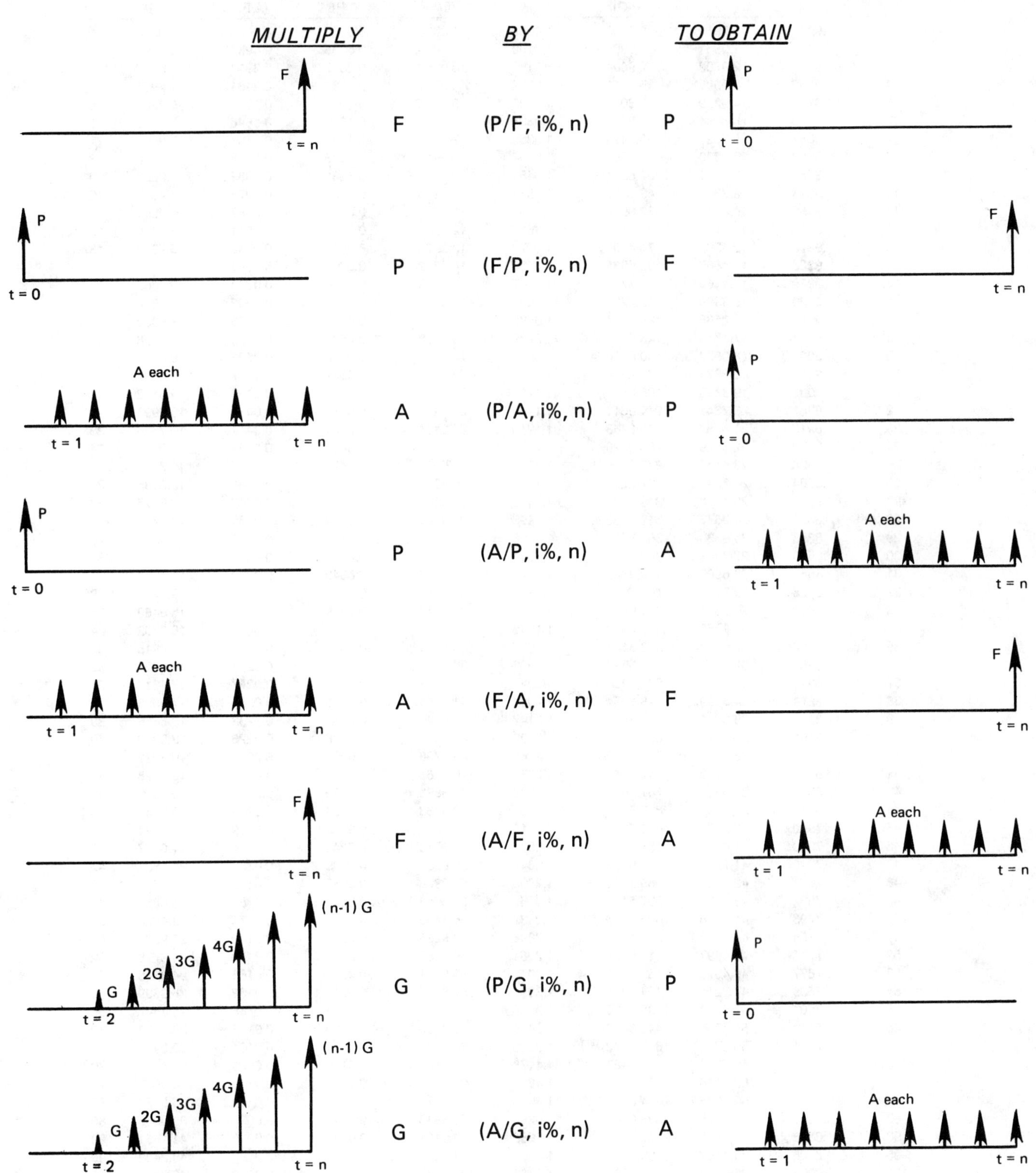

**I = 0.50 %**

| N | (P/F) | (P/A) | (P/G) | (F/P) | (F/A) | (A/P) | (A/F) | (A/G) | N |
|---|---|---|---|---|---|---|---|---|---|
| 1 | .9950 | 0.9950 | −0.0000 | 1.0050 | 1.0000 | 1.0050 | 1.0000 | −0.0000 | 1 |
| 2 | .9901 | 1.9851 | 0.9901 | 1.0100 | 2.0050 | 0.5038 | 0.4988 | 0.4988 | 2 |
| 3 | .9851 | 2.9702 | 2.9604 | 1.0151 | 3.0150 | 0.3367 | 0.3317 | 0.9967 | 3 |
| 4 | .9802 | 3.9505 | 5.9011 | 1.0202 | 4.0301 | 0.2531 | 0.2481 | 1.4938 | 4 |
| 5 | .9754 | 4.9259 | 9.8026 | 1.0253 | 5.0503 | 0.2030 | 0.1980 | 1.9900 | 5 |
| 6 | .9705 | 5.8964 | 14.6552 | 1.0304 | 6.0755 | 0.1696 | 0.1646 | 2.4855 | 6 |
| 7 | .9657 | 6.8621 | 20.4493 | 1.0355 | 7.1059 | 0.1457 | 0.1407 | 2.9801 | 7 |
| 8 | .9609 | 7.8230 | 27.1755 | 1.0407 | 8.1414 | 0.1278 | 0.1228 | 3.4738 | 8 |
| 9 | .9561 | 8.7791 | 34.8244 | 1.0459 | 9.1821 | 0.1139 | 0.1089 | 3.9668 | 9 |
| 10 | .9513 | 9.7304 | 43.3865 | 1.0511 | 10.2280 | 0.1028 | 0.0978 | 4.4589 | 10 |
| 11 | .9466 | 10.6770 | 52.8526 | 1.0564 | 11.2792 | 0.0937 | 0.0887 | 4.9501 | 11 |
| 12 | .9419 | 11.6189 | 63.2136 | 1.0617 | 12.3356 | 0.0861 | 0.0811 | 5.4406 | 12 |
| 13 | .9372 | 12.5562 | 74.4602 | 1.0670 | 13.3972 | 0.0796 | 0.0746 | 5.9302 | 13 |
| 14 | .9326 | 13.4887 | 86.5835 | 1.0723 | 14.4642 | 0.0741 | 0.0691 | 6.4190 | 14 |
| 15 | .9279 | 14.4166 | 99.5743 | 1.0777 | 15.5365 | 0.0694 | 0.0644 | 6.9069 | 15 |
| 16 | .9233 | 15.3399 | 113.4238 | 1.0831 | 16.6142 | 0.0652 | 0.0602 | 7.3940 | 16 |
| 17 | .9187 | 16.2586 | 128.1231 | 1.0885 | 17.6973 | 0.0615 | 0.0565 | 7.8803 | 17 |
| 18 | .9141 | 17.1728 | 143.6634 | 1.0939 | 18.7858 | 0.0582 | 0.0532 | 8.3658 | 18 |
| 19 | .9096 | 18.0824 | 160.0360 | 1.0994 | 19.8797 | 0.0553 | 0.0503 | 8.8504 | 19 |
| 20 | .9051 | 18.9874 | 177.2322 | 1.1049 | 20.9791 | 0.0527 | 0.0477 | 9.3342 | 20 |
| 21 | .9006 | 19.8880 | 195.2434 | 1.1104 | 22.0840 | 0.0503 | 0.0453 | 9.8172 | 21 |
| 22 | .8961 | 20.7841 | 214.0611 | 1.1160 | 23.1944 | 0.0481 | 0.0431 | 10.2993 | 22 |
| 23 | .8916 | 21.6757 | 233.6768 | 1.1216 | 24.3104 | 0.0461 | 0.0411 | 10.7806 | 23 |
| 24 | .8872 | 22.5629 | 254.0820 | 1.1272 | 25.4320 | 0.0443 | 0.0393 | 11.2611 | 24 |
| 25 | .8828 | 23.4456 | 275.2686 | 1.1328 | 26.5591 | 0.0427 | 0.0377 | 11.7407 | 25 |
| 26 | .8784 | 24.3240 | 297.2281 | 1.1385 | 27.6919 | 0.0411 | 0.0361 | 12.2195 | 26 |
| 27 | .8740 | 25.1980 | 319.9523 | 1.1442 | 28.8304 | 0.0397 | 0.0347 | 12.6975 | 27 |
| 28 | .8697 | 26.0677 | 343.4332 | 1.1499 | 29.9745 | 0.0384 | 0.0334 | 13.1747 | 28 |
| 29 | .8653 | 26.9330 | 367.6625 | 1.1556 | 31.1244 | 0.0371 | 0.0321 | 13.6510 | 29 |
| 30 | .8610 | 27.7941 | 392.6324 | 1.1614 | 32.2800 | 0.0360 | 0.0310 | 14.1265 | 30 |
| 31 | .8567 | 28.6508 | 418.3348 | 1.1672 | 33.4414 | 0.0349 | 0.0299 | 14.6012 | 31 |
| 32 | .8525 | 29.5033 | 444.7618 | 1.1730 | 34.6086 | 0.0339 | 0.0289 | 15.0750 | 32 |
| 33 | .8482 | 30.3515 | 471.9055 | 1.1789 | 35.7817 | 0.0329 | 0.0279 | 15.5480 | 33 |
| 34 | .8440 | 31.1955 | 499.7583 | 1.1848 | 36.9606 | 0.0321 | 0.0271 | 16.0202 | 34 |
| 35 | .8398 | 32.0354 | 528.3123 | 1.1907 | 38.1454 | 0.0312 | 0.0262 | 16.4915 | 35 |
| 36 | .8356 | 32.8710 | 557.5598 | 1.1967 | 39.3361 | 0.0304 | 0.0254 | 16.9621 | 36 |
| 37 | .8315 | 33.7025 | 587.4934 | 1.2027 | 40.5328 | 0.0297 | 0.0247 | 17.4317 | 37 |
| 38 | .8274 | 34.5299 | 618.1054 | 1.2087 | 41.7354 | 0.0290 | 0.0240 | 17.9006 | 38 |
| 39 | .8232 | 35.3531 | 649.3883 | 1.2147 | 42.9441 | 0.0283 | 0.0233 | 18.3686 | 39 |
| 40 | .8191 | 36.1722 | 681.3347 | 1.2208 | 44.1588 | 0.0276 | 0.0226 | 18.8359 | 40 |
| 41 | .8151 | 36.9873 | 713.9372 | 1.2269 | 45.3796 | 0.0270 | 0.0220 | 19.3022 | 41 |
| 42 | .8110 | 37.7983 | 747.1886 | 1.2330 | 46.6065 | 0.0265 | 0.0215 | 19.7678 | 42 |
| 43 | .8070 | 38.6053 | 781.0815 | 1.2392 | 47.8396 | 0.0259 | 0.0209 | 20.2325 | 43 |
| 44 | .8030 | 39.4082 | 815.6087 | 1.2454 | 49.0788 | 0.0254 | 0.0204 | 20.6964 | 44 |
| 45 | .7990 | 40.2072 | 850.7631 | 1.2516 | 50.3242 | 0.0249 | 0.0199 | 21.1595 | 45 |
| 46 | .7950 | 41.0022 | 886.5376 | 1.2579 | 51.5758 | 0.0244 | 0.0194 | 21.6217 | 46 |
| 47 | .7910 | 41.7932 | 922.9252 | 1.2642 | 52.8337 | 0.0239 | 0.0189 | 22.0831 | 47 |
| 48 | .7871 | 42.5803 | 959.9188 | 1.2705 | 54.0978 | 0.0235 | 0.0185 | 22.5437 | 48 |
| 49 | .7832 | 43.3635 | 997.5116 | 1.2768 | 55.3683 | 0.0231 | 0.0181 | 23.0035 | 49 |
| 50 | .7793 | 44.1428 | 1035.6966 | 1.2832 | 56.6452 | 0.0227 | 0.0177 | 23.4624 | 50 |
| 51 | .7754 | 44.9182 | 1074.4670 | 1.2896 | 57.9284 | 0.0223 | 0.0173 | 23.9205 | 51 |
| 52 | .7716 | 45.6897 | 1113.8162 | 1.2961 | 59.2180 | 0.0219 | 0.0169 | 24.3778 | 52 |
| 53 | .7677 | 46.4575 | 1153.7372 | 1.3026 | 60.5141 | 0.0215 | 0.0165 | 24.8343 | 53 |
| 54 | .7639 | 47.2214 | 1194.2236 | 1.3091 | 61.8167 | 0.0212 | 0.0162 | 25.2899 | 54 |
| 55 | .7601 | 47.9814 | 1235.2686 | 1.3156 | 63.1258 | 0.0208 | 0.0158 | 25.7447 | 55 |
| 60 | .7414 | 51.7256 | 1448.6458 | 1.3489 | 69.7700 | 0.0193 | 0.0143 | 28.0064 | 60 |
| 65 | .7231 | 55.3775 | 1675.0272 | 1.3829 | 76.5821 | 0.0181 | 0.0131 | 30.2475 | 65 |
| 70 | .7053 | 58.9394 | 1913.6427 | 1.4178 | 83.5661 | 0.0170 | 0.0120 | 32.4680 | 70 |
| 75 | .6879 | 62.4136 | 2163.7525 | 1.4536 | 90.7265 | 0.0160 | 0.0110 | 34.6679 | 75 |
| 80 | .6710 | 65.8023 | 2424.6455 | 1.4903 | 98.0677 | 0.0152 | 0.0102 | 36.8474 | 80 |
| 85 | .6545 | 69.1075 | 2695.6389 | 1.5280 | 105.5943 | 0.0145 | 0.0095 | 39.0065 | 85 |
| 90 | .6383 | 72.3313 | 2976.0769 | 1.5666 | 113.3109 | 0.0138 | 0.0088 | 41.1451 | 90 |
| 95 | .6226 | 75.4757 | 3265.3298 | 1.6061 | 121.2224 | 0.0132 | 0.0082 | 43.2633 | 95 |
| 100 | .6073 | 78.5426 | 3562.7934 | 1.6467 | 129.3337 | 0.0127 | 0.0077 | 45.3613 | 100 |

**I = 0.75 %**

| N | (P/F) | (P/A) | (P/G) | (F/P) | (F/A) | (A/P) | (A/F) | (A/G) | N |
|---|---|---|---|---|---|---|---|---|---|
| 1 | .9926 | 0.9926 | -0.0000 | 1.0075 | 1.0000 | 1.0075 | 1.0000 | -0.0000 | 1 |
| 2 | .9852 | 1.9777 | 0.9852 | 1.0151 | 2.0075 | 0.5056 | 0.4981 | 0.4981 | 2 |
| 3 | .9778 | 2.9556 | 2.9408 | 1.0227 | 3.0226 | 0.3383 | 0.3308 | 0.9950 | 3 |
| 4 | .9706 | 3.9261 | 5.8525 | 1.0303 | 4.0452 | 0.2547 | 0.2472 | 1.4907 | 4 |
| 5 | .9633 | 4.8894 | 9.7058 | 1.0381 | 5.0756 | 0.2045 | 0.1970 | 1.9851 | 5 |
| 6 | .9562 | 5.8456 | 14.4866 | 1.0459 | 6.1136 | 0.1711 | 0.1636 | 2.4782 | 6 |
| 7 | .9490 | 6.7946 | 20.1808 | 1.0537 | 7.1595 | 0.1472 | 0.1397 | 2.9701 | 7 |
| 8 | .9420 | 7.7366 | 26.7747 | 1.0616 | 8.2132 | 0.1293 | 0.1218 | 3.4608 | 8 |
| 9 | .9350 | 8.6716 | 34.2544 | 1.0696 | 9.2748 | 0.1153 | 0.1078 | 3.9502 | 9 |
| 10 | .9280 | 9.5996 | 42.6064 | 1.0776 | 10.3443 | 0.1042 | 0.0967 | 4.4384 | 10 |
| 11 | .9211 | 10.5207 | 51.8174 | 1.0857 | 11.4219 | 0.0951 | 0.0876 | 4.9253 | 11 |
| 12 | .9142 | 11.4349 | 61.8740 | 1.0938 | 12.5076 | 0.0875 | 0.0800 | 5.4110 | 12 |
| 13 | .9074 | 12.3423 | 72.7632 | 1.1020 | 13.6014 | 0.0810 | 0.0735 | 5.8954 | 13 |
| 14 | .9007 | 13.2430 | 84.4720 | 1.1103 | 14.7034 | 0.0755 | 0.0680 | 6.3786 | 14 |
| 15 | .8940 | 14.1370 | 96.9876 | 1.1186 | 15.8137 | 0.0707 | 0.0632 | 6.8606 | 15 |
| 16 | .8873 | 15.0243 | 110.2973 | 1.1270 | 16.9323 | 0.0666 | 0.0591 | 7.3413 | 16 |
| 17 | .8807 | 15.9050 | 124.3887 | 1.1354 | 18.0593 | 0.0629 | 0.0554 | 7.8207 | 17 |
| 18 | .8742 | 16.7792 | 139.2494 | 1.1440 | 19.1947 | 0.0596 | 0.0521 | 8.2989 | 18 |
| 19 | .8676 | 17.6468 | 154.8671 | 1.1525 | 20.3387 | 0.0567 | 0.0492 | 8.7759 | 19 |
| 20 | .8612 | 18.5080 | 171.2297 | 1.1612 | 21.4912 | 0.0540 | 0.0465 | 9.2516 | 20 |
| 21 | .8548 | 19.3628 | 188.3253 | 1.1699 | 22.6524 | 0.0516 | 0.0441 | 9.7261 | 21 |
| 22 | .8484 | 20.2112 | 206.1420 | 1.1787 | 23.8223 | 0.0495 | 0.0420 | 10.1994 | 22 |
| 23 | .8421 | 21.0533 | 224.6682 | 1.1875 | 25.0010 | 0.0475 | 0.0400 | 10.6714 | 23 |
| 24 | .8358 | 21.8891 | 243.8923 | 1.1964 | 26.1885 | 0.0457 | 0.0382 | 11.1422 | 24 |
| 25 | .8296 | 22.7188 | 263.8029 | 1.2054 | 27.3849 | 0.0440 | 0.0365 | 11.6117 | 25 |
| 26 | .8234 | 23.5422 | 284.3888 | 1.2144 | 28.5903 | 0.0425 | 0.0350 | 12.0800 | 26 |
| 27 | .8173 | 24.3595 | 305.6387 | 1.2235 | 29.8047 | 0.0411 | 0.0336 | 12.5470 | 27 |
| 28 | .8112 | 25.1707 | 327.5416 | 1.2327 | 31.0282 | 0.0397 | 0.0322 | 13.0128 | 28 |
| 29 | .8052 | 25.9759 | 350.0867 | 1.2420 | 32.2609 | 0.0385 | 0.0310 | 13.4774 | 29 |
| 30 | .7992 | 26.7751 | 373.2631 | 1.2513 | 33.5029 | 0.0373 | 0.0298 | 13.9407 | 30 |
| 31 | .7932 | 27.5683 | 397.0602 | 1.2607 | 34.7542 | 0.0363 | 0.0288 | 14.4028 | 31 |
| 32 | .7873 | 28.3557 | 421.4675 | 1.2701 | 36.0148 | 0.0353 | 0.0278 | 14.8636 | 32 |
| 33 | .7815 | 29.1371 | 446.4746 | 1.2796 | 37.2849 | 0.0343 | 0.0268 | 15.3232 | 33 |
| 34 | .7757 | 29.9128 | 472.0712 | 1.2892 | 38.5646 | 0.0334 | 0.0259 | 15.7816 | 34 |
| 35 | .7699 | 30.6827 | 498.2471 | 1.2989 | 39.8538 | 0.0326 | 0.0251 | 16.2387 | 35 |
| 36 | .7641 | 31.4468 | 524.9924 | 1.3086 | 41.1527 | 0.0318 | 0.0243 | 16.6946 | 36 |
| 37 | .7585 | 32.2053 | 552.2969 | 1.3185 | 42.4614 | 0.0311 | 0.0236 | 17.1493 | 37 |
| 38 | .7528 | 32.9581 | 580.1511 | 1.3283 | 43.7798 | 0.0303 | 0.0228 | 17.6027 | 38 |
| 39 | .7472 | 33.7053 | 608.5451 | 1.3383 | 45.1082 | 0.0297 | 0.0222 | 18.0549 | 39 |
| 40 | .7416 | 34.4469 | 637.4693 | 1.3483 | 46.4465 | 0.0290 | 0.0215 | 18.5058 | 40 |
| 41 | .7361 | 35.1831 | 666.9144 | 1.3585 | 47.7948 | 0.0284 | 0.0209 | 18.9556 | 41 |
| 42 | .7306 | 35.9137 | 696.8709 | 1.3686 | 49.1533 | 0.0278 | 0.0203 | 19.4040 | 42 |
| 43 | .7252 | 36.6389 | 727.3297 | 1.3789 | 50.5219 | 0.0273 | 0.0198 | 19.8513 | 43 |
| 44 | .7198 | 37.3587 | 758.2815 | 1.3893 | 51.9009 | 0.0268 | 0.0193 | 20.2973 | 44 |
| 45 | .7145 | 38.0732 | 789.7173 | 1.3997 | 53.2901 | 0.0263 | 0.0188 | 20.7421 | 45 |
| 46 | .7091 | 38.7823 | 821.6283 | 1.4102 | 54.6898 | 0.0258 | 0.0183 | 21.1856 | 46 |
| 47 | .7039 | 39.4862 | 854.0056 | 1.4207 | 56.1000 | 0.0253 | 0.0178 | 21.6280 | 47 |
| 48 | .6986 | 40.1848 | 886.8404 | 1.4314 | 57.5207 | 0.0249 | 0.0174 | 22.0691 | 48 |
| 49 | .6934 | 40.8782 | 920.1243 | 1.4421 | 58.9521 | 0.0245 | 0.0170 | 22.5089 | 49 |
| 50 | .6883 | 41.5664 | 953.8486 | 1.4530 | 60.3943 | 0.0241 | 0.0166 | 22.9476 | 50 |
| 51 | .6831 | 42.2496 | 988.0050 | 1.4639 | 61.8472 | 0.0237 | 0.0162 | 23.3850 | 51 |
| 52 | .6780 | 42.9276 | 1022.5852 | 1.4748 | 63.3111 | 0.0233 | 0.0158 | 23.8211 | 52 |
| 53 | .6730 | 43.6006 | 1057.5810 | 1.4859 | 64.7859 | 0.0229 | 0.0154 | 24.2561 | 53 |
| 54 | .6680 | 44.2686 | 1092.9842 | 1.4970 | 66.2718 | 0.0226 | 0.0151 | 24.6898 | 54 |
| 55 | .6630 | 44.9316 | 1128.7869 | 1.5083 | 67.7688 | 0.0223 | 0.0148 | 25.1223 | 55 |
| 60 | .6387 | 48.1734 | 1313.5189 | 1.5657 | 75.4241 | 0.0208 | 0.0133 | 27.2665 | 60 |
| 65 | .6153 | 51.2963 | 1507.0910 | 1.6253 | 83.3709 | 0.0195 | 0.0120 | 29.3801 | 65 |
| 70 | .5927 | 54.3046 | 1708.6065 | 1.6872 | 91.6201 | 0.0184 | 0.0109 | 31.4634 | 70 |
| 75 | .5710 | 57.2027 | 1917.2225 | 1.7514 | 100.1833 | 0.0175 | 0.0100 | 33.5163 | 75 |
| 80 | .5500 | 59.9944 | 2132.1472 | 1.8180 | 109.0725 | 0.0167 | 0.0092 | 35.5391 | 80 |
| 85 | .5299 | 62.6838 | 2352.6375 | 1.8873 | 118.3001 | 0.0160 | 0.0085 | 37.5318 | 85 |
| 90 | .5104 | 65.2746 | 2577.9961 | 1.9591 | 127.8790 | 0.0153 | 0.0078 | 39.4946 | 90 |
| 95 | .4917 | 67.7704 | 2807.5694 | 2.0337 | 137.8225 | 0.0148 | 0.0073 | 41.4277 | 95 |
| 100 | .4737 | 70.1746 | 3040.7453 | 2.1111 | 148.1445 | 0.0143 | 0.0068 | 43.3311 | 100 |

**I = 1.00 %**

| N | (P/F) | (P/A) | (P/G) | (F/P) | (F/A) | (A/P) | (A/F) | (A/G) | N |
|---|---|---|---|---|---|---|---|---|---|
| 1 | .9901 | 0.9901 | -0.0000 | 1.0100 | 1.0000 | 1.0100 | 1.0000 | -0.0000 | 1 |
| 2 | .9803 | 1.9704 | 0.9803 | 1.0201 | 2.0100 | 0.5075 | 0.4975 | 0.4975 | 2 |
| 3 | .9706 | 2.9410 | 2.9215 | 1.0303 | 3.0301 | 0.3400 | 0.3300 | 0.9934 | 3 |
| 4 | .9610 | 3.9020 | 5.8044 | 1.0406 | 4.0604 | 0.2563 | 0.2463 | 1.4876 | 4 |
| 5 | .9515 | 4.8534 | 9.6103 | 1.0510 | 5.1010 | 0.2060 | 0.1960 | 1.9801 | 5 |
| 6 | .9420 | 5.7955 | 14.3205 | 1.0615 | 6.1520 | 0.1725 | 0.1625 | 2.4710 | 6 |
| 7 | .9327 | 6.7282 | 19.9168 | 1.0721 | 7.2135 | 0.1486 | 0.1386 | 2.9602 | 7 |
| 8 | .9235 | 7.6517 | 26.3812 | 1.0829 | 8.2857 | 0.1307 | 0.1207 | 3.4478 | 8 |
| 9 | .9143 | 8.5660 | 33.6959 | 1.0937 | 9.3685 | 0.1167 | 0.1067 | 3.9337 | 9 |
| 10 | .9053 | 9.4713 | 41.8435 | 1.1046 | 10.4622 | 0.1056 | 0.0956 | 4.4179 | 10 |
| 11 | .8963 | 10.3676 | 50.8067 | 1.1157 | 11.5668 | 0.0965 | 0.0865 | 4.9005 | 11 |
| 12 | .8874 | 11.2551 | 60.5687 | 1.1268 | 12.6825 | 0.0888 | 0.0788 | 5.3815 | 12 |
| 13 | .8787 | 12.1337 | 71.1126 | 1.1381 | 13.8093 | 0.0824 | 0.0724 | 5.8607 | 13 |
| 14 | .8700 | 13.0037 | 82.4221 | 1.1495 | 14.9474 | 0.0769 | 0.0669 | 6.3384 | 14 |
| 15 | .8613 | 13.8651 | 94.4810 | 1.1610 | 16.0969 | 0.0721 | 0.0621 | 6.8143 | 15 |
| 16 | .8528 | 14.7179 | 107.2734 | 1.1726 | 17.2579 | 0.0679 | 0.0579 | 7.2886 | 16 |
| 17 | .8444 | 15.5623 | 120.7834 | 1.1843 | 18.4304 | 0.0643 | 0.0543 | 7.7613 | 17 |
| 18 | .8360 | 16.3983 | 134.9957 | 1.1961 | 19.6147 | 0.0610 | 0.0510 | 8.2323 | 18 |
| 19 | .8277 | 17.2260 | 149.8950 | 1.2081 | 20.8109 | 0.0581 | 0.0481 | 8.7017 | 19 |
| 20 | .8195 | 18.0456 | 165.4664 | 1.2202 | 22.0190 | 0.0554 | 0.0454 | 9.1694 | 20 |
| 21 | .8114 | 18.8570 | 181.6950 | 1.2324 | 23.2392 | 0.0530 | 0.0430 | 9.6354 | 21 |
| 22 | .8034 | 19.6604 | 198.5663 | 1.2447 | 24.4716 | 0.0509 | 0.0409 | 10.0998 | 22 |
| 23 | .7954 | 20.4558 | 216.0660 | 1.2572 | 25.7163 | 0.0489 | 0.0389 | 10.5626 | 23 |
| 24 | .7876 | 21.2434 | 234.1800 | 1.2697 | 26.9735 | 0.0471 | 0.0371 | 11.0237 | 24 |
| 25 | .7798 | 22.0232 | 252.8945 | 1.2824 | 28.2432 | 0.0454 | 0.0354 | 11.4831 | 25 |
| 26 | .7720 | 22.7952 | 272.1957 | 1.2953 | 29.5256 | 0.0439 | 0.0339 | 11.9409 | 26 |
| 27 | .7644 | 23.5596 | 292.0702 | 1.3082 | 30.8209 | 0.0424 | 0.0324 | 12.3971 | 27 |
| 28 | .7568 | 24.3164 | 312.5047 | 1.3213 | 32.1291 | 0.0411 | 0.0311 | 12.8516 | 28 |
| 29 | .7493 | 25.0658 | 333.4863 | 1.3345 | 33.4504 | 0.0399 | 0.0299 | 13.3044 | 29 |
| 30 | .7419 | 25.8077 | 355.0021 | 1.3478 | 34.7849 | 0.0387 | 0.0287 | 13.7557 | 30 |
| 31 | .7346 | 26.5423 | 377.0394 | 1.3613 | 36.1327 | 0.0377 | 0.0277 | 14.2052 | 31 |
| 32 | .7273 | 27.2696 | 399.5858 | 1.3749 | 37.4941 | 0.0367 | 0.0267 | 14.6532 | 32 |
| 33 | .7201 | 27.9897 | 422.6291 | 1.3887 | 38.8690 | 0.0357 | 0.0257 | 15.0995 | 33 |
| 34 | .7130 | 28.7027 | 446.1572 | 1.4026 | 40.2577 | 0.0348 | 0.0248 | 15.5441 | 34 |
| 35 | .7059 | 29.4086 | 470.1583 | 1.4166 | 41.6603 | 0.0340 | 0.0240 | 15.9871 | 35 |
| 36 | .6989 | 30.1075 | 494.6207 | 1.4308 | 43.0769 | 0.0332 | 0.0232 | 16.4285 | 36 |
| 37 | .6920 | 30.7995 | 519.5329 | 1.4451 | 44.5076 | 0.0325 | 0.0225 | 16.8682 | 37 |
| 38 | .6852 | 31.4847 | 544.8835 | 1.4595 | 45.9527 | 0.0318 | 0.0218 | 17.3063 | 38 |
| 39 | .6784 | 32.1630 | 570.6616 | 1.4741 | 47.4123 | 0.0311 | 0.0211 | 17.7428 | 39 |
| 40 | .6717 | 32.8347 | 596.8561 | 1.4889 | 48.8864 | 0.0305 | 0.0205 | 18.1776 | 40 |
| 41 | .6650 | 33.4997 | 623.4562 | 1.5038 | 50.3752 | 0.0299 | 0.0199 | 18.6108 | 41 |
| 42 | .6584 | 34.1581 | 650.4514 | 1.5188 | 51.8790 | 0.0293 | 0.0193 | 19.0424 | 42 |
| 43 | .6519 | 34.8100 | 677.8312 | 1.5340 | 53.3978 | 0.0287 | 0.0187 | 19.4723 | 43 |
| 44 | .6454 | 35.4555 | 705.5853 | 1.5493 | 54.9318 | 0.0282 | 0.0182 | 19.9006 | 44 |
| 45 | .6391 | 36.0945 | 733.7037 | 1.5648 | 56.4811 | 0.0277 | 0.0177 | 20.3273 | 45 |
| 46 | .6327 | 36.7272 | 762.1765 | 1.5805 | 58.0459 | 0.0272 | 0.0172 | 20.7524 | 46 |
| 47 | .6265 | 37.3537 | 790.9938 | 1.5963 | 59.6263 | 0.0268 | 0.0168 | 21.1758 | 47 |
| 48 | .6203 | 37.9740 | 820.1460 | 1.6122 | 61.2226 | 0.0263 | 0.0163 | 21.5976 | 48 |
| 49 | .6141 | 38.5881 | 849.6237 | 1.6283 | 62.8348 | 0.0259 | 0.0159 | 22.0178 | 49 |
| 50 | .6080 | 39.1961 | 879.4176 | 1.6446 | 64.4632 | 0.0255 | 0.0155 | 22.4363 | 50 |
| 51 | .6020 | 39.7981 | 909.5186 | 1.6611 | 66.1078 | 0.0251 | 0.0151 | 22.8533 | 51 |
| 52 | .5961 | 40.3942 | 939.9175 | 1.6777 | 67.7689 | 0.0248 | 0.0148 | 23.2686 | 52 |
| 53 | .5902 | 40.9844 | 970.6057 | 1.6945 | 69.4466 | 0.0244 | 0.0144 | 23.6823 | 53 |
| 54 | .5843 | 41.5687 | 1001.5743 | 1.7114 | 71.1410 | 0.0241 | 0.0141 | 24.0945 | 54 |
| 55 | .5785 | 42.1472 | 1032.8148 | 1.7285 | 72.8525 | 0.0237 | 0.0137 | 24.5049 | 55 |
| 60 | .5504 | 44.9550 | 1192.8061 | 1.8167 | 81.6697 | 0.0222 | 0.0122 | 26.5333 | 60 |
| 65 | .5237 | 47.6266 | 1358.3903 | 1.9094 | 90.9366 | 0.0210 | 0.0110 | 28.5217 | 65 |
| 70 | .4983 | 50.1685 | 1528.6474 | 2.0068 | 100.6763 | 0.0199 | 0.0099 | 30.4703 | 70 |
| 75 | .4741 | 52.5871 | 1702.7340 | 2.1091 | 110.9128 | 0.0190 | 0.0090 | 32.3793 | 75 |
| 80 | .4511 | 54.8882 | 1879.8771 | 2.2167 | 121.6715 | 0.0182 | 0.0082 | 34.2492 | 80 |
| 85 | .4292 | 57.0777 | 2059.3701 | 2.3298 | 132.9790 | 0.0175 | 0.0075 | 36.0801 | 85 |
| 90 | .4084 | 59.1609 | 2240.5675 | 2.4486 | 144.8633 | 0.0169 | 0.0069 | 37.8724 | 90 |
| 95 | .3886 | 61.1430 | 2422.8811 | 2.5735 | 157.3538 | 0.0164 | 0.0064 | 39.6265 | 95 |
| 100 | .3697 | 63.0289 | 2605.7758 | 2.7048 | 170.4814 | 0.0159 | 0.0059 | 41.3426 | 100 |

**I = 1.50 %**

| N | (P/F) | (P/A) | (P/G) | (F/P) | (F/A) | (A/P) | (A/F) | (A/G) | N |
|---|---|---|---|---|---|---|---|---|---|
| 1 | .9852 | 0.9852 | -0.0000 | 1.0150 | 1.0000 | 1.0150 | 1.0000 | -0.0000 | 1 |
| 2 | .9707 | 1.9559 | 0.9707 | 1.0302 | 2.0150 | 0.5113 | 0.4963 | 0.4963 | 2 |
| 3 | .9563 | 2.9122 | 2.8833 | 1.0457 | 3.0452 | 0.3434 | 0.3284 | 0.9901 | 3 |
| 4 | .9422 | 3.8544 | 5.7098 | 1.0614 | 4.0909 | 0.2594 | 0.2444 | 1.4814 | 4 |
| 5 | .9283 | 4.7826 | 9.4229 | 1.0773 | 5.1523 | 0.2091 | 0.1941 | 1.9702 | 5 |
| 6 | .9145 | 5.6972 | 13.9956 | 1.0934 | 6.2296 | 0.1755 | 0.1605 | 2.4566 | 6 |
| 7 | .9010 | 6.5982 | 19.4018 | 1.1098 | 7.3230 | 0.1516 | 0.1366 | 2.9405 | 7 |
| 8 | .8877 | 7.4859 | 25.6157 | 1.1265 | 8.4328 | 0.1336 | 0.1186 | 3.4219 | 8 |
| 9 | .8746 | 8.3605 | 32.6125 | 1.1434 | 9.5593 | 0.1196 | 0.1046 | 3.9008 | 9 |
| 10 | .8617 | 9.2222 | 40.3675 | 1.1605 | 10.7027 | 0.1084 | 0.0934 | 4.3772 | 10 |
| 11 | .8489 | 10.0711 | 48.8568 | 1.1779 | 11.8633 | 0.0993 | 0.0843 | 4.8512 | 11 |
| 12 | .8364 | 10.9075 | 58.0571 | 1.1956 | 13.0412 | 0.0917 | 0.0767 | 5.3227 | 12 |
| 13 | .8240 | 11.7315 | 67.9454 | 1.2136 | 14.2368 | 0.0852 | 0.0702 | 5.7917 | 13 |
| 14 | .8118 | 12.5434 | 78.4994 | 1.2318 | 15.4504 | 0.0797 | 0.0647 | 6.2582 | 14 |
| 15 | .7999 | 13.3432 | 89.6974 | 1.2502 | 16.6821 | 0.0749 | 0.0599 | 6.7223 | 15 |
| 16 | .7880 | 14.1313 | 101.5178 | 1.2690 | 17.9324 | 0.0708 | 0.0558 | 7.1839 | 16 |
| 17 | .7764 | 14.9076 | 113.9400 | 1.2880 | 19.2014 | 0.0671 | 0.0521 | 7.6431 | 17 |
| 18 | .7649 | 15.6726 | 126.9435 | 1.3073 | 20.4894 | 0.0638 | 0.0488 | 8.0997 | 18 |
| 19 | .7536 | 16.4262 | 140.5084 | 1.3270 | 21.7967 | 0.0609 | 0.0459 | 8.5539 | 19 |
| 20 | .7425 | 17.1686 | 154.6154 | 1.3469 | 23.1237 | 0.0582 | 0.0432 | 9.0057 | 20 |
| 21 | .7315 | 17.9001 | 169.2453 | 1.3671 | 24.4705 | 0.0559 | 0.0409 | 9.4550 | 21 |
| 22 | .7207 | 18.6208 | 184.3798 | 1.3876 | 25.8376 | 0.0537 | 0.0387 | 9.9018 | 22 |
| 23 | .7100 | 19.3309 | 200.0006 | 1.4084 | 27.2251 | 0.0517 | 0.0367 | 10.3462 | 23 |
| 24 | .6995 | 20.0304 | 216.0901 | 1.4295 | 28.6335 | 0.0499 | 0.0349 | 10.7881 | 24 |
| 25 | .6892 | 20.7196 | 232.6310 | 1.4509 | 30.0630 | 0.0483 | 0.0333 | 11.2276 | 25 |
| 26 | .6790 | 21.3986 | 249.6065 | 1.4727 | 31.5140 | 0.0467 | 0.0317 | 11.6646 | 26 |
| 27 | .6690 | 22.0676 | 267.0002 | 1.4948 | 32.9867 | 0.0453 | 0.0303 | 12.0992 | 27 |
| 28 | .6591 | 22.7267 | 284.7958 | 1.5172 | 34.4815 | 0.0440 | 0.0290 | 12.5313 | 28 |
| 29 | .6494 | 23.3761 | 302.9779 | 1.5400 | 35.9987 | 0.0428 | 0.0278 | 12.9610 | 29 |
| 30 | .6398 | 24.0158 | 321.5310 | 1.5631 | 37.5387 | 0.0416 | 0.0266 | 13.3883 | 30 |
| 31 | .6303 | 24.6461 | 340.4402 | 1.5865 | 39.1018 | 0.0406 | 0.0256 | 13.8131 | 31 |
| 32 | .6210 | 25.2671 | 359.6910 | 1.6103 | 40.6883 | 0.0396 | 0.0246 | 14.2355 | 32 |
| 33 | .6118 | 25.8790 | 379.2691 | 1.6345 | 42.2986 | 0.0386 | 0.0236 | 14.6555 | 33 |
| 34 | .6028 | 26.4817 | 399.1607 | 1.6590 | 43.9331 | 0.0378 | 0.0228 | 15.0731 | 34 |
| 35 | .5939 | 27.0756 | 419.3521 | 1.6839 | 45.5921 | 0.0369 | 0.0219 | 15.4882 | 35 |
| 36 | .5851 | 27.6607 | 439.8303 | 1.7091 | 47.2760 | 0.0362 | 0.0212 | 15.9009 | 36 |
| 37 | .5764 | 28.2371 | 460.5822 | 1.7348 | 48.9851 | 0.0354 | 0.0204 | 16.3112 | 37 |
| 38 | .5679 | 28.8051 | 481.5954 | 1.7608 | 50.7199 | 0.0347 | 0.0197 | 16.7191 | 38 |
| 39 | .5595 | 29.3646 | 502.8576 | 1.7872 | 52.4807 | 0.0341 | 0.0191 | 17.1246 | 39 |
| 40 | .5513 | 29.9158 | 524.3568 | 1.8140 | 54.2679 | 0.0334 | 0.0184 | 17.5277 | 40 |
| 41 | .5431 | 30.4590 | 546.0814 | 1.8412 | 56.0819 | 0.0328 | 0.0178 | 17.9284 | 41 |
| 42 | .5351 | 30.9941 | 568.0201 | 1.8688 | 57.9231 | 0.0323 | 0.0173 | 18.3267 | 42 |
| 43 | .5272 | 31.5212 | 590.1617 | 1.8969 | 59.7920 | 0.0317 | 0.0167 | 18.7227 | 43 |
| 44 | .5194 | 32.0406 | 612.4955 | 1.9253 | 61.6889 | 0.0312 | 0.0162 | 19.1162 | 44 |
| 45 | .5117 | 32.5523 | 635.0110 | 1.9542 | 63.6142 | 0.0307 | 0.0157 | 19.5074 | 45 |
| 46 | .5042 | 33.0565 | 657.6979 | 1.9835 | 65.5684 | 0.0303 | 0.0153 | 19.8962 | 46 |
| 47 | .4967 | 33.5532 | 680.5462 | 2.0133 | 67.5519 | 0.0298 | 0.0148 | 20.2826 | 47 |
| 48 | .4894 | 34.0426 | 703.5462 | 2.0435 | 69.5652 | 0.0294 | 0.0144 | 20.6667 | 48 |
| 49 | .4821 | 34.5247 | 726.6884 | 2.0741 | 71.6087 | 0.0290 | 0.0140 | 21.0484 | 49 |
| 50 | .4750 | 34.9997 | 749.9636 | 2.1052 | 73.6828 | 0.0286 | 0.0136 | 21.4277 | 50 |
| 51 | .4680 | 35.4677 | 773.3629 | 2.1368 | 75.7881 | 0.0282 | 0.0132 | 21.8047 | 51 |
| 52 | .4611 | 35.9287 | 796.8774 | 2.1689 | 77.9249 | 0.0278 | 0.0128 | 22.1794 | 52 |
| 53 | .4543 | 36.3830 | 820.4986 | 2.2014 | 80.0938 | 0.0275 | 0.0125 | 22.5517 | 53 |
| 54 | .4475 | 36.8305 | 844.2184 | 2.2344 | 82.2952 | 0.0272 | 0.0122 | 22.9217 | 54 |
| 55 | .4409 | 37.2715 | 868.0285 | 2.2679 | 84.5296 | 0.0268 | 0.0118 | 23.2894 | 55 |
| 60 | .4093 | 39.3803 | 988.1674 | 2.4432 | 96.2147 | 0.0254 | 0.0104 | 25.0930 | 60 |
| 65 | .3799 | 41.3378 | 1109.4752 | 2.6320 | 108.8028 | 0.0242 | 0.0092 | 26.8393 | 65 |
| 70 | .3527 | 43.1549 | 1231.1658 | 2.8355 | 122.3638 | 0.0232 | 0.0082 | 28.5290 | 70 |
| 75 | .3274 | 44.8416 | 1352.5600 | 3.0546 | 136.9728 | 0.0223 | 0.0073 | 30.1631 | 75 |
| 80 | .3039 | 46.4073 | 1473.0741 | 3.2907 | 152.7109 | 0.0215 | 0.0065 | 31.7423 | 80 |
| 85 | .2821 | 47.8607 | 1592.2095 | 3.5450 | 169.6652 | 0.0209 | 0.0059 | 33.2676 | 85 |
| 90 | .2619 | 49.2099 | 1709.5439 | 3.8189 | 187.9299 | 0.0203 | 0.0053 | 34.7399 | 90 |
| 95 | .2431 | 50.4622 | 1824.7224 | 4.1141 | 207.6061 | 0.0198 | 0.0048 | 36.1602 | 95 |
| 100 | .2256 | 51.6247 | 1937.4506 | 4.4320 | 228.8030 | 0.0194 | 0.0044 | 37.5295 | 100 |

I = 2.00 %

| N | (P/F) | (P/A) | (P/G) | (F/P) | (F/A) | (A/P) | (A/F) | (A/G) | N |
|---|---|---|---|---|---|---|---|---|---|
| 1 | .9804 | 0.9804 | -0.0000 | 1.0200 | 1.0000 | 1.0200 | 1.0000 | -0.0000 | 1 |
| 2 | .9612 | 1.9416 | 0.9612 | 1.0404 | 2.0200 | 0.5150 | 0.4950 | 0.4950 | 2 |
| 3 | .9423 | 2.8839 | 2.8458 | 1.0612 | 3.0604 | 0.3468 | 0.3268 | 0.9868 | 3 |
| 4 | .9238 | 3.8077 | 5.6173 | 1.0824 | 4.1216 | 0.2626 | 0.2426 | 1.4752 | 4 |
| 5 | .9057 | 4.7135 | 9.2403 | 1.1041 | 5.2040 | 0.2122 | 0.1922 | 1.9604 | 5 |
| 6 | .8880 | 5.6014 | 13.6801 | 1.1262 | 6.3081 | 0.1785 | 0.1585 | 2.4423 | 6 |
| 7 | .8706 | 6.4720 | 18.9035 | 1.1487 | 7.4343 | 0.1545 | 0.1345 | 2.9208 | 7 |
| 8 | .8535 | 7.3255 | 24.8779 | 1.1717 | 8.5830 | 0.1365 | 0.1165 | 3.3961 | 8 |
| 9 | .8368 | 8.1622 | 31.5720 | 1.1951 | 9.7546 | 0.1225 | 0.1025 | 3.8681 | 9 |
| 10 | .8203 | 8.9826 | 38.9551 | 1.2190 | 10.9497 | 0.1113 | 0.0913 | 4.3367 | 10 |
| 11 | .8043 | 9.7868 | 46.9977 | 1.2434 | 12.1687 | 0.1022 | 0.0822 | 4.8021 | 11 |
| 12 | .7885 | 10.5753 | 55.6712 | 1.2682 | 13.4121 | 0.0946 | 0.0746 | 5.2642 | 12 |
| 13 | .7730 | 11.3484 | 64.9475 | 1.2936 | 14.6803 | 0.0881 | 0.0681 | 5.7231 | 13 |
| 14 | .7579 | 12.1062 | 74.7999 | 1.3195 | 15.9739 | 0.0826 | 0.0626 | 6.1786 | 14 |
| 15 | .7430 | 12.8493 | 85.2021 | 1.3459 | 17.2934 | 0.0778 | 0.0578 | 6.6309 | 15 |
| 16 | .7284 | 13.5777 | 96.1288 | 1.3728 | 18.6393 | 0.0737 | 0.0537 | 7.0799 | 16 |
| 17 | .7142 | 14.2919 | 107.5554 | 1.4002 | 20.0121 | 0.0700 | 0.0500 | 7.5256 | 17 |
| 18 | .7002 | 14.9920 | 119.4581 | 1.4282 | 21.4123 | 0.0667 | 0.0467 | 7.9681 | 18 |
| 19 | .6864 | 15.6785 | 131.8139 | 1.4568 | 22.8406 | 0.0638 | 0.0438 | 8.4073 | 19 |
| 20 | .6730 | 16.3514 | 144.6003 | 1.4859 | 24.2974 | 0.0612 | 0.0412 | 8.8433 | 20 |
| 21 | .6598 | 17.0112 | 157.7959 | 1.5157 | 25.7833 | 0.0588 | 0.0388 | 9.2760 | 21 |
| 22 | .6468 | 17.6580 | 171.3795 | 1.5460 | 27.2990 | 0.0566 | 0.0366 | 9.7055 | 22 |
| 23 | .6342 | 18.2922 | 185.3309 | 1.5769 | 28.8450 | 0.0547 | 0.0347 | 10.1317 | 23 |
| 24 | .6217 | 18.9139 | 199.6305 | 1.6084 | 30.4219 | 0.0529 | 0.0329 | 10.5547 | 24 |
| 25 | .6095 | 19.5235 | 214.2592 | 1.6406 | 32.0303 | 0.0512 | 0.0312 | 10.9745 | 25 |
| 26 | .5976 | 20.1210 | 229.1987 | 1.6734 | 33.6709 | 0.0497 | 0.0297 | 11.3910 | 26 |
| 27 | .5859 | 20.7069 | 244.4311 | 1.7069 | 35.3443 | 0.0483 | 0.0283 | 11.8043 | 27 |
| 28 | .5744 | 21.2813 | 259.9392 | 1.7410 | 37.0512 | 0.0470 | 0.0270 | 12.2145 | 28 |
| 29 | .5631 | 21.8444 | 275.7064 | 1.7758 | 38.7922 | 0.0458 | 0.0258 | 12.6214 | 29 |
| 30 | .5521 | 22.3965 | 291.7164 | 1.8114 | 40.5681 | 0.0446 | 0.0246 | 13.0251 | 30 |
| 31 | .5412 | 22.9377 | 307.9538 | 1.8476 | 42.3794 | 0.0436 | 0.0236 | 13.4257 | 31 |
| 32 | .5306 | 23.4683 | 324.4035 | 1.8845 | 44.2270 | 0.0426 | 0.0226 | 13.8230 | 32 |
| 33 | .5202 | 23.9886 | 341.0508 | 1.9222 | 46.1116 | 0.0417 | 0.0217 | 14.2172 | 33 |
| 34 | .5100 | 24.4986 | 357.8817 | 1.9607 | 48.0338 | 0.0408 | 0.0208 | 14.6083 | 34 |
| 35 | .5000 | 24.9986 | 374.8826 | 1.9999 | 49.9945 | 0.0400 | 0.0200 | 14.9961 | 35 |
| 36 | .4902 | 25.4888 | 392.0405 | 2.0399 | 51.9944 | 0.0392 | 0.0192 | 15.3809 | 36 |
| 37 | .4806 | 25.9695 | 409.3424 | 2.0807 | 54.0343 | 0.0385 | 0.0185 | 15.7625 | 37 |
| 38 | .4712 | 26.4406 | 426.7764 | 2.1223 | 56.1149 | 0.0378 | 0.0178 | 16.1409 | 38 |
| 39 | .4619 | 26.9026 | 444.3304 | 2.1647 | 58.2372 | 0.0372 | 0.0172 | 16.5163 | 39 |
| 40 | .4529 | 27.3555 | 461.9931 | 2.2080 | 60.4020 | 0.0366 | 0.0166 | 16.8885 | 40 |
| 41 | .4440 | 27.7995 | 479.7535 | 2.2522 | 62.6100 | 0.0360 | 0.0160 | 17.2576 | 41 |
| 42 | .4353 | 28.2348 | 497.6010 | 2.2972 | 64.8622 | 0.0354 | 0.0154 | 17.6237 | 42 |
| 43 | .4268 | 28.6616 | 515.5253 | 2.3432 | 67.1595 | 0.0349 | 0.0149 | 17.9866 | 43 |
| 44 | .4184 | 29.0800 | 533.5165 | 2.3901 | 69.5027 | 0.0344 | 0.0144 | 18.3465 | 44 |
| 45 | .4102 | 29.4902 | 551.5652 | 2.4379 | 71.8927 | 0.0339 | 0.0139 | 18.7034 | 45 |
| 46 | .4022 | 29.8923 | 569.6621 | 2.4866 | 74.3306 | 0.0335 | 0.0135 | 19.0571 | 46 |
| 47 | .3943 | 30.2866 | 587.7985 | 2.5363 | 76.8172 | 0.0330 | 0.0130 | 19.4079 | 47 |
| 48 | .3865 | 30.6731 | 605.9657 | 2.5871 | 79.3535 | 0.0326 | 0.0126 | 19.7556 | 48 |
| 49 | .3790 | 31.0521 | 624.1557 | 2.6388 | 81.9406 | 0.0322 | 0.0122 | 20.1003 | 49 |
| 50 | .3715 | 31.4236 | 642.3606 | 2.6916 | 84.5794 | 0.0318 | 0.0118 | 20.4420 | 50 |
| 51 | .3642 | 31.7878 | 660.5727 | 2.7454 | 87.2710 | 0.0315 | 0.0115 | 20.7807 | 51 |
| 52 | .3571 | 32.1449 | 678.7849 | 2.8003 | 90.0164 | 0.0311 | 0.0111 | 21.1164 | 52 |
| 53 | .3501 | 32.4950 | 696.9900 | 2.8563 | 92.8167 | 0.0308 | 0.0108 | 21.4491 | 53 |
| 54 | .3432 | 32.8383 | 715.1815 | 2.9135 | 95.6731 | 0.0305 | 0.0105 | 21.7789 | 54 |
| 55 | .3365 | 33.1748 | 733.3527 | 2.9717 | 98.5865 | 0.0301 | 0.0101 | 22.1057 | 55 |
| 60 | .3048 | 34.7609 | 823.6975 | 3.2810 | 114.0515 | 0.0288 | 0.0088 | 23.6961 | 60 |
| 65 | .2761 | 36.1975 | 912.7085 | 3.6225 | 131.1262 | 0.0276 | 0.0076 | 25.2147 | 65 |
| 70 | .2500 | 37.4986 | 999.8343 | 3.9996 | 149.9779 | 0.0267 | 0.0067 | 26.6632 | 70 |
| 75 | .2265 | 38.6771 | 1084.6393 | 4.4158 | 170.7918 | 0.0259 | 0.0059 | 28.0434 | 75 |
| 80 | .2051 | 39.7445 | 1166.7868 | 4.8754 | 193.7720 | 0.0252 | 0.0052 | 29.3572 | 80 |
| 85 | .1858 | 40.7113 | 1246.0241 | 5.3829 | 219.1439 | 0.0246 | 0.0046 | 30.6064 | 85 |
| 90 | .1683 | 41.5869 | 1322.1701 | 5.9431 | 247.1567 | 0.0240 | 0.0040 | 31.7929 | 90 |
| 95 | .1524 | 42.3800 | 1395.1033 | 6.5617 | 278.0850 | 0.0236 | 0.0036 | 32.9189 | 95 |
| 100 | .1380 | 43.0984 | 1464.7527 | 7.2446 | 312.2323 | 0.0232 | 0.0032 | 33.9863 | 100 |

**I = 3.00 %**

| N | (P/F) | (P/A) | (P/G) | (F/P) | (F/A) | (A/P) | (A/F) | (A/G) | N |
|---|---|---|---|---|---|---|---|---|---|
| 1 | .9709 | 0.9709 | -0.0000 | 1.0300 | 1.0000 | 1.0300 | 1.0000 | -0.0000 | 1 |
| 2 | .9426 | 1.9135 | 0.9426 | 1.0609 | 2.0300 | 0.5226 | 0.4926 | 0.4926 | 2 |
| 3 | .9151 | 2.8286 | 2.7729 | 1.0927 | 3.0909 | 0.3535 | 0.3235 | 0.9803 | 3 |
| 4 | .8885 | 3.7171 | 5.4383 | 1.1255 | 4.1836 | 0.2690 | 0.2390 | 1.4631 | 4 |
| 5 | .8626 | 4.5797 | 8.8888 | 1.1593 | 5.3091 | 0.2184 | 0.1884 | 1.9409 | 5 |
| 6 | .8375 | 5.4172 | 13.0762 | 1.1941 | 6.4684 | 0.1846 | 0.1546 | 2.4138 | 6 |
| 7 | .8131 | 6.2303 | 17.9547 | 1.2299 | 7.6625 | 0.1605 | 0.1305 | 2.8819 | 7 |
| 8 | .7894 | 7.0197 | 23.4806 | 1.2668 | 8.8923 | 0.1425 | 0.1125 | 3.3450 | 8 |
| 9 | .7664 | 7.7861 | 29.6119 | 1.3048 | 10.1591 | 0.1284 | 0.0984 | 3.8032 | 9 |
| 10 | .7441 | 8.5302 | 36.3088 | 1.3439 | 11.4639 | 0.1172 | 0.0872 | 4.2565 | 10 |
| 11 | .7224 | 9.2526 | 43.5330 | 1.3842 | 12.8078 | 0.1081 | 0.0781 | 4.7049 | 11 |
| 12 | .7014 | 9.9540 | 51.2482 | 1.4258 | 14.1920 | 0.1005 | 0.0705 | 5.1485 | 12 |
| 13 | .6810 | 10.6350 | 59.4196 | 1.4685 | 15.6178 | 0.0940 | 0.0640 | 5.5872 | 13 |
| 14 | .6611 | 11.2961 | 68.0141 | 1.5126 | 17.0863 | 0.0885 | 0.0585 | 6.0210 | 14 |
| 15 | .6419 | 11.9379 | 77.0002 | 1.5580 | 18.5989 | 0.0838 | 0.0538 | 6.4500 | 15 |
| 16 | .6232 | 12.5611 | 86.3477 | 1.6047 | 20.1569 | 0.0796 | 0.0496 | 6.8742 | 16 |
| 17 | .6050 | 13.1661 | 96.0280 | 1.6528 | 21.7616 | 0.0760 | 0.0460 | 7.2936 | 17 |
| 18 | .5874 | 13.7535 | 106.0137 | 1.7024 | 23.4144 | 0.0727 | 0.0427 | 7.7081 | 18 |
| 19 | .5703 | 14.3238 | 116.2788 | 1.7535 | 25.1169 | 0.0698 | 0.0398 | 8.1179 | 19 |
| 20 | .5537 | 14.8775 | 126.7987 | 1.8061 | 26.8704 | 0.0672 | 0.0372 | 8.5229 | 20 |
| 21 | .5375 | 15.4150 | 137.5496 | 1.8603 | 28.6765 | 0.0649 | 0.0349 | 8.9231 | 21 |
| 22 | .5219 | 15.9369 | 148.5094 | 1.9161 | 30.5368 | 0.0627 | 0.0327 | 9.3186 | 22 |
| 23 | .5067 | 16.4436 | 159.6566 | 1.9736 | 32.4529 | 0.0608 | 0.0308 | 9.7093 | 23 |
| 24 | .4919 | 16.9355 | 170.9711 | 2.0328 | 34.4265 | 0.0590 | 0.0290 | 10.0954 | 24 |
| 25 | .4776 | 17.4131 | 182.4336 | 2.0938 | 36.4593 | 0.0574 | 0.0274 | 10.4768 | 25 |
| 26 | .4637 | 17.8768 | 194.0260 | 2.1566 | 38.5530 | 0.0559 | 0.0259 | 10.8535 | 26 |
| 27 | .4502 | 18.3270 | 205.7309 | 2.2213 | 40.7096 | 0.0546 | 0.0246 | 11.2255 | 27 |
| 28 | .4371 | 18.7641 | 217.5320 | 2.2879 | 42.9309 | 0.0533 | 0.0233 | 11.5930 | 28 |
| 29 | .4243 | 19.1885 | 229.4137 | 2.3566 | 45.2189 | 0.0521 | 0.0221 | 11.9558 | 29 |
| 30 | .4120 | 19.6004 | 241.3613 | 2.4273 | 47.5754 | 0.0510 | 0.0210 | 12.3141 | 30 |
| 31 | .4000 | 20.0004 | 253.3609 | 2.5001 | 50.0027 | 0.0500 | 0.0200 | 12.6678 | 31 |
| 32 | .3883 | 20.3888 | 265.3993 | 2.5751 | 52.5028 | 0.0490 | 0.0190 | 13.0169 | 32 |
| 33 | .3770 | 20.7658 | 277.4642 | 2.6523 | 55.0778 | 0.0482 | 0.0182 | 13.3616 | 33 |
| 34 | .3660 | 21.1318 | 289.5437 | 2.7319 | 57.7302 | 0.0473 | 0.0173 | 13.7018 | 34 |
| 35 | .3554 | 21.4872 | 301.6267 | 2.8139 | 60.4621 | 0.0465 | 0.0165 | 14.0375 | 35 |
| 36 | .3450 | 21.8323 | 313.7028 | 2.8983 | 63.2759 | 0.0458 | 0.0158 | 14.3688 | 36 |
| 37 | .3350 | 22.1672 | 325.7622 | 2.9852 | 66.1742 | 0.0451 | 0.0151 | 14.6957 | 37 |
| 38 | .3252 | 22.4925 | 337.7956 | 3.0748 | 69.1594 | 0.0445 | 0.0145 | 15.0182 | 38 |
| 39 | .3158 | 22.8082 | 349.7942 | 3.1670 | 72.2342 | 0.0438 | 0.0138 | 15.3363 | 39 |
| 40 | .3066 | 23.1148 | 361.7499 | 3.2620 | 75.4013 | 0.0433 | 0.0133 | 15.6502 | 40 |
| 41 | .2976 | 23.4124 | 373.6551 | 3.3599 | 78.6633 | 0.0427 | 0.0127 | 15.9597 | 41 |
| 42 | .2890 | 23.7014 | 385.5024 | 3.4607 | 82.0232 | 0.0422 | 0.0122 | 16.2650 | 42 |
| 43 | .2805 | 23.9819 | 397.2852 | 3.5645 | 85.4839 | 0.0417 | 0.0117 | 16.5660 | 43 |
| 44 | .2724 | 24.2543 | 408.9972 | 3.6715 | 89.0484 | 0.0412 | 0.0112 | 16.8629 | 44 |
| 45 | .2644 | 24.5187 | 420.6325 | 3.7816 | 92.7199 | 0.0408 | 0.0108 | 17.1556 | 45 |
| 46 | .2567 | 24.7754 | 432.1856 | 3.8950 | 96.5015 | 0.0404 | 0.0104 | 17.4441 | 46 |
| 47 | .2493 | 25.0247 | 443.6515 | 4.0119 | 100.3965 | 0.0400 | 0.0100 | 17.7285 | 47 |
| 48 | .2420 | 25.2667 | 455.0255 | 4.1323 | 104.4084 | 0.0396 | 0.0096 | 18.0089 | 48 |
| 49 | .2350 | 25.5017 | 466.3031 | 4.2562 | 108.5406 | 0.0392 | 0.0092 | 18.2852 | 49 |
| 50 | .2281 | 25.7298 | 477.4803 | 4.3839 | 112.7969 | 0.0389 | 0.0089 | 18.5575 | 50 |
| 51 | .2215 | 25.9512 | 488.5535 | 4.5154 | 117.1808 | 0.0385 | 0.0085 | 18.8258 | 51 |
| 52 | .2150 | 26.1662 | 499.5191 | 4.6509 | 121.6962 | 0.0382 | 0.0082 | 19.0902 | 52 |
| 53 | .2088 | 26.3750 | 510.3742 | 4.7904 | 126.3471 | 0.0379 | 0.0079 | 19.3507 | 53 |
| 54 | .2027 | 26.5777 | 521.1157 | 4.9341 | 131.1375 | 0.0376 | 0.0076 | 19.6073 | 54 |
| 55 | .1968 | 26.7744 | 531.7411 | 5.0821 | 136.0716 | 0.0373 | 0.0073 | 19.8600 | 55 |
| 60 | .1697 | 27.6756 | 583.0526 | 5.8916 | 163.0534 | 0.0361 | 0.0061 | 21.0674 | 60 |
| 65 | .1464 | 28.4529 | 631.2010 | 6.8300 | 194.3328 | 0.0351 | 0.0051 | 22.1841 | 65 |
| 70 | .1263 | 29.1234 | 676.0869 | 7.9178 | 230.5941 | 0.0343 | 0.0043 | 23.2145 | 70 |
| 75 | .1089 | 29.7018 | 717.6978 | 9.1789 | 272.6309 | 0.0337 | 0.0037 | 24.1634 | 75 |
| 80 | .0940 | 30.2008 | 756.0865 | 10.6409 | 321.3630 | 0.0331 | 0.0031 | 25.0353 | 80 |
| 85 | .0811 | 30.6312 | 791.3529 | 12.3357 | 377.8570 | 0.0326 | 0.0026 | 25.8349 | 85 |
| 90 | .0699 | 31.0024 | 823.6302 | 14.3005 | 443.3489 | 0.0323 | 0.0023 | 26.5667 | 90 |
| 95 | .0603 | 31.3227 | 853.0742 | 16.5782 | 519.2720 | 0.0319 | 0.0019 | 27.2351 | 95 |
| 100 | .0520 | 31.5989 | 879.8540 | 19.2186 | 607.2877 | 0.0316 | 0.0016 | 27.8444 | 100 |

**I = 4.00 %**

| N | (P/F) | (P/A) | (P/G) | (F/P) | (F/A) | (A/P) | (A/F) | (A/G) | N |
|---|---|---|---|---|---|---|---|---|---|
| 1 | .9615 | 0.9615 | -0.0000 | 1.0400 | 1.0000 | 1.0400 | 1.0000 | -0.0000 | 1 |
| 2 | .9246 | 1.8861 | 0.9246 | 1.0816 | 2.0400 | 0.5302 | 0.4902 | 0.4902 | 2 |
| 3 | .8890 | 2.7751 | 2.7025 | 1.1249 | 3.1216 | 0.3603 | 0.3203 | 0.9739 | 3 |
| 4 | .8548 | 3.6299 | 5.2670 | 1.1699 | 4.2465 | 0.2755 | 0.2355 | 1.4510 | 4 |
| 5 | .8219 | 4.4518 | 8.5547 | 1.2167 | 5.4163 | 0.2246 | 0.1846 | 1.9216 | 5 |
| 6 | .7903 | 5.2421 | 12.5062 | 1.2653 | 6.6330 | 0.1908 | 0.1508 | 2.3857 | 6 |
| 7 | .7599 | 6.0021 | 17.0657 | 1.3159 | 7.8983 | 0.1666 | 0.1266 | 2.8433 | 7 |
| 8 | .7307 | 6.7327 | 22.1806 | 1.3686 | 9.2142 | 0.1485 | 0.1085 | 3.2944 | 8 |
| 9 | .7026 | 7.4353 | 27.8013 | 1.4233 | 10.5828 | 0.1345 | 0.0945 | 3.7391 | 9 |
| 10 | .6756 | 8.1109 | 33.8814 | 1.4802 | 12.0061 | 0.1233 | 0.0833 | 4.1773 | 10 |
| 11 | .6496 | 8.7605 | 40.3772 | 1.5395 | 13.4864 | 0.1141 | 0.0741 | 4.6090 | 11 |
| 12 | .6246 | 9.3851 | 47.2477 | 1.6010 | 15.0258 | 0.1066 | 0.0666 | 5.0343 | 12 |
| 13 | .6006 | 9.9856 | 54.4546 | 1.6651 | 16.6268 | 0.1001 | 0.0601 | 5.4533 | 13 |
| 14 | .5775 | 10.5631 | 61.9618 | 1.7317 | 18.2919 | 0.0947 | 0.0547 | 5.8659 | 14 |
| 15 | .5553 | 11.1184 | 69.7355 | 1.8009 | 20.0236 | 0.0899 | 0.0499 | 6.2721 | 15 |
| 16 | .5339 | 11.6523 | 77.7441 | 1.8730 | 21.8245 | 0.0858 | 0.0458 | 6.6720 | 16 |
| 17 | .5134 | 12.1657 | 85.9581 | 1.9479 | 23.6975 | 0.0822 | 0.0422 | 7.0656 | 17 |
| 18 | .4936 | 12.6593 | 94.3498 | 2.0258 | 25.6454 | 0.0790 | 0.0390 | 7.4530 | 18 |
| 19 | .4746 | 13.1339 | 102.8933 | 2.1068 | 27.6712 | 0.0761 | 0.0361 | 7.8342 | 19 |
| 20 | .4564 | 13.5903 | 111.5647 | 2.1911 | 29.7781 | 0.0736 | 0.0336 | 8.2091 | 20 |
| 21 | .4388 | 14.0292 | 120.3414 | 2.2788 | 31.9692 | 0.0713 | 0.0313 | 8.5779 | 21 |
| 22 | .4220 | 14.4511 | 129.2024 | 2.3699 | 34.2480 | 0.0692 | 0.0292 | 8.9407 | 22 |
| 23 | .4057 | 14.8568 | 138.1284 | 2.4647 | 36.6179 | 0.0673 | 0.0273 | 9.2973 | 23 |
| 24 | .3901 | 15.2470 | 147.1012 | 2.5633 | 39.0826 | 0.0656 | 0.0256 | 9.6479 | 24 |
| 25 | .3751 | 15.6221 | 156.1040 | 2.6658 | 41.6459 | 0.0640 | 0.0240 | 9.9925 | 25 |
| 26 | .3607 | 15.9828 | 165.1212 | 2.7725 | 44.3117 | 0.0626 | 0.0226 | 10.3312 | 26 |
| 27 | .3468 | 16.3296 | 174.1385 | 2.8834 | 47.0842 | 0.0612 | 0.0212 | 10.6640 | 27 |
| 28 | .3335 | 16.6631 | 183.1424 | 2.9987 | 49.9676 | 0.0600 | 0.0200 | 10.9909 | 28 |
| 29 | .3207 | 16.9837 | 192.1206 | 3.1187 | 52.9663 | 0.0589 | 0.0189 | 11.3120 | 29 |
| 30 | .3083 | 17.2920 | 201.0618 | 3.2434 | 56.0849 | 0.0578 | 0.0178 | 11.6274 | 30 |
| 31 | .2965 | 17.5885 | 209.9556 | 3.3731 | 59.3283 | 0.0569 | 0.0169 | 11.9371 | 31 |
| 32 | .2851 | 17.8736 | 218.7924 | 3.5081 | 62.7015 | 0.0559 | 0.0159 | 12.2411 | 32 |
| 33 | .2741 | 18.1476 | 227.5634 | 3.6484 | 66.2095 | 0.0551 | 0.0151 | 12.5396 | 33 |
| 34 | .2636 | 18.4112 | 236.2607 | 3.7943 | 69.8579 | 0.0543 | 0.0143 | 12.8324 | 34 |
| 35 | .2534 | 18.6646 | 244.8768 | 3.9461 | 73.6522 | 0.0536 | 0.0136 | 13.1198 | 35 |
| 36 | .2437 | 18.9083 | 253.4052 | 4.1039 | 77.5983 | 0.0529 | 0.0129 | 13.4018 | 36 |
| 37 | .2343 | 19.1426 | 261.8399 | 4.2681 | 81.7022 | 0.0522 | 0.0122 | 13.6784 | 37 |
| 38 | .2253 | 19.3679 | 270.1754 | 4.4388 | 85.9703 | 0.0516 | 0.0116 | 13.9497 | 38 |
| 39 | .2166 | 19.5845 | 278.4070 | 4.6164 | 90.4091 | 0.0511 | 0.0111 | 14.2157 | 39 |
| 40 | .2083 | 19.7928 | 286.5303 | 4.8010 | 95.0255 | 0.0505 | 0.0105 | 14.4765 | 40 |
| 41 | .2003 | 19.9931 | 294.5414 | 4.9931 | 99.8265 | 0.0500 | 0.0100 | 14.7322 | 41 |
| 42 | .1926 | 20.1856 | 302.4370 | 5.1928 | 104.8196 | 0.0495 | 0.0095 | 14.9828 | 42 |
| 43 | .1852 | 20.3708 | 310.2141 | 5.4005 | 110.0124 | 0.0491 | 0.0091 | 15.2284 | 43 |
| 44 | .1780 | 20.5488 | 317.8700 | 5.6165 | 115.4129 | 0.0487 | 0.0087 | 15.4690 | 44 |
| 45 | .1712 | 20.7200 | 325.4028 | 5.8412 | 121.0294 | 0.0483 | 0.0083 | 15.7047 | 45 |
| 46 | .1646 | 20.8847 | 332.8104 | 6.0748 | 126.8706 | 0.0479 | 0.0079 | 15.9356 | 46 |
| 47 | .1583 | 21.0429 | 340.0914 | 6.3178 | 132.9454 | 0.0475 | 0.0075 | 16.1618 | 47 |
| 48 | .1522 | 21.1951 | 347.2446 | 6.5705 | 139.2632 | 0.0472 | 0.0072 | 16.3832 | 48 |
| 49 | .1463 | 21.3415 | 354.2689 | 6.8333 | 145.8337 | 0.0469 | 0.0069 | 16.6000 | 49 |
| 50 | .1407 | 21.4822 | 361.1638 | 7.1067 | 152.6671 | 0.0466 | 0.0066 | 16.8122 | 50 |
| 51 | .1353 | 21.6175 | 367.9289 | 7.3910 | 159.7738 | 0.0463 | 0.0063 | 17.0200 | 51 |
| 52 | .1301 | 21.7476 | 374.5638 | 7.6866 | 167.1647 | 0.0460 | 0.0060 | 17.2232 | 52 |
| 53 | .1251 | 21.8727 | 381.0686 | 7.9941 | 174.8513 | 0.0457 | 0.0057 | 17.4221 | 53 |
| 54 | .1203 | 21.9930 | 387.4436 | 8.3138 | 182.8454 | 0.0455 | 0.0055 | 17.6167 | 54 |
| 55 | .1157 | 22.1086 | 393.6890 | 8.6464 | 191.1592 | 0.0452 | 0.0052 | 17.8070 | 55 |
| 60 | .0951 | 22.6235 | 422.9966 | 10.5196 | 237.9907 | 0.0442 | 0.0042 | 18.6972 | 60 |
| 65 | .0781 | 23.0467 | 449.2014 | 12.7987 | 294.9684 | 0.0434 | 0.0034 | 19.4909 | 65 |
| 70 | .0642 | 23.3945 | 472.4789 | 15.5716 | 364.2905 | 0.0427 | 0.0027 | 20.1961 | 70 |
| 75 | .0528 | 23.6804 | 493.0408 | 18.9453 | 448.6314 | 0.0422 | 0.0022 | 20.8206 | 75 |
| 80 | .0434 | 23.9154 | 511.1161 | 23.0498 | 551.2450 | 0.0418 | 0.0018 | 21.3718 | 80 |
| 85 | .0357 | 24.1085 | 526.9384 | 28.0436 | 676.0901 | 0.0415 | 0.0015 | 21.8569 | 85 |
| 90 | .0293 | 24.2673 | 540.7369 | 34.1193 | 827.9833 | 0.0412 | 0.0012 | 22.2826 | 90 |
| 95 | .0241 | 24.3978 | 552.7307 | 41.5114 | 1012.7846 | 0.0410 | 0.0010 | 22.6550 | 95 |
| 100 | .0198 | 24.5050 | 563.1249 | 50.5049 | 1237.6237 | 0.0408 | 0.0008 | 22.9800 | 100 |

**I = 5.00 %**

| N | (P/F) | (P/A) | (P/G) | (F/P) | (F/A) | (A/P) | (A/F) | (A/G) | N |
|---|---|---|---|---|---|---|---|---|---|
| 1 | .9524 | 0.9524 | -0.0000 | 1.0500 | 1.0000 | 1.0500 | 1.0000 | -0.0000 | 1 |
| 2 | .9070 | 1.8594 | 0.9070 | 1.1025 | 2.0500 | 0.5378 | 0.4878 | 0.4878 | 2 |
| 3 | .8638 | 2.7232 | 2.6347 | 1.1576 | 3.1525 | 0.3672 | 0.3172 | 0.9675 | 3 |
| 4 | .8227 | 3.5460 | 5.1028 | 1.2155 | 4.3101 | 0.2820 | 0.2320 | 1.4391 | 4 |
| 5 | .7835 | 4.3295 | 8.2369 | 1.2763 | 5.5256 | 0.2310 | 0.1810 | 1.9025 | 5 |
| 6 | .7462 | 5.0757 | 11.9680 | 1.3401 | 6.8019 | 0.1970 | 0.1470 | 2.3579 | 6 |
| 7 | .7107 | 5.7864 | 16.2321 | 1.4071 | 8.1420 | 0.1728 | 0.1228 | 2.8052 | 7 |
| 8 | .6768 | 6.4632 | 20.9700 | 1.4775 | 9.5491 | 0.1547 | 0.1047 | 3.2445 | 8 |
| 9 | .6446 | 7.1078 | 26.1268 | 1.5513 | 11.0266 | 0.1407 | 0.0907 | 3.6758 | 9 |
| 10 | .6139 | 7.7217 | 31.6520 | 1.6289 | 12.5779 | 0.1295 | 0.0795 | 4.0991 | 10 |
| 11 | .5847 | 8.3064 | 37.4988 | 1.7103 | 14.2068 | 0.1204 | 0.0704 | 4.5144 | 11 |
| 12 | .5568 | 8.8633 | 43.6241 | 1.7959 | 15.9171 | 0.1128 | 0.0628 | 4.9219 | 12 |
| 13 | .5303 | 9.3936 | 49.9879 | 1.8856 | 17.7130 | 0.1065 | 0.0565 | 5.3215 | 13 |
| 14 | .5051 | 9.8986 | 56.5538 | 1.9799 | 19.5986 | 0.1010 | 0.0510 | 5.7133 | 14 |
| 15 | .4810 | 10.3797 | 63.2880 | 2.0789 | 21.5786 | 0.0963 | 0.0463 | 6.0973 | 15 |
| 16 | .4581 | 10.8378 | 70.1597 | 2.1829 | 23.6575 | 0.0923 | 0.0423 | 6.4736 | 16 |
| 17 | .4363 | 11.2741 | 77.1405 | 2.2920 | 25.8404 | 0.0887 | 0.0387 | 6.8423 | 17 |
| 18 | .4155 | 11.6896 | 84.2043 | 2.4066 | 28.1324 | 0.0855 | 0.0355 | 7.2034 | 18 |
| 19 | .3957 | 12.0853 | 91.3275 | 2.5270 | 30.5390 | 0.0827 | 0.0327 | 7.5569 | 19 |
| 20 | .3769 | 12.4622 | 98.4884 | 2.6533 | 33.0660 | 0.0802 | 0.0302 | 7.9030 | 20 |
| 21 | .3589 | 12.8212 | 105.6673 | 2.7860 | 35.7193 | 0.0780 | 0.0280 | 8.2416 | 21 |
| 22 | .3418 | 13.1630 | 112.8461 | 2.9253 | 38.5052 | 0.0760 | 0.0260 | 8.5730 | 22 |
| 23 | .3256 | 13.4886 | 120.0087 | 3.0715 | 41.4305 | 0.0741 | 0.0241 | 8.8971 | 23 |
| 24 | .3101 | 13.7986 | 127.1402 | 3.2251 | 44.5020 | 0.0725 | 0.0225 | 9.2140 | 24 |
| 25 | .2953 | 14.0939 | 134.2275 | 3.3864 | 47.7271 | 0.0710 | 0.0210 | 9.5238 | 25 |
| 26 | .2812 | 14.3752 | 141.2585 | 3.5557 | 51.1135 | 0.0696 | 0.0196 | 9.8266 | 26 |
| 27 | .2678 | 14.6430 | 148.2226 | 3.7335 | 54.6691 | 0.0683 | 0.0183 | 10.1224 | 27 |
| 28 | .2551 | 14.8981 | 155.1101 | 3.9201 | 58.4026 | 0.0671 | 0.0171 | 10.4114 | 28 |
| 29 | .2429 | 15.1411 | 161.9126 | 4.1161 | 62.3227 | 0.0660 | 0.0160 | 10.6936 | 29 |
| 30 | .2314 | 15.3725 | 168.6226 | 4.3219 | 66.4388 | 0.0651 | 0.0151 | 10.9691 | 30 |
| 31 | .2204 | 15.5928 | 175.2333 | 4.5380 | 70.7608 | 0.0641 | 0.0141 | 11.2381 | 31 |
| 32 | .2099 | 15.8027 | 181.7392 | 4.7649 | 75.2988 | 0.0633 | 0.0133 | 11.5005 | 32 |
| 33 | .1999 | 16.0025 | 188.1351 | 5.0032 | 80.0638 | 0.0625 | 0.0125 | 11.7566 | 33 |
| 34 | .1904 | 16.1929 | 194.4168 | 5.2533 | 85.0670 | 0.0618 | 0.0118 | 12.0063 | 34 |
| 35 | .1813 | 16.3742 | 200.5807 | 5.5160 | 90.3203 | 0.0611 | 0.0111 | 12.2498 | 35 |
| 36 | .1727 | 16.5469 | 206.6237 | 5.7918 | 95.8363 | 0.0604 | 0.0104 | 12.4872 | 36 |
| 37 | .1644 | 16.7113 | 212.5434 | 6.0814 | 101.6281 | 0.0598 | 0.0098 | 12.7186 | 37 |
| 38 | .1566 | 16.8679 | 218.3378 | 6.3855 | 107.7095 | 0.0593 | 0.0093 | 12.9440 | 38 |
| 39 | .1491 | 17.0170 | 224.0054 | 6.7048 | 114.0950 | 0.0588 | 0.0088 | 13.1636 | 39 |
| 40 | .1420 | 17.1591 | 229.5452 | 7.0400 | 120.7998 | 0.0583 | 0.0083 | 13.3775 | 40 |
| 41 | .1353 | 17.2944 | 234.9564 | 7.3920 | 127.8398 | 0.0578 | 0.0078 | 13.5857 | 41 |
| 42 | .1288 | 17.4232 | 240.2389 | 7.7616 | 135.2318 | 0.0574 | 0.0074 | 13.7884 | 42 |
| 43 | .1227 | 17.5459 | 245.3925 | 8.1497 | 142.9933 | 0.0570 | 0.0070 | 13.9857 | 43 |
| 44 | .1169 | 17.6628 | 250.4175 | 8.5572 | 151.1430 | 0.0566 | 0.0066 | 14.1777 | 44 |
| 45 | .1113 | 17.7741 | 255.3145 | 8.9850 | 159.7002 | 0.0563 | 0.0063 | 14.3644 | 45 |
| 46 | .1060 | 17.8801 | 260.0844 | 9.4343 | 168.6852 | 0.0559 | 0.0059 | 14.5461 | 46 |
| 47 | .1009 | 17.9810 | 264.7281 | 9.9060 | 178.1194 | 0.0556 | 0.0056 | 14.7226 | 47 |
| 48 | .0961 | 18.0772 | 269.2467 | 10.4013 | 188.0254 | 0.0553 | 0.0053 | 14.8943 | 48 |
| 49 | .0916 | 18.1687 | 273.6418 | 10.9213 | 198.4267 | 0.0550 | 0.0050 | 15.0611 | 49 |
| 50 | .0872 | 18.2559 | 277.9148 | 11.4674 | 209.3480 | 0.0548 | 0.0048 | 15.2233 | 50 |
| 51 | .0831 | 18.3390 | 282.0673 | 12.0408 | 220.8154 | 0.0545 | 0.0045 | 15.3808 | 51 |
| 52 | .0791 | 18.4181 | 286.1013 | 12.6428 | 232.8562 | 0.0543 | 0.0043 | 15.5337 | 52 |
| 53 | .0753 | 18.4934 | 290.0184 | 13.2749 | 245.4990 | 0.0541 | 0.0041 | 15.6823 | 53 |
| 54 | .0717 | 18.5651 | 293.8208 | 13.9387 | 258.7739 | 0.0539 | 0.0039 | 15.8265 | 54 |
| 55 | 0683 | 18.6335 | 297.5104 | 14.6356 | 272.7126 | 0.0537 | 0.0037 | 15.9664 | 55 |
| 60 | .0535 | 18.9293 | 314.3432 | 18.6792 | 353.5837 | 0.0528 | 0.0028 | 16.6062 | 60 |
| 65 | .0419 | 19.1611 | 328.6910 | 23.8399 | 456.7980 | 0.0522 | 0.0022 | 17.1541 | 65 |
| 70 | .0329 | 19.3427 | 340.8409 | 30.4264 | 588.5285 | 0.0517 | 0.0017 | 17.6212 | 70 |
| 75 | .0258 | 19.4850 | 351.0721 | 38.8327 | 756.6537 | 0.0513 | 0.0013 | 18.0176 | 75 |
| 80 | .0202 | 19.5965 | 359.6460 | 49.5614 | 971.2288 | 0.0510 | 0.0010 | 18.3526 | 80 |
| 85 | .0158 | 19.6838 | 366.8007 | 63.2544 | 1245.0871 | 0.0508 | 0.0008 | 18.6346 | 85 |
| 90 | .0124 | 19.7523 | 372.7488 | 80.7304 | 1594.6073 | 0.0506 | 0.0006 | 18.8712 | 90 |
| 95 | .0097 | 19.8059 | 377.6774 | 103.0347 | 2040.6935 | 0.0505 | 0.0005 | 19.0689 | 95 |
| 100 | .0076 | 19.8479 | 381.7492 | 131.5013 | 2610.0252 | 0.0504 | 0.0004 | 19.2337 | 100 |

**I = 6.00 %**

| N | (P/F) | (P/A) | (P/G) | (F/P) | (F/A) | (A/P) | (A/F) | (A/G) | N |
|---|---|---|---|---|---|---|---|---|---|
| 1 | .9434 | 0.9434 | -0.0000 | 1.0600 | 1.0000 | 1.0600 | 1.0000 | -0.0000 | 1 |
| 2 | .8900 | 1.8334 | 0.8900 | 1.1236 | 2.0600 | 0.5454 | 0.4854 | 0.4854 | 2 |
| 3 | .8396 | 2.6730 | 2.5692 | 1.1910 | 3.1836 | 0.3741 | 0.3141 | 0.9612 | 3 |
| 4 | .7921 | 3.4651 | 4.9455 | 1.2625 | 4.3746 | 0.2886 | 0.2286 | 1.4272 | 4 |
| 5 | .7473 | 4.2124 | 7.9345 | 1.3382 | 5.6371 | 0.2374 | 0.1774 | 1.8836 | 5 |
| 6 | .7050 | 4.9173 | 11.4594 | 1.4185 | 6.9753 | 0.2034 | 0.1434 | 2.3304 | 6 |
| 7 | .6651 | 5.5824 | 15.4497 | 1.5036 | 8.3938 | 0.1791 | 0.1191 | 2.7676 | 7 |
| 8 | .6274 | 6.2098 | 19.8416 | 1.5938 | 9.8975 | 0.1610 | 0.1010 | 3.1952 | 8 |
| 9 | .5919 | 6.8017 | 24.5768 | 1.6895 | 11.4913 | 0.1470 | 0.0870 | 3.6133 | 9 |
| 10 | .5584 | 7.3601 | 29.6023 | 1.7908 | 13.1808 | 0.1359 | 0.0759 | 4.0220 | 10 |
| 11 | .5268 | 7.8869 | 34.8702 | 1.8983 | 14.9716 | 0.1268 | 0.0668 | 4.4213 | 11 |
| 12 | .4970 | 8.3838 | 40.3369 | 2.0122 | 16.8699 | 0.1193 | 0.0593 | 4.8113 | 12 |
| 13 | .4688 | 8.8527 | 45.9629 | 2.1329 | 18.8821 | 0.1130 | 0.0530 | 5.1920 | 13 |
| 14 | .4423 | 9.2950 | 51.7128 | 2.2609 | 21.0151 | 0.1076 | 0.0476 | 5.5635 | 14 |
| 15 | .4173 | 9.7122 | 57.5546 | 2.3966 | 23.2760 | 0.1030 | 0.0430 | 5.9260 | 15 |
| 16 | .3936 | 10.1059 | 63.4592 | 2.5404 | 25.6725 | 0.0990 | 0.0390 | 6.2794 | 16 |
| 17 | .3714 | 10.4773 | 69.4011 | 2.6928 | 28.2129 | 0.0954 | 0.0354 | 6.6240 | 17 |
| 18 | .3503 | 10.8276 | 75.3569 | 2.8543 | 30.9057 | 0.0924 | 0.0324 | 6.9597 | 18 |
| 19 | .3305 | 11.1581 | 81.3062 | 3.0256 | 33.7600 | 0.0896 | 0.0296 | 7.2867 | 19 |
| 20 | .3118 | 11.4699 | 87.2304 | 3.2071 | 36.7856 | 0.0872 | 0.0272 | 7.6051 | 20 |
| 21 | .2942 | 11.7641 | 93.1136 | 3.3996 | 39.9927 | 0.0850 | 0.0250 | 7.9151 | 21 |
| 22 | .2775 | 12.0416 | 98.9412 | 3.6035 | 43.3923 | 0.0830 | 0.0230 | 8.2166 | 22 |
| 23 | .2618 | 12.3034 | 104.7007 | 3.8197 | 46.9958 | 0.0813 | 0.0213 | 8.5099 | 23 |
| 24 | .2470 | 12.5504 | 110.3812 | 4.0489 | 50.8156 | 0.0797 | 0.0197 | 8.7951 | 24 |
| 25 | .2330 | 12.7834 | 115.9732 | 4.2919 | 54.8645 | 0.0782 | 0.0182 | 9.0722 | 25 |
| 26 | .2198 | 13.0032 | 121.4684 | 4.5494 | 59.1564 | 0.0769 | 0.0169 | 9.3414 | 26 |
| 27 | .2074 | 13.2105 | 126.8600 | 4.8223 | 63.7058 | 0.0757 | 0.0157 | 9.6029 | 27 |
| 28 | .1956 | 13.4062 | 132.1420 | 5.1117 | 68.5281 | 0.0746 | 0.0146 | 9.8568 | 28 |
| 29 | .1846 | 13.5907 | 137.3096 | 5.4184 | 73.6398 | 0.0736 | 0.0136 | 10.1032 | 29 |
| 30 | .1741 | 13.7648 | 142.3588 | 5.7435 | 79.0582 | 0.0726 | 0.0126 | 10.3422 | 30 |
| 31 | .1643 | 13.9291 | 147.2864 | 6.0881 | 84.8017 | 0.0718 | 0.0118 | 10.5740 | 31 |
| 32 | .1550 | 14.0840 | 152.0901 | 6.4534 | 90.8898 | 0.0710 | 0.0110 | 10.7988 | 32 |
| 33 | .1462 | 14.2302 | 156.7681 | 6.8406 | 97.3432 | 0.0703 | 0.0103 | 11.0166 | 33 |
| 34 | .1379 | 14.3681 | 161.3192 | 7.2510 | 104.1838 | 0.0696 | 0.0096 | 11.2276 | 34 |
| 35 | .1301 | 14.4982 | 165.7427 | 7.6861 | 111.4348 | 0.0690 | 0.0090 | 11.4319 | 35 |
| 36 | .1227 | 14.6210 | 170.0387 | 8.1473 | 119.1209 | 0.0684 | 0.0084 | 11.6298 | 36 |
| 37 | .1158 | 14.7368 | 174.2072 | 8.6361 | 127.2681 | 0.0679 | 0.0079 | 11.8213 | 37 |
| 38 | .1092 | 14.8460 | 178.2490 | 9.1543 | 135.9042 | 0.0674 | 0.0074 | 12.0065 | 38 |
| 39 | .1031 | 14.9491 | 182.1652 | 9.7035 | 145.0585 | 0.0669 | 0.0069 | 12.1857 | 39 |
| 40 | .0972 | 15.0463 | 185.9568 | 10.2857 | 154.7620 | 0.0665 | 0.0065 | 12.3590 | 40 |
| 41 | .0917 | 15.1380 | 189.6256 | 10.9029 | 165.0477 | 0.0661 | 0.0061 | 12.5264 | 41 |
| 42 | .0865 | 15.2245 | 193.1732 | 11.5570 | 175.9505 | 0.0657 | 0.0057 | 12.6883 | 42 |
| 43 | .0816 | 15.3062 | 196.6017 | 12.2505 | 187.5076 | 0.0653 | 0.0053 | 12.8446 | 43 |
| 44 | .0770 | 15.3832 | 199.9130 | 12.9855 | 199.7580 | 0.0650 | 0.0050 | 12.9956 | 44 |
| 45 | .0727 | 15.4558 | 203.1096 | 13.7646 | 212.7435 | 0.0647 | 0.0047 | 13.1413 | 45 |
| 46 | .0685 | 15.5244 | 206.1938 | 14.5905 | 226.5081 | 0.0644 | 0.0044 | 13.2819 | 46 |
| 47 | .0647 | 15.5890 | 209.1681 | 15.4659 | 241.0986 | 0.0641 | 0.0041 | 13.4177 | 47 |
| 48 | .0610 | 15.6500 | 212.0351 | 16.3939 | 256.5645 | 0.0639 | 0.0039 | 13.5485 | 48 |
| 49 | .0575 | 15.7076 | 214.7972 | 17.3775 | 272.9584 | 0.0637 | 0.0037 | 13.6748 | 49 |
| 50 | .0543 | 15.7619 | 217.4574 | 18.4202 | 290.3359 | 0.0634 | 0.0034 | 13.7964 | 50 |
| 51 | .0512 | 15.8131 | 220.0181 | 19.5254 | 308.7561 | 0.0632 | 0.0032 | 13.9137 | 51 |
| 52 | .0483 | 15.8614 | 222.4823 | 20.6969 | 328.2814 | 0.0630 | 0.0030 | 14.0267 | 52 |
| 53 | .0456 | 15.9070 | 224.8525 | 21.9387 | 348.9783 | 0.0629 | 0.0029 | 14.1355 | 53 |
| 54 | .0430 | 15.9500 | 227.1316 | 23.2550 | 370.9170 | 0.0627 | 0.0027 | 14.2402 | 54 |
| 55 | .0406 | 15.9905 | 229.3222 | 24.6503 | 394.1720 | 0.0625 | 0.0025 | 14.3411 | 55 |
| 60 | .0303 | 16.1614 | 239.0428 | 32.9877 | 533.1282 | 0.0619 | 0.0019 | 14.7909 | 60 |
| 65 | .0227 | 16.2891 | 246.9450 | 44.1450 | 719.0829 | 0.0614 | 0.0014 | 15.1601 | 65 |
| 70 | .0169 | 16.3845 | 253.3271 | 59.0759 | 967.9322 | 0.0610 | 0.0010 | 15.4613 | 70 |
| 75 | .0126 | 16.4558 | 258.4527 | 79.0569 | 1300.9487 | 0.0608 | 0.0008 | 15.7058 | 75 |
| 80 | .0095 | 16.5091 | 262.5493 | 105.7960 | 1746.5999 | 0.0606 | 0.0006 | 15.9033 | 80 |
| 85 | .0071 | 16.5489 | 265.8096 | 141.5789 | 2342.9817 | 0.0604 | 0.0004 | 16.0620 | 85 |
| 90 | .0053 | 16.5787 | 268.3946 | 189.4645 | 3141.0752 | 0.0603 | 0.0003 | 16.1891 | 90 |
| 95 | .0039 | 16.6009 | 270.4375 | 253.5463 | 4209.1042 | 0.0602 | 0.0002 | 16.2905 | 95 |
| 100 | .0029 | 16.6175 | 272.0471 | 339.3021 | 5638.3681 | 0.0602 | 0.0002 | 16.3711 | 100 |

**I = 7.00 %**

| N | (P/F) | (P/A) | (P/G) | (F/P) | (F/A) | (A/P) | (A/F) | (A/G) | N |
|---|---|---|---|---|---|---|---|---|---|
| 1 | .9346 | 0.9346 | -0.0000 | 1.0700 | 1.0000 | 1.0700 | 1.0000 | -0.0000 | 1 |
| 2 | .8734 | 1.8080 | 0.8734 | 1.1449 | 2.0700 | 0.5531 | 0.4831 | 0.4831 | 2 |
| 3 | .8163 | 2.6243 | 2.5060 | 1.2250 | 3.2149 | 0.3811 | 0.3111 | 0.9549 | 3 |
| 4 | .7629 | 3.3872 | 4.7947 | 1.3108 | 4.4399 | 0.2952 | 0.2252 | 1.4155 | 4 |
| 5 | .7130 | 4.1002 | 7.6467 | 1.4026 | 5.7507 | 0.2439 | 0.1739 | 1.8650 | 5 |
| 6 | .6663 | 4.7665 | 10.9784 | 1.5007 | 7.1533 | 0.2098 | 0.1398 | 2.3032 | 6 |
| 7 | .6227 | 5.3893 | 14.7149 | 1.6058 | 8.6540 | 0.1856 | 0.1156 | 2.7304 | 7 |
| 8 | .5820 | 5.9713 | 18.7889 | 1.7182 | 10.2598 | 0.1675 | 0.0975 | 3.1465 | 8 |
| 9 | .5439 | 6.5152 | 23.1404 | 1.8385 | 11.9780 | 0.1535 | 0.0835 | 3.5517 | 9 |
| 10 | .5083 | 7.0236 | 27.7156 | 1.9672 | 13.8164 | 0.1424 | 0.0724 | 3.9461 | 10 |
| 11 | .4751 | 7.4987 | 32.4665 | 2.1049 | 15.7836 | 0.1334 | 0.0634 | 4.3296 | 11 |
| 12 | .4440 | 7.9427 | 37.3506 | 2.2522 | 17.8885 | 0.1259 | 0.0559 | 4.7025 | 12 |
| 13 | .4150 | 8.3577 | 42.3302 | 2.4098 | 20.1406 | 0.1197 | 0.0497 | 5.0648 | 13 |
| 14 | .3878 | 8.7455 | 47.3718 | 2.5785 | 22.5505 | 0.1143 | 0.0443 | 5.4167 | 14 |
| 15 | .3624 | 9.1079 | 52.4461 | 2.7590 | 25.1290 | 0.1098 | 0.0398 | 5.7583 | 15 |
| 16 | .3387 | 9.4466 | 57.5271 | 2.9522 | 27.8881 | 0.1059 | 0.0359 | 6.0897 | 16 |
| 17 | .3166 | 9.7632 | 62.5923 | 3.1588 | 30.8402 | 0.1024 | 0.0324 | 6.4110 | 17 |
| 18 | .2959 | 10.0591 | 67.6219 | 3.3799 | 33.9990 | 0.0994 | 0.0294 | 6.7225 | 18 |
| 19 | .2765 | 10.3356 | 72.5991 | 3.6165 | 37.3790 | 0.0968 | 0.0268 | 7.0242 | 19 |
| 20 | .2584 | 10.5940 | 77.5091 | 3.8697 | 40.9955 | 0.0944 | 0.0244 | 7.3163 | 20 |
| 21 | .2415 | 10.8355 | 82.3393 | 4.1406 | 44.8652 | 0.0923 | 0.0223 | 7.5990 | 21 |
| 22 | .2257 | 11.0612 | 87.0793 | 4.4304 | 49.0057 | 0.0904 | 0.0204 | 7.8725 | 22 |
| 23 | .2109 | 11.2722 | 91.7201 | 4.7405 | 53.4361 | 0.0887 | 0.0187 | 8.1369 | 23 |
| 24 | .1971 | 11.4693 | 96.2545 | 5.0724 | 58.1767 | 0.0872 | 0.0172 | 8.3923 | 24 |
| 25 | .1842 | 11.6536 | 100.6765 | 5.4274 | 63.2490 | 0.0858 | 0.0158 | 8.6391 | 25 |
| 26 | .1722 | 11.8258 | 104.9814 | 5.8074 | 68.6765 | 0.0846 | 0.0146 | 8.8773 | 26 |
| 27 | .1609 | 11.9867 | 109.1656 | 6.2139 | 74.4838 | 0.0834 | 0.0134 | 9.1072 | 27 |
| 28 | .1504 | 12.1371 | 113.2264 | 6.6488 | 80.6977 | 0.0824 | 0.0124 | 9.3289 | 28 |
| 29 | .1406 | 12.2777 | 117.1622 | 7.1143 | 87.3465 | 0.0814 | 0.0114 | 9.5427 | 29 |
| 30 | .1314 | 12.4090 | 120.9718 | 7.6123 | 94.4608 | 0.0806 | 0.0106 | 9.7487 | 30 |
| 31 | .1228 | 12.5318 | 124.6550 | 8.1451 | 102.0730 | 0.0798 | 0.0098 | 9.9471 | 31 |
| 32 | .1147 | 12.6466 | 128.2120 | 8.7153 | 110.2182 | 0.0791 | 0.0091 | 10.1381 | 32 |
| 33 | .1072 | 12.7538 | 131.6435 | 9.3253 | 118.9334 | 0.0784 | 0.0084 | 10.3219 | 33 |
| 34 | .1002 | 12.8540 | 134.9507 | 9.9781 | 128.2588 | 0.0778 | 0.0078 | 10.4987 | 34 |
| 35 | .0937 | 12.9477 | 138.1353 | 10.6766 | 138.2369 | 0.0772 | 0.0072 | 10.6687 | 35 |
| 36 | .0875 | 13.0352 | 141.1990 | 11.4239 | 148.9135 | 0.0767 | 0.0067 | 10.8321 | 36 |
| 37 | .0818 | 13.1170 | 144.1441 | 12.2236 | 160.3374 | 0.0762 | 0.0062 | 10.9891 | 37 |
| 38 | .0765 | 13.1935 | 146.9730 | 13.0793 | 172.5610 | 0.0758 | 0.0058 | 11.1398 | 38 |
| 39 | .0715 | 13.2649 | 149.6883 | 13.9948 | 185.6403 | 0.0754 | 0.0054 | 11.2845 | 39 |
| 40 | .0668 | 13.3317 | 152.2928 | 14.9745 | 199.6351 | 0.0750 | 0.0050 | 11.4233 | 40 |
| 41 | .0624 | 13.3941 | 154.7892 | 16.0227 | 214.6096 | 0.0747 | 0.0047 | 11.5565 | 41 |
| 42 | .0583 | 13.4524 | 157.1807 | 17.1443 | 230.6322 | 0.0743 | 0.0043 | 11.6842 | 42 |
| 43 | .0545 | 13.5070 | 159.4702 | 18.3444 | 247.7765 | 0.0740 | 0.0040 | 11.8065 | 43 |
| 44 | .0509 | 13.5579 | 161.6609 | 19.6285 | 266.1209 | 0.0738 | 0.0038 | 11.9237 | 44 |
| 45 | .0476 | 13.6055 | 163.7559 | 21.0025 | 285.7493 | 0.0735 | 0.0035 | 12.0360 | 45 |
| 46 | .0445 | 13.6500 | 165.7584 | 22.4726 | 306.7518 | 0.0733 | 0.0033 | 12.1435 | 46 |
| 47 | .0416 | 13.6916 | 167.6714 | 24.0457 | 329.2244 | 0.0730 | 0.0030 | 12.2463 | 47 |
| 48 | .0389 | 13.7305 | 169.4981 | 25.7289 | 353.2701 | 0.0728 | 0.0028 | 12.3447 | 48 |
| 49 | .0363 | 13.7668 | 171.2417 | 27.5299 | 378.9990 | 0.0726 | 0.0026 | 12.4387 | 49 |
| 50 | .0339 | 13.8007 | 172.9051 | 29.4570 | 406.5289 | 0.0725 | 0.0025 | 12.5287 | 50 |
| 51 | .0317 | 13.8325 | 174.4915 | 31.5190 | 435.9860 | 0.0723 | 0.0023 | 12.6146 | 51 |
| 52 | .0297 | 13.8621 | 176.0037 | 33.7253 | 467.5050 | 0.0721 | 0.0021 | 12.6967 | 52 |
| 53 | .0277 | 13.8898 | 177.4447 | 36.0861 | 501.2303 | 0.0720 | 0.0020 | 12.7751 | 53 |
| 54 | .0259 | 13.9157 | 178.8173 | 38.6122 | 537.3164 | 0.0719 | 0.0019 | 12.8500 | 54 |
| 55 | .0242 | 13.9399 | 180.1243 | 41.3150 | 575.9286 | 0.0717 | 0.0017 | 12.9215 | 55 |
| 60 | .0173 | 14.0392 | 185.7677 | 57.9464 | 813.5204 | 0.0712 | 0.0012 | 13.2321 | 60 |
| 65 | .0123 | 14.1099 | 190.1452 | 81.2729 | 1146.7552 | 0.0709 | 0.0009 | 13.4760 | 65 |
| 70 | .0088 | 14.1604 | 193.5185 | 113.9894 | 1614.1342 | 0.0706 | 0.0006 | 13.6662 | 70 |
| 75 | .0063 | 14.1964 | 196.1035 | 159.8760 | 2269.6574 | 0.0704 | 0.0004 | 13.8136 | 75 |
| 80 | .0045 | 14.2220 | 198.0748 | 224.2344 | 3189.0627 | 0.0703 | 0.0003 | 13.9273 | 80 |
| 85 | .0032 | 14.2403 | 199.5717 | 314.5003 | 4478.5761 | 0.0702 | 0.0002 | 14.0146 | 85 |
| 90 | .0023 | 14.2533 | 200.7042 | 441.1030 | 6287.1854 | 0.0702 | 0.0002 | 14.0812 | 90 |
| 95 | .0016 | 14.2626 | 201.5581 | 618.6697 | 8823.8535 | 0.0701 | 0.0001 | 14.1319 | 95 |
| 100 | .0012 | 14.2693 | 202.2001 | 867.7163 | 12381.6618 | 0.0701 | 0.0001 | 14.1703 | 100 |

**I = 8.00 %**

| N | (P/F) | (P/A) | (P/G) | (F/P) | (F/A) | (A/P) | (A/F) | (A/G) | N |
|---|---|---|---|---|---|---|---|---|---|
| 1 | .9259 | 0.9259 | -0.0000 | 1.0800 | 1.0000 | 1.0800 | 1.0000 | -0.0000 | 1 |
| 2 | .8573 | 1.7833 | 0.8573 | 1.1664 | 2.0800 | 0.5608 | 0.4808 | 0.4808 | 2 |
| 3 | .7938 | 2.5771 | 2.4450 | 1.2597 | 3.2464 | 0.3880 | 0.3080 | 0.9487 | 3 |
| 4 | .7350 | 3.3121 | 4.6501 | 1.3605 | 4.5061 | 0.3019 | 0.2219 | 1.4040 | 4 |
| 5 | .6806 | 3.9927 | 7.3724 | 1.4693 | 5.8666 | 0.2505 | 0.1705 | 1.8465 | 5 |
| 6 | .6302 | 4.6229 | 10.5233 | 1.5869 | 7.3359 | 0.2163 | 0.1363 | 2.2763 | 6 |
| 7 | .5835 | 5.2064 | 14.0242 | 1.7138 | 8.9228 | 0.1921 | 0.1121 | 2.6937 | 7 |
| 8 | .5403 | 5.7466 | 17.8061 | 1.8509 | 10.6366 | 0.1740 | 0.0940 | 3.0985 | 8 |
| 9 | .5002 | 6.2469 | 21.8081 | 1.9990 | 12.4876 | 0.1601 | 0.0801 | 3.4910 | 9 |
| 10 | .4632 | 6.7101 | 25.9768 | 2.1589 | 14.4866 | 0.1490 | 0.0690 | 3.8713 | 10 |
| 11 | .4289 | 7.1390 | 30.2657 | 2.3316 | 16.6455 | 0.1401 | 0.0601 | 4.2395 | 11 |
| 12 | .3971 | 7.5361 | 34.6339 | 2.5182 | 18.9771 | 0.1327 | 0.0527 | 4.5957 | 12 |
| 13 | .3677 | 7.9038 | 39.0463 | 2.7196 | 21.4953 | 0.1265 | 0.0465 | 4.9402 | 13 |
| 14 | .3405 | 8.2442 | 43.4723 | 2.9372 | 24.2149 | 0.1213 | 0.0413 | 5.2731 | 14 |
| 15 | .3152 | 8.5595 | 47.8857 | 3.1722 | 27.1521 | 0.1168 | 0.0368 | 5.5945 | 15 |
| 16 | .2919 | 8.8514 | 52.2640 | 3.4259 | 30.3243 | 0.1130 | 0.0330 | 5.9046 | 16 |
| 17 | .2703 | 9.1216 | 56.5883 | 3.7000 | 33.7502 | 0.1096 | 0.0296 | 6.2037 | 17 |
| 18 | .2502 | 9.3719 | 60.8426 | 3.9960 | 37.4502 | 0.1067 | 0.0267 | 6.4920 | 18 |
| 19 | .2317 | 9.6036 | 65.0134 | 4.3157 | 41.4463 | 0.1041 | 0.0241 | 6.7697 | 19 |
| 20 | .2145 | 9.8181 | 69.0898 | 4.6610 | 45.7620 | 0.1019 | 0.0219 | 7.0369 | 20 |
| 21 | .1987 | 10.0168 | 73.0629 | 5.0338 | 50.4229 | 0.0998 | 0.0198 | 7.2940 | 21 |
| 22 | .1839 | 10.2007 | 76.9257 | 5.4365 | 55.4568 | 0.0980 | 0.0180 | 7.5412 | 22 |
| 23 | .1703 | 10.3711 | 80.6726 | 5.8715 | 60.8933 | 0.0964 | 0.0164 | 7.7786 | 23 |
| 24 | .1577 | 10.5288 | 84.2997 | 6.3412 | 66.7648 | 0.0950 | 0.0150 | 8.0066 | 24 |
| 25 | .1460 | 10.6748 | 87.8041 | 6.8485 | 73.1059 | 0.0937 | 0.0137 | 8.2254 | 25 |
| 26 | .1352 | 10.8100 | 91.1842 | 7.3964 | 79.9544 | 0.0925 | 0.0125 | 8.4352 | 26 |
| 27 | .1252 | 10.9352 | 94.4390 | 7.9881 | 87.3508 | 0.0914 | 0.0114 | 8.6363 | 27 |
| 28 | .1159 | 11.0511 | 97.5687 | 8.6271 | 95.3388 | 0.0905 | 0.0105 | 8.8289 | 28 |
| 29 | .1073 | 11.1584 | 100.5738 | 9.3173 | 103.9659 | 0.0896 | 0.0096 | 9.0133 | 29 |
| 30 | .0994 | 11.2578 | 103.4558 | 10.0627 | 113.2832 | 0.0888 | 0.0088 | 9.1897 | 30 |
| 31 | .0920 | 11.3498 | 106.2163 | 10.8677 | 123.3459 | 0.0881 | 0.0081 | 9.3584 | 31 |
| 32 | .0852 | 11.4350 | 108.8575 | 11.7371 | 134.2135 | 0.0875 | 0.0075 | 9.5197 | 32 |
| 33 | .0789 | 11.5139 | 111.3819 | 12.6760 | 145.9506 | 0.0869 | 0.0069 | 9.6737 | 33 |
| 34 | .0730 | 11.5869 | 113.7924 | 13.6901 | 158.6267 | 0.0863 | 0.0063 | 9.8208 | 34 |
| 35 | .0676 | 11.6546 | 116.0920 | 14.7853 | 172.3168 | 0.0858 | 0.0058 | 9.9611 | 35 |
| 36 | .0626 | 11.7172 | 118.2839 | 15.9682 | 187.1021 | 0.0853 | 0.0053 | 10.0949 | 36 |
| 37 | .0580 | 11.7752 | 120.3713 | 17.2456 | 203.0703 | 0.0849 | 0.0049 | 10.2225 | 37 |
| 38 | .0537 | 11.8289 | 122.3579 | 18.6253 | 220.3159 | 0.0845 | 0.0045 | 10.3440 | 38 |
| 39 | .0497 | 11.8786 | 124.2470 | 20.1153 | 238.9412 | 0.0842 | 0.0042 | 10.4597 | 39 |
| 40 | .0460 | 11.9246 | 126.0422 | 21.7245 | 259.0565 | 0.0839 | 0.0039 | 10.5699 | 40 |
| 41 | .0426 | 11.9672 | 127.7470 | 23.4625 | 280.7810 | 0.0836 | 0.0036 | 10.6747 | 41 |
| 42 | .0395 | 12.0067 | 129.3651 | 25.3395 | 304.2435 | 0.0833 | 0.0033 | 10.7744 | 42 |
| 43 | .0365 | 12.0432 | 130.8998 | 27.3666 | 329.5830 | 0.0830 | 0.0030 | 10.8692 | 43 |
| 44 | .0338 | 12.0771 | 132.3547 | 29.5560 | 356.9496 | 0.0828 | 0.0028 | 10.9592 | 44 |
| 45 | .0313 | 12.1084 | 133.7331 | 31.9204 | 386.5056 | 0.0826 | 0.0026 | 11.0447 | 45 |
| 46 | .0290 | 12.1374 | 135.0384 | 34.4741 | 418.4261 | 0.0824 | 0.0024 | 11.1258 | 46 |
| 47 | .0269 | 12.1643 | 136.2739 | 37.2320 | 452.9002 | 0.0822 | 0.0022 | 11.2028 | 47 |
| 48 | .0249 | 12.1891 | 137.4428 | 40.2106 | 490.1322 | 0.0820 | 0.0020 | 11.2758 | 48 |
| 49 | .0230 | 12.2122 | 138.5480 | 43.4274 | 530.3427 | 0.0819 | 0.0019 | 11.3451 | 49 |
| 50 | .0213 | 12.2335 | 139.5928 | 46.9016 | 573.7702 | 0.0817 | 0.0017 | 11.4107 | 50 |
| 51 | .0197 | 12.2532 | 140.5799 | 50.6537 | 620.6718 | 0.0816 | 0.0016 | 11.4729 | 51 |
| 52 | .0183 | 12.2715 | 141.5121 | 54.7060 | 671.3255 | 0.0815 | 0.0015 | 11.5318 | 52 |
| 53 | .0169 | 12.2884 | 142.3923 | 59.0825 | 726.0316 | 0.0814 | 0.0014 | 11.5875 | 53 |
| 54 | .0157 | 12.3041 | 143.2229 | 63.8091 | 785.1141 | 0.0813 | 0.0013 | 11.6403 | 54 |
| 55 | .0145 | 12.3186 | 144.0065 | 68.9139 | 848.9232 | 0.0812 | 0.0012 | 11.6902 | 55 |
| 60 | .0099 | 12.3766 | 147.3000 | 101.2571 | 1253.2133 | 0.0808 | 0.0008 | 11.9015 | 60 |
| 65 | .0067 | 12.4160 | 149.7387 | 148.7798 | 1847.2481 | 0.0805 | 0.0005 | 12.0602 | 65 |
| 70 | .0046 | 12.4428 | 151.5326 | 218.6064 | 2720.0801 | 0.0804 | 0.0004 | 12.1783 | 70 |
| 75 | .0031 | 12.4611 | 152.8448 | 321.2045 | 4002.5566 | 0.0802 | 0.0002 | 12.2658 | 75 |
| 80 | .0021 | 12.4735 | 153.8001 | 471.9548 | 5886.9354 | 0.0802 | 0.0002 | 12.3301 | 80 |
| 85 | .0014 | 12.4820 | 154.4925 | 693.4565 | 8655.7061 | 0.0801 | 0.0001 | 12.3772 | 85 |
| 90 | .0010 | 12.4877 | 154.9925 | 1018.9151 | 12723.9386 | 0.0801 | 0.0001 | 12.4116 | 90 |
| 95 | .0007 | 12.4917 | 155.3524 | 1497.1205 | 18701.5069 | 0.0801 | 0.0001 | 12.4365 | 95 |
| 100 | .0005 | 12.4943 | 155.6107 | 2199.7613 | 27484.5157 | 0.0800 | 0.0000 | 12.4545 | 100 |

**I = 9.00 %**

| N | (P/F) | (P/A) | (P/G) | (F/P) | (F/A) | (A/P) | (A/F) | (A/G) | N |
|---|---|---|---|---|---|---|---|---|---|
| 1 | .9174 | 0.9174 | -0.0000 | 1.0900 | 1.0000 | 1.0900 | 1.0000 | -0.0000 | 1 |
| 2 | .8417 | 1.7591 | 0.8417 | 1.1881 | 2.0900 | 0.5685 | 0.4785 | 0.4785 | 2 |
| 3 | .7722 | 2.5313 | 2.3860 | 1.2950 | 3.2781 | 0.3951 | 0.3051 | 0.9426 | 3 |
| 4 | .7084 | 3.2397 | 4.5113 | 1.4116 | 4.5731 | 0.3087 | 0.2187 | 1.3925 | 4 |
| 5 | .6499 | 3.8897 | 7.1110 | 1.5386 | 5.9847 | 0.2571 | 0.1671 | 1.8282 | 5 |
| 6 | .5963 | 4.4859 | 10.0924 | 1.6771 | 7.5233 | 0.2229 | 0.1329 | 2.2498 | 6 |
| 7 | .5470 | 5.0330 | 13.3746 | 1.8280 | 9.2004 | 0.1987 | 0.1087 | 2.6574 | 7 |
| 8 | .5019 | 5.5348 | 16.8877 | 1.9926 | 11.0285 | 0.1807 | 0.0907 | 3.0512 | 8 |
| 9 | .4604 | 5.9952 | 20.5711 | 2.1719 | 13.0210 | 0.1668 | 0.0768 | 3.4312 | 9 |
| 10 | .4224 | 6.4177 | 24.3728 | 2.3674 | 15.1929 | 0.1558 | 0.0658 | 3.7978 | 10 |
| 11 | .3875 | 6.8052 | 28.2481 | 2.5804 | 17.5603 | 0.1469 | 0.0569 | 4.1510 | 11 |
| 12 | .3555 | 7.1607 | 32.1590 | 2.8127 | 20.1407 | 0.1397 | 0.0497 | 4.4910 | 12 |
| 13 | .3262 | 7.4869 | 36.0731 | 3.0658 | 22.9534 | 0.1336 | 0.0436 | 4.8182 | 13 |
| 14 | .2992 | 7.7862 | 39.9633 | 3.3417 | 26.0192 | 0.1284 | 0.0384 | 5.1326 | 14 |
| 15 | .2745 | 8.0607 | 43.8069 | 3.6425 | 29.3609 | 0.1241 | 0.0341 | 5.4346 | 15 |
| 16 | .2519 | 8.3126 | 47.5849 | 3.9703 | 33.0034 | 0.1203 | 0.0303 | 5.7245 | 16 |
| 17 | .2311 | 8.5436 | 51.2821 | 4.3276 | 36.9737 | 0.1170 | 0.0270 | 6.0024 | 17 |
| 18 | .2120 | 8.7556 | 54.8860 | 4.7171 | 41.3013 | 0.1142 | 0.0242 | 6.2687 | 18 |
| 19 | .1945 | 8.9501 | 58.3868 | 5.1417 | 46.0185 | 0.1117 | 0.0217 | 6.5236 | 19 |
| 20 | .1784 | 9.1285 | 61.7770 | 5.6044 | 51.1601 | 0.1095 | 0.0195 | 6.7674 | 20 |
| 21 | .1637 | 9.2922 | 65.0509 | 6.1088 | 56.7645 | 0.1076 | 0.0176 | 7.0006 | 21 |
| 22 | .1502 | 9.4424 | 68.2048 | 6.6586 | 62.8733 | 0.1059 | 0.0159 | 7.2232 | 22 |
| 23 | .1378 | 9.5802 | 71.2359 | 7.2579 | 69.5319 | 0.1044 | 0.0144 | 7.4357 | 23 |
| 24 | .1264 | 9.7066 | 74.1433 | 7.9111 | 76.7898 | 0.1030 | 0.0130 | 7.6384 | 24 |
| 25 | .1160 | 9.8226 | 76.9265 | 8.6231 | 84.7009 | 0.1018 | 0.0118 | 7.8316 | 25 |
| 26 | .1064 | 9.9290 | 79.5863 | 9.3992 | 93.3240 | 0.1007 | 0.0107 | 8.0156 | 26 |
| 27 | .0976 | 10.0266 | 82.1241 | 10.2451 | 102.7231 | 0.0997 | 0.0097 | 8.1906 | 27 |
| 28 | .0895 | 10.1161 | 84.5419 | 11.1671 | 112.9682 | 0.0989 | 0.0089 | 8.3571 | 28 |
| 29 | .0822 | 10.1983 | 86.8422 | 12.1722 | 124.1354 | 0.0981 | 0.0081 | 8.5154 | 29 |
| 30 | .0754 | 10.2737 | 89.0280 | 13.2677 | 136.3075 | 0.0973 | 0.0073 | 8.6657 | 30 |
| 31 | .0691 | 10.3428 | 91.1024 | 14.4618 | 149.5752 | 0.0967 | 0.0067 | 8.8083 | 31 |
| 32 | .0634 | 10.4062 | 93.0690 | 15.7633 | 164.0370 | 0.0961 | 0.0061 | 8.9436 | 32 |
| 33 | .0582 | 10.4644 | 94.9314 | 17.1820 | 179.8003 | 0.0956 | 0.0056 | 9.0718 | 33 |
| 34 | .0534 | 10.5178 | 96.6935 | 18.7284 | 196.9823 | 0.0951 | 0.0051 | 9.1933 | 34 |
| 35 | .0490 | 10.5668 | 98.3590 | 20.4140 | 215.7108 | 0.0946 | 0.0046 | 9.3083 | 35 |
| 36 | .0449 | 10.6118 | 99.9319 | 22.2512 | 236.1247 | 0.0942 | 0.0042 | 9.4171 | 36 |
| 37 | .0412 | 10.6530 | 101.4162 | 24.2538 | 258.3759 | 0.0939 | 0.0039 | 9.5200 | 37 |
| 38 | .0378 | 10.6908 | 102.8158 | 26.4367 | 282.6298 | 0.0935 | 0.0035 | 9.6172 | 38 |
| 39 | .0347 | 10.7255 | 104.1345 | 28.8160 | 309.0665 | 0.0932 | 0.0032 | 9.7090 | 39 |
| 40 | .0318 | 10.7574 | 105.3762 | 31.4094 | 337.8824 | 0.0930 | 0.0030 | 9.7957 | 40 |
| 41 | .0292 | 10.7866 | 106.5445 | 34.2363 | 369.2919 | 0.0927 | 0.0027 | 9.8775 | 41 |
| 42 | .0268 | 10.8134 | 107.6432 | 37.3175 | 403.5281 | 0.0925 | 0.0025 | 9.9546 | 42 |
| 43 | .0246 | 10.8380 | 108.6758 | 40.6761 | 440.8457 | 0.0923 | 0.0023 | 10.0273 | 43 |
| 44 | .0226 | 10.8605 | 109.6456 | 44.3370 | 481.5218 | 0.0921 | 0.0021 | 10.0958 | 44 |
| 45 | .0207 | 10.8812 | 110.5561 | 48.3273 | 525.8587 | 0.0919 | 0.0019 | 10.1603 | 45 |
| 46 | .0190 | 10.9002 | 111.4103 | 52.6767 | 574.1860 | 0.0917 | 0.0017 | 10.2210 | 46 |
| 47 | .0174 | 10.9176 | 112.2115 | 57.4176 | 626.8628 | 0.0916 | 0.0016 | 10.2780 | 47 |
| 48 | .0160 | 10.9336 | 112.9625 | 62.5852 | 684.2804 | 0.0915 | 0.0015 | 10.3317 | 48 |
| 49 | .0147 | 10.9482 | 113.6661 | 68.2179 | 746.8656 | 0.0913 | 0.0013 | 10.3821 | 49 |
| 50 | .0134 | 10.9617 | 114.3251 | 74.3575 | 815.0836 | 0.0912 | 0.0012 | 10.4295 | 50 |
| 51 | .0123 | 10.9740 | 114.9420 | 81.0497 | 889.4411 | 0.0911 | 0.0011 | 10.4740 | 51 |
| 52 | .0113 | 10.9853 | 115.5193 | 88.3442 | 970.4908 | 0.0910 | 0.0010 | 10.5158 | 52 |
| 53 | .0104 | 10.9957 | 116.0593 | 96.2951 | 1058.8349 | 0.0909 | 0.0009 | 10.5549 | 53 |
| 54 | .0095 | 11.0053 | 116.5642 | 104.9617 | 1155.1301 | 0.0909 | 0.0009 | 10.5917 | 54 |
| 55 | .0087 | 11.0140 | 117.0362 | 114.4083 | 1260.0918 | 0.0908 | 0.0008 | 10.6261 | 55 |
| 60 | .0057 | 11.0480 | 118.9683 | 176.0313 | 1944.7921 | 0.0905 | 0.0005 | 10.7683 | 60 |
| 65 | .0037 | 11.0701 | 120.3344 | 270.8460 | 2998.2885 | 0.0903 | 0.0003 | 10.8702 | 65 |
| 70 | .0024 | 11.0844 | 121.2942 | 416.7301 | 4619.2232 | 0.0902 | 0.0002 | 10.9427 | 70 |
| 75 | .0016 | 11.0938 | 121.9646 | 641.1909 | 7113.2321 | 0.0901 | 0.0001 | 10.9940 | 75 |
| 80 | .0010 | 11.0998 | 122.4306 | 986.5517 | 10950.5741 | 0.0901 | 0.0001 | 11.0299 | 80 |
| 85 | .0007 | 11.1038 | 122.7533 | 1517.9320 | 16854.8003 | 0.0901 | 0.0001 | 11.0551 | 85 |
| 90 | .0004 | 11.1064 | 122.9758 | 2335.5266 | 25939.1842 | 0.0900 | 0.0000 | 11.0726 | 90 |
| 95 | .0003 | 11.1080 | 123.1287 | 3593.4971 | 39916.6350 | 0.0900 | 0.0000 | 11.0847 | 95 |
| 100 | .0002 | 11.1091 | 123.2335 | 5529.0408 | 61422.6755 | 0.0900 | 0.0000 | 11.0930 | 100 |

**I = 10.00 %**

| N | (P/F) | (P/A) | (P/G) | (F/P) | (F/A) | (A/P) | (A/F) | (A/G) | N |
|---|---|---|---|---|---|---|---|---|---|
| 1 | .9091 | 0.9091 | −0.0000 | 1.1000 | 1.0000 | 1.1000 | 1.0000 | −0.0000 | 1 |
| 2 | .8264 | 1.7355 | 0.8264 | 1.2100 | 2.1000 | 0.5762 | 0.4762 | 0.4762 | 2 |
| 3 | .7513 | 2.4869 | 2.3291 | 1.3310 | 3.3100 | 0.4021 | 0.3021 | 0.9366 | 3 |
| 4 | .6830 | 3.1699 | 4.3781 | 1.4641 | 4.6410 | 0.3155 | 0.2155 | 1.3812 | 4 |
| 5 | .6209 | 3.7908 | 6.8618 | 1.6105 | 6.1051 | 0.2638 | 0.1638 | 1.8101 | 5 |
| 6 | .5645 | 4.3553 | 9.6842 | 1.7716 | 7.7156 | 0.2296 | 0.1296 | 2.2236 | 6 |
| 7 | .5132 | 4.8684 | 12.7631 | 1.9487 | 9.4872 | 0.2054 | 0.1054 | 2.6216 | 7 |
| 8 | .4665 | 5.3349 | 16.0287 | 2.1436 | 11.4359 | 0.1874 | 0.0874 | 3.0045 | 8 |
| 9 | .4241 | 5.7590 | 19.4215 | 2.3579 | 13.5795 | 0.1736 | 0.0736 | 3.3724 | 9 |
| 10 | .3855 | 6.1446 | 22.8913 | 2.5937 | 15.9374 | 0.1627 | 0.0627 | 3.7255 | 10 |
| 11 | .3505 | 6.4951 | 26.3963 | 2.8531 | 18.5312 | 0.1540 | 0.0540 | 4.0641 | 11 |
| 12 | .3186 | 6.8137 | 29.9012 | 3.1384 | 21.3843 | 0.1468 | 0.0468 | 4.3884 | 12 |
| 13 | .2897 | 7.1034 | 33.3772 | 3.4523 | 24.5227 | 0.1408 | 0.0408 | 4.6988 | 13 |
| 14 | .2633 | 7.3667 | 36.8005 | 3.7975 | 27.9750 | 0.1357 | 0.0357 | 4.9955 | 14 |
| 15 | .2394 | 7.6061 | 40.1520 | 4.1772 | 31.7725 | 0.1315 | 0.0315 | 5.2789 | 15 |
| 16 | .2176 | 7.8237 | 43.4164 | 4.5950 | 35.9497 | 0.1278 | 0.0278 | 5.5493 | 16 |
| 17 | .1978 | 8.0216 | 46.5819 | 5.0545 | 40.5447 | 0.1247 | 0.0247 | 5.8071 | 17 |
| 18 | .1799 | 8.2014 | 49.6395 | 5.5599 | 45.5992 | 0.1219 | 0.0219 | 6.0526 | 18 |
| 19 | .1635 | 8.3649 | 52.5827 | 6.1159 | 51.1591 | 0.1195 | 0.0195 | 6.2861 | 19 |
| 20 | .1486 | 8.5136 | 55.4069 | 6.7275 | 57.2750 | 0.1175 | 0.0175 | 6.5081 | 20 |
| 21 | .1351 | 8.6487 | 58.1095 | 7.4002 | 64.0025 | 0.1156 | 0.0156 | 6.7189 | 21 |
| 22 | .1228 | 8.7715 | 60.6893 | 8.1403 | 71.4027 | 0.1140 | 0.0140 | 6.9189 | 22 |
| 23 | .1117 | 8.8832 | 63.1462 | 8.9543 | 79.5430 | 0.1126 | 0.0126 | 7.1085 | 23 |
| 24 | .1015 | 8.9847 | 65.4813 | 9.8497 | 88.4973 | 0.1113 | 0.0113 | 7.2881 | 24 |
| 25 | .0923 | 9.0770 | 67.6964 | 10.8347 | 98.3471 | 0.1102 | 0.0102 | 7.4580 | 25 |
| 26 | .0839 | 9.1609 | 69.7940 | 11.9182 | 109.1818 | 0.1092 | 0.0092 | 7.6186 | 26 |
| 27 | .0763 | 9.2372 | 71.7773 | 13.1100 | 121.0999 | 0.1083 | 0.0083 | 7.7704 | 27 |
| 28 | .0693 | 9.3066 | 73.6495 | 14.4210 | 134.2099 | 0.1075 | 0.0075 | 7.9137 | 28 |
| 29 | .0630 | 9.3696 | 75.4146 | 15.8631 | 148.6309 | 0.1067 | 0.0067 | 8.0489 | 29 |
| 30 | .0573 | 9.4269 | 77.0766 | 17.4494 | 164.4940 | 0.1061 | 0.0061 | 8.1762 | 30 |
| 31 | .0521 | 9.4790 | 78.6395 | 19.1943 | 181.9434 | 0.1055 | 0.0055 | 8.2962 | 31 |
| 32 | .0474 | 9.5264 | 80.1078 | 21.1138 | 201.1378 | 0.1050 | 0.0050 | 8.4091 | 32 |
| 33 | .0431 | 9.5694 | 81.4856 | 23.2252 | 222.2515 | 0.1045 | 0.0045 | 8.5152 | 33 |
| 34 | .0391 | 9.6086 | 82.7773 | 25.5477 | 245.4767 | 0.1041 | 0.0041 | 8.6149 | 34 |
| 35 | .0356 | 9.6442 | 83.9872 | 28.1024 | 271.0244 | 0.1037 | 0.0037 | 8.7086 | 35 |
| 36 | .0323 | 9.6765 | 85.1194 | 30.9127 | 299.1268 | 0.1033 | 0.0033 | 8.7965 | 36 |
| 37 | .0294 | 9.7059 | 86.1781 | 34.0039 | 330.0395 | 0.1030 | 0.0030 | 8.8789 | 37 |
| 38 | .0267 | 9.7327 | 87.1673 | 37.4043 | 364.0434 | 0.1027 | 0.0027 | 8.9562 | 38 |
| 39 | .0243 | 9.7570 | 88.0908 | 41.1448 | 401.4478 | 0.1025 | 0.0025 | 9.0285 | 39 |
| 40 | .0221 | 9.7791 | 88.9525 | 45.2593 | 442.5926 | 0.1023 | 0.0023 | 9.0962 | 40 |
| 41 | .0201 | 9.7991 | 89.7560 | 49.7852 | 487.8518 | 0.1020 | 0.0020 | 9.1596 | 41 |
| 42 | .0183 | 9.8174 | 90.5047 | 54.7637 | 537.6370 | 0.1019 | 0.0019 | 9.2188 | 42 |
| 43 | .0166 | 9.8340 | 91.2019 | 60.2401 | 592.4007 | 0.1017 | 0.0017 | 9.2741 | 43 |
| 44 | .0151 | 9.8491 | 91.8508 | 66.2641 | 652.6408 | 0.1015 | 0.0015 | 9.3258 | 44 |
| 45 | .0137 | 9.8628 | 92.4544 | 72.8905 | 718.9048 | 0.1014 | 0.0014 | 9.3740 | 45 |
| 46 | .0125 | 9.8753 | 93.0157 | 80.1795 | 791.7953 | 0.1013 | 0.0013 | 9.4190 | 46 |
| 47 | .0113 | 9.8866 | 93.5372 | 88.1975 | 871.9749 | 0.1011 | 0.0011 | 9.4610 | 47 |
| 48 | .0103 | 9.8969 | 94.0217 | 97.0172 | 960.1723 | 0.1010 | 0.0010 | 9.5001 | 48 |
| 49 | .0094 | 9.9063 | 94.4715 | 106.7190 | 1057.1896 | 0.1009 | 0.0009 | 9.5365 | 49 |
| 50 | .0085 | 9.9148 | 94.8889 | 117.3909 | 1163.9085 | 0.1009 | 0.0009 | 9.5704 | 50 |
| 51 | .0077 | 9.9226 | 95.2761 | 129.1299 | 1281.2994 | 0.1008 | 0.0008 | 9.6020 | 51 |
| 52 | .0070 | 9.9296 | 95.6351 | 142.0429 | 1410.4293 | 0.1007 | 0.0007 | 9.6313 | 52 |
| 53 | .0064 | 9.9360 | 95.9679 | 156.2472 | 1552.4723 | 0.1006 | 0.0006 | 9.6586 | 53 |
| 54 | .0058 | 9.9418 | 96.2763 | 171.8719 | 1708.7195 | 0.1006 | 0.0006 | 9.6840 | 54 |
| 55 | .0053 | 9.9471 | 96.5619 | 189.0591 | 1880.5914 | 0.1005 | 0.0005 | 9.7075 | 55 |
| 60 | .0033 | 9.9672 | 97.7010 | 304.4816 | 3034.8164 | 0.1003 | 0.0003 | 9.8023 | 60 |
| 65 | .0020 | 9.9796 | 98.4705 | 490.3707 | 4893.7073 | 0.1002 | 0.0002 | 9.8672 | 65 |
| 70 | .0013 | 9.9873 | 98.9870 | 789.7470 | 7887.4696 | 0.1001 | 0.0001 | 9.9113 | 70 |
| 75 | .0008 | 9.9921 | 99.3317 | 1271.8954 | 12708.9537 | 0.1001 | 0.0001 | 9.9410 | 75 |
| 80 | .0005 | 9.9951 | 99.5606 | 2048.4002 | 20474.0021 | 0.1000 | 0.0000 | 9.9609 | 80 |
| 85 | .0003 | 9.9970 | 99.7120 | 3298.9690 | 32979.6903 | 0.1000 | 0.0000 | 9.9742 | 85 |
| 90 | .0002 | 9.9981 | 99.8118 | 5313.0226 | 53120.2261 | 0.1000 | 0.0000 | 9.9831 | 90 |
| 95 | .0001 | 9.9988 | 99.8773 | 8556.6760 | 85556.7605 | 0.1000 | 0.0000 | 9.9889 | 95 |
| 100 | .0001 | 9.9993 | 99.9202 | 13780.6123 | 137796.1234 | 0.1000 | 0.0000 | 9.9927 | 100 |

**I = 12.00 %**

| N | (P/F) | (P/A) | (P/G) | (F/P) | (F/A) | (A/P) | (A/F) | (A/G) | N |
|---|---|---|---|---|---|---|---|---|---|
| 1 | .8929 | 0.8929 | -0.0000 | 1.1200 | 1.0000 | 1.1200 | 1.0000 | -0.0000 | 1 |
| 2 | .7972 | 1.6901 | 0.7972 | 1.2544 | 2.1200 | 0.5917 | 0.4717 | 0.4717 | 2 |
| 3 | .7118 | 2.4018 | 2.2208 | 1.4049 | 3.3744 | 0.4163 | 0.2963 | 0.9246 | 3 |
| 4 | .6355 | 3.0373 | 4.1273 | 1.5735 | 4.7793 | 0.3292 | 0.2092 | 1.3589 | 4 |
| 5 | .5674 | 3.6048 | 6.3970 | 1.7623 | 6.3528 | 0.2774 | 0.1574 | 1.7746 | 5 |
| 6 | .5066 | 4.1114 | 8.9302 | 1.9738 | 8.1152 | 0.2432 | 0.1232 | 2.1720 | 6 |
| 7 | .4523 | 4.5638 | 11.6443 | 2.2107 | 10.0890 | 0.2191 | 0.0991 | 2.5515 | 7 |
| 8 | .4039 | 4.9676 | 14.4714 | 2.4760 | 12.2997 | 0.2013 | 0.0813 | 2.9131 | 8 |
| 9 | .3606 | 5.3282 | 17.3563 | 2.7731 | 14.7757 | 0.1877 | 0.0677 | 3.2574 | 9 |
| 10 | .3220 | 5.6502 | 20.2541 | 3.1058 | 17.5487 | 0.1770 | 0.0570 | 3.5847 | 10 |
| 11 | .2875 | 5.9377 | 23.1288 | 3.4785 | 20.6546 | 0.1684 | 0.0484 | 3.8953 | 11 |
| 12 | .2567 | 6.1944 | 25.9523 | 3.8960 | 24.1331 | 0.1614 | 0.0414 | 4.1897 | 12 |
| 13 | .2292 | 6.4235 | 28.7024 | 4.3635 | 28.0291 | 0.1557 | 0.0357 | 4.4683 | 13 |
| 14 | .2046 | 6.6282 | 31.3624 | 4.8871 | 32.3926 | 0.1509 | 0.0309 | 4.7317 | 14 |
| 15 | .1827 | 6.8109 | 33.9202 | 5.4736 | 37.2797 | 0.1468 | 0.0268 | 4.9803 | 15 |
| 16 | .1631 | 6.9740 | 36.3670 | 6.1304 | 42.7533 | 0.1434 | 0.0234 | 5.2147 | 16 |
| 17 | .1456 | 7.1196 | 38.6973 | 6.8660 | 48.8837 | 0.1405 | 0.0205 | 5.4353 | 17 |
| 18 | .1300 | 7.2497 | 40.9080 | 7.6900 | 55.7497 | 0.1379 | 0.0179 | 5.6427 | 18 |
| 19 | .1161 | 7.3658 | 42.9979 | 8.6128 | 63.4397 | 0.1358 | 0.0158 | 5.8375 | 19 |
| 20 | .1037 | 7.4694 | 44.9676 | 9.6463 | 72.0524 | 0.1339 | 0.0139 | 6.0202 | 20 |
| 21 | .0926 | 7.5620 | 46.8188 | 10.8038 | 81.6987 | 0.1322 | 0.0122 | 6.1913 | 21 |
| 22 | .0826 | 7.6446 | 48.5543 | 12.1003 | 92.5026 | 0.1308 | 0.0108 | 6.3514 | 22 |
| 23 | .0738 | 7.7184 | 50.1776 | 13.5523 | 104.6029 | 0.1296 | 0.0096 | 6.5010 | 23 |
| 24 | .0659 | 7.7843 | 51.6929 | 15.1786 | 118.1552 | 0.1285 | 0.0085 | 6.6406 | 24 |
| 25 | .0588 | 7.8431 | 53.1046 | 17.0001 | 133.3339 | 0.1275 | 0.0075 | 6.7708 | 25 |
| 26 | .0525 | 7.8957 | 54.4177 | 19.0401 | 150.3339 | 0.1267 | 0.0067 | 6.8921 | 26 |
| 27 | .0469 | 7.9426 | 55.6369 | 21.3249 | 169.3740 | 0.1259 | 0.0059 | 7.0049 | 27 |
| 28 | .0419 | 7.9844 | 56.7674 | 23.8839 | 190.6989 | 0.1252 | 0.0052 | 7.1098 | 28 |
| 29 | .0374 | 8.0218 | 57.8141 | 26.7499 | 214.5828 | 0.1247 | 0.0047 | 7.2071 | 29 |
| 30 | .0334 | 8.0552 | 58.7821 | 29.9599 | 241.3327 | 0.1241 | 0.0041 | 7.2974 | 30 |
| 31 | .0298 | 8.0850 | 59.6761 | 33.5551 | 271.2926 | 0.1237 | 0.0037 | 7.3811 | 31 |
| 32 | .0266 | 8.1116 | 60.5010 | 37.5817 | 304.8477 | 0.1233 | 0.0033 | 7.4586 | 32 |
| 33 | .0238 | 8.1354 | 61.2612 | 42.0915 | 342.4294 | 0.1229 | 0.0029 | 7.5302 | 33 |
| 34 | .0212 | 8.1566 | 61.9612 | 47.1425 | 384.5210 | 0.1226 | 0.0026 | 7.5965 | 34 |
| 35 | .0189 | 8.1755 | 62.6052 | 52.7996 | 431.6635 | 0.1223 | 0.0023 | 7.6577 | 35 |
| 36 | .0169 | 8.1924 | 63.1970 | 59.1356 | 484.4631 | 0.1221 | 0.0021 | 7.7141 | 36 |
| 37 | .0151 | 8.2075 | 63.7406 | 66.2318 | 543.5987 | 0.1218 | 0.0018 | 7.7661 | 37 |
| 38 | .0135 | 8.2210 | 64.2394 | 74.1797 | 609.8305 | 0.1216 | 0.0016 | 7.8141 | 38 |
| 39 | .0120 | 8.2330 | 64.6967 | 83.0812 | 684.0102 | 0.1215 | 0.0015 | 7.8582 | 39 |
| 40 | .0107 | 8.2438 | 65.1159 | 93.0510 | 767.0914 | 0.1213 | 0.0013 | 7.8988 | 40 |
| 41 | .0096 | 8.2534 | 65.4997 | 104.2171 | 860.1424 | 0.1212 | 0.0012 | 7.9361 | 41 |
| 42 | .0086 | 8.2619 | 65.8509 | 116.7231 | 964.3595 | 0.1210 | 0.0010 | 7.9704 | 42 |
| 43 | .0076 | 8.2696 | 66.1722 | 130.7299 | 1081.0826 | 0.1209 | 0.0009 | 8.0019 | 43 |
| 44 | .0068 | 8.2764 | 66.4659 | 146.4175 | 1211.8125 | 0.1208 | 0.0008 | 8.0308 | 44 |
| 45 | .0061 | 8.2825 | 66.7342 | 163.9876 | 1358.2300 | 0.1207 | 0.0007 | 8.0572 | 45 |
| 46 | .0054 | 8.2880 | 66.9792 | 183.6661 | 1522.2176 | 0.1207 | 0.0007 | 8.0815 | 46 |
| 47 | .0049 | 8.2928 | 67.2028 | 205.7061 | 1705.8838 | 0.1206 | 0.0006 | 8.1037 | 47 |
| 48 | .0043 | 8.2972 | 67.4068 | 230.3908 | 1911.5898 | 0.1205 | 0.0005 | 8.1241 | 48 |
| 49 | .0039 | 8.3010 | 67.5929 | 258.0377 | 2141.9806 | 0.1205 | 0.0005 | 8.1427 | 49 |
| 50 | .0035 | 8.3045 | 67.7624 | 289.0022 | 2400.0182 | 0.1204 | 0.0004 | 8.1597 | 50 |
| 51 | .0031 | 8.3076 | 67.9169 | 323.6825 | 2689.0204 | 0.1204 | 0.0004 | 8.1753 | 51 |
| 52 | .0028 | 8.3103 | 68.0576 | 362.5243 | 3012.7029 | 0.1203 | 0.0003 | 8.1895 | 52 |
| 53 | .0025 | 8.3128 | 68.1856 | 406.0273 | 3375.2272 | 0.1203 | 0.0003 | 8.2025 | 53 |
| 54 | .0022 | 8.3150 | 68.3022 | 454.7505 | 3781.2545 | 0.1203 | 0.0003 | 8.2143 | 54 |
| 55 | .0020 | 8.3170 | 68.4082 | 509.3206 | 4236.0050 | 0.1202 | 0.0002 | 8.2251 | 55 |
| 60 | .0011 | 8.3240 | 68.8100 | 897.5969 | 7471.6411 | 0.1201 | 0.0001 | 8.2664 | 60 |
| 65 | .0006 | 8.3281 | 69.0581 | 1581.8725 | 13173.9374 | 0.1201 | 0.0001 | 8.2922 | 65 |
| 70 | .0004 | 8.3303 | 69.2103 | 2787.7998 | 23223.3319 | 0.1200 | 0.0000 | 8.3082 | 70 |
| 75 | .0002 | 8.3316 | 69.3031 | 4913.0558 | 40933.7987 | 0.1200 | 0.0000 | 8.3181 | 75 |
| 80 | .0001 | 8.3324 | 69.3594 | 8658.4831 | 72145.6925 | 0.1200 | 0.0000 | 8.3241 | 80 |
| 85 | .0001 | 8.3328 | 69.3935 | 15259.2057 | 127151.7140 | 0.1200 | 0.0000 | 8.3278 | 85 |
| 90 | .0000 | 8.3330 | 69.4140 | 26891.9342 | 224091.1185 | 0.1200 | 0.0000 | 8.3300 | 90 |
| 95 | .0000 | 8.3332 | 69.4263 | 47392.7766 | 394931.4719 | 0.1200 | 0.0000 | 8.3313 | 95 |
| 100 | .0000 | 8.3332 | 69.4336 | 83522.2657 | 696010.5477 | 0.1200 | 0.0000 | 8.3321 | 100 |

**I = 15.00 %**

| N | (P/F) | (P/A) | (P/G) | (F/P) | (F/A) | (A/P) | (A/F) | (A/G) | N |
|---|---|---|---|---|---|---|---|---|---|
| 1 | .8696 | 0.8696 | -0.0000 | 1.1500 | 1.0000 | 1.1500 | 1.0000 | -0.0000 | 1 |
| 2 | .7561 | 1.6257 | 0.7561 | 1.3225 | 2.1500 | 0.6151 | 0.4651 | 0.4651 | 2 |
| 3 | .6575 | 2.2832 | 2.0712 | 1.5209 | 3.4725 | 0.4380 | 0.2880 | 0.9071 | 3 |
| 4 | .5718 | 2.8550 | 3.7864 | 1.7490 | 4.9934 | 0.3503 | 0.2003 | 1.3263 | 4 |
| 5 | .4972 | 3.3522 | 5.7751 | 2.0114 | 6.7424 | 0.2983 | 0.1483 | 1.7228 | 5 |
| 6 | .4323 | 3.7845 | 7.9368 | 2.3131 | 8.7537 | 0.2642 | 0.1142 | 2.0972 | 6 |
| 7 | .3759 | 4.1604 | 10.1924 | 2.6600 | 11.0668 | 0.2404 | 0.0904 | 2.4498 | 7 |
| 8 | .3269 | 4.4873 | 12.4807 | 3.0590 | 13.7268 | 0.2229 | 0.0729 | 2.7813 | 8 |
| 9 | .2843 | 4.7716 | 14.7548 | 3.5179 | 16.7858 | 0.2096 | 0.0596 | 3.0922 | 9 |
| 10 | .2472 | 5.0188 | 16.9795 | 4.0456 | 20.3037 | 0.1993 | 0.0493 | 3.3832 | 10 |
| 11 | .2149 | 5.2337 | 19.1289 | 4.6524 | 24.3493 | 0.1911 | 0.0411 | 3.6549 | 11 |
| 12 | .1869 | 5.4206 | 21.1849 | 5.3503 | 29.0017 | 0.1845 | 0.0345 | 3.9082 | 12 |
| 13 | .1625 | 5.5831 | 23.1352 | 6.1528 | 34.3519 | 0.1791 | 0.0291 | 4.1438 | 13 |
| 14 | .1413 | 5.7245 | 24.9725 | 7.0757 | 40.5047 | 0.1747 | 0.0247 | 4.3624 | 14 |
| 15 | .1229 | 5.8474 | 26.6930 | 8.1371 | 47.5804 | 0.1710 | 0.0210 | 4.5650 | 15 |
| 16 | .1069 | 5.9542 | 28.2960 | 9.3576 | 55.7175 | 0.1679 | 0.0179 | 4.7522 | 16 |
| 17 | .0929 | 6.0472 | 29.7828 | 10.7613 | 65.0751 | 0.1654 | 0.0154 | 4.9251 | 17 |
| 18 | .0808 | 6.1280 | 31.1565 | 12.3755 | 75.8364 | 0.1632 | 0.0132 | 5.0843 | 18 |
| 19 | .0703 | 6.1982 | 32.4213 | 14.2318 | 88.2118 | 0.1613 | 0.0113 | 5.2307 | 19 |
| 20 | .0611 | 6.2593 | 33.5822 | 16.3665 | 102.4436 | 0.1598 | 0.0098 | 5.3651 | 20 |
| 21 | .0531 | 6.3125 | 34.6448 | 18.8215 | 118.8101 | 0.1584 | 0.0084 | 5.4883 | 21 |
| 22 | .0462 | 6.3587 | 35.6150 | 21.6447 | 137.6316 | 0.1573 | 0.0073 | 5.6010 | 22 |
| 23 | .0402 | 6.3988 | 36.4988 | 24.8915 | 159.2764 | 0.1563 | 0.0063 | 5.7040 | 23 |
| 24 | .0349 | 6.4338 | 37.3023 | 28.6252 | 184.1678 | 0.1554 | 0.0054 | 5.7979 | 24 |
| 25 | .0304 | 6.4641 | 38.0314 | 32.9190 | 212.7930 | 0.1547 | 0.0047 | 5.8834 | 25 |
| 26 | .0264 | 6.4906 | 38.6918 | 37.8568 | 245.7120 | 0.1541 | 0.0041 | 5.9612 | 26 |
| 27 | .0230 | 6.5135 | 39.2890 | 43.5353 | 283.5688 | 0.1535 | 0.0035 | 6.0319 | 27 |
| 28 | .0200 | 6.5335 | 39.8283 | 50.0656 | 327.1041 | 0.1531 | 0.0031 | 6.0960 | 28 |
| 29 | .0174 | 6.5509 | 40.3146 | 57.5755 | 377.1697 | 0.1527 | 0.0027 | 6.1541 | 29 |
| 30 | .0151 | 6.5660 | 40.7526 | 66.2118 | 434.7451 | 0.1523 | 0.0023 | 6.2066 | 30 |
| 31 | .0131 | 6.5791 | 41.1466 | 76.1435 | 500.9569 | 0.1520 | 0.0020 | 6.2541 | 31 |
| 32 | .0114 | 6.5905 | 41.5006 | 87.5651 | 577.1005 | 0.1517 | 0.0017 | 6.2970 | 32 |
| 33 | .0099 | 6.6005 | 41.8184 | 100.6998 | 664.6655 | 0.1515 | 0.0015 | 6.3357 | 33 |
| 34 | .0086 | 6.6091 | 42.1033 | 115.8048 | 765.3654 | 0.1513 | 0.0013 | 6.3705 | 34 |
| 35 | .0075 | 6.6166 | 42.3586 | 133.1755 | 881.1702 | 0.1511 | 0.0011 | 6.4019 | 35 |
| 36 | .0065 | 6.6231 | 42.5872 | 153.1519 | 1014.3457 | 0.1510 | 0.0010 | 6.4301 | 36 |
| 37 | .0057 | 6.6288 | 42.7916 | 176.1246 | 1167.4975 | 0.1509 | 0.0009 | 6.4554 | 37 |
| 38 | .0049 | 6.6338 | 42.9743 | 202.5433 | 1343.6222 | 0.1507 | 0.0007 | 6.4781 | 38 |
| 39 | .0043 | 6.6380 | 43.1374 | 232.9248 | 1546.1655 | 0.1506 | 0.0006 | 6.4985 | 39 |
| 40 | .0037 | 6.6418 | 43.2830 | 267.8635 | 1779.0903 | 0.1506 | 0.0006 | 6.5168 | 40 |
| 41 | .0032 | 6.6450 | 43.4128 | 308.0431 | 2046.9539 | 0.1505 | 0.0005 | 6.5331 | 41 |
| 42 | .0028 | 6.6478 | 43.5286 | 354.2495 | 2354.9969 | 0.1504 | 0.0004 | 6.5478 | 42 |
| 43 | .0025 | 6.6503 | 43.6317 | 407.3870 | 2709.2465 | 0.1504 | 0.0004 | 6.5609 | 43 |
| 44 | .0021 | 6.6524 | 43.7235 | 468.4950 | 3116.6334 | 0.1503 | 0.0003 | 6.5725 | 44 |
| 45 | .0019 | 6.6543 | 43.8051 | 538.7693 | 3585.1285 | 0.1503 | 0.0003 | 6.5830 | 45 |
| 46 | .0016 | 6.6559 | 43.8778 | 619.5847 | 4123.8977 | 0.1502 | 0.0002 | 6.5923 | 46 |
| 47 | .0014 | 6.6573 | 43.9423 | 712.5224 | 4743.4824 | 0.1502 | 0.0002 | 6.6006 | 47 |
| 48 | .0012 | 6.6585 | 43.9997 | 819.4007 | 5456.0047 | 0.1502 | 0.0002 | 6.6080 | 48 |
| 49 | .0011 | 6.6596 | 44.0506 | 942.3108 | 6275.4055 | 0.1502 | 0.0002 | 6.6146 | 49 |
| 50 | .0009 | 6.6605 | 44.0958 | 1083.6574 | 7217.7163 | 0.1501 | 0.0001 | 6.6205 | 50 |
| 51 | .0008 | 6.6613 | 44.1360 | 1246.2061 | 8301.3737 | 0.1501 | 0.0001 | 6.6257 | 51 |
| 52 | .0007 | 6.6620 | 44.1715 | 1433.1370 | 9547.5798 | 0.1501 | 0.0001 | 6.6304 | 52 |
| 53 | .0006 | 6.6626 | 44.2031 | 1648.1075 | 10980.7167 | 0.1501 | 0.0001 | 6.6345 | 53 |
| 54 | .0005 | 6.6631 | 44.2311 | 1895.3236 | 12628.8243 | 0.1501 | 0.0001 | 6.6382 | 54 |
| 55 | .0005 | 6.6636 | 44.2558 | 2179.6222 | 14524.1479 | 0.1501 | 0.0001 | 6.6414 | 55 |
| 60 | .0002 | 6.6651 | 44.3431 | 4383.9987 | 29219.9916 | 0.1500 | 0.0000 | 6.6530 | 60 |
| 65 | .0001 | 6.6659 | 44.3903 | 8817.7874 | 58778.5826 | 0.1500 | 0.0000 | 6.6593 | 65 |
| 70 | .0001 | 6.6663 | 44.4156 | 17735.7200 | 118231.4669 | 0.1500 | 0.0000 | 6.6627 | 70 |
| 75 | .0000 | 6.6665 | 44.4292 | 35672.8680 | 237812.4532 | 0.1500 | 0.0000 | 6.6646 | 75 |
| 80 | .0000 | 6.6666 | 44.4364 | 71750.8794 | 478332.5293 | 0.1500 | 0.0000 | 6.6656 | 80 |
| 85 | .0000 | 6.6666 | 44.4402 | 144316.6470 | 962104.3133 | 0.1500 | 0.0000 | 6.6661 | 85 |
| 90 | .0000 | 6.6666 | 44.4422 | 290272.3252 | 1935142.1680 | 0.1500 | 0.0000 | 6.6664 | 90 |
| 95 | .0000 | 6.6667 | 44.4433 | 583841.3276 | 3892268.8509 | 0.1500 | 0.0000 | 6.6665 | 95 |
| 100 | .0000 | 6.6667 | 44.4438 | 1174313.4507 | 7828749.6713 | 0.1500 | 0.0000 | 6.6666 | 100 |

**I = 20.00 %**

| N | (P/F) | (P/A) | (P/G) | (F/P) | (F/A) | (A/P) | (A/F) | (A/G) | N |
|---|---|---|---|---|---|---|---|---|---|
| 1 | .8333 | 0.8333 | -0.0000 | 1.2000 | 1.0000 | 1.2000 | 1.0000 | -0.0000 | 1 |
| 2 | .6944 | 1.5278 | 0.6944 | 1.4400 | 2.2000 | 0.6545 | 0.4545 | 0.4545 | 2 |
| 3 | .5787 | 2.1065 | 1.8519 | 1.7280 | 3.6400 | 0.4747 | 0.2747 | 0.8791 | 3 |
| 4 | .4823 | 2.5887 | 3.2986 | 2.0736 | 5.3680 | 0.3863 | 0.1863 | 1.2742 | 4 |
| 5 | .4019 | 2.9906 | 4.9061 | 2.4883 | 7.4416 | 0.3344 | 0.1344 | 1.6405 | 5 |
| 6 | .3349 | 3.3255 | 6.5806 | 2.9860 | 9.9299 | 0.3007 | 0.1007 | 1.9788 | 6 |
| 7 | .2791 | 3.6046 | 8.2551 | 3.5832 | 12.9159 | 0.2774 | 0.0774 | 2.2902 | 7 |
| 8 | .2326 | 3.8372 | 9.8831 | 4.2998 | 16.4991 | 0.2606 | 0.0606 | 2.5756 | 8 |
| 9 | .1938 | 4.0310 | 11.4335 | 5.1598 | 20.7989 | 0.2481 | 0.0481 | 2.8364 | 9 |
| 10 | .1615 | 4.1925 | 12.8871 | 6.1917 | 25.9587 | 0.2385 | 0.0385 | 3.0739 | 10 |
| 11 | .1346 | 4.3271 | 14.2330 | 7.4301 | 32.1504 | 0.2311 | 0.0311 | 3.2893 | 11 |
| 12 | .1122 | 4.4392 | 15.4667 | 8.9161 | 39.5805 | 0.2253 | 0.0253 | 3.4841 | 12 |
| 13 | .0935 | 4.5327 | 16.5883 | 10.6993 | 48.4966 | 0.2206 | 0.0206 | 3.6597 | 13 |
| 14 | .0779 | 4.6106 | 17.6008 | 12.8392 | 59.1959 | 0.2169 | 0.0169 | 3.8175 | 14 |
| 15 | .0649 | 4.6755 | 18.5095 | 15.4070 | 72.0351 | 0.2139 | 0.0139 | 3.9588 | 15 |
| 16 | .0541 | 4.7296 | 19.3208 | 18.4884 | 87.4421 | 0.2114 | 0.0114 | 4.0851 | 16 |
| 17 | .0451 | 4.7746 | 20.0419 | 22.1861 | 105.9306 | 0.2094 | 0.0094 | 4.1976 | 17 |
| 18 | .0376 | 4.8122 | 20.6805 | 26.6233 | 128.1167 | 0.2078 | 0.0078 | 4.2975 | 18 |
| 19 | .0313 | 4.8435 | 21.2439 | 31.9480 | 154.7400 | 0.2065 | 0.0065 | 4.3861 | 19 |
| 20 | .0261 | 4.8696 | 21.7395 | 38.3376 | 186.6880 | 0.2054 | 0.0054 | 4.4643 | 20 |
| 21 | .0217 | 4.8913 | 22.1742 | 46.0051 | 225.0256 | 0.2044 | 0.0044 | 4.5334 | 21 |
| 22 | .0181 | 4.9094 | 22.5546 | 55.2061 | 271.0307 | 0.2037 | 0.0037 | 4.5941 | 22 |
| 23 | .0151 | 4.9245 | 22.8867 | 66.2474 | 326.2369 | 0.2031 | 0.0031 | 4.6475 | 23 |
| 24 | .0126 | 4.9371 | 23.1760 | 79.4968 | 392.4842 | 0.2025 | 0.0025 | 4.6943 | 24 |
| 25 | .0105 | 4.9476 | 23.4276 | 95.3962 | 471.9811 | 0.2021 | 0.0021 | 4.7352 | 25 |
| 26 | .0087 | 4.9563 | 23.6460 | 114.4755 | 567.3773 | 0.2018 | 0.0018 | 4.7709 | 26 |
| 27 | .0073 | 4.9636 | 23.8353 | 137.3706 | 681.8528 | 0.2015 | 0.0015 | 4.8020 | 27 |
| 28 | .0061 | 4.9697 | 23.9991 | 164.8447 | 819.2233 | 0.2012 | 0.0012 | 4.8291 | 28 |
| 29 | .0051 | 4.9747 | 24.1406 | 197.8136 | 984.0680 | 0.2010 | 0.0010 | 4.8527 | 29 |
| 30 | .0042 | 4.9789 | 24.2628 | 237.3763 | 1181.8816 | 0.2008 | 0.0008 | 4.8731 | 30 |
| 31 | .0035 | 4.9824 | 24.3681 | 284.8516 | 1419.2579 | 0.2007 | 0.0007 | 4.8908 | 31 |
| 32 | .0029 | 4.9854 | 24.4588 | 341.8219 | 1704.1095 | 0.2006 | 0.0006 | 4.9061 | 32 |
| 33 | .0024 | 4.9878 | 24.5368 | 410.1863 | 2045.9314 | 0.2005 | 0.0005 | 4.9194 | 33 |
| 34 | .0020 | 4.9898 | 24.6038 | 492.2235 | 2456.1176 | 0.2004 | 0.0004 | 4.9308 | 34 |
| 35 | .0017 | 4.9915 | 24.6614 | 590.6682 | 2948.3411 | 0.2003 | 0.0003 | 4.9406 | 35 |
| 36 | .0014 | 4.9929 | 24.7108 | 708.8019 | 3539.0094 | 0.2003 | 0.0003 | 4.9491 | 36 |
| 37 | .0012 | 4.9941 | 24.7531 | 850.5622 | 4247.8112 | 0.2002 | 0.0002 | 4.9564 | 37 |
| 38 | .0010 | 4.9951 | 24.7894 | 1020.6747 | 5098.3735 | 0.2002 | 0.0002 | 4.9627 | 38 |
| 39 | .0008 | 4.9959 | 24.8204 | 1224.8096 | 6119.0482 | 0.2002 | 0.0002 | 4.9681 | 39 |
| 40 | .0007 | 4.9966 | 24.8469 | 1469.7716 | 7343.8578 | 0.2001 | 0.0001 | 4.9728 | 40 |
| 41 | .0006 | 4.9972 | 24.8696 | 1763.7259 | 8813.6294 | 0.2001 | 0.0001 | 4.9767 | 41 |
| 42 | .0005 | 4.9976 | 24.8890 | 2116.4711 | 10577.3553 | 0.2001 | 0.0001 | 4.9801 | 42 |
| 43 | .0004 | 4.9980 | 24.9055 | 2539.7653 | 12693.8263 | 0.2001 | 0.0001 | 4.9831 | 43 |
| 44 | .0003 | 4.9984 | 24.9196 | 3047.7183 | 15233.5916 | 0.2001 | 0.0001 | 4.9856 | 44 |
| 45 | .0003 | 4.9986 | 24.9316 | 3657.2620 | 18281.3099 | 0.2001 | 0.0001 | 4.9877 | 45 |
| 46 | .0002 | 4.9989 | 24.9419 | 4388.7144 | 21938.5719 | 0.2000 | 0.0000 | 4.9895 | 46 |
| 47 | .0002 | 4.9991 | 24.9506 | 5266.4573 | 26327.2863 | 0.2000 | 0.0000 | 4.9911 | 47 |
| 48 | .0002 | 4.9992 | 24.9581 | 6319.7487 | 31593.7436 | 0.2000 | 0.0000 | 4.9924 | 48 |
| 49 | .0001 | 4.9993 | 24.9644 | 7583.6985 | 37913.4923 | 0.2000 | 0.0000 | 4.9935 | 49 |
| 50 | .0001 | 4.9995 | 24.9698 | 9100.4382 | 45497.1908 | 0.2000 | 0.0000 | 4.9945 | 50 |
| 51 | .0001 | 4.9995 | 24.9744 | 10920.5258 | 54597.6289 | 0.2000 | 0.0000 | 4.9953 | 51 |
| 52 | .0001 | 4.9996 | 24.9783 | 13104.6309 | 65518.1547 | 0.2000 | 0.0000 | 4.9960 | 52 |
| 53 | .0001 | 4.9997 | 24.9816 | 15725.5571 | 78622.7856 | 0.2000 | 0.0000 | 4.9966 | 53 |
| 54 | .0001 | 4.9997 | 24.9844 | 18870.6685 | 94348.3427 | 0.2000 | 0.0000 | 4.9971 | 54 |
| 55 | .0000 | 4.9998 | 24.9868 | 22644.8023 | 113219.0113 | 0.2000 | 0.0000 | 4.9976 | 55 |
| 60 | .0000 | 4.9999 | 24.9942 | 56347.5144 | 281732.5718 | 0.2000 | 0.0000 | 4.9989 | 60 |
| 65 | .0000 | 5.0000 | 24.9975 | 140210.6469 | 701048.2346 | 0.2000 | 0.0000 | 4.9995 | 65 |
| 70 | .0000 | 5.0000 | 24.9989 | 348888.9569 | 1744439.7847 | 0.2000 | 0.0000 | 4.9998 | 70 |
| 75 | .0000 | 5.0000 | 24.9995 | 868147.3693 | 4340731.8466 | 0.2000 | 0.0000 | 4.9999 | 75 |

**I = 25.00 %**

| N | (P/F) | (P/A) | (P/G) | (F/P) | (F/A) | (A/P) | (A/F) | (A/G) | N |
|---|---|---|---|---|---|---|---|---|---|
| 1 | .8000 | 0.8000 | 0.0 | 1.2500 | 1.0000 | 1.2500 | 1.0000 | 0.0 | 1 |
| 2 | .6400 | 1.4400 | 0.6400 | 1.5625 | 2.2500 | 0.6944 | 0.4444 | 0.4444 | 2 |
| 3 | .5120 | 1.9520 | 1.6640 | 1.9531 | 3.8125 | 0.5123 | 0.2623 | 0.8525 | 3 |
| 4 | .4096 | 2.3616 | 2.8928 | 2.4414 | 5.7656 | 0.4234 | 0.1734 | 1.2249 | 4 |
| 5 | .3277 | 2.6893 | 4.2035 | 3.0518 | 8.2070 | 0.3718 | 0.1218 | 1.5631 | 5 |
| 6 | .2621 | 2.9514 | 5.5142 | 3.8147 | 11.2588 | 0.3388 | 0.0888 | 1.8683 | 6 |
| 7 | .2097 | 3.1611 | 6.7725 | 4.7684 | 15.0735 | 0.3163 | 0.0663 | 2.1424 | 7 |
| 8 | .1678 | 3.3289 | 7.9469 | 5.9605 | 19.8419 | 0.3004 | 0.0504 | 2.3872 | 8 |
| 9 | .1342 | 3.4631 | 9.0207 | 7.4506 | 25.8023 | 0.2888 | 0.0388 | 2.6048 | 9 |
| 10 | .1074 | 3.5705 | 9.9870 | 9.3132 | 33.2529 | 0.2801 | 0.0301 | 2.7971 | 10 |
| 11 | .0859 | 3.6564 | 10.8460 | 11.6415 | 42.5661 | 0.2735 | 0.0235 | 2.9663 | 11 |
| 12 | .0687 | 3.7251 | 11.6020 | 14.5519 | 54.2077 | 0.2684 | 0.0184 | 3.1145 | 12 |
| 13 | .0550 | 3.7801 | 12.2617 | 18.1899 | 68.7596 | 0.2645 | 0.0145 | 3.2437 | 13 |
| 14 | .0440 | 3.8241 | 12.8334 | 22.7374 | 86.9495 | 0.2615 | 0.0115 | 3.3559 | 14 |
| 15 | .0352 | 3.8593 | 13.3260 | 28.4217 | 109.6868 | 0.2591 | 0.0091 | 3.4530 | 15 |
| 16 | .0281 | 3.8874 | 13.7482 | 35.5271 | 138.1085 | 0.2572 | 0.0072 | 3.5366 | 16 |
| 17 | .0225 | 3.9099 | 14.1085 | 44.4089 | 173.6357 | 0.2558 | 0.0058 | 3.6084 | 17 |
| 18 | .0180 | 3.9279 | 14.4147 | 55.5112 | 218.0446 | 0.2546 | 0.0046 | 3.6698 | 18 |
| 19 | .0144 | 3.9424 | 14.6741 | 69.3889 | 273.5558 | 0.2537 | 0.0037 | 3.7222 | 19 |
| 20 | .0115 | 3.9539 | 14.8932 | 86.7362 | 342.9447 | 0.2529 | 0.0029 | 3.7667 | 20 |
| 21 | .0092 | 3.9631 | 15.0777 | 108.4202 | 429.6809 | 0.2523 | 0.0023 | 3.8045 | 21 |
| 22 | .0074 | 3.9705 | 15.2326 | 135.5253 | 538.1011 | 0.2519 | 0.0019 | 3.8365 | 22 |
| 23 | .0059 | 3.9764 | 15.3625 | 169.4066 | 673.6264 | 0.2515 | 0.0015 | 3.8634 | 23 |
| 24 | .0047 | 3.9811 | 15.4711 | 211.7582 | 843.0329 | 0.2512 | 0.0012 | 3.8861 | 24 |
| 25 | .0038 | 3.9849 | 15.5618 | 264.6978 | 1054.7912 | 0.2509 | 0.0009 | 3.9052 | 25 |
| 26 | .0030 | 3.9879 | 15.6373 | 330.8722 | 1319.4890 | 0.2508 | 0.0008 | 3.9212 | 26 |
| 27 | .0024 | 3.9903 | 15.7002 | 413.5903 | 1650.3612 | 0.2506 | 0.0006 | 3.9346 | 27 |
| 28 | .0019 | 3.9923 | 15.7524 | 516.9879 | 2063.9515 | 0.2505 | 0.0005 | 3.9457 | 28 |
| 29 | .0015 | 3.9938 | 15.7957 | 646.2349 | 2580.9394 | 0.2504 | 0.0004 | 3.9551 | 29 |
| 30 | .0012 | 3.9950 | 15.8316 | 807.7936 | 3227.1743 | 0.2503 | 0.0003 | 3.9628 | 30 |
| 31 | .0010 | 3.9960 | 15.8614 | 1009.7420 | 4034.9678 | 0.2502 | 0.0002 | 3.9693 | 31 |
| 32 | .0008 | 3.9968 | 15.8859 | 1262.1774 | 5044.7098 | 0.2502 | 0.0002 | 3.9746 | 32 |
| 33 | .0006 | 3.9975 | 15.9062 | 1577.7218 | 6306.8872 | 0.2502 | 0.0002 | 3.9791 | 33 |
| 34 | .0005 | 3.9980 | 15.9229 | 1972.1523 | 7884.6091 | 0.2501 | 0.0001 | 3.9828 | 34 |
| 35 | .0004 | 3.9984 | 15.9367 | 2465.1903 | 9856.7613 | 0.2501 | 0.0001 | 3.9858 | 35 |
| 36 | .0003 | 3.9987 | 15.9481 | 3081.4879 | 12321.9516 | 0.2501 | 0.0001 | 3.9883 | 36 |
| 37 | .0003 | 3.9990 | 15.9574 | 3851.8599 | 15403.4396 | 0.2501 | 0.0001 | 3.9904 | 37 |
| 38 | .0002 | 3.9992 | 15.9651 | 4814.8249 | 19255.2994 | 0.2501 | 0.0001 | 3.9921 | 38 |
| 39 | .0002 | 3.9993 | 15.9714 | 6018.5311 | 24070.1243 | 0.2500 | 0.0000 | 3.9935 | 39 |
| 40 | .0001 | 3.9995 | 15.9766 | 7523.1638 | 30088.6554 | 0.2500 | 0.0000 | 3.9947 | 40 |
| 41 | .0001 | 3.9996 | 15.9809 | 9403.9548 | 37611.8192 | 0.2500 | 0.0000 | 3.9956 | 41 |
| 42 | .0001 | 3.9997 | 15.9843 | 11754.9435 | 47015.7740 | 0.2500 | 0.0000 | 3.9964 | 42 |
| 43 | .0001 | 3.9997 | 15.9872 | 14693.6794 | 58770.7175 | 0.2500 | 0.0000 | 3.9971 | 43 |
| 44 | .0001 | 3.9998 | 15.9895 | 18367.0992 | 73464.3969 | 0.2500 | 0.0000 | 3.9976 | 44 |
| 45 | .0000 | 3.9998 | 15.9915 | 22958.8740 | 91831.4962 | 0.2500 | 0.0000 | 3.9980 | 45 |
| 46 | .0000 | 3.9999 | 15.9930 | 28698.5925 | 114790.3702 | 0.2500 | 0.0000 | 3.9984 | 46 |
| 47 | .0000 | 3.9999 | 15.9943 | 35873.2407 | 143488.9627 | 0.2500 | 0.0000 | 3.9987 | 47 |
| 48 | .0000 | 3.9999 | 15.9954 | 44841.5509 | 179362.2034 | 0.2500 | 0.0000 | 3.9989 | 48 |
| 49 | .0000 | 3.9999 | 15.9962 | 56051.9386 | 224203.7543 | 0.2500 | 0.0000 | 3.9991 | 49 |
| 50 | .0000 | 3.9999 | 15.9969 | 70064.9232 | 280255.6929 | 0.2500 | 0.0000 | 3.9993 | 50 |
| 51 | .0000 | 4.0000 | 15.9975 | 87581.1540 | 350320.6161 | 0.2500 | 0.0000 | 3.9994 | 51 |
| 52 | .0000 | 4.0000 | 15.9980 | 109476.4425 | 437901.7701 | 0.2500 | 0.0000 | 3.9995 | 52 |
| 53 | .0000 | 4.0000 | 15.9983 | 136845.5532 | 547378.2126 | 0.2500 | 0.0000 | 3.9996 | 53 |
| 54 | .0000 | 4.0000 | 15.9986 | 171056.9414 | 684223.7658 | 0.2500 | 0.0000 | 3.9997 | 54 |
| 55 | .0000 | 4.0000 | 15.9989 | 213821.1768 | 855280.7072 | 0.2500 | 0.0000 | 3.9997 | 55 |
| 60 | .0000 | 4.0000 | 15.9996 | 652530.4468 | 2610117.7872 | 0.2500 | 0.0000 | 3.9999 | 60 |

**Practice Problems: ENGINEERING ECONOMY**

Untimed

1. A structure costing $10,000 has the operating costs and salvage values given. (a) What is the economic life of the structure? (b) Assuming that the structure has been owned and operated for 4 years, what is the cost of owning the structure for exactly one more year? Use 20% as the interest rate.

Year 1: maintenance $2000, salvage $8000
Year 2: maintenance $3000, salvage $7000
Year 3: maintenance $4000, salvage $6000
Year 4: maintenance $5000, salvage $5000
Year 5: maintenance $6000, salvage $4000

2. A man purchases a car for $5000 for personal use, intending to drive 15,000 miles per year. It costs him $200 per year for insurance and $150 per year for maintenance. He gets 15 mpg and gasoline costs $.60 per gallon. The resale value after 5 years is $1000. Because of unexpected business driving (5000 miles per year extra), his insurance is increased to $300 per year and maintenance to $200. Salvage is reduced to $500. Use 10% to answer the following questions. (a) The man's company offers $.10 per mile reimbursement. Is that adequate? (b) How many miles must be driven per year at $.10 per mile to justify the company buying a car for its use? The cost would be $5000, but insurance, maintenance, and salvage would be $250, $200, and $800 respectively.

3. A shredder installed at the entrance to a sewage treatment plant can remove 7 pounds per hour of debris from the incoming flow. The economic life of this shredder is 20 years. Any debris left in the flow will cause $25,000 damage to the wet-well pumps. Several investments are available to increase the capacity of the shredder. At 10% interest, what should be done?

| debris rate (pounds per hour) | probability of exceeding debris rate | required investment to meet debris rate |
|---|---|---|
| 7 | .15 per year | no cost |
| 8 | .10 per year | $15,000 |
| 9 | .07 per year | $20,000 |
| 10 | .03 per year | $30,000 |

4. A new machine will cost $17,000 and will have a value of $14,000 in 5 years. Special tooling will cost $5000 and it will have a resale value of $2500 after 5 years. Maintenance will be $200 per year. What will be the average cost of ownership during the next 5 years if interest is at 6%?

5. An old highway bridge can be strengthened at a cost of $9000, or it can be replaced for $40,000. The present salvage value of the old bridge is $13,000. It is estimated that the reinforced bridge will last for 20 years with an annual cost of $500 and will have a salvage value of $10,000 at the end of 20 years. The estimated salvage of the new bridge after 25 years is $15,000. The maintenance for the new b $100 annually. Which is the best altern interest?

6. A firm expects to receive $32,000 each year for 15 years from the sale of a product. It will require an initial investment of $150,000. Expenses will run $7530 per year. Salvage is zero and straight-line depreciation is used. The tax rate is 48%. What is the after-tax rate of return?

7. A public works project has initial costs of $1,000,000, benefits of $1,500,000, and disbenefits of $300,000. (a) What is the benefit/cost ratio? (b) What is the excess of benefits over costs?

8. An apartment complex is purchased for $500,000. What is the depreciation in each of the first 3 years if the salvage value is $100,000 in 25 years? Use (a) straight-line, (b) sum-of-the-year's digits, and (c) double declining balance depreciations.

9. Equipment is purchased for $12,000 which is expected to be sold after 10 years for $2000. The estimated maintenance is $1000 the first year, but is expected to increase $200 each year thereafter. Using 10%, find the present worth and the annual cost.

10. One of 5 grades of pipe with average lives (in years) and costs (in dollars) of (9,1500), (14,1600), (30,1750), (52,1900), and (86,2100) is to be chosen for a 20-year project. A failure of the pipe at any time during the project will result in a cost equal to 35% of the original cost. Annual costs are 4% of the initial cost, and the pipes are not recoverable. At 6%, which pipe is superior? HINT: The lives are average lives, not absolute replacement times. Assume a rectangular failure distribution, with probability of failure in any year inversely proportional to the average life.

11. Make a recommendation to your client to accept one of the following alternatives. Use the present worth comparison method. (Initial costs are the same.)

- A 25 year annuity paying $4800 at the end of each year, where the interest rate is a nominal 12% per annum.
- A 25 year annuity paying $1200 every quarter at 12% nominal annual interest.

12. A firm has two alternatives for improvement of its existing production line. The data are as follows:

| | A | B |
|---|---|---|
| initial installment cost | 1500 | 2500 |
| annual operating cost | 800 | 650 |
| service life | 5 years | 8 years |
| salvage value | 0 | 0 |

Determine the best alternative using an interest rate of 15%.

13. Two mutually exclusive alternatives requiring different investments are being considered. The life of both alternatives is estimated at 20 years with no salvage values. The minimum rate of return that is considered acceptable is 4%. Which alternative is best?

| | A | B |
|---|---|---|
| investment required | 70,000 | 40,000 |
| net income per year | 5620 | 4075 |
| rate of return on total investment | 5% | 8% |

14. Compare the costs of two plant renovation schemes A and B. Assume equal lives of 25 years, no salvage values, and interest at 25%. Make the comparison on the basis of (a) present worth, (b) capitalized cost, and (c) annual cost.

| | A | B |
|---|---|---|
| first cost | $20,000 | $25,000 |
| annual expenditure | $ 3000 | $ 2500 |

15. With interest at 8%, obtain the solutions to the following to the nearest dollar. (a) A machine costs $18,000 and has a salvage value of $2000. It has a useful life of 8 years. What is its book value at the end of 5 years using straight-line depreciation? (b) Using data from part (a), find the depreciation in the first three years using the sinking fund method. (c) Repeat part (a) using double declining-balance depreciation to find the first five years' depreciation.

16. A chemical pump motor unit is purchased for $14,000. The estimated life is 8 years, after which it will be sold for $1800. Find the depreciation in the first two years by the sum-of-the-year's digits method. Calculate the after-tax depreciation recovery using 15% interest with 52% income tax.

17. A soda ash plant has the water effluent from the processing equipment treated in a large settling basin. The settling basin eventually discharges into a river that runs alongside the basin. Recently enacted environmental regulations require all rainfall on the plant to be diverted and treated in the settling basin. A heavy rainfall will cause the entire basin to overflow. An uncontrolled overflow will cause environmental damage and heavy fines. The construction of additional height on the existing basin walls is under consideration.

Data on the costs of construction and expected costs for environmental clean up and fines are shown. Data on 50 typical winter storms has been collected. The soda ash plant management considers 12% to be their minimum rate of return and it is felt that after 15 years the plant will be closed. The company wants to select the alternative that minimizes its total expected costs.

| additional basin height | number of rainfalls causing overflow of basin | expense for environmental clean up per year | construction cost |
|---|---|---|---|
| 0 feet | 24 | | 0 |
| 5 | 14 | $600,000 | $ 600,000 |
| 10 | 8 | $650,000 | $ 710,000 |
| 15 | 3 | $700,000 | $ 900,000 |
| 20 | 1 | $800,000 | $1,000,000 |
| | 50 | | |

18. A wood processing plant installed a waste gas scrubber at a cost of $30,000 to remove pollutants from the exhaust discharged into the atmosphere. The scrubber has no salvage value and will cost $18,700 to operate next year, with operating costs expected to increase at the rate of $1200 per year thereafter. When should the company consider replacing the scrubber? Money can be borrowed at 12%.

19. Two alternative piping schemes are being considered by a water treatment facility. On the basis of a 10 year life and an interest rate of 12%, determine the number of hours of operation for which the two installations will break even.

| | A | B |
|---|---|---|
| pipe diameter | 4 in | 6 in |
| head loss for required flow | 48 ft | 26 ft |
| size motor required | 20 hp | 7 hp |
| energy cost per hour operation | $ .30 | $ .10 |
| cost of motor installed | $3600 | $2800 |
| cost of pipes and fittings | $3050 | $5010 |
| salvage value at end of 10 years | $ 200 | $ 280 |

20. An 88% learning curve is used with an item whose first production time was 6 weeks. How long will it take to produce the 4th item? How long will it take to produce the 6th through 14th items?

## Timed

1. A company is considering two alternatives, only one of which can be selected.

| alternative | initial investment | salvage value | annual net profit | life |
|---|---|---|---|---|
| A | 120,000 | 15,000 | 57,000 | 5 yrs |
| B | 170,000 | 20,000 | 67,000 | 5 yrs |

The net profit is after operating and maintenance costs, but before taxes. The company pays 45% of its year-end profit as income taxes. Use straight-line depreciation. Do not use investment tax credit. Find the best alternative if the company's marginal return on investment is 15%.

2. A company is considering the purchase of equipment to expand its capacity. The equipment cost is $300,000. The equipment is needed for 5 years, after which it will be sold for $50,000. The company's before-tax cash flow will be improved $90,000 annually by the purchase of the asset.

The corporate tax rate is 48% and straight-line depreciation will be used. The company will take the investment tax credit of 6.67%. What is the after-tax rate of return associated with this equipment purchase?

3. A 120-room hotel is purchased for $2,500,000. A 25-year loan is available for 12%. A study was conducted to determine the various occupancy rates.

| occupancy | probability |
|---|---|
| 65% full | .40 |
| 70% | .30 |
| 75% | .20 |
| 80% | .10 |

The operating costs of the hotel are:

| | |
|---|---|
| taxes & insurance | $ 20,000 annually |
| maintenance | 50,000 annually |
| operating | 200,000 annually |

The life of the hotel is figured to be 25 years when operating 365 days per year. The salvage after 25 years is $500,000.

Neglect tax credit and income taxes. Determine the average rate which should be charged per room per night to return 15% of the initial cost each year.

4. A company is insured for $3,500,000 against fire. The insurance rate is $0.69/1000. The insurance company will decrease the rate to $0.47/1000 if fire sprinklers are installed. The initial cost of the sprinklers is $7500. Annual costs are $200; additional taxes are $100 annually. The system life is 25 years. What is the rate of return on this investment?

5. Heat losses through the walls in an existing building cost a company $1,300,000 per year. This amount is considered excessive, and two alternatives are being evaluated. None of the alternatives will increase the life of the existing building beyond the current expected life of 6 years. None of the alternatives will produce a salvage value.

Alternative A: Do nothing. Continue with current losses.

Alternative B: Spend $2,000,000 immediately to upgrade the building and reduce the loss by 80%. This alternative will require annual maintenance of $150,000.

Alternative C: Spend $1,200,000 immediately. Repeat the $1,200,000 expenditure 3 years from now. Heat loss the first year will be reduced 80%. Due to deterioration, the reduction will be 55% and 20% in the second and third years. (The pattern is repeated starting after the second expenditure.) There are no maintenance costs.

All energy and maintenance costs are regarded as expenses for tax purposes. The company's tax rate is 48%, and straight-line depreciation is used. 15% is regarded as the effective annual interest rate. Evaluate each alternative on an after-tax basis, and recommend the best alternative.

6. You have been asked to determine if a 7-year-old machine should be replaced. Give a full explanation for your recommendation. Base your decision on a before-tax interest rate of 15%.

The existing machine is presumed to have a 10-year life. It has been depreciated on a straight-line basis from its original value of $1,250,000 to a current book value of $620,000. Its ultimate salvage value was assumed to be $350,000 for purposes of depreciation. Its present salvage value is estimated at $400,000, and this is not expected to change over the next 3 years. The current operating costs are not expected to change from $200,000 per year.

A new machine costs $800,000, with operating costs of $40,000 the first year, and increasing by $30,000 each year thereafter. The new machine has an expected life of 10 years. The salvage value depends on the year the new machine is retired:

| year retired | salvage |
|---|---|
| 1 | $600,000 |
| 2 | $500,000 |
| 3 | $450,000 |
| 4 | $400,000 |
| 5 | $350,000 |
| 6 | $300,000 |
| 7 | $250,000 |
| 8 | $200,000 |
| 9 | $150,000 |
| 10 | $100,000 |

RESERVED FOR FUTURE USE

# 3 FLUID STATICS AND DYNAMICS

*Nomenclature*

| | | |
|---|---|---|
| a | acceleration | ft/sec$^2$ |
| bhp | brake horsepower | hp |
| A | area | ft$^2$ |
| c | speed of sound in fluid | ft/sec |
| C | compressibility, Hazen-Williams constant, or coefficient | ft$^2$/lbf, –, – |
| d | depth, diameter | ft |
| D | diameter, drag | ft, lbf |
| ehp | electrical horsepower | hp |
| E | bulk modulus, energy | lbf/ft$^2$, ft-lbf |
| f | Darcy friction factor | – |
| fhp | friction horsepower | hp |
| F | force | lbf |
| $F_{va}$ | velocity of approach factor | – |
| g | local gravitational acceleration | ft/sec$^2$ |
| $g_c$ | gravitational constant (32.2) | lbm-ft/lbf-sec$^2$ |
| G | mass flow rate per unit area | lbm/sec-ft$^2$ |
| h | fluid height, head, depth | ft |
| H | total head | ft |
| I | moment of inertia | ft$^4$ |
| k | ratio of specific heats | – |
| K | minor loss coefficient | – |
| L | length of pipe, lift | ft, lbf |
| m | mass | lbm |
| $\dot{m}$ | mass flow rate | lbm/sec |
| n | rotational speed | rpm |
| $n_s$ | specific speed | – |
| $N_{Fr}$ | Froude number | |
| $N_{Re}$ | Reynolds number | – |
| $N_W$ | Weber number | – |
| p | pressure | lbf/ft$^2$ |
| P | power | ft-lbf/sec |
| Q | flow rate | gpm |
| r | radius | ft |
| $r_h$ | hydraulic radius | ft |
| R | specific gas content | ft-lbf/lbm-°R |
| s | length | ft |
| S.G. | specific gravity | – |
| t | time | sec |
| T | absolute temperature | °R |
| v | velocity | ft/sec |
| V | volume | ft$^3$ |
| $\dot{V}$ | volumetric flow rate | ft$^3$/sec |
| w | weight | lbf |
| whp | water horsepower | hp |
| x | x-coordinate | ft |
| y | distance, y-coordinate | ft |
| z | height above datum | ft |

*Symbols*

| | | |
|---|---|---|
| $\beta$ | contact angle, beta ratio | ° |
| $\gamma$ | specific weight | lbf/ft$^3$ |
| $\epsilon$ | specific roughness | ft |
| $\eta$ | efficiency | – |
| $\theta$ | angle | ° |
| $\mu$ | absolute viscosity | lbf-sec/ft$^2$ |
| $\nu$ | kinematic viscosity | ft$^2$/sec |
| $\rho$ | density | lbm/ft$^3$ |
| $\tau$ | shear stress | lbf/ft$^2$ |
| $T$ | surface tension | lbf/ft |
| $\upsilon$ | specific volume | ft$^3$/lbm |
| $\phi$ | angle, deflection angle | ° |
| $\omega$ | rotational speed | rad/sec |

*Subscripts*

| | |
|---|---|
| a | atmospheric |
| A | added |
| b | blade |
| c | centroid, contraction |
| d | discharge |
| D | drag |
| e | equivalent, entrance |
| E | English, extracted |
| f | friction, flow |
| i | inside, inlet |

| | |
|---|---|
| j | jet |
| k | kinetic |
| L | lift |
| m | manometer fluid, metacentric, model, motor |
| M | metric |
| n | nozzle |
| o | outside, outlet |
| p | static pressure, pump |
| r | ratio |
| R | resultant |
| s | stagnation |
| STP | standard temperature and pressure |
| t | total, tank, true |
| v | velocity |
| vp | vapor pressure |

# PART 1: Fluid Properties

Fluids are generally divided into two categories: ideal and real. *Ideal fluids* are those which have zero viscosity and shearing forces, are incompressible, and have uniform velocity distributions when flowing.

*Real fluids* are divided into Newtonian and non-Newtonian fluids. *Newtonian fluids* are typified by gases, thin liquids, and most fluids having simple chemical formulas. *Non-Newtonian fluids* are typified by gels, emulsions, and suspensions. Both Newtonian and non-Newtonian fluids exhibit finite viscosities and nonuniform velocity distributions. However, Newtonian fluids exhibit viscosities which are independent of the rate of change of shear stress, while non-Newtonian fluids exhibit viscosities dependent on the rate of change of shear stress.

Most fluid problems assume Newtonian fluid characteristics.

## 1 FLUID DENSITY

Most fluid flow calculations are based on an inconsistent system of units which measures density in pounds-mass per cubic foot. That convention is followed in this chapter when the symbol $\rho$ is employed.

$$\rho = \text{fluid density in lbm/ft}^3 \tag{3.1}$$

Hydrostatic pressure and energy conservation equations given in this chapter require a standard local gravity of 32.2 ft/sec$^2$. However, a short discussion of situations with non-standard gravity is given at the ends of parts 2 and 4 in this chapter.

The density of a fluid in liquid form is usually given, known in advance, or easily obtained from a table (see end of this chapter). The density of a gas can be found from the following formula, which has been derived from the ideal gas law:

$$\rho = \frac{p}{RT} \tag{3.2}$$

## 2 SPECIFIC VOLUME

Specific volume is the volume occupied by a pound of fluid. It is the reciprocal of the density.

$$v = \frac{1}{\rho} \tag{3.3}$$

## 3 SPECIFIC GRAVITY

Specific gravity is the ratio of a fluid's density to some specified reference density. For liquids, the reference density is the density of pure water. There is some confusion about this reference since the density of water varies with temperature, and various reference temperatures have been used (e.g., 39°F, 60°F, 70°F, etc.).

Strictly speaking, specific gravity of a liquid cannot be given without specifying the reference temperature at which the water's density was evaluated. However, the reference temperature is often omitted since water's density is fairly constant over the normal ambient temperature range. Using three significant digits, this reference density is 62.4 lbm/ft$^3$.

$$S.G._{\text{liquid}} = \frac{\rho}{62.4} \tag{3.4}$$

Specific gravities of petroleum products and aqueous acid solutions can be found from hydrometer readings. There are two basic hydrometer scales. The Baumé scale has been used widely in the past. Now, however, the API (American Petroleum Institute) scale is recommended for use with all liquids.

For liquids lighter than water, specific gravity may be found from the Baumé hydrometer reading:

$$S.G. = \frac{140.0}{130.0 + \text{°Baumé}} \tag{3.5}$$

For liquids heavier than water, specific gravity may be found from the Baumé hydrometer reading:

$$S.G. = \frac{145.0}{145.0 - \text{°Baumé}} \tag{3.6}$$

The modern API scale may be used with all liquids:

$$S.G. = \frac{141.5}{131.5 + \text{°API}} \tag{3.7}$$

Specific gravities also can be given for gases. The reference density is the density of air at specified conditions of pressure and temperature. The density of air evaluated at STP is approximately 0.075 lbm/ft$^3$. Therefore,

$$S.G._{\text{gas}} = \frac{\rho_{\text{STP}}}{0.075} \tag{3.8}$$

If the gas and air densities both are evaluated at the same temperature and pressure, the specific gravity is the inverse ratio of specific gas constants.

$$S.G._{\text{gas}} = \frac{R_{\text{air}}}{R_{\text{gas}}} = \frac{53.3}{R_{\text{gas}}} \tag{3.9}$$

*Example 3.1*

Determine the specific gravity of carbon dioxide (150°F, 20 psia) using STP air as a reference.

The specific gas constant for carbon dioxide is approximately 35.1 ft-lbf/lbm-°R. Using equation 3.2 and converting temperature to °R, the density is

$$\rho = \frac{(20)(144)}{(35.1)(150+460)} = 0.135 \text{ lbm/ft}^3$$

From equation 3.8,

$$S.G. = \frac{0.135}{0.075} = 1.8$$

## 4 VISCOSITY

Viscosity of a fluid is a measure of its resistance to flow. Consider two plates separated by a viscous fluid with thickness equal to $y$. The bottom plate is fixed. The top plate is kept in motion at a constant velocity $v$ by a constant force $F$.

Experiments with Newtonian fluids have shown that the force required to maintain the velocity is proportional to the velocity and inversely proportional to the separation of the plates. That is,

$$\frac{F}{A} \propto \frac{dv}{dy} \tag{3.10}$$

The constant of proportionality is known as the *absolute*[1] *viscosity*. Recognizing that the quantity $(F/A)$ is the *fluid shear stress* allows the following equation to be written.

$$\tau = \mu \frac{dv}{dy} \tag{3.11}$$

[1] Another name for *absolute* viscosity is *dynamic* viscosity. The term *absolute* is preferred.

Another quantity using the name *viscosity* is the combination of units given by equation 3.12. This combination of units, known as the *kinematic viscosity*, appears sufficiently often in fluids problems to warrant its own symbol and name.

$$\nu = \frac{\mu g_c}{\rho} \tag{3.12}$$

There are a number of different units used to measure viscosity. Table 3.1 lists the most commonly used units in the English and SI systems.

**Table 3.1**
Typical Viscosity Units

| | Absolute | Kinematic |
|---|---|---|
| English | lbf-sec/ft$^2$ (slug/ft-sec) | ft$^2$/sec |
| Conventional Metric | dyne-sec/cm$^2$ (poise) | cm$^2$/sec (stoke) |
| SI | Pascal-second (N-s/m$^2$) | m$^2$/s |

Conversions between the two types of viscosities and between the English and various metric systems can be accomplished with table 3.2.

*Example 3.2*

Water at 60°F has a specific gravity of 0.999 and a kinematic viscosity of 1.12 centistokes. What is the absolute viscosity in lbf-sec/ft$^2$?

$$\nu_M = \frac{1.12}{100} = 0.0112 \text{ stokes}$$
$$\mu_M = (0.0112)(0.999) = 0.01119 \text{ poise}$$
$$\mu_E = \frac{0.01119}{478.8} = 2.34 \text{ EE} - 5 \text{ lbf-sec/ft}^2$$

Viscosity also can be measured by a viscometer. A viscometer essentially is a container which allows the fluid to leak out through a small hole. The more viscous the fluid, the more time will be required to leak out a given quantity. Viscosity measured in this indirect manner has the units of seconds. Seconds Saybolt Universal (SSU) and Seconds Saybolt Furol (SSF) are two systems of indirect viscosity measurement.

In liquids, molecular cohesion is the dominating cause of viscosity. As the temperature of a liquid increases, these cohesive forces decrease, resulting in an absolute viscosity decrease.

In gases, the dominating cause of viscosity is random collisions between gas molecules. This molecular agita-

**Table 3.2**
Viscosity Conversions

| to obtain | multiply | by | and divide by |
|---|---|---|---|
| $ft^2$/sec | lbf-sec/$ft^2$ | 32.2 | density |
| $ft^2$/sec | stokes | 1.076 EE–3 | 1 |
| lbf-sec/$ft^2$ | $ft^2$/sec | density | 32.2 |
| lbf-sec/$ft^2$ | poise | 1 | 478.8 |
| $m^2$/s | centistokes | 1 EE–6 | 1 |
| $m^2$/s | stokes | 1 EE–4 | 1 |
| $m^2$/s | $ft^2$/sec | 9.29 EE–2 | 1 |
| pascal-sec | centipoise | 1 EE–3 | 1 |
| pascal-sec | lbm/ft-sec | 1.488 | 1 |
| pascal-sec | lbf-sec/$ft^2$ | 47.88 | 1 |
| pascal-sec | poise | 0.1 | 1 |
| pascal-sec | slug/ft-sec | 47.88 | 1 |
| poise | lbf-sec/$ft^2$ | 478.8 | 1 |
| poise | stokes | specific gravity | 1 |
| stokes | $ft^2$/sec | 929 | 1 |
| stokes | poise | 1 | specific gravity |

tion increases with increases in temperature. Therefore, viscosity in gases increases with temperature.

The absolute viscosity of both gases and liquids is independent of changes in pressure. Of course, kinematic viscosity greatly depends on both temperature and pressure since these variables affect density.

## 5 VAPOR PRESSURE

Molecular activity in a liquid tends to free some surface molecules. This tendency toward vaporization is dependent on temperature. The partial pressure exerted at the surface by these free molecules is known as the *vapor pressure.* Boiling occurs when the vapor pressure is increased (by increasing the fluid temperature) to the local ambient pressure. Thus, a liquid's boiling point depends on both the temperature and the external pressure. Liquids with low vapor pressures are used in accurate barometers.

Vapor pressure is a function of temperature only. Typical values are given in table 3.3. Appendix A lists values for water.

**Table 3.3**
Typical Vapor Pressures at 68°F

| Fluid | Vapor Pressure |
|---|---|
| Ethyl alcohol | 122.4 psf |
| Turpentine | 1.115 |
| Water | 48.9 |
| Ether | 1231. |
| Mercury | 0.00362 |

## 6 SURFACE TENSION

The skin which seems to form on the free surface of a fluid is due to the intermolecular cohesive and adhesive forces known as *surface tension.* Surface tension is the amount of work required to form a new unit of surface area. The units, therefore, are ft-lbf/$ft^2$ or just lbf/ft.

Surface tension can be measured as the tension between two points on the surface separted by a foot. It decreases with temperature increases and depends on the gas contacting the free surface. Surface tension values usually are quoted for air contact.

**Table 3.4**
Typical Surface Tensions
(68°F, air contact)

| Fluid | T |
|---|---|
| ethyl alcohol | 0.001527 lbf/ft |
| turpentine | 0.001857 |
| water | 0.004985 |
| mercury | 0.03562 |
| n-octane | 0.00144 |
| acetone | 0.00192 |
| benzene | 0.00192 |
| carbon tetrachloride | 0.00180 |

The relationship between surface tension and the pressure in a bubble surrounded by gas is given by equation 3.13. $r$ is the radius of the bubble.

$$T = \frac{1}{4}r\,(p_{\text{inside}} - p_{\text{outside}}) \qquad 3.13$$

The surface tension in a full spherical droplet or in a bubble in a liquid is given by equation 3.14.

$$T = \frac{1}{2} r \left(p_{\text{inside}} - p_{\text{outside}}\right) \qquad 3.14$$

## 7 CAPILLARITY

Surface tension is the cause of capillarity which occurs whenever a liquid comes into contact with a vertical solid surface. In water, adhesive forces dominate. They cause water to attach itself readily to a vertical surface, to climb the wall. In a thin-bore tube, water will rise above the general level as it tries to wet the interior surface.

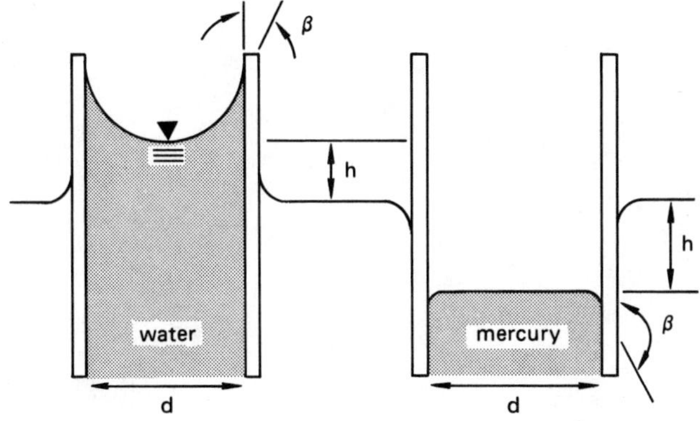


**Figure 3.1** Capillarity in Thin-Wall Tubes

On the other hand, cohesive forces dominate in mercury, since mercury molecules have a great affinity for each other. The curved surface called the *meniscus* formed inside a thin-bore tube inserted into a container of mercury will be below the general level.

Whether adhesive or cohesive forces dominate can be determined by the *angle of contact*, $\beta$, as shown in figure 3.1. For a contact angle less than 90°, adhesive forces dominate. Typical values of $\beta$ are given in table 3.5.

If the tube diameter is less than 0.1″, the surface tension inside a circular capillary tube can be approximated by equation 3.15. The meniscus is assumed spherical with radius $r$. Equation 3.15 also can be used to estimate the capillary rise in a capillary tube, as illustrated in figure 3.1.

$$T = \frac{h\rho d}{4\cos\beta} \qquad 3.15$$

$$r = \frac{d}{2\cos\beta} \qquad 3.16$$

Table 3.5 can be used to determine T and $\beta$ for various combinations of contacting liquids.

## 8 COMPRESSIBILITY

Usually, fluids are considered to be incompressible. Actually, fluids are somewhat compressible. Compressibility is the percentage change in a unit volume per unit change in pressure.

$$C = \frac{\frac{\Delta V}{V}}{\Delta p} \qquad 3.17$$

## 9 BULK MODULUS

The bulk modulus of a liquid is the reciprocal of the compressibility.

$$E = \frac{1}{C} \qquad 3.18$$

The bulk modulus of an ideal gas is given by equation 3.19, where $k$ is the ratio of specific heats. $k$ is equal to 1.4 for air.

$$E = kp \qquad 3.19$$

## 10 SPEED OF SOUND

The speed of sound in a pure liquid or gas is given by equations 3.20 and 3.21.

$$c_{\text{liquid}} = \sqrt{\frac{Eg_c}{\rho}} = \sqrt{\frac{g_c}{C\rho}} \qquad 3.20$$

$$c_{\text{gas}} = \sqrt{kg_cRT} = \sqrt{\frac{kpg_c}{\rho}} \qquad 3.21$$

**Table 3.5**
Capillary Constants

| combination | surface tension, $T$ | contact angle, $\beta$ |
|---|---|---|
| Mercury-vacuum-glass | 3.29 EE–2 lbf/ft | 140° |
| Mercury-air-glass | 2.02 EE–2 | 140° |
| Mercury-water-glass | 2.60 EE–2 | 140° |
| Water-air-glass | 5.00 EE–3 | 0° |

The temperature term in equation 3.21 must be in degrees absolute.

*Example 3.3*

What is the velocity of sound in 150°F water?

From the tables at the end of this chapter, the density of water at 150°F is 61.2 lbm/ft$^3$. Similarly, the bulk modulus is 328 EE3 psi. From equation 3.20,

$$c = \sqrt{\frac{(328\text{ EE3})(144)(32.2)}{61.2}} = 4985\text{ ft/sec}$$

*Example 3.4*

What is the velocity of sound in 150°F air at atmospheric pressure?

The specific gas constant for air is 53.3 ft-lbf/lbm-°R. Using equation 3.21,

$$\begin{aligned} c &= \sqrt{(1.4)(32.2)(53.3)(150+460)} \\ &= 1210.7\text{ ft/sec} \end{aligned}$$

# PART 2: Fluid Statics

## 1 MEASURING PRESSURES

The value of pressure, regardless of the device used to measure it, is dependent on the reference point chosen. Two such reference points exist: zero absolute pressure and standard atmospheric pressure.

If standard atmospheric pressure (approximately 14.7 psia) is chosen as the reference, pressures are known as *gage* pressures. Positive gage pressures always are pressures above atmospheric pressure. Vacuum (negative gage pressure) is the pressure below atmospheric. Maximum vacuum, according to this convention, is −14.7 psig. The term *gage* is somewhat misleading, as a mechanical gauge may not be used to measure gage pressures.

If zero absolute pressure is chosen as the reference, the pressures are known as *absolute* pressures. The barometer is a common device for measuring the absolute pressure of the atmosphere. It is constructed by filling a long, hollow tube, open at one end, with mercury, and inverting it such that the open end is below the level of a mercury-filled container. If the vapor pressure is neglected, the mercury will be supported only by the atmospheric pressure transmitted through the container fluid at the lower, open end. The equation balancing the weight of the fluid against the atmospheric force is:

$$p_a = 0.491(h)(144) \qquad 3.22$$

$h$ is the height of the mercury column in inches, and 0.491 is the density of mercury in pounds per cubic inch.

Any fluid can be used to measure atmospheric pressure, although vapor pressure may be significant. For any fluid used in a barometer,

$$p_a = [(0.0361)(S.G.)(h) + p_v](144) \qquad 3.23$$

0.0361 is the density of water in pounds per cubic inch. $p_v$ should be given in psi in equation 3.23.

*Example 3.5*

A vacuum pump is used to drain a flooded mine shaft of 68°F water. The pump is incapable of lifting the water beyond 400 inches. What is the atmospheric pressure?

From table 3.3, the vapor pressure of 68°F water is

$$\frac{48.9}{144} = 0.34 \text{ psi}$$

Then, the atmospheric pressure is

$$p_a = [(0.0361)(1)(400) + 0.34](144) = 2128.3 \text{ psf}$$

This is 14.78 psia.

## 2 MANOMETERS

Manometers are used frequently to measure pressure differentials. Figure 3.2 shows a simple U-tube manometer whose ends are connected to two pressure vessels. Often, one end will be open to the atmosphere, which then determines that end's pressure.

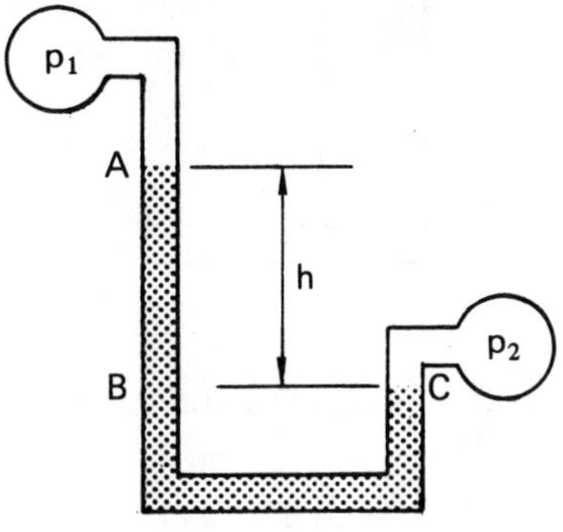


**Figure 3.2** A Simple Manometer

Since the pressure at point B is the same as at point C, the pressure differential produces the fluid column of height $h$.

$$\Delta p = p_2 - p_1 = \rho_m h \qquad 3.24$$

Equation 3.24 assumes that the manometer is small and that only low density gases fill the tubes above the measuring fluid. If a high density fluid (such as water) is present above the measuring fluid, or if the gas columns $h_1$ or $h_2$ are very long, corrections must be made:

$$\Delta p = \rho_m h + \rho_1 h_1 - \rho_2 h_2 \qquad 3.25$$

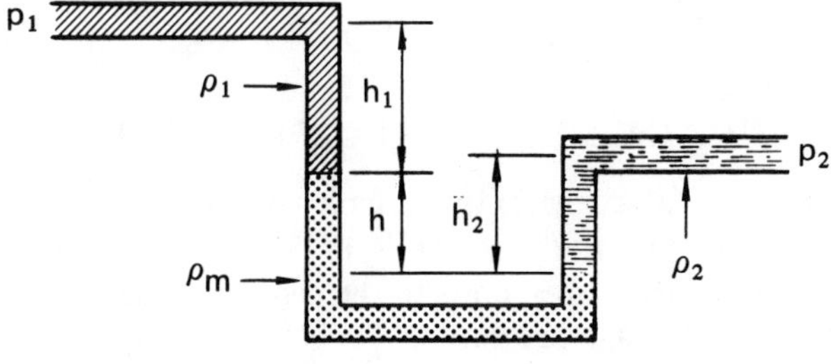


**Figure 3.3** A Manometer Requiring Corrections

Corrections for capillarity are seldom needed since manometer tubes generally are large in diameter.

*Example 3.6*

What is the pressure at the bottom of the water tank?

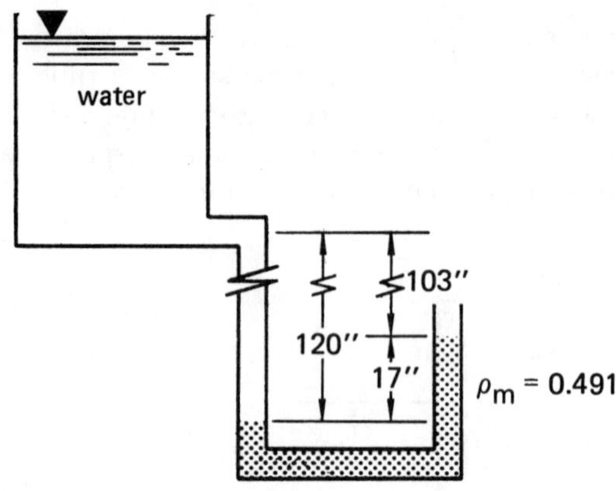


Using equation 3.25, the pressure differential is

$$\Delta p = p_{\text{tank bottom}} - p_a = (0.491)(17) - (0.0361)(120)$$
$$= 8.347 - 4.332$$
$$= 4.015 \text{ psig}$$

The third term in equation 3.25 was omitted because the density of air is much smaller than that of water or mercury.

## 3 HYDROSTATIC PRESSURE DUE TO INCOMPRESSIBLE FLUIDS

Hydrostatic pressure is the pressure which a fluid exerts on an object or container walls. It always acts through the center of pressure and is normal to the exposed surface, regardless of the object's orientation or shape. It varies linearly with depth and is a function of depth and density only.

### A. FORCE ON A HORIZONTAL PLANE SURFACE

In the case of a horizontal surface, such as the bottom of a container, the pressure is uniform, and the center of pressure corresponds to the centroid of the plane surface. The gage pressure is

$$p = \rho h \qquad 3.26$$

The total vertical force on the horizontal plane is

$$R = pA \qquad 3.27$$

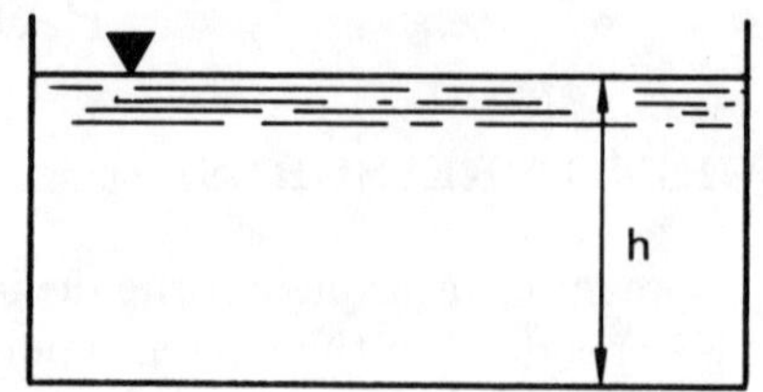


**Figure 3.4** Horizontal Plane Surface

### B. VERTICAL AND INCLINED RECTANGULAR PLANE SURFACES

If a rectangular plate is vertical or inclined within a fluid body, the linear variation in pressure with depth is maintained. The pressures at the top and bottom of the plate are

$$p_1 = \rho h_1 \sin\theta \qquad 3.28$$

$$p_2 = \rho h_2 \sin\theta \qquad 3.29$$

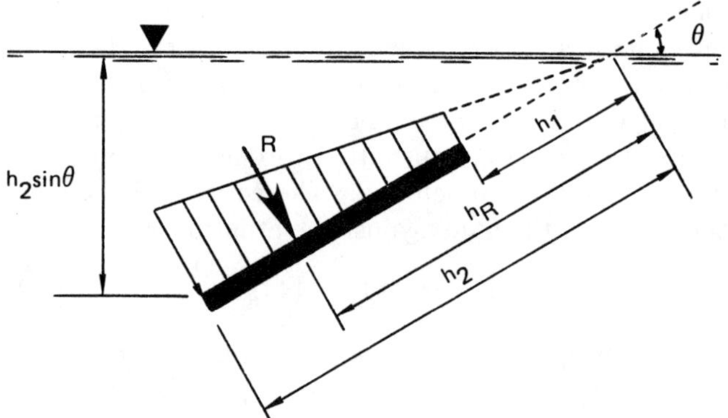


**Figure 3.5** Inclined Rectangular Plate

The average pressure occurs at the average depth $(\frac{1}{2})(h_1 + h_2)\sin\theta$. The average pressure over the entire vertical or inclined surface is

$$\overline{p} = \frac{1}{2}\rho(h_1 + h_2)\sin\theta \qquad 3.30$$

The total resultant force on the inclined plane is

$$R = \overline{p}A \qquad 3.31$$

The center of pressure is not located at the average depth but is located at the centroid of the triangular or trapezoidal pressure distribution. That depth is

$$h_R = \frac{2}{3}\left[h_1 + h_2 - \frac{h_1 h_2}{(h_1 + h_2)}\right] \qquad 3.32$$

If the object is inclined, $h_R$ must be measured parallel to the object's surface (e.g., an inclined length).

*Example 3.7*

The tank shown is filled with water. What is the force on a one-foot width of the inclined portion of the wall? Where is the resultant located on the inclined section?

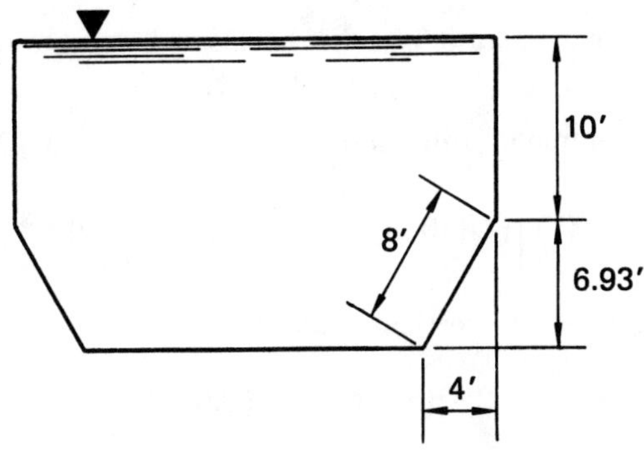


The $\sin\theta$ terms in equations 3.28, 3.29, and 3.30 convert the inclined distances to vertical distances. Therefore, the $\sin\theta$ terms may be omitted if the vertical distances are known. The average pressure in the inclined section is

$$\bar{p} = \frac{1}{2}(62.4)(10 + 16.93) = 840.2 \text{ psf}$$

The total force is

$$R = (840.2)(8)(1) = 6721.6 \text{ lbf}$$

To determine $h_R$, $\theta$ must be known in order to calculate $h_1$ and $h_2$.

$$\theta = \arctan\left(\frac{6.93}{4}\right) = 60°$$

$$h_1 = \frac{10}{\sin\ 60°} = 11.55 \text{ ft}$$

$$h_2 = \frac{16.93}{\sin\ 60°} = 19.55 \text{ ft}$$

$h_R$ can be calculated from equation 3.32 by substituting the inclined distances.

$$h_R = \frac{2}{3}\left[11.55 + 19.55 - \frac{(11.55)(19.55)}{11.55 + 19.55}\right]$$
$$= 15.89 \text{ ft}$$

## C. GENERAL PLANE SURFACE

For any non-rectangular plane surface, the average pressure depends on the location of the surface's centroid, $h_c$.

$$\bar{p} = \rho h_c \sin\theta \qquad 3.33$$

$$R = \bar{p}A \qquad 3.34$$

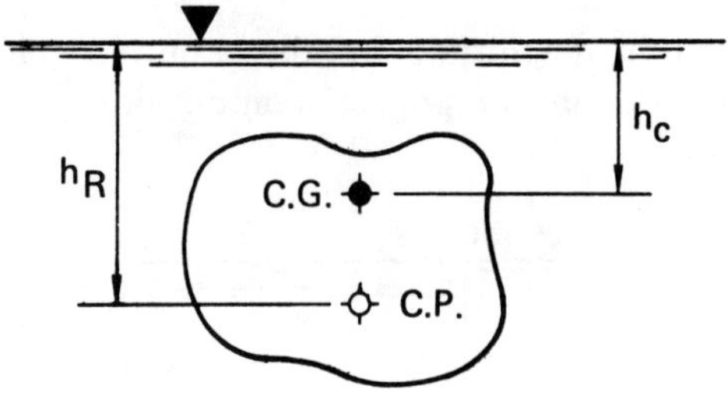


**Figure 3.6** General Plane Surface

The resultant is normal to the surface, acting at depth $h_R$.

$$h_R = h_c + \frac{I_c}{Ah_c} \qquad 3.35$$

$I_c$ is the moment of inertia about an axis parallel to the surface through the area's centroid. As with the previous case, $h_c$ and $h_r$ must be measured parallel to the area's surface. That is, if the plane is inclined, $h_c$ and $h_R$ also must be the inclined distances.

*Example 3.8*

What is the force on a one-foot diameter circular sight-hole whose top edge is located 4′ below the water surface? Where does the resultant act?

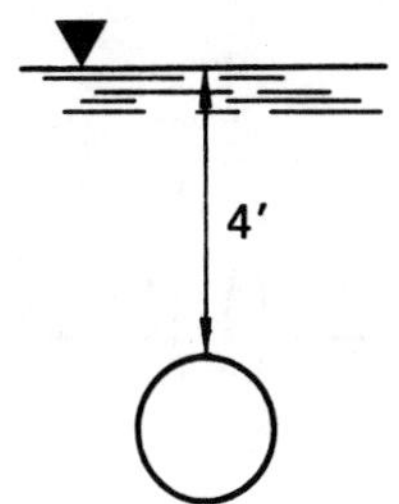


$$h_c = 4.5 \text{ ft}$$

$$A = \frac{1}{4}\pi(1)^2 = 0.7854 \text{ ft}^2$$

$$I_c = \frac{1}{4}\pi r^4 = 0.049 \text{ ft}^4$$

$$\bar{p} = (62.4)(4.5) = 280.8 \text{ psf}$$

$$R = (280.8)(0.7854) = 220.5 \text{ lb}$$

$$h_R = 4.5 + \frac{0.049}{(0.7854)(4.5)} = 4.514 \text{ ft}$$

## D. CURVED SURFACES

The horizontal component of the resultant force acting on a curved surface can be found by the same method used for a vertical plane surface. The vertical component of force on an area usually will equal the weight of the liquid above it. In figure 3.7, the vertical force on length AB is the weight of area ABCD, with a line of action passing through the centroid of the area ABCD.

The resultant magnitude and direction may be found from conventional component composition.

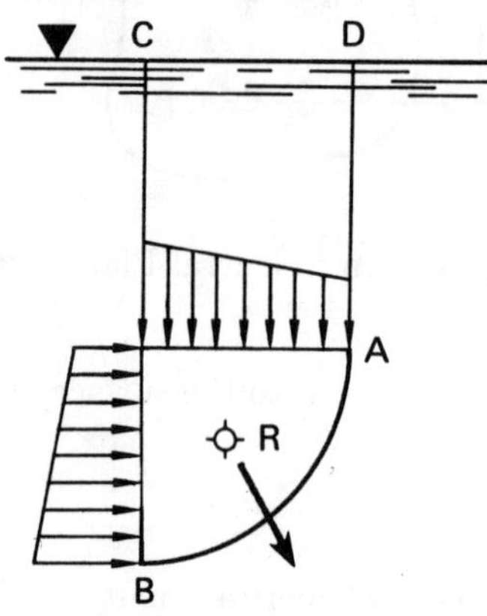


**Figure 3.7** Forces On a Curved Surface

*Example 3.9*

What is the total force on a one-foot section of the wall described in example 3.7?

The centroid of section ABCD (with point B serving as the reference) is

$$\begin{aligned}\overline{x} &= \frac{\sum A_i\overline{x}_i}{\sum A_i}\\ &= \frac{(4)(10)(2) + \left(\frac{1}{2}\right)(4)(6.93)\left(\frac{1}{3}\right)(4)}{40 + 13.86}\\ &= 1.83\end{aligned}$$

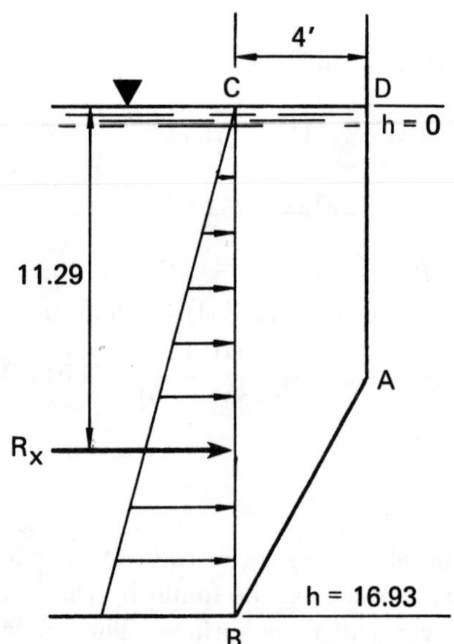


The average depth is $\frac{1}{2}(0 + 16.93) = 8.465$

Using equations 3.30 and 3.31, the average pressure and horizontal component of the resultant are

$$\overline{p} = 62.4(8.465) = 528.2 \text{ psf}$$
$$R_x = (16.93)(1)(528.2) = 8942.4 \text{ lbf}$$

From equation 3.32, the horizontal component acts $(\frac{2}{3})(16.93) = 11.29$ ft from the top.

The volume of a one foot section of area ABCD is

$$(1)\left[(4)(10) + \frac{1}{2}(4)(6.93)\right] = 53.86 \text{ ft}^3$$

Therefore, the vertical component is

$$R_y = (62.4)(53.86) = 3360.9 \text{ lbf}$$

The resultant of $R_x$ and $R_y$ is

$$R = \sqrt{(8942.4)^2 + (3360.9)^2} = 9553.1 \text{ lbf}$$
$$\phi = \arctan\left(\frac{3360.9}{8942.4}\right) = 20.6°$$

In general, it is not correct to calculate the vertical component of force on a submerged surface as being the weight of the fluid above it, as was done in example 3.9. This procedure is valid only when there is no change in the cross section of the tank area.

The *hydrostatic paradox* is illustrated by figure 3.8. The pressure anywhere on the bottom of either container is the same. This pressure is dependent on only the maximum height of the fluid, not the volume.

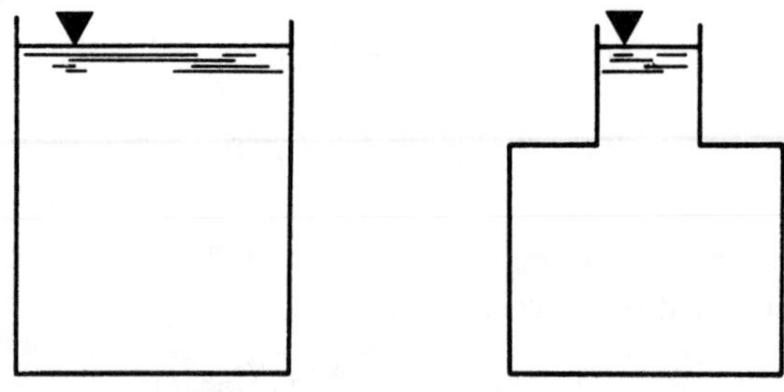

**Figure 3.8** Hydrostatic Paradox

## 4 HYDROSTATIC PRESSURE DUE TO COMPRESSIBLE FLUIDS

Equation 3.26 is a special case of a more general equation known as the *Fundamental Equation of Fluid Statics*, presented as equation 3.36. As defined in the nomenclature, $h$ is a variable representing height, and it is assumed that $h_2$ is greater than $h_1$. The minus sign in equation 3.36 indicates that pressure decreases when height increases.

$$\int_1^2 \frac{dp}{\rho} = -(h_2 - h_1) \qquad 3.36$$

If the fluid is a compressible layer of perfect gas, and if compression is assumed to be isothermal, equation 3.36 becomes

$$h_2 - h_1 = RT\ ln\left(\frac{p_1}{p_2}\right) \qquad 3.37$$

The pressure at height $h_2$ in a layer of gas which has been isothermally compressed is

$$p_2 = p_1\left[e^{\frac{h_1-h_2}{RT}}\right] \qquad 3.38$$

The following relationships assume a polytropic compression of the gas layer. These three relationships can be used for adiabatic compression by substituting $k$ for $n$.

$$h_2 - h_1 = \frac{n}{n-1}RT_1\left[1-\left(\frac{p_2}{p_1}\right)^{\frac{n-1}{n}}\right] \qquad 3.39$$

$$p_2 = p_1\left[1-\left(\frac{n-1}{n}\right)\left(\frac{h_2-h_1}{RT_1}\right)\right]^{\frac{n}{n-1}} \qquad 3.40$$

$$T_2 = T_1\left[1-\left(\frac{n-1}{n}\right)\left(\frac{h_2-h_1}{RT_1}\right)\right] \qquad 3.41$$

*Example 3.10*

The pressure at sea level is 14.7 psia. Assume 70°F isothermal compression, and calculate the pressure at 5000 feet altitude.

$R$ = 53.3 ft-lbf/lbm-°R for air. $T$ = (70 + 460) = 530 °R.

From equation 3.38,

$$p_{5000} = 14.7\left[e^{\frac{0-5000}{(53.3)(530)}}\right] = 12.32 \text{ psia}$$

## 5 BUOYANCY

The buoyancy theorem, also known as *Archimedes' principle*, states that the upward force on an immersed object is equal to the weight of the displaced fluid. A buoyant force due to displaced air also is relevant in the case of partially-submerged objects. For lighter-than-air crafts, the buoyant force results entirely from displaced air.

$$F_{\text{buoyant}} = \left(\begin{array}{c}\text{displaced}\\ \text{volume}\end{array}\right)\left(\begin{array}{c}\text{density of}\\ \text{displaced fluid}\end{array}\right) \qquad 3.42$$

In the case of floating or submerged objects not moving vertically, the buoyant force and weight are equal. If the forces are not in equilibrium, the object will rise or fall until some equilibrium is reached. The object will sink until it is supported by the bottom or until the density of the supporting fluid increases sufficiently. It will rise until the weight of the displaced fluid is reduced, either by a decrease in the fluid density or by breaking the surface.

*Example 3.11*

An empty polyethylene telemetry balloon with payload has a mass of 500 pounds. It is charged with helium when the atmospheric conditions are 60°F and 14.8 psia. What volume of helium is required for liftoff from a sea level platform? The specific gas constant of helium is 386.3 ft-lbf/lbm-°R.

Using equation 3.2 and converting temperature to °R, the gas densities are

$$\rho_{\text{air}} = \frac{p}{RT} = \frac{(14.8)(144)}{(53.3)(60+460)} = 0.07689 \text{ lbm/ft}^3$$

$$\rho_{\text{helium}} = \frac{(14.8)(144)}{(386.3)(60+460)} = 0.01061 \text{ lbm/ft}^3$$

The total mass of the balloon, payload, and helium is

$$m = 500 + (0.01061)\left(\begin{array}{c}\text{helium}\\ \text{volume}\end{array}\right)$$

The buoyant force is the weight of the displaced air.

$$F = (0.07689)\left(\begin{array}{c}\text{helium}\\ \text{volume}\end{array}\right)$$

Equating $F$ and $m$ results in a helium volume of 7544 ft$^3$.

*Example 3.12*

A 6-foot diameter sphere floats half-submerged in sea water. How much concrete (in lbm) is required as an external anchor to just submerge the sphere completely?

Assume the densities are 64.0 lbm/ft$^3$ for sea water, 150 lbm/ft$^3$ for concrete, and 0.075 lbm/ft$^3$ for air.

The weight of the sphere can be calculated from the buoyant force required to support it when half-submerged. Both the displaced sea water and the displaced air contribute to the buoyant force.

$$V_{\text{sphere}} = \frac{4\pi(3)^3}{3} = 113.1 \text{ ft}^3$$

$$w_{\text{sphere}} = \left(\frac{1}{2}\right)(113.1)(64) + \left(\frac{1}{2}\right)(113.1)(0.075)$$
$$= 3623.4 \text{ lbf}$$

The bouyant force equation for a fully-submerged sphere and anchor can be solved for the concrete volume.

$$w_{\text{sphere}} + w_{\text{concrete}} = (V_{\text{sphere}} + V_{\text{concrete}})\ (64.0)$$
$$3623.4 + 150V_{\text{concrete}} = (113.1 + V_{\text{concrete}})\ (64.0)$$
$$V_{\text{concrete}} = 42.0 \text{ ft}^3$$
$$m_{\text{concrete}} = (42)\ (150) = 6305 \text{ lbm}$$

## 6 STABILITY OF FLOATING OBJECTS

The bouyant force on a floating object acts upward through the center of gravity of the displaced *volume*, known as the *center of buoyancy.* The weight acts downward through the center of gravity of the *object.* For totally submerged objects such as balloons and submarines, the center of buoyancy must be above the center of gravity for stability. For partially submerged vessels, the metacenter must be above the center of gravity. Stability exists because a righting moment is created if the vessel heels over, since the center of buoyancy moves outboard of the center of gravity.

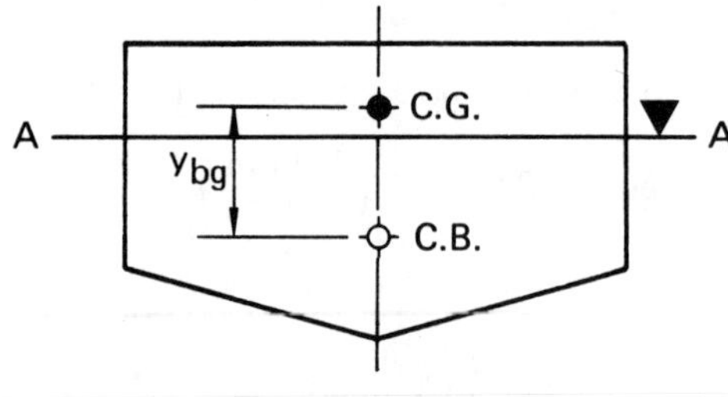


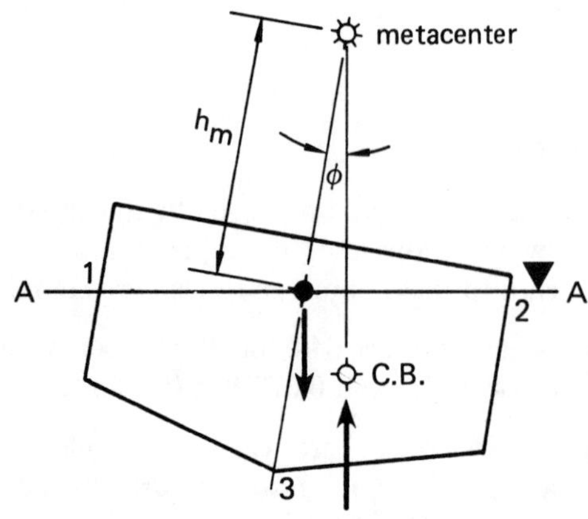


**Figure 3.9** Locating the Metacenter

Refer to figure 3.9. If the floating body heels through an angle $\phi$, the location of the center of gravity does not change. However, the center of buoyancy will shift to the center of gravity of the new submerged section 123. The center of buoyancy and gravity are no longer in line. The couple thus formed tends to resist further overturning.

This righting couple exists when the extension of the buoyant force F intersects line O-O above the center of gravity at M, the *metacenter.* If M lies below the center of gravity, an overturning couple will exist. The distance between the center of gravity and the metacenter is called the *metacentric height*, and it is reasonably constant for heel angles less than 10 degrees.

The metacentric height, $h_m$, can be found from equation 3.43 where $I$ is the moment of inertia of the submerged portion about the line A-A, and $V$ is the displaced volume.

$$h_m = \frac{I}{V} \pm y_{bg} \tag{3.43}$$

## 7 FLUID MASSES UNDER ACCELERATION

The pressures obtained thus far have assumed that the fluid is subjected only to gravitational acceleration. As soon as the fluid is subjected to any other acceleration, additional forces which change hydrostatic pressures are imposed.

If the fluid is subjected to constant accelerations in the vertical and/or horizontal directions, the fluid behavior will be given by equations 3.44 and 3.45.

$$\theta = \arctan\left[\frac{a_x}{a_y + g}\right] \tag{3.44}$$

$$p_h = \rho h\left(1 + \frac{a_y}{g}\right) \tag{3.45}$$

$a_y$ is negative if the acceleration is downward. Notice that a plane of equal pressure also is inclined if the fluid mass experiences a horizontal acceleration.

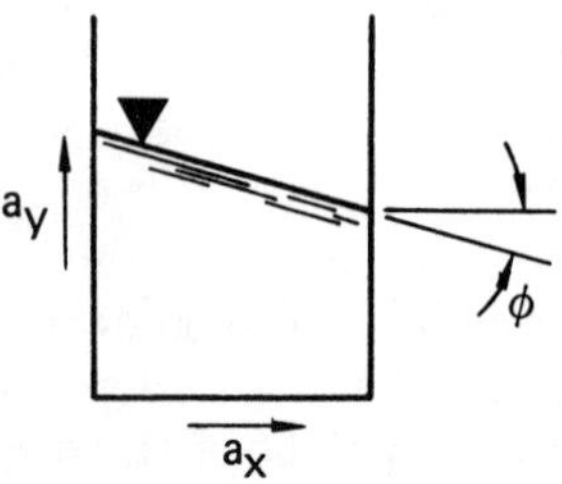


**Figure 3.10** Constant Linear Acceleration

If the fluid mass is rotated about a vertical axis, a parabolic fluid surface will result. The distance $h$ is measured from the lowermost part of the fluid during rotation. $h$ is not measured from the original level of the stationary fluid. $h$ is the height of the fluid at a distance $r$ from the center of rotation.

$$\theta = \arctan\left(\frac{\omega^2 r}{g}\right) \quad 3.46$$

$$h = \frac{(\omega r)^2}{2g} = \frac{v^2}{2g} \quad 3.47$$

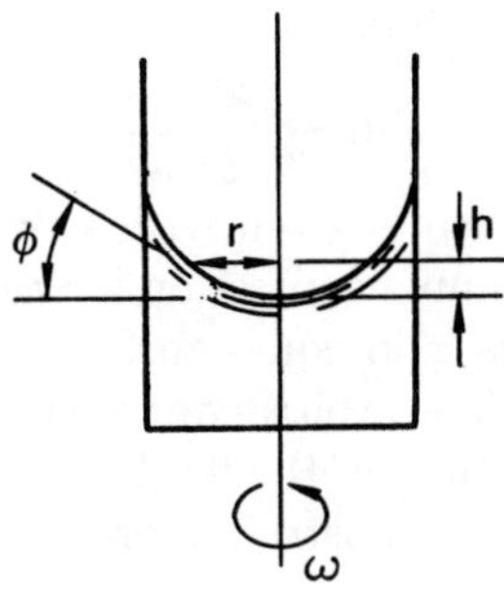


**Figure 3.11** Constant Rotational Acceleration

## 8 MODIFICATION FOR OTHER GRAVITIES

All of the equations in Part 2 of this chapter can be used in a standard gravitational field of 32.2 ft/sec$^2$. In such a standard field, the terms *lbm* and *lbf* can be cancelled freely. However, a modification to the formulas is needed if the local gravity deviates from standard.

The modification that must be made is that of converting the mass density in lbm/ft$^3$ to a *specific weight* in lbf/ft$^3$. This can be done through the use of equation 3.48.

$$\gamma = \frac{\rho g_{\text{local}}}{g_c} \quad 3.48$$

$g_{\text{local}}$ is the actual gravitational acceleration in ft/sec$^2$. $g_c$ is the dimensional conversion factor presented in chapter 22. $g_c$ is approximately equal to 32.2 lbm-ft/lbf-sec$^2$.

The various hydrostatic pressure formulas can be used in non-standard gravitational fields by substituting $\gamma$ for $\rho$.

*Example 3.13*

What is the maximum height that a vacuum pump can lift 60°F water if the atmospheric pressure is 14.6 psia and the local gravity is 28 ft/sec$^2$?

From appendix A, the vapor pressure head at 60°F is approximately 0.59 feet, corresponding to 0.26 psi. The effective pressure which can be used to lift water is (14.6 − 0.26) = 14.34 psi.

$$p = \left(\frac{g}{g_c}\right)\rho h$$

$$h = \frac{(14.34)(144)\left(\frac{32.2}{28.0}\right)}{62.4} = 38.06 \text{ ft}$$

# PART 3: Fluid Flow Parameters

## 1 INTRODUCTION

Pressure commonly is measured in pounds per square inch (psi) or pounds per square foot (psf). However, pressure may be changed into a new variable called *head* by dividing by the density of the fluid. This operation does more than just scale down the pressure by a factor equal to the reciprocal of the density. Since density itself possesses dimensional units, the units of head are not the same as the units of pressure.

$$(h \text{ in ft}) = \frac{(p \text{ in lbf/ft}^2)}{(\rho \text{ in lbm/ft}^3)} \qquad 3.49$$

Equation 3.49 is, of course, the same as equation 3.26. As long as the fluid density and local gravitational acceleration remain constant, there is complete interchangeability between the variables of pressure and head.

When Bernoulli's equation is introduced in this chapter, head also will be used as a measure of energy. Actually, head is used as a measure of *specific* energy. This is commonly justified by equation 3.50.

$$(h \text{ in ft}) = \frac{(E \text{ in ft-lbf})}{(m \text{ in lbm})} \qquad 3.50$$

A certain amount of care in the use of equations 3.49 and 3.50 is required since lb$f$ is being cancelled completely by lb$m$. The actual operation being performed is given by equation 3.51.

$$(h \text{ in ft}) = \frac{\left(g_c \text{ in } \frac{\text{lbm-ft}}{\text{lbf-sec}^2}\right)(p \text{ in lbf/ft}^2)}{\left(g \text{ in } \frac{\text{ft}}{\text{sec}^2}\right)\left(\rho \text{ in } \frac{\text{lbm}}{\text{ft}^3}\right)} \qquad 3.51$$

As $g_c$ always equals 32.2, it can be seen from equation 3.51 that equations 3.49 and 3.50 will give the correct numerical value for head as long as the local gravitational acceleration is 32.2 ft/sec$^2$.

## 2 FLUID ENERGY

A fluid can possess energy in three forms[2]—as pressure, kinetic, or potential energies. The energy (work) that must be put into a fluid to raise its pressure is known as the *pressure energy* or *static energy.* (This form of energy also is known as *flow work* or *flow energy*). In keeping with the common convention to put fluid energy terms into units of feet, the *pressure head* or *static head* is defined by equation 3.52.

$$h_p = \frac{p}{\rho} \qquad 3.52$$

Energy also is required to accelerate fluid to velocity $v$. The specific kinetic energy with units of feet is known as the *velocity head* or *dynamic head.* Although the units in equation 3.53 actually do yield feet, you should remember that specific energy is in foot-pounds per pound mass of fluid.

$$h_v = \frac{v^2}{2g_c} \qquad 3.53$$

Potential energy also is given with units of feet. This results in a very simple expression for *potential head* or *gravitational head.* $z$ in equation 3.54 is the height of the fluid above some arbitrary reference point. Cancelling the weight (lbf) of the fluid by its mass (lbm) is acceptable under the constraints previously given.

$$z = \frac{wz}{m} \qquad 3.54$$

## 3 BERNOULLI'S EQUATION

The Bernoulli equation is an energy conservation equation. It states that the total energy of a fluid flowing without losses in a pipe cannot change. The total energy possessed by a fluid is the sum of its pressure, kinetic, and potential energies.

$$\frac{p_1}{\rho} + \frac{v_1^2}{2g_c} + z_1 = \frac{p_2}{\rho} + \frac{v_2^2}{2g_c} + z_2 \qquad 3.55$$

Equation 3.55 is valid for laminar and turbulent flow. It can be used for gases as well as liquids if the gases are incompressible.[3] It is assumed that the flow between points 1 and 2 is frictionless and adiabatic.

The sum of the three head terms is known as the *total head, H*. The total pressure can be calculated from the total head.

$$H = \frac{p}{\rho} + \frac{v^2}{2g_c} + z \qquad 3.56$$

$$p_t = \rho H \qquad 3.57$$

The total energy of the fluid stream has two definitions. The *total specific energy* is the same as total head, $H$. The total energy of all fluid flowing can be calculated from equation 3.58.

$$E_t = mH \qquad 3.58$$

A graph of the total specific energy versus distance along a pipe is known as a *total energy line.* In a friction-

[2] Another important energy form is *thermal* energy. Thermal energy terms (internal energy and enthalpy) are not included in this analysis since it is assumed that the temperature of the fluid remains constant.

[3] A gas can be considered to be incompressible as long as its pressure does not change more than 10% between points 1 and 2, and its velocity is less than Mach 0.3 everywhere.

less pipe without pumps or turbines, the total specific energy will remain constant. Total specific energy will decrease if fluid friction is present.

*Example 3.14*

A pipe takes water from the reservoir as shown and discharges it freely 100 feet below. The flow is frictionless. (a) What is the total specific energy at point B? (b) What is the velocity at point C?

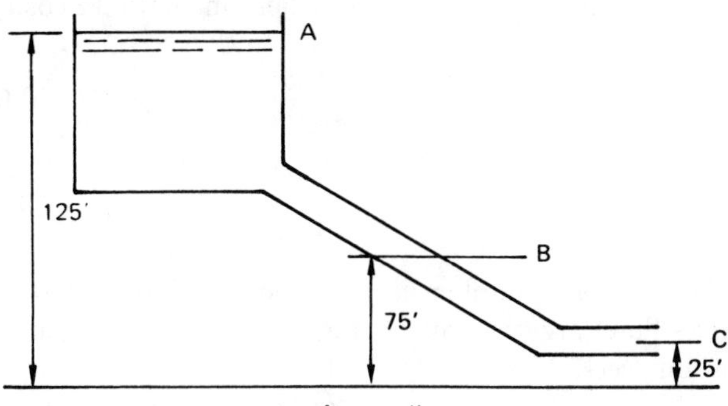


(a) At point A, the velocity and the gage pressure both are zero, so the total specific energy with respect to the reference line is

$$H_A = 0 + 0 + 125 = 125 \text{ ft}$$

At point B, the fluid is moving and possesses kinetic energy. The fluid is also under hydrostatic pressure. However, the flow is frictionless, and the total specific energy is constant (see equation 3.56). The velocity and static heads have increased at the expense of the potential head. Therefore,

$$H_B = H_A = 125 \text{ ft}$$

(b) At point C, the pressure head is again zero since the discharge is at atmospheric pressure. The potential head with respect to the energy reference line is 25 ft. From equation 3.56,

$$125 = 0 + \frac{v^2}{2g_c} + 25$$

$$v^2 = (2)(32.2)(100)$$

$$v = 80.2 \text{ ft/sec}$$

*Example 3.15*

Water is pumped at a rate of 3 cfs through the piping system illustrated. If the pump has a discharge pressure of 150 psig, to what elevation can the tank be raised? Assume the head loss due to friction is 10 feet.

$$h_{p,1} = \frac{(150)(144)}{62.4} = 346.15 \text{ ft}$$

From appendix F, the internal area of 4″ pipe is 0.0884 ft$^2$.

$$v_1 = \frac{3}{0.0884} = 33.94 \text{ fps}$$

$$h_{v,1} = \frac{(33.94)^2}{(2)(32.2)} = 17.89 \text{ ft}$$

$$z_1 = 0$$

$$h_{p,2} = 0 \text{ (at free surface)}$$

$$v_2 = 0 \text{ (at free surface)}$$

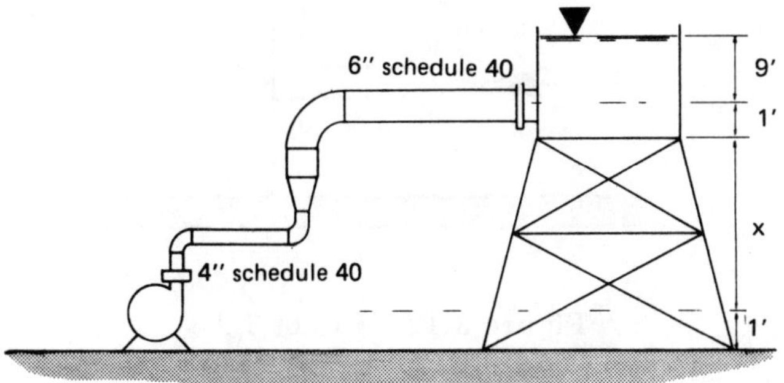


From equation 3.55,

$$346.15 + 17.89 = z_2 + 10$$

$$z_2 = 354.0 \text{ ft}$$

The tank bottom can be raised to $(354.0 - 10 + 1) = 345$ feet above the ground.

## 4 IMPACT ENERGY

Impact energy (also known as stagnation or total[4] energy) is the sum of the kinetic and pressure energies. The impact head is

$$h_s = \frac{p}{\rho} + \frac{v^2}{2g_c} = h_p + h_v \qquad 3.59$$

Impact head represents the effective head in a fluid which has been brought to rest (stagnated) in an adiabatic and reversible manner. Equation 3.59 can be used with a gas as long as the velocity is low—less than 400 ft/sec.

Impact head can be measured directly by using a pitot tube. This is illustrated in figure 3.12. Equation 3.59 can be used with a pitot tube.

A mercury manometer must be used if the stagnation properties of a gas or high-pressure liquid are being measured. Measurement of stagnation properties is covered in greater detail in part 4 of this chapter.

[4] There is confusion about *total head* as defined by equations 3.56 and 3.59. The effective pressure in a fluid which has been brought to rest adiabatically does not depend on the potential energy term, $z$. The application will determine which definition of total head is intended.

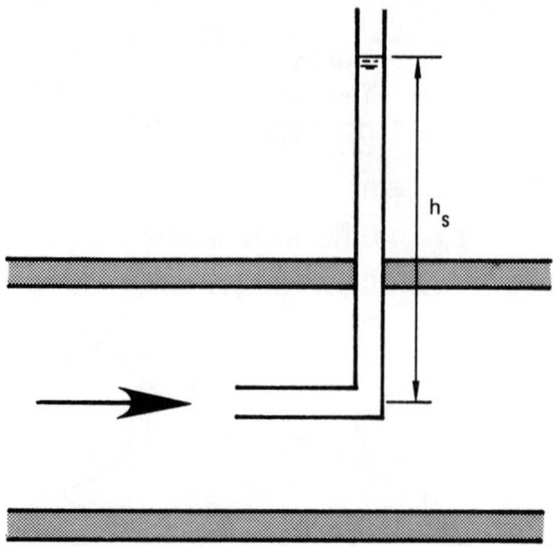


**Figure 3.12** A Pitot Tube

*Example 3.16*

The static pressure of air ($\rho = 0.075$ lbm/ft$^3$) flowing in a pipe is measured by a precision gage to be 10.0 psig. A pitot tube manometer indicates 20.6 inches of mercury. What is the velocity of the air in the pipe?

The pitot tube measures stagnation pressure. From equation 3.24, using 0.491 lbm/in$^3$ as the density of mercury,

$$p_s = (20.6)(0.491) = 10.11 \text{ psig}$$

Since stagnation pressure is the sum of static and velocity pressures, the velocity pressure is

$$p_v = p_s - p_p = 10.11 - 10.0 = 0.11 \text{ psig}$$

The velocity head is

$$h_v = \frac{p_v}{\rho} = \frac{(0.11)(144)}{0.075} = 211.1 \text{ ft}$$

From equation 3.53,

$$v = \sqrt{(2)(32.2)(211.2)} = 116.6 \text{ ft/sec}$$

## 5 HYDRAULIC GRADE LINE

The hydraulic grade line is a graphical representation of the sum of the static and potential heads versus position along the pipeline

$$\text{hydraulic grade} = z + h_p \qquad 3.60$$

Since the pressure head can increase at the expense of the velocity head, the hydraulic grade line can increase if an increase in flow area is encountered.

## 6 REYNOLDS' NUMBER

The Reynolds number is a dimensionless ratio of the inertial flow forces to the viscous forces within the fluid. Two expressions for Reynolds' number are used, one requiring absolute viscosity, the other kinematic viscosity:

$$N_{Re} = \frac{D_e v \rho}{\mu g_c} \qquad 3.61$$

$$= \frac{D_e v}{\nu} \qquad 3.62$$

The Reynolds number also can be calculated from the mass flow rate per unit area, $G$. $G$ must have the units of lbm/sec-ft$^2$.

$$N_{Re} = \frac{D_e G}{\mu g_c} \qquad 3.63$$

The Reynolds number is an important indicator in many types of problems. In addition to being used quantitatively in many equations, the Reynolds number also is used to determine whether fluid flow is laminar or turbulent.

A Reynolds number of 2000 or less indicates *laminar flow.* Fluid particles in laminar flow move in straight paths parallel to the flow direction. Viscous effects are dominant, resulting in a parabolic velocity distribution with a maximum velocity along the fluid flow centerline.

The fluid is said to be *turbulent* if the Reynolds number is greater than 2000.[5] Turbulent flow is characterized by random movement of fluid particles. The velocity distribution is essentially uniform with turbulent flow.

## 7 EQUIVALENT DIAMETER

The equivalent diameter, $D_e$, used in equations 3.61 and 3.62, is equal to the inside diameter of a circular pipe. The equivalent diameters of other cross sections in flow are given by table 3.6.

*Example 3.17*

Determine the equivalent diameter of the open trapezoidal channel shown.

[5] The beginning of the turbulent region is difficult to predict. There actually is a transition region between Reynolds numbers 2000 to 4000. In most fluid problems, however, flow is well within the turbulent region.

**Table 3.6**
Equivalent Diameters

| conduit cross section | $D_e$ |
|---|---|
| flowing full | |
| annulus | $D_o - D_i$ |
| square | $L$ |
| rectangle | $\frac{2L_1L_2}{L_1+L_2}$ |
| flowing partially full | |
| half-filled circle | $D$ |
| rectangle ($h$ deep, $L$ wide) | $\frac{4hL}{L+2h}$ |
| wide, shallow stream ( $h$ deep) | $4h$ |
| triangle ($h$ deep, $L$ broad, $s$ side) | $\frac{hL}{s}$ |
| trapezoid ($h$ deep, $a$ wide at top, $b$ wide at bottom, $s$ side) | $\frac{2h(a+b)}{b+2s}$ |

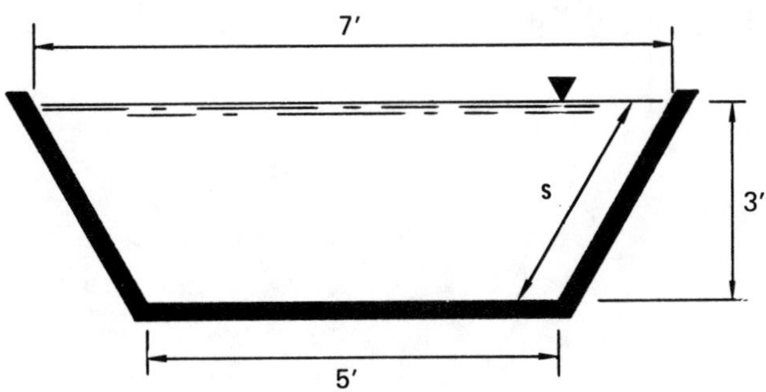


$$s = \sqrt{3^2 + 1^2} = 3.16 \text{ feet}$$

$$D_e = \frac{2(3)(7+5)}{5+2(3.16)} = 6.36 \text{ feet}$$

## 8 HYDRAULIC RADIUS

The equivalent diameter also can be found from the hydraulic radius, which is defined as the area in flow divided by the wetted perimeter. The wetted perimeter does not include free fluid surface.

$$D_e = 4r_h \qquad 3.64$$

$$r_h = \frac{\text{area in flow}}{\text{wetted perimeter}} \qquad 3.65$$

Consider a circular pipe flowing full. The area in flow is $\pi r^2$. The wetted perimeter is the entire circumference, $2\pi r$. The hydraulic radius is

$$\left(\frac{\pi r^2}{2\pi r}\right) = \frac{1}{2}r$$

Therefore, the hydraulic radius and the pipe radius are not the same. (The hydraulic radius of a pipe flowing half full is also $\frac{1}{2}r$, as the flow area and the wetted perimeter both are halved.)

The hydraulic radius of a pipe flowing less than full can be found from table 3.7 or appendix E.

*Example 3.18*

What is the hydraulic radius of the trapezoidal channel described in example 3.17?

From equation 3.65,

$$r_h = \frac{(5)(3)+(3)(1)}{3.16+5+3.16} = 1.59 \text{ feet}$$

Using the results of the previous example and equation 3.64,

$$r_h = \frac{6.36}{4} = 1.59 \text{ feet}$$

**Table 3.7**
Hydraulic Radius
of Partially Filled Circular Pipes
(Also see appendix E)

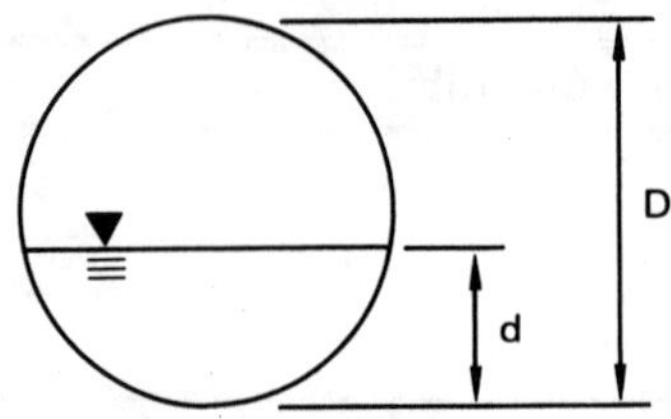


| $\frac{d}{D}$ | $\frac{\text{hyd. rad.}}{D}$ | $\frac{d}{D}$ | $\frac{\text{hyd. rad.}}{D}$ |
|---|---|---|---|
| 0.05 | 0.0326 | 0.55 | 0.2649 |
| 0.10 | 0.0635 | 0.60 | 0.2776 |
| 0.15 | 0.0929 | 0.65 | 0.2881 |
| 0.20 | 0.1206 | 0.70 | 0.2962 |
| 0.25 | 0.1466 | 0.75 | 0.3017 |
| 0.30 | 0.1709 | 0.80 | 0.3042 |
| 0.35 | 0.1935 | 0.85 | 0.3033 |
| 0.40 | 0.2142 | 0.90 | 0.2980 |
| 0.45 | 0.2331 | 0.95 | 0.2864 |
| 0.50 | 0.2500 | 1.00 | 0.2500 |

# PART 4: Fluid Dynamics

## 1 FLUID CONSERVATION LAWS

Many fluid flow problems can be solved by using the principles of conservation of mass and energy.

When applied to fluid flow, the principle of mass conservation is known as the *continuity equation*:

$$\rho_1 A_1 v_1 = \rho_2 A_2 v_2 \qquad 3.66$$

$$\dot{m}_1 = \dot{m}_2 \qquad 3.67$$

If the fluid is incompressible, $\rho_1 = \rho_2$, so

$$A_1 v_1 = A_2 v_2 \qquad 3.68$$

$$\dot{V}_1 = \dot{V}_2 \qquad 3.69$$

The energy conservation principle is based on the Bernoulli equation. However, terms for friction loss and hydraulic machines must be included.

$$\left(\frac{p_1}{\rho} + \frac{v_1^2}{2g_c} + z_1\right) + h_A = \left(\frac{p_2}{\rho} + \frac{v_2^2}{2g_c} + z_2\right) + h_E + h_f \qquad 3.70$$

## 2 HEAD LOSS DUE TO FRICTION

The most common expression for calculating head loss due to friction ($h_f$) is the *Darcy formula*:

$$h_f = \frac{fLv^2}{2Dg_c} \qquad 3.71$$

The *Moody friction factor chart* (figure 3.13) probably is the most convenient method of determining the friction factor, $f$.

The basic parameter required to use the Moody friction factor chart is the Reynolds number. If the Reynolds number is less than 2000, the friction factor is given by equation 3.72.

$$f = \frac{64}{N_{Re}} \qquad 3.72$$

For turbulent flow ($N_{Re} > 2000$), the friction factor depends on the relative roughness of the pipe. This roughness is expressed by the ratio $\frac{\epsilon}{D}$, where $\epsilon$ is the specific surface roughness and $D$ is the inside diameter. Values of $\epsilon$ for various types of pipe are found in table 3.8.

Another method for finding the friction head loss is the *Hazen-Williams formula.* The Hazen-Williams formula gives good results for liquids that have kinematic viscosities around 1.2 EE−5 ft$^2$/sec (corresponding to 60°F water). At extremely high and low temperatures, the Hazen-Williams formula can be as much as 20% in error for water. The Hazen-Williams formula should be used only for turbulent flow.

The Hazen-Williams head loss is

$$h_f = \frac{(3.022)(v)^{1.85}L}{(C)^{1.85}(D)^{1.165}} \qquad 3.73$$

Or, in terms of other units,

$$h_f = (10.44)(L)\frac{(\text{gpm})^{1.85}}{(C)^{1.85}(d_{\text{inches}})^{4.8655}} \qquad 3.74$$

Use of these formulas requires a knowledge of the Hazen-Williams coefficient, $C$, which is assumed to be independent of the Reynolds number. Table 3.8 gives values of $C$ for various types of pipe.

Values of $f$ and $h_f$ are appropriate for clean, new pipe. As some pipes age, it is not uncommon for scale build-up to decrease the equivalent flow diameter. This diameter decrease produces a dramatic increase in the friction loss.

$$\frac{h_{f,\text{scaled}}}{h_{f,\text{new}}} = \left(\frac{D_{\text{new}}}{D_{\text{scaled}}}\right)^5 \qquad 3.75$$

Because of this scale effect, an uprating factor of 10-30% is commonly applied to $f$ or $h_f$ in anticipation of future service conditions.

*Example 3.19*

50°F water is pumped through 4″ schedule 40 welded steel pipe ($\epsilon = 0.0002$) at the rate of 300 gpm. What is the friction head loss calculated by the Darcy formula for 1000 feet of pipe?

First, it is necessary to collect data on the pipe and water. The fluid viscosity and pipe dimensions can be found from tables at the end of the chapter.

kinematic viscosity = 1.41 EE − 5 ft$^2$/sec

inside diameter = 0.3355 ft

flow area = 0.0884 ft$^2$

The flow quantity is

$$(300)(0.002228) = 0.6684 \text{ cfs}$$

The velocity is

$$v = \frac{\dot{V}}{A} = \frac{0.6684}{0.0884} = 7.56 \text{ fps}$$

The Reynolds number is

$$N_{Re} = \frac{(0.3355)(7.56)}{1.41 \text{ EE} - 5} = 1.8 \text{ EE5}$$

**Table 3.8**
Specific Roughness and Hazen-Williams Constants for Various Pipe Materials

| | $\epsilon$(ft) | | $C$ | | |
|---|---|---|---|---|---|
| type of pipe or surface | range | design | range | clean | design |
| STEEL | | | | | |
| welded and seamless | 0.0001–0.0003 | 0.0002 | 150–80 | 140 | 100 |
| interior riveted, no projecting rivets | | | | 139 | 100 |
| projecting girth rivets | | | | 130 | 100 |
| projecting girth and horizontal rivets | | | | 115 | 100 |
| vitrified, spiral-riveted, flow with lap | | | | 110 | 100 |
| vitrified, spiral-riveted, flow against lap | | | | 100 | 90 |
| corrugated | | | | 60 | 60 |
| MINERAL | | | | | |
| concrete | 0.001–0.01 | 0.004 | 152–85 | 120 | 100 |
| cement-asbestos | | | 160–140 | 150 | 140 |
| vitrified clays | | | | | 110 |
| brick sewer | | | | | 100 |
| IRON | | | | | |
| cast, plain | 0.0004–0.002 | 0.0008 | 150–80 | 130 | 100 |
| cast, tar (asphalt) coated | 0.0002–0.0006 | 0.0004 | 145–50 | 130 | 100 |
| cast, cement lined | 0.000008 | 0.000008 | | 150 | 140 |
| cast, bituminous lined | 0.000008 | 0.000008 | 160–130 | 148 | 140 |
| cast, centrifugally spun | 0.00001 | 0.00001 | | | |
| galvanized, plain | 0.0002–0.0008 | 0.0005 | | | |
| wrought, plain | 0.0001–0.0003 | 0.0002 | 150–80 | 130 | 100 |
| MISCELLANEOUS | | | | | |
| fiber | | | | 150 | 140 |
| copper and brass | 0.000005 | 0.000005 | 150–120 | 140 | 130 |
| wood stave | 0.0006–0.003 | 0.002 | 145–110 | 120 | 110 |
| transite | 0.000008 | 0.000008 | | | |
| lead, tin, glass | | 0.000005 | 150–120 | 140 | 130 |
| plastic (PVC and ABS) | | 0.000005 | 150–120 | 140 | 130 |

The relative roughness is

$$\frac{\epsilon}{D} = \frac{0.0002}{0.3355} = 0.0006$$

From the Moody friction factor chart, $f = 0.0195$.

From equation 3.71,

$$h_f = \frac{(0.0195)(1000)(7.56)^2}{(2)(0.3355)(32.2)} = 51.6 \text{ ft}$$

*Example 3.20*

Repeat example 3.19 using the Hazen-Williams formula. Assume $C = 100$.

Using equation 3.73,

$$h_f = \frac{(3.022)(7.56)^{1.85}(1000)}{(100)^{1.85}(0.3355)^{1.165}} = 90.8 \text{ ft}$$

Using equation 3.74,

$$h_f = (10.44)(1000)\frac{(300)^{1.85}}{(100)^{1.85}(4.026)^{4.8655}} = 90.9 \text{ ft}$$

## 3 MINOR LOSSES

In addition to the head loss caused by friction between the fluid and the pipe wall, losses also are caused by obstructions in the line, changes in direction, and changes in flow area. These losses are named *minor losses* because they are much smaller in magnitude than the $h_f$ term. Two methods are used to determine these losses: the method of equivalent lengths and the method of loss coefficients.

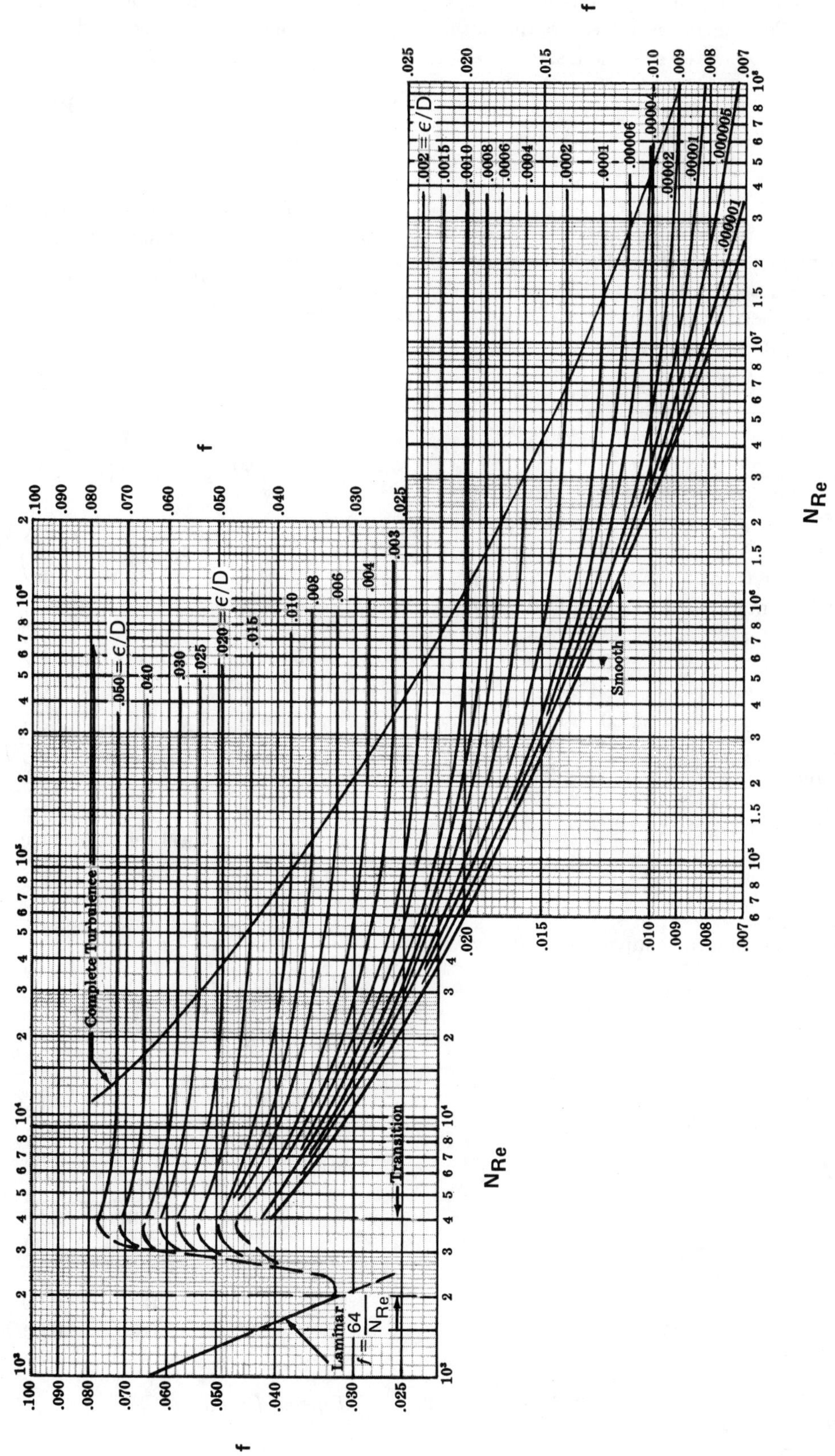

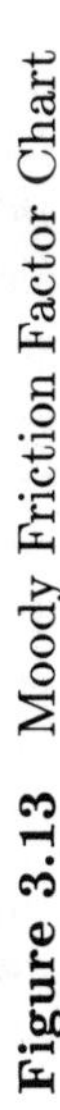

**Figure 3.13** Moody Friction Factor Chart

hod of *equivalent lengths* uses a table to convert each valve and fitting into an equivalent length of straight pipe. This length is added to the actual pipeline length and substituted into the Darcy equation for $L_e$.

$$h_f = \frac{f L_e v^2}{2 D g_c} \qquad 3.76$$

**Table 3.9**
Typical Equivalent Lengths of Schedule 40 Straight Pipe
For Screwed Steel Fittings and Valves
(Also, see appendix L)
(For any fluid in turbulent flow)

| | equivalent length, ft | | |
|---|---|---|---|
| | pipe size | | |
| fitting type | 1″ | 2″ | 4″ |
| short radius, regular 90° elbow | 5.2 | 8.5 | 13.0 |
| long radius 90° elbow | 2.7 | 3.6 | 4.6 |
| regular 45° elbow | 1.3 | 2.7 | 5.5 |
| tee, flow through line (run) | 3.2 | 7.7 | 17.0 |
| tee, flow through stem | 6.6 | 12.0 | 21.0 |
| 180° return bend | 5.2 | 8.5 | 13.0 |
| globe valve | 29.0 | 54.0 | 110.0 |
| gate valve | 0.84 | 1.5 | 2.5 |
| angle valve | 17.0 | 18.0 | 18.0 |
| swing check valve | 11.0 | 19.0 | 38.0 |
| coupling or union | 0.29 | 0.45 | 0.65 |

*Example 3.21*

Using table 3.9, determine the equivalent length of the piping network shown.

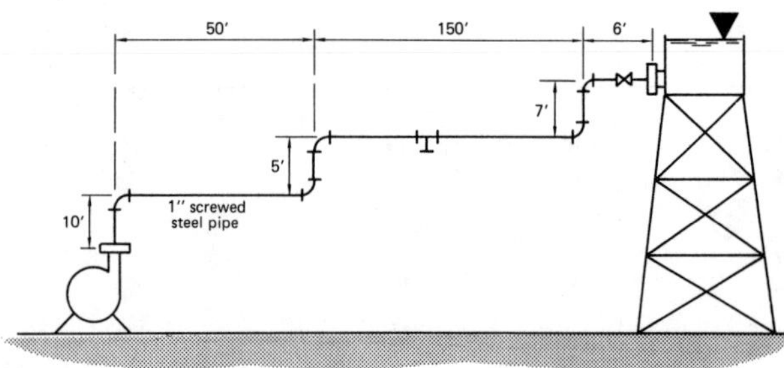


The line consists of:

| | |
|---|---|
| 1 gate valve | 0.84 |
| 5 90° standard elbows | 5.2×5 |
| 1 tee run | 3.2 |
| straight pipe | 228 |
| $L_e$ = | 258 feet |

The alternative is to use a loss coefficient, $K$. This loss coefficient, when multiplied by the velocity head, will give the head loss in feet. This method must be used to find exit and entrance losses.

$$h_f = K \frac{v^2}{2 g_c} \qquad 3.77$$

Values of $K$ are widely tabulated, but they also can be calculated from the following formulas.

*Valves and Fittings*: Refer to the manufacturer's data, or calculate from the equivalent length.

$$K = \frac{f L_e}{D} \qquad 3.78$$

*Sudden Enlargements*: $(D_1 < D_2)$

$$K = \left[1 - \left(\frac{D_1}{D_2}\right)^2\right]^2 \qquad 3.79$$

*Sudden Contractions*: $(D_1 < D_2)$

$$K = \frac{1}{2}\left[1 - \left(\frac{D_1}{D_2}\right)^2\right] \qquad 3.80$$

*Pipe Exit*: (projecting exit, sharp-edged, and rounded)

$$K = 1.0$$

*Pipe Entrance*:

*Reentrant*: $K = 0.78$

*Sharp edged*: $K = 0.5$

*Rounded*:

| $\frac{r}{D}$ | $K$ |
|---|---|
| 0.02 | 0.28 |
| 0.04 | 0.24 |
| 0.06 | 0.15 |
| 0.10 | 0.09 |
| 0.15 | 0.04 |

*Tapered Diameter Changes*:

$$\beta = \frac{\text{small diameter}}{\text{large diameter}}$$

$$\phi = \text{wall-to-horizontal angle}$$

| | Gradual, $\phi < 22°$ | Sudden, $\phi > 22°$ | |
|---|---|---|---|
| Enlargement | $2.6(\sin\phi)(1-\beta^2)^2$ | $(1-\beta^2)^2$ | 3.81 |
| Contraction | $0.8(\sin\phi)(1-\beta^2)$ | $\frac{1}{2}(1-\beta^2)\sqrt{\sin\phi}$ | 3.82 |

## 4 HEAD ADDITIONS/EXTRACTIONS

A pump adds head (energy) to the fluid stream. A turbine extracts head from the fluid stream. The amount of head added or extracted can be found by evaluating Bernoulli's equation (equation 3.55) on both sides of the device.

$$h_A = (H_2 - H_1) \quad \text{(pumps)} \qquad 3.83$$

$$h_E = (H_1 - H_2) \quad \text{(turbines)} \qquad 3.84$$

The head increase from a pump is given by equation 3.85.

$$h_A = \frac{(550)(\text{pump input horsepower})\eta_{\text{pump}}}{\dot{m}} \qquad 3.85$$

Bernoulli's equation also can be used to calculate the power available to a turbine in a fluid stream by multiplying the total energy by the mass flow rate. This is called the *water horsepower*.

$$P = \dot{m}H = \dot{m}\left(\frac{p}{\rho} + \frac{v^2}{2g_c} + z\right) \qquad 3.86$$

$$\dot{m} = \rho A v \qquad 3.87$$

$$whp = \frac{P}{550} \qquad 3.88$$

Pumps and turbines are covered in greater detail in chapter 4.

## 5 DISCHARGE FROM TANKS

Flow from a tank discharging liquid to the atmosphere through an opening in the tank wall (figure 3.14) is affected by both the area and the shape of the opening. At the orifice, the total head of the fluid is converted into kinetic energy according to equation 3.89.[6]

$$v_o = C_v\sqrt{2gh} \qquad 3.89$$

[6] Although the term $g_c$ appears in the equation for velocity head (equation 3.53), here it is the local gravity, $g$, which appears in equation 3.89. An analysis of equation 3.197 will show you why this is so.

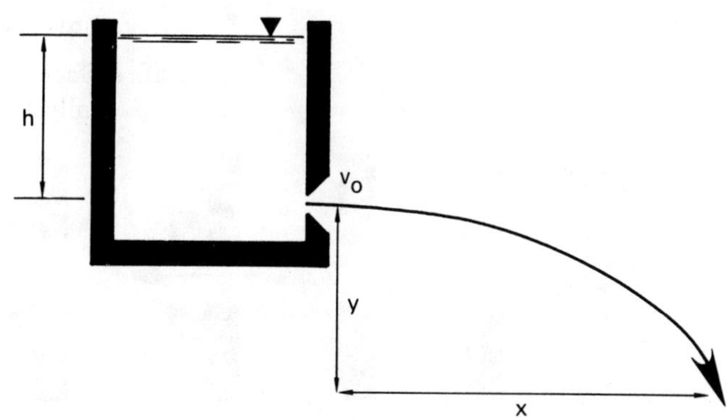


**Figure 3.14** Discharge From a Tank

$C_v$ is the *coefficient of velocity* which can be calculated from the *coefficients of discharge* and *contraction*. Typical values of $C_v$, $C_d$, and $C_c$ are given in table 3.10.

$$C_v = \frac{C_d}{C_c} \qquad 3.90$$

The discharge from the orifice is

$$\dot{V} = (C_c A_o)v_o = C_c A_o C_v\sqrt{2gh} = C_d A_o\sqrt{2gh} \qquad 3.91$$

The head loss due to turbulence at the orifice is

$$h_f = \left(\frac{1}{C_v^2} - 1\right)\frac{v_o^2}{2g_c} \qquad 3.92$$

The discharge stream coordinates (see figure 3.14) are

$$x = v_o t = v_o\sqrt{\frac{2y}{g}} = 2C_v\sqrt{hy} \qquad 3.93$$

$$y = \frac{gt^2}{2} = \frac{g}{2}\left(\frac{x}{v_o}\right)^2 \qquad 3.94$$

The fluid velocity at a point downstream of the orifice is

$$v_x = v_o \qquad 3.95$$

$$v_y = gt \qquad 3.96$$

If the liquid in a tank is not being replenished constantly, the static head forcing discharge through the orifice will decrease. For a tank with a constant cross-sectional area, the time required to lower the fluid level from level $h_1$ to $h_2$ is calculated from equation 3.97.

$$t = \frac{2A_t(\sqrt{h_1} - \sqrt{h_2})}{C_d A_o\sqrt{2g}} \qquad 3.97$$

If the tank has a varying cross section, the following basic relationship holds.

$$Q\,dt = -A_t\,dh \qquad 3.98$$

**Table 3.10**
**Orifice Coefficients for Water**
(fully turbulent)

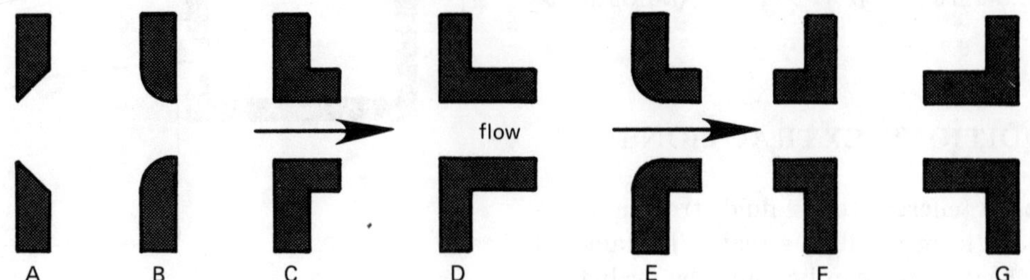


| illustration | description | $C_d$ | $C_c$ | $C_v$ |
|---|---|---|---|---|
| A | sharp-edged | 0.62 | 0.63 | 0.98 |
| B | round-edged | 0.98 | 1.00 | 0.98 |
| C | short tube (fluid separates from walls) | 0.61 | 1.00 | 0.61 |
| D | short tube (no separation) | 0.82 | 1.00 | 0.82 |
| E | short tube with rounded entrance | 0.97 | 0.99 | 0.98 |
| F | reentrant tube, length less than one-half of pipe diameter | 0.54 | 0.55 | 0.99 |
| G | reentrant tube, length 2 to 3 pipe diameters | 0.72 | 1.00 | 0.72 |
| not shown | smooth, well-tapered nozzle | 0.98 | 0.99 | 0.99 |

An expression for the tank area, $A_t$, as a function of $h$, must be determined. Then, the time to empty the tank from height $h_1$ to lower height $h_2$ is

$$t = \int_{h_1}^{h_2} \frac{A_t\, dh}{C_d A_o \sqrt{2gh}} \qquad 3.99$$

For a tank being fed at a rate, $\dot{V}_{\text{in}}$, which is less than the discharge through the orifice, the time to empty expression is

$$t = \int_{h_1}^{h_2} \frac{A_t dh}{(C_d A_o \sqrt{2gh}) - \dot{V}_{\text{in}}} \qquad 3.100$$

When a tank is being fed at a rate greater than the discharge, equation 3.100 will become positive, indicating a rising head. $t$ then will be the time it takes to raise the fluid level from $h_1$ to $h_2$.

The preceding discussion has assumed that the tank has been open or vented to the atmosphere. If the fluid is discharging from a pressurized tank, the total head will be increased by the gage pressure converted to head of fluid by means of equation 3.52.

*Example 3.22*

A 15′ diameter tank discharges 150°F water through a sharp edged 1″ diameter orifice. If the original water depth is 12′ and the tank is continually pressurized to 50 psig, find the time to empty the tank.

At 150°F, $\rho = 61.20 \text{ lbm/ft}^3$

For the orifice,

$$A_o = 0.00545 \text{ ft}^2,\ C_d = 0.62$$

$$h_1 = 12 + \frac{(50)(144)}{61.2} = 129.65 \text{ ft}$$

$$h_2 = \frac{(50)(144)}{61.20} = 117.65 \text{ ft}$$

From equation 3.97,

$$t = \frac{2\,[\pi(7.5)^2](\sqrt{129.65} - \sqrt{117.65})}{(0.62)(0.00545)\sqrt{(2)(32.2)}} = 7035 \text{ seconds}$$

## 6 CULVERTS AND SIPHONS

A culvert is a water path used to drain runoff from an obstructing geographical feature. Most culvert designs are empirical. However, if the entrance and exit of the culvert both are submerged, the discharge will be independent of the barrel slope. (Chapter 5 treats this subject in greater detail.) In that case, equation 3.101 can be used to evaluate the discharge. $h$ is the difference in surface levels of the headwater and tailwater.

$$\dot{V} = C_d A \sqrt{2gh} \qquad 3.101$$

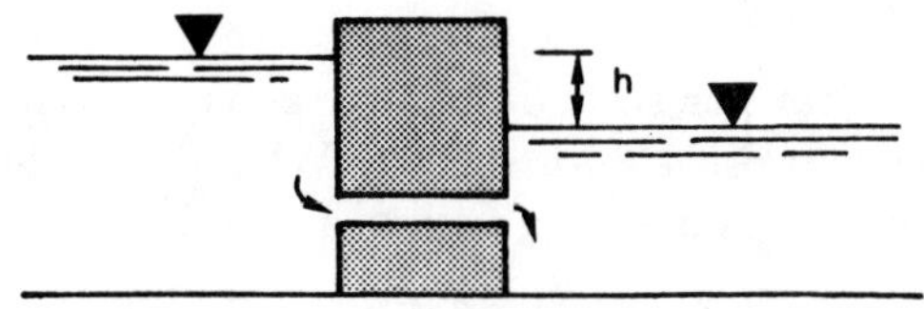


**Figure 3.15** A Simple Pipe Culvert

If the culvert length is greater than 50 feet or if the entrance is not smooth, the available energy will be divided between friction and velocity heads. The effective head to be used in equation 3.101 is:

$$h' = h - h_{\text{entrance}} - h_f \qquad 3.102$$

The entrance head loss is calculated using loss coefficients:

$$h_{\text{entrance}} = K_e \left(\frac{v^2}{2g_c}\right) \qquad 3.103$$

Typical values of $K_e$ are:

0.08 for a smooth and tapered entrance

0.10 for a flush concrete groove or bell design

0.15 for a projecting concrete groove or bell design

0.50 for a flush square-edged entrance

0.90 for a projecting square-edged entrance

The friction loss, $h_f$, can be found in the usual manner, either from the Darcy equation and Moody friction factor chart or from the Hazen-Williams equation. A trial and error solution may be necessary since $v$ is not known, but is needed to find the friction factor.

## 7 MULTIPLE PIPE SYSTEMS

### A. SERIES PIPE SYSTEMS

A series pipe system has one or more diameters along its run. If $Q$ or $v$ is known in any part of the system, the friction loss can be found easily as the sum of the friction losses in the sections.

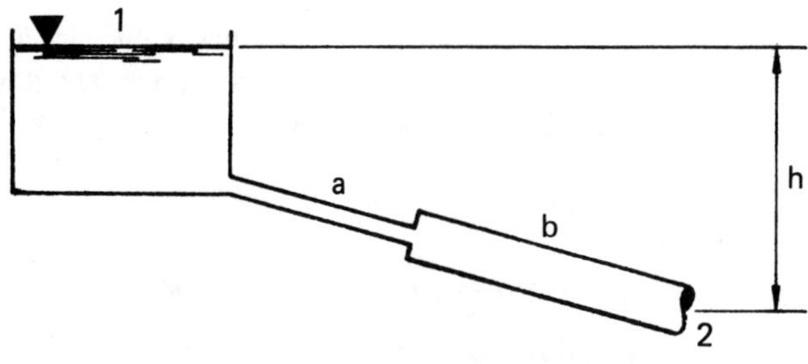


**Figure 3.16** A Series Pipe System

If both $v$ and $Q$ are unknown, a trial and error solution method is required. The following procedure can be used with the Darcy friction factor.

*step 1*: Using the Moody diagram with $\epsilon_a$, $\epsilon_b$, $D_a$, and $D_b$, find $f_a$ and $f_b$ for fully turbulent flow.

*step 2*: Write all of the velocities in terms of one unknown velocity.

$$v_a = v_a \qquad 3.104$$

$$v_b = \left(\frac{A_a}{A_b}\right) v_a \qquad 3.105$$

*step 3*: Write the friction loss in terms of the unknown velocity.

$$h_{f',\text{ total}} = \frac{f_a L_a v_a^2}{2D_a g_c} + \frac{f_b L_b}{2D_b g_c}\left(\frac{A_a}{A_b}\right)^2 v_a^2 \qquad 3.106$$

$$= \frac{v_a^2}{2g_c}\left[\frac{f_a L_a}{D_a} + \frac{f_b L_b}{D_b}\left(\frac{A_a}{A_b}\right)^2\right] \qquad 3.107$$

*step 4*: Solve for the unknown velocity using Bernoulli's equation between points 1 and 2. Include the pipe friction but, for convenience, ignore minor losses.

$$h = \frac{v_b^2}{2g_c} + h_f \qquad 3.108$$

$$= \frac{v_a^2}{2g_c}\left[\left(\frac{A_a}{A_b}\right)^2\left(1 + \frac{f_b L_b}{D}\right) + \frac{f_a L_a}{D_a}\right] \qquad 3.109$$

*step 5*: Using the values of $v_a$ and $v_b$ found from step 4, check the values of $f_a$ and $f_b$. Repeat steps 3 and 4 using the new values of $f_a$ and $f_b$ if necessary.

If the Hazen-Williams coefficients are given for the pipe sections, the procedure for finding unknown velocities and flow quantities is similar although considerably more difficult since the $v^2$ and $v^{1.85}$ terms cannot be combined. A first approximation, however, can be obtained by replacing $v^{1.85}$ with $v^2$ in the Hazen-Williams formula for friction loss. A trial and error method can then be used to find $v$.

### B. PARALLEL PIPE SYSTEMS

A common method of increasing the capacity of an existing line is to install a second line parallel to the first.

If that is done, the flow will divide in such a manner as to make the friction loss the same in both branches.

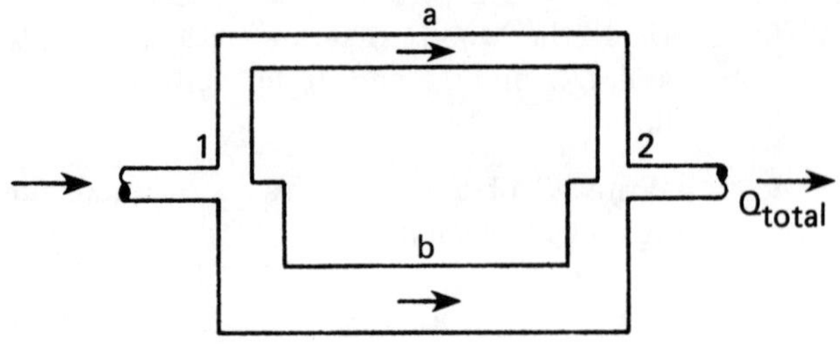


**Figure 3.17** A Parallel Pipe System

If the parallel system has only two branches, a simultaneous solution approach can be taken.

$$h_{f,a} = h_{f,b} = \frac{f_a L_a v_a^2}{2D_a g_c} = \frac{f_b L_b v_b^2}{2D_b g_c} \qquad 3.110$$

$$Q_a + Q_b = Q_c \qquad 3.111$$

$$\frac{1}{4}\pi\left(D_a^2 v_a + D_b^2 v_b\right) = Q_c \qquad 3.112$$

However, if the parallel system has three or more branches, it is easier to use the following procedure, which can also be modified for use with the Hazen-Williams loss formula.

*step 1*: Write $h_f = \frac{fLv^2}{2Dg_c}$ for each branch. Both $h_f$ and $v$ will be unknown.

*step 2*: Solve for $v$ for each branch.

$$v = \sqrt{\frac{2Dg}{fL}h_f} \qquad 3.113$$

*step 3*: Solve for $Q$ for each branch. (There will be a different value of $K'$ for each branch.)

$$Q = Av = A\sqrt{\frac{2Dg}{fL}h_f} = K'\sqrt{h_f} \qquad 3.114$$

*step 4*:

$$\begin{aligned} Q_{\text{total}} &= Q_1 + Q_2 + Q_3 \\ &= (K_1' + K_2' + K_3')\sqrt{h_f} \end{aligned} \qquad 3.115$$

Since $Q_{\text{total}}$, $K_1'$, $K_2'$, and $K_3'$ are known, it is possible to solve for the friction loss.

*step 5*: Check the values of $f$ and repeat as necessary.

*Example 3.23*

3 cubic feet per second of water enter the schedule 40 piping network shown below. What is the head loss between the connecting points $A$ and $B$?

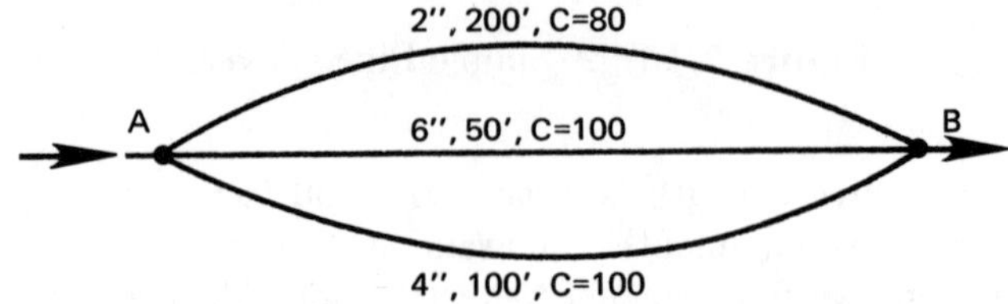


The pipe dimensions are determined from a table of schedule 40 pipe. (See appendix F.)

| | 2″ | 4″ | 6″ |
|---|---|---|---|
| flow area | 0.0233 | 0.0884 | 0.2006 |
| diameter | 0.1723 | 0.3355 | 0.5054 |

*step 1*: Since the Hazen-Williams loss coefficients are given for each branch, the Hazen-Williams friction loss equation will be used.

$$h_f = \frac{(3.012)(v)^{1.85}L}{(C)^{1.85}(D)^{1.165}}$$

*step 2*: The Hazen-Williams friction loss equation can be solved for the velocity term.

$$V = \frac{(0.551)(C)(D)^{0.63}}{L^{0.54}}(h_f)^{0.54}$$

The velocity in the 2″ pipe section can be written in terms of the constant friction loss term.

$$v_{2''} = \frac{(0.551)(80)(0.1723)^{0.63}}{(200)^{0.54}}(h_f)^{0.54} = 0.833(h_f)^{0.54}$$

Similarly,

$$v_{6''} = 4.335(h_f)^{0.54}$$
$$v_{4''} = 2.303(h_f)^{0.54}$$

*step 3*: Since the pipe areas and flow velocities (in terms of the friction loss) are known, the flow quantities can be calculated.

$$Q = Av$$
$$Q_{2''} = (0.0233)(0.833)(h_f)^{0.54} = 0.0194(h_f)^{0.54}$$
$$Q_{6''} = (0.2006)(4.335)(h_f)^{0.54} = 0.8696(h_f)^{0.54}$$
$$Q_{4''} = (0.0884)(2.303)(h_f)^{0.54} = 0.2036(h_f)^{0.54}$$

*step 4*: The three flow quantities must total the quantity flowing through the system, 3 cfs. Since the friction loss is the same across all of the branches, it can be calculated directly.

$$Q_{\text{total}} = Q_{2''} + Q_{6''} + Q_{4''}$$
$$3 = (0.0194 + 0.8696 + 0.2036)(h_f)^{0.54}$$
$$h_f = \left(\frac{3}{1.0926}\right)^{\frac{1}{0.54}} = (2.746)^{1.85} = 6.48 \text{ ft}$$

## C. RESERVOIR BRANCHING SYSTEMS

The three-reservoir problem requires a trial and error solution method which is easily proceduralized. Although there are many possible choices for the unknown variable (e.g., pipe length, diameter, head, flow rate, etc.), there are three common types of problems.

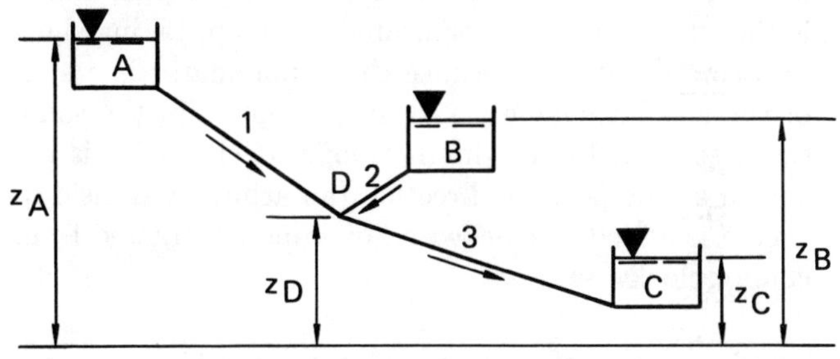


**Figure 3.18** 3-Reservoir System

*Case 1*: Given: all lengths, diameters, and elevations

Find: $Q_1$, $Q_2$, $Q_3$

Although an analytical solution method is possible, this problem is best solved by trial and error.

*step 1*: Assume $Q_1$ and use the Bernoulli equation to find the assumed pressure at point $D$. Ignore minor losses and velocity head. The friction term depends on the assumed value of $Q_1$.

$$v_1 = \frac{Q_1}{A_1} \qquad 3.116$$
$$z_A = z_D + (p_D/\rho) + h_{f,1} \qquad 3.117$$

*step 2*: Once the assumed $p_D$ is known, use it to find $Q_2$.

$$z_B = z_D + (p_D/\rho) + h_{f,2} \qquad 3.118$$
$$Q_2 = v_2 A_2 \qquad 3.119$$

If $z_D + (p_D/\rho)$ is greater than $z_B$, flow will be into reservoir $B$. In that case, the friction term should be subtracted, not added.

*step 3*: Find $Q_3$.

$$z_C = z_D + (p_D/\rho) - h_{f,3} \qquad 3.120$$
$$Q_3 = v_3 A_3 \qquad 3.121$$

*step 4*: Check that $Q_1 + Q_2 = Q_3$. If it does not, return to step 1. After two iterations, plot $Q_1 + Q_2 - Q_3$ versus $Q$. Estimate $Q_1$ by interpolation or extrapolation.

*Alternative Solution Method*: Assume a value of $p_D$ and solve for the flow quantities. Repeat with different values of $p_D$.

*Case 2*: Given: $Q_1$, all lengths, diameters, $z_A$, and $z_B$

Find: $z_C$

*step 1*: $v_1 = Q_1/A_1$ 3.122

*step 2*: Solve for $p_D$ from

$$z_A = z_D + (p_D/\rho) + h_{f,1} \qquad 3.123$$

*step 3*: Solve for $v_2$ from

$$z_B = z_D + (p_D/\rho) + h_{f,2} \qquad 3.124$$

If $z_D + (p_D/\rho)$ is greater than $z_B$, the flow will be into reservoir $B$. In that case, the friction $h_{f,2}$ should be subtracted, not added.

*step 4*: $Q_2 = A_2 v_2$ 3.125

*step 5*: $Q_3 = Q_1 \pm Q_2$ 3.126

*step 6*: $v_3 = Q_3/A_3$ 3.127

*step 7*: Calculate $h_{f,3}$ from $v_3$, $L_3$, and $D_3$.

*step 8*: Find $z_C$ from

$$z_C = z_D + (p_D/\rho) - h_{f,3} \qquad 3.128$$

*Case 3*: Given: All lengths, elevations, $Q_1$, and diameters $D_1$ and $D_2$

Find: $D_3$

*step 1–5*: Repeat steps 1–5 from case 2.

*step 6*: Find $h_{f,3}$ from

$$z_C = z_D + (p_D/\rho) - h_{f,3} \qquad 3.129$$

*step 7*: Find $D_3$ from $h_{f,3}$.

## D. PIPE NETWORKS

Network flows in multi-loop systems can be determined with the Hardy Cross method when a manual solution is necessary. This is a systematic trial-and-error method which first assumes flows and then adds consecutive adjustments to the assumed flows. The Hardy Cross method is easy to apply. It is based on the following principles:

- The flows entering a junction must equal the flows leaving the junction.
- The algebraic sum of friction losses around any closed loop is zero.

If $Q_a$ is the assumed flow in a pipe, and $Q_t$ is the true flow, the difference is $\delta$, where

$$\delta = Q_t - Q_a \qquad 3.130$$

The true flow can be written in terms of the assumed flow and the correction:

$$Q_t = Q_a + \delta \qquad 3.131$$

The friction loss in the pipe has the form of $h_f = K'Q_t^n$, where $n = 2$ if the Darcy equation is used, and $n = 1.85$ if the Hazen-Williams equation is used.

- For $Q$ in cfs, $L$ in feet, and $D$ in feet, the Hazen-Williams friction coefficient is

$$K' = \frac{(4.727)L}{D^{4.87}C^{1.85}} \qquad 3.132$$

- For $Q$ in gpm, $L$ in feet, and $d$ in inches, the Hazen-Williams friction coefficient is

$$K' = \frac{(10.44)L}{C^{1.85}d^{4.87}} \qquad 3.133$$

- For $Q$ in cfs, $L$ in feet, and $D$ in feet, the Darcy friction coefficient is

$$K' = \frac{(0.0252)fL}{D^5} \qquad 3.134$$

  $f$ is usually assumed to be the same (such as 0.02) in all parts of the network.

- For $Q$ in gpm, $L$ in feet, and $D$ in feet, the Darcy friction coefficient is

$$K' = (1.251\,\text{EE--}7)\frac{fL}{D^5} \qquad 3.135$$

Combining and expanding the friction loss as a series,

$$h_f = K'(Q_a+\delta)^n \simeq K'Q_a^n + nK'\delta Q_a^{n-1} + \cdots \qquad 3.136$$

Subsequent higher order terms can be omitted because it is assumed that the correction is small.

Around a complete loop in a network, the sum of the friction drops is zero. Therefore,

$$\sum h_f = \sum K'Q_a^n + \delta \sum nK'Q_a^{n-1} = 0 \qquad 3.137$$

$\delta$ has been taken outside of the summation because all branches in the loop have the same correction. If $n$ is the same for all pipes, it can be taken out of the summation also.

This equation can be solved for $\delta$.

$$\delta = \frac{-\sum K'Q_a^n}{n\sum\left|K'Q_a^{n-1}\right|} = \frac{-\sum h_f}{n(\sum h_f/Q_a)} \qquad 3.138$$

To use this equation, it is necessary to first assume the flow directions as well as the flow rates. The numerator is the sum of head losses around the loop, taking signs into consideration. Because the denominator is a sum of the absolute values, $\delta$ must be applied in the same sense to each branch in the loop. If clockwise is assumed as the positive direction (an arbitrary decision), then $\delta$ is added to clockwise flows and subtracted from counterclockwise flows.

The application of the Hardy Cross method is as follows:

*step 1*: Choose between the Darcy and Hazen-Williams friction loss equations. The Darcy equation results in an easier expression to evaluate because the exponent $(n-1)$ is 1.

*step 2*: Choose a positive direction (e.g., clockwise).

*step 3*: Number all pipes in the network or identify all nodes.

*step 4*: Divide the network into independent loops such that each branch is included in at least one loop.

*step 5*: Calculate $K'$ for each pipe in the network.

*step 6*: Assume flow rates and directions. This may seem like a difficult step, but it is not. Most inaccurate first assumptions yield good results after several iterations.

*step 7*: Calculate $\delta$ for each independent loop.

*step 8*: Apply $\delta$ to each pipe in its loop using the previously mentioned sign convention.

*step 9*: Return to step 7.

*Example 3.24*

Use a Moody friction factor of $f = 0.02$ to calculate the flow in each pipe in the network shown. (Pipes are shown as straight lines for convenience only.)

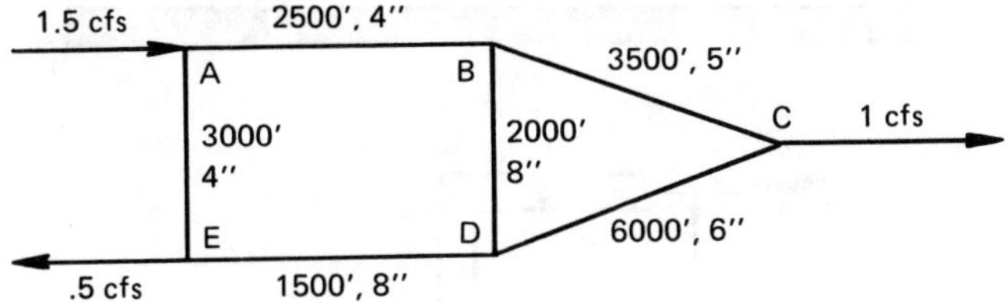


*step 1*: The Moody friction factor is given, so the Darcy friction loss equation will be used.

*step 2*: Choose clockwise as the positive direction.

*step 3*: Use the identification system shown on the network.

*step 4*: Work with loops $ABDE$ and $BCD$. Notice that loop $ABCDE$ is not independent if the other two loops are used.

*step 5*:

$$\text{pipe } AB : D = (4/12) = 0.3333$$

$$K' = \frac{(0.0252)(0.02)(L)}{D^5}$$

$$= \frac{(0.0252)(0.02)(2500)}{(0.3333)^5}$$

$$= 306.2$$

$$\text{pipe } BC : K' = 140.5$$

$$\text{pipe } DC : K' = 96.8$$

$$\text{pipe } BD : K' = 7.7$$

$$\text{pipe } ED : K' = 5.7$$

$$\text{pipe } AE : K' = 367.4$$

*step 6*: Assume the directions and flows shown in the figure.

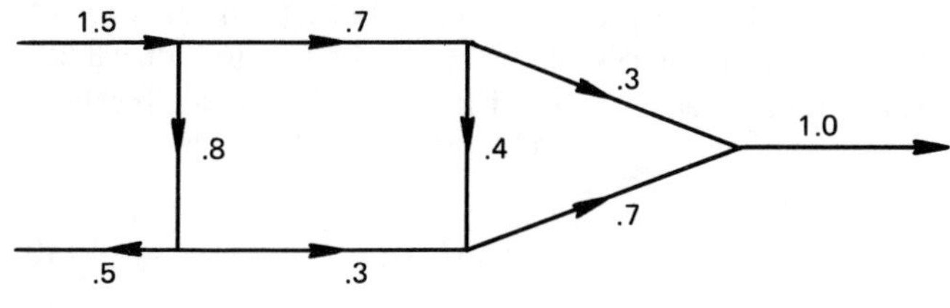


*step 7*:

$$\delta_{ABDE} = \frac{-[(306.2)(0.7)^2+(7.7)(0.4)^2-(5.7)(0.3)^2-(367.4)(0.8)^2]}{2[(306.2)(0.7)+(7.7)(0.4)+(5.7)(0.3)+(367.4)(0.8)]} = +0.08$$

$$\delta_{BCD} = \frac{-[(140.5)(0.3)^2-(96.8)(0.7)^2-(7.7)(0.4)]}{2[(140.5)(0.3)+(96.8)(0.7)+(7.7)(0.4)]} = +0.16$$

*step 8*: The corrected flows are:

$$\text{pipe } AB : \ 0.7 + (0.08) = 0.78$$

$$\text{pipe } BC : \ 0.3 + (0.16) = 0.46$$

$$\text{pipe } DC : \ 0.7 - (0.16) = 0.54$$

$$\text{pipe } BD : \ 0.4 + (0.08) - (0.16) = 0.32$$

$$\text{pipe } ED : \ 0.3 - (0.08) = 0.22$$

$$\text{pipe } AE : \ 0.8 - (0.08) = 0.72$$

*step 9*: The procedure is repeated using the corrected flows.

## 8 FLOW MEASURING DEVICES

The total energy in a fluid flow is the sum of pressure head, velocity head, and gravitational head.

$$H = \frac{p}{\rho} + \frac{v^2}{2g_c} + z \qquad 3.139$$

Change in gravitational head within a flow-measuring instrument is negligible. Therefore, if two of the three remaining variables ($H$, $p$, or $v$) are known, the third can be found from subtraction. The flow measuring devices discussed in this section are capable of measuring total head ($H$) or pressure head ($p$).

### A. VELOCITY MEASUREMENT

Velocity of a fluid stream is determined by measuring the difference between the static and the stagnation pressures, then solving for the velocity head.

A *piezometer tap* can be used to measure the pressure head directly in feet of fluid.

$$h_p = \frac{p}{\rho} \qquad 3.140$$

For liquids with pressures higher than the capability of the direct reading tap, a manometer can be used with a piezometer tap or with a *static probe* as shown in figure 3.20.

For either configuration of figure 3.20, the static pressure is

$$p = \rho_m \Delta h_m - \rho y \qquad 3.141$$

$$h_p = \frac{\rho_m \Delta h_m}{\rho} - y \qquad 3.142$$

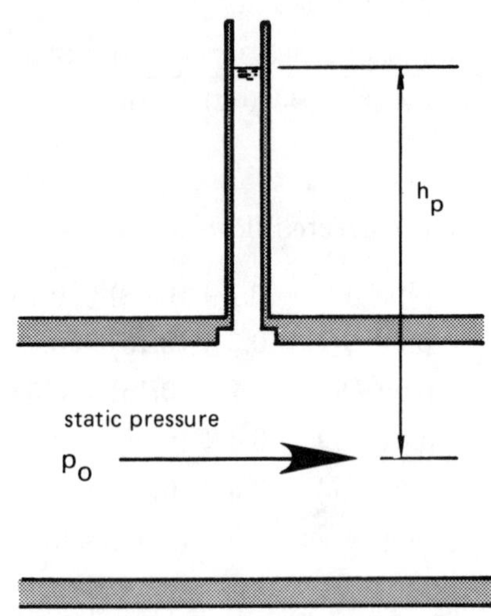


**Figure 3.19** Piezometer Tap

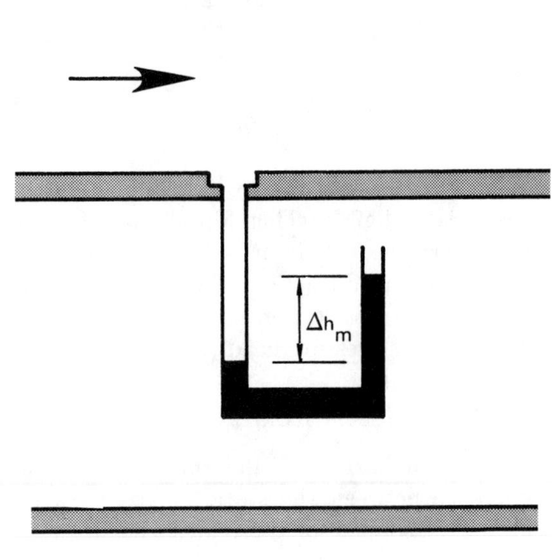


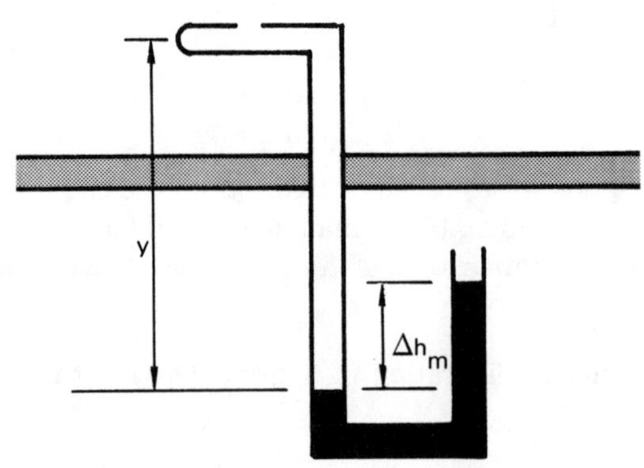


**Figure 3.20** Use of Manometers to Measure Static Pressure

*Stagnation pressure*, also known as *total pressure* or *impact pressure*, can be measured directly in feet of fluid by using a pitot tube as shown in figure 3.12.

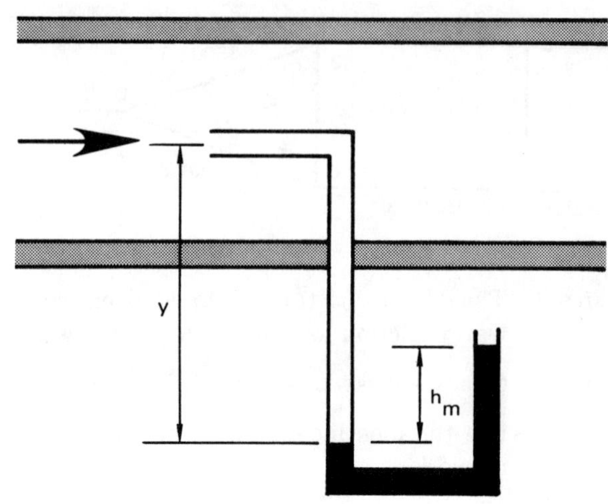


**Figure 3.21** Use of Manometer to Measure Total Pressure

Using the results of the measurements above, the velocity head can be calculated from equation 3.145.

For high-pressure fluids, a manometer must be used to measure stagnation pressure:

$$p_s = \rho h_s = \rho_m \Delta h_m - \rho y \qquad 3.143$$

$$h_s = \frac{p}{\rho} + \frac{v^2}{2g_c} = \frac{\rho_m \Delta h_m}{\rho} - y \qquad 3.144$$

$$v = \sqrt{2g(h_s - h_p)} = \sqrt{\frac{2g(p_s - p)}{\rho}} \qquad 3.145$$

If the piezometer tap of figure 3.19 and the pitot tube of figure 3.12 are placed at the same point as in figure 3.22, the velocity head in feet of fluid can be read directly.

$$\frac{v^2}{2g_c} = \Delta h \qquad 3.146$$

$$v = \sqrt{2g\Delta h} \qquad 3.147$$

The instrumentation arrangement of figures 3.20 and 3.21 can be combined into a single instrument to provide a measurement of velocity head as shown in figure 3.23.

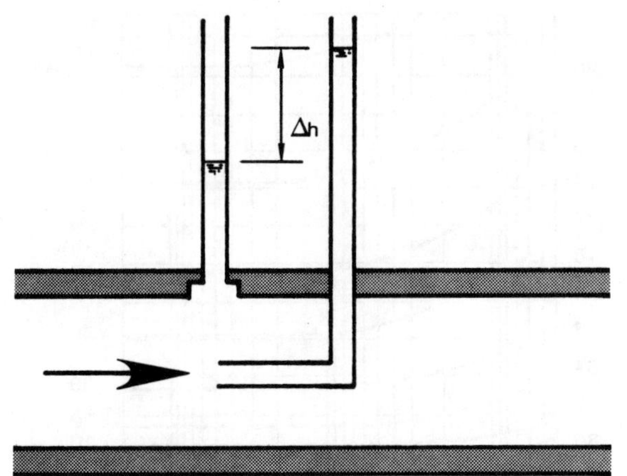


**Figure 3.22** Comparative Velocity Head Measurement

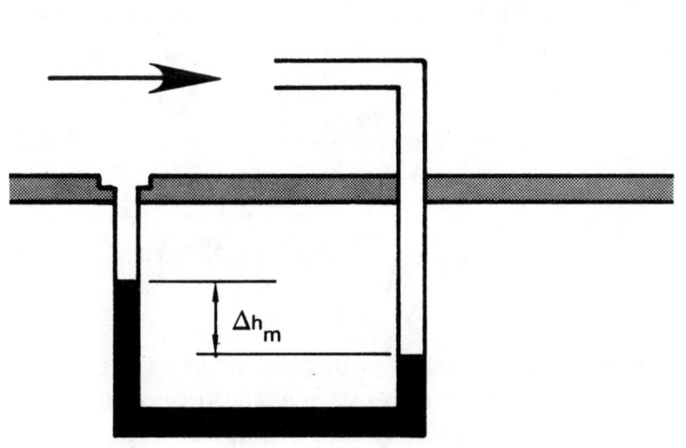


**Figure 3.23** Velocity Head Measurement

$$\frac{v^2}{2g_c} = \frac{\Delta h_m(\rho_m - \rho)}{\rho} \qquad 3.148$$

$$v = \sqrt{\frac{2g(\rho_m - \rho)\Delta h_m}{\rho}} \qquad 3.149$$

*Example 3.25*

50°F water is flowing through a pipe. A pitot-static gage registers a 3″ deflection of mercury. What is the velocity within the pipe? (The density of mercury is 848.6 pcf.)

Using equation 3.149,

$$v = \sqrt{\frac{2\,(32.2)(848.6 - 62.4)\,\left(\frac{3}{12}\right)}{62.4}} = 14.24 \text{ fps}$$

## B. FLOW MEASUREMENT

Using the techniques described in the preceding section, the flow rate in a line can be determined by measuring the pressure drop across a restriction. Once the geometry of the restriction is known, the Bernoulli equation, along with empirically determined correction coefficients, can be applied to obtain an expression directly relating flow rate with pressure drop.

If potential head is neglected, Bernoulli's equation becomes

$$\frac{p_1}{\rho} + \frac{v_1^2}{2g_c} = \frac{p_2}{\rho} + \frac{v_2^2}{2g_c} \qquad 3.150$$

But, $v_1$ and $v_2$ are related. From equation 3.68,

$$v_1 = v_2\left(\frac{A_2}{A_1}\right) \qquad 3.151$$

Combining equations 3.150 ar.d 3.151 yields the standard flow measurement equation.

$$v_2 = \frac{\sqrt{2g\left(\frac{p_1 - p_2}{\rho}\right)}}{\sqrt{1 - \left(\frac{A_2}{A_1}\right)^2}} \qquad 3.152$$

The flow quantity can be found from

$$\dot{V} = v_2 A_2 \qquad 3.153$$

The reciprocal of the denominator of equation 3.152 is known as the *velocity of approach factor*, $F_{va}$. The *beta ratio* can be incorporated into the formula for $F_{va}$.

$$F_{va} = \frac{1}{\sqrt{1 - \beta^4}} \qquad 3.154$$

$$\beta = \frac{D_2}{D_1} \qquad 3.155$$

The simplest fluid flow measuring device is the *orifice plate*. This consists of a thin plate or diaphragm with a central hole through which the fluid flows.

Equations 3.156 and 3.157 are the governing orifice plate equations for liquid flow.

$$\dot{V} = F_{va}C_dA_o\sqrt{\frac{2g(p_1 - p_2)}{\rho}} \qquad 3.156$$

$$= F_{va}C_dA_o\sqrt{\frac{2g(\rho_m - \rho)\Delta h_m}{\rho}} \qquad 3.157$$

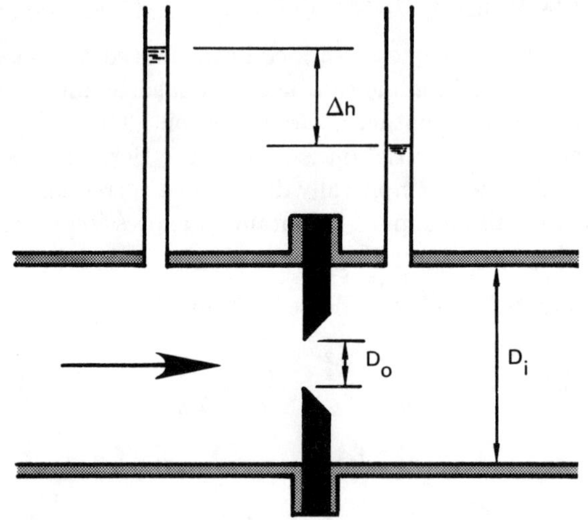


**Figure 3.24** Comparative Reading Orifice Plate

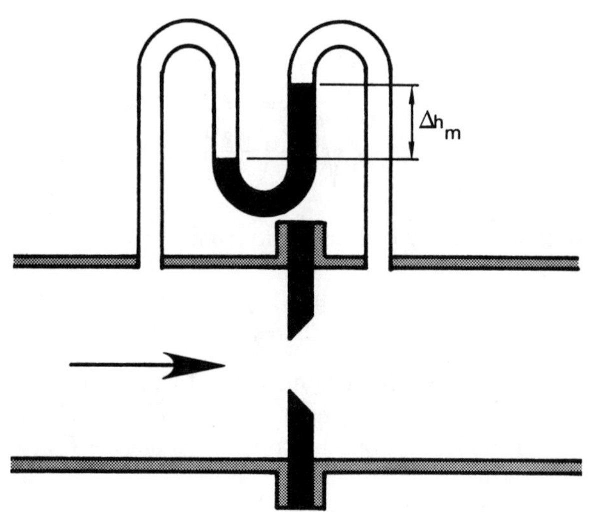


**Figure 3.25** Direct Reading Orifice Plate

The definition of the velocity of approach factor is modified slightly for the orifice plate.

$$F_{va} = \frac{1}{\sqrt{1 - \left(\frac{C_c A_o}{A_i}\right)^2}} \quad 3.158$$

The *flow coefficient* depends on the velocity of approach factor and the discharge coefficient. It also can be obtained from figure 3.26.

$$C_f = F_{va} C_d \quad 3.159$$

$$C_d = C_v C_c \quad 3.160$$

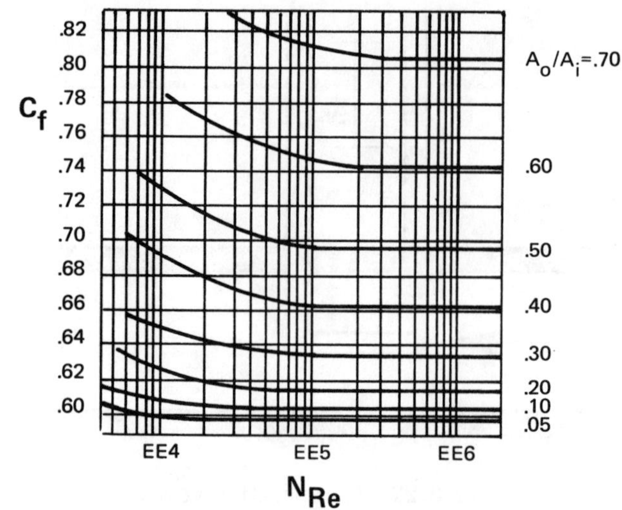


**Figure 3.26** Flow Coefficients for I.S.A. Orifice Plates

The flow coefficients can be used to rewrite equations 3.156 and 3.157.

$$\dot{V} = C_f A_o \sqrt{\frac{2g(p_1 - p_2)}{\rho}} \quad 3.161$$

$$= C_f A_o \sqrt{\frac{2g(\rho_m - \rho)\Delta h_m}{\rho}} \quad 3.162$$

Operating on the same principles as the orifice plate, the *venturi meter* induces a smaller pressure drop. It is, however, mechanically more complex, as shown by figure 3.27.

The governing equations are similar to those for orifice plates. $C_c$ usually is 1.0 for venturi meters.

$$v_2 = F_{va} \sqrt{\frac{2g(p_1 - p_2)}{\rho}} \quad 3.163$$

$$\dot{V} = F_{va} C_d A_2 \sqrt{\frac{2g(p_1 - p_2)}{\rho}}$$

$$= C_f A_2 \sqrt{\frac{2g(p_1 - p_2)}{\rho}} \quad 3.164$$

$$F_{va} = \frac{1}{\sqrt{1 - \left(\frac{A_2}{A_1}\right)^2}} \quad 3.165$$

$$C_d = C_v C_c \quad 3.166$$

$$C_f = F_{va} C_d \quad 3.167$$

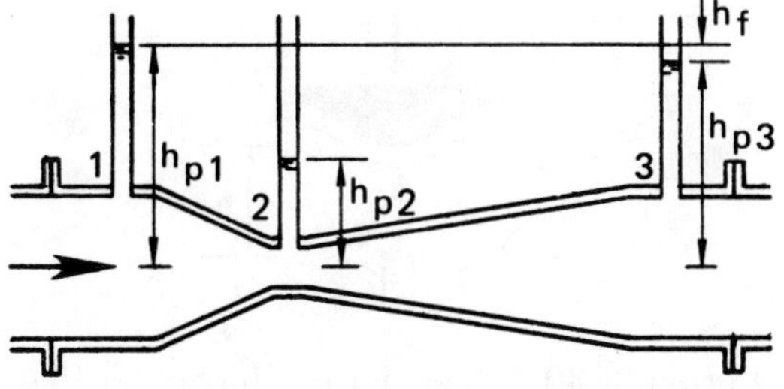


**Figure 3.27** Venturi Meter with Wall Taps

**Table 3.11**
$C_d$ for Venturi Meters

| $2 < (A_1/A_2) < 3$ | |
|---|---|
| $C_d$ | $N_{Re}$ |
| 0.94 | 6000 |
| 0.95 | 10,000 |
| 0.96 | 20,000 |
| 0.97 | 50,000 |
| 0.98 | 200,000 |
| 0.99 | 2000,000 |

*Example 3.26*

150°F water is flowing in an 8″ schedule 40 steel pipe at 2.23 cfs. If a 7 inch sharp edged orifice plate is bolted across the line, what manometer deflection in inches of mercury would be expected? (Mercury has a density of 848.6 pcf.) Neglect the correction for water in the manometer tubes.

| | 7″ orifice | 8″ schedule 40 |
|---|---|---|
| flow area | 0.267 ft$^2$ | 0.3474 ft$^2$ |
| diameter | 0.583 ft | 0.6651 ft |

From table 3.10 for the orifice: $C_c = 0.63$, $C_d = 0.62$.

Using equation 3.158,

$$F_{va} = \left[1 - \frac{C_c A_o}{A_i}\right]^{-\frac{1}{2}}$$

$$= \left[1 - \left(\frac{(0.63)(0.267)}{0.3474}\right)^2\right]^{-\frac{1}{2}} = 1.14$$

From equation 3.157,

$$\Delta h_m = \left(\frac{\dot{V}}{F_{va} C_d A_o}\right)^2 \frac{\rho}{2g(\rho_m - \rho)}$$

$$= \left(\frac{2.23}{(1.14)(0.62)(0.267)}\right)^2 \times \frac{61.2}{(2)(32.2)(848.6 - 61.2)} \times 12\frac{\text{in}}{\text{ft}}$$

$$= 2.02''$$

## 9 THE IMPULSE/MOMENTUM PRINCIPLE

A force is required to cause a direction or velocity change in a flowing fluid. Conventions necessary to determine such a force are:

1. $\Delta v = v_2 - v_1$
2. A positive $\Delta v$ indicates an increase in velocity. A negative $\Delta v$ indicates a decrease in velocity.
3. $F$ and $x$ are positive to the right. $F$ and $y$ are positive upward.
4. $F$ is the force on the fluid. The force on the walls or support has the same magnitude but opposite direction.
5. The fluid is assumed to flow horizontally from left to right and is assumed to possess no $y$-component of velocity.

The *momentum* possessed by a moving fluid is defined as the product of mass (in slugs) and velocity (in ft/sec).

The $g_c$ term in equation 3.168 is needed to convert pounds-mass into slugs.

$$\text{momentum} = \frac{mv}{g_c} \tag{3.168}$$

*Impulse* is defined as the product of a force and the length of time the force is applied.

$$\text{impulse} = F\Delta t \tag{3.169}$$

The *impulse-momentum principle* states that the impulse applied to a moving body is equal to the change in momentum. This is expressed by equation 3.170.

$$F\Delta t = \frac{m\Delta v}{g_c} \tag{3.170}$$

Solving for $F$ and combining $m$ and $\Delta t$ yields equation 3.171.

$$F = \frac{m\Delta v}{g_c \Delta t} = \frac{\dot{m}\Delta v}{g_c} \tag{3.171}$$

Since $F$ is a vector, it can be broken into its components

$$F_x = \frac{\dot{m}\Delta v_x}{g_c} \tag{3.172}$$

$$F_y = \frac{\dot{m}\Delta v_y}{g_c} \tag{3.173}$$

If the fluid flow is directed through an angle $\phi$,

$$\Delta v_x = v(\cos\phi - 1) \tag{3.174}$$

$$\Delta v_y = v \sin\phi \tag{3.175}$$

There are several fluid applications of the impulse-momentum principle.

A. JET PROPULSION

$$\dot{m}_2 = \dot{m}_1 + \dot{m}_{\text{fuel}} = \dot{V}_1\rho_1 + \dot{V}_{\text{fuel}}\rho_{\text{fuel}} \qquad 3.176$$

$$F_x = \frac{\dot{V}_2\rho_2 v_{2x} - \dot{V}_1\rho_1 v_{1x}}{g_c} \qquad 3.177$$

$$F_y = \frac{\dot{V}_2\rho_2 v_{2y} - \dot{V}_1\rho_1 v_{1y}}{g_c} \qquad 3.178$$

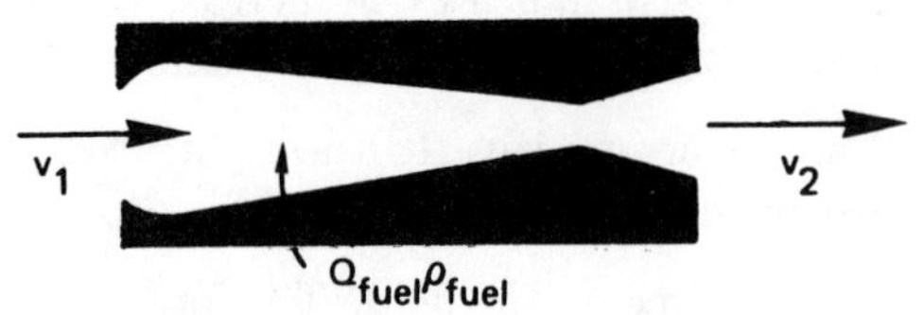


**Figure 3.28** Jet Propulsion

B. OPEN JET ON VERTICAL FLAT PLATE

$$\Delta v_y = 0 \qquad 3.179$$

$$\Delta v_x = -v \qquad 3.180$$

$$F_x = \frac{-\dot{m}v}{g_c} = \frac{-\dot{V}\rho v}{g_c} \qquad 3.181$$

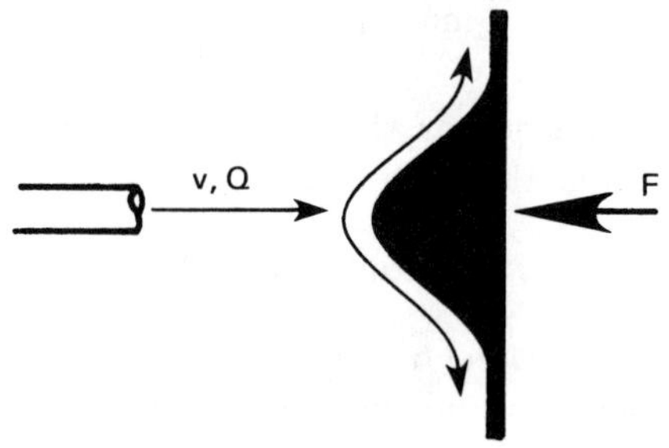


**Figure 3.29** Open Jet on Vertical Plate

C. OPEN JET ON HORIZONTAL FLAT PLATE

As the jet travels upwards, its velocity decreases since gravity is working against it. By the time the liquid has reached the plate, the velocity has become

$$v_y = \sqrt{v_0^2 - 2gh} \qquad 3.182$$

$$\Delta v_x = 0 \qquad 3.183$$

$$\Delta v_y = -\sqrt{v_0^2 - 2gh} \qquad 3.184$$

$$F = \left(\frac{-\dot{m}}{g_c}\right)\sqrt{v_o^2 - 2gh}$$

$$= \left(\frac{\dot{V}\rho}{g_c}\right)\sqrt{v_o^2 - 2gh} \qquad 3.185$$

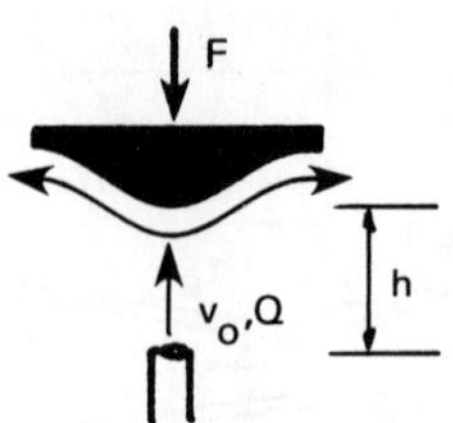


**Figure 3.30** Open Jet on Horizontal Plate

D. OPEN JET ON SINGLE STATIONARY BLADE

$v_2$ may not be the same as $v_1$ if friction is present. If no information is given, assume that $v_2 = v_1$.

$$\Delta v_x = v_2\cos\phi - v_1 \qquad 3.186$$

$$\Delta v_y = v_2\sin\phi \qquad 3.187$$

$$F_x = \left(\frac{\dot{m}}{g_c}\right)(v_2\cos\phi - v_1)$$

$$= \left(\frac{\dot{V}\rho}{g_c}\right)(v_2\cos\phi - v_1) \qquad 3.188$$

$$F_y = \left(\frac{\dot{m}}{g_c}\right)(v_2\sin\phi) = \left(\frac{\dot{V}\rho}{g_c}\right)(v_2\sin\phi) \qquad 3.189$$

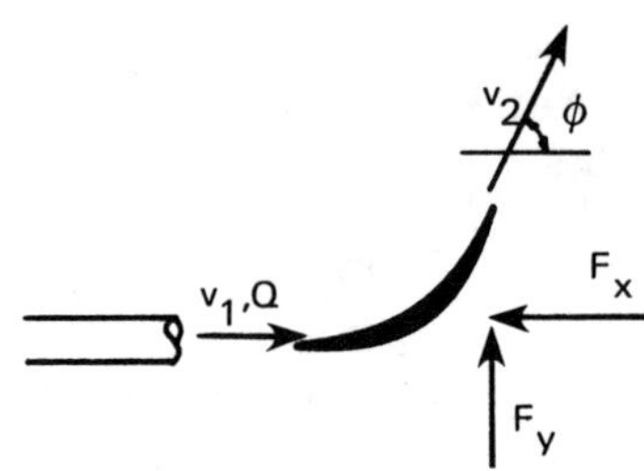


**Figure 3.31** Open Jet on Stationary Blade

E. OPEN JET ON SINGLE MOVING BLADE

$v_b$ is the blade velocity. For simplicity, friction is ignored. The discharge overtaking the moving blade is $\dot{V}'$.

$$\dot{V}' = \left(\frac{v - v_b}{v}\right)\dot{V} \qquad 3.190$$

$$\Delta v_x = (v - v_b)(\cos\phi - 1) \qquad 3.191$$

$$\Delta v_y = (v - v_b)(\sin\phi) \qquad 3.192$$

$$F_x = \frac{\dot{m}'\Delta v_x}{g_c} = \left(\frac{\dot{V}'\rho}{g_c}\right)\Delta v_x \qquad 3.193$$

$$F_y = \frac{\dot{m}'\Delta v_y}{g_c} = \left(\frac{\dot{V}'\rho}{g_c}\right)\Delta v_y \qquad 3.194$$

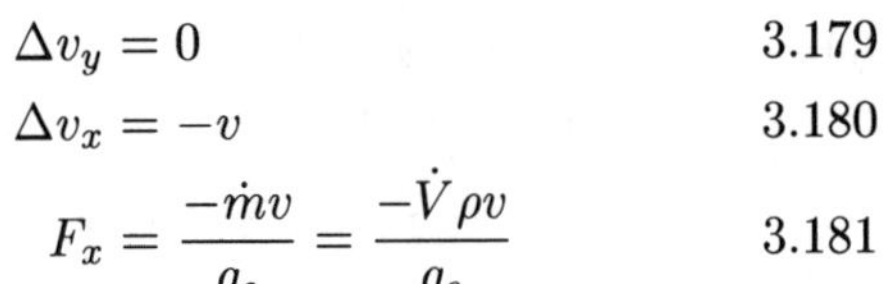

The power transferred to the blade is given by equation 3.195. Power is maximized when $\phi = 180°$ and $v_b = \frac{1}{2}v$.

$$P = F_x v_b \qquad 3.195$$

Equations 3.193 and 3.194 can be used with a *multiple-bladed wheel* by using the full $\dot{V}$ instead of $\dot{V}'$.

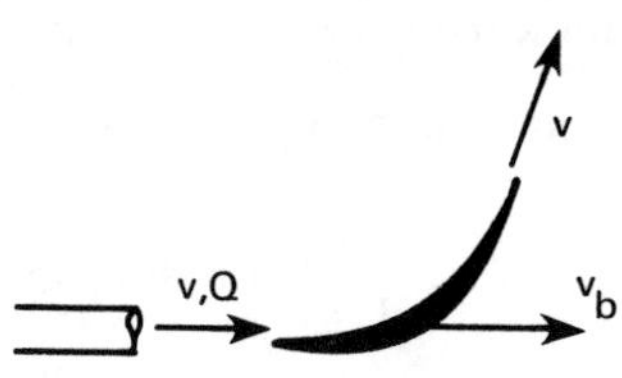


**Figure 3.32** Open Jet on Moving Blade

F. CONFINED STREAMS IN PIPE BENDS

Since the fluid is confined, the forces caused by static pressure must be included along with the force from momentum changes. Using gage pressures and neglecting the fluid weight,

$$F_x = p_2 A_2 \cos\phi - p_1 A_1 + \left(\frac{\dot{V}\rho}{g_c}\right)(v_2 \cos\phi - v_1) \quad 3.196$$

$$F_y = \left[p_2 A_2 + \frac{\dot{V}\rho v_2}{g}\right] \sin\phi \qquad 3.197$$

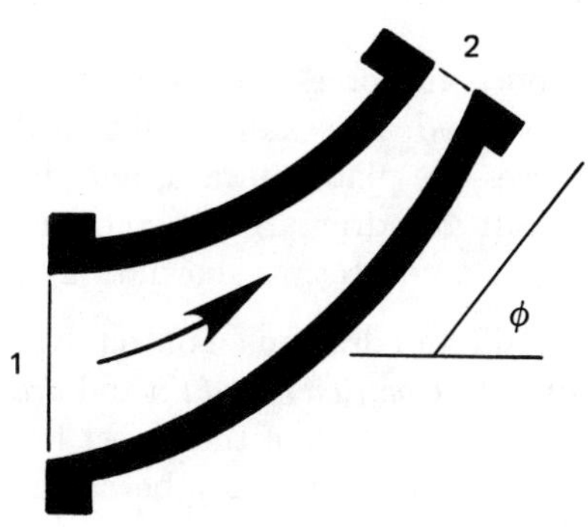


**Figure 3.33** A Pipe Bend

G. WATER HAMMER

Water hammer is an increase in pressure in a pipe caused by a sudden velocity decrease. The sudden velocity decrease usually will be caused by a valve's closing.

Assuming the pipe material is inelastic, the time required for the water hammer shock wave to travel from a valve to the end of a pipe and back is given by

$$t = \frac{2L}{c} \qquad 3.198$$

The fluid pressure increase resulting from this shock wave is

$$\Delta p = \frac{\rho c \Delta v}{g_c} \qquad 3.199$$

Water hammer or *surge* can be handled by one or more of the following methods:

- Moderating the valve closure time by either a manual or automatic valve controller
- A surge tank with a free water surface
- An air chamber on the discharge line
- A surge surpressor
- A surge relief valve

*Example 3.27*

60°F water at 40 psig flowing at 8 ft/sec enters a 12″×8″ reducing elbow as shown and is turned 30°. (a) What is the resultant force on the water? (b) What other forces should be considered in the design of supports for the fitting?

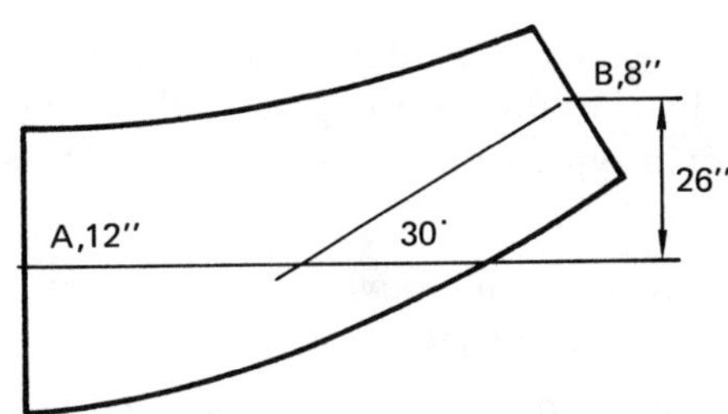


(a) The total head at point A is

$$\frac{(40)(144)}{(62.4)} + \frac{(8)^2}{(2)(32.2)} + 0 = 93.3 \text{ ft}$$

At point B, the velocity is

$$(8)\left(\frac{12}{8}\right)^2 = 18 \text{ ft/sec}$$

The pressure at B can be found from Bernoulli's equation.

$$93.3 = \frac{p_B(144)}{62.4} + \frac{(18)^2}{(2)(32.2)} + \frac{26}{12}$$

So, $p_B = 37.3$ psig

$$\dot{V} = vA = (8)\left(\frac{1}{4}\right)\pi\left(\frac{12}{12}\right)^2 = 6.28 \text{ cfs}$$

From equation 3.196,

$$\begin{aligned} F_x &= +(37.3)(144)\left(\frac{1}{4}\right)\pi\left(\frac{8}{12}\right)^2 \cos 30^\circ \\ &\quad - (40)(144)\left(\frac{1}{4}\right)\pi\left(\frac{12}{12}\right)^2 \\ &\quad + \left(\frac{(6.28)(62.4)}{32.2}\right)[(18)(\cos 30^\circ) - 8] \\ &= -2808 \text{ lbf} \end{aligned}$$

From equation 3.197,

$$\begin{aligned} F_y &= \left[(37.3)(144)\left(\frac{1}{4}\right)\pi\left(\frac{8}{12}\right)^2 \right. \\ &\quad \left. + \left(\frac{(6.28)(62.4)(18)}{32.2}\right)\right] \sin 30^\circ \\ &= 1047 \text{ lbf} \end{aligned}$$

The resultant force on the water is

$$R = \sqrt{(-2808)^2 + (1047)^2} = 2997 \text{ lbf}$$

(b) The support also should be designed to carry the weight of the water in the pipe and the bend, and the weight of the pipe and the bend itself.

*Example 3.28*

40°F water is flowing at 10 ft/sec through a 4″ schedule 40 welded steel pipe. A valve suddenly is closed. Assuming rigid pipe, what increase in fluid pressure will occur?

Assume that the closing valve completely stops the flow. Therefore, $\Delta v$ is 10 ft/sec.

At 40°F, $E$ = 294 EE3 psi, and $\rho$ = 62.43 lbm/ft$^3$. From equation 3.20, the speed of sound in the water is

$$c = \sqrt{\frac{(294 \text{ EE3})(144)(32.2)}{62.43}} = 4673 \text{ fps}$$

From equation 3.199,

$$\Delta p = \frac{(62.43)(4673)(10)}{32.2} = 90{,}600 \text{ psf}$$

## H. OPEN JET ON INCLINED PLATE

An open jet will be diverted both up and down a stationary, inclined flat plate. The velocity in each diverted flow will be $v$, the same as in the approaching jet. The fractions $f_1$ and $f_2$ of the jet which are diverted up and down can be found from equations 3.200 and 3.201.

$$f_1 = \frac{1+\cos\phi}{2} \qquad 3.200$$

$$f_2 = \frac{1-\cos\phi}{2} \qquad 3.201$$

$$f_1 - f_2 = \cos\phi \qquad 3.202$$

$$f_1 + f_2 = 1 \qquad 3.203$$

As the flow along the plate is assumed to be frictionless, there will be no force component parallel to the plate.

The force perpendicular to the plate is

$$F = \left(\frac{\dot{V}\rho}{g_c}\right) v \sin\phi \qquad 3.204$$

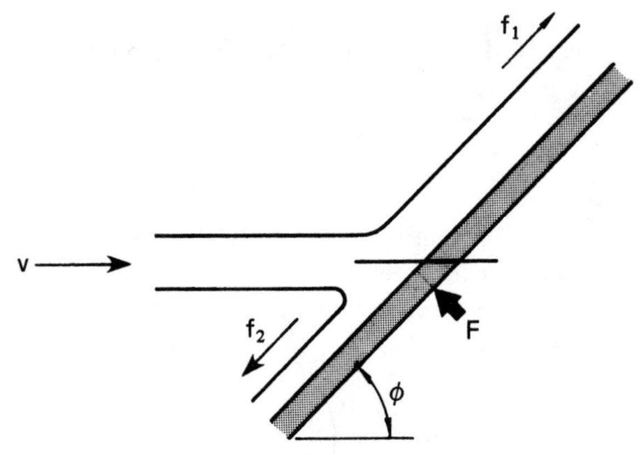


**Figure 3.34** Open Jet on Inclined Plate

## 10 LIFT AND DRAG

Lift and drag both are forces exerted on an object as it passes through a fluid. For example, lift on the wing of an airplane forces the plane upward, and drag tries to slow it down. Lift and drag are the components of the resultant force on an object, as shown in figure 3.35.

The amounts of lift and drag on an object depend on the shape of the object. *Coefficients of lift* and *drag* are used to measure the effectiveness of the object in producing lift and drag. Lift and drag may be calculated from equations 3.205 and 3.206.

$$L = \frac{C_L A \rho v^2}{2g_c} \qquad 3.205$$

$$D = \frac{C_D A \rho v^2}{2g_c} \qquad 3.206$$

$A$, in equation 3.205, is the object's area projected onto a plane parallel to the direction of motion. In equation 3.206, $A$ is the area projected onto a plane normal to the direction of motion.

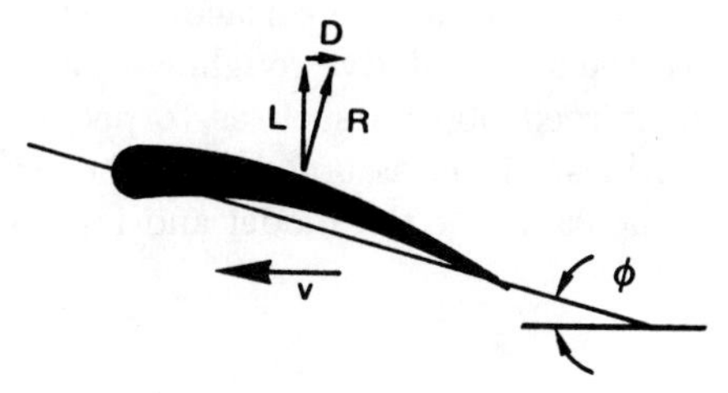


**Figure 3.35** Lift and Drag on an Airfoil

Values of $C_L$ for various airfoil sections have been correlated with $N_{Re}$. No simple relationship can be given for airfoils in general. However, the theoretical relationship for a thin flat plate inclined at an angle $\phi$ is

$$C_L = 2\pi \sin \phi \qquad 3.207$$

The drag coefficient for a sphere moving with $N_{Re}$ less than 0.4 is predicted by *Stokes' law*, equation 3.208. The same equation may be used for circular disks. Values of $C_D$ for other shapes are given in table 3.12.

$$C_D = \frac{24}{N_{Re}} \qquad 3.208$$

**Table 3.12**
Approximate Drag Coefficients

(Do not interpolate between $N_{Re}$ = EE5 and $N_{Re}$ = EE6)

Reynolds Number, $N_{Re}$

| body shape, and (characteristic dimension) | EE0 | EE1 | EE2 | EE3 | EE4 | EE5 | EE6 | fully turbulent EE6-EE7 |
|---|---|---|---|---|---|---|---|---|
| sphere (diameter) | (a) | 4 | 1.0 | 0.45 | 0.40 | 0.55 | 0.25 | 0.2 |
| flat disk (diameter) | (a) | 4 | 1.5 | 1.9(b) | 1.1 | 1.1 | 1.1 | 1.1 |
| flat plate, normal to flow, (short side) | | | | | | | | |
| length/breadth = 1 | | | | (b) | 1.16 | 1.16 | 1.16 | 1.16 |
| 4 | | | | (b) | 1.17 | 1.17 | 1.17 | 1.17 |
| 8 | | | | (b) | 1.23 | 1.23 | 1.23 | 1.23 |
| 12.5 | | | | (b) | 1.34 | 1.34 | 1.34 | 1.34 |
| 20 | | | | (b) | 1.50 | 1.50 | 1.50 | 1.50 |
| 25 | | | | (b) | 1.57 | 1.57 | 1.57 | 1.57 |
| 50 | | | | (b) | 1.76 | 1.76 | 1.76 | 1.76 |
| ∞ | | | | (b) | 2.0 | 2.0 | 2.0 | 2.0 |
| circular cylinder, axis normal to flow (diameter) | | | | | | | | |
| length/diameter = 1 | | | | 0.6 | 0.6 | 0.6 | | 0.35 |
| 5 | | | | 0.7 | 0.9 | 0.9 | | |
| 20 | | | | 0.9 | 0.9 | 0.9 | | |
| ∞ | 10 | 2.5 | 1.3 | 0.9 | 1.1 | 1.4 | 0.37 | 0.33 |
| circular cylinder, axis parallel to flow (diameter) | | | | | | | | |
| length/diameter = 1 | | | | 0.91 | 0.91 | 0.91 | | |
| 2 | | | | 0.85 | 0.85 | 0.85 | | |
| 4 | | | | 0.87 | 0.87 | 0.87 | | |
| 7 | | | | 0.99 | 0.99 | 0.99 | | |

Note a: Use Stokes' law, equation 3.208    Note b: Becomes fully turbulent at $N_{Re}$ = 3 EE3

## 11 SIMILARITY

*Similarity* between a model (subscript $m$) and a full-sized object (subscript $t$) implies that the model can be used to predict the performance of the full-sized object. Such a model is said to be *mechanically similar* to the full-sized object.

Complete mechanical similarity requires geometric and dynamic similarity.[7] *Geometric similarity* means that the model is true to scale in length, area, and volume. *Dynamic similarity* means that the ratios of all types of forces are equal. These forces result from inertia, gravity, viscosity, elasticity (fluid compressibility), surface tension, and pressure.

The *model scale* or *length ratio* is

$$L_r = \frac{\text{size of model}}{\text{full size}} \qquad 3.209$$

The area and volume ratios are based on the length ratio.

$$\frac{A_m}{A_t} = (L_r)^2 \qquad 3.210$$

$$\frac{V_m}{V_t} = (L_r)^3 \qquad 3.211$$

The number of possible ratios of forces is large. Fortunately, some force ratios may be ignored because the forces are negligible or self-canceling. Three important cases where the analysis can be simplified are dominant viscous and inertial forces, dominant inertial and gravitational forces, and dominant surface tension.

### A. VISCOUS AND INERTIAL FORCES DOMINATE

Consider the testing of a completely submerged object such as a submarine. Surface tension effects are negligible. The fluid is assumed incompressible for low velocities. Gravity does not change the path of the fluid particles significantly during the time the submarine is near.

Only inertial, viscous, and pressure forces are significant. Being the only significant ones, these three forces are in equilibrium. Since they are in equilibrium, knowing any two will define the third completely. Since it is not an independent force, pressure is omitted from the similarity analysis.

The ratio of the inertial forces to the viscous forces is the Reynolds number. Setting the model and full-size Reynolds numbers equal will ensure similarity. That is,

$$(N_{Re})_m = (N_{Re})_t \qquad 3.212$$

This approach works for problems involving fans, pumps, turbines, drainage through holes in tanks, closed-pipe flow with no free surfaces (in the turbulent region with the same relative roughness), and for completely submerged objects such as torpedoes, airfoils, and submarines. It is assumed that the drag coefficients are the same for the model and for the full-size object.

[7] Complete mechanical similarity also requires kinematic and thermal similarity, which are not discussed in this book.

*Example 3.29*

A 1/30th size model is tested in a wind tunnel at 120 mph. The wind tunnel conditions are 50 psia and 100°F. What would be the equivalent speed of a prototype traveling at 14.0 psia, 40°F still air?

Start by setting the Reynolds number of the model and its prototype.

$$\frac{v_m L_m}{\nu_m} = \frac{v_p L_p}{\nu_p}$$

$$v_p = v_m \left(\frac{L_m}{L_p}\right)\left(\frac{\nu_p}{\nu_m}\right) = (120)\left(\frac{1}{30}\right)\left(\frac{\nu_p}{\nu_m}\right)$$

Air viscosity terms must be evaluated at the respective temperatures and pressures. As tables of viscosities are not readily available, the viscosities must be calculated.

Absolute viscosity essentially is independent of pressure. In Appendix B, the absolute viscosity of air is

$$\mu_p \,@\, 40° = 3.62 \text{ EE}-7$$

$$\mu_m \,@\, 100°F = 3.96 \text{ EE}-7$$

The density of air at the two conditions is

$$\rho_p = \frac{(14.0)(144)}{(53.3)(460+40)} = 0.0756 \text{ lbm/ft}^3$$

$$\rho_m = \frac{(50)(144)}{(53.3)(460+100)} = 0.2412$$

The kinematic viscosity can be calculated from the absolute viscosity. (The $g_c$ terms are omitted as they ultimately cancel out.)

$$\nu = \frac{\mu g_c}{\rho}$$

$$\nu_p = \frac{3.62 \text{ EE}-7}{0.0756} = 4.79 \text{ EE}-6$$

$$\nu_m = \frac{3.96 \text{ EE}-7}{0.2412} = 1.64 \text{ EE}-6$$

Then, the prototype velocity is

$$v_p = (120)\left(\frac{1}{30}\right)\left(\frac{4.79}{1.64}\right) = 11.7 \text{ mph}$$

B. INERTIAL AND GRAVITATIONAL FORCES DOMINATE

Elasticity and surface tension can be neglected in the analysis of large surface vessels. This leaves pressure, inertia, viscosity, and gravity. Pressure, again, is omitted as being dependent.

There are only two possible combinations of the remaining three forces. The ratio of inertial and viscous forces is recognized again as the Reynolds number. The ratio of the inertial forces to the gravitational forces is known as the *Froude number.*

$$N_{Fr} = \frac{v^2}{Lg} \tag{3.213}$$

Similarity is ensured when equations 3.214 and 3.215 are satisfied.

$$(N_{Re})_m = (N_{Re})_t \tag{3.214}$$

$$(N_{Fr})_m = (N_{Fr})_t \tag{3.215}$$

As an alternative, equations 3.213 and 3.62 can be solved simultaneously. This results in the following requirement for complete similarity.

$$\frac{\nu_m}{\nu_t} = \left(\frac{L_m}{L_t}\right)^{3/2} = (L_r)^{3/2} \tag{3.216}$$

Sometimes it is not possible to satisfy equation 3.215 or 3.216. This occurs when a model viscosity that is not available is called for. If only equation 3.214 is satisfied, the model is said to be *partially similar.*

This analysis is valid for surface ships, seaplane hulls, and open channels with varying surface levels such as weirs and spillways.

C. SURFACE TENSION DOMINATES

Problems involving waves, droplets, bubbles, and air entrainment can be solved by setting the *Weber numbers* equal.

$$N_W = \frac{v^2 L\rho}{T} \tag{3.217}$$

$$(N_W)_m = (N_W)_t \tag{3.218}$$

## 12 EFFECTS OF NON-STANDARD GRAVITY

Most of the equations in part 4 are based on Bernoulli's equation. This equation can be modified to allow for non-standard gravities. Assuming an incompressible fluid, Bernoulli's equation becomes

$$\frac{p_1}{\rho} + \frac{v_1^2}{2g_c} + \frac{gz_1}{g_c} = \frac{p_2}{\rho} + \frac{v_2^2}{2g_c} + \frac{gz_2}{g_c} \tag{3.219}$$

## 13 CHOICE OF PIPING MATERIALS

Steel and copper are commonly used in pressure piping systems. Each material is available in several configurations. For example, steel can be uncoated or galvanized. Copper tubing can be hard or soft. Table 3.13 can be used to select an appropriate pipe material.

Steel pipe is specified by its nominal size and schedule. In the past, steel pipe was designated as *standard* (S), *extra-strong* (X), and *double extra-strong* (XX). However, these designations have been replaced by a numerical rating. For example, schedule 40 now corresponds to a standard wall steel pipe in most cases.

The approximate schedule required can be found from equation 3.220.

$$\text{schedule} \approx \frac{(1000)(p)}{SE} \tag{3.220}$$

$p$ is the operating pressure in psig; $S$ is the allowable material stress in psi; $E$ is the joint efficiency. A value of 6500 psi can be used for the product $SE$ with low carbon steel in butt-welded lines and temperatures less than 650°F.

When copper is used as a pipe material, there is a potential for confusion, as there are two different sets of dimensions for copper pipe. Copper pipe in the K, L, and M categories is available in both annealed rolls ("tubing") and hardened straight lengths. Dimensions for such copper pipe, commonly referred to as *copper water tubing*, are given in Appendix G.

Type DWV copper drainage tube also is available. It is recommended for sanitary drainage installations above ground. The tube walls are thinner than type M, making it lighter and less expensive. It is strictly for non-pressure applications.

Copper and brass can also be formed into pipe with the dimensions given in Appendix H. Since the term "copper pipe" is ambiguous, the application must be used to determine the correct dimensions. Type L tubing in straight lengths is used principally in domestic and commercial plumbing because of its cost and the availability of soldered fittings. However, brass piping may be used with high-temperature water and corrosive fluids.

## 14 FLOW OF COMPRESSIBLE FLUIDS AND STEAM

Under certain conditions, Compressible fluids can be handled as incompressible flow. Specifically, a compressible gas, such as air or steam, can be treated as

f the pressure drop along the pipe run

lrop, based on the entrance pressure, is he fluid properties can be evaluated at any known point along the pipe run. If the pressure drop is greater than 10% but less than 40%, use of the mid-point properties will yield reasonably close friction loss calculations. If the pressure drop is greater than 40%, exact compressible gas dynamics equations should be used.

Since the pressure drop is being used to determine if the pressure drop is excessive, several iterations may be necessary to determine the pressures by trial and error.

**Table 3.13**
Recommended Pipe Materials for Various Services

| SERVICE | | PIPE |
|---|---|---|
| REFRIGERANTS 12, 22, 500 and 502 | Suction Line | Hard copper tubing, Type L*<br>Steel pipe, standard wall Lap welded or seamless |
| | Liquid Line | Hard copper tubing, Type L*<br>Steel pipe, standard wall Lap welded or seamless |
| | Hot Gas Line | Hard copper tubing, Type L*<br>Steel pipe, standard wall Welded or seamless |
| CHILLED WATER | | Plain or Galvanized steel pipe†<br>Hard copper tubing† |
| CONDENSER OR MAKE-UP WATER | | Galvanized steel pipe†<br>Hard copper tubing† |
| DRAIN OR CONDENSATE LINES | | Galvanized steel pipe†<br>Hard copper tubing† |
| STEAM OR CONDENSATE | | Steel pipe†<br>Hard copper tubing† |
| HOT WATER | | Steel pipe<br>Hard copper tubing† |

* Except for sizes 1/4″ and 3/8″ OD where wall thicknesses of 0.30 and 0.32 in. are required. Soft copper refrigeration tubing may be used for sizes 1 3/8″ OD and smaller. Mechanical joints must not be used with soft copper tubing in sizes larger than 7/8″ OD.

† Normally, standard wall steel pipe or Type M hard copper tubing is satisfactory for air conditioning applications. However, the piping material selected should be checked for the design temperature-pressure ratings.

# Appendix A: Properties of Water at Atmospheric Pressure

| temp. °F | density $lbm/ft^3$ | absolute viscosity $lbf\text{-}sec/ft^2$ | kinematic viscosity $ft^2/sec$ | surface tension lbf/ft | vapor pressure head ft | bulk modulus $lbf/in^2$ |
|---|---|---|---|---|---|---|
| 32 | 62.42 | 3.746 EE−5 | 1.931 EE−5 | 0.518 EE−2 | 0.20 | 293 EE3 |
| 40 | 62.43 | 3.229 EE−5 | 1.664 EE−5 | 0.514 EE−2 | 0.28 | 294 EE3 |
| 50 | 62.41 | 2.735 EE−5 | 1.410 EE−5 | 0.509 EE−2 | 0.41 | 305 EE3 |
| 60 | 62.37 | 2.359 EE−5 | 1.217 EE−5 | 0.504 EE−2 | 0.59 | 311 EE3 |
| 70 | 62.30 | 2.050 EE−5 | 1.059 EE−5 | 0.500 EE−2 | 0.84 | 320 EE3 |
| 80 | 62.22 | 1.799 EE−5 | 0.930 EE−5 | 0.492 EE−2 | 1.17 | 322 EE3 |
| 90 | 62.11 | 1.595 EE−5 | 0.826 EE−5 | 0.486 EE−2 | 1.61 | 323 EE3 |
| 100 | 62.00 | 1.424 EE−5 | 0.739 EE−5 | 0.480 EE−2 | 2.19 | 327 EE3 |
| 110 | 61.86 | 1.284 EE−5 | 0.667 EE−5 | 0.473 EE−2 | 2.95 | 331 EE3 |
| 120 | 61.71 | 1.168 EE−5 | 0.609 EE−5 | 0.465 EE−2 | 3.91 | 333 EE3 |
| 130 | 61.55 | 1.069 EE−5 | 0.558 EE−5 | 0.460 EE−2 | 5.13 | 334 EE3 |
| 140 | 61.38 | 0.981 EE−5 | 0.514 EE−5 | 0.454 EE−2 | 6.67 | 330 EE3 |
| 150 | 61.20 | 0.905 EE−5 | 0.476 EE−5 | 0.447 EE−2 | 8.58 | 328 EE3 |
| 160 | 61.00 | 0.838 EE−5 | 0.442 EE−5 | 0.441 EE−2 | 10.95 | 326 EE3 |
| 170 | 60.80 | 0.780 EE−5 | 0.413 EE−5 | 0.433 EE−2 | 13.83 | 322 EE3 |
| 180 | 60.58 | 0.726 EE−5 | 0.385 EE−5 | 0.426 EE−2 | 17.33 | 313 EE3 |
| 190 | 60.36 | 0.678 EE−5 | 0.362 EE−5 | 0.419 EE−2 | 21.55 | 313 EE3 |
| 200 | 60.12 | 0.637 EE−5 | 0.341 EE−5 | 0.412 EE−2 | 26.59 | 308 EE3 |
| 212 | 59.83 | 0.593 EE−5 | 0.319 EE−5 | 0.404 EE−2 | 33.90 | 300 EE3 |

# Appendix B: Properties of Air at Atmospheric Pressure

| temp. °F | density $lbm/ft^3$ | kinematic viscosity $ft^2/sec$ | absolute viscosity $lbf\text{-}sec/ft^2$ |
|---|---|---|---|
| 0 | 0.0862 | 12.6 EE−5 | 3.28 EE−7 |
| 20 | 0.0827 | 13.6 EE−5 | 3.50 EE−7 |
| 40 | 0.0794 | 14.6 EE−5 | 3.62 EE−7 |
| 60 | 0.0763 | 15.8 EE−5 | 3.74 EE−7 |
| 68 | 0.0752 | 16.0 EE−5 | 3.75 EE−7 |
| 80 | 0.0735 | 16.9 EE−5 | 3.85 EE−7 |
| 100 | 0.0709 | 18.0 EE−5 | 3.96 EE−7 |
| 120 | 0.0684 | 18.9 EE−5 | 4.07 EE−7 |
| 250 | 0.0559 | 27.3 EE−5 | 4.74 EE−7 |

# Appendix C: Viscosity of Water

| | absolute viscosity | kinematic viscosity | | |
|---|---|---|---|---|
| temperature (°F) | centipoise | centistokes | SSU | $ft^2$/sec |
| 32 | 1.79 | 1.79 | 33.0 | 0.00001931 |
| 50 | 1.31 | 1.31 | 31.6 | 0.00001410 |
| 60 | 1.12 | 1.12 | 31.2 | 0.00001217 |
| 70 | 0.98 | 0.98 | 30.9 | 0.00001059 |
| 80 | 0.86 | 0.86 | 30.6 | 0.00000930 |
| 85 | 0.81 | 0.81 | 30.4 | 0.00000869 |
| 100 | 0.68 | 0.69 | 30.2 | 0.00000739 |
| 120 | 0.56 | 0.57 | 30.0 | 0.00000609 |
| 140 | 0.47 | 0.48 | 29.7 | 0.00000514 |
| 160 | 0.40 | 0.41 | 29.6 | 0.00000442 |
| 180 | 0.35 | 0.36 | 29.5 | 0.00000385 |
| 212 | 0.28 | 0.29 | 29.3 | 0.00000319 |

# Appendix D: Important Fluid Conversions

| multiply | by | to obtain |
|---|---|---|
| cubic feet | 7.4805 | gallons |
| cfs | 448.83 | gpm |
| cfs | 0.64632 | MGD |
| gallons | 0.1337 | cubic feet |
| gpm | 0.002228 | cfs |
| inches of mercury | 0.491 | psi |
| inches of mercury | 70.7 | psf |
| inches of mercury | 13.60 | inches of water |
| inches of water | 5.199 | psf |
| inches of water | 0.0361 | psi |
| inches of water | 0.0735 | inches of mercury |
| psi | 144 | psf |
| psi | 2.308 | feet of water |
| psi | 27.7 | inches of water |
| psi | 2.037 | inches of mercury |
| psf | 0.006944 | psi |

# Appendix E: Area, Wetted Perimeter and Hydraulic Radius of Partially Filled Circular Pipes

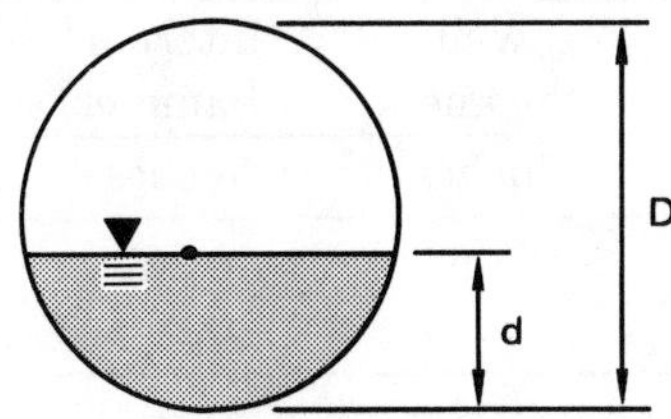


| $\frac{d}{D}$ | $\frac{area}{D^2}$ | $\frac{wet.\ per.}{D}$ | $\frac{hyd.\ rad.}{D}$ | $\frac{d}{D}$ | $\frac{area}{D^2}$ | $\frac{wet.\ per.}{D}$ | $\frac{hyd.\ rad.}{D}$ |
|---|---|---|---|---|---|---|---|
| 0.01 | 0.0013 | 0.2003 | 0.0066 | 0.51 | 0.4027 | 1.5908 | 0.2531 |
| 0.02 | 0.0037 | 0.2838 | 0.0132 | 0.52 | 0.4127 | 1.6108 | 0.2561 |
| 0.03 | 0.0069 | 0.3482 | 0.0197 | 0.53 | 0.4227 | 1.6308 | 0.2591 |
| 0.04 | 0.0105 | 0.4027 | 0.0262 | 0.54 | 0.4327 | 1.6509 | 0.2620 |
| 0.05 | 0.0147 | 0.4510 | 0.0326 | 0.55 | 0.4426 | 1.6710 | 0.2649 |
| 0.06 | 0.0192 | 0.4949 | 0.0389 | 0.56 | 0.4526 | 1.6911 | 0.2676 |
| 0.07 | 0.0242 | 0.5355 | 0.0451 | 0.57 | 0.4625 | 1.7113 | 0.2703 |
| 0.08 | 0.0294 | 0.5735 | 0.0513 | 0.58 | 0.4723 | 1.7315 | 0.2728 |
| 0.09 | 0.0350 | 0.6094 | 0.0574 | 0.59 | 0.4822 | 1.7518 | 0.2753 |
| 0.10 | 0.0409 | 0.6435 | 0.0635 | 0.60 | 0.4920 | 1.7722 | 0.2776 |
| 0.11 | 0.0470 | 0.6761 | 0.0695 | 0.61 | 0.5018 | 1.7926 | 0.2797 |
| 0.12 | 0.0534 | 0.7075 | 0.0754 | 0.62 | 0.5115 | 1.8132 | 0.2818 |
| 0.13 | 0.0600 | 0.7377 | 0.0813 | 0.63 | 0.5212 | 1.8338 | 0.2839 |
| 0.14 | 0.0688 | 0.7670 | 0.0871 | 0.64 | 0.5308 | 1.8546 | 0.2860 |
| 0.15 | 0.0739 | 0.7954 | 0.0929 | 0.65 | 0.5404 | 1.8755 | 0.2881 |
| 0.16 | 0.0811 | 0.8230 | 0.0986 | 0.66 | 0.5499 | 1.8965 | 0.2899 |
| 0.17 | 0.0885 | 0.8500 | 0.1042 | 0.67 | 0.5594 | 1.9177 | 0.2917 |
| 0.18 | 0.0961 | 0.8763 | 0.1097 | 0.68 | 0.5687 | 1.9391 | 0.2935 |
| 0.19 | 0.1039 | 0.9020 | 0.1152 | 0.69 | 0.5780 | 1.9606 | 0.2950 |
| 0.20 | 0.1118 | 0.9273 | 0.1206 | 0.70 | 0.5872 | 1.9823 | 0.2962 |
| 0.21 | 0.1199 | 0.9521 | 0.1259 | 0.71 | 0.5964 | 2.0042 | 0.2973 |
| 0.22 | 0.1281 | 0.9764 | 0.1312 | 0.72 | 0.6054 | 2.0264 | 0.2984 |
| 0.23 | 0.1365 | 1.0003 | 0.1364 | 0.73 | 0.6143 | 2.0488 | 0.2995 |
| 0.24 | 0.1449 | 1.0239 | 0.1416 | 0.74 | 0.6231 | 2.0714 | 0.3006 |
| 0.25 | 0.1535 | 1.0472 | 0.1466 | 0.75 | 0.6318 | 2.0944 | 0.3017 |
| 0.26 | 0.1623 | 1.0701 | 0.1516 | 0.76 | 0.6404 | 2.1176 | 0.3025 |
| 0.27 | 0.1711 | 1.0928 | 0.1566 | 0.77 | 0.6489 | 2.1412 | 0.3032 |
| 0.28 | 0.1800 | 1.1152 | 0.1614 | 0.78 | 0.6573 | 2.1652 | 0.3037 |
| 0.29 | 0.1890 | 1.1373 | 0.1662 | 0.79 | 0.6655 | 2.1895 | 0.3040 |
| 0.30 | 0.1982 | 1.1593 | 0.1709 | 0.80 | 0.6736 | 2.2143 | 0.3042 |
| 0.31 | 0.2074 | 1.1810 | 0.1755 | 0.81 | 0.6815 | 2.2395 | 0.3044 |
| 0.32 | 0.2167 | 1.2025 | 0.1801 | 0.82 | 0.6893 | 2.2653 | 0.3043 |
| 0.33 | 0.2260 | 1.2239 | 0.1848 | 0.83 | 0.6969 | 2.2916 | 0.3041 |
| 0.34 | 0.2355 | 1.2451 | 0.1891 | 0.84 | 0.7043 | 2.3186 | 0.3038 |
| 0.35 | 0.2450 | 1.2661 | 0.1935 | 0.85 | 0.7115 | 2.3462 | 0.3033 |
| 0.36 | 0.2546 | 1.2870 | 0.1978 | 0.86 | 0.7186 | 2.3746 | 0.3026 |
| 0.37 | 0.2642 | 1.3078 | 0.2020 | 0.87 | 0.7254 | 2.4038 | 0.3017 |
| 0.38 | 0.2739 | 1.3284 | 0.2061 | 0.88 | 0.7320 | 2.4341 | 0.3008 |
| 0.39 | 0.2836 | 1.3490 | 0.2102 | 0.89 | 0.7384 | 2.4655 | 0.2996 |
| 0.40 | 0.2934 | 1.3694 | 0.2142 | 0.90 | 0.7445 | 2.4981 | 0.2980 |
| 0.41 | 0.3032 | 1.3898 | 0.2181 | 0.91 | 0.7504 | 2.5322 | 0.2963 |
| 0.42 | 0.3130 | 1.4101 | 0.2220 | 0.92 | 0.7560 | 2.5681 | 0.2944 |
| 0.43 | 0.3229 | 1.4303 | 0.2257 | 0.93 | 0.7612 | 2.6061 | 0.2922 |
| 0.44 | 0.3328 | 1.4505 | 0.2294 | 0.94 | 0.7662 | 2.6467 | 0.2896 |
| 0.45 | 0.3428 | 1.4706 | 0.2331 | 0.95 | 0.7707 | 2.6906 | 0.2864 |
| 0.46 | 0.3527 | 1.4907 | 0.2366 | 0.96 | 0.7749 | 2.7389 | 0.2830 |
| 0.47 | 0.3627 | 1.5108 | 0.2400 | 0.97 | 0.7785 | 2.7934 | 0.2787 |
| 0.48 | 0.3727 | 1.5308 | 0.2434 | 0.98 | 0.7816 | 2.8578 | 0.2735 |
| 0.49 | 0.3827 | 1.5508 | 0.2467 | 0.99 | 0.7841 | 2.9412 | 0.2665 |
| 0.50 | 0.3927 | 1.5708 | 0.2500 | 1.00 | 0.7854 | 3.1416 | 0.2500 |

# Appendix F: Dimensions of Welded and Seamless Steel Pipe

| nominal diameter inches | schedule | outside diameter inches | wall thickness inches | internal diameter inches | internal area sq inches | internal diameter feet | internal area sq feet |
|---|---|---|---|---|---|---|---|
| $\frac{1}{8}$ | 40 (S) | 0.405 | 0.068 | 0.269 | 0.0568 | 0.0224 | 0.00039 |
| | 80 (X) | | 0.095 | 0.215 | 0.0363 | 0.0179 | 0.00025 |
| $\frac{1}{4}$ | 40 (S) | 0.540 | 0.088 | 0.364 | 0.1041 | 0.0303 | 0.00072 |
| | 80 (X) | | 0.119 | 0.302 | 0.0716 | 0.0252 | 0.00050 |
| $\frac{3}{8}$ | 40 (S) | 0.675 | 0.091 | 0.493 | 0.1909 | 0.0411 | 0.00133 |
| | 80 (X) | | 0.126 | 0.423 | 0.1405 | 0.0353 | 0.00098 |
| | 40 (S) | 0.840 | 0.109 | 0.622 | 0.3039 | 0.0518 | 0.00211 |
| $\frac{1}{2}$ | 80 (X) | | 0.147 | 0.546 | 0.2341 | 0.0455 | 0.00163 |
| | 160 | | 0.187 | 0.466 | 0.1706 | 0.0388 | 0.00118 |
| | (XX) | | 0.294 | 0.252 | 0.499 | 0.0210 | 0.00035 |
| | 40 (S) | 1.050 | 0.113 | 0.824 | 0.5333 | 0.0687 | 0.00370 |
| $\frac{3}{4}$ | 80 (X) | | 0.154 | 0.742 | 0.4324 | 0.0618 | 0.00300 |
| | 160 | | 0.219 | 0.612 | 0.2942 | 0.0510 | 0.00204 |
| | (XX) | | 0.308 | 0.434 | 0.1479 | 0.0362 | 0.00103 |
| | 40 (S) | 1.315 | 0.133 | 1.049 | 0.8643 | 0.0874 | 0.00600 |
| 1 | 80 (X) | | 0.179 | 0.957 | 0.7193 | 0.0798 | 0.00500 |
| | 160 | | 0.250 | 0.815 | 0.5217 | 0.0679 | 0.00362 |
| | (XX) | | 0.358 | 0.599 | 0.2818 | 0.0499 | 0.00196 |
| | 40 (S) | 1.660 | 0.140 | 1.380 | 1.496 | 0.1150 | 0.01039 |
| $1\frac{1}{4}$ | 80 (X) | | 0.191 | 1.278 | 1.283 | 0.1065 | 0.00890 |
| | 160 | | 0.250 | 1.160 | 1.057 | 0.0967 | 0.00734 |
| | (XX) | | 0.382 | 0.896 | 0.6305 | 0.0747 | 0.00438 |
| | 40 (S) | 1.900 | 0.145 | 1.610 | 2.036 | 0.1342 | 0.01414 |
| $1\frac{1}{2}$ | 80 (X) | | 0.200 | 1.500 | 1.767 | 0.1250 | 0.01227 |
| | 160 | | 0.281 | 1.338 | 1.406 | 0.1115 | 0.00976 |
| | (XX) | | 0.400 | 1.100 | 0.9503 | 0.0917 | 0.00660 |
| | 40 (S) | 2.375 | 0.154 | 2.067 | 3.356 | 0.1723 | 0.02330 |
| 2 | 80 (X) | | 0.218 | 1.939 | 2.953 | 0.1616 | 0.02051 |
| | 160 | | 0.344 | 1.687 | 2.235 | 0.1406 | 0.01552 |
| | (XX) | | 0.436 | 1.503 | 1.774 | 0.1253 | 0.01232 |
| | 40 (S) | 2.875 | 0.203 | 2.469 | 4.788 | 0.2058 | 0.03325 |
| $2\frac{1}{2}$ | 80 (X) | | 0.276 | 2.323 | 4.238 | 0.1936 | 0.02943 |
| | 160 | | 0.375 | 2.125 | 3.547 | 0.1771 | 0.02463 |
| | (XX) | | 0.552 | 1.771 | 2.464 | 0.1476 | 0.01711 |
| | 40 (S) | 3.500 | 0.216 | 3.068 | 7.393 | 0.2557 | 0.05134 |
| 3 | 80 (X) | | 0.300 | 2.900 | 6.605 | 0.2417 | 0.04587 |
| | 160 | | 0.438 | 2.624 | 5.408 | 0.2187 | 0.03755 |
| | (XX) | | 0.600 | 2.300 | 4.155 | 0.1917 | 0.02885 |
| $3\frac{1}{2}$ | 40 (S) | 4.000 | 0.226 | 3.548 | 9.887 | 0.2957 | 0.06866 |
| | 80 (X) | | 0.318 | 3.364 | 8.888 | 0.2803 | 0.06172 |

| nominal diameter | | outside diameter | wall thickness | internal diameter | internal area | internal diameter | internal area |
|---|---|---|---|---|---|---|---|
| inches | schedule | inches | inches | inches | sq inches | feet | sq feet |
| | 40 (S) | 4.500 | 0.237 | 4.026 | 12.73 | 0.3355 | 0.08841 |
| 4 | 80 (X) | | 0.337 | 3.826 | 11.50 | 0.3188 | 0.07984 |
| | 120 | | 0.438 | 3.624 | 10.32 | 0.3020 | 0.07163 |
| | 160 | | 0.531 | 3.438 | 9.283 | 0.2865 | 0.06447 |
| | (XX) | | 0.674 | 3.152 | 7.803 | 0.2627 | 0.05419 |
| 5 | 40 (S) | 5.563 | 0.258 | 5.047 | 20.01 | 0.4206 | 0.1389 |
| | 80 (X) | | 0.375 | 4.813 | 18.19 | 0.4011 | 0.1263 |
| | 120 | | 0.500 | 4.563 | 16.35 | 0.3803 | 0.1136 |
| | 160 | | 0.625 | 4.313 | 14.61 | 0.3594 | 0.1015 |
| | (XX) | | 0.750 | 4.063 | 12.97 | 0.3386 | 0.09004 |
| 6 | 40 (S) | 6.625 | 0.280 | 6.065 | 28.89 | 0.5054 | 0.2006 |
| | 80 (X) | | 0.432 | 5.761 | 26.07 | 0.4801 | 0.1810 |
| | 120 | | 0.562 | 5.501 | 23.77 | 0.4584 | 0.1650 |
| | 160 | | 0.719 | 5.187 | 21.13 | 0.4323 | 0.1467 |
| | (XX) | | 0.864 | 4.897 | 18.83 | 0.4081 | 0.1308 |
| 8 | 20 | 8.625 | 0.250 | 8.125 | 51.85 | 0.6771 | 0.3601 |
| | 30 | | 0.277 | 8.071 | 51.16 | 0.6726 | 0.3553 |
| | 40 (S) | | 0.322 | 7.981 | 50.03 | 0.6651 | 0.3474 |
| | 60 | | 0.406 | 7.813 | 47.94 | 0.6511 | 0.3329 |
| | 80 (X) | | 0.500 | 7.625 | 45.66 | 0.6354 | 0.3171 |
| | 100 | | 0.594 | 7.437 | 43.44 | 0.6198 | 0.3017 |
| | 120 | | 0.719 | 7.187 | 40.57 | 0.5989 | 0.2817 |
| | 140 | | 0.812 | 7.001 | 38.50 | 0.5834 | 0.2673 |
| | (XX) | | 0.875 | 6.875 | 37.12 | 0.5729 | 0.2578 |
| | 160 | | 0.906 | 6.813 | 36.46 | 0.5678 | 0.2532 |
| 10 | 20 | 10.75 | 0.250 | 10.250 | 82.52 | 0.85417 | 0.5730 |
| | 30 | | 0.307 | 10.136 | 80.69 | 0.84467 | 0.5604 |
| | 40 (S) | | 0.365 | 10.020 | 78.85 | 0.83500 | 0.5476 |
| | 60 (X) | | 0.500 | 9.750 | 74.66 | 0.8125 | 0.5185 |
| | 80 | | 0.594 | 9.562 | 71.81 | 0.7968 | 0.4987 |
| | 100 | | 0.719 | 9.312 | 68.11 | 0.7760 | 0.4730 |
| | 120 | | 0.844 | 9.062 | 64.50 | 0.7552 | 0.4479 |
| | 140 (XX) | | 1.000 | 8.750 | 60.13 | 0.7292 | 0.4176 |
| | 160 | | 1.125 | 8.500 | 56.75 | 0.7083 | 0.3941 |
| 12 | 20 | 12.75 | 0.250 | 12.250 | 117.86 | 1.0208 | 0.8185 |
| | 30 | | 0.330 | 12.090 | 114.80 | 1.0075 | 0.7972 |
| | (S) | | 0.375 | 12.000 | 113.10 | 1.0000 | 0.7854 |
| | 40 | | 0.406 | 11.938 | 111.93 | 0.99483 | 0.7773 |
| | (X) | | 0.500 | 11.750 | 108.43 | 0.97917 | 0.7530 |
| | 60 | | 0.562 | 11.626 | 106.16 | 0.96883 | 0.7372 |
| | 80 | | 0.688 | 11.374 | 101.61 | 0.94783 | 0.7056 |
| | 100 | | 0.844 | 11.062 | 96.11 | 0.92183 | 0.6674 |
| | 120 (XX) | | 1.000 | 10.750 | 90.76 | 0.89583 | 0.6303 |
| | 140 | | 1.125 | 10.500 | 86.59 | 0.87500 | 0.6013 |
| | 160 | | 1.312 | 10.126 | 80.53 | 0.84383 | 0.5592 |

| nominal diameter inches | schedule | outside diameter inches | wall thickness inches | internal diameter inches | internal area sq inches | internal diameter feet | internal area sq feet |
|---|---|---|---|---|---|---|---|
| 14 OD | 10 | 14.00 | 0.250 | 13.500 | 143.14 | 1.1250 | 0.9940 |
| | 20 | | 0.312 | 13.376 | 140.52 | 1.1147 | 0.9758 |
| | 30 (S) | | 0.375 | 13.250 | 137.89 | 1.1042 | 0.9575 |
| | 40 | | 0.438 | 13.124 | 135.28 | 1.0937 | 0.9394 |
| | (X) | | 0.500 | 13.000 | 132.67 | 1.0833 | 0.9213 |
| | 60 | | 0.594 | 12.812 | 128.92 | 1.0677 | 0.8953 |
| | 80 | | 0.750 | 12.500 | 122.72 | 1.0417 | 0.8522 |
| | 100 | | 0.938 | 12.124 | 115.45 | 1.0104 | 0.8017 |
| | 120 | | 1.094 | 11.812 | 109.58 | 0.98433 | 0.7610 |
| | 140 | | 1.250 | 11.500 | 103.87 | 0.95833 | 0.7213 |
| | 160 | | 1.406 | 11.188 | 98.31 | 0.93233 | 0.6827 |
| 16 OD | 10 | 16.00 | 0.250 | 15.500 | 188.69 | 1.2917 | 1.3104 |
| | 20 | | 0.312 | 15.376 | 185.69 | 1.2813 | 1.2895 |
| | 30 (S) | | 0.375 | 15.250 | 182.65 | 1.2708 | 1.2684 |
| | 40 (X) | | 0.500 | 15.000 | 176.72 | 1.2500 | 1.2272 |
| | 60 | | 0.656 | 14.688 | 169.44 | 1.2240 | 1.1767 |
| | 80 | | 0.844 | 14.312 | 160.88 | 1.1927 | 1.1172 |
| | 100 | | 1.031 | 13.938 | 152.58 | 1.1615 | 1.0596 |
| | 120 | | 1.219 | 13.562 | 144.46 | 1.1302 | 1.0032 |
| | 140 | | 1.438 | 13.124 | 135.28 | 1.0937 | 0.9394 |
| | 160 | | 1.594 | 12.812 | 128.92 | 1.0677 | 0.8953 |
| 18 OD | 10 | 18.00 | 0.250 | 17.500 | 240.53 | 1.4583 | 1.6703 |
| | 20 | | 0.312 | 17.376 | 237.13 | 1.4480 | 1.6467 |
| | (S) | | 0.375 | 17.250 | 233.71 | 1.4375 | 1.6230 |
| | 30 | | 0.438 | 17.124 | 230.00 | 1.4270 | 1.5993 |
| | (X) | | 0.500 | 17.000 | 226.98 | 1.4167 | 1.5762 |
| | 40 | | 0.562 | 16.876 | 223.68 | 1.4063 | 1.5533 |
| | 60 | | 0.750 | 16.500 | 213.83 | 1.3750 | 1.4849 |
| | 80 | | 0.938 | 16.124 | 204.19 | 1.3437 | 1.4180 |
| | 100 | | 1.156 | 15.688 | 193.30 | 1.3073 | 1.3423 |
| | 120 | | 1.375 | 15.250 | 182.65 | 1.2708 | 1.2684 |
| | 140 | | 1.562 | 14.876 | 173.81 | 1.2397 | 1.2070 |
| | 160 | | 1.781 | 14.438 | 163.72 | 1.2032 | 1.1370 |
| 20 OD | 10 | 20.00 | 0.250 | 19.500 | 298.65 | 1.6250 | 2.0739 |
| | 20 (S) | | 0.375 | 19.250 | 291.04 | 1.6042 | 2.0211 |
| | 30 (X) | | 0.500 | 19.000 | 283.53 | 1.5833 | 1.9689 |
| | 40 | | 0.594 | 18.812 | 277.95 | 1.5677 | 1.9302 |
| | 60 | | 0.812 | 18.376 | 265.21 | 1.5313 | 1.8417 |
| | 80 | | 1.031 | 17.938 | 252.72 | 1.4948 | 1.7550 |
| | 100 | | 1.281 | 17.438 | 238.83 | 1.4532 | 1.6585 |
| | 120 | | 1.500 | 17.000 | 226.98 | 1.4167 | 1.5762 |
| | 140 | | 1.750 | 16.500 | 213.83 | 1.3750 | 1.4849 |
| | 160 | | 1.969 | 16.062 | 202.62 | 1.3385 | 1.4071 |

| nominal diameter inches | schedule | outside diameter inches | wall thickness inches | internal diameter inches | internal area sq inches | internal diameter feet | internal area sq feet |
|---|---|---|---|---|---|---|---|
| 24 OD | 10 | 24.00 | 0.250 | 23.500 | 433.74 | 1.9583 | 3.0121 |
| | 20 (S) | | 0.375 | 23.250 | 424.56 | 1.9375 | 2.9483 |
| | (X) | | 0.500 | 23.000 | 415.48 | 1.9167 | 2.8852 |
| | 30 | | 0.562 | 22.876 | 411.01 | 1.9063 | 2.8542 |
| | 40 | | 0.688 | 22.624 | 402.00 | 1.8853 | 2.7917 |
| | 60 | | 0.969 | 22.062 | 382.28 | 1.8385 | 2.6547 |
| | 80 | | 1.219 | 21.562 | 365.15 | 1.7802 | 2.5358 |
| | 100 | | 1.531 | 20.938 | 344.32 | 1.7448 | 2.3911 |
| | 120 | | 1.812 | 20.376 | 326.92 | 1.6980 | 2.2645 |
| | 140 | | 2.062 | 19.876 | 310.28 | 1.6563 | 2.1547 |
| | 160 | | 2.344 | 19.312 | 292.92 | 1.6093 | 2.0342 |
| 30 OD | 10 | 30.00 | 0.312 | 29.376 | 677.76 | 2.4480 | 4.7067 |
| | (S) | | 0.375 | 29.250 | 671.62 | 2.4375 | 4.6640 |
| | 20 (X) | | 0.500 | 29.000 | 660.52 | 2.4167 | 4.5869 |
| | 30 | | 0.625 | 28.750 | 649.18 | 2.3958 | 4.5082 |

S = Wall thickness, formerly designated "standard weight"

X = Wall thickness, formerly designated "extra strong"

XX = Wall thickness, formerly designated "double extra strong"

Actual wall thickness may vary slightly.

Extracted from American Standard Wrought Steel and Wrought Iron Pipe (ASA B36, 10—1959), The American Society of Mechanical Engineers.

# Appendix G: Dimensions of Copper Water Tubing

| CLASSIFICATION | NOM. TUBE SIZE (in.) | OUTSIDE DIAM (in.) | WALL THICK-NESS (in.) | INSIDE DIAM (in.) | TRANS-VERSE AREA (sq. in.) | SAFE WORKING PRESSURE (psi) |
|---|---|---|---|---|---|---|
| HARD | 1/4 | 3/8 | .025 | .325 | .083 | 1000 |
| | 3/8 | 1/2 | .025 | .450 | .159 | 1000 |
| | 1/2 | 5/8 | .028 | .569 | .254 | 890 |
| | 3/4 | 7/8 | .032 | .811 | .516 | 710 |
| | 1 | 1 1/8 | .035 | 1.055 | .874 | 600 |
| | 1 1/4 | 1 3/8 | .042 | 1.291 | 1.309 | 590 |
| Type | 1 1/2 | 1 5/8 | .049 | 1.527 | 1.831 | 580 |
| "M" | 2 | 2 1/8 | .058 | 2.009 | 3.17 | 520 |
| 250 psi | 2 1/2 | 2 5/8 | .065 | 2.495 | 4.89 | 470 |
| Working | 3 | 3 1/8 | .072 | 2.981 | 6.98 | 440 |
| Pressure | 3 1/2 | 3 5/8 | .083 | 3.459 | 9.40 | 430 |
| | 4 | 4 1/8 | .095 | 3.935 | 12.16 | 430 |
| | 5 | 5 1/8 | .109 | 4.907 | 18.91 | 400 |
| | 6 | 6 1/8 | .122 | 5.881 | 27.16 | 375 |
| | 8 | 8 1/8 | .170 | 7.785 | 47.6 | 375 |
| HARD | 3/8 | 1/2 | .035 | .430 | .146 | 1000 |
| | 1/2 | 5/8 | .040 | .545 | .233 | 1000 |
| | 3/4 | 7/8 | .045 | .785 | .484 | 1000 |
| | 1 | 1 1/8 | .050 | 1.025 | .825 | 880 |
| | 1 1/4 | 1 3/8 | .055 | 1.265 | 1.256 | 780 |
| Type | 1 1/2 | 1 5/8 | .060 | 1.505 | 1.78 | 720 |
| "L" | 2 | 2 1/8 | .070 | 1.985 | 3.094 | 640 |
| 250 psi | 2 1/2 | 2 5/8 | .080 | 2.465 | 4.77 | 580 |
| Working | 3 | 3 1/8 | .090 | 2.945 | 6.812 | 550 |
| Pressure | 3 1/2 | 3 5/8 | .100 | 3.425 | 9.213 | 530 |
| | 4 | 4 1/8 | .110 | 3.905 | 11.97 | 510 |
| | 5 | 5 1/8 | .125 | 4.875 | 18.67 | 460 |
| | 6 | 6 1/8 | .140 | 5.845 | 26.83 | 430 |
| HARD | 1/4 | 3/8 | .032 | .311 | .076 | 1000 |
| | 3/8 | 1/2 | .049 | .402 | .127 | 1000 |
| | 1/2 | 5/8 | .049 | .527 | .218 | 1000 |
| | 3/4 | 7/8 | .065 | .745 | .436 | 1000 |
| | 1 | 1 1/8 | .065 | .995 | .778 | 780 |
| Type | 1 1/4 | 1 3/8 | .065 | 1.245 | 1.217 | 630 |
| "K" | 1 1/2 | 1 5/8 | .072 | 1.481 | 1.722 | 580 |
| 400 psi | 2 | 2 1/8 | .083 | 1.959 | 3.014 | 510 |
| Working | 2 1/2 | 2 5/8 | .095 | 2.435 | 4.656 | 470 |
| Pressure | 3 | 3 1/8 | .109 | 2.907 | 6.637 | 450 |
| | 3 1/2 | 3 5/8 | .120 | 3.385 | 8.999 | 430 |
| | 4 | 4 1/8 | .134 | 3.857 | 11.68 | 420 |
| | 5 | 5 1/8 | .160 | 4.805 | 18.13 | 400 |
| | 6 | 6 1/8 | .192 | 5.741 | 25.88 | 400 |
| SOFT | 1/4 | 3/8 | .032 | .311 | .076 | 1000 |
| | 3/8 | 1/2 | .049 | .402 | .127 | 1000 |
| | 1/2 | 5/8 | .049 | .527 | .218 | 1000 |
| | 3/4 | 7/8 | .065 | .745 | .436 | 1000 |
| | 1 | 1 1/8 | .065 | .995 | .778 | 780 |
| Type | 1 1/4 | 1 3/8 | .065 | 1.245 | 1.217 | 630 |
| "K" | 1 1/2 | 1 5/8 | .072 | 1.481 | 1.722 | 580 |
| 250 psi | 2 | 2 1/8 | .083 | 1.959 | 3.014 | 510 |
| Working | 2 1/2 | 2 5/8 | .095 | 2.435 | 4.656 | 470 |
| Pressure | 3 | 3 1/8 | .109 | 2.907 | 6.637 | 450 |
| | 3 1/2 | 2 5/8 | .120 | 3.385 | 8.999 | 430 |
| | 4 | 4 1/8 | .134 | 3.857 | 11.68 | 420 |
| | 5 | 5 1/8 | .160 | 4.805 | 18.13 | 400 |
| | 6 | 6 1/8 | .192 | 5.741 | 25.88 | 400 |

# Appendix H: Dimensions of Brass and Copper Tubing

**regular**

| pipe size in. | nominal dimensions in. | | | cross sectional area of bore sq. in. | lb per ft | |
|---|---|---|---|---|---|---|
| | O.D. | I.D. | wall | | red brass | copper |
| 1/8 | .405 | .281 | .062 | .062 | .253 | .259 |
| 1/4 | .540 | .376 | .082 | .110 | .447 | .457 |
| 3/8 | .675 | .495 | .090 | .192 | .627 | .641 |
| 1/2 | .840 | .626 | .107 | .307 | .934 | .955 |
| 3/4 | 1.050 | .822 | .114 | .531 | 1.270 | 1.300 |
| 1 | 1.315 | 1.063 | .126 | .887 | 1.780 | 1.820 |
| 1 1/4 | 1.660 | 1.368 | .146 | 1.470 | 2.630 | 2.690 |
| 1 1/2 | 1.900 | 1.600 | .150 | 2.010 | 3.130 | 3.200 |
| 2 | 2.375 | 2.063 | .156 | 3.340 | 4.120 | 4.220 |
| 2 1/2 | 2.875 | 2.501 | .187 | 4.910 | 5.990 | 6.120 |
| 3 | 3.500 | 3.062 | .219 | 7.370 | 8.560 | 8.750 |
| 3 1/2 | 4.000 | 3.500 | .250 | 9.620 | 11.200 | 11.400 |
| 4 | 4.500 | 4.000 | .250 | 12.600 | 12.700 | 12.900 |
| 5 | 5.562 | 5.062 | .250 | 20.100 | 15.800 | 16.200 |
| 6 | 6.625 | 6.125 | .250 | 29.500 | 19.000 | 19.400 |
| 8 | 8.625 | 8.001 | .312 | 50.300 | 30.900 | 31.600 |
| 10 | 10.750 | 10.020 | .365 | 78.800 | 45.200 | 46.200 |
| 12 | 12.750 | 12.000 | .375 | 113.000 | 55.300 | 56.500 |

**extra strong**

| pipe size in. | nominal dimensions in. | | | cross sectional area of bore sq. in. | lb per ft | |
|---|---|---|---|---|---|---|
| | O.D. | I.D. | wall | | red brass | copper |
| 1/8 | .405 | .205 | .100 | .033 | .363 | .371 |
| 1/4 | .540 | .294 | .123 | .068 | .611 | .625 |
| 3/8 | .675 | .421 | .127 | .139 | .829 | .847 |
| 1/2 | .840 | .542 | .149 | .231 | 1.230 | 1.250 |
| 3/4 | 1.050 | .736 | .157 | .425 | 1.670 | 1.710 |
| 1 | 1.315 | .951 | .182 | .710 | 2.460 | 2.510 |
| 1 1/4 | 1.660 | 1.272 | .194 | 1.270 | 3.390 | 3.460 |
| 1 1/2 | 1.900 | 1.494 | .203 | 1.750 | 4.100 | 4.190 |
| 2 | 2.375 | 1.933 | .221 | 2.94 | 5.670 | 5.800 |
| 2 1/2 | 2.875 | 2.315 | .280 | 4.21 | 8.660 | 8.850 |
| 3 | 3.500 | 2.892 | .304 | 6.57 | 11.600 | 11.800 |
| 3 1/2 | 4.000 | 3.358 | .321 | 8.86 | 14.100 | 14.400 |
| 4 | 4.500 | 3.818 | .341 | 11.50 | 16.900 | 17.300 |
| 5 | 5.562 | 4.812 | .375 | 18.20 | 23.200 | 23.700 |
| 6 | 6.625 | 5.751 | .437 | 26.00 | 32.200 | 32.900 |
| 8 | 8.625 | 7.625 | .500 | 45.70 | 48.400 | 49.500 |
| 10 | 10.750 | 9.750 | .500 | 74.70 | 61.100 | 62.400 |

# Appendix I: Typical Dimensions and Weights of Concrete Sewer Pipe

(All dimensions in inches)

(Weights given are for reinforced tongue and groove pipe. Reinforced bell and spigot pipe is heavier. Weights are based on 150 pcf concrete.)

| 3000 psi | | | 3500 psi | | | 4000 psi | | |
|---|---|---|---|---|---|---|---|---|
| internal diameter, inches | minimum shell thickness, inches | weight per foot, in pounds | internal diameter, inches | minimum shell thickness, inches | weight per foot, in pounds | internal diameter, inches | minimum shell thickness, inches | weight per foot, in pounds |
| 12 | 2 | 93 | 12 | 13/4 | 79 | 12 | | |
| 15 | 2 1/4 | 127 | 15 | 2 | 111 | 15 | | |
| 18 | 2 1/2 | 168 | 18 | | | 18 | 2 | 131 |
| 21 | 2 3/4 | 214 | 21 | | | 21 | 2 1/4 | 171 |
| 24 | 3 | 264 | 24 | 2 5/8 | 229 | 24 | 2 1/2 | 217 |
| 27 | 3 | 295 | 27 | 2 3/4 | 268 | 27 | 2 5/8 | 255 |
| 30 | 3 1/2 | 384 | 30 | 3 | 324 | 30 | 2 3/4 | 295 |
| 33 | 3 3/4 | 451 | 33 | 3 1/4 | 396 | 33 | 2 3/4 | 322 |
| 36 | 4 | 524 | 36 | 3 3/8 | 435 | 36 | 3 | 383 |
| 42 | 4 1/2 | 686 | 42 | 3 3/4 | 561 | 42 | 3 3/8 | 500 |
| 48 | 5 | 867 | 48 | 4 1/4 | 727 | 48 | 3 3/4 | 635 |
| 54 | 5 1/2 | 1068 | 54 | 4 5/8 | 887 | 54 | 4 1/4 | 810 |
| 60 | 6 | 1295 | 60 | 5 | 1064 | 60 | 4 1/2 | 950 |
| 66 | 6 1/2 | 1542 | 66 | 5 3/8 | 1256 | 66 | 4 3/4 | 1100 |
| 72 | 7 | 1811 | 72 | 5 3/4 | 1463 | 72 | 5 | 1260 |
| 78 | 7 1/2 | 2100 | | | | | | |
| 84 | 8 | 2409 | | | | | | |
| 90 | 8 | 2565 | | | | | | |
| 96 | 8 1/2 | 2906 | | | | | | |
| 108 | 9 | 3446 | | | | | | |

# Appendix J: Cast Iron Pipe Dimensions

(all dimensions in inches)

| nominal diameter | class A 100 foot head 43 pounds pressure | | class B 200 foot head 86 pounds pressure | | class C 300 foot head 130 pounds pressure | | class D 400 foot head 173 pounds pressure | |
|---|---|---|---|---|---|---|---|---|
| | outside diameter | inside diameter | outside diameter | inside diameter | outside diameter | inside diameter | outside diameter | inside diameter |
| 3 | 3.80 | 3.02 | 3.96 | 3.12 | 3.96 | 3.06 | 3.96 | 3.00 |
| 4 | 4.80 | 3.96 | 5.00 | 4.10 | 5.00 | 4.04 | 5.00 | 3.96 |
| 6 | 6.90 | 6.02 | 7.10 | 6.14 | 7.10 | 6.08 | 7.10 | 6.00 |
| 8 | 9.05 | 8.13 | 9.05 | 8.03 | 9.30 | 8.18 | 9.30 | 8.10 |
| 10 | 11.10 | 10.10 | 11.10 | 9.96 | 11.40 | 10.16 | 11.40 | 10.04 |
| 12 | 13.20 | 12.12 | 13.20 | 11.96 | 13.50 | 12.14 | 13.50 | 12.00 |
| 14 | 15.30 | 14.16 | 15.30 | 13.98 | 15.65 | 14.17 | 15.65 | 14.01 |
| 16 | 17.40 | 16.20 | 17.40 | 16.00 | 17.80 | 16.20 | 17.80 | 16.02 |
| 18 | 19.50 | 18.22 | 19.50 | 18.00 | 19.92 | 18.18 | 19.92 | 18.00 |
| 20 | 21.60 | 20.26 | 21.60 | 20.00 | 22.06 | 20.22 | 22.06 | 20.00 |
| 24 | 25.80 | 24.28 | 25.80 | 24.02 | 26.32 | 24.22 | 26.32 | 24.00 |
| 30 | 31.74 | 29.98 | 32.00 | 29.94 | 32.40 | 30.00 | 32.74 | 30.00 |
| 36 | 37.96 | 35.98 | 38.30 | 36.00 | 38.70 | 39.98 | 39.16 | 36.00 |
| 42 | 44.20 | 42.00 | 44.50 | 41.94 | 45.10 | 42.02 | 45.58 | 42.02 |
| 48 | 50.50 | 47.98 | 50.80 | 47.96 | 51.40 | 47.98 | 51.98 | 48.06 |
| 54 | 56.66 | 53.96 | 57.10 | 54.00 | 57.80 | 54.00 | 58.40 | 53.94 |
| 60 | 62.80 | 60.02 | 63.40 | 60.06 | 64.20 | 60.20 | 64.82 | 60.06 |
| 72 | 75.34 | 72.10 | 76.00 | 72.10 | 76.88 | 72.10 | | |
| 84 | 87.54 | 84.10 | 88.54 | 84.10 | | | | |

| nominal diameter | class E 500 foot head 217 pounds pressure | | class F 600 foot head 260 pounds pressure | | class G 700 foot head 304 pounds pressure | | class H 800 foot head 347 pounds pressure | |
|---|---|---|---|---|---|---|---|---|
| | outside diameter | inside diameter | outside diameter | inside diameter | outside diameter | inside diameter | outside diameter | inside diameter |
| 6 | 7.22 | 6.06 | 7.22 | 6.00 | 7.38 | 6.08 | 7.38 | 6.00 |
| 8 | 9.42 | 8.10 | 9.42 | 8.00 | 9.60 | 8.10 | 9.60 | 8.00 |
| 10 | 11.60 | 10.12 | 11.60 | 10.00 | 11.84 | 10.12 | 11.84 | 10.00 |
| 12 | 13.78 | 12.14 | 13.78 | 12.00 | 14.08 | 12.14 | 14.08 | 12.00 |
| 14 | 15.98 | 14.18 | 15.98 | 14.00 | 16.32 | 14.18 | 16.32 | 14.00 |
| 16 | 18.16 | 16.20 | 18.16 | 16.00 | 18.54 | 16.18 | 18.54 | 16.00 |
| 18 | 20.34 | 18.20 | 20.34 | 18.00 | 20.78 | 18.22 | 20.78 | 18.00 |
| 20 | 22.54 | 20.24 | 22.54 | 20.00 | 23.02 | 20.24 | 23.02 | 20.00 |
| 24 | 26.90 | 24.28 | 26.90 | 24.00 | 27.76 | 24.26 | 27.76 | 24.00 |
| 30 | 33.10 | 30.00 | 33.46 | 30.00 | | | | |
| 36 | 39.60 | 36.00 | 40.04 | 36.00 | | | | |

# Appendix K: American Standard Piping Symbols

| | Flanged | Screwed | Bell & Spigot | Welded | Soldered |
|---|---|---|---|---|---|
| Joint | | | | | |
| Elbow—90° | | | | | |
| Elbow—45° | | | | | |
| Elbow—Turned Up | | | | | |
| Elbow—Turned Down | | | | | |
| Elbow—Long Radius | | | | | |
| Reducing Elbow | | | | | |
| Tee | | | | | |
| Tee—Outlet Up | | | | | |
| Tee—Outlet Down | | | | | |
| Side Outlet Tee—Outlet Up | | | | | |
| Cross | | | | | |
| Reducer—Concentric | | | | | |
| Reducer—Eccentric | | | | | |
| Lateral | | | | | |
| Gate Valve | | | | | |
| Globe Valve | | | | | |
| Check Valve | | | | | |
| Stop Cock | | | | | |
| Safety Valve | | | | | |
| Expansion Joint | | | | | |
| Union | | | | | |
| Sleeve | | | | | |
| Bushing | | | | | |

# Appendix L: Equivalent Length of Straight Pipe for Various Fittings (feet)

(turbulent flow only, for any fluid)
c.i. = cast iron

| fittings | | | 1/4 | 3/8 | 1/2 | 3/4 | 1 | 1 1/4 | 1 1/2 | 2 | 2 1/2 | 3 | 4 | 5 | 6 | 8 | 10 | 12 | 14 | 16 | 18 | 20 | 24 |
|---|---|---|---|---|---|---|---|---|---|---|---|---|---|---|---|---|---|---|---|---|---|---|---|
| | | | pipe size | | | | | | | | | | | | | | | | | | | | |
| regular 90°ell | screwed | steel | 2.3 | 3.1 | 3.6 | 4.4 | 5.2 | 6.6 | 7.4 | 8.5 | 9.3 | 11.0 | 13.0 | | | | | | | | | | |
| | screwed | c.i. | | | | | | | | | | 9.0 | 11.0 | | | | | | | | | | |
| | flanged | steel | | | 0.92 | 1.2 | 1.6 | 2.1 | 2.4 | 3.1 | 3.6 | 4.4 | 5.9 | 7.3 | 8.9 | 12.0 | 14.0 | 17.0 | 18.0 | 21.0 | 23.0 | 25.0 | 30.0 |
| | flanged | c.i. | | | | | | | | | | 3.6 | 4.8 | | 7.2 | 9.8 | 12.0 | 15.0 | 17.0 | 19.0 | 22.0 | 24.0 | 28.0 |
| long radius 90°ell | screwed | steel | 1.5 | 2.0 | 2.2 | 2.3 | 2.7 | 3.2 | 3.4 | 3.6 | 3.6 | 4.0 | 4.6 | | | | | | | | | | |
| | screwed | c.i. | | | | | | | | | | 3.3 | 3.7 | | | | | | | | | | |
| | flanged | steel | | | 1.1 | 1.3 | 1.6 | 2.0 | 2.3 | 2.7 | 2.9 | 3.4 | 4.2 | 5.0 | 5.7 | 7.0 | 8.0 | 9.0 | 9.4 | 10.0 | 11.0 | 12.0 | 14.0 |
| | flanged | c.i. | | | | | | | | | | 2.8 | 3.4 | | 4.7 | 5.7 | 6.8 | 7.8 | 8.6 | 9.6 | 11.0 | 11.0 | 13.0 |
| regular 45°ell | screwed | steel | 0.34 | 0.52 | 0.71 | 0.92 | 1.3 | 1.7 | 2.1 | 2.7 | 3.2 | 4.0 | 5.5 | | | | | | | | | | |
| | screwed | c.i. | | | | | | | | | | 3.3 | 4.5 | | | | | | | | | | |
| | flanged | steel | | | 0.45 | 0.59 | 0.81 | 1.1 | 1.3 | 1.7 | 2.0 | 2.6 | 3.5 | 4.5 | 5.6 | 7.7 | 9.0 | 11.0 | 13.0 | 15.0 | 16.0 | 18.0 | 22.0 |
| | flanged | c.i. | | | | | | | | | | 2.1 | 2.9 | | 4.5 | 6.3 | 8.1 | 9.7 | 12.0 | 13.0 | 15.0 | 17.0 | 20.0 |
| tee-line flow | screwed | steel | 0.79 | 1.2 | 1.7 | 2.4 | 3.2 | 4.6 | 5.6 | 7.7 | 9.3 | 12.0 | 17.0 | | | | | | | | | | |
| | screwed | c.i. | | | | | | | | | | 9.9 | 14.0 | | | | | | | | | | |
| | flanged | steel | | | 0.69 | 0.82 | 1.0 | 1.3 | 1.5 | 1.8 | 1.9 | 2.2 | 2.8 | 3.3 | 3.8 | 4.7 | 5.2 | 6.0 | 6.4 | 7.2 | 7.6 | 8.2 | 9.6 |
| | flanged | c.i. | | | | | | | | | | 1.9 | 2.2 | | 3.1 | 3.9 | 4.6 | 5.2 | 5.9 | 6.5 | 7.2 | 7.7 | 8.8 |
| tee-branch flow | screwed | steel | 2.4 | 3.5 | 4.2 | 5.3 | 6.6 | 8.7 | 9.9 | 12.0 | 13.0 | 17.0 | 21.0 | | | | | | | | | | |
| | screwed | c.i. | | | | | | | | | | 14.0 | 17.0 | | | | | | | | | | |
| | flanged | steel | | | 2.0 | 2.6 | 3.3 | 4.4 | 5.2 | 6.6 | 7.5 | 9.4 | 12.0 | 15.0 | 18.0 | 24.0 | 30.0 | 34.0 | 37.0 | 43.0 | 47.0 | 52.0 | 62.0 |
| | flanged | c.i. | | | | | | | | | | 7.7 | 10.0 | | 15.0 | 20.0 | 25.0 | 30.0 | 35.0 | 39.0 | 44.0 | 49.0 | 57.0 |
| 180° return bend | screwed | steel | 2.3 | 3.1 | 3.6 | 4.4 | 5.2 | 6.6 | 7.4 | 8.5 | 9.3 | 11.0 | 13.0 | | | | | | | | | | |
| | screwed | c.i. | | | | | | | | | | 9.0 | 11.0 | | | | | | | | | | |
| | reg. flanged | steel | | | 0.92 | 1.2 | 1.6 | 2.1 | 2.4 | 3.1 | 3.6 | 4.4 | 5.9 | 7.3 | 8.9 | 12.0 | 14.0 | 17.0 | 18.0 | 21.0 | 23.0 | 25.0 | 30.0 |
| | reg. flanged | c.i. | | | | | | | | | | 3.6 | 4.8 | | 7.2 | 9.8 | 12.0 | 15.0 | 17.0 | 19.0 | 22.0 | 24.0 | 28.0 |
| | long rad. flanged | steel | | | 1.1 | 1.3 | 1.6 | 2.0 | 2.3 | 2.7 | 2.9 | 3.4 | 4.2 | 5.0 | 5.7 | 7.0 | 8.0 | 9.0 | 9.4 | 10.0 | 11.0 | 12.0 | 14.0 |
| | long rad. flanged | c.i. | | | | | | | | | | 2.8 | 3.4 | | 4.7 | 5.7 | 6.8 | 7.8 | 8.6 | 9.6 | 11.0 | 11.0 | 13.0 |
| globe valve | screwed | steel | 21.0 | 22.0 | 22.0 | 24.0 | 29.0 | 37.0 | 42.0 | 54.0 | 62.0 | 79.0 | 110.0 | | | | | | | | | | |
| | screwed | c.i. | | | | | | | | | | 65.0 | 86.0 | | | | | | | | | | |
| | flanged | steel | | | 38.0 | 40.0 | 45.0 | 54.0 | 59.0 | 70.0 | 77.0 | 94.0 | 120.0 | 150.0 | 190.0 | 260.0 | 310.0 | 390.0 | | | | | |
| | flanged | c.i. | | | | | | | | | | 77.0 | 99.0 | | 150.0 | 210.0 | 270.0 | 330.0 | | | | | |
| gate valve | screwed | steel | 0.32 | 0.45 | 0.56 | 0.67 | 0.84 | 1.1 | 1.2 | 1.5 | 1.7 | 1.9 | 2.5 | | | | | | | | | | |
| | screwed | c.i. | | | | | | | | | | 1.6 | 2.0 | | | | | | | | | | |
| | flanged | steel | | | | | | | | 2.6 | 2.7 | 2.8 | 2.9 | 3.1 | 3.2 | 3.2 | 3.2 | 3.2 | 3.2 | 3.2 | 3.2 | 3.2 | 3.2 |
| | flanged | c.i. | | | | | | | | | | 2.3 | 2.4 | | 2.6 | 2.7 | 2.8 | 2.9 | 2.9 | 3.0 | 3.0 | 3.0 | 3.0 |
| angle valve | screwed | steel | 12.8 | 15.0 | 15.0 | 15.0 | 17.0 | 18.0 | 18.0 | 18.0 | 18.0 | 18.0 | 18.0 | | | | | | | | | | |
| | screwed | c.i. | | | | | | | | | | 15.0 | 15.0 | | | | | | | | | | |
| | flanged | steel | | | 15.0 | 15.0 | 17.0 | 18.0 | 18.0 | 21.0 | 22.0 | 28.0 | 38.0 | 50.0 | 63.0 | 90.0 | 120.0 | 140.0 | 160.0 | 190.0 | 210.0 | 240.0 | 300.0 |
| | flanged | c.i. | | | | | | | | | | 23.0 | 31.0 | | 52.0 | 74.0 | 98.0 | 120.0 | 150.0 | 170.0 | 200.0 | 230.0 | 280.0 |
| swing check valve | screwed | steel | 7.2 | 7.3 | 8.0 | 8.8 | 11.0 | 13.0 | 15.0 | 19.0 | 22.0 | 27.0 | 38.0 | | | | | | | | | | |
| | screwed | c.i. | | | | | | | | | | 22.0 | 31.0 | | | | | | | | | | |
| | flanged | steel | | | 3.8 | 5.3 | 7.2 | 10.0 | 12.0 | 17.0 | 21.0 | 27.0 | 38.0 | 50.0 | 63.0 | 90.0 | 120.0 | 140.0 | | | | | |
| | flanged | c.i. | | | | | | | | | | 22.0 | 31.0 | | 52.0 | 74.0 | 98.0 | 120.0 | | | | | |
| coupling or union | screwed | steel | 0.14 | 0.18 | 0.21 | 0.24 | 0.29 | 0.36 | 0.39 | 0.45 | 0.47 | 0.53 | 0.65 | | | | | | | | | | |
| | screwed | c.i. | | | | | | | | | | 0.44 | 0.52 | | | | | | | | | | |
| | bell mouth inlet | steel | 0.04 | 0.07 | 0.10 | 0.13 | 0.18 | 0.26 | 0.31 | 0.43 | 0.52 | 0.67 | 0.95 | 1.3 | 1.6 | 2.3 | 2.9 | 3.5 | 4.0 | 4.7 | 5.3 | 6.1 | 7.6 |
| | bell mouth inlet | c.i. | | | | | | | | | | 0.55 | 0.77 | | 1.3 | 1.9 | 2.4 | 3.0 | 3.6 | 4.3 | 5.0 | 5.7 | 7.0 |
| | square mouth inlet | steel | 0.44 | 0.68 | 0.96 | 1.3 | 1.8 | 2.6 | 3.1 | 4.3 | 5.2 | 6.7 | 9.5 | 13.0 | 16.0 | 23.0 | 29.0 | 35.0 | 40.0 | 47.0 | 53.0 | 61.0 | 76.0 |
| | square mouth inlet | c.i. | | | | | | | | | | 5.5 | 7.7 | | 13.0 | 19.0 | 24.0 | 30.0 | 36.0 | 43.0 | 50.0 | 57.0 | 70.0 |
| | re-entrant pipe | steel | 0.88 | 1.4 | 1.9 | 2.6 | 3.6 | 5.1 | 6.2 | 8.5 | 10.0 | 13.0 | 19.0 | 25.0 | 32.0 | 45.0 | 58.0 | 70.0 | 80.0 | 95.0 | 110.0 | 120.0 | 150.0 |
| | re-entrant pipe | c.i. | | | | | | | | | | 11.0 | 15.0 | | 26.0 | 37.0 | 49.0 | 61.0 | 73.0 | 86.0 | 100.0 | 110.0 | 140.0 |

# Appendix M: Hazen-Williams Nomograph

$(C = 100)$

For values of $C$ other than 100, multiply the nomograph values for head loss by $\left(\frac{100}{C}\right)^{1.85}$

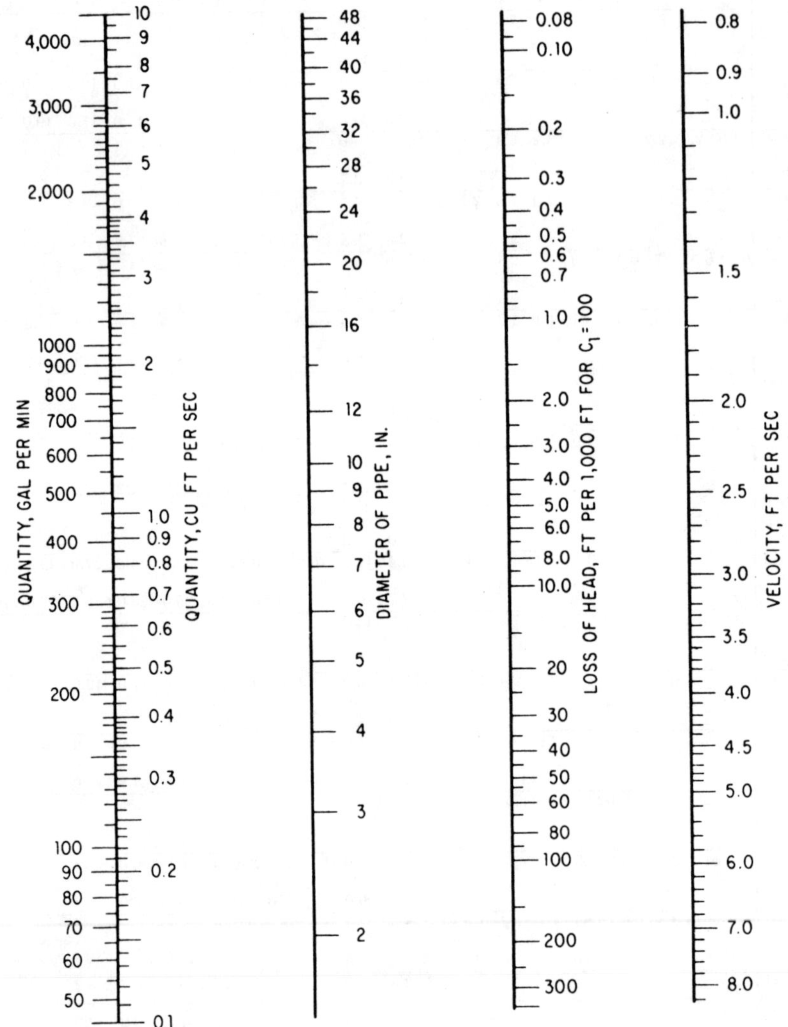

Fig. 1. Panoramic view of the Twisp River Bridge, Okanogan County, Washington State (picture taken in February 2002). Courtesy: Alex Young, WSDOT Bridge Architect.

A plan and elevation of the Twisp River Bridge is shown in Fig. 3. The structure replaces a four-span cast-in-place concrete T-beam bridge, built in 1935, that had become functionally obsolete (see Fig. 4). The bridge crosses the Twisp River, which flows into the Methow River.

The Twisp and Methow rivers are home to several endangered fish species: the Upper Columbia Steelhead, the Upper Columbia Chinook Salmon, and the Bull Trout. Under normal circumstances, WSDOT would have designed a two-span prestressed concrete girder superstructure with an intermediate pier in the river. Environmental restrictions, however, allowed work below the ordinary high water line only during the months of July and August, which make the installation of an intermediate pier difficult.

This obstacle meant that WSDOT would have to come up with a new innovative solution. Prior to the availability of the W95PTMG sections, the likely scheme would have been to span the river with steel plate girders. But in this case, the solution was a single-span bridge using the new deep WSDOT girder sections, which would eliminate construction in the river during the September to June fish closure.

WSDOT received $500,000 of supplementary funding from the Federal Highway Administration's (FHWA) Innovative Bridge Research and Construction Program for the new design features and construction techniques used to advance the state of practice in transportation structures.

## BACKGROUND

Currently, there are roughly 3000 bridges on the Washington State highway system. Approximately 87 percent of these bridges are concrete structures, about half of which are prestressed. Today, WSDOT uses precast, pretensioned concrete girders with cast-in-place decking as its standard concrete bridge.

For many years, the deepest standard girder was the W74MG, which is 1867 mm (73.5 in.) deep and weighs 12.1 kN/m (0.83 kips per ft). These girders are typically used for bridges in the 36.6 to 42.7 m (120 to 140 ft) span range, although spans of up to 48.8 m (160 ft) are possible.

Fig. 2. Closeup of bridge showing new 2.4 m (7.87 ft) deep, W95PTMG girder.

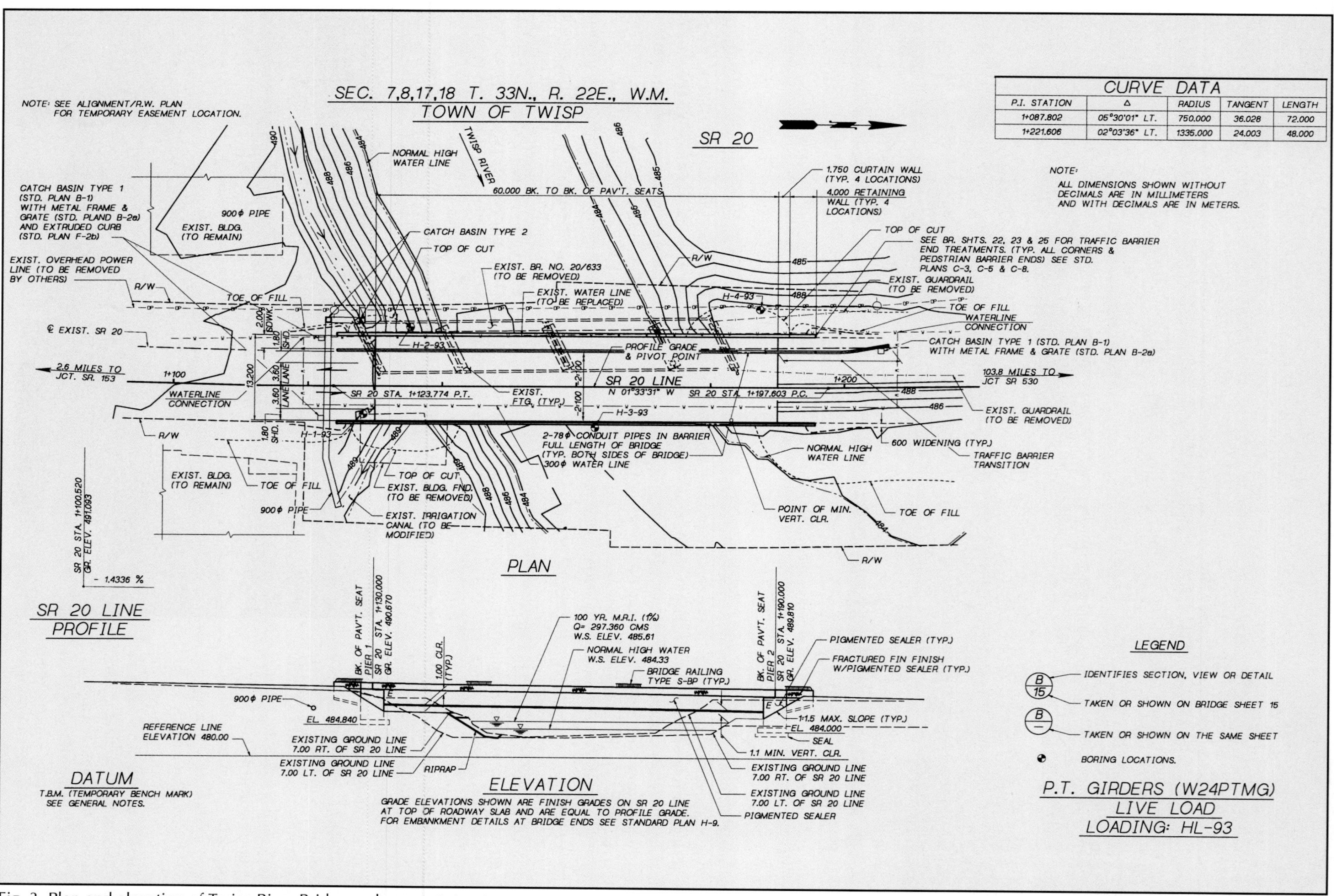


CURVE DATA

| P.I. STATION | Δ | RADIUS | TANGENT | LENGTH |
|---|---|---|---|---|
| 1+087.802 | 05°30'01" LT. | 750.000 | 36.028 | 72.000 |
| 1+221.606 | 02°03'36" LT. | 1335.000 | 24.003 | 48.000 |

Fig. 3. Plan and elevation of Twisp River Bridge replacement structure.

# Appendix N: Manning Nomograph

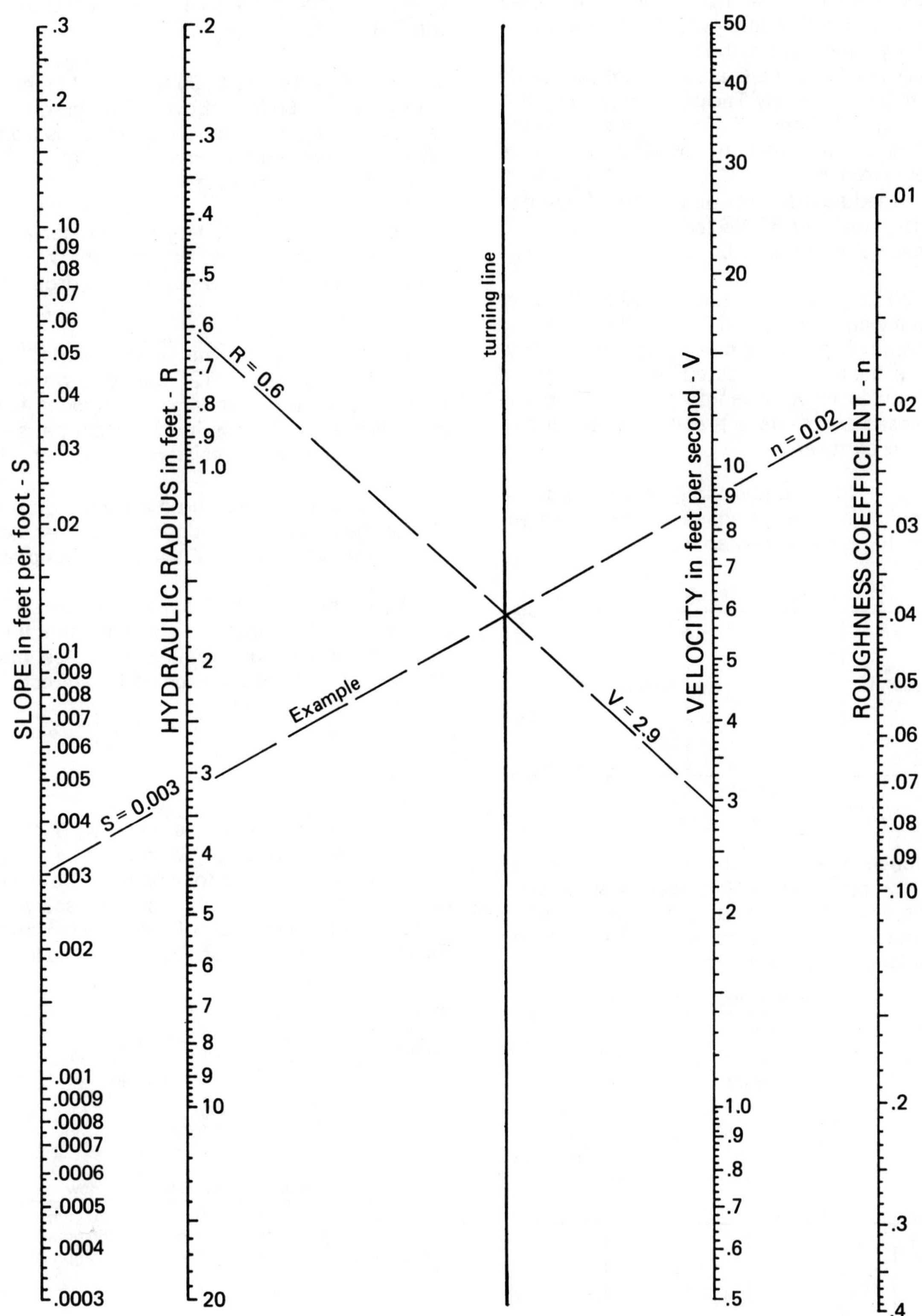

## Practice Problems: HYDRAULICS

Untimed

1. Three reservoirs (A, B, and C) are interconnected with a common junction at elevation 25 feet above some arbitrary reference point. The water surface levels for reservoirs A, B, and C are at elevations of 50, 40, and 22 feet respectively. The pipe from reservoir A to the junction is 800 feet of 3" steel pipe. The pipe from reservoir B to the junction is 500 feet of 10" steel pipe. The pipe from reservoir C to the junction is 1000 feet of 4" schedule-40 steel pipe. Is the flow into or out of the reservoir B? Neglect minor losses and velocity heads. Assume f = .02.

2. A class 20 cast iron pipe has an inside diameter of 24" and a wall thickness of 1.16". Water is flowing at 6 fps. (a) If a valve is closed instantaneously, what will be the pressure created? (b) If the pipe is 500 feet long, over what length of time must the valve be closed to create a pressure equivalent to instantaneous closure?

3. Assume C = 100 and find the flow in each of the pipes in the distribution system shown below. An accuracy of ± 10 gpm is acceptable.

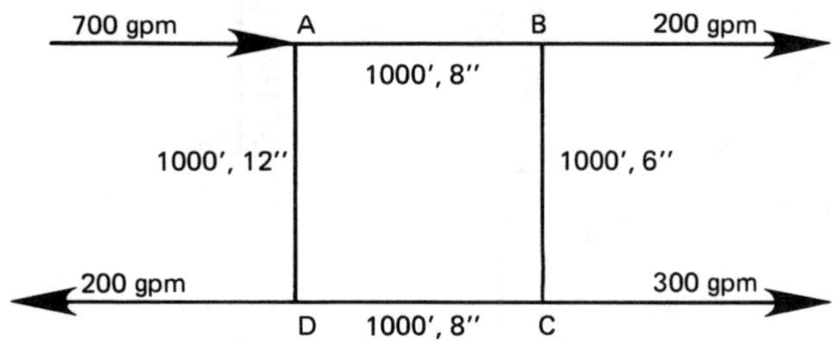


4. The distance between each labeled point on the distribution graph below is 1000 feet. Assume C = 100 for each pipe section. What is the pressure at all labeled points? If the pump receives 20 psig water, what horsepower is required?

| point | pressure | elevation |
|---|---|---|
| A | | 200 ft |
| B | | 150 |
| C | 40 psig | 300 |
| D | | 150 |
| E | | 200 |
| F | | 150 |

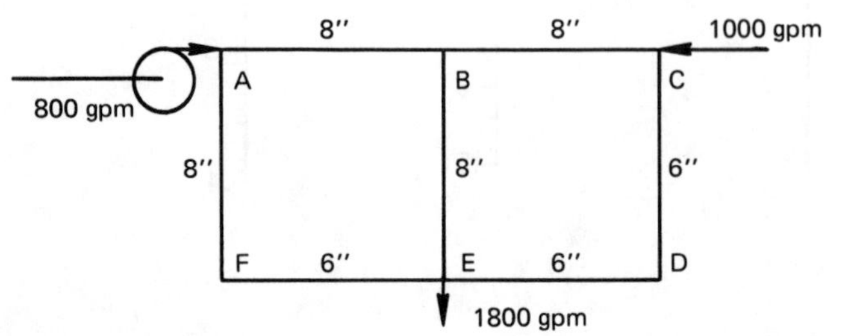


5. 70°F water is carried in a schedule-40 steel pipe which changes gradually from 6" at point A to 18" at point B. B is 15 feet higher than A. 5 cfs are flowing and the respective pressures at A and B are 10 psia and 7 psia. What is the direction of the flow?

6. Points A and B are 3000 feet apart along a new 6" steel pipe. B is 60 feet above A. 750 gpm of 60°F water flow in the pipe. The flow direction is from A to B. What pressure must be maintained at A if the pressure at B is to be 50 psig?

7. A cylindrical tank 20 feet in diameter and 40 feet high has a 4" hole in the bottom with $C_d$ = .98. How long will it take for the water level to drop from 40' to 20'?

8. A venturi water meter with an 8" diameter throat is installed in a 12" diameter water line. Assume the venturi discharge coefficient is equal to one. What is the flow in cubic feet per second if a mercury manometer registers a 4" differential?

9. What will be the measured pressure drop across a .2 foot diameter sharp-edged orifice installed in a 1 foot diameter pipe if 70°F water is flowing at 2 fps?

10. A pipe necks down from 24" at point A to 12" at point B. The discharge is 8 cfs in the direction of A to B. The pressure head at A is 20 feet. Assume no friction. Find the resultant force and its direction on the fluid if water is flowing.

Timed

1. A pipe network connects points A, B, C, and D as shown in the figure below. Water can be added or removed at any of these four points. The flow is from point A to point D. The minimum pressure is 20 psi. (a) What is the elevation of the gradient at point A? (b) What are the pressures (in psi) for the four points?

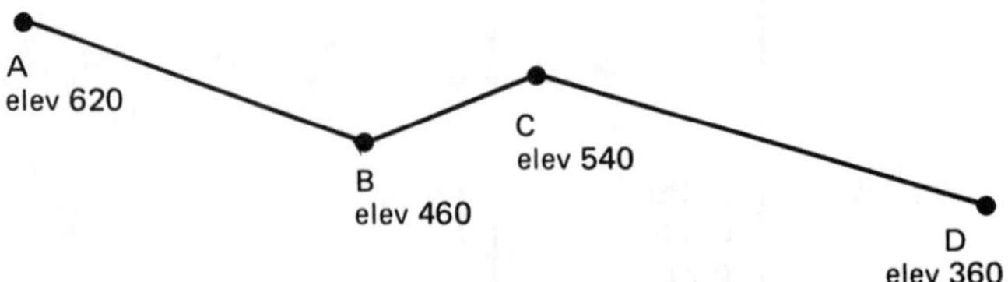


| pipe section | length | diameter | flow | C |
|---|---|---|---|---|
| A to B | 20,000 ft | 6 in. | 120 gpm | 150 |
| B to C | 10,000 ft | 6 in. | 160 gpm | 150 |
| C to D | 30,000 ft | 4 in. | 120 gpm | 150 |

# 4 HYDRAULIC MACHINES

*Nomenclature*

| | | |
|---|---|---|
| bhp | brake horsepower | hp |
| $C_v$ | coefficient of velocity | – |
| d | diameter | inches |
| D | diameter | ft |
| ehp | electrical horsepower | hp |
| E | energy | ft-lbf/lbm |
| f | Darcy friction factor, or frequency | –, hertz |
| fhp | friction horsepower | hp |
| g | acceleration due to gravity (32.2) | $\text{ft/sec}^2$ |
| $g_c$ | gravitational constant (32.2) | $\frac{\text{lbm-ft}}{\text{lbf-sec}^2}$ |
| h | head | ft |
| H | dynamic head | ft |
| L | pipe length | ft |
| m | mass | lbm |
| n | rotational speed | rpm |
| $n_s$ | specific speed | rpm |
| $n_{ss}$ | suction specific speed | rpm |
| NPSHA | net positive suction head available | ft |
| NPSHR | net positive suction head required | ft |
| p | pressure | psf |
| Q | flow quantity | gpm |
| s | slip | – |
| S.G. | specific gravity | – |
| T | temperature | °F |
| v | velocity | ft/sec |
| V | volume | $\text{ft}^3$ |
| w | weight | lbf |
| whp | hydraulic ("water") horsepower | hp |
| z | height above datum | ft |

*Symbols*

| | | |
|---|---|---|
| $\eta$ | efficiency | – |
| $\epsilon$ | specific pipe roughness | ft |
| $\rho$ | density | $\text{lbm/ft}^3$ |
| $\beta$ | blade angle | ° |
| $\sigma$ | cavitation coefficient | – |

*Subscripts*

| | |
|---|---|
| a | atmospheric |
| A | added by pump |
| d | discharge |
| f | friction |
| i | intake |
| j | jet |
| m | motor |
| n | nozzle |
| p | pump or pressure |
| s | suction |
| sd | static discharge |
| th | theoretical |
| ts | total static |
| T | turbine |
| v | velocity |
| vp | vapor pressure |

## 1 INTRODUCTION

Pumps and turbines are the two types of hydraulic machines discussed in this chapter. Pumps convert mechanical energy into fluid energy. Turbines convert fluid energy into mechanical energy.

## 2 TYPES OF PUMPS

Pumps can be classified in several ways. The clearest categorization is based on the method by which pumping energy is transmitted to the fluid. This approach separates pumps into *kinetic pumps* and *positive displacement pumps.*

The two most common forms of positive displacement pumps are *reciprocating action pumps* (which use pistons, plungers, diaphragms, or bellows) and *rotary action pumps* (using vanes, screws, lobes, or progressing cavities). Such pumps discharge a given volume for each stroke or revolution. Energy is added intermittently to the fluid flow.

Kinetic pumps rely on a transformation of kinetic energy to static pressure. Jet and ejector pumps fall into this category, but centrifugal pumps are the primary examples of kinetic pumps. Pumps covered in this chapter are assumed to be centrifugal pumps.

**Table 4.1**
Characteristics of Kinetic and Displacement Pumps

| characteristic | displacement | kinetic |
|---|---|---|
| flow rate | low | high |
| pressure rise per stage | high | low |
| constant variable over operating range | flow rate | pressure rise |
| self-priming | yes | no |
| outlet stream | pulsing | steady |
| works with high-viscosity fluids | yes | no |

Liquid flowing into the *suction side* (the *inlet*) of a centrifugal pump is captured by the *impeller* and thrown to the outside of the pump casing. Within the casing, the velocity imparted to the fluid by the impeller is converted into pressure energy.

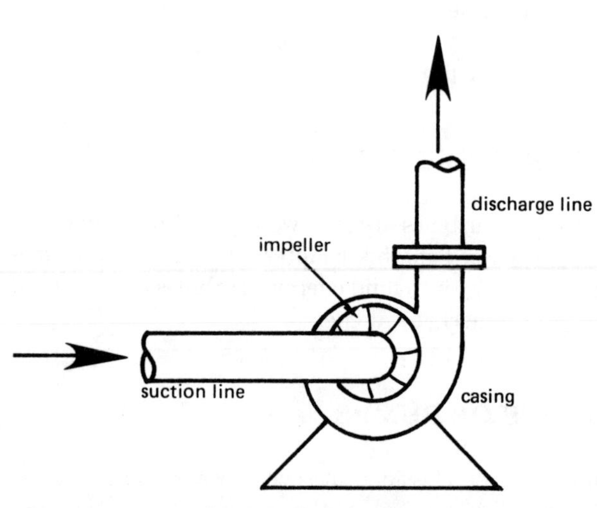


**Figure 4.1** A Centrifugal Pump

## 3 TYPES OF CENTRIFUGAL PUMPS

Centrifugal pumps can be classified into three general categories according to the way the impeller imparts energy to the fluid. Each of these categories has its own range of specific speeds and appropriate applications.

*Radial flow impellers* impart energy primarily by centrifugal force. Liquid enters the impeller at the hub and flows radially to the outside of the casing. Single suction impellers have a specific speed less than 5000. Double suction impellers have a specific speed less than 6000.

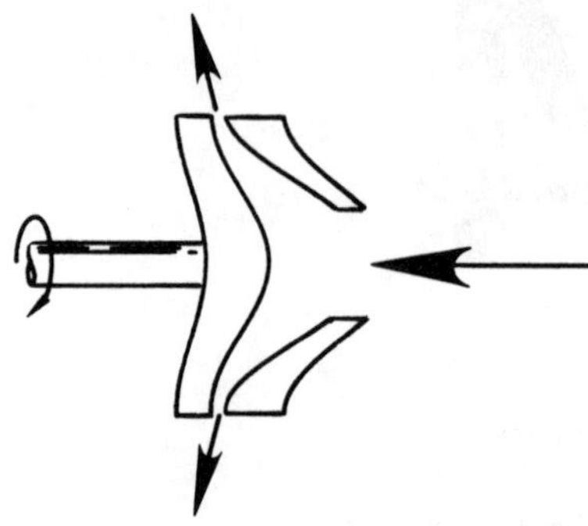

**Figure 4.2** Centrifugal (Radial) Flow Pump

*Mixed flow impellers* impart energy partially by centrifugal force and partially by axial force, since the vanes act partially as an axial compressor. Liquid enters the impeller at the hub and flows both radially and axially to discharge. Specific speeds of mixed flow pumps range from 4200 to 9000.

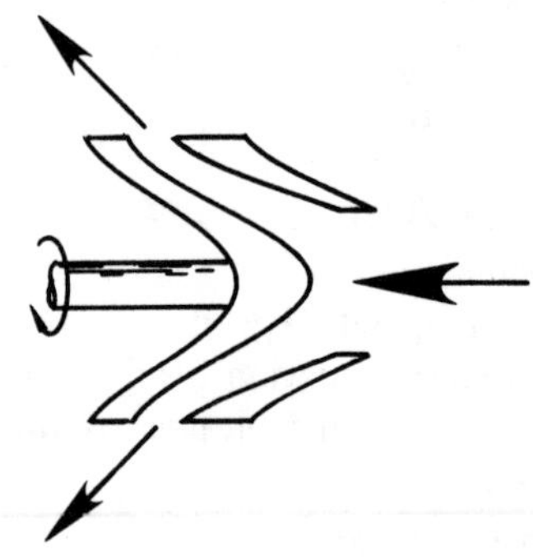

**Figure 4.3** Mixed Flow Pump

*Axial flow impellers* impart energy to the fluid by acting as axial flow compressors. Fluid enters and exits along the axis of rotation. Specific speed is greater than 9000.

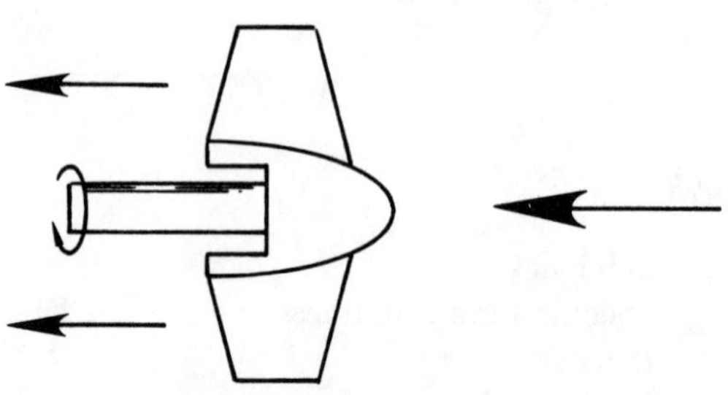

**Figure 4.4** Axial Flow Pump

Radial flow and mixed flow centrifugal pumps can be designed for either single or double suction operation. In a *single suction pump*, fluid enters only one side of the impeller. In a *double suction pump*, fluid enters both sides of the impeller. Thus, for an impeller with a given specific speed, a greater flow rate can be expected from a double suction pump. In addition, a double suction pump has a lower NPSHR for a given flow than a single suction pump.

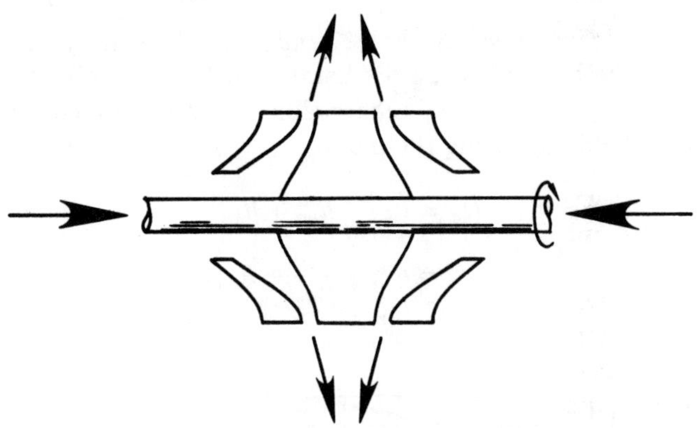

**Figure 4.5** Radial Flow Pump (Double Suction)

A *multiple stage pump* consists of two or more impellers within a single casing. The discharge of one stage is the input of the next stage. In this manner, higher heads are achieved than would be possible with a single impeller.

## 4 PUMP AND HEAD TERMINOLOGY

Like most specialized subjects, the centrifugal pump field has developed its own terminology. It is essential that this terminology be understood, since its interpretation will often affect an installation's physical configuration.

All of the terms which follow are *head terms*, and as such, have units of feet. Of course, any head term can be converted to pressure by using equation 4.1.

$$p = \rho h \tag{4.1}$$

*Friction head* ($h_f$): The head required to overcome resistance to flow in the pipe, fittings, valves, entrances, and exits.

$$h_f = \frac{fL_e v^2}{2Dg_c} \tag{4.2}$$

*Velocity head* ($h_v$): The head of a fluid as a result of its kinetic energy.

$$h_v = \frac{v^2}{2g_c} \tag{4.3}$$

*Atmospheric head* ($h_a$): Atmospheric pressure converted to feet of fluid being pumped.

$$h_a = \frac{p_a}{\rho} \tag{4.4}$$

*Pressure head* ($h_p$): Pressure converted to feet of fluid being pumped.

$$h_p = \frac{p}{\rho} \tag{4.5}$$

*Vapor pressure head* ($h_{vp}$): Fluid vapor pressure converted to feet of fluid being pumped. Steam tables can be used to evaluate the vapor pressure of water. Figure 4.9 can be used with hydrocarbons.

$$h_{vp} = \frac{p_{vp}}{\rho} \tag{4.6}$$

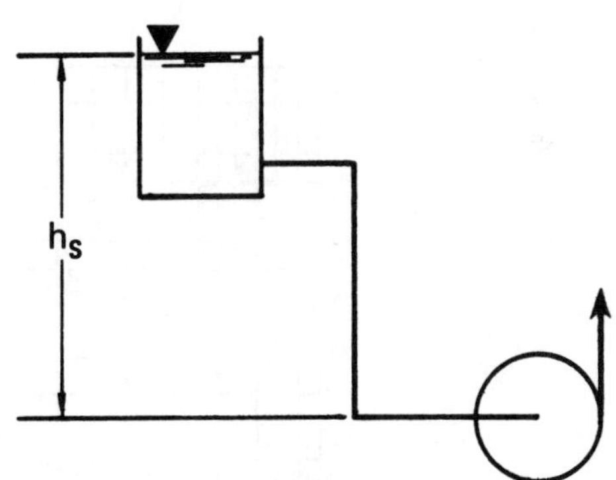


**Figure 4.6** Static Suction Head

*Static suction head* ($h_s$): The vertical distance in feet above the centerline of the inlet to the free level of the fluid source. If the free level of the fluid source is below the inlet, $h_s$ will be negative. In this case, $h_s$ is known as *static suction lift*.

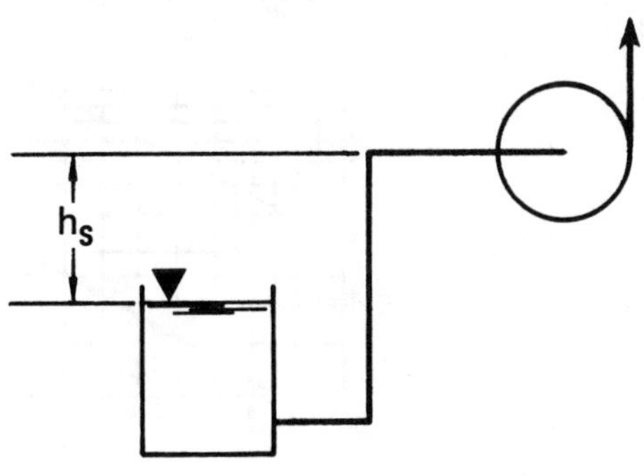


**Figure 4.7** Static Suction Lift

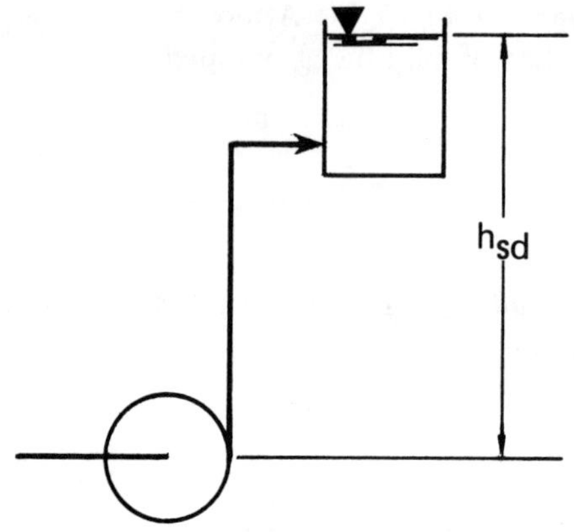


Figure 4.8 Static Discharge Head

*Total (dynamic) suction head* ($H_s$): The static suction head minus the friction head in the suction line (i.e., the total energy of the fluid entering the impeller). If $h_s$ is negative (i.e., the free level of the fluid source is below the inlet), then $H_s$ will be negative. In this case, $H_s$ is known as *total (dynamic) suction lift.*

$$H_s = h_s - h_{f\,(s)} \tag{4.7}$$

*Static discharge head* ($h_{sd}$): The vertical distance in feet above the pump inlet centerline to the free level of the discharge tank or point of free discharge.

*Total (dynamic) discharge head* ($H_d$): The static discharge head plus the discharge velocity head plus the friction head in the discharge line (i.e., the total energy of the fluid leaving the pump).

$$H_d = h_{sd} + h_{vd} + h_{f\,(d)} \tag{4.8}$$

$$= h_{sd} + \frac{v_d^2}{2g_c} + h_{f\,(d)} \tag{4.9}$$

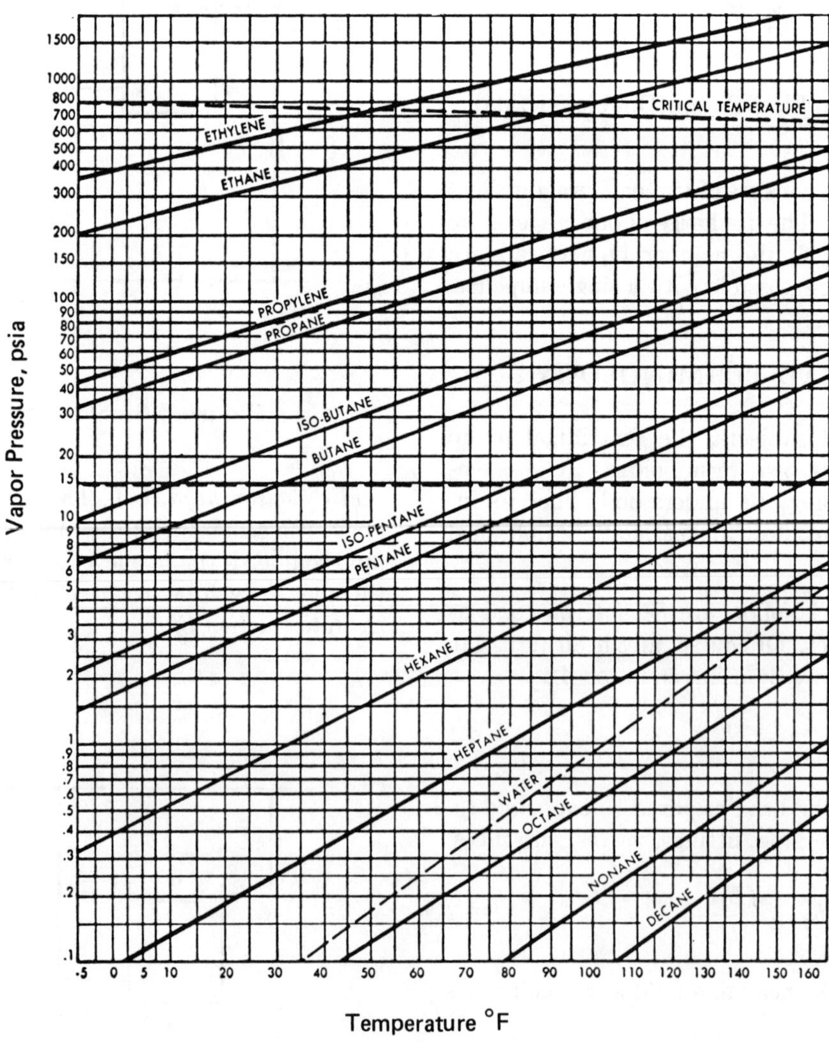


Figure 4.9 Vapor Pressure of Hydrocarbons

*Total static head* ($h_{ts}$): The vertical distance in feet between the free level of the supply and either the point of free discharge or the free level of the discharge tank.

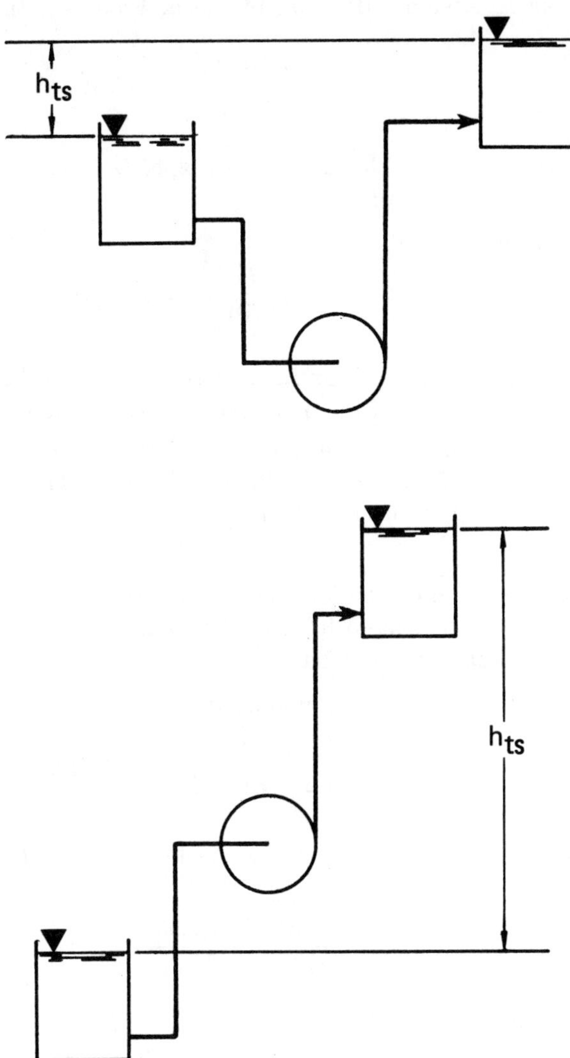


**Figure 4.10** Total Static Head

*Total (dynamic) head* ($H$): The total discharge head less the total suction head.

$$H = H_d - H_s \quad 4.10$$

$$= h_{sd} - h_s + \frac{v_d^2}{2g_c} + h_{f\,(s)} + h_{f\,(d)} \quad 4.11$$

## 5 NET POSITIVE SUCTION HEAD AND CAVITATION

Liquid is not sucked into a pump. A positive head (normally atmospheric pressure) must push the liquid into the impeller (i.e., "flood" the impeller). *Net Positive Suction Head Required* (*NPSHR*) is the minimum fluid energy required at the inlet by the pump for satisfactory operation. NPSHR is usually specified by the pump manufacturer.[1] *Net Positive Suction Head Available* (*NPSHA*) is the actual fluid energy at the inlet.[2]

$$\text{NPSHA} = h_a + h_s - h_{f\,(s)} - h_{vp} \quad 4.12$$

$$= h_{p\,(i)} + h_{v\,(i)} - h_{vp} \quad 4.13$$

If NPSHA is less than NPSHR, the fluid will cavitate. *Cavitation* is the vaporization of fluid within the casing or suction line. If the fluid pressure is less than the vapor pressure, pockets of vapor will form. As vapor pockets reach the surface of the impeller, the local high fluid pressure will collapse them, causing noise, vibration, and possible structural damage to the pump.

Cavitation may be caused by any of the following conditions:

1. Discharge heads far below the pump's calibrated head at peak efficiency.
2. Suction lift higher or suction head lower than the manufacturer's recommendation.
3. Speeds higher than the manufacturer's recommendation.
4. Liquid temperatures (thus, vapor pressures) higher than that for which the system was designed.

The following steps can be used to check for cavitation:

*step 1*: Determine the minimum NPSHR for the given pump. This should be given as part of the performance data. NPSHR follows the $Q^2$ law. If NPSHR is known for one flow rate, it can be determined for another flow rate from equation 4.14.

$$\frac{\text{NPSHR}_2}{\text{NPSHR}_1} = \left(\frac{Q_2}{Q_1}\right)^2 \quad 4.14$$

*step 2*: Calculate NPSHA from either equation 4.12 or 4.13.

*step 3*: If NPSHA is greater than NPSHR, cavitation will not occur. A good safety margin is

[1] If NPSHR is multiplied by the fluid density, it is known as *NIPR*, the *Net Inlet Pressure Required*. Similarly, NPSHA can be converted to *NIPA*.

[2] Equations 4.12 and 4.13 represent two totally different methods, both of which are correct, for calculating NPSHA. Equation 4.12 is based on the conditions at the fluid surface at the top of a tank. There is potential energy (the $h_s$ term) but no kinetic energy. Equation 4.13 is based on the conditions at the immediate entrance to the pump. At that point, some of the potential head has been converted to velocity head.

2–3 feet of fluid. If NPSHA is insufficient, it should be increased or the NPSHR should be decreased. NPSHA can be increased by:

a. increasing the height of the free fluid level of the supply tank

b. reducing the distance and minor losses between the supply tank and the pump, or by using a larger pipe size

c. reducing the temperature of the fluid

d. pressurizing the supply tank

e. reducing the flow rate or velocity

NPSHR can be reduced by:

a. placing a throttling valve in the discharge line (i.e., this will increase the total head, thereby reducing the capacity of the pump and driving its operating point into a region of lower NPSHR)

b. using a double suction pump

Applications which require very high NPSHR, such as boiler feed pumps needing 150 to 250 feet, should use booster pumps in front of the high NPSHR pumps. Such booster pumps are typically single stage, double suction pumps running at low speed. Their NPSHR can be 25 feet or less.

It is important to note that throttling the input line to a pump and venting or evacuating the receiving tank both work to increase cavitation. Throttling the input line increases the friction head term and decreases NPSHA. Evacuating the receiving tank increases the flow rate, which also increases the friction head term.

*Example 4.1*

2 cfs of water are pumped from a feed tank mounted on a platform to an open reservoir through 6″ schedule 40 ($\epsilon/D = 0.000293$) steel pipe. Determine the NPSHA.

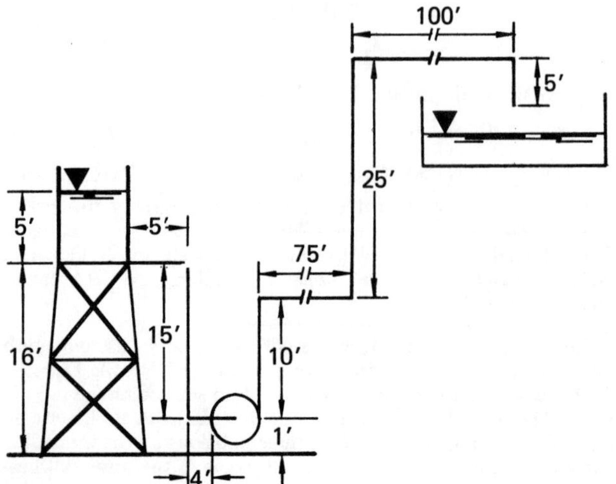


*step 1*: Assume 60°F and 14.7 psia. From equation 4.4,

$$h_a = \frac{(14.7)(144)}{62.4} = 33.9 \text{ ft}$$

*step 2*: For 6″ schedule 40 steel pipe, $D = 0.505$ ft, $A = 0.201 \text{ ft}^2$.

*step 3*:

$$v = \frac{Q}{A} = \frac{2}{0.201} = 9.95 \text{ ft/sec}$$

*step 4*: From appendix L, chapter 3, the equivalent lengths of the pipe and flanged fittings are:

| | |
|---|---|
| square entrance loss | $1 \times 16 = 16$ |
| 90° long radius elbows | $2 \times 5.7 = 11.4$ |
| pipe run $(5 + 15 + 4)$ | 24 |
| | 51.4 ft |

*step 5*: At 60°, the kinematic viscosity of water is 1.217 EE−5 ft²/sec. The vapor pressure is 0.6 feet of water.

*step 6*: The Reynolds number is

$$N_{Re} = \frac{(0.505)(9.95)}{1.217 \text{ EE} - 5} = 4.13 \text{ EE5}$$

*step 7*: From the Moody friction factor chart, $f = 0.0165$, so

$$h_f = \frac{(0.0165)(51.4)(9.95)^2}{(2)(0.505)(32.2)} = 2.6 \text{ ft}$$

*step 8*: From equation 4.12,

$$\text{NPSHA} = 33.9 + 20 - 2.6 - 0.6 = 50.7 \text{ ft}$$

## 6 PUMPING HYDROCARBONS AND OTHER LIQUIDS

NPSHR is specified by pump manufacturers for use with cold (85°F) water.[3] Minor variations in the water temperature do not change the NPSHR appreciably. Experiments have shown, however, that the NPSHR can be reduced from the cold water values when hydrocarbons are pumped. This reduction apparently is due to the slow vapor release of complex organic liquids. Figure 4.11 gives the percentage correction to be applied to NPSHR values for cold water. Notice that the vapor pressure at the pumping temperature must be known. This can be obtained from figure 4.9.

[3] The term "cold *clear* water" is frequently used. This chapter omits the *clear* qualification.

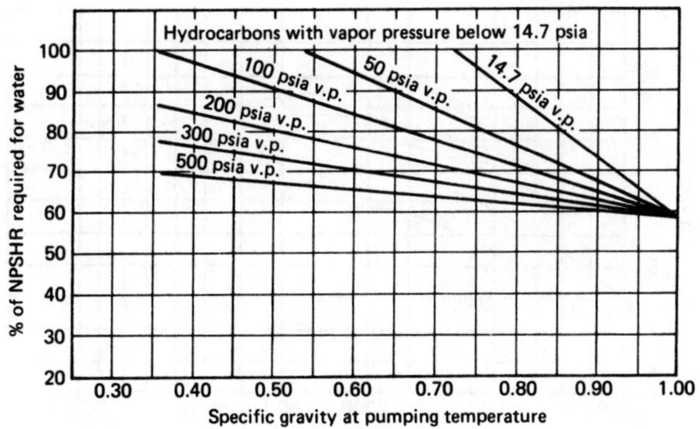


**Figure 4.11** Hydrocarbon NPSHR Correction Factor

Head developed by a pump is independent of the liquid being pumped, although the required horsepower is dependent on the fluid specific gravity. Because of this independence, pump performance curves from water tests can be used with other Newtonian fluids (e.g., gasoline, alcohol, and saline solutions).

High viscosity fluids, however, result in decreased head and capacity, as well as in an increase in input horsepower when compared to the pumping of water. Therefore, corrections are required when using water-based pump curves with viscous fluids. The parameters that would be used with water curves are calculated from equations 4.15 through 4.17.

$$H_{\text{water}} = \frac{H_{\text{viscous}}}{C_H} \qquad 4.15$$

$$Q_{\text{water}} = \frac{Q_{\text{viscous}}}{C_Q} \qquad 4.16$$

$$\eta_{\text{water}} = \frac{\eta_{\text{viscous}}}{C_E} \qquad 4.17$$

No exact method for determining the correction factors, other than actual tests of an installation with both fluids, exists. Nevertheless, several sources have produced correction factor charts based on experiments in limited viscosity and size ranges. One such chart is reproduced as Appendix E of this chapter.

*Example 4.2*

10°F liquid iso-butane (vapor pressure = 15 psia, specific gravity = 0.58) is to be used with a centrifugal pump whose NPSHR is 12 psia with cold water. What NPSHR should be used with the butane?

From figure 4.11, the intersection of 0.58 specific gravity and 15 psia is above the horizontal line. Therefore, the NPSHR is a full 12 psia.

## 7 RECIRCULATION

Cavitation is a high-volume problem. If a pump is forced to operate a flow rate higher than originally intended, one of the results could be cavitation.

Operating a pump at a flow rate much less than it was designed for, on the other hand, can result in *recirculation.* Recirculation of the fluid at both the impeller inlet and the pump outlet is possible. Such recirculation produces characteristics similar to cavitation. Specifically, vibration and noise are produced when the fluid energy is reduced through internal friction and fluid shear.

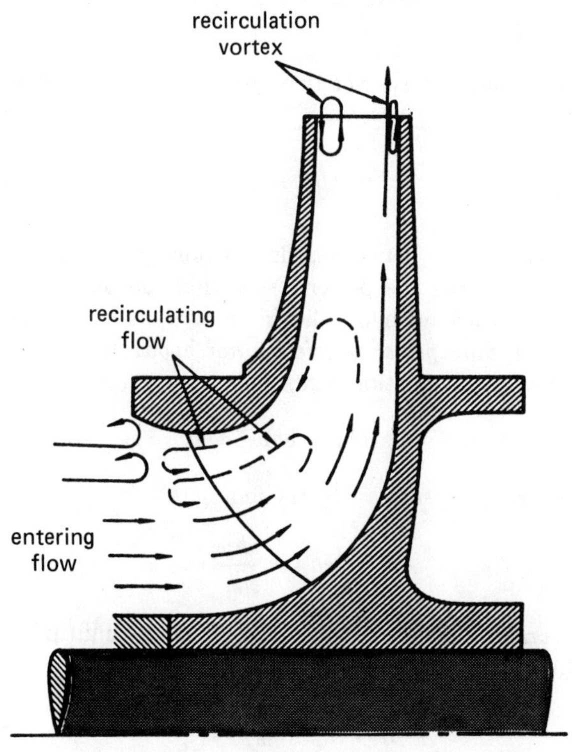


**Figure 4.12** Low-Volume Recirculation

## 8 PUMPING POWER AND EFFICIENCY

The energy (head) added by a pump can be determined by evaluating Bernoulli's equation on either side of the pump. Writing Bernoulli's equation for the discharge and inlet conditions produces equation 4.18.

$$h_A = \frac{p_d}{\rho} - \frac{p_i}{\rho} + \frac{v_d^2}{2g_c} - \frac{v_i^2}{2g_c} + z_d - z_i \qquad 4.18$$

The work performed by a pump is a function of the total head and the mass of the liquid pumped in a given time. Pump output is measured in *hydraulic horsepower, whp.*[4] Relationships for finding the hydraulic horsepower are given in table 4.2.

**Table 4.2**
Hydraulic Horsepower Equations

| | $Q$ in gpm | $\dot{m}$ in lbm/sec | $\dot{V}$ in cfs |
|---|---|---|---|
| $h_A$ is added head in feet | $\frac{h_A Q\,(S.G.)}{3956}$ | $\frac{h_A \dot{m}}{550}$ | $\frac{h_A \dot{V}\,(S.G.)}{8.814}$ |
| p is added head in psf | $\frac{pQ}{2.468\text{EE}5}$ | $\frac{p\dot{m}}{(34320)(S.G.)}$ | $\frac{p\dot{V}}{550}$ |

The input horsepower delivered to the pump shaft is the *brake horsepower.*

$$bhp = \frac{whp}{\eta_p} \qquad 4.19$$

The difference between hydraulic horsepower and brake horsepower is the power lost within the pump due to mechanical and hydraulic friction. This is referred to as *heat horsepower* or *friction horsepower* and is determined from equation 4.20.

$$fhp = bhp - whp \qquad 4.20$$

*Electrical horsepower* to the motor is

$$ehp = \frac{bhp}{\eta_m} \qquad 4.21$$

*Overall efficiency* is the pump efficiency multiplied by the motor efficiency:

$$\eta = (\eta_p)(\eta_m) = \frac{whp}{ehp} \qquad 4.22$$

Ideal pump efficiency is a function of the flow rate and specific speed. Figure 4.13 can be used if both quantities are known.

Larger horsepower pumps usually are driven by three-phase *induction motors.* The *synchronous speed* of such a motor is the speed of the rotating field.

$$n = \frac{(120)(f)}{\text{no. of poles}} \qquad 4.23$$

The frequency, $f$, is either 60 Hz (cycles per second) or 50 Hz, depending on the location of the installation.[5] The number of poles can be two or usually some multiple of four, but it must be an even number.

[4] This term also is known as *water horsepower.*

[5] 50 Hz is used in Europe.

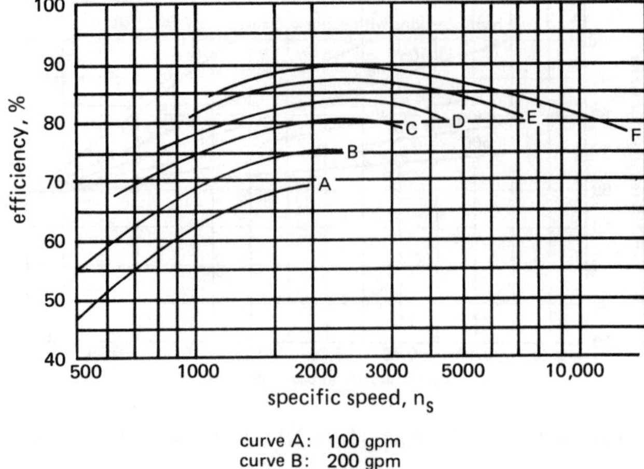


**Figure 4.13** Efficiency versus Specific Speed

**Table 4.3**
Synchronous Speeds

| Number of Poles | Synchronous Speed |
|---|---|
| 2 | 3600 |
| 4 | 1800 |
| 6 | 1200 |
| 8 | 900 |
| 12 | 600 |
| 18 | 400 |
| 24 | 300 |
| 48 | 150 |

Induction motors do not run at their synchronous speeds. Rather, they run at slightly less than synchronous speed. The percentage deviation is known as the *slip.* Slip is seldom greater than 10%, and it is usually much less than that. 4% is a good estimate for evaluation studies.

$$s = \frac{n_{\text{synchronous}} - n_{\text{actual}}}{n_{\text{synchronous}}} \qquad 4.24$$

Of course, special gear or belt drives can be used with various reduction ratios to obtain any required operating speed. For belt drive applications, most motors are of the 1800 and 1200 rpm varieties. Running speeds of 1750 and 1150 rpm are typically assumed for these motors when fully loaded.

Table 4.4 lists common motor sizes in horsepower. It is always best to specify standard motor sizes when possible.

**Table 4.4**
Standard Motor Sizes (BHP)

| |
|---|
| $\frac{1}{8}, \frac{1}{6}, \frac{1}{4}, \frac{1}{3}$ |
| $0.5, 0.75, 1, 1.5, 2, 3, 5, 7.5$ |
| $10, 15, 20, 25, 30, 40, 50, 60$ |
| $75, 100, 125, 150, 200, 250$ |

Induction motors are usually specified in terms of their KVA (kilo-volt-amp) ratings. KVA is not the same as the motor power in kilowatts, although one can be derived from the other if the motor's power factor is known. Such power factors can range from 0.8 to 0.9, depending on the installation and motor size.

$$\text{KVA rating} = \frac{\text{kilowatt power}}{\text{power factor}} \qquad 4.25$$

$$\text{kilowatt power} = (0.7457)\,ehp \qquad 4.26$$

If exact flow is not critical when changes in system and pump conditions occur, a constant speed drive can be used. Integral gear motors and motor-reducer drives are rugged, self-contained drives generally using 1800 rpm (typically taken as 1750 rpm to account for slip), three-phase induction motors and helical gear reducers. Horsepowers up to 50 are commonly available. With the gear reductions, the following approximate stock speeds are obtained: 37, 45, 56, 68, 84, 100, 125, 155, 180, 230, 280, 350, 420, 520, and 640 rpm.

V-belts are usually the lowest initial cost constant speed drive. The V-belt drive provides some flexibility for changing pump speeds through changes in sheave size. Using readily available standard motors of 1200 and 1800 (taken as 1150 and 1750) rpm, a range of pump speeds is possible. Due to sheave size and space limitations, the useful range of pump speeds is generally 200 to 600 rpm.

One disadvantage of a V-belt drive is the side load or overhung load it puts on both the pump and motor shafts and bearings. This problem is particularly significant with low speeds and high horsepowers. If it is determined that the overhung load is excessive, it will be necessary to use a jack drive or outboard bearing installation.

The integral gear motor is generally more compact, lower in cost, and easier to install. However, the motor and separate reducer is sometimes preferred because of its flexibility, especially in changing standard motors for maintenance.

If variable flow will require variations in motor speed, there are many types of packaged variable speed drives available. Such drives offer the ability to adjust pump speed to control flow, as well as adjusting for changes in the system and eventual pump wear. It is important to match the torque capacity of the drive with the required pump torque with all types of variable speed drives.

Belt type variable speed drives are available in a wide choice of horsepower and speed ranges, and they provide a compact drive at reasonable cost. Traction type drives are infinitely variable from zero speed, and they are reversible. Electronic variable speed drives, using both DC and AC motors with variable voltage or frequency to vary speed, generally require a speed reducer to obtain the required torque at lower pump speeds. Hydraulic drives, either packaged or custom designed, have excellent high torque capacity over a broad speed range.

When the calculated speed and required horsepower are known, a conservative approach is to select the next lower stock speed and a stock horsepower equal to or greater than the requirement. If a minimum flow must be maintained, even with system changes and pump wear, the next higher speed may be needed. In this case, the system should be recalculated, as the higher speed and resulting higher flow and pressure drop will require more horsepower. The motor/drive selected must be able to supply this horsepower.

*Example 4.3*

Recommend a 6-pole induction motor size for the pump in example 4.1. The friction loss in the discharge line is 12.0 feet. Neglect the electrical motor efficiency. Assume the pump is primed.

The Bernoulli equation can be used to find the head added by the pump.

$$0 + 0 + 20 + h_A = 0 + \frac{(9.95)^2}{2\,(32.2)} + 30 + 2.6 + 12$$

$$h_A = 26.14 \text{ ft}$$

From table 4.2,

$$whp = \frac{(26.14)(2)(1)}{8.814} = 5.93 \text{ hp}$$

The flow rate is

$$\frac{(2)}{(0.002228)} = 897.7 \text{ gpm}$$

Assuming a motor speed of 1150 rpm, equation 4.27 gives the specific speed

$$n_s = \frac{1150\sqrt{897.7}}{(26.14)^{0.75}} = 2980 \text{ rpm}$$

From figure 4.13, a pump efficiency of about 82% can be expected. So, the minimum motor horsepower would be

$$\frac{5.93}{0.82} = 7.23$$

From table 4.4, choose a 7.5 hp or larger motor.

## 9 SPECIFIC SPEED

The capacity or flow rate of a centrifugal pump is governed by the impeller thickness. For a given impeller diameter, the deeper the vanes, the greater the capacity of the pump.

For a desired flow rate or a desired discharge head, there will be one optimum impeller design. The impeller that is best for developing a high discharge pressure will have different proportions from an impeller designed to produce a high flow rate. The quantitative index of this optimization is called *specific speed* ($n_s$).

Specific speed is a function of the a pump's capacity, head, and rotational speed at peak efficiency. For a given pump and impeller configuration, the specific speed remains essentially constant over a range of flow rates and heads. Theoretically, specific speed is the speed in rpm at which a homologous pump would have to turn in order to put out 1 gpm at 1 foot total head. (For double suction pumps, $Q$ in equation 4.27 is not divided by 2.)

$$n_s = \frac{n\sqrt{Q}}{(h_A)^{0.75}} \qquad 4.27$$

Specific speed is used as a guide to selecting the most efficient pump type. Given a desired flow rate, pipeline geometry, and motor speed, $n_s$ is calculated from equation 4.27. The type of impeller is chosen from table 4.5

**Table 4.5**
Specific Speed versus Impeller Types

| approximate range of specific speed (rpm) | impeller type |
|---|---|
| 500–1000 | radial vane |
| 2000–3000 | Francis (mixed) vane |
| 4000–7000 | mixed flow |
| 9000 and above | axial flow |

Highest heads per stage are developed at low specific speeds. However, for best efficiency, a centrifugal pump's specific speed should be greater than 650 at its operating point. At low specific speeds, the impeller diameter is large with high mechanical friction and hydraulic losses. If the specific speed for a given set of conditions drops below 650, a multiple stage pump should be selected.[6]

As the specific speed increases, the ratio of the impeller diameter to the inlet diameter decreases. As this ratio decreases, the pump is capable of developing less head. Best efficiencies are usually obtained from pumps with specific speeds between 1500 and 3000. At specific speeds of 10,000 or higher, the pump is suitable for high flow rates but low discharge heads.

Other uses for specific speed are:

- If the pump and impeller are known, a maximum specific speed can be determined from table 4.5. This maximum specific speed can be translated into maximum values of rpm and flow rate, as well as a minimum value of total head added.
- If specific speed is known, an approximate pump efficiency can be found from figure 4.13.
- Specific speed limits have been established by the Hydraulic Institute.[7] These limits are presented in graphical form at the end of this chapter.

If NPSHR is substituted for total head in the expression for specific speed, a formula for *suction specific speed* results.

Suction specific speed is an index of the suction characteristics of the impeller.

$$n_{ss} = \frac{n\sqrt{Q}}{\text{NPSHR}^{0.75}} \qquad 4.28$$

Ideally, $n_{ss}$ should be approximately 7900 for single suction pumps and 11,200 for double suction pumps. (The value of $Q$ in equation 4.28 should be halved for double suction pumps.) If these ideal values are assumed, equation 4.28 can be solved for approximate values of NPSHR.

$$\text{NPSHR}_{\text{single suction}} \approx (6.36\ \text{EE}-6)\, n^{1.33} Q^{0.67} \qquad 4.29$$

$$\text{NPSHR}_{\text{double suction}} \approx (3.99\ \text{EE}-6)\, n^{1.33} Q^{0.67} \qquad 4.30$$

*Example 4.4*

A 3600 rpm pump ($Q = 150$ gpm) is used to increase fluid pressure from 35 psi to 220 psi. The pump adds negligible velocity or potential energy. If efficiency is to

[6] Relatively recent advances in *partial emission, forced vortex centrifugal pumps* allow operation down to specific speeds of 150. Such partial emission pumps use radial vanes and a single tangential discharge. Discharge is through a conical diffuser in which the energy conversion occurs. Partial emission pumps have been used for low flow, high head applications, such as petrochemical high-pressure cracking processes.

[7] Hydraulic Institute, 712 Lakewood Center North, 14600 Detroit Avenue, Cleveland, OH 44107.

be maximized, how many stages should be used? Assume that each stage adds its proportionate share of head.

The head added is

$$h_A = \frac{(220 - 35)(144)}{62.4} = 426.9 \text{ ft}$$

From equation 4.27, the specific speed is

$$n_s = \frac{(3600)\sqrt{150}}{(426.9)^{0.75}} = 470$$

From figure 4.13, the approximate efficiency is 45%.

Try a two stage pump. The head is split between the stages.

$$n_s = \frac{(3600)\sqrt{150}}{\left(\frac{426.9}{2}\right)^{0.75}} = 790$$

From figure 4.13, the efficiency is approximately 60%.

This process continues until the specific speed reaches 2000, at which time the number of stages will be approximately seven. It is questionable, however, whether the cost of multi-staging is worthwhile in this low-volume application.

*Example 4.5*

A direct driven pump is to discharge 150 gpm against a 300 foot total head when turning at the fully-loaded speed of 3500 rpm. What type of pump should be selected?

Calculate the specific speed from equation 4.27.

$$n_s = \frac{3500\sqrt{150}}{(300)^{0.75}} = 594.7 \text{ rpm}$$

From table 4.5, the pump should be a radial vane type. However, pumps with best efficiencies have $n_s$ greater than 650. To increase the specific speed, the rotational speed can be increased or the total head can be decreased. Since 3600 rpm is the maximum practical speed for induction motors, the better choice would be to divide the total head between two stages (or to use two pumps in a series).

In a two stage system, the specific speed would be:

$$n_s = \frac{3500\sqrt{150}}{(150)^{0.75}} = 1000$$

This is satisfactory for a radial vane pump.

## 10 AFFINITY LAWS—CENTRIFUGAL PUMPS

Most parameters (impeller diameter, speed, and flow rate) determining a pump's performance can vary. If the impeller diameter is held constant and the speed varied, the following ratios are maintained with no change of efficiency:

$$\frac{Q_2}{Q_1} = \frac{n_2}{n_1} \qquad 4.31$$

$$\frac{h_2}{h_1} = \left(\frac{n_2}{n_1}\right)^2 = \left(\frac{Q_2}{Q_1}\right)^2 \qquad 4.32$$

$$\frac{bhp_2}{bhp_1} = \left(\frac{n_2}{n_1}\right)^3 = \left(\frac{Q_2}{Q_1}\right)^3 \qquad 4.33$$

These relationships assume that the efficiencies of the larger and smaller pumps are the same. In reality, larger pumps will be more efficient than smaller pumps. Therefore, extrapolations to much larger or much smaller sizes should be avoided.

Equation 4.34 can be used to predict the efficiency of a larger or smaller pump based on homologous data.

$$\frac{1 - \eta_{\text{smaller}}}{1 - \eta_{\text{larger}}} = \left(\frac{d_{\text{larger}}}{d_{\text{smaller}}}\right)^{0.2} \qquad 4.34$$

If the speed is held constant and the impeller size varied,

$$\frac{Q_2}{Q_1} = \frac{d_2}{d_1} \qquad 4.35$$

$$\frac{h_2}{h_1} = \left(\frac{d_2}{d_1}\right)^2 \qquad 4.36$$

$$\frac{bhp_2}{bhp_1} = \left(\frac{d_2}{d_1}\right)^3 \qquad 4.37$$

*Example 4.6*

A pump delivers 500 gpm against a total head of 200 feet operating at 1770 rpm. Changes have increased the total head to 375 feet. At what rpm should this pump be operated to achieve this new head at the same efficiency?

From equation 4.32,

$$n_2 = 1770\sqrt{\frac{375}{200}} = 2424 \text{ rpm}$$

## 11 PUMP SIMILARITY

The performance of one pump can be used to predict the performance of a *dynamically similar* (*homologous*) *pump*. This can be done by using equations 4.38 through 4.40.

$$\frac{n_1 d_1}{\sqrt{h_1}} = \frac{n_2 d_2}{\sqrt{h_2}} \quad 4.38$$

$$\frac{Q_1}{d_1^2 \sqrt{h_1}} = \frac{Q_2}{d_2^2 \sqrt{h_2}} \quad 4.39$$

$$\frac{bhp_1}{\rho_1 d_1^2 h_1^{1.5}} = \frac{bhp_2}{\rho_2 d_2^2 h_2^{1.5}} \quad 4.40$$

$$\frac{Q_1}{n_1 d_1^3} = \frac{Q_2}{n_2 d_2^3} \quad 4.41$$

$$\frac{bhp_1}{\rho_1 n_1^3 d_1^5} = \frac{bhp_2}{\rho_2 n_2^3 d_2^5} \quad 4.42$$

$$\frac{n_1 \sqrt{Q_1}}{(h_1)^{0.75}} = \frac{n_2 \sqrt{Q_2}}{(h_2)^{0.75}} \quad 4.43$$

These so-called *similarity laws* assume that both pumps

- operate in the turbulent region.
- have the same operating efficiency.
- operate with the same percentage of wide-open flow.

Similar pumps also will have the same specific speed and cavitation number.

*Example 4.7*

A 6″ pump operating at 1770 rpm discharges 1500 gpm of cold water (S.G. = 1.0) against an 80 foot head at 80% efficiency. A homologous 8″ pump operating at 1170 rpm is being considered as a replacement. What capacity and total head can be expected from the new pump? What would be the new power requirement?

From equation 4.38,

$$H_2 = \left[\frac{(8)(1170)}{(6)(1770)}\right]^2 (80) = 62.14 \text{ ft}$$

From equation 4.41,

$$Q_2 = \left[\frac{(1170)(8)^3}{(1770)(6)^3}\right] (1500) = 2350.3 \text{ gpm}$$

$$whp_2 = \frac{(2350.3)(62.14)(1.0)}{3956} = 36.92 \text{ hp}$$

$$bhp_2 = \frac{36.92}{0.8} = 46.2 \text{ hp}$$

## 12 THE CAVITATION NUMBER

Although it is difficult to predict when cavitation will occur, the *cavitation number* (*cavitation coefficient*) can be used in modeling and in comparing experimental results. The actual cavitation number, given by equation 4.44, is compared with the critical cavitation number from experimental results. If the critical cavitation number is larger, it is concluded that cavitation will result.

$$\sigma = \frac{p - p_{vp}}{\frac{\rho v^2}{2g_c}} = \frac{\text{NPSHA}}{h_A \text{ per stage}} \quad 4.44$$

The two forms of equation 4.44 yield slightly different results. The first form is essentially the ratio of the net pressure available for collapsing a vapor bubble to the velocity pressure creating the vapor. The first form is useful in model experiments, whereas the second form is applicable to tests of production model pumps.

## 13 PUMP PERFORMANCE CURVES

Evaluating the performance of a pump is often simplified by examining a graphical representation of its operating characteristics. For a given impeller diameter and constant speed, the head added by a centrifugal pump will decrease as the flow rate increases. This is illustrated by figure 4.14. Other operating characteristics also vary with the flow rate. These can be presented on individual graphs. However, since the independent variable (flow rate) is the same for all, common practice is to plot all charcteristics together on a single graph. A pump *performance curve* is illustrated in figure 4.15.

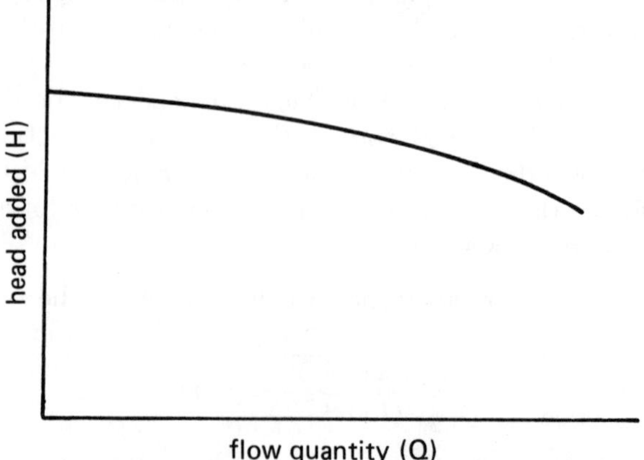


**Figure 4.14** Head versus Flow Rate

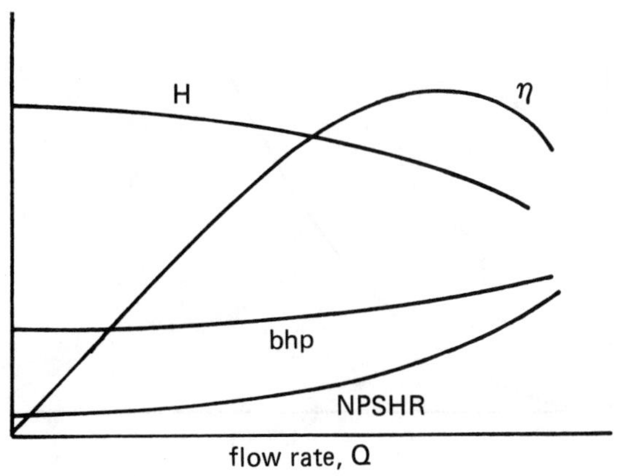


**Figure 4.15** Pump Performance Curves

Figures 4.14 and 4.15 are for a pump with a fixed impeller diameter and rotational speed. The characteristics of a pump operated over a range of speeds are illustrated in figure 4.16. For maximum efficiency, the operating point should fall along the dotted line.

Manufacturers' performance curves show pump performance at a limited number of calibration speeds. The desired operating point can be outside the range of the published curves. It is then necessary to estimate a speed at which the pump would give the required performance. This is done by using the affinity laws, as illustrated in example 4.8.

*Example 4.8*

A pump with the 1750 rpm performance curve shown is required to pump 500 gpm at 425 feet total head. At what speed must this pump be driven to achieve the desired performance with no change in efficiency or impeller size?

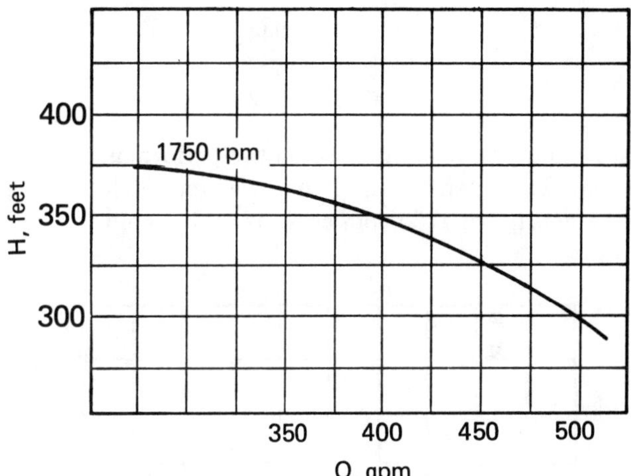


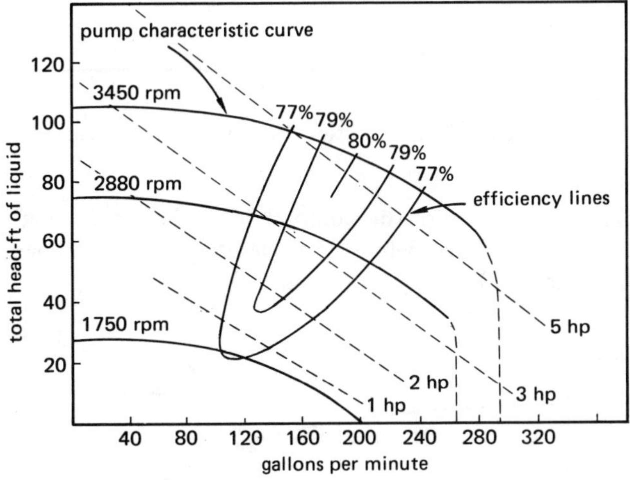


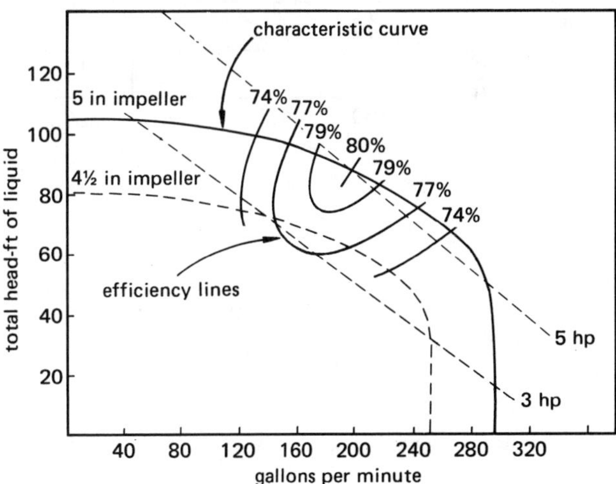


**Figure 4.16** Two Types of Characteristic Curves

From equation 4.32, the quantity $(H/Q^2)$ is constant for a pump with a given impeller size. In this case,

$$\frac{425}{(500)^2} = 1.7 \text{ EE} - 3$$

In order to apply an affinity equation, it is necessary to know the operating point on the 1750 rpm curve. To find the operating point, choose random values of Q and solve for H such that $(H/Q^2) = 1.7 \text{ EE} - 3$.

| Q | H |
|---|---|
| 475 | 383 |
| 450 | 344 |
| 425 | 307 |
| 400 | 272 |

These four points are plotted on the performance curve graph. The intersection of the constant efficiency line and the original 1750 rpm curve is at 440 gpm.

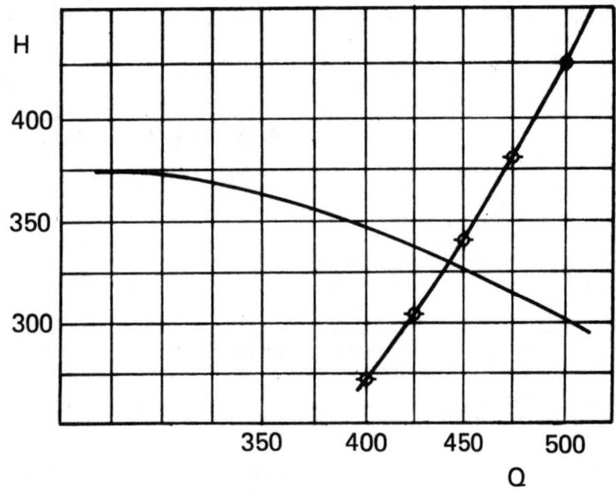


Then, from equation 4.31,

$$n_2 = 1750 \left(\frac{500}{440}\right) = 1989 \text{ rpm}$$

## 14 SYSTEM CURVES

A *system curve* graph can also be made from the resistance to flow of the piping system. This resistance varies with the square of the flow rate since $h_f$ varies with $v^2$ in the Darcy friction formula.

$$\frac{H_1}{Q_1^2} = \frac{H_2}{Q_2^2} \qquad 4.45$$

Equation 4.45 is illustrated by figure 4.17, in which there is no static head ($h_{ts}$) to overcome.

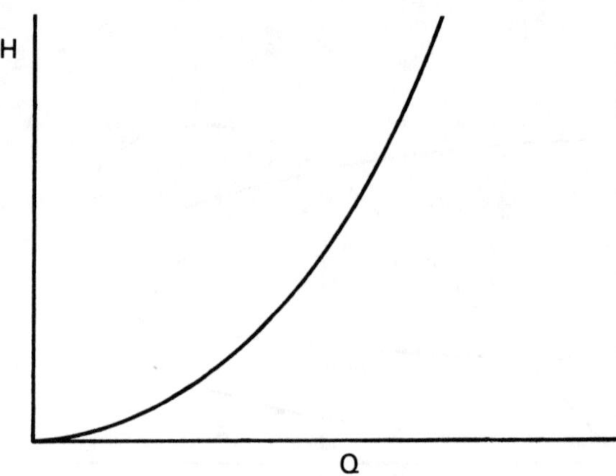


**Figure 4.17** System Performance Curve (Dynamic Losses Only)

When a static head ($h_{ts}$) exists in a system, the loss curve is displaced upward an amount equal to the static head. This is illustrated in figure 4.18.

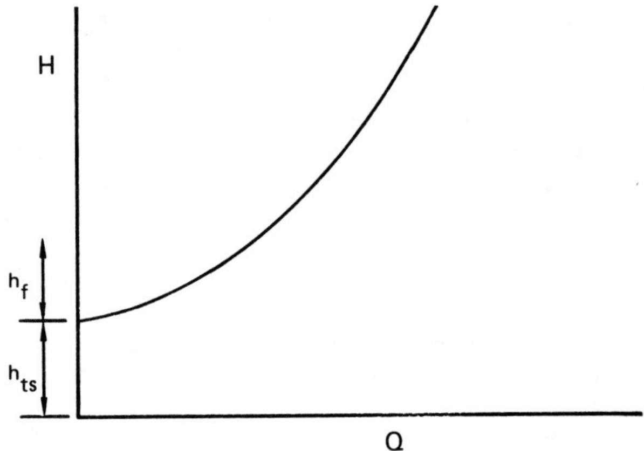


**Figure 4.18** System Performance Curve

The intersection of the pump characteristic curve with the system curve defines the *operating point* as shown in figure 4.19.

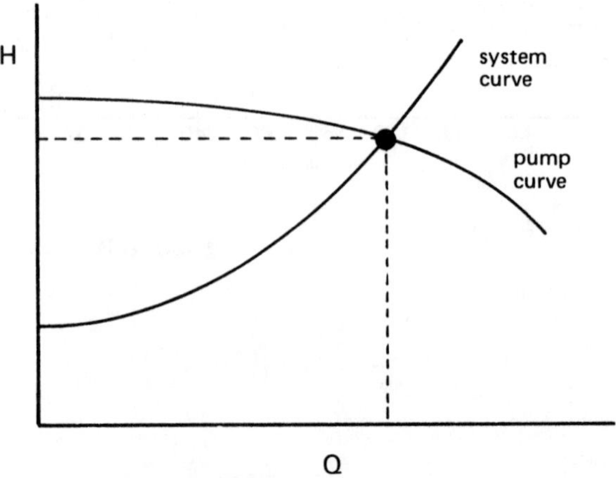


**Figure 4.19** Operating Point

After a pump is installed, it may be desired to vary the pump's performance. If a valve is placed in the discharge line, the operating point may be moved along the performance curve by opening or closing the valve. This is illustrated in figure 4.20. (A throttling valve should never be placed in the intake line since that would reduce NPSHA.)

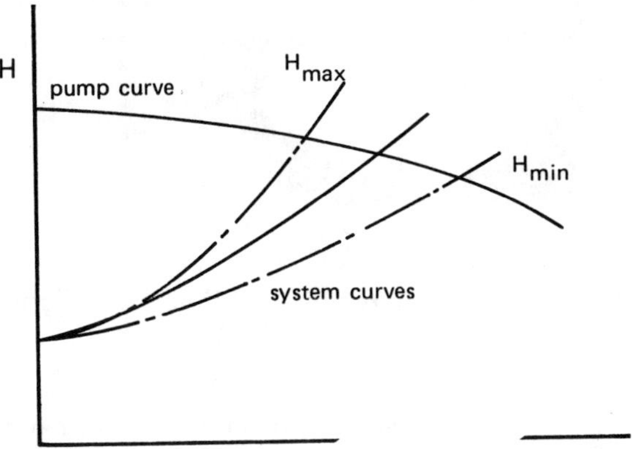


**Figure 4.20** Effe ... ttling the Discharge

In most systems, the static head will vary as the feed tank is drained or as the discharge tank fills. The system head is then defined by a pair of parallel curves intersecting the performance curve. The two intercept points are the maximum and the minimum capacity requirements.

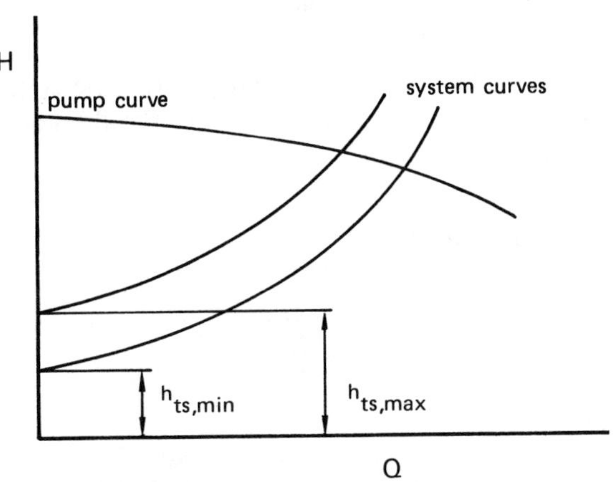


**Figure 4.21** Extreme Operating Points

## 15 PUMPS IN SERIES AND PARALLEL

Parallel operation is obtained by having two pumps discharging into a common header. This type of connection is advantageous when the system demand varies greatly. A single pump providing total flow would have to operate far from its optimum efficiency at one point or another. With two pumps in parallel, one can be shut down during low demand. This allows the remaining pump to operate close to its optimum efficiency point.

Figure 4.22 illustrates that parallel operation increases the capacity of the system while maintaining the same total head.

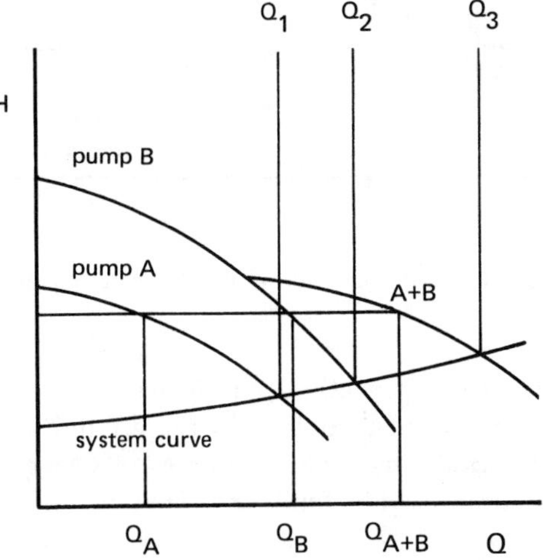


**Figure 4.22** Pumps Operating in Parallel

The performance curve for a set of pumps in parallel can be plotted by adding the capacities of the two pumps at various heads. Capacity does not increase at heads above the maximum head of the smaller pump. Furthermore, a second pump will operate only when its discharge head is greater than the discharge head of the pump already running.

When the parallel performance curve is plotted with the system head curve, the operating point is the intersection of the system curve with the $A + B$ curve. With pump A operating alone, the capacity is given by $Q_1$. When pump B is added, the capacity increases to $Q_3$ with a slight increase in total head.

Series operation is achieved by having one pump discharge into the suction of the next. This arrangement is used primarily to increase the discharge head, although a small increase in capacity also results.

The performance curve for a set of pumps in series can be plotted by adding the heads of the two pumps at various capacities.

## 16 CONSIDERATION FOR WASTEWATER PLANTS

The primary consideration in choosing a pump to lift sewage is the pump's tendency to clog. Centrifugal pumps for sewage and liquids with large solids should always be of the single-suction type with non-clog, open

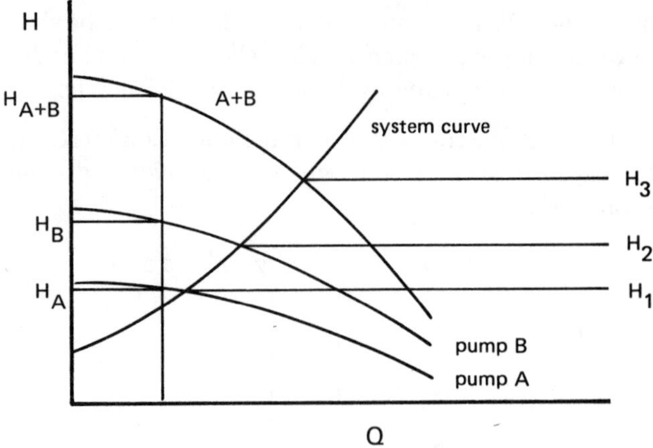


**Figure 4.23** Pumps Operating in Series

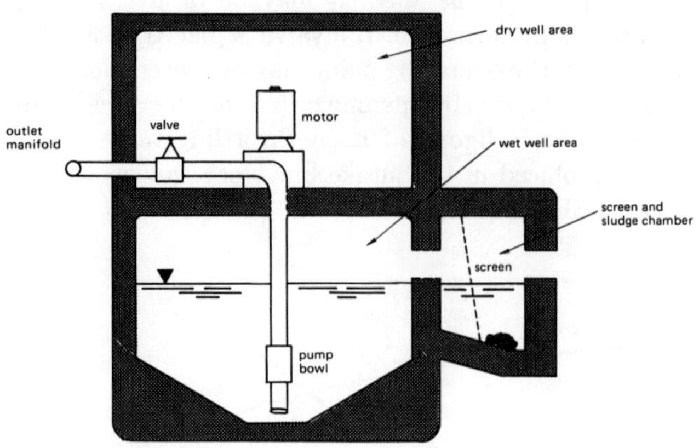


**Figure 4.24** Typical Wastewater Pump Installation (greatly simplified)

impellers. (Double suction pumps are prone to clogging because rags will catch and wrap around the shaft which extends through the impeller eye.) Clogging can be further minimized by limiting the number of impeller vanes to two or three, providing for large passageways, and using a bar screen ahead of the pump.

Non-clog pumps are of heavy construction but are constructed for ease of cleaning and repair. Horizontal pumps usually have a split casing, one-half of which can be removed for maintenance. A hand-sized cleanout opening may also be built into the casing. Although designed for long life, a sewage pump should normally be used with a grit chamber for prolonged bearing life.

The solid-handling capacity of a pump may be given in terms of the largest sphere which can pass through it without clogging. For example, a wastewater pump with a 6″ inlet should be able to pass a 4″ sphere. Of course, the pump should also be capable of handling spheres with diameters slightly larger than the bar screen spacing.

Figure 4.24 shows a simplified wastewater pump installation. Not shown are instrumentation and water level measurement devices, baffles, lighting, drains for the dry well, electrical power, pump lubrication equipment, and access holes. In addition, the multiplicity of pumping equipment is not apparent from the figure. (Totally submerged pumps do not require dry wells. However, such pumps may be more difficult to access, maintain, and repair.)

The number of pumps used in a wastewater installation is largely dependent on expected demand, pump capacity, and design criteria for backup operation. Although there may be state and federal regulations affecting the design, it is considered good practice to install pumps in sets, with a backup pump being available for each set of pumps that performs the same function. The number of pumps and their capacities should be able to handle the peak flow with one pump in the set out of service.

## 17 IMPULSE TURBINES

As shown in figure 4.25, an *impulse turbine* converts the energy of a fluid stream into kinetic energy by use of a nozzle which directs the stream jet against the turbine blades. Impulse turbines are generally employed where the available head exceeds 800 feet. Their efficiencies are typically in the 80% to 90% range.

The *total head available* to an impulse turbine is given by equation 4.46. ($p$ is the pressure of the fluid at the nozzle entrance.)

$$H' = H - h_n = \frac{p}{\rho} + \frac{v^2}{2g_c} - h_n \tag{4.46}$$

$$= h_s - \frac{fL_e v^2}{2Dg_c} - h_n \tag{4.47}$$

The *nozzle loss* is

$$h_n = \left(\frac{p}{\rho} + \frac{v^2}{2g_c}\right)(1 - C_v^2) \tag{4.48}$$

The velocity of the fluid jet is

$$v_j = \sqrt{2gH'} = C_v\sqrt{2gH} \tag{4.49}$$

The energy transmitted by each pound of fluid to the turbine runner is given by equation 4.50.

$$E = \frac{v_T(v_j - v_T)}{g_c}(1 - \cos\beta) \tag{4.50}$$

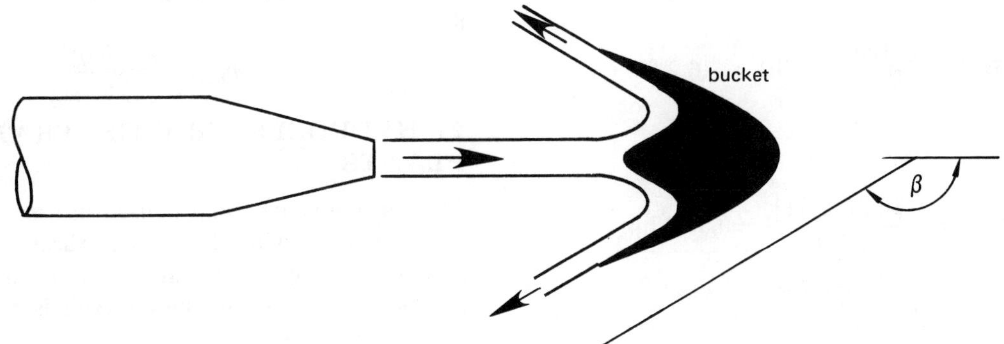


**Figure 4.25** A Simple Impulse Turbine

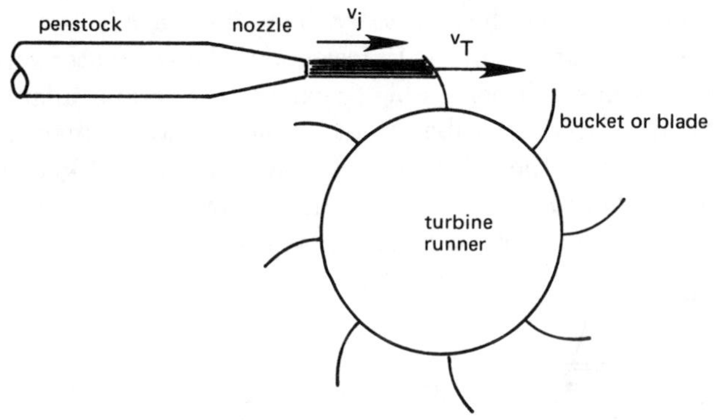


**Figure 4.26** Turbine Blade Geometry

Multiplying the energy by the fluid flow rate gives an expression for the theoretical horsepower output of the turbine.

$$bhp_{th} = \frac{Q\rho\,(v_j - v_T)\,v_T\,(1 - \cos\beta)}{(2.47\ \text{EE5})g_c} \qquad 4.51$$

The actual output will be less than the theoretical output. Typical efficiencies range from 80% to 90%.

*Example 4.9*

A Pelton wheel impulse turbine developing 100 bhp (net) is driven by a water stream from an 8″ schedule 40 penstock. Total head (before nozzle loss) is 200 feet. If the turbine runner is rotating at 500 rpm and its efficiency is 80%, determine the area of the jet, the flow rate, and the pressure head at the nozzle entrance. ($C_v = 0.95$)

From equation 4.49, the jet velocity is

$$v_j = 0.95\sqrt{2\,(32.2)\,(200)} = 107.8\ \text{ft/sec}$$

From equation 4.48, the nozzle loss is

$$h_n = 200[1 - (0.95)^2] = 19.5\ \text{ft}$$

Using table 4.2, the flow rate is

$$\dot{V} = \frac{(8.814)\,(100)}{(200 - 19.5)\,(1)\,(0.8)} = 6.10\ \text{cfs}$$

The jet area is

$$A_j = \frac{6.10}{107.8} = 0.0566\ \text{ft}^2$$

The flow area of 8″ schedule 40 pipe is 0.3474 ft$^2$.

The velocity at the nozzle entrance is

$$\frac{6.10}{0.3474} = 17.56\ \text{ft/sec}$$

The pressure head at the nozzle entrance is

$$200 - \frac{(17.56)^2}{2(32.2)} = 195.2\ \text{ft}$$

## 18 REACTION TURBINES

*Reaction turbines* are essentially centrifugal pumps in reverse. They are used when the total head is small, typically below 800 feet. However, their energy conversion efficiency is higher than for impulse turbines, typically 90% to 95%.

Since reaction turbines are centrifugal pumps in reverse, all of the affinity and similarity relationships (equations 4.31 through 4.42) can be used when comparing homologous turbines.

*Example 4.10*

A reaction turbine develops 500 bhp. Flow through the turbine is 50 cfs. Water enters at 20 fps with a 100′ pressure head. Elevation of the turbine above tailwater level is 10′. Find the effective head and turbine efficiency.

The effective (total) fluid head is

$$H = 100 + \frac{(20)^2}{2\,(32.2)} + 10 = 116.2 \text{ ft}$$

From table 4.2,

$$(whp)_{\text{in}} = \frac{(116.2)\,(50)\,(1)}{8.814} = 659.2 \text{ hp}$$

$$\eta_T = \frac{500}{659.2} = 0.758$$

## 19 TURBINE SPECIFIC SPEED

Like centrifugal pumps, turbines are classified according to the manner in which the impeller extracts energy from the fluid flow. This is measured by the turbine specific speed equation, which is different from the equation used to calculate specific speed for pumps.

$$n_s = \frac{n\sqrt{bhp}}{H^{1.25}} \qquad 4.52$$

Each of the three types of turbines is associated with a range of specific speeds.

- Axial-flow turbines are used for low heads, high rotational speeds, and large flow rates. The propeller turbines operate with specific speeds in the 80 to 200 range. Their best efficiencies, however, are produced between 120 and 160.
- For reaction turbines, the specific speed varies from 10 to 100. Best efficiencies are found in the 40 to 60 range with heads between 80 and 600 feet.
- Radial flow turbines have the lowest specific speeds. These impulse wheels have specific speeds below 5.

## 20 TURBINE EFFICIENCY

A turbine extracts energy from fluid. Its efficiency is the ratio of actual energy extracted to ideal energy extracted—the opposite of the definition of pump efficiency.

$$\eta_{\text{turbine}} = \frac{bhp}{whp} \qquad 4.53$$

## 21 HYDROELECTRIC GENERATING PLANTS

The turbine is generally housed in a *powerhouse*, with water conducted to the turbine through the *penstock* piping. Water originates in a reservoir, dam, or *forebay* (in the instance where the reservoir is a large distance from the turbine).

After the water passes through the turbine, it is discharged through the draft tube to the receiving reservoir, known as the *tail water*. The *draft tube* is used to keep the turbine up to 15 feet above the tail water surface, while still being able to extract the total available head. If a draft tube is not employed, water may be returned to the tail water by way of a channel known as the *tail race*. The turbine, draft tube, and all related parts comprise what is known as the *setting*.

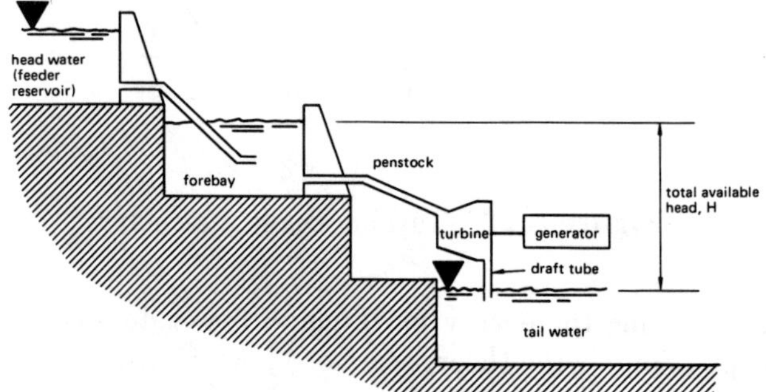


**Figure 4.27** A Typical Hydroelectric Plant

When a forebay is not part of the generating plant's design, it will be desirable to provide a *surge chamber* in order to relieve the effects of rapid changes in flow rate. In the case of a sudden power demand, the surge chamber would provide an immediate source of water, without waiting for a contribution from the feeder reservoir. Similarly, in the case of a sudden decrease in discharge through the turbine, the excess water would surge back into the surge chamber.

# Appendix A: Atmospheric Pressure versus Altitude

| altitude ft | pressure psia | altitude ft | pressure psia |
|---|---|---|---|
| 0 | 14.696 | 33000 | 3.797 |
| 1000 | 14.175 | 34000 | 3.625 |
| 2000 | 13.664 | 35000 | 3.458 |
| 3000 | 13.168 | 36000 | 3.296 |
| 4000 | 12.692 | 37000 | 3.143 |
| 5000 | 12.225 | 38000 | 2.996 |
| 6000 | 11.778 | 39000 | 2.854 |
| 7000 | 11.341 | 40000 | 2.721 |
| 8000 | 10.914 | 41000 | 2.593 |
| 9000 | 10.501 | 42000 | 2.475 |
| 10000 | 10.108 | 43000 | 2.358 |
| 11000 | 9.720 | 44000 | 2.250 |
| 12000 | 9.347 | 45000 | 2.141 |
| 13000 | 8.983 | 46000 | 2.043 |
| 14000 | 8.630 | 47000 | 1.950 |
| 15000 | 8.291 | 48000 | 1.857 |
| 16000 | 7.962 | 49000 | 1.768 |
| 17000 | 7.642 | 50000 | 1.690 |
| 18000 | 7.338 | 51000 | 1.611 |
| 19000 | 7.038 | 52000 | 1.532 |
| 20000 | 6.753 | 53000 | 1.464 |
| 21000 | 6.473 | 54000 | 1.395 |
| 22000 | 6.2 | 55000 | 1.331 |
| 23000 | 5.943 | 56000 | 1.267 |
| 24000 | 5.693 | 57000 | 1.208 |
| 25000 | 5.452 | 58000 | 1.154 |
| 26000 | 5.216 | 59000 | 1.100 |
| 27000 | 4.990 | 60000 | 1.046 |
| 28000 | 4.774 | 61000 | 0.997 |
| 29000 | 4.563 | 62000 | 0.953 |
| 30000 | 4.362 | 63000 | 0.909 |
| 31000 | 4.165 | 64000 | 0.864 |
| 32000 | 3.978 | 65000 | 0.825 |

# Appendix B: Upper Limits of Specific Speeds

Single Stage, Single and Double Suction Pumps
Handling Clear Water at 85°F at Sea Level

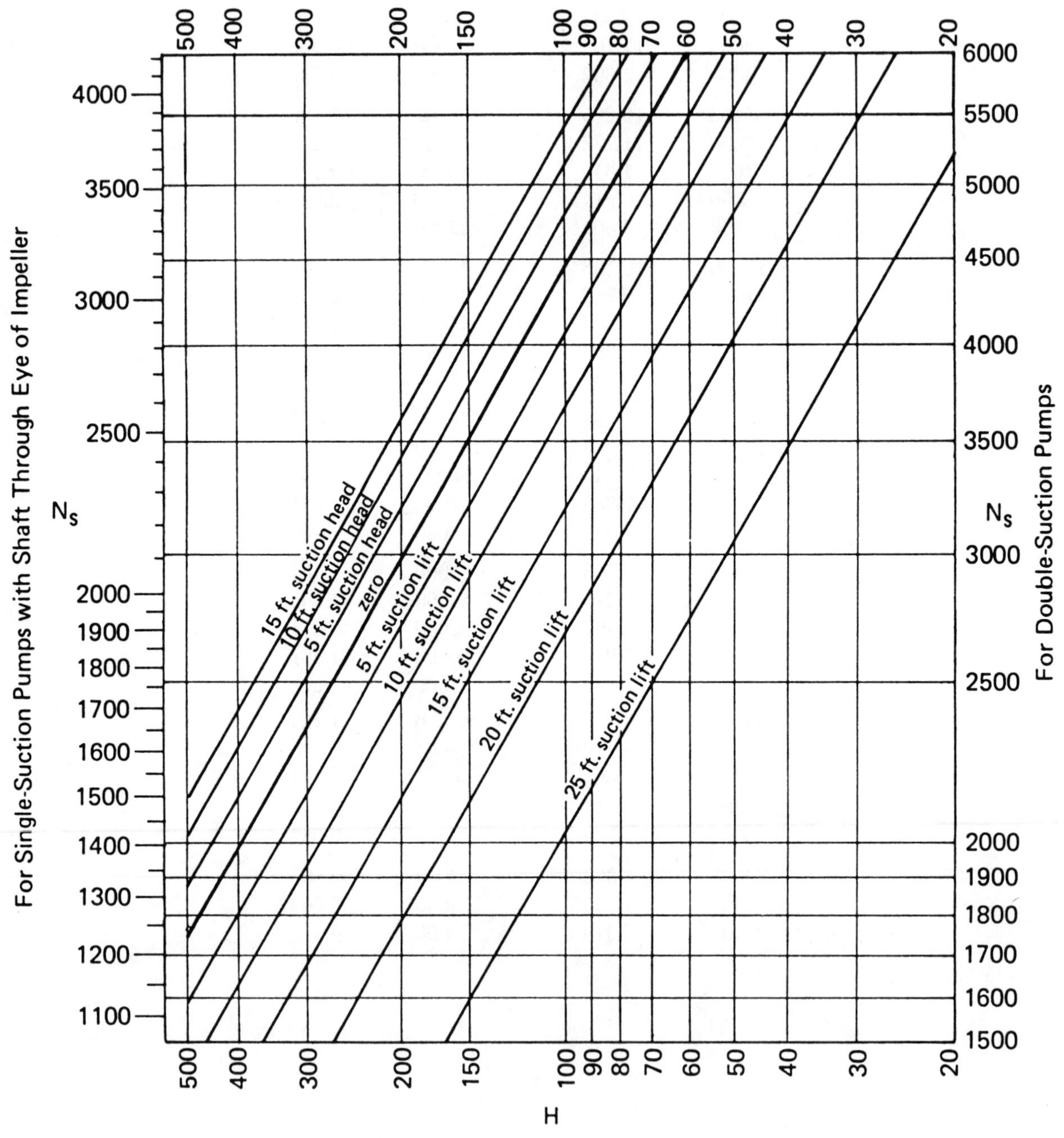

# Appendix C: Upper Limits of Specific Speeds

Single Stage, Single Suction, Mixed and Axial Flow Pumps
Handling Clear Water at 85°F at Sea Level

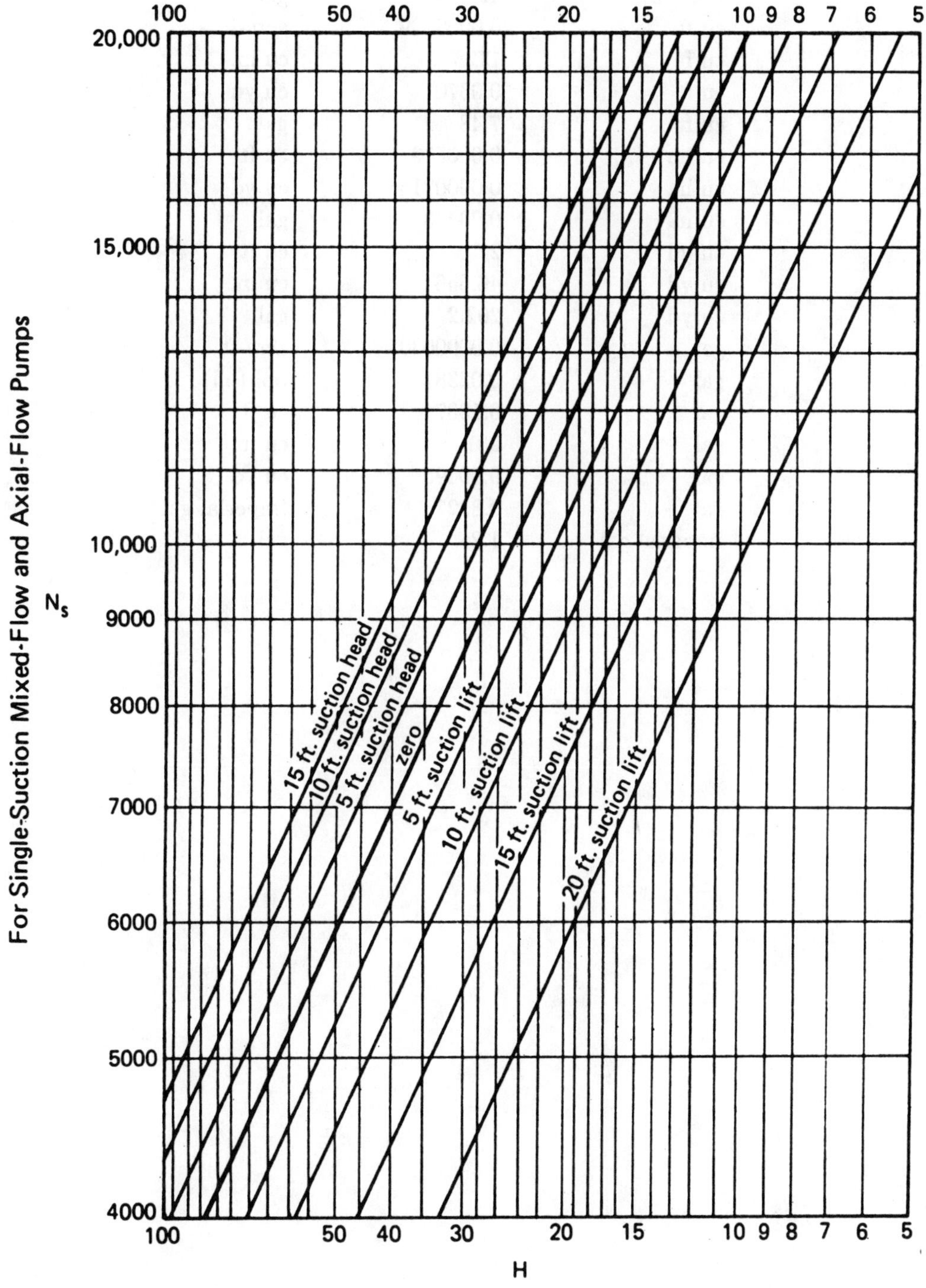

# Appendix D: Volumetric Conversion Factors

| multiply | by | to obtain |
|---|---|---|
| acre-ft | 43,560 | cu ft |
| acre-ft | 325,851 | gal |
| bbl (oil) | 42 | gal |
| cu ft | 0.0000229 | acre-ft |
| cu ft | 1728 | cu in. |
| cu ft | 0.0370 | cu yd |
| cu ft | 7.48 | gal |
| cu in. | 0.000579 | cu ft |
| cu in. | 0.0000214 | cu yd |
| cu in. | 0.00433 | gal |
| cu yd | 27 | cu ft |
| cu yd | 46,656 | cu in. |
| cu yd | 202.2 | gal |
| gal | 0.00000307 | acre-ft |
| gal | 0.0238 | bbl (oil) |
| gal | 0.1337 | cu ft |
| gal | 231 | cu in. |
| gal | 0.00495 | cu yd |
| gal | 0.8327 | Imperial gal |
| Imperial gal | 1.2 | gal |

# Appendix E: Pump Performance Correction Factor Chart

(Prepared from tests on 1″ and smaller pumps)

*Instructions for use*: Start with the actual fluid flow rate on the bottom horizontal scale. Move vertically until the required head is reached on the diagonal head scale. Move horizontally to the left until the fluid viscosity is reached on the diagonal viscosity scale. Move upward and read the correction factors.

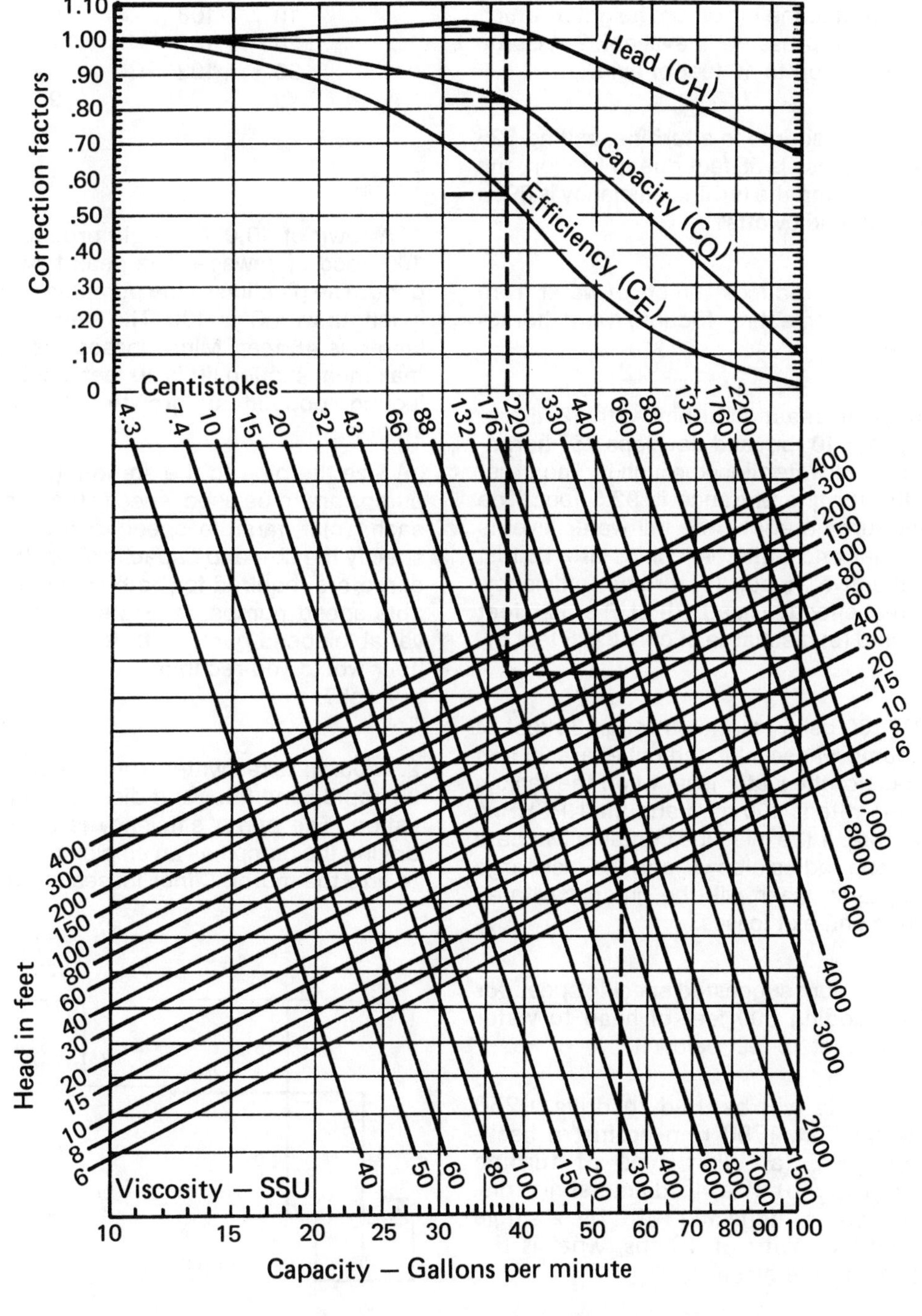

**Practice Problems: HYDRAULIC MACHINES**

Untimed

1. A sludge slurry with a specific gravity of 1.2 is pumped at the rate of 2000 gpm through an inlet of 12″ and out an 8″ discharge at the same level. The inlet gage reads 8″ of mercury below atmospheric. The discharge gage reads 20 psig and is located 4 feet above the centerline of the pump outlet. If the pump efficiency is 85%, what is the input power?

2. A pump discharges water at 12 fps through a 6″ line. The inlet is a section of 8″ line. Suction is 5 psig below atmospheric. If the pump is 20 horsepower and is 70% efficient, what is the maximum height at which water at atmospheric pressure is available? Assume all friction losses add up to 10 feet of fluid.

3. Water flows from a source to a turbine, exiting 625 feet lower. The head loss is 58 feet due to friction, the flow rate is 1000 cfs, and the turbine efficiency is 89%. What is the output in kilowatts?

4. A horizontal turbine reduces 100 cfs of water from 30 psia to 5 psia. Neglecting friction, what horsepower is generated?

5. A Francis hydraulic reaction turbine with 22″ diameter blades runs at 610 rpm and develops 250 horsepower when 25 cfs of water flow through it. The pressure head at the turbine entrance is 92.5 feet. The elevation of the turbine above the tail water level is 5.26 feet. The inlet and outlet velocities are 12 fps. Find the (a) effective head, (b) turbine efficiency, (c) rpm at 225 feet effective head, (d) BHP at 225 feet effective head, and (e) discharge in cfs at 225 feet effective head.

6. Water (180°F, 80 psia) empties through 30 feet of 1½″ pipe by a pump whose inlet and outlet are 20 feet below the surface of the water level when the tank is full. The pumping rate is 100 gpm and the NPSHR is 10 feet for that rate. If the inlet line contains two gate valves and two long radius elbows, and the discharge is into a 2 psig tank, when will the pump cavitate? Neglect entrance and exit losses.

7. What is the maximum suggested specific speed for a 2-stage pump adding 300 feet of head to water pulled through an inlet 10 feet below it?

8. Water at 500 psig will be used to drive a 250 horsepower turbine at 1750 rpm against a backpressure of 30 psig. (a) What type of turbine would you suggest? (b) If a 4″ diameter jet discharging at 35 fps is deflected 80° by a single moving vane with velocity of 10 fps, what is the total force acting on the blade?

9. A 1750 rpm pump is normally splined to a ½ horsepower motor. What horsepower is required if the pump is to run at 2000 rpm?

10. The inlet of a centrifugal pump is 7 feet above a water surface level. The inlet is 12 feet of 2″ pipe and contains one long-radius elbow and one check valve. The 2″ outlet contains two long-radius elbows in its 80 feet of length. The discharge is 20 feet above the surface. The following pump curve data is available. What is the flow if water is at 70°F?

| gpm | head | gpm | head |
|---|---|---|---|
| 0 | 110 | 50 | 93 |
| 10 | 108 | 60 | 87 |
| 20 | 105 | 70 | 79 |
| 30 | 102 | 80 | 66 |
| 40 | 98 | 90 | 50 |

Timed

1. A town of 10,000 people produces an average of 100 gpcd of sewage. The peak flow, however, is 250 gpcd. The pipeline to the pumping station is 5000 feet in length with C = 130. The elevation drop along the length is 48 feet. Minor losses are insignificant. The maximum suction lift is 10 feet inside the pump. Neglecting population growth, design the pumping station.

(a) Size the pipe to the station. (b) If constant speed pumps are to be used, specify the size and capacity of each. (c) If variable speed pumps are to be used, specify the size and capacity of each. (d) Find the size of motors required for both constant speed and variable speed pumps. (e) State six ways of controlling variable speed pumps. (f) How many air changes per hour would you recommend for the wet well? For the dry well?

2. A pump takes water from the clear well of a 10′ × 20′ rapid sand filter and discharges it at a higher elevation. The pump efficiency is 85%, and the motor driving the pump has an efficiency of 90%. It is desired to size the motor. Minor losses are to be ignored.

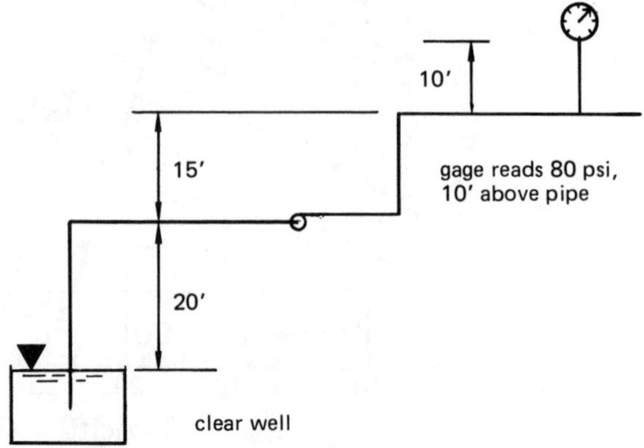

(a) What is the static suction lift? (b) What is the static discharge head? (c) What is the total dynamic head? (d) Determine the motor horsepower required to maintain the clear well at its current level if the filter output is 3.5 MGD. (e) If the filter output is increased 25%, how much surplus will there be in 8 hours?

3. A water distribution network is 10–15 years old, and is constructed of cast iron pipe. There is a requirement for 40 psi minimum pressure at point D. The inlet is at point A and the outlet is at point B.

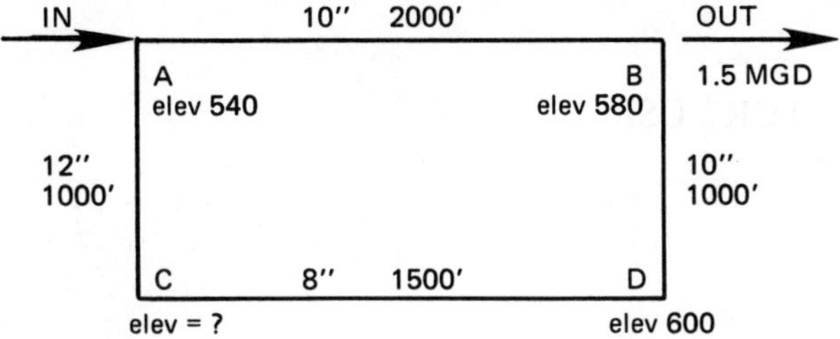


(a) Determine the flow rates into the two branches. (b) Determine the pressures at points A and B.

RESERVED FOR FUTURE USE

# 5 OPEN CHANNEL FLOW

*Nomenclature*

| | | |
|---|---|---|
| A | area | $ft^2$ |
| b | weir width | ft |
| C | coefficient | $ft^{\frac{1}{2}}/sec$ |
| d | depth | ft |
| $d_H$ | hydraulic diameter | ft |
| D | pipe diameter | ft |
| E | specific energy | ft-lbf/lbm |
| f | Darcy friction factor | – |
| g | acceleration due to gravity (32.2) | $ft/sec^2$ |
| $g_c$ | gravitational constant | $\frac{\text{lbm-ft}}{\text{lbf-sec}^2}$ |
| h | head | ft |
| H | total hydraulic head | ft |
| k | minor loss coefficient | – |
| K | conveyance | $ft^3/sec$ |
| L | channel length | ft |
| n | Manning roughness coefficient | – |
| N | number of end contractions | – |
| p | pressure | $lbf/ft^2$ |
| P | wetted perimeter, or weir height | ft, ft |
| Q | flow quantity | $ft^3/sec$ |
| $r_H$ | hydraulic radius | ft |
| S | slope of energy line (energy gradient) | (decimal) |
| $S_o$ | channel slope | (decimal) |
| v | velocity | ft/sec |
| w | channel width | ft |
| Y | weir height | ft |
| z | height above datum | ft |

*Symbols*

| | | |
|---|---|---|
| $\rho$ | density | $lbm/ft^3$ |

*Subscripts*

| | |
|---|---|
| b | brink |
| c | critical, or composite |
| e | equivalent, or entrance |
| f | friction |
| H | hydraulic |
| n | normal |
| o | uniform, or culvert barrel |
| s | spillway |
| t | total |
| w | weir |

## 1 INTRODUCTION

An open channel is a fluid passageway which allows part of the fluid to be exposed to the atmosphere. This type of channel includes natural waterways, canals, culverts, flumes, and pipes flowing under the influence of gravity (as opposed to pressure conduits which always flow full).

There are difficulties in evaluating open channel flow. The unlimited geometric cross sections and variations in roughness have contributed to a small number of scientific observations upon which to estimate the required coefficients and exponents. Therefore, the analysis of open channel flow is more empirical and less exact than that of pressure conduit flow. This lack of precision, however, is more than offset by the percentage error in runoff calculations that generally precede the channel calculations.

Flow in open channels is almost always turbulent. However, within that are many somewhat confusing categories of flow. Flow can also be categorized on the basis of the channel material. Except for a short discussion of erodible canals, this chapter assumes the channel is non-erodible.

Flow can be a function of time and location. If the flow quantity is invariant, it is said to be *steady.* If the flow cross section does not depend on the location along the channel, the flow is said to be *uniform.* Steady flow can be *non-uniform,* as in the case of a river with a varying cross section or on a steep slope. Furthermore, uniform channel construction does not ensure uniform flow, as will be seen in the case of hydraulic jumps.

Due to the adhesion between the wetted surface of the channel and the water, the velocity will not be uniform across the area in flow. The velocity term used in this chapter is the *mean velocity.* The mean velocity, when multiplied by the flow area, gives the flow quantity.

$$Q = Av \tag{5.1}$$

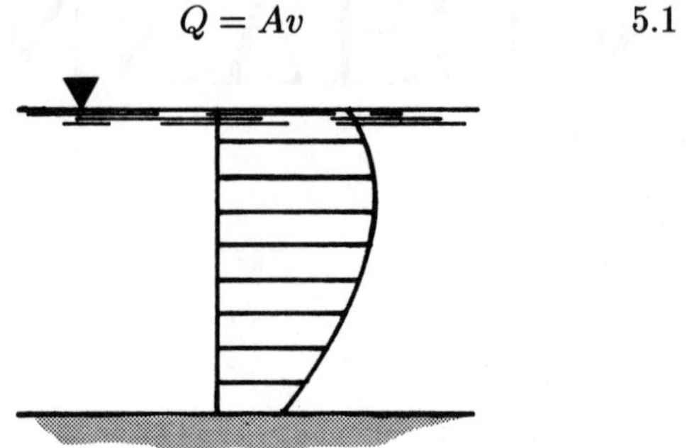

**Figure 5.1** Distribution of Velocities in a Channel

The location of the mean velocity depends on the distribution of velocities in the waterway, which is generally quite complex. The procedure for measuring the velocity of a channel (called *stream gaging*) involves measuring the average channel velocity at multiple locations across the channel width. These sub-average velocities are then themselves averaged to give a grand average (mean) flow velocity.

## 2 DEFINITIONS

Accelerated Flow: A form of varied flow in which the velocity is increasing and the depth is decreasing.

Apron: An underwater 'floor' constructed along the channel bottom to prevent scour. Aprons are almost always extensions of spillways and culverts.

Backwater: Water upstream from a dam or other obstruction which is deeper than it would normally be without the obstruction.

Backwater Curve: A plot of depth versus location along the channel containing backwater.

Check: A short section of built-up channel placed in a canal or irrigation ditch and provided with gates or flashboards to control flow or raise upstream level for diversion.

Colloid: An extremely fine sediment which will not easily settle out.

Conjugate Depths: The depths on either side of a hydraulic jump.

Contraction: A decrease in the width or depth of flow caused by the geometry of a weir, orifice, or obstruction.

Critical Flow: Flow at the critical depth and velocity. Critical flow minimizes the specific energy and maximizes discharge.

Critical Depth: The depth which minimizes the specific energy of flow.

Critical Slope: The slope which produces critical flow.

Critical Velocity: The velocity which minimizes specific energy. When water is moving at its critical velocity, a disturbance wave cannot move upstream since it moves at the critical velocity.

Downpull: A force on a gate, typically less at lower depths than at upper depths due to increased velocity, when the gate is partially open.

Energy Gradient: The slope of the specific energy line (i.e., the sum of the potential and velocity heads).

Flume: An open channel constructed above the earth's surface, usually supported on a trestle or on piers.

Forebay: A reservoir holding water for use after it has been discharged from a dam.

Freeboard: The height of the channel side above the water level.

Gradient: See 'Slope.'

Headwall: Entrance to a culvert or sluiceway.

Hydraulic Jump: A spontaneous increase in flow depth from a velocity higher than critical to a velocity lower than critical.

Hydraulic Mean Depth: Same as hydraulic radius.

Normal Depth: The depth of uniform flow. This is a unique depth of flow for any combination of channel conditions. Normal depth is found from the Chezy-Manning equation.

Overchute: A flume passing over a canal to carry floodwaters away without contaminating the canal water. An elevated culvert.

$q$-curve: A plot of depth of flow versus quantity flowing for a channel with a constant specific energy.

Rapid Flow: Flow at less than the critical depth, as typically occurs on steep slopes.

Rating Curve: A plot of quantity flowing versus depth for a natural watercourse.

Reach: A section of channel.

Retarded Flow: A form of varied flow in which the velocity is decreasing and the depth increasing.

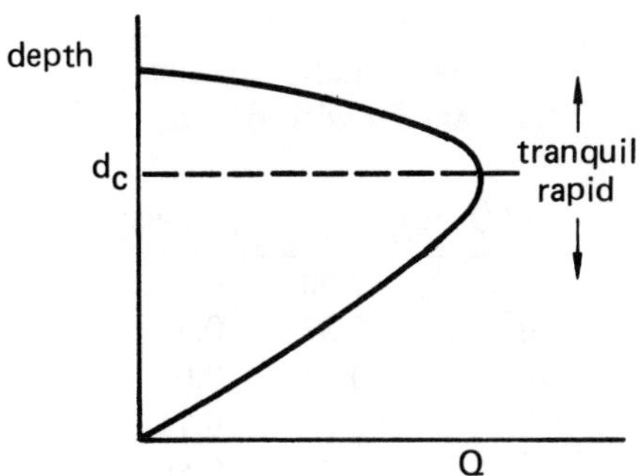


**Figure 5.2** A $q$-Curve

Sand Trap: A section constructed deeper than the rest of the channel to allow sediment to settle out.

Scour: Erosion at the exit of an open channel or toe of a spillway.

Settling Basin: A large, shallow basin through which water passes at low velocity, causing most of the suspended sediment to settle out.

Shooting Flow: See 'Rapid Flow.'

Sill: A submerged wall or weir.

Slope: The head loss per foot. For almost-level channels in uniform flow, the slope is equal to the tangent of the angle made by the channel bottom.

Standing Wave: A stationary wave caused by an obstruction in a water-course. The wave cannot move (propagate) because the water is flowing at its critical speed.

Steady Flow: Flow which does not vary with time.

Stilling Basin: An excavated pool downstream from a spillway used to decrease tailwater depth and to ensure an energy-dissipating hydraulic jump.

Stream Gaging: A method of determining the velocity in an open channel.

Subcritical Flow: Flow at greater than the critical depth (less than the critical velocity).

Supercritical Flow: Flow at less than the critical depth (greater than the critical velocity).

Tail Race: An open waterway leading water out of a dam spillway back to a natural channel.

Tail Water: The water into which a spillway or outfall discharges.

Tranquil Flow: Flow at greater than the critical depth.

Turnout: A pipe placed through a canal embankment to carry water from the canal for other uses.

Uniform Flow: Flow which has a constant depth, volume, and shape along its course.

Varied Flow: Flow that has a changing depth along the water course. The variation is with respect to location, not time.

Wasteway: A canal or pipe which returns excess irrigation water back to the main channel.

Wetted Perimeter: The length of the channel which has water contact. The air-water interface is not included in the wetted perimeter.

## 3 PARAMETERS USED IN OPEN CHANNEL FLOW

The *hydraulic radius* is the ratio of area in flow to wetted perimeter.

$$r_H = \frac{A}{P} \qquad 5.2$$

For a circular channel flowing either full or half-full, the hydraulic radius is $(D/4)$. Hydraulic radii of other channel shapes are easily calculated from the basic definition.

The *hydraulic depth* is the ratio of area in flow to the width of the channel at the fluid surface.[1]

$$d_H = \frac{A}{w} \qquad 5.3$$

The *slope*, $S$, in open channel equations is the slope of the energy line. If the flow is uniform, the slope of the energy line will parallel the water surface and channel bottom, $S_o$. In general, the slope can be calculated from the Bernoulli equation as the energy loss per unit length of channel.

$$S = \frac{dH}{dL} \qquad 5.4$$

## 4 GOVERNING EQUATIONS FOR UNIFORM FLOW

Since water is incompressible, the continuity equation is

$$A_1 v_1 = A_2 v_2 \qquad 5.5$$

The more common equation used to calculate the flow velocity in open channels is the *Chezy equation*.[2]

$$v = C\sqrt{r_H S} \qquad 5.6$$

Various equations for evaluating the coefficient $C$ have been proposed. If it is assumed that the channel is

[1] The *hydraulic mean depth*, another name for the hydraulic radius, is not the same as the hydraulic depth, however.

[2] Pronounced "Shay'-zee".

large, then the friction loss will not depend so much on the Reynolds number as on the channel roughness. The *Manning formula* is frequently used to evaluate the constant $C$. Notice that the value of $C$ depends only on the channel roughness and geometry.[3]

$$C = \frac{1.49}{n}(r_H)^{1/6} \qquad 5.7$$

$n$ is the Manning roughness constant. Combining equations 5.6 and 5.7 produces the *Chezy-Manning equation*, applicable when the slope is less than 0.10.

$$v = \frac{1.49}{n}(r_H)^{2/3}\sqrt{S} \qquad 5.8$$

All of the coefficients and constants in the Chezy-Manning equation may be combined into the *conveyance*, $K$.

$$Q = vA = \frac{1.49}{n}(A)(r_H)^{2/3}\sqrt{S} = K\sqrt{S} \qquad 5.9$$

*Example 5.1*

A rectangular channel on a 0.002 slope is constructed of finished concrete and is 8 feet wide. What is the uniform flow if water is at a depth of 5 feet?

The hydraulic radius is: $r_H = \frac{(8)(5)}{5+8+5} = 2.22$ ft

From Appendix A, the roughness coefficient for finished concrete is 0.012. The Manning coefficient is

$$C = \frac{1.49}{0.012}(2.22)^{1/6} = 141.8$$

The discharge from equation 5.9 is

$$Q = (141.8)(8)(5)\sqrt{(2.22)(0.002)} = 377.9 \text{ cfs}$$

## 5 VARIATIONS IN THE MANNING ROUGHNESS CONSTANT, $n$

For most calculations, $n$ is assumed to be constant. The accuracy of other parameters used in open flow calculations often does not warrant considering the variation of $n$ with depth, and the choice to use varying $n$ values is left to the individual designer.

If it is desired to acknowledge variations in $n$ with respect to depth, it is expedient to make use of tables or graphs of hydraulic elements. Table 5.1 lists such hydraulic elements under the assumption that $n$ varies. (Appendix C can be used for both varying and non-varying $n$.)

**Table 5.1**
Circular Channel Ratios
($n$ varying with depth)

| $d/D$ | $Q/Q_{full}$ | $v/v_{full}$ |
|---|---|---|
| 0.1 | 0.02 | 0.31 |
| 0.2 | 0.07 | 0.48 |
| 0.3 | 0.14 | 0.61 |
| 0.4 | 0.26 | 0.71 |
| 0.5 | 0.41 | 0.80 |
| 0.6 | 0.56 | 0.88 |
| 0.7 | 0.72 | 0.95 |
| 0.8 | 0.87 | 1.01 |
| 0.9 | 0.99 | 1.04 |
| 0.95 | 1.02 | 1.03 |
| 1.00 | 1.00 | 1.00 |

*Example 5.2*

2.5 cfs of water are in uniform flow in a 20″ sewer line ($n = 0.015$, $S = 0.001$). What are the depth and velocity? (Assume $n$ varies with depth.)

If the pipe were to flow full, it would carry $Q_{full}$.

$$D = 20/12 = 1.667 \text{ ft}$$

$$r_H = \frac{1}{4}D = \frac{1}{4}(1.667) = 0.417 \text{ ft}$$

$$Q_{full} = \frac{1}{4}\pi(1.667)^2\left(\frac{1.49}{0.015}\right)(0.417)^{2/3}\sqrt{0.001} = 3.83 \text{ cfs}$$

$$v_{full} = \frac{3.83}{\frac{1}{4}\pi(1.667)^2} = 1.75 \text{ ft/sec}$$

$Q/Q_{full} = 2.5/3.83 = 0.65$. From Appendix C, $(d/D) = 0.66$ and $(v/v_{full}) = 0.92$. So,

$$v = (0.92)(1.75) = 1.61 \text{ ft/sec}$$

$$d = (0.66)(20) = 13.2 \text{ inches}$$

## 6 NORMAL DEPTH

When the depth of flow is constant along the length of the channel (i.e., depth is neither increasing nor decreasing), the flow is said to be *uniform*. The depth of flow in that case is known as the *normal depth*, $d_o$. If the normal depth is known or can be calculated, it can be compared with the actual depth of flow to determine if the flow is uniform.[4]

[3] Originally proposed in 1868 by an investigator with the name of Gaukler. In Europe, the Manning equation may be known as *Strickler's equation.*

[4] In reality, there are two normal depths for any given discharge. This is apparent from the specific energy curves presented elsewhere in this chapter.

The difficulty or ease with which the normal depth is calculated depends on the cross section of the channel. If the width is very large compared to the depth, the Chezy-Manning equation can be used. (Equation 5.10 assumes that the hydraulic radius equals the normal depth.)

$$d_o = 0.79\left(\frac{nQ}{\sqrt{S}w}\right)^{3/5} \qquad 5.10$$

Normal depth in circular channels can be calculated directly only under limited conditions. If the circular channel is flowing full, the normal depth is the inside pipe diameter.

$$D = d_o = 1.33\left(\frac{nQ}{\sqrt{S}}\right)^{3/8} \qquad 5.11$$

If a circular channel is flowing half-full, the normal depth is half the inside pipe diameter.

$$D = 2d_o = 1.73\left(\frac{nQ}{\sqrt{S}}\right)^{3/8} \qquad 5.12$$

For other cases of uniform flow, it is more difficult to determine normal depth. Various researchers have prepared tables and figures to assist in the calculations. For example, table 5.1 is derived from Appendix C, and can be used for circular channels flowing other than full or half-full.

In the absence of tables or figures, trial and error solutions are required. The appropriate expression for the hydraulic radius is used in the Chezy-Manning equation. Trial values are used in conjunction with graphical techniques or linear interpolation to determine the normal depth. For example, for a rectangular channel whose width is not large compared to the depth, the hydraulic radius depends on the normal depth.

$$r_H = \frac{wd_o}{w + 2d_o} \qquad 5.13$$

The Chezy-Manning equation in terms of the normal depth can then be compared to the actual known flow quantity.

$$Q = \frac{1.49}{n}(wd_o)\left(\frac{wd_o}{w+2d_o}\right)^{2/3}\sqrt{S} \qquad 5.14$$

## 7 ENERGY AND FRICTION RELATIONSHIPS

The Bernoulli equation can be written for two points along the bottom of an open channel experiencing uniform flow.

$$\frac{p_1}{\rho} + \frac{v_1^2}{2g_c} + z_1 = \frac{p_2}{\rho} + \frac{v_2^2}{2g_c} + z_2 + h_f \qquad 5.15$$

However, $(p/\rho) = d$. And since $d_1 = d_2$ and $v_1 = v_2$ for uniform flow,

$$h_f = z_1 - z_2 \qquad 5.16$$

For small slopes typical of almost all natural waterways, the channel length and horizontal run are essentially identical. Then, the hydraulic slope is

$$S = \frac{z_1 - z_2}{L} \qquad 5.17$$

Therefore, the friction loss in a length of channel is

$$h_f = LS \qquad 5.18$$

The friction loss can also be calculated from equation 5.19.

$$h_f = \frac{Ln^2v^2}{2.21(r_H)^{4/3}} \qquad 5.19$$

*Example 5.3*

Both $v_2$ and $v_1$ are unknown in the sluice gate shown. What is $v_2$?

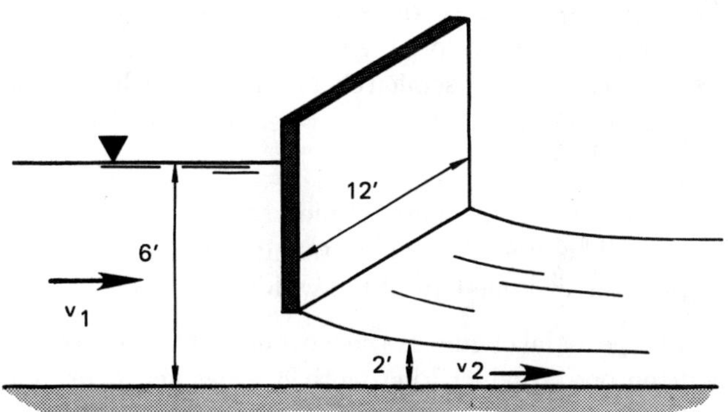


This problem can be solved using Bernoulli's equation and the continuity of flow equation. Since the channel bottom is essentially level, $z_1 = z_2$. Bernoulli's equation reduces to

$$6 + \frac{v_1^2}{2g_c} = 2 + \frac{v_2^2}{2g_c}$$

$v_1$ and $v_2$ are related by continuity.

$$Q_1 = Q_2$$
$$A_1v_1 = A_2v_2$$
$$(6)(12)v_1 = (2)(12)v_2$$
$$v_1 = \frac{1}{3}v_2$$

Substituting the expression for $v_1$ into the Bernoulli equation,

$$6 + \frac{v_2^2}{(3)^2(2)(32.2)} = 2 + \frac{v_2^2}{(2)(32.2)}$$
$$6 + 0.00173\,v_2^2 = 2 + 0.0155\,v_2^2$$
$$v_2 = 17.0 \text{ fps}$$

*Example 5.4*

In example 5.1, an open channel in normal flow had the following characteristics: $S = 0.002$, $n = 0.012$, $v = 9.447$ ft/sec, $r_H = 2.22$ ft. What is the energy loss per 1000 feet?

From equation 5.18, $h_f = (1000)(0.002) = 2$ feet

From equation 5.19, $h_f = \dfrac{1000(0.012)^2(9.447)^2}{2.21(2.22)^{4/3}} = 2$ feet

## 8 MOST EFFICIENT CROSS SECTION

The most efficient cross section (from an open channel standpoint) is the one which has the maximum hydraulic radius, and therefore, the maximum discharge for a given slope, area, and roughness. Wetted perimeter will be at a minimum (to minimize friction) when the flow is maximum. Furthermore, construction cost will be minimized.

Semicircular cross sections have the smallest wetted perimeter, and therefore the cross section with the highest efficiency is the semicircle. Although such a shape can be constructed with concrete, it cannot be used with earth channels.

Rectangular channels are frequently used with wooden flumes. The most efficient rectangle is one which has a depth equal to one-half of the width.

For trapezoidal channels, the most efficient cross section will be one in which the depth is twice the hydraulic radius. The sides of such a trapezoid will be inclined at 30° from the horizontal, and the flow area is half a hexagon.

A semicircle with its center at the middle of the water surface can always be inscribed in a cross section with maximum efficiency.

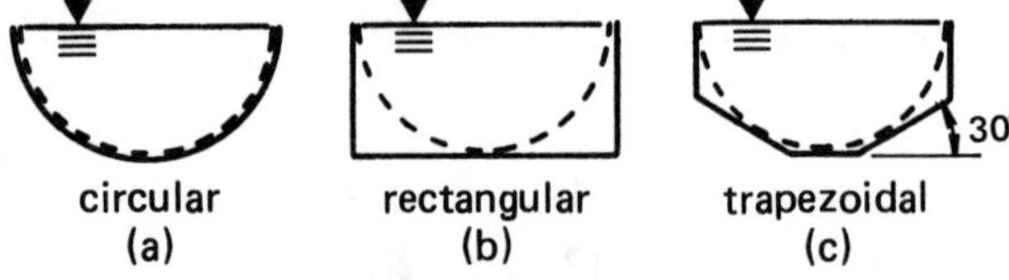


**Figure 5.3** Circles Inscribed in Efficient Channels

*Example 5.5*

A rubble masonry open channel is being designed to carry 500 cfs of water on a 0.0001 slope. Using $n = 0.017$, find the most efficient dimensions for a rectangular channel.

Let the depth and width be $d$ and $w$ respectively. For an efficient rectangle, $d = \frac{1}{2}w$.

$$A = dw = \frac{1}{2}w^2$$

$$P = d + w + d = 2w$$

$$r_H = \frac{\frac{1}{2}w^2}{2w} = \frac{1}{4}w$$

Using equation 5.9,

$$500 = \left(\frac{1}{2}w^2\right)\left(\frac{1.49}{0.017}\right)\left(\frac{1}{4}w\right)^{2/3}(0.0001)^{1/2}$$

$$500 = (0.1739)w^{8/3}$$

$$w = 19.82$$

$$d = \frac{1}{2}w = 9.91 \text{ ft}$$

## 9 ANALYSIS OF NATURAL WATERCOURSES

Natural watercourses do not have uniform paths or cross sections. This complicates their analysis considerably. Frequently, analyzing the flow from a river is a case of doing the best you can. Some types of problems can be solved with a reasonable amount of error.

As was seen in equation 5.19, the friction loss (and hence the hydraulic gradient) depends on the square of the roughness coefficient. Therefore, an attempt must be made to evaluate $n$ as accurately as possible. If the channel consists of a river with overbank *flood plains*, it should be treated as parallel channels. The flow from each subdivision can be calculated independently, and the separate values added to obtain the total flow. Alternatively, a composite value of the roughness coefficient, $n_c$, can be approximated from the individual values of $n$ and the corresponding wetted perimeters.

$$n_c = \left[\frac{\sum P_i(n_i)^{3/2}}{\sum P_i}\right]^{2/3} \qquad 5.20$$

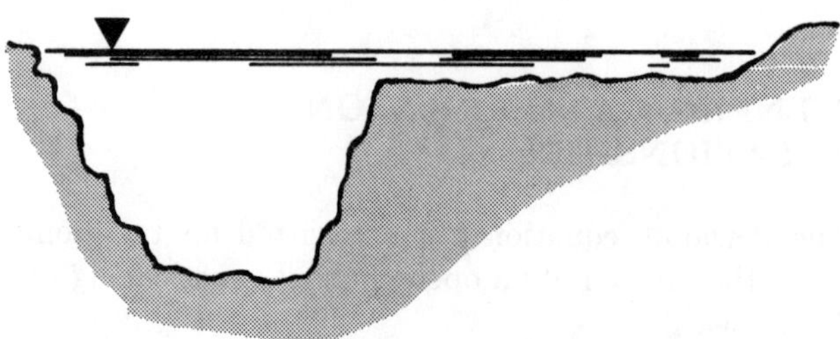

**Figure 5.4** River with Flood Plain

If the channel is divided by an island into two channels (figure 5.5), $Q$ will usually be known. It may be necessary to calculate $Q_1$ and $Q_2$ in that case, or, if $Q_1$ and $Q_2$ are known, it may be necessary to find the slope.

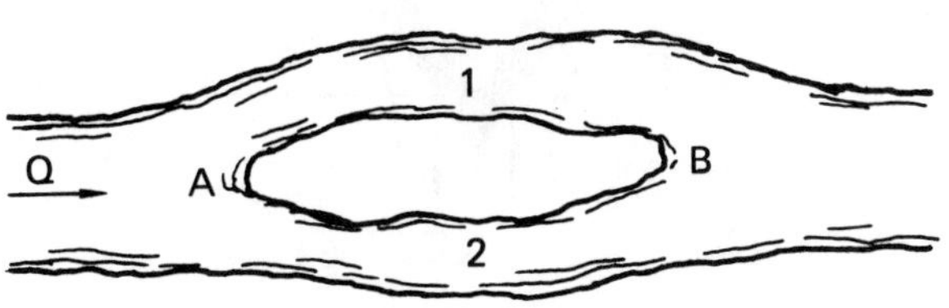


**Figure 5.5** Divided Channel

Since the drop $(z_B - z_A)$ between points $A$ and $B$ is the same regardless of flow path,

$$S_1 = \frac{z_B - z_A}{L_1} \quad 5.21$$

$$S_2 = \frac{z_B - z_A}{L_2} \quad 5.22$$

Once the slopes are known, $Q_1$ and $Q_2$ can be found from equation 5.9. The sum of $Q_1$ and $Q_2$ will probably not be the same as the given flow quantity, $Q$. In that case, $Q$ should be prorated according to the ratios of $Q_1$ and $Q_2$ to $(Q_1 + Q_2)$.

If the lengths $L_1$ and $L_2$ are the same or almost so, the Chezy-Manning equation may be solved for the slope by writing equation 5.23.

$$Q = Q_1 + Q_2 = 1.49\left[\frac{A_1}{n_1}(r_{H,1})^{2/3} + \frac{A_2}{n_2}(r_{H,2})^{2/3}\right]\sqrt{S} \quad 5.23$$

## 10 FLOW MEASUREMENT WITH WEIRS

A *weir* is an obstruction in an open channel over which flow occurs. Although a dam spillway is a specific type of weir, most weirs are designed for flow measurement. These weirs consist of a vertical flat plate with sharpened edges. Because of their construction, they are called *sharp-crested weirs.*

Sharp-crested weirs are most frequently rectangular, consisting of a straight, horizontal crest. However, weirs may also have trapezoidal and triangular openings.

If a rectangular weir is constructed with an opening width less than the channel width, the overfalling liquid sheet (called the *nappe*) decreases in width as it falls. This *contraction* of the nappe causes these weirs to be called *contracted weirs*, although it is the nappe that is actually contracted. If the opening of the weir extends the full channel width, the weir is called a *suppressed weir*, since the contractions are suppressed.

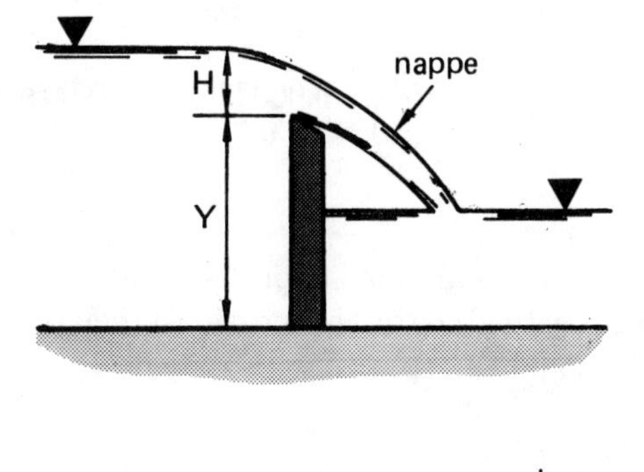


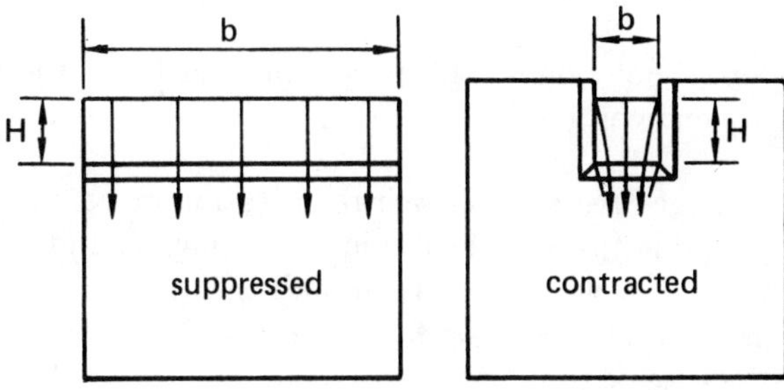


**Figure 5.6** Contracted and Suppressed Weirs

The derivation of the basic weir equation is not particularly difficult, but it is dependent on many simplifying assumptions. The basic weir equation (equation 5.24 or 5.25) is, therefore, an approximate result requiring correction by the inclusion of experimental coefficients.

If it is assumed that the contractions are suppressed, upstream velocity is uniform, flow is laminar over the crest, nappe pressure is zero, the nappe is fully ventilated, and viscosity, turbulence, and surface tension effects are negligible, then the following equation may be derived from the Bernoulli equation:

$$Q = \frac{2}{3}b\sqrt{2g}\left[\left(H + \frac{v_1^2}{2g}\right)^{3/2} - \left(\frac{v_1^2}{2g}\right)^{3/2}\right] \quad 5.24$$

If $v_1$ is negligible, then

$$Q = \frac{2}{3}b\sqrt{2g}(H)^{3/2} \quad 5.25$$

Equation 5.25 must be corrected for all of the assumptions made. This is done by introducing a coefficient, $C_1$, to account primarily for a non-uniform velocity distribution.

$$Q = \frac{2}{3}(C_1)b\sqrt{2g}(H)^{3/2} \quad 5.26$$

A number of investigations have been done to evaluate $C_1$. Perhaps the most widely known is the coefficient formula developed by *Rehbock:*[5]

$$C_1 = \left[0.6035+0.0813\left(\frac{H}{Y}\right)+\frac{0.000295}{Y}\right]\left[1+\frac{0.00361}{H}\right]^{3/2} \quad 5.27$$

If the contractions are not suppressed (i.e., one or both sides do not extend to the channel sides) then the actual width, $b$, should be replaced with the *effective width.*

$$b_{\text{effective}} = b_{\text{actual}} - (0.1)(N)(H) \quad 5.28$$

$N$ is one if one side is contracted, and $N$ is two if there are two end contractions.

A submerged rectangular weir requires a more complex analysis, due to the difficulty in measuring $H$, and because the discharge depends on both the upstream and downstream depths. The following equation, however, may be used with little difficulty.

$$Q_{\text{submerged}} = Q_{\text{free flow}}\left[1-\left(\frac{H_{\text{downstream}}}{H_{\text{upstream}}}\right)^{3/2}\right]^{0.385} \quad 5.29$$

Equation 5.29 is used by first finding $Q$ from equation 5.26 and then correcting it with the bracketed quantity.

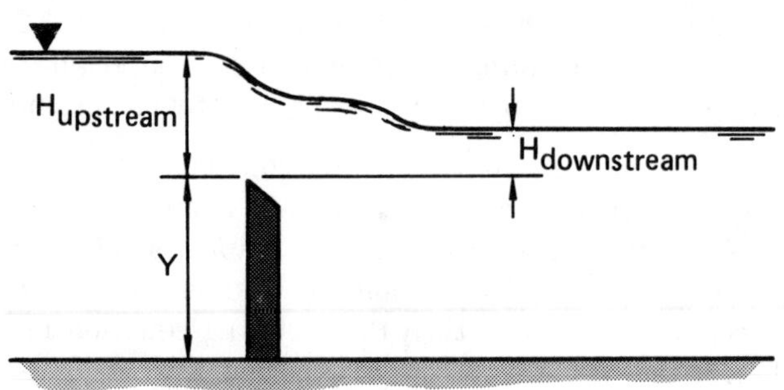


**Figure 5.7** Submerged Weir

*Triangular (V-notch) weirs* should be used when small flow rates are to be measured. The flow over a triangular weir depends on the notch angle, $\theta$. For a 90° weir, $C_2 \approx 0.593$.

$$Q = C_2\left(\frac{8}{15}\right)\tan\left(\frac{1}{2}\theta\right)\sqrt{2g}(H)^{5/2} \quad 5.30$$

$$Q \approx 2.5H^{2.5}(90° \text{ weir}) \quad 5.31$$

[5] There is much variation in how different investigators calculate the discharge coefficient, $C_1$. For ratios of $H/b$ less than 5, $C_1 = 0.622$ gives a reasonable value. With the questionable accuracy of some of the other variables used in open channel flow problems, the pursuit of greater accuracy is of dubious value.

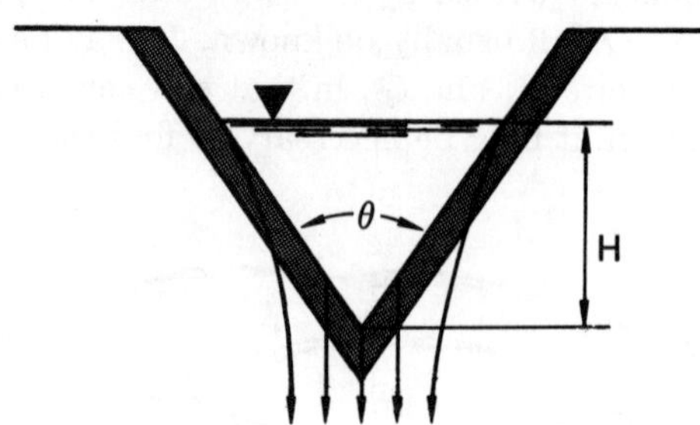


**Figure 5.8** Triangular Weir

A *trapezoidal weir* is essentially a rectangular weir with a triangular weir on either side. If the angle of the sides from the vertical is approximately 14° (i.e., 4 vertical and 1 horizontal) the weir is known as a *Cipoletti weir.* The discharge from the triangular ends of a Cipoletti weir approximately make up for the contractions that reduce rectangular flow. Therefore, no correction is theoretically necessary. The discharge from a Cipoletti weir is given by equation 5.32.

$$Q = 3.367(b)(H)^{3/2} \quad 5.32$$

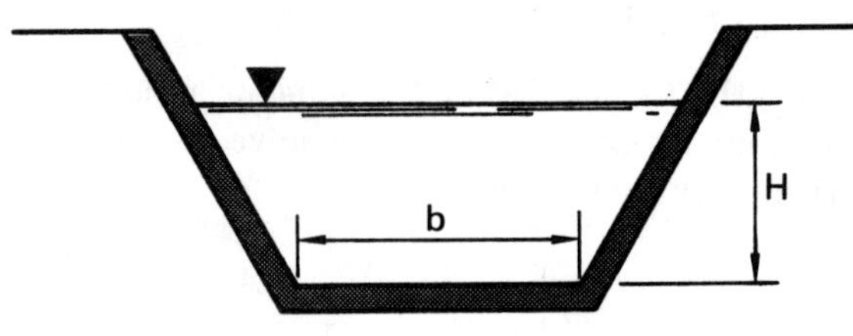


**Figure 5.9** Trapezoidal Weir

Equation 5.26 can also be used for *broad-crested weirs* ($C = 0.5$ to $0.57$) and *ogee spillways* ($C = 0.60$ to $0.75$).

*Example 5.6*

A sharp-crested, rectangular weir with two contractions is $2\frac{1}{2}$ feet high and 4 feet long. A 4″ head exists upstream from the weir. What is the velocity of approach?

$$H = 4/12 = 0.333 \text{ ft}$$

From equation 5.28, $N = 2$ and the effective width is

$$b_{\text{effective}} = 4 - (0.1)(2)(0.333) = 3.93$$

The Rehbock coefficient (from equation 5.27) is

$$\begin{aligned} C_1 &= \left(0.6035 + 0.0813\left(\frac{0.333}{2.5}\right) \right. \\ &\quad \left. + \frac{0.000295}{2.5}\right)\left(1 + \frac{0.00361}{0.333}\right)^{3/2} \\ &= 0.624 \end{aligned}$$

From equation 5.26, the flow is

$$\begin{aligned} Q &= \frac{2}{3}(0.624)(3.93)\sqrt{(2)(32.2)}(0.333)^{3/2} \\ &= 2.52 \text{ cfs} \\ v &= \frac{Q}{A} = \frac{2.52}{(4)(2.5 + 0.333)} = 0.222 \text{ ft/sec} \end{aligned}$$

## 11 FLOW MEASUREMENT WITH PARSHALL FLUMES

The Parshall flume is one of the most widely used devices for measuring open channel wastewater flows. It performs well even in instances where head loss must be kept to a minimum or when there is a high concentration of suspended solids.

The Parshall flume is constructed with a converging upstream section, a throat, and a diverging downstream section. The walls of the flume are vertical, and the floor of the throat section drops. The length, width, and height of the flume are essentially predefined by the flow rate anticipated.[6]

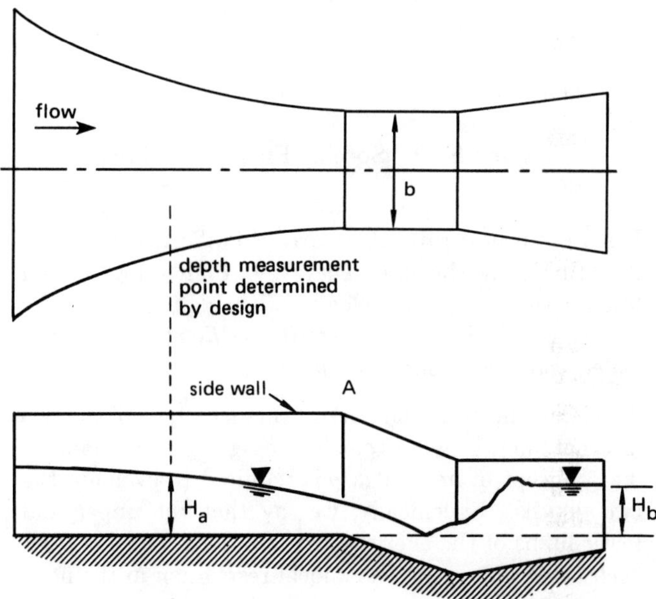


**Figure 5.10** The Parshall Flume

[6] This chapter does not attempt to design the Parshall flume, only to predict flow rates through its use.

The throat geometry in a Parshall flume has been chosen so as to force the occurrence of a critical flow there. Following the critical section is a short length of supercritical flow followed by a hydraulic jump. This construction does not produce a dead water region where debris and silt can accumulate (as is the case with broad crested and other flat-top weirs).

The discharge relationship for a Parshall flume is given by equation 5.33 for submergence ratios of $H_b/H_a$ up to 0.7. Above 0.7, the true discharge is less than predicted by the equation. Values of $K$ are given in the table 5.2, although using a value of 4.0 is accurate for most purposes.

$$Q = Kb(H_a)^n \qquad 5.33$$

$$n = 1.522(b)^{0.026} \qquad 5.34$$

**Table 5.2**
$K$ values for the Parshall Flume

| $b$ | $K$ |
|---|---|
| 0.25 ft | 3.97 |
| 0.50 | 4.12 |
| 0.75 | 4.09 |
| 1.0 | 4.00 |
| 1.5 | 4.00 |
| 2.0 | 4.00 |
| 3.0 | 4.00 |
| 4.0 | 4.00 |

## 12 STEADY FLOW

Steady open flow is one of constant-volume flow. However, the flow may be uniform or non-uniform. Figure 5.11 illustrates that three definitions of "slope" exist for open channel flow. These three slopes are the slope of the channel bottom, the slope of the water surface, and the slope of the energy gradient line.

Under conditions of uniform flow, all of these three slopes are equal, since the flow quantity and flow depth are constant along the length of flow.[7] With non-uniform flow, however, the flow velocity and depth vary along the length of channel, and the three slopes are not necessarily equal.

[7] As a simplification, this chapter deals only with channels of constant width. If the width is varied, changes in flow depth may not coincide with changes in flow quantity.

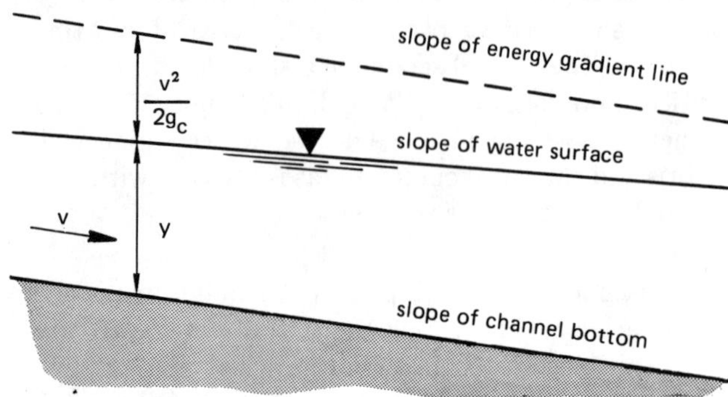


**Figure 5.11** Slopes Used in Open Channel Flow

If water is introduced down a path with a steep slope (as down a spillway), the effect of gravity will be to cause an increasing velocity. This velocity will be opposed by friction. Since the gravitational force is constant but friction varies as the square of velocity, these two forces eventually become equal. When they become equal, the velocity stops increasing, the depth stops decreasing, and the flow becomes uniform. Until they become equal, however, the flow is non-uniform (varied).

The total head is given by the Bernoulli equation.

$$H_t = z + \frac{p}{\rho} + \frac{v^2}{2g_c} \qquad 5.35$$

*Specific energy* is the total head with respect to the channel bottom. In this case, $z = 0$ and $(p/\rho) = d$.

$$E = d + \frac{v^2}{2g_c} \qquad 5.36$$

Equation 5.36 is not meant to imply that the potential energy is not an important factor in open channel flow problems. The concept of specific energy is used for convenience only, and it should be clear that the Bernoulli equation is still valid. Indeed, for problems in which there is a step in the channel bottom, the Bernoulli equation written in terms of the specific energy is invaluable in perceiving the behavior of the flow.

$$E_1 + z_1 = E_2 + z_2 \qquad 5.37$$

$$E_1 - E_2 = z_2 - z_1 \qquad 5.38$$

In uniform flow, total head decreases due to the frictional effects, but specific energy is constant. In non-uniform flow, total head decreases, but specific energy may increase or decrease.

Since $v = Q/A$, equation 5.36 can be written

$$E = d + \frac{Q^2}{2g_c A^2} \qquad 5.39$$

For a rectangular channel, the velocity can be written in terms of the width and flow depth.

$$v = \frac{Q}{A} = \frac{Q}{wd} \qquad 5.40$$

The specific energy equation for a rectangular channel is, then,

$$E = d + \frac{Q^2}{2g_c(wd)^2} \qquad 5.41$$

Solving for $d$ from equation 5.41 requires working with a cubic equation. There are three values of $d$ which will satisfy equation 5.41. One of them is negative, as figure 5.12 shows. Since depth cannot be negative, that value can be discarded. The two remaining values are known as *alternate depths.*

Since the area depends on the depth, fixing the channel shape and slope and assuming a depth will determine $Q$. This also will determine the specific energy, as illustrated in the *specific energy diagram.*

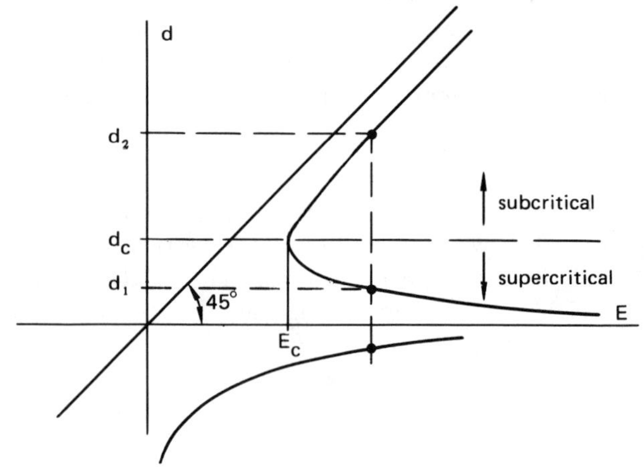


**Figure 5.12** Specific Energy Diagram

For a given flow rate, there are two different depths of flow that have the same energy—a high velocity with low depth and a low velocity with high depth. The former is called *rapid (supercritical) flow*; the latter is called *tranquil (subcritical) flow.*

The Bernoulli equation doesn't predict which of the two alternate depths will occur for any given flow quantity. The concept of *accessibility* is required to evaluate the two depths. Specifically, we say that the upper and lower limbs of the energy curve are not accessible from each other unless there is a local restriction in the flow.

Energy curves can be drawn for different flow quantities, as shown in figure 5.13 for flow quantities $Q_A$ and $Q_B$. Suppose that flow is initially at point 1. (Since the flow is on the upper limb, the flow is initially subcritical.) If there is a step up in the channel bottom, equation 5.38 predicts that the specific energy will decrease.

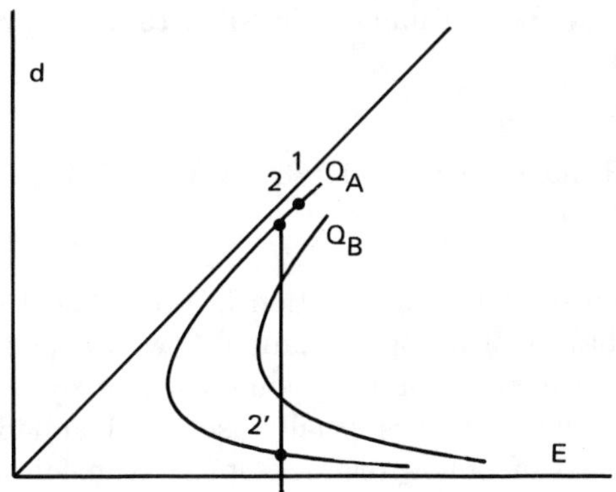


**Figure 5.13** Specific Energy Curve Families

However, the flow cannot arrive at point 2′ without changing flow quantity (i.e., going through a specific energy curve for a different flow quantity). Therefore, we say that point 2′ is not accessible from point 1 without going through point 2 first.[8]

*Example 5.7*

An open channel flow is initially flowing at 4 fps in a 7 foot wide channel 6 feet deep. The flow encounters a 1.0 foot step in the channel bottom. What is the depth of flow above the step?

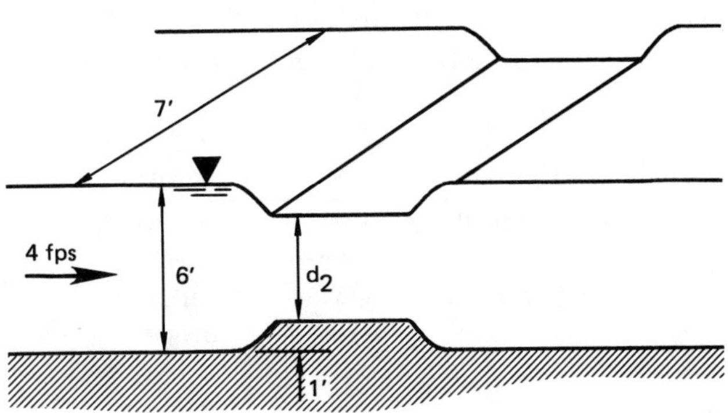


The initial specific energy is found from equation 5.36.

$$E_1 = 6 + \frac{(4)^2}{(2)(32.2)} = 6.25$$

From equation 5.38, the specific energy over the step is

$$\begin{aligned} E_2 &= E_1 + z_1 - z_2 \\ &= 6.25 + 0 - 1 = 5.25 \end{aligned}$$

The flow quantity is

$$Q = Av = (4)(6)(7) = 168 \text{ cfs}$$

Substituting $Q$ into equation 5.41,

$$5.25 = d_2 + \frac{(168)^2}{(2)(32.2)(7)^2(d_2)^2}$$

By trial and error,

$$d_2 = 1.55, 4.9 \quad \text{(alternate depths)}$$

Since the 1.55 foot depth is not accessible from the initial depth of 6 feet, the depth over the step is 4.9 feet. Notice that the drop in the water level is $6-(4.9+1) = 0.1$ feet.

Since the water level dropped only slightly, it is apparent that the flow is well up on the top limb of the specific energy curve. Since the upper limb is asymptotic to a 45° line, any change in specific energy will result in almost the same change in depth.[9] That is,

$$\Delta d \approx \Delta E \qquad 5.42$$

Therefore, the surface level will remain almost the same.

However, if the initial point is close to the critical point, then a small change in the specific energy (such as might be caused by a small variation in the channel floor) will cause a large change in depth. That is why there commonly is severe turbulence near critical flow.

## 13 CRITICAL FLOW IN RECTANGULAR CHANNELS

There is one depth, known as the *critical depth*, which minimizes the energy of flow. (The depth is not minimized, however.) The critical depth for a given flow depends on the shape of the channel.

For a rectangular channel, if equation 5.41 is differentiated with respect to depth in order to minimize the specific energy, equation 5.43 results.

$$d_c^3 = \frac{Q^2}{gw^2} \qquad 5.43$$

[8] Actually, specific energy curves are typically plotted for $q = Q/w$. If that is the case, a jump from one limb to the other could take place if the width was allowed to change as well as depth. However, the width is constant in most open channel flow problems.

[9] A rise in the channel bottom does not always produce a drop in the water surface. Only if the flow is initially subcritical will the water surface drop upon encountering a step. The water surface will rise if the flow is initially supercritical.

Geometrical and analytical methods can be used to correlate the critical depth and the minimum specific energy.

$$d_c = \frac{2}{3}E_c \quad 5.44$$

For a rectangular channel, $Q = d_c w v_c$. Substituting this into equation 5.43 produces an equation for the *critical velocity.*

$$v_c = \sqrt{g d_c} \quad 5.45$$

The expression for critical velocity also coincides with the expression for the velocity of a *surge (surface) wave* moving in liquid of depth $d_c$. Since surface disturbances are transmitted as ripples upstream (and downstream) at velocity $v_c$, it is apparent that a surge wave will be stationary in a channel moving at the critical velocity. Such a motionless wave is known as a *standing wave.*

If the flow velocity is less than the surge wave velocity (for the actual depth), then a ripple can make its way upstream. If the flow velocity exceeds the surge wave velocity, the ripple will be swept downstream.

## 14 CRITICAL FLOW IN NON-RECTANGULAR CHANNELS

For non-rectangular shapes, the critical depth can be found by trial and error from the following equation in which $b$ is the surface width. To use equation 5.46, assume trial values of the critical depth, use it to calculate dependent quantities in the equation, and then verify the equality.

$$\frac{Q^2}{g} = \frac{A^3}{b} \quad 5.46$$

Equation 5.46 is particularly difficult to use with circular channels. Appendix D is a convenient method of determining critical depth in that case.

## 15 THE FROUDE NUMBER

The Froude number, when sufficient information is available to calculate it, can be used to determine if the flow is subcritical or supercritical. This dimensionless number can be calculated from equation 5.47.

$$N_{Fr} = \frac{v}{\sqrt{gd}} \quad 5.47$$

When the Froude number is less than one, the flow is subcritical. That is, the depth of flow is greater than the critical depth, and the velocity is less than critical velocity.

When the Froude number is greater than one, the flow is supercritical. The depth is less than critical depth, and the flow velocity is greater than critical velocity.

When the Froude number is equal to one, the flow is critical.[10]

## 16 PREDICTING OPEN CHANNEL FLOW BEHAVIOR

Upon encountering a variation in the channel bottom, the behavior of an open channel flow is dependent on whether the flow is initially subcritical or supercritical. The Froude number is usually used as the quantitative determination of flow regime. Open channel flow is governed by the following equation, in which the Froude number is the primary independent variable.

$$\frac{dd}{dx}\left[1 - N_{Fr}^2\right] + \frac{dz}{dx} = 0 \quad 5.48$$

The quantity $\frac{dd}{dx}$ is the slope of the surface. (That is, it is the derivative of the depth with respect to the channel length.) The quantity $\frac{dz}{dx}$ is the slope of the channel bottom.

For an upward step, $\frac{dz}{dx} > 0$. If the flow is initially subcritical (i.e., $N_{FR} < 1$), then equation 5.48 requires that $\frac{dd}{dx} < 0$, a drop in depth. This logic can be repeated for other combinations of the terms.

Table 5.3 lists the various behaviors of open channel flow surface levels based on equation 5.48.

**Table 5.3**
Surface Level Change Behavior

| initial flow | step up | step down |
|---|---|---|
| subcritical | surface drops | surface rises |
| supercritical | surface rises | surface drops |

If $\frac{dz}{dx} = 0$ (i.e., a horizontal slope), then either the depth must be constant, or the Froude number must be unity. The former case is obvious. The latter case predicts critical flow. Such critical flow actually occurs where the slope is horizontal over broad crested weirs (see figure 5.16) and at the top of a rounded spillway. Since broad crested weirs and spillways produce critical flow, they represent a class of *controls on flow.*

[10] The similarity of the Froude number to the Mach number used to classify gas flows is more than coincidental. Both bodies of knowledge employ similar concepts.

## 17 OCCURRENCES OF CRITICAL FLOW

For any given discharge and cross section, there is a unique slope that will produce and maintain flow at critical depth. Once $d_c$ is known, this critical slope can be found from the Chezy-Manning equation.

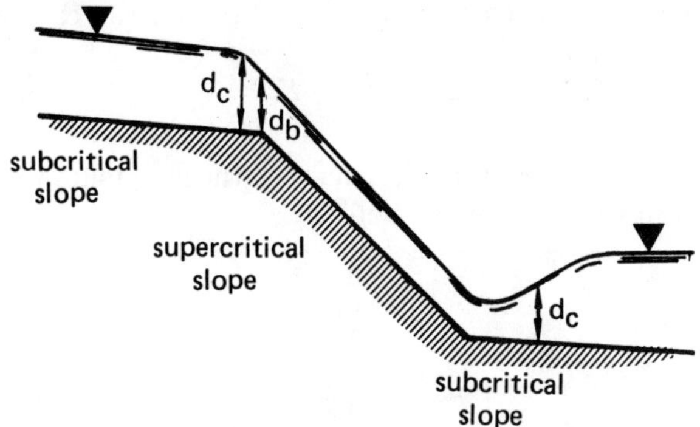


**Figure 5.14** Occurrence of Critical Depth

Critical depth occurs at free outfall from a channel of mild slope. The occurrence is at the point of curvature inversion, just upstream from the brink. For mild slopes, the brink depth is approximately

$$d_b = (0.715)d_c \tag{5.49}$$

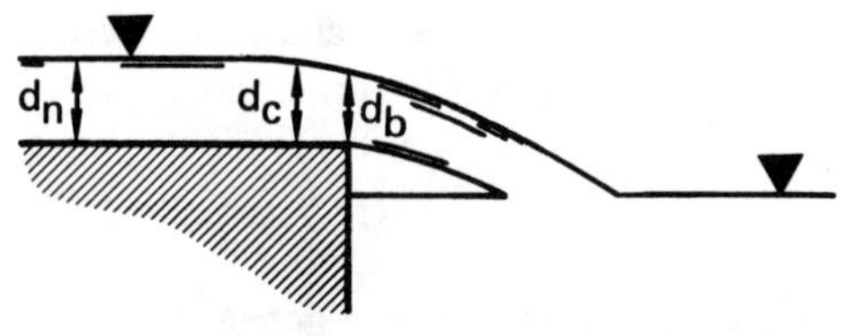


**Figure 5.15** Free Outfall

Critical flow can occur across a broad-crested weir.[11] With no obstruction to hold the water, it falls from the normal depth to the critical depth, but it can fall no more than that because there is no source to increase the specific energy (to increase the velocity). This is not a contradiction of the previous free outfall case where the brink depth is less than the critical depth. The flow curvatures in free outfall are a result of the constant gravitational acceleration.

[11] Figure 5.16 is an example of a *hydraulic drop*, the opposite of a hydraulic jump. A hydraulic drop can be recognized by the sudden decrease in depth over a short length of channel.

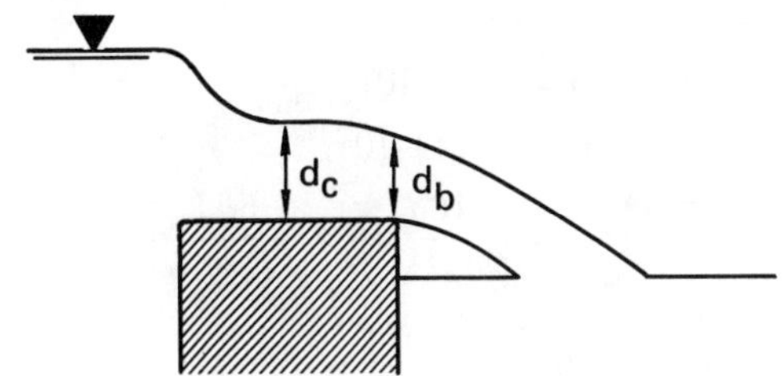


**Figure 5.16** Broad-Crested Weir

Critical depth can also occur when a channel bottom has been raised to choke the flow. A raised channel bottom is essentially a broad-crested weir.

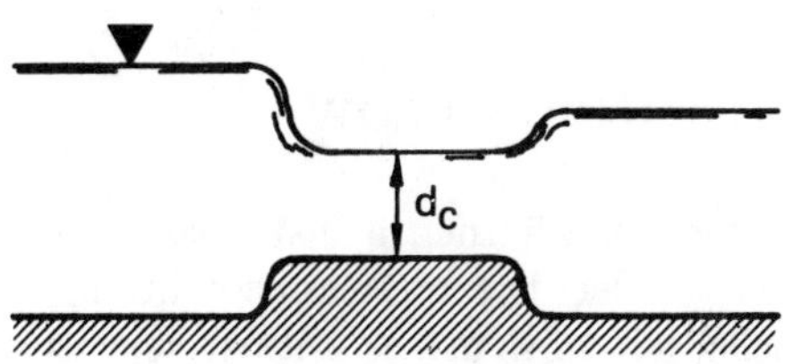


**Figure 5.17** Raised Channel Bottom with Choked Flow

The critical depth is important, not because it minimizes the energy of flow, but because it maximizes the quantity flowing for a given cross section and slope. Critical flow is generally quite turbulent because of the large changes in energy that occur with small elevation and depth changes. Critical depth flow is characterized by water surface undulations.

In all of the previous instances of critical depth, equation 5.41 can be used to calculate the actual velocity.

*Example 5.8*

At a particular point in an open rectangular channel ($n = 0.013$, $S = 0.002$, $w = 10$ feet) the flow is 250 cfs and the depth is 4.2 feet.

(a) Is the flow tranquil, critical, or rapid?

(b) What is the normal depth?

(c) If the flow ends in a free outfall, what is the brink depth?

(a) From equation 5.43, the critical depth is

$$d_c = \sqrt[3]{(250)^2/(32.2)(10)^2} = 2.69$$

Since the actual depth exceeds the critical depth, the flow is tranquil.

(b)

$$A = (d_n)(10)$$
$$P = 2d_n + 10$$
$$r_H = \frac{10d_n}{2d_n + 10} = \frac{5d_n}{d_n + 5}$$

From equation 5.9,

$$250 = (10)(d_n)\left(\frac{1.49}{0.013}\right)\left(\frac{5d_n}{d_n + 5}\right)^{2/3}\sqrt{0.002}$$

By trial and error, $d_n = 3.1$ feet. Since the actual and normal depths are different, the flow is non-uniform.

(c) $d_b = 0.715(2.69) = 1.92$ ft

## 18 CONTROLS ON FLOW

If flow is subcritical, then a disturbance downstream will be able to affect the upstream conditions. Since the flow velocity is less than the critical velocity, a ripple will be able to propagate upstream to signal a change in the downstream conditions. Any object downstream which affects the flow rate, velocity, or depth upstream is known as a *downstream control.*

If a flow is supercritical, then a downstream obstruction will have no effect, since disturbances cannot propagate upstream faster than the flow velocity. The only effect on supercritical flow is from an upstream obstruction. Such an obstruction is said to be an *upstream control.*

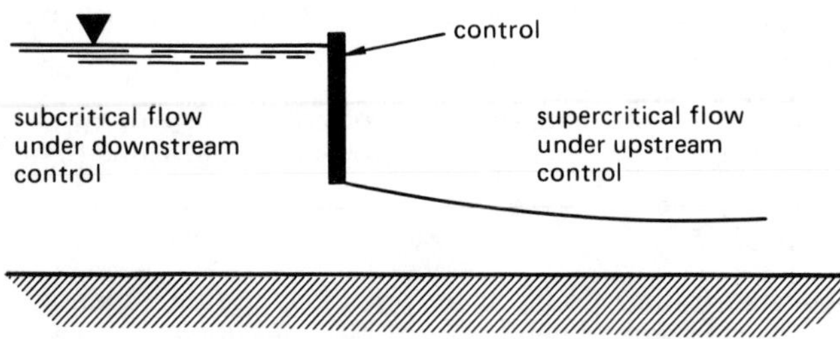


**Figure 5.18** A Control on Flow

A downstream control may also be an upstream control, as figure 5.18 shows.

In general, any feature which determines depth and discharge rate is known as a *control on flow.* Controls may consist of control structures (weirs, gates, sluices, etc.), forced flow through critical depth (as in a free outfall), sudden changes of slope (which forces a hydraulic jump or hydraulic drop to the new normal depth), or free flow between reservoirs of different surface elevations.

## 19 FLOW CHOKING

A control which is severe enough to influence the upstream flow is known as a *choke*, and the corresponding flow through the control (and any flow thereafter) is known as *choked flow.* Choked flow will occur, in the case of upward or downward steps in the channel bottom, when the step size becomes equal to the difference between the upstream specific energy and the critical flow energy.

$$\Delta z = E_1 - E_c \qquad 5.50$$

In the case of a rectangular channel, the maximum variation in channel bottom will be

$$\Delta z = E_1 - \left(d_c + \frac{v_c^2}{2g_c}\right)$$
$$= E_1 - \frac{3}{2}d_c \qquad 5.51$$

If critical depth is achieved on the step, the flow further downstream can be subcritical or supercritical depending on the downstream conditions. If there is a downstream control, such as a sluice gate, the flow will be subcritical. If there is no downstream control, then the step will serve as an upstream control, and the flow will be supercritical.

## 20 VARIED FLOW CALCULATIONS

*Accelerated flow* occurs in any channel where the actual slope exceeds the friction loss per foot.

$$S_o > h_f/L \qquad 5.52$$

*Retarded flow* occurs when the actual slope is less than the unit friction loss.

$$S_o < h_f/L \qquad 5.53$$

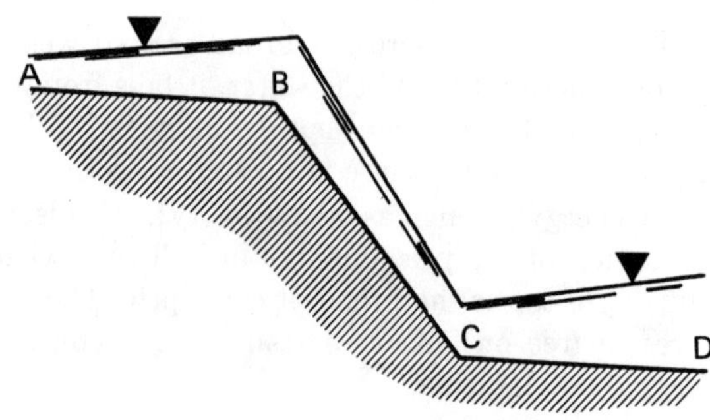


**Figure 5.19** Varied Flow

In sections $AB$ and $CD$ of figure 5.19, the slopes are less than the energy gradient, so the flows are retarded. In section $BC$, the slope is greater than the energy gradient, and the velocity increases (i.e., the flow is accelerated). If section $BC$ were long enough, the friction loss would eventually become equal to the accelerating energy, and the flow would become uniform.

Cases of accelerated and retarded flow (except for hydraulic jump) can be evaluated from the following procedure which will give the distance between points of two known or assumed depths. Assuming that the friction losses are the same for varied flow as for uniform flow, the following equations are needed:

$$S = \left(\frac{n v_{ave}}{1.486 (r_{H,ave})^{2/3}}\right)^2 \qquad 5.54$$

$$v_{ave} = \frac{1.486}{n}(r_{H,ave})^{2/3}(S)^{1/2} \qquad 5.55$$

$S$ is the slope of the energy gradient from equation 5.54, not the channel slope $S_o$. The usual method of finding the depth profile is to start at a point in the channel where $d_2$ and $v_2$ are known. Then, assume a depth $d_1$, find $v_1$ and $S$, and solve equations 5.56 for $L$.

$$L = \frac{\left(d_1 + \frac{v_1^2}{2g_c}\right) - \left(d_2 + \frac{v_2^2}{2g_c}\right)}{S - S_o} = \frac{E_1 - E_2}{S - S_o} \qquad 5.56$$

*Example 5.9*

How far from the point in example 5.8 will the depth be 4 feet?

The difference between 4 feet and 4.2 feet is small, so a one-step calculation will probably be sufficient.

$$d_1 = 4 \text{ feet}$$

$$v_1 = Q/A = \frac{250}{4(10)} = 6.25 \text{ ft/sec}$$

$$E_1 = 4 + \frac{(6.25)^2}{(2)(32.2)} = 4.607 \text{ ft}$$

$$r_H = \frac{(4)(10)}{4 + 10 + 4} = 2.22$$

$$d_2 = 4.2$$

$$v_2 = \frac{250}{(4.2)(10)} = 5.95$$

$$E_2 = 4.2 + \frac{(5.95)^2}{(2)(32.2)} = 4.75$$

$$r_H = \frac{(4.2)(10)}{4.2 + 10 + 4.2} = 2.28$$

$$v_{ave} = \frac{1}{2}(6.25 + 5.95) = 6.1$$

$$r_{H,ave} = \frac{1}{2}(2.22 + 2.28) = 2.25$$

From equation 5.54,

$$S = \left(\frac{(0.013)(6.1)}{1.486(2.25)^{0.667}}\right)^2 = 0.000965$$

From equation 5.56,

$$L = \frac{4.607 - 4.75}{0.000965 - 0.002} = 138 \text{ ft}$$

## 21 HYDRAULIC JUMP

If water is introduced at high (supercritical) velocity to a section of slow-moving (subcritical) flow (as in section C in figure 5.19), the velocity will be reduced through a hydraulic jump. A hydraulic jump is an abrupt rise in the water surface. The increase in depth is always from below the critical depth to above the critical depth.[12]

The hydraulic jump has practical applications in the design of stilling basins where supercritical velocities are reduced to slower velocities by having the flow cross a series of baffles on the channel bottom. Although energy is lost across the baffles, momentum is conserved. Momentum considerations can be used to predict the downstream depth.

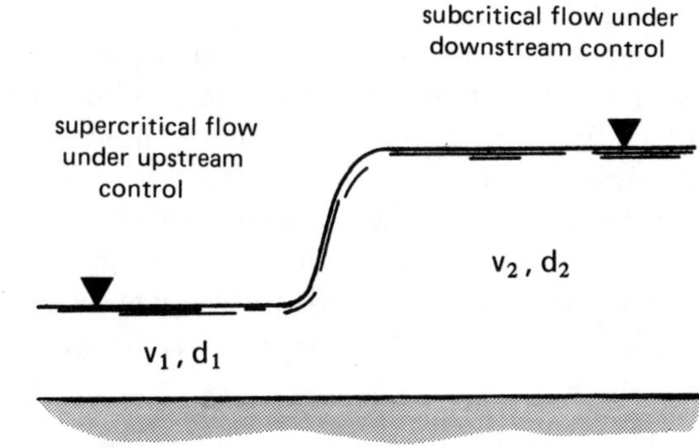


**Figure 5.20** Conjugate Depths

[12] This gives us a way to determine if a hydraulic jump can occur in a channel. If the original depth is above the critical depth, the flow is already subcritical. Therefore, a hydraulic jump cannot form.

If the depths $d_1$ and $d_2$ are known, then the velocity $v_1$ can be found from equation 5.57.

$$v_1^2 = \frac{g\,d_2}{2d_1}(d_1 + d_2) \qquad 5.57$$

$d_1$ and $d_2$ are known as *conjugate depths*, because they occur on either side of the jump. These depths are:

$$d_1 = -\frac{1}{2}d_2 + \sqrt{\frac{2v_2^2 d_2}{g} + \frac{d_2^2}{4}} \qquad 5.58$$

$$d_2 = -\frac{1}{2}d_1 + \sqrt{\frac{2v_1^2 d_1}{g} + \frac{d_1^2}{4}} \qquad 5.59$$

The specific energy lost in the jump is the energy lost per pound of water flowing.

$$\Delta E = \left[\left(d_1 + \frac{v_1^2}{2g}\right) - \left(d_2 + \frac{v_2^2}{2g}\right)\right] \times \frac{g}{g_c} \qquad 5.60$$

In the case of an apron at the bottom of a dam spillway, the apron is usually insufficient to overcome friction, and the water depth will gradually increase.

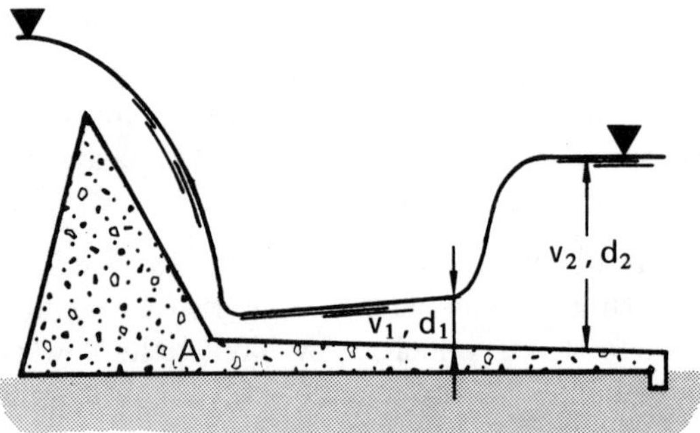


**Figure 5.21** Hydraulic Jump to Reach Tailwater Level

If the tailwater depth $d_2$ is less than critical, no hydraulic jump will occur. If the tailwater depth at the toe is less than the conjugate depth corresponding to $d_2$ but greater than the critical depth, flow will continue until the depth increases to $d_1$, and then a hydraulic jump will form. If the tailwater depth is equal to the conjugate depth at the toe, the jump will occur at the toe. If the tailwater depth exceeds the conjugate depth, the hydraulic jump may occur up on the spillway, or it may be completely submerged.

*Example 5.10*

A hydraulic jump is produced at a point in a 10 foot wide channel where the depth is 1 foot and the flow is 200 cfs. (a) What is the depth after the jump? (b) What is the total power dissipated?

(a) $v_1 = Q/A = 200/(10)(1) = 20 \text{ ft/sec}$

From equation 5.59,

$$d_2 = -\frac{1}{2}(1) + \sqrt{\frac{(2)(20)^2(1)}{32.2} + \frac{(1)^2}{4}} = 4.51$$

(b) The flow rate is

$$(200)(62.4) = 12{,}480 \text{ lbm/sec}$$

The velocity after the jump is

$$v_2 = 200/(10)(4.51) = 4.43$$

From equation 5.60, the change in specific energy is

$$\left(1 + \frac{(20)^2}{2(32.2)}\right) - \left(4.51 + \frac{(4.43)^2}{2(32.2)}\right) = 2.4 \text{ feet}$$

The total power dissipated is

$$(12{,}480)(2.4) = 29{,}952 \ \frac{\text{ft-lbf}}{\text{sec}}$$

## 22 LENGTH OF HYDRAULIC JUMP

The length of a hydraulic jump is somewhat difficult to measure (and investigate) due to the difficulty in defining the downstream termination of the jump. The majority of the evidence collected indicates that the length of the jump varies within the limits of $5 < L/d_2 < 6.5$, in which $L$ is the length of the jump. Where greater accuracy is warranted, table 5.4 can be used. This table correlates the length of the jump to the upstream Froude number.

**Table 5.4**
Approximate Lengths of Hydraulic Jumps

| $N_{Fr,1}$ | $L/d_2$ |
|---|---|
| 3 | 5.25 |
| 4 | 5.8 |
| 5 | 6.0 |
| 6 | 6.1 |
| 7 | 6.15 |
| 8 | 6.15 |

## 23 SPILLWAYS

A spillway is designed for a capacity based on the dam's inflow hydrograph, turbine capacity, and storage capacity. *Overflow spillways* frequently have a cross section known as an *ogee*, which closely approximates the underside of a nappe from a sharp-crested weir. This cross section reduces cavitation which is likely to occur when the water surface breaks contact with the spillway at heads higher than were designed for.[13]

Discharge from an overflow spillway is the same as for a weir:

$$Q = C_s b(H)^{3/2} \qquad 5.61$$

$C_s$ is the spillway coefficient, which varies from about 3 to 4 for an ogee spillway. (Use 3.97 in the absence of other information.) $C_s$ is dependent on $H$, the upstream design head above the spillway top. $C_s$ increases as $H$ increases.

If the velocity of approach is significant, the flow quantity is

$$Q = C_s b\left(H + \frac{v^2}{2g_c}\right)^{3/2} \qquad 5.62$$

*Scour protection* is usually needed at the toe of a spillway to protect the area exposed to a hydraulic jump. This protection usually takes the form of an extended horizontal or sloping apron. Other measures, however, are needed if the tailwater exhibits large variations in depth.

## 24 SLUICEWAYS

A sluiceway carries water away from a dam or reservoir, and is usually constructed in such a manner as to allow withdrawal at various reservoir levels. Its intake is submerged, but the inlet level is seldom below the minimum reservoir level. Sluiceways are generally round or square in shape.

Analysis and design of sluiceways is the same as for culverts.

## 25 ERODIBLE CANALS

Design of erodible channels is similar to that of concrete or pipe channels, except for the added considerations of maximum velocities and permissible side slopes.

The sides of the channel should not be slopes exceeding the natural *angle of repose* for the material used. Although there are other factors which determine the maximum permissible side slope, table 5.5 lists some guidelines.

**Table 5.5**
Recommended Side Slopes

| Type of Channel | (horizontal : vertical) |
|---|---|
| firm rock | vertical to $\frac{1}{4}:1$ |
| concrete lined stiff clay | $\frac{1}{2}:1$ |
| fissured rock | $\frac{1}{2}:1$ |
| firm earth with stone lining | $1:1$ |
| firm earth, large channels | $1:1$ |
| firm earth, small channels | $1\frac{1}{2}:1$ |
| loose, sandy earth | $2:1$ |
| sandy, porous loam | $3:1$ |

Maximum velocities that should be used with erodible channels are given in table 5.6.

## 26 APPLICATIONS OF OPEN CHANNEL FLOW TO CULVERT DESIGN[14]

Culverts are classified as having either *inlet* or *outlet control.* Specifically, either the inlet or outlet will control the discharge capacity. The most important feature of a culvert, particularly a culvert under inlet control, is whether or not the culvert flows full. Since length of culvert is one of the most important factors in determining the degree of filling, a culvert may be known as "hydraulically long" if it runs full and "hydraulically short" if it does not.[15]

Due to the numerous variables involved, no single formula can be given to design a culvert. Culvert design is often an empirical, trial and error process. Figure 5.22 illustrates some of the important variables which affect culvert performance.

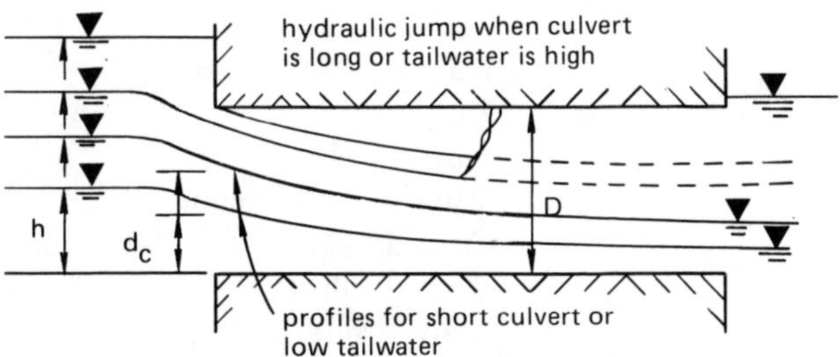


**Figure 5.22** Flow Profiles in Culvert Design

13 No cavitation or separation is expected as long as the actual head, $H$, is less than twice the design value. The shape of the ogee spillway will be a function of the design head.

14 The methods of culvert flow analysis in this chapter are based on "Measurement of Peak Discharge at Culverts by Indirect Methods," U.S. Department of the Interior, 1968.

15 Proper design of culvert entrances can reduce the importance of length on culvert filling.

The culvert can operate with its entrance totally submerged or partially submerged. Similarly, the exit can be totally submerged, partially submerged, or free outfall. The upstream head, $h$, is the water surface level above the lowest part of the culvert barrel, known as the *invert.*[16]

In figure 5.22, the three lowermost profiles are of the type which would be produced with conditions of *inlet control.* Such a situation could occur when the culvert is short and steep. Flow at the entrance would be critical (hence, the inlet control) as the water falls over the brink.

If the tailwater covers the culvert completely (i.e., a submerged exit), the culvert will be full at that point, even though the inlet control forces the culvert to be only partially full at the inlet. The transition from partially full to full occurs in a hydraulic jump, the location of which depends on the flow resistance and water levels. If the flow resistance is very high, or if the headwater and tailwater levels are high enough, the jump will occur close to or at the entrance.

[16] The highest part of the culvert barrel is known as the *soffit* or *crown.*

If the flow in a culvert is full for its entire length, then the flow is said to be under *outlet control.* The discharge will be a function of the differences in tailwater and headwater levels, as well as the flow resistance along the barrel length.

For convenience, culvert flow is classified into six different types on the basis of the type of control and the relative tailwater and headwater heights.[18] These six types are illustrated in figure 5.23.

## 27 DETERMINING TYPE OF CULVERT FLOW

The types of flow are categorized according to the steepness of the barrel, the elevations of the tailwater and headwater, and in some cases, the relationship between critical depth and culvert size. These parameters are quantified through the use of the ratios in table 5.7.

[18] It should be cautioned that the six cases which are presented here do not exhaust the various possibilities for entrance and exit control. Culvert design is complicated by this multiplicity of possible flows. Each situation needs to be carefully evaluated, since only the easiest problems can be immediately categorized as one of the six cases which follow.

**Table 5.6**
Suggested Maximum Velocities[17]

| Soil type or lining (earth; no vegetation) | Maximum permissible velocities (fps) | | |
|---|---|---|---|
| | clear water | water carrying fine silts | water carrying sand and gravel |
| Fine sand (noncolloidal) | 1.5 | 2.5 | 1.5 |
| Sandy loam (noncolloidal) | 1.7 | 2.5 | 2.0 |
| Silt loam (noncolloidal) | 2.0 | 3.0 | 2.0 |
| Ordinary firm loam | 2.5 | 3.5 | 2.2 |
| Volcanic ash | 2.5 | 3.5 | 2.0 |
| Fine gravel | 2.5 | 5.0 | 3.7 |
| Stiff clay (very colloidal) | 3.7 | 5.0 | 3.0 |
| Graded, loam to cobbles (noncolloidal) | 3.7 | 5.0 | 5.0 |
| Graded, silt to cobbles (colloidal) | 4.0 | 5.5 | 5.0 |
| Alluvial silts (noncolloidal) | 2.0 | 3.5 | 2.0 |
| Alluvial silts (colloidal) | 3.7 | 5.0 | 3.0 |
| Coarse gravel (noncolloidal) | 4.0 | 6.0 | 6.5 |
| Cobbles and shingles | 5.0 | 5.5 | 6.5 |
| Shales and hard pans | 6.0 | 6.0 | 5.0 |

[17] As recommended by Special Committee on Irrigation Research, ASCE, 1926.

**Table 5.7**
Flow Type Classification Parameters

| Flow type | $\frac{h_1-z}{D}$ | $\frac{h_4}{h_c}$ | $\frac{h_4}{D}$ | culvert slope | barrel flow | location of control | kind of control |
|---|---|---|---|---|---|---|---|
| 1 | $< 1.5$ | $< 1.0$ | $\leq 1.0$ | steep | partial | inlet | critical depth |
| 2 | $< 1.5$ | $< 1.0$ | $\leq 1.0$ | mild | partial | outlet | critical depth |
| 3 | $< 1.5$ | $> 1.0$ | $\leq 1.0$ | mild | partial | outlet | backwater |
| 4 | $> 1.0$ | | $> 1.0$ | any | full | outlet | backwater |
| 5 | $\geq 1.5$ | | $\leq 1.0$ | any | partial | inlet | entrance geometry |
| 6 | $\geq 1.5$ | | $\leq 1.0$ | any | full | outlet | entrance and barrel geometry |

Identification of the type of flow beyond the guidelines in table 5.7 requires a trial and error procedure.

In the following cases, several variables appear repeatedly. $C_d$ is the *discharge coefficient*, a function of the barrel inlet geometry. Use orifice data if specific information is unavailable. $v_1$ is the average velocity of the water approaching the culvert entrance, often an insignificant quantity. Not all of the kinetic energy of the approaching water survives the inlet transition, and the *velocity-head coefficient*, $\alpha$, for the approach section accounts for the loss. $d_c$ is the critical depth, which may not correspond to the actual depth of flow. (It will have to be calculated from the flow conditions.) Finally, $h_f$ is the friction loss in the identified section. The friction head loss between sections 1 and 2, for example, can be calculated from equation 5.63.

$$h_{f,1-2} = \frac{LQ^2}{K_1 K_2} \qquad 5.63$$

$$K = \frac{1.486}{n}(r_H)^{2/3} A \qquad 5.64$$

The friction loss can be found in the usual manner, from the Darcy equation and the Moody friction factor chart. The *Manning equation* can also be used, and it is particularly useful since it eliminates the need for trial and error solutions.

$$h_f = \frac{v^2 n^2 L}{(2.21)(r_H)^{4/3}} \qquad 5.65$$

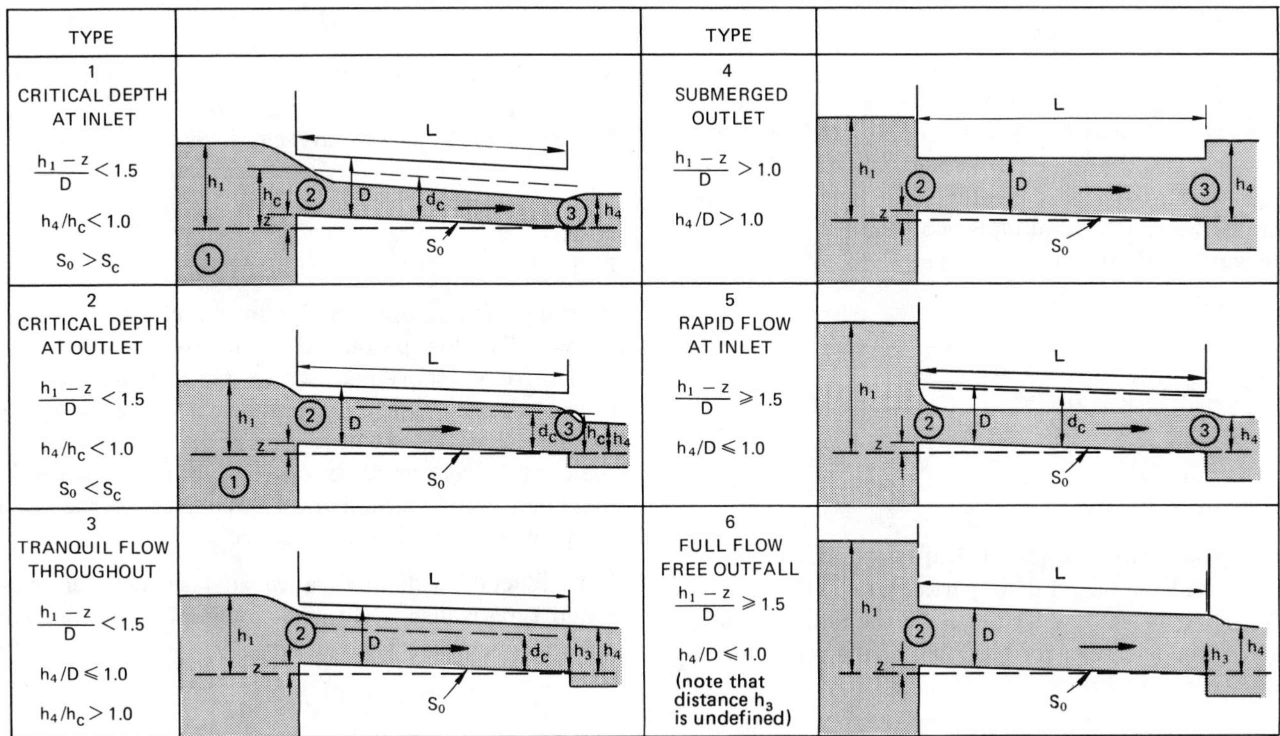


**Figure 5.23** Culvert Flow Classifications

For example, suppose the total hydraulic head available was $H$. This head would be divided between the velocity head in the culvert, the entrance loss (if considered), and the friction.

$$H = \frac{v^2}{2g_c} + k_e\left(\frac{v^2}{2g_c}\right) + \frac{v^2n^2L}{(2.21)(r_H)^{4/3}} \quad 5.66$$

This can be solved directly for the velocity.

$$v = \sqrt{\frac{H}{\frac{(1+k_e)}{2g} + \frac{n^2L}{(2.21)(r_H)^{4/3}}}} \quad 5.67$$

**Table 5.8**

Minor Entrance Loss Coefficients

| $k_e$ | condition of entrance |
|---|---|
| 0.08 | smooth, tapered |
| 0.10 | flush concrete groove |
| 0.10 | flush concrete bell |
| 0.15 | projecting concrete groove |
| 0.15 | projecting concrete bell |
| 0.50 | flush, square-edged |
| 0.90 | projecting, squared-edged |

The area, $A$, used in the discharge equations which follow, is not always the culvert area, particularly when the culvert does not flow full. $A_c$ is the area in flow at the critical section, while $A_i$ is the area in flow at numbered section i.

A. TYPE 1 FLOW

Water passes through the critical depth near the culvert entrance, and the culvert flows partially full. The slope of the culvert barrel is greater than the critical slope, and thc tailwater elevation is less than the elevation of the water surface at the control section.

The discharge is

$$Q=C_dA_c\sqrt{2g\left(h_1 - z + \frac{\alpha_1 v_1^2}{2g} - d_c - h_{f1-2} - h_{f2-3}\right)} \quad 5.68$$

B. TYPE 2 FLOW

Flow passes through the critical depth at the culvert outlet, and the barrel flows partially full. The slope of the culvert is less than critical, and the tailwater elevation does not exceed the elevation of the water surface at the control section.

$$Q = C_dA_c\sqrt{2g\left(h_1 + \frac{\alpha_1 v_1^2}{2g} - d_c - h_{f1-2} - h_{f2-3}\right)} \quad 5.69$$

C. TYPE 3 FLOW

When backwater is the controlling factor in culvert flow, the critical depth cannot occur. The upstream water-surface elevation for a given discharge is a function of the height of the tailwater.

For type 3 flow, flow is subcritical for the entire length of the culvert, with the flow being partial. The outlet is not submerged, but the tailwater elevation does exceed the elevation of critical depth at the terminal section.

$$Q = C_dA_3\sqrt{2g\left(h_1 + \frac{\alpha_1 v_1^2}{2g} - h_3 - h_{f1-2} - h_{f2-3}\right)} \quad 5.70$$

D. TYPE 4 FLOW

As in type 3 flow, the backwater elevation is the controlling factor in this case. Critical depth cannot occur, and the upstream water surface elevation for a given discharge is a function of the tailwater elevation. Discharge is independent of barrel slope.

The culvert is submerged at both the headwater and tailwater. No differentiation between low-head and high-head is made for this case. If the velocity head at section 1, the entrance friction loss and the exit friction loss are neglected, the discharge can be calculated.

$$Q = C_dA_o\sqrt{\frac{2g(h_1 - h_4)}{1 + \frac{29C_d^2n^2L}{(r_H)^{4/3}}}} \quad 5.71$$

The complicated term in the denominator corrects for friction. For rough estimates and for culverts less than 50 feet long, the friction loss can be ignored.

$$Q = C_dA_o\sqrt{2g(h_1 - h_4)} \quad 5.72$$

E. TYPE 5 FLOW

Partially-full flow under a high head is classified as type 5 flow. The flow pattern is similar to the flow downstream from a sluice gate with rapid flow near the entrance. Usually, type 5 flow requires a relatively square entrance that causes contraction of the flow area to less than the culvert area. In addition, the barrel length, roughness, and bed slope must be sufficient to maintain the flow area less than the culvert area.

It is difficult to distinguish in advance between type 5 and type 6 flow. Within a range of the important parameters, either flow can occur.[19]

$$Q = C_dA_o\sqrt{2g(h_1 - z)} \quad 5.73$$

[19] If the water surface ever touches the top of the culvert, the passage of air to the culvert will be prevented, and the culvert will flow full everywhere. This is type 6 flow.

F. TYPE 6 FLOW

Type 6 flow, like type 5, is considered a high-head flow. The culvert is full under pressure with free outfall. The discharge is

$$Q = C_d A_o \sqrt{2g(h_1 - h_3 - h_{f2-3})} \qquad 5.74$$

Equation 5.74 is inconvenient because $h_3$ (the true piezometric head at the outfall) is difficult to evaluate without special graphical aids. The actual hydraulic head driving the culvert flow is a function of the Froude number. For conservative first approximations, $h_3$ can be taken as the barrel diameter. In reality, it varies from somewhat less than half the barrel diameter to the full diameter. This will give the minimum hydraulic head.

If $h_3$ is taken as the barrel diameter, the total hydraulic head ($H = h_1 - h_3$) will be split between the velocity head and friction. In that case, equation 5.67 can be used to calculate the velocity. The discharge is easily calculated from equation 5.75.[20]

$$Q = A_o v \qquad 5.75$$

*Example 5.11*

Size a square culvert which has the following characteristics:

$$\text{slope} = 0.01$$
$$\text{length} = 250\,\text{feet}$$
$$\text{capacity} = 45\,\text{cfs}$$
$$n = 0.013$$

entrance fluid level 5 feet above barrel top

free exit

Since the $h_1$ dimension in all six cases is measured from the culvert invert, it is difficult to classify the type of flow at this point. However, either type 5 or type 6 is likely.

*step 1*: Assume a trial culvert size. Select a square opening with 1.0 foot sides.

*step 2*: Calculate the flow assuming case 5. With entrance control, the culvert will not flow full. The entrance will act like an orifice.

$$A_o = (1)(1) = 1\,\text{ft}^2$$
$$H = h_1 - z = 5 + 1 = 6\,\text{ft}$$
$$C_d = 0.62 \text{ for square-edged openings}$$
$$Q = (0.62)(1)\sqrt{(2)(32.2)(6)} = 12.2\ \text{cfs}$$

Since this size has insufficient capacity, try a larger culvert. Choose 2.0 foot sides.

$$A_o = (2)(2) = 4$$
$$H = 5 + 2 = 7$$
$$Q = (0.62)(4)\sqrt{(2)(32.2)(7)} = 52.7\,\text{cfs}$$

*step 3*: Begin checking the entrance control assumption by calculating the maximum hydraulic radius. Because the flow is entrance controlled, the upper surface of the culvert is not wetted. The hydraulic radius is maximum at the entrance.

$$r_H = \frac{A_o}{p_w} = \frac{4}{2+2+2} = 0.667\ \text{ft}$$

*step 4*: Calculate the velocity using the Manning equation for open channel flow. Since the hydraulic radius is maximum, the velocity will also be maximum.

$$v = \frac{1.486}{0.013}(0.667)^{2/3}\sqrt{0.01}$$
$$= 8.72\ \text{ft/sec}$$

*step 5*: Calculate the normal depth, $d_n$.

$$d_n = \frac{Q}{(v)(w)} = \frac{45}{(8.72)(2)} = 2.58$$

Since the normal depth is greater than the culvert size, the discharge will be full pipe flow under pressure. (It was not necessary to calculate the critical depth since the flow is implicitly subcritical.) The entrance control assumption was, therefore, not valid for this size culvert.[21] At this point, two things can be done. A larger culvert can be chosen if entrance control is desired. Or, the solution can continue by checking to see if the culvert has the required capacity as a pressure conduit.

*step 6*: Check the capacity as a pressure conduit. $H$ is the total available head.

$$H = h_1 - h_3$$
$$= [5 + 2 + (0.01)(250)] - 2$$
$$= 7.5\ \text{ft}$$

[20] Equation 5.75 does not include the discharge coefficient. $v$, when calculated from equation 5.67, is implicitly the velocity in the barrel.

[21] If the normal depth had been less than the barrel diameter, it would still be necessary to determine the critical depth of flow. If the normal depth was less than the critical depth, the entrance control assumption would have been valid.

*step 7*: Since the pipe is flowing full, the hydraulic radius is

$$r_H = \frac{4}{8} = 0.5$$

*step 8*: Equation 5.67 can be used to calculate the flow velocity. Since the culvert has a square-edged entrance, a loss coefficient of $k_e = 0.5$ is used. However, this doesn't affect the velocity greatly.

$$v = \sqrt{\frac{7.5}{\frac{(1+0.5)}{2(32.2)} + \frac{(0.013)^2(250)}{(2.21)(0.5)^{4/3}}}}$$
$$= 10.24\ \text{ft/sec}$$

*step 9*: Check the capacity.

$$Q = vA_o = (10.24)(4) = 40.96\ \text{cfs}$$

The culvert size is not acceptable since its discharge under the maximum head does not have a capacity of 45 cfs.

*step 10*: Repeat from step 2, trying a larger size culvert. With a 2.5 foot side, the following values are obtained:

$$A_o = (2.5)(2.5) = 6.25\ \text{ft}^2$$
$$H = 5 + 2.5 = 7.5\ \text{ft}$$
$$Q = (0.62)(6.25)\sqrt{(2)(32.2)(7.5)}$$
$$= 85.2\ \text{cfs}$$
$$r_H = \frac{6.25}{7.5} = 0.833\ \text{ft}$$
$$v = \frac{1.486}{0.013}(0.833)^{2/3}\sqrt{0.01} = 10.12\ \text{ft/sec}$$
$$d_n = \frac{45}{(10.12)(2.5)} = 1.78$$

*step 11*: Calculate the critical depth. For rectangular channels, equation 5.43 can be used.

$$d_c = \sqrt[3]{\frac{Q^2}{gw^2}} = \sqrt[3]{\frac{(45)^2}{32.2(2.5)^2}} = 2.16\ \text{ft}$$

Since the normal depth is less than the critical depth, the flow is supercritical. The entrance control assumption was correct for the culvert. The culvert has sufficient capacity to carry 45 cfs.

# Appendix A: Design Use Values of Manning's $n$

| channel material | $n$ |
|---|---|
| clean, uncoated cast iron | 0.013–0.015 |
| clean, coated cast iron | 0.012–0.014 |
| dirty, tuberculated cast iron | 0.015–0.035 |
| riveted steel | 0.015–0.017 |
| lock-bar and welded | 0.012–0.013 |
| galvanized iron | 0.015–0.017 |
| brass and glass | 0.009–0.013 |
| wood stave | |
| small diameter | 0.011–0.012 |
| large diameter | 0.012–0.013 |
| concrete | |
| with rough joints | 0.016–0.017 |
| dry mix, rough forms | 0.015–0.016 |
| wet mix, steel forms | 0.012–0.014 |
| very smooth, finished | 0.011–0.012 |
| vitrified sewer | 0.013–0.015 |
| common-clay drainage tile | 0.012–0.014 |
| asbestos | 0.011 |
| planed timber | 0.011 |
| canvas | 0.012 |
| unplaned timber | 0.014 |
| brick | 0.016 |
| rubble masonry | 0.017 |
| smooth earth | 0.018 |
| firm gravel | 0.023 |
| corrugated metal pipe | 0.022 |
| natural channels, good condition | 0.025 |
| natural channels with stones and weeds | 0.035 |
| very poor natural channels | 0.060 |

# Appendix B: Manning Nomograph

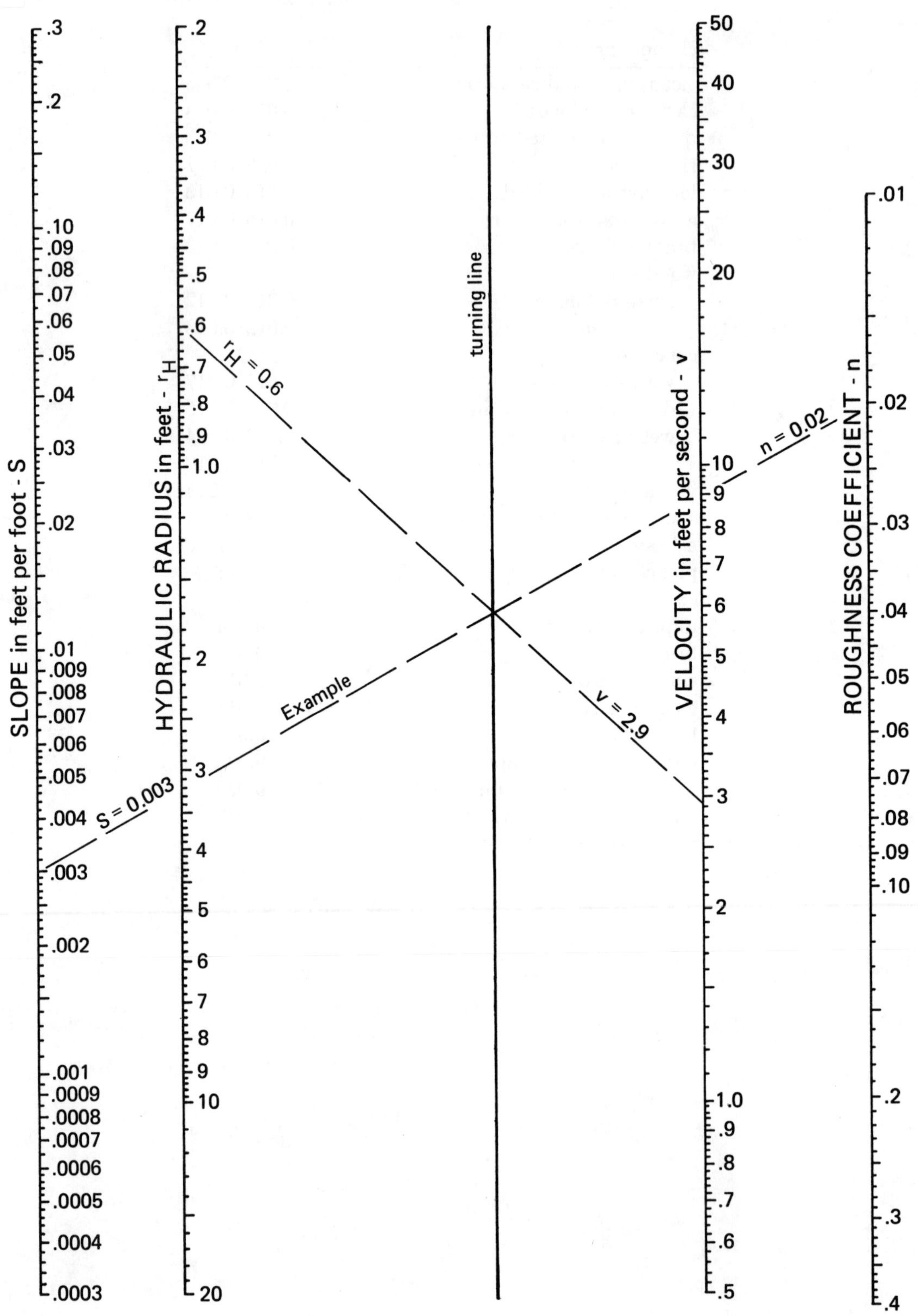

# Appendix C: Circular Channel Ratios

Experiments have shown that $n$ varies slightly with depth. This figure gives velocity and flow rate ratios for varying $n$ (solid line) and constant $n$ (broken line) assumptions.

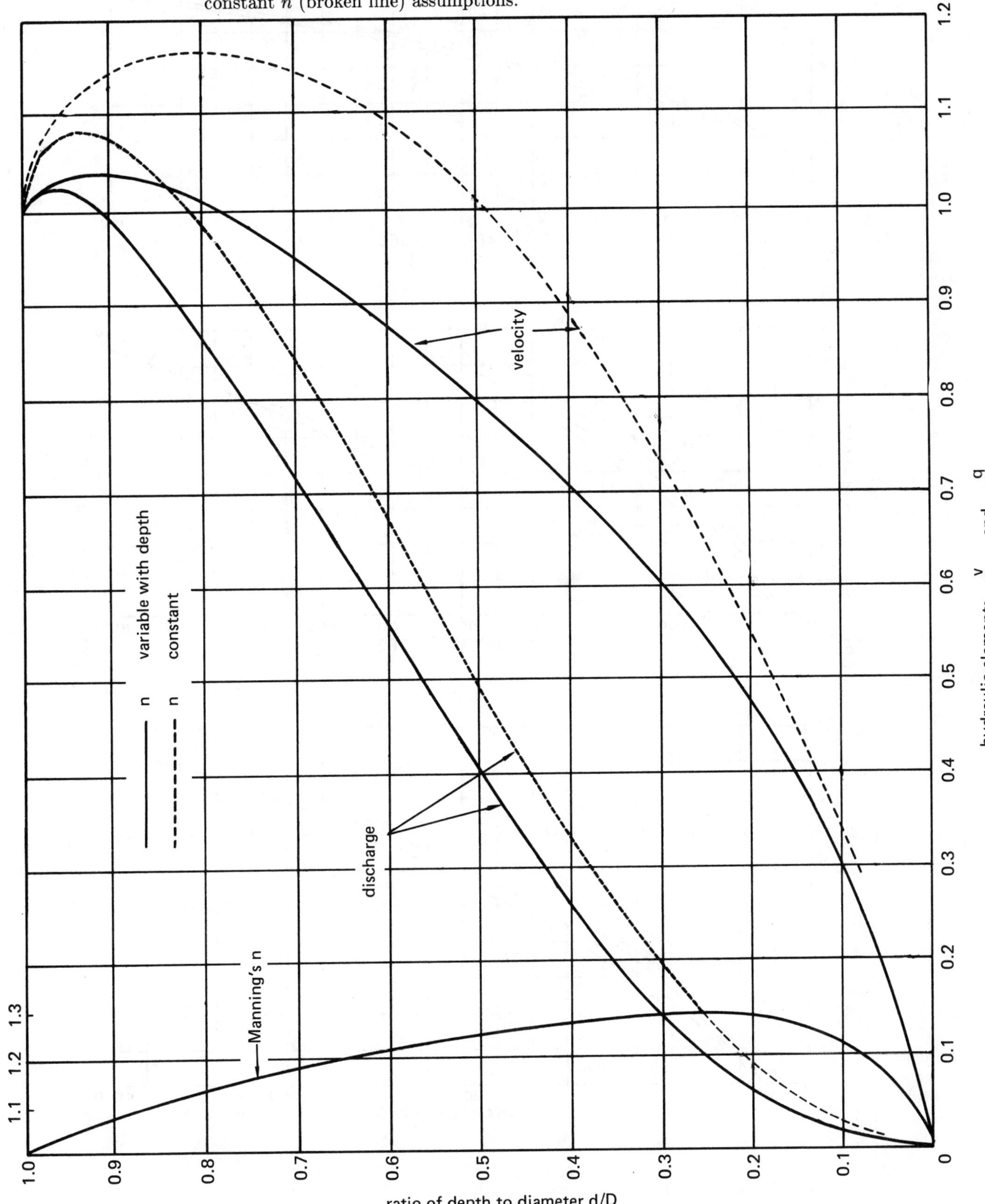

# Appendix D: Critical Depths in Circular Channels

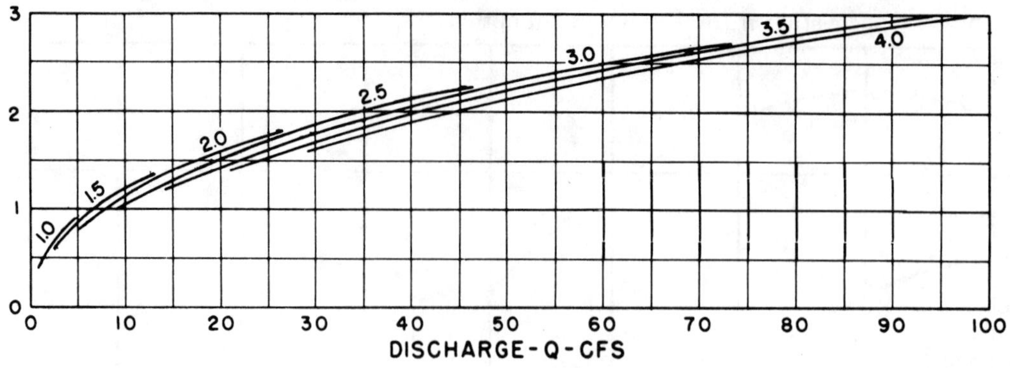


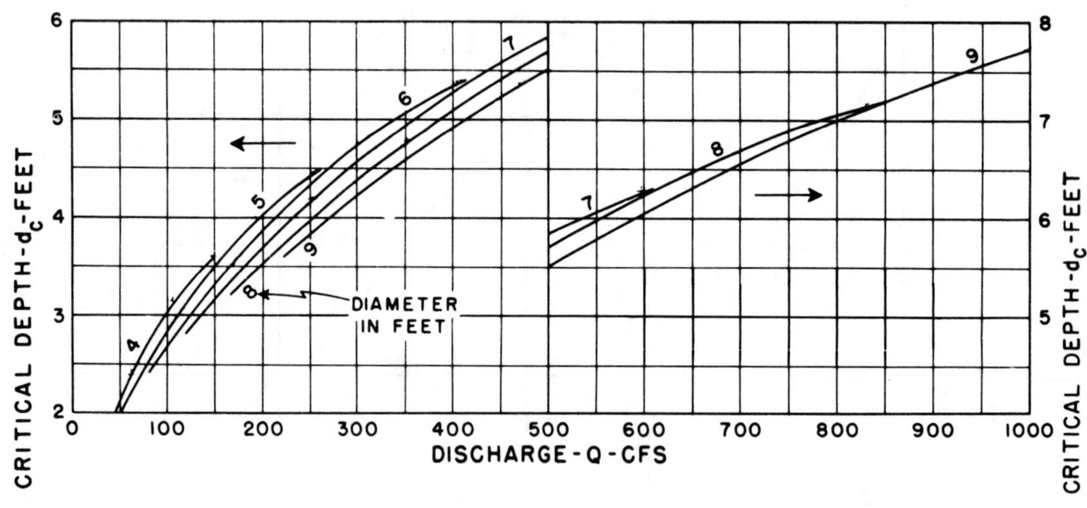


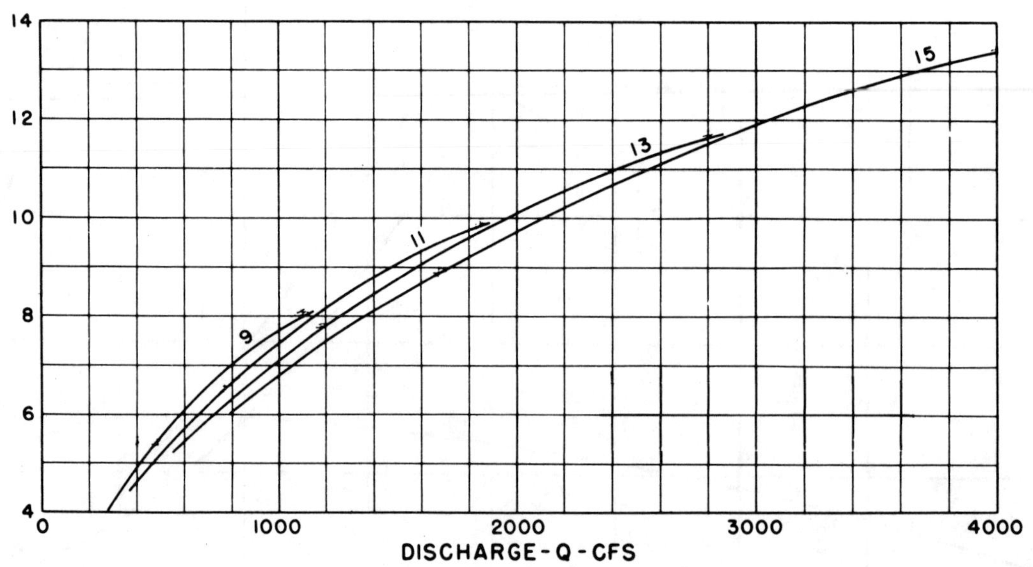

## Practice Problems: OPEN CHANNEL FLOW

Untimed

1. 30 years ago, a 24 inch diameter pipe ($n = .013$) was installed on a .001 slope. Recent tests indicate that the full-flow capacity of the pipe is 6.0 cfs. Find (a) the original velocity when full, (b) the present velocity when full, (c) the present value of $n$, and (d) the original capacity.

2. A sewer is to be installed on a 1% grade. Its roughness coefficient is $n = .013$. Maximum full capacity is to be 3.5 cfs. (a) What size pipe would you recommend? (b) What is the capacity of the pipe size you have chosen when flowing full? (c) What is the velocity when flowing full? (d) What is the depth of flow when the flow is .7 cfs? (e) What minimum velocity will prevent solids settling out in the sewer?

3. The depth of flow upstream from a hydraulic jump is 1 foot. The depth of flow after the jump is 2.4 feet. The channel is rectangular with a 5 foot width. What is the discharge rate of the channel?

4. A wooden flume ($n = .012$) of rectangular cross section is 2 feet wide. The flume carries 3 cfs of water on a 1% slope. What is the depth?

5. A 4-foot diameter concrete storm drain ($n = .013$, slope $= .02$) carries water at a depth of 1.5 feet. (a) What is the velocity of the water in the pipe? (b) What is the maximum velocity of water flowing in the pipe? (c) What is the maximum capacity of the pipe?

6. A hydraulic jump forms at the toe of a spillway. The depths of flow are 0.2 feet and 6 feet on either side of the jump. The velocity before the jump is 54.7 fps. What is the energy loss in the jump?

7. A spillway operates with 2 feet of head. The toe of the spillway is 40 feet below the top of the spillway. Use a spillway coefficient of 3.5. (a) What is the discharge per foot of crest? (b) What is the depth of flow at the toe?

8. An ogee spillway operates with $C_S = 3.5$ and a head of 5 feet. The spillway crest is 10 feet above the toe. What is the discharge per foot of crest?

9. 10,000 cfs of water flow down a 100-foot wide spillway placed on a 5% grade. The spillway surface has a roughness coefficient of $n = .012$. Neglect sidewall effects. (a) What is the depth of flow (normal depth) down the spillway? (b) What is the critical depth? (c) Is the flow tranquil or shooting? (d) A hydraulic jump forms at the junction of the 5% slope and a horizontal toe. What is the depth after the jump?

10. A sharp-crested rectangular weir with two end contractions is 5 feet wide. The weir height is 6 feet. What is the flow rate if the head is measured as .43 feet?

11. The trapezoidal channel shown is laid at a slope of .002. The depth of flow is 2.0 ft. Determine the flow rate. Use $n = .013$.

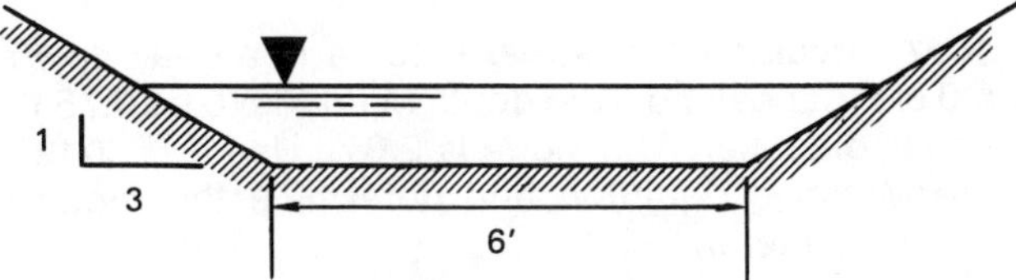


12. The spillway shown has a crest length of 60.0 ft. The stilling basin is 60 ft wide with a level bottom. The head loss in the chute is 20% of the difference in level between the reservoir surface and the stilling basin bottom. Use a spillway coefficient of 3.7. (a) Determine the flow rate. (b) Determine the depth of flow at the toe of the spillway. (c) What tailwater depth is required to cause a hydraulic jump at the toe? (d) What is the energy loss in the hydraulic jump?

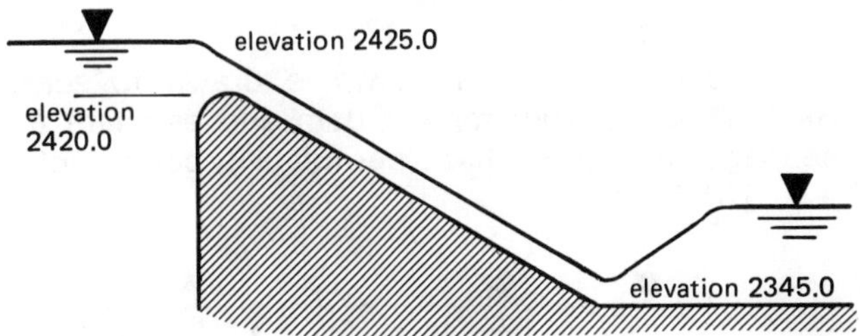


13. A Parshall flume has a throat width of 6.0 ft. The upstream head measured from the throat floor is 18.0 in. (a) What is the flow rate? (b) What are the advantages of the Parshall flume?

14. The weir shown has an 18.0 in. base and a 4-to-1 side slope. The depth of flow over the weir is 9.0 in. What is the rate of discharge?

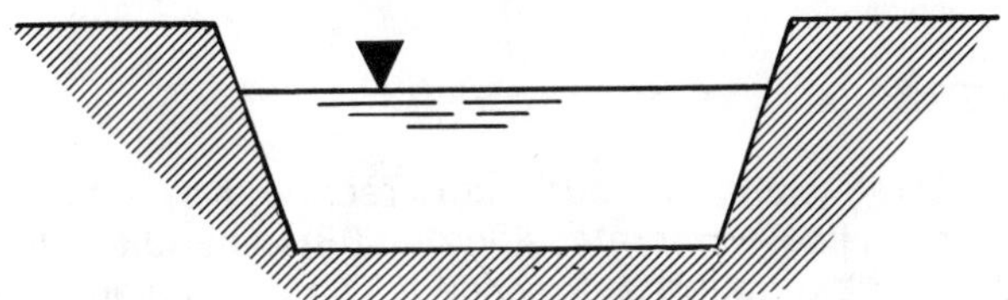

15. Water is flowing at the rate of 50 cfs and a depth of 3.0 ft in a 6.0 ft wide rectangular channel. How high a hump must be placed in the channel to produce critical flow over the hump?

16. A 42 in. diameter concrete culvert is 250 ft long and laid at a slope of .006. The culvert entrance is flush and square-edged. Determine the capacity of the culvert when the outlet is submerged to the crown of the barrel and the headwater is 5.0 ft above the crown of the culvert's inlet.

17. A rectangular channel 8.0 ft wide carries a flow of 150 cfs. The channel slope is .0015 and the Manning coefficient is $n = .015$. A weir installed across the channel raises the depth at the weir to 6.0 ft. (a) Determine the normal depth of flow. (b) Draw the backwater curve.

18. A circular storm sewer is to carry a peak flow of 5.0 cfs. At peak flow the depth of flow is to be 0.75 the sewer diameter. The slope is 2.0%. Use $n = .012$ in the Manning equation and determine the required sewer diameter.

19. Water is flowing in a 6.0 ft wide rectangular channel. A hydraulic jump takes place as shown. (a) Determine the flow rate. (b) What is the energy loss in the jump?

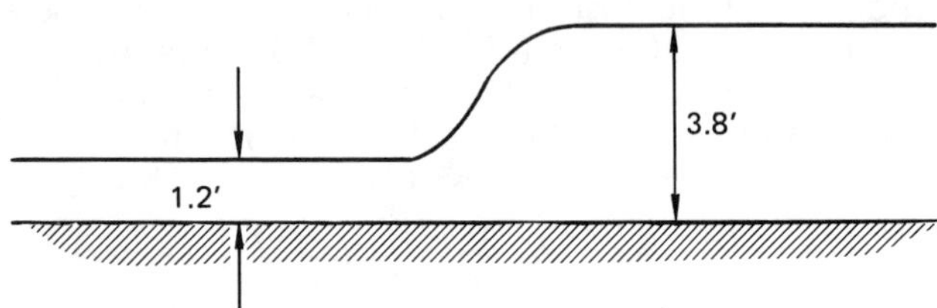


20. The masonry dam shown is drawn to scale. Sketch the flow net for the dam and estimate the seepage rate. The silt has a permeability coefficient of $k = 0.15$ ft/hr.

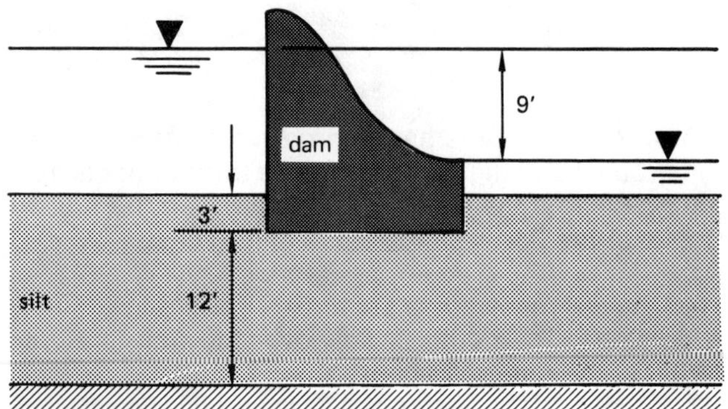


## Timed

1. Dimension an optimum rectangular channel (smooth lined concrete, slope = .08) to handle a flow of 17 cubic meters per second. The dimensions can be in units of feet.

2. An overflow structure has three inlets sized 1′ high by 2′ wide as shown. The coefficient of discharge for each orifice is 0.7. The water standing in the intake structure is 4 feet above the top of the box culvert. The coefficient of discharge of the box culvert entrance is 2/3. Assume $n$ = 0.013 (Manning's coefficient) and slope = 0.05 for the culvert which is 100 feet long. Size the square box culvert to handle the flow. Assume the water level behind the dam is constant.

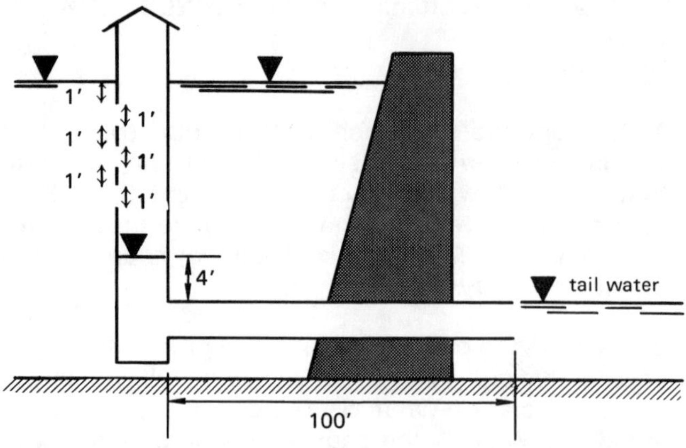


3. A dam spills 70,000 cfs of water over its crest as shown. The width of the dam is 500 feet. The depth at point B is 2 feet. (a) What is the friction loss between points A and B? (b) What is the depth of water at point C? (c) Sketch the energy gradient line from point A to point B. (d) What else needs to be calculated to determine if the hydraulic jump occurs at the toe (point B)?

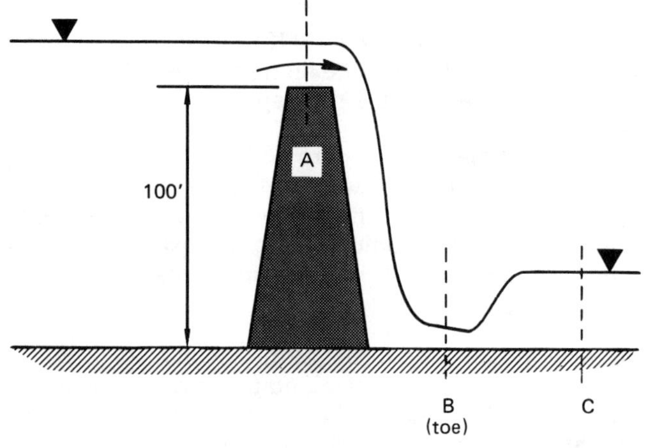


4. Flow in a 6′ diameter, newly-formed concrete culvert is 150 cfs at a point where the depth of flow is 3′. (a) What is the hydraulic radius of the flow? (b) What is the slope of the pipe? (c) Is the flow subcritical, critical, or supercritical? (d) What would be the flow if the culvert were full? (e) Assuming that you know the entrance type and discharge rate, list 3 factors to be considered in selecting a culvert size.

5. A 17′ × 20′ rectangular tank is fed from the bottom and overflows through a double-constricted rectangular weir (to be designed) into a trough.

The following specifications are known:

- discharge rate is 2 MGD minimum
  4 MGD average
  8 MGD maximum
- minimum freeboard in tank = 4″
- minimum head (tank to channel) = 10″
- maximum channel elevation = 590′
- elevation of tank bottom = 580′

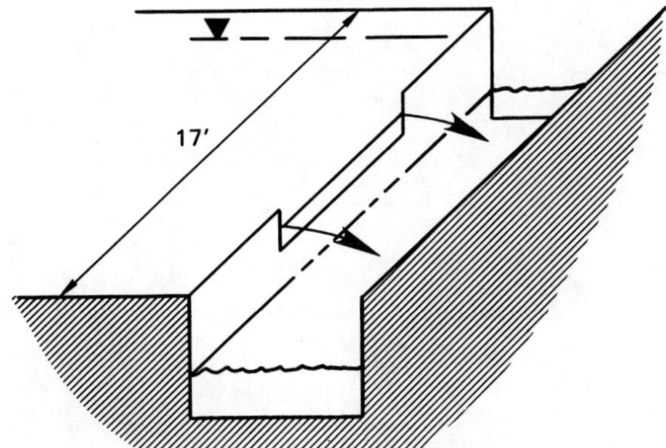


(a) Specify the length of the weir opening. (b) What is the surface elevation at maximum flow? (c) What is the water surface elevation at minimum flow?

RESERVED FOR FUTURE USE

# 6 HYDROLOGY

*Nomenclature*

| | | |
|---|---|---|
| A | area | $ft^2$ |
| $A_d$ | drainage area | acres |
| B | aquifer width | ft |
| C | rational runoff coefficient, or other coefficient | – |
| $C_p$ | pan coefficient | – |
| $C_r$ | retardance coefficient | – |
| CN | curve number | – |
| d | drawdown, or distance between stations | ft, or miles |
| E | evaporation | in/day |
| F | frequency of occurrence, or infiltration | –/yrs, inches |
| $G_H$ | hydraulic gradient | ft/ft |
| H | total hydraulic head | ft |
| I | rainfall intensity | in/hr |
| $I_a$ | initial abstraction | inches |
| $K_p$ | constant of permeability | $ft^3$/day-$ft^2$ |
| $L_c$ | centroidal stream length | miles |
| $L_o$ | overland flow path length | ft |
| $L_s$ | main stream length | miles |
| N | average precipitation per year, or time from peak to end of runoff | inches, or hours |
| P | precipitation over a short period | inches |
| Q | flow quantity | cfs |
| r | radial distance from well | ft |
| R | regional factor | – |
| S | slope of the surface, or storage | –, inches |
| $t_c$ | storm duration (time of concentration) | minutes |
| $t_p$ | time from start of storm to peak runoff | hrs |
| $t_r$ | rain storm duration | hrs |
| T | transmissivity | $ft^3$/day-ft |
| v | flow velocity | ft/sec |
| V | volume | $ft^3$ |
| y | aquifer thickness after drawdown | ft |
| Y | original aquifer thickness | ft |

*Subscripts*

| | |
|---|---|
| f | flow |
| g | gross |
| N | net |
| o | at well |
| p | peak, pan, or equipotential |
| r | radius r, rainfall |
| R | reservoir |
| t | total |
| u | unit |
| x | unknown |
| y | years |

## 1 IMPORTANT CONVERSIONS

| Multiply | By | To Get |
|---|---|---|
| acre | 43,560 | $ft^2$ |
| acre-ft | 43,560 | $ft^3$ |
| acre-ft | 325,851 | gallons (U.S.) |
| acre-inches/hr | 1.008 | cfs |
| cubic feet | 7.4805 | gallons |
| cubic feet/sec | 1.9834 | acre-ft/day |
| cubic feet/sec | 448.83 | gpm |
| cubic feet/sec | 0.64632 | MGD |
| darcy | 1.062 EE–11 | $ft^2$ |
| gallons (U.S.) | 0.1337 | $ft^3$ |
| gallons | 3.07 EE–6 | acre-ft |
| gallons/day | 1.547 EE–6 | $ft^3$/sec |
| gpm | 1440 | gallon/day |
| gpm | 0.002228 | cfs |
| gpm | 192.5 | $ft^3$/day |
| hectare | 2.471 | acres |
| horsepower | 0.7457 | kw |
| horsepower | 550 | ft-lbf/sec |
| inch of runoff | 53.3 | acre-ft/sq. mile |
| inch of runoff | 2.323 EE6 | $ft^3$/sq. mile |
| Meinzer unit | 1.00 | gal/day-$ft^2$ |
| MGD | 1 EE6 | gallon/day |

| | | |
|---|---|---|
| sq. mile | 640 | acres |
| sq. mile | 2.788 EE7 | $ft^2$ |
| sq. mile-inch | 53.3 | acre-ft |
| sq. mile-inch/day | 26.88 | cfs |

## 2 DEFINITIONS

Anabranch: The intertwining channels of a braided stream.

Anticlinal spring: A portion of an exposed aquifer (usually on a slope) between two impervious layers.

Aquiclude: An underground source of water with insufficient porosity to support any sufficient removal rate.

Aquifer: An underground source of water capable of supplying a well or other use.

Aquifuge: An underground geological formation which has no porosity or openings at all through which water can enter or be removed.

Artesian well: A spring in which water flows naturally out of the earth's surface due to pressure placed on the water by an impervious overburden and hydrostatic head.

Artesian formation: An aquifer in which the piezometric height is greater than the aquifer thickness.

Base flow: Runoff which percolates down to the water table and then discharges into a stream. Up to 2 years may elapse between precipitation and discharge.

Bifurcation ratio: The average number of streams feeding into the next side (order) waterway. The range is usually 2 to 4.

Blind drainage: Geographically large (with respect to the drainage basin) depressions which store water during a storm and, therefore, stop it from contributing to surface runoff.

Braided stream: A wide, shallow stream with many anabranches.

Capillary water: Water just above the water table which is drawn up out of an aquifer due to capillary action of the soil.

Cone of depression: The shape of the water table around a well during and immediately after use. The cone's apex differs from the original water table by the well's drawdown.

Confined water: Artesian water overlaid with an impervious layer, usually under pressure.

Connate water: Water, frequently saline, present in rock at its formation.

Depression storage: Initial storage of rain in small surface puddles.

Depth-Area-Duration analysis: A study made to determine the maximum amounts of rain within a given time period over a given area.

Dimple spring: A depression in the earth below the water table.

Drainage density: The total length of streams in a watershed divided by the drainage area.

Drawdown: The difference in water table level at a well head and far from it.

Dry weather flow: See 'Base flow.'

Effluent stream: A stream which intersects the water table and receives groundwater. Effluent streams seldom go completely dry during the rainless periods.

Ephemeral stream: A stream which goes dry during rainless periods.

Evapotranspiration: Evaporation of water from a study area due to all sources including water, soil, snow, ice, vegetation, and transpiration.

Flowing well: A well which flows on its own accord to the surface. See also 'Artesian well.'

Forebay: An area which recharges an aquifer.

Gravitational water: Water in transit downward through the earth.

Groundwater: Subsurface water flowing in an aquifer towards a stream. Groundwater is not water flowing on the ground. It is water flowing underground.

Hydrological cycle: The cycle experienced by water in its travel from the ocean, through evaporation and precipitation, percolation, runoff, and return to the ocean.

Hydrometeor: Any form of water falling from the sky.

Hygroscopic water: Moisture adhering in a thin film to soil grains.

Impervious layer: A geologic layer through which no water can pass.

Infiltration: The movement of water through the upper soil.

Influent stream: A stream above the water table. Influent streams may go dry during the rainless season.

Initial loss: The sum of interception and depression loss, but excluding blind drainage.

Interception: Rain which falls on vegetation and other impervious objects and which evaporates without contributing to runoff.

Interflow: Infiltrated subsurface water which travels to a stream without percolating down to the water level.

Juvenile water: Water formed chemically within the earth.

Lysimeter: A soil container used to observe and measure evaptotranspiration.

Meandering stream: A stream which flows in large loops, not in a straight line.

Meteoric water: See 'Hydrometeor.'

Negative boundaries: A fault or similar geologic structure.

Net rain: Rain which contributes to surface runoff.

Overland flow: Water which travels over the ground surface to a stream.

Pan: A container used to measure surface evaporation rates.

Perched spring: A localized saturated area which occurs above an impervious layer on a slope.

Percolation: The travel of water down through the soil to the water table.

Phreatic zone: The layer below the water table down to an impervious layer.

Phreatophytes: Trees with root systems which extend into the water table.

Piezometric level: The level to which water will rise in a pipe due to its own pressure.

Plat: A small plot of land.

Porosity: The ratio of pore volume to total formation volume.

Probable maximum rainfall: The rainfall corresponding to some given probability (e.g., 1 in 100 years).

Safe yield: The maximum rate of water withdrawal which is economically and ecologically feasible.

Seep: See 'Spring.'

Sinuosity: The stream length divided by the valley length.

Specific yield: The ratio of water volume which will drain freely from a substance sample to the total volume. Specific yield is always less than porosity. Specific yield is also defined as the volume of water obtained by lowering one square foot of the water table by one foot.

Spring: A place where the earth surface and aquifer coincide.

Stream order: An artificial categorization of stream geneology. Small streams are 1st order. 2nd order streams are fed by 1st order streams, 3rd order streams are fed by 2nd order streams, etc.

Subsurface runoff: See 'Interflow.'

Surface detention: Rain water which collects as a film and runs off of the saturated surface during a storm.

Surface retention: That part of a storm which does not immediately appear as infiltration or surface runoff. Retention is made up of depression storage, interception, and evaporation.

Surface runoff: Water flow over the surface which reaches a stream after a storm.

Time of concentration: The time required for water to flow from the most distant point on a runoff area to the measurement or collection point.

Transpiration: The process in which plants give off internal moisture to the atmosphere.

Unit stream power: The product of velocity and slope, representing the rate of energy expenditure per pound of water.

Vaclose water: All water above the water table, including soil water, gravitational water, and capillary water.

Vadose zone: A zone above the water table containing both saturated and empty soil pores.

Water table: The top level of an aquifer, defined as the locus of points where the water pressure is equal to the atmospheric pressure.

Xerophytes: Drought-resistant plants, typically existing with root systems well above the water table.

Zone of aeration: See 'Vadose zone.'

Zone of saturation: See 'Phreatic zone.'

## 3 PRECIPITATION

Although the word *precipitation* encompasses all hydrometeoric forms, it is often applied only to rainfall in the liquid form. Precipitation data may be collected in a number of ways, but the open 8-inch precipitation rain gage is quite common.

If a rain measurement is lost or is not available, it can be estimated by one of the following procedures:

*Method 1*: Choose three stations evenly spaced around and close to the location which has missing data. If the normal annual precipitations at the three sites do not vary more than 10% from the missing station's normal annual precipitation, the rainfall is estimated as the arithmetic mean of the three neighboring stations' precipitations for the period in question.

*Method 2*: If the difference is more than 10%, the Normal-Ratio Method can be used:

$$P_x = \left(\frac{1}{3}\right)\left[\left(\frac{N_x}{N_A}\right)P_A + \left(\frac{N_x}{N_B}\right)P_B + \left(\frac{N_x}{N_C}\right)P_C\right] \qquad 6.1$$

*Method 3*: A method used by the U.S. National Weather Service for river forecasting is to use data from stations in the four nearest quadrants (North, South, East, and West of the unknown station) and to weight the data with the square of the distance between stations.

$$P_x = \frac{d_{A-x}^2 P_A + d_{B-x}^2 P_B + d_{C-x}^2 P_C + d_{D-x}^2 P_D}{d_{A-x}^2 + d_{B-x}^2 + d_{C-d}^2 + d_{D-x}^2} \qquad 6.2$$

The average precipitation over a specific area or time basis can be found from station data in several ways.

*Method 1*: If the stations are uniformly distributed over a flat site, their precipitations can be averaged. This also requires that the individual precipitation records not vary too much from the mean.

*Method 2*: The *Thiessen method* calculates the average by weighting station measurements by the area of the assumed basin for each station. These assumed basin areas are found by drawing dotted lines between all stations and bisecting these dotted lines with solid lines (which are extended outward until they connect with other solid lines). The solid lines will form a polygon whose area is the assumed basin area.

*Method 3*: The most accurate method is the *Isohyetal method.* This method requires plotting lines of constant precipitation (isohyets) and weighting the isohyet values by the areas enclosed by the lines. Station data is used to draw isohyets, but not in the calculation of average rainfall.

Effective design of a structure will depend on the geographical location and degree of protection required. Once the location of a structure has been chosen, it is up to the engineer to design for the area's most probable maximum rainfall or probable maximum flood. Both of these require some judgment since the maximum is not a deterministic number.

*Rainfall intensity* is the rate of precipitation per hour. Intensity will be low for most storms, but it will be high for some storms. These high intensity storms can be expected very infrequently—say every 20, 50, or 100 years. The average number of years between storms of a given magnitude is known as the *design-storm frequency of occurrence* or *recurrence interval.*

The intensity of a storm can be calculated from equation 6.3.

$$I = \frac{K'(F)^a}{(t_c + b)^c} \qquad 6.3$$

$K'$, $a$, $b$, and $c$ are constants which depend on the conditions and location.

The *Steel formula* is a simplification of equation 6.3.

$$I = \frac{K}{t_c + b} \qquad 6.4$$

$K$ and $b$ are dependent on the storm frequency and geographical location. $t_c$ can be either the time of concentration (in runoff studies) or the rainfall duration.

Values of the constants $K$ and $b$ in equation 6.4 are not difficult to obtain once the intensity-duration-frequency curve is established. Although a logarithmic transformation could be used to obtain the data in straight line form, an easier method exists. This method starts by taking the reciprocal of equation 6.4, converting the equation to a straight line.

$$\frac{1}{I} = \frac{t_c + b}{K} = \frac{t_c}{K} + \frac{b}{K} = C_1 t_c + C_2 \qquad 6.5$$

As equation 6.5 shows, it is possible to use linear regression to determine the relationship between $1/I$ and $t_c$. Once $C_1$ and $C_2$ have been found, $K$ and $b$ can be calculated.

$$K = \frac{1}{C_1} \qquad 6.6$$

$$b = \frac{C_2}{C_1} \qquad 6.7$$

For very rough estimates, published values of $K$ and $b$ can be obtained from the geographical location. Table 6.1 is typical of some of this general data.

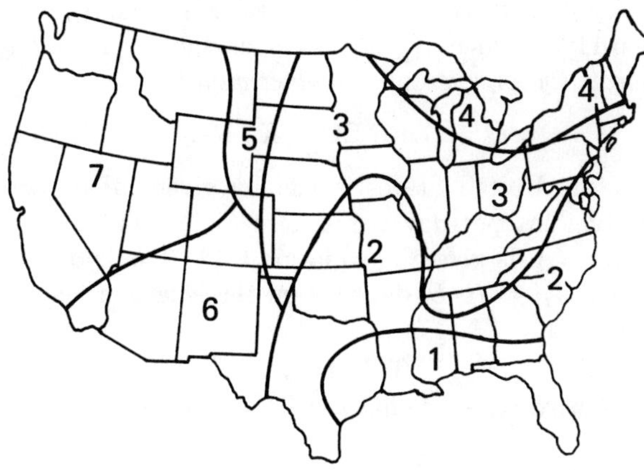


**Figure 6.1** Steel Formula Rainfall Regions

**Table 6.1**
Steel Formula Coefficients

| frequency in years | coefficients | region 1 | 2 | 3 | 4 | 5 | 6 | 7 |
|---|---|---|---|---|---|---|---|---|
| 2 | K | 206 | 140 | 106 | 70 | 70 | 68 | 32 |
| | b | 30 | 21 | 17 | 13 | 16 | 14 | 11 |
| 4 | K | 247 | 190 | 131 | 97 | 81 | 75 | 48 |
| | b | 29 | 25 | 19 | 16 | 13 | 12 | 12 |
| 10 | K | 300 | 230 | 170 | 111 | 111 | 122 | 60 |
| | b | 36 | 29 | 23 | 16 | 17 | 23 | 13 |
| 25 | K | 327 | 260 | 230 | 170 | 130 | 155 | 67 |
| | b | 33 | 32 | 30 | 27 | 17 | 26 | 10 |
| 50 | K | 315 | 350 | 250 | 187 | 187 | 160 | 65 |
| | b | 28 | 38 | 27 | 24 | 25 | 21 | 8 |
| 100 | K | 367 | 375 | 290 | 220 | 240 | 210 | 77 |
| | b | 33 | 36 | 31 | 28 | 29 | 26 | 10 |

Since the intensity, $I$, is the average intensity, the total rainfall (in inches) can be calculated from the storm duration.

$$P = It_c \qquad 6.8$$

*Example 6.1*

A storm has an intensity given by the equation:

$$I = \frac{100}{t_c + 10}$$

15 minutes are required for runoff from the farthest corner of a 5-acre plat to reach a discharge culvert. What is the design intensity?

$$I = \frac{100}{15 + 10} = 4\text{ inches/hour}$$

## 4 FLOOD CONSIDERATIONS

A flood occurs when more water than a river can carry is introduced. When a water course is too small to contain the flow, the water overflows the banks. The results of this overflow may produce nuisance flooding, damaging flooding, or devastating flooding.

*Nuisance flooding* may result in inconveniences such as wet feet, tire spray, and soggy lawns. *Damaging floods* go on to soak flooring, carpeting, and first-floor furniture. *Devastating floods* can wash buildings and vehicles downstream, as well as take lives.

Although rain causes flooding, large rain storms do not always create flooding. The size of a flood depends not only on the amount of rainfall, but also on the conditions within the watershed before and during the storm. When rain falls on a very wet watershed that is unable to absorb more water, or when a very large amount of rain falls on a dry watershed faster than it can be absorbed, the water runs off.

It is not economically justifiable to attempt to provide protection against the largest flood that could occur. Such protection would be too costly for such a rare event. Therefore, governmental and private institutions have agreed that a *one percent flood* is the most infrequent event to be protected against. A one percent flood, also called a *hundred year flood*, is a flood that would be exceeded in severity only once every hundred years on the average.

Planning for the one percent flood has proved to be a good compromise between not doing enough and spending too much. The U.S. Army Corps of Engineers, Soil Conservation Service, Department of Transportation, and others use the one percent flood to design channels and bridges and to protect facilities in which they have an interest.

Specific terms are sometimes used to designate the degree of protection required. For example, the *probable maximum flood* (*PMF*) is a hypothetical flood which can be expected to occur as a result of the most severe combination of critical meteorlogic and hydrologic conditions possible within a region. Since many structures are built with a limited life expectancy, it may be extremely uneconomical to design against the PMF.

The *standard project flood* (*SPF*) is a flood which can be selected from the most severe combinations of meteorological and hydrological conditions reasonably characteristic of the region, excluding extremely rare combinations of events. SPF volumes are commonly 40% to 60% of the PMF volumes.

The *design basis flood* (*DBF*) depends on the site. It is the flood that is adopted as the basis for design of a particular project. The DBF is usually determined from economic considerations, or is specified as part of the contract document.

Although the one percent flood is a common choice for the design basis flood, shorter recurrence intervals are often used, particularly in low value areas such as cropland. For example, 5-year storm curves are used in residential areas, 10-year curves for business sections, and 15-year frequencies for high-value districts where flooding will result in more extensive damage. The ultimate choice of a recurrence interval, however, must be made on the basis of economic considerations and tradeoffs.

The probability that a flooding event in any year will equal or exceed the design basis flood based on a recurrence interval or frequency, $F$, is

$$p\{F \text{ event in one year}\} = \frac{1}{F} \qquad 6.9$$

The probability of an $F$ event occurring in $n$ years is

$$p\{F \text{ event in } n \text{ years}\} = 1 - \left(1 - \frac{1}{F}\right)^n \qquad 6.10$$

*Example 6.2*

A wastewater treatment plant has been designed to be in use for 40 years. What is the probability that a one percent flood will occur within the useful lifetime of the plant?

$$F = 100$$

$$p\{100 \text{ year flood in 40 years}\} = 1 - \left(1 - \frac{1}{100}\right)^{40} = 0.33 \quad (33\%)$$

## 5 SUBSURFACE WATER

Subsurface water is a major source of all water used in the United States. In dry areas, it may be the only source of water used for domestic and irrigation uses. Subsurface zones are divided into two parts by the water table. The *vadose zone* exists above the water table, and pores in the vadose zone may be either empty or full. Below the water table is the *phreatic zone*, whose pores are always full.

Soil moisture content is measured in pounds per cubic foot. Soil moisture is usually determined by oven drying a sample of soil and measuring the weight loss. The moisture content can also be determined with a tensiometer, which measures the vapor pressure of the moisture in the soil.

Movement of water through an aquifer is given by equation 6.11 in which $K_p$ is known as the *coefficient of permeability* (or *hydraulic conductivity*).[1]

$$Q = (1.157\,\text{EE–5})K_p A G_H \qquad 6.11$$

$K_p$ in the United States is usually given in *Meinzer units* (gallons per day per square foot). The aquifer flow area, $A$, is given by equation 6.13. The hydraulic gradient $G_H = H/L$, depends on the length, $L$, of the aquifer.

**Table 6.2**

Approximate Coefficients of Permeability, $K_p$

| material | ft/day or ft³/day-ft² | gal/day-ft² | darcys |
|---|---|---|---|
| Clay | 1.3 EE–3 | EE–2 | EE–3 |
| Sand | 1.3 EE2 | EE3 | EE2 |
| Gravel | 1.3 EE4 | EE5 | EE4 |
| Gravel/Sand | 1.3 EE3 | EE4 | EE3 |
| Sandstone | 1.3 EE1 | EE2 | EE1 |
| Dense shale & limestone | 1.3 EE–1 | 1.0 | EE–1 |
| Quartzite & granite | 1.3 EE–3 | EE–2 | EE–3 |

The *transmissivity* of flow from a saturated aquifer of thickness $Y$ and width $B$ is given by equation 6.12.

$$T = K_p Y \qquad 6.12$$

The cross sectional area in flow depends on the aquifer dimensions.[2]

$$A = BY = \frac{BT}{K_p} \qquad 6.13$$

Combining equation 6.11 with equation 6.13,

$$Q = (1.157\,\text{EE–5})BTG_H \qquad 6.14$$

## 6 WELL DRAWDOWN IN AQUIFERS

A virgin aquifer with a well is shown in figure 6.2 (a). Once pumping of the well starts, the water table will be lowered in the vicinity of the well, and the resulting water table surface is known as the *cone of depression.* The decrease in water level at the well is known as the *drawdown*, $d_o$. The drawdown at some distance $r$ from the well is $d_r$.

If the drawdown is small with respect to the aquifer thickness, $Y$, and the well completely penetrates the aquifer, then the equilibrium (steady state) well dis-

[1] This is one form of *Darcy's law.* The 1.157 EE–5 is a conversion factor from cubic feet per day to cfs.

[2] Since water can only flow between rock, not all of the aquifer area contributes to flow. It is important to know that the area, $A$, and the permeability, $K_p$, are consistent. If the permeability value is based on the clear flow area only, it will be necessary to multiply the aquifer area by the porosity to reduce the area in flow.

charge is given by the *Druit equation*, equation 6.15. The 86,400 converts cubic feet per day to cubic feet per second.

$$Q = \frac{\pi K_p (y_1^2 - y_2^2)}{86{,}400\, ln(r_1/r_2)} \qquad 6.15$$

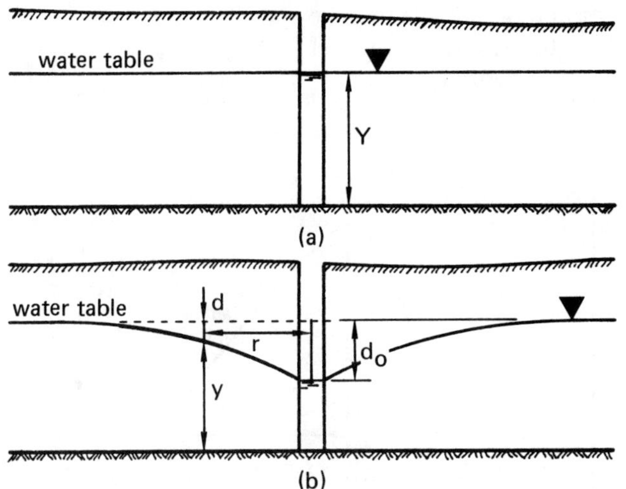


**Figure 6.2** Well Drawdown

In equation 6.15, $y_1$ and $y_2$ are the aquifer depths at any two radii, $r_1$ and $r_2$, respectively. $y_1$ can also be taken as the original aquifer depth, $Y$, if $r_1$ is the well's *radius of influence.*

For an artesian well fed by a confined aquifer of thickness $Y$, the discharge is

$$Q = \frac{2\pi K_p (y_1 - y_2) Y}{86{,}400\, ln(r_1/r_2)} \qquad 6.16$$

When pumping first begins, the removed water also comes from the aquifer above the equilibrium cone of depression. Therefore, equation 6.15 cannot be used, and a non-equilibrium analysis is required.

*Example 6.3*

A 9″ diameter well is pumped at the rate of 50 gpm. The aquifer is 100 feet thick. The well sides cave in and are replaced with an 8″ diameter tube. Assuming a 6 foot drawdown, what will be the steady flow from the new well? Assume the water table recovers its original thickness 2500 feet from the well.

$$(50)\text{gpm}(0.002228)\frac{\text{cfs}}{\text{gpm}} = 0.1114\,\text{cfs}$$

From equation 6.15,

$$y_2 = 100 - 6 = 94\,\text{ft}$$

$$r_2 = 9/(2)(12) = 0.375\,\text{ft}$$

$$0.1114 = \frac{\pi K_p((100)^2 - (94)^2)}{(86{,}400) ln(2500/0.375)}$$

$$K_p = 23.17\,\text{ft}^3/\text{day-ft}^2$$

For an 8″ (r = 0.333 ft) pipe,

$$Q = \frac{\pi(23.17)((100)^2 - (94)^2)}{(86{,}400)\, ln(2500/0.333)} = 0.110\,\text{cfs}$$

## 7 SEEPAGE AND FLOW NETS

Groundwater flows from locations of high hydraulic head to locations of lower hydraulic head. Problems requiring the actual calculation of the flow quantity are best solved with the flow net concept, particularly when a manual solution is required.

*Flow nets* are constructed from streamlines and equipotential lines. *Streamlines* show the path taken by the seepage. *Equipotential lines* connect points of constant pressure. The flow net concept is limited to cases where the flow is steady, two-dimensional, incompressible, through a homogeneous media, and where the liquid has a constant viscosity.

The flow net is constructed according to the following rules.

- Streamlines enter and leave pervious surfaces perpendicular to those surfaces.
- Streamlines approach the *line of seepage* (above which there is no hydrostatic pressure) asymptotically to that surface.
- Streamlines are parallel to impervious surfaces.
- Streamlines are parallel to the flow direction.
- Equipotential lines are drawn perpendicular to streamlines, such that the resulting cells are square and the intersections are 90° angles.
- Equipotential lines enter and leave impervious surfaces perpendicular to those surfaces.

The size of the cells formed by the intersection of the streamlines and equipotential lines is not important, although the more cells formed, the greater will be the accuracy.

The object of a graphical solution is to construct a network of flow paths (outlined by the streamlines) and equal pressure drops (bordered by equipotential lines). No fluid flows across streamlines, and a constant amount of fluid flows between any two streamlines.

Figure 6.3 shows flow nets for several common situations. A careful study of the flow nets will help to clarify the rules and conventions previously listed. In drawing flow nets, remember that the number of streamlines and equipotential lines drawn is up to the investigator. There are many flow nets that can be drawn, all more or less correct, differing only in their accuracy.

Once the flow net is drawn, it can be used to calculate the seepage. First, the number of flow channels, $N_f$, between the streamlines is counted. Then, the number of equipotential drops, $N_p$, between equipotential lines is counted. The total hydraulic head is determined as a function of the water surface levels.

$$Q = K_p H \left(\frac{N_f}{N_p}\right) \qquad 6.17$$

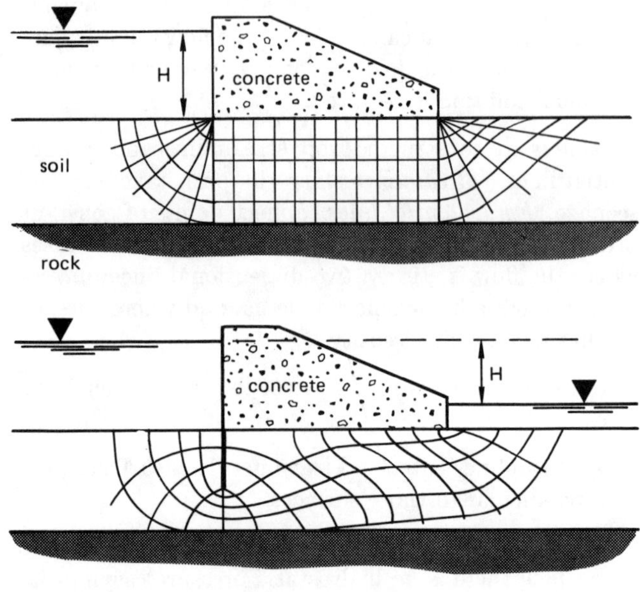


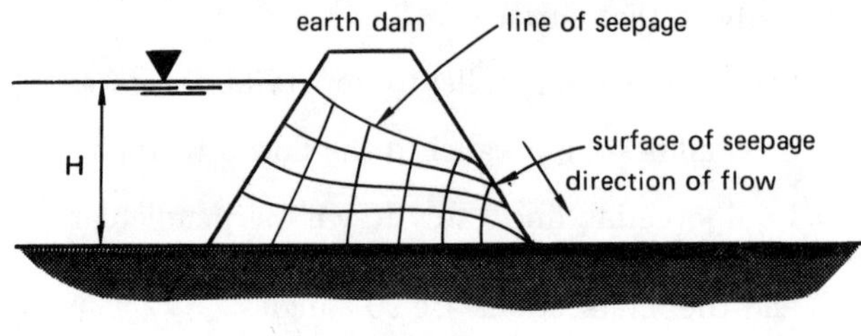


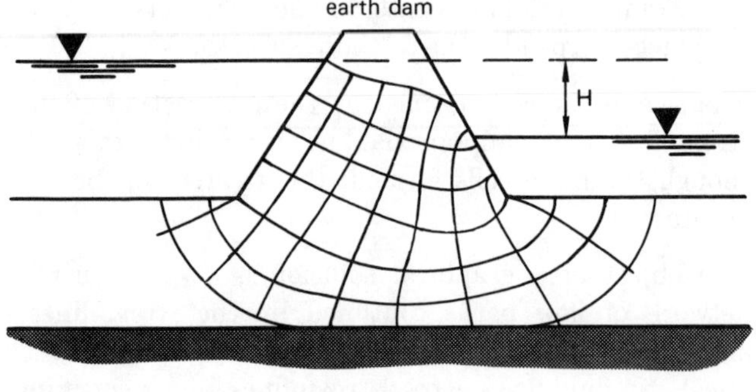


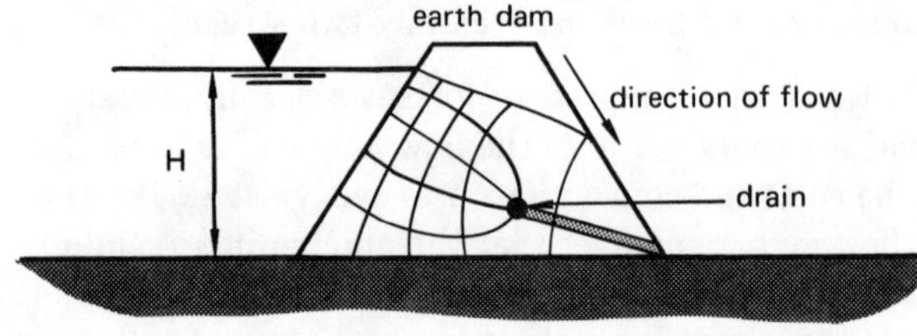


**Figure 6.3** Typical Flow Nets

## 8 TOTAL SURFACE RUNOFF FROM STREAM HYDROGRAPHS

After a rain, water makes its way to a stream. A plot of stream discharge versus time is known as a *hydrograph.* Hydrograph periods may be very long (such as a year) down to very short (hours). A typical hydrograph is shown in figure 6.4.

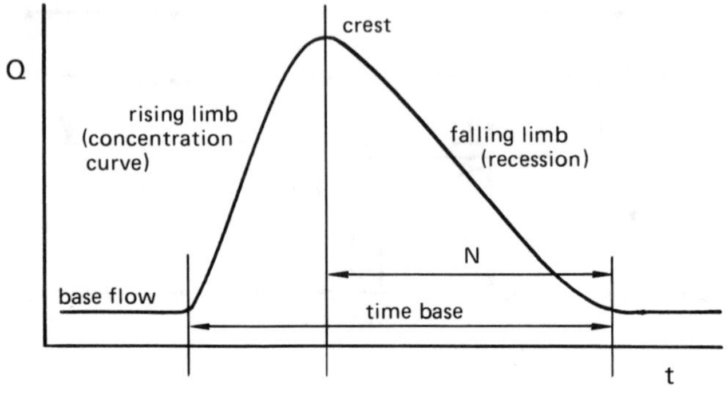


**Figure 6.4** A Stream Hydrograph

In figure 6.4, the portion of the hydrograph to the left of the crest is known as the *rising limb.* To the right of the crest, the curve is known as the *recession* or *falling limb.*

The stream discharge is assumed to consist of *overland flow* and *groundwater flow.* Since culverts do not have to be designed to carry groundwater, a procedure called *hydrograph separation* or *hydrograph analysis* is necessary to separate surface and groundwater.[3]

There are several methods of separating baseflow from overland flow. All of the methods are somewhat arbitrary. Three of the methods easily carried out manually are presented here.

*Method 1*: (Straight Line Method): A horizontal line is drawn from the start of the rising limb to the falling limb. All of the flow under the horizontal line is considered base flow. This is illustrated in figure 6.5.

[3] The total rain dropped by the storm is the *gross rain.* The rain which actually appears as immediate runoff is known as *surface runoff, overland flow, surface flow,* and *net rain.* The water which is absorbed by the soil and which does not contribute to the surface runoff is known as *base flow, groundwater, infiltration,* and *dry weather flow.*

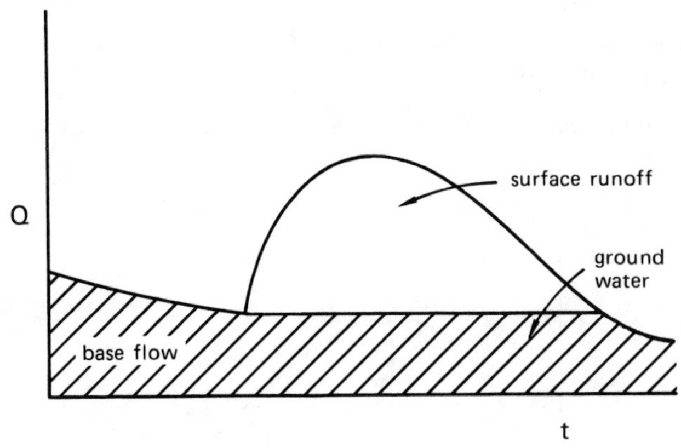


**Figure 6.5** Hydrograph Separation Method 1

*Method 2*: (Fixed Base Method): The base flow existing before the storm is projected or continued down to a point directly under the crest of the hydrograph. Then, a straight line is used to connect the projection to the falling limb $N$ days later. $N$ is determined by inspection or it can be calculated from the rule-of-thumb equation 6.18.

$$\begin{aligned} N &= (\text{area in sq. mi})^{0.2} \text{ in days} \\ &= 6.59\,(\text{area in acres})^{0.2} \text{ in hours} \end{aligned} \qquad 6.18$$

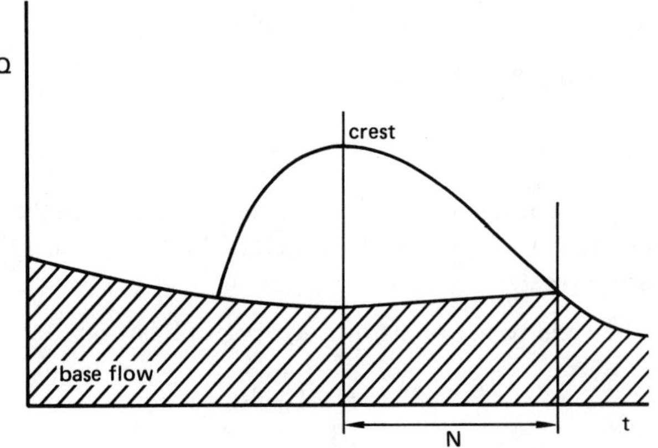


**Figure 6.6** Hydrograph Separation Method 2

*Method 3*: (Variable Slope Method): This method recognizes that the shape of the base flow curve before the storm will probably match the shape of the base flow curve after the storm. The groundwater curve after the storm is projected back under the hydrograph to a point under the inflection point of the falling limb. The separation line under the rising limb is drawn arbitrarily.

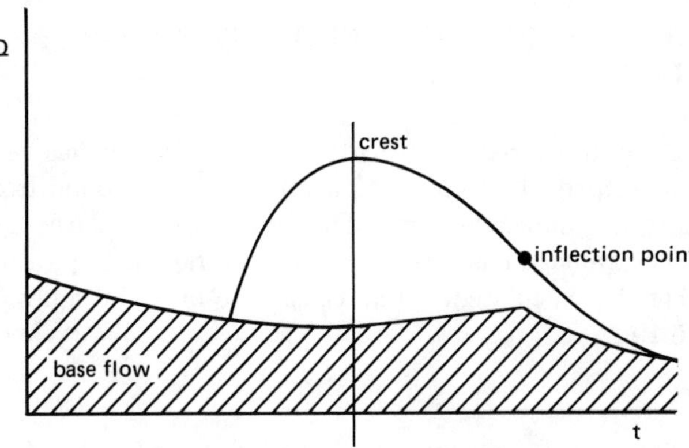


**Figure 6.7** Hydrograph Separation Method 3

Once the base flow is separated out, the hydrograph of overland flow will take on the appearance of figure 6.8.

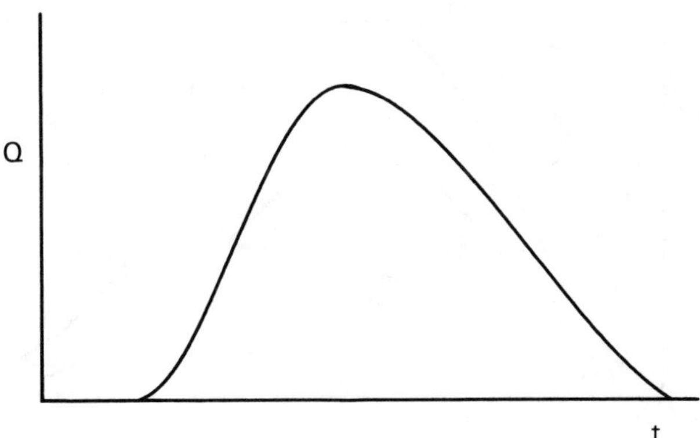


**Figure 6.8** Overland Flow Hydrograph

The total volume of the storm can be found from the area under the hydrograph. Although this can be accomplished by integration or planimetry, it is often sufficiently accurate to approximate the hydrograph with a histogram.

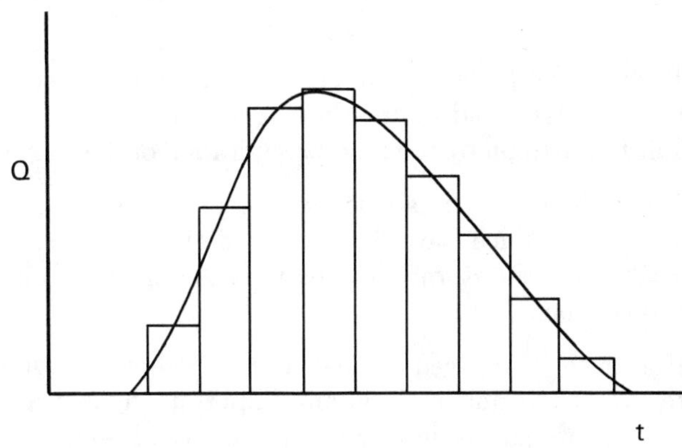


**Figure 6.9** Hydrograph Histogram

## 9 PEAK RUNOFFS FROM THE UNIT HYDROGRAPH

Once the overland flow hydrograph for a storm has been developed, the total rainfall volume can be found from the area under the curve. Furthermore, since the area of the drainage basin is known, the average precipitation can be calculated. (Use consistent units in equation 6.19.)

$$P = \frac{V}{A_d} \qquad 6.19$$

This precipitation will be some number of inches (e.g., 1.7″ or 0.92″, etc.). If every point on the hydrograph is divided by this average precipitation, a unit hydrograph will be derived. This is a hydrograph of a storm dropping 1″ of rain on the entire basin. Figure 6.10 shows how a unit hydrograph compares to a similar storm dropping an average of 1.7″ of rain.

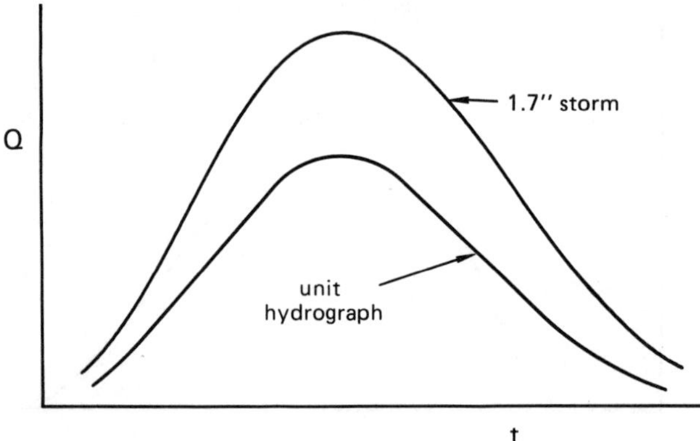


**Figure 6.10** A Storm and Unit Hydrograph

Direct use of a unit hydrograph assumes that:

- All storms in that basin have the same duration.
- The time base, $N$, is constant for all storms.
- The shape of the distribution is the same for all storms.
- Only the total amount of rainfall varies from storm to storm.

The hydrograph of a storm producing more or less than 1″ of rain is found by multiplying all ordinates on the unit hydrograph by the total precipitation of the storm.

A unit hydrograph can be used to predict the runoff for storms which have durations differing as much as ±25% from the storm duration used to derive the unit hydrograph.

If a basin is ungaged so that no records are available to produce a unit hydrograph, important hydrograph parameters can be derived analytically (with some success). Knowing these parameters and recognizing that the total precipitation from a unit hydrograph must be one inch permits sketching a rough approximation to a unit hydrograph.

The *Snyder synthetic hydrograph* is shown in figure 6.11.

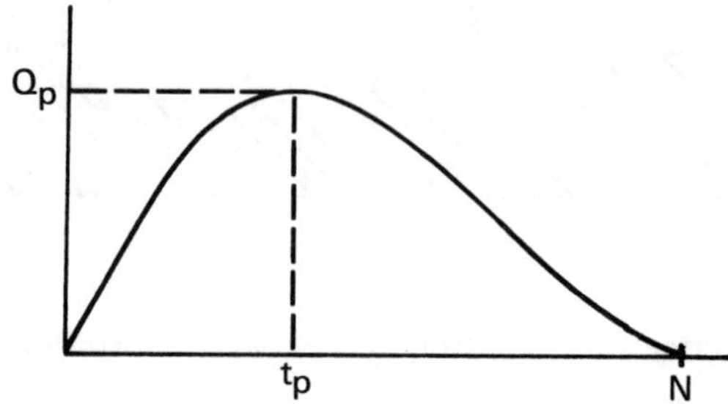


**Figure 6.11** Snyder Synthetic Hydrograph

An estimate for the time to peak runoff for the Snyder hydrograph is

$$t_p = C_t(L_s L_c)^{0.3} \qquad 6.20$$

$C_t$ is a coefficient which varies from 1.8 to 2.2, depending on slope, averaging 2 for steep slopes. $L_s$ is the main stream length from outlet to divide in miles. $L_c$ is the distance in miles from the outlet to a point on the stream nearest the basin centroid. Equation 6.20 is valid for basin areas of 10 to 10,000 square miles.

The peak flow for the unit hydrograph is

$$Q_p = C_p A_d / t_p \qquad 6.21$$

$C_p$ is a coefficient which varies between 0.4 and 0.8 depending on slope and other factors, but values near 0.60 are typical.

The time base for the unit hydrograph is

$$N = 72 + 3t_p \qquad 6.22$$

Since the value of $N$ can never be less than 72 hours, the Snyder hydrograph is not suitable for small basins which reach their peaks in a matter of hours.

Equations 6.21 and 6.22 are valid if the storm rain duration is

$$t_r = t_p/5.5 \qquad 6.23$$

If the duration is known to be otherwise, say $t'_r$, the actual time to peak runoff that should be used in equations 6.21 and 6.22, is $t'_p$.

$$t'_p = t_p + \frac{t'_r - \frac{t_p}{5.5}}{4} \qquad 6.24$$

**Table 6.3**
Representative Snyder Hydrograph Constants

| location | $C_t$ | $C_p$ |
|---|---|---|
| Southern California | 0.4 | 0.94 |
| Appalachian highlands | 2.0 | 0.63 |
| Gulf of Mexico eastern portions | 8.0 | 0.61 |

*Example 6.4*

After a 2-hour storm, a station downstream from a 45 square mile drainage basin measures 9400 cfs as a peak discharge and 3300 acre-feet as total runoff. Find the 2-hour unit hydrograph peak discharge. What would be the peak runoff and design flood volume if a 2-hour storm dropped $2\frac{1}{2}$ inches net precipitation?

1 inch of runoff from 45 square miles is

$$V = (45)\,\text{mile}^2\,(1)\,\text{inch}\,(53.3)\frac{\text{acre-ft}}{\text{sq. mile-in}}$$
$$= 2399\,\text{acre-feet}$$

The runoff ratio is $3300/2399 = 1.38$.

The unit hydrograph peak discharge is $9400/1.38 = 6812$ cfs.

For a $2\frac{1}{2}''$ storm, peak runoff would be $(2.5)(6812) = 17{,}030$ cfs.

The design flood would be $(2.5)(2399) = 5998$ acre-feet.

*Example 6.5*

A storm drops its significant rainfall in 6 hours on a 25 square mile basin. The resulting surface runoff is listed. (a) Construct the unit hydrograph of this 6-hour storm. (b) Find the runoff at $t = 15$ hours from a two storm system (both at 6 hours duration) if the first storm drops 2″ net starting at $t = 0$ and the second storm drops 5″ net starting at $t = 12$ hours.

| hours after rainfall starts | runoff (cfs) | hours after rainfall starts | runoff (cfs) |
|---|---|---|---|
| 0 | 0 | 21 | 600 |
| 3 | 400 | 24 | 400 |
| 6 | 1300 | 27 | 300 |
| 9 | 2500 | 30 | 200 |
| 12 | 1700 | 33 | 100 |
| 15 | 1200 | 36 | 0 |
| 18 | 800 | | |
| | | TOTAL | 9500 |

(a) The actual runoff is plotted below:

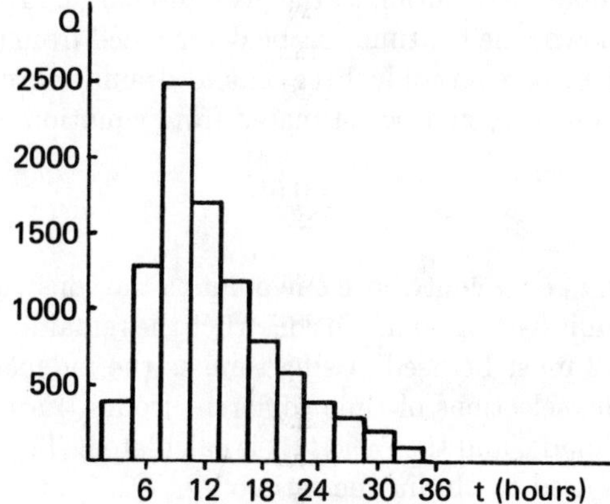


Finding the total area under the curve is equivalent to the following calculation:

$$V = (9500)\,\text{cfs}(3)\,\text{hours}(3600)\frac{\text{sec}}{\text{hr}}$$
$$= 1.026\,\text{EE8 ft}^3\,\text{runoff}$$

The basin area is

$$(25)\,\text{miles}^2\,(5280)^2\frac{\text{feet}}{\text{mile}} = 6.97\,\text{EE8 ft}^2$$

The net precipitation is

$$\frac{(1.026\,\text{EE8})\text{ft}^3\,(12)\text{in/ft}}{(6.97\,\text{EE8})\,\text{ft}^2} = 1.767\,\text{inches}$$

The unit hydrograph has the same shape as the actual hydrograph, with all ordinates reduced by 1.767.

(b) The flow at $t = 15 - 12 = 3$ is 400 cfs. The runoff at $t = 15$ is

$$2\left(\frac{1200}{1.767}\right) + 5\left(\frac{400}{1.767}\right) = 2490\,\text{cfs}$$

## 10 THE SCS UNIT HYDROGRAPH (1957)

The U.S. Soil Conservation Service has developed a synthetic unit hydrograph based on dimensionless variables.[4] In order to use this method, it is necessary to know the time to peak flow ($t_p$) and the peak discharge ($Q_p$). Provisions for calculating both of these important parameters are included in the method.

$$t_r = 0.133t_c \qquad 6.25$$
$$t_p = 0.5t_r + t_1 \qquad 6.26$$
$$Q_p = \frac{0.756A_d}{t_p} \qquad 6.27$$

[4] A hydrograph can also be constructed from the SCS curve number (CN) method used to calculate peak flow. This method is described elsewhere in this chapter.

$t_1$ in equation 6.26 is the lag time from the centroid of the rainfall distribution to the peak discharge, in hours. If unknown, the lag time can be determined from correlations with geographical region and drainage area. As a last resort, $t_1$ can be estimated from equation 6.28.

$$t_1 = 0.6t_c \qquad 6.28$$

$Q_p$ and $t_p$ only contribute one point to the construction of the unit hydrograph. To construct the remainder, figure 6.12 must be used. Using time as the independent variable, selections of time (different from $t_p$) are arbitrarily made, and the ratio $t/t_p$ is calculated. The curve then is used to obtain the ratio of $Q_t/Q_p$.

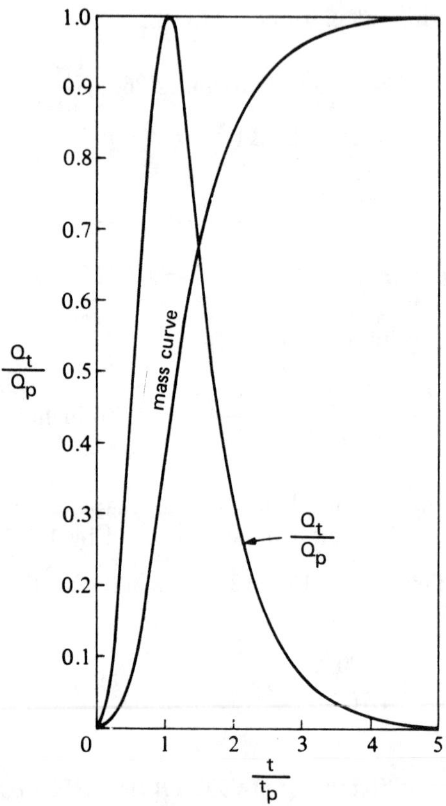


**Figure 6.12** Dimensionless Unit Hydrograph

## 11 HYDROGRAPH SYNTHESIS

If a storm's duration is not the same as the hydrograph's base length, the unit hydrograph cannot be used to predict runoff. For example, the runoff from a 6-hour storm cannot be predicted from a unit hydrograph derived from a 2-hour storm. However, the technique of hydrograph synthesis can be used to construct the hydrograph of the longer storm from the unit hydrograph of a shorter storm.

### A. LAGGING STORM METHOD

If a unit hydrograph for a storm of duration $t_r$ is available, it can be used to construct the hydrograph of a storm whose duration is a whole multiple of $t_r$.[5] For example, a 6-hour storm hydrograph can be constructed from a 2-hour unit hydrograph since $6/2 = 3$ is a whole number.

Let the whole multiple number be $N$. To construct the longer hydrograph, draw $N$ unit hydrographs, each separated by time $t_r$. Then add the ordinates to obtain a hydrograph for an $Nt_r$ duration storm. Since the total rainfall from this new hydrograph is $N$ inches (having been constructed from $N$ unit hydrographs), the curve will have to be reduced (i.e., divided) by $N$ everywhere to produce a unit hydrograph.

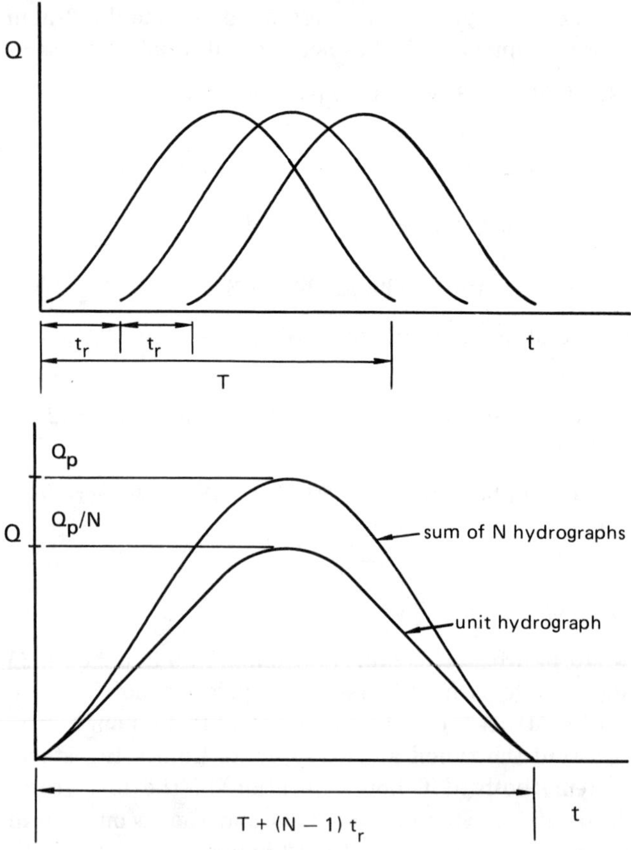


**Figure 6.13** Hydrograph Synthesis by the Lagging Storm Method

### B. S-CURVE METHOD

The S-curve method can be used to construct hydrographs from unit hydrographs with longer or shorter durations, and when the storm durations are not multiples. This method begins by adding the ordinates of

[5] Distinction is made between the *storm duration*, $t_r$ and the *time base* of the hydrograph. A 2-hour storm may produce a hydrograph with runoff continuing over a time base of many days.

many unit hydrographs, each lagging the other by time $t$, the duration of the storm which produced the unit hydrograph. After a sufficient number of lagging unit hydrographs have been added together, the accumulation will level off and remain constant. At that point, the lagging can be stopped. The resulting accumulation is known as an *S-curve.*

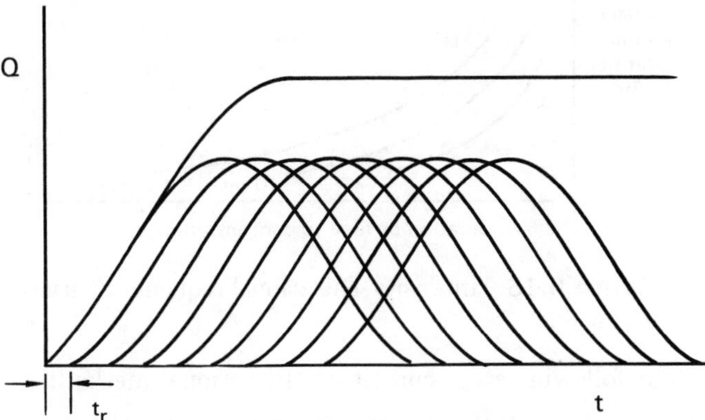


**Figure 6.14** Constructing the S-Curve

If two S-curves are drawn, one lagging the other by time $t'$, the area between the two curves represents a hydrograph area for a storm of duration $t'_r$. The differences between the two curves can be plotted and scaled to a unit hydrograph by multiplying by the ratio of $(t_r/t'_r)$.

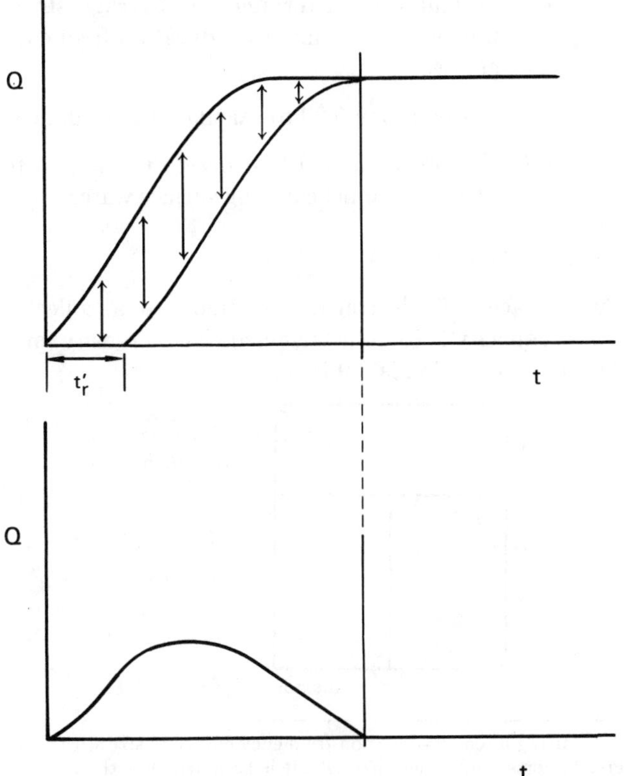


**Figure 6.15** Using S-Curves to Construct a $t'$ hydrograph

## 12 PEAK RUNOFF FROM THE RATIONAL METHOD

Although total runoff data is required for reservoir and dam design, the instantaneous peak runoff is needed to size culverts and storm drains.

In the closed-form equations used to find peak runoff, it is usually assumed that the rainfall is applied to a surface at a constant rate. If this assumption is true and the surface is largely impervious, the runoff will eventually equal the rate of rainfall. The time between the start of rainfall and the time of peak flow is known as the *time of concentration*, $t_c$. Typical values of $t_c$ for areas less than 50 acres range from 5 to 30 minutes.

The *rational formula* is based on the assumptions given and has been in widespread use for small areas (i.e., less than 100 acres or so) despite its serious deficiencies for larger areas. Values of $C$ are found in appendix A.

$$Q_p = CIA_d \tag{6.29}$$

Strictly speaking, $Q_p$ is in acre-inches/hour, but it is typically taken as cubic feet per second since the conversion between these two units is 1.008. For a small drainage area, $t_c$ is taken as the largest combination of overland flow time and channel time. *Channel time* is found by dividing the channel length by the (usually assumed) channel velocity.[6] Assuming laminar flow, the overland flow time for small areas without defined channels can be found from the *Izzard formula.* Equation 6.30 should be used only if the product $(IL_o)$ is less than 500.

$$t = \frac{(41)(b)(L_o)^{1/3}}{(CI)^{2/3}} \tag{6.30}$$

$$b = \frac{0.0007(I) + C_r}{S^{1/3}} \tag{6.31}$$

The retardance coefficient, $C_r$, is given in table 6.4.

**Table 6.4**
Retardance Coefficient

| Type of Surface | $C_r$ |
|---|---|
| smooth asphalt | 0.007 |
| concrete pavement | 0.012 |
| tar & gravel pavement | 0.017 |
| closely clipped lawn | 0.046 |
| dense bluegrass turf | 0.060 |

[6] If the pipe or channel size is known, the velocity can be found from $Q = Av$. If the pipe size is not known, then $A$ will have to be estimated. In that case, one might as well estimate $v$ instead. 5 fps is a reasonable velocity. The minimum velocity for a self-cleansing pipe is 2 fps.

The overland flow time has been predicted by the Federal Aviation Administration after analyses of airport drainage areas.

$$t = \frac{1.8(1.1 - C)\sqrt{L_o}}{S^{1/3}} \qquad 6.32$$

The distance $L_o$ in equations 6.30 and 6.32 is the longest distance to the collection point, as shown in figure 6.16.

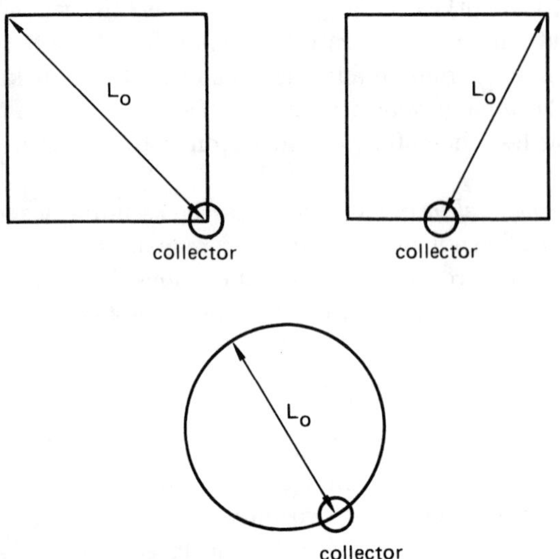


**Figure 6.16** Overland Flow Distances

For irregularly-shaped drainage areas, it may be necessary to evaluate several alternative overland flow distances. For example, figure 6.17 shows a drainage area with a long tongue. Although the tongue area contributes to the drainage area, it does lengthen the overland flow time. Depending on the intensity-duration-frequency curve, the longer overland flow time (resulting in a lower rainfall intensity) may offset the increase in area due to the tongue. Therefore, two runoffs need to be compared, one ignoring and the other including the tongue, and taking the larger.

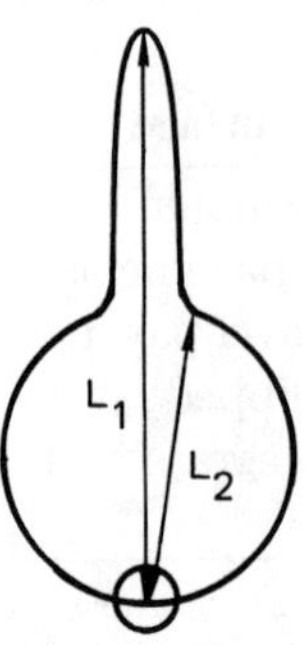


**Figure 6.17** An Irregular Drainage Area

The most important part of equation 6.29 is the rainfall intensity. Rainfall data can be compiled into intensity-duration-frequency curves similar to those in figure 6.18. The intensity used in equation 6.29 will depend on the time of concentration and the degree of protection desired.[7]

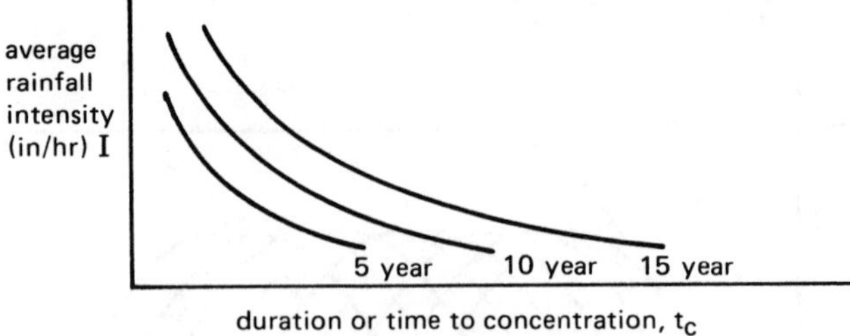


**Figure 6.18** Intensity-Duration-Frequency Curves

The following steps constitute the rational method.

*step 1*: Estimate $t_c$. This is the sum of the overland flow and conduit flow times. $t_c$ will increase the farther you get from the drainage area.

*step 2*: Choose a value of $C$. If more than one area contributes to the runoff, $C$ is weighted by the areas.

*step 3*: Select a frequency or return period for the storm.

*step 4*: Calculate or determine the average storm intensity from intensity-duration-frequency curves.

*step 5*: Use equation 6.29 to calculate the peak flow.

*step 6*: Use open channel flow design techniques to size the channel carrying surface water away.

*Example 6.6*

Two adjacent fields contribute runoff to a collector whose capacity is to be determined. The intensity for a 25 minute duration is 3.9 in/hr.

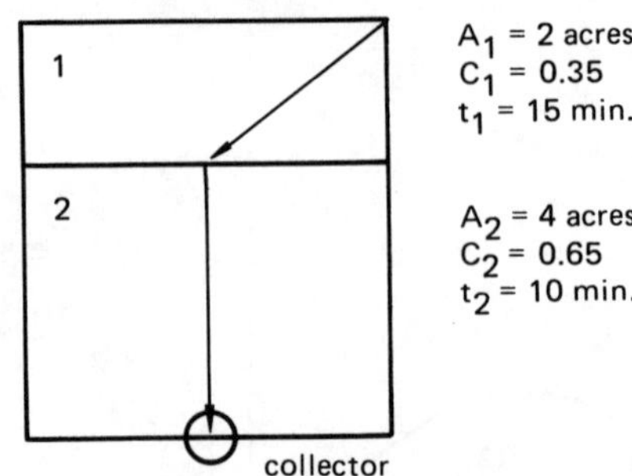


[7] By using intensity-duration-frequency curves to size storm sewers, culverts, and other channels, it is assumed that the frequencies and probabilities of flood damage and storms are identical. This is not generally true, but the assumption is made when other supporting data is not available.

*step 1*: The overland flow time is given for both areas. The time for water from the farthest corner to reach the collector is

$$t_c = 15 + 10 = 25\,\text{minutes}$$

*step 2*: The runoff coefficients are given for each area. Since we want to size the pipe carrying the total runoff, the coefficients are weighted by their respective contributing areas.

$$C = \frac{(2)(0.35) + (4)(0.65)}{2+4} = 0.55$$

*step 3 & 4*: The intensity for a 25 minute duration was given as 3.9 in/hr.

*step 5*: The total area is $4 + 2 = 6$ acres. The peak flow is found from equation 6.29.

$$Q_p = (0.55)(3.9)(6) = 12.9\,\text{cfs}$$

*Example 6.7*

A drainage area has the following characteristics:

| area | size (acres) | overland flow time (min) | $C$ |
|---|---|---|---|
| A | 10 | 20 | 0.3 |
| B | 2 | 5 | 0.7 |
| C | 15 | 25 | 0.4 |

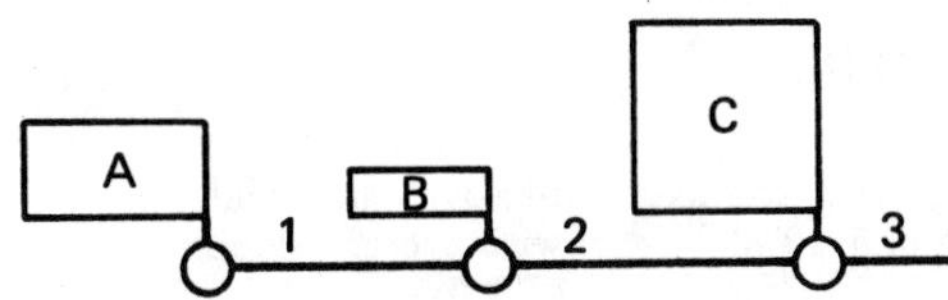


The rainfall intensity for the area is given by

$$I = \frac{115}{t_c + 15}$$

The manholes are 300 feet apart. The pipe slope is 0.009. Manning's roughness is 0.015. Assume the flow velocity is 5 ft/sec in all sections. What should be the pipe size in section 2? What is the maximum flow through section 3?

For area $A$:

$$t_c = 20$$

$$I = \frac{115}{20+15} = 3.29\,\text{in/hr}$$

$$Q = (0.3)(10)(3.29) = 9.87\,\text{cfs}$$

This value of $Q$ (9.87 cfs) should be used to size line #1.

To find the flow time between manholes, a flow velocity is needed. This would normally be calculated from $Q$ and $A$ obtained from equation 5.11. However, v is given in this example.

Using a flow velocity of 5 ft/sec, the flow time between drainage inlets is

$$t = \frac{300}{5} = 60\,\text{seconds} = 1\,\text{minute}$$

For area $B$:

At $t = 5$, $I = 5.75$ in/hr. The run-off from 2 acres is $(0.7)(2)(5.75)$ or 8.05 cfs. However, at $t = (20+1)$, the flow from area $A$ will reach the second manhole. At $t = 21$, $I = 3.19$ in/hr.

The sum of $CA$ values is $(0.3)(10) + (0.7)(2) = 4.4$

$$Q = (4.4)(3.19) = 14.0\,\text{cfs}$$

This value of $Q$ should be used to design section 2 of the pipe. If the pipe is assumed to flow full, the required diameter is found from open channel flow considerations.

$$d = (1.33)\left[\frac{(14.0)(0.015)}{\sqrt{0.009}}\right]^{3/8} = 1.79\,\text{ft}\ \ (\text{round to } 2.0\,\text{ft})$$

For area $C$:

Assuming 5 fps as the flow velocity in the pipe, the time from the start of the storm for water from plat $A$ to reach the 3rd manhole is

$$20 + 1 + 1 = 22$$

Since 25 is larger than 22, the maximum runoff will occur 25 minutes after the start of the storm. The 22 minute datum is not used.

$$I = \frac{115}{25+15} = 2.875$$

The sum of $CA$ values is $(0.3)(10) + (0.7)(2) + (0.4)(15) = 10.4$.

$$Q = (10.4)(2.875) = 29.9\,\text{cfs}$$

## 13 PEAK RUNOFF BY SCS METHODS

Several methods of calculating peak runoff have been suggested by the U.S. Soil Conservation Service. These methods have generally been well correlated with actual experience, but the classification of terrain and soil conditions into SCS categories is difficult. Correct application of these SCS methods is dependent on being able to determine the various coefficients and parameters used. SCS literature should be consulted in most cases.

A. THE COOK EQUATION (1940)

The Cook equation is used for watersheds of less than 2500 acres.

*step 1*: Determine the frequency factor, $F$. For 50-year protection, use $F = 1.00$; for 25-year storms, use $F = 0.83$; for 10-year storms, use $F = 0.71$.

*step 2*: Determine the rainfall factor, $R$, from the geology of the region. This factor varies from 0.6 to 1.1.

*step 3*: Determine the watershed characteristics. Based on the relief, soil infiltration, vegetation, and surface storage conditions, calculate the total watershed coefficient, $W = \sum w_i$. Various values of the weights, $w_i$, are listed in table 6.5. $W$ varies between 25 and 100.

*step 4*: Determine $Q_{p,50}$, the standard 50-year peak flow, from figure 6.19.

*step 5*: Calculate the peak discharge associated with the given return period and geological region.

$$Q_p = Q_{p,50}RF \qquad 6.33$$

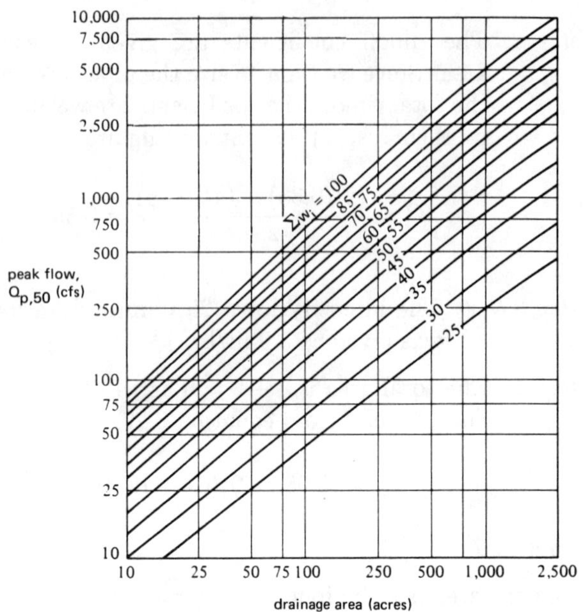


**Figure 6.19** Standard 50-year Peak Flows

**Table 6.5**
Cook Equation Watershed Weights ($w_i$)

| | | | | |
|---|---|---|---|---|
| Relief | (40) Steep, rugged, terrain with average slopes generally above 30% | (30) Hilly, with average slopes of 10–30% | (20) Rolling, with average slopes of 5–10% | (10) Relatively flat land, with average slopes of 0–5% |
| Soil infiltration | (20) No effective soil cover; either rock or thin soil mantle of negligible infiltration capacity | (15) Slow to take up water; clay or other soil of low infiltration capacity, such as heavy gumbo. | (10) Normal, deep loam with infiltration about equal to that of typical prairie soil | (5) High; deep sand or other soil that takes up water readily and rapidly |
| Vegetal cover | (20) No effective plant cover; bare except for very sparse cover | (15) Poor to fair; clean-cultivated crops or poor natural cover; less than 10% of drainage area under good cover | (10) Fair to good; about 50% of drainage area in good grassland, woodland, or equivalent cover; not more than 50% of area in clean-cultivated crops | (5) Good to excellent, about 90% of drainage area in good grassland, woodland, or equivalent cover |
| Surface storage | (20) Negligible; surface depressions are few and shallow; drainage-ways steep and small; no ponds or marshes | (15) Low; well-defined system of small drainage-ways; no ponds or marshes | (10) Normal; considerable surface depression storage; drainage system similar to that of typical prairie lands; lakes, ponds and marshes less than 2% of drainage area | (5) High; surface-depression storage high; drainage system not sharply defined; large flood plain storage or a large number of lakes, ponds or marshes |

## B. THE SCS DRAINAGE COEFFICIENT METHOD (1939)

A simple method used by the Soil Conservation Service for surface drainage from flatlands depends on the characteristics of the watershed to be protected. This method applies to areas where the natural land slopes are about 1 percent or less. The formula can be used for minor portions of steeper land in a watershed which is predominantly flatland. Flow from uplands in the watershed should be computed by other procedures and added to the flatlands flow.

$$Q_p = C\left(\frac{A_d}{640}\right)^{5/6} \qquad 6.34$$

The 640 term in equation 6.34 converts the drainage area from acres to square miles. The coefficient $C$ depends on use of the flatland. Table 6.6 lists typical values of the coefficient $C$.

**Table 6.6**
SCS Drainage Coefficients

| use and location | $C$ |
|---|---|
| SE coastal plain, cultivated | 45 |
| SE delta, cultivated | 40 |
| NE and cornbelt, cultivated | 37 |
| SE coastal plain, pasture | 30 |
| NE and cornbelt, pasture | 25 |
| SE delta and SE coastal ricelands | 22.5 |
| Red River Valley, cultivated | 20 |
| SW range | 15 |
| SE coastal plain, woodland | 10 |

## C. CURVE NUMBER METHOD (1975)

This SCS method is dependent on being able to classify the land use and soil type by a single parameter called the *curve number*, $CN$. This method can be used for any size homogeneous watershed with a known percentage of imperviousness. If the watershed varies in soil type or in cover, it should be divided into regions to be analyzed separately.[8] This method uses precipitation records and an assumed distribution of rainfall to construct a *synthetic storm*.

*step 1*: Classify the soil according to *runoff potential.* Soil is classified into types $A$ (low runoff potential) through $D$ (high runoff potential).

- Type A: High infiltration rates (0.30–0.45 in/hr) even if thoroughly saturated; chiefly sands and gravels with good drainage and high moisture transmission.
- Type B: Moderate infiltration rates if thoroughly wetted (0.15–0.30 in/hr), moderate rates of moisture transmission, and consisting chiefly of coarse to moderately fine textures.
- Type C: Slow infiltration rates (0.05–0.15 in/hr) if thoroughly wetted, and slow moisture transmission; soils which have moderately fine to fine textures, or which impede the downward movement of water.
- Type D: Very slow infiltration rates (less than 0.05 in/hr) if thoroughly wetted, very slow water transmission, and consisting primarily of clay soils with high potential for swelling, soils with permanent high water tables, or soils with an impervious layer near the surface.

*step 2*: Determine the pre-existing soil conditions. The soil condition is classified into *antecedent moisture conditions* (*AMC*) I through III.

- AMC I: Dry soils, as prior to or after plowing or cultivation.
- AMC II: Typical condition existing before maximum annual flood.
- AMC III: Saturated soil due to heavy rainfall (or light rainfall with low temperatures) during 5 days prior to storm.

*step 3*: Classify the hydrologic condition of the soil-cover complex. The condition is *good* if it is lightly grazed or has plant cover over 75% or more of its area. The condition is *fair* if plant coverage is 50%–75% or not heavily grazed. The condition is *poor* if the area is heavily grazed, has no mulch, or has plant cover over less than 50% of the area. These definitions are applicable to pasture or range, and may be difficult to apply to row crops and grains.

*step 4*: Use table 6.7 to determine the curve number, $CN$, corresponding to the soil classification for AMC II.

*step 5*: If the soil is AMC I or AMC III, convert the curve number from step 4 using table 6.8.

*step 6*: If any significant fraction of the watershed is impervious (i.e., $CN = 100$), or if the watershed consists of areas with different curve numbers, calculate the composite curve number by weighting by the runoff areas.

[8] If the net rain and time to concentration are known, it may be possible to jump immediately to step 11 in this procedure.

**Table 6.7**
Runoff Curve Numbers for Soil Use and Condition (AMC II)

| use of land | treatment | hydrologic condition | runoff potential *A* | *B* | *C* | *D* |
|---|---|---|---|---|---|---|
| fallow | straight row | – | 77 | 86 | 91 | 94 |
| row crops | straight row | poor | 72 | 81 | 88 | 91 |
| | straight row | good | 67 | 78 | 85 | 89 |
| | contoured | poor | 70 | 79 | 84 | 88 |
| | contoured | good | 65 | 75 | 82 | 86 |
| | contoured and terraced | poor | 66 | 74 | 80 | 82 |
| | contoured and terraced | good | 62 | 71 | 78 | 81 |
| small grain | straight row | poor | 65 | 76 | 84 | 88 |
| | | good | 63 | 75 | 83 | 87 |
| | contoured | poor | 63 | 74 | 82 | 85 |
| | | good | 61 | 73 | 81 | 84 |
| | contoured and terraced | poor | 61 | 72 | 79 | 82 |
| | | good | 59 | 70 | 78 | 81 |
| close-seeded legumes | straight row | poor | 66 | 77 | 85 | 89 |
| or rotation meadow | straight row | good | 58 | 72 | 81 | 85 |
| | contoured | poor | 64 | 75 | 83 | 85 |
| | contoured | good | 55 | 69 | 78 | 83 |
| | contoured and terraced | poor | 63 | 73 | 80 | 83 |
| | contoured and terraced | good | 51 | 67 | 76 | 80 |
| pasture or range | | poor | 68 | 79 | 86 | 89 |
| | | fair | 49 | 69 | 79 | 84 |
| | | good | 39 | 61 | 74 | 80 |
| | contoured | poor | 47 | 67 | 81 | 88 |
| | contoured | fair | 25 | 59 | 75 | 83 |
| | contoured | good | 6 | 35 | 70 | 79 |
| meadow | | good | 30 | 58 | 71 | 78 |
| woods | | poor | 45 | 66 | 77 | 83 |
| | | fair | 36 | 60 | 73 | 79 |
| | | good | 25 | 55 | 70 | 77 |
| farmsteads | | – | 59 | 74 | 82 | 86 |
| roads(dirt) | | – | 72 | 82 | 87 | 89 |
| (hard surface) | | – | 74 | 84 | 90 | 92 |

**Table 6.8**
Curve Numbers for AMC I and AMC III

| *CN* for AMC II | Corresponding *CN*'s AMC I | AMC II |
|---|---|---|
| 100 | 100 | 100 |
| 95 | 87 | 98 |
| 90 | 78 | 96 |
| 85 | 70 | 94 |
| 80 | 63 | 91 |
| 75 | 57 | 88 |
| 70 | 51 | 85 |
| 65 | 45 | 82 |
| 60 | 40 | 78 |
| 55 | 35 | 74 |
| 50 | 31 | 70 |
| 45 | 26 | 65 |
| 40 | 22 | 60 |
| 35 | 18 | 55 |
| 30 | 15 | 50 |
| 25 | 12 | 43 |
| 20 | 9 | 37 |
| 15 | 6 | 30 |
| 10 | 4 | 22 |
| 5 | 2 | 13 |

*step 7*: Estimate the time to concentration of the watershed.

*step 8*: Determine the gross rain from the storm. To do this, it is necessary to assume the storm length and frequency. A 24-hour storm is typically chosen, probably because of the 24-hour thunderstorms experienced in all but the Pacific Coast states. Maps from the U.S. Weather Bureau can be used to read gross point rainfalls for storms with durations from 30 minutes to 24 hours, and with frequencies from 1 to 100 years.[9]

*step 9*: Multiply the gross rain point value from step 8 by a factor from figure 6.20 to make the gross rain representative of larger areas. This is the *areal gross rain.*

*step 10*: Calculate the *net rain* (*precipitation excess*) from the *gross rain.* Subtract losses from interception, storm period evaporation, depression storage, and infiltration from the gross rain to obtain the net rain.

This step is difficult. The SCS procedure is to first determine the distribution of the storm. Type I storms are applicable to Hawaii, Alaska, and the coastal side of the Sierra Nevada and Cascade Mountains in California, Oregon, and Washington. Type II distributions are typical of the rest of the United States, Puerto Rico, and the Virgin Islands. Table 6.9 shows the cumulative rainfall over 24 hours for type I and II storms.

9 "Rainfall Frequency Atlas of the United States," U.S. Weather Bureau, Technical Paper 40.

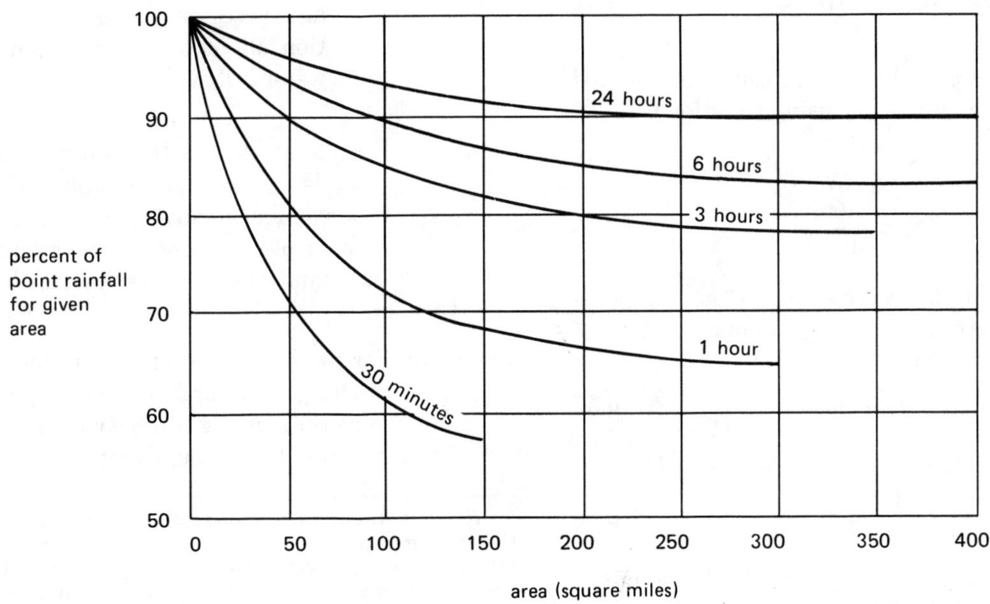


**Figure 6.20** Point to Areal Depth Conversion Factors

**Table 6.9**
Cumulative Distribution of Type I and II 24-hour Storms

| time(hours) | type I | type II |
|---|---|---|
| 0 | 0 | 0 |
| 2.0 | 0.035 | 0.022 |
| 4.0 | 0.076 | 0.048 |
| 6.0 | 0.125 | 0.080 |
| 7.0 | 0.156 | – |
| 8.0 | 0.194 | 0.120 |
| 8.5 | 0.219 | – |
| 9.0 | 0.254 | 0.147 |
| 9.5 | 0.303 | 0.163 |
| 9.75 | 0.362 | – |
| 10.0 | 0.515 | 0.181 |
| 10.5 | 0.583 | 0.204 |
| 11.0 | 0.624 | 0.235 |
| 11.5 | 0.654 | 0.283 |
| 11.75 | – | 0.387 |
| 12.0 | 0.682 | 0.663 |
| 12.5 | – | 0.735 |
| 13.0 | 0.727 | 0.772 |
| 13.5 | – | 0.799 |
| 14.0 | 0.767 | 0.820 |
| 16.0 | 0.830 | 0.880 |
| 20.0 | 0.926 | 0.952 |
| 24.0 | 1.000 | 1.000 |

Next, the soil complex curve number ($CN$) is used to calculate the excess of rain over infiltration. The following assumptions are common:

- Infiltration follows an exponential decay curve with time.
- *Storage capacity* of the soil, $S$, can be calculated from the curve number by using equation 6.35.

$$S = \frac{1000}{CN} - 10 \qquad 6.35$$

$$CN = \frac{1000}{10 + S} \qquad 6.36$$

- *Initial abstraction* (depression storage, evaporation, and interception losses) is equal to 20% of the storage capacity.

$$I_a = 0.2S \qquad 6.37$$

- The gross rain equals or exceeds the initial abstraction.

$$P \geq I_a \qquad 6.38$$

- The storage capacity equals or exceeds the initial abstraction plus the infiltration.

$$S \geq I_a + F \qquad 6.39$$

The net rain is then calculated from the areal gross rain, $P_g$, by using equation 6.40.

$$P_N = \frac{(P_g - 0.2S)^2}{P_g + 0.8S} \qquad 6.40$$

If it is necessary to draw a stream flow hydrograph, calculate the increments in accumulated rainfall excess.[10]

*step 11*: Use figure 6.21 to determine the peak discharge. Notice that the peak discharge is given per inch of net rain with units of cubic feet per second per square mile of drainage area. Figure 6.21 is valid only for type II, 24-hour storms. Refer to SCS literature for other storm types, or determine the peak flow directly from the hydrograph drawn from the incremental data in step 10.

## 14. RESERVOIR YIELD AND RESERVOIR SIZING

The reservoir yield problem is an accounting problem (i.e., keeping track of what comes in and what goes out). The purpose of the reservoir yield analysis is to determine the proper size of a dam or reservoir, or to evaluate the ability of an existing dam to meet water demands. There are three basic methods of solving the reservoir yield problem.

*Method #1*: Tabular simulation is the easiest method to apply, although its validity is dependent on choosing time increments as small as possible. It is also necessary to make some assumption about the distribution of annual water inflows.

*step 1*: Determine the starting storage volume, $V_n$. If $V_n$ is zero or considerably different than the average steady-state storage, a large number of iterations will be required before the simulation reaches its steady-state results.

*step 2*: For the next iteration, determine the inflow, discharge, evaporation, and seepage. Determine the starting storage volume for the next iteration by solving equation 6.41.

[10] In effect, the runoff hydrograph is simulated. For example, the areal gross rain is multiplied consecutively by the factors in table 6.9 to give the accumulated areal gross rainfall at the various times. The accumulated excess (net) is calculated from equation 6.40. The consecutive excesses are subtracted to get the incremental excesses. These incremental excesses can be used to draw the hydrograph.

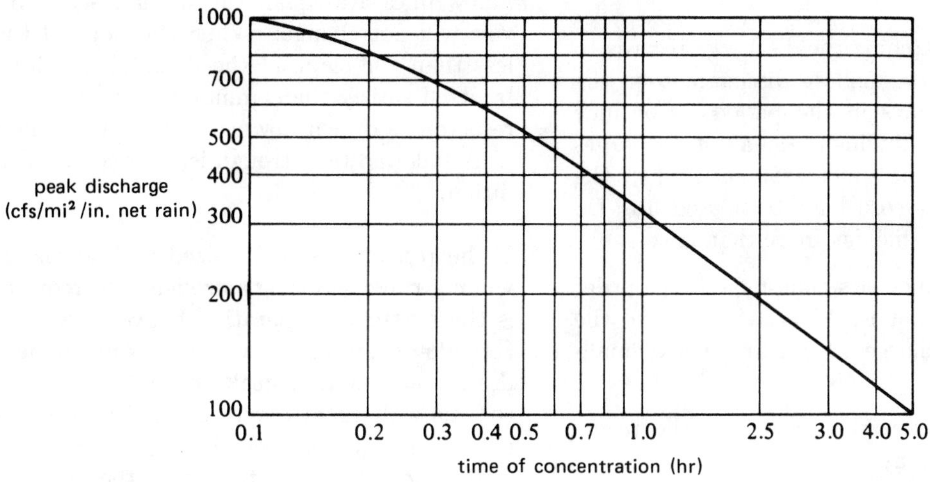


**Figure 6.21** Peak Runoff versus Time of Concentration

$$V_{n+1} = V_n + (\text{inflow})_n - (\text{discharge})_n - (\text{seepage})_n - (\text{evaporation})_n \qquad 6.41$$

Repeat step 2 as many times as necessary.

Reservoir seepage is generally very small compared to inflow and discharge. It is, therefore often neglected. Reservoir evaporation can be estimated from analytical relationships or by evaluating data from *evaporation pans.* Pan data is extended to reservoir evaporation by the pan coefficient formula. In equation 6.42, the summation is taken over the number of days in the simulation period.

$$E_R = \sum C_p E_p \qquad 6.42$$

Units of $E$ are typically inches per day. A typical value of $C_p$ is 0.7.

Inflow can be taken as actual past history. However, since it is not very likely that history will repeat, the next method is preferred.

*Method #2*: This method is the same as method #1 except for its method of determining the inflow. Method #2 uses a *Monte Carlo simulation* which is dependent on enough historical data to establish a cumulative inflow distribution. A Monte Carlo simulation is suitable if long periods are to be simulated. If short periods are to be simulated, the simulation should be performed several times and the results averaged.

*step 1*: Tabulate or otherwise determine a frequency distribution of inflow quantities.

*step 2*: Form a cumulative distribution of inflow quantities.

*step 3*: Multiply the cumulative x-axis (which runs from 0 to 1) by 100 or 1000, depending on the accuracy needed.

*step 4*: Generate random numbers between 0 and 100 (or 0 and 1000). Use of a random number table (appendix B) is adequate for hand simulation.

*step 5*: Locate the inflow quantity corresponding to the random number from the cumulative distribution x-axis.

*Method #3*: The *non-sequential drought method* is more complex, but it has the advantage of giving an estimate of the required reservoir size, rather than evaluating a trial size as the first two methods do.

In the absence of synthetic drought information, it is first necessary to develop intensity-duration-frequency curves from stream flow records.

*step 1*: Choose a duration. Usually, the first duration used will be 7 days, although choosing 15 days will not introduce too much error.

*step 2*: Search the streamflow records to find the smallest flow during the duration chosen. (The first time through, for example, find the smallest discharge totaled over any 7 days.) The days do not have to be sequential.

*step 3*: Continue searching the discharge records to find the next smallest discharge over the number of days in the period. Continue searching and finding the smallest discharges (which gradually increase) until all of the days in the record have been used up. Do not use the same day more than once.

*step 4*: Give the values of smallest discharge order numbers. That is, $M = 1$ is given to the smallest discharge, $M = 2$ to the next smallest, etc.

*step 5*: For each observation, calculate the recurrence interval as

$$F = n_y/M \qquad 6.43$$

$n_y$ is the number of years of streamflow data that was searched to find the smallest discharges.

*step 6*: Plot the points as discharge on the $y$-axis versus $F$ in years on the $x$-axis. Draw a reasonably continuous curve through the points.

*step 7*: Return to step 1 for the next duration. Repeat for all of the following durations: 7, 15, 30, 60, 120, 183, and 365 days.

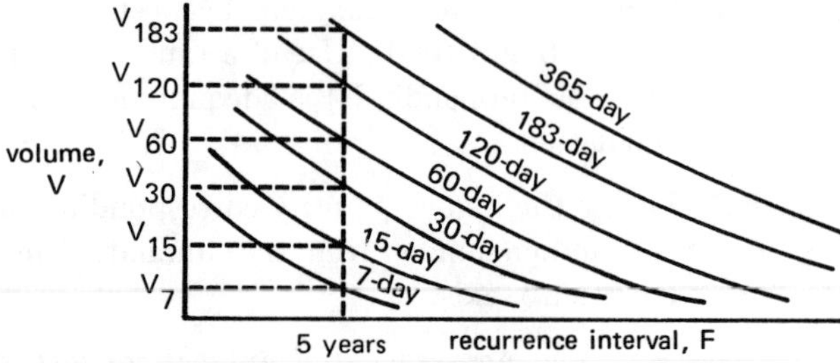


**Figure 6.22** Sample Family of Synthetic Inflow Curves

A synthetic drought can be constructed for any recurrence interval. For example, if a 5-year drought is to be planned for, the discharges $V_7, V_{15}, V_{30}, \ldots, V_{365}$ are read from the appropriate curves for $F = 5$ years.

The next step is to plot the *mass diagram* (also known as a *Rippl diagram*) for the reservoir. This is a simultaneous plot of the cumulative demand (known as *draft*) and cumulative inflow. The mass diagram is used to graphically determine the reservoir storage requirements (i.e., size).

As long as the slopes of the cumulative demand and inflow lines are equal, the water reserve in the reservoir will not change. When the slope of the inflow is less than the slope of the demand, the inflow cannot by itself satisfy the community's water needs, and the reservoir is drawn down to make up the difference. A peak followed by a trough is, therefore, a drought condition.

If the reservoir is to be sized so that the community will not run dry during a drought, the required capacity is the maximum separation between two parallel lines (pseudo-demand lines with slopes of the demand rate) drawn tangent to a peak and a subsequent trough. If the mass diagram covers enough time so that multiple droughts are present, the largest separation between peaks and subsequent troughs is the capacity.

In order for the reservoir to supply enough water during a drought condition, the reservoir must be full prior to the start of the drought. This fact is not represented when the mass diagram is drawn, hence the need to draw a pseudo-demand line parallel to the peak.

After a drought equal to the capacity of the reservoir, the reservoir will be empty. At the trough, however, the reservoir will begin to fill up again. When the cumulative excess exceeds the reservoir capacity, the reservoir will have to spill (i.e., release) water. This occurs when the cumulative inflow line crosses the prior peak's pseudo-demand line as shown in figure 6.23.

A *flood-control dam* is built to keep water in and must be sized so that water is not spilled. The mass diagram can still be used, but the maximum separation between troughs and subsequent peaks (not peaks followed by troughs) becomes the capacity.

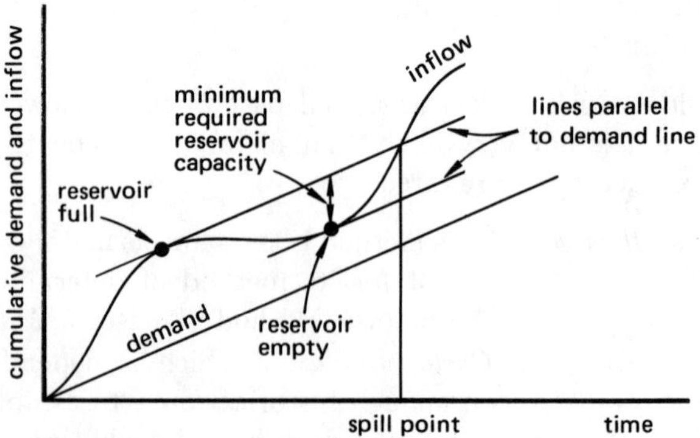


**Figure 6.23** Reservoir Mass Diagram (Rippl Diagram)

*Example 6.8*

A well-monitored stream has been observed for 50 years and has the following frequency distribution of total annual discharges.

| discharge (units) | frequency (years) | % of time |
|---|---|---|
| 0 to 0.5 | 5 | 0.10 |
| 0.5 to 1.0 | 21 | 0.42 |
| 1.0 to 1.5 | 17 | 0.34 |
| 1.5 to 2.0 | 7 | 0.14 |
| | 50 | 1.00 total |

It is proposed to dam the stream and create a reservoir with a capacity of 1.8 units. The reservoir is to support a town which will draw 1.2 units per year. Simulate 10 years of reservoir operation assuming it starts with 1.5 units.

*step 1*: The frequency distribution was given.

*steps 2 and 3*: The cumulative distribution is

| discharge | cumulative frequency | cumulative frequency × 100 |
|---|---|---|
| 0 to 0.5 | 0.10 | 10 |
| 0.5 to 1.0 | 0.52 | 52 |
| 1.0 to 1.5 | 0.86 | 86 |
| 1.5 to 2.0 | 1.00 | 100 |

*step 4*: From appendix B, choose ten 2-digit numbers. Their choice is arbitrary, but they must come sequentially from a row or column. Use the first row for this simulation:

$$78, 46, 68, 33, 26, 96, 58, 98, 87, 27$$

*step 5*: For the first year, the random number is 78. Since 78 is greater than 52 but less than 86, the inflow is in the 1.0 to 1.5 unit range. The mid-point of this range is taken as the inflow which would be 1.25. The reservoir volume after the first year would be $1.5 + 1.25 - 1.20 = 1.55$. The remaining years can be similarly simulated.

No shortages are experienced; one spill is required.

| year | starting volume | + inflow | − usage | = ending volume | + spill |
|---|---|---|---|---|---|
| 1 | 1.5 | 1.25 | 1.2 | 1.55 | |
| 2 | 1.55 | 0.75 | 1.2 | 1.1 | |
| 3 | 1.11 | 1.25 | 1.2 | 1.15 | |
| 4 | 1.15 | 0.75 | 1.2 | 0.7 | |
| 5 | 0.7 | 0.75 | 1.2 | 0.25 | |
| 6 | 0.25 | 1.75 | 1.2 | 0.8 | |
| 7 | 0.8 | 1.25 | 1.2 | 0.85 | |
| 8 | 0.85 | 1.75 | 1.2 | 1.4 | |
| 9 | 1.4 | 1.75 | 1.2 | 1.95 | 0.15 |
| 10 | 1.8 | 0.75 | 1.2 | 1.35 | |

# Appendix A: Rational Method Runoff Coefficients

| categorized by surface | |
|---|---|
| forested | 0.059–0.2 |
| asphalt | 0.7–0.95 |
| brick | 0.7–0.85 |
| concrete | 0.8–0.95 |
| shingle roof | 0.75–0.95 |
| lawns, well drained (sandy soil) | |
| up to 2% slope | 0.05–0.1 |
| 2% to 7% slope | 0.10–0.15 |
| over 7% slope | 0.15–0.2 |
| lawns, poor drainage (clay soil) | |
| up to 2% slope | 0.13–0.17 |
| 2% to 7% slope | 0.18–0.22 |
| over 7% slope | 0.25–0.35 |
| driveways, walkways | 0.75–0.85 |

| categorized by use | |
|---|---|
| farmland | 0.05–0.3 |
| pasture | 0.05–0.3 |
| unimproved | 0.1–0.3 |
| parks | 0.1–0.25 |
| cemeteries | 0.1–0.25 |
| railroad yard | 0.2–0.40 |
| playgrounds (except asphalt or concrete) | 0.2–0.35 |
| business districts | |
| neighborhood | 0.5–0.7 |
| city (downtown) | 0.7–0.95 |
| residential | |
| single family | 0.3–0.5 |
| multi-plexes, detached | 0.4–0.6 |
| multi-plexes, attached | 0.6–0.75 |
| suburban | 0.25–0.4 |
| apartments, condominiums | 0.5–0.7 |
| industrial | |
| light | 0.5–0.8 |
| heavy | 0.6–0.9 |

# Appendix B: Random Numbers

| 78466 83326 | 96589 88727 | 72655 49682 | 82338 28583 | 01522 11248 |
|---|---|---|---|---|
| 78722 47603 | 03477 29528 | 63956 01255 | 29840 32370 | 18032 82051 |
| 06401 87397 | 72898 32441 | 88861 71803 | 55626 77847 | 29925 76106 |
| 04754 14489 | 39420 94211 | 58042 43184 | 60977 74801 | 05931 73822 |
| 97118 06774 | 87743 60156 | 38037 16201 | 35137 54513 | 68023 34380 |
| 71923 49313 | 59713 95710 | 05975 64982 | 79253 93876 | 33707 84956 |
| 78870 77328 | 09637 67080 | 49168 75290 | 50175 34312 | 82593 76606 |
| 61208 17172 | 33187 92523 | 69895 28284 | 77956 45877 | 08044 58292 |
| 05033 24214 | 74232 33769 | 06304 54676 | 70026 41957 | 40112 66451 |
| 95983 13391 | 30369 51035 | 17042 11729 | 88647 70541 | 36026 23113 |
| 19946 55448 | 75049 24541 | 43007 11975 | 31797 05373 | 45893 25665 |
| 03580 67206 | 09635 84610 | 62611 86724 | 77411 99415 | 58901 86160 |
| 56823 49819 | 20283 22272 | 00114 92007 | 24369 00543 | 05417 92251 |
| 87633 31761 | 99865 31488 | 49947 06060 | 32083 47944 | 00449 06550 |
| 95152 10133 | 52693 22480 | 50336 49502 | 06296 76414 | 18358 05313 |
| 05639 24175 | 79438 92151 | 57602 03590 | 25465 54780 | 79098 73594 |
| 65927 55525 | 67270 22907 | 55097 63177 | 34119 94216 | 84861 10457 |
| 59005 29000 | 38395 80367 | 34112 41866 | 30170 84658 | 84441 03926 |
| 06626 42682 | 91522 45955 | 23263 09764 | 26824 82936 | 16813 13878 |
| 11306 02732 | 34189 04228 | 58541 72573 | 89071 58066 | 67159 29633 |
| 45143 56545 | 94617 42752 | 31209 14380 | 81477 36952 | 44934 97435 |
| 97612 87175 | 22613 84175 | 96413 83336 | 12408 89318 | 41713 90669 |
| 97035 62442 | 06940 45719 | 39918 60274 | 54353 54497 | 29789 82928 |
| 62498 00257 | 19179 06313 | 07900 46733 | 21413 63627 | 48734 92174 |
| 80306 19257 | 18690 54653 | 07263 19894 | 89909 76415 | 57246 02621 |
| 84114 84884 | 50129 68942 | 93264 72344 | 98794 16791 | 83861 32007 |
| 58437 88807 | 92141 88677 | 02864 02052 | 62843 21692 | 21373 29408 |
| 15702 53457 | 54258 47485 | 23399 71692 | 56806 70801 | 41548 94809 |
| 59966 41287 | 87001 26462 | 94000 28457 | 09469 80416 | 05897 87970 |
| 43641 05920 | 81346 02507 | 25349 93370 | 02064 62719 | 45740 62080 |
| 25501 50113 | 44600 87433 | 00683 79107 | 22315 42162 | 25516 98434 |
| 98294 08491 | 25251 26737 | 00071 45090 | 68628 64390 | 42684 94956 |
| 52582 89985 | 37863 60788 | 27412 47502 | 71577 13542 | 31077 13353 |
| 26510 83622 | 12546 00489 | 89304 15550 | 09482 07504 | 64588 92562 |
| 24755 71543 | 31667 83624 | 27085 65905 | 32386 30775 | 19689 41437 |
| 38399 88796 | 58856 18220 | 51056 04976 | 54062 49109 | 95563 48244 |
| 18889 87814 | 52232 58244 | 95206 05947 | 26622 01381 | 28744 38374 |
| 51774 89694 | 02654 63161 | 54622 31113 | 51160 29015 | 64730 07750 |
| 88375 37710 | 61619 69820 | 13131 90406 | 45206 06386 | 06398 68652 |
| 10416 70345 | 93307 87360 | 53452 61179 | 46845 91521 | 32430 74795 |

## Appendix C: Map of the U.S. Showing Average Annual Precipitation in Inches for the Period 1889–1938

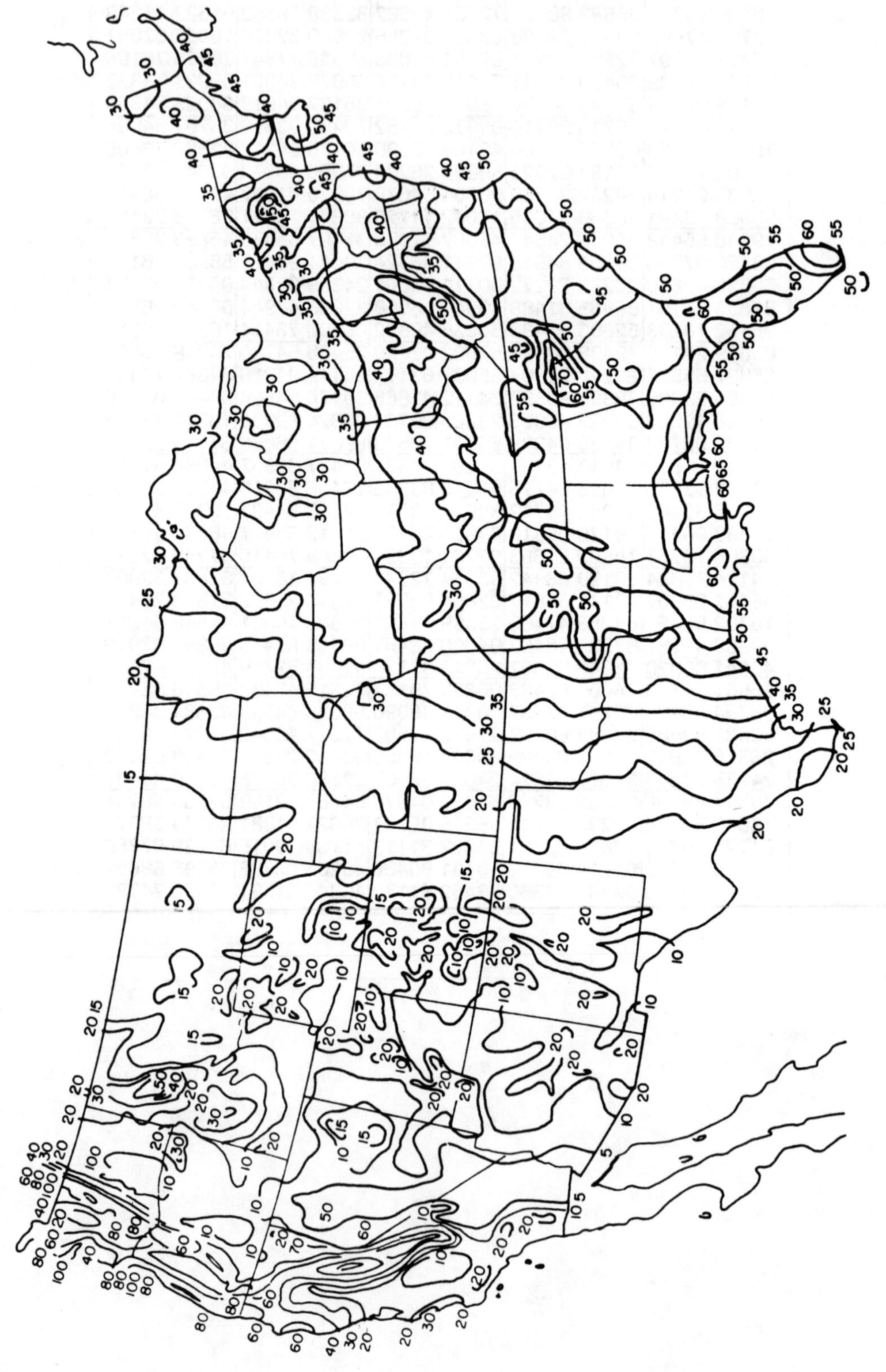

## Practice Problems: HYDROLOGY

Untimed

1. Four 5-acre areas are served by a 1200 foot storm drain ($n = .013$ and slope $= .005$). Inlets to the storm drain are placed every 300 feet along the storm drain. The inlet time for each area served by an inlet is 15 minutes, and the run-off coefficient is 0.55. A storm to be used for design purposes has the following characteristics:

$$I = \frac{100}{t_c + 10}$$

I is in inches/hr
t is in minutes

What is the size of the last section of storm drain assuming that all flows are maximum, and all pipe sizes are available?

2. An aquifer has a water table level of 100 feet below the ground surface. An 18-inch diameter well extends 200 feet into the aquifer, for a total depth of 300 feet. The aquifer transmissivity is 10,000 gallons/day-foot. The well's radius of influence is 900 feet with a 20-foot drawdown at the well. (a) What steady discharge is possible? (b) What horsepower motor is required to achieve the steady discharge?

3. A 2-hour storm over a 43 square mile area produced a flood volume of 3300 acre-feet with a peak discharge of 9300 cfs. (a) What is the unit hydrograph peak discharge? (b) If a 2-hour storm producing 2.5 inches of runoff is to be used to design a culvert, what is the design flood hydrograph volume? (c) What is the design discharge?

4. A 0.5 square mile drainage area has a suggested run-off coefficient of 0.6 and a time of concentration of 60 minutes. The drainage area is in Steel region #3, and a 10-year storm is to be used for design purposes. What is the run-off?

5. A well extends from the ground surface (elevation 383 feet) through a gravel bed to a layer of bedrock at elevation 289 feet. The screened well is 1500 feet from a river whose surface level is 363 feet. The well is pumped by a 10″ schedule 40 steel pipe which draws 120,000 gallons per day. The permeability of the well is 1600 gallons/day-ft$^2$. The pump discharges into a piping network whose friction head is 100 feet. What net horsepower is required for steady flow?

6. A measurement station on a stream recorded the following discharges from a 1.2 square mile drainage area.

| hour | cfs | hour | cfs | hour | cfs |
|---|---|---|---|---|---|
| 0 | 102 | 6 | 455 | 12 | 55 |
| 1 | 99 | 7 | 325 | 13 | 49 |
| 2 | 101 | 8 | 205 | 14 | 43 |
| 3 | 215 | 9 | 145 | 15 | 38 |
| 4 | 507 | 10 | 100 | | |
| 5 | 625 | 11 | 70 | | |

(a) Draw the actual hydrograph. (b) Draw the unit hydrograph. (c) Determine the time base, N, for direct run-off. (d) Separate the groundwater and surface water.

7. A watershed has an area of 100 square miles. The length of the main stream channel is 20 miles and the distance to a point opposite the centroid is 11 miles. Assume that $C_t = 1.8$ and find (a) the time lag, (b) the duration of the synthetic hydrograph, and (c) the peak discharge.

8. A reservoir has a total capacity of 7 units. At the beginning of a study, the reservoir contains 5.5 units. The monthly demand on the reservoir from a nearby city is 0.7 units. The monthly inflow to the reservoir is normally distributed with a mean of 0.9 units and standard deviation of 0.2 units. Simulate one year of reservoir operation.

9. Repeat problem 8 assuming that the monthly demand on the reservoir is normally distributed with a mean of 0.7 units and a standard deviation of 0.2 units.

10. A class A pan located near a reservoir shows an evaporation loss of 0.8 inches in one day. If the pan coefficient is 0.7, what is the approximate evaporation loss in the reservoir?

Timed

1. A reservoir is needed to provide 20 acre-ft/month of water over a 13-month period. The inflow for each of the 13 months is given below. Size the reservoir by whatever means you choose.

| month | F | M | A | M | J | J | A | S | O | N | D | J | F |
|---|---|---|---|---|---|---|---|---|---|---|---|---|---|
| inflow (acre-ft) | 30 | 60 | 20 | 10 | 5 | 10 | 5 | 10 | 20 | 90 | 85 | 75 | 50 |

2. An impervious concrete dam is shown. The dam reduces seepage by use of two impervious sheets extending 15 feet below the dam bottom. (a) Sketch the flow net. (b) Determine the seepage in cubic feet per minute. (c) What is the uplift on the dam at point A, midway between the left and right edges?

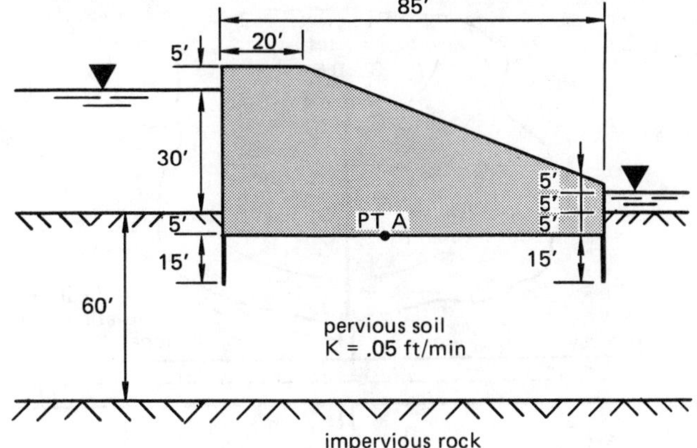

3. A standard 4-hour storm produces 2 inches net of runoff. A stream gauging report is produced from successive sampling every few hours. Two weeks later, the first four hours of an 8-hour storm over the same basin produces 1 inch of runoff. The second four hours produce 2 inches of runoff. Neglecting ground water, draw a hydrograph of the 8-hour storm.

| hour | flow (cfs) |
|---|---|
| 0 | 0 |
| 2 | 100 |
| 4 | 350 |
| 6 | 600 |
| 8 | 420 |
| 10 | 300 |
| 12 | 250 |
| 14 | 150 |
| 16 | 100 |
| 18 | — |
| 20 | 50 |
| 22 | — |
| 24 | 0 |

4. A reservoir with constant draft of 240 MG/-mi$^2$-yr is being designed. Given the rainfall distribution, what should be the minimum reservoir size? When would the reservoir start to spill?

| month | inflow (MG/mi$^2$) |
|---|---|
| J | 20 |
| F | 30 |
| M | 45 |
| A | 30 |
| M | 40 |
| J | 30 |
| J | 15 |
| A | 5 |
| S | 15 |
| O | 60 |
| N | 90 |
| D | 40 |

5. A 75-acre urbanized section of land drains into a rectangular 5′ x 7′ channel which directs runoff through a round culvert under a roadway. The culvert is concrete, and 60″ in diameter. It is 36′ long and placed on a 1% slope.

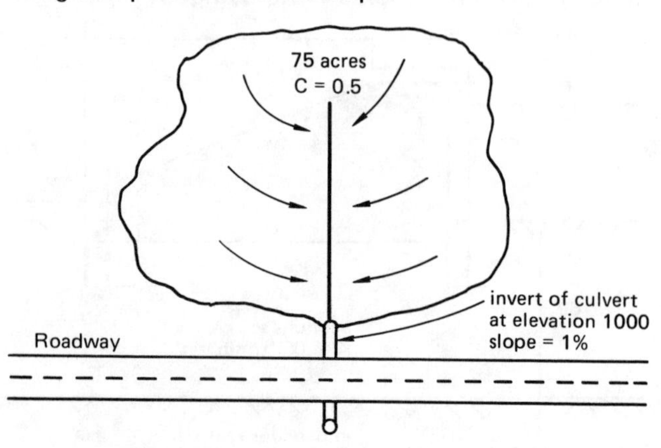


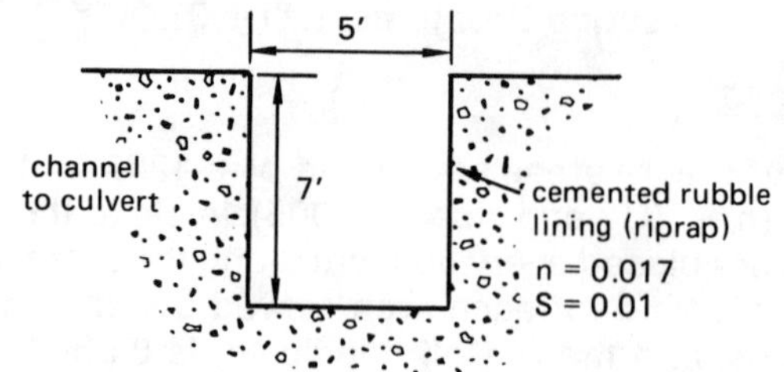


It is desired to evaluate the culvert design based on a 50-year storm. It is known that the time of concentration to the head of the culvert is 30 minutes.

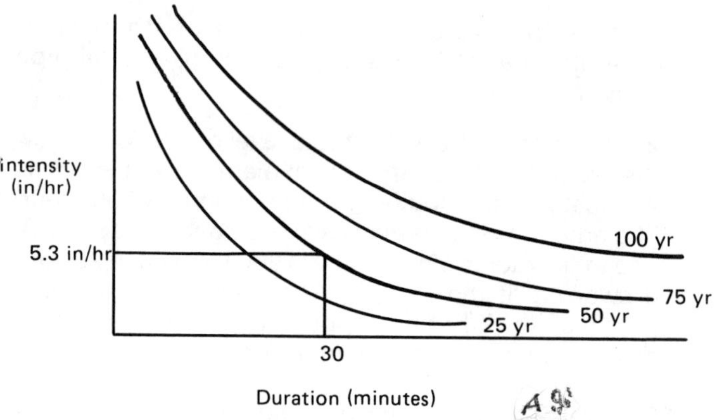


(a) If the minimum road surface elevation is 1010′, will the road surface be flooded? (b) What is the water elevation upstream from the culvert entrance after 30 minutes? (c) What is the water elevation downstream of the culvert exit after 30 minutes?

6. Draw the flow net for the coffer-dam shown. Sheet piles extend below the mud line, and the coffer-dam floor is unlined. The figure is drawn to scale.

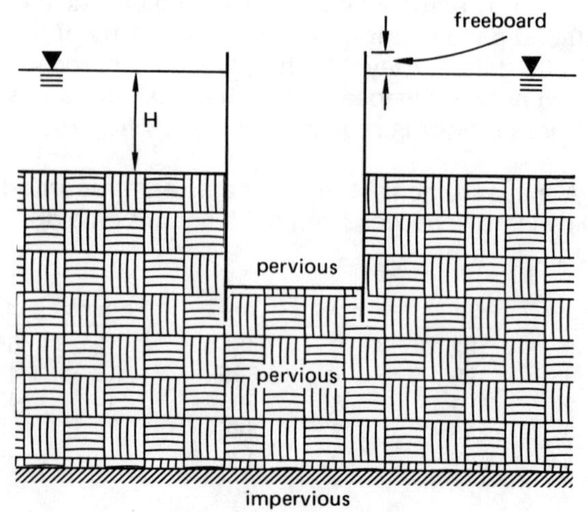

# 7 WATER SUPPLY ENGINEERING

*Nomenclature*

| | | |
|---|---|---|
| A | area | $ft^2$ |
| B | width | ft |
| C | constant, or concentration | –, or mg/l |
| $C_D$ | drag coefficient | – |
| d | particle diameter | ft |
| D | outside pipe diameter, or drag force | ft, lbf |
| f | change in charge | – |
| $F_I$ | impact factor | – |
| g | acceleration due to gravity (32.2) | $ft/sec^2$ |
| $g_c$ | gravitational constant | $ft\text{-}lbm/sec^2\text{-}lbf$ |
| G | mean velocity gradient | 1/sec |
| h | height or depth | ft |
| H | height | ft |
| K | rate constant | 1/sec |
| $K_{eq}$ | equilibrium constant | – |
| $K_{sp}$ | solubility product | – |
| L | length | ft |
| LF | load factor | – |
| $N_{Re}$ | Reynolds number | – |
| p | pressure | psf |
| P | population, force, or power | 1000's of people, pounds, or ft-lbf/sec |
| Q | flow | cfs |
| t | time | seconds |
| v | velocity, or rate of reaction | ft/sec,– |
| v* | surface loading | ft/sec |
| V | volume | $ft^3$ |
| w | pipe load | lbf/ft |
| x | mole fraction | – |
| z | diagonal distance | ft |

*Symbols*

| | | |
|---|---|---|
| $\rho$ | density | $lbm/ft^3$ |
| $\nu$ | kinematic viscosity | $ft^2/sec$ |
| $\mu$ | absolute viscosity | $lbf\text{-}sec/ft^2$ |
| $\eta$ | efficiency | – |

*Subscripts*

| | |
|---|---|
| d | detention |
| DL | dead load |
| f | flow-through |
| i | incoming |
| o | outgoing |
| s | settling |

## 1 CONVERSIONS

| multiply | by | to obtain |
|---|---|---|
| Clark degrees | 1 | grains/Imperial gallon |
| cubic feet | 7.481 | gallons |
| cubic feet | 28.32 | liters |
| cubic feet/sec | 0.6463 | million gallons/day |
| cubic feet/sec | 448.8 | gpm |
| feet | 0.3048 | meters |
| gallons | 3785 | cubic centimeters |
| gallons | 0.1337 | cubic feet |
| gallons | 3.785 | liter |
| gpm | 0.002228 | cfs |
| grains | 1.429 EE–4 | pounds |
| grains/gallon | 142.9 | pounds/million gallons |
| grains/gallon | 17.1 | ppm |
| grains/gallon | 17.1 | mg/l |
| grains/gallon | 1.2 | Clark degrees |
| grams | 0.03527 | ounces |
| grams | 0.002205 | pounds |
| Imperial gallons | 1.2 | U.S. gallons |
| inches | 2.540 | centimeters |
| kilograms | 2.205 | pounds |
| $kg/m^3$ | 0.06243 | $pounds/ft^3$ |
| liters | 1000 | cubic centimeters |
| liters | 0.03532 | cubic feet |
| liters | 0.2642 | gallons |
| liter/sec | 15.85 | gpm |
| meters | 3.281 | feet |
| meters | 39.37 | inches |
| microns | 0.001 | millimeters |
| MGD | 1.547 | cubic feet/sec |
| mg/l | 1.0 | ppm |

| | | |
|---|---|---|
| mg/l | 0.0583 | grains/gallon |
| mg/l | 8.345 | pounds/million gallons |
| pounds | 7000 | grains |
| pounds | 453.6 | grams |
| ppm | 0.0583 | grains/gallon |
| ppm | 0.07 | Clark degrees |

## 2 DEFINITIONS

Acid: A compound containing hydrogen ions ($H^+$ or $H_3O^+$) in an aqueous solution. Acids have a sour taste, conduct electricity, turn blue litmus paper red, and have a pH between 0 and 7.

Aeration: Mixing water with air, either by spraying water or diffusing air through it.

Aerobic: Requiring oxygen.

Air break: A means by which drinking water can be used for fire fighting without contaminating the drinking supply. Drinking water is freely discharged into the top of a tank which feeds the fire main.

Algae: One-celled plant life.

Altitude valve: A valve that automatically opens to prevent overflow in storage tanks.

AMU: Atomic mass unit. One AMU is 1/12th the atomic weight of carbon.

Anaerobic: Not requiring oxygen.

Anion: Negative ions that migrate to the positive electrode in an electrolytic solution.

Atomic number: The number of protons in the nucleus of an atom.

Atomic weight: Approximately the number of protons and neutrons in the nucleus of an element.

Avogadro's law: A gram-mole of any substance contains 6.023 EE23 molecules.

B. coli: Bacteria coli. See 'E. coli.'

Backfill: The soil that is used to cover a pipe put into a trench.

Base: A compound containing hydroxide ions ($OH^-$) combined with alkali metals and alkali earths. Bases in aqueous solutions have a bitter taste, conduct electricity, turn red litmus paper blue, and have a pH between 7 and 14.

Belt-line layout: See 'Grid iron layout.'

Breakpoint chlorination: Application of chlorine which results in a minimum of chloramine residuals. No free chlorine residuals are produced unless the breakpoint is exceeded.

Capita: Person.

Carbonate hardness: Hardness caused by bicarbonates.

Cation: A positive ion that migrates to the negative electrode in an electrolytic solution.

Chloramine: Compounds of chlorine and ammonia (e.g., $NH_2Cl$, $NHCl_2$, or $NCl_3$).

Chlorine demand: The difference between applied chlorine and the chlorine residual. Chlorine demand is chlorine that has been reduced in chemical reactions and is no longer available for purification.

Clear well: Storage in a water treatment plant which normally takes water from the filters.

Coagulation: Floc formation as the result of adding coagulating chemicals. Coagulants destabilize (by reducing repulsive forces) suspended particles, allowing them to agglomerate.

Coliform: See 'Colon bacilli.'

Colloid: A fine particle ranging in size from 1 to 500 millimicrons. Colloids cause turbidity.

Colon bacilli: Bacteria residing normally in the intestinal tract. These are not necessarily dangerous if present in drinking water, but they are indicators of other possible pathogens.

Combined residuals: Compounds of an additive (such as chlorine) that have combined with something else. Chloramines are combined residuals.

Compound: A homogeneous substance composed of two or more elements that can be decomposed by chemical changes only.

Confirmed test: A second test used if the presumptive test for coliforms is positive.

Cross connection: Connecting fire and drinking water supplies together.

Detention time: The average time spent by water in a settling basin.

Detritus: See 'Grit.'

Distillation: Salt removal in water by boiling and condensation.

Domestic use: Water use by the public (home use).

Double main system: Separate water mains for domestic and fire fighting use.

E. coli: See 'Escherichia coli.'

Element: A pure substance that cannot be decomposed by chemical means.

Electrodialysis: A method of using induced currents

and direction-selective membranes to remove dissolved salts from water.

Electrolysis: The production of an oxidation-reduction reaction by a D.C. current.

Enteric: Intestinal.

Equivalent weight: The molecular weight divided by the change in charge or oxidation number.

Escherichia coli: A common bacterium found in the digestive tract.

Eutrophication: Aging of a lake due to plant growth and sedimentation.

Facultative: Able to live under different or changing conditions.

Floc: Agglomerated colloidal particles.

Flush hydrant: Hydrants located in pits below street level.

Free residuals: Ions or compounds not combined or reduced. The presence of free residuals signifies excess dosage.

Fungus: Multi-cellular plant growth common to humid areas.

Gravity distribution: A water supply that uses natural flow from an elevated tank or mountain reservoir to supply pressure.

Grid-iron layout: A system of distribution pipes in which there are alternative paths through which water can flow in case one path is disturbed.

Grit: Sand-like particles mixed with mud and other debris.

Hard water: Water containing bicarbonates of calcium and magnesium, as well as chlorides and sulfates.

Hydrogen ion: Positively charged combination of a proton and water molecule ($H_3O^+$).

Hydrophilic: Seeking or liking water.

Hydrophobic: Disliking water.

Hydronium ion: See 'hydrogen ion.'

Ions: Atoms that have lost or gained one or more electrons, giving them a charge.

Intrinsic water: Ultra-pure water with essentially no mineral or ion content. Typically used in electronic industries.

Isotopes: Atoms of the same atomic number but with different atomic weights due to a variable number of neutrons.

Mixture: A heterogeneous physical combination of two or more substances, each of which retains its identity and specific properties.

Mole: A quantity of substance equal to its molecular weight in grams (*gmole* or *gram-mole*) or in pounds (*pmole* or *pound-mole*).

Molecule: The smallest division of an element or compound.

Molecular weight: The sum of the atomic weights of all atoms in the molecule.

Non-pathogenic: Not biologically harmful.

Osmosis: The flow of a solvent through a permeable membrane separating two solutions of different concentrations.

Oxidation: The loss of electrons.

Oxidation number: An arbitrarily assigned number used in redox calculations to balance molecular charges.

Pathogenic: Biologically harmful.

Permanent hardness: Hardness that cannot be removed by heating.

pH: A measure of a solution's hydrogen ion concentration (acidity).

pOH: A measure of a solution's hydroxyl ion concentration (alkalinity).

Polished water: See 'Intrinsic water.'

Post hydrant: A fire hydrant that rises from the sidewalk surface.

Potable: Drinkable.

Presumptive test: A first stage test which is inconclusive if positive, but is conclusive if negative.

Protozoa: Single-celled aquatic animals.

Radical: A charged group of atoms that combines as a single element.

Recarbonation: Addition of carbon dioxide to water to neutralize excess lime.

Redox reaction: A reaction in which oxidation and reduction occur.

Reduction: A gain in electrons.

Residual: Additive that is left over after some of the additive has been combined or inactivated.

Retention period: See 'Detention time.'

Reverse osmosis: A process that uses pressure to force a solvent to flow against the ionization gradient.

Salt: An ionic compound formed by direct union of elements, reactions between acids and bases, reactions of acids and salts, and reactions between different salts.

Single main system: One main supplies both potable and fire fighting water.

Solution: A homogeneous mixture of solute and solvent.

Stoichiometry: The study of how elements combine in predetermined quantities to form compounds.

Superchlorination: Chlorination past the breakpoint.

Syndets: Synthetic detergents containing phosphates.

Temporary hardness: Hardness that can be removed by heating water.

Thermocline: A layer of water in a lake or reservoir where temperature declines rapidly with depth.

Turbidity: Cloudiness of water.

Valence: The relative combining capacity of an atom or group of atoms compared to that of the standard hydrogen atom. Valence of an ion is the same as the ion's charge.

Zeolite: A natural or synthetic resin that has an affinity for ions.

## 3 CHEMISTRY REVIEW

### A. VALENCE

The charge of any element in its free state is zero. A charged condition will occur when the element has lost or gained one or more electrons. The *valence* of an ion is equal to its charge.

Common elements with one valence number are:

| cations | | | anions | | |
|---|---|---|---|---|---|
| +1 | +2 | +3 | −1 | −2 | ±4 |
| H | Mg | Al | F | O | C |
| Li | Ca | B | Cl | S | Si |
| Na | Sr | | Br | | |
| K | Ba | | I | | |
| Ag | Zn | | | | |
| | Cd | | | | |

Some elements have two valence numbers. The lower valence is associated with the generic ending 'ous'. The higher valence is associated with the ending 'ic'.

| +1/+2 | +2/+3 | +2/+4 | +3/+5 |
|---|---|---|---|
| Cu | Fe | Pb | As |
| Hg | Ni | Sn | Sb |
| | Co | | Bi |

The following elements have more than two valence numbers:

| | | | | | | |
|---|---|---|---|---|---|---|
| Cr: | +2, | +3, | +6 | | | |
| Mn: | +2, | +3, | +4, | +6, | +7 | |
| S: | −2 | +4, | +6 | | | |
| N: | −3, | +1, | +2, | +3, | +4, | +5 |
| P: | −3, | +3, | +5 | | | |

Common radicals are:

| | | |
|---|---|---|
| $NH_4$ | Ammonium | +1 |
| $ClO_3$ | Chlorate | −1 |
| $ClO_2$ | Chlorite | −1 |
| $ClO$ | Hypochlorite | −1 |
| $NO_3$ | Nitrate | −1 |
| $NO_2$ | Nitrite | −1 |
| $C_2H_3O_2$ | Acetate | −1 |
| $MnO_4$ | Permanganate | −1 |
| $OH$ | Hydroxide | −1 |
| $HSO_3$ | Bisulfite | −1 |
| $HSO_4$ | Bisulfate | −1 |
| $HCO_3$ | Bicarbonate | −1 |
| $CO_3$ | Carbonate | −2 |
| $SO_4$ | Sulfate | −2 |
| $SO_3$ | Sulfite | −2 |
| $CrO_4$ | Chromate | −2 |
| $Cr_2O_7$ | Dichromate | −2 |
| $BO_3$ | Borate | −3 |
| $PO_4$ | Phosphate | −3 |
| $Fe(CN)_6$ | Ferricyanide | −3 |
| $Fe(CN)_6$ | Ferrocyanide | −4 |

Compounds always form in such a manner as to obtain a neutral charge. For example, $MgNO_3$ would not be allowed because the magnesium has a charge of +2 and the nitrate ion has a charge of −1. However, $Mg(NO_3)_2$ is a valid compound.

### B. CHEMICAL REACTIONS

During chemical changes, bonds between atoms are broken and new bonds are formed. Reactants are either converted to simpler products or are synthesized into more complex compounds. There are five common types of chemical reactions.

- **Direct combination or synthesis**. This is the simplest type of reaction where two elements or compounds combine directly to form a compound.

$$2H_2 + O_2 \rightarrow 2H_2O$$

$$SO_2 + H_2O \rightarrow H_2SO_3$$

- **Decomposition**. Bonds uniting a compound are disrupted by heat or other energy source to yield simpler compounds or elements.

$$2HgO \rightarrow 2Hg + O_2$$
$$H_2CO_3 \rightarrow H_2O + CO_2$$

- **Single displacements**. This type of reaction is identified by one element and one compound as reactants.

$$2Na + 2H_2O \rightarrow 2NaOH + H_2$$
$$2KI + Cl_2 \rightarrow 2KCl + I_2$$

- **Double decomposition**. These are reactions characterized by having two compounds as reactants and forming two new compounds.

$$AgNO_3 + NaCl \rightarrow AgCl + NaNO_3$$
$$H_2SO_4 + ZnS \rightarrow H_2S + ZnSO_4$$

- **Oxidation-Reduction (Redox)**. These reactions involve oxidation of one substance and reduction of another. In the example, calcium loses electrons and is oxidized; oxygen gains electrons and is reduced.

$$2Ca + O_2 \rightarrow 2CaO$$

Balancing chemical equations is largely a matter of deductive trial and error. The coefficients in front of each compound can be thought of as the number of molecules or moles taking part in the reaction. The number of atoms for each element must be equal on both sides of the equation. Total atomic weights must also be equal on both sides.

*Example 7.1*

Balance the following reaction equation:

$$Al + H_2SO_4 \rightarrow Al_2(SO_4)_3 + H_2$$

As written, the reaction is not balanced. For example, there is one aluminum on the left, but there are two on the right. The starting element in the balancing procedure is chosen somewhat arbitrarily.

*step 1*: Since there are two aluminums on the right, multiply (Al) by 2.

$$2Al + H_2SO_4 \rightarrow Al_2(SO_4)_3 + H_2$$

*step 2*: Since there are three sulfate radicals ($SO_4$) on the right, multiply ($H_2SO_4$) by 3.

$$2Al + 3H_2SO_4 \rightarrow Al_2(SO_4)_3 + H_2$$

*step 3*: Now there are six hydrogens on the left, so multiply $H_2$ by 3 to balance the equation.

$$2Al + 3H_2SO_4 \rightarrow Al_2(SO_4)_3 + 3H_2$$

## C. STOICHIOMETRY

Stoichiometric problems are known as 'weight and proportion' problems. Their solutions use simple ratios to determine the weight of reactants required to produce some given amount of products. The procedure for solving these problems is essentially the same regardless of the complexity of the reaction.

*step 1*: Write and balance the chemical equation.

*step 2*: Calculate the molecular weight of each compound or element in the equation.

*step 3*: Multiply the molecular weights by their respective coefficients and write the products under the formulas.

*step 4*: Write the given weight data under the molecular weights calculated from step 3.

*step 5*: Fill in missing information by using simple ratios.

*Example 7.2*

Caustic soda (NaOH) is made from sodium carbonate ($Na_2CO_3$) and slaked lime ($Ca(OH)_2$). How many pounds of caustic soda can be made from one ton of sodium carbonate?

The balanced chemical equation is

| | $Na_2CO_3$ | + $Ca(OH)_2$ | $\rightarrow$ 2NaOH | + $CaCO_3$ |
|---|---|---|---|---|
| MW's: | 106 | 74 | 80 | 100 |
| given: | 2000 | | X | |

The ratio used is

$$\frac{NaOH}{Na_2CO_3} = \frac{80}{106} = \frac{X}{2000}$$

$$X = 1509\,\text{pounds}$$

## D. EQUIVALENT WEIGHTS

The *equivalent weight* of a molecule that takes place in a chemical reaction is the molecular weight divided by the change in charge (or oxidation number) which is experienced by the molecule or ion.

*Example 7.3*

What are the equivalent weights of the following compounds?

(a) Al in the reaction: $Al^{+++} + 3e^- \rightarrow Al$

(b) $H_2SO_4$ in the reaction: $H_2SO_4 + H_2O \rightarrow 2H^+ + SO_4^{--} + H_2O$

(c) NaOH in the reaction: $NaOH + H_2O \rightarrow Na^+ + OH^- + H_2O$

(a) The atomic weight of aluminum is 27. Since the change in charge is three, the equivalent weight is $27/3 = 9$.

(b) The molecular weight of sulfuric acid is 98.1. Since it went from neutral to ions with 2 charges each, the equivalent weight is $98.1/2 = 49.05$.

(c) Sodium hydroxide has a molecular weight of 40. The molecule went from neutral to a singly-charged state. So, the equivalent weight is $40/1 = 40$.

E. SOLUTIONS OF SOLIDS IN LIQUIDS

Various methods exist for measuring the strengths of solutions.

N – **Normality**: The number of gram equivalent weights of solute per liter of solution. A solution is 'normal' if there is exactly one gram equivalent weight per liter.

M – **Molarity**: The number of gram moles of solute per liter of solution. A 'molar' solution contains one gram mole per liter of solution. Hydrated water adds to molecular weight.

**Formality:** The number of gram formula weights (i.e., molecular weights in grams) per liter of solution. Hydrated water does not add to formula weight.

m – **Molality**: The number of gram moles of solute per 1000 grams of solvent. A 'molal' solution contains 1 gmole per 1000 grams.

x – **Mole fraction**: The number of moles of solute divided by the number of moles of solvent and all solutes.

mg/l – **milligrams per liter**: The number of milligrams per liter. Same as ppm for solutions of water.

ppm – **parts per million**: The number of pounds (or grams) of solute per million pounds (or grams) of solution. Same as mg/l for solutions of water.

meq/l – **milligram equivalent weights** of solute per liter of solution: Calculated by multiplying normality by 1000.

*Example 7.4*

A solution is made by dissolving 0.353 grams of $Al_2(SO_4)_3$ in 730 grams of water. Assuming 100% ionization, what is the concentration expressed as normality, molarity, and mg/l?

The molecular weight of $Al_2(SO_4)_3$ is

$$MW = (2)(26.98) + 3[32.06 + (4)(16)] = 342.14$$

The equivalent weight is $342.14/6 = 57.02$

The number of gram equivalent weights used is $0.353/57.02 = 6.19$ EE–3.

The number of liters of solution (same as the solvent volume if the small amount of solute is neglected) is 0.73.

The normality is

$$N = \frac{6.19\,\text{EE–3}}{0.73} = 8.48\,\text{EE–3}$$

The number of moles of solute used is $0.353/342.14 = 1.03$ EE–3.

The molarity is

$$M = \frac{1.03\,\text{EE–3}}{0.73} = 1.41\,\text{EE–3}$$

The number of milligrams is $0.353/0.001 = 353$.

$$\text{mg/l} = \frac{353}{0.73} = 483.6$$

For the purpose of water supply calculations, however, the most useful measure of solution strength is the *$CaCO_3$ equivalent* measurement. In this method, substance quantities are reported in mg/l *as $CaCO_3$*, even when $CaCO_3$ is unrelated to the substance or reaction that produced the substance.

Actual gravimetric amounts of a substance can be converted to amounts as $CaCO_3$ by use of the conversion factors in appendix A. These factors are easily derived from stoichiometric principles.

The value of converting all substance quantities to amounts as $CaCO_3$ is that equal $CaCO_3$ amounts constitute stoichiometric reaction quantities. For example, 100 mg/l *as $CaCO_3$* of sodium ion ($Na^+$) reacts with 100 mg/l *as $CaCO_3$* of chloride ion ($Cl^-$) to produce 100 mg/l *as $CaCO_3$* of salt (NaCl), even though the gravimetric quantities differ and $CaCO_3$ is not part of the reaction.

*Example 7.5*

Lime is added to water to remove carbon dioxide gas.

$$CO_2 + Ca(OH)_2 \rightarrow CaCO_3 \downarrow + H_2O$$

If water contains 5 mg/l of $CO_2$, how much lime is required for its removal?

From appendix A, the factor which converts $CO_2$ *as substance* to $CO_2$ *as* $CaCO_3$ is 2.27.

$$\begin{aligned} CO_2 \text{ as } CaCO_3 \text{ equivalent} &= (2.27)(5) \\ &= 11.35 \text{ mg/l as } CaCO_3 \end{aligned}$$

Therefore, the $CaCO_3$ equivalent of lime required will also be 11.35 mg/l.

From appendix A, the factor which converts lime as $CaCO_3$ to lime as substance is (1/1.35).

$$Ca(OH)_2 \text{ substance} = \frac{11.35}{1.35} = 8.41 \text{ mg/l as substance}$$

## F. SOLUTIONS OF GASES IN LIQUIDS

*Henry's law* states that the amount of gas dissolved in a liquid is proportional to the partial pressure of the gas. This applies separately to each gas to which the liquid is exposed.

*Example 7.6*

At 20 °C and 760 mm Hg, one liter of water can absorb 0.043 grams of oxygen gas or 0.19 grams of nitrogen. Find the weight of oxygen and nitrogen in one liter of water exposed to air at 20 °C and 760 mm Hg total pressure.

Partial pressure is volumetrically weighted. Air is 20.9% oxygen by volume. The remainder is taken as nitrogen.

$$\begin{aligned} \text{oxygen dissolved} &= (0.209)(0.043) = 0.009 \text{ g/l} \\ \text{nitrogen dissolved} &= (0.791)(0.19) = 0.150 \text{ g/l} \end{aligned}$$

## G. ACIDS AND BASES

An *acid* is any substance that dissociates in water into $H^+$ (or $H_3O^+$). A *base* dissociates in water and gives up $OH^-$. A measure of the strength of the acid or base is the number of these hydrogen or hydroxide ions in a liter of solution. Since these are large (small) numbers, a logarithmic scale is used.

$$pH = -\log_{10}[H^+] \qquad 7.1$$

$$pOH = -\log_{10}[OH^-] \qquad 7.2$$

The following relationship always exists between pH and pOH:

$$pH + pOH = 14 \qquad 7.3$$

[X] is defined as the *ionic concentration* in moles per liter. The number of moles can be calculated from Avogadro's law by dividing the actual number of ions by 6.023 EE23. An easier method is to use equation 7.4.

$$[X] = (\text{fraction ionized})(\text{molarity}) \qquad 7.4$$

*Example 7.7*

Calculate the concentrations of $H^+$, $OH^-$, pH, and pOH in 4.2% ionized 0.010M ammonia solution prepared from ammonium hydroxide ($NH_4OH$).

$$\begin{aligned} [OH^-] &= (0.042)(0.010) = 4.2\,\text{EE–4 moles/liter} \\ pOH &= -\log(4.2\,\text{EE–4}) = 3.38 \\ pH &= 14 - 3.38 = 10.62 \\ [H^+] &= 10^{-10.62} = 2.4\,\text{EE–11 moles/liter} \end{aligned}$$

Since $H^+ + OH^- \rightarrow H_2O$, acids and bases neutralize each other by forming water. The volumes required for complete neutralization are:

$$\begin{aligned} &(\text{vol. base})(\text{normality of base}) \\ &\qquad = (\text{vol. acid})(\text{normality of acid}) \qquad 7.5 \end{aligned}$$

If the concentrations are expressed as molarities,

$$\begin{aligned} &(\text{vol. base})(\text{base molarity})(\text{change in base charge}) = \\ &(\text{vol. acid})(\text{acid molarity})(\text{change in acid charge}) \qquad 7.6 \end{aligned}$$

Both equation 7.5 and 7.6 assume 100% ionization of the solute.

## H. REVERSIBLE REACTIONS

Some reactions are capable of going in either direction. Such reactions are called *reversible reactions* and are characterized by the presence of all reactants and all products simultaneously.

$$N_2 + 3H_2 \rightleftharpoons 2NH_3 + 24{,}500 \text{ calories}$$

*Le Chatelier's principle* predicts the direction the reaction will go when some property or condition is changed. That principle says that when a reversible reaction has resulted in an equilibrium condition, and the reactants and products are further stressed by changing the pressure, temperature, or concentration, a new equilibrium will be formed in the direction that reduces the stress.

Consider the reaction between nitrogen and hydrogen. When the reaction proceeds in the forward direction, heat is given off (an *exothermic reaction*). If the reaction proceeds in the reverse direction, heat is absorbed (*endothermic reaction*). If the system is stressed by increasing the temperature, the reaction will proceed in the reverse direction because that direction will absorb heat and reduce the temperature.

For reactions that involve gases, the coefficients in front of the molecules can be interpreted as volumes. In the nitrogen-hydrogen reaction, 4 volumes combine to form

2 volumes. If the equilibrium system is stressed by increasing the pressure, then the forward reaction will occur because this direction reduces the volume and thereby reduces the pressure.

If the concentration of any substance is increased, the reaction proceeds in a direction away from the substance with the increased concentration.

If a *catalyst* is introduced, the equilibrium will not be changed. However, the reaction speed is increased, and equilibrium is reached more quickly. This reaction speed is known as the *rate of reaction* or *reaction velocity*.

The *law of mass action* says that the reaction speed is proportional to the concentrations of reactants. Consider the following reaction.

$$A + B \rightleftharpoons C + D$$

The rates of reaction can be calculated as the products of the reactants' concentrations.

$$v_{\text{forward}} = C_1[A][B] \qquad 7.7$$

$$v_{\text{reverse}} = C_2[C][D] \qquad 7.8$$

The constants, $C_1$ and $C_2$, are needed to obtain the proper units. For reversible reactions, the *equilibrium constant* is defined by equation 7.9. This equilibrium constant is essentially independent of pressure, but will depend on temperature.

$$K_{\text{eq}} = \frac{[C][D]}{[A][B]} \qquad 7.9$$

Consider the following complex reaction,

$$aA + bB \rightleftharpoons cC + dD$$

The equilibrium constant is calculated from the *mass action equation.*

$$K_{\text{eq}} = \frac{[C]^c[D]^d}{[A]^a[B]^b} \qquad 7.10$$

The only special rule to be observed is that if any of $A, B, C$, or $D$ are pure solids or liquids, then their concentrations are omitted in the calculation of the equilibrium constant.

*Example 7.8*

Acetic acid dissociates according to the following equation. What is the equilibrium constant if the ion concentrations are as given?

$$HC_2H_3O_2 + H_2O \rightleftharpoons H_3O^+ + C_2H_3O_2^-$$

$$[HC_2H_3O_2] = 0.09866 \text{ moles/liter}$$

$$[H_2O] = 55.5555 \text{ moles/liter}$$

$$[H_3O^+] = 0.00134 \text{ moles/liter}$$

$$[C_2H_3O_2^-] = 0.00134 \text{ moles/liter}$$

From equation 7.9, omitting the water's concentration,

$$K_{eq} = \frac{(0.00134)(0.00134)}{0.09866} = 1.82 \text{ EE}{-5}$$

For weak aqueous solutions, the concentration of water is very large and essentially constant. Therefore, it is omitted. The new constant is known as the *ionization constant* to distinquish it from the equilibrium constant.

$$K_{\text{ionization}} = K_{\text{equilibrium}} \qquad 7.11$$

Pure water is a very weak electrolyte and it ionizes only slightly by itself.

$$2H_2O \rightleftharpoons H_3O^+ + OH^- \qquad 7.12$$

At equilibrium, the ion concentrations can be measured to be

$$[H_3O^+] = \text{EE–7}$$

$$[OH^-] = \text{EE–7}$$

The ionization constant for pure water is

$$K_{\text{ionization}} = [H_3O^+][OH^-] = (\text{EE–7})(\text{EE–7}) = \text{EE–14} \qquad 7.13$$

Notice that the concentration of water was omitted because the ionization constant, not the equilibrium constant, was calculated.

Taking logs of both sides of equation 7.13 will derive equation 7.3.

*Example 7.9*

A 0.1M acetic acid solution is 1.34% ionized. Find the

(a) hydrogen ion concentration

(b) acetate ion concentration

(c) un-ionized acid concentration

(d) ionization constant

(a) From equation 7.4,

$$[H_3O^+] = (0.0134)(0.1) = 0.00134 \text{ moles/liter.}$$

(b) Since every hydrogen ion has a corresponding acetate ion (see example 7.8), the acetate ion concentration is also 0.00134 moles per liter.

(c) The concentration of un-ionized acid can be derived from equation 7.4:

$$[HC_2H_3O_2] = \text{(fraction not ionized)(molarity)}$$
$$= (1 - 0.0134)(0.1) = 0.09866 \text{ moles/liter}$$

(d) The ionization constant is

$$K_{\text{ionization}} = \frac{(0.00134)(0.00134)}{0.09866} = 1.82 \text{ EE}{-5}$$

A faster way to find the ionization constant is to use equation 7.14.

$$K_{\text{ionization}} = \frac{\text{(molarity)(fraction ionized)}^2}{1 - \text{fraction ionized}} \quad 7.14$$

*Example 7.10*

Find the concentration of the hydrogen ion for a 0.2M acetic acid solution if $K_{\text{ionization}} = 1.8$ EE–5.

From equation 7.14, letting $X$ be the fraction ionized,

$$1.8\,\text{EE–5} = \frac{(0.2)(X)^2}{1 - X}$$

If $X$ is small, then $(1 - X) \approx 1$. So, $X = 9.49$ EE–3.

From equation 7.4, the concentration is

$$[H_3O^+] = (9.49\,\text{EE–3})(0.2) = 1.9\,\text{EE–3}$$

The *common ion effect law* is a form of Le Chatelier's law. If a salt containing a common ion is added to a weak acid or base solution, ionization will be suppressed. This is a consequence of the need to have an unchanged ionization constant.

*Example 7.11*

What is the hydrogen ion concentration of a solution with 0.1 gmole of 80% ionized ammonium acetate in one liter of 0.1M acetic acid?

As before, $1.8 \text{ EE–5} = \frac{[H_3O^+][C_2H_3O_2^-]}{[HC_2H_3O_2]}$.

Let $X = [H_3O^+]$.

Then, $[HC_2H_3O_2] = 0.1 - X \approx 0.1$

$$[C_2H_3O_2^-] = X + (0.8)(0.1) \approx (0.8)(0.1) = 0.08$$

So, from the ionization constant equation,

$$1.8\,\text{EE–5} = \frac{(X)(0.08)}{0.1}$$
$$X = 2.2\,\text{EE–5}$$

I. SOLUBILITY PRODUCT

When an ionic solid is dissolved in a solvent, it dissociates and ionizes.

$$AgCl \rightleftharpoons Ag^+ + Cl^- \quad \text{(in water)}$$

When the equation for the equilibrium constant is written, the terms for solid components are omitted. When the equation for the ionization constant is written, the term for water concentration is also omitted. Thus, when an ionic solid is placed in water, the ionization constant will consist only of the ion concentrations. This ionization constant is known as the *solubility product*.

$$K_{sp} = [Ag^+][Cl^-]$$

As with the general case of ionization constants, the solubility product is essentially constant for slightly soluble solutes. Any time that the product of terms exceeds the standard value of the solubility product, solute will

**Table 7.1**
Approximate Ionization Constants (moles/liter)

| substance | 0°C | 5°C | 10°C | 15°C | 20°C | 25°C |
|---|---|---|---|---|---|---|
| HClO | 2.0 EE–8 | 2.3 EE–8 | 2.6 EE–8 | 3.0 EE–8 | 3.3 EE–8 | 3.7 EE–8 |
| $HC_2H_3O_2$ | 1.67 EE–5 | 1.70 EE–5 | 1.73 EE–5 | 1.75 EE–5 | 1.75 EE–5 | 1.75 EE–5 |
| HBrO | | | | | ≈ 2 EE–9 | |
| $H_2CO_3$ ($K_1$) | 2.6 EE–7 | 3.04 EE–7 | 3.44 EE–7 | 3.81 EE–7 | 4.16 EE–7 | 4.45 EE–7 |
| $HClO_2$ | | | | | ≈ 1.1 EE–2 | |
| $NH_3$ | 1.37 EE–5 | 1.48 EE–5 | 1.57 EE–5 | 1.65 EE–5 | 1.71 EE–5 | 1.77 EE–5 |
| $NH_4OH$ | | | | | | 1.79 EE–5 |
| $Ca(OH)_2$ | | | | | | 3.74 EE–3 |
| water* | 14.9435 | 14.7338 | 14.5346 | 14.3463 | 14.1669 | 13.9965 |

*$-\log_{10}(K)$ given.

precipitate out until the product of the remaining ion concentrations attains the standard value. If the product is less than the standard value, the solution is not saturated.

**Table 7.2**
Approximate Solubility Products

(moles/liter)

| substance | 12°F | 15°C | 18°C | 25°C |
|---|---|---|---|---|
| $Al(OH)_3$ | | 4 EE–13 | 1.1 EE–15 | 3.7 EE–15 |
| $CaCO_3$ | | 0.99 EE–8 | | 0.87 EE–8 |
| $CaF_2$ | | | 3.4 EE–11 | 4.0 EE–11 |
| $Fe(OH)_3$ | | | 1.1 EE–36 | |
| $Mg(OH)_2$ | | | 1.2 EE–11 | 1 EE–11 |
| $MgCO_3$ | 2.6 EE–5 | | | 1 EE–5 |
| $BaCO_3$ | | 7 EE–9 | | 8.1 EE–9 |
| $BaSO_4$ | | | 0.87 EE–10 | 1.1 EE–10 |
| $CaSO_4$ | 2 EE–4 | | | |
| $Fe(OH)_2$ | | | 1.6 EE–14 | 8 EE–6 |
| $MnCO_3$ | | | | 1.8 EE–11 |
| $Mn(OH)_2$ | | | 4 EE–14 | |
| $MgF_2$ | | | 7.1 EE–9 | 6.6 EE–9 |

*Example 7.12*

What is the solubility product of lead sulfate ($PbSO_4$) if its solubility is 38 mg/l?

$$PbSO_4 \rightleftharpoons Pb^{++} + SO_4^{--} \quad \text{(in water)}$$

The molecular weight of $PbSO_4$ is 207.19 + 32.06 + (4)(16) = 303.25

The number of moles of $PbSO_4$ in a liter of saturated solution is

$$0.038/303.25 = 1.25\,\text{EE–4}$$

This is also the number of moles of $Pb^{++}$ and $SO_4^{--}$ that will form in the solution. Therefore,

$$K_{sp} = [Pb^{++}][SO_4^{--}] = (1.25\,\text{EE–4})^2 = 1.56\,\text{EE–8}$$

The method used in example 7.12 to find the solubility product works well with chromates ($CrO_4^{--}$), halides ($F^-$, $Cl^-$, $Br^-$, $I^-$), sulfates ($SO_4^{--}$), and iodates ($IO_3^-$). However, sulfides ($S^{--}$), carbonates ($CO_3^{--}$), phosphates ($PO_4^{---}$), and the salts of transition elements (such as iron) hydrolyze and must be treated differently.

## 4 QUALITIES OF SUPPLY WATER

### A. ACIDITY AND ALKALINITY

*Acidity* is a measure of acids in solution. Acidity in surface water is caused by formation of *carbonic acid* ($H_2CO_3$) from carbon dioxide in the air.[1]

$$CO_2 + H_2O \rightarrow H_2CO_3 \qquad 7.15$$

$$H_2CO_3 + H_2O \rightarrow HCO_3^- + H_3O^+ \ (\text{pH} > 4.5) \qquad 7.16$$

$$HCO_3^- + H_2O \rightarrow CO_3^{--} + H_3O^+ \ (\text{pH} > 8.3) \qquad 7.17$$

Measurement of acidity is done by titration with a standard basic measuring solution. Acidity in water is typically given in terms of the $CaCO_3$ equivalent that would neutralize the acid.

$$\text{acidity (mg/l of } CaCO_3) = \frac{(\text{vol. titrant})(\text{titrant normality})(50{,}000)}{(\text{vol. sample})} \qquad 7.18$$

*Alkalinity* is a measure of the amount of negative ions in the water. Specifically, $OH^-$, $CO_3^{--}$, and $HCO_3^-$ all contribute to alkalinity.[2] The measure of alkalinity is the sum of concentrations of each of the substances measured as $CaCO_3$.

*Example 7.13*

Water from a city well is analyzed and is found to carry 20 mg/l as substance of $HCO_3^-$ and 40 mg/l as substance of $CO_3^{--}$. What is the alkalinity of this water?

From appendix A, the factors converting $HCO_3^-$ and $CO_3^{--}$ ions to $CaCO_3$ equivalents are 0.82 and 1.67 respectively.

$$\begin{aligned}\text{alkalinity} &= (0.82)(20) + (1.67)(40) \\ &= 83.2\,\text{mg/l as } CaCO_3\end{aligned}$$

Alkalinity can also be found by using an acidic titrant.

$$\text{alkalinity (mg/l of } CaCO_3) = \frac{(\text{vol. titrant})(\text{titrant normality})(50{,}000)}{(\text{vol. sample})} \qquad 7.19$$

Alkalinity and acidity of a titrated sample is determined from color changes in indicators added to the titrant. Table 7.3 lists indicators that are commonly used.

[1] Carbonic acid is very aggressive and must be neutralized to eliminate the cause of water pipe corrosion. If the pH of water is greater than 4.5, carbonic acid ionizes to form bicarbonate (equation 7.16). If the pH is greater than 8.3, carbonate ions form which cause water hardness by combining with calcium. (See equation 7.17.)

[2] Other radicals, such as $NO_3^-$, also contribute to alkalinity, but their presence is rare. If detected, they should be included in the calculation of alkalinity.

**Table 7.3**
Commonly Used Indicator Solutions

| common name | pH visual transition interval | color acidic | basic |
|---|---|---|---|
| cresol red | 0.2–1.8 | red | yellow |
| thymol blue | 1.2–2.8 | red | yellow |
| methyl yellow | 2.4–4.0 | red | yellow |
| bromophenol blue | 3.0–4.6 | yellow | blue |
| methyl orange | 3.2–4.4 | red | yellow-orange |
| methyl orange + xylene cyanole FF, 40 : 56 | (3.8–4.1)[a] | violet | green |
| bromocresol green | 3.9–5.4 | yellow | blue |
| methyl red | 4.2–6.2 | pink | yellow |
| methyl red + methylene blue, 1 : 1 | (~ 5.3)[a] | red-violet | green |
| bromocresol purple | 5.2–6.8 | yellow | purple |
| bromothymol blue | 6.0–7.6 | yellow | blue |
| cresol red | 7.2–8.8 | yellow | red |
| phenol red | 6.8–8.2 | yellow | red |
| thymol blue | 8.0–9.2 | yellow | blue |
| phenol-phthalein | (8.0–9.8)[b] | colorless | red-violet |
| phenol-phthalein + methylene green, 1 : 2 | (8.8)[c] | green | violet |
| Thymol-phthalein | (9.0–10.5)[b] | colorless | blue |
| eriochrome black T | 7–10 | blue | wine-red |
| alizarin yellow | 10.1–12 | yellow | red |

[a] Screened indicator, neutral gray at stated pH.

[b] Based on addition of 1 or 2 drops of a 0.1% indicator solution to 10 ml of aqueous solution.

[c] Screened indicator, pale blue at stated pH.

For strongly basic samples (pH > 8.3), consecutive titration with both phenolphthalein and methyl orange is often done. The alkalinity is the sum total of the *phenolphthalein alkalinity* and the *methyl orange alkalinity.* (Note: Phenolphthalein alkalinity is not the same as carbonate alkalinity since $CO_3^-$ has been converted to $HCO_3^-$ but not neutralized. See equation 7.17. The *carbonate alkalinity* is twice the phenolphthalein alkalinity.)

*Example 7.14*

0.02N sulfuric acid is used to titrate 110 ml of water. 3.3 ml of titrant is needed to reach the phenolphthalein point, and 13.2 ml is needed to reach the methyl orange point. What are the total and phenolphthalein alkalinities?

From equation 7.19, the phenolphthalein alkalinity is

$$(3.3)(0.02)(50{,}000)/(110) = 30\,\text{mg/l as } CaCO_3$$

The total alkalinity is

$$(13.2 + 3.3)(0.02)(50{,}000)/(110) = 150\,\text{mg/l as } CaCO_3$$

The alkalinity of 150 mg/l is caused by carbonates (2 × 30 = 60 mg/l) and bicarbonates (150 – 60 = 90 mg/l).

## B. HARDNESS

Water hardness is caused by multi-valent (doubly-charged, triply-charged, etc., but not singly-charged) positive metallic ions such as calcium, magnesium, iron, and manganese. (Iron and manganese are not as common, however.) Hardness reacts with soap to reduce its cleansing effectiveness, and to form scum on the water surface and ring around the bathtub.

Water containing bicarbonate ($HCO_3^-$) ions can be heated to precipitate a carbonate molecule.[3] This hardness is known as *temporary hardness* or *carbonate hardness.*[4]

$$Ca^{++} + 2HCO_3^- + \text{heat} \rightarrow CaCO_3 \downarrow + CO_2 + H_2O \qquad 7.20$$

$$Mg^{++} + 2HCO_3^- + \text{heat} \rightarrow MgCO_3 \downarrow + CO_2 + H_2O \qquad 7.21$$

Remaining hardness due to sulfates, chlorides, and nitrates is known as *permanent hardness* or *non-carbonate hardness* because it cannot be removed by heating. Permanent hardness can be calculated numerically by causing precipitation, drying, and then weighing the precipitate.

$$Ca^{++} + SO_4^{--} + Na_2CO_3 \rightarrow 2Na^+ + SO_4^{--} + CaCO_3 \downarrow \qquad 7.22$$

$$Mg^{++} + 2Cl^- + 2NaOH \rightarrow 2Na^+ + 2Cl^- + Mg(OH)_2 \downarrow \qquad 7.23$$

***Total hardness*** is the sum of temporary and permanent hardnesses, both expressed in mg/l as $CaCO_3$.

Hardness can also be measured by the titration method using a titrant (complexione, versene, EDTA, or BDH)

[3] Hard water forms scale when heated. This scale, if it forms in pipes, eventually restricts water flow. Even in small quantities, the scale insulates boiler tubes. Therefore, water used in steam-producing equipment must be essentially hardness-free.

[4] The hardness is known as *carbonate* hardness even though it is caused by *bicarbonate* radicals, not carbonate radicals.

and an indicator (such as eriochrome black T). The standard hardness reagent used for titration has an equivalent hardness of 1 mg/l per ml used.

The hardness of water can be classified according to table 7.4. Although high values of hardness are not organically dangerous, public acceptance of the water supply requires a hardness of well below 150 mg/l. Except for special industrial uses, potable water should have the carbonate hardness reduced to at least 40 mg/l, and the total hardness should be below 75 mg/l. Where it is economically feasible, the carbonate hardness should be reduced to 25 mg/l.

**Table 7.4**
Hardness Classifications

| class | type | hardness |
|---|---|---|
| A | soft | below 60 mg/l |
| B | medium hard | 60–120 |
| C | hard | 120–180 |
| D | very hard | 180–350 |
| E | saline, brackish | above 350 |

*Example 7.15*

A 75 ml water sample required 8.1 ml of EDTA. What is the hardness?

$$\text{hardness} = \frac{(8.1)\,\text{ml}(1)\,\text{mg/l}}{(75/1000)} = 108\,\text{mg/l}$$

*Example 7.16*

Water is found to contain sodium ($Na^{+}$, 15 mg/l), magnesium ($Mg^{++}$, 70 mg/l), and calcium ($Ca^{++}$, 40 mg/l). What is the hardness?

Sodium is singly-charged, so it does not contribute to hardness. The approximate equivalent weights of the relevant compounds and elements are:

$$\text{Mg} : 12 \quad \text{Ca} : 20 \quad \text{CaCO}_3 : 50$$

The equivalent hardness is

$$(70)\left(\frac{50}{12}\right) + (40)\left(\frac{50}{20}\right) = 392\,\text{mg/l as CaCO}_3$$

Alternatively, appendix A could have been used to convert the ionic concentrations to $CaCO_3$ equivalents.

$$(70)(4.10) + (40)(2.50) = 387\,\text{mg/l as CaCO}_3$$

## C. IRON CONTENT

Even in low concentrations, iron is objectionable because it stains bathroom fixtures, causes a brown color in laundered clothing, and affects taste. Water originally pumped from anaerobic sources may contain ($Fe^{++}$) ferrous ions which are invisible and soluble. When exposed to oxygen, insoluble ($Fe^{+++}$) ferric oxides form which give water the rust coloration.

Iron is measured optically by comparing the color of a sample with standard colors. The comparison can be made by eye or with a photoelectric *colorimeter.* Iron concentrations greater than 0.3 mg/l are undesirable.

## D. MANGANESE CONTENT

Manganese ions are similar in effect, detection, and measurement to iron ions. Manganous manganese ($Mn^{++}$) oxidizes to manganic manganese ($Mn^{++++}$) to give water a rust color. An undesirable concentration is 0.05 mg/l.

## E. FLUORIDE CONTENT

An optimum concentration of fluoride in the form of a fluoride ion, $F^{-}$, is between 0.8 mg/l for hot climates (80°F–90°F average) to 1.2 mg/l for cold climates (50°F average). These amounts reduce the population cavity rate to a minimum without producing significant fluorosis (staining) of the teeth. The actual amount of fluoridation depends on the average outside temperature since the temperature affects the amount of water that is ingested by the population.

**Table 7.5**
Maximum Fluoride Concentrations

(Note that the 1974 Safe Drinking Water Act and its 1986 amendments set the maximum fluoride concentration at 4 mg/l for all temperatures.)

| 5-year average of maximum daily air temperatures | fluoride concentrations |
|---|---|
| (deg F) | (mg/l) |
| 50.0–53.7 | 2.4 |
| 53.8–58.3 | 2.2 |
| 58.4–63.8 | 2.0 |
| 63.9–70.6 | 1.8 |
| 70.7–79.2 | 1.6 |
| 79.3–90.5 | 1.4 |

Fluoridation can be obtained by the readily dissociating compounds in table 7.6.

**Table 7.6**
Fluoridation Chemicals

| | |
|---|---|
| $(NH_4)_2SiF_6$ | (ammonium silicofluoride) |
| $CaF_2$ | (calcium fluoride) |
| $H_2SiF_6$ | (fluosilic acid) |
| NaF | (sodium fluoride) |
| $Na_2SiF_6$ | (sodium silicofluoride) |

Fluoride content is measured by colorimetric and electrical methods.

## F. CHLORIDE CONTENT

Chlorine is used as a disinfectant for water, but its strong oxidation potential allows it also to be used to remove iron and manganese ions. Chlorine gas in water forms hypochlorous and hydrochloric acids.

$$Cl_2 + H_2O \underset{pH<4}{\overset{pH\geq 4}{\rightleftharpoons}} HCl + HOCl \underset{pH<9}{\overset{pH\geq 9}{\rightleftharpoons}} H^+ + OCl^- \qquad 7.24$$

Free chlorine, hypochlorous acid, and hypochlorite ions are known as *free chlorine residuals.* Hypochlorous acid reacts with ammonia (if it is present) to form mono-, di-, and trichloramines. Chloramines are known as *combined residuals.* Chloramines are more stable than free residuals, but their disinfecting ability is weaker. Their action may extend for a considerable distance into the distribution system.

The amount of chlorine to be added depends on the organic and inorganic matter present in the water. However, most waters are effectively treated within 10 minutes if a free residual of 0.2 mg/l is maintained. Larger residual concentrations may cause objectional odor and taste.

If the water contains phenol, it and the chlorine will form chlorophenol compounds which produce an objectionable taste. This may be stopped by adding ammonia to the water before chlorination.

Both free and combined residual chlorine can be detected by color comparison. However, organic matter in waste water makes it necessary to use a test based on water conductivity. The color comparison test with supply water, however, is adequate.

## G. PHOSPHORUS CONTENT

*Orthophosphates* ($H_2PO_4^-$, $HPO_4^{--}$, and $PO_4^{---}$) and *polyphosphates* (such as $Na_3(PO_3)_6$) result from the use of *synthetic detergents* (*syndets*). Phosphate content is more of a concern in waste water than in supply water. Excessive phosphate discharge contributes to aquatic plant growth and subsequent *eutrophication.*

Phosphates are measured by a variety of means, including colorimetry and filtered precipitation analysis. Care should be taken not to confuse phosphates with phosphorus. Multiply mg/l of phosphate by 0.326 to obtain mg/l of phosphorus.

## H. NITROGEN CONTENT

Nitrogen is present in water in many forms, including organic (protein), ammonia, nitrate, and gaseous ammonia. As with phosphates, nitrogen contamination is more of a problem with waste water than with supply water. Nitrogen pollution promotes algae growth. Ammonia is toxic to fish.

Drinking water is typically tested only for nitrates. The following tests are used:

**Table 7.7**
Tests for Nitrogen

| to test for | procedure |
|---|---|
| ammonia | distillation |
| organic nitrogen | digestion with distillation |
| nitrate, nitrite | colorimetry |

Gaseous nitrogen is of little concern since it is not normally metabolized by plants and it is of no danger to animal or human life.

## I. COLOR

Color in domestic water is undesirable aesthetically, and it may dull the color of clothes and stain bathroom fixtures. Some industries (such as beverage production, dairy, food processing, paper manufacturing, and textile production) also have strict water color standards.

Water color is measured with a colorimeter or comparitively with tubes containing standard platinum/cobalt solutions. Color is graded on a scale of 0 (clear) to 70 color units.

## J. TURBIDITY

Turbidity is a measure of the insoluble solids (soil, organics, and microorganisms) in water which impede light passage. Completely clean water measures 0 *turbidity units* (NTU). 5 NTU is noticeable to an average consumer, and this is a practical upper limit for drinking water. Muddy water exceeds 100 NTU. A TU is equivalent to 1 mg/l of silica in suspension.

Turbidity is usually measured with a *nephelometer*, *Jackson candle apparatus*, or *Baylis turbidimeter*. Units can be NTU (nephelometer turbidity units) or JTU (Jackson turbidity units).

### K. SUSPENDED AND DISSOLVED SOLIDS

Solids present in a sample of drinking water can be divided into several categories, not all of which are mutually exclusive.

- **Suspended solids**: Suspended solids, the same as *filterable solids*, are measured by filtering a sample of water and weighing the residue.
- **Dissolved solids**: Dissolved solids, same as *non-filterable solids*, are measured as the difference between total solids and suspended solids.
- **Total solids**: Total solids are made up of suspended and dissolved solids. They are measured by drying a sample of water and weighing the residue.
- **Volatile solids**: Volatile solids are measured as the decrease in weight of total solids which have been ignited in an electric furnace.
- **Fixed solids**: The fixed solids can be found as the difference between total solids and volatile solids.
- **Settleable solids**: The volume (ml/l) of settleable solids is measured by allowing a sample to stand for one hour in a graduated conical container (*Imhoff cone*).

An upper limit of 500 mg/l of total solids is recommended.

**Table 7.8**
Water-Borne Organisms

| this organism | causes this disease |
|---|---|
| BACTERIA | |
| Salmonella typhosa | typhoid fever |
| Vibrio comma | cholera |
| Shigella dysenteriae | dysentery |
| Escherichia coli | enteric problems |
| fecal streptococci | enteric problems |
| VIRUSES | |
| Poliomyelitis | polio |
| Infectious hepatitis | hepatitis |
| PROTOZOA | |
| Entamoeba histolytica | dysentery |
| PARASITES | |
| flatworms | Schistosomiasis |
| flatworms | Bilharziasis |

### L. WATER-BORNE DISEASES

Organisms that are present in water consist of bacteria, fungi, viruses, algae, protozoa, and multicellular animals. Not all of these are dangerous, but some are. Important organisms are listed in table 7.8.

## 5 TRIHALOMETHANES

Trihalomethanes (THM's) are organic chemicals produced during the disinfection of water. The chemically active elements of chlorine, iodine, and bromine react with various organic precursors to produce THM's. However, iodine is seldom used in disinfection. Therefore, only four THM's are found in significant quantities:[5]

- $CHCl_3$ – trichloromethane (chloroform)
- $CHBr_3$ – tribromomethane (bromoform)
- $CHBrCl_2$ – bromodichloromethane
- $CHBr_2Cl$ – dibromochloromethane

The *organic precursors* which react with chlorine to produce THM's tend to be naturally occurring. For example, decaying vegetation produces humic and fulvic acids which are natural precursors. The precursors in themselves are not harmful, but the THM's produced from them have been shown to be carcinogenic.[6]

Table 7.9 lists the maximum contaminant level (MCL) for THM's in drinking water. Communities with populations of 10,000 and above are covered by the MCL if they add a disinfectant to their drinking water supply. The MCL reported in table 7.9 is for total THM (TTHM), not just chloroform. Precursors are not limited or monitored.

When THM levels need to be reduced, several options are available. These options fall into two categories, depending on whether the precursors are removed prior to chlorination, or a disinfectant is chosen that does not produce THM's. The first category includes the following options:

- Using *granular activated carbon* (GAC), other adsorbents, or filters, including weak-base resins, to remove precursors
- Selecting a water source with fewer precursors

[5] Bromine can be present in gaseous chlorine as an impurity. Bromine also results from reacting chlorine with the bromide present in high-salinity water.

[6] Actually, tests have shown that only chloroform in high doses is carcinogenic to rats and mice. The other THM's are considered carcinogenic by association.

- Moving the chlorination point to the end of the treatment process so that most precursors are removed prior to disinfection (70–75% TTHM reduction)
- Optimizing coagulation and settling processes to improve precursor removal

Within the second category are the following options:

- Using ozone, chlorine dioxide, or potassium permanganate to disinfect without THM formation (60–90% TTHM reduction)
- Dechlorination using *sodium metabisulfate* or other methods after chlorination to prevent the reaction of chlorine with precursors
- Adding ammonia to water prior to discharge to induce chloramine formation, since chloramines suppress the formation of THM's

25–60% TTHM's can also be removed after formation by contacting with granular activated carbon.

Caution is required when changing to alternate disinfectants. The disinfectant and its byproducts should be evaluated to determine disinfecting power, residual power, toxicity, and other health effects.[7]

Costs of operation will increase when alternate disinfectants are used. Moving the point of application may not result in any significant operating costs after a modest capital expenditure is made. Cost of using ozone as an alternative disinfectant is often less than using chlorine dioxide, but it is more than using chloramines or changing the points of chlorine application.

[7] Ozone does not form any potentially dangerous byproducts.

## 6 COMPARISON OF ALKALINITY AND HARDNESS

Hardness measures the presence of $Mg^{++}$, $Ca^{++}$, $Fe^{++}$, and other multi-valent ions. Alkalinity measures the presence of $HCO_3^-$, $SO_4^{--}$, $Cl^-$, $NO_3^-$, and $OH^-$ ions. Both positive and negative ions can exist side by side, so an alkaline water can also be hard.

If certain assumptions are made, then it is possible to draw conclusions about the water composition. For example, $Fe^{++}$ is an unlikely ion in most water supplies, and it is often neglected in comparing alkalinity and hardness.

Figure 7.1 gives an easy method of comparing hardness and alkalinity, and using the comparison to deduce other compounds in the water.

If hardness (as $CaCO_3$) and alkalinity (also as $CaCO_3$) are the same, then there are no $SO_4^{--}$, $Cl^-$, or $NO_3^-$ ions present. (That is, there is no non-carbonate, permanent hardness.) If hardness is greater than alkalinity, however, then non-carbonate, permanent hardness is present, and the temporary carbonate hardness is equal to the alkalinity. If hardness is less than alkalinity, then all hardness is carbonate, temporary hardness, and the extra $HCO_3^-$ comes from other sources (such as $NaHCO_3$).

M = alkalinity
H = total hardness
Ca = calcium
O = hydroxides
S = sulfate hardness
L = free lime

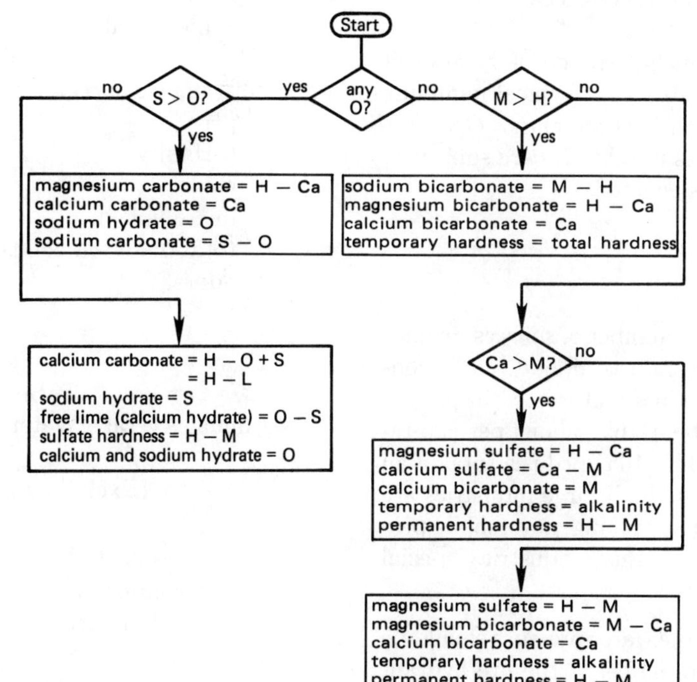


**Figure 7.1** Hardness and Alkalinity
(All results expressed as $CaCO_3$)

*Example 7.17*

A sample of water has been found to contain the following:

alkalinity: 220 mg/l as $CaCO_3$

hardness: 180 mg/l as $CaCO_3$

calcium ($Ca^{++}$): 140 mg/l as $CaCO_3$

(a) What is the non-carbonate hardness?

(b) What is the $Mg^{++}$ content in mg/l as substance?

To use figure 7.1, the absence of any significant hydroxides must be assumed. Since the alkalinity is greater than the hardness, the figure indicates the following compounds in the water:

$$NaHCO_3 = 220 - 180 = 40 \text{ mg/l as } CaCO_3$$
$$Mg(HCO_3)_2 = 180 - 140 = 40 \text{ mg/l as } CaCO_3$$
$$Ca(HCO_3)_2 = 140 \text{ mg/l}$$

There is no non-carbonate hardness in the water.

The $Mg^{++}$ ion content in $CaCO_3$ is equal to the $Mg(HCO_3)_2$ content as $CaCO_3$. Appendix A can be used to convert $CaCO_3$ equivalents to amounts as substance.

$$Mg^{++} = \frac{40}{4.1} = 9.6 \text{ mg/l as substance}$$

## 7 WATER QUALITY STANDARDS

Minimum drinking water quality standards have been set by the Safe Drinking Water Act. Typical minimum standards are given in table 7.9 as *Maximum Contaminant Levels (MCL's)*. Values in table 7.9 are subject to change as new legislation is enacted.

**Table 7.9**
Typical Water Quality Standards
(subject to change)

| contaminant/quality | MCL |
|---|---|
| inorganic compounds | |
| arsenic | 0.05 mg/l |
| barium | 1.0 mg/l |
| cadmium | 0.01 mg/l |
| chloride | 250 mg/l |
| chromium | 0.05 mg/l |
| copper | 1.0 mg/l |
| cyanide | 0.005 mg/l |
| iron | 0.3 mg/l |
| lead | 0.05 mg/l |
| manganese | 0.05 mg/l |
| mercury | 0.002 mg/l |
| nitrate | 10 mg/l |
| selenium | 0.01 mg/l |
| silver | 0.05 mg/l |
| sulfate | 250 mg/l |
| zinc | 5.0 mg/l |
| organic compounds | |
| trihalomethanes (total) | 0.1 mg/l |
| organic pesticides | |
| endrin | 0.0002 mg/l |
| lindane | 0.004 mg/l |
| methoxychlor | 0.1 mg/l |
| toxaphene | 0.005 mg/l |
| 2,4-D | 0.1 mg/l |
| 2,4,5-TP(silvex) | 0.01 mg/l |
| miscellaneous regulations | |
| pH | 6.5–8.5 |
| turbidity | 1 NTU |
| color | 15 units |
| microbiological | 1 coliform/100 ml |
| total dissolved solids | 500 mg/l |
| odor | 3 tons |

## 8 WATER DEMAND

Water demand comes from a number of sources, including residential, commercial, industrial, and public consumers, as well unavoidable loss and waste. In project planning, a minimum of about 165 gallons per capita-day should be considered. This 165 gpcd is a total of all demands, as given in table 7.10. If large industries are present (such as canning, steel making, automobile production, electronics, etc.), then those industries' special needs must also be considered.

For ordinary domestic use, the water pressure should be 25 to 40 psi. A minimum of 60 psi at the fire hydrant is usually adequate, since that allows for up to 20 psi pressure drop in fire hoses. 75 psi and higher is common in commercial and industrial districts.

**Table 7.10**
Annual Average Water Requirements (gpcd)

(Excluding fire fighting)

| | |
|---|---|
| residential | 75–130 |
| commercial & industrial | 70–100 |
| public | 10– 20 |
| loss & waste | 10– 20 |
| | 165–270 total |

Variations can be expected with the time of day and season. If the average daily demand is to be used to estimate peak demands or fluctuation, then table 7.11 lists some multipliers. These multipliers are to be used against the 165 gpcd (or whatever other average is available).

**Table 7.11**
Demand Multipliers For Peak Periods

| consumption time/period | multiplier |
|---|---|
| winter | 0.80 |
| summer | 1.30 |
| maximum daily | 1.50–1.80 |
| maximum hourly | 2.00–3.00 |
| early morning | 0.25–0.40 |
| noon | 1.50–2.0 |

Water demand (in gpcd) must be multiplied by the population to obtain the total demand. Since a population changes, a supply system must be designed to handle demands through a reasonable time into the future. Several methods exist for estimating future demand.

- Uniform growth rate
- Constant percentage growth rate (e.g., 40% per decade)
- Decreasing growth rate (e.g., 6% each decade)
- Straight-line graphical extension
- Comparison with neighboring cities

Typical growth factors are 1.25 for large systems and 1.50 for small systems. Economic aspects of the project dictate the number of years into the future which should be designed for.

Since the maximum demand can be up to 3 times the average daily demand (table 7.11), the design rate should be 3 times the average daily rate plus an allowance for fire fighting. If the water treatment plant's capacity is fixed, the distribution system should be able to handle the plant's capacity plus an allowance for fire fighting (which can be passed around the treatment plant).

The requirements for fire fighting at any point will vary between 500 gpm (a minimum) to 12,000 gpm for a single fire.[8] Multiple fires will place a greater demand on the distribution system. A municipality must continue to serve its domestic, commercial, and industrial customers during a fire, however. The Insurance Services Office recommends that the fire system be able to operate with the remainder of the potable water system operating at the maximum daily rate, as taken over all 24 hour periods within the last three years.

Recommended fire flow in a neighborhood will depend on construction type, occupancy, and floor area. An estimate for a neighborhood can be found from equation 7.25, as proposed by the Insurance Services Office.

$$Q = (0.04)C\sqrt{A} \qquad 7.25$$

$C$ is a constant which depends on construction: 1.5 for wood frame, 1.0 for ordinary construction, 0.8 for noncombustible construction, and 0.6 for fire resistant construction. $A$ is the area (in square feet) of all stories in the building, except for basements. Special rules are used to find $A$ for multi-story fire-resistant structures, buildings with various fire loadings, or buildings with sprinkler systems. $Q$ is rounded to the nearest 250 gpm, but it should not be less than 500 gpm or more than 8000 gpm for a single building.

In estimating the water requirements for fire fighting on a population basis, the American Insurance Association has recommended the following formula:

$$\begin{aligned} Q &= 2.27\sqrt{P}(1 - 0.01\sqrt{P})\,\text{cfs} \\ &= 1020\sqrt{P}(1 - 0.01\sqrt{P})\,\text{gpm} \qquad 7.26 \end{aligned}$$

(P in thousands of people)

Most insurance requirements will be met if the flow rate can be maintained for $T$ hours, where $T$ is the flow rate in 1000's of gallons per minute, with a maximum of 10 hours.

Fire hydrants are spaced at a maximum distance of 500 feet. They are ordinarily located at street corners where use from four directions is possible. The actual separation of hydrants can be calculated from standards presented by the Insurance Services Offices. These standards are presented in table 7.12.

The Insurance Services Office also suggests durations that various fire flow rates must be able to be maintained. Critical equipment such as power sources, pumps, supply mains, and treatment processes should be able to provide several days of peak load in addition to the flow for fire fighting for the duration specified.[9]

## 9 METHODS OF WATER DISTRIBUTION

Several methods are used to distribute water, depending on terrain, economics, and other local conditions.

*Gravity distribution* is available when a lake or reservoir is located significantly higher in elevation than the

[8] A standard 1.125″ smooth nozzle discharges approximately 250 gpm of water.

[9] It should be possible to maintain peak flow for 2–5 days, with the actual duration depending on the component, redundancy, and expected repair time.

**Table 7.12**
Standard Fire Hydrant Distribution

| fire flow required (gpm) | minimum average area per hydrant (sq ft) |
|---|---|
| 1000 or less | 160,000 |
| 1500 | 150,000 |
| 2000 | 140,000 |
| 2500 | 130,000 |
| 3000 | 120,000 |
| 3500 | 110,000 |
| 4000 | 100,000 |
| 4500 | 95,000 |
| 5000 | 90,000 |
| 5500 | 85,000 |
| 6000 | 80,000 |
| 6500 | 75,000 |
| 7000 | 70,000 |
| 7500 | 65,000 |
| 8000 | 60,000 |
| 8500 | 57,500 |
| 9000 | 55,000 |
| 10,000 | 50,000 |
| 11,000 | 45,000 |
| 12,000 | 40,000 |

**Table 7.13**
Required Durations for Fire Fighting

| required fire flow (gpm) | required duration (hr) |
|---|---|
| 10,000 and greater | 10 |
| 9500 | 9 |
| 9000 | 9 |
| 8500 | 8 |
| 8000 | 8 |
| 7500 | 7 |
| 7000 | 7 |
| 6500 | 6 |
| 6000 | 6 |
| 5500 | 5 |
| 5000 | 5 |
| 4500 | 4 |
| 4000 | 4 |
| 3500 | 3 |
| 3000 | 3 |
| 2500 or less | 2 |

city. The high hydraulic head that results can be easily maintained for domestic and fire fighting mains.

Distribution by pumping with storage is the most desirable option if gravity distribution is not used. Excess water is pumped into elevated storage during periods of low consumption. During periods of high consumption, water is drawn from the storage to augment the pumped water. More uniform pumping rates result, and the pumps are able to run near their rated capacity most of the time.

Using pumps without storage to force water directly into mains is the least desirable option. Pumps must be able to maintain the flow during peak consumption, and pumps will not always run in their economical capacity ranges. In the event of a power outage, all water supply will be lost unless backup power is available.

Water mains in a distribution system should be at least 6 inches in diameter.

## 10 STORAGE OF WATER

Water is stored to equalize pumping rates, to equalize supply and demand over periods of high consumption, and to furnish extraordinary volumes during emergencies such as fires.

To equalize the pumping rate during the day will ordinarily require storage oı 15 to 30 percent of the maximum daily use. Storage for emergencies is more difficult to determine, and is dictated by economic benefits to the public. Fire insurance rates are generally lower the greater the emergency storage capacity is.

## 11 PIPE MATERIALS

Many types of pipe are available. Successful pipe materials used for distribution must have adequate strength to withstand external loads from backfill, traffic, and earth movement; high burst strength to withstand water pressure; smooth interior surfaces; corrosion resistant exteriors; and tight joints.

The types of pipes listed in table 7.14 are suitable for use in the water distribution system.

## 12 LOADS ON BURIED PIPES

If a pipe is buried (placed in an excavated trench and backfilled), it must support an external vertical load in addition to its internal pressure load. The magnitude of the load depends on the amount of backfill, type of soil, and type of pipe. For rigid pipes (concrete, cast iron, and clay) which cannot deform and which are placed in narrow trenches (2 or 3 diameters), the load in pounds per foot of pipe is given by equation 7.27.[10]

$$w = C\rho B^2 \qquad 7.27$$

[10] Equation 7.27 is known as *Marston's formula.*

**Table 7.14**
Water Pipe Materials

| type | comments |
|---|---|
| ductile and gray cast iron | Long life, strong, impervious. High cost and heavy. May be coated to resist exterior and interior corrosion. Available in 4″ to 54″ standard sizes. 350 psi working pressure. |
| asbestos-cement | Immune to electrolysis and corrosion. Low flexural strength. Smooth interior surface. Available 4″ to 42″ diameter. Up to 200 psi working pressure. |
| concrete | Durable, watertight, low maintenance, smooth interior surface. Diameters 16″ to 144″, 50 psi (plain) and 250 psi (reinforced) working pressures. |
| steel | High strength, good yielding and shock resistance, but susceptible to corrosion. Exterior may be tarred, painted, or wrapped. Interior may have enamel or cement mortar lining. Smooth interior surface. 16″ to 120″ diameters, 250 psi pressures. |
| plastic | Chemically inert, corrosion resistant, smooth interior. PVC most popular. PVC available in rating to 315 psi, diameters 1/2″ to 16″. |

Typical values of $C$ and $\rho$ are given in table 7.15.[11] $B$ is the trench width in feet at the top of the pipe. (A minimum trench width to allow working room is commonly estimated at 4/3 times the pipe diameter plus 8″.)

[11] There is considerable literature on the coefficients used in equations 7.27, 7.29, and 7.30. The values listed in this chapter are representative, but are not intended to cover every case. In most instances, other factors may be necessary to correctly select the coefficients.

The dead load pressure is simply the experienced load divided by the trench width.

$$p_{DL} = \frac{w}{B} \qquad 7.28$$

*Flexible pipes* (steel, plastic, copper) are sufficiently flexible to develop horizontal restraining pressures equal to the vertical pressures if the backfill is well compacted.

$$w = C\rho BD \qquad 7.29$$

**Table 7.15**
Approximate Pipe Load Correction Coefficients

| backfill material | cohesionless granular material | sand & gravel | saturated topsoil | clay | saturated clay |
|---|---|---|---|---|---|
| density | 100 | 100 | 100 | 120 | 130 |
| h/B | | | values of C | | |
| 1 | 0.82 | 0.84 | 0.86 | 0.88 | 0.90 |
| 2 | 1.40 | 1.45 | 1.50 | 1.55 | 1.62 |
| 3 | 1.80 | 1.90 | 2.00 | 2.10 | 2.20 |
| 4 | 2.05 | 2.22 | 2.33 | 2.49 | 2.65 |
| 5 | 2.20 | 2.45 | 2.60 | 2.80 | 3.03 |
| 6 | 2.35 | 2.60 | 2.78 | 3.04 | 3.33 |
| 7 | 2.45 | 2.75 | 2.95 | 3.23 | 3.57 |
| 8 | 2.50 | 2.80 | 3.03 | 3.37 | 3.76 |
| 10 | 2.55 | 2.92 | 3.17 | 3.56 | 4.04 |
| 12 | 2.60 | 2.97 | 3.24 | 3.68 | 4.22 |
| ∞ | 2.60 | 3.00 | 3.25 | 3.80 | 4.60 |

If a pipe is placed on undisturbed ground and covered with fill (*broad fill* or *embankment fill*) the load is

$$w = C_p \rho D^2 \tag{7.30}$$

**Table 7.16**
Representative Values of $C_p$

| h/D | rigid pipe, rigid surface, noncohesive backfill | flexible pipe, average conditions |
|---|---|---|
| 1 | 1.2 | 1.1 |
| 2 | 2.8 | 2.6 |
| 3 | 4.7 | 4.0 |
| 4 | 6.7 | 5.4 |
| 6 | 11.0 | 8.2 |
| 8 | 16.0 | 11.0 |

Equation 7.27 shows that the trench width is an important parameter in determining whether or not the pipe is overloaded. There is a depth for each conduit size beyond which no additional load is transmitted to the conduit, regardless of trench width. This limiting value is known as the *transition width.* Specifically, the load on a rigid pipe (e.g., cast iron, concrete, ductile iron) can never exceed the value calculated from equation 7.30. The transition width can be calculated by equating equations 7.27 and 7.30.

$$B_{\text{transition}} = D\sqrt{\frac{C_p}{C}} \tag{7.31}$$

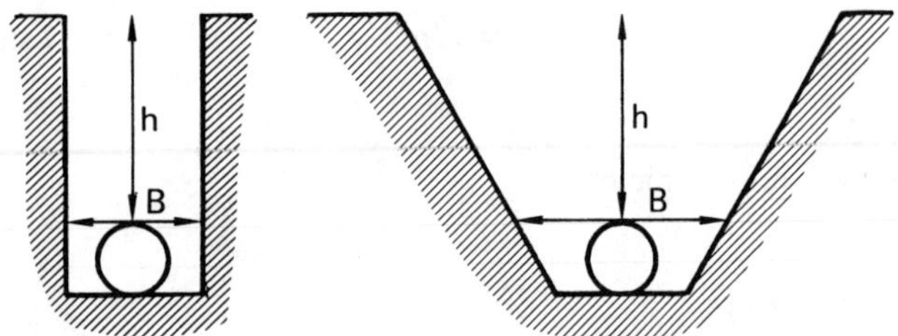


**Figure 7.2** Backfilled Trenches

*Boussinesq's equation* should be used to calculate the load on a pipe due to a superimposed line load, $P$, at the surface. This load should be added to the loadings calculated from equations 7.27, 7.29, and 7.30.

$$p = \frac{3h^3P}{2\pi z^5} = \frac{3P}{2\pi h^2}\left[\frac{1}{1+\left(\frac{r}{h}\right)^2}\right]^{5/2} \tag{7.32}$$

$$w = Dp \tag{7.33}$$

If the pipe has less than 3 feet of cover, a multiplicative impact factor should also be used.

$$w = F_I D p \tag{7.34}$$

**Table 7.17**
Impact Factors

| | |
|---|---|
| BY DEPTH (general use) | |
| less than 1 foot | 1.3 |
| 1 to 2 feet | 1.2 |
| 2 to 3 feet | 1.1 |
| more than 3 feet | 1.00 |
| BY USE | |
| highway | 1.50 |
| railway | 1.75 |
| airfield runway | 1.00 |
| airfield taxiway, apron | 1.50 |

**Figure 7.3** External Loads on Buried Pipes

Concrete pipes are tested in a 3-edge bearing mechanism, as shown in figure 7.4, to determine the *crushing strength.* Crushing strength (taken as the load which produces a 0.01″ crack) is given in pounds per foot of pipe per foot of inside diameter. Therefore, crushing strength (in pounds per linear foot of pipe) is

$$\text{crushing strength} = (\text{load per unit diameter})(\text{diameter}) \tag{7.35}$$

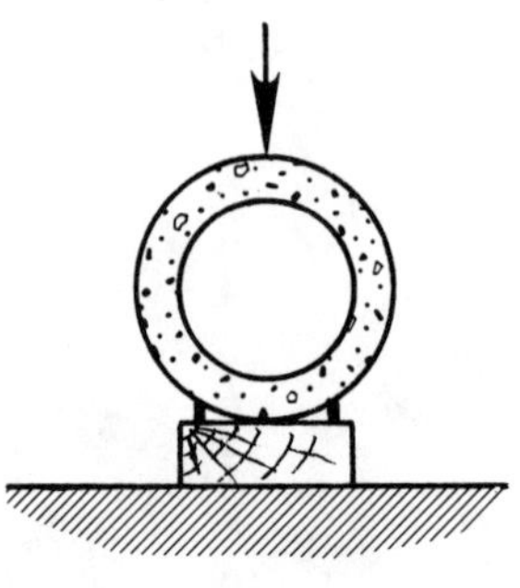

**Figure 7.4** Three-Edge Bearing Test

The load on the pipe can be decreased below the crushing strength by proper bedding. Different bedding methods and their load factors are given in figure 7.5. The allowable load per foot of pipe is given by equation 7.36. The factor of safety varies from 1.25 for flexible pipe to 1.50 for rigid pipe. For reinforced concrete, the factor of safety is 1.0.

$$\text{allowable load per foot of pipe} = \frac{(\text{crushing strength})(\text{load factor, LF})}{\text{factor of safety}} \quad 7.36$$

## 13 WATER SUPPLY TREATMENT METHODS

### A. AERATION

Aeration can be used where there is a high concentration of carbon dioxide, where tastes and odors are objectionable, and where iron and manganese are present in amounts above 0.3 ppm.

Typical aerating devices are listed in table 7.18. Greatest efficiencies can be achieved by designing for increased water surface exposed to air, rapid change of air in contact with the water, and increased aeration periods.

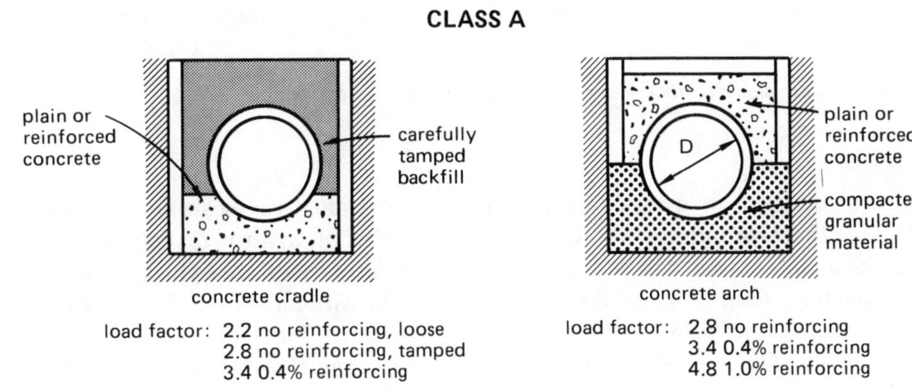


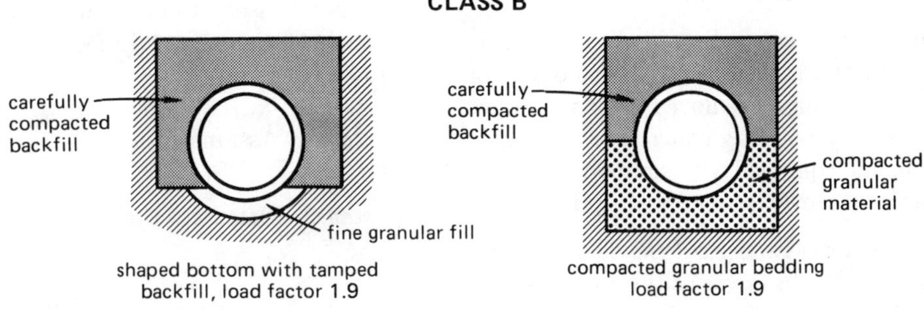


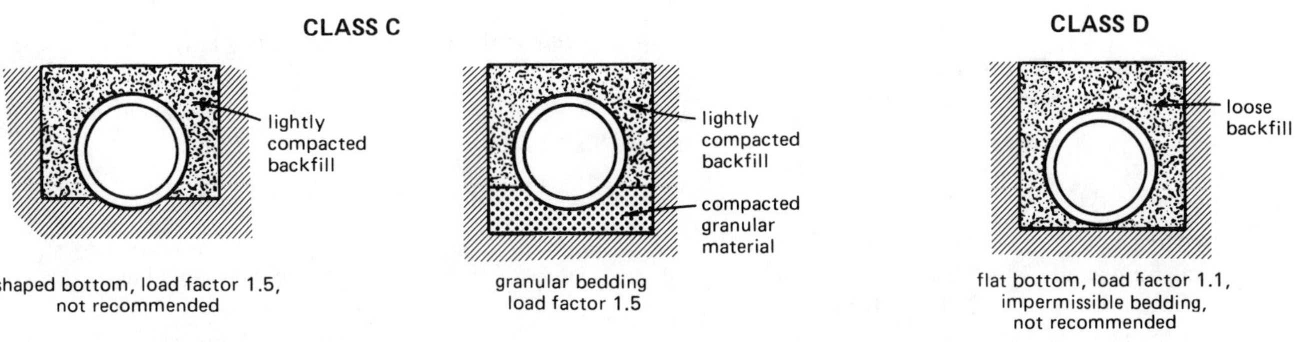


**Figure 7.5** Bedding Classes and Load Factors

**Table 7.18**
Characteristics of Aerators

| type | operating head (ft) | loading capacity | efficiency in $CO_2$ removal | remarks |
|---|---|---|---|---|
| spray | 8–28 | 4–180 gpm nozzle | high | requires protection from loss of water by wind; ice hazard in cold climates |
| cascade | 3–10 | 20–50 gpm/sq ft | low to | requires large space |
| perforated tray | | | fair | requires larger space and higher head than coke tray |
| coke tray | 6–10 | <35 gpm sq ft | high | used also for iron and manganese removal |
| forced draft | 10–25 | 16–18 gpm sq ft | high | compact but more complex than above types |
| diffused air | 5–10 psi[1] | 0.02–0.2 cfm/gpm[2] | high | requires compressed air; most complex |

[1]Air pressure depends upon water depth and pipe friction losses.

[2]Air requirement.

Transfer efficiencies of diffused air systems vary with depth and bubble size. If coarse bubbles are produced, only 4% to 8% of the available oxygen will be transferred to the water. With medium-sized bubbles, the *transfer efficiency* varies between 6% and 15%. Fine bubble systems are capable of transferring 10% to 30% of the supplied oxygen to the water.

## B. PLAIN SEDIMENTATION (CLARIFICATION)

Water contaminated with sand, dirt, mud, etc., can be treated in a sedimentation basin or tank. Up to 80% of the incoming sediment can be removed in this manner. Sedimentation basins are usually concrete, rectangular or circular in plan, and equipped with scrapers or raking arms to periodically remove accumulated sludge.

Settlement of water-borne particles depends on the water temperature (which affects viscosity), particle size, and particle specific gravity. Typical specific gravities are given in table 7.19.

**Table 7.19**
Properties of Particles

| particle type | specific gravity | particle size, mm |
|---|---|---|
| coarse sand | 2.65 | 0.50–2.00 |
| medium sand | 2.65 | 0.25–0.50 |
| fine sand | 2.65 | 0.10–0.25 |
| very fine sand | 2.65 | 0.05–0.10 |
| silt | 2.65 | 0.005–0.05 |
| clay | – | 0.001–0.005 |
| flocculated mud | 1.03 | – |

Settlement time can be calculated from the settling velocity, flow-through velocity, and depth of the tank. Tank depths are typically 6 to 15 feet. The settling velocities for spherical particles in 68°F water are given in figure 7.6. Of course, settling velocities will be much less than those shown in figure 7.6 because actual sediment particles are not spherical.

If it is necessary to calculate the settling velocity of a particle of diameter $d$ (in feet), the following procedure can be used.

*step 1*: Assume $v_s$.

*step 2*: Calculate the Reynolds number.

$$N_{Re} = \frac{v_s d}{\nu} \tag{7.37}$$

*step 3*: If $N_{Re} < 1$, use Stoke's law.

$$v_s = \frac{(\rho_{\text{particle}} - \rho_{\text{water}})d^2 g}{18\mu g_c} \tag{7.38}$$

If $1 < N_{Re} < 2000$, use figure 7.6.

If $N_{Re} > 2000$, use Newton's law.

$$v_s = \sqrt{\frac{4g(\rho_{\text{particle}} - \rho_{\text{water}})d}{3(\rho_{\text{water}})C_D}} \tag{7.39}$$

Values of $C_D$ are given in table 7.20.

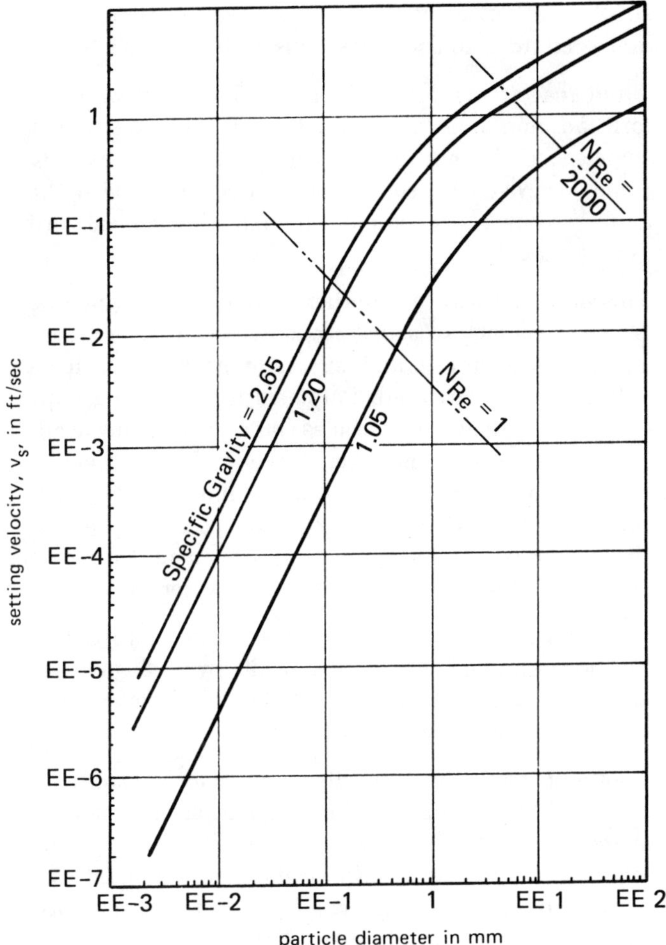


**Figure 7.6** Settling Velocities (spherical particles, 68°F water)

**Table 7.20**
Approximate Drag Coefficients for Spheres

| $N_{Re}$ | $C_D$ |
|---|---|
| 2000 | 0.4 |
| 10,000 | 0.4 |
| 50,000 | 0.5 |
| 100,000 | 0.5 |
| 200,000 | 0.4 |

If it is assumed that the water velocity is a uniform $v_f$, then all particles with $v_s > v^*$ will be removed. $v^*$ is known as the *overflow rate* (*surface loading* or *critical velocity*) and typically has a value of 600 to 1000 gpd/ft$^2$ for rectangular basins. For square and circular basins, the surface loading should be within 500–750 gpd/ft$^2$. $B$ is the tank width, typically 30 to 40 feet, and $L$ is the length, typically 100 to 200 feet. (For radial flow circular basins, a typical diameter is 100 feet.)

$$v^* = \frac{Q}{A_{\text{surface}}} = \frac{Q}{BL} \qquad 7.40$$

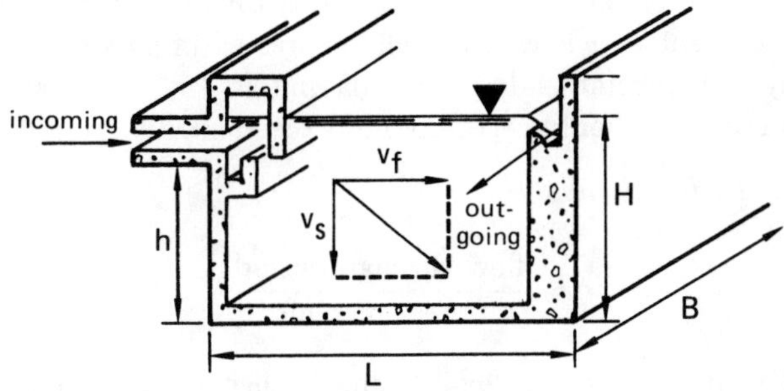


**Figure 7.7** Rectangular Settling Basin

If water enters at some level other than the surface, such as at level $h$ in figure 7.7, all particles will be removed that have

$$v_s > \frac{hQ}{HBL} \qquad 7.41$$

Rectangular basins are preferred, even if sludge removal means emptying and taking them out of service. Rectangular basins should be constructed with aspect ratios of greater than 3:1, and preferably in excess of 4:1. Slope the bottom toward the drain at no less than 1 percent. Use multiple inlets along the entire inlet wall, if possible. If fewer than four ports are used, an inlet baffle should be provided.

In the case of square or circular basins, the slope toward the drain should be greater, typically on the order of 1 : 12. A baffled center inlet should be provided. Square and circular basins are appropriate when space is limited. Otherwise, rectangular basins or solid contact units are preferred.

In using equation 7.40, divide the total flow to be treated into at least two basins. That way, one basin can be out of service for cleaning without interruption of operation.

Basins should be constructed from concrete for all permanent installations. Steel should be used only for small or temporary installations. Where steel parts are unavoidable, as in the case of internal parts of rotors, adequate corrosion resistance is necessary.

The time spent by water in the basin is known as the *detention* (or *retention*) *time* (or *period*). The detention time is given by equation 7.42. A minimum time recommended is 3–4 hours, although periods from 1 to 10 hours are used.[12]

$$t = \frac{\text{tank volume}}{Q} \qquad 7.42$$

[12] Long detention times, up to 12 hours, are required to remove fine particles.

The *weir loading* is the daily flow rate divided by the total effluent weir length, usually expressed in gallons/ft-day. A recommended weir loading is 15,000–20,000 gpd/ft, but should certainly not exceed 50,000 gpd/ft.

The *basin efficiency* is

$$\eta = \frac{\text{flow-through period}}{\text{detention time}} \tag{7.43}$$

The *flow-through period* is found by using color dye in the basin. The flow velocity should not exceed 1 fpm.

*Sludge* needs to be removed according to the demand placed on the basin. Removal is called for when the sludge has reached approximately 25 mg/l or is organic. Various methods of removing sludge are available, including scrapers and pumps. A linear velocity less than 15 fpm should be used with sludge scrapers.

## C. MIXING AND FLOCCULATION

Chemicals can be added to obtain a desired water quality. These chemicals are added to the water in mixing basins. There are two types of mixing basins: complete mixing and plug flow mixing.

*Complete mixing basins* dispense the chemical immediately throughout the volume by using mixing paddles. If the volume of water being mixed is small, the tank is known as a *flash* or *quick mixer*. Quick mixing detention time is often less than 60 seconds.

The retention time required for complete mixing in a tank of volume $V$ depends on the rate constant, $K$, and is given by equation 7.44.

$$t = \frac{V}{Q} = \frac{1}{K}\left(\frac{C_i}{C_o} - 1\right) \tag{7.44}$$

*Plug flow mixing* is accomplished by allowing the water and additive to flow through a long chamber at a uniform rate without mechanical agitation. The retention time in a plug flow mixer of length $L$ is given by equation 7.45.

$$t = \frac{V}{Q} = \frac{L}{v} = \frac{1}{K} ln(C_i/C_o) \tag{7.45}$$

If the mixer adds coagulant for the removal of colloidal sediment, it is known as a *flocculator*. *Floc* is a precipitate that forms when the coagulant allows the colloidal particles to agglomerate. Flocculation is enhanced by gentle agitation, but the floc disintegrates with violent agitation.

Common coagulants among the *hydrolyzing metal ions* are aluminium sulfate ('alum,' $Al_2(SO_4)_3 \cdot 14H_2O$), ferrous sulfate ('copperas,' $FeSO_4 \cdot 7H_20$), ferric chloride ($FeCl_3$), or chlorinated copperas (a mixture of ferrous sulfate and ferric chloride). The most common coagulant is alum. The usual dosage is 10 to 40 mg/l.

Alum reacts with alkalinity in the incoming water to form an aluminum hydroxide floc. If the water is not sufficiently alkaline, an auxiliary chemical such as $CaO$ (lime) or $Na_2CO_3$ (soda ash) is used along with the alum.[13] Lime is also used as an auxiliary chemical with $FeSO_4$ coagulant.

*Polymers* constitute another group of coagulants. *Organic polymers* (*polyelectrolytes*) such as starches and polysaccharides and *synthetic polymers* such as polysacylimides are used. Polymers are chains or groups of repeating identical molecules (*mers*) with many available active adsorption sites. Their molecular weights range from a few hundred to a few million. Polymers can be positively charged (*cationic polymers*), negatively charged (*anionic polymers*), or neutral (*nonionic polymers*). The charge can vary with water pH.

Polymers are effective in a narrow range of turbidity. Turbidity can be increased artificially by adding clay or alum (to produce $Al(OH)_3$ floc). Polymers also require an incoming water with alkalinity.

Another category of chemicals are *flocculation additives*. These additives improve the coagulation efficiency by increasing or decreasing the floc size. *Weighting agents* (e.g., bentonite clays), *adsorbents* (e.g., activated carbon), and *oxidants* (e.g., chlorine, ozone, and potassium permanganate) are used. Polymers are also used in conjunction with metallic ion coagulants.

After flash mixing, a 20- to 60-minute period of gentle mixing is used to permit flocculation. During this period, the flow-through velocity should be between 0.5 and 1.5 ft/min. The peripheral speed of mixing paddles should range from 0.5 ft/sec for fragile, cold water floc to 3.0 ft/sec for tough, warm water floc. The flocculation is followed by sedimentation for 2 to 8 hours in a low-velocity basin.

The *drag force* on a paddle is given by the standard fluid drag force equation. For flat plates, $C_D \approx 1.8$.

$$D = \frac{C_D A \rho v^2}{2g_c} \tag{7.46}$$

The power requirement is easily calculated from the drag force and mixing velocity. (The *mixing velocity*, also known as the *relative paddle velocity*, is approximately 0.7 to 0.8 times the tip speed.)

$$P = Dv \tag{7.47}$$

$$\text{horsepower} = \frac{P}{550} \tag{7.48}$$

13 One mg/l of alum requires $\frac{1}{2}$ mg/l of alkalinity.

The *mean velocity gradient*, $G$, varies from 20 to 75 $sec^{-1}$ for a 15 to 30 minute mixing period. However, $G$ is often used in conjunction with the power requirement and detention time.[14]

$$G = \sqrt{\frac{P}{\mu V_{\text{tank}}}} \quad 7.49$$

$$P = \mu G^2 V_{\text{tank}} \quad 7.50$$

$$Gt_d = \frac{V_{\text{tank}}}{Q}\sqrt{\frac{P}{\mu V_{\text{tank}}}} = \frac{1}{Q}\sqrt{\frac{PV_{\text{tank}}}{\mu}} \quad 7.51$$

Typical values of $Gt_d$ range from $10^4$ to $10^5$. $Gt_d$ is the *mixing opportunity parameter*.

## D. CLARIFICATION WITH FLOCCULATION

The *flocculation clarifier* combines mixing, flocculation, and sedimentation into a single tank. These units are called *solid contact units* and *upflow tanks*. They are generally round in construction, with mixing and flocculation taking place near the central hub, and sedimentation occurring at the periphery. Flocculation clarifiers are most suitable when combined with softening since the precipitated solids help seed the floc.

[14] For mixing units, such as rapid-mix flash units, which do not have to preserve the integrity of floc, a mean velocity gradient in the range of 1000–3000 1/sec can be achieved. For gentle mixing in a flocculator, however, the mean velocity gradient must be much less.

The following general guidelines can be used to design a flocculation clarifier.

| | |
|---|---|
| • minimum flocculation and mixing time | 30 minutes |
| • minimum retention time | 1.5 hours (2.0 hours typical) |
| • maximum weir loading | 10 gpm/ft |
| • upflow rate | 0.8–1.7 $gpm/ft^2$ (1.0 typical) |
| • maximum sludge formation rate | 5% of water flow |

## E. FILTRATION

Nonsettling floc can be removed by filtering. The *rapid sand (gravity) filter* is the most common filter for this use. Rapid sand filters are essentially beds of granulated gravel and sand. Although the box depth can be 10 feet, the sand bed will be only about 3 feet deep. Since the water has been coagulant treated, it may be passed through the filter quickly—hence the name 'rapid.' Rapid filters are usually square or nearly square in design, and operate with a one to eight foot hydraulic head.

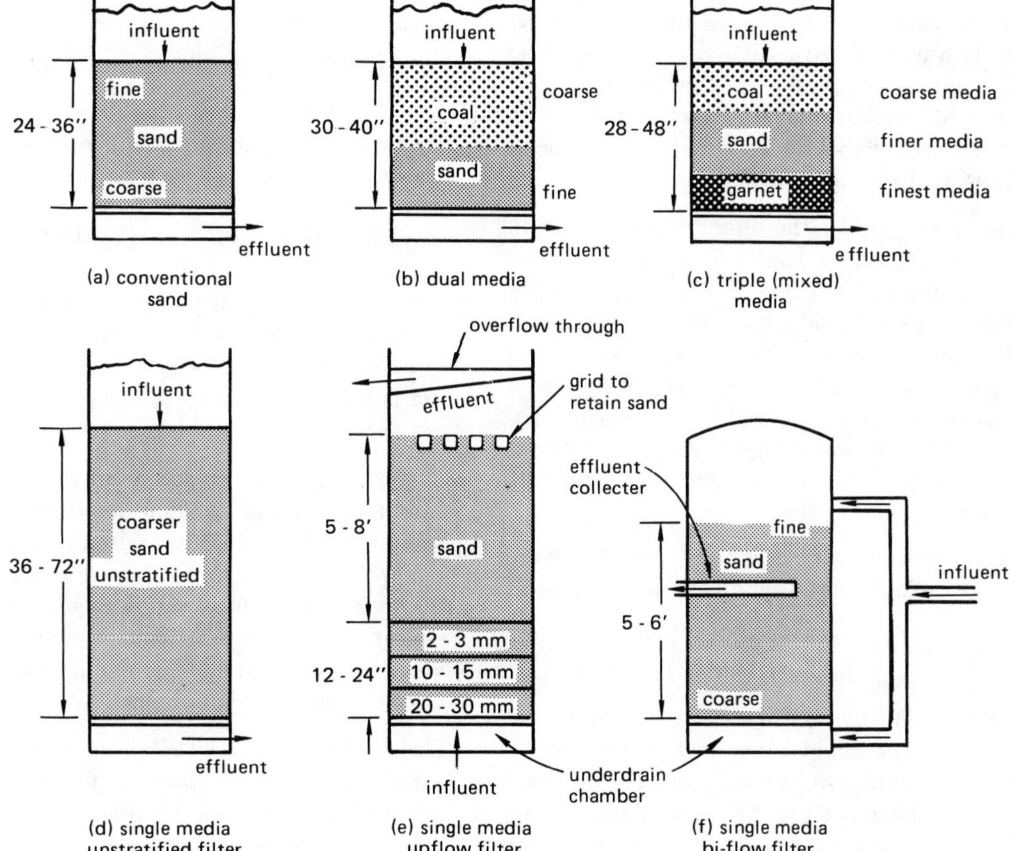


**Figure 7.8** Granular Media Filters

Optimum filter operation will occur when the top layer of sand is slightly more coarse than the rest of the sand. During backwashing, however, the finest sand rises to the top. Various designs using coal and garnet layers in conjunction with sand layers (i.e., *dual-* and *triple-media* filters) overcome this difficulty due to the differences in specific gravity.

Historically, the flow rate has been 2 gpm/ft$^2$ in rapid sand filter design, although some current filters operate at 8 gpm/ft$^2$. 4 gpm/ft$^2$ is a reasonable rate for modern designs.

A water treatment plant should have at least three filters so that two can be in operation when one is being cleaned. Typical total through-puts per filter range from 350 gpm for small plants to 3500 gpm for large plants.

Optimum design of a filtering system includes discharge into a *clearwell.* Clearwells are storage reservoirs with capacities of 30% to 60% of the daily output with a minimum of 12 hours of maximum daily consumption. Demand can be satisfied by the clearwell if one or more of the filters is serviced.

The most common type of service needed by filters is *backwashing.* Filters require backwashing when the pores between sand particles clog up. Typically, this occurs after 1 to 3 days of operation. Backwashing is done when the head loss through the filter bed reaches approximately 8 feet. Backwashing with filtered water expands the sand layer up to 50%, which dislodges the trapped material. Backwashing for 3 to 5 minutes at 8–15 gpm/ft$^2$ is a reasonable design standard. The head loss is reduced to 1 foot after washing.

Water is pumped through the filter from the bottom during backwashing. The rate at which the water rises in the filter housing varies between 12 and 36 inches per minute. This rise rate should not exceed the settling velocity of the smallest particle which is to be retained in the filter. Backwashing usually takes between 3 and 5 minutes. Water which is collected in troughs for disposal and used for backwashing, constitutes between 1% and 5% of the total processed water (approximately 75–100 gal/ft$^2$ total).

The actual amount of backwash water can be found from equation 7.52. Be sure to use consistent units.

$$\text{water volume} = \begin{pmatrix}\text{backwash}\\ \text{time}\end{pmatrix}\begin{pmatrix}\text{filter}\\ \text{area}\end{pmatrix}\begin{pmatrix}\text{rise}\\ \text{rate}\end{pmatrix} \qquad 7.52$$

*Slow sand filters* are similar in design to rapid sand filters, except that the sand layer is thicker (24″ to 48″), the gravel layer is thinner (6″ to 12″), and the flow rate is much lower (0.05 to 0.1 gpm/ft$^2$). Slow sand filters are limited to low-turbidity applications not requiring chemical treatment. Cleaning is usually accomplished by removing a few inches of sand. Slow sand filters operate with a 0.2 to 4.0 foot head loss.

Other types of filters are seeing limited use.

- **Pressure filters**: Similar to rapid sand filters except incoming water is pressurized up to 25 feet (hydraulic). Filter rates of 2 to 4 gpm/ft$^2$. Not used in large installations.
- **Diatomaceous earth filters**: 1 to 3 gpm/ft$^2$; short (20-hour) cycle life.
- **Microstrainers**: Woven stainless steel fabric, usually mounted on a rotating drum.

## F. DISINFECTION

Chlorination is used for disinfection and oxidation. As a disinfectant, chlorine destroys bacteria and microorganisms. As an oxidant, it removes iron, manganese, and ammonia nitrogen.

Chlorine can be added as a gas or a solid. (If it is added to the water as a gas, it is stored as a liquid which vaporizes around −35°C.) Liquid chlorine is the predominant form since it is cheaper than hypochlorite solid ($Ca(OCl)_2$). If chlorine liquid or gas is added to water, the following reaction occurs to form hypochlorous acid, which itself ionizes to hypochlorite and hydrogen ions.

$$Cl_2 + H_2O \rightarrow HCl + HOCl \underset{pH<7}{\overset{pH>8}{\rightleftharpoons}} H^+ + OCL^- + HCL \qquad 7.53$$

If calcium hypochlorite solid is added to water, the ionization follows immediately.

$$Ca(OCl)_2 + H_2O \longrightarrow Ca^{++} + 2OCl^- + H_2O \qquad 7.54$$

Chlorine existing in water as hypochlorous acid and hypochlorite ions is known as *free available chlorine* (*free residuals*). Chlorine in combination with ammonia is known as *combined available chlorine* (*combined residuals*).

The average chlorine dose is in the 1 to 2 mg/l range. Minimum chlorine residuals for 70°F water are given in table 7.21. Reliable chlorination requires a pH of water below 9.0. However, inactivated viruses (such as might be present in surface water) require a heavier chlorine concentration. Since treatment of water is by both free and combined residuals, ammonia can be added to the water to produce chloramines.

Excess chlorine can be removed with a reducing agent, usually called a *dechlor.* Sulfur dioxide and sodium bisulfate (sodium metabisulfate) are used in this manner. Aeration also reduces chlorine content, as does passing the water through an activated charcoal filter.

**Table 7.21**
Minimum Chlorine Residuals (mg/l)

| pH Value | Free residuals after 10–20 minutes | Combined residuals after 1–2 hours |
|---|---|---|
| 6.0 | 0.2 | 1.0 |
| 7.0 | 0.2 | 1.5 |
| 8.0 | 0.4 | 1.8 |
| 9.0 | 0.8 | 3.0* |
| 10.0 | 0.8 | 3.0* |

*not recommended

Alternatives to chlorination have become popular since THM's were traced to the chlorination process. Chlorine dioxide can be used in place of chlorine, but it is expensive. In high dosages, chlorine dioxide is also thought to produce its own toxic byproducts. Ozone is a more powerful disinfectant than chlorine, but it is expensive to generate and requires costly contact chambers. Ozone, which is used extensively in Western Europe, Canada, the USSR, and Japan, is generated onsite by running high voltage electrical currents through dry air or pure oxygen.

## G. FLUORIDATION

Fluoridation can occur any time after filtering. Smaller utilities almost always choose liquid solution and a volumetric feeding mechanism, with solutions being manually prepared. Larger utilities use gravimetric dry feeders with sodium silicofluoride or solution feeders with fluorsilic acid. The characteristics and dose rates of common fluorine compounds are given in table 7.22.

**Table 7.22**
Dose Rates for Fluorine Compounds

| Formula | $H_2SiF_6$ | NaF | $Na_2SiF_6$ |
|---|---|---|---|
| Form | liquid | solid | solid |
| Typical purity | 22–30% | 90–98% | 98–99% |
| Dose to obtain 1.0 mg/l (pounds per million gallons) | 35.2 (with 30% purity) | 18.8 (with 98% purity) | 14.0 (with 98.5% purity) |

*Defluoridation* (required if the fluoride exceeds 1.5 mg/l) can be achieved with calcined alumina or bone char (tricalcium phosphate). Softening using lime can also be used when waters contain smaller amounts of fluoride. Each 45 to 65 mg/l reduction in magnesium will result in a 1.0 mg/l reduction in fluoride.

## H. IRON AND MANGANESE REMOVAL

Several methods of removing iron exist. (Manganese is not easily removed by aeration alone. However, the remaining methods work.)

- Aeration, followed by sedimentation and filtration

$$Fe^{++} + O_x \rightarrow FeO_x \downarrow \qquad 7.55$$

- Aeration, followed by chemical oxidation, sedimentation, and filtration. Chlorine or potassium permanganate may be used as an oxidizer.
- Manganese zeolite process: Manganese dioxide removes soluble iron ions.
- Lime water softening

Table 7.23 lists the characteristics of iron and manganese removal processes.

## I. WATER SOFTENING

- Lime and Soda Ash Softening

Water softening can be accomplished with lime and soda ash to precipitate calcium and magnesium ions from the solution. Lime treatment has added benefits of disinfection, iron removal, and clarification.

*Lime* (CaO) is available as *granular quicklime* (90% CaO) or *hydrated lime* (93% $Ca(OH)_2$). Both forms are slaked prior to use, which means that water is added to form a lime slurry in an exothermic reaction.

$$CaO + H_2O \rightarrow Ca(OH)_2 + \text{heat} \qquad 7.56$$

*Soda ash* is usually available as 98% pure sodium carbonate ($Na_2CO_3$).

**FIRST STAGE TREATMENT**: In the first stage treatment, lime added to water reacts with free carbon dioxide to form calcium carbonate precipitate.

$$CO_2 + Ca(OH)_2 \rightarrow CaCO_3 \downarrow + H_2O \qquad 7.57$$

Next, the lime reacts with calcium bicarbonate.

$$Ca(HCO_3)_2 + Ca(OH)_2 \rightarrow 2CaCO_3 \downarrow + 2H_2O \qquad 7.58$$

Any magnesium hardness is also removed at this time.

$$Mg(HCO_3)_2 + Ca(OH)_2 \rightarrow CaCO_3 \downarrow + 2H_2O + MgCO_3 \qquad 7.59$$

To remove the soluble $MgCO_3$, the pH must be above 10.8. This is accomplished by adding an excess of approximately 35 mg/l of CaO or 50 mg/l of $Ca(OH)_2$ plus lime to satisfy equation 7.60.

$$MgCO_3 + Ca(OH)_2 \rightarrow CaCO_3 \downarrow + Mg(OH)_2 \downarrow \qquad 7.60$$

**Table 7.23**
Iron/Manganese Removal Processes

| processes | iron and/or manganese removed | pH required | remarks |
|---|---|---|---|
| aeration, settling, and filtration | ferrous bicarbonate | 7.5 | provide aeration unless incoming water contains adequate dissolved oxygen |
| | ferrous sulfate | 8.0 | |
| | manganous bicarbonate | 10.3 | |
| | manganous sulfate | 10.0 | |
| aeration, free residual chlorination, settling, and filtration | ferrous bicarbonate | 5.0 | provide aeration unless incoming water contains adequate dissolved oxygen |
| | manganous bicarbonate | 9.0 | |
| aeration, lime softening, settling, and filtration | ferrous bicarbonate | 8.5–9.6 | |
| | manganous bicarbonate | | |
| aeration, coagulation, lime softening, settling, and filtration | colloidal or organic iron | 8.5–9.6 | require lime, and alum or iron coagulant |
| | colloidal or organic manganese | 10.0 | |
| ion exchange | ferrous bicarbonate | 6.5± | water must be devoid of oxygen |
| | manganous bicarbonate | | iron and manganese in raw water not to exceed 2.0 mg/l |
| | | | consult manufacturers for type of ion exchange resin to be used |

**FIRST STAGE RECARBONATION**: Lime added to precipitate hardness removes itself. This is desirable because any calcium that remains in the water has the potential for forming scale. Further stabilization can be achieved by *recarbonation* (treatment with carbon dioxide).

$$Ca(OH)_2 + CO_2 \rightarrow CaCO_3 \downarrow + H_2O \qquad 7.61$$

Excess recarbonation should be avoided. If the pH is allowed to drop below 9.5, then carbonate hardness reappears.

$$CaCO_3 + CO_2 + H_2O \rightarrow Ca(HCO_3)_2 \qquad 7.62$$

At this time, any unsettled $Mg(OH)_2$ can be returned to a soluble state.

$$2Mg(OH)_2 + 2CO_2 \rightarrow 2MgCO_3 + 2H_2O \qquad 7.63$$

**SECOND STAGE TREATMENT**: The second stage treatment removes calcium noncarbonate hardness (sulfates and chlorides) which needs soda ash for precipitation.

$$CaSO_4 + Na_2CO_3 \rightarrow CaCO_3 \downarrow + Na_2SO_4 \qquad 7.64$$

Magnesium noncarbonate hardness needs both lime and soda ash.

$$MgSO_4 + Ca(OH)_2 \rightarrow Mg(OH)_2 \downarrow + CaSO_4 \qquad 7.65$$

$$CaSO_4 + Na_2CO_3 \rightarrow CaCO_3 \downarrow + Na_2SO_4 \qquad 7.66$$

Excess soda ash leaves sodium ions in the water. However, noncarbonate hardness is a small part of total hardness. Soda ash is also costly, so the actual dose might be slightly reduced from what is needed.

**SECOND STAGE RECARBONATION**: Second stage recarbonation is needed to remove $CaCO_3$. $CO_2$ is added until the pH is about 8.6, at which time no further precipitation will occur because $[Ca^{++}][CO_3^{=}] < K_{sp}$ of $CaCO_3$.

Sodium polyphosphate can be added at this time to inhibit crusting on filter sand and scale formation in pipes.

A *split process* can be used to reduce the amount of lime that is neutralized by recarbonation (and is wasted). Excess lime is added in the first stage. This forces precipitation of magnesium in the first stage instead of in the second stage. Excess lime reacts with calcium hardness in the second stage. The amount of bypass depends on the allowable hardness of water leaving the plant. A typical split process is shown in figure 7.9.

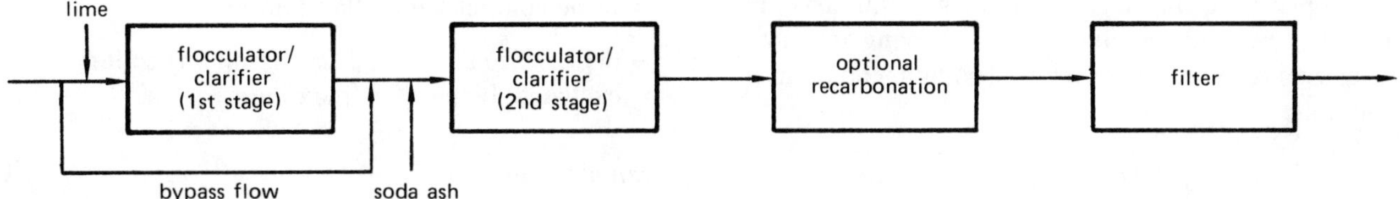


**Figure 7.9** A Split (Bypass) Process

Practical limits of *precipitation softening* by the lime process are 30 to 35 mg/l of $CaCO_3$ and 8 to 10 mg/l of $Mg(OH)_2$ as $CaCO_3$ due to intrinsic solubilities. Water softened with this process usually leaves the apparatus with a hardness of between 50 and 80.

*Example 7.18*

Water contains 130 mg/l as $CaCO_3$ of $Ca(HCO_3)_2$. How much slaked lime ($Ca(OH)_2$) is required to remove the hardness?

Since the $Ca(HCO_3)_2$ is given in $CaCO_3$ equivalents, 130 mg/l of lime (as $CaCO_3$) is implicitly required. It only remains to convert the $CaCO_3$ equivalent to a substance measurement using appendix A.

$$Ca(OH)_2 = \frac{130}{1.35} = 96.3 \text{ mg/l as substance}$$

*Example 7.19*

How much slaked lime (90% pure), soda ash, and carbon dioxide are required to reduce the hardness of the water evaluated below to zero using the lime-soda ash process? Neglect the fact that this process cannot really produce zero hardness, and base your answer on stoichiometric considerations.

| | |
|---|---|
| total hardness: | 250 mg/l as $CaCO_3$ |
| alkalinity: | 150 mg/l as $CaCO_3$ |
| carbon dioxide: | 5 mg/l |

Using appendix A, the $CO_2$ is first converted to its $CaCO_3$ equivalent.

$$(2.27)(5) = 11.35 \text{ mg/l as } CaCO_3$$

The alkalinity of 150 mg/l is already in $CaCO_3$ equivalent form. Therefore, the total $CaCO_3$ equivalent from substances requiring lime for neutralization is $11.35 + 150 = 161.35$ mg/l as $CaCO_3$.

From appendix A, the amount of 90% pure slaked lime ($Ca(OH)_2$) is

$$\frac{161.35}{(1.35)(0.90)} = 132.8 \text{ mg/l as substance}$$

50 mg/l of lime is arbitrarily added to raise the pH above 10.8. The total lime requirement is then

$$132.8 + \frac{50}{0.90} = 188.4 \text{ mg/l as substance}$$

The noncarbonate hardness is $250 - 150 = 100$ mg/l as $CaCO_3$. The soda ash ($Na_2CO_3$, 98% pure) requirement is

$$\frac{100}{(0.94)(0.98)} = 108.6 \text{ mg/l as substance}$$

The first stage recarbonation $CO_2$ requirement depends on the excess lime added.

$$\frac{(50)(1.35)}{(2.27)} = 29.7 \text{ mg/l as substance}$$

• Ion Exchange Method

In the ion exchange process (also known as *zeolite process*, *resin exchange process*, or *ion exchange method*), water is passed through a filter bed of exchange material. This exchange material is known as *zeolite*. Ions in the insoluble exchange material are displaced by ions in the water. When the exchange material is spent, it is regenerated with a rejuvenating solution such as sodium chloride (salt), or, in the case of common cationic resins, sulfuric and hydrochloric acids are used as *regenerants*. Soda ash is used as a regenerant in weakly-basic exchangers.

The processed water will have a zero hardness. However, since there is no need for water with zero hardness, some water is usually bypassed around the unit.

There are three types of ion exchange materials. *Greensand* (*glauconite*) is a natural substance that is mined and treated with manganese dioxide. *Siliceous-gel zeolite* is an artificial solid used in small volume deionizer columns. *Polystyrene resins* are also synthetic. Polystyrene resins currently dominate the softening field.[15]

[15] Differences in the polymerization step can result in polymers with gel or macroporous structures. *Gel polymers* have low cross linking, high capacity, and fast reaction kinetics. *Macroporous polymers* have high cross linking, reduced capacity, and lower kinetics. Gel resins have historically been used in water softening. However, the chemical resistance of macroporous forms is advantageous in special applications.

During operation, the calcium and magnesium ions are removed in reactions similar to the following reaction. Z is the zeolite anion. The resulting sodium compounds are soluble.

$$\left\{\begin{matrix} Ca \\ Mg \end{matrix}\right\}\left\{\begin{matrix} (HCO_3)_2 \\ SO_4 \\ Cl_2 \end{matrix}\right\} + Na_2Z$$

$$\rightarrow \left\{\begin{matrix} 2NaHCO_3 \\ Na_2SO_4 \\ 2NaCl \end{matrix}\right\} + \left\{\begin{matrix} Ca \\ Mg \end{matrix}\right\} Z \qquad 7.67$$

Typical characteristics of an ion exchange unit are expressed per 1000 grains of hardness removed.[16]

- exchange capacity: 3000 grains hardness/ft$^3$ zeolite for natural; 5000–30,000 (20,000 typical) for synthetic.
- flow rate: 2 to 6 gpm/ft$^3$ (2 gpm/ft$^3$ standard)

  6 gpm/ft$^2$ of filter bed
- backwash flow: 5 to 6 gpm/ft$^2$
- salt dosage: 5 to 20 pounds/ft$^3$. Alternatively, 0.3 to 0.7 pound of salt per 1000 grains of hardness removed
- brine contact time: 25 to 45 minutes
- depth of ion exchange bed: 2 ft (minimum) to 9 ft (maximum)

16 1000 grains of hardness is also known as a *kilograin*.

*Example 7.20*

A municipal plant receives water with a total hardness of 200 mg/l. The designed discharge hardness is 50 mg/l. If an ion exchange unit is used, what is the bypass factor?

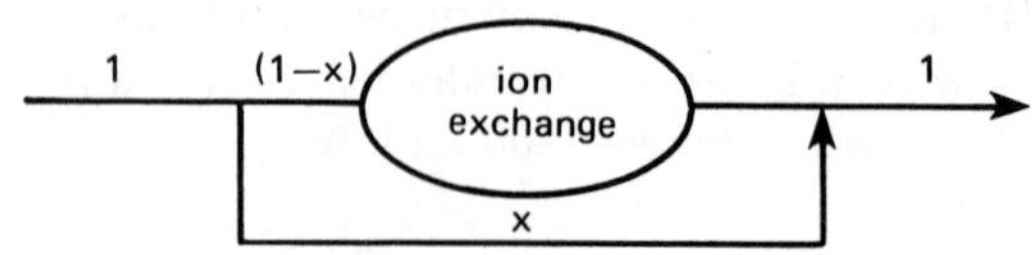


Let $x$ be the bypass factor. Since the water passing through the ion exchange unit is reduced to zero hardness,

$$(1-x)0 + x(200) = 50$$
$$x = 0.25$$

J. TURBIDITY REMOVAL

*Coagulants* can be based on aluminum (e.g., aluminum sulfate, sodium aluminate, potash alum, or ammonia alum) or iron (e.g., ferric sulfate, ferrous sulfate, chlorinated copperas, or ferric chloride). If significant hydrolysis of iron and aluminum salts is ignored, the re-

**Table 7.24**
Types of Synthetic Exchange Materials

| type of resin | drained density lbm/ft$^3$ | operating pH range | regeneration | characteristics |
|---|---|---|---|---|
| strong acid | 49–53 | 0–14 | excess strong acid | high exchange rates; are stable; low swelling; long life, up to 20 years or more; can split strong and weak salts |
| weak acid | 45 | 7–14 | weak or strong acid | capacities double of strong acid; resistant to chlorine and other oxidants; high (90%) swell; not effective for electrolytic salt cations |
| strong base | 45 | 0–14 | excess strong base | irreversibly fouled by humic acids from decaying vegetation; can split strong or weak salts; less stable than cation resins (life less than 3 years); can remove silica; often used with food processing |
| weak base | 32 | 0–6 | weak or strong base | resistant to organic fouling; does not remove $CO_2$ or silica; capacity double of strong base; can remove color |
| intermediate base | 43 | 0–14 | strong base | can absorb $CO_2$ silica and phenol. Useful as substitutes for weak base resins in multiple-bed processes |

lationships in this section can be used to calculate the approximate stoichiometric quantities.

The most-used coagulant is aluminum sulfate ($Al_2(SO_4)_3 \cdot 14H_2O$). Filter alum is about 17% soluble material. The hydrolysis of the aluminum ion is complex. Assuming that the aluminum floc is $Al(OH)_3$ and the water pH is near neutral, then 1 mg/l of alum with a molecular weight of 600 removes the following quantities:

0.5 mg/l ($CaCO_3$) of natural alkalinity

0.39 mg/l of 95% hydrated lime ($Ca(OH)_2$)

0.33 mg/l of 85% quicklime ($CaO$)

0.53 mg/l of soda ash ($Na_2CO_3$)

If the alum has a molecular weight that is different than 600 (due to the variation in the number of waters of hydration), multiply the above quantities by (600/actual molecular weight).

Typical doses of alum are 5 to 50 mg/l, depending on turbidity. Alum flocculation is effective within pH limits of 5.5 to 8.0.

Ferrous sulfate ($FeSO_4 \cdot 7H_2O$), also known as *copperas*, reacts with lime ($Ca(OH)_2$) to flocculate ferric hydroxide ($Fe(OH)_3$). This is an effective method of clarifying turbid waters at higher pH, as in lime softening. 1 mg/l of ferrous sulfate with a molecular weight of 278 will react with 0.27 mg/l of lime.

Ferric sulfate ($Fe_2(SO_4)_3$) reacts with natural alkalinity or lime to create floc. 1 mg/l of ferric sulfate will react with

1.22 mg/l of $Ca(HCO_3)_2$

0.56 mg/l of $Ca(OH)_2$

0.62 mg/l of natural alkalinity (as $CaCO_3$)

Ferric sulfate can be used for color removal at low pH; at high pH, it is useful for iron and manganese removal, as well as a coagulant with precipitation softening.

## K. TASTE AND ODOR CONTROL

- **Copper Sulfate Treatment**. This treatment is used in impounding reservoirs, lakes, storage reservoirs, and occasionally in settling basins or treated water, to prevent biological growths. Dosages may vary from 0.5 to 2.0 milligrams per liter; the lower dosage ordinarily suffices for soft water. For very hard water, a dosage above 2.0 milligrams per liter may be used after laboratory tests to determine the necessary algicidal dose. Effects on fish life should be monitored.

- **Aeration**. This process can be used to improve tastes and odors in water where the cause is hydrogen sulfide or the absence of dissolved oxygen. This method has little effect on most tastes and odors.

- **Activated Carbon**. This material removes most tastes and odors. Dosages may vary from 0.5 to 200 milligrams per liter, ordinarily ranging from 2 to 10 milligrams per liter.

- **Superchlorination and Dechlorination**. This treatment will improve tastes and odors caused by organic matter and industrial wastes, especially phenolic wastes. Normally, the dosage required will be several times greater than those for ordinary disinfection (as determined by testing). Provide chlorinating equipment capable of dosing at these high values; allow a minimum of 20 minutes contact time; furnish equipment for dechlorinating with sulfur dioxide or other reducing agent.

- **Chlorine-Ammonia Treatment**. Where chlorosubstitution products cause tastes and odors, the chlorine-ammonia treatment can be used to prevent them. It can also be used for maintaining the combined residual chlorine for an extended period as, for example, in reservoirs or distribution systems.

    - Chloramines are less active disinfectants than free chlorine and, therefore, may not be substituted where adequate disinfection requires free residual chlorine.

    - The ratio of chlorine to ammonia required for disinfection varies from 3:1 to 7:1.

    - Periodic laboratory tests should be conducted to determine the proper dosage. Apply chlorine after ammonia has been properly dispersed in the water.

- **Free Residual Chlorination**. Use this method before filtration to reduce tastes and odors caused by organic matter at locations where experience shows it to be effective and acceptable. Increase the chlorine dosage until the residual consists solely of free available chlorine.

- **Chlorine Dioxide**. In some cases, this chemical can be used to destroy phenolic and other organic tastes and odors in raw water. The dosage varies from 0.2 to 0.3 milligram per liter, as determined by testing.

- **Microstraining**. This method is used as a means of reducing the number of algae and other organisms in the water, and thus reduces the subsequent production of tastes and odors. The microstrainer removes no dissolved or colloidal organic matter. It utilizes monel metal cloth with 35 micron (0.0014 inch) openings. Finer mesh can be obtained.

### L. DEMINERALIZATION/DESALINATION

If dissolved salts are to be removed, one of the following methods must be used.

- **Distillation**: The water is vaporized, leaving the salt behind. The vapor is reclaimed by condensation.
- **Electrodialysis**: Positive and negative ions flow through selective membranes under the influence of an induced electrical current.
- **Ion exchange**: This is the same process as described for water softening.
- **Reverse osmosis**: This is the least expensive method of demineralization. In operation, a thin membrane of cellulose acetate plastic separates two salt solutions of different concentrations. Although ions would normally flow through the membrane into the solution with the lower concentration, the migration direction can be reversed by applying pressure to the low concentration fluid. Typical reverse osmosis units operate at 400 psi and produce about 2 gallons per day of fresh water for each square foot of surface.

## 14 TYPICAL MUNICIPAL SYSTEMS

The processes employed in treating incoming water will depend on the characteristics of the water. However, some sequences work better than others due to the physical and chemical nature of the processes. Listed in this section are some typical sequences. Not present in the lists are the usual system hardware items such as intake screens, pumps, pipes, hydrants, reservoirs, and holding basins.

Table 7.25 provides guidelines for choosing processes required to achieve satisfactory water quality. This table bases the required processes on the incoming water quality.

Additives and chemicals can be applied to the water supply at various points along the treatment path. Figure 7.10 indicates typical application points.

- **For Well Ground Water** (typically cleaner than surface water)

sequence #1:
intake
chlorination
fluoridation

sequence #2:
intake
aeration
oxidation(chlorine or potassium permanganate)
settling
filtering
chlorination
fluoridation

sequence #3:
intake
aeration
lime addition
soda ash addition
rapid mix
flocculation
settling
recarbonation
filtering
chlorination
fluoridation

- **For Lake or Surface Water** (typically turbid, and carrying odor and color)

sequence #1:
intake
chlorination
coagulation
rapid mixing
flocculation
optional chlorination
addition of activated carbon
settling
addition of activated carbon
filtering
chlorination
fluoridation

- **For River Surface Water** (very turbid)

sequence #1:
intake
presedimentation (holding basin)
chlorination
coagulation
rapid mix
flocculation
settling
coagulation
rapid mix
flocculation
addition of activated carbon
settling
addition of activated carbon
filtering
chlorination
fluoridation

**Table 7.25**
Applicability of Treatment Methods

| water quality | | | pretreatment | | | treatment | | | | | special treatments | | | |
|---|---|---|---|---|---|---|---|---|---|---|---|---|---|---|
| constituents | concentration, mg/l | screeening | prechlorination | plain settling | aeration | lime softening | coagulation and sedimentation | rapid sand filtration | slow sand filtration | postchlorination | superchlorination[1] or chlor-ammoniation | active carbon | special chemical treatment | salt water conversion[2] |
| coliform monthly avg mpn/100 ml[5] | 0–20 | | | | | | | | | E | | | | |
| | 20–100 | | | O | | | O | O | O | E | | | | |
| | 100–5000 | | E | | | | E | E | O | E | | | | |
| | >5000 | | E | O[3] | | | E | E | | E | O | | | |
| suspended solids | 0–100 | O | | | | | | | O | | | | | |
| | 100–200 | O | | | | | E | E | | | | | | |
| | > 200 | O | | O[4] | | | E | E | | | | | | |
| color, mg/l | 20–70 | | | | | | O | O | | | O | | | |
| | > 70 | | | | | | E | E | | | O | | | |
| tastes and odors | noticeable | | O | | O | | | | O | | O | E | | |
| $CaCO_3$, mg/l | > 200 | | | | | E | E | E | | | | | E | |
| pH | < 5.0–9.0< | | | | | | | | | | | | | |
| iron and manganese mg/l | ≤ 0.3 | | O | O | | | | | | | | | | |
| | 0.3–1.0 | | | | O | | E | E | O | | | | | |
| | > 1.0 | | E | | E | | E | E | O | | | | O | |
| chloride, mg/l | 0–250 | | | | | | | | | | | | | |
| | 250–500 | | | | | | | | | | | | | O |
| | 500 + | | | | | | | | | | | | | E |
| phenolic compounds, mg/l | 0–0.005 | | | | | | O | O | | | O | O | | |
| | > 0.005 | | | | | | E | E | | | O | E | O | |
| toxic chemicals | | | | | | | E | E | | | | E | O | |
| less critical chemicals | | | | | | | O | O | | | | O | O | |

Note: E = essential, O = optional
[1]Superchlorination shall be followed by dechlorination.
[2]As alternate, dilute with low chloride water.
[3]Double settling shall be provided for coliform exceeding 20,000 mpn/100 ml[5].
[4]For extremely muddy water, presedimentation by plain settling may be provided.
[5]mpn = most probable number

- **For Hard Water**

sequence #1:
intake (bypass to second flocculator)
lime addition
alum addition
rapid mixing
flocculation
sedimentation
oxidation (chlorine or potassium permanganate)
second flocculation
sedimentation
filtering
fluoridation
chlorination

sequence #2:
intake
presedimentation (in a basin)
chlorination
mixing
addition of activated carbon
lime addition
alum addition
flocculation
sedimentation
addition of activated carbon
mixing
filtering
fluoridation
chlorination
soda ash addition

sequence #3: intake
lime addition
alum addition
addition of activated carbon
mixing
flocculation
chlorination
sedimentation
recarbonation
filtering (bypass to discharge)
zeolite treatment

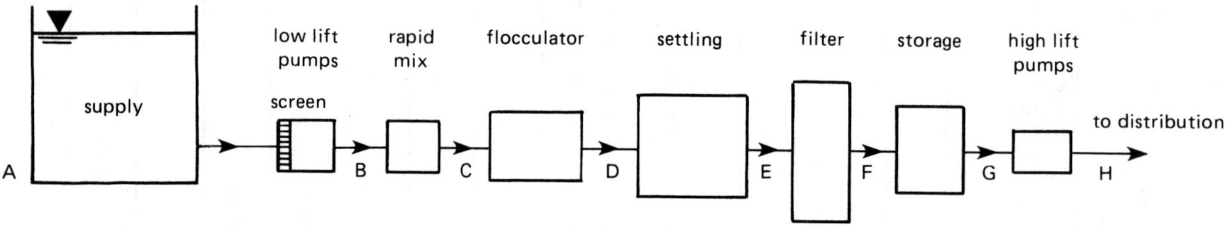


Typical flow diagram of water treatment plant

| category of chemicals | possible points of application | | | | | | | |
|---|---|---|---|---|---|---|---|---|
| | A | B | C | D | E | F | G | H |
| algicide | X | | | | X | | | |
| disinfectant | | X | X | | X | X | X | X |
| activated carbon | | X | X | X | X | | | |
| coagulants | | X | X | | | | | |
| coagulation aids | | X | X | | X | | | |
| alkali: | | | | | | | | |
| for flocculation | | | X | | | | | |
| for corrosion control | | | | | | X | | |
| for softening | | | X | | | | | |
| acidifier | | | X | | | X | | |
| fluoride | | | | | | X | | |
| cupric-chloramine | | | | | | X | | |
| dechlorinating agent | | | | | | X | | X |

Note: With solids contact reactors, point C is same as point D.

**Figure 7.10** Application Points for Chemicals

# Appendix A: Conversions from mg/l as a Substance to mg/l as $CaCO_3$

Multiply the mg/l of the substances listed below by the corresponding factors to obtain mg/l as $CaCO_3$. For example, 70 mg/l of $Mg^{++}$ would be (70)(4.10) = 287 mg/l as $CaCO_3$.

| Substance | Factor | Substance | Factor |
|---|---|---|---|
| $Al^{+++}$ | 5.56 | $HCO_3^-$ | 0.82 |
| $Al_2(SO_4)_3$* | 0.88 | $K^+$ | 1.28 |
| $AlCl_3$ | 1.13 | KCl | 0.67 |
| $Al(OH)_3$ | 1.92 | $K_2CO_3$ | 0.72 |
| $Ba^{++}$ | 0.73 | $Mg^{++}$ | 4.10 |
| $Ba(OH)_2$ | 0.59 | $MgCl_2$ | 1.05 |
| $BaSO_4$ | 0.43 | $MgCO_3$ | 1.19 |
| $Ca^{++}$ | 2.50 | $Mg(HCO_3)_2$ | 0.68 |
| $CaCl_2$ | 0.90 | MgO | 2.48 |
| $CaCO_3$ | 1.00 | $Mg(OH)_2$ | 1.71 |
| $Ca(HCO_3)_2$ | 0.62 | $Mg(NO_3)_2$ | 0.67 |
| CaO | 1.79 | $MgSO_4$ | 0.83 |
| $Ca(OH)_2$ | 1.35 | $Mn^{++}$ | 1.82 |
| $CaSO_4$* | 0.74 | $Na^+$ | 2.18 |
| $Cl^-$ | 1.41 | NaCl | 0.85 |
| $CO_2$ | 2.27 | $Na_2CO_3$ | 0.94 |
| $CO_3^{--}$ | 1.67 | $NaHCO_3$ | 0.60 |
| $Cu^{++}$ | 1.57 | $NaNO_3$ | 0.59 |
| $Cu^{+++}$ | 2.36 | NaOH | 1.25 |
| $CuSO_4$ | 0.63 | $Na_2SO_4$* | 0.70 |
| $F^-$ | 2.66 | $NH_3$ | 2.94 |
| $Fe^{++}$ | 1.79 | $NH_4^+$ | 2.78 |
| $Fe^{+++}$ | 2.69 | $NH_4OH$ | 1.43 |
| $Fe(OH)_3$ | 1.41 | $(NH_4)_2SO_4$ | 0.76 |
| $FeSO_4$* | 0.66 | $NO_3^-$ | 0.81 |
| $Fe_2(SO_4)_3$ | 0.75 | $OH^-$ | 2.94 |
| $FeCl_3$ | 0.93 | $PO_4^{---}$ | 1.58 |
| $H^+$ | 50.0 | $SO_4^{--}$ | 1.04 |
| | | $Zn^{++}$ | 1.54 |

* anhydrous

# Appendix B: Atomic Weights of Elements Referred to Carbon (12)

| Element | Symbol | Atomic Weight |
|---|---|---|
| Actinium | Ac | (227) |
| Aluminum | Al | 26.9815 |
| Americium | Am | (243) |
| Antimony | Sb | 121.75 |
| Argon | Ar | 39.948 |
| Arsenic | As | 74.9216 |
| Astatine | At | (210) |
| Barium | Ba | 137.34 |
| Berkelium | Bk | (249) |
| Beryllium | Be | 9.0122 |
| Bismuth | Bi | 208.980 |
| Boron | B | 10.811 |
| Bromine | Br | 79.909 |
| Cadmium | Cd | 112.40 |
| Calcium | Ca | 40.08 |
| Californium | Cf | (251) |
| Carbon | C | 12.01115 |
| Cerium | Ce | 140.12 |
| Cesium | Cs | 132.905 |
| Chlorine | Cl | 35.453 |
| Chromium | Cr | 51.996 |
| Cobalt | Co | 58.9332 |
| Copper | Cu | 63.54 |
| Curium | Cm | (247) |
| Dysprosium | Dy | 162.50 |
| Einsteinium | Es | (254) |
| Erbium | Er | 167.26 |
| Europium | Eu | 151.96 |
| Fermium | Fm | (253) |
| Fluorine | F | 18.9984 |
| Francium | Fr | (223) |
| Gadolinium | Gd | 157.25 |
| Gallium | Ga | 69.72 |
| Germanium | Ge | 72.59 |
| Gold | Au | 196.967 |
| Hafnium | Hf | 178.49 |
| Helium | He | 4.0026 |
| Holmium | Ho | 164.930 |
| Hydrogen | H | 1.00797 |
| Indium | In | 114.82 |
| Iodine | I | 126.9044 |
| Iridium | Ir | 192.2 |
| Iron | Fe | 55.847 |
| Krypton | Kr | 83.80 |
| Lanthanum | La | 138.91 |
| Lead | Pb | 207.19 |
| Lithium | Li | 6.939 |
| Lutetium | Lu | 174.97 |
| Magnesium | Mg | 24.312 |
| Manganese | Mn | 54.9380 |
| Mendelevium | Md | (256) |
| Mercury | Hg | 200.59 |
| Molybdenum | Mo | 95.94 |
| Neodymium | Nd | 144.24 |
| Neon | Ne | 20.183 |
| Neptunium | Np | (237) |
| Nickel | Ni | 58.71 |
| Niobium | Nb | 92.906 |
| Nitrogen | N | 14.0067 |
| Osmium | Os | 190.2 |
| Oxygen | O | 15.9994 |
| Palladium | Pd | 106.4 |
| Phosphorus | P | 30.9738 |
| Platinum | Pt | 195.09 |
| Plutonium | Pu | (242) |
| Polonium | Po | (210) |
| Potassium | K | 39.102 |
| Praseodymium | Pr | 140.907 |
| Promethium | Pm | (145) |
| Protactinium | Pa | (231) |
| Radium | Ra | (226) |
| Radon | Rn | (222) |
| Rhenium | Re | 186.2 |
| Rhodium | Rh | 102.905 |
| Rubidium | Rb | 85.47 |
| Ruthenium | Ru | 101.07 |
| Samarium | Sm | 150.35 |
| Scandium | Sc | 44.956 |
| Selenium | Se | 78.96 |
| Silicon | Si | 28.086 |
| Silver | Ag | 107.870 |
| Sodium | Na | 22.9898 |
| Strontium | Sr | 87.62 |
| Sulfur | S | 32.064 |
| Tantalum | Ta | 180.948 |
| Technetium | Tc | (99) |
| Tellurium | Te | 127.60 |
| Terbium | Tb | 158.924 |
| Thallium | Tl | 204.37 |
| Thorium | Th | 232.038 |
| Thulium | Tm | 168.934 |
| Tin | Sn | 118.69 |
| Titanium | Ti | 47.90 |
| Tungsten | W | 183.85 |
| Uranium | U | 238.03 |
| Vanadium | V | 50.942 |
| Xenon | Xe | 131.30 |
| Ytterbium | Yb | 173.04 |
| Yttrium | Y | 88.905 |
| Zinc | Zn | 65.37 |
| Zirconium | Zr | 91.22 |

# Appendix C: Inorganic Chemicals Used in Water Treatment

| Chemical Name | Formula | Use | Molecular Weight | Equivalent Weight |
|---|---|---|---|---|
| Activated carbon | C | Taste and odor control | 12.0 | ---- |
| Aluminum sulfate (filter alum) | $Al_2(SO_4)_3{\cdot}14.3H_2O$ | Coagulation | 600 | 100 |
| Aluminum hydroxide | $Al(OH)_3$ | (Hypothetical combination) | 78.0 | 26.0 |
| Ammonia | $NH_3$ | Chloramine disinfection | 17.0 | ---- |
| Ammonium fluosilicate | $(NH_4)_2SiF_6$ | Fluoridation | 178 | ---- |
| Ammonium sulfate | $(NH_4)_2SO_4$ | Coagulation | 132 | 66.1 |
| Calcium bicarbonate | $Ca(HCO_3)_2$ | (Hypothetical combination) | 162 | 81.0 |
| Calcium carbonate | $CaCO_3$ | Corrosion control | 100 | 50.0 |
| Calcium fluoride | $CaF_2$ | Fluoridation | 78.1 | ---- |
| Calcium hydroxide | $Ca(OH)_2$ | Softening | 74.1 | 37.0 |
| Calcium hypochlorite | $Ca(ClO)_2{\cdot}2H_2O$ | Disinfection | 179 | ---- |
| Calcium oxide (lime) | CaO | Softening | 56.1 | 28.0 |
| Carbon dioxide | $CO_2$ | Recarbonation | 44.0 | 22.0 |
| Chlorine | $Cl_2$ | Disinfection | 71.0 | ---- |
| Chlorine dioxide | $ClO_2$ | Taste and odor control | 67.0 | ---- |
| Copper sulfate | $CuSO_4$ | Algae control | 160 | 79.8 |
| Ferric chloride | $FeCl_3$ | Coagulation | 162 | 54.1 |
| Ferric hydroxide | $Fe(OH)_3$ | (Hypothetical combination) | 107 | 35.6 |
| Ferric sulfate | $Fe_2(SO_4)_3$ | Coagulation | 400 | 66.7 |
| Ferrous sulfate (copperas) | $FeSO_4{\cdot}7H_2O$ | Coagulation | 278 | 139 |
| Fluosilicic acid | $H_2SiF_6$ | Fluoridation | 144 | ---- |
| Hydrochloric acid | HCl | pH adjustment | 36.5 | 36.5 |
| Magnesium hydroxide | $Mg(OH)_2$ | Defluoridation | 58.3 | 29.2 |
| Oxygen | $O_2$ | Aeration | 32.0 | 16.0 |
| Potassium permanganate | $KMnO_4$ | Oxidation | 158 | ---- |
| Sodium aluminate | $NaAlO_2$ | Coagulation | 82.0 | ---- |
| Sodium bicarbonate (baking soda) | $NaHCO_3$ | pH adjustment | 84.0 | 84.0 |
| Sodium carbonate (soda ash) | $Na_2CO_3$ | Softening | 106 | 53.0 |
| Sodium chloride (common salt) | NaCl | Ion-exchange regeneration | 58.4 | 58.4 |
| Sodium fluoride | NaF | Fluoridation | 42.0 | ---- |
| Sodium hexametaphosphate | $(NaPO_3)_n$ | Corrosion control | ---- | ---- |
| Sodium hydroxide | NaOH | pH adjustment | 40.0 | 40.0 |
| Sodium hypochlorite | NaClO | Disinfection | 74.4 | ---- |
| Sodium silicate | $Na_4SiO_4$ | Coagulation aid | 184 | ---- |
| Sodium fluosilicate | $Na_2SiF_6$ | Fluoridation | 188 | ---- |
| Sodium thiosulfate | $Na_2S_2O_3$ | Dechlorination | 158 | ---- |
| Sulfur dioxide | $SO_2$ | Dechlorination | 64.1 | ---- |
| Sulfuric acid | $H_2SO_4$ | pH adjustment | 98.1 | 49.0 |
| Water | $H_2O$ | ---- | 18.0 | ---- |

**Practice Problems: WATER SUPPLY ENGINEERING**

Untimed

1. A water treatment plant has four rapid sand filters, each of which has a capacity of 600,000 gallons per day. Each filter is backwashed once a day for eight minutes. (a) Determine the inside dimensions of the sand filter. (b) What percentage of the filtered water is used for backwashing?

2. Design a circular, mechanically-cleaned clarifier using the following specifications:

| | |
|---|---|
| flow rate: | 2.8 million gallons/day |
| detention period: | 2 hours |
| surface loading: | 700 gallons/ft$^2$-day |

If the initial flow rate is only 1.1 million gallons per day, what are the surface loading and average detention periods?

3. A town's water supply is to be taken from a river with the following quality characteristics:

| | |
|---|---|
| turbidity: | varies between 20 and 100 units |
| total hardness: | less than 60 mg/l (as $CaCO_3$) |
| coliform count: | varies between 200 and 1000 per 100 ml |

The town has a design population of 15,000 people and an average consumption of 110 gpcd. (a) What rate (gpm) should the distribution be designed to carry? (b) What total filter area would you recommend? (c) Is softening required? (d) If 2 mg/l of chlorine are required to obtain the necessary chlorine residual, how many pounds per 24 hours of chlorine are required?

4. A town's water supply has the following hypothetical ion concentrations. (a) What is the total hardness in mg/l (as $CaCO_3$)? (b) How much lime ($Ca(OH)_2$) and soda ash are required to react with the carbonate hardness?

| | | | |
|---|---|---|---|
| $Ca^{++}$ | 80.2 mg/l | $CO_3^{--}$ | 0 |
| $Na^{+}$ | 46.0 mg/l | $Mg^{++}$ | 24.3 mg/l |
| $NO_3^{-}$ | 0 | $Fl^{-}$ | 0 |
| $Cl^{-}$ | 85.9 mg/l | $SO_4^{--}$ | 125 mg/l |
| $CO_2$ | 19 mg/l | $Fe^{++}$ | 1.0 mg/l |
| $Al^{+++}$ | 0.5 mg/l | $HCO_3^{-}$ | 185 mg/l |

5. The following concentrations of inorganic compounds are found during a routine analysis of a city's water supply.

| | |
|---|---|
| $Ca(HCO_3)_2$ | 137 mg/l (as $CaCO_3$) |
| $CO_2$ | 0 mg/l |
| $MgSO_4$ | 72 mg/l (as $CaCO_3$) |

(a) How many pounds of lime ($Ca(OH)_2$) and soda ash ($Na_2CO_3$) are required to soften one million gallons of this water to 100 mg/l if 30 mg/l excess lime is required for a complete reaction? (b) How many pounds of salt would be required if a zeolite process is used with the following characteristics:

| | |
|---|---|
| exchange capacity: | 10,000 grains hardness/ft$^3$ |
| salt requirement: | .5 pound/1000 grains hardness removed |

6. A water treatment plant has five square rapid sand filters, each of which has a capacity of one million gallons per day. (a) What are the recommended dimensions for the filters? (b) If each filter is backwashed each day for 5 minutes, what percentage of the plant's filtered water is used for backwashing?

7. Water from an underground aquifer is to be reduced from 245 mg/l hardness to 80 mg/l hardness by the zeolite process. (a) Draw a line schematic of the process used to accomplish this reduction. (b) What is the time between regenerations of the softener if the exchanger has the following characteristics:

| | |
|---|---|
| flow volume: | 20,000 gallons per day |
| exchanger resin volume: | 2 cubic feet |
| resin exchange capacity: | 20,000 grains per cubic foot |

8. A 12″ standard strength clay sewer pipe is to be installed under a backfill of 11 feet of saturated topsoil which has a density of 120 pounds per cubic foot. The pipe strength is 1,500 pounds per foot. Design a bedding using a safety factor of 1.5.

9. A settling tank has an overflow rate of 100,000 gal/ft$^2$-day. Water carrying grit of various sizes is introduced. The grit has the following distribution of settling velocities:

| settling velocity (fpm) | weight fraction remaining |
|---|---|
| 10.0 | .54 |
| 5.0 | .45 |
| 2.0 | .35 |
| 1.0 | .20 |
| .75 | .10 |
| .50 | .03 |

What is the percentage by weight of the grit removed?

10. What is the settling velocity of a spherical sand particle which has a specific gravity of 2.6 and a diameter of 1 millimeter?

Timed

1. A flocculator tank with a volume of 200,000 ft$^3$ uses a paddle wheel to disperse chemicals throughout the mixture. Use the values listed to determine (a) the required paddle area, (b) the drag force on the paddle, and (c) the theoretical power requirement to drive the paddle.

| | |
|---|---|
| mean velocity gradient: | 45 1/sec |
| water temperature: | 60°F |
| drag coefficient: | 1.75 |
| paddle-tip velocity: | 2 ft/sec |
| relative water/paddle velocity: | 1.5 ft/sec |

2. The figure shows the cumulative per capita water demand for a peak day in an area expected to attain a population of 40,000 after 20 years.

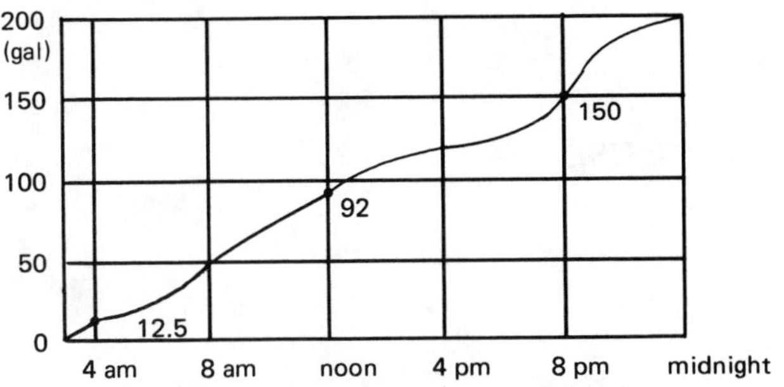


(a) Determine the daily per capita demand for a peak day. (b) Assuming uniform operation, what storage volume is required in the treatment plant for all uses, including fire fighting demand? Assume 24 hour per day operation. (c) Assume that the pumping station only runs from 4 AM to 8 AM. If pumping is uniform during this period, what storage is required to meet all uses including fire fighting?

3. You are to determine whether two sedimentation basins have been correctly designed. The current design features are:

| | |
|---|---|
| design average daily flow | 1.5 MGD (2 basins) |
| configuration | 2 basins, 90′ × 16′ × 12′ deep |
| total weir length per basin | 48 ft |
| 3-month sustained average low | .7 (design average daily flow) |
| 3-month sustained average high | 2.0 (design average daily flow) |

The basins must meet the following government standards.

| | |
|---|---|
| minimum retention time | 4.0 hours |
| maximum weir load | 20,000 gpd/ft |
| maximum velocity | .5 ft/min |

4. Two boreholes are located at points *A* and *B*. The soil between them has a permeability of 3 EE-5 ft/sec and a porosity of .4. Water flows from *B* to *A* in the water table. A 500′ × 100′ hazardous waste containment cell will be built between points *A* and *B* as indicated. The bottom of the containment cell must be at least 5 feet above the water table at all points. (a) What is the hydraulic (ground water) gradient between points *A* and *B*? (b) What is the minimum elevation of the containment cell? (c) Assume that the protective lining of the cell fails and contaminant reaches the groundwater. How long will it take for the contaminated water to reach point *A*?

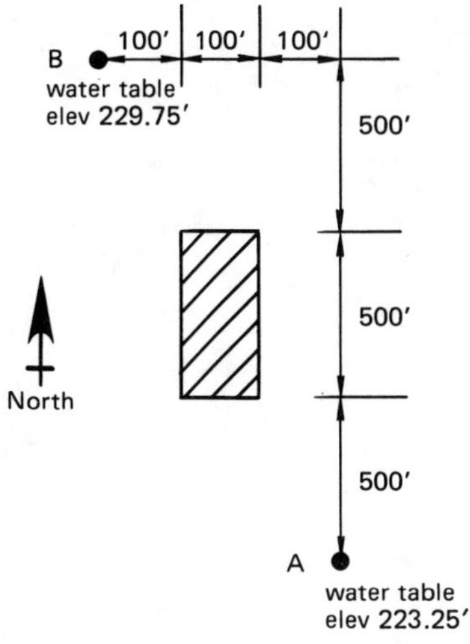

RESERVED FOR FUTURE USE

# 8 WASTE-WATER ENGINEERING

*Nomenclature*

| | | |
|---|---|---|
| A | area | $ft^2$ |
| BOD | biochemical oxygen demand | mg/l |
| C | concentration | mg/l |
| COD | chemical oxygen demand | mg/l |
| d | stream flow depth, or particle diameter | ft, or mm |
| D | oxygen deficit | mg/l |
| DO | dissolved oxygen | mg/l |
| f | Darcy friction factor | – |
| F | effective number of passes | – |
| g | acceleration due to gravity (32.2) | $ft/sec^2$ |
| k | rate constant | –/days |
| $K_D$ | deoxygenation coefficient | –/days |
| $K_R$ | reoxygenation coefficient | –/days |
| $K_t$ | oxygen transfer coefficient | –/hrs |
| L | loading | varies |
| MLSS | mixed liquor suspended solids | mg/l |
| P | population | 1000's of people |
| Q | flow quantity | gallons/day |
| Q′ | flow quantity | cfs |
| R | ratio | – |
| ROT | rate of oxygen transfer | mg/l-hr |
| s | sludge suspended solids | decimal |
| SA | sludge age | days |
| SG | specific gravity | – |
| SS | suspended solids | mg/l |
| SVI | sludge volume index | |
| t | time | days |
| T | temperature | °C |
| v | velocity | ft/sec |
| V | volume | ml |
| w | weighting factor | – |
| W | sludge removal rate | lbm/day (dry) |
| x | a fraction | decimal |

*Symbols*

| | | |
|---|---|---|
| $\eta$ | efficiency | – |
| $\beta$ | oxygen saturation coefficient | – |

*Subscripts*

| | |
|---|---|
| a | aeration |
| as | activated sludge |
| c | critical |
| d | discharge or detention |
| D | deoxygenation |
| e | equivalent |
| f | final |
| F–M | food to microorganism |
| H | hydraulic |
| i | initial |
| ML | mixed liquor |
| o | immediately after mixing, or original |
| p | particle, or primary |
| R | recirculation, return, or reoxygenation |
| RS | return sludge |
| req | required at discharge |
| s | standard 5-day, or secondary |
| sat | saturated |
| ss | suspended solids |
| t | at time t |
| T | at temperature T |
| u | ultimate carbonaceous |
| w | raw wastewater |

## 1 CONVERSIONS

| multiply | by | to obtain |
|---|---|---|
| acre-feet | 43.56 | 1000's of cubic feet |
| cubic feet | 7.48 | gallons |
| cubic feet/sec (cfs) | 0.6463 | MGD |
| cubic feet/sec (cfs) | 448.8 | gpm |
| gallons | 0.1337 | cubic feet |

| | | |
|---|---|---|
| gallons/day (gpd) | 1.547 EE–6 | cfs |
| gallons/min (gpm) | 0.002228 | cfs |
| gallons/acre-day (gad) | 2.296 EE–5 | gallons/day-$ft^2$ |
| gallons/$ft^2$-day ($gpd/ft^2$) | 0.04356 | million gallons/ acre-day |
| million gallons/ acre-day (mgad) | 22.96 | $gpd/ft^2$ |
| million gallons/ day (MGD) | 1.547 | cfs |
| milligrams (mg) | 2.205 EE–6 | pounds |
| milligrams/liter (mg/l) | 8.345 | pounds/million gallons |
| meters | 3.281 | feet |
| millimeters/meter | 0.012 | in/foot |
| $m^3/m^2$-day | 24.54 | gallons/$ft^2$-day |
| $m^3$/m-day | 80.52 | gallons/ft-day |
| miles per hour (mph) | 1.4667 | ft/sec |
| pounds | 4.536 EE5 | milligrams |
| pounds/acre-ft-day | 0.02296 | 1bm/1000 $ft^3$-day |
| pounds/1000 $ft^3$-day | 43.56 | 1bm/acre-ft-day |
| pounds/1000 $ft^3$-day | 133.7 | 1bm/million gal-day |
| pounds/million gallons | 0.1198 | mg/l |
| pounds/million gallons-day | 0.00748 | lbm/1000 $ft^3$-day |

## 2 DEFINITIONS

Activated sludge: Solids from aerated settling tanks which are rich in bacteria.

Aerated lagoon: A holding basin into which air is mechanically introduced to speed up aerobic decomposition.

Appurtenance: A thing which belongs with (or is designed to complement) something else. For example, a manhole is a sewer appurtenance.

Bioactivation process: A process using sedimentation, trickling filter, and secondary sedimentation before adding activated sludge. Aeration and final sedimentation are the follow-up processes.

Biosorption process: A process which mixes raw sewage and sludge which have been pre-aerated in a separate tank.

Biota: The flora and fauna of a region, process, or tank.

Branch sewer: A sewer off the main sewer.

Bulking: See 'Sludge bulking.'

Carbonaceous demand: Oxygen demand due to biological activity in a water sample.

Chemical precipitation: Causing suspended solids to settle out by adding coagulating chemicals.

Clean-out: A pipe through which snakes can be pushed to unplug a sewer.

Combined system: A system using a single sewer for domestic waste and storm water.

Comminutor: A device which cuts solid waste into small pieces.

Complete mixing: Mixing accomplished by mechanical means (stirring).

Cunette: A small channel in the invert of a large combined sewer for dry weather flow.

Deoxygenation: The act of removing dissolved oxygen from water.

Dewatering: Removal of excess moisture from sludge waste.

Digestion: Conversion of sludge solids to gas.

Dilution disposal: Relying on a large water volume (lake or stream) to dilute waste to an acceptable concentration.

Domestic waste: Waste which originates from households.

Effluent: That which flows out of a process.

Elutriation: A counter-current sludge washing process used to remove dissolved salts.

First-stage demand: See 'Carbonaceous demand.'

Floatation: Adding chemicals or bubbling air through waste to get solids to float to the top as scum.

Force main: A sewer line which is pressurized.

Humus: A greyish brown sludge consisting of relatively large particle biological debris, as is the material sloughed off from a trickling filter.

Infiltration: Ground water which enters sewer pipes through cracks and joints.

Influent: Flow entering a process.

Inverted siphon: A sewer line which drops below the hydraulic gradient.

Kraus process: Mixing raw sewage, activated sludge, and material from sludge digesters.

Lamp holes: Sewer inspection holes large enough to lower a lamp into but too small for a man.

Lateral: A sewer line which goes off at right angles to another.

Main: A large sewer from which all other branches originate.

Malodorous: Offensive smelling.

Mesophilic bacteria: Bacteria growing between 10 and 40°C , with an optimum temperature of 37°C. 40°C is, therefore, the upper limit for most wastewater processes.

Mohlman index: Same as the 'Sludge volume index.'

Nitrogenous demand: Oxygen demand from nitrogen-consuming bacteria.

Outfall: The pipe which discharges completely treated wastewater into a lake, stream, or ocean.

Partial treatment: Primary treatment only.

Post-chlorination: Addition of chlorine after all other processes have been completed.

Pre-chlorination: Addition of chlorine prior to sedimentation to help control odors and to aid in grease removal.

Putrefaction: Anaerobic decomposition of organic matter with accompanying foul odors.

Refractory: Dissolved organic materials which are biologically resistant and difficult to remove.

Regulator: A device or weir which deflects large volume flows into a special high-capacity sewer.

Sag pipe: See 'Inverted siphon.'

Second stage demand: See 'Nitrogenous demand.'

Seed: The activated sludge initially taken from the secondary settling tank and returned to the aeration tank to start the activated sludge process.

Separate system: Separate sewers for domestic and storm waste water.

Septic: Produced by putrefaction.

Sludge bulking: Failure of suspended solids to completely settle out.

Split chlorination: Addition of chlorine prior to sedimentation and after final processing.

Submain: See 'Branch.'

Supernatant: The clarified liquid floating on top of a digesting sludge layer.

Thermophilic bacteria: Bacteria which thrive in the 45°C to 75°C range (optimum near 55°C).

Volatile solid: Solid material in a water sample or in sludge which can be burned or vaporized at high temperature.

Wet well: A short-term storage tank containing a pump or pump entrance, and into which the raw influent is brought.

Zooglea: The gelatinous film of aerobic organisms which cover the rocks in a trickling filter.

## 3 WASTEWATER QUALITY CHARACTERISTICS

### A. DISSOLVED OXYGEN

Fish and most aquatic life require oxygen.[1] The biological decomposition of organic solids is also dependent on oxygen. If the dissolved oxygen content of water is less than the saturated values given in appendix B, there is good reason to believe that the water is organically polluted. Other reasons for measuring the dissolved oxygen concentration are for aerobic treatment monitoring, aeration process monitoring, BOD testing, and pipe corrosion studies.

The difference between the saturated and actual dissolved oxygen concentrations is known as the *oxygen deficit.*

$$D = DO_{\text{sat}} - DO \qquad 8.1$$

The oxygen deficit is reduced by aerating the water (i.e., the dissolved oxygen concentration is increased). An exponential decay is traditionally used to predict the oxygen deficit as a function of time. Equation 8.2 assumes that oxygen is not being depleted during the reoxygenation process.

$$D_t = D_o 10^{-K_R t} \qquad 8.2$$

$K_R$ is the *reoxygenation coefficient*, which depends on the type of flow and temperature.[2] Reoxygenation coefficients are also given for use with a different logarithmic base.

$$D_t = D_o e^{-K'_R t} \qquad 8.3$$

[1] 4–6 mg/l is the generally accepted range of dissolved oxygen required to support fish populations. 5 mg/l is adequate, as is verifiable from high-altitude trout lakes. However, 6 mg/l is preferable, particularly for large fish populations.

[2] $K_R$ may be written as $K_1$ in the literature.

The constants $K_R$ and $K'_R$ are not the same, but they are related.[3]

$$K'_R = 2.3K_R \qquad 8.4$$

Table 8.12 lists representative values of $K_R$.

## B. BIOCHEMICAL OXYGEN DEMAND

When oxidizing organic waste material in water, biological organisms remove oxygen from the water. Therefore, oxygen use is an indication of the organic waste content. The biochemical oxygen demand (BOD) of a biologically active sample is given by equation 8.5:

$$\text{BOD}_s = \frac{DO_i - DO_f}{\frac{V_{\text{sample}}}{V_{\text{sample}} + V_{\text{dilution}}}} \qquad 8.5$$

BOD is determined by adding a measured amount of wastewater (which supplies the organic material) to a measured amount of dilution water (which reduces toxicity and supplies dissolved oxygen). An oxygen use curve similar to that in figure 8.1 will result. (More than one identical sample must be prepared in order to determine initial and final concentrations of dissolved oxygen.)

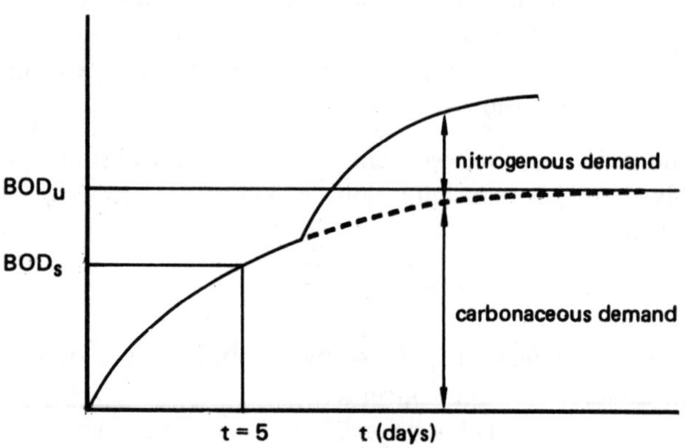


**Figure 8.1** BOD Time Curve

The deviation from the expected exponential growth curve in figure 8.1 is due to *nitrification* or *nitrogenous demand.* Nitrification is the use of oxygen by *autotrophic bacteria.*[4] Such bacteria use fixed carbon as food. (For example, the carbon in carbon dioxide is used by autotrophic bacteria.) Autotrophic bacteria oxidize ammonia to nitrites and nitrates. However, the number of autotrophic bacteria is small. Generally, six to ten days are required for the autotrophic population to become sufficiently large enough to affect a BOD test. Therefore, the standard BOD test is terminated before the autotrophic contribution to BOD becomes significant.

The standard BOD test typically calls for a 5-day incubation period at 20°C. The BOD at any time can be found from equation 8.6.

$$\text{BOD}_t = \text{BOD}_u(1 - 10^{-K_D t}) \qquad 8.6$$

$K_D$ is the *deoxygenation rate constant*, typically taken as 0.1. The ultimate BOD cannot be found from long term studies due to the effect of nitrogen-consuming bacteria in the sample. However, if $K_D$ is 0.1, the ultimate BOD can be found from equation 8.7.

$$\text{BOD}_u \approx 1.47\,\text{BOD}_s \qquad 8.7$$

$K_D$ for other temperatures can be found from equation 8.8. (The 1.047 constant is often quoted in literature. Recent research suggests 1.135 for 4°C to 20°C, and 1.056 for 20°C to 30°C.)

$$K_{D,T} = (1.047)^{T-20} K_{D,20°\text{C}} \qquad 8.8$$

The variation in BOD with temperature is given by equation 8.9.

$$\text{BOD}_T = \text{BOD}_{20°\text{C}}(0.02T + 0.6) \qquad 8.9$$

**Table 8.1**
Typical Values of $K_D$

| | |
|---|---|
| treatment plant effluents | 0.05–0.10 |
| highly polluted shallow streams | 0.25 |

*Example 8.1*

Ten 5-ml samples of wastewater are placed in 300 ml BOD bottles. Half of the bottles are titrated immediately with an average initial concentration of dissolved oxygen of 7.9 mg/l. The remaining bottles are incubated for 5 days, after which the average dissolved oxygen is determined to be 4.5 mg/l. What is the standard BOD and ultimate carbonaceous BOD assuming $K_D = 0.13$?

From equation 8.5:

$$\text{BOD}_s = \frac{7.9 - 4.5}{\frac{5}{300}} = 204\ \text{mg/l}$$

[3] $K'_R$ may be written as $K_R$ (*base e*) in the literature.

[4] Most bacteria in wastewater are heterotrophic. *Heterotrophic bacteria* use organic carbon as food.

From equation 8.6, the ultimate BOD is

$$\text{BOD}_u = \frac{204}{1 - 10^{(-0.13)(5)}} = 263 \text{ mg/l}$$

If a sample of industrial wastewater is taken, it will probably lack sufficient microorganisms to metabolize the organic matter. In such a case, seed organisms must be added. The BOD for seeded experiments is found by measuring dissolved oxygen in the seeded sample after 15 minutes ($DO_i$) and after 5 days ($DO_f$), as well as the dissolved oxygen of the seed material itself after 15 minutes ($DO_i^*$) and after 5 days ($DO_f^*$).

$$\text{BOD}_s = \frac{DO_i - DO_f - x[DO_i^* - DO_f^*]}{\frac{V_{\text{sample}}}{V_{\text{sample}} + V_{\text{dilution}}}} \qquad 8.10$$

$$x = \frac{\text{volume of seed added to sample}}{\text{volume of seed used to find } DO^*} \qquad 8.11$$

The BOD of domestic waste is typically taken as 0.17 to 0.20 pounds per capita-day, excluding industrial wastes. This makes it possible to calculate the *population equivalent* of any BOD loading.

$$P_e = \frac{(\text{BOD})(Q)(8.345\ \text{EE--9})}{(0.17)} \quad \text{(in 1000's of people)} \qquad 8.12$$

Values of BOD for various industrial wastewaters are given in table 8.2.

BOD of 100 mg/l is considered a *weak wastewater*; BOD of 200 to 250 mg/l is considered a *medium strength wastewater*; above 300 mg/l, it is considered to be a *strong wastewater*.

## C. RELATIVE STABILITY

The relative stability test is much easier to perform than the BOD test, although it is much less accurate. The relative stability of an effluent is defined as the percent of initial BOD that has been satisfied. The test consists of taking a sample of effluent and adding a small amount of methylene blue dye. When all oxygen has been removed from the water, anaerobic bacteria start to remove the dye. The time for the color to start degrading is known as the *stabilization time* or *decoloration time*.

The relative stability can be found from the stabilization time by using table 8.3.

**Table 8.3**
Relative Stability (at 20°C)

| stabilization time (days) | relative stability % | stabilization time (days) | relative stability % |
|---|---|---|---|
| 1/2 | 11 | 8 | 84 |
| 1 | 21 | 9 | 87 |
| 1 1/2 | 30 | 10 | 90 |
| 2 | 37 | 11 | 92 |
| 2 1/2 | 44 | 12 | 94 |
| 3 | 50 | 13 | 95 |
| 4 | 60 | 14 | 96 |
| 5 | 68 | 16 | 97 |
| 6 | 75 | 18 | 98 |
| 7 | 80 | 20 | 99 |

**Table 8.2**
Typical BOD and COD of Industrial Effluents

| industry/type of waste | BOD | COD |
|---|---|---|
| canning | | |
| corn | 19.5 lbm/ton corn | |
| tomatoes | 8.4 lbm/ton tomatoes | |
| dairy milk processing | 1150 lbm/ton raw milk<br>1000 mg/l | 1900 mg/l |
| beer brewing | 1.2 lbm/barrel beer | |
| commercial laundry | 1250 lbm/1000 pounds dry<br>700 mg/l | 2400 mg/l |
| slaughterhouse (meat packing) | 7.7 lbm/animal<br>1400 mg/l | 2100 mg/l |
| papermill | 121 lbm/ton pulp | |
| synthetic textile | 1500 mg/l | 3300 mg/l |
| chlorophenolic manufacturing | 4300 mg/l | 5400 mg/l |
| milk bottling | 230 mg/l | 420 mg/l |
| cheese production | 3200 mg/l | 5600 mg/l |
| candy production | 1600 mg/l | 3000 mg/l |

*Example 8.2*

A sample treatment plant effluent begins to clarify after 13 days. What percent of the original BOD remains unsatisfied?

From table 8.3, the relative stability is 95%. Therefore, only 5% of the initial BOD remains unsatisfied.

## D. CHEMICAL OXYGEN DEMAND

Unlike BOD, which is a measure of oxygen removed by biological organisms, chemical oxygen demand (COD) is a measure of maximum oxidizable substances. Therefore, COD is an excellent measure of *effluent strength.*

COD testing is required in environments of chemical pollution. In such environments, the organisms necessary to metabolize organic compounds may not exist. Furthermore, the toxicity of the water may make the standard BOD test impossible to carry out. The COD test also produces results faster than the BOD test. COD test results are usually available in a matter of hours.

If the toxicity is low, BOD and COD test results can be correlated. The $BOD_s$/COD ratio typically varies from 0.4 to 0.8. This is a wide range, but for any given treatment plant and waste type, the correlation is essentially constant. The correlation can, however, vary along the treatment path.

## E. CHLORINE DEMAND

Chlorination destroys bacteria, hydrogen sulfide, and other noxious substances. For example, hydrogen sulfide is oxidized according to equation 8.13.

$$H_2S + 4H_2O + 4Cl_2 \rightarrow H_2SO_4 + 8HCl \qquad 8.13$$

*Chlorine demand* is the amount of chlorine (or its chloramine or hypochlorite equivalent) required to give a 0.5 mg/l residual after 15 minutes of contact time. 15 minutes is the recommended contact and mixing time prior to discharge since this period will kill nearly all pathogenic bacteria in the water. Typical doses for wastewater effluent are given in table 8.4.

**Table 8.4**
Typical Chlorine Doses

| final process | dose (mg/l) |
|---|---|
| no treatment (straight discharge) | 10–30 |
| secondary sand filter | 2–6 |
| secondary activated sludge | 2–8 |
| secondary trickling filter | 3–15 |
| primary sedimentation | 5–25 |

In actuality, the chlorine dose needs to be determined by careful monitoring of coliform counts and free residuals, since there are several ways that chlorine can be used up without producing significant disinfection. Only after uncombined (free) chlorine starts showing up is it assumed that all chemical reactions and disinfection are complete.[5]

Because of their reactivity, chlorine is initially used up in the neutralization of hydrogen sulfide and the rare ferrous and manganous ($Fe^{++}$ and $Mn^{++}$) ions. The resulting HCl, $FeCl_2$, and $MnCl_2$ ions do not contribute to disinfection. They are known as *unavailable combined residuals.*

Plants and animals use nitrogen. Bacterial decomposition and the hydrolysis of urea produces ammonia, $NH_3$. This ammonia, once it enters the wastewater stream, forms ammonium ion, $NH_4^+$, also known as *ammonia nitrogen.*

Ammonia nitrogen combines with chlorine to form the family of *chloramines.* Depending on the water pH, *monochloramines* ($NH_2Cl$), *dichloramines* ($NHCl_2$), or *trichloramines* (nitrogen trichloride, $NCl_3$) may form.[6] Chloramines have long-term disinfection capabilities, and chloramines are therefore known as *available combined residuals.* Equation 8.14 is a typical chloramine formation reaction.

$$NH_4^+ + HOCl \rightleftharpoons NH_2Cl + H_2O + H^+ \qquad 8.14$$

The continued addition of chlorine after chloramine formation changes the pH, and *chloramine destruction* begins. Chloramines are converted to nitrogen gas ($N_2$) and nitrous oxide ($N_2O$). The destruction of chloramines continues with the continued application of chlorine, until no ammonia remains in the water. The point at which all ammonia has been removed is known as the *breakpoint.* Equation 8.15 is a typical chloramine destruction reaction.

$$2NH_2Cl + HOCl \rightleftharpoons N_2 + 3HCl + H_2O \qquad 8.15$$

In the *breakpoint chlorination* method, additional chlorine is added after the breakpoint in order to obtain free chlorine residuals. The free residuals have a high disinfection capacity. Typical free residuals are free chlorine ($Cl_2$), hypochlorous acid (HOCl), and hypochlorite ions. Equations 8.16 and 8.17 illustrate the formation of these free residuals.

$$Cl_2 + H_2O \rightarrow HCl + HOCl \qquad 8.16$$

$$HOCl \rightarrow H^+ + ClO^- \qquad 8.17$$

[5] Chlorine kills most bacteria, but many viruses are resistant.

[6] Lower pH favors the formation of di- and trichloramines.

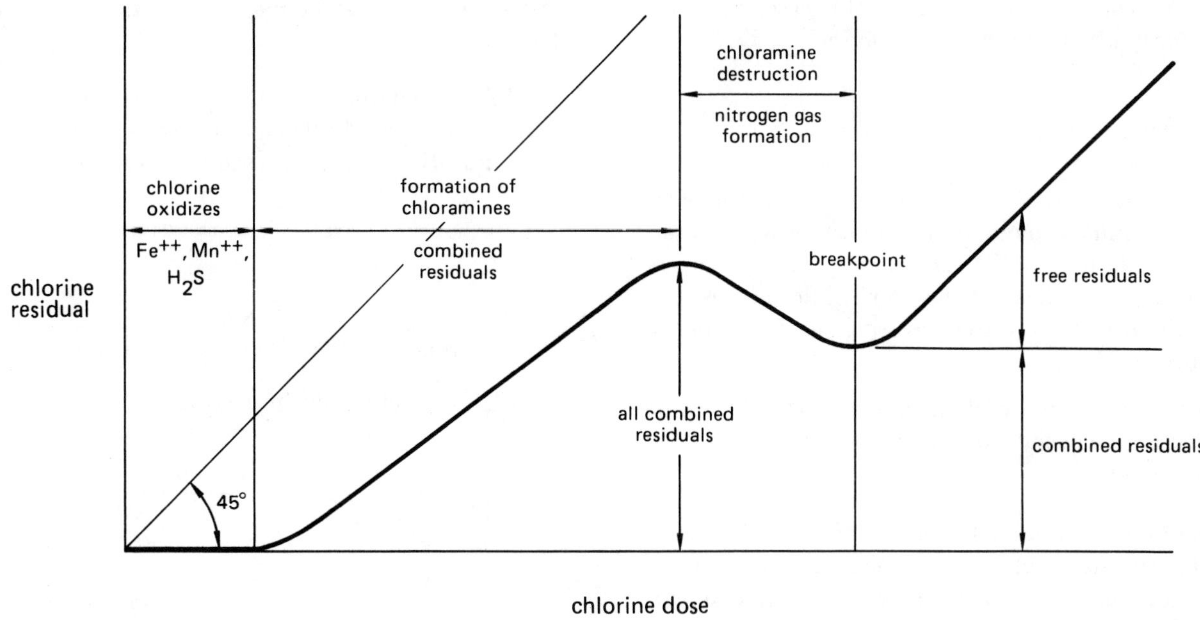


**Figure 8.2** Breakpoint Chlorination

There are several problems associated with breakpoint chlorination.

- It may not be economical to use breakpoint chlorination unless the ammonia nitrogen has been reduced.
- Free chlorine residuals favor the formation of trihalomethanes. Where free residuals are not permitted, the water may need to be dechlorinated using sulfur dioxide gas or sodium bisulfate. (Where small concentrations of free residuals are permitted, dechlorination may be needed only during the dry months. During winter storm months, the chlorine residuals may be adequately diluted with rain water.)

F. GREASE

Greases are organic substances including fats, vegetable and mineral oils, waxes, fatty acids from soaps, and other hydrocarbons. Grease's low solubility causes adhesion problems in pipes and tanks, reduces contact area during various filtering processes, and produces sludge which is difficult to dispose of.

G. VOLATILE ACIDS

Volatile acids (acetic, propionic, and butyric) occur in anaeorobically digested sludge. These acids can be used to indicate the completion of a sludge digestion process. Acid content is given in mg/l as acetic acid.

H. SUSPENDED SOLIDS

Suspended solids, as in water supply engineering, can be categorized in several ways. Generally, suspended solids constitute only a small amount of the incoming flow, less than 1/10%. Together with *dissolved solids*, suspended solids constitute *total solids*. Figure 8.3 illustrates the relationships between the various solids categories. A further division of each category into organic and inorganic solids is possible.

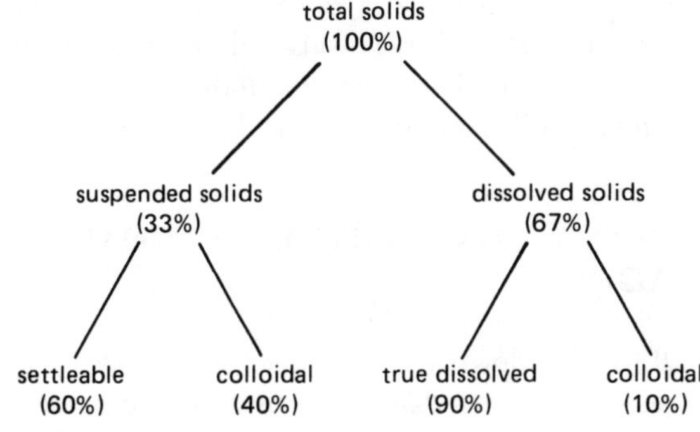


**Figure 8.3** Family of Solids
(Typical percentages given.)

The term *volatile solids* can be used as a measure of organic pollutants capable of affecting the oxygen content of the flow. Volatile solids, as in water supply testing, are measured by igniting filtered solids, and measuring the decrease in mass.

*Refractory pollutants* are solids which are difficult to remove by common processes. In this case, the term "refractory" is used to mean "stubborn."

## 4 DISINFECTION

Chlorine gas is the least expensive, and therefore the most common, method of disinfecting wastewater. However, chlorine gas is toxic, corrosive, and displaces oxygen since it is heavier than air. Chlorine gas also lowers the pH of the water, favoring the formation of combined residuals.

Because of these disadvantages, alternatives to chlorine gas need to be considered when recommending the disinfection method.

- **Hypochlorites**: Both sodium and calcium hypochlorite are solids that dissolve in water. They have a limited shelf life, and are susceptible to photodecomposition. Hypochlorites are less effective and slightly more expensive than chlorine gas.
- **Chlorine Dioxide Gas**: $Cl_2O$ is explosive and must be generated on-site. It reacts with many compounds, requiring larger doses than chlorine gas. However, it combines with organics without combining with ammonia.
- **Ozone Gas**: Ozone is one of the most effective oxidizing agents. In addition to its disinfection capabilities, ozone also increases the dissolved oxygen content of water. Ozone is toxic and corrosive. It must be generated at the point of application. Because of its very short half-life, step feeding is required to obtain the necessary contact period.
- **Exotics**: Other methods exist, but are typically high in cost. These exotic alternatives include bromine (bromine chloride), iodine, silver oxide, gamma radiation, and ultraviolet radiation.

## 5 TYPICAL COMPOSITION OF DOMESTIC SEWAGE

Not all sewage flows are the same. Some sewages are stronger than others. Table 8.5 lists typical values for strong and weak domestic sewages. A medium classification would be approximately midway between the values for strong and weak.

## 6 WASTEWATER QUALITY STANDARDS

Applicable wastewater quality standards have been set by both the Water Pollution Control Act and the Environmental Protection Agency. General standards set by the Water Pollution Control Act are given in table 8.6. These standards must be met by facilities that receive federal funding.

The EPA's standards for secondary treatment are given in terms of 5-day BOD, suspended solids, coliform count, and pH. Table 8.7 presents typical values.

**Table 8.5**
Strong and Weak Domestic Sewages

(All concentrations in mg/l unless noted.)

| constituent | strong | weak |
|---|---|---|
| solids, total | 1200 | 350 |
| dissolved, total | 850 | 250 |
| fixed | 525 | 145 |
| volatile | 325 | 105 |
| suspended, total | 350 | 100 |
| fixed | 75 | 30 |
| volatile | 275 | 70 |
| settleable solids, (ml/liter) | 20 | 5 |
| biochemical oxygen demand, 5-day, 20°C | 300 | 100 |
| total organic carbon | 300 | 100 |
| chemical oxygen demand | 1000 | 250 |
| nitrogen, (total as N) | 85 | 20 |
| organic | 35 | 8 |
| free ammonia | 50 | 12 |
| nitrites | 0 | 0 |
| nitrates | 0 | 0 |
| phosphorus (total as P) | 20 | 6 |
| organic | 5 | 2 |
| inorganic | 15 | 4 |
| chlorides | 100 | 30 |
| alkalinity (as $CaCO_3$) | 200 | 50 |
| grease | 150 | 50 |

## 7 DESIGN FLOW QUANTITY

Approximately 70 to 80% of a community's domestic and industrial water use will return as wastewater. This water is discharged into the sewer systems, which may be different or the same as storm drains. Therefore, the nature of the return system must be known before sizing can occur.

*Sanitary sewer* sizing can often be based on an average of 100–125 gpcd. There will be variations with time in the flow, although the variations are not as pronounced as they are for water supply.

**Table 8.6**
Typical Surface Water Standards

| water use | minimum dissolved oxygen (mg/l) | maximum dissolved solids | maximum coliforms per 100 ml |
|---|---|---|---|
| domestic use (food preparation) | 6 | * | none |
| water contact recreation | 4 to 5 | * | 1000 total ave.<br>200 fecal ave.<br>not more than 10% exceeding 400 fecal (2000 total) |
| fisheries | 4 to 6 | * | 5000 ave. |
| industrial supply | 3 to 5 | 750 to 1500 mg/l | – |
| agricultural irrigation | 3 to 5 | 750 to 1500 mg/l | – |
| shellfish harvesting | 4 to 6 | * | 70 total ave.<br>not more than 10% exceeding 230 |

* No floating solids or settling solids that form deposits.

Design codes frequently specify a design loading of 400 gpcd (laterals and submains) and 250 gpcd (mains, trunks, and outfall). Both of these include the effect of *infiltration.* Infiltration due to cracks and poor joints is limited by many municipal codes to 500 gallons per day per mile of pipe per inch of diameter. Modern piping materials and joints should be able to reduce this quantity to 200 gpd/inch-mile. Infiltration may also be roughly estimated at 3%–5% of the peak hourly domestic rate, or as 10% of the average rate.

## 8 COLLECTION SYSTEMS

### A. STORM DRAINS AND INLETS

Curb inlets to storm drains should be placed no more than 600 feet apart, and a limit of 300 feet is advisable. Inlets are required at all low points where pondage could occur. A common practice is to install 3 inlets in a sag vertical curve—one at the lowest point and one on each side with an elevation of 0.2 feet above the center inlet. Openings may be of the covered grate type or the curb inlet type.

**Table 8.7**
Typical Secondary Effluent Standards

| quality | average over | discharge maximum |
|---|---|---|
| BOD (5-day) | 30 days | 30 mg/l |
| | 7 days | 45 mg/l |
| | 30 days | 15% of incoming BOD |
| suspended solids | 30 days | 30 mg/l |
| | 7 days | 45 mg/l |
| | 30 days | 15% of incoming SS |
| fecal coliforms** | 30 days | 200 per 100 ml |
| | 7 days | 400 per 100 ml |
| pH | at all times | within 6 to 9 |

**A geometric mean is used, not arithmetic

**Table 8.8**
Variations in Wastewater Flow
(based on the average daily flow)

| description | when/where | variation |
|---|---|---|
| daily peak | 10–12 a.m. (residential) | 225% |
| | constant during day (commercial) | 150% |
| | 12 noon at the outfall | 150% |
| daily minimum | 4–5 a.m. | 40% |
| seasonal peak | late summer | 125% |
| seasonal minimum | winter's end | 90% |
| seasonal average | May, June | 100% |
| maximum peaks | in laterals | 300% |
| | treatment plant influent | 200% |

The capacity of a curb-type opening which diverts 100% of gutter flow is given by equation 8.18. A typical curb depression is 5 inches.

$$Q' = (0.7)\begin{pmatrix}\text{curb}\\ \text{opening}\\ \text{length, ft}\end{pmatrix}\begin{pmatrix}\text{inlet flow} \\ \text{depth, ft}\end{pmatrix} + \begin{pmatrix}\text{curb inlet}\\ \text{depression, ft}\end{pmatrix}^{3/2} \qquad 8.18$$

Grate inlets accepting flows less than 0.4 feet deep have a capacity given by equation 8.19. The bars should be parallel to the flow and at least 18 inches long. Equation 8.19 should also be used for combined curb-grate inlets.

$$Q' = 3\begin{pmatrix}\text{grate}\\ \text{perimeter, ft}\end{pmatrix}\begin{pmatrix}\text{inlet flow}\\ \text{depth, ft}\end{pmatrix}^{3/2} \qquad 8.19$$

B. MANHOLES

Manholes should be provided at junctions and at changes in elevation, direction, size, diameter, and slope of sewers. If the sewer is too small for a man to enter, manholes should be placed every 400 feet to allow for cleaning. A maximum recommended spacing is every 700 feet.

**Table 8.9**
Recommended Manhole Spacing

| pipe diameter | spacing |
|---|---|
| less than 18″ | 400 ft |
| 18″–48″ | 500 |
| more than 48″ | 600 |

C. PIPES

Concrete pipe is commonly used for storm sewers. Circular pipe is used in most applications, although special shapes (arch, egg, elliptical, etc.) are available at extra cost. Concrete pipe in diameters up to 24″ are usually not reinforced, and are available in standard 3 and 4 foot lengths. Reinforced pipe in diameters ranging from 12″ to 144″ is available in lengths from 4 to 12 feet.

Sewer pipes are constructed from clay, concrete, asbestos-cement, steel, cast iron, and plastic derivatives. *Vitrified clay pipe* is especially resistant to acids, alkalines, hydrogen sulfide (septic sewage), erosion, and scour. Clay is typically used for diameters less than 36″. Clay pipe is available in standard diameters of 4″ and 6″ in 2 foot lengths; 8″, 10″, 12″, 15″, 18″, 21″, and 24″ diameters in $2\frac{1}{2}$ foot lengths; and 27″, 30″, 33″, and 36″ diameters in 3 foot lengths.

Two strengths of clay pipe are available. The standard strength is suitable for pipes less than 12″ in diameter for any depth of cover if the '4/3D + 8″' trench width rule is observed. Double strength pipe is recommended for large pipe deeply trenched.

*Asbestos-cement pipe* can be used for both gravity and pressure sewers carrying non-septic and non-corrosive wastes through non-corrosive soils. Light weight and longer laying lengths are the inherent advantages of asbestos-cement pipe.

*Concrete pipe* is primarily used for large diameter (16 inches or larger) trunk and interceptor sewers. In some geographical regions, concrete pipe is used for domestic sewers in smaller pipe sizes. However, concrete domestic lines should be selected only where stale or septic sewage is not anticipated. Concrete pipe can be used for gravity as well as pressure mains. However, it should not be used with corrosive wastes or soils.

*Cast iron pipe* is particularly suited to installations where scour, high velocity waste, and high external loads are anticipated. It can be used for domestic connections, although it is more expensive than clay pipe. Special linings, coatings, wrappings, or encasements are required for corrosive wastes and soils.

*Polyvinyl chloride* (*PVC*) and *acrylonitrile-butadiene-styrene* (*ABS*) are two plastic compositions that can be used for normal domestic sewage and industrial wastewaters. They have excellent resistance to corrosive soils. However, special care must be given to trench loadings and pipe beddings.

ABS plastic may also be combined with concrete reinforcement for collector lines for corrosive domestic sewage and industrial wastes. Such pipe is known as *truss pipe* due to its construction.[7]

For pressure lines, welded steel pipe with an epoxy liner, and cement-lined and coated steel pipe are also used occasionally.

In general, sewers in the collection system (including laterals, interceptors, trunks, and mains) should be at least 8″ in diameter. Building service connections can be as small as 4″ in diameter.

## 9 PIPE FLOW VELOCITIES

Sewer flow velocities greater than 15 ft/sec require special provisions to protect against erosion and momentum effects. 2 ft/sec is often quoted as the minimum *self-cleansing velocity*.[8] However, the minimum design velocity depends on the particulate matter size. Ta-

[7] The plastic is extruded with inner and outer web-connected pipe walls. The voids between the inner and outer walls are filled with lightweight concrete.

[8] Even 1.5 ft/sec is acceptable if the main is occasionally flushed out by peak flow.

ble 8.10 can be used to select velocity for the collection system as well as for the plant treatment system.

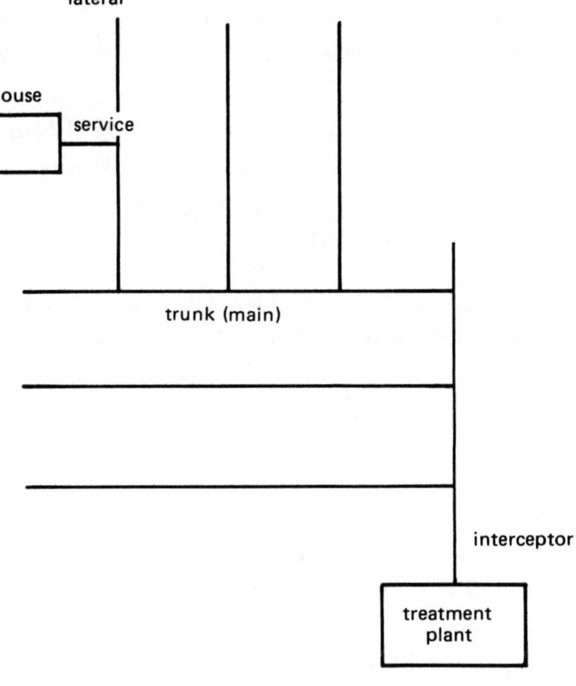


**Figure 8.4** Types of Sewer Lines

**Table 8.10**
Minimum Flow Velocities

| pipe carrying | minimum velocity to keep particles in suspension | minimum resuspension velocity |
|---|---|---|
| raw sewage | 2.5 | 3.5 |
| grit tank effluent | 2 | 2.5 |
| primary settling tank effluent | 1.5 | 2 |
| mixed liquor | 1.5 | 2 |
| trickling filter effluent | 1.5 | 2 |
| secondary settling tank effluent | 0.5 | 1 |

## 10 PUMPS USED IN WASTEWATER PLANTS

Wastewater plant flows should be gravity fed wherever possible. However, there are still many instances where pumping assistance is required. Table 8.11 lists pump types as a function of type of material to be pumped.

**Table 8.11**
Pumps Used in Wastewater Plants

| type of material being pumped | flow rate (gpm) | pump type |
|---|---|---|
| raw sewage | 0 to 50 | pneumatic ejector |
| | 50 to 200 | submersible or end-suction non-clog centrifugal |
| | 200 up | end-suction non-clog centrifugal |
| settled sewage | 0 to 500 | end-suction non-clog centrifugal |
| | 500 up | vertical axial or mixed flow centrifugal |
| sludge (primary, thickened, or digested) | | plunger pump |
| secondary sludge | | end-suction non-clog centrifugal |
| scum | | plunger pump or recessed impeller |
| grit | | recessed impeller centrifugal, pneumatic ejector, or conveyor rake |

## 11 DILUTION PURIFICATION

Dilution purification (also known as *self purification*) refers to discharge of partially treated sewage into a body of water such as a stream or river. If the body is large and is adequately oxygenated, the sewage's BOD may be satisfied without putrefaction. Other conditions which must be monitored besides BOD are oxygen content and suspended solids.

Equation 8.20 can be used to calculate the final concentration of BOD, oxygen, and sediment when the two flows are mixed. Dilution requirements may be expressed in terms of ratios (e.g., 23 stream volumes per discharge volume) or absolute flow quantities (e.g., 4 to 7 cfs per 1000 population).

$$C_1Q_1 + C_2Q_2 = C_f(Q_1 + Q_2) \qquad 8.20$$

*Example 8.3*

Wastewater ($DO$ = 0.9 mg/l, 6 MGD) is discharged into a 50°F stream flowing at 40 cfs. Assuming the stream is saturated with oxygen, what is the oxygen content of the stream immediately after mixing?

From appendix B, the saturated oxygen content at 50°F (10°C) is 11.3 mg/l.

$$(6)\text{MGD}(1.547)\frac{\text{cfs}}{\text{MGD}} = 9.28\,\text{cfs}$$

$$C = \frac{(0.9)(9.28) + (11.3)(40)}{(9.28) + (40)} = 9.34\,\text{mg/l}$$

The *oxygen deficit* is the difference between actual and saturated oxygen concentrations. Since reoxygenation and deoxygenation of a polluted river occur simultaneously, an oxygen deficit will occur only if the reoxygenation rate is less than the deoxygenation rate. If the oxygen content goes to zero, anaerobic decomposition and putrefaction will occur.

The oxygen deficit at any time $t$ is given by the *Streeter-Phelps equation*:

$$D_t = DO_{\text{sat}} - DO_t$$

$$= \frac{K_D\text{BOD}_u}{K_R - K_D}\left(10^{-K_Dt} - 10^{-K_Rt}\right)$$

$$+ D_o\left(10^{-K_Rt}\right) \qquad 8.21$$

$D_t$ is the dissolved oxygen deficit, $t$ is in days, and $\text{BOD}_u$ is the ultimate carbonaceous BOD of the stream immediately after mixing. $K_D$ and $K_R$ are the deoxygenation and reoxygenation rate constants respectively, and $D_o$ is the dissolved oxygen deficit immediately after mixing.[9]

$K_R$ can be approximated by equation 8.22 if field test data is not available.[10]

$$K_{R,20^\circ\text{C}} \approx \frac{3.3v}{d^{1.33}} \qquad 8.22$$

$K_R$ for different temperatures is given by equation 8.23. Typical values of $K_R$ are given in table 8.12.

$$K_{R,T} = (1.016)^{T-20}K_{R,20^\circ\text{C}} \qquad 8.23$$

**Table 8.12**
Typical Reoxygenation Constants (base 10, 1/days)

| | |
|---|---|
| white water | 0.5 and above |
| swiftly flowing | 0.3 to 0.5 |
| large streams | 0.15 to 0.3 |
| large lakes | 0.10 to 0.15 |
| sluggish streams | 0.10 to 0.15 |
| small ponds | 0.05 to 0.10 |

Equations 8.6 and 8.21 can be plotted simultaneously as shown in figure 8.5. The plot of equation 8.21 is known as the *oxygen sag curve*. The difference between the two curves is the effect of reoxygenation.

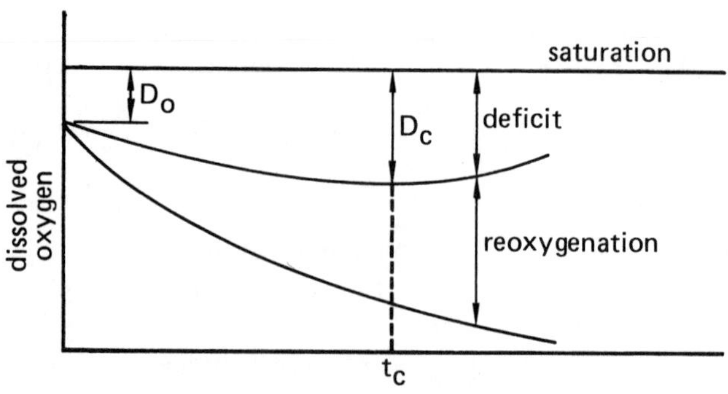


**Figure 8.5** The Oxygen Sag Curve

[9] $K_R$ and $K_D$ may be written as $K_1$ and $K_2$ by other authors.

[10] Equation 8.22 is the O'Connor and Dobbins formula for natural streams.

The time to the minimum or *critical point* on the sag curve is given by equation 8.24.

$$t_c = \frac{1}{K_R - K_D} \times \log_{10}\left[\left(\frac{K_D\text{BOD}_u - K_R D_o + K_D D_o}{K_D\text{BOD}_u}\right)\left(\frac{K_R}{K_D}\right)\right] \qquad 8.24$$

The ratio of $K_R/K_D$ is known as the *self-purification coefficient.* The *critical oxygen deficit* is given by equation 8.25.

$$D_c = \left(\frac{K_D\text{BOD}_u}{K_R}\right)10^{-K_D t_c} \qquad 8.25$$

Knowing $t_c$ and the stream flow velocity will locate the point where the oxygen level is the lowest.

*Example 8.4*

A treatment plant discharge has the following characteristics:

15 cfs
45 mg/l BOD (5 day, 20°C)
2.9 mg/l DO
24°C
$K_{D,20°C} = 0.1$ per day (when mixed with river water)

The outfall is located in a river with the following characteristics:

0.55 ft/sec velocity
4.0 feet average depth
120 cfs
4 mg/l BOD (5 day, 20°C)
8.3 mg/l DO
16°C

Determine the distance downstream where the oxygen level is minimum, and predict if the river can support fish life at that point.

*step 1*: Find the river conditions immediately after mixing. Use equation 8.20 three times.

$$\text{BOD} = \frac{(15)(45) + (120)(4)}{15 + 120} = 8.56\text{ mg/l}$$

$$DO = \frac{(15)(2.9) + (120)(8.3)}{135} = 7.7\text{ mg/l}$$

$$T = \frac{(15)(24) + (120)(16)}{135} = 16.89°\text{C}$$

*step 2*: Calculate the rate constants. From equation 8.8,

$$K_{D,16.89°C} = (0.1)(1.047)^{16.89-20} = 0.0867$$

From equation 8.22,

$$K_{R,20°C} \approx 3.3(0.55/(4)^{1.33}) = 0.287$$

From equation 8.23,

$$K_{R,16.89°C} = (0.287)(1.016)^{16.89-20} = 0.275$$

*step 3*: Estimate $\text{BOD}_u$. Using equation 8.6,

$$\text{BOD}_u = \frac{8.56}{1 - (10)^{-(0.0867)(5)}} = 13.56$$

*step 4*: Calculate $D_o$. From appendix B, the saturated oxygen concentration at 16.89°C is approximately 9.7 mg/l. So, $D_o = 9.7 - 7.7 = 2.0$.

*step 5*: Calculate $t_c$ from equation 8.24.

$$t_c = \frac{1}{0.275 - 0.0867} \times \log\left[\frac{(0.0867)(13.56) - (0.275)(2) + (0.0867)(2)}{0.0867(13.56)} \times \left(\frac{0.275}{0.0867}\right)\right] = 1.77\text{ days}$$

*step 6*: The distance downstream is

$$\frac{(1.77)\text{ days}(0.55)\text{ ft/sec}(86{,}400)\text{ sec/day}}{(5280)\text{ ft/mile}} = 15.9\text{ miles}$$

*step 7*: The critical oxygen deficit is found from equation 8.25.

$$D_c = \frac{(0.0867)(13.56)}{0.275}(10)^{-(0.0867)(1.77)} = 3.0$$

*step 8*: If the temperature 15.9 miles downstream is 16°C, the saturated oxygen content is 10 mg/l. Since the critical deficit is 3, the minimum oxygen content is 7 mg/l. This is adequate for fish life.

## 12 SMALL VOLUME DISPOSAL

### A. CESSPOOLS

A cesspool is a form of sub-surface disposal consisting of a lined and covered underground cavern into which sewage is discharged. Cesspools may be watertight if they are for temporary storage only. *Leaching cesspools* allow seepage of sewage into the ground. Cesspools are acceptable only for small volumes (1 or 2 families) of sewage.

### B. SEPTIC TANK

A septic tank is a simple tank which allows both sedimentation and digestion to occur. Typical detention times are 8 to 24 hours. Effluent is discharged into underground tile fields which allow the water to percolate into the soil. Only 30–50% of the suspended solids are removed by septic tanks. The remaining solids eventually clog the tank and must be mechanically removed.

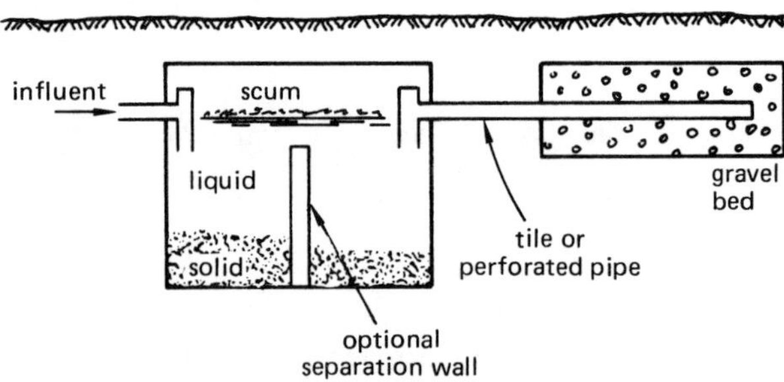


**Figure 8.6** Typical Septic Tank

Typical design parameters of a domestic septic tank are:

| | |
|---|---|
| • minimum capacity below flow line | 300 gal. for 5 persons or less; 500 gal. preferred. No garbage disposals. Add 30 gal. per additional person. |
| • plan aspect ratio | 1:2 |
| • minimum depth below flow line | 3 to 4 feet |
| • minimum freeboard | 1 foot |
| • tank burial depth | 1 to 2 feet |
| • tile length | 30 feet per person |
| • maximum tile run length | 60 feet |
| • minimum tile depth | 15 inches (30″ preferred) |
| • lateral line spacing | 6 feet |
| • gravel bed size | 4 inch radius around tile, 12″ below |
| • minimum soil layer below tile bed | 10 feet |

Municipal septic tanks should be designed to hold 12–24 hours of flow plus stored sludge. A general rule is to allow at least 25 gallons per person served by the tank.

### C. IMHOFF TANK

An Imhoff tank is similar to a septic tank in that sedimentation and sludge digestion both occur. However, these two processes occur in different parts of the tank, and Imhoff vessels are larger in capacity than simple domestic septic tanks. Wastewater enters the tank at the top where sedimentation occurs. The sediment slides down the sloped inner baffles.

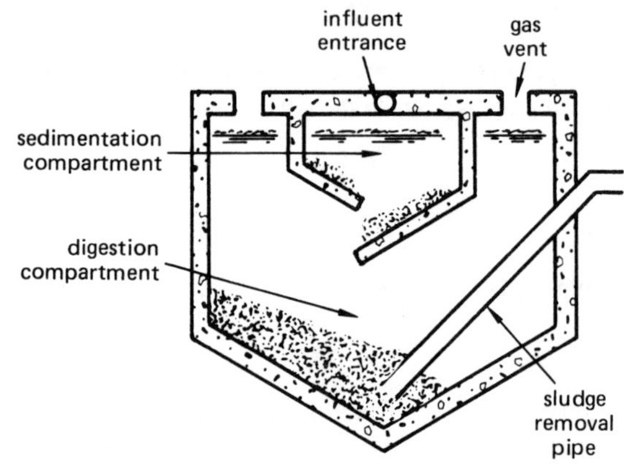


**Figure 8.7** Simple Imhoff Tank

One of the inner baffles extends past the other so that gas produced in the digester chamber will not enter the sedimentation chamber.

Imhoff tanks are usually very large (i.e., 2 stories high) and have been used in the past where the loading is between 250,000 gpd and 1,000,000 gpd. Imhoff tanks can remove up to 60% of the suspended solids during the 1 to 2 hour retention time. They are more efficient than septic tanks, but require very frequent (up to hourly) attention.

Typical characteristics of an Imhoff tank are:

| | |
|---|---|
| • sludge chamber capacity | 2.5 $ft^3$/person |
| • slope of inner baffles | 2 vertical: 1 horizontal |
| • total depth | 15 feet minimum |
| • gas vent area | 15% to 25% of top area |
| • fall-through slot width | 8″ minimum |
| • baffle overhang | 8″ |
| • sludge pipe diameter | 8″ minimum |
| • distance from slot to sludge | 18″minimum |

## 13 WASTEWATER PLANT SITING CONSIDERATIONS

Wastewater plants should be located as far as possible from inhabited areas. A minimum distance of 1000 feet for uncovered plants is desired. Uncovered plants should be located downwind when a definite wind direction prevails. Foundation conditions need to be evaluated, as does the elevation of the water table. Elevation in relationship to the need for sewage pumping (and for dikes around the site) is relevant. Furthermore, the plant must be protected against flooding. 100-year storms are typically chosen as the design flood when designing dikes and similar facilities. Distance to the outfall and possible effluent pumping need to be considered.

Table 8.13 lists the approximate acreage for preliminary engineering estimates. Of course, an estimate of expansion is proper when evaluating acreage requirements.

**Table 8.13**
Treatment Plant Acreage Requirements

| type of treatment | acres per MGD |
|---|---|
| activated sludge plants | 2 |
| trickling filter plants | 3 |
| aerated lagoons | 16 |
| stabilization basins | 20 |
| physical-chemical plants | 1.5 |

## 14 PRETREATMENT OF INDUSTRIAL WASTES

Industrial wastes that would harm collection or treatment facilities or upset subsequent biological processes need to be pretreated. The guidelines which follow should be evaluated for applicability and conformance to local, state, and federal codes. If possible, eliminate the contaminants at their sources.

- **chromium removal**: If hexavalent chromium is greater than 2 mg/l in the influent, use chemical reduction followed by chemical precipitation.

- **heavy metal removal**: Use chemical precipitation if total heavy metals exceed 1 mg/l. If recovery of the metals is desired, use ion exchange methods.

- **cyanide removal**: Use chemical oxidation if the concentration exceeds 2 mg/l. Use electrolysis for high-strength, low-flow waste streams.

- **phenol removal**: Biochemical oxidation and chemical oxidation can both be used, although separate biological treatment for phenol may be uneconomical. Maximum concentration needs to be determined empirically.

- **pH adjustment**: The pH of water entering biological treatment should be between 6.0 and 9.0. Neutralize with acid or alkalai additives.

- **emulsified oil removal**: Use coagulation and flotation adsorption on activated carbon.

- **hydrogen sulfide removal**: Preaerate if sulfides exceed 50 mg/l.

- **oil separation**: Use gravity separation.

## 15 WASTEWATER PROCESSES

### A. PRELIMINARY TREATMENT

Preliminary preparation of the wastewater stream is essentially a mechanical process. It removes large objects, rags, and wood from the flow. Heavy solids and excessive oils and grease are also eliminated. Damage to pumps and other equipment would be expected without preliminary treatment.

**Screens**: Trash racks or coarse screens with openings 2 inches or larger should precede pumps to prevent clogging. Screenings usually consist of paper, wood, and rags. Medium screens ($\frac{1}{2}''$ to $1\frac{1}{2}''$ openings) and fine screens ($\frac{1}{16}''$ to $\frac{1}{8}''$) are also used to relieve the load on grit chambers and sedimentation basins.[11] Screens are cleaned by automatic scraping arms. Screen capacities and head losses are specified by the manufacturer. In general, however, flow through screens should be limited to 3 fps or less.

**Grit Chambers**: Grit is an abrasive that wears pumps, clogs pipes, and accumulates in excessive volumes. A grit chamber (also known as *grit clarifier* or *detritus tank*) slows the wastewater down to approximately 1 ft/sec. This velocity allows the grit to settle out but moves the organic matter through. The grit can be manually or mechanically removed with buckets or screw conveyors.

[11] Fine screens are rare except when used with some industrial waste processing plants.

Typical design standards for grit chambers are:

- grit removal rate — 1 to 5 $ft^3$/MG
- grit size — 0.2 mm or larger
- grit specific gravity — 2.65
- depth — 3 to 4 feet
- length (width not critical) — 40 to 100 feet
- detention time — 45 to 90 seconds
- horizontal velocity — 0.75 to 1.25 ft/sec

In actual practice, grit chambers are designed to keep the flow velocity as close to 1.0 ft/sec as possible. If an analytical design based on settling velocity is required, the *scouring velocity* should not be exceeded. Scouring of the minimum-sized particles which have already settled will be prevented if the velocity is kept below that in equation 8.26. ($SG_p$ is the specific gravity of the particle. $d_p$ is the particle diameter in millimeters.) The friction factor is approximately 0.03 for grit chambers. The scouring velocity has been converted to ft/sec even though d is in mm.[12]

$$v = 2.2\sqrt{\frac{gd_p}{f}(SG_p - 1)} \approx 1.3\sqrt{d_p(SG_p - 1)} \qquad 8.26$$

**Aerated Grit Chambers**: In smaller plants, the grit chamber may be a hopper-bottomed tank with a small detention time. Diffused aeration from one side of the tank rolls the water and is employed to keep the organics in suspension while the grit settles out. Influent enters on one side of the tank, and an effluent weir on the opposite side removes the degritted wastewater. The water spirals or rolls through the tank.

Solids are removed by airlift pump, screw conveyer, bucket elevator, or gravity flow. However, since the scouring velocity is not maintained, the grit will have a significant organic content. A grit washer or cyclone separator may be used to clean the grit.

Typical characteristics of aerated grit chambers are:

- detention time — 2 to 5 minutes at peak flow (3 typical)
- air supply:
  - shallow tanks — 1.5 to 5.0 cfm/ft length (3.0 typical)
  - deep tanks — 3.0 to 8.0 cfm/ft length (5.0 typical)
- grit and scum quantities — 0.5 to 25.0 $ft^3$/MG (2.0 typical)
- length-width ratio — 2.5:1 to 5:1 (3:1 typical)
- depth — 6 to 15 feet
- length — 20 to 60 feet
- width — 7 to 20 feet

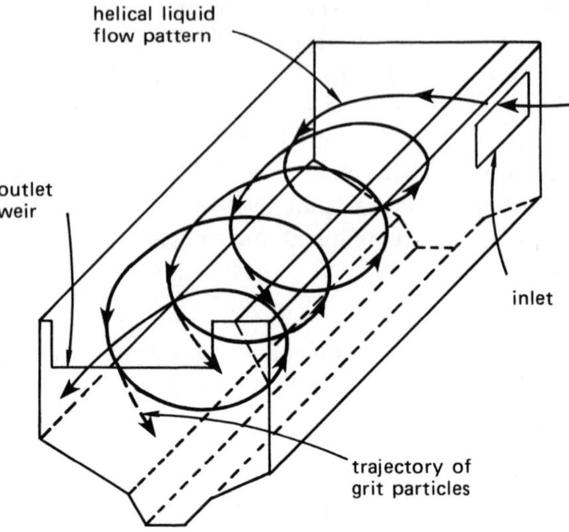


**Figure 8.8** Basic Aerated Grit Chamber

**Skimming Tanks**: If the sewage has excessive grease or oil, a basin 8 to 10 feet deep providing 5 to 15 minutes of detention time will allow the grease to rise to the surface.[13] An aerating device below will help coagulate and float grease to the surface. Approximately 0.01 to 0.1 cubic feet of 40 to 80 psig air per gallon of influent should be used. (The actual air pressure will depend on the tank depth.) Surface grease can be mechanically removed by skimming troughs. The tank outlet is submerged, and it is lower than the inlet.

If the skimming tank is enclosed and the air evacuated to approximately 9″ of mercury, rising bubbles in the sewage will expand and help float grease upwards without the need for mechanical aeration.

A small fraction (e.g., 30%) of the influent may be recycled in some cases.

**Shredders**: Shredders (also called *comminutors*) cut waste solids to approximately 1/4″ in size. They reduce the amount of screenings which must be disposed of. Shreddings generally stay with the flow for later settling.

**Other Pretreatment Processes**: Odor control through chlorination, aeration, or ozonation, freshening of septic waste by aeration, and flow equalization

[12] Equation 8.26 is known as the *Camp formula*.

[13] 50 mg/l or more of total floatables should be considered as requiring skimming.

in holding basins can also be loosely categorized as pretreatment processes.

## B. PRIMARY TREATMENT

Primary treatment is a mechanical (settling) process used to remove most of the settleable solids. A 25 to 35% reduction in BOD is also achieved, but BOD reduction is not the goal of primary treatment.

**Plain Sedimentation**: Plain sedimentation basins are described in chapter 7. Design characteristics for wastewater treatment are:

| | |
|---|---|
| • BOD reduction | 20% to 40% |
| • total suspended solids reduction | 35% to 65% |
| • bacteria reduction | 50% to 60% |
| • organic content of settled solids | 50% to 75% |
| • specific gravity of settled solids | 1.2 or less |
| • typical settling velocity | above 4 feet/hr |
| • plan shape | rectangular or circular |
| • basin depth | 6 to 15 feet (12 typical) |
| • basin width | 10 to 50 feet |
| • minimum freeboard | 18 inches |
| • minimum hopper wall angle | 60° |
| • aspect ratio (rectangular) | 3:1 to 5:1 |
| • detention time | 1.5 to 2.5 hours |
| • circular diameter | 30 to 150 feet (100 common) |
| • flow-through velocity | 0.005 ft/sec |
| • flow-through time | at least 30% of detention time |
| • overflow rate | 400 to 2000 gpd/ft$^2$ (800 to 1200 typical) |
| • bottom | slight slope (8%) towards hopper |
| • inlet | baffled for uniform velocity |
| • scum removal | mechanical or manual |
| • weir loading | 10,000 to 20,000 gpd/ft |

**Chemical Sedimentation**: Chemical flocculation (*clarification* or *coagulation*) is similar to that described in chapter 7 except that the coagulant doses are greater. Typically, the most economical coagulant used is ferric chloride. Lime and sulfuric acid may be used to adjust the pH for proper coagulation. Chemical precipitation is used when the stream into which the outfall discharges is running low, when there is a large increase in sewage flow, and generally when plain sedimentation is insufficient.

## C. SECONDARY TREATMENT

Secondary treatment is a biological treatment. It became mandatory for all publicly owned water treatment plants as of July 1977 under the Federal Water Pollution Control Act ammendments of 1972.

**Trickling Filters**: Trickling filters (also known as *biological beds*) consist of beds of 2″ to 5″ rocks up to 9 feet thick (6 feet typical) over which influent is sprayed. The biological and microbial slime growth attached to the rocks purify the wastewater as it trickles through the rocks. The water is introduced into the filter by rotating arms which move by virtue of the spray reaction. The clarified water is collected by an underdrain system. Some water may be returned to the filter for a longer contact time.

On the average, one acre of standard filter area is needed for each 20,000 people served. Trickling filters can remove 70% to 90% of the suspended solids, 65% to 85% of the BOD, and 70% to 95% of the bacteria. Most of the reduction occurs in the first few feet of bed, and organisms in the lower part of the bed may be in a near-starvation condition. The bed will periodically slough off (unload) parts of its slime coating, and sedimentation after filtering is necessary.

Since there are limits to the heights of trickling filters, longer contact times can be achieved by returning some of the collected filter water to the filter. This is known as *recirculation.* Recirculation is also used to keep the filter medium from drying out and to smooth out fluctuations in the hydraulic loading.

*High rate filters* are now in use by most modern facilities. The higher hydraulic loading flushes the rockpile and inhibits excess biological growth. High rate filters may be only 3 to 4 feet deep. The high rate is possible because much of the filter discharge is recirculated.

The *hydraulic loading* of a trickling filter is the water flow divided by the plan area. Typical values of hydraulic loading are 25 to 100 gpd/ft$^2$ for standard filters, and up to 1000 gpd/ft$^2$ for high-rate filters.

$$L_H = \frac{Q_w + Q_R}{A_{\text{filter}}} = \frac{Q_w + R_R Q_w}{A_{\text{filter}}} = \frac{Q_w(1+R_R)}{A_{\text{filter}}} \quad 8.27$$

The *recirculation ratio* is given by equation 8.28. It can be as high as 3 for high rate filters, although it is zero for standard low-rate filters.

$$R_R = Q_R/Q_w = \frac{L_H A_{\text{filter}}}{Q_w} - 1 \quad 8.28$$

The *BOD loading* (same as *organic loading*) is calculated without considering any recirculated flow. BOD

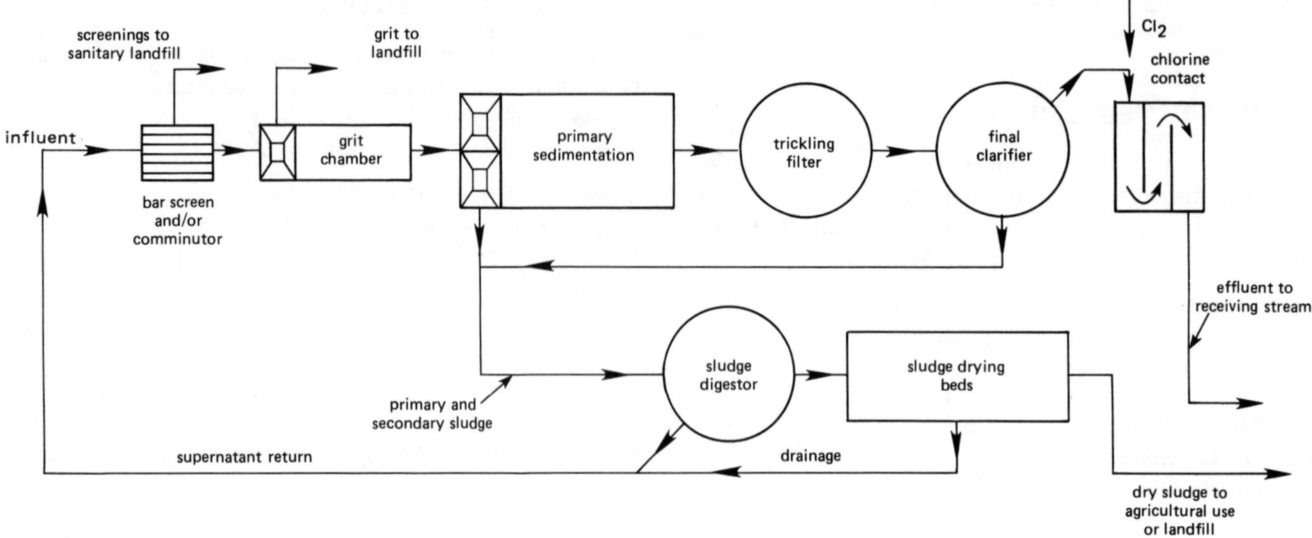


**Figure 8.9** A Typical Trickling Filter Plant

loading for the filter/clarifier combination is essentially the BOD of the applied wastewater divided by the filter volume.

$$L_{\mathrm{BOD}} = \frac{(Q_w)(\mathrm{BOD}_{s,i})(8.345\ \mathrm{EE}-3)}{\text{filter volume in ft}^3} \qquad 8.29$$

BOD loading is given in pounds per 1000 cubic feet per day. Typical values are 5 to 25 lbm/1000 cubic feet-day (low rate) and 30 to 90 lbm/1000 cubic feet-day (high rate.)

Significant reduction in BOD occurs in a trickling filter. Standard rate filters produce an 80%–85% reduction. Because they offer less contact area and time, high rate filters only remove 65%–80% of the BOD.

If it is assumed that the biological layer and hydraulic loading are uniform, the water is at 20°C, and the filter is single-stage rock followed by a settling tank, then the following equation developed by the National Research Council can be used to calculate the BOD removal efficiency of the filter/clarifier combination.[14] Equation 8.30 is easily solved from figure 8.10.[15]

$$\eta = \frac{\mathrm{BOD}_{\mathrm{removed}}}{\mathrm{BOD}_{\mathrm{entering}}} = \frac{1}{1+0.0561\sqrt{L_{\mathrm{BOD}}/F}} \qquad 8.30$$

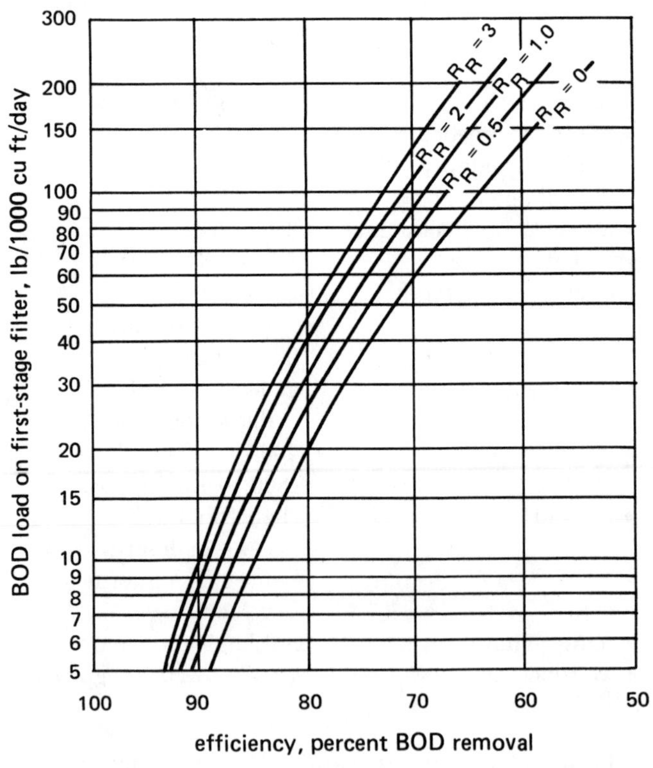


**Figure 8.10** Solution of Equation 8.30

For other temperatures, calculate the efficiency assuming 20°C, and then correct the efficiency using equation 8.31.

$$\eta_T = \eta_{20^\circ\mathrm{C}}(1.01)^{T-20} \qquad 8.31$$

[14] The National Research Council did studies in 1946 on sewage treatment plants at military installations. It concluded that the organic loading had a greater effect on removal efficiency than did volumetric loading.

[15] The constant 0.0561 in equation 8.30 is also reported as 0.0085 in the literature. However, that value is for use with media volumes expressed in acre-feet, not 1000's of ft$^3$.

If the incoming and required BOD are known, the approximate BOD loading required is given by equation 8.32.

$$L_{BOD} = 317.7 \left( \frac{BOD_{req}}{BOD_i - BOD_{req}} \right)^2 \qquad 8.32$$

The BOD loading for a high rate, single stage filter with recirculation is given by equation 8.34, where $F$ is the effective number of passes through the filter. The *weighting factor*, $w$, is typically assigned a value of 0.1.

$$F = \frac{1 + R_R}{(1 + wR_R)^2} \qquad 8.33$$

$$L_{BOD} = (317.7)F \left( \frac{BOD_{req}}{BOD_i - BOD_{req}} \right)^2 \qquad 8.34$$

There are a number of ways to recirculate water back to the filter. Some methods are shown in figure 8.11. The variations in performance are not significant. Equation 8.34 should be used only with the first four recirculation schemes shown in figure 8.11.

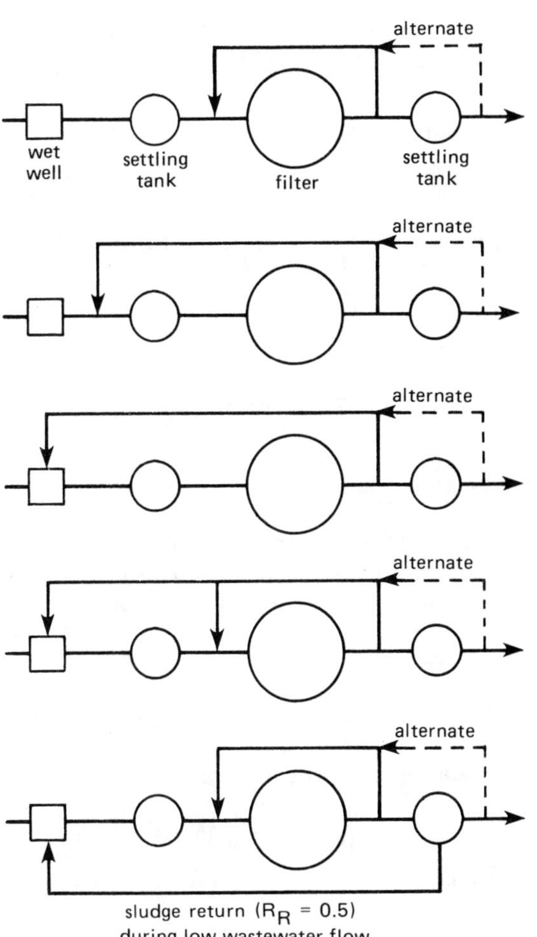


**Figure 8.11** Typical One-Stage Recirculation Methods

If even more BOD and solids removal is needed, two filters can be connected in series to form a two-stage filter system. A two-stage filter system is shown in figure 8.12(a) with an optional intermediate settling tank.

Typical two-stage performance is:

- BOD loading 45 to 70 lbm/1000 $ft^3$-day
- hydraulic loading 0.16 to 0.48 gpm/$ft^2$
- recirculation ratio 0.5 to 4.0
- final effluent BOD 30 mg/l

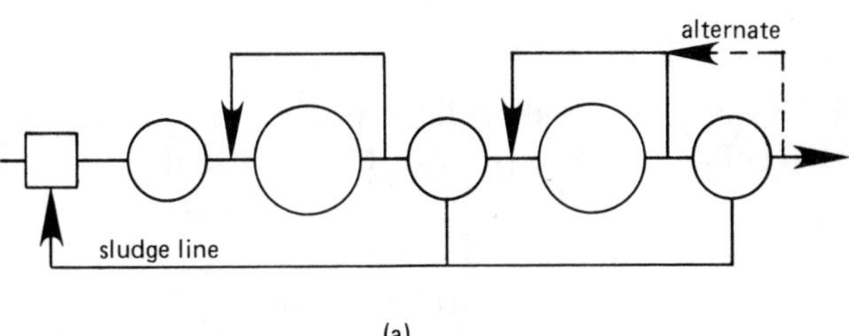


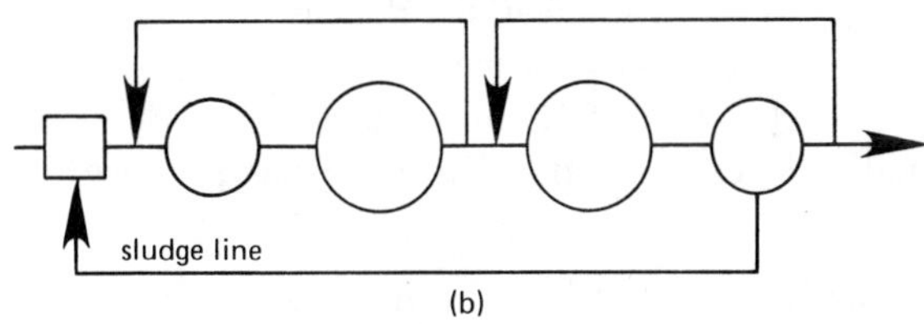


**Figure 8.12** Typical Two-Stage Filter Systems

The approximate BOD loading that is required to accomplish a given reduction in BOD is given by equation 8.35.

$$L_{BOD} = (317.7)F_{2nd\ stage} \times \left( \frac{BOD_{out,1st}}{BOD_{in,1st}} \right)^2 \left( \frac{BOD_{out,2nd}}{BOD_{in,2nd} - BOD_{out,2nd}} \right)^2 \qquad 8.35$$

The efficiency of a second-stage filter is considerably less than that of a first-stage filter because much of the biological food has been removed. This lowered efficiency can be considered as an increase in BOD loading, as given by equation 8.36.

$$L_{BOD,\ adjusted\ 2nd\ stage} = \frac{L_{BOD,\ actual\ second\ stage}}{[1 - \text{first stage efficiency}]^2} \qquad 8.36$$

The actual second stage load is calculated from equation 8.29 using the incoming BOD from the intermediate clarifier. Equation 8.30 is then used to find the 2nd stage efficiency.

The overall BOD efficiency of the two-stage system is:

$$\eta_{\text{overall}} = \frac{\text{BOD}_{\text{in}} - \text{BOD}_{\text{out}}}{\text{BOD}_{\text{in}}} = 1 - (1 - \eta_{\text{settling basin}})(1 - \eta_{\text{1st stage}})(1 - \eta_{\text{2nd stage}}) \quad 8.37$$

For temperatures other than 20°C, equation 8.38 can be used.

$$\eta_T = \eta_{20°\text{C}}(1.035)^{T-20} \quad 8.38$$

*Example 8.5*

A single-stage trickling filter plant is to process 1.4 MGD of domestic waste with 170 mg/l BOD.

| | |
|---|---|
| primary clarifier | 50 feet diameter |
| | 7 feet wet depth |
| | peripheral weir |
| trickling filter | 90 feet diameter |
| | 7 feet wet depth |
| | rock media |
| | 50% recirculation |
| final clarifier | same dimensions as primary |

Determine if the units have been sized correctly, and estimate the final BOD if the plant operates at 16°C.

Primary Clarifier

circumference: $\pi(50) = 157\,\text{ft}$

surface area: $\frac{1}{4}\pi(50)^2 = 1963\,\text{ft}^2$

volume: $(1963)(7) = 13{,}740\,\text{ft}^3$

The surface loading is

$$\frac{1.4\,\text{EE6}}{1963} = 713\,\text{gpd/ft}^2 \text{ (ok)}$$

The retention time is

$$t = \frac{V}{Q} = \frac{(13{,}740)\,\text{ft}^3\,(24)\,\text{hr/day}}{(1.4\,\text{EE6})\,\text{gal/day}(0.1337)\,\text{ft}^3/\text{gal}} = 1.76\,\text{hrs (ok)}$$

The weir loading is $\frac{1.4\,\text{EE6}}{157} = 8917$ gpd/ft (ok)

Assume a 30% reduction in BOD at 20°C. The effluent BOD is

$$\text{BOD} = (0.7)(170) = 119\,\text{mg/l}$$

Trickling Filter

area: $\frac{1}{4}\pi(90)^2 = 6362\,\text{ft}^2$

volume: $(6362)(7) = 44{,}534\,\text{ft}^3$

From equation 8.27, the hydraulic load on the filter is

$$\frac{(1.5)(1.4\,\text{EE6})}{6362} = 330\,\text{gpd/ft}^2 \quad \text{(ok for high-rate filter)}$$

The BOD loading is

$$\frac{(1.4)\text{MGD}(119)\,\frac{\text{mg}}{\text{l}}(8.345)\frac{\text{lbm-l}}{\text{MG-mg}}\frac{1000\,\text{ft}^3}{1000\,\text{ft}^3}}{(44{,}534)\,\text{ft}^3} = 31.2\,\text{lbm/day-1000 ft}^3 \text{ (ok for high-rate filter)}$$

From equation 8.33,

$$F = \frac{1 + 0.5}{(1 + (0.1)(0.5))^2} = 1.36$$

From equation 8.30, the approximate BOD removal efficiency is

$$\eta = \frac{1}{1 + 0.0561\sqrt{31.2/1.36}} = 0.788 \text{ at } 20°\text{C}$$

The final clarifier effluent BOD is

$$(1 - 0.788)(119) = 25.2\,\text{mg/l}$$

Final Settling Tank

Same area, volume, surface loading, weir loading, and retention time as the primary clarifier. The BOD removal effect is included in equation 8.30.

16°C Performance

The 20°C overall efficiency is $\frac{170 - 25.2}{170} = 0.852$

From equation 8.38, the efficiency at 16°C is

$$\eta_{16°\text{C}} = 0.852(1.035)^{16-20} = 0.742$$
$$\text{BOD}_{16°\text{C}} = (1 - 0.742)(170) = 43.9\,\text{mg/l}$$

**Rotating Biological Contactors**: *Rotating biological contactors* (also known as *rotating biological reactors*) consist of large-diameter plastic disks mounted on a shaft which turns slowly. The rotation progressively wets the disks, on which a microorganism population grows. This biomass population, since it is well oxygenated, is efficient at removing organic solids from the wastewater.

Several stages of RBC's typically constitute the process. The process can be placed in series or parallel with existing trickling filter or activated sludge processes. Recirculation is not common with RBC processes.

The primary design criterion is hydraulic loading, not organic (BOD) loading. For a specific hydraulic loading, the BOD removal efficiency will be essentially constant, regardless of variations in BOD. Other design criteria are listed here:

- plant design flows — all
- hydraulic loading:
  - (secondary treatment) — 2 to 4 gpd/ft$^2$ of plan area
  - (denitrification) — 0.75 to 2 gpd/ft$^2$
- optimum peripheral rotational velocity — 1 ft/sec
- tank volume — 0.12 gal/ft$^2$ of biomass area
- temperature — 55°–90°F
- BOD removed — 70–80%
- retention time — fixed by tank volume and hydraulic loading
- number of stages — 2–4
- diameter of disks — 10–12 feet
- immersion — 40% at any instant

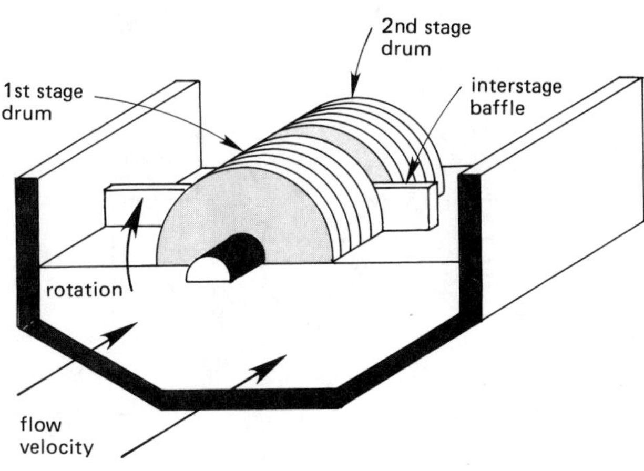


**Figure 8.13** Rotating Biological Contactor

**Intermittent Sand Filters**: For small populations, a *slow sand filter* or *intermittent sand filter* can be used. Because of the lower flow rate, the filter area per person is higher than for a trickling filter. Roughly one acre is needed for a population of 1000. The filter is constructed as a sand bed 2 to 3 feet deep over a 6″ to 12″ gravel bed. Application rates are usually 2 to $2\frac{1}{2}$ gpd/ft$^2$. The filter is cleaned by removing the top layer of clogged sand.

The filter is alternately exposed to water from a settling tank and to air. Straining and aerobic decomposition cleans the water. If the water is applied continuously as a final process from a secondary treatment plant, the filter is known as a *polishing filter*. The water rate of a polishing filter may be as high as 10 gpd/ft$^2$. Up to 95% of the BOD can be satisfied in an intermittent sand filter.

**Stabilization Ponds**: A stabilization pond (also known as an *oxidation pond*) holds partially treated water for 3 to 6 weeks (up to 6 months in cold weather) at a depth of 2 to 6 feet (4 typical). Aquatic plants, weeds, algae, and microorganisms are used to stabilize the organic matter. The algae gives off oxygen from growth in sunlight. The oxygen is used by microorganisms to digest organic matter. The microorganisms give off $CO_2$, ammonia, and phosphates which the algae use.

Areas required are large—up to one acre per 50 pounds BOD per day (about one acre per 300 people) for warm southern states. In cold climates, twice this area may be required. The BOD loading is 15 to 35 lbm/acre-day.

**Aerated Lagoons**: If a stabilization pond has air added mechanically, it is known as an *aerated lagoon*. Such a lagoon typically is deeper and has a shorter detention time than a stabilization pond. With floating aerators, one acre can support 500 to 1000 pounds of BOD per day. In cold climates, twice this area is required.

The lagoon volume must be sufficient to hold the design flow during the detention period. (Since $V/Q = t_d$, equation 8.39 can be used to determine $t_d$ if the BOD removal efficiency is known.)

$$V = Qt_d \tag{8.39}$$

The BOD reduction can be calculated if the overall, first-order BOD *removal rate constant* is known. Typical values of $k$ are 0.25 to 1.0 $\frac{1}{\text{day}}$ (base-10).

$$\frac{\text{BOD}_f}{\text{BOD}_i} = \frac{1}{1 + \dfrac{kV}{Q}} \tag{8.40}$$

If the removal rate constant is quoted for a different temperature than that at which the lagoon is to operate, it must be corrected using equation 8.41.

$$k_T = k_{20°\text{C}}(1.06)^{\text{T}-20} \tag{8.41}$$

Other common design characteristics of a mechanically aerated lagoon are:

| | |
|---|---|
| • aspect ratio | less than 3:1 |
| • depth | 10 to 12 feet |
| • detention time | 4 to 10 days |
| • BOD loading | 20–400 lbm/day-acre (220 typical) |
| • temperature range | 32° to 100°F (70°F optimum) |
| • typical effluent BOD | 20 to 70 mg/l |
| • oxygen requirements | 0.7 to 1.4 times BOD removed |

**Activated Sludge Processes**: Sludge produced during the oxidation process has an extremely high concentration of active aerobic bacteria. For this reason, partially oxidized sludge is called *activated sludge.* Purification of raw sewage can be speeded up considerably if the raw sewage is mixed (seeded) with activated sludge. The mixture of raw sewage and activated sludge is known as *mixed liquor* (*ML*). The biological systems in the mixed liquor are known as *mixed liquor suspended solids* (*MLSS*).

In operation, an activated sludge process takes raw water and allows it to settle. The settled effluent is mixed with activated sludge in the approximate ratio of 1 part sludge per 3 or 4 parts effluent. Mechanical aeration is used. The effluent is then settled in a second sedimentation tank, chlorinated, and discharged. Settled sludge from this last tank supplies the continuous seed for the activation.

Activated sludge processes are highly efficient, with the following typical characteristics for a conventional system. (Also, see table 8.14).

| | |
|---|---|
| • BOD reduction | 90 to 95% |
| • BOD loading | 0.25 to 1 lbm/lbm MLSS |
| • maximum aeration chamber volume | 5000 $ft^3$ |
| • aeration chamber depth | 10 to 15 feet |
| • aeration chamber width | 20 ft |
| • air rate | 1/2 to 2 $ft^3$/gal raw sewage |
| • minimum dissolved oxygen | 2 mg/l |
| • biological mass density | 1000 to 4000 mg/l |
| • sedimentation basin depth | 15 ft |
| • sedimentation basin detention time | 2 hours |
| • basin overflow rate | 400 to 2000 gpd/$ft^2$ (1000 typical) |
| • % sludge returned | 20 to 30% |
| • frequency of sludge transfer | once each hour |
| • activated sludge volume index | 50 to 150 |
| • weir loading | 10,000 gpd/ft |
| • maximum tank volume | 2500 $ft^2$ |

The *rate of oxygen transfer* (ROT) from the air to the mixed liquor during aeration is given by equation 8.42.

$$\text{ROT} = K_t D \qquad 8.42$$

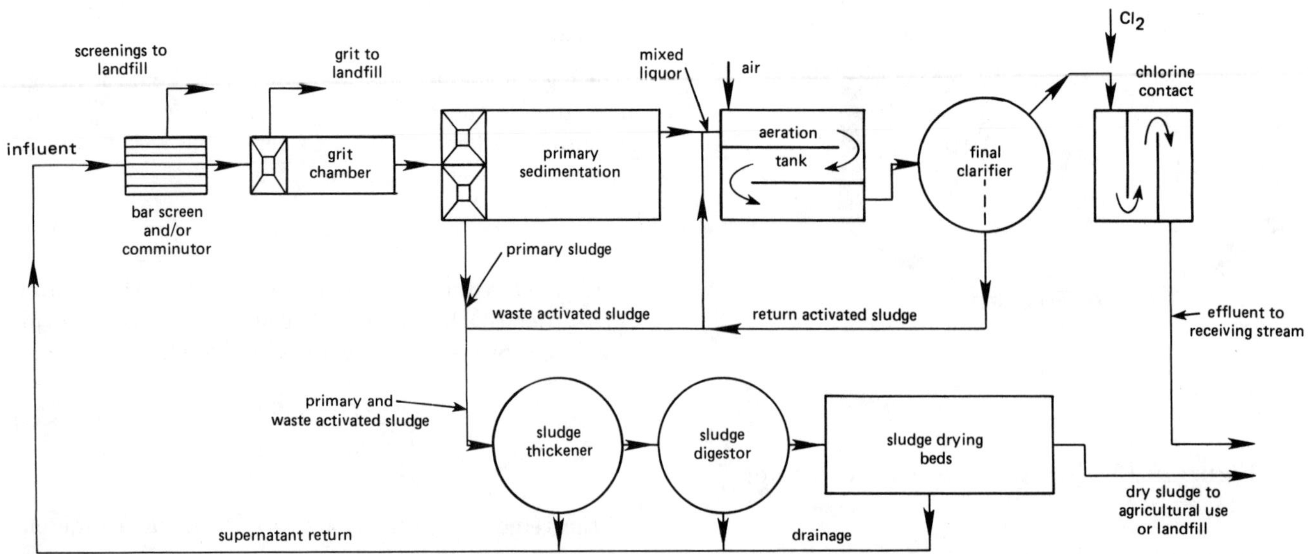


**Figure 8.14** A Typical Activated Sludge Plant

$K_t$ is a transfer coefficient which depends on the equipment and waste characteristics. The oxygen deficit, $D$, is given by equation 8.43.

$$D = \beta DO_{\text{saturated}} - DO_{\text{actual ML}} \qquad 8.43$$

$\beta$ is the water's *oxygen saturation coefficient*, usually 0.8 or 0.9.

*Example 8.6*

20°C wastewater ($\beta = 0.9$) is processed in an aerating system with an oxygen transfer coefficient of 2.7 per hour. If the wastewater dissolved oxygen is 3 mg/l, what is the rate of oxygen transfer?

At 20°C, the saturated oxygen content is 9.2 mg/l. So, the oxygen deficit is

$$(0.9)(9.2) - 3 = 5.28\,\text{mg/l}$$

From equation 8.42,

$$\text{ROT} = (2.7)(5.28) = 14.3\,\text{mg/l-hr}$$

The *aerating period* is calculated from equation 8.44. Recirculation is not a factor in calculating the aerating period.

$$t_A = \frac{\text{aeration tank volume, gallons}}{Q_w} \qquad 8.44$$

The BOD loading on the aeration tank is

$$L_{\text{BOD}} = \frac{(Q_w)(\text{BOD}_w)(0.0624)}{\text{aeration tank volume, gallons}} \qquad 8.45$$

The *food-to-microorganism ratio* in pounds of BOD per day per pound of MLSS is defined by equation 8.46.

$$R_{F-M} = \frac{(Q_w)(\text{BOD}_w)}{(MLSS)(\text{aeration tank volume, gallons})} \qquad 8.46$$

The *rate of return sludge* is

$$R_{RS} = Q_{RS}/Q_w \qquad 8.47$$

The BOD efficiency of an activated sludge process is

$$\eta_{\text{BOD}} = \frac{\text{BOD}_w - \text{BOD}_{\text{after settling}}}{\text{BOD}_w} \qquad 8.48$$

*Sludge bulking* refers to a condition in which the sludge will not settle out. Since the solids do not settle, they leave the sedimentation tank and cause problems in subsequent processes. The *sludge volume index* (also known as the *Mohlman index*) can be calculated by taking one liter of mixed liquor and measuring the volume of settled solids after 30 minutes. Then, the sludge volume index (SVI) is

$$\text{SVI} = \frac{(\text{settled volume in ml})(1000)}{\text{MLSS}} \qquad 8.49$$

If SVI is above 150, the sludge is bulking. Remedies include the addition of lime, chlorine, more aeration, and the reduction in MLSS.

SVI is related to the concentration of suspended solids in the activated sludge by equation 8.50.

$$SS_{as} = \frac{1{,}000{,}000}{\text{SVI}} \qquad 8.50$$

The theoretical quantity of return sludge required can be calculated from the SVI test results:

$$Q_R/Q_w = \frac{(\text{settled volume in ml/l})}{(1000 - \text{settled volume in ml/l})} \qquad 8.51$$

Another important parameter is *sludge age*. Although the water passes through the system in a matter of hours, the sludge is recycled continuously and has an average stay much longer in duration. Two measures of sludge age are used: age of the suspended solids and age of BOD. Sludge age ($SA_{ss}$) is typically 3 to 5 days.

$$SA_{ss} = \frac{\text{pounds MLSS in aerating basin}}{\text{pounds SS in effluent and waste sludge per day}} \qquad 8.52$$

$$SA_{\text{BOD}} = 1/R_{\text{F-M}} = \frac{\text{pounds MLSS in aerating basin}}{\text{pounds BOD applied to basin per day}} \qquad 8.53$$

*Example 8.7*

One liter of liquid is taken from an aerating lagoon near its discharge point. After settling for 30 minutes in a one liter graduated cylinder, 250 ml of solids have settled out. A second water sample is taken and the suspended solids concentration determined to be 2300 mg/l. Find the SVI and percentage of required return sludge.

From equation 8.49, $\text{SVI} = \dfrac{(250)(1000)}{2300} = 109$

From equation 8.51, $Q_R/Q_w = \dfrac{250}{1000 - 250} = 0.33$

**Activated Sludge Aeration Methods:** Various methods of aeration are used, each having its own characteristic ranges of operating parameters. Table 8.14 lists typical parameters for selected aeration methods.

*Extended Aeration*: Small flows can be treated with the extended aeration method. This method uses mechanical floating or fixed sub-surface aerators to oxygenate the mixed liquor for 24 to 36 hours. There is no primary clarification and there are generally no sludge wasting facilities. Sludge is allowed to accumulate at the bottom of the lagoon for several months. Then, the system is shut down and the lagoon is pumped out.

Sedimentation basins are sized very small with low overflow rates (200 to 600 gpd/ft$^2$), and they have long retention times. All sludge is returned to the aerating basin.

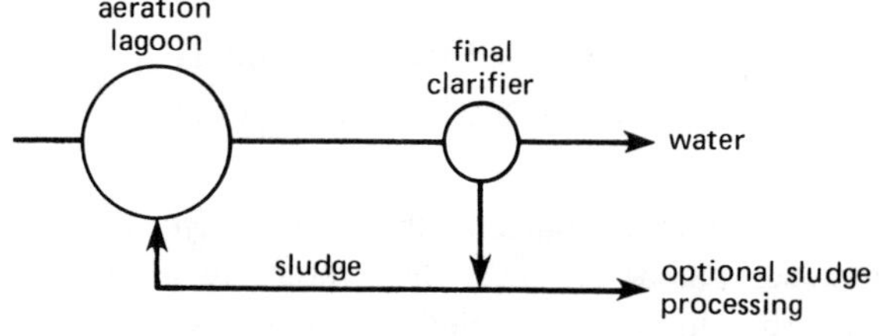


**Figure 8.15** Extended Aeration

*Conventional Aeration*: In this method, (also known as *plug flow*), the influent is taken from a primary clarifier and then aerated. The amount of aeration may be decreased (i.e., *tapered aeration*) as the wastewater travels through the circuitous route since the BOD also decreases along the route.

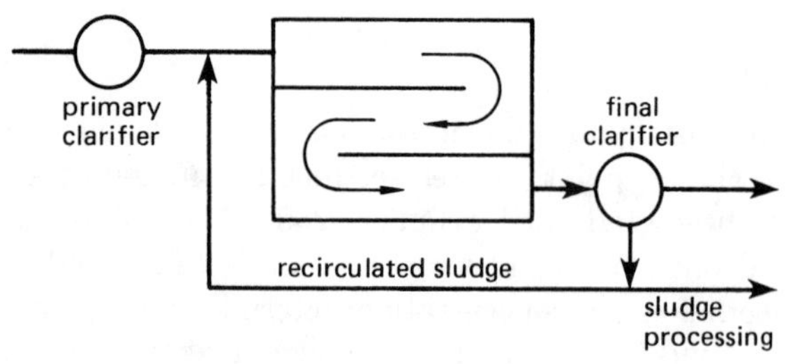


**Figure 8.16** Conventional Aeration

*Step Flow*: In step flow, aeration is constant along the length of the aeration path, but influent is introduced at various points along the path.

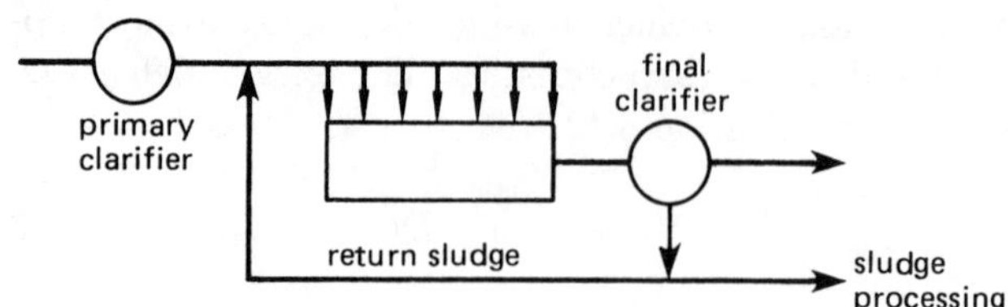


**Figure 8.17** Step Aeration

*Complete Mix*: Waste is added uniformly and the mixed liquor is removed uniformly over the length of the tank. This method is often used for industrial waste processing.

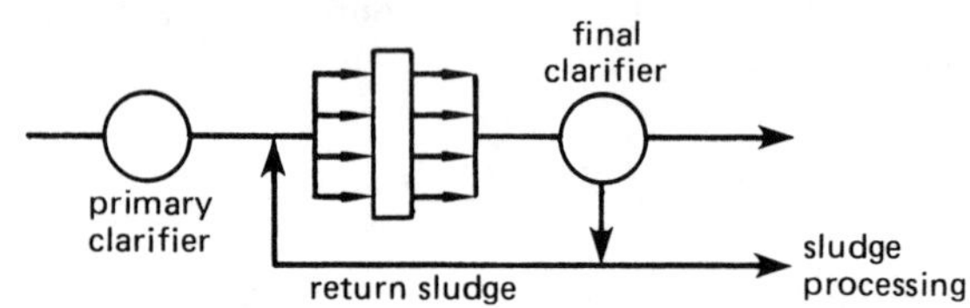


**Figure 8.18** Complete Mixing

*Contact Stabilization*: Units for the contact stabilization (*biosorption*) method are typically factory-built and erected on a site. They are compact, but not as economical or efficient as a regular plant. The units may be constructed as concentric compartmentalized cylinders. This method is designed primarily to handle colloidal wastes.

In this method, the aeration tank is called a *contact tank*. The stabilization tank takes the sludge from the clarifier and aerates it. Less time and space is required for this process because the sludge stabilization is done when the sludge is still concentrated.

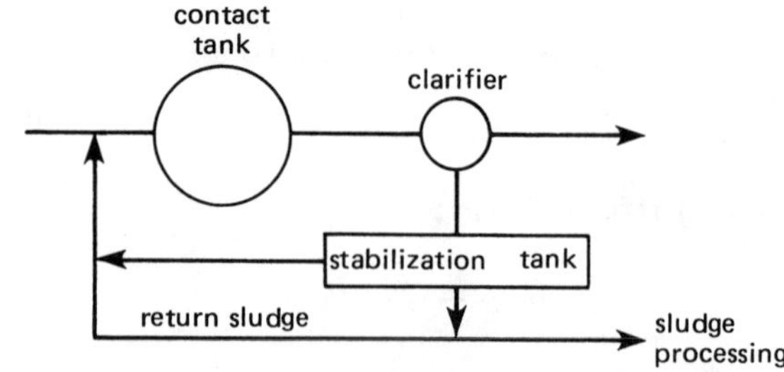


**Figure 8.19** Contact Stabilization

*High Rate Aeration*: This method uses mechanical mixing along with aeration to decrease the aeration period and to increase the BOD load per unit volume.

*High Purity Oxygen Aeration*: This method requires the use of bottled or manufactured oxygen which is introduced to closed aerating tanks in place of atmospheric air. Mechanical mixing is needed to take full advantage of the oxygen.

**Intermediate Clarifiers**: Sedimentation tanks located between trickling filter stages (see figure 8.12) or between a filter and subsequent aeration are known as *intermediate clarifiers*. Recommended standards are:

- maximum overflow rate 1000 gpd/ft$^2$
- minimum water depth 7 feet
- maximum weir loading 10,000 gpd/ft (plants 1 MGD or less); 15,000 gpd/ft (plants over 1 MGD)

**Final Clarifiers**: Final sedimentation in secondary treatment is done in final clarifiers. The purpose of final clarifiers is to collect sloughed off filter material (trickling filter processes) or to collect sludge and return it for aeration (activated sludge processes).

General characteristics for clarifiers following trickling filters are:

- BOD removal — See equation 8.30
- minimum depth — 7 feet
- maximum overflow rate — 800 gpd/ft$^2$
- maximum weir loading — Same as for intermediate clarifiers, but lower preferred.

If the final clarifier follows an activated sludge process, the sludge should be removed rapidly from the entire bottom of the clarifier. Characteristics of clarifiers following an activated sludge process are given in table 8.15.

## D. ADVANCED TERTIARY TREATMENT

*Suspended Solids*: Suspended solids are removed by microstrainers or polishing filter beds.

*Phosphorus Removal*: Phosphorus can be removed by chemical precipitation. Aluminum and iron coagulants, as well as lime, are effective in removing phosphates.

*Nitrogen Conversion and Removal*: In the *ammonia stripping* (*air stripping*) method, lime is added to water to increase its pH to above 10. The water is then passed through a packed tower into which air is blown. The air (at the rate of approximately 400 ft$^3$/gallon) strips the ammonia out of the water. Recarbonation follows to remove the excess lime.

$$NH_4 + OH^- \underset{pH<10}{\overset{pH\geq 11}{\rightleftharpoons}} NH_4OH \xrightarrow{\text{air}} H_2O + NH_3 \qquad 8.54$$

In the *nitrification and denitrification process*, bacteria oxidize ammonium ions to nitrate and nitrite in an aeration tank kept at low BOD. Nitrate and nitrite ions do not absorb further oxygen and may be discharged.

$$NH_3^+ \underset{\text{oxygen}}{\overset{\text{bacteria}}{\longrightarrow}} NO_2^- + NO_3^- \qquad 8.55$$

**Table 8.14**
Representative Operating Conditions for Aeration

| type of aeration | plant flow rate (MGD) | $t_A$ (hrs) | oxygen required (lbm/lbm BOD removed) | waste sludge (lbm/lbm BOD removed) | total plant BOD load (lbm/day) | aerator BOD load, $L_{BOD}$ (lbm/1000 ft$^3$-day) | $R_{F/M}$ (lbm/lbm) | MLSS (mg/l) | $R_R$ (%) | $\eta_{BOD}$ (%) |
|---|---|---|---|---|---|---|---|---|---|---|
| conventional | 0–0.5 | 7.5 | 0.8–1.1 | 0.4–0.6 | 0–1000 | 30 | 0.2–0.5 | 1500–3000 | 30 | 90–95 |
| | 0.5–1.5 | 7.5–6.0 | | | 1000–3000 | 30–40 | | | | |
| | 1.5 up | 6.0 | | | 3000 up | 40 | | | | |
| contact stabilization | 0–0.5 | 3.0* | 0.8–1.1 | 0.4–0.6 | 0–1000 | 30 | 0.2–0.5 | 1000–3000* | 100 | 85–90 |
| | 0.5–1.5 | 3.0–2.0* | 0.4–0.6 | | 1000–3000 | 30–50 | | | | |
| | 1.5 up | 1.5–2.0* | 0.4–0.6 | | 3000 up | 50 | | | | |
| extended | 0–0.05 | 24 | 1.4–1.6 | 0.15–0.3 | all | 10.0 | 0.05–0.1 | 3000–6000 | 100 | 85-95 |
| | 0.05–0.15 | 20 | | | | 12.5 | | | | |
| | 0.15 up | 16 | | | | 15.0 | | | | |
| high rate | 0–0.5 | 4.0 | 0.7–0.9 | 0.5–0.7 | 2000 up | 100 | 1.0 or less | 4000–10,000 | 100 | 80–85 |
| | 0.5–1.5 | 3.0 | | | | | | | | |
| | 1.5 up | 2.0 | | | | | | | | |
| step aeration | 0–0.5 | 7.5 | | | 0–1000 | 30 | 0.2–0.5 | 2000–3500 | 50 | 85–95 |
| | 0.5–1.5 | 7.5–5.0 | | | 1000–3000 | 30-50 | | | | |
| | 1.5 up | 5.0 | | | 3000 up | 50 | | | | |
| high purity oxygen | | 1.0–3.0 | | | | above 120 | 0.6–1.5 | 6000–8000 | 50 | 90–95 |

* in contact unit only

**Table 8.15**
Final Clarifiers for Activated Sludge Processes

| type of aeration | design flow (MGD) | minimum detention time (hr) | maximum overflow rate (gpd/ft$^2$) |
|---|---|---|---|
| conventional, high rate, and step | $< 0.5$ | 3.0 | 600 |
| | 0.5 to 1.5 | 2.5 | 700 |
| | $> 1.5$ | 2.0 | 800 |
| contact stabilization | $< 0.5$ | 3.6 | 500 |
| | 0.5 to 1.5 | 3.0 | 600 |
| | $> 1.5$ | 2.5 | 700 |
| extended aeration | $< 0.05$ | 4.0 | 300 |
| | 0.05 to 0.15 | 3.6 | 300 |
| | $> 0.15$ | 3.0 | 600 |

Following sedimentation and sludge recirculation, the effluent may be treated to convert nitrates and nitrites to nitrogen gas. Methanol supplies the energy required by the denitrification bacteria.

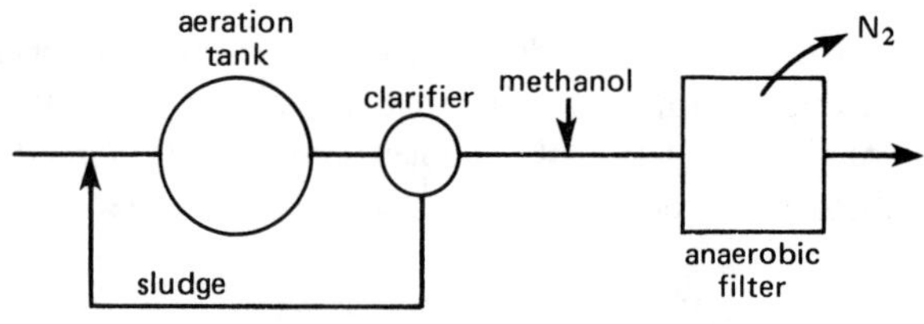


**Figure 8.20** Nitrification and Denitrification Process

Ammonia can also be removed by breakpoint chlorination. Other methods of nitrogen/ammonia removal include anion ion exchange and algae ponds.

*Inorganic Salt Removal*: Ions from inorganic salts can be economically removed by electrodialysis and reverse osmosis.

*Dissolved Solids Removal*: The so-called *trace organics* or *refractory substances* are dissolved organic solids that are biologically resistant. They can be removed by filtering through activated carbon or ozonation.

## 16 SLUDGE DISPOSAL

### A. SLUDGE QUANTITIES

Sludge removed from sedimentation basins is 95% to 99% moisture. With primary treatment only, about 0.1 pounds of dried sludge can be expected per capita day.

With secondary treatment, the total sludge load will be about 0.2 pounds per capita day (when dried). This amounts to approximately 2 quarts of sludge per 100 gallons of wastewater processed.

**Table 8.16**
Typical Characteristics of Domestic Sewage Sludge
(also, see table 8.18)

| origin of sludge | solids content of wet sludge (s in percent) | dry solids (lbm/day/capita) |
|---|---|---|
| primary settling tank | 6 | 0.12 |
| trickling filter secondary | 4 | 0.04 |
| mixed primary and trickling filter secondary | 5 | 0.16 |
| high rate activated sludge secondary | 2.5–5 | 0.06 |
| mixed primary and high rate activated sludge secondary | 5 | 0.18 |
| conventional activated sludge secondary | 0.5–1 | 0.07 |
| mixed primary and conventional activated sludge secondary | 2–3 | 0.19 |
| extended aeration secondary | 2 | 0.02 |

The dry weight of solids from primary settling basins is:

$$W_p = (\text{decrease in SS})(Q)(8.345\,\text{EE–6}) \qquad 8.56$$

The dry weight of solids from secondary aeration lagoons and biological filters is:

$$W_s = K(\text{BOD})_{\text{removed}}(Q)(8.345\,\text{EE–6}) \qquad 8.57$$

$K$ in equation 8.57 depends on the food-to-microorganism ratio, as given in table 8.17. $K$ is the fraction of BOD that appears as excess biological solids. For trickling filters and extended aeration, $K$ ranges from 0.2 to 0.33. For conventional and step aeration, $K$ ranges from 0.33 to 0.42. $K$ is known as the *cell yield* or *yield coefficient.*

**Table 8.17**
Cell Yield (Yield Coefficient)

| $R_{\text{F-M}}$ | $K$ |
|---|---|
| 0.05 | 0.2 |
| 0.07 | 0.21 |
| 0.1 | 0.24 |
| 0.15 | 0.28 |
| 0.2 | 0.33 |
| 0.3 | 0.37 |
| 0.4 | 0.4 |
| 0.5 | 0.43 |

Equations 8.56 and 8.57 give dry weight of sludge. Assuming the sludge specific gravity is near 1, the volume of wet sludge with a solids concentration, $s$, can be found from the dry weight by using equation 8.58.

$$\begin{array}{c}\text{gallons of}\\ \text{sludge/day}\end{array} = \frac{\text{dried weight per day, lbm}}{(s)(8.345)} \qquad 8.58$$

Typical values of $s$ are given in table 8.18.

**Table 8.18**
Total Sludge Solids, s
(also, see table 8.16)

| source or type of sludge | s, as fraction |
|---|---|
| primary settling tank sludge | 0.06 to 0.08 |
| primary settling tank sludge mixed with filter sludge | 0.04 to 0.06 |
| primary settling tank sludge mixed with activated sludge from aeration lagoons | 0.03 to 0.04 |
| excess activated sludge | 0.005 to 0.02 |
| filter backwashing water | 0.01 to 0.1 |
| softening sludge | 0.02 to 0.15 |

The specific gravity of the wet sludge is:

$$\frac{1}{(SG)_{\text{total}}} = \frac{\text{fraction moisture}}{1.0} + \frac{\text{fraction solids}}{(SG)_{\text{solids}}} \qquad 8.59$$

The volume of sludge is:

$$V = \frac{W}{(SG)_{\text{total}}(62.4)} \text{ in ft}^3\text{/day} \qquad 8.60$$

*Example 8.8*

A trickling filter plant processes domestic waste with the following characteristics: 190 mg/l BOD, 230 mg/l SS, and 4,000,000 gpd.

(a) What is the wet sludge volume from the primary sedimentation tank and trickling filter? Assume the combined sludge solids content is 5%.

(b) What is the approximate weight of dry solids produced per person-day?

*step 1*: Find the weight of the dry solids obtained from the primary settling basin. From equation 8.56, assuming 50% of solids can be removed,

$$\begin{aligned} W_p &= (0.5)(230)(4\,\text{EE}6)(8.345\,\text{EE–}6) \\ &= 3839\,\text{pounds/day} \end{aligned}$$

Assume a 30% BOD reduction, so the BOD leaving the basin is $(0.7)(190) = 133$ mg/l.

*step 2*: Assume a cell yield value of 0.25. Then, the weight of dry solids from the filter is given by equation 8.57.

$$\begin{aligned} W_s &= (0.25)(133)(4\,\text{EE}6)(8.345\,\text{EE–}6) \\ &= 1110\,\text{pounds/day} \end{aligned}$$

*step 3*: The wet sludge volume can be found from equation 8.58.

$$\frac{3839 + 1110}{(0.05)(8.345)} = 11{,}860\,\text{gallons/day}$$

*step 4*: The equivalent population is given by equation 8.12.

$$P_e = \frac{(190)(8.345)(4)}{0.17} = 37{,}310\,\text{people}$$

*step 5*: The per capita dried solids rate is

$$\frac{3839 + 1110}{37{,}310} = 0.13\frac{\text{lbm dry solids}}{\text{person-day}}$$

## B. SLUDGE THICKENING

Since the volume of wet sludge is inversely proportional to its solids content (equation 8.58), thickening of sludge is desirable. Thickening is required to at least 4% solids if dewatering is to be feasible. *Gravity thickening* uses a stirred sedimentation tank into which sludge is fed. A doubling of solids content is usually possible with a gravity thickener.

Thickening can also be accomplished through the *dissolved air flotation* method. Air is bubbled through a tank containing sludge. The solid particles adhere to the air bubbles, float to the surface, and are skimmed away. The skimmed scum has a solids content of approximately 4%. Up to 85% of the total solids may be recovered, although chemical flocculants may be used to increase this to 95%. Two to four pounds of solids are obtained each hour for each square foot of surface area.

Floatation thickening, as well as centrifuge thickening, of activated sludge may be desirable where unusually large quantities are produced in proportion to primary sludge. Gravity thickening is usually best for primary or mixtures of primary and secondary sludges.

The following criteria can be used to design gravity thickening tanks.

- required overflow rate, gpd per sq ft — 600 to 800
- maximum solids loading, lbm of dry solids/sq ft per day:
  - primary sludge — 22
  - primary plus trickling filter sludge — 15
  - primary plus modified aeration activated sludge — 12
  - primary plus conventional activated sludge — 8
  - waste activated sludge — 4
- shape — circular
- minimum side water depth, feet — 10
- minimum detention time, hours — 6
- minimum floor slope, vertical to horizontal — 2.75 to 12
- minimum number of tanks (unless alternate means is available for thickening or storing sludge) — 2
- minimum angle between hopper wall and horizontal plane — 60°
- minimum freeboard — 18 in
- sludge collection and stirring mechanism — rotary scraper collector

## C. SLUDGE DEWATERING

Once the sludge has been thickened, it can be digested or dewatered prior to disposal. Several methods of dewatering are available, including vacuum filtration, pressure filtration, centrifugation, drying beds of sand or gravel, and lagooning.

A common method of dewatering is vacuum filtration in a *rotary drum filter*. Suction is applied from within the drum to attract solids to the filter and to extract moisture. The dried cake is scraped off of the drum in the discharge section of the device. Chemical flocculants are used to collect fine particles on the filter drum. A final solids content of 20% to 25% is attained. This is sufficient for sanitary landfill. A solids content of 30% is needed, however, for direct incineration.

Sand drying beds are preferable when digested sludge is to be disposed of in a landfill. Sand beds produce cake with a high dryness. Generally, vacuum filtration and centrifugation are used with undigested sludge. (Gritty sludge should not be centrifuged.) Vacuum filtering presents more odor and cleanliness problems than centrifugation. Lagooning of sludge prior to landfilling should be considered only in isolated areas where the average temperature is above 60°F.

Sludge drying beds can be designed according to the following criteria:

- depth of application — 8″
- number of applications — 8/year
- area required (open beds):
  - primary — 1 $ft^2$/cap
  - trickling filter — 1.5
  - activated sludge — 2
  - chemical coagulation — 2
- typical dimensions — 20 to 30 ft wide, 50 to 100 ft long
- enclosing wall height — 18 inches

## D. DIGESTION AND STABILIZATION

Much of the organic material in sludge is easily digested (i.e., "stabilized") by anaerobic microbes. Solids which are capable of being digested are known as *volatile solids*. Digestion of volatile solids results in methane, carbon dioxide, and hydrogen sulfide gases.

Anaerobic digestion is more complex and more easily upset than aerobic digestion. However, it has a lower operating cost. Aerobic digestion is preferable with stabilized primary and mixed primary-secondary sludge. Chloride concentrations must be monitored with both types of digestion.

A third alternative, that of *wet oxidation* with low pressure and low temperature, conditions sludge as well as stabilizes it. However, below plant flows of 2 MGD, this process may be too complex. Wet oxidation is not subject to toxic upsets as are anaerobic digesters. This method, as well as anaerobic digestion, yields a byproduct liquor with a high BOD and ammonia content. Separate treatment or recirculation may be required. Where there are nitrogenous oxygen demand or nutrient restrictions on the plant effluent, wet oxidation (and thermal conditioning and anaerobic digestion) may be impractical.

**Anaerobic Digestion**: If the digestion takes place in the absense of oxygen, it is known as *anaerobic digestion.* Two types of bacteria are involved: acid forming and acid splitting. The pH must be kept above 6.5 for the methane producing bacteria to function.

In a single-stage, floating-cover digester (figure 8.21) raw sludge is brought into the tank at the cover and top of the dome. The contents of the digester stratify into four layers: scum on top, *supernatant*, a layer of actively digesting sludge, and a bottom layer of concentrated sludge. Some sludge may be withdrawn, heated, and returned to keep the temperature up.

Supernatant is removed along the periphery of the digester and returned to the inlet of the processing plant. Digested sludge is removed from the bottom and is then dewatered. Gas is removed from the gas dome and is burned. The heat from the burning methane can be used to warm the sludge that is withdrawn.

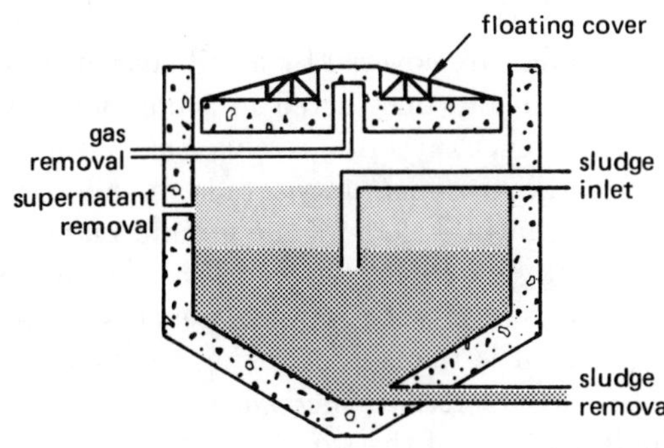


**Figure 8.21** Floating Cover Digester

The following characteristics are relevant to a single-stage digester:

- sludge loading — 0.13–0.2 lbm of volatile solids/ft$^3$-day
- optimum temperature — 95 to 98°F
- optimum pH — 6.7 to 7.8 (7.0 to 7.1 preferred)
- gas production — 7 to 10 ft$^3$/lbm volatile solids added; 0.5 to 1.0 ft$^3$/capita day
- gas composition — 65% methane, 35% $CO_2$
- retention time — 30 to 90 days (conventional); 15 to 25 days (high rate)
- final moisture content — 90 to 95%
- gas heat content — 600 BTU/ft$^3$
- depth — 20 to 45 feet
- depth to diameter ratio — use manufacturer's recommendations
- minimum freeboard — 2 feet
- slope of tank bottom — 1:12 to 1:4
- minimum number of tanks — 2
- typical diameter — 20 to 115 feet

A single stage digester performs the functions of digestion, gravity thickening, and storage in one tank. In a two-stage process, two digesters in series are used. Heating and mechanical mixing occur in the first digester. Since the sludge is continually mixed, it will not settle. Settling and further digestion occur in the unheated second tank.

Historically, rules of thumb were used to size digesters. For single-stage, heated digesters taking primary waste from a trickling filter, 3 to 4 cubic feet of digester volume per equivalent capita is required. For primary and secondary waste, 6 cubic feet are required. Equation 8.61 can also be used to size the digester.

$$\text{digester volume} = \tfrac{1}{2}(Q_{\text{raw sludge}} + Q_{\text{digested sludge}})t_{\text{digestion}} + (Q_{\text{digested sludge}})t_{\text{storage}} \qquad 8.61$$

**Aerobic Digestion**: Aerobic digestion in a holding tank or digester uses mechanical aerators in a manner similar to aerated lagoons. Construction of an aerobic digester is similar to that of an aerated lagoon.

Aerobic digesters have the ability to significantly reduce volatile solids. Typical operating and physical characteristics are:

- sizing — 2 to 3 ft$^3$/capita
- aeration period:
  - only activated sludge — 10 to 15 days
  - activated sludge without primary settling — 12 to 18
  - mixture of primary sludge and activated or trickle filter sludge — 15 to 20
- maximum volatile solids loading — 0.1 to 0.3
- sludge age:
  - primary sludge — 25 to 30 days
  - activated sludge — 15 to 20
- maximum tank depth — 15 feet
- oxygen required — 1.5 to 1.9 lbm $O_2$/lbm BOD removed
- volatile solids reduction — 35 to 45% typical; 45 to 70% maximum
- dissolved oxygen level — 1 to 2 mg/l
- mixing energy required — 0.75 to 1.5 hp/1000 ft$^3$

### E. ADDITIONAL SLUDGE PROCESSING METHODS

Sludge may be heated when an economical source of combustion heat is available. Even then, other methods may be more economical. Incineration produces the smallest quantity of inert residue. Sludge storage is not really a process at all, but it can be considered for short-term solutions prior to mechanical dewatering.

### F. RESIDUAL DISPOSAL

Stabilized sludge and clean grit may be eligible for ocean dumping. However, this disposal method is discouraged. Where isolated land is available, land surface spreading is an option. Water pollution from sludge must be prevented. (Harrowing sludge into the soil and other runoff-erosion methods must be used.) Sludge must also be free of heavy metals and other toxic materials in excess of concentrations permitted.

Stabilized sludge and clean grit can be disposed of by lagooning, and restrictions to those for land surface spreading apply. In addition, lagooning should be considered only where eventual restoration of the lagoon area to other uses is possible.

Where a satisfactory landfill site is available, sludge cake, grit, screenings, and scum can be disposed of. However the possible effects on leachate quality will affect the treatment methods and sludge quality.

Incineration is the most expensive treatment, requiring the near total destruction of volatile solids in sludge, scum, and grit. Land disposal of ash should be considered only when lower-cost options are unavailable.

## 17 EFFLUENT DISPOSAL[16]

Wherever possible, effluent should be discharged to receiving waters. Flowing surface waters (such as streams, estuaries, and oceans) are desired. Disposal to lakes and reservoirs should be avoided.

Wastes from very small installations (up to about 30 people) in rural or remote locations can be discharged to groundwater through soil absorption systems. *Deep well injection* can be used for difficult-to-treat industrial wastes if permitted by regulatory agencies and if soil strata are suitable for the discharge type.

If economical, treated effluent can be used for industrial purposes, irrigation, flushing, or (with sufficient treatment) for potable water.

[16] Regulatory agencies should be consulted to determine the applicability of any of these disposal methods.

## 18 SANITARY LANDFILLS

### A. INTRODUCTION

Sanitary landfilling is the disposing of waste solids by thin layering, compacting to minimum volume, and daily soil covering. There is no burning of waste solids.

Sanitary landfill sites need to be selected on the basis of many relevant characteristics. Some of these characteristics are:

- economics and availability of cheap land
- location, including ease of access, distance, acceptance by the population, and aesthetics
- availability of cover soil, if required
- wind direction and wind speed, including odor, dust, and erosion considerations
- flat topography
- dry climate, or low water table and low infiltration rates
- low risk of acquifer contamination
- type and permeability of underlying strata
- absence of winter freezing

Each day's solids are covered by a soil layer, producing a *cell.* Cells can be constructed in a number of ways, but the area, trench, and progressive methods are most common. In the *area method,* waste is spread and compacted on the natural slope of the ground before being covered. With the *trench method,* waste is placed in a trench and covered with its own trench soil. A new trench extension is then dug for the subsequent day's waste. In the *progressive method,* cover soil is taken from the front (toe) of the working face.

Soils and cover soil of clay are preferred, since clay is both hard and strong when dry.

The cell height including the soil cover is known as the *lift.* Lift and cell thickness are illustrated in figure 8.22.

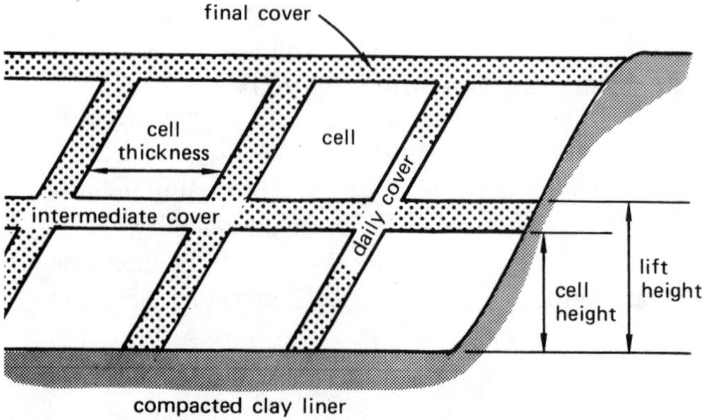


**Figure 8.22** Cells and Landfilling Terms

The daily, intermediate, and final soil covers are essential to proper landfill operation. The cover prevents fly emergence, discourages rats and rodents, and controls entering moisture. The soil cover also serves a purpose when gases of decomposition are to be controlled or collected. In the event of ignition, the soil covers form fire stops within the landfill.

Once covered, the generous food supply within a cell hosts considerable biological decomposition activity. The waste exhausts the oxygen from the cell quickly. Digestion continues anaerobically, producing methane ($CH_4$) gas.[17] The temperature may increase to up to 150°F.

Once closed, the landfill can be used for grassed green areas, shallow-rooted agriculture, and recreational areas (golf, ball fields, etc.). Some light construction uses are also possible, although there may be problems with gas accumulation, corrosion of pipe and foundation piles, and settlement.

### B. DESIGN OF SANITARY LANDFILLS

Sanitary landfills should be designed for a 5-year minimum life, during which time the design volume will be 5 to 8 lbm/capita-day. (5 lbm/capita-day is a typical conservative estimate of incoming volume which is commonly used.)

The waste is placed in layers typically 2 to 3 feet thick. If the cell is sloped, the slope will be less than 40°, typically 20 to 30°. The daily soil cover can be assumed to be 6 inches; the intermediate soil cover will be approximately 1 foot; and the final cover of 2 feet should be applied and compacted in 4 to 6 inch layers. (However, the cover may be up to 4 feet thick, with compaction in 1 foot layers.) When the landfill layer is complete, the ratio of solid waste volume to cover volume will be between 4:1 and 3:1.

Although the *cell height* can be taken as 8 feet for design studies, it can actually be up to 30 feet. The height should be chosen to minimize the amount of cover material required.

Once placed, the solid waste can be compacted to a density of 400 to 1500 lbm/cu. yd., although 1000 lbm/cu. yd. is typical. Generally, a density of 800 lbm/cu. yd. is the minimum acceptable density. Compaction is achieved by 2 to 5 passes by a tracked bulldozer. A 50% reduction in volume can be expected.

The daily increase in landfill volume can be predicted by equation 8.62. The loading factor is 1.25 for a 4:1 ratio of solid waste to cover volumes. It is 1.33 for a 3:1 ratio. (The compacted density can be calculated as a weighted average of soil and solid waste densities. Fine grained soils have densities in the 70 to 120 $\text{lbm/ft}^3$ range; coarse grained soils have densities in the 100 to 135 $\text{lbm/ft}^3$ range.)[18]

$$\frac{\text{cubic yards}}{\text{day}} = \frac{\begin{pmatrix}\text{population}\\ \text{size}\end{pmatrix}\begin{pmatrix}\text{waste in}\\ \text{lbm/capita-day}\end{pmatrix}\begin{pmatrix}\text{loading}\\ \text{factor}\end{pmatrix}}{\text{compacted density, lbm/cu. yd.}} \qquad 8.62$$

Once a landfill site has been closed, settlement can be expected. In areas with high rainfall, there will be greater waste decomposition, with up to 20% settlement based on the overall landfill height. In dry areas, the settlement may be limited to 2% or 3%. Little information is available on bearing capacity of the landfill. Some studies have suggested capacities in the 500 to 800 $\text{lbf/ft}^2$ range.

### C. GAS PRODUCTION

Little or no gas will be produced during the first year. However, the peak gas production rate will be reached quickly within 12 to 18 months after closing. Steady gas production can be expected for up to 5 years. Over the gas-producing life of the landfill, 40 cubic feet of gas per cubic yard of landfill can be expected. The gas will be approximately 50% methane.

Gas can be allowed to escape through a semi-permeable cover material such as sand or gravel. It can also be vented through a gravel trench or inclined gravel vent. Vent pipes with collection laterals are expensive, as is vacuum collection.

Compacted clay liners and clay trenches outside the landfill should be placed to prevent the widespread lateral distribution of gas. At least 5 feet should separate the water table and the bottom clay liner. This amount of separation will be sufficient to remove decomposed organisms and living bacteria from the leachate before it reaches the groundwater.

## 19 HAZARDOUS WASTE DISPOSAL

### A. INTRODUCTION

Hazardous wastes can be handled by disposal, dispersal, encapsulation, treatment, or recovery and recycling.[19]

[17] The products of *aerobic decomposition* are $CO_2$, $H_2O$, and nitrates. The products of *anaerobic decomposition* include $CH_4$, CO and $CO_2$, $H_2O$, $N_2$, ammonia, organic acids, and various metallic sulfides.

[18] Multiply cubic yards by $\frac{27}{43{,}560}$ to get acre-feet.

[19] The 1984 amendments to the *Resource Conservation Recovery Act* (*RCRA*) of 1976 eliminate land disposal for many wastes and keep such disposal at a minimum for many other wastes. The types of disposal methods prohibited unless approved by the Environmental Protection Agency include landfills, surface impoundments, waste piles, injection wells, land treatment facilities, salt dome formations, salt bed formations, and underground mines.

Disposal can be at landfills, impoundment of liquids at the surface, underground injection wells, aboveground piles, and storage in containers and tanks. Dispersal (dilution) is suitable only for degradable substances. Land spreading and ocean dumping are examples of dispersal.

Encapsulation is commonly used with disposal. Hazardous substances can be encapsulated with cement or potted in plastic (polymerization).

Treatment is a method of modifying the hazard. Organic substances can be incinerated or treated by high temperature decomposition. Chemical reduction in tanks or biological processes (digestion) in biocontactors can also be used.

Recovery/recycling methods include filtration, electrolysis, distillation/evaporation, sedimentation/precipitation, ion exchange and reverse osmosis, and adsorption.

## B. HAZARDOUS WASTE LANDFILLS

Hazardous waste disposal sites must be designed to hold wastes so that they do not adversely affect the air and water quality. Primarily, this means that free liquids within the landfill must be controlled against migration.

Hazardous waste disposal sites can be designed for almost any type of waste, but they are best at controlling soluble and volatile wastes. The most difficult wastes are those that are soluble, toxic, and persistent.[20]

A disposal site's performance will be a function of the reliability and longevity of the leachate collection systems, the hydrogeological characteristics of the site, the type of waste, and the site use, management, and monitoring.

## C. HYDROGEOLOGICAL SITE CHARACTERISTICS

An ideal disposal site would have unfractured bedrock topped with many feet of native, low-permeability clay. The rate of evaporation for the area would exceed precipitation, and the water table would be low or nonexistent. The disposal site would not extend over acquifers which serve as a sole source water supply. The site would not be within the 100-year flood plain.

## D. PRETREATMENT REQUIRED

Ignitable, reactive, and corrosive wastes must be pretreated to reduce those characteristics. When placed in a disposal site, the wastes must be compatible. Containerized liquids should be rendered non-free flowing. Serious thought should be given to disposing of toxic, persistent, and highly mobile wastes when allowed.

## E. LINER AND DISPOSAL SITE DESIGNS

Two alternative liner schemes can be used, both of which require a double bottom liner. The top liner for both alternatives is a synthetic membrane with minimum thickness of 30 mils. The two designs have different bottom liners, however. For the first, the bottom is of compacted clay. The other design is preferred by the EPA. It has a synthetic sheet above the clay liner.

Both designs require two leachate collection systems. One of the collection systems is within the landfill itself to limit the hydraulic head of liquid that has reached the bottom of the landfill. The other collection system catches and recycles leachate which has passed through the top liner.

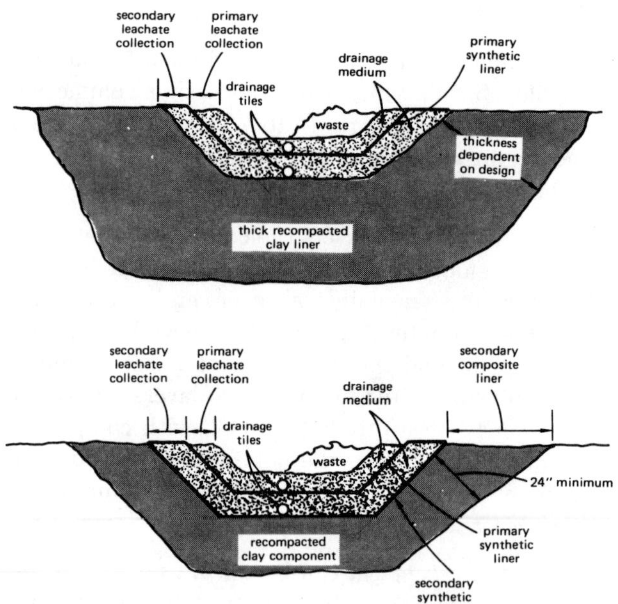


**Figure 8.23** Alternative Liner Designs

## F. LINERS FOR DISPOSAL SITES

Bottom liners are designed to reduce the rate of leachate migration into the subsoil. Clay liners 2 to 20 feet thick have been used frequently in the past. Synthetic *flexible membrane liners* (*FML's*) 30 to 80 mils thick are now required. Other liner materials that have been used include soil cement, asphalt concrete, sprayed-on asphalt and urethane membranes, and soil sealants. Liners fail by faulty installation, structural failure due to hydro-

[20] Arsenic, chromium, spent solvents, dioxins, and other highly toxic wastes may not be disposed of in landfills.

static pressure, settling of the soil, and chemical decomposition of the liner from wastes. Clay liners are particularly susceptible to dessication from concentrated organic chemical such as solvents.

Migration through undamaged liners is proportional to the conductivity and pressure. The pressure is a function of the leachate depth. Conductivities of clay liners vary between EE–9 and EE–11 $m^3/m^2$-sec. For synthetic liners, the conductivity ranges between EE–11 and EE–14 $m^3/m^2$-sec.

Double liner systems require ground water monitoring with monitoring wells extending past the bottom of the disposal site into the acquifer.

### G. COVERS FOR DISPOSAL SITES

Covers are required to prevent infiltration of rain into disposal cells. Clay or synthetic materials, or combinations of the two, can be used. (Liners are effective at diverting water off of mounded clay covers.) Covers fail by dessication (drying out), penetration by roots, animal activity, freezing and thawing cycles, erosion, settling, and breakdown of synthetic liners from sunlight exposure. Failure will also occur when the cover collapses into the disposal site due to open voids below or excess loading (rainwater ponding) above.

### H. LEACHATE COLLECTION/RECOVERY SYSTEMS

Leachate will be prevented from leaving the landfill by a leachate collection system. By removing the leachate above the first liner, the hydrostatic pressure on the liner will be reduced. A pump is used to raise collected leachate to the surface once a predetermined level has been reached. Tracer compounds (e.g., lithium compounds or radioactive hydrogen) can be buried with the wastes to signal migration and leakage.[21]

Leachate collection and recovery systems fail by clogging drainage layers and pipes, crushing of the collection pipes due to waste load, and pump failures.

## 20 INJECTION WELLS

Liquid or low-viscous wastes can be injected under high pressure into appropriate strata 2000 to 6000 feet below the surface. Wastes will displace native fluids under pressure. The injection well is capped to maintain the pressure.

[21] Tritium has a half-life of 12.3 years. It is essentially completely decayed after 120 years. It is easily detected, and is not affected chemically by other wastes.

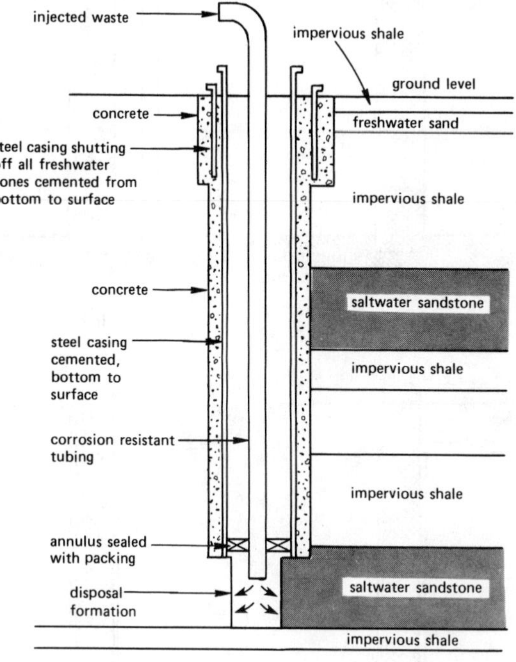


**Figure 8.24** A Typical Injection Well

Injection wells fail primarily by waste plumes (capillary action) through fractures, shrinkage cracks, pressure fractures, dissolution channels, fault slips in strata, and seepage around the borehole.

## 21 SLURRY TRENCH CONTAINMENT

Slurry trenches 4 to 12 inches wide are opened hydraulically using a bentonite slurry trenching fluid. (The slurry contains *bentonite* 1 to 6% by weight.) When the slurry hardens on the walls of the trench, the trench is backfilled. If backfilled with soil, the trench is known as a *soil bentonite* trench. If the trench is filled with bentonite slurry as the trench is opened, it is known as a *pure bentonite* or *cement bentonite trench.*

Bentonite is not very chemical resistant, particularly when it is fully hydrated. Furthermore, it's permeability is on the order of only EE–10 $m^3/m^2$-sec. So, a synthetic liner is required with exposure to hazardous wastes. Asphaltic cutoffs can also be used.

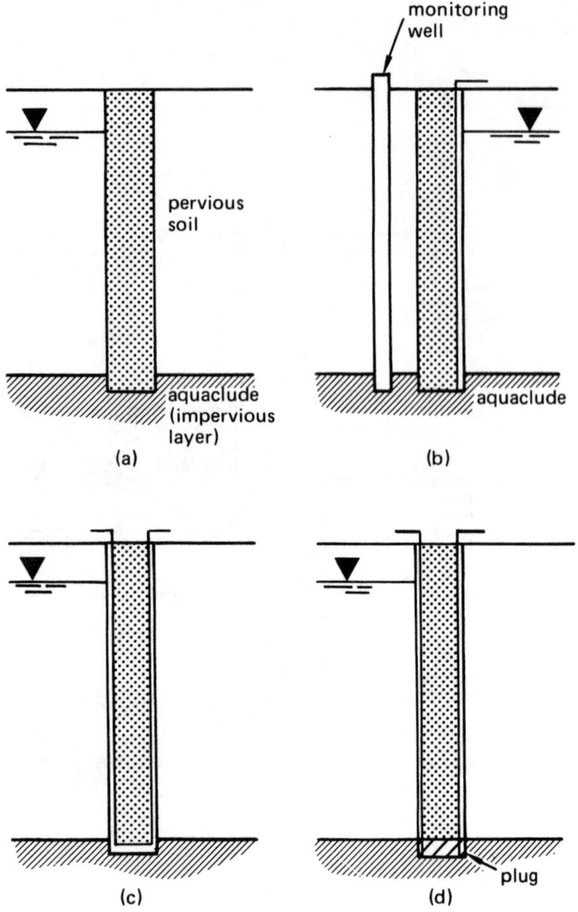


(a) standard trench
(b) trench with single sheet and monitoring well
(c) envelope design double sheet
(d) double sheet with bottom plug

**Figure 8.25** Slurry Trench Containment Walls

## 22 STEPS IN CLEANING UP HAZARDOUS WASTE SPILLS

The steps required to clean up a hazardous waste spill depend on the type of waste spilled, the weather, environment, geology, proximity to population, and other factors. Also, some steps will need to be performed as quickly as possible, while others are long term. Not all of the steps listed here will be appropriate for every case.

- immediate confinement: sandbag dams, bentonite clay linings, slurry walls
- physical removal: removal of solids, liquids, bulk containers, and contaminated soil and vegetation to hazardous waste disposal sites
- pumping through purge wells: removing contaminated water from the acquifer
- holding contaminated water in basins for later treatment
- dilution: mixing clean or processed water with contaminated water through spraying, injection into the soil or acquifer, or deep well disposal injection.
- adsorption: mixing surface soil with granular carbon, soda ash, and other chemicals to raise pH, elevate temperature, and initiate decay and decomposition of remaining waste
- coverage: covering the contaminated area by a surface layer to decrease rain water forming leachate and driving contaminant further down

# Appendix A: Selected "10-States' Standards"

The following selected standards are derived from "Recommended Standards for Sewage Works", originally developed by the Great Lakes-Upper Mississippi River Board of State Sanitary Engineers. Since there are ten states in this board, the publication is commonly referred to as "10-States' Standards."

The standards are subject to change, and this summary is not complete. The items listed here are merely meant to be representative of good waste water plant design.

- **Pumps**: At least 2 pumps are required, and 3 are required when the design flow exceeds 1 MGD. Both pumps should have the same capacity if only 2 pumps are used. This capacity must exceed the design flow for each pump. If 3 or more pumps are used, the capacities may vary, but capacity pumping must be possible with one pump out of service.

- **Racks and Bar Screens**: All racks and screens shall have openings less than 1.75″ wide. The smallest opening for manually-cleaned screens is 1.0″. The smallest opening for automatically-cleaned screens is 0.625″. Flow velocity should be 1.25 to 3.0 fps.

- **Grinders and Shredders**: Grinder/shredders are required if there is no primary sedimentation or fine screens. Gravel traps or grit-removal equipment should precede comminutors.

- **Grit Chambers**: Grit chambers are required when combined storm and sanitary sewers are used. A minimum of two grit chambers in parallel should be used, with a provision for bypassing. The optimum velocity is 1.0 fps throughout. The detention time is dependent on particle sizes to be removed.

- **Plain Sedimentation Tanks**: Multiple units are desirable and must be provided if the flow exceeds 100,000 gpd. For primary settling, the depth should be 7.0′ or greater. The maximum peak overflow rate is 1500 gpd/ft$^2$.[22] 15,000 gpd/ft is the maximum weir loading. If the flow rate is less than 1.0 MGD, the weir loading should be reduced to 10,000 gpd/ft.

- **Trickle Filters**: Rock media should have a depth of 5′ to 10′. Manufactured media should have a depth of 10′ to 30′. The rock media should be 1″ to $4\frac{1}{2}$″ in size, with no fines. Freeboard of 4′ or more is required. The drain should slope at 1% or more.

- **Activated Sludge Processes**: For sedimentation basins, the following hydraulic loading maximums are specified: 1200 gpd/ft$^2$ for conventional, step, and contact units; 1000 gpd/ft$^2$ for extended aeration units; 800 gpd/ft$^2$ for separate nitrification units. The maximum BOD loading shall be: 40 lbm/day-1000 ft$^3$ for conventional, step, and complete mix units: 50 lbm/day-1000 ft$^3$ for contact stabilization units; 15 lbm/day-1000 ft$^3$ for extended aeration units.

  Aeration tank depths should be between 10′ and 30′. At least two aeration tanks should be used. The dissolved oxygen content should not be allowed to drop below 2.0 mg/l at any time. The aeration rate should be 1500 ft$^3$ oxygen per pound of $BOD_s$. For extended aeration, the rate should be 2000 ft$^3$ oxygen per pound of $BOD_s$.

- **Final Clarifiers**: Maximum surface settling rate is 800 gpd/ft$^2$ for separate nitrification stages, 1000 gpd/ft$^2$ for extended aeration, and 1200 gpd/ft$^2$ for all other cases, including fixed film biological processes.

- **Lagoons**: Maximum BOD application is 15 to 35 lbm $BOD_s$ per acre-day for both controlled-discharge and flow-through stabilization ponds.

- **Chlorination**: Requires a 15 minute contact period at peak flow.

- **Anaerobic Digesters**: For completely mixed digesters, up to 80 lbm of volatile solids per day per 1000 ft$^3$ of digester. For moderately mixed digesters, the limit is 40 lbm/day-1000 ft$^3$. Multiple units. Minimum 20 feet sidewater depth.

- **Sludge Drying Beds**: Requires 2 ft$^2$/capita-day, if drying beds are the primary dewatering method, and 1 ft$^2$/capita-day if beds are a back-up dewatering method.

[22] The basin size shall also be calculated based on the average design flow rate and a maximum overflow rate of 1000 gpd/ft$^2$. The larger of the two sizes shall be used.

# Appendix B: Saturated Oxygen Concentrations

To convert °C to °F:

$$°F = \frac{9}{5}°C + 32°$$

To convert °F to °C:

$$°C = \frac{5}{9}(°F - 32°)$$

(1 atmosphere)

| temperature °C | dissolved oxygen, mg/l | subtract for each 100 mg/l chloride |
|---|---|---|
| 0 | 14.6 | 0.017 |
| 1 | 14.2 | 0.016 |
| 2 | 13.8 | 0.015 |
| 3 | 13.5 | 0.015 |
| 4 | 13.1 | 0.014 |
| 5 | 12.8 | 0.014 |
| 6 | 12.5 | 0.014 |
| 7 | 12.2 | 0.013 |
| 8 | 11.9 | 0.013 |
| 9 | 11.6 | 0.012 |
| 10 | 11.3 | 0.012 |
| 11 | 11.1 | 0.011 |
| 12 | 10.8 | 0.011 |
| 13 | 10.6 | 0.011 |
| 14 | 10.4 | 0.010 |
| 15 | 10.2 | 0.010 |
| 16 | 10.0 | 0.010 |
| 17 | 9.7 | 0.010 |
| 18 | 9.5 | 0.009 |
| 19 | 9.4 | 0.009 |
| 20 | 9.2 | 0.009 |
| 21 | 9.0 | 0.009 |
| 22 | 8.8 | 0.008 |
| 23 | 8.7 | 0.008 |
| 24 | 8.5 | 0.008 |
| 25 | 8.4 | 0.008 |
| 26 | 8.2 | 0.008 |
| 27 | 8.1 | 0.008 |
| 28 | 7.9 | 0.008 |
| 29 | 7.8 | 0.008 |
| 30 | 7.6 | 0.008 |

# Appendix C: Typical Sequences Used in Wastewater Plants

The following partial sequences are used to construct a complete treatment plant:

| | |
|---|---|
| I | intake and preconditioning |
| P | primary treatment |
| S | secondary treatment |
| T | tertiary treatment |
| D | discharge |
| SP | sludge processing |
| SD | sludge disposal |

### I: Intake and Preconditioning

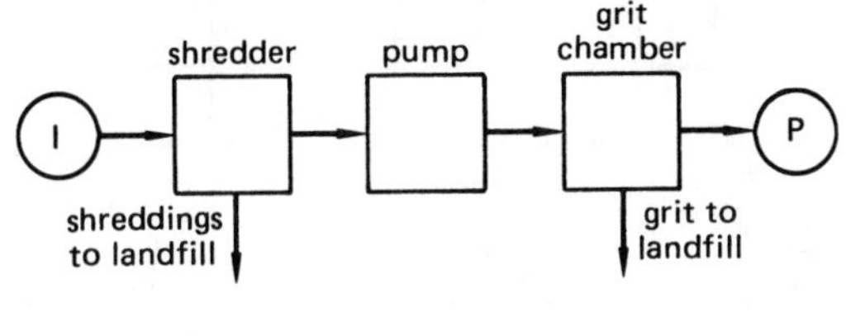


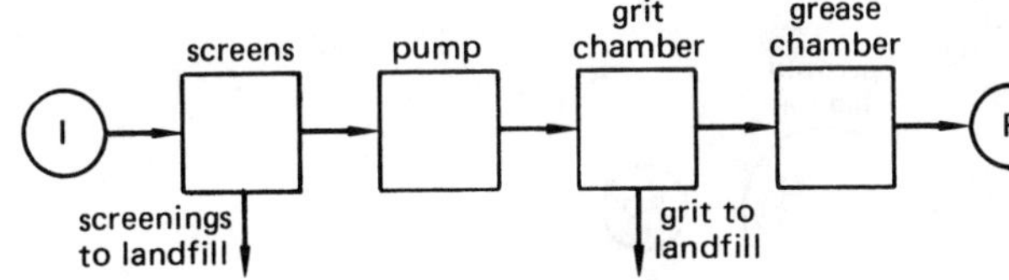


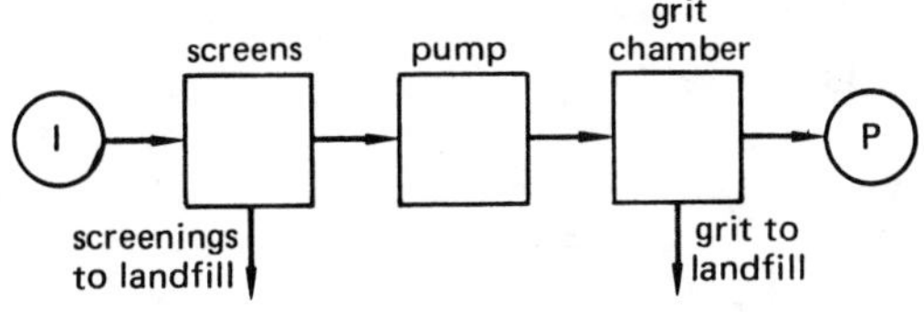


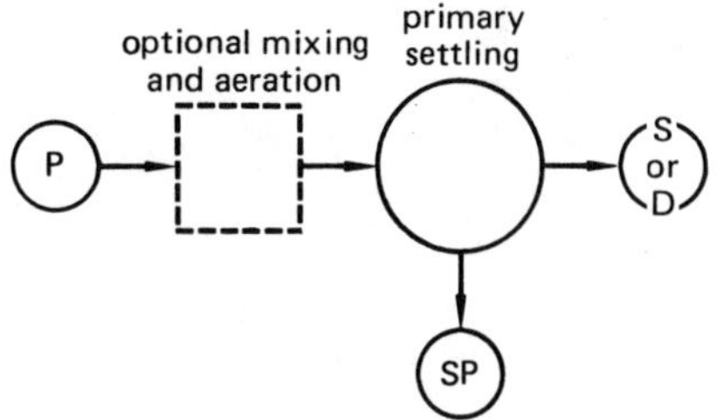


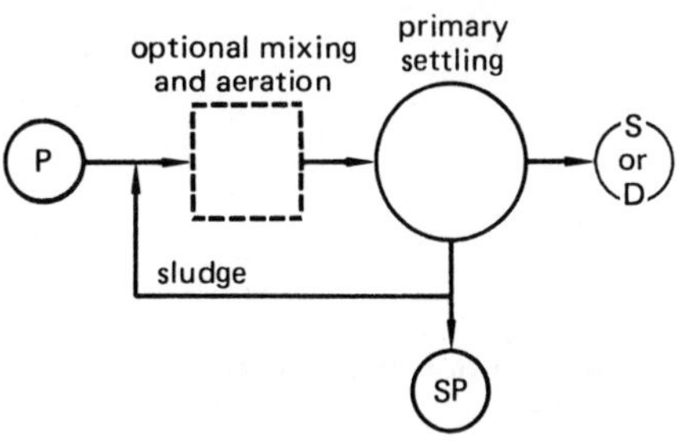


### P: Primary Treatment

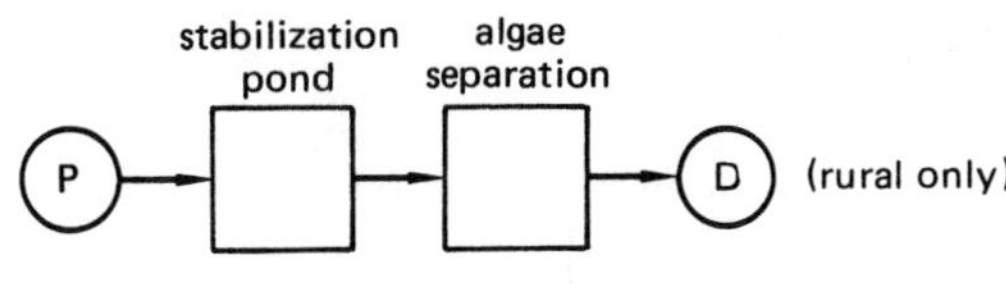


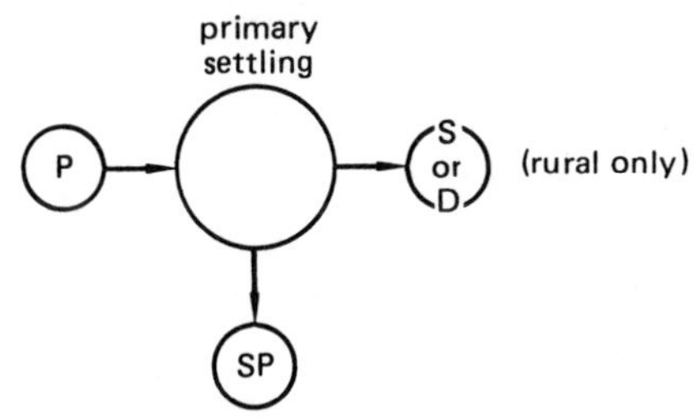


### S: Secondary Treatment

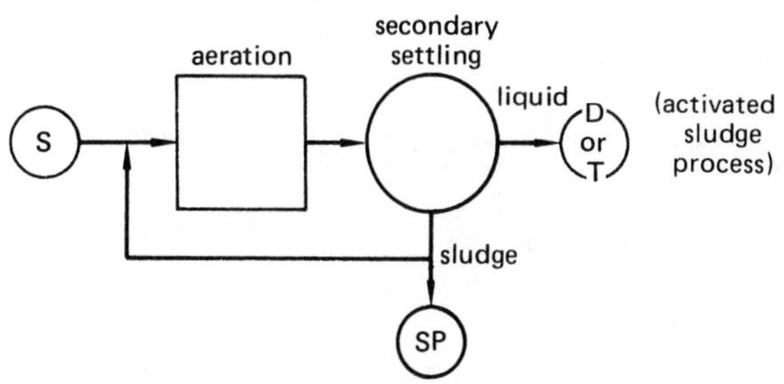


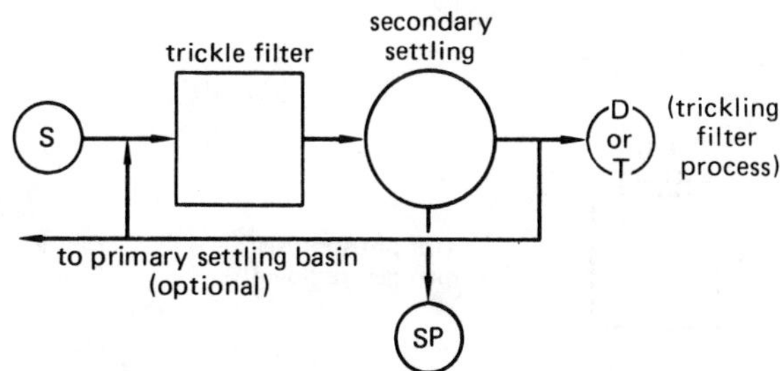

## T: Tertiary Treatment

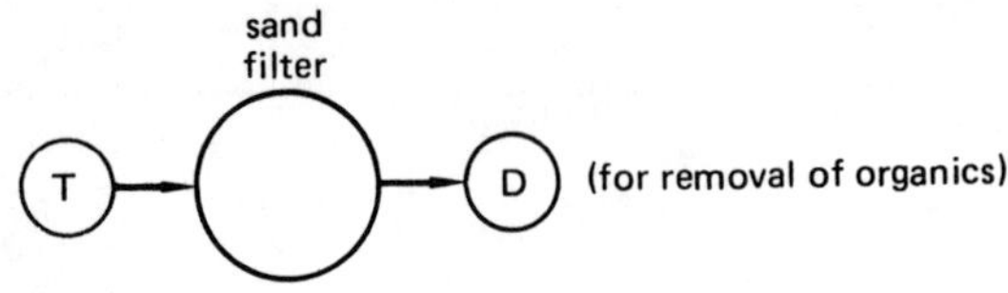


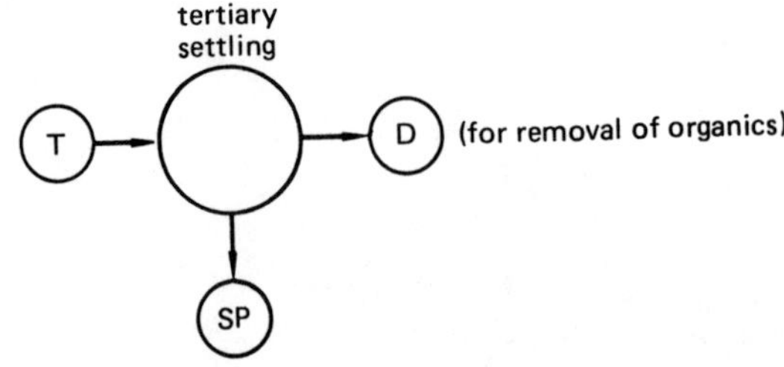


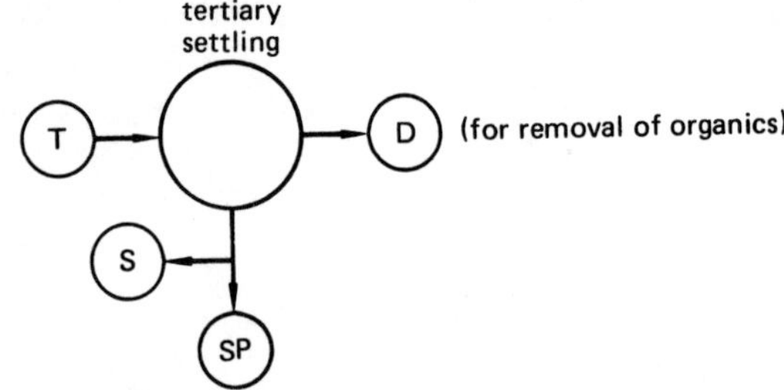


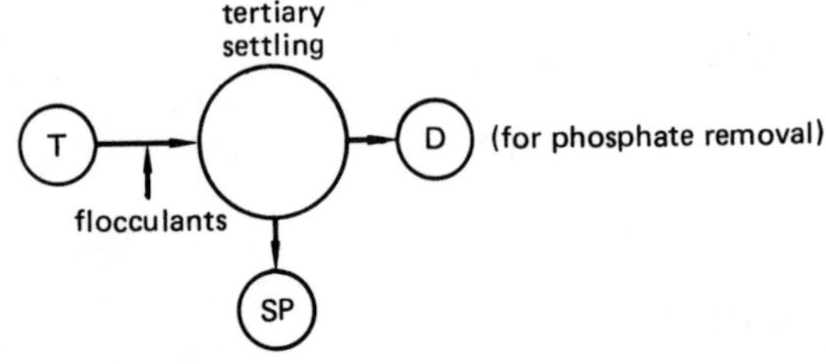


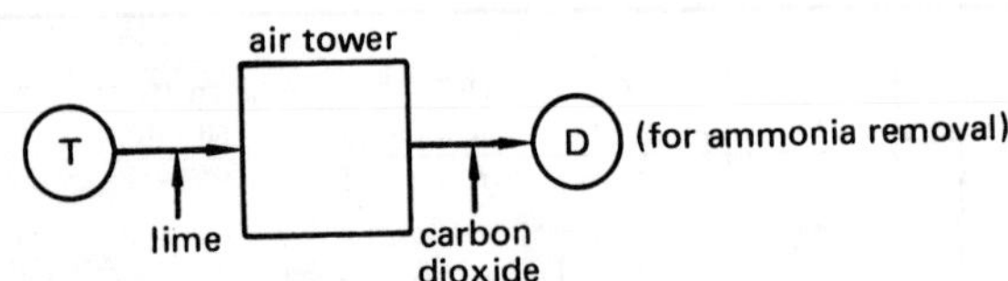


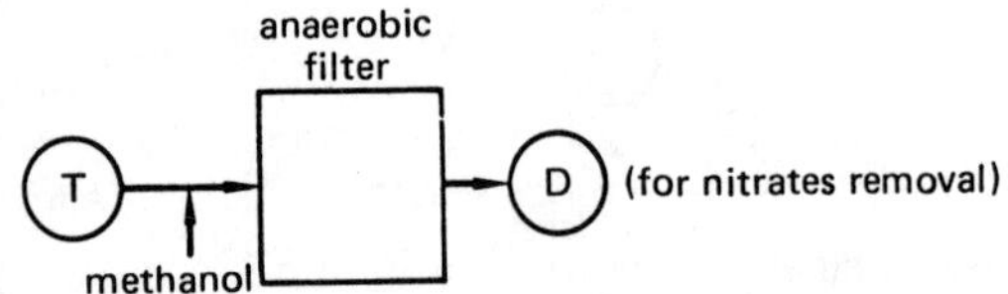


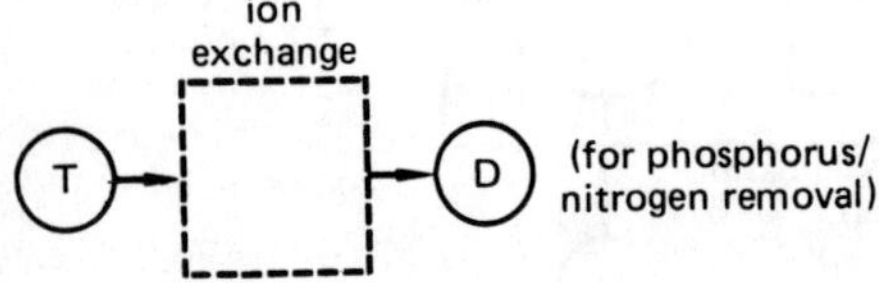


## SP: Sludge Processing

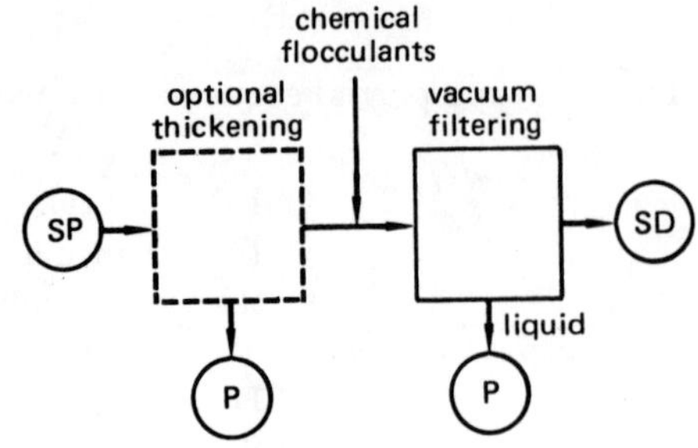


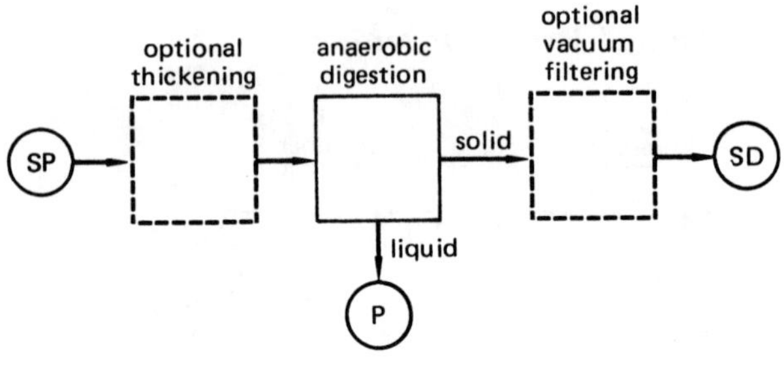


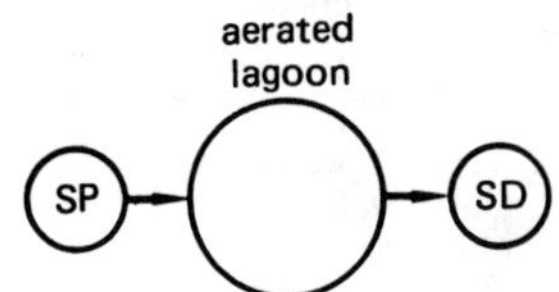


## SD: Solids Disposal

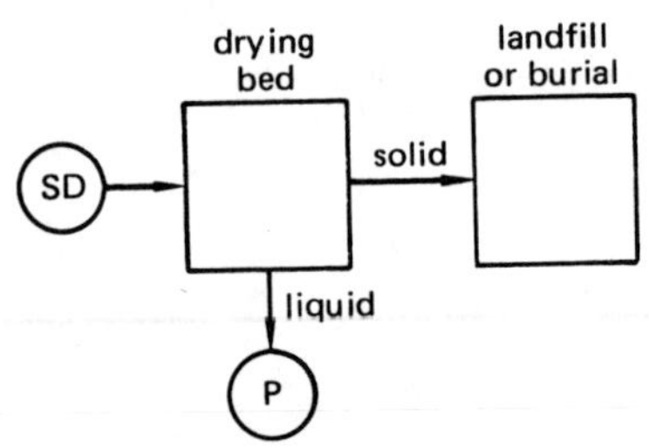


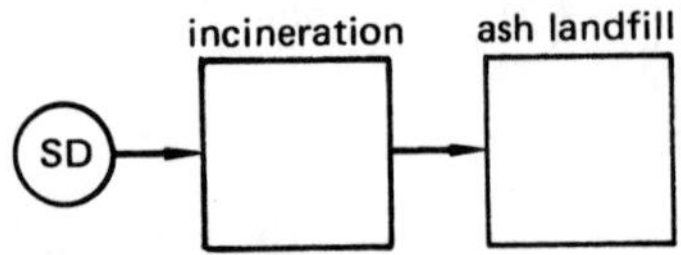


## D: Discharge

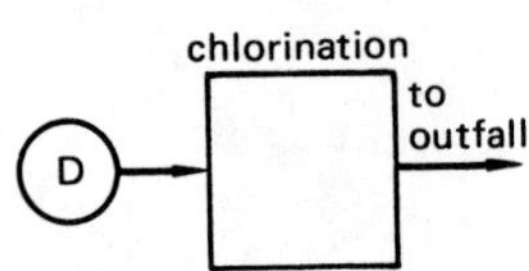

**Practice Problems: WASTE WATER ENGINEERING**

Untimed

1. It is estimated that the BOD of raw sewage received at a treatment plant you are designing will be 300 mg/l from a population of 20,000. A single-stage, high-rate filter is to be used to reduce the plant effluent to 50 mg/l. 30% of the raw sewage BOD is removed by settling. Recirculation is from the filter effluent to the primary settling influent. Use '10-State Standards'. (a) What is the design flow rate? (b) If a round filter is used, what should be its depth and diameter? (c) What volume should be recirculated? (d) Draw a flow diagram of the process. (e) What is the overall plant efficiency?

2. The average waste-water flow from a community of 20,000 is 125 gpcd. The 5-day, 20°C BOD is 250 mg/l. The suspended solids content is 300 mg/l. The final plant effluent is to be 50 mg/l BOD through use of settling tanks and trickling filters. The settling tanks are to have a surface settling rate of 1000 gpd/ft². The trickling filters are to be 6 feet deep and used without recirculation. (a) Design circular settling tanks for this plant. (b) Estimate the BOD removal in the settling tanks. (c) Determine the size of the trickling filters.

3. A small town of 10,000 people discharges its wastes directly into a stream which has the following characteristics:

| | |
|---|---|
| minimum flow rate: | 120 cfs |
| minimum dissolved oxygen: (at 15°C) | 7.5 mg/l |
| velocity: | 3 mph |
| temperature: | 15°C |
| reaeration coefficient of stream and wastes: | 0.2 @ 20°C |
| BOD reaction coefficient of stream and wastes: | 0.1 @ 20°C |

The town's waste consists of the following:

| | volume | BOD @ 20°C | temperature |
|---|---|---|---|
| domestic | 122 gpcd | 0.191 lb/cd | 64°F |
| infiltration | 116,000 gpd | | 51°F |
| industrial #1 | 180,000 gpd | 800 mg/l | 95°F |
| industrial #2 | 76,000 gpd | 1700 mg/l | 84°F |

(a) What is the domestic waste BOD in mg/l? (b) What is the overall waste BOD in mg/l just before discharge into the stream? (c) What is the temperature of the waste just before discharge into the stream? (d) What is the theoretical minimum dissolved oxygen concentration in the stream? (e) How far downstream would you expect the minimum dissolved oxygen concentration to occur?

4. A sewage treatment plant is being designed to handle both domestic and industrial waste with the following characteristics:

| | |
|---|---|
| industrial #1: | 1.3 MGD; 1100 mg/l BOD |
| industrial #2: | 1.0 MGD; 500 mg/l BOD |
| domestic: | 100 gpcd; 0.18 lb/cd BOD |

The city has a population of 20,000, and an excess capacity factor of 15% is to be used. (a) What is the design population equivalent for the plant? (b) What is the plant's organic loading? (c) What is the plant's hydraulic loading?

5. Sewage from a city of 40,000 has an average daily flow of 4.4 MGD. An analysis of the raw sewage shows the following characteristics:

| | |
|---|---|
| pH | 7.8 |
| suspended solids | 180 mg/l |
| 5-day BOD | 160 mg/l |
| COD | 800 mg/l |
| total solids | 900 mg/l |
| volatile solids | 320 mg/l |
| settleable solids | 8 mg/l |

(a) What are the dimensions of a circular primary sedimentation basin that would remove about 30% of the BOD? (b) How many basins are required? (c) What are the dimensions of square sedimentation basins that would remove about 30% of the BOD? (d) How many square basins would be required? (e) Design a single-stage trickling filter and final sedimentation basin (circular) which would produce, in combination with the primary unit, a final effluent having a 5-day BOD of not more than 20 mg/l.

6. A cheese factory discharges 35,000 gpd with the following effluent characteristics: 10,000 gpd with 1000 mg/l BOD; 25,000 gpd with 250 mg/l BOD. (a) Assuming an average depth of 4 feet, what lagoon size is required for stabilization? (b) What is the detention time?

7. An activated sludge plant processes 10 MGD of influent with 240 mg/l BOD and 225 mg/l of suspended solids. The discharge from the final clarifier contains 15 mg/l BOD and 20 mg/l suspended solids. The following assumptions may be used:

- Primary clarification removes 60% of the suspended solids and 35% of the BOD.
- Sludge has a specific gravity of 1.02 and consists of 6% solids by weight.
- The cell yield (BOD conversion to biological solids) in the aeration basin is 60%.
- The final clarifier does not reduce BOD.

(a) What is the daily weight of sludge produced? (b) Assuming the sludge is completely dried prior to disposal, determine the daily sludge volume.

Timed

1. Your client has just completed a subdivision and his sewage hooks up to a collector which goes into a trunk. The problem is that the first manhole up from the trunk on the collector pipe overflows periodically. An industrial plant is hooked directly into the trunk and it is the plant's volume which is making your cli-

ent's line back up. The subdivision is in a flood plain and the sewer lines are very flat and cannot be steepened. The rim on the manhole cannot be raised. (a) Give two solutions to this problem. (b) Sketch a plan view showing what your solutions would look like.

2. A high-rate, two-stage trickle filter system processes 5.2 MGD of sewage with a 5-day BOD of 320 mg/l. The effluent is to have a 5-day BOD of 28 mg/l, within a ± 20 mg/l range. Determine the required filter size (diameters). Both filters have the same size.

hydraulic load to filter: 32 mgad including recycle
30% BOD removal in primary clarifier
filter depth: 5 feet for both filters
Use NRC standards.

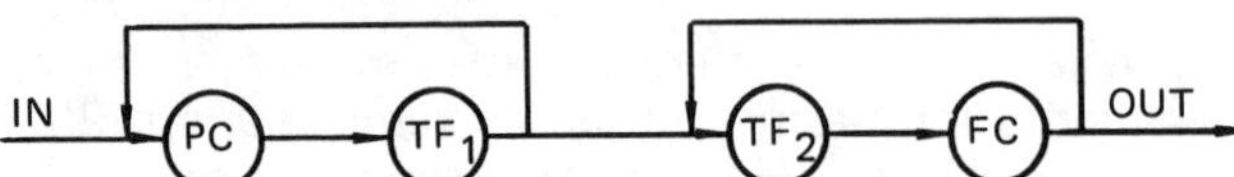


3. A town has a current population of 10,000 and is expected to double in size in 15 years. The town has plans to deposit its solid waste in a 30-acre landfill that will be used as a park in 20 years. (a) Describe the general requirements for a sanitary landfill. (b) What factors should be considered in selecting a site? (c) If the average person generates 5 pounds of solid waste per day, how long will it take to produce a 6-foot lift? Hint: The town continues to grow after 15 years.

4. A town of 10,000 people has selected a 50 acre site to deposit its solid waste. The minimum side borders are 50 feet. The maximum trench depth is ± 20 feet. There is a minimum requirement of 10 feet of earth cover for the lift. (a) What is the service life of the disposal site? (b) What is the daily annual solid waste volume? (c) What is the volumetric capacity of the disposal site? (d) Discuss the environmental considerations relating to traffic, pollution, aesthetics, and other factors.

5. A town of 10,000 people has its own primary treatment plant. (a) What mass of total solids (in pounds per day) should the treatment plant expect? (b) If the town is 4 miles from the treatment plant and 400 feet above it in elevation, what size pipe should be used between the town and the plant?

6. A small community has a projected flow of 1 MGD. Incoming wastewater has the following properties:

BOD of 250 mg/l
grit specific gravity of 2.65
total suspended solids of 400 mg/l

The community wishes to have a wastewater treatment system consisting of an aerated grit chamber, primary clarification, two trickling filters, and a secondary clarifier. The final effluent is to have a final BOD of 30 mg/l. The recirculation ratio of the system is 100%. (a) Size the aerated grit chamber. (b) Determine the air requirements for grit chamber in order to capture 95% of the grit. (c) Size the primary clarifier. (d) Size the trickling filters in acre-feet. (e) Size the secondary clarifier.

7. Wastewater enters the recirculating biological contactor (RBC) treatment process shown with 250 mg/l BOD at the rate of 1.5 MGD. The RBC removes a fraction of the incoming BOD.

$$\text{Fraction BOD removal in RBC (expressed as a decimal)} = \frac{1}{\left[1 + \frac{kA}{Q}\right]^3}$$

$$k = 2.45\ \frac{\text{gpd}}{\text{ft}^2}$$

Q does not include recirculation

(a) Find the area of the RBC such that $BOD_{out}$ is 30 mg/l. (b) Find the recirculation ratio such that the $BOD_{out}$ is 20 mg/l. Keep the RBC area the same as in (a). (c) Find the sludge volume produced from the clarifiers if the yield is 0.4 lbs/lb BOD removed. S.G. of raw sludge – 1.0. (d) What is the organic loading to the RBC under the flow scheme outlined in part (a)?

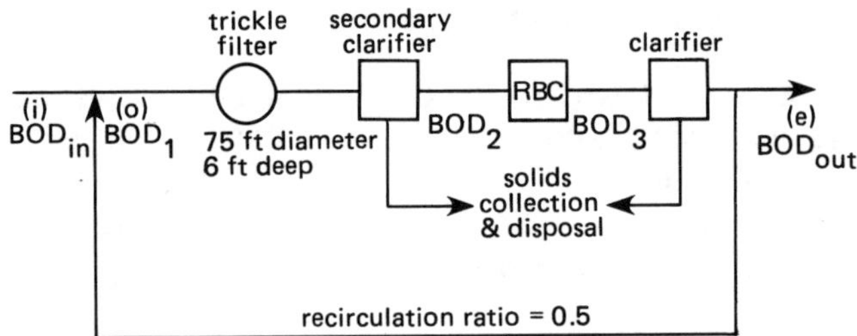

# 9 SOILS

*Nomenclature*

| | | |
|---|---|---|
| A | area | various |
| c | cohesion | psf |
| $C_c$ | compression index | – |
| $C_u$ | uniformity coefficient | – |
| $C_z$ | coefficient of curvature | – |
| CBR | California bearing ratio | – |
| D | diameter | mm |
| e | void ratio | – |
| F | percent passing through the sieve, or shape factor | –, various |
| $G_H$ | hydraulic gradient | – |
| h | head | cm |
| $I_d$ | density index | – |
| $I_g$ | group index | – |
| $I_l$ | liquidity index | – |
| $I_p$ | plasticity index | – |
| k | coefficient of permeability | cm/sec |
| L | flow path length | cm |
| n | porosity | – |
| N | number of blows | – |
| p | pressure | psi |
| P | load | lbf |
| PPS | percent pore space | – |
| Q | flow quantity | $cm^3$/sec |
| r | radius | various |
| R | overconsolidation ratio, or Hveem's resistance | –, – |
| s | degree of saturation | – |
| S | strength | psi |
| SG | specific gravity[1] | – |
| t | time | seconds |
| v | velocity | cm/sec |
| V | volume | $cm^3$ |
| w | water content | – |
| W | weight | grams |
| $w_l$ | liquid limit | – |
| $w_p$ | plastic limit | – |

*Symbols*

| | | |
|---|---|---|
| $\rho$ | density | $g/cm^3$ or $lbm/ft^3$ |
| $\sigma$ | normal stress | psi |
| $\phi$ | angle of internal friction | degrees |
| $\tau$ | shear stress | psf |
| $\theta$ | angle of principal stress plane | |

*Subscripts*

| | |
|---|---|
| A | axial |
| B | borrow |
| c | compressive |
| d | dry |
| eq | equilibrium |
| f | final |
| F | fill |
| g | air |
| i | ith component, or initial |
| n | unconfined |
| o | consolidated |
| R | radial |
| s | soil |
| sat | saturated |
| t | total |
| u | ultimate |
| us | ultimate shear |
| v | void |
| w | water |
| z | zero air voids |

## 1 CONVERSIONS

| multiply | by | to obtain |
|---|---|---|
| centimeters | 0.3937 | inches |
| centimeters squared | 0.155 | square inches |
| cubic yards | 27 | cubic feet |
| cubic yards | 202.2 | gallons |
| dynes | EE–5 | newtons |
| cubic feet | 7.48 | gallons |
| cubic feet | 0.03704 | cubic yards |

[1] As a peculiarity of soils engineering, the specific gravity is usually given the symbol *G*, as opposed to this book which uses *SG* throughout.

| | | |
|---|---|---|
| feet/min | 0.508 | cm/sec |
| gallons | 0.1337 | cubic feet |
| gallons | 4.95 EE–3 | cubic yards |
| gallons of water | 8.345 | pounds of water |
| grams/cubic centimeter | 62.428 | pounds/cubic foot |
| inches | 2.54 | centimeters |
| square inches | 6.4516 | square centimeters |
| kilograms | 2.20462 | pounds |
| newtons | 0.22481 | pounds |
| newtons | EE5 | dynes |
| pascals | 0.145 EE–3 | psi |
| pounds | 0.4536 | kilograms |
| pounds | 453.59 | grams |
| pounds | 4.448 | newtons |
| pounds/square inch | 144 | psf |
| pounds/square inch | 6.894 EE3 | pascals |
| pounds/cubic foot | 0.01602 | grams/cubic centimeter |

## 2 DEFINITIONS

Admixture: Material added to soil to increase its workability, strength, or imperviousness, or to lower its freezing point.

Adsorption: Absorption characterized by a higher concentration of water at the surface of the solid than throughout.

Adsorbed water: Water held near the surface of a material by electrochemical forces.

Aggregate: A mixture of various soil components (e.g., sand, gravel, and silt).

Bentonite: A volcanic clay which exhibits extremely large volume changes with moisture content changes.

Caisson: An air- and water-tight chamber used to work or excavate below the water level.

Catena: A group of different soils which frequently occur together.

Cation: Positively charged ion.

Dilatancy: Property of increasing in volume when changing shape.

Fine: Combined silt and clay.

Friable: Easily crumbled.

Frost susceptibility: Susceptible to having water continually drawn up from the water table by capillary action, forming ice crystals below the surface (but above the frost line).

Gap graded: Having large particles and small particles, but no medium-sized particles.

Glacial till: Soil resulting from a receding glacier, consisting of mixed clay, sand, gravel, and boulders.

Gumbo: Silty soil that becomes soapy, sticky, or waxy when wet.

Horizon: Dividing line between layers of soil with different colors or compositions.

Loess: A deposit of wind-blown silt.

Normally loaded soil: A soil which has never been loaded to a greater extent than at present.

Pycnometer: A stoppered flask with graduations.

Pedology: The study of the soil constituting the upper 4 or 5 feet of the earth's crust.

Rock flour: Fine grained, rounded quartz grains with little plasticity, characteristic of glacial activity.

Stratum: Layer.

Thixotropic: Gradually increasing in strength as absorbed water distributes itself through the soil.

Till: See 'Glacial till'.

## 3 SOIL TYPES

Soil is an aggregate of loose mineral and organic particles. This definition distinguishes soil from rock, which exhibits strong and permanent cohesive forces between the mineral particles.

Proper calculations for foundations and retaining walls require that the nature of the soil be known. This can be done either quantitatively or qualitatively. The primary components of any soil are gravel, sand, silt, and clay. Organic material can also be present in surface samples. If the soil is a mixture of two or more of these components, the soil is given the name of the constituent that has the greatest influence on its mechanical behavior (e.g., silty clay or sandy loam).

Particle size limits for defining gravel and sand have been suggested, and these are given in table 9.1. Whereas sand and gravel are classified as coarse grained soils, inorganic silt and clay are classified as fine grained soils. The clay-silt distinction cannot be made on the basis of size alone. Silt possesses little plasticity and cohesion. Clay, on the other hand, is very plastic and cohesive.

*Clays* may be distinguished from *silts* by using the following simple tests:

**Table 9.1**
Soil Classification by Particle Size

| system | date | sizes(mm) gravel | sand | silt | clay |
|---|---|---|---|---|---|
| Bureau of Soils | 1890 | 1–100 | 0.05–1 | 0.005–0.05 | < 0.005 |
| Atterberg | 1905 | 2–100 | 0.2–2 | 0.002–0.2 | < 0.002 |
| MIT | 1931 | 2–100 | 0.06–2 | 0.002–0.06 | < 0.002 |
| USDA | 1938 | 2–100 | 0.05–2 | 0.002–0.05 | < 0.002 |
| Unified (or AC) | 1953 | 4.75–75 | 0.075–4.75 | < 0.075 | |
| ASTM | 1967 | 4.75–75 | 0.075–4.75 | < 0.075 | |
| AASHTO | 1970 | 2–75 | 0.075–2 | 0.002–0.075 | 0.001–0.002 |

Dry Strength Test: Mold a small brick of soil and allow it to air dry. Break the brick and place a small (1/8″) fragment between thumb and finger. A silt fragment will break easily, whereas clay will not.

Dilatancy Test: Mix a small sample with water to form a thick slurry. When the sample is squeezed, water will flow back into a silty sample quickly. The return rate will be much lower for clay.

Plasticity Test: Roll a moist soil sample into a thin (1/8″) thread. As the thread dries, silt will be weak and friable, but clay will be tough.

Dispersion Test: Disperse a sample of soil in water. Measure the time for the particles to settle. Sand will settle in 30 to 60 seconds. Silt will settle in 15 to 60 minutes, and clay will remain in suspension for a long time.

*Organic matter* can also be present in soil, and this presence can have a significant effect on the mechanical properties of the soil. Organic material is classified into *organic silt* and *organic clay*. Generally, the greater the organic content, the darker will be the soil color.

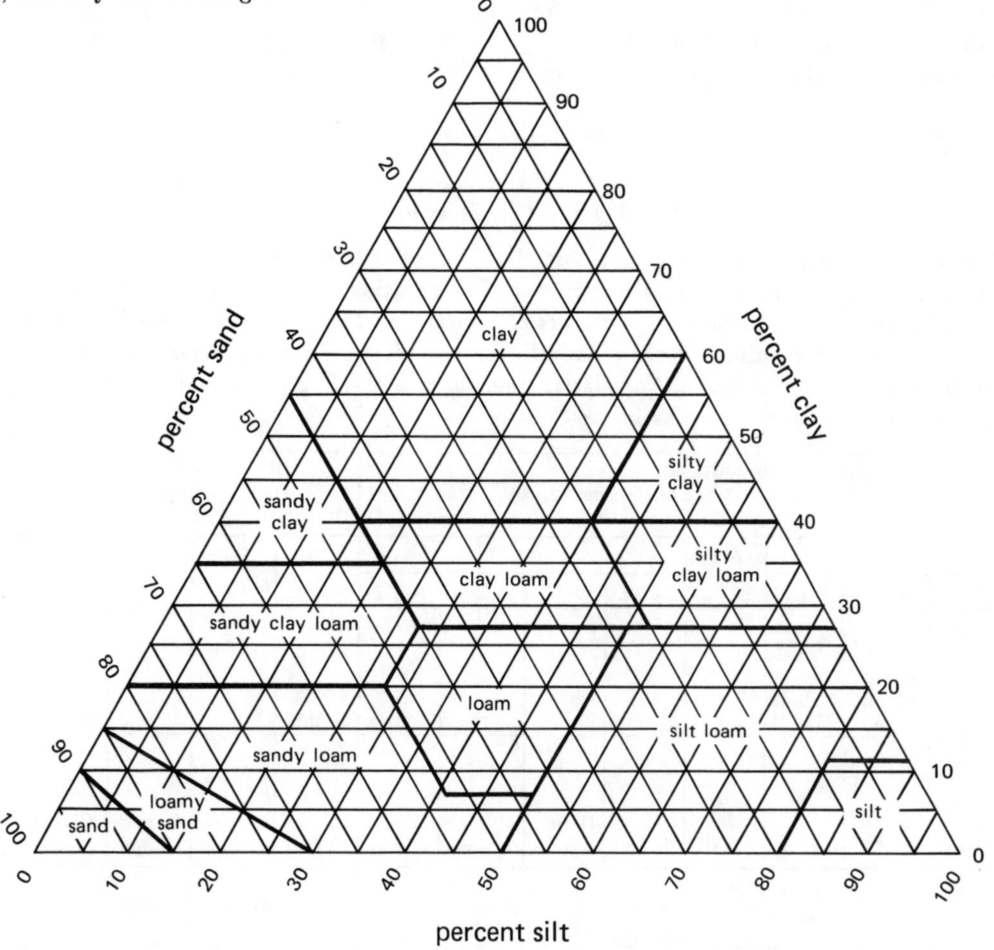


**Figure 9.1** USDA Triangle Chart

General soil classifications have been established by a number of organizations. The most common single index used in classification is particle size. The various classification schemes are presented in table 9.1. The actual classification of a soil will depend on the percentages of each constituent. For example, a soil might be classified as "4% gravel, 45% sand, 15% silt, and 36% clay (MIT)."

One method of giving qualitative descriptions to the soil is by using the *USDA triangular chart*, figure 9.1. This classification ignores the presence of any gravel, although the adjective 'stony' can be used in conjunction with the chart classifications.

Since the qualitative description obtained from the USDA classification chart does not necessarily reflect the mechanical properties of the soil, other systems have been developed. The American Association of State Highway Transportation Officials (AASHTO) has developed a system based on the sieve analysis, liquid limit, and plasticity index.

Soils excellent for roadway subgrade construction are classified as A-1. Highly organic soils not suitable for roadway subgrade construction are classified as A-8. Subgroup classifications are also used, as well as group indexes for fine grained soil. The *group index* of a fine grained soil is given by equation 9.1. $w_l - 40$ cannot be less than zero.

$$I_g = (F_{200} - 35)[0.2 + 0.005(w_l - 40)] + 0.01(F_{200} - 15)(I_p - 10) \qquad 9.1$$

$$(I_p = w_l - w_p)$$

$F_{200}$ is the percentage of soil that passes through a #200 sieve. The AASHTO classification system is given in table 9.2. A group index of zero is a good subgrade material. Group indexes of 20 or higher represent poor subgrade materials.

*Example 9.1*

Determine the AASHTO classification of an inorganic soil with the following characteristics:

| soil size | fraction retained on sieve | |
|---|---|---|
| $< 0.002$ | 0.19 | $w_l = 0.53$ |
| 0.002–0.005 | 0.12 | |
| 0.005–0.05 | 0.36 | $w_p = 0.22$ |
| 0.05–0.075 | 0.04 | $F_{200} = 0.04 + 0.36 + 0.12 + 0.19 = 0.71$ |
| 0.075–2.0 | 0.29 | |
| $> 2.0$ | 0 | |

**Table 9.2**
AASHTO Soil Classification System

**Classification procedure: Using the test data, proceed from left to right in the chart. The correct group will be found by process of elimination. The first group from the left consistent with the test data is the correct classification. The A-7 group is subdivided into A-7-5 or A-7-6 depending on the plastic limit. For plastic limit $w_p = w_L - I_p$ less than 30, the classification is A-7-6. For plastic limit $w_p = w_L - I_p$ greater than or equal to 30, it is A-7-5. NP means non-plastic.**

| | granular materials (35% or less passing no. 200 sieve) | | | | | | | silt-clay materials (more than 35% passing no. 200 sieve) | | | | |
|---|---|---|---|---|---|---|---|---|---|---|---|---|
| | A-1 | | A-3 | A-2 | | | | A-4 | A-5 | A-6 | A-7 | A-8 |
| | A-1-a | A-1-b | | A-2-4 | A-2-5 | A-2-6 | A-2-7 | | | | A-7-5 or A-7-6 | |
| sieve analysis: % passing | | | | | | | | | | | | |
| no. 10 | 50 max | | | | | | | | | | | |
| no. 40 | 30 max | 50 max | 51 min | | | | | | | | | |
| no. 200 | 15 max | 25 max | 10 max | 35 max | 35 max | 35 max | 35 max | 36 min | 36 min | 36 min | 36 min | |
| characteristics of fraction passing no. 40: | | | | | | | | | | | | |
| $w_l$: liquid limit | | | | 40 max | 41 min | 40 max | 41 min | 40 max | 41 min | 40 max | 41 min | |
| $I_p$: plasticity index | 6 max | | NP | 10 max | 10 max | 11 min | 11 min | 10 max | 10 max | 11 min | 11 min | |
| usual types of significant constituents | stone fragments gravel and sand | | fine sand | silty or clayey gravel and sand | | | | silty soils | | clayey soils | | peat, highly organic soils |
| general subgrade rating | excellent to good | | | | | | | fair to poor | | | | unsatisfactory |

From table 9.2, the classification is A-7-5 or A-7-6. Since the plastic limit is 22 (less than 30), the classification is A-7-6. The group index is

$$\begin{aligned} I_g &= (71-35)[0.2+0.005(53-40)] \\ &\quad +0.01(71-15)(31-10) \\ &= 21.3 \end{aligned}$$

The soil would be classified as A-7-6 (21).

ASTM standards also provide a method of classifying soils based on qualitative descriptions. Coarse-grained soils are divided into two categories: gravel soils (symbol $G$) and sand soils (symbol $S$). Sands and gravels are further subdivided into 4 subcategories:

symbol $W$: well-graded, fairly clean

symbol $C$: well-graded, with excellent clay binder

symbol $P$: poorly graded, fairly clean

symbol $M$: coarse materials with fines, not in preceding 3 groups

Fine-grained soils are divided into three categories: inorganic silty and very fine sandy soils (symbol $M$), inorganic clays (symbol $C$), and organic silts and clays (symbol $O$). These three are subdivided into two subcategories:

symbol $L$: low compressibilities ($w_l$ 50 or less)

symbol $H$: high compressibilities ($w_l$ greater than 50)

Table 9.7 illustrates the use of some of these classification symbols.

## 4 SOIL INDEXING

Indexing of the soil is necessary in order to apply some of the quantitative property relationships contained in this chapter. Indexing is accomplished by performing various classification tests on the soil. Index properties are of two types: grain properties and aggregate properties.

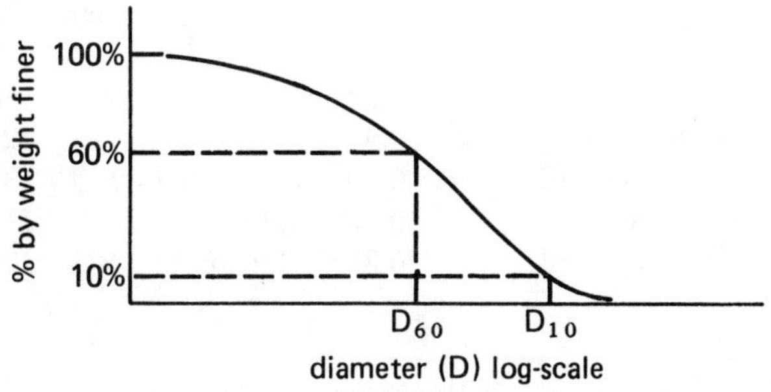


**Figure 9.2** Particle Size Distribution

Soil grain properties include particle size distribution, density, and mineral composition. Density is found from a hydrometer test. Particle size distribution is found from a sieve test for coarse soils, and from a dispersion test for fine soils.

**Table 9.3**
Sieve Sizes

| this size sieve | has this size openings |
|---|---|
| 4″ | 100 mm |
| 3″ | 75 |
| 2″ | 50 |
| $1\frac{1}{2}$″ | 37.5 |
| 1″ | 25 |
| $\frac{3}{4}$″ | 19 |
| $\frac{1}{2}$″ | 12.5 |
| $\frac{3}{8}$″ | 9.5 |
| no.4 | 4.75 |
| 8 | 2.36 |
| 10 | 2.00 |
| 16 | 1.18 |
| 20 | 0.850 |
| 30 | 0.600 |
| 40 | 0.425 (425 μm) |
| 50 | 0.300 |
| 60 | 0.250 |
| 70 | 0.212 |
| 100 | 0.150 |
| 140 | 0.106 |
| 200 | 0.075 (75 μm) |

The results of particle size tests are graphed as a *particle size distribution.*

The *effective grain size* is defined as the diameter for which only 10% of the particles are finer ($D_{10}$). The *Hazen uniformity coefficient* is given by equation 9.2.

$$C_u = \frac{D_{60}}{D_{10}} \qquad 9.2$$

If the uniformity coefficient is less than 4 or 5, the soil is considered uniform in particle size. Well graded soils have uniformity coefficients greater than 10.

The *coefficient of curvature* is

$$C_z = \frac{D_{30}^2}{D_{10}\,D_{60}} \qquad 9.3$$

**Table 9.4**
Typical Soil Coefficients

| soil | $C_u$ | $C_z$ |
|---|---|---|
| gravel | $>4$ | 1–3 |
| fine sand | 5–10 | 1–3 |
| coarse sand | 4–6 | |
| mixture of silty sand and gravel | 15–300 | |
| mixture of clay, sand, silt, and gravel | 25–1000 | |

## 5 AGGREGATE SOIL PROPERTIES

The aggregate index properties are essentially weight-volume relationships. In any sample of soil, there will be some air-filled voids, water-filled voids, and solid material. The percentages of these constituents (by both volume and weight) are used to calculate the aggregate properties.

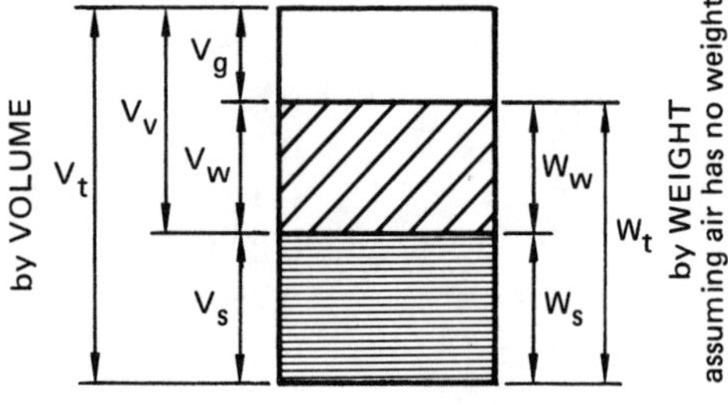


**Figure 9.3** Soil Sample Constituents

The *porosity* is

$$n = \frac{V_v}{V_t} = \frac{V_g + V_w}{V_g + V_w + V_s} \tag{9.4}$$

The *void ratio* is

$$e = \frac{V_v}{V_s} = \frac{V_g + V_w}{V_s} \tag{9.5}$$

The *water content* is

$$w = \frac{W_w}{W_s} \tag{9.6}$$

The volume of the sample will decrease as the water content is reduced down to the *shrinkage limit*, $w = SL$. Below the shrinkage limit, air enters the voids and the water content decreases are not accompanied by decreases in volume.

The *degree of saturation* is

$$s = \frac{V_w}{V_v} = \frac{V_w}{V_g + V_w} \tag{9.7}$$

The *soil density* is[2]

$$\rho = \frac{W_t}{V_t} = \frac{W_w + W_s}{V_g + V_w + V_s} \tag{9.8}$$

The *dry density* is

$$\rho_d = \frac{W_s}{V_t} = \frac{W_s}{V_g + V_w + V_s} \tag{9.9}$$

If the water content is known, the dry density (also known as the *bulk density*) of a moist sample can be found from equation 9.10.

$$\rho_d = \frac{W_t}{(1+w)V_t} = \frac{\rho}{1+w} \tag{9.10}$$

The *compacted density* (*zero air voids density*) is

$$\rho_z = \frac{W_s}{V_w + V_s} \tag{9.11}$$

The density of the solid constituents is

$$\rho_s = \frac{W_s}{V_s} \tag{9.12}$$

The *percent pore space* is

$$PPS = V_v/V_t = 1 - (V_s/V_t) = 1 - (\rho_d/\rho_s) \tag{9.13}$$

The specific gravity of the solid constituents is given by equation 9.14. The specific gravity of sand is approximately 2.65, and for clay it ranges from 2.5 to 2.9 with an average around 2.7.

$$SG_s = \frac{\rho_s}{\rho_w} = \frac{\rho_s}{62.4} \tag{9.14}$$

Typical values of these soil parameters are given in table 9.6

**Table 9.6**
Typical Soil Indexes

| description | n | e | $w_{sat}$ | $\rho_d$ | $\rho_{sat}$ |
|---|---|---|---|---|---|
| sand, loose and uniform | 0.46 | 0.85 | 0.32 | 90 | 118 |
| sand, dense and uniform | 0.34 | 0.51 | 0.19 | 109 | 130 |
| sand, loose and mixed | 0.40 | 0.67 | 0.25 | 99 | 124 |
| sand, dense and mixed | 0.30 | 0.43 | 0.16 | 116 | 135 |
| glacial clay, soft | 0.55 | 1.20 | 0.45 | 76 | 110 |
| glacial clay, stiff | 0.37 | 0.60 | 0.22 | 106 | 129 |

[2] Soils engineering uses the symbol $\gamma$ and the term *unit weight.* This chapter is consistent with the rest of the book in specifying mass density with symbol $\rho$.

**Table 9.5**
Consolidated Soil Indexing Formulas
(Specific Gravity = G)

| | | property | saturated sample ($W_s, W_w, G$, are known) | unsaturated sample ($W_s, W_w, G, V$ are known) | supplementary formulas relating measured and computed factors | | | |
|---|---|---|---|---|---|---|---|---|
| volume components | $V_s$ | volume of solids | | $\frac{W_s}{G\rho_w}$ | $V-(V_a+V_w)$ | $V(1-n)$ | $\frac{V}{(1+e)}$ | $\frac{V_v}{e}$ |
| | $V_w$ | volume of water | | $\frac{W_w}{\rho_w{}^*}$ | $V_v-V_a$ | $SV_v$ | $\frac{SVe}{(1+e)}$ | $Sv_se$ |
| | $V_a$ | volume of air or gas | zero | $V-(V_s+V_w)$ | $V_v-V_w$ | $(1-S)V_v$ | $\frac{(1-S)Ve}{(1+e)}$ | $(1-S)V_se$ |
| | $V_v$ | volume of voids | $\frac{W_w}{\rho_w{}^*}$ | $V-\frac{W_s}{G\rho_w}$ | $V-V_s$ | $\frac{V_sn}{1-n}$ | $\frac{Ve}{(1+e)}$ | $V_se$ |
| | $V$ | total volume of sample | $V_s+V_w$ | measured | $V_s+V_a+V_w$ | $\frac{V_s}{1-n}$ | $V_s(1+e)$ | $\frac{V_v(1+e)}{e}$ |
| | $n$ | porosity | | $\frac{V_v}{V}$ | $1-\frac{V_s}{V}$ | $1-\frac{W_s}{GV\rho_w}$ | $\frac{e}{1+e}$ | |
| | $e$ | void ratio | | $\frac{V_v}{V_s}$ | $\frac{V}{V_s}-1$ | $\frac{GV\rho_w}{W_s}-1$ | $\frac{W_wG}{W_sS}$ | $\frac{n}{1-n}$ \| $\frac{wG}{S}$ |
| weights for specific sample | $W_s$ | weight of solids | | measured | $\frac{W_T}{(1+w)}$ | $GV\rho_w(1-n)$ | $\frac{W_wG}{eS}$ | |
| | $W_w$ | weight of water | | measured | $wW_s$ | $S\rho_wV_v$ | $\frac{eW_sS}{G}$ | $V\cdot\rho_D\cdot w$ |
| | $W_t$ | total weight of sample | | $W_s+W_w$ | $W_s(1+w)$ | | | |
| weights for sample of unit volume | $\rho_D$ | dry unit weight | $\frac{W_s}{V_s+V_w}$ | $\frac{W_s}{V}$ | $\frac{W_t}{V(1+w)}$ | $\frac{G\rho_w}{(1+e)}$ | $\frac{G\rho_w}{1+wG/S}$ | |
| | $\rho_T$ | wet unit weight | $\frac{W_s+W_w}{V_s+V_w}$ | $\frac{W_s+W_w}{V}$ | $\frac{W_T}{V}$ | $\frac{(G+Se)\rho_w}{(1+e)}$ | $\frac{(1+w)\rho_w}{w/S+1/G}$ | $\rho_D(1+w)$ |
| | $\rho_{SAT}$ | saturated unit weight | $\frac{W_s+W_w}{V_s+V_w}$ | $\frac{W_s+V_v\rho_w}{V}$ | $\frac{W_s}{V}+\left(\frac{e}{1+e}\right)\rho_w$ | $\frac{(G+e)\rho_w}{(1+e)}$ | $\frac{(1+w)\rho_w}{w+1/G}$ | |
| | $\rho_{SUB}$ | submerged (buoyant) unit weight | | $\rho_{SAT}-\rho_w{}^*$ | $\frac{W_s}{V}-\left(\frac{1}{1+e}\right)\rho_w{}^*$ | $\left(\frac{G+e}{1+e}-1\right)\rho_w{}^*$ | $\left(\frac{1-1/G}{w+1/G}\right)\rho_w{}^*$ | |
| combined relations | $w$ | moisture content | | $\frac{W_w}{W_s}$ | $\frac{W_t}{W_s}-1$ | $\frac{Se}{G}$ | $S\left[\frac{\rho_w{}^*}{\rho_D}-\frac{1}{G}\right]$ | |
| | $S$ | degree of saturation | 1.00 | $\frac{V_w}{V_v}$ | $\frac{W_w}{V_v\rho_w{}^*}$ | $\frac{wG}{e}$ | $\frac{w}{\left[\frac{\rho_w{}^*}{\rho_D}-\frac{1}{G}\right]}$ | |
| | $G$ | specific gravity | | $\frac{W_s}{V_s\rho_w}$ | $\frac{Se}{w}$ | | | |

$\rho_w$ is the density of water. Where noted with an asterisk (*) use the actual density of water. In other cases, use 62.4 lbm/ft$^3$.

The *density index* (also known as *relative density*) is a measure of the tendency or ability to compact during loading. The density index is equal to 1 for a very dense soil; it is equal to 0 for a very loose soil.

$$I_d = \frac{e_{max} - e}{e_{max} - e_{min}} \quad 9.15$$

The many relationships between the above soil indexes and parameters are listed in table 9.5.

*Example 9.2*

What is the degree of saturation for a sand sample with $SG = 2.65$, $\rho = 115$ lbm/ft$^3$, and $w = 0.17$?

In these problems it is always a good idea to keep track of the various weight and volume phases on a phase diagram.

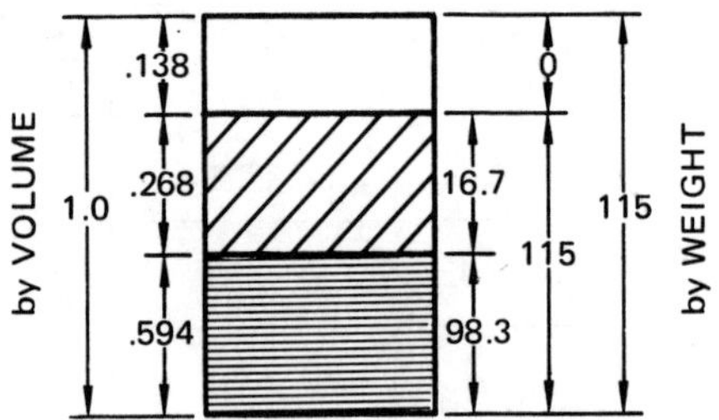


From equation 9.6,

$$W_w = wW_s = (0.17)(W_s)$$

But since $W_w + W_s = 115$,

$$(0.17)W_s + W_s = 115$$
$$W_s = 98.3$$
$$W_w = 16.7$$

The solids volume is given by equation 9.12:

$$V_s = W_s/\rho_s = W_s/(SG)_s\rho_w = \frac{98.3}{(2.65)(62.4)} = 0.594\,\text{ft}^3$$

Similarly, the water volume is

$$V_w = \frac{16.7}{62.4} = 0.268$$

The air volume is $1 - 0.594 - 0.268 = 0.138$.

The degree of saturation is

$$s = \frac{0.268}{0.268 + 0.138} = 0.66$$

*Example 9.3*

Borrow soil is used to fill a 100,000 cubic yard depression. The borrow soil has the following characteristics: density = 96.0 lbm/ft$^3$; water content = 8%; specific gravity of the solids = 2.66. The final in-place dry density should be 112.0 lbm/ft$^3$, and the final water content should be 13% (dry basis).

(a) How many cubic yards of borrow soil are needed? (b) Assuming no evaporation loss, how many pounds of water are needed to achieve 13% moisture? (c) What will be the density of the in-place fill after a long rain?

The first step in borrow problems is to draw the phase diagrams for both the borrow and compacted fill soils. Use subscript $B$ for borrow soil and $F$ for fill soil, and work with 1 cubic foot of fill material.

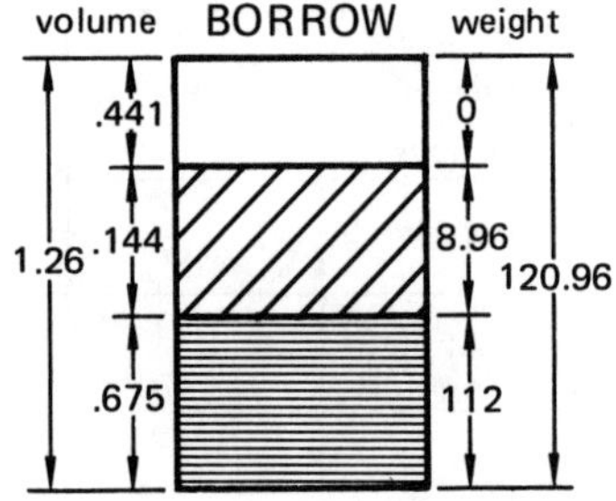


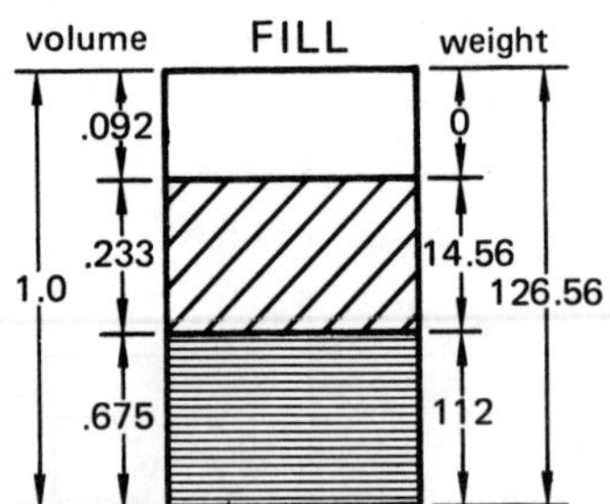


*step 1*: The air has no weight. Dry density precludes water. The soil content (weight) is the same at both locations. (That is, getting 112 pounds of soil in the fill requires taking 112 pounds of borrow soil.) Per cubic foot of fill,

$$W_{sB} = W_{sF} = 112$$

*step 2*: The weight of the water in one cubic foot of fill is

$$W_{wF} = wW_s = (0.13)(112) = 14.56$$

*step 3*: Total weight and density of the fill are

$$W_{tF} = 112 + 14.56 = 126.56$$

$$\rho_F = 126.56/1 = 126.56 \text{ lbm/ft}^3 \text{ fill}$$

*step 4*: The solid volume of the solids in the fill and borrow is

$$V_{sF} = \frac{112}{(2.66)(62.4)} = 0.675 \text{ ft}^3$$

*step 5*: The volume of the water in the fill is

$$V_{wF} = \frac{14.56}{62.4} = 0.233 \text{ ft}^3$$

*step 6*: The air volume in the fill is

$$1 - 0.233 - 0.675 = 0.092 \text{ ft}^3$$

*step 7*: The weight of the water (per cubic foot of fill) in the borrow soil is

$$W_{wB} = wW_{sB} = (0.08)(112) = 8.96$$

(Note that the weight of water per cubic foot of borrow soil is 8.96/1.26.)

*step 8*: The total weight of the borrow soil per cubic foot of fill is

$$W_{tB} = 112 + 8.96 = 120.96$$

*step 9*: The total volume of the borrow soil per cubic foot of fill is

$$V_{tB} = \frac{120.96}{96} = 1.26$$

*step 10*: From step 4, the volume of solids in the borrow soil is

$$V_{sB} = V_{sF} = 0.675$$

*step 11*: The volume of water in the borrow soil per cubic foot of fill is

$$V_{wB} = \frac{8.96}{62.4} = 0.144$$

*step 12*: The air volume in the borrow soil per cubic foot of fill is

$$1.26 - 0.144 - 0.675 = 0.441$$

*step 13* (a):

$$V_{\text{required},B} = \left(\frac{1.26}{1}\right)(100{,}000) = 126{,}000 \text{ cubic yards}$$

(b): The actual moisture in the co[illegible] borrow soil is

$$\frac{(126{,}000)(27)(8.96)}{1.26} = 2.42 \text{ EE7 pounds}$$

The required moisture in the fill soil is

$$(100{,}000)(27)(14.56) = 3.93 \text{ EE7 pounds}$$

The required additional moisture is

$$3.93 \text{ EE7} - 2.42 \text{ EE7} = 1.51 \text{ EE7 pounds}$$

(c): The saturated density is

$$(0.233 + 0.092)(62.4) + 112 = 132.3 \text{ lbm/ft}^3$$

## 6 SOIL TESTING AND MECHANICAL PROPERTIES

### A. PENETRATION RESISTANCE TEST

The most common test is the *standard penetration test* (SPT), which measures resistance to the penetration of a standard split spoon sampler.[3] The number of blows required to drive the sampler a distance of 12 inches (after an initial penetration of 6 inches) is referred to as the *N-value*, in blows per foot.

The N-value has been correlated with many other mechanical properties, including shear modulus, unconfined compressive strength, and effective vertical stress. Figure 9.4 relates $N$ and $S_{nc}$.

### B. MOISTURE-DENSITY RELATIONSHIPS

Soils are compacted to increase their stability, decrease permeability, enhance resistance to erosion, and decrease compressibility. The laboratory test to determine the optimum moisture content and dry density in clay soils is known as the (modified) *Proctor test.* (Nuclear gauges are used in the field to measure density and moisture content in situ.)

A soil sample is compacted in 3 layers by a specific number of hammer blows. The actual density is then given by equation 9.8. The dry density of the sample can be found from equation 9.10. This procedure is repeated for various water contents, and a graph similar to figure 9.5 is obtained. $\rho_d^*$ is known as the *maximum dry density*, or *density at 100% compaction.* $w^*$ is known as the *optimum water content.*

[3] *Cone penetrometer* tests are also performed.

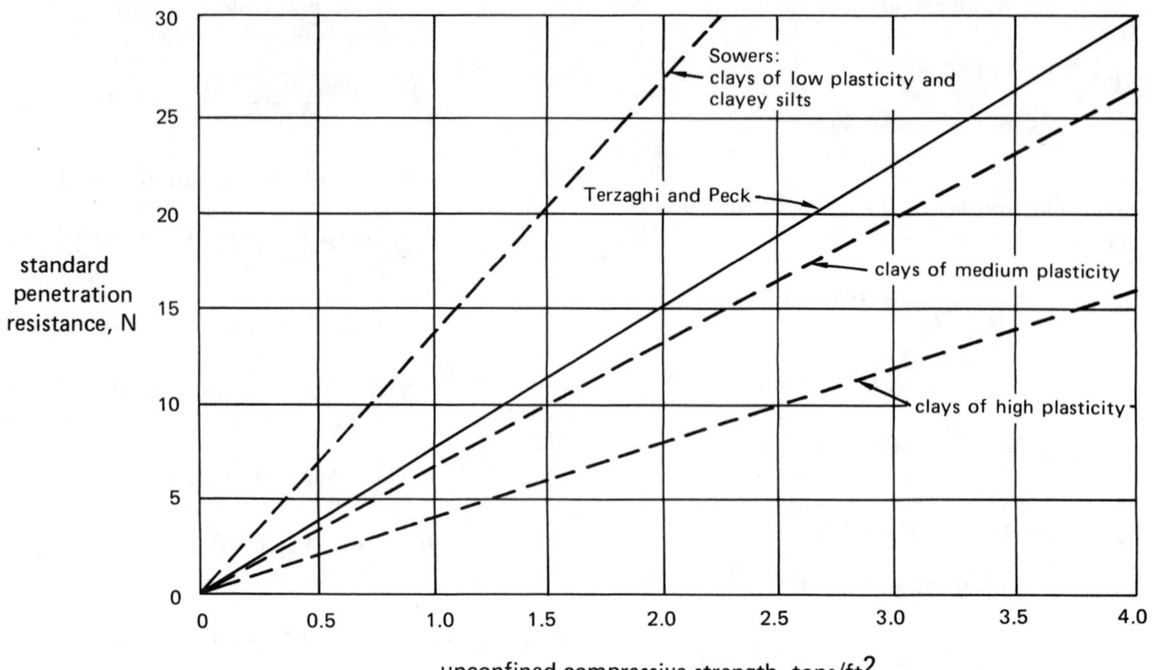


**Figure 9.4** Approximate Relationships Between N and the Unconfined Compressive Strength for Clay

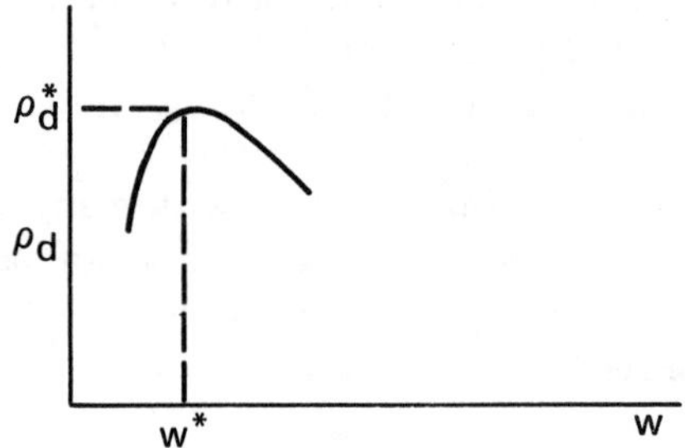


**Figure 9.5** Proctor Test Results

Since the actual compacted density will usually be below (or above) the maximum dry density, the *percentage of compaction* is defined as $\rho_d/\rho_d^*$.

It is not usually feasible to compact soil to the optimum value derived from the Proctor test. Construction compaction methods do not parallel the compaction method used in the test. Usually, some percentage of the maximum Proctor dry density is specified. Table 9.7 lists typical values of optimum moisture content for various soil types, as well as the suggested degree of compaction.

The degree of compaction suggested in table 9.7 depends on the category of soil use. Class 1 uses include the upper 9 feet of fills supporting 1- or 2-story buildings, the upper 3 feet of subgrade under pavements, and the upper 1 foot of subgrade under floors. Class 2 uses include deeper parts of fills under buildings and pavements, as well as earth dams. All other fills requiring some degree of strength or incompressibility are classified as class 3.

For a given water content, saturation will result from perfect compaction, since all air will be removed. The densities resulting from saturation at each water content can be plotted versus water content, and the result is known as a *zero air voids curve.*[4] The theoretical maximum density of the zero air voids curve is calculated from equation 9.16.

$$\rho_z = \frac{\rho_w}{w + \left(\frac{1}{SG}\right)} = \frac{62.4}{w + \left(\frac{1}{SG}\right)} \qquad 9.16$$

The maximum value of the zero air voids density occurs at $w = 0$. At that point, the maximum dry zero air voids density is equal to the density of the solid itself (as calculated from the solid specific gravity).

$$\rho_{zd,\max} = (62.4)(SG) \qquad 9.17$$

$\rho_{zd,\max}$ and $\rho_d^*$ are not the same, however, since air voids exist in the $\rho_d^*$ case.

[4] The zero air voids curve always lies above the Proctor test curve, since that test cannot expel all air.

**Table 9.7**
Typical Values of Optimum Moisture Content and Suggested Compactions (Based on Standard Proctor Test)

| class group symbol | description | range of maximum dry densities lbm/ft$^3$ | range of optimum moisture content % | recommended percentage of Proctor maximum class 1 | 2 | 3 |
|---|---|---|---|---|---|---|
| GW | well-graded, clean gravels, gravel-sand mixtures | 125–135 | 11–8 | 97 | 94 | 90 |
| GP | poorly-graded clean gravels, gravel-sand mixtures | 115–125 | 14–11 | 97 | 94 | 90 |
| GM | silty gravels, poorly graded gravel-sand silt | 120–135 | 12–8 | 98 | 94 | 90 |
| GC | clayey gravels, poorly-graded gravel-sand-clay | 115–130 | 14–9 | 98 | 94 | 90 |
| SW | well-graded clean sands, gravely sands | 110–130 | 16–9 | 97 | 95 | 91 |
| SP | poorly-graded clean sands, sand-gravel mix | 100–120 | 21–12 | 98 | 95 | 91 |
| SM | silty sands, poorly-graded sand-silt mix | 110–125 | 16–11 | 98 | 95 | 91 |
| SM-SC | sand-silt-clay mix with slightly plastic fines | 110–130 | 15–11 | 99 | 96 | 92 |
| SC | clayey sands, poorly-graded sand-clay mix | 105–125 | 19–11 | 99 | 96 | 92 |
| ML | inorganic silts and clayey silts | 95–120 | 24–12 | 100 | 96 | 92 |
| ML-CL | mixture of organic silt and clay | 100–120 | 22–12 | 100 | 96 | 92 |
| CL | inorganic clays of low-to-medium plasticity | 95–120 | 24–12 | 100 | 96 | 92 |
| OL | organic silts and silt-clays, low plasticity | 80–100 | 33–21 | – | 96 | 93 |
| MH | inorganic clayey silts, elastic silts | 70–95 | 40–24 | – | 97 | 93 |
| CH | inorganic clays of high plasticity | 75–105 | 36–19 | – | – | 93 |
| OH | organic and silty clays | 65–100 | 45–21 | – | 97 | 93 |

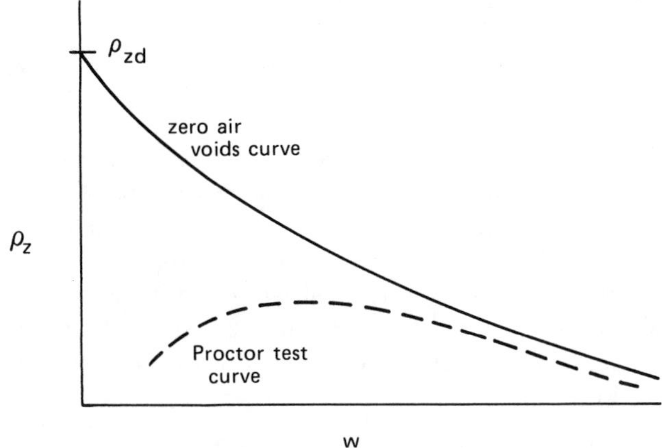


**Figure 9.6** A Typical Zero Air Voids Curve

*Example 9.4*

A Proctor test using a 1/30 ft$^3$ mold is performed on a sample of soil.

| test no. | sample net mass (lbm) | water content (%) |
|---|---|---|
| 1 | 4.28 | 7.3 |
| 2 | 4.52 | 9.7 |
| 3 | 4.60 | 11.0 |
| 4 | 4.55 | 12.8 |
| 5 | 4.50 | 14.4 |

If 0.032 cubic feet of compacted soil tested at a construction site weighed 3.87 pounds wet and 3.74 pounds dry, what is the percent of compaction?

The actual density of sample 1 is (4.28)(30) = 128.4 lbm/ft$^3$. From equation 9.10, the dry density is

$$\rho_d = \frac{100(128.4)}{100 + 7.3} = 119.7\,\text{lbm/ft}^3$$

The following table is constructed from the results of all 5 tests.

| test no. | dry density |
|---|---|
| 1 | 119.7 |
| 2 | 123.6 |
| 3 | 124.3 |
| 4 | 121.0 |
| 5 | 118.0 |

If this data is graphed, the following figure results. The peak is near point 3, so take the maximum density to be 124.3 lbm/ft$^3$. The sample density is 3.74/0.032 = 116.9, so the percentage of compaction is 116.9/124.3 = 0.94.

C. MODIFIED PROCTOR TEST

This test is similar to the standard Proctor test except that the soil is compacted in 5 layers with a heavier hammer falling a greater distance. The result is a denser soil which is more representative of compaction densities available from modern equipment. Table 9.7 can be used by adding 10 to 20 lbm/ft$^3$ to the densities and taking 3 to 10% from the moisture contents.

D. IN-PLACE DENSITY TEST

This test, also known as the *field density test*, starts by compacting soil in the field and digging a 3″ to 5″ deep hole with smooth sides. All soil taken from the hole is saved and weighed before the water content can change. The hole volume is determined by filling the hole with sand or a water-filled rubber balloon. The required densities are given by equations 9.8 and 9.10.

E. UNCONFINED COMPRESSIVE STRENGTH TEST

A cylinder of cohesive soil (usually clay) is loaded to compressive failure. (Failure of elastic soils is taken as a 20% strain.) The unconfined compressive strength is given by equation 9.18. The ultimate shear strength is taken as one half of the unconfined compressive strength.

$$S_{nc} = P/A \qquad 9.18$$

$$S_{us} = \frac{S_{nc}}{2} \qquad 9.19$$

F. SENSITIVITY TESTS

Clay will become softer as it is worked, and clay soils may turn into viscous liquids during construction. This tendency is determined by measuring the ultimate strength of two unconfined samples, one of which has been packed and extruded.

$$\text{sensitivity} = \frac{S_{nc,\text{undisturbed}}}{S_{nc,\text{remolded}}} \qquad 9.20$$

**Table 9.8**
Sensitivity Classifications

| sensitivity | class |
|---|---|
| 1–8 | natural clays |
| 4–8 | sensitive |
| 8–15 | extra sensitive |
| > 15 | quick |

G. ATTERBERG LIMIT TESTS (CONSISTENCY TESTS)

Clay soils can be either solid, plastic, or liquid depending on the water content. The water contents corresponding to the transitions from solid to plastic or plastic to liquid are known as the *Atterberg limits.* These transitions are called the *plastic limit* ($w_p$) and *liquid limit* ($w_l$) respectively.

When a soil has a liquid limit of 100, the weight of moisture equals the weight of the dry soil (i.e., $w = 1$). Alternatively, at the liquid limit, the soil is half water and half solids. A liquid limit of 50 means that the soil at the liquid limit is two-thirds soil and one-third water.

Sandy soils have low liquid limits—on the order of 20. In such soils, the test is of little significance in judging load carrying capacities. Silts and clays can have significant liquid limits—as high as 100. Most clays, however, have liquid limits between 40 and 60. High liquid limits indicate high clay content and low load carrying capacity.

The plastic limit depends on the clay content. Some silt and sand soils have no plastic limit at all. They are known as *non-plastic soils.* The test is of no value in judging the relative load carrying capacities of such soils.

The difference between the liquid and plastic limits is known as the *plasticity index.*

$$I_p = w_l - w_p \qquad 9.21$$

The plasticity index gives the range in moisture content over which the soil is in a plastic condition. A small plasticity index shows that a small change in moisture content will change the soil from a semisolid to a liquid condition. Such a soil is sensitive to moisture. A large plasticity index (i.e., greater than 20) shows that considerable water can be added before the soil becomes liquid.

*Atterberg limits* vary with the clay content, type of clay, and the ions (cations) contained in the clay.

The *liquidity index* of a clay soil is

$$I_l = \frac{w - w_p}{I_p} \qquad 9.22$$

The Atterberg *liquid limit* is found by taking a soil sample and placing it in a shallow container. The sample is parted in half with a special grooving tool. The container is dropped 25 times. At the liquid limit, the sample will have rejoined for a length of $\frac{1}{2}''$.

The *plastic limit* test consists of rolling a soil sample

into a 1/8″ thread. The sample will crumble when it is at the plastic limit when rolled to that diameter.

*Example 9.5*

A clay has the following Atterberg limits: liquid limit = 60%; plastic limit = 40%; shrinkage limit = 25%. The clay shrinks from 15 cubic centimeters to 9.57 cubic centimeters when the moisture content is decreased from the liquid limit to the shrinkage limit in the Atterberg tests. What is the clay's specific gravity (dry)?

The water reduction is 15 − 9.57 = 5.43 cubic centimeters. Since 1 cubic centimeter of water weighs 1 gram, the weight loss is 5.43 grams. The percentage weight loss (dry basis) is 60% − 25% = 35%. Therefore, from equation 9.6, the solid weight is

$$W_s = \frac{\Delta W_w}{\Delta w} = \frac{5.43}{0.35} = 15.5 \text{ g}$$

The water volume at the shrinkage limit is

$$V_w = (0.25)(15.5) = 3.875$$

Since at and above the shrinkage limit there are no air voids, the volume of solid at the shrinkage limit is

$$9.57 - 3.875 = 5.695$$

The density of the solid is

$$\rho = \frac{15.5}{5.695} = 2.72 \text{ g/cm}^3$$

$$SG = 2.72$$

## H. PERMEABILITY TESTS

Permeability of a soil is a measure of continuous voids. A permeable material permits a significant flow of water. The flow of water through a permeable acquifer or soil is given by equation 9.23, known as *Darcy's law.*

$$v = kG_H/n \qquad 9.23$$

$$Q = nAv \qquad 9.24$$

The area $A$ in equation 9.24 is the cross sectional area of the aquifer, not the actual area in flow. Water can only flow through the area between the solids. This open area is $nA$.

Typical values of the coefficient of permeability, $k$, are given in table 9.9. Soils with permeabilities of less than EE–6 are essentially impervious. The soil is considered pervious if $k$ is greater than EE–4.

**Table 9.9**
Typical Permeabilities

| group symbol | typical coefficient of permeability (cm/sec) |
|---|---|
| GW | 2.5 EE–2 |
| GP | 5 EE–2 |
| GM | > 5 EE–7 |
| GC | > 5 EE–8 |
| SW | > 5 EE–4 |
| SP | > 5 EE–4 |
| SM | > 2.5 EE–5 |
| SM-SC | > EE–6 |
| SC | > 2.5 EE–7 |
| ML | > 5 EE–6 |
| ML-CL | > 2.5 EE–7 |
| CL | > 5 EE–8 |
| OL | – |
| MH | > 2.5 EE–7 |
| CH | > 5 EE–8 |
| OH | – |

For loose filter sands, $k$ is given approximately by equation 9.25.

$$k \approx 100(D_{10})^2 \qquad 9.25$$

Actual numerical values can be calculated from controlled permeability tests using constant- or falling-head *permeators* (figure 9.7). For *constant-head tests,* $k$ can be found from equation 9.26. ($V$ is the water volume.)

$$k = \frac{VL}{hAt} \qquad 9.26$$

For *falling-head tests,* $k$ can be found from equation 9.27.

$$k = \frac{A'L}{At} ln(h_i/h_f) \qquad 9.27$$

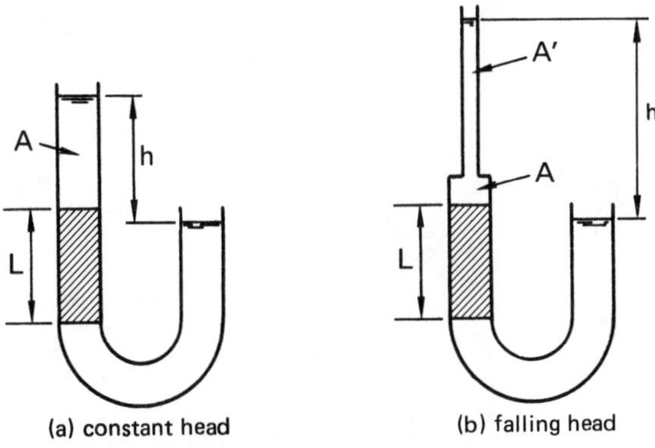


**Figure 9.7** Permeators

For the *auger-hole method* (i.e., in-field, falling-head tests), the combination of area and length variables may be known as the *shape factor* or *conductivity coefficient*, $F$. For example, for a cased hole below the water table of length $L$ and radius $r$ whose impervious casing extends all the way to the hole bottom and whose liquid level rises from $h_i$ to $h_f$ in time $t$ (approaching the water table level), the shape factor and permeability are:

$$F = \frac{11r}{2} \quad 9.28$$

$$k = \frac{\pi r^2}{Ft} ln\left(\frac{h_i}{h_f}\right) = \frac{2\pi r}{11t} ln\left(\frac{h_i}{h_f}\right) \quad 9.29$$

Other in-field tests can use cased holes with constant head, or uncased holes with constant and variable head. The shape factors for these tests are not the same as that in equation 9.28.

*Example 9.6*

The permeability of a semi-impervious soil was evaluated in a falling-head permeator whose head decreased from 100 to 40 cm in 5 minutes. The body diameter was 13 cm; the standpipe diameter was 0.3 cm; and the sample length was 8 cm. What was the permeability of the soil?

From equation 9.27:

$$k = \frac{\frac{1}{4}\pi(0.3)^2(8)}{\frac{1}{4}\pi(13)^2(5)(60)} ln(100/40) = 1.3\,\text{EE–5 cm/sec}$$

## I. CONSOLIDATION TESTS

*Consolidation tests* (also known as *confined compression* and *oedometer tests*) start with a disc of cohesive soil (usually clay) confined by a metal ring. The faces of the disc are covered with porous plates. The disc sandwich is loaded in a water tank. The testing time is very long, since the water out-seepage is very slow. The load versus the void ratio is plotted as an *e-log p curve.*

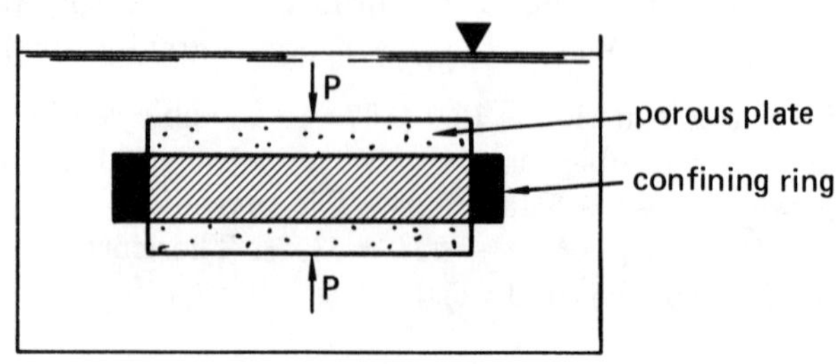


**Figure 9.8** Consolidation Test

Figure 9.9 shows an e-log p curve for a soil sample from which the load has been removed at $m$ allowing the clay to recover.

The line segment $m$–$r$ is known as the *virgin branch* or *virgin consolidation line.* This type of behavior is typical of *normally consolidated clay.* Normally consolidated clay can either be virgin, previously unloaded clay, or it can be clay carrying a load which has never been removed or exceeded.

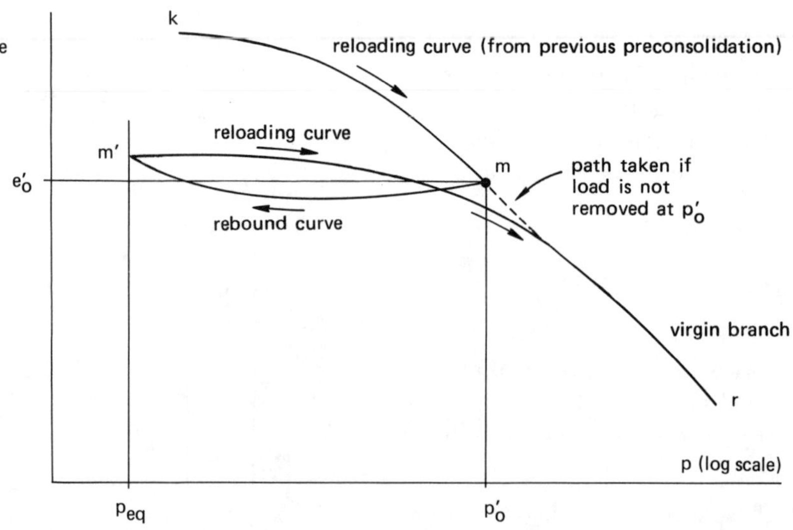


**Figure 9.9** e-log p Curve

Line $m$–$m'$ is a *rebound curve*. Line $m'$–$r$ is known as a *reloading curve*. Such curves result when a normally loaded clay is relaxed and restressed.

Notice that point $m'$ can only be reached by loading the soil to a pressure of $p'_o$ and then removing the pressure. This clay is said to have been *preloaded* or *overconsolidated*.[5] Although the pressure of the clay is essentially the same as when it started, its void ratio has been reduced. The *overconsolidation ratio* is defined by equation 9.30.

$$R_o = p'_o/p_{eq} \qquad 9.30$$

The *overconsolidation pressure*, $p'_o$, can be estimated by eye as a point slightly above the point of maximum curvature. Graphical means are also used.

The shape of the e-log p curve will depend on the degree of previous overconsolidation, as shown in figure 9.10.

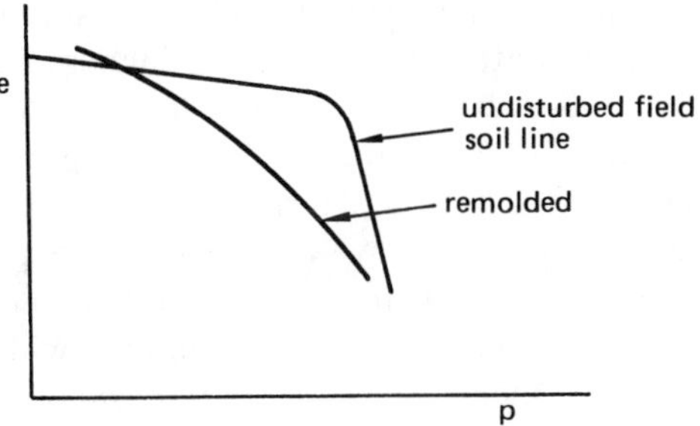


**Figure 9.10** Consolidation of Various Soils

Laboratory results can be used to find the *preconsolidation pressure* (point $m$ in figure 9.9). This is illustrated in the following procedure.

> Draw 2 lines—a tangent line and a horizontal line—through the point of maximum curvature. Bisect the resulting angle. Draw a tangent to the tail of the field soil line. The intersection of this tangent and the bisection line defines $e_o$ and $p_o$.[6]

Line $k$ can be used to predict consolidation of the soil under various loadings. The *compression index* is the (negative of the) logarithmic slope of line $k$ and is given by equation 9.31, where points 1 and 2 correspond to any two points on line $k$.

$$C_c = \frac{e_2 - e_1}{\log_{10}(p_1/p_2)} = \frac{e_o - e_1}{\log_{10}(p_1/p_o)} \qquad 9.31$$

If the clay is soft and near its liquid limit, the compression index can be approximated by equation 9.32.

(In equation 9.32, $w_l$ is a whole number, not a decimal percentage.)

$$C_c \approx 0.009(w_l - 10) \qquad 9.32$$

The *reconsolidation index* and *swelling index* can be found from the (negative of the) logarithmic slopes of the rebound and reloading curves, respectively.

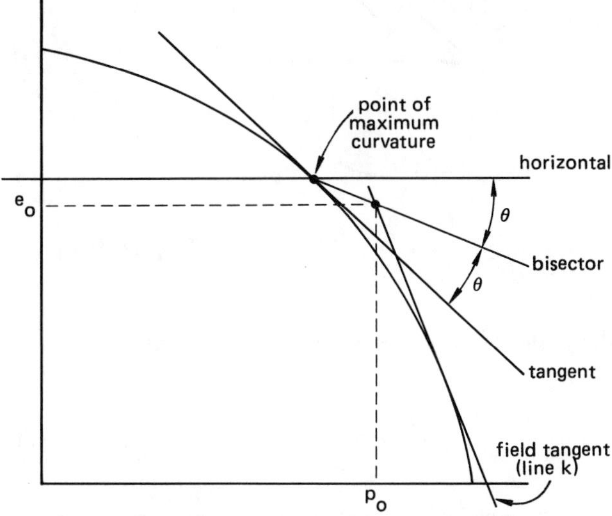


**Figure 9.11** Casagrande Method

## J. TRIAXIAL STRESS TESTS

In a triaxial stress test, a cylindrical sample is loaded on both ends and all around its surface. Usually, the radial stress ($\sigma_R$) is kept constant and the axial stress ($\sigma_A$) is varied. The normal and shear stresses on a plane of any angle can be found from the combined stress equations. (Consider compression positive.)

$$\sigma_\theta = \tfrac{1}{2}(\sigma_A + \sigma_R) + \tfrac{1}{2}(\sigma_A - \sigma_R)\cos 2\theta \qquad 9.33$$

$$\tau_\theta = +\tfrac{1}{2}(\sigma_A - \sigma_R)\sin 2\theta \qquad 9.34$$

These equations represent points on Mohr's circle, which can easily be constructed once $\sigma_A$ and $\sigma_R$ are known. (Care must be taken when plotting this graph. The sample is usually exposed to a pressure $p_R$ over all of its surface, including the ends. Thus, $p_R$ is equal to $\sigma_R$. The pressure applied to the ends, $p_A$, is in addition to radial pressure. Therefore, $\sigma_A = p_R + p_A$.) Test results are shown in figure 9.12 for two different samples which were both tested to failure. The ultimate shear strength, $S_{us}$, can be read directly from the y-axis.[7]

The equation for the *rupture line* (also known as the *envelope of rupture*) is given by Coulomb's equation.[8]

$$\tau = S_{us} = c + \sigma(\tan\phi) \qquad 9.35$$

[5] *Preloaded clay* is also known as *preconsolidated clay*, as well as *overconsolidated clay*.

[6] This method of finding the preconsolidation pressure is known as the *Casagrande method*.

[7] The ultimate shear strength is given the symbol $s$ in most soils books.

[8] Equation 9.35 is also known as the *Mohr-Coulomb equation*.

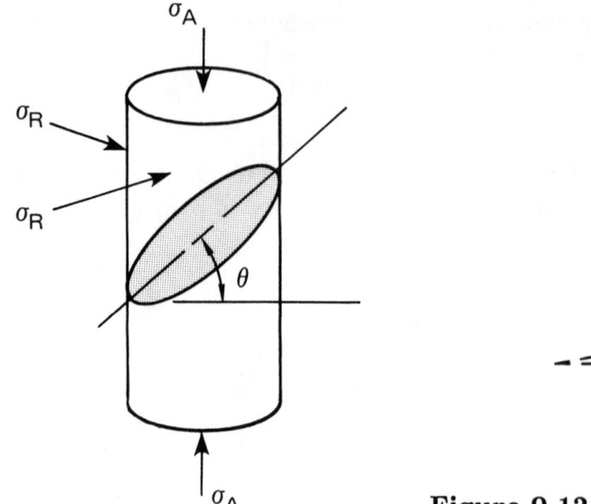


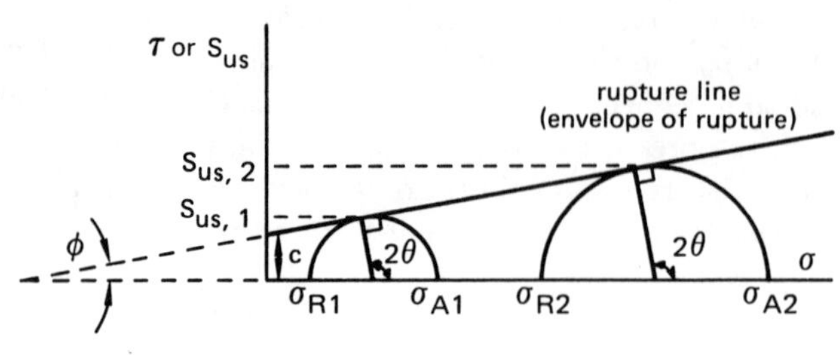


**Figure 9.12** Mohr's Circle of Stress

The *plane of failure* is inclined at the angle

$$\theta = 45^\circ + \tfrac{1}{2}\phi \qquad 9.36$$

For slow shear of drained sands and gravels, the *cohesion*, $c$, is zero. Therefore, it is possible to draw the rupture line with only one test. Typically, $c$ varies from 200 to 2000 psf for very soft and very stiff clays, respectively.

For saturated clays in quick shear, it is commonly assumed that $\phi = 0$. This would be represented as a horizontal rupture line.

Representative values of $\phi$ are given in table 9.10. $\phi$ is known as the *angle of internal friction*.[9]

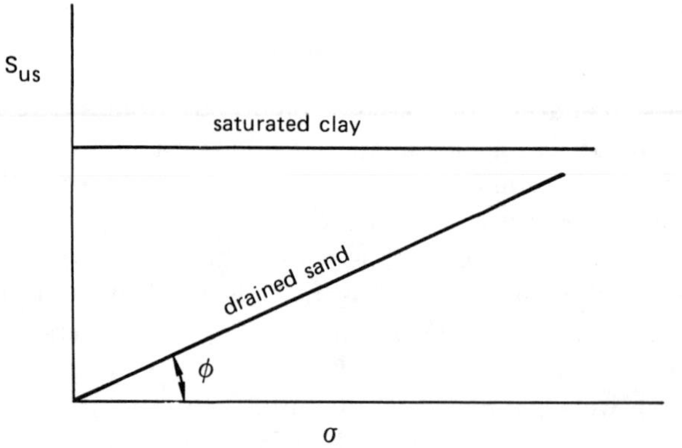


**Figure 9.13** Rupture Lines for Ideal Sand and Saturated Clay

[9] For cohesionless soils, the angle of internal friction is the angle from the horizontal naturally formed by a pile. For example, sand makes a pile with a slope of approximately 30°. Saturated clay, on the other hand, ideally behaves like a liquid, with $\phi = 0$. For most soils, the natural angle of repose will not be the same as the angle of internal friction, due to the effects of cohesion.

Presence of water in the pores of a sample will not affect these results much if the triaxial test is conducted in such a manner as to allow pore water pressure to dissipate (i.e., pore water to flow freely). Such triaxial tests are known as *S-tests* or *consolidated-drained tests.* However, testing of slow-draining soils may require several weeks time. If the test is peformed quickly so that the pore pressure does not have a chance to dissipate, the test is known as an *R-test* or *consolidated-undrained test.* In such a case, much of the axial load will be carried by the pore moisture. In the *Q-test* (*quick test*), the water content of the specimen is not allowed to change. Such a test is justified only with low permeability (e.g., EE-3 cm/sec) soils.

The *effective soil pressure* is the pressure that soil grains exert on each other. This pressure provides the shear strength of granular materials. It can be calculated from equation 9.37, where $c'$ and $\phi'$ are the *effective stress parameters.*

$$S_{us} = c' + \sigma' \tan\phi' \qquad 9.37$$

The *total pressure* also includes the *pore pressure*, $\mu$. The pore pressure can be found from the rise in a capillary tube, or it can be measured directly in a triaxial shear test.

$$\sigma' = \sigma - \mu \qquad 9.38$$

R-tests are used to determine the effective stress parameters, $c'$ and $\phi'$. In the absence of pore pressure measurements, R-tests can only record the total stress parameters $c$ and $\phi$.[10]

[10] If a soil is always going to be saturated, the total stress parameters can be used for foundation design. In cases where the soil is not always saturated, only the effective stress parameters should be used.

**Table 9.10**
Typical Strength Characteristics

| group symbol | cohesion (as compacted) psf c | cohesion (saturated) psf $c_{sat}$ | effective stress envelope degrees $\phi$ |
|---|---|---|---|
| GW | 0 | 0 | > 38 |
| GP | 0 | 0 | > 37 |
| GM | – | – | > 34 |
| GC | – | – | > 31 |
| SW | 0 | 0 | 38 |
| SP | 0 | 0 | 37 |
| SM | 1050 | 420 | 34 |
| SM-SC | 1050 | 300 | 33 |
| SC | 1550 | 230 | 31 |
| ML | 1400 | 190 | 32 |
| ML-CL | 1350 | 460 | 32 |
| CL | 1800 | 270 | 28 |
| OL | – | – | – |
| MH | 1500 | 420 | 25 |
| CH | 2150 | 230 | 19 |
| OH | – | – | – |

*Example 9.7*

A sample of dry sand is taken and a triaxial test performed. The added axial stress causing failure was 5.43 tons/ft$^2$ when the radial stress was 1.5 tons/ft$^2$. What is the angle of internal friction? What is the angle of the failure plane?

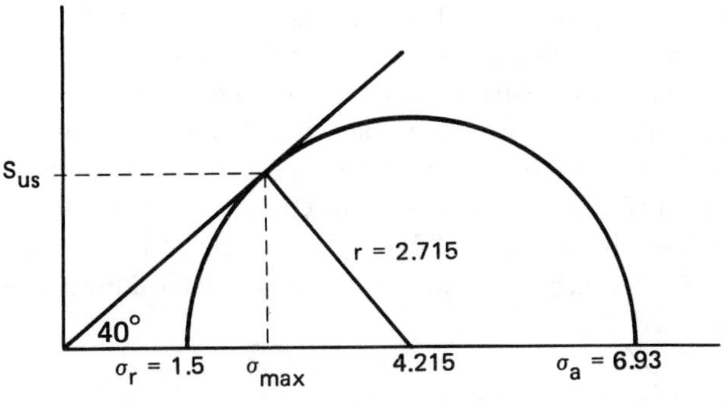


$$\phi = 40°$$

$$\theta = 45° + \frac{1}{2}(40) = 65°$$

For any given radial pressure, a stress-strain curve can be plotted. This is illustrated in figure 9.14. The strain is volumetric strain due to the axial load only. The stress is the difference between the axial and radial stresses. The ultimate compressive stress ($S_{uc}$) can be read directly from the chart. $S_{uc}$ is usually taken as the stress difference for which the strain is 20%. The initial slope of the line is the *elastic modulus.*

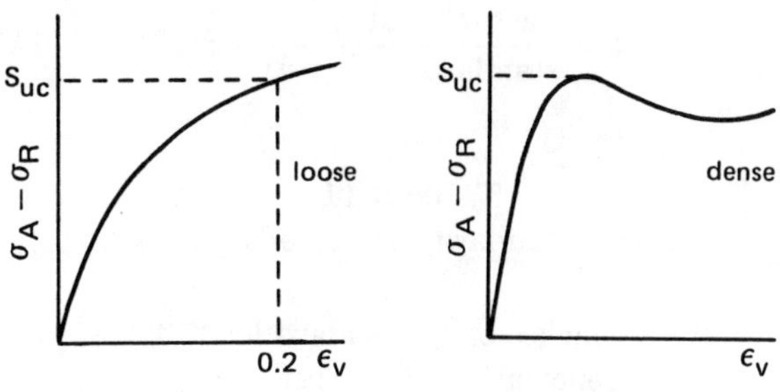


**Figure 9.14** Stress Strain Curves

## K. CALIFORNIA BEARING RATIO TEST: SHEARING RESISTANCE[11]

The *California Bearing Ratio* (*CBR*) test consists of measuring the relative load required to cause a standard (3 square inches) plunger to penetrate a water-saturated soil specimen at a specific rate to a specific depth. The word 'relative' is used because the actual load is compared to a standard load derived from a sample of crushed stone. The ratio is multiplied by 100 and the percent omitted.

The resulting data will be in the form of inches of penetration versus load. This data can be plotted as shown in figure 9.15. If the plot is concave upward (curve B), the steepest slope is extended downward to the x-axis. This point is taken as the zero penetration point and all penetration values adjusted accordingly.

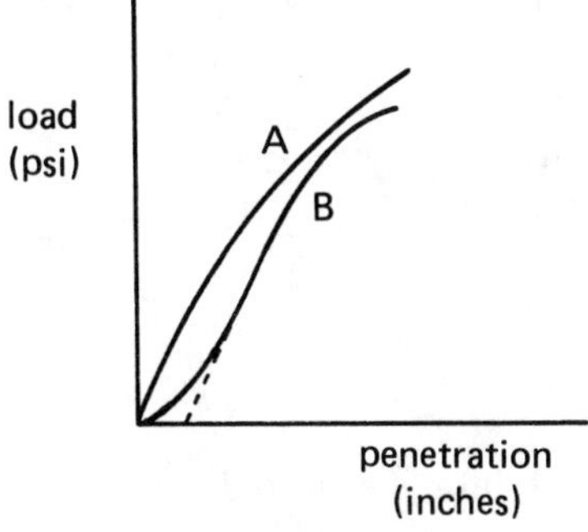


**Figure 9.15** Plotting CBR Test Data

Standard loads for crushed stone are given in table 9.11. For a plunger of 3 square inches, the CBR is the ratio of the load for a 0.1 inch penetration divided by 1000 psi. The CBR for 0.2 inches should also be calculated.

[11] California's Department of Transportation was the first to make use of the CBR test. However, other states and the Corps of Engineers have adopted CBR testing techniques. These states have, generally, retained the California Bearing Ratio name.

uld be repeated if $CBR_{0.2} > CBR_{0.1}$. If re similar, use $CBR_{0.2}$.

$$\text{BR} = \frac{\text{actual load (psi)}}{\text{standard load (psi)}} \times (100) \qquad 9.39$$

**Table 9.11**
Standard CBR Loads

| inches of penetration | standard load (psi) |
|---|---|
| 0.1 | 1000 |
| 0.2 | 1500 |
| 0.3 | 1900 |
| 0.4 | 2300 |
| 0.5 | 2600 |

**Table 9.12**
Typical CBR Values

| group symbol | range of CBR values |
|---|---|
| GW | 40 to 80 |
| GP | 30 to 60 |
| GM | 20 to 60 |
| GC | 20 to 40 |
| SW | 20 to 40 |
| SP | 10 to 40 |
| SM | 10 to 40 |
| SM-SC | 5 to 30 |
| SC | 5 to 20 |
| ML | ≤ 15 |
| ML-CL | – |
| CL | ≤ 15 |
| OL | ≤ 5 |
| MH | ≤ 10 |
| CH | ≤ 15 |
| OH | ≤ 5 |

*Example 9.8*

The following load data is collected for a 3 square inch plunger test.

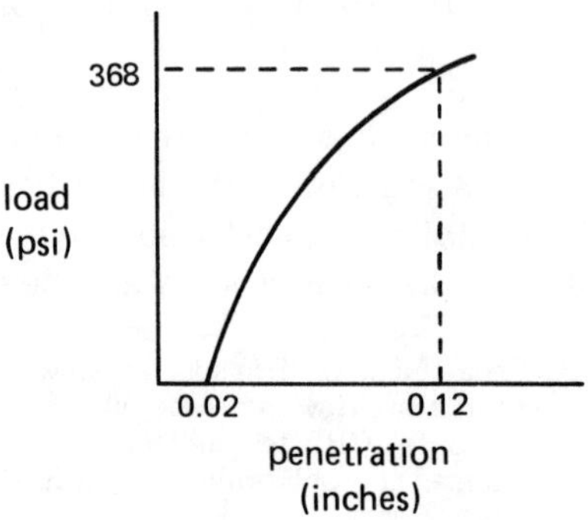


| penetration (inches) | load (psi) |
|---|---|
| 0.025 | 20 |
| 0.050 | 130 |
| 0.075 | 230 |
| 0.100 | 320 |
| 0.125 | 380 |
| 0.150 | 470 |
| 0.175 | 530 |
| 0.200 | 600 |
| 0.250 | 700 |
| 0.300 | 830 |

Upon graphing the data, it is apparent that a 0.02 inch correction is required. Therefore, the 0.1″ load is read from the graph as a 0.12 inch load.

$$CBR_{0.1} = \frac{(368)(100)}{1000} = 36.8 \text{ (percent omitted)}$$

$$CBR_{0.2} = \frac{(645)(100)}{1500} = 43$$

Since $CBR_{0.2}$ is greater than $CBR_{0.1}$, the test should be repeated.

## L. PLATE BEARING VALUE TEST: THE SUBGRADE MODULUS

A standard diameter round steel plate is set over soil on a bed consisting of fine sand and/or plaster of paris. Smaller diameter plates are placed on top of the bottom plate to ensure rigidity. After the plate is seated by a quick but temporary load, it is loaded to a deflection of about 0.04 inches. This load is maintained until the deflection rate decreases to 0.01 inch/minute. Then the load is released. The deflection prior to loading, the final deflection, and the deflection each minute are recorded.

The test is repeated 10 times. For each repetition of each load, the *end-point deflection* is found for which the deflection rate is exactly 0.001 inch/minute. The loads are then corrected for dead weights of jacks, plates, etc.

The corrected load versus the corrected deflection is graphed for the 10th repetition. The *bearing value* is the interpolated load which would produce a deflection of 0.5 inches. Figure 9.16 can be used to find the *subgrade modulus*, or *modulus of subgrade reaction, k*, which is the slope of the line (in psi/inch) in the loading range encountered by the soil.

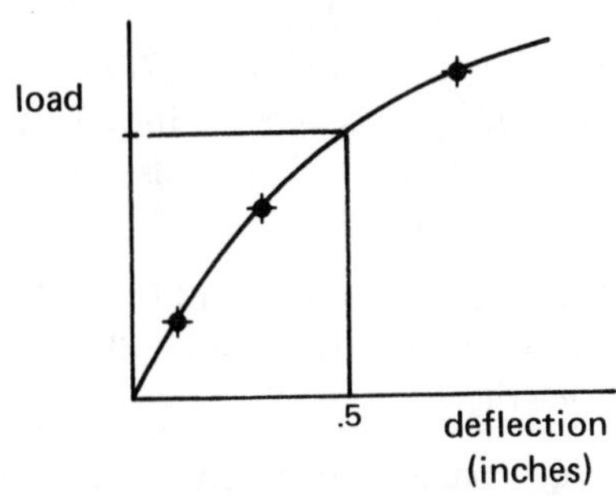


**Figure 9.16** 10th Repetition Bearing Load

M. HVEEM'S RESISTANCE VALUE TEST: THE R-VALUE

The term 'resistance' refers to the ability of a soil to resist lateral deformation when a vertical load acts upon it. When displacement does occur, the soil moves out and away from the applied load.

Measuring the *R-value* of a soil is done with a *stabilometer test.* The R-value will range from zero (the resistance of water) to 100 (the approximate resistance of steel). R-values of soil and aggregate usually range from 5 to 85.

**Table 9.13**
Typical Values of Subgrade Modulus

| group symbol | range of subgrade modulus $k$ 1 psi/in |
|---|---|
| GW | 300–500 |
| GP | 250–400 |
| GM | 100–400 |
| GC | 100–300 |
| SW | 200–300 |
| SP | 200–300 |
| SM | 100–300 |
| SM-SC | 100–300 |
| SC | 100–300 |
| ML | 100–200 |
| ML-CL | – |
| CL | 50–200 |
| OL | 50–100 |
| MH | 50–100 |
| CH | 50–150 |
| OH | 25–100 |

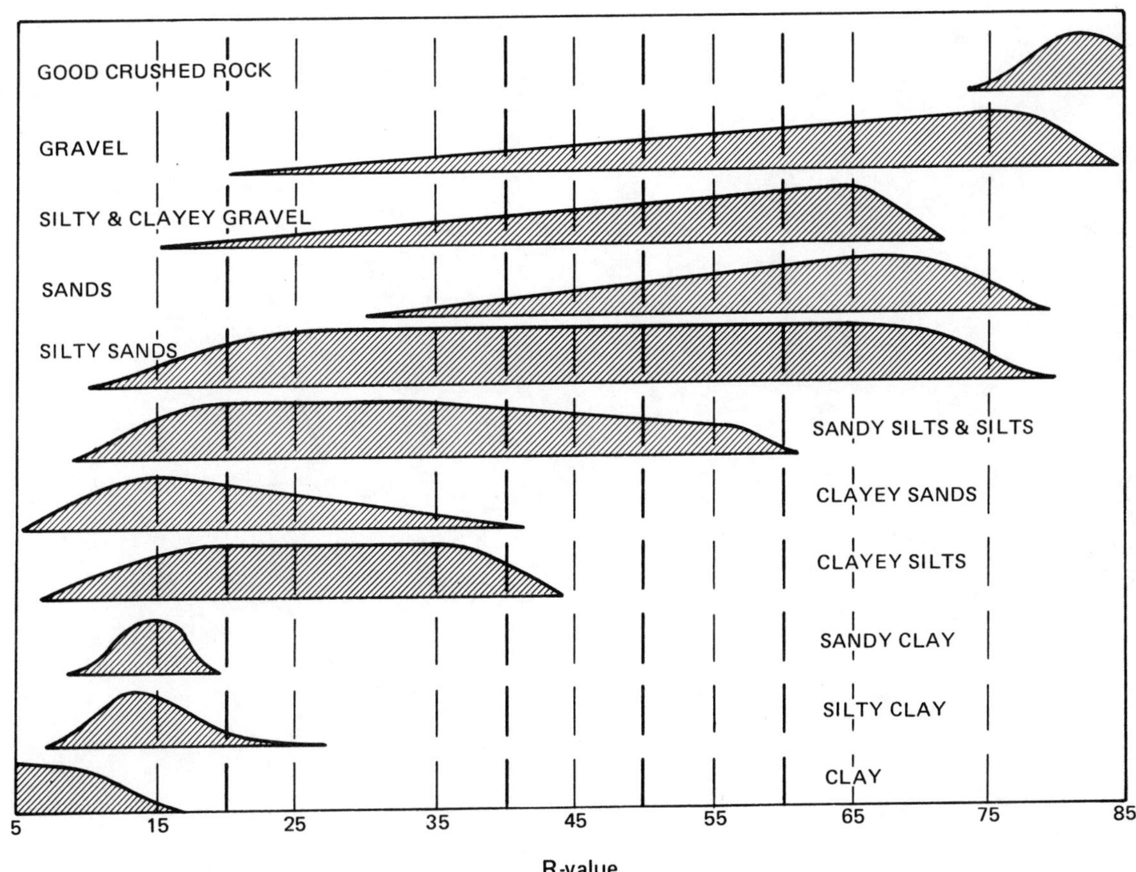


**Figure 9.17** R-Values of Various Soils

The R-value is determined using soil samples which are compacted as they would be during normal construction. They are tested as near to saturation as possible to give the lowest expected R-value. Thus, the R-value represents the worst possible state the soil might attain during use.

The procedure also takes into account the fact that some soils are expansive. When a compacted soil expands due to the absorption of water, the R-value also decreases. The test procedure accounts for this by lowering the R-value.

In the absence of stabilometer testing, rough estimates of the R-value can be made using simple soil classification tests. The soil type can be found from sieve analyses, hydrometer tests, or figure 9.1.

After the soil has been classified, figure 9.17 can be used to determine an approximate R-value. It will be seen that each soil type covers an R-value range. The curves respresenting the various soils are approximate frequency distributions.

For fine-grained soils, the upper tail (high R-value) represents a lower plasticity; the lower tail represents soils with higher plasticity.

The curves for coarse-grained materials are affected in the same manner. Lower tails represent materials with either more clay or clay with a higher activity. In coarse-grained materials with little or no clay, the lower tail represents hard, smooth-surfaced and uniformly sized material. The upper tail represents rough-surfaced material with a distribution of sizes.

## Practice Problems: SOILS

Untimed

1. A sample of moist soil was found to have the following characteristics:

| | |
|---|---|
| volume: | .01456 $m^3$ (as sampled) |
| mass: | 25.74 kg (as sampled) |
| | 22.10 kg (after oven drying) |
| specific gravity of solids: | 2.69 |

Find the density, unit weight, void ratio, porosity, and degree of saturation for the soil.

2. A Proctor test was performed on a soil which has a specific gravity for its solids of 2.71. For the data below, (a) plot the moisture-dry density curve. (b) Find the maximum density and optimum moisture. (c) What range of moisture is permitted if a contractor must achieve 90% compaction?

| water content % | actual density pcf | water content % | actual density pcf |
|---|---|---|---|
| 10 | 98 | 20 | 129 |
| 13 | 106 | 22 | 128 |
| 16 | 119 | 25 | 123 |
| 18 | 125 | | |

3. For the soil in problem #2, how many gallons of water need to be added to obtain 1 $yd^3$ of soil at the maximum density if the soil is originally at 10% water content (dry basis)?

4. A triaxial shear test is performed on a well-drained sand sample. Failure occurred when the normal stress was 6260 psf and the shear stress was 4175 psf. What is the angle of internal friction? What are the principal stresses?

5. The results of a sieve test below give the percentage passing through the sieve. (a) Plot the particle size distribution. (b) Calculate the uniformity coefficient. (c) Calculate the coefficient of curvature.

| sieve | % finer by weight |
|---|---|
| ½″ | 52 |
| #4 | 37 |
| #10 | 32 |
| #20 | 23 |
| #40 | 11 |
| #60 | 7 |
| #100 | 4 |

6. A permeability test is conducted with a sample of soil which is 60 mm in diameter and 120 mm long. The head is kept constant at 225 mm. The flow is 1.5 ml in 6.5 minutes. What is the coefficient of permeability in units of meter/year?

7. A consolidation test is performed on a soil with the following results:

| pressure (psf) | e | pressure (psf) | e |
|---|---|---|---|
| 250 | .755 | 8350 | .724 |
| 520 | .754 | 16,700 | .704 |
| 1040 | .753 | 33,400 | .684 |
| 2090 | .750 | 8350 | .691 |
| 4180 | .740 | 250 | .710 |

(a) Graph the curve of stress versus void ratio. (Hint: use log or semi-log paper.) (b) What is the compression index? (c) If the initial pressure on the soil layer is 1400 psf, how much stress can a 10 foot thickness of this soil carry without settling more than 1.0 inch?

8. A sample of soil has the following characteristics:

| | |
|---|---|
| % passing #40 screen: | 95 |
| % passing #200 screen: | 57 |
| liquid limit of 40 fraction: | 37 |
| plasticity index of 40 fraction: | 18 |

Use the AASHTO system to classify this soil. Include the group index number.

9. A sample of sand has a relative density of 40% with a solids specific gravity of 2.65. The minimum void ratio is .45; the maximum void ratio is .97. (a) What is the density of this sand in a saturated condition? (b) If the sand is compacted to a relative density of 65%, what will be the decrease in thickness of a 4 foot thick layer?

10. Specifications on a job require a fill using borrow soil to be compacted to 95% of its standard Proctor maximum dry density. Tests indicate that this maximum is 124.0 pcf when dry. The soil now has 12% moisture. The borrow material has a void ratio of .60 and a solid specific gravity of 2.65. What is the minimum volume of borrow soil required to fill 1.0 cubic foot?

Timed

1. Two choices for borrow soil are available.

| borrow A | | borrow B |
|---|---|---|
| 115 lb/ft³ | density in place | 120 |
| ? | density in transport | 95 |
| .92 | void ratio in transport | ? |
| 25% | water content in place | 20% |
| \$.20/yd³ | cost to excavate | \$.10/yd³ |
| \$.30/yd³ | cost to haul | \$.40/yd³ |
| 2.7 | S.G. of solids | 2.7 |
| (112) lb/ft³ | max Proctor dry density | (110) |

It will be necessary to fill a 200,000 yd³ depression, and the fill material must be compacted to 95% of the standard Proctor (maximum) density. A final 10% moisture content (dry basis) is desired in either case. (a) What soil would be cheaper to use? (b) What is the volume of borrow from each site? (c) What is the minimum quantity (volume) of material to haul?

2. Two series of triaxial shear tests on a soil were performed with the following results:

| undrained series | | drained series | |
|---|---|---|---|
| confining pressure | total axial pressure | confining pressure | total axial pressure |
| 0 psi | 60 psi | 50 psi | 250 psi |
| 50 | 110 | 100 | 400 |
| 100 | 160 | 150 | 550 |

(a) Determine the angle of internal friction for both series. (b) What is the cohesion for both series? (c) What is the angle of the failure plane (with respect to the horizontal axis) for both series? (d) Given a fourth test of the drained sample with radial confining pressure = 300 psi, what is the expected axial load at failure?

# 10 FOUNDATIONS AND RETAINING WALLS

*Nomenclature*

| | | |
|---|---|---|
| $a_v$ | coefficient of compressibility | $ft^2/lbf$ |
| A | area | $ft^2$ |
| B | width or diameter | ft |
| c | cohesion | $lbf/ft^2$ |
| $c_a$ | adhesion | $lbf/ft^2$ |
| C | multiplicative correction factor | – |
| $C_c$ | compression index | – |
| $C_v$ | coefficient of consolidation | $ft^2/sec$ |
| d | depth factor | – |
| D | depth | ft |
| e | void ratio | – |
| $f_o$ | skin friction coefficient | – |
| F | factor of safety | – |
| h | depth | ft |
| H | soil layer thickness or depth | ft |
| k | permeability coefficient, or a constant | ft/sec,– |
| $k_o$ | coefficient of earth pressure at rest | – |
| L | length | ft |
| M | moment | ft-lbf |
| N | capacity factor, or number of blows | –, – |
| $N_o$ | stability number | – |
| p | pressure | $lbf/ft^2$ |
| P | load or force | lbf |
| q | uniform surcharge | lbf/ft, or $lbf/ft^2$ |
| r | distance (moment arm) | ft |
| R | force (resistance) | lbf/ft of wall |
| S | strength or settlement | $lbf/ft^2$, or ft |
| SG | specific gravity | – |
| t | time | various |
| $T_v$ | time factor | – |
| $U_z$ | percent of total consolidation | – |
| w | water content | – |
| $w_l$ | liquid limit | – |
| W | weight (mass) | lbm |
| z | depth | ft |

*Symbols*

| | | |
|---|---|---|
| $\phi$ | angle of internal friction | degrees |
| $\delta$ | angle of wall friction | degrees |
| $\alpha$ | secondary compression index | – |
| $\rho$ | density | $lbm/ft^3$ |
| $\mu$ | pore pressure | $lbf/ft^2$ |
| $\epsilon$ | eccentricity | ft |
| $\eta$ | efficiency | – |
| $\beta$ | cut angle | degrees |

*Subscripts*

| | |
|---|---|
| a | allowable |
| A | active |
| b | below mudline |
| c | compressive |
| f | footing |
| g | gross |
| h | horizontal |
| i | the ith component |
| n | unconfined |
| P | passive |
| q | surcharge |
| s | shear, or sliding |
| v | vertical |
| w | water |
| $\gamma$ | density (as a subscript) |

## 1 CONVERSIONS

| multiply | by | to obtain |
|---|---|---|
| kips | 1000 | pounds |
| pounds | 5 EE–4 | tons |
| pounds | 0.001 | kips |
| pounds/square foot | 5 EE–4 | $tons/ft^2$ |
| pounds/square inch | 0.072 | $tons/ft^2$ |
| tons | 2000 | pounds |
| tons/square foot | 2000 | $pounds/ft^2$ |
| tons/square foot | 13.889 | $pounds/inch^2$ |
| Newtons/square meter | 0.021 | $pounds/ft^2$ |

## 2 DEFINITIONS

Abutment: A retaining wall which also supports a vertical load.

Active pressure: Pressure causing a wall to move away from the soil.

Batter pile: A pile inclined from the vertical.

Bell: An enlarged section at the base of a pile or pier used as an anchor.

Berm: A shelf, ledge, or pile.

Cased hole: An excavation whose sides are lined or sheeted.

Dead load: An inert, inactive load, primarily due to the structure's own weight.

Dredge level: See 'Mud line.'

Freeze (of piles): A large increase in the ultimate capacity (and required driving energy) of a pile after it has been driven some distance.

Grillage: A footing or part of a footing consisting of horizontally laid timbers or steel beams.

Lagging: Heavy planking used to construct walls in excavations and braced cuts.

Live load: The weight of all non-permanent objects in a structure, including people and furniture. Live load does not include seismic or wind loading.

Mud line: The lower surface of an excavation or braced cut.

Passive pressure: A pressure acting to counteract active pressure.

Pier shaft: The part of a pier structure which is supported by the pier foundation.

Ranger: See 'Wale.'

Rip rap: Pieces of broken stone used to protect the sides of waterways from erosion.

Sheeted pit: See 'Cased hole.'

Slickenside: A surface (plane) in stiff clay which is a potential slip plane.

Soldier pile: An upright pile used to hold lagging.

Stringer: See 'Wale.'

Surcharge: A surface loading in addition to the soil load behind a retaining wall.

Wale: A horizontal brace used to hold timbers in place against the sides of an excavation, or to transmit the braced loads to the lagging.

## 3 COMPARISON OF SAND AND CLAY AS FOUNDATION MATERIALS

Ordinarily, sand makes a good foundation material. It doesn't settle after its initial loading. It drains quickly. However, it behaves poorly in excavations. When sand is fine and saturated, it can become quick, and a major loss in supporting strength occurs.

Care must be taken when distinguishing between "moist" and "saturated" sands. Sand which has been allowed to drain may be "moist" in the normal sense of the word. However, if the water is not captive, pore pressure will not develop, and the sand can be considered dry. However, special considerations are required if the sand is below the water table. Such sand is saturated, not moist.

Clay, on the other hand, is good in excavations, but is poor for foundations. It continues to settle indefinitely. It retains water for a long time, and large volume changes can result when large changes in moisture content occur.

## 4 GENERAL CONSIDERATIONS FOR FOOTINGS

A *footing* is an enlargement at the base of a load-supporting column designed to transmit forces to the soil. The area of the footing will depend on the load and the soil characteristics. The following types of footings are used.

- spread footing: A footing used to support a single column. This is also known as an *individual column footing* and *isolated footing.*
- continuous footing: A long footing supporting a continuous wall. Also known as *wall footing.*
- combined footing: A footing carrying more than one column.
- cantilever footing: A combined footing that supports a column and an exterior wall or column.

If possible, footings should be designed according to the following general considerations:

- The footing should be located below the frost line and below the level which is affected by moisture content changes.
- Footings need not be any lower than the highest-adequate stratum.
- The centroid of the footing should coincide with the centroid of the applied load.
- Allowable soil pressures should not be exceeded.

- Below-grade footings should be equipped with a drainage system.
- Footings on fill over loose sand should be densified with piles.
- If possible, footings should be placed in excavations made in compacted fill. They should not be put in place prior to compaction.
- Size footings to the nearest 3″ above or equal to the theoretical size.

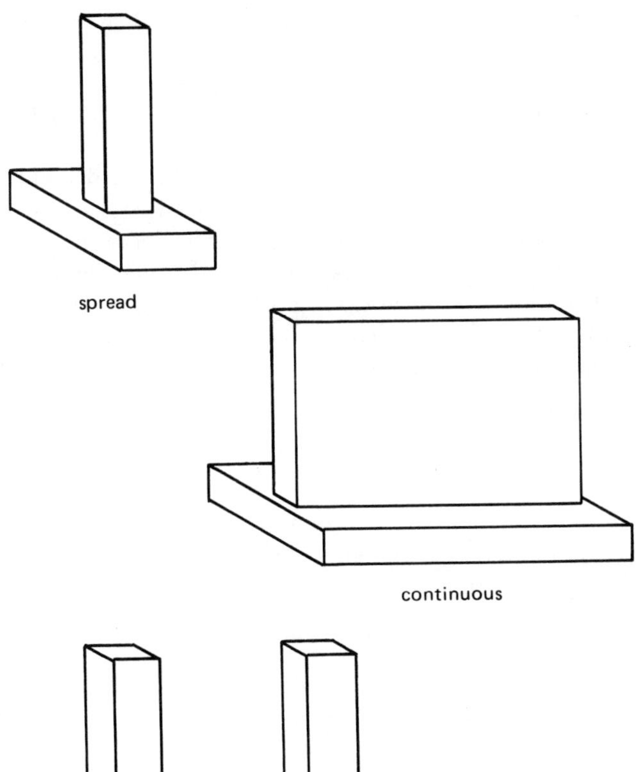


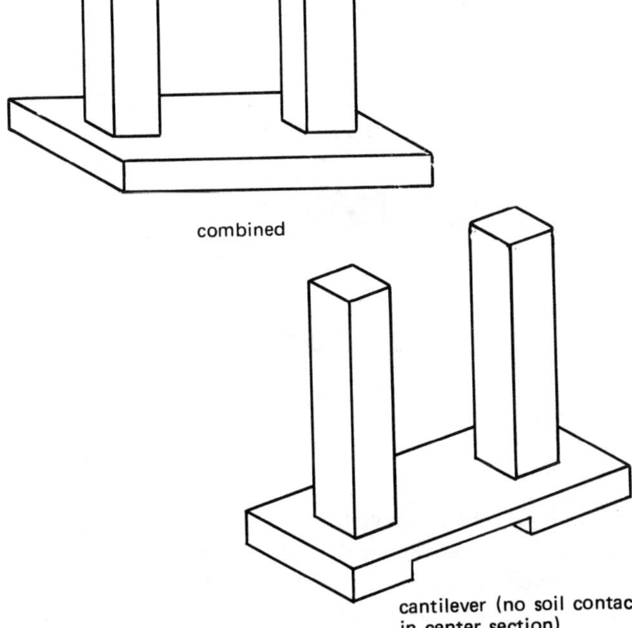


**Figure 10.1** Types of Footings

## 5 ALLOWABLE SOIL PRESSURES

When data from soil tests is unavailable, table 10.1 can be used for preliminary calculations.

**Table 10.1**
Typical Allowable Soil Bearing Pressures[1]

| type of soil | allowable pressure |
|---|---|
| massive crystalline bedrock | 4000 $\text{lbf/ft}^2$ |
| sedimentary and foliated rock | 2000 |
| sandy gravel and/or gravel (GW and GP) | 2000 |
| sand, silty sand, clayey sand, silty gravel, and clayey gravel (SW, SP, SM, SC, GM, GC) | 1500 |
| clay, sandy clay, silty clay, and clayey silt (CL, ML, MH, and CH) | 1000 |

## 6 GENERAL FOOTING DESIGN EQUATION

The *gross* (or *ultimate*) *bearing capacity* or *gross pressure* for a soil is given by equation 10.1, which is known as the *Terzaghi-Meyerhoff equation.* The equation is good for both sandy and clayey soils. It is specifically valid for continuous wall footings. ($p_q$ is a surface surcharge.)

$$p_g = \frac{1}{2}\rho B N_\gamma + c N_c + (p_q + \rho D_f) N_q \qquad 10.1$$

Various researchers have made improvements on this theory, leading to somewhat different terms and sophistication in evaluating $N_\gamma$, $N_c$, and $N_q$.[2] However, the general form remains valid for design, with corrections for various footing geometries.

Figure 10.2 and table 10.3 can be used to evaluate the capacity factors $N_\gamma$, $N_c$, and $N_q$ in equation 10.1.

**Table 10.2**
$N_c$ Bearing Capacity Factor Multipliers for Various Values of B/L (See figure 10.3)

| B/L | multiplier |
|---|---|
| 1 (square) | 1.25 |
| 0.5 | 1.12 |
| 0.2 | 1.05 |
| 0.0 | 1.00 |
| 1 (circular) | 1.20 |

[1] As in the definition of $p_a$, the term 'allowable' implies that a factor of safety has already been applied.

[2] Differences in reported values of $N_\gamma$, $N_c$, and $N_q$ may also be due to the different units used by researchers.

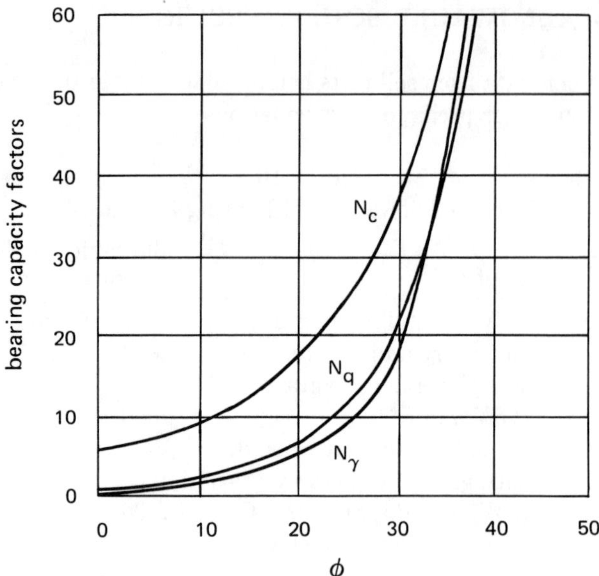


**Figure 10.2** Bearing Capacity Factors

**Table 10.3**
Terzaghi Bearing Capacity Factors for General Shear[3]

| $\phi$ | $N_c$ | $N_q$ | $N_\gamma$ |
|---|---|---|---|
| 0 | 5.7 | 1.0 | 0.0 |
| 5 | 7.3 | 1.6 | 0.5 |
| 10 | 9.6 | 2.7 | 1.2 |
| 15 | 12.9 | 4.4 | 2.5 |
| 20 | 17.7 | 7.4 | 5.0 |
| 25 | 25.1 | 12.7 | 9.7 |
| 30 | 37.2 | 22.5 | 19.7 |
| 34 | 52.6 | 36.5 | 35.0 |
| 35 | 57.8 | 41.4 | 42.4 |
| 40 | 95.7 | 81.3 | 100.4 |
| 45 | 172.3 | 173.3 | 297.5 |
| 48 | 258.3 | 287.9 | 780.1 |
| 50 | 347.5 | 415.1 | 1153.2 |

**Table 10.4**
$N_\gamma$ Multipliers for Various Values of B/L
(See figure 10.3)

| B/L | multiplier |
|---|---|
| 1.0 (square) | 0.85 |
| 1.0 (circular) | 0.70 |
| 0.5 | 0.90 |
| 0.2 | 0.95 |
| 0.0 | 1.0 |

[3] In *general shear*, the soil resists an increased load until failure is sudden. There is another case, that of *local shear*, which results with looser soil. However, it is unlikely that foundations would be designed for loose soil without compaction. With compaction, the general shear case holds.

Once a gross pressure is determined, it is corrected by the *overburden*, giving the *net soil pressure.*

$$p_{\text{net}} = p_g - \rho D_f \qquad 10.2$$

The *allowable soil pressure* is determined by dividing the net pressure by a factor of safety.[4] A safety factor of 3 (based on $p_{\text{net}}$) should be used for average conditions. Exceptional loadings and improbable combinations of snow, wind, and seismic forces may be allowed to reduce the safety factor to 2.

$$p_a = \frac{p_{\text{net}}}{F} \qquad 10.3$$

## 7 FOOTINGS ON CLAY AND PLASTIC SILT

Clay is normally soft, fairly impermeable, and highly preloaded. When loads are first applied to saturated clay, the pore pressure increases. For a short time, this pore pressure does not dissipate and the angle of internal friction should be taken as $\phi = 0°$. This is known as the $\phi = 0°$ or *undrained case.* The undrained clay shear strength is one-half of the unconfined strength.

$$c = \frac{1}{2} S_{nc} \qquad 10.4$$

If $\phi = 0°$, then $N_\gamma = 0$ and $N_q = 1$. If there is no surface surcharge, the gross bearing capacity is given by equation 10.5

$$p_g = cN_c + \rho D_f \qquad 10.5$$

$$p_{\text{net}} = p_g - \rho D_f = cN_c \qquad 10.6$$

$$p_a = \frac{p_{\text{net}}}{F} \qquad 10.7$$

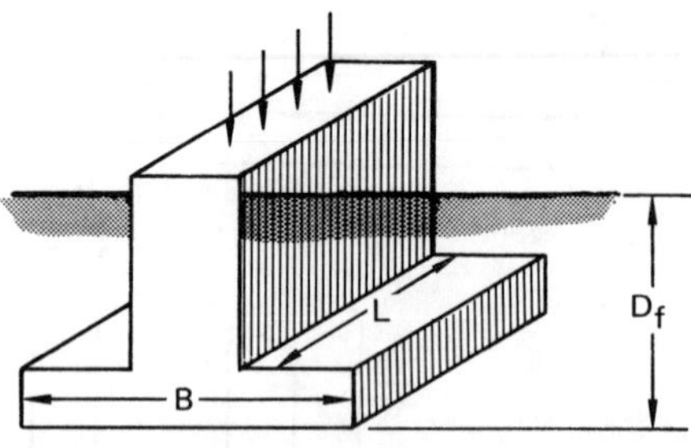


**Figure 10.3** A Spread Footing

*Example 10.1*

An individual square column footing carries an 83,800 pound dead load and a 75,400 pound live load. The unconfined compressive strength of the supporting clay is 0.84 tons/ft$^2$ and its density is 115 lbm/ft$^3$. The

[4] The term 'allowable' implies that the safety factor has been included. This pressure is also known as "safe" and "net allowable" pressure.

footing is covered by a 6″ basement slab. The footing thickness is initially unknown. What size footing is required? Neglect depth correction factors. Do not design the structural steel.

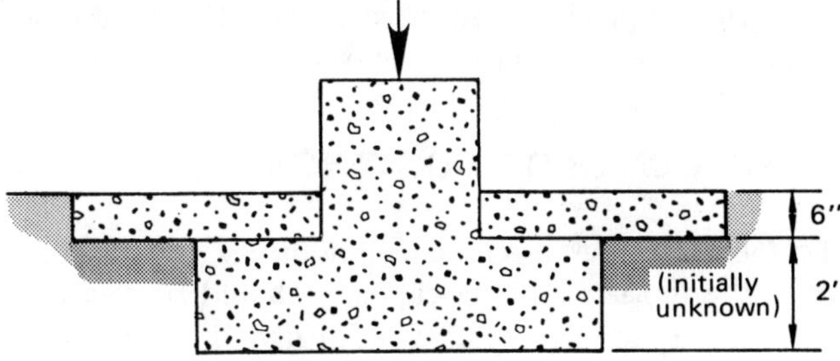


The total load on the column is

$$\frac{83{,}800 + 75{,}400}{2000} = 79.6 \text{ tons}$$

From tables 10.2 and 10.3, for square footings,

$$N_c \approx (1.25)(5.7) = 7.1$$

The cohesion is estimated from the unconfined compressive strength and equation 10.4.

$$c = \frac{0.84}{2} = 0.42 \text{ tons/ft}^2$$

Using a factor of safety of 3, the allowable pressure is

$$p_a = \frac{(0.42)(7.1)}{3} = 0.99 \text{ tons/ft}^2$$

The approximate area required is

$$A = \frac{79.6}{0.99} = 80.4 \text{ ft}^2$$

So, try a 9′3″ square footing (area = 85.6 ft$^2$). (At this point, a footing thickness would be determined based on concrete design considerations.) Assume a 2.0 foot footing thickness.

The actual pressure under the footing due to applied load is

$$p = \frac{79.6}{85.6} = 0.93 \text{ tons/ft}^2$$

This first iteration did not consider the concrete weight. The concrete density is approximately 150 lbm/ft$^3$. Therefore, the pressure surcharge due to one square foot of concrete floor is

$$p_q = \frac{1 \times 1 \times \frac{6}{12} \times 150}{2000} = 0.04 \text{ tons/ft}^2$$

Similarly, the footing itself has weight. The footing extends 2 feet down.

$$p_f = \frac{1 \times 1 \times 2 \times 150}{2000} = 0.15 \text{ tons/ft}^2$$

Therefore, the total pressure under the footing is

$$p_{\text{total}} = 0.93 + 0.04 + 0.15 = 1.12 \text{ tons/ft}^2$$

Equation 10.2 gives the allowable load in excess of the soil surcharge. The footing bottom is 2.5 feet below the original grade, so the soil surcharge is[5]

$$\frac{(2.5)(115)}{2000} = 0.14 \text{ tons/ft}^2$$

The net actual pressure to be compared to the allowable pressure is

$$p = 1.12 - 0.14 = 0.98$$

This is essentially the same as $p_a$.

## 8 FOOTINGS ON SAND

The cohesion, $c$, is zero in sand. The gross ultimate bearing capacity can be derived from equation 10.1 by setting $c = 0$.

$$p_g = \frac{1}{2}B\rho N_\gamma + (p_q + \rho D_f)N_q \qquad 10.8$$

The net ultimate bearing capacity when there is no surface surcharge (i.e., $p_q = 0$) is

$$p_{\text{net}} = p_g - \rho D_f = \frac{1}{2}B\rho N_\gamma + \rho D_f(N_q - 1) \qquad 10.9$$

If the water table level is above the footing face (*submerged condition*), $p_{\text{net}}$ should be reduced by 50%.

The allowable sand loading is based on a factor of safety, which is typically taken as 2 for sand.

$$p_a = \frac{p_{\text{net}}}{F} = \frac{B}{F}\left[\frac{1}{2}\rho N_\gamma + \rho(N_q - 1)\frac{D_f}{B}\right] \qquad 10.10$$

Since sand is permeable and rapidly adjusts to changes in loading, design the footing based on the maximum instantaneous load. Determine the allowable soil pressure based on the footing with the maximum load, smallest $N$, deepest (highest) surface water, etc. Use this soil pressure for all footings in the building foundation.

[5] A depth of 2 feet could also be used if the basement slab was constructed on the original grade. This calculation assumes the slab is poured 6″ below the original grade.

Since the quantity in brackets in equation 10.10 is constant for specific $D_f/B$ ratios, and since the bearing capacity factors depend on $\phi$ (which can be correlated to $N$), equation 10.11 can be derived.[6] This equation assumes $F = 2$, $\rho = 100\ \text{lbm/ft}^3$, and $D_f < B$.

$$p_a = (0.11)C_nN \qquad 10.11$$

No correction is usually made if the density is different from 100 lbm/ft$^3$. However, the equation assumes that the overburden load ($D_f\rho$) is approximately 1 ton/ft$^2$. This means that the $N$ values were derived from data corresponding to depths of 10 to 15 feet below the original surface (not the basement surface). If the footing is to be installed close to the original surface, then a correction factor is required.[7]

**Table 10.5**
Overburden Corrections

| overburden | $C_n$ |
|---|---|
| 0 tons/ft$^2$ | 2 |
| 0.25 | 1.45 |
| 0.5 | 1.21 |
| 1.0 | 1.00 |
| 1.5 | 0.87 |
| 2.0 | 0.77 |
| 2.5 | 0.70 |
| 3.0 | 0.63 |
| 3.5 | 0.58 |
| 4.0 | 0.54 |
| 4.5 | 0.50 |
| 5.0 | 0.46 |

For a given sand settlement, the soil pressure will be greatest in intermediate width ($B$ = 2 to 4 feet) footings. This is illustrated in figure 10.4. Equation 10.11 should not be used for small-width footings, since bearing pressure governs. For wide footings, (i.e., B > 2 to 4 feet), settlement governs.

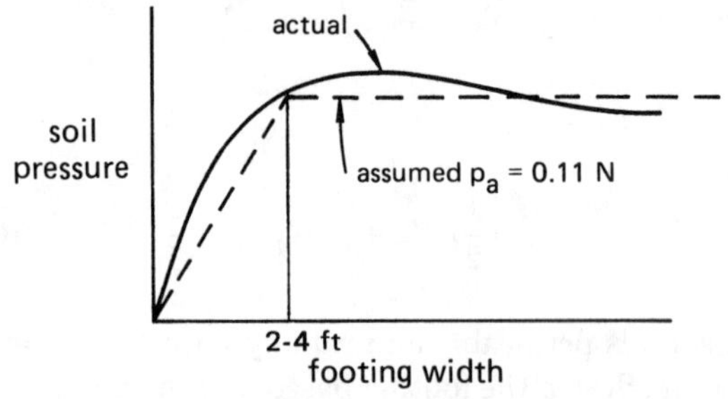


**Figure 10.4** Soil Pressure on Sand with Constant Settlement

[6] $N$ is the number of blows per foot from a standard penetration test.

[7] The correction is actually a correction for $N$. If corrected $N$ values are known, $C_n$ may be neglected.

## 9 FOOTINGS ON ROCK

If bedrock can be reached by excavation, the allowable pressure is likely to be determined by local codes. A safety factor of 5 based on the unconfined compressive strength is typical. For most rock beds, the design will be based on settlement characteristics, not strength.

## 10 MOMENTS ON FOOTINGS

If a footing carries a moment in addition to its vertical load, the footing bearing capacity should be analyzed assuming a smaller area.[8] Specifically, the size should be reduced by twice the eccentricity.[9]

$$\epsilon_B = \frac{M_B}{P}; \quad \epsilon_L = \frac{M_L}{P} \qquad 10.12$$

$$L' = L - 2\epsilon_L; \quad B' = B - 2\epsilon_B \qquad 10.13$$

$$A' = L'B' \qquad 10.14$$

This area reduction places the equivalent force at the centroid of the reduced area. The actual value of $B$ should be used in calculating $D_f/B$ ratios used in finding capacity and depth factors. However, $B'$ should be used in equation 10.1 and in other equations where $B$ appears by itself. The footing bearing capacity is reduced in two ways. First, a smaller $B$ in equation 10.1 results in a smaller $p_g$. Second, a smaller $p_g$ results in a smaller allowable load (i.e., $P = p_aB'L'$).

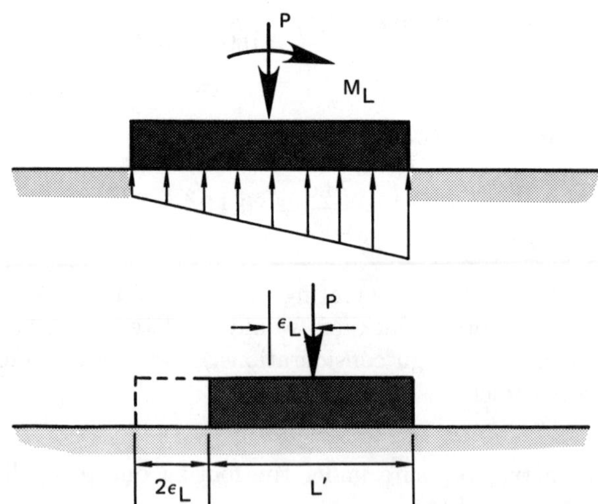


**Figure 10.5** A Footing with an Overturning Moment

Although the eccentricity is independent of the footing dimensions, a trial and error solution may be necessary when designing footings. Trial and error is not required when analyzing a footing of known dimensions.

[8] Usually, there will be no $M_B$ moment.

[9] This discussion is for rectangular footings. It is much more difficult to construct an equivalent footing for circular shapes.

Assuming $M_B = 0$, $\epsilon_L = \epsilon$, and disregarding the concrete and overburden weights, the actual soil pressure distribution is given by equation 10.15. $B'$ and $L'$ should not be used in equation 10.15 because these variables place the load at the centroid of the reduced area, producing a uniform pressure distribution.

$$p_{\text{max}}, p_{\text{min}} = \frac{P}{BL}\left(1 \pm \frac{6\epsilon}{L}\right) \qquad 10.15$$

If the eccentricity, $\epsilon$, is sufficiently large, a negative soil pressure will result. Since soil cannot carry a tensile stress, such stresses are neglected. This results in a reduced area to carry the load.

If the resultant force is within the middle third of the footing, all of the footing will contribute. That is, the maximum eccentricity without incurring a reduction footing area will be L/6.

## 11 GENERAL CONSIDERATIONS FOR RAFTS

A *raft* or *mat* is a combined footing-slab that covers the entire area beneath a building and supports all walls and columns. A raft foundation should be used (at least for economic reasons) any time the individual footings would constitute half or more of the area beneath a building.

## 12 RAFTS ON CLAY

The net ultimate bearing pressure for rafts on clay can be found in the same manner as for footings. Since the size of the raft is essentially fixed by the building size (plus or minus a few feet), the only method available to increase the loading is to lower the elevation (increase $D_f$) of the raft.

The factor of safety produced by a raft construction is given by equation 10.16, which can also be solved to give the required $D_f$ if the factor of safety is known. The factor of safety should be at least 3 for normal loadings, but may be reduced to 2 during temporary extreme loading.

$$F = \frac{cN_c}{\frac{\text{total load}}{\text{raft area}} - \rho D_f} \qquad 10.16$$

If the denominator in equation 10.16 is small, the factor of safety is very large. If the denominator is zero, the raft is said to be a *fully compensated foundation.* For $D_f$ less than the fully-compensated depth, the raft is said to be *partially compensated.*

*Example 10.2*

A raft foundation is to be designed for a $120' \times 200'$ building with a total loading of 5.66 EE7 pounds. The clay density is 115 lbm/ft$^3$, and the clay has an average unconfined compressive strength of 0.3 tons/ft$^2$. (a) What should be the raft depth, $D_f$, for full compensation? (b) What should be the raft depth for a factor of safety of 3? Neglect depth correction factors.

The loading pressure is

$$p_{\text{load}} = \frac{5.66\,\text{EE7}}{(120)(200)} = 2.36\,\text{EE3}\,\text{lbf/ft}^2$$

(a) For full compensation, $p_{\text{load}} = \rho D_f$.

$$D_f = \frac{2.36\,\text{EE3}}{115} = 20.5\,\text{ft}$$

(b) From table 10.3, $N_c = 5.7$. Since $B/L = 120/200 = 0.6$, use an $N_c$ multiplier of 1.15.

$$3 = \frac{\left(\frac{1}{2}\right)(0.3)(2000)(1.15)(5.7)}{2.36\,\text{EE3} - (115)D_f}$$

$$D_f = 14.8\text{ ft}$$

## 13 RAFTS ON SAND

Rafts on sand are always well protected against bearing capacity failure. Therefore, settlement will govern the design. Since differential settlement will be much smaller for various locations on the raft (due to the raft's rigidity), the allowable soil pressure may be doubled.

$$p_a = 0.22 C_n N \ (\text{tons/ft}^2) \qquad 10.17$$

$N$ should always be at least 5 after correcting for overburden. Otherwise, the sand should be compacted or a pier/pile foundation used.

The net soil pressure should be compared with the allowable pressure. The net soil pressure is

$$p_{\text{net}} = \frac{\text{total load}}{\text{raft area}} - \rho D_f \qquad 10.18$$

## 14 GENERAL CONSIDERATIONS FOR PIERS

A pier is a large underground structure with a length (depth) greater than its width (diameter). It differs from a pile in its diameter, load carrying capacity, and installation method. A pier is usually constructed within an excavation.

## 15 PIERS IN CLAY

If $D_f/B > 4$, then $N_c$ is constant (i.e., depth factors do not affect $N_c$). The design of a pier foundation is similar in other respects to a footing design. A factor of safety of 3 is usually used.

$$p_{\text{net}} = p_g - \rho_{\text{clay}} D_f - \rho_w h \qquad 10.19$$

$$p_a = \frac{cN_c}{F} \qquad 10.20$$

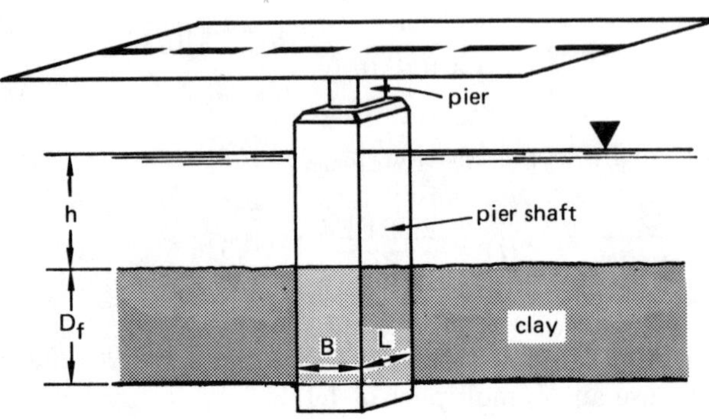


**Figure 10.6** A Pier in Saturated Clay

Piers derive additional supporting strength from skin friction. Skin friction strength is 0.3 to 0.5 times the average undrained shear strength.[10] The additional load that the pier can support is the skin friction times the surface area of the pier shaft. If the pier is belled, only the straight part of the pier is used to calculate the skin friction capacity.

If the skin friction is used to support any of the applied load, $D_f$ should be taken as zero in finding $N_c$.

## 16 PIERS IN SAND

Piers in sand are designed similarly to footings, since skin friction is insignificant. A conservative estimate of bearing capacity can be found from equation 10.21.

$$p_{\text{net}} = \frac{1}{2} B \rho N_\gamma + \rho D_f (N_q - 1) \qquad 10.21$$

$$p_a = (0.11) C_w (C_n N) \qquad 10.22$$

$C_w$ is a correction for water table height, which is neglected when $D_w \geq D_f + B$.

$$C_w = 0.5 + 0.5 \left( \frac{D_w}{D_f + B} \right) \qquad 10.23$$

[10] The skin friction strength should not be greater than 1 ton/ft$^2$. If it is, use 1 ton/ft$^2$.

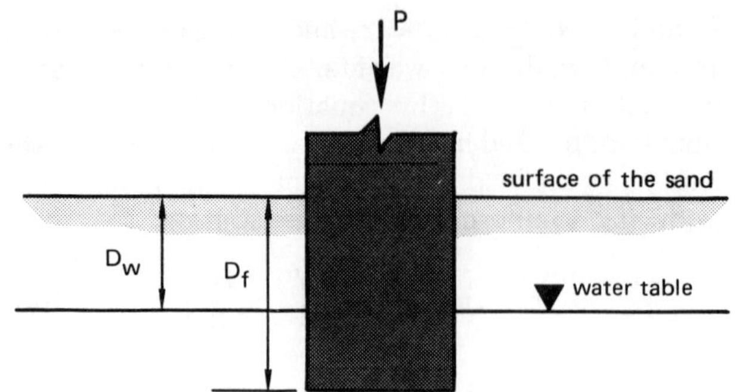


**Figure 10.7** A Pier in Partially Drained Sand

## 17 GENERAL CONSIDERATIONS FOR PILES

Piles are small-area members that are usually hammered or vibrated into place. They provide strength to soils that are too weak or compressible to otherwise support a foundation. Piles are often grouped together to provide the required strength to support a column or wall.

Two major pile classifications exist: friction and point-bearing piles. *Friction piles* derive their load-bearing ability from the friction between the soil and pile. *Point-bearing piles* derive their strength from the support of the soil near the point.

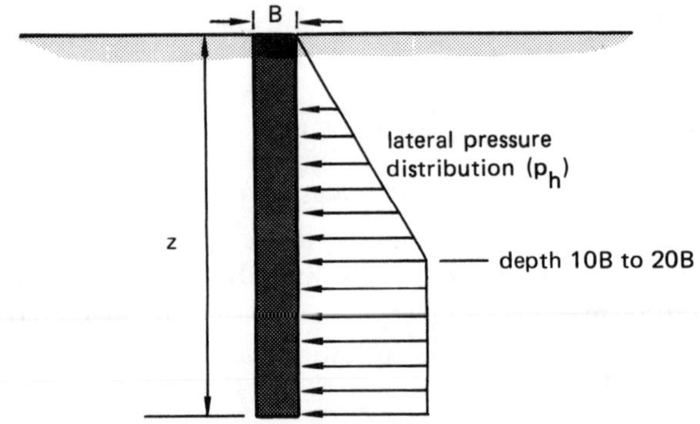


**Figure 10.8** Pressure Distribution on a Typical Pile

### A. CAPACITY OF INDIVIDUAL PILES

In reality, friction bearing capacity and end bearing capacity are both present in friction piles and point-bearing piles. However, one mode of bearing capacity may be predominant. The capacity which a pile possesses is

$$\begin{aligned} W &= [\text{end capacity}] + [\text{friction capacity}] \\ &= \left[ \left( \frac{\pi B^2}{4} \right) \left( \frac{1}{2} \rho B N_\gamma + cN_c + \rho z (N_q - 1) \right) \right] \\ &\quad + [\pi B z f_o] \end{aligned} \qquad 10.24$$

If the pile diameter, $B$, is small, the $\frac{1}{2}\rho BN_\gamma$ term can be omitted. There is also some evidence that the $\rho z(N_q-1)$ term does not increase without bound, but rather, has as upper limit of $N_q \tan\phi$.

The *skin friction coefficient*, $f_o$, includes both cohesive and adhesive terms. In evaluating $f_o$ and the bearing capacity factors, the friction angle $\phi$ should be increased by 2° to 5° for piles driven into sand. For drilled or jetted piles, no increase is necessary.

$$f_o = \text{smaller of} \begin{Bmatrix} c + p_h \tan\phi \\ c_a + p_h \tan\delta \end{Bmatrix} \qquad 10.25$$

The friction angle, $\delta$, can be obtained from table 10.6. The lateral earth pressure depends on the depth, down to a *critical depth*, after which it is essentially constant.[11]

$$p_h = k(\rho z - \mu) \qquad 10.26$$

[11] Between relative densities of 30% and 70%, the critical depth can be interpolated between 10 and 20 diameters.

$$z_{\text{critical}} = \begin{Bmatrix} 10B \text{ for relative density} < 30\% \\ 20B \text{ for relative density} > 70\% \end{Bmatrix} \qquad 10.27$$

The *adhesion*, $c_a$, should be obtained from testing. In the absence of such tests, it can be approximated as a fraction of the cohesion. For rough concrete, rusty steel, and corrugated metal, $c_a = c$. For wood, $0.9c \leq c_a \leq c$. For smooth concrete, $0.8c \leq c_a \leq c$. For clean steel, $0.5c \leq c_a \leq 0.9c$.

For driven piles, the *coefficient of lateral earth pressure at failure*, $k$, also depends on the relative density. For loose sands (relative density < 30%), $2 \leq k \leq 3$. For driven piles in dense sand (relative density > 70%), $3 \leq k \leq 4$. For drilled piles, the coefficients of lateral earth pressure are approximately 50% of the values for driven piles. For jetted piles, the coefficients are approximately 25% of the driven values.

Of course, the pore pressure will not develop in drained sandy soils. For sand below the water table, the pore pressure will be

$$\mu = 62.4 \times \text{depth} \qquad 10.28$$

**Table 10.6**
Friction Angles

| interface materials* | friction angle, $\delta$, degrees |
|---|---|
| concrete or masonry on the following foundation materials: | |
| clean, sound rock | 35 |
| clean gravel, gravel-sand mixtures, and coarse sand | 29–31 |
| clean fine to medium sand, silty medium to coarse sand, and silty or clayey gravel | 24–29 |
| clean fine sand, and silty or clayey fine to medium sand | 19–24 |
| fine sandy silt, and non-plastic silt | 17–19 |
| very stiff clay, and hard residual or preconsolidated clay | 22–26 |
| medium stiff clay, stiff clay, and silty clay | 17–19 |
| steel sheet piles against the following soils: | |
| clean gravel, gravel-sand mixtures, and well-graded rock fill with spalls | 22 |
| clean sand, silty sand-gravel mixtures, and single-size hard rock fill | 17 |
| silty sand, gravel or sand mixed with silt or clay | 14 |
| fine sandy silt, and non-plastic silt | 11 |
| formed concrete or concrete sheet piling against the following soils: | |
| clean gravel, gravel-sand mixtures, and well-graded rock fill with spalls | 22–26 |
| clean sand, silty sand-gravel mixtures, and single-size hard rock fill | 17–22 |
| silty sand, and gravel or sand mixed with silt or clay | 17 |
| fine sandy silt, and non-plastic silt | 14 |
| miscellaneous combinations of structural materials: | |
| masonry on masonry, igneous and metamorphic rocks: | |
| dressed soft rock on dressed soft rock | 35 |
| dressed hard rock on dressed soft rock | 33 |
| dressed hard rock on dressed hard rock | 29 |
| masonry on wood (cross grain) | 26 |
| steel on steel at sheet-steel interlocks | 17 |

* Angles given are ultimate values. Sufficient movement is required before failure will occur.

*Example 10.3*

An 11″, smooth concrete pile with a blunt end is driven 60 feet into clay. The clay's cohesion and density are 1400 psf and 120 lbm/ft$^3$ respectively. The water table extends to the ground surface. What is the ultimate bearing capacity of the pile?

*step 1*: The pile diameter and areas are:

$$B = \frac{11}{12} = 0.917\,\text{ft}$$
$$A_{\text{end}} = \frac{\pi}{4}(0.917)^2 = 0.66\,\text{ft}^2$$
$$A_{\text{surface}} = \pi(0.917)(60) = 172.9\,\text{ft}^2$$

*step 2*: Assume $\phi = 0°$ for saturated clay. From table 10.3, $N_q = 1$. From table 10.2 and table 10.3, $N_c = (1.2)(5.7) = 6.8$. (The contribution of $N_\gamma$ is ignored since $B$ is small.)

*step 3*: The point bearing capacity is

$$(0.66)(1400 \times 6.8) = 6283 \text{ lbf}$$

*step 4*: The lateral earth pressure is disregarded in evaluating the friction capacity, since $\phi = 0°$. Estimate $c_a = 0.9c$ for smooth concrete. Then, the friction capacity is

$$(172.9)(0.9)(1400) = 217{,}850 \text{ lbf}$$

*step 5*: The total capacity is 6283 + 217,850 = 224,133 lbf.

B. CAPACITY OF PILE GROUPS

The capacity of a pile group will generally be more or less than the sum of the individual piles.[12] The *pile group efficiency* is the ratio of actual capacity to the sum of individual capacities.

$$\eta = \frac{\text{group capacity}}{\sum \text{individual capacities}} \qquad 10.29$$

The group capacity of a large number of piles can be approximated by assuming that the piles form a large footing. This large footing extends from the surface to the depth of the pile points. The length and width of the large footing are the length and width of the pile group. The group bearing capacity is computed from the general bearing capacity equation, equation 10.1. The shearing-related capacity includes the effects of cohesion (or adhesion), but the increase in friction capacity due to lateral soil pressure is disregarded.

[12] For sand, the efficiency is maximized ($\eta$ = 200%) with pile spacings of approximately 2B center-to-center. For clay, the efficiency is less than 100% up to a spacing of 2B, after which the efficiency of 100% is reached and maintained for all reasonable pile spacings above 2B.

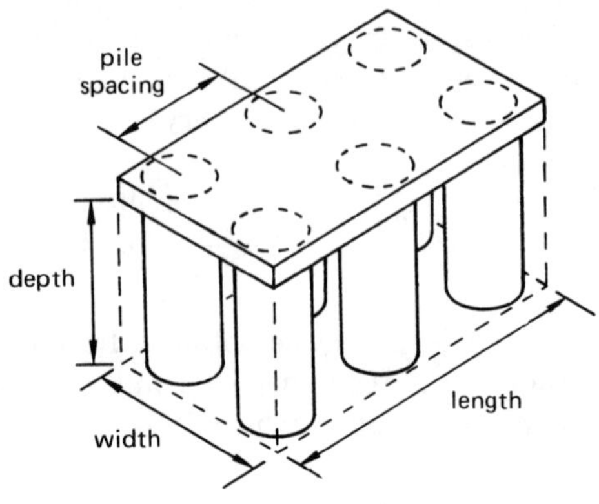


**Figure 10.9** A Pile Group

## 18 SOIL PRESSURES DUE TO APPLIED LOADS

A. BOUSSINESQ'S EQUATION

Figure 10.10 illustrates a load applied through a footing to a soil below. The increase in pressure, $p_v$, due to the application of the building load can be found from *Boussinesq's equation.* This equation requires the footing width to be small compared to the depth, $h$, at which the increase in pressure is desired (i.e., $h > 2B$).

$$p_v = \frac{3h^3P}{2\pi z^5} = \frac{3P}{2\pi(h^2)}\left[\frac{1}{1+(r/h)^2}\right]^{5/2} \qquad 10.30$$

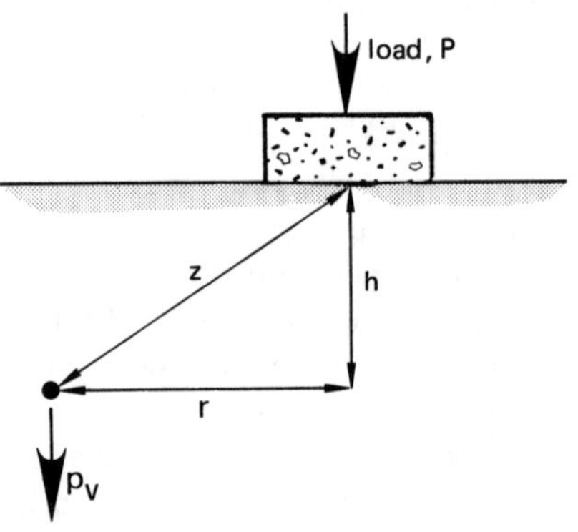


**Figure 10.10** Application of Boussinesq's Equation

## B. INFLUENCE CHART METHOD

If a footing or mat foundation is large compared to the depth where the pressure is wanted, the vertical pressure can be found by use of an *influence chart*, similar to figure 10.11. This chart is used in the following manner.

Let the distance A–B on the chart correspond to the depth at which the pressure is wanted. Using this scale, draw a plan view of the footing on a piece of tracing paper.

Place the tracing paper over the influence chart. Locate the footing tracing so that the center of the chart coincides with the location under the footing where the pressure is wanted.

Count the number of squares seen under the footing drawing. Count partial squares as fractions. Count the pie-shaped areas in the center circle as squares.

Calculate the pressure from equation 10.31[13]

$$p = (\#\,\text{squares})(0.005)(\text{applied pressure}) \qquad 10.31$$

[13] Equation 10.31 assumes the influence chart's *influence value* is 0.005, as it is in figure 10.11. Other charts may have other influence values.

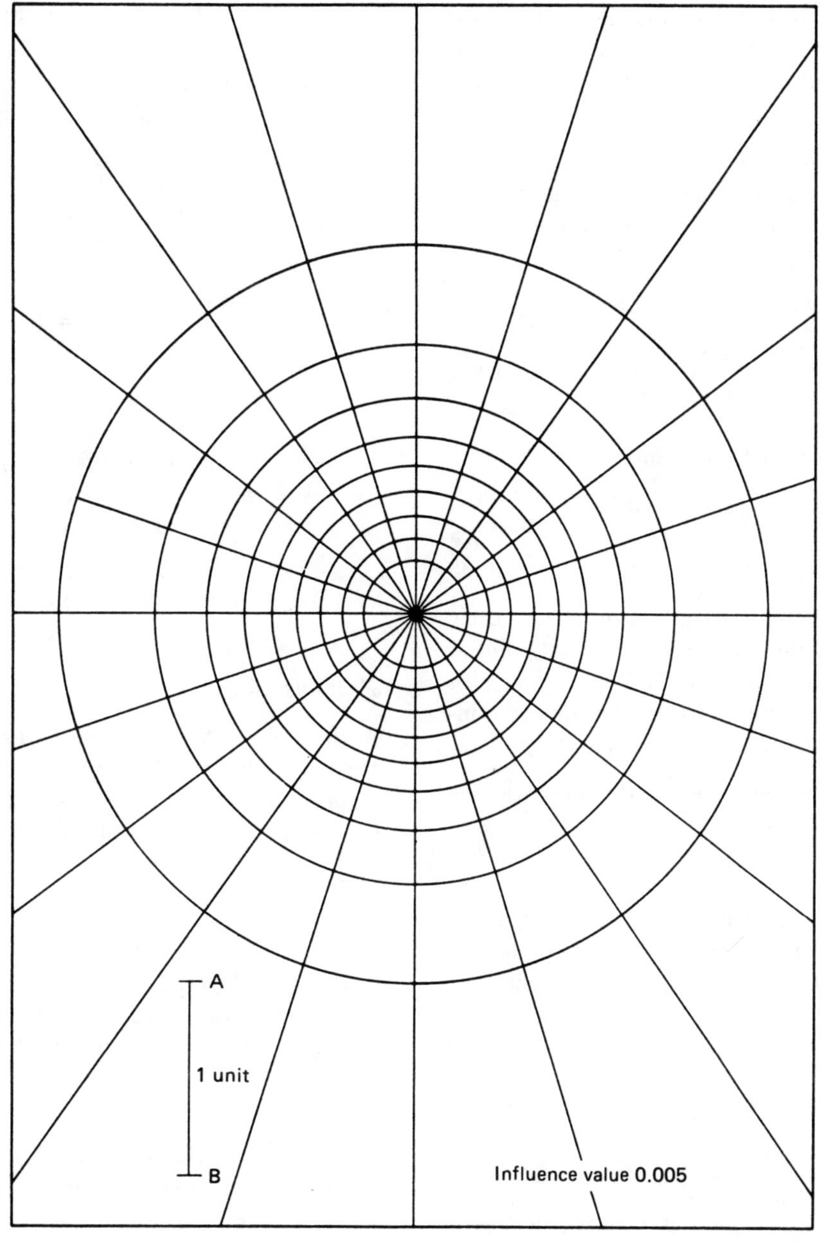


**Figure 10.11** Influence Chart

C. STRESS CONTOUR CHARTS

If a soil is assumed to be semi-infinite, elastic, isotropic, and homogeneous, stress contour charts based on the *Boussinesq case* can be used to obtain subsurface pressures.[14] Such charts exist for a variety of surface loadings, including point loads, square or rectangular footings, circular footings, corners of mats, etc. Appendixes C and D are typical of such contour charts.

## 19 PRIMARY SETTLEMENT IN CLAY

The purpose of pre-construction settlement calculations is to determine the magnitude of expected settlement due to an increase in surface loading. However, these calculations can also be used to find the settlement due to change in any variable, such as a drop in the water table.

*step 1*: Find the original *effective pressure*, $p_o$, at the mid-height of the clay layer. The average effective pressure is the sum of the following items:

- For layers above the water table:

$$p = \begin{pmatrix}\text{layer}\\ \text{thickness}\end{pmatrix}\begin{pmatrix}\text{layer}\\ \text{density}\end{pmatrix} \qquad 10.32$$

- For layers below the water table:

$$p = \begin{pmatrix}\text{layer}\\ \text{thickness}\end{pmatrix}\begin{pmatrix}\text{saturated layer}\\ \text{density}\end{pmatrix} - 62.4 \qquad 10.33$$

*step 2*: Find the increase in pressure, $p_v$, directly below the foundation and at the midpoint of the clay layer due to the building load. This can be done using Boussinesq's equation, influence charts, or stress contour charts.

*step 3*: Estimate the unconsolidated voids ratio, $e_o$.

$$e_o \approx w(SG) \qquad 10.34$$

*step 4*: Estimate the *compression index* (*coefficient of consolidation*) of the soil. This is the logarithmic slope of the primary consolidation curve.

$$C_c = \frac{e_2 - e_1}{\log_{10}\left(\frac{p_1}{p_2}\right)} \qquad 10.35$$

If necessary, the compression index can be estimated from other soil parameters. For inorganic soils with sensitivities less than 4, equation 10.36 applies.

$$C_c = 0.009(w_l - 10) \qquad 10.36$$

For organic soils, such as peat, the compression index can be estimated from the natural moisture content.

$$C_c = 0.0155 w_n \qquad 10.37$$

For clays, a general expression is

$$C_c = 1.15(e_o - 0.35) \qquad 10.38$$

*step 5*: Calculate the settlement. For any clay (normally loaded or pre-loaded) for which the original void ratio and change in void ratio are known, equation 10.39 should be used. $H$ is the thickness of the clay layer, regardless of any overlying layers or surcharges.

$$S = H\left(\frac{\Delta e}{1 + e_o}\right) \qquad 10.39$$

For normally loaded clays only, equation 10.40 can be used. $p_v$ is the change in vertical pressure due to applied loads. Therefore, $p_o + p_v$ is the total pressure after load application or removal.[15]

$$S = \left(\frac{C_c}{1 + e_o}\right) H \log_{10}\left(\frac{p_o + p_v}{p_o}\right) \qquad 10.40$$

*Example 10.4*

A $40' \times 60'$ raft is constructed as shown. The building rests on sand which has already settled. What long-term settlement can be expected in the clay (a) at the center of the raft? (b) at a corner of the raft?

*step 1*:

| | | |
|---|---|---|
| silt layer: | (5)(90) | = 450 lbf/ft² |
| dry sand layer: | (14)(120) | = 1680 |
| wet sand layer: | (22)(120 − 62.4) | = 1270 |
| clay layer: | $\frac{1}{2}$(14)(110 − 62.4) | = 330 |
| | $p_o$ | = 3730 lbf/ft² |

*step 2*: Use the influence chart (figure 10.11) to calculate the pressure increase due to the building. The distance from the bottom of the

[14] The Boussinesq case is not the only case possible. The *Westergaard case* assumes layered or anisotropic foundation soil, consisting of alternating layers of soft and stiff materials. The effect of such layering is to reduce the stresses substantially below those obtained from the Boussinesq case. The Westergaard case is typically used in the analysis of wheel loads over multi-layered highway pavement sections.

[15] $p_o$ in equation 10.40 is the *effective pressure*, excluding the pore pressure. Experience has indicated that effective stress alone can cause consolidation.

raft to the mid-point of the clay layer is $36 - 3 + 7 = 40$ feet. Using a scale of 1 inch = 40 feet, the raft is $1'' \times 1\frac{1}{2}''$. 86 squares are covered.

The net pressure at the base of the raft is the applied pressure minus the overburden pressure due to the excavated sand and silt.

$$p_{\text{net}} = 2400 - 5(90) - 3(120) = 1590\,\text{lbf/ft}^2$$

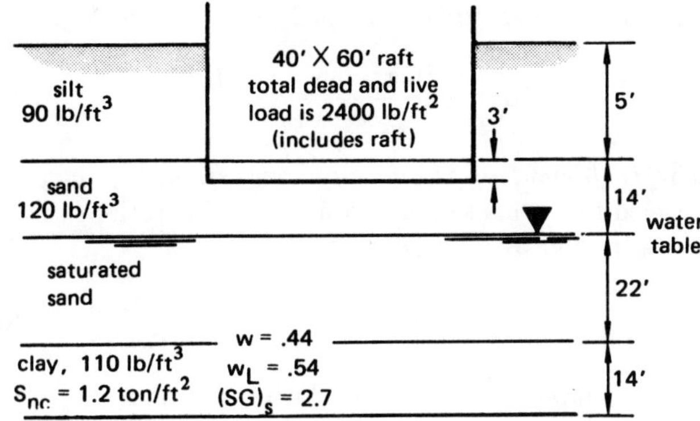


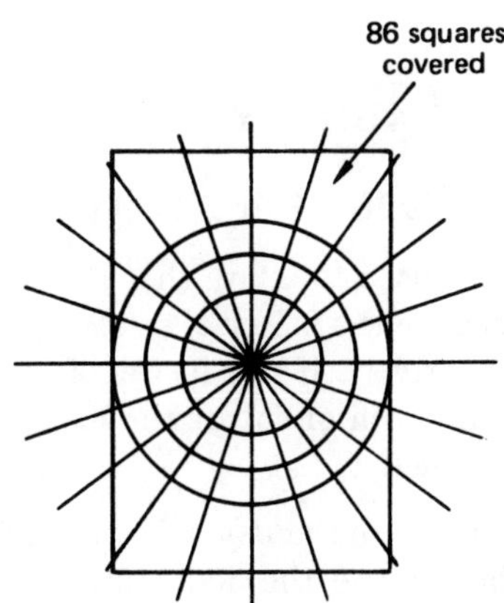


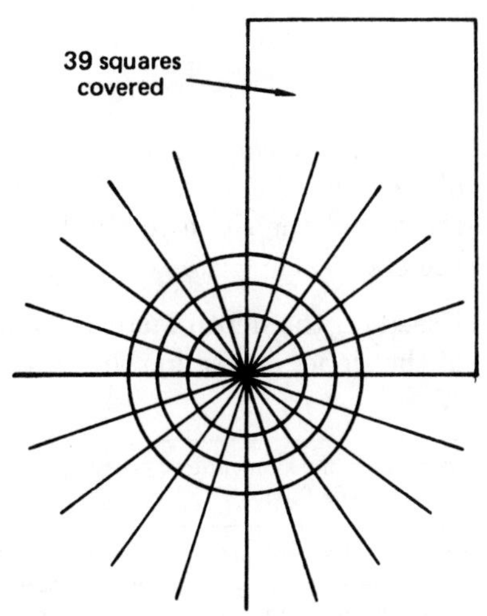


The mid-point pressure below the centroid of the raft is

$$p_v = (1590)(86)(0.005) = 680\,\text{lbf/ft}^2$$

Similarly, the pressure at the corner is

$$p_v = (1590)(39)(0.005) = 310\,\text{lbf/ft}^2$$

*step 3*: The original void content of the clay is found from equation 10.34.

$$e_o = (0.44)(2.7) = 1.188$$

*step 4*: The compression index is estimated from equation 10.36.

$$C_c = 0.009(54 - 10) = 0.396$$

*step 5*: The settlements are

$$S_{\text{center}} = \left(\frac{0.396}{1 + 1.188}\right)(14)\log\left(\frac{3730 + 680}{3730}\right) = 0.184\,\text{ft}$$

$$S_{\text{corner}} = \left(\frac{0.396}{1 + 1.188}\right)(14)\log\left(\frac{3730 + 310}{3730}\right) = 0.087\,\text{ft}$$

## 20 TIME RATE OF PRIMARY CONSOLIDATION

Settling in clay is a continuous process. The time to reach a specific settlement is given by equation 10.41. $z$ is the layer's *half-thickness* if drainage is through the top and bottom surfaces (i.e., *two-way drainage*). If drainage is from one surface only (i.e., *one-way drainage*), $z$ is the layer's full thickness. Units of $t$ will depend on units of $C_v$.

$$t = \frac{T_v z^2}{C_v} \qquad 10.41$$

The *coefficient of consolidation* is assumed to remain constant over small variations in the void ratio, $e$.

$$C_v = \frac{k(1 + e)}{62.4(a_v)} \qquad 10.42$$

The *coefficient of compressibility*, $a_v$, can be found from the void ratio and effective stress for any two different loadings.

$$a_v = \frac{e_1 - e_2}{p_2 - p_1} \qquad 10.43$$

$T_v$ is a dimensionless number known as the *time factor.* $T_v$ depends on the *degree of consolidation,* $U_z$. $U_z$ is the percent of the total consolidation (settlement) expected. For $U_z$ less than 0.60 (60%), $T_v$ is given by equation 10.44.

$$T_v = \frac{1}{4}\pi U_z^2 \qquad 10.44$$

Table 10.7 should be used to find $T_v$ for larger values of $U_z$.

**Table 10.7**
Approximate Time Factors

| $U_z$ | $T_v$ |
|---|---|
| 0.65 | 0.34 |
| 0.70 | 0.40 |
| 0.75 | 0.48 |
| 0.80 | 0.55 |
| 0.85 | 0.70 |
| 0.90 | 0.85 |
| 0.95 | 1.3 |
| 1.0 | $\infty$ |

Following the end of *primary consolidation* (approximately $T_v = 1$ in equation 10.44), the rate of consolidation will decrease considerably. The continued consolidation is known as *secondary consolidation.*

## 21 SECONDARY CONSOLIDATION

*Secondary consolidation* is a gradual consolidation which continues long after the majority of the initial consolidation has occurred. Secondary consolidation may not occur at all, as in the case of granular soils. However, secondary consolidation may be a major factor for inorganic clays and silts, as well as for highly-organic soils.

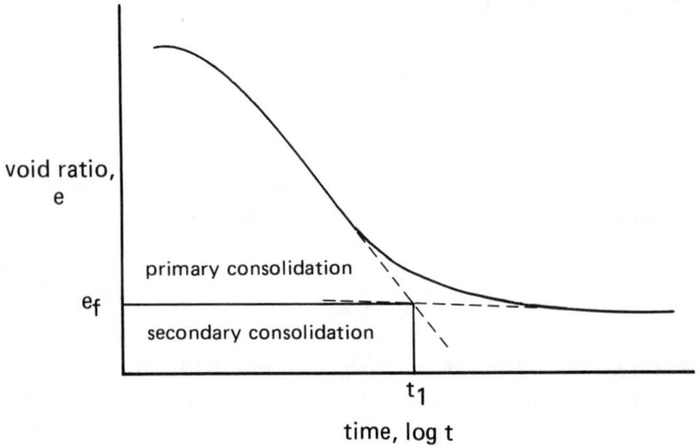


**Figure 10.12** Primary and Secondary Consolidation

Secondary consolidation can be identified by a plot of void ratio (or settlement) versus time on a logarithmic scale. The region of secondary consolidation is characterized by a slope reduction on the plot. The plot can be used to obtain important parameters necessary to calculate the magnitude and progression of secondary consolidation.

The final void ratio, $e_f$, at the end of the primary consolidation period is read from the intersection of the projections of the primary and secondary curves. The logarithmic slope of the secondary compression line is known as the *secondary compression index,* $\alpha$.[16]

$$\alpha = \frac{-(\log_{10} t_2 - \log_{10} t_1)}{e_2 - e_1} \qquad 10.45$$

The *coefficient of secondary compression,* $C_\alpha$, can be derived from this slope. The initial voids ratio, $e_o$, can be estimated from equation 10.34.

$$C_\alpha = \frac{\alpha}{1 + e_o} \qquad 10.46$$

The secondary consolidation during any period $t_2 - t_1$ is

$$S_{\text{secondary}} = C_\alpha H \log\left(\frac{t_2}{t_1}\right) \qquad 10.47$$

## 22 SLOPE STABILITY

### A. HOMOGENEOUS, SOFT CLAY ($\phi = 0°$)

For homogeneous, soft clay, the *Taylor chart* can be used to determine the factor of safety against slope failure.[17] Alternatively, if the factor of safety is known, the maximum depth of cut or the maximum cut angle can be determined.

*step 1:* Calculate the *depth factor,* $d$, from the slope height and the depth from the slope toe to the lowest point on the slip circle.

$$d = \frac{D}{H} \qquad 10.48$$

The *slope height,* $H$, is essentially the depth of the cut. $D$ is the vertical distance from the toe of the slope to the firm base below the clay.

*step 2:* Based on the depth factor, $d$, and the angle of the slope, determine the *stability number,* $N_o$.

[16] The secondary compression index generally ranges from 0 to 0.03, and seldom exceeds 0.04.

[17] The *circular arc method* and *method of slices* can also be used to analyze a particular failure surface. However, if the failure surface is unknown, these methods may require trial and error solutions to locate the critical failure plane.

*step 3*: Calculate the factor of safety, $F$, from equation 10.49. For submerged clays, the buoyant force should be subtracted from the clay density.

$$F = \frac{N_o c}{\rho_{\text{eff}} H} \qquad 10.49$$

$$\rho_{\text{eff}} = \rho_{\text{sat}} - 62.4 \text{ (if submerged)} \qquad 10.50$$

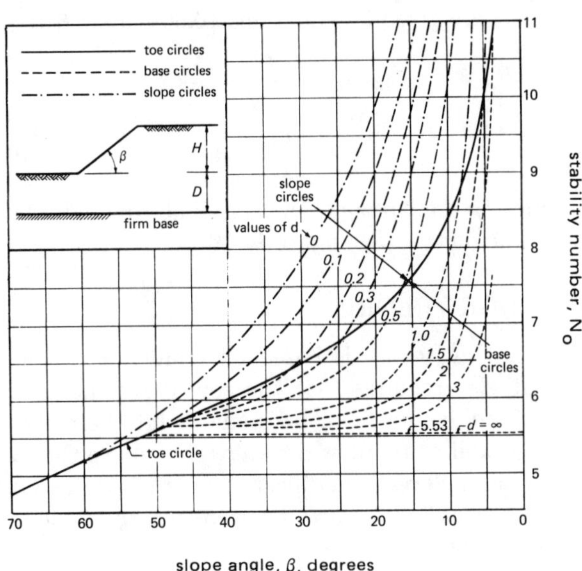


**Figure 10.13** The Taylor Chart for Slope Stability ($\phi = 0$)

Figure 10.13 shows that *toe circle failures* occur for all slopes steeper than 53°. For slopes less than 53°, *slope circle failure*, toe circle failure, or *base circle failure* may occur. These possibilities are illustrated in figure 10.14.

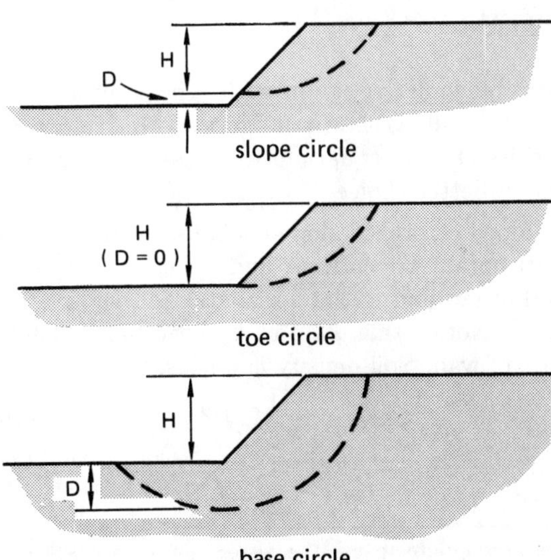


**Figure 10.14** Failure Modes in Clay

*Example 10.5*

An underwater trench is excavated in soft bay mud. The walls of the trench are sloped at 4.5 vertical:3 horizontal. The mud has a saturated density of 100 lbm/ft$^3$ and a cohesion of 400 lbf/ft$^2$. There is a layer of dense sand and gravel 59 feet below the surface of the mud. What is the depth of cut that can be used while maintaining a 1.5 safety factor?

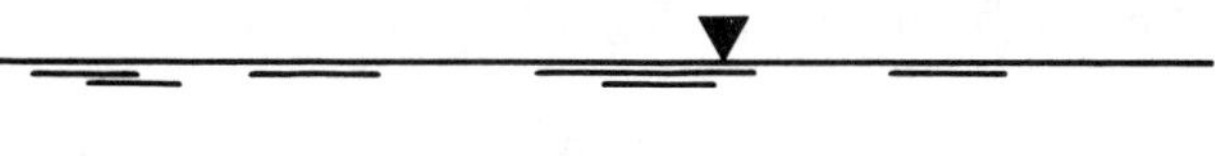

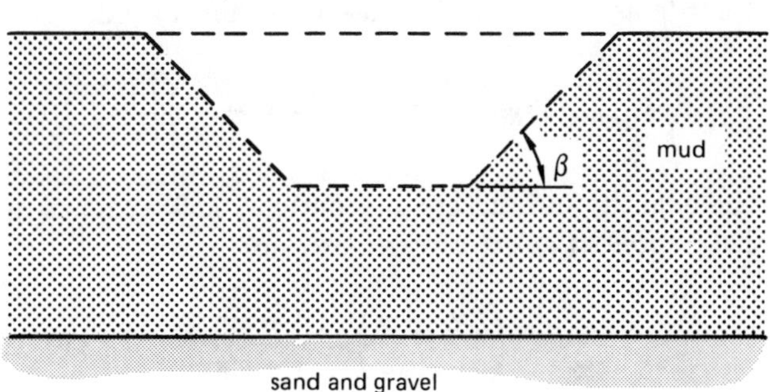


The cut angle is

$$\beta = \arctan\left(\frac{4.5}{3}\right) = 56°$$

From figure 10.13, the stability number for $\beta = 56°$ is approximately 5.4.

Since the clay is submerged, the effective unit weight is $(100 - 62.4) = 37.6$.

From equation 10.49, the maximum cut depth is

$$H = \frac{N_o c}{F\rho} = \frac{(5.4)(400)}{(1.5)(37.6)} = 38.3 \text{ ft}$$

## B. HOMOGENEOUS, COHESIVE SOIL ($c > 0$, $\phi > 0°$)

For cohesive, non-granular soils, the failure will be through the slope toe for $\phi > 5°$.

C. COHESIONLESS SAND ($c = 0$)

The maximum slope angle for cohesionless sand is the angle of internal friction, $\phi$.

## 23 EARTH PRESSURE THEORIES

The general equation for horizontal *active earth pressure* is:

$$p_{\text{horizontal}} = p_{\text{vertical}} \tan^2\left(45° - \frac{\phi}{2}\right) - 2c\tan\left(45° - \frac{\phi}{2}\right) \quad 10.51$$

$c$ in equation 10.51 is the soil's cohesion.

$p_{\text{vertical}}$ can be due to surcharge, externally applied loads, or the soil's own mass.

If $\phi = 0°$, as in the limiting case for saturated clay, then

$$p_{\text{horizontal}} = p_{\text{vertical}} - 2c \quad 10.52$$

If $c = 0$, as in the limiting case for drained sand,

$$p_{\text{horizontal}} = p_{\text{vertical}}\left[\tan^2\left(45° - \frac{\phi}{2}\right)\right] \quad 10.53$$

The quantity in brackets in equation 10.53 is known as the *coefficient of active earth pressure.*

$$k_A = \tan^2\left(45° - \frac{\phi}{2}\right) = \frac{1 - \sin\phi}{1 + \sin\phi} \quad 10.54$$

The general equation for horizontal *passive earth pressure* is:

$$p_{\text{horizontal}} = p_{\text{vertical}} \tan^2\left(45° + \frac{\phi}{2}\right) + 2c\tan\left(45° + \frac{\phi}{2}\right) \quad 10.55$$

The *coefficient of passive earth pressure* for sand is

$$k_P = \frac{1}{k_A} = \tan^2\left(45° + \frac{\phi}{2}\right) = \frac{1 + \sin\phi}{1 - \sin\phi} \quad 10.56$$

A. THE RANKINE THEORY

If it is assumed that the backfill soil is dry, cohesionless sand, then the *Rankine theory* can be used. At any depth, $H$, the vertical pressure is

$$p_{\text{vertical}} = \rho H \quad 10.57$$

The horizontal pressure depends on the *coefficient of earth pressure at rest*, $k_o$, which varies from 0.4 to 0.5 for untamped sand.[18]

$$p_{\text{horizontal}} = k_o \rho H \quad 10.58$$

$$k_O \approx 1 - \sin\phi \quad 10.59$$

$$R_o = \tfrac{1}{2} k_o \rho H^2$$

Equations 10.57 and 10.58 apply only to a sand deposit of infinite depth and extent. For sand that is compressed or tensioned (as in around a retaining wall) the reactions are given by equations 10.60 and 10.61.

$$R_A = (p_{\text{horizontal}})\left(\frac{H}{2}\right) = \frac{1}{2} k_A \rho H^2 \quad 10.60$$

$$R_P = \frac{1}{2} k_P \rho H^2 \quad 10.61$$

$R_A$ and $R_P$ are horizontal if the soil above the heel and toe is horizontal. (See figure 10.19.)

B. WEDGE THEORIES

The Rankine theory is based on infinite, cohesionless soil. It also requires that the soil above the heel be level. Modifications can be made to lift these restrictions, as well as to allow a water table above the foundation base. Several modifications are known as *wedge theories. Coulomb's earth-pressure theory* is one such wedge theory.

The wedge methods are based on the observation that retaining walls fail when the active soil shears. Although the shear plane is actually a slightly curved surface, it is assumed to be linear (line ∗–∗ in figure 10.15). However, since the actual shear plane is not known in advance, several trial planes need to be taken. This is known as the *trial wedge method.*

## 24 SLOPED AND BROKEN SLOPE BACKFILL

It is possible to derive equations for the active force with a sloped backfill, as shown in figure 10.16. However, the complexity of these equations usually makes a graphical solution a better choice.

With sloped or broken slope backfill, the active force is not horizontal. Appendix A and appendix B provide a method of evaluating the horizontal and vertical earth pressure. Notice that $k_h$ and $k_v$ have units of lbf/ft$^2$ per foot of wall. Soil density is not used.

$$R_{A,h} = \frac{1}{2} k_h H^2 \quad 10.62$$

$$R_{A,v} = \frac{1}{2} k_v H^2 \quad 10.63$$

[18] It is appropriate to use the at-rest soil case whenever the wall does not move. Bridge abutments and basement walls are examples where movement is essentially nonexistent.

$$R_A = \sqrt{R_h^2 + R_v^2} \qquad 10.64$$

$$\theta = \arctan\left(\frac{R_v}{R_h}\right) = \beta \qquad 10.65$$

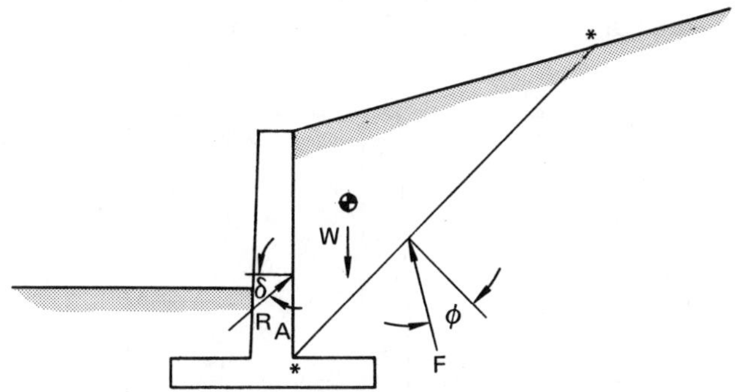


**Figure 10.15** Failure Wedge

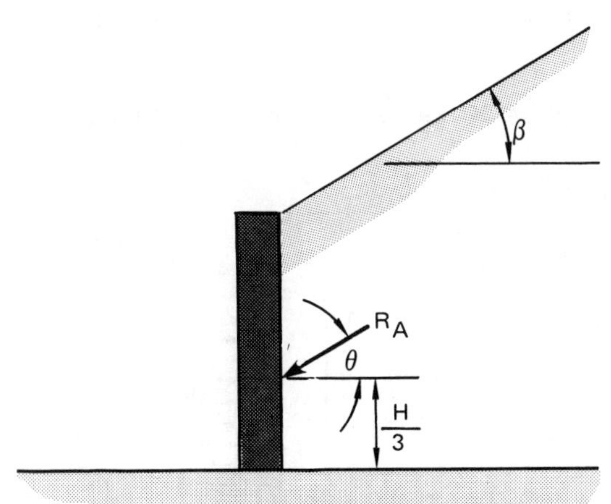


**Figure 10.16** Sloped Backfill

## 25 SURCHARGE LOADING

### A. UNIFORM SURCHARGE

If there is a uniform surcharge load of $q$ lbf/ft$^2$ above the backfill, there will be an additional active reaction, $R_q$. $R_q$ acts at $H/2$ above the base. This reaction is in addition to the regular active reaction which acts at $H/3$.

$$R_q = k_A q H \times \text{wall width} \qquad 10.66$$

### B. POINT LOAD

If a point load is applied a distance $x$ back from the wall face, as in figure 10.18, the distribution of pressure behind the wall can be found from equations 10.67 through 10.70.[19]

[19] These equations are based on elastic theories with a Poisson's ratio of $\mu = 0.5$. The coefficients have been adjusted to bring the theory into agreement with observed values.

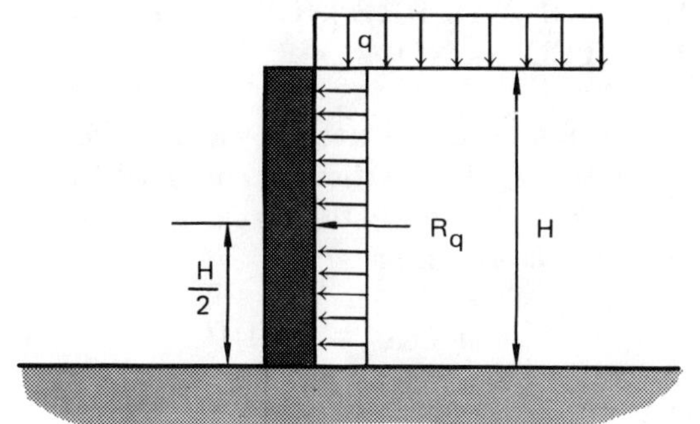


**Figure 10.17** A Uniform Surcharge

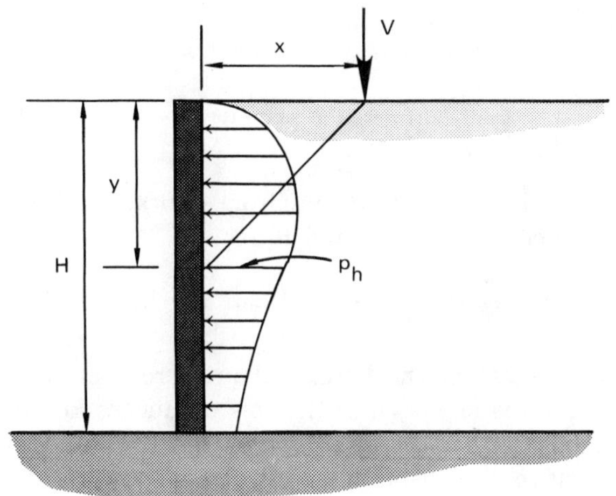


**Figure 10.18** A Point Load Distribution

$$m = \frac{x}{H} \qquad 10.67$$

$$n = \frac{y}{H} \qquad 10.68$$

$$p_h = \frac{1.77V}{H^2}\frac{m^2n^2}{(m^2+n^2)^3} \quad (m > 0.4) \qquad 10.69$$

$$p_h = \frac{0.28V}{H^2}\frac{n^2}{(0.16+n^2)^3} \quad (m \le 0.4) \qquad 10.70$$

### C. LINE LOAD

For a line load of $q$ lbf/ft, the distribution of pressure behind the wall is given by equations 10.71 and 10.72. Figure 10.18 applies if $V$ is replaced with $q$.

$$p_h = \frac{4}{\pi}\frac{q}{H}\left(\frac{m^2 n}{(m^2+n^2)^2}\right) \quad (m > 0.4) \qquad 10.71$$

$$p_h = \frac{q}{H}\frac{0.203n}{(0.16+n^2)^2} \quad (m \le 0.4) \qquad 10.72$$

## 26 HORIZONTAL PRESSURES FROM SATURATED SAND

Sand, being porous, allows trapped water to exert a horizontal pressure against vertical retaining walls. Therefore, the sand and water both exert a horizontal force. The hydrostatic pressure is

$$p_{\text{hydrostatic}} = (62.4)H \qquad 10.73$$

However, there is also a buoyant force, since each sand particle is submerged below the water table. The soil pressure depends on its saturated density and the buoyant force.

$$p_{\text{soil, vertical}} = (\rho_{\text{sat}} - 62.4)H \qquad 10.74$$

$$p_{\text{soil, horizontal}} = k_A(\rho_{\text{sat}} - 62.4)H \qquad 10.75$$

The total horizontal pressure from saturated soil is the sum of equations 10.73 and 10.75.

$$p_{\text{horizontal}} = [62.4 + k_A(\rho_{\text{sat}} - 62.4)]H \qquad 10.76$$

The increase in the horizontal pressure over the dry condition is $(1 - k_A)(62.4)$. This product is known as the *effective hydrostatic loading* of the fluid. If it is convenient to do so, equation 10.76 can be interpreted as a pressure from a fluid with *effective hydrostatic density* $\rho_{\text{eff}}$.[20]

$$\rho_{\text{eff}} = 62.4(1 - k_A) \qquad 10.77$$

$$p_{\text{horizontal}} = [k_A\rho_{\text{sat}} + \rho_{\text{eff}}]H \qquad 10.78$$

The saturated soil density used in the above equations can be calculated if the dry soil density and either the porosity or void ratio is known.

$$\rho_{\text{sat}} = \rho_{\text{dry}} + 62.4n = \rho_{\text{dry}} + 62.4\left(\frac{e}{1+e}\right) \qquad 10.79$$

## 27 RETAINING WALLS

Retaining walls must be safe against settlement. In this regard, their design is similar to footings. They must have sufficient resistance against overturning and sliding. Retaining walls must also possess adequate structural strength. The method of meeting these requirements is one of trial and error.

The analysis of a retaining wall's stability requires knowledge of at least six different distributed force systems. These forces result from the distributions shown in figure 10.20. They are the forward (active or tensioned) earth reaction, $R_A$, the backward (passive or compressed) earth reaction, $R_P$, the soil force, $R_v$, the shear resistance, $R_s$, the weights of the earth masses, $R_T$ and $R_H$, and the weight of the wall itself.

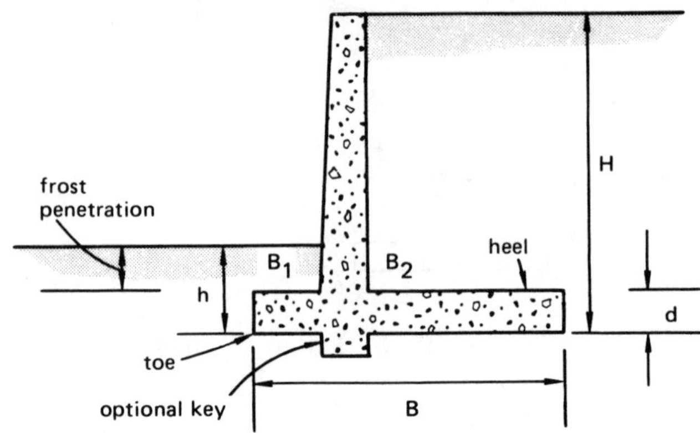


**Figure 10.19** A Cantilever Retaining Wall

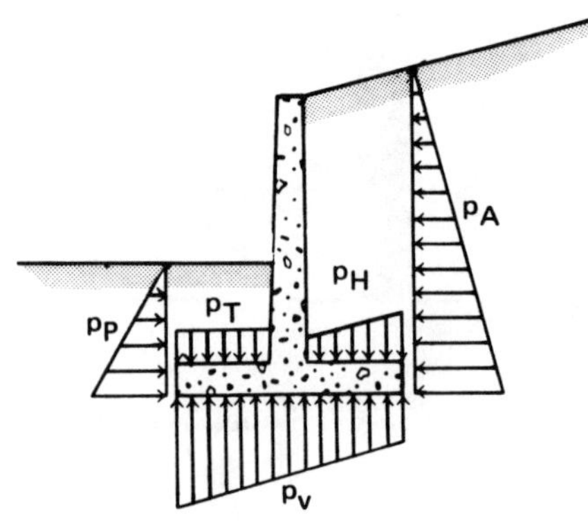


**Figure 10.20** Pressure Distributions on a Retaining Wall

The important factors to be considered when evaluating the design of a retaining wall are factor of safety against overturning, the maximum soil pressure under the base, and the factor of safety against sliding.

The following procedure can be used to analyze a retaining wall.

*step 1*: Determine the active reaction and its point of application. Include the reactions from all point, line, and distributed surcharges.[21]

[20] 45 $lbm/ft^3$ is typically taken as the effective hydrostatic density of the fluid behind a retaining wall.

[21] Retaining walls should be analyzed and designed for a minimum density of 30 $lbm/ft^3$ of fill, regardless of the actual load.

*step 2*: Determine the passive reaction and its point of application.[22]

*step 3*: Find the vertical forces against the base. These forces are the weights, $W_i$, of the areas shown in figure 10.21. Find the centroid of each area and the moment arm, $r_i$, from the centroid of the $i$th area to point $G$.[23]

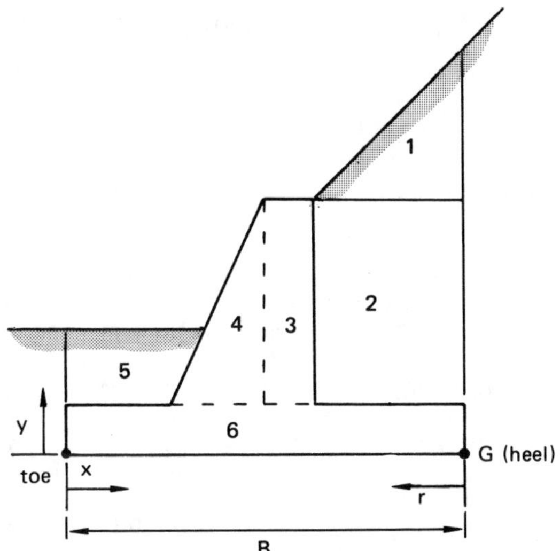


**Figure 10.21** Elements Contributing to Vertical Force

*step 4*: Find the moment about point $G$ of all the vertical forces and the active pressure.

$$M_G = \sum W_i r_i + R_{A,h} r_A \qquad 10.80$$

*step 5*: Find the location and eccentricity of the vertical force resultant. The eccentricity is the distance from the center of the base to the vertical force resultant. Eccentricity should be less than $B/6$ for the entire base to be in compression.

$$r_R = \frac{M_G}{\sum W_i + R_{A,v}} \qquad 10.81$$

$$\epsilon = r_R - \frac{1}{2}B \qquad 10.82$$

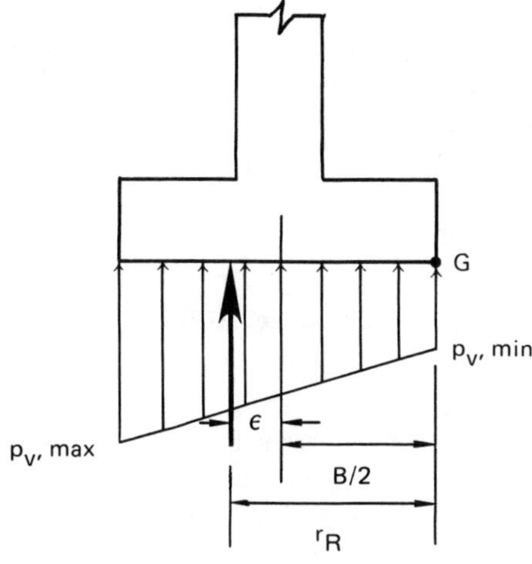


**Figure 10.22** Resultant Forces on Base

*step 6*: Check the *factor of safety against overturning* by summing moments about the toe. The moment arms $x$ and $y$ must be measured from the toe.[24] The factor of safety should generally exceed 1.5 for cohesionless soils, and 2.0 for cohesive soils.

$$F_{\text{overturning}} = \frac{\sum W_i x_i + R_{A,v} x_{A,v}}{R_{A,h} y_{A,h}} \qquad 10.83$$

*step 7*: Find the maximum (at the toe) and minimum (at the heel) foundation pressure on the base.

$$p_{v,\text{max}}, p_{v,\text{min}} = \frac{\sum W_i + R_{A,v}}{B}\left(1 \pm \frac{6\epsilon}{B}\right) \qquad 10.84$$

The maximum pressure should not exceed the allowable soil pressure.

*step 8*: Calculate the *resistance against sliding.* Disregarding the passive pressure, the active pressure must be resisted by the shearing strength of the soil or the friction between the base and the soil. Equation 10.85 is for use when the wall has a key, and then only for the soil to the left of the key. Equation 10.86 is for use with the soil to the right of a key, and for flat-bottomed walls.

$$R_s = \left(\sum W_i + R_{A,v}\right) \tan\phi + c_a B \qquad 10.85$$

$$R_s = \left(\sum W_i + R_{A,v}\right) \tan\delta + c_a B \qquad 10.86$$

[22] The passive pressure is usually disregarded on the assumption that the backfill will be in place prior to the front fill, or that the front fill will be removed at some future date for repairs to the wall.

[23] Moments can also be taken about the toe.

[24] Other interpretations may include $R_{A,v}$ as a negative term in the denominator.

In the absense of friction coefficient data, the resistance can be found from equation 10.87 and table 10.8.

$$R_s = k_s\left(\sum W_i + R_{A,v}\right) + c_a B \qquad 10.87$$

**Table 10.8**
Values of $k_s$

| | |
|---|---|
| coarse grained soil without silt | 0.55 |
| coarse grained soil with silt | 0.45 |
| silt | 0.35 |

*step 9*: Calculate the *factor of safety against sliding*. In cases where the passive pressure is disregarded, a factor of safety in excess of 1.5 is desired. If passive pressure is included in the resistance against sliding, the factor of safety should exceed 2.0. (Neglecting the passive force will give a minimum factor of safety.)

$$F_{\text{sliding}} = \frac{R_s}{R_{A,h}} \qquad 10.88$$

If the factor of safety is too low, increase the base length, $B$, or include a key.

*Example 10.6*

A retaining wall is being designed for a soil with $\phi = 30°$, $\delta = 17°$, and a maximum allowable pressure of 3000 psf. The backfill is coarse-grained sand with silt, with a density of 125 lbm/ft$^3$. Using an adhesion of 950 psi, check the tentative design for stability against sliding. Do not check for factor of safety against overturning.

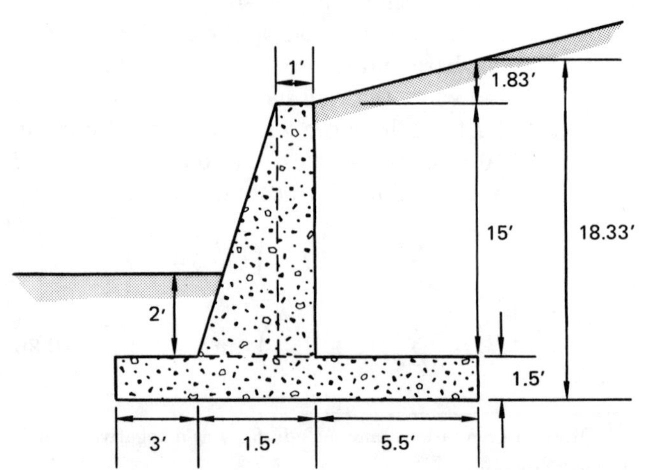


*step 1*: The backfill is sloped. Therefore, assume a type-2 fill and use appendix A.

$$\beta = \arctan\left(\frac{1.83}{5.5}\right) = 18.4° \quad (3:1)$$

From appendix A, $k_v \approx 10$ and $k_h \approx 40$.

$$R_{A,v} = \frac{1}{2}(10)(18.33)^2 = 1680\,\text{lbf}$$

$$R_{A,h} = \frac{1}{2}(40)(18.33)^2 = 6720\,\text{lbf}$$

*step 2*: Assume $R_P = 0$.

*step 3*:

| $i$ | area | $\rho$ | $W_i$(lbf) | $r_i$ | moment | |
|---|---|---|---|---|---|---|
| 1 | $\frac{1}{2}$(5.5)(1.83)=5.03 | 125 | 629 | 1.83 | 1151 | |
| 2 | 5.5(15)=82.5 | 125 | 10,313 | 2.75 | 28,361 | |
| 3 | 1(15)=15 | 150 | 2250 | 6.0 | 13,500 | |
| 4 | $\frac{1}{2}$(0.5)(15)=3.75 | 150 | 563 | 6.67 | 3755 | |
| 6 | 1.5(10)=15 | 150 | 2250 | 5.0 | 11,250 | |
| 5 | 2(3)=6 | 125 | 750 | 8.5 | 6375 | |
| | | | 16,755 | | 64,392 | totals |
| $R_{A,v}$ | | | 1680 | 0 | 0 | |
| $R_{A,h}$ | | | 6720 | 6.11 | 41,060 | |

(Notice $R_{A,v}$ goes through point G.)

*step 4*: $M_G = 64{,}392 + 41{,}060 = 105{,}450$ ft-lbf

*step 5*:

$$r = \frac{105{,}450}{16{,}755 + 1680} = 5.72\,\text{ft}$$

$$\epsilon = 5.72 - \frac{1}{2}(10) = 0.72\,\text{ft}$$

Since 0.72 is less than 10/6, the base is in compression everywhere.

*step 6*: Skipped.

*step 7*: The maximum pressure at the base is

$$p_{\max} = \left(\frac{16,755 + 1680}{10}\right)\left(1 + \frac{6(0.72)}{10}\right) = 2640\,\text{lbf/ft}^2$$

This is less than 3000 lbf/ft$^2$.

*step 8*: The resisting force against sliding is

$$R_s = (16,755 + 1680)\tan 17° + (950)(10) = 15,136\text{ lbf}$$

*step 9*: From equation 10.88,

$$F_{s,\min} = \frac{15,136}{6720} = 2.25\ \text{(O.K.)}$$

## 28 FLEXIBLE ANCHORED BULKHEADS

*Anchored bulkheads* are supported at their bases by having been sunk into the ground. They are anchored further up with rods projecting back into the soil. These rods can terminate at deadmen, piles, walls, or beams.

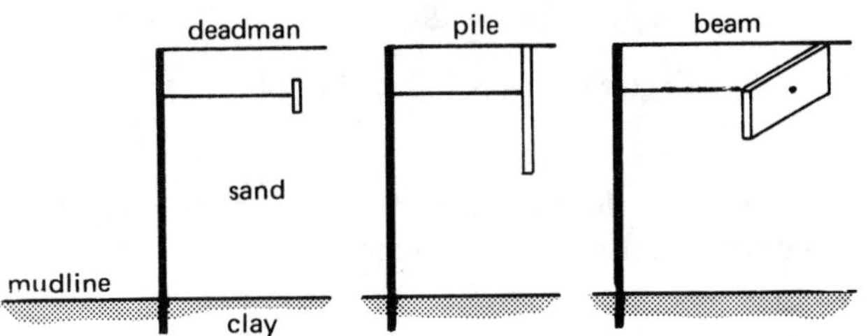


**Figure 10.23** Anchored Bulkheads

Anchored bulkheads can fail in the following ways:

- The base clay can fail due to inadequate bearing capacity. The clay will shear along a circular arc passing under the bulkhead.
- The anchorage can fail. There are several ways for this to occur, including rod tension failure and deadman movement.
- The toe embedment at the mud line can fail.
- The sheeting can fail, although this is rare.

The depth of embedment that is required is found by taking moments about the anchor attachment point on the bulkhead. This anchor pull is found by summation of all horizontal loads on the bulkhead.

The maximum bending moment in the bulkhead itself should be found by taking moments about a point of counterflexure listed in table 10.9.

**Table 10.9**
Points of Counterflexure

| embedment material | point |
|---|---|
| firm and dense | mud line |
| loose and weak | 1 or 2 feet below mud line |
| soft over hard layer | at hard layer depth |

## 29 BRACED CUTS

### A. INTRODUCTION

A braced cut is an excavation in which the load from one bulkhead is used to support the opposite bulkhead's load. Failure in dry soils above the water line generally occurs by wale crippling followed by strut buckling. Planned excavations below the water line should be dewatered prior to cutting. The analysis of braced cuts is approximate due to extensive bending of the sheeting.

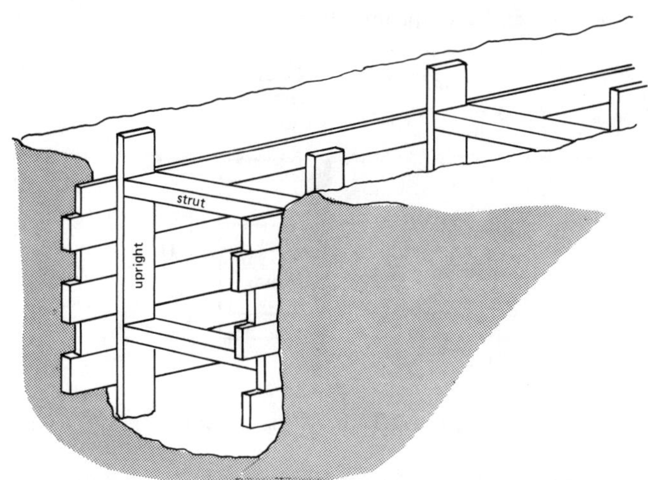


**Figure 10.24** A Braced Cut with Box Shoring

Since the struts are installed as the excavation goes down, the upper part of the wall deflects very little due to the strut restraints. Therefore, the final pressure on the upper part of the wall will be considerably higher than would be normally predicted by the active pressure equation. These larger than expected loads near the top have resulted in failure of the upper strut in braced cuts.

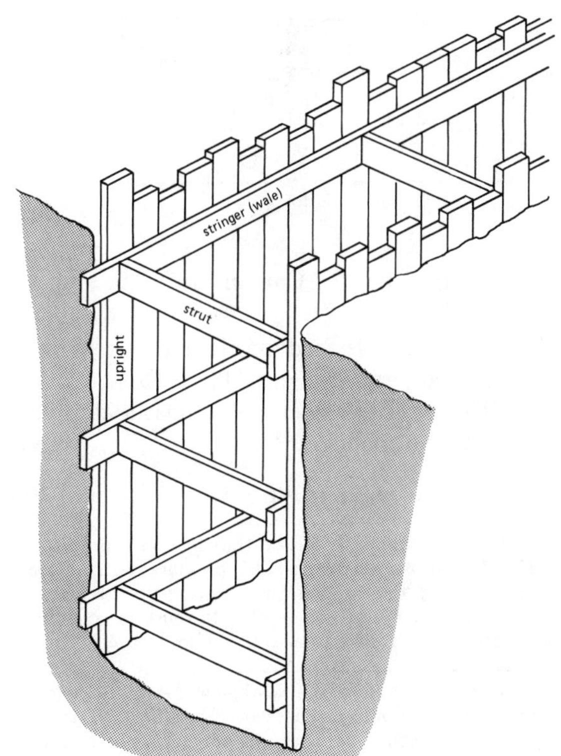


**Figure 10.25** A Braced Cut with Close Sheeting

## B. PRESSURE DISTRIBUTIONS IN BRACED CUTS

The pressure distribution depends on the cut material. For dry or drained sand, the pressure distribution is similar to that in figure 10.26.[25] The maximum pressure is given by equation 10.89.

$$p_{\text{max}} = (0.65)k_A \rho H \qquad 10.89$$

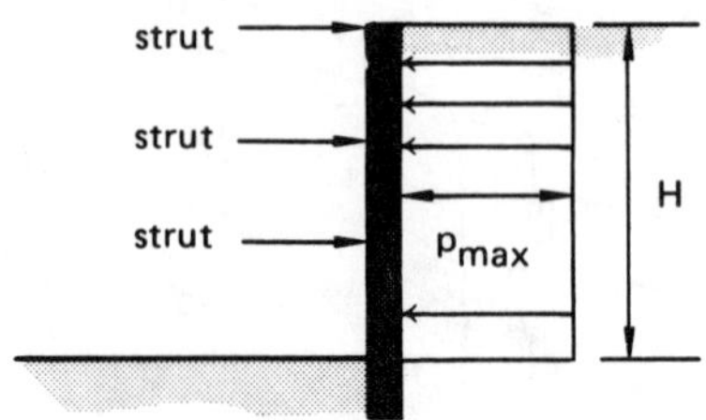


**Figure 10.26** Cuts in Sand

For undrained clay (typical of cuts made rapidly in comparison to drainage times) where $\phi = 0°$, the pressure envelope depends on the average undrained shear strength of the clay in the cut zone. If $\rho H/c \leq 4$, (e.g., stiff clay), the pressure envelope is shown by figure 10.27.

$$0.2\rho H \leq p_{\text{max}} \leq 0.4\rho H \qquad 10.90$$

Use the lower values of $p_{\text{max}}$ when movement is minimal or when the construction period is short.

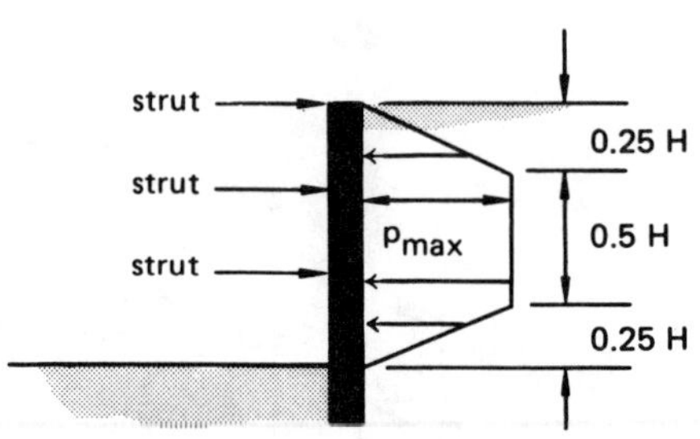


**Figure 10.27** Cuts in Stiff Clay

If $\rho H/c \geq 6$ (e.g., soft to medium clay), the pressure distribution is as appears in figure 10.28. Except for cuts underlain by deep, soft, normally consolidated clays, equation 10.91 can be used.[26]

$$p_{\text{max}} = \rho H - 4c \qquad 10.91$$

[25] This is not the only distribution that has been proposed for sand. The Tschebotarioff trapezoidal pressure distribution increases (starting at the surface and working down) from 0 to 0.8 $k_A \rho H$ in the first $0.1H$ of depth. It remains constant for $0.7H$, and decreases linearly to zero in the lower $0.2H$.

[26] If the shearing strength of the clay below the cut, $S_b$, is known, the quantity $\rho H/S_b$ should be checked. If it is below 6, the bearing capacity of the soil is sufficient to prevent shearing and upward heave. Simple braced cuts should not be attempted if this quantity exceeds 8.

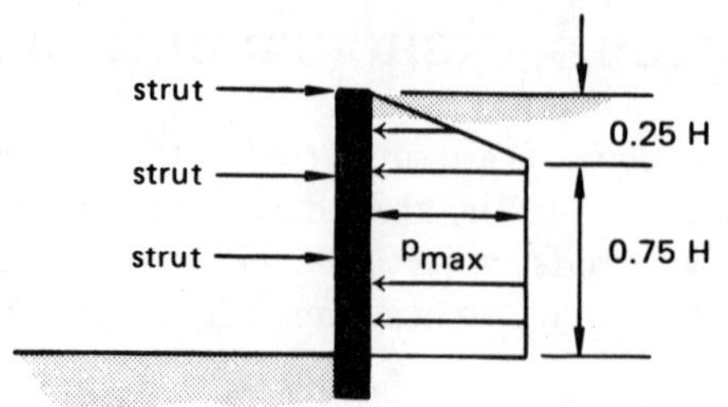


**Figure 10.28** Cuts in Soft Clay

If $4 < \rho H/c < 6$, use either the soft or stiff clay case, whichever results in the stronger design.

## C. DESIGN OF BRACED CUTS

The first step in the design of braced cuts is to determine the pressure distribution based on equations 10.60 and 10.76. The passive pressure reduces the active pressure below the mudline. The point at which the active pressure distribution is zero is taken as the hinge point. The passive pressure below the hinge point is disregarded.[27]

If there are no struts, the wall is designed as a cantilever beam.

If there is one strut, moments are taken about the hinge point to determine the strut load.

With multiple struts, the moment distribution method can be used to determine strut loads. An easier method is to assign portions of the active distribution to the struts based on areas between the mid-points of the strut separation distances.[28]

Once a tentative design has been reached, the design should be checked using the distributions presented in the previous section. Strut loads are computed by tributary areas. Vertical members are designed as beams on unyielding supports.

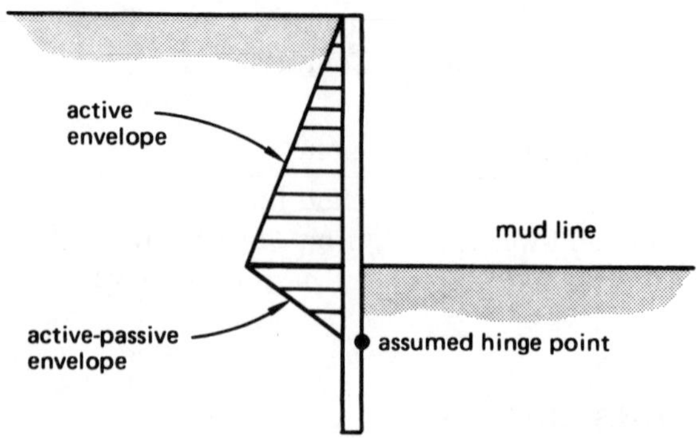


**Figure 10.29** Preliminary Design of Braced Cuts

[27] Passive pressures greater than the active pressure are disregarded.

[28] That is, the sheeting is assumed to be hinged at the strut points.

# Appendix A: Active Components for Retaining Walls with Straight Slope Backfill

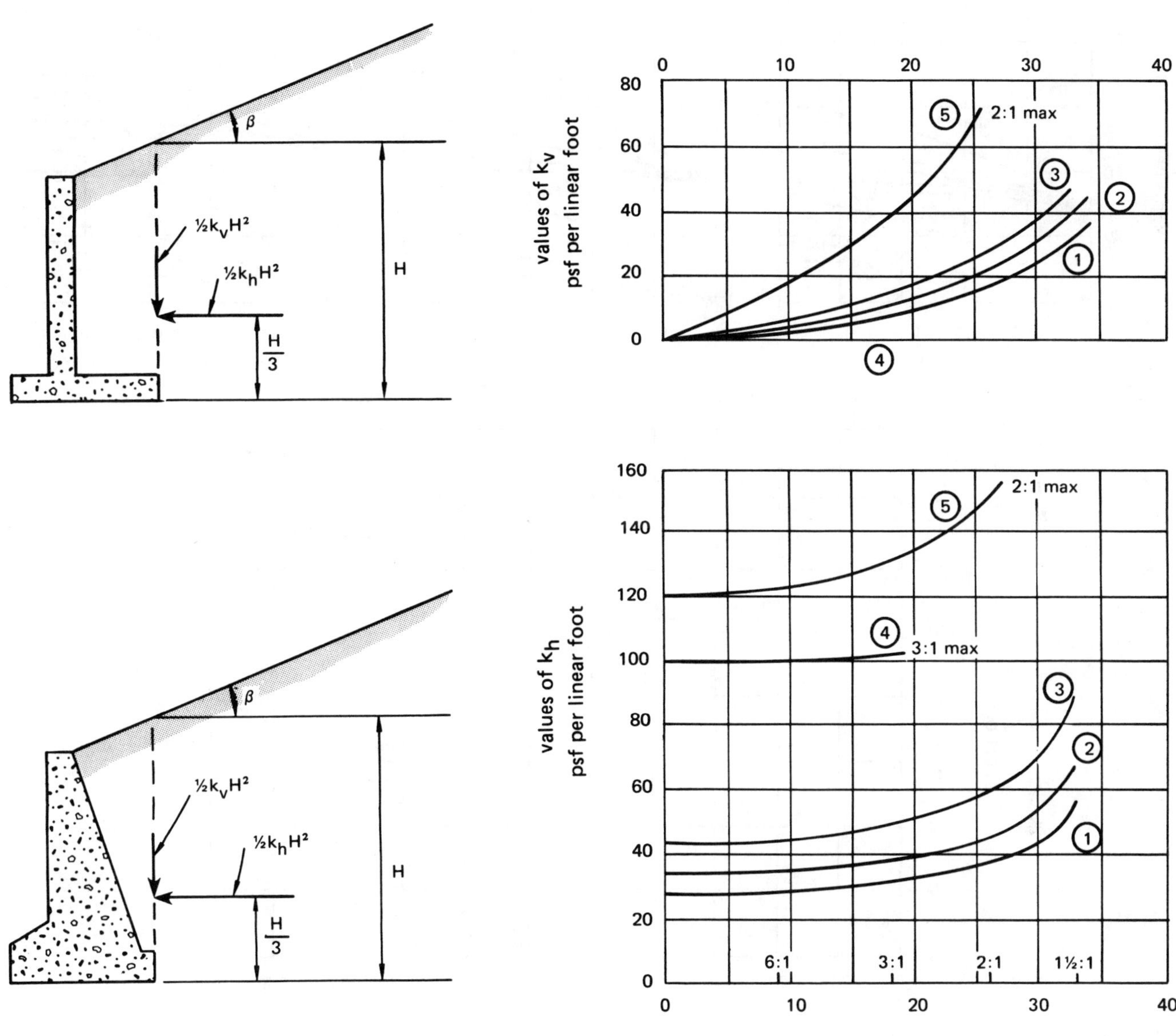


Circled numbers indicate the following soil types.

1. Clean sand and gravel: GW, GP, SW, SP.
2. Dirty sand and gravel of restricted permeability: GM, GM–GP, SM, SM–SP.
3. Stiff residual silts and clays, silty fine sands, and clayey sands and gravels: CL, ML, CH, MH, SM, SC, GC.
4. Very soft to soft clay, silty clay, organic silt and clay: CL, ML, OL, CH, MH, OH.
5. Medium to stiff clay deposited in chunks and protected from infiltration: CL, CH.

For type 5 material, H is reduced by 4 feet. The resultant is assumed to act (H–4)/3 above the bottom of the base.

# Appendix B: Active Components for Retaining Walls with Broken Slope Backfill

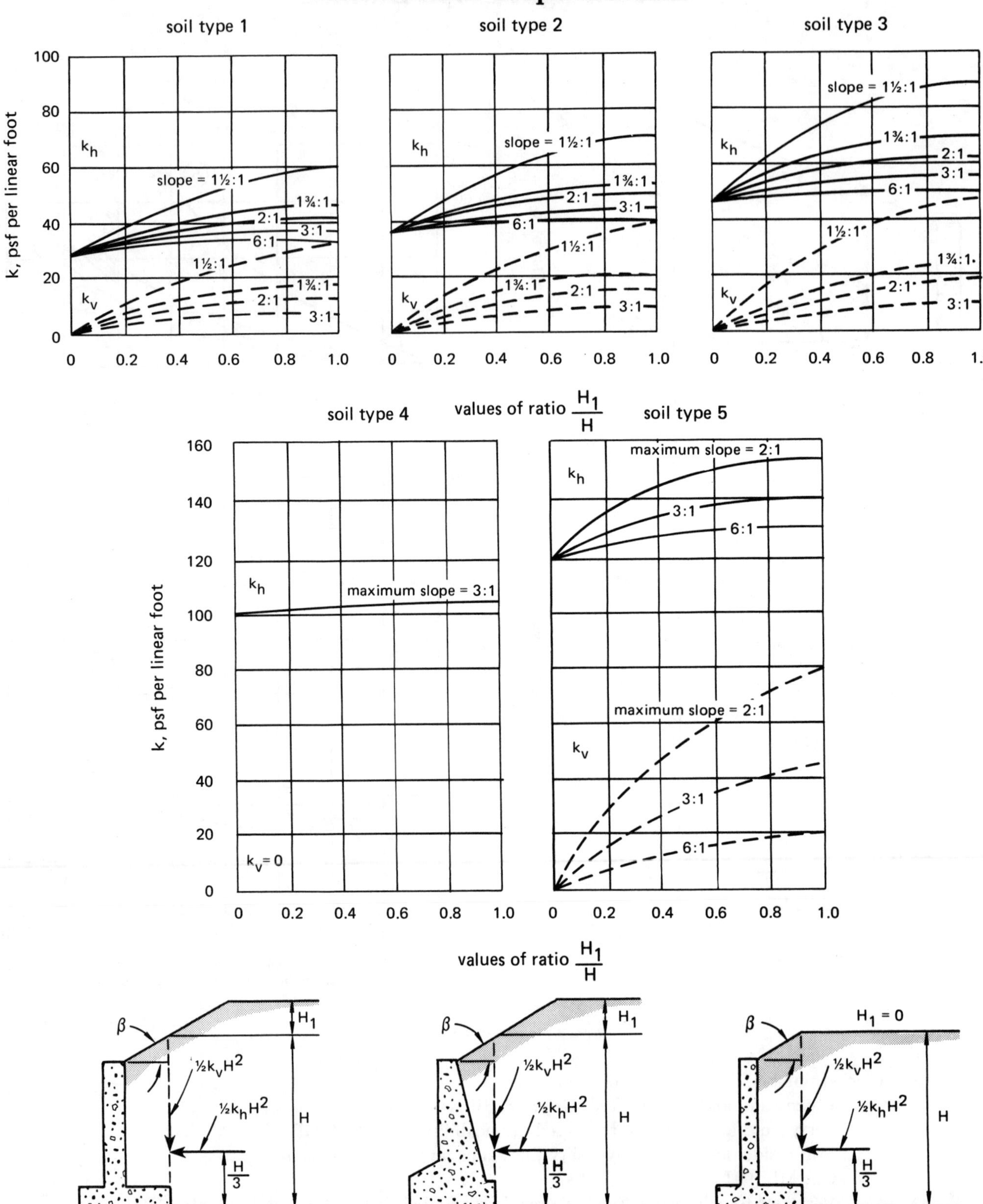


for type 5 material H is reduced by 4 feet, resultant acts at a height of $\frac{(H - 4)}{3}$ above base

# Appendix C: Boussinesq Stress Contour Charts Infinitely Long and Square Footings

$p$ = uniform foundation pressure

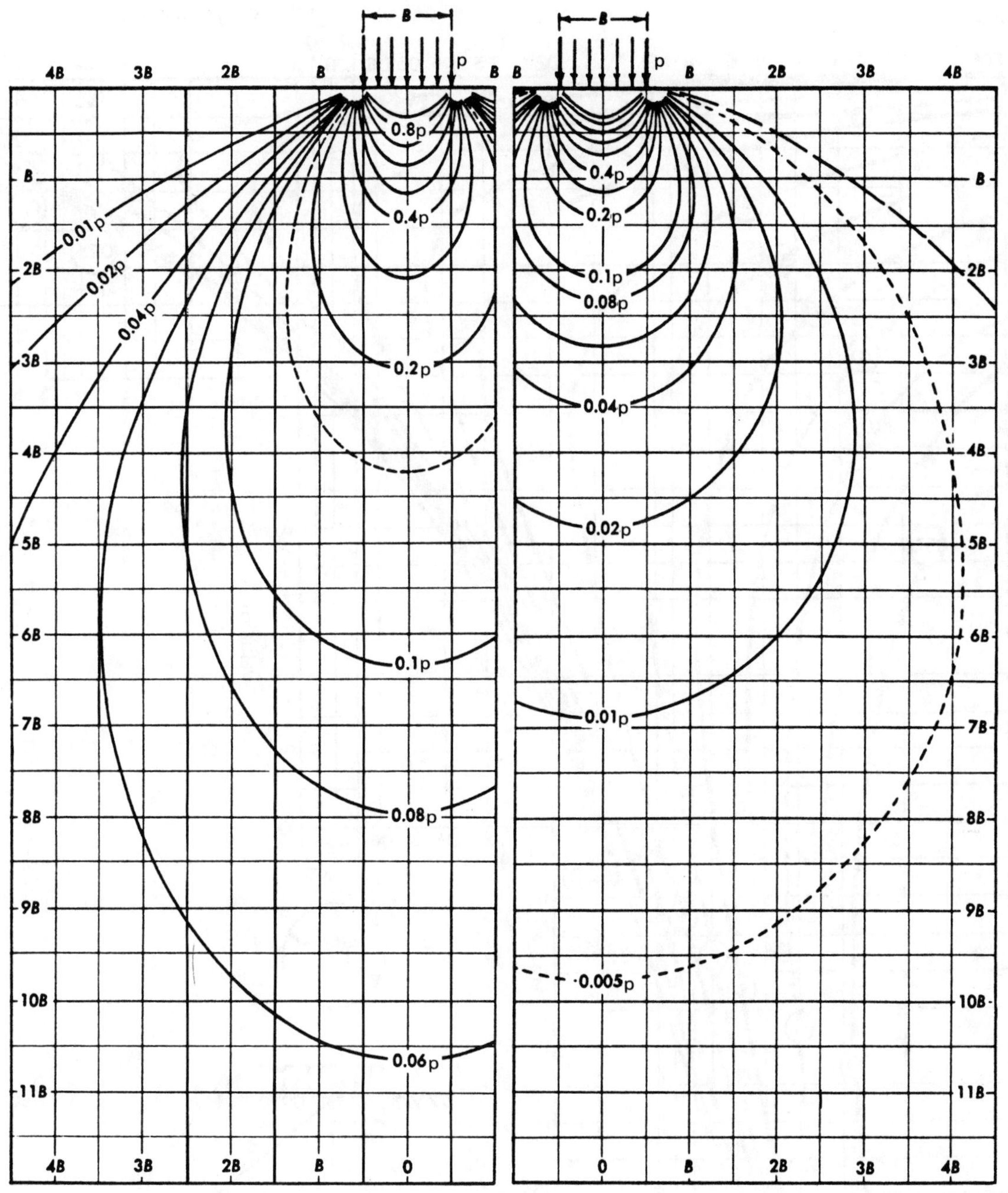


infinitely long foundation

square foundation

# Appendix D: Boussinesq Stress Contour Chart Uniformly Loaded Circular Footings

$p$ = uniform foundation pressure

$\sigma_{z,\text{vertical}} = p \times \text{chart value}$

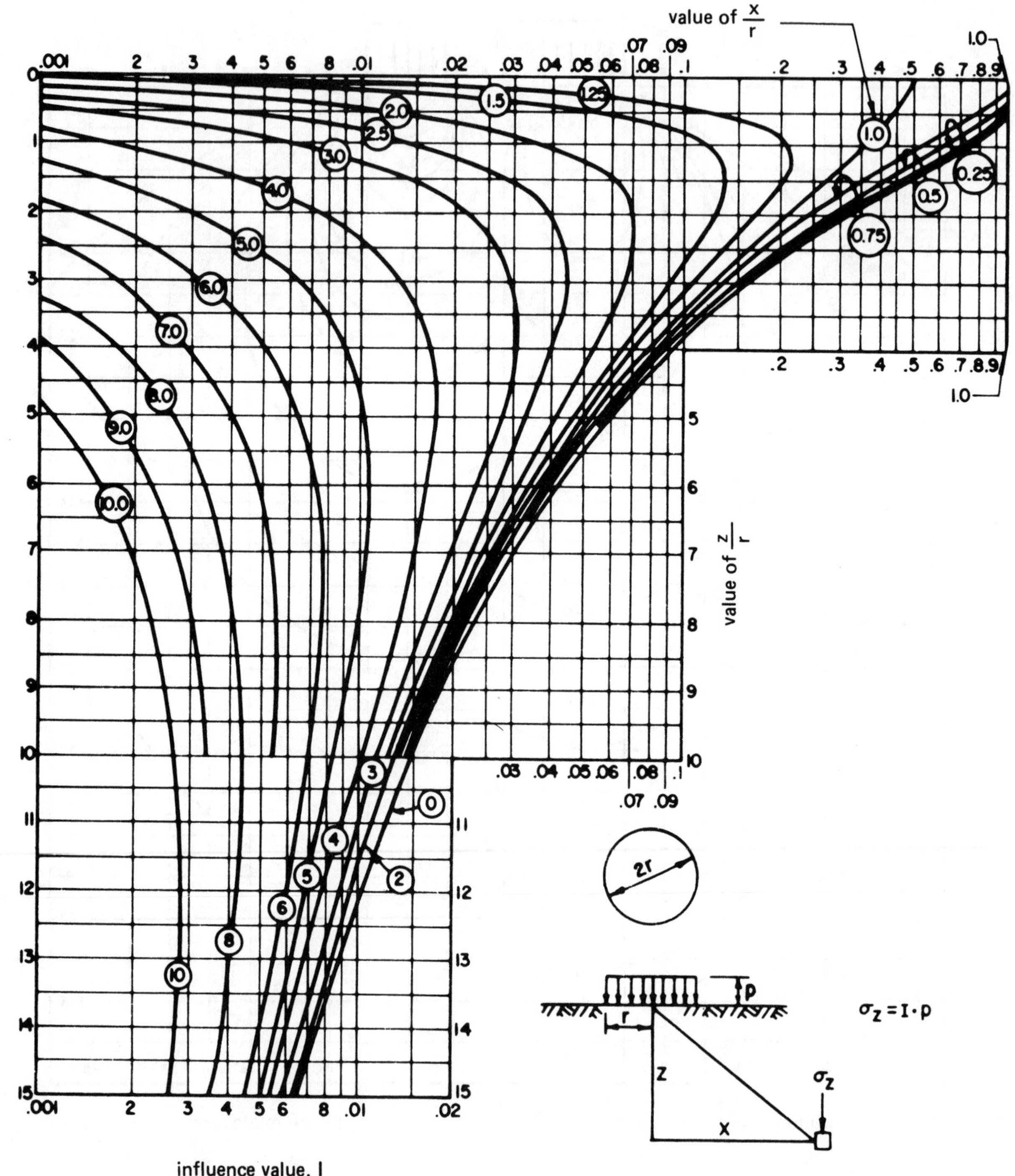

## Practice Problems: FOUNDATIONS AND RETAINING WALLS

Untimed

1. A large manufacturing plant with a floor load of 500 psf is supported on a large mat foundation. A 38 foot thick layer of silty soil is underlain by sand and gravel. The plant was constructed on the silty soil 10 feet above the water table. After a number of years and after all settlement of the building had stopped, a series of wells were drilled which dropped the water table from 10 feet below the surface to 18 feet below the surface. The unit weight of the soil above the water table is 100 pcf. It is 120 pcf below the water table. The silt compression index is .02. The void ratio before the water table was lowered was 0.6. How much can the building be expected to settle?

2. 360 kips are to be supported by a square footing. The footing is to rest directly on sand which has a density of 121 pcf and an angle of internal friction of 38°. (a) Size the footing. (b) Size the footing if it is to be placed 4 feet below the surface.

3. A reinforced concrete retaining wall is to be 14 feet high. The top surface is horizontal but supports a surcharge of 500 psf. The soil has the following characteristics:

density: 130 pcf (sandy soil)
angle of internal friction: 35°
coefficient of friction with concrete: .5
allowable soil pressure: 4500 psf
frost line: 4 feet below grade

Use a factor of safety of 1.5 against sliding and overturning to design the dimensions of the retaining wall. Neglect structural details.

4. A 26 foot high wall holds back sand with a 96 pcf dry density. The water table is permanently 10 feet below the top of the wall. The saturated density of the sand is 121 pcf. The angle of internal friction is estimated to be 36°. (a) Disregarding capillary rise, calculate the active earth resultant. (b) Where is the resultant active earth resultant located? (c) Assuming the water table level could be dropped 16 feet to the bottom of the wall, what would be the reduction in overturning moment?

5. The compression index of a soil is 0.31. When the effective stress on the soil is 2600 psf, the void ratio is 1.04 and the permeability is 4 EE-7 mm/sec. The stress is increased gradually to 3900 psf. (a) What is the change in the void ratio? (b) Compute the settlement of a 16 foot layer. (c) How much time is required for the settlement to be 75% complete?

6. A 30 foot deep excavation is being planned for sand ($\phi = 40°$, density of 121 pcf) with a drained water table during construction. The excavation will be 40 feet square. The bracing is to consist of horizontal lagging supported by 8-inch H-beam soldier piles. (a) What is the pressure diagram? (b) If the soldier piles are 8 feet apart, what bending moment is placed on the lagging?

7. A 2:1 (horizontal:vertical) slope is cut in homogeneous, saturated clay which has a density of 112 pcf and a shear strength of 1.1 EE3 psf. The cut is 43 feet deep, and the clay extends 15 feet below the cut bottom to a rock layer. Compute the safety of this slope.

8. What is the increase in stress at a depth of 10 feet and 8 feet from the center of a 10 foot square footing that exerts a pressure of 3000 psf on the soil?

9. A concentrated vertical load of 6000 pounds is applied at the ground surface. What is the vertical pressure caused by this load at a point 3.5 feet below the surface and 4 feet from the action line of the force?

10. A 10.75" O.D. steel pile is driven 65 feet into a stiff, insensitive clay with a shear strength of 1.3 EE3 psf. The pile has a flat-plate, closed end. The soil density is 115 pcf, and the water table is at the ground surface. The relative density is 75%. What is the ultimate bearing capacity?

Timed

1. Use Terzaghi's equations to size the three foundations illustrated below. The soil is sandy clay with the following characteristics: density – 108 lbm/ft³; angle of internal friction – 25°; cohesion – 400 psf. The water table is 35 feet below the ground surface. Use a safety factor of 2.5. Neglect footing weight.

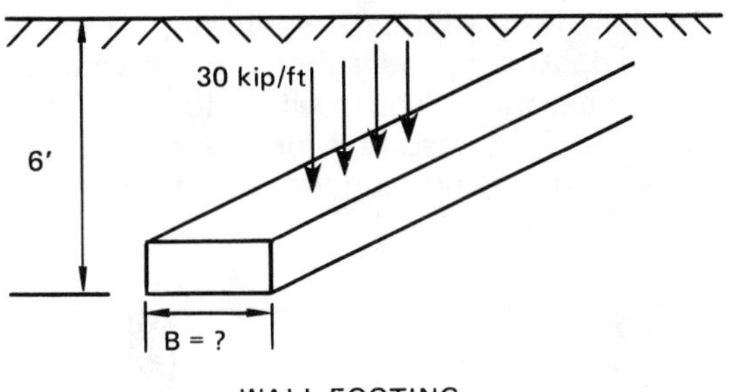


WALL FOOTING

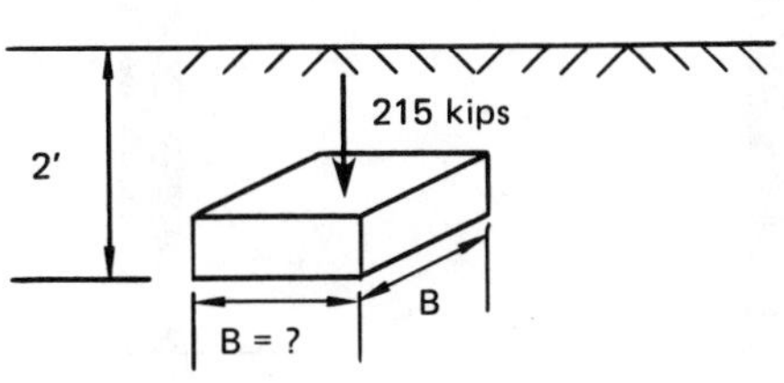


SQUARE FOOTING

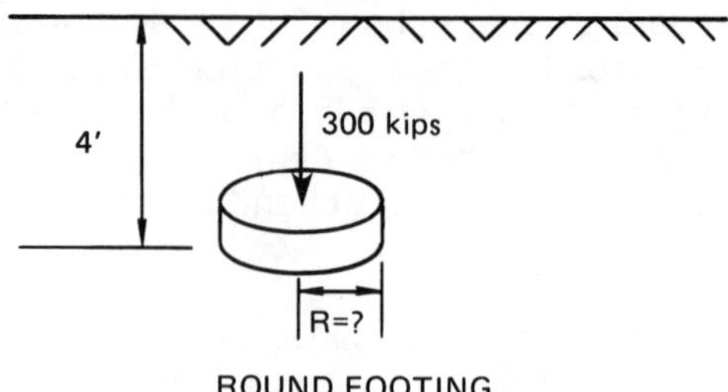

ROUND FOOTING

2. A construction firm wants to consolidate a clay layer at a building site. The soil consists of 8 feet of soft clay on top of a rock layer. The clay is covered with 15 feet of silty sand. The water table is 18 feet above the rock layer. It is proposed to consolidate the clay layer by surcharging the site with 10 more feet of sandy fill (density = 110 pcf) and by dewatering the sand 5 feet (i.e., lowering the water table 5 feet). What is the settlement caused by the surcharging and dewatering?

SITE AS FOUND

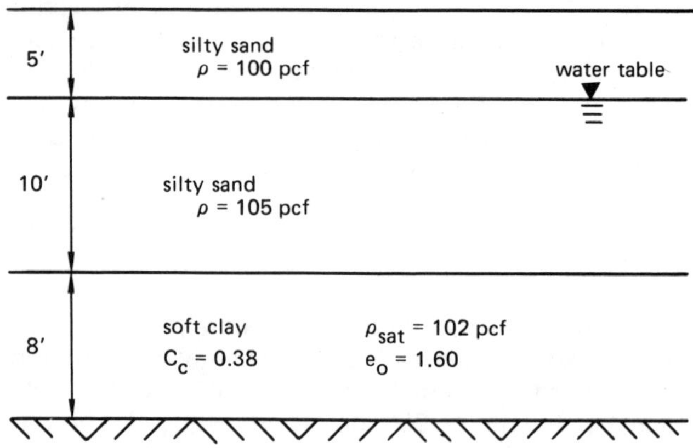


3. A retaining wall is designed for free-draining granular backfill with adequate subdrains. After several years of operation, the subdrains become plugged and the water table rises to within 10 feet of the top of the wall. Find the resultant force and its location for both the drained and plugged conditions.

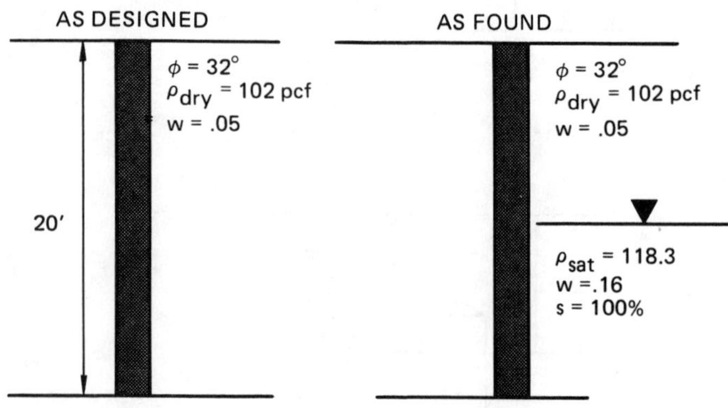


4. A sheet pile is driven through 10 feet of clay to support a 25 foot vertical cut through sand. The total height of the sheet pile is 35 feet, and bedrock is located below that depth. No penetration of the bedrock is made. A tieback is located 8 feet below the surface, and it terminates at a deadman. There is no significant water table. What is the resistive force in the tie rod?

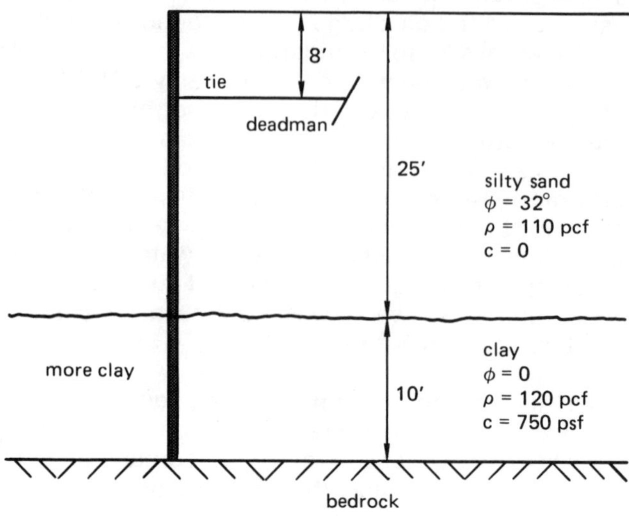


5. A test has shown that a typical pile in the group of 36 piles shown has an uplift capacity of 150 tons and a compressive capacity of 500 tons. The pile grouping is capped by a thick concrete slab producing an axial load of 600 tons. (a) Find the maximum moment that the pile can take in the x- and y-directions. (b) If the concrete slab is separated along the y-axis into two separate pieces, and then reattached by a steel plate, what is the maximum moment that the pile group can take about the y-axis?

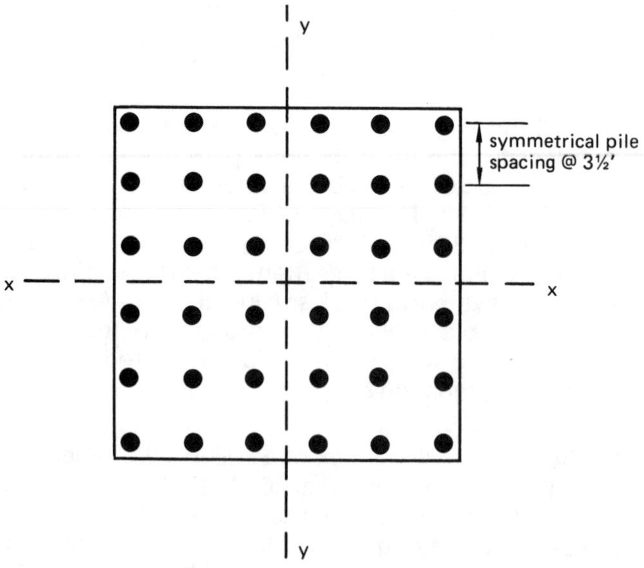


6. A small PVC pipe (2" diameter) is to be buried in a trench 18 inches wide and 3½ feet deep. The angle of internal friction of the cohesive soil is 18°, cohesion is 200 psf. The moisture content is 20%, dry density is 115 pcf. (a) Will the soil stand or

slide during trenching if the spoils are placed directly alongside the trench sides? (b) How far away from the trench must the spoils be placed to satisfy OSHA? (c) When the spoils are kept at the distance given in (b), will the soil remain stable? (d) Recommend initial bedding and backfill materials. (e) What additional precautions would you recommend to protect the workers?

7. A building weighing 20 tons is placed on a basement slab as shown. The soil below the basement is dense sand of 100 feet thickness. Find the pressure produced by the building weight (a) at a depth 30 feet below point A, and (b) at a depth of 45 feet below point B.

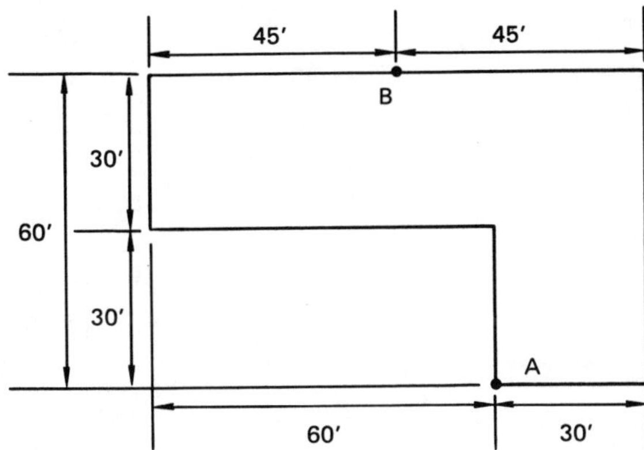

RESERVED FOR FUTURE USE

# 11 STATICS

*Nomenclature*

| | | |
|---|---|---|
| a | horizontal distance from point of maximum sag | ft |
| A | area | $ft^2$ |
| c | parameter in catenary equations | ft |
| d | distance | ft |
| E | modulus of elasticity | $lbf/ft^2$ |
| f | coefficient of friction | – |
| F | vertical force | lbf |
| H | horizontal component of tension | lbf |
| I | moment of inertia | $ft^4$ |
| J | polar moment of inertia | $ft^4$ |
| L | length, cable length from point of maximum sag | ft |
| m | mass | slugs |
| M | moment | ft-lbf |
| N | normal force | lbf |
| p | pressure | $lbf/ft^2$ |
| P | load, or product of inertia | lbf, $ft^4$ |
| r | radius of gyration | ft |
| S | maximum cable sag | ft |
| T | tension, or temperature | lbf, °F |
| w | load per unit weight, or weight | lbf/ft, lbf |

*Symbols*

| | | |
|---|---|---|
| $\alpha$ | coefficient of linear thermal expansion | 1/°F |
| $\delta$ | deflection | ft |
| $\rho$ | density | $lbm/ft^3$ |

*Subscripts*

| | |
|---|---|
| c | centroidal, or concrete |
| i | the ith component |
| R | resultant |
| s | steel |

## 1 CONCENTRATED FORCES AND MOMENTS

Forces are vector quantities having magnitude, direction, and location in 3-dimensional space. The direction of a force **F** is given by its *direction cosines*, which are cosines of the true angles made by the force vector with the $x$, $y$, and $z$ axes. The components of the force are given by equations 11.1, 11.2, and 11.3.

$$\mathbf{F}_x = \mathbf{F}\,(\cos\theta_x) \tag{11.1}$$

$$\mathbf{F}_y = \mathbf{F}\,(\cos\theta_y) \tag{11.2}$$

$$\mathbf{F}_z = \mathbf{F}\,(\cos\theta_z) \tag{11.3}$$

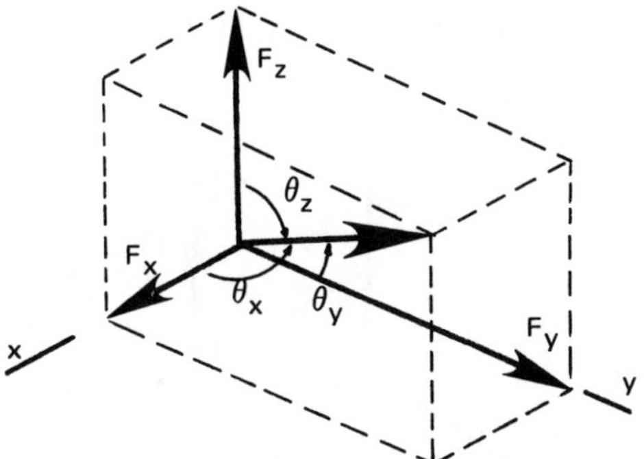


**Figure 11.1** Components of a Force **F**

A force which would cause an object to rotate is said to contribute a *moment* to the object. The magnitude of a moment can be found by multiplying the magnitude of the force times the appropriate moment arm. That is, **M**=**F**·d.

The *moment arm* is a perpendicular distance from the force's line of application to some arbitrary reference point. This reference point should be chosen to eliminate one or more unknowns. This can be done by choosing the reference as a point at which unknown reactions are applied.

Moments also can be treated as vector quantities, and they are shown as double-headed arrows. Using the *right-hand rule* as shown, the direction cosines again are used to give the $x$, $y$, and $z$ components of a moment vector.[1]

$$\mathbf{M}_x = \mathbf{M}(\cos\theta_x) \quad 11.4$$

$$\mathbf{M}_y = \mathbf{M}(\cos\theta_y) \quad 11.5$$

$$\mathbf{M}_z = \mathbf{M}(\cos\theta_z) \quad 11.6$$

$$|\mathbf{M}| = \sqrt{\mathbf{M}_x^2 + \mathbf{M}_y^2 + \mathbf{M}_z^2} \quad 11.7$$

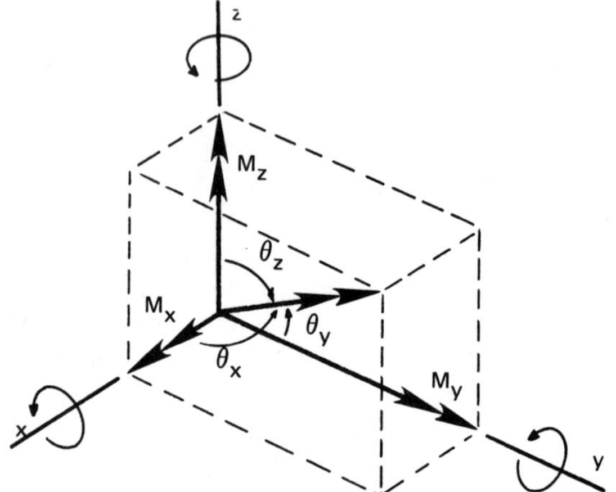


**Figure 11.2** Components of a Moment **M**

Moment vectors have the properties of magnitude and direction, but not of location (point of application). Moment vectors can be moved from one location to another without affecting the equilibrium of solid bodies.

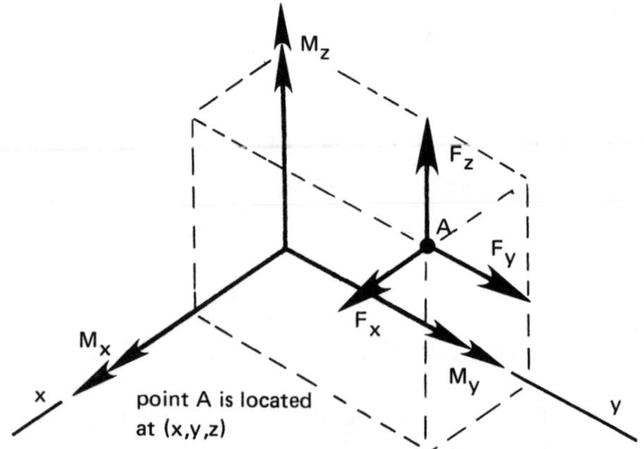


**Figure 11.3** Coordinates of a Point A

If a force is not parallel to an axis, it produces a moment around that axis. The moment is evaluated by finding the components of the force and their respective distances to the axis. In figure 11.3, a force acts through point A located at $(x, y, z)$ and produces moments given by equations 11.8, 11.9, and 11.10.

$$\mathbf{M}_x = y\mathbf{F}_z - z\mathbf{F}_y \quad 11.8$$

$$\mathbf{M}_y = z\mathbf{F}_x - x\mathbf{F}_z \quad 11.9$$

$$\mathbf{M}_z = x\mathbf{F}_y - y\mathbf{F}_x \quad 11.10$$

Any two equal, opposite, and parallel forces constitute a *couple*. A couple is statically equivalent to a single moment vector. In figure 11.4, the two forces, $\mathbf{F}_1$ and $\mathbf{F}_2$, of equal magnitude produce a moment vector $\mathbf{M}_z$ of magnitude **F**y. The two forces can be replaced by this moment vector which then can be moved to any location on the object.

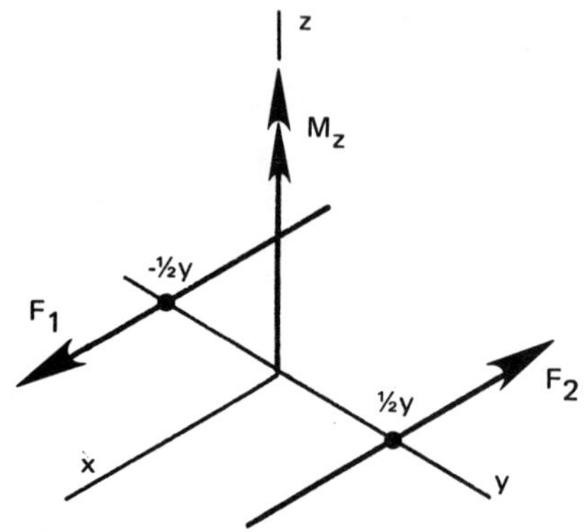


**Figure 11.4** A Couple

## 2 DISTRIBUTED LOADS

If an object is loaded by its own weight or by another type of continuous loading, it is said to be subjected to a *distributed load.* Provided that the load per unit length, $w$, is acting in the same direction everywhere, the statically equivalent concentrated load can be found from equation 11.11 by integrating over the line of application.

$$\mathbf{F}_R = \int w\,dx \quad 11.11$$

The location of the resultant is given by equation 11.12.

$$\overline{x} = \frac{\int (wx)\,dx}{\mathbf{F}_R} \quad 11.12$$

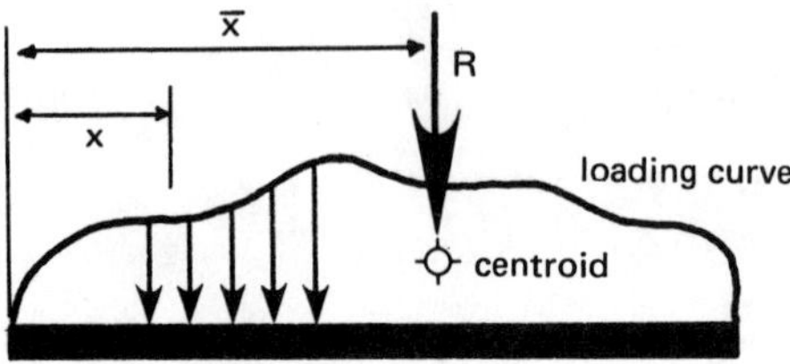


**Figure 11.5** A Distributed Load and Resultant

[1] The right-hand rule: Close your right hand in such a way that your fingers curl in the direction of the force. Your thumb will point in the direction of the moment.

In the case of a straight beam under *transverse loading*, the magnitude of $\mathbf{F}$ equals the area under the loading curve.[2] The location of the resultant force coincides with the centroid of that area. If the distributed load is uniform so that $w$ is constant along the beam,

$$\mathbf{F} = wL \tag{11.13}$$

$$\overline{x} = \tfrac{1}{2}L \tag{11.14}$$

If the distribution is triangular and increases to a maximum of $w$ pounds per unit length as $x$ increases,

$$\mathbf{F} = \tfrac{1}{2}wL \tag{11.15}$$

$$\overline{x} = \tfrac{2}{3}L \tag{11.16}$$

*Example 11.1*

Find the magnitude and the location of the resultant of the distributed loads on each span of the beam shown.

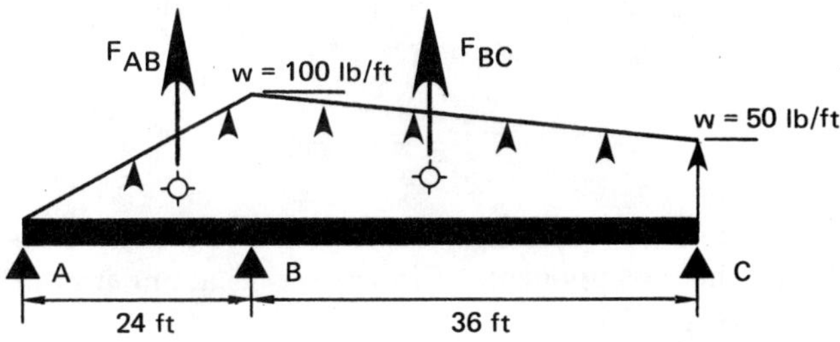


*For span A-B*: The area under the loading curve is $\frac{1}{2}(100)(24) = 1200$ pounds. The centroid of the loading triangle is $\left(\frac{2}{3}\right)(24) = 16$ feet from point $A$. Therefore, the triangular load on the span A-B can be replaced (for the purposes of statics) with a concentrated load of 1200 pounds located 16 feet from the left end.

*For span B-C*: The area under the loading curve is

$$(50)(36) + \tfrac{1}{2}(50)(36) = 2700 \text{ lbs}$$

The centroid of the trapezoid is

$$\frac{36\left[(2)(100) + 50\right]}{3\,(100 + 50)} = 20 \text{ ft from pt } C$$

Therefore, the distributed load on span B-C can be replaced (for the purposes of statics) with a concentrated load of 2700 pounds located 20 feet to the left of point $C$.

[2] Loading is said to be *transverse* if its line of action is perpendicular to the length of the beam.

## 3 PRESSURE LOADS

Hydrostatic pressure is an example of a pressure load that is distributed over an area. The pressure is denoted as $p$ pounds per unit area of surface. It is normal to the surface at every point. If the surface is plane, the statically equivalent concentrated load can be found by integrating over the area. The resultant is numerically equal to the average pressure times the area. The point of application will be the centroid of the area over which the integration was performed.

## 4 RESOLUTION OF FORCES AND MOMENTS

Any system (collection) of forces and moments is statically equivalent to a single resultant force vector plus a single resultant moment vector in 3-dimensional space. Either or both of these resultants may be zero.

The $x$-component of the resultant force is the sum of all the $x$-components of the individual forces, and similarly for the $y$- and the $z$-components of the resultant force.

$$\mathbf{F}_{Rx} = \sum \mathbf{F}_i(\cos\theta_{x,i}) \tag{11.17}$$

$$\mathbf{F}_{Ry} = \sum \mathbf{F}_i(\cos\theta_{y,i}) \tag{11.18}$$

$$\mathbf{F}_{Rz} = \sum \mathbf{F}_i(\cos\theta_{z,i}) \tag{11.19}$$

The determination of the resultant moment vector is more complex. The resultant moment vector includes the moments of all system forces around the reference axes plus the components of all system moments.

$$\mathbf{M}_{Rx} = \sum(y_i\mathbf{F}_{z,i} - z_i\mathbf{F}_{y,i}) + \sum \mathbf{M}_i(\cos\theta_{x,i}) \tag{11.20}$$

$$\mathbf{M}_{Ry} = \sum(z_i\mathbf{F}_{x,i} - x_i\mathbf{F}_{z,i}) + \sum \mathbf{M}_i(\cos\theta_{y,i}) \tag{11.21}$$

$$\mathbf{M}_{Rz} = \sum(x_i\mathbf{F}_{y,i} - y_i\mathbf{F}_{x,i}) + \sum \mathbf{M}_i(\cos\theta_{z,i}) \tag{11.22}$$

## 5 CONDITIONS OF EQUILIBRIUM

An object which is not moving is said to be static. All forces on a static object are in equilibrium. For an object to be in equilibrium, it is necessary that the resultant force vector and the resultant moment vectors be equal to zero.

$$\mathbf{F}_R = \sqrt{\mathbf{F}_{Rx}^2 + \mathbf{F}_{Ry}^2 + \mathbf{F}_{Rz}^2} = 0 \tag{11.23}$$

$$\mathbf{M}_R = \sqrt{\mathbf{M}_{Rx}^2 + \mathbf{M}_{Ry}^2 + \mathbf{M}_{Rz}^2} = 0 \tag{11.24}$$

Since the square of any quantity cannot be negative, equations 11.25 through 11.30 follow directly from equations 11.23 and 11.24.

$$\mathbf{F}_{Rx} = \sum \mathbf{F}_x = 0 \qquad 11.25$$

$$\mathbf{F}_{Ry} = \sum \mathbf{F}_y = 0 \qquad 11.26$$

$$\mathbf{F}_{Rz} = \sum \mathbf{F}_z = 0 \qquad 11.27$$

$$\mathbf{M}_{Rx} = \sum \mathbf{M}_x = 0 \qquad 11.28$$

$$\mathbf{M}_{Ry} = \sum \mathbf{M}_y = 0 \qquad 11.29$$

$$\mathbf{M}_{Rz} = \sum \mathbf{M}_z = 0 \qquad 11.30$$

## 6 FREE-BODY DIAGRAMS

A *free-body diagram* is a representation of an object in equilibrium, showing all external forces, moments, and support reactions. Since the object is in equilibrium, the resultant of all forces and moments on the free-body is zero.

If any part of the object is removed and replaced by the forces and moments which are exerted on the cut surface, a free-body of the remaining structure is obtained, and the conditions of equilibrium will be satisfied by the new free-body.

By dividing the object into a sufficient number of free-bodies, the internal forces and moments can be found at all points of interest, providing that the conditions of equilibrium are sufficient to give a static solution.

## 7 REACTIONS

A typical first step in solving statics problems is to determine the supporting reaction forces. The manner in which the structure is supported will determine the type, the location, and the direction of the reactions. Conventional symbols can be used to define the type of reactions which occur at each point of support. Some examples are shown in table 11.1.

*Example 11.2*

Find the reactions $\mathbf{R}_1$ and $\mathbf{R}_2$.

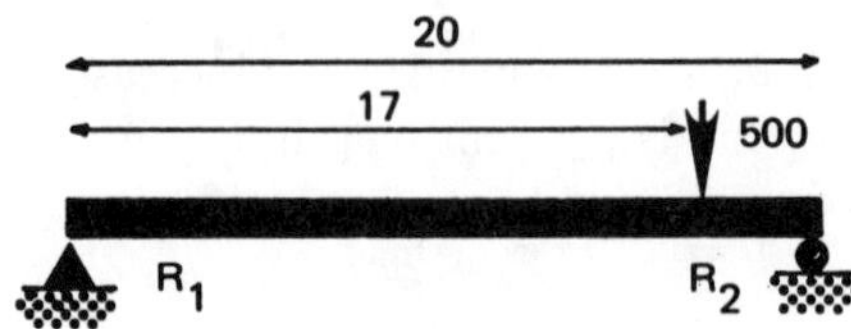


Since the left support is a simple support, its reaction can have any direction. $\mathbf{R}_1$ can, therefore, be written in terms of its $x$ and $y$ components, $\mathbf{R}_{1,x}$ and $\mathbf{R}_{1,y}$, respectively. $\mathbf{R}_2$ is a roller support which cannot sustain an $x$ component.

From equation 11.25, choosing forces to the right as positive, $\mathbf{R}_{1,x}$ is found to be zero.

$$\sum \mathbf{F}_x = \mathbf{R}_{1,x} = 0$$

Equation 11.26 can be used to obtain a relationship between the $y$ components of force. Forces acting upward are considered positive.

$$\sum \mathbf{F}_y = \mathbf{R}_{1,y} + \mathbf{R}_2 - 500 = 0$$

Since both $\mathbf{R}_{1,y}$ and $\mathbf{R}_2$ are unknown, a second equation is needed. Equation 11.30 is used. The reference point is chosen as the left end to make the moment arm for $\mathbf{R}_{1,y}$ equal to zero. This eliminates $\mathbf{R}_{1,y}$ as an unknown, allowing $\mathbf{R}_2$ to be found directly. Clockwise moments are considered positive.

$$\sum \mathbf{M}_{\text{left end}} = (500)(17) - (\mathbf{R}_2)(20) = 0$$

$$\mathbf{R}_2 = 425$$

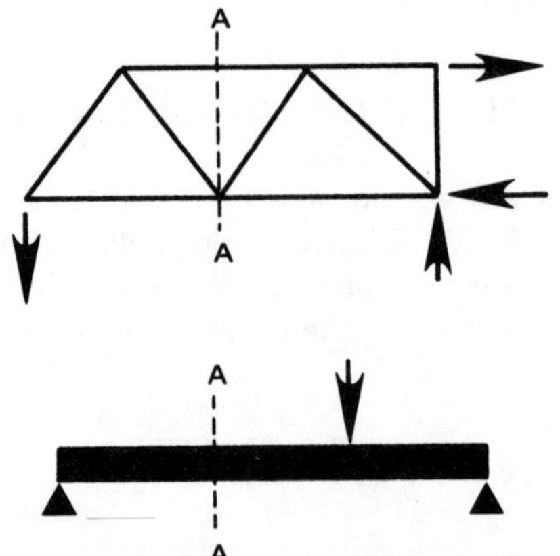


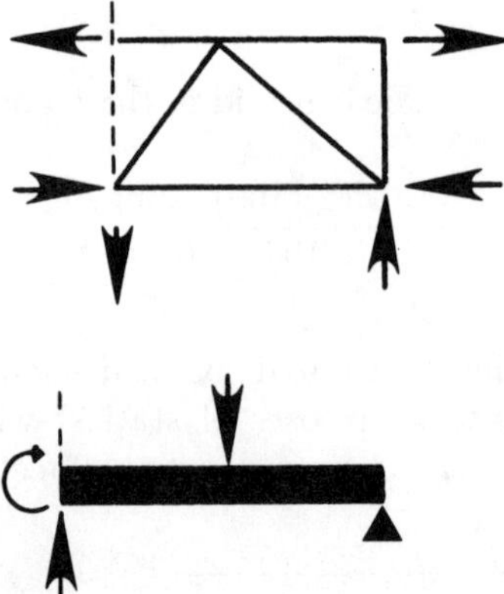

**Figure 11.6** Original and Cut Free-bodies

**Table 11.1**
Common Support Symbols

| type of support | symbol | characteristics |
|---|---|---|
| Built-in | | Moments and forces in any direction |
| Simple | | Load in any direction; no moment |
| Roller | | Load normal to surface only; no moment |
| Cable | | Load in cable direction; no moment |
| Guide | | No load or moment in guide direction |
| Hinge | | Load in any direction; no moment |

Once $\mathbf{R}_2$ is known, $\mathbf{R}_{1,y}$ is found easily from equation 11.26.

$$\sum \mathbf{F}_y = \mathbf{R}_{1,y} + 425 - 500 = 0$$

$$\mathbf{R}_{1,y} = 75$$

## 8 INFLUENCE LINES FOR REACTIONS

An *influence line* (*influence graph*) is an $x$-$y$ plot of the magnitude of a reaction (any reaction on the object) as it would vary as the load is placed at different points on the object. The x-axis corresponds to the location along the object (as along the length of a beam); the y-axis corresponds to the magnitude of the reaction. For uniformity, the load is taken as 1 unit. Therefore, for an actual load of $\mathbf{P}$ units, the actual reaction would be given by equation 11.31.

$$\text{actual reaction} = \mathbf{P}\begin{bmatrix}\text{influence graph}\\ \text{ordinate}\end{bmatrix} \qquad 11.31$$

*Example 11.3*

Draw the influence graphs of the left and right reactions for the beam shown.

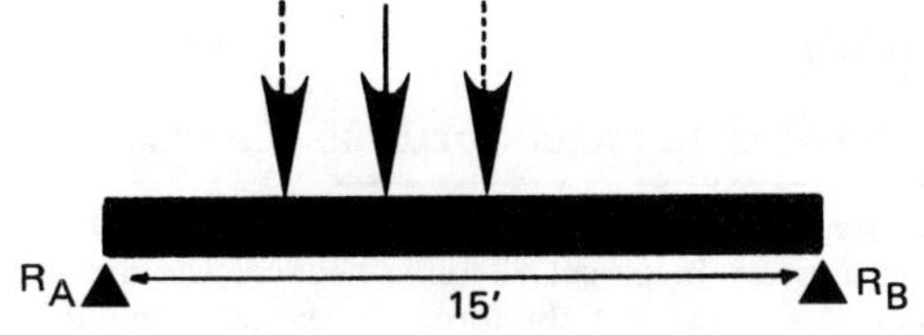


If a unit load is at the left end, reaction $\mathbf{R}_A$ will be equal to 1. If the unit load is at the right end, it will be supported entirely by $\mathbf{R}_B$, so $\mathbf{R}_A$ will be zero. The influence line for $\mathbf{R}_A$ is

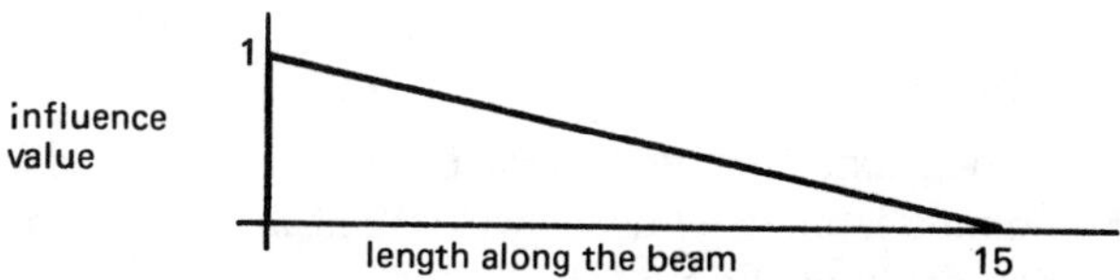


The influence line for $\mathbf{R}_B$ is found similarly.

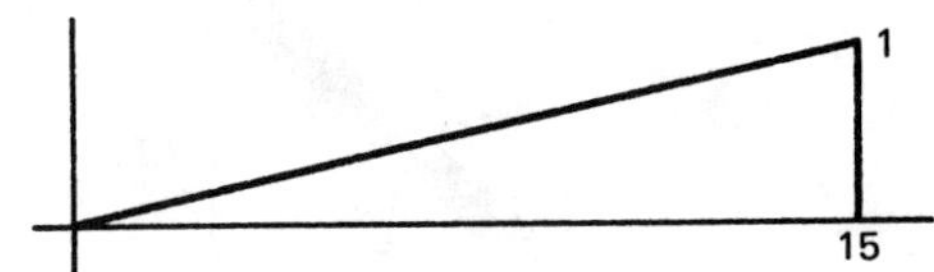


## 9 AXIAL MEMBERS

A member which is in equilibrium when acted upon by forces at each end and by no other forces or moments is an *axial member*. For equilibrium to exist, the resultant forces at the ends must be equal, opposite, and collinear. In an actual truss, this type of loading can be approached through the use of frictionless bearings or pins at the ends of the axial members. In simple truss analysis, the members are assumed to be axial members, regardless of the end conditions.

A typical inclined axial member is illustrated in figure 11.7. For that member to be in equilibrium, the following equations must hold.

$$\mathbf{F}_{Rx} = \mathbf{F}_{Bx} - \mathbf{F}_{Ax} = 0 \qquad 11.32$$

$$\mathbf{F}_{Ry} = \mathbf{F}_{By} - \mathbf{F}_{Ay} = 0 \qquad 11.33$$

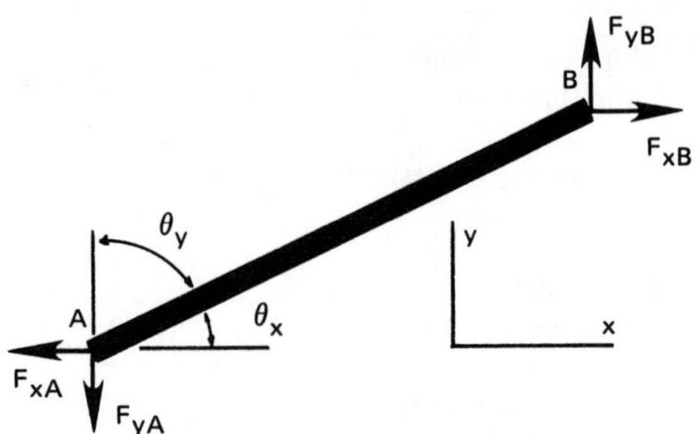


**Figure 11.7** An Axial Member

The resultant force, $\mathbf{F}_R$, can be derived from the components by trigonometry and direction cosines.

$$\mathbf{F}_{Rx} = \mathbf{F}_R \cos\theta_x \qquad 11.34$$

$$\mathbf{F}_{Ry} = \mathbf{F}_R \cos\theta_y = \mathbf{F}_R \sin\theta_x \qquad 11.35$$

If, however, the geometry of the axial member is known, similar triangles can be used to find the resultant and/or the components. This is illustrated in example 11.4.

*Example 11.4*

A 12′ long axial member carrying an internal load of 180 pounds is inclined as shown. What are the $x$- and $y$-components of the load?

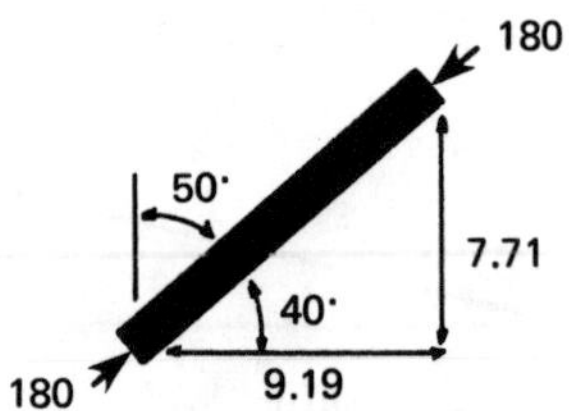


*method 1: Direction Cosines*

$$\mathbf{F}_x = 180\,(\cos\ 40°) = 137.9$$

$$\mathbf{F}_y = 180\,(\cos\ 50°) = 115.7$$

*method 2: Similar Triangles*

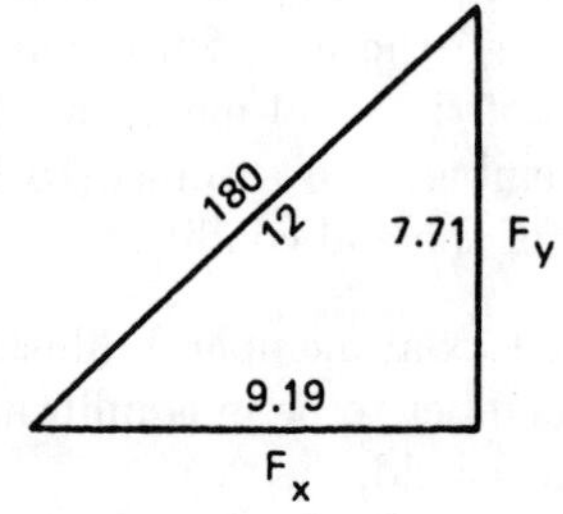


$$\mathbf{F}_x = \left(\frac{9.19}{12}\right)(180) = 137.9$$

$$\mathbf{F}_y = \left(\frac{7.71}{12}\right)(180) = 115.7$$

## 10 TRUSSES

This discussion is directed toward 2-dimensional trusses. The loads in truss members are represented by arrows pulling away from the joints for tension, and by arrows pushing toward the joints for compression.[3]

The equations of equilibrium can be used to find the external reactions on a truss. To find the internal resultants in each axial member, three methods can be used. These methods are *method of joints*, *cut-and-sum*, and *method of sections*.

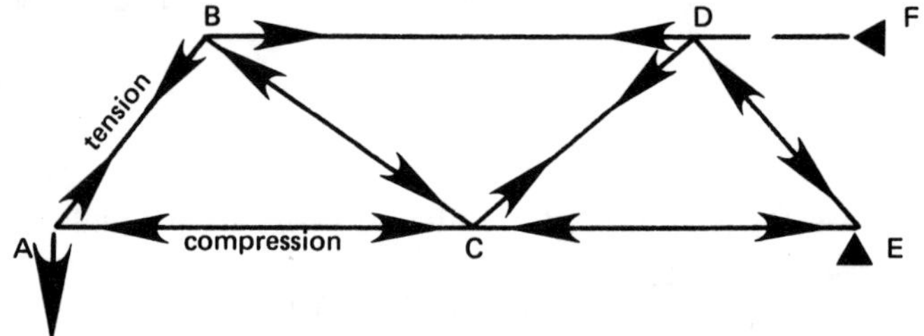


**Figure 11.8** Truss Notation

All joints in a truss will be *determinate*, and all member loads can be found if equation 11.36 holds.

$$\#\text{ truss members} = 2\,(\#\text{ joints}) - 3 \qquad 11.36$$

If the left-hand side is greater than the right-hand side, indeterminate methods must be used to solve the truss. If the left-hand side is less than the right-hand side, the truss is not rigid and will collapse under certain types of loading.

### A. METHOD OF JOINTS

The *method of joints* is a direct application of equations 11.25 and 11.26. The sums of forces in the $x$- and $y$-directions are taken at consecutive joints in the truss. At each joint, there may be up to two unknown axial forces, each of which may have two components. Since there are two equations of equilibrium, a joint with two unknown forces will be determinate.

*Example 11.5*

Find the force in member **BD** in the truss shown.

[3] The method of showing tension and compression on a truss drawing appears incorrect. This is because the arrows show the forces on the joints, not the forces in the axial members.

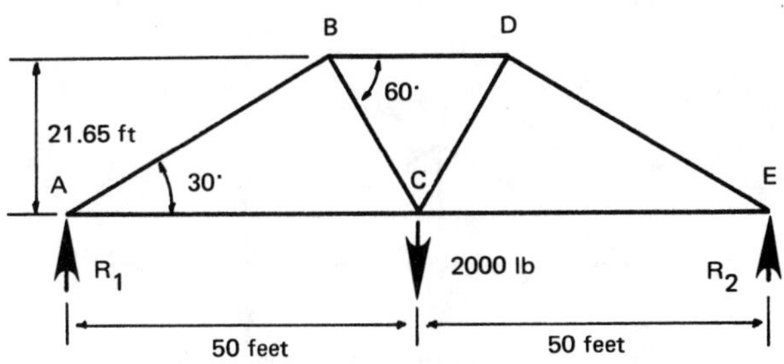


*step 1*: Find the reactions $\mathbf{R}_1$ and $\mathbf{R}_2$. From equation 11.29, the sum of moments must be zero. Taking moments (counterclockwise as positive) about point A gives $\mathbf{R}_2$.

$$\sum \mathbf{M}_A = 100\,(\mathbf{R}_2) - 2000\,(50) = 0$$
$$\mathbf{R}_2 = 1000$$

From equation 11.26, the sum of the forces (vertical positive) in the $y$ direction must be zero.

$$\sum \mathbf{F}_y = \mathbf{R}_1 - 2000 + 1000 = 0$$
$$\mathbf{R}_1 = 1000$$

*step 2*: Although we want the force in member **BD**, there are three unknowns at joints B and D. Therefore, start with joint A where there are only two unknowns (**AB** and **AC**). The free-body of joint A is shown below. The direction of $\mathbf{R}_1$ is known. However, the directions of the member forces usually are not known and need to be found by inspection or assumption. If an incorrect direction is assumed, the force will show up with a negative sign in later calculations.

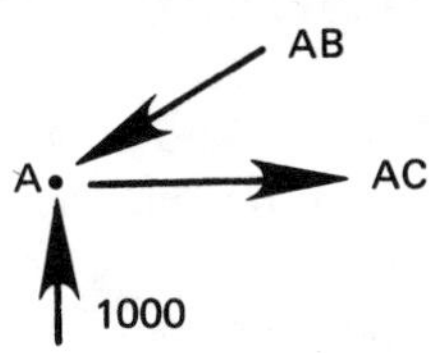


*step 3*: Resolve all inclined forces on joint A into horizontal and vertical components using trigonometry or similar triangles. $\mathbf{R}_1$ and **AC** already are parallel to the $y$ and $x$ axes, respectively. Only **AB** needs to be resolved into components. By observation, it is clear that $\mathbf{AB}_y = 1000$. If this were not true, equation 11.26 would not hold.

$$\mathbf{AB}_y = \mathbf{AB}\,(\sin\ 30°)$$
$$1000 = \mathbf{AB}\,(0.5) \text{ or } \mathbf{AB} = 2000$$
$$\mathbf{AB}_x = \mathbf{AB}\,(\cos\ 30°) = 1732$$

*step 4*: Draw the free-body diagram of joint B. Notice that the direction of force **AB** is toward the joint, just as it was for joint A. The direction of load **BC** is chosen to counteract the vertical component of load **AB**. The direction of load **BD** is chosen to counteract the horizontal components of loads **AB** and **BC**.

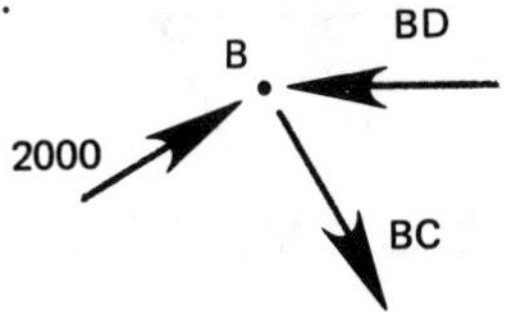


*step 5*: Resolve all inclined forces into horizontal and vertical components.

$$\mathbf{AB}_x = 1732$$
$$\mathbf{AB}_y = 1000$$
$$\mathbf{BC}_x = \mathbf{BC}\,(\sin\ 30°) = 0.5\mathbf{BC}$$
$$\mathbf{BC}_y = \mathbf{BC}\,(\cos\ 30°) = 0.866\mathbf{BC}$$

*step 6*: Write the equations of equilibrium for joint B.

$$\sum \mathbf{F}_x = 1732 + 0.5\mathbf{BC} - \mathbf{BD} = 0$$
$$\sum \mathbf{F}_y = 1000 - 0.866\mathbf{BC} = 0$$

**BC** from the second equation is found to be 1155. Substituting 1155 into the first equilibrium condition equation gives

$$1732 + 0.5\,(1155) - \mathbf{BD} = 0$$
$$\mathbf{BD} = 2310$$

Since **BD** turned out to be positive, its direction was chosen correctly. The direction of the arrow indicates that the member is compressing the pin joint. Consequently, the pin is compressing the member, and member **BD** is in compression.

## B. CUT-AND-SUM METHOD

The *cut-and-sum method* can be used if a load in an inclined member in the middle of a truss is wanted. The method is strictly an application of the equilibrium condition requiring the sum of forces in the vertical direction to be zero. The method is illustrated in example 11.6.

*Example 11.6*

Find the force in member **BC** for the truss shown in example 11.5.

*step 1*: Find the external reactions. This is the same step as in example 11.5. $\mathbf{R}_1 = \mathbf{R}_2 = 1000$.

*step 2*: Cut the truss through, making sure that the cut goes through only one member with a vertical component. In this case, that member is **BC**.

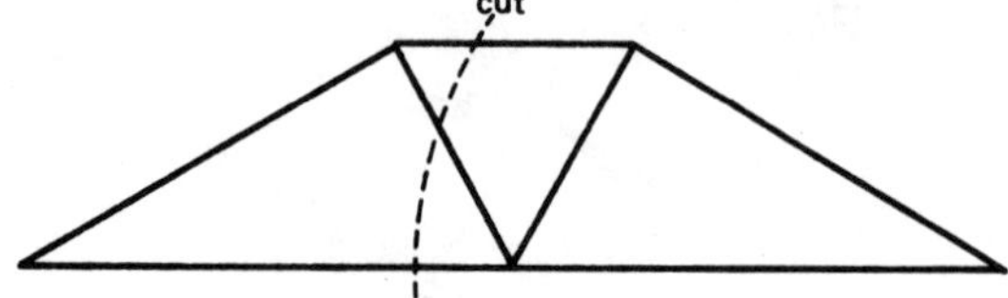


*step 3*: Draw the free-body of either part of the remaining truss.

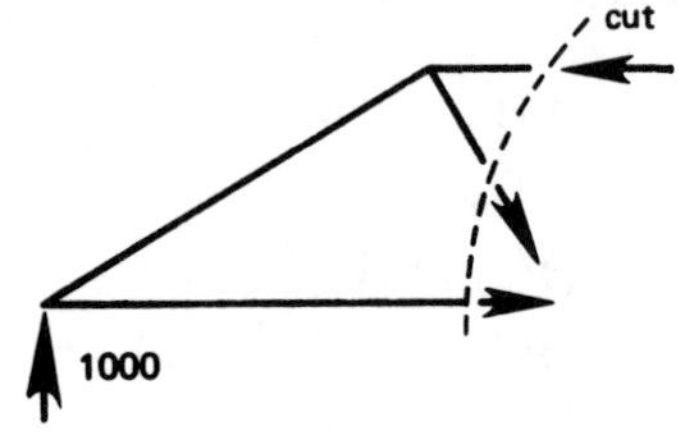


*step 4*: Resolve the unknown inclined force into vertical and horizontal components.

$$\mathbf{BC}_x = 0.5\,(\mathbf{BC})$$
$$\mathbf{BC}_y = 0.866\,(\mathbf{BC})$$

*step 5*: Sum forces in the $y$ direction for the entire free-body.

$$\sum \mathbf{F}_y = 1000 - 0.866\ (\mathbf{BC}) = 0$$
$$\mathbf{BC} = 1155$$

C. METHOD OF SECTIONS

The cut-and-sum method will work only if it is possible to cut the truss without going through two members with vertical components.

The *method of sections* is a direct approach for finding member loads at any point in a truss. In this method, the truss is cut at an appropriate section, and the conditions of equilibrium are applied to the resulting free-body. This is illustrated in example 11.7.

*Example 11.7*

For the truss shown, find the load in members **CE** and **CD**.

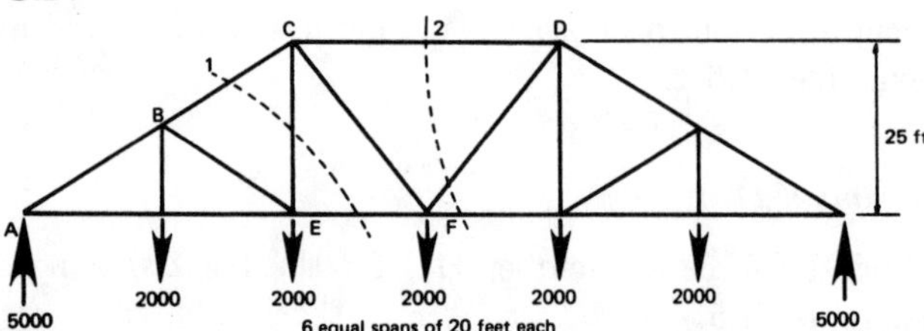


For member force **CE**, the truss is cut at section 1.

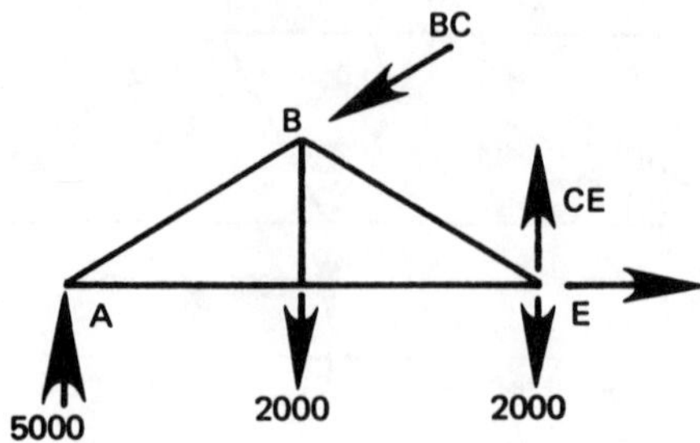


Taking moments about A will eliminate all unknowns except force **CE**.

$$\sum \mathbf{M}_A = \mathbf{CE}\,(40) - 2000\,(20) - 2000\,(40) = 0$$
$$\mathbf{CE} = 3000$$

For member **CD**, the truss is cut at section 2. Taking moments about point F will eliminate all unknowns except **CD**.

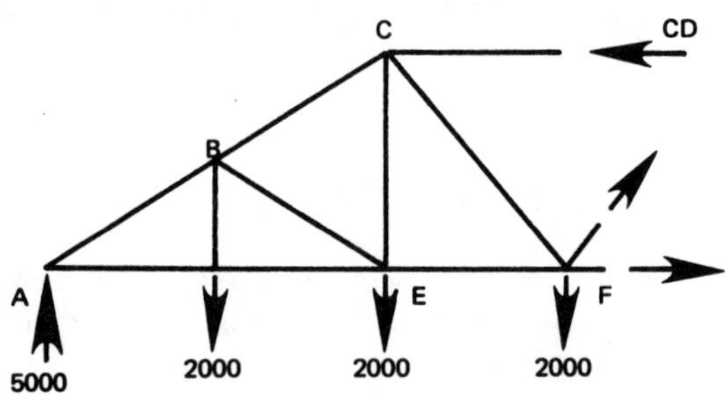


$$\sum \mathbf{M}_F = \mathbf{CD}(25) + 2000(20) + 2000(40) - 5000(60) = 0$$
$$\mathbf{CD} = 7200$$

## 11 SUPERPOSITION OF LOADINGS

For any group of forces and moments which satisfies the conditions of equilibrium, the resultant force and moment vectors are zero. The resultant of two zero vectors is another zero vector. Therefore, any number of such equilibrium systems can be combined without disturbing the equilibrium.

Superposition methods must be used with discretion in working with actual structures since some structures change shape significantly under load. If the actual structure were to deflect so that the points of application of loads were quite different from those in the undeflected structure, superposition would not be applicable.

In simple truss analysis, the change of shape under load is neglected when finding the member loads. Superposition, therefore, can be assumed to apply.

## 12 CABLES

### A. CABLES UNDER CONCENTRATED LOADS

An ideal cable is assumed to be completely flexible. It acts as an axial member in tension between any two points of concentrated load application.

The method of joints and sections used in truss analysis applies equally well to cables under concentrated loads. However, no compression members will be found. As in truss analysis, if the cable loads are unknown, some information concerning the geometry of the cable must be known in order to solve for the axial tension in the segments.

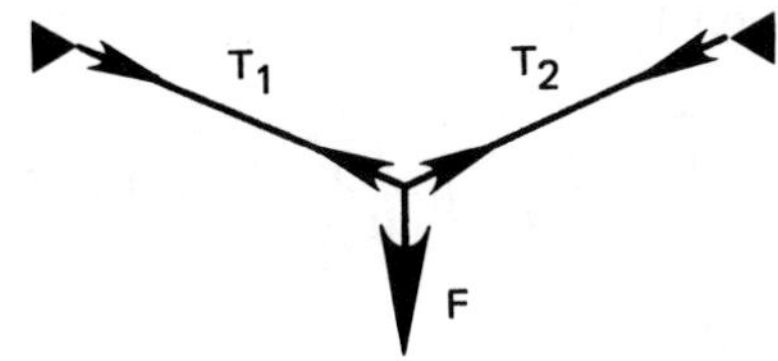


**Figure 11.9** Cable Under Transverse Loading

*Example 11.8*

Find the tension $\mathbf{T}_2$ between points B and C.

*step 1*: Take moments about point A to find $\mathbf{T}_3$.

$$\sum \mathbf{M}_A = a\mathbf{F}_1 + b\mathbf{F}_2 + d\mathbf{T}_3 \cos\theta_3 - c\mathbf{T}_3 \sin\theta_3 = 0$$

*step 2*: Sum forces in the $x$ direction to find $\mathbf{T}_1$.

$$\sum \mathbf{F}_x = \mathbf{T}_1 \cos\theta_1 - \mathbf{T}_3 \cos\theta_3 = 0$$

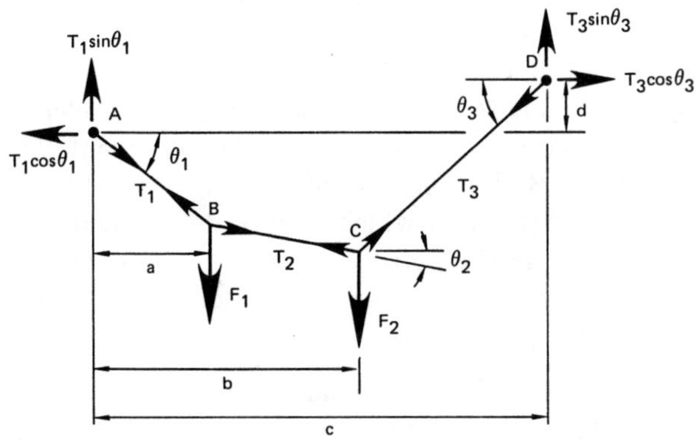


*step 3*: Sum forces in the $x$ direction at point B to find $\mathbf{T}_2$.

$$\sum \mathbf{F}_x = \mathbf{T}_2 \cos\theta_2 - \mathbf{T}_1 \cos\theta_1 = 0$$

### B. CABLES UNDER DISTRIBUTED LOADS

An idealized tension cable under a distributed load is similar to a linkage made up of a very large number of axial members. The cable is an axial member in the sense that the internal tension acts in a direction which is along the centerline of the cable everywhere.

Figure 11.10 illustrates a cable under a unidirectionally distributed load. A free-body diagram of segment B-C of the cable also is known. **F** is the vertical resultant of the distributed load on the segment.

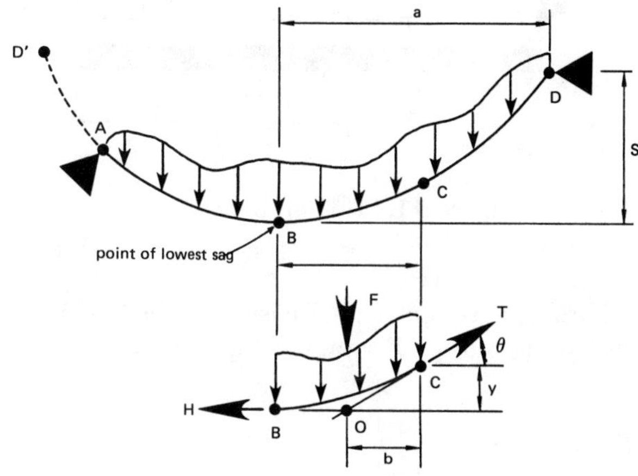


**Figure 11.10** Cable Under Distributed Load

**T** is the cable tension at point C, and **H** is the cable tension at the point of lowest sag. From the conditions of equilibrium for free-body B-C, it is apparent that the three forces, **H**, **F**, and **T**, must be concurrent at point O. Taking moments about point C, the following equations are obtained.

$$\sum \mathbf{M}_C = \mathbf{F}b - \mathbf{H}y = 0 \quad 11.37$$

$$\mathbf{H} = \frac{\mathbf{F}b}{y} \quad 11.38$$

But $\tan\theta = \left(\frac{y}{b}\right)$. So,

$$\mathbf{H} = \frac{\mathbf{F}}{\tan\theta} \quad 11.39$$

From the summation of forces in the vertical and horizontal directions,

$$\mathbf{T}\cos\theta = \mathbf{H} \quad 11.40$$

$$\mathbf{T}\sin\theta = \mathbf{F} \quad 11.41$$

$$\mathbf{T} = \sqrt{\mathbf{H}^2 + \mathbf{F}^2} \quad 11.42$$

The shape of the cable is a function of the relative amount of sag at point B and the relative distribution (not the absolute magnitude) of the applied running load.

## C. PARABOLIC CABLES

If the distribution load per unit length, $w$, is constant with respect to a horizontal line (as is the load from a bridge floor), the cable will be parabolic in shape. This is illustrated in figure 11.11.

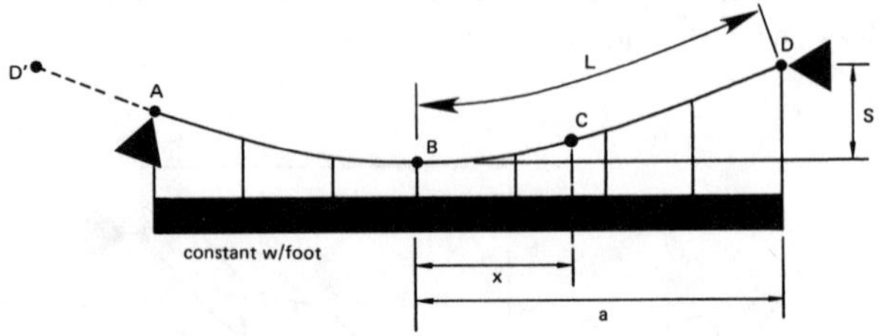


**Figure 11.11** Parabolic Cable

The horizontal component of tension can be found from equation 11.38 using **F**=wa, b= $\frac{1}{2}$a, and y=S.

$$\mathbf{H} = \frac{\mathbf{F}b}{y} = \frac{wa^2}{2S} \qquad 11.43$$

$$\mathbf{T} = \sqrt{\mathbf{H}^2 + \mathbf{F}^2} = \sqrt{\left(\frac{wa^2}{2S}\right)^2 + (wx)^2} \qquad 11.44$$

$$= w\sqrt{x^2 + \left(\frac{a^2}{2S}\right)^2} \qquad 11.45$$

The shape of the cable is given by equation 11.46.

$$y = \frac{wx^2}{2\mathbf{H}} \qquad 11.46$$

The approximate length of the cable from the lowest point to the support is given by equation 11.47.

$$L \approx (a)\left[1 + \tfrac{2}{3}\left(\frac{S}{a}\right)^2 - \tfrac{2}{5}\left(\frac{S}{a}\right)^4\right] \qquad 11.47$$

*Example 11.9*

A pedestrian bridge has two suspension cables and a flexible floor. The floor weighs 28 pounds per foot. The span of the bridge is 100 feet between the two end supports. When the bridge is empty, the tension at point A is 1500 pounds. What is the cable sag, S, at the center? What is the approximate cable length?

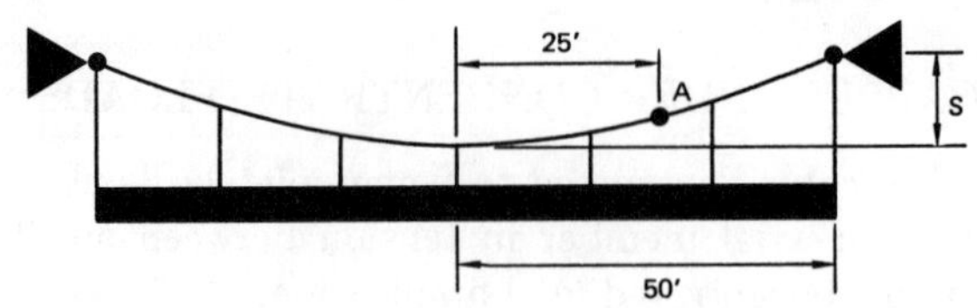


The floor weight per cable is 28/2= 14 lb/ft. From equation 11.45,

$$1500 = 14\sqrt{[25]^2 + \left[\frac{(50)^2}{2S}\right]^2}$$

$$S = 12 \text{ feet}$$

From equation 11.47,

$$L = 50\left[1 + \left(\frac{2}{3}\right)\left(\frac{12}{50}\right)^2 - \left(\frac{2}{5}\right)\left(\frac{12}{50}\right)^4\right]$$

$$= 51.9 \text{ feet}$$

The cable length is 2 × 51.9 = 103.8 ft.

## D. THE CATENARY

If the distributed load, $w$, is constant along the length of the cable (as in the case of a cable loaded by its own weight), the cable will have the shape of a *catenary.* This is illustrated in figure 11.12.

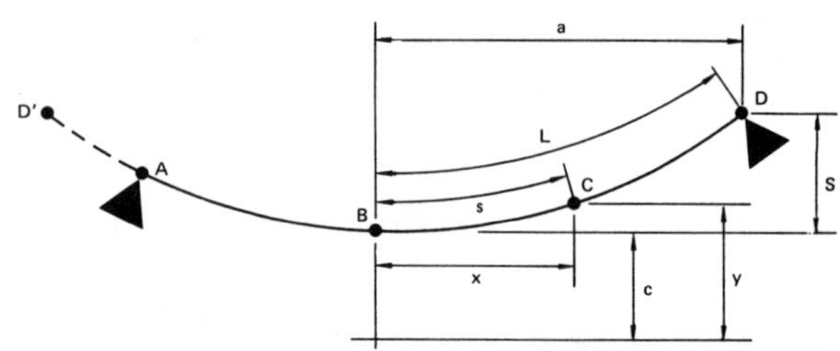


**Figure 11.12** A Catenary

As shown in figure 11.12, $y$ is measured from a reference plane located a distance c below the lowest point of the cable, point B. The location of this reference plane is a parameter of the cable which must be determined before equations 11.48 through 11.53 are used. The value of $c$ does not correspond to any physical distance, nor does the reference plane correspond to the ground.

The equations of the catenary are presented below. Some judgment usually is necessary to determine which equations should be used and in which order. To define cable shape, it is necessary to have some initial information which can be entered into the equations. For example, if $a$ and the sag $S$ are given, equation 11.51 can be solved by trial and error to obtain $c$. Once $c$

is known, the cable geometry and forces are defined by the remaining equations.

$$y = c\left[\cosh\left(\frac{x}{c}\right)\right] \quad 11.48$$

$$s = c\left[\sinh\left(\frac{x}{c}\right)\right] \quad 11.49$$

$$y = \sqrt{s^2 + c^2} \quad 11.50$$

$$S = c\left[\cosh\left(\frac{a}{c}\right) - 1\right] \quad 11.51$$

$$\tan\theta = \frac{s}{c} \quad 11.52$$

$$\mathbf{H} = wc \quad 11.53$$

$$\mathbf{F} = ws \quad 11.54$$

$$\mathbf{T} = wy \quad 11.55$$

$$\tan\theta = \frac{ws}{\mathbf{H}} \quad 11.56$$

$$\cos\theta = \frac{\mathbf{H}}{\mathbf{T}} \quad 11.57$$

*Example 11.10*

A cable 100′ long is loaded by its own weight. The sag is 25′, and the supports are on the same level. What is the distance between the supports?

From equation 11.50 at point $D$, with $S = 25$,

$$c + S = \sqrt{s^2 + c^2}$$

$$c + 25 = \sqrt{(50)^2 + c^2}$$

$$c = 37.5$$

From equation 11.49,

$$50 = 37.5\left[\sinh\left(\frac{a}{37.5}\right)\right]$$

$$a = 41.2 \text{ feet}$$

The distance between supports is

$$(2a) = (2)(41.2) = 82.4 \text{ feet}$$

Providing that the lowest point, B, is known or can be found, the location of the cable supports at different levels does not significantly affect the analysis of cables. The same procedure is used in proceeding from point B to either support. In fact, once the theoretical shape of the cable has been determined, the supports can be relocated anywhere along the cable line without affecting the equilibrium of the supporting segment.

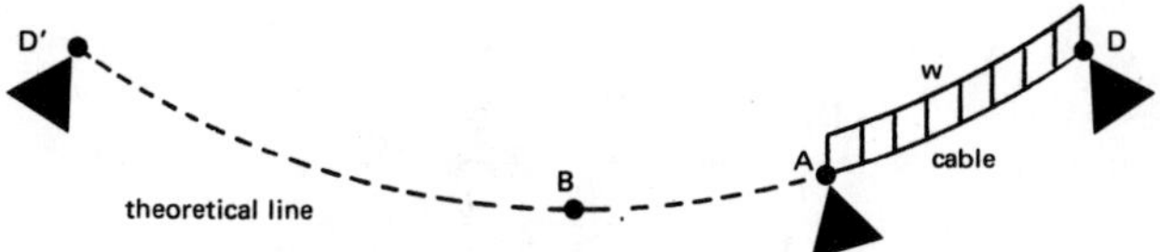


**Figure 11.13** Non-Symmetrical Segment of Symmetrical Cable

## 13 3-DIMENSIONAL STRUCTURES

The static analysis of 3-dimensional structures usually requires the following steps.

*step 1*: Determine the components of all loads and reactions. This usually is accomplished by finding the $x$, $y$, and $z$ coordinates of all points and then using direction cosines.

*step 2*: Draw three free-bodies of the structure—one each for the $x$, $y$, and $z$ components of loads and reactions.

*step 3*: Solve for unknowns using $\sum \mathbf{F} = 0$ and $\sum \mathbf{M} = 0$.

*Example 11.11*

Beam ABC is supported by the two cables as shown. The connection at A is pinned (hinged). Find the cable tensions $\mathbf{T}_1$ and $\mathbf{T}_2$.

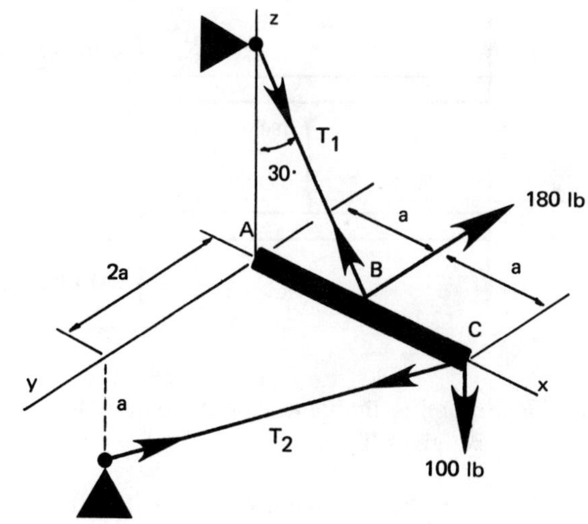


*step 1*: For the 180 pound load,

$$\mathbf{F}_x = 0$$

$$\mathbf{F}_y = -180$$

$$\mathbf{F}_z = 0$$

For the 100 pound load,

$$\mathbf{F}_x = 0$$

$$\mathbf{F}_y = 0$$

$$\mathbf{F}_z = -100$$

For cable 1:

$$\cos\theta_x = \cos\ 120^\circ = -0.5$$

$$\cos\theta_y = 0$$

$$\cos\theta_z = \cos\ 30^\circ = 0.866$$

$$\mathbf{T}_{1x} = -0.5\mathbf{T}_1$$

$$\mathbf{T}_{1y} = 0$$

$$\mathbf{T}_{1z} = 0.866\mathbf{T}_1$$

For cable 2: The length of the cable is

$$L = \sqrt{(2a)^2 + (-2a)^2 + (-a)^2} = 3a$$

$$\cos\theta_x = \frac{-2a}{3a} = -0.667$$

$$\cos\theta_y = \frac{2a}{3a} = 0.667$$

$$\cos\theta_z = \frac{-a}{3a} = -0.333$$

$$\mathbf{T}_{2x} = -0.667\mathbf{T}_2$$

$$\mathbf{T}_{2y} = 0.667\mathbf{T}_2$$

$$\mathbf{T}_{2z} = -0.333\mathbf{T}_2$$

*step 2:*

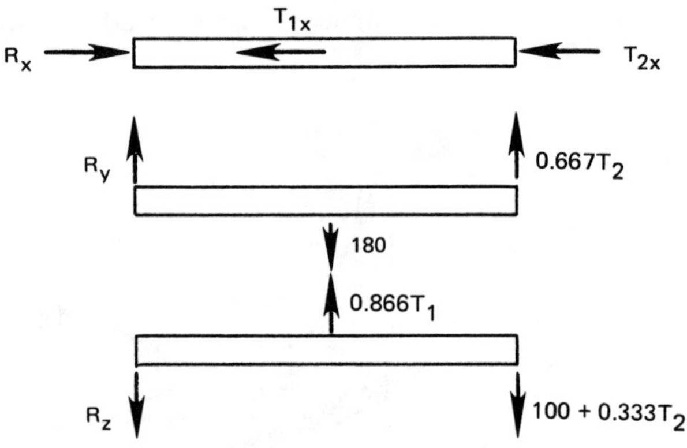


*step 3:* Summing moments about point A for the $y$ case gives $\mathbf{T}_2$.

$$\sum \mathbf{M}_{Az} = 0.667\mathbf{T}_2(2a) - 180(a) = 0$$

$$\mathbf{T}_2 = 135$$

Summing moments about point A for the z case gives $\mathbf{T}_1$.

$$\sum \mathbf{M}_{Ay} = 0.866\mathbf{T}_1(a) - 0.333(135)(2a) - 100(2a) = 0$$

$$\mathbf{T}_1 = 335$$

## 14 GENERAL TRIPOD SOLUTION

The procedure given in the preceding section will work with a tripod consisting of three axial pin-ended members with a load in any direction applied at the apex. However, the tripod problem occurs frequently enough to develop a specialized procedure for solution.

*step 1:* Use the direction cosines of the force, $\mathbf{F}$, to find its components.

$$\mathbf{F}_x = \mathbf{F}(\cos\theta_x) \qquad 11.58$$

$$\mathbf{F}_y = \mathbf{F}(\cos\theta_y) \qquad 11.59$$

$$\mathbf{F}_z = \mathbf{F}(\cos\theta_z) \qquad 11.60$$

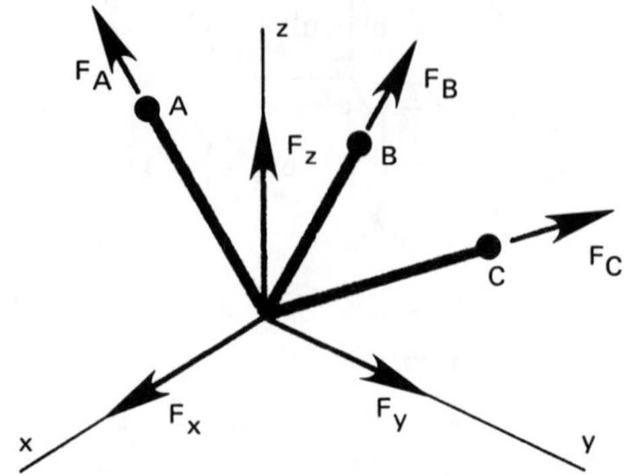


**Figure 11.14** A General Tripod

*step 2:* Using the $x$, $y$, and $z$ coordinates of points A, B, and C (taking the origin at the apex), find the direction cosines for the legs. Repeat the following four equations for each member, observing algebraic signs of $x$, $y$, and $z$.

$$L^2 = x^2 + y^2 + z^2 \qquad 11.61$$

$$\cos\theta_x = \frac{x}{L} \qquad 11.62$$

$$\cos\theta_y = \frac{y}{L} \qquad 11.63$$

$$\cos\theta_z = \frac{z}{L} \qquad 11.64$$

*step 3:* Write the equations of equilibrium for joint O. The following simultaneous equations assume tension in all three members. A minus sign in the solution for any member indicates compression instead of tension.

$$\mathbf{F}_A\cos\theta_{xA} + \mathbf{F}_B\cos\theta_{xB} + \mathbf{F}_C\cos\theta_{xC} + \mathbf{F}_x = 0 \qquad 11.65$$

$$\mathbf{F}_A\cos\theta_{yA} + \mathbf{F}_B\cos\theta_{yB} + \mathbf{F}_C\cos\theta_{yC} + \mathbf{F}_y = 0 \qquad 11.66$$

$$\mathbf{F}_A\cos\theta_{zA} + \mathbf{F}_B\cos\theta_{zB} + \mathbf{F}_C\cos\theta_{zC} + \mathbf{F}_z = 0 \qquad 11.67$$

*Example 11.12*

Find the load on each leg of the tripod shown.

| Point | x | y | z |
|---|---|---|---|
| O | 0 | 0 | 6 |
| A | 2 | −2 | 0 |
| B | 0 | 3 | −4 |
| C | −3 | 3 | 2 |

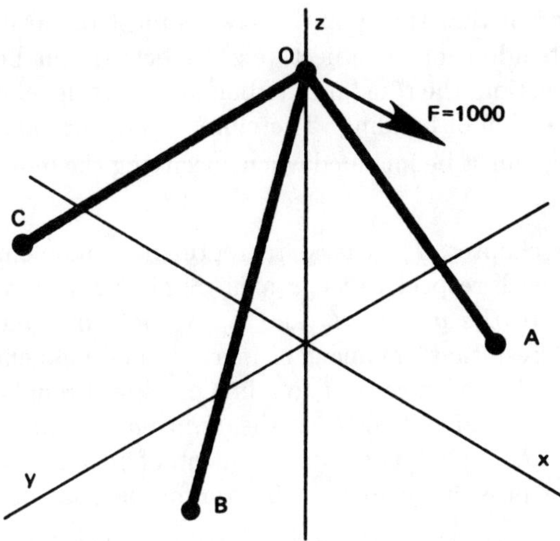


Since the apex is not at point (0, 0, 0), it is necessary to transfer the origin to the apex. This is done by the following equations. Only the $z$ values are affected.

$$x' = x - x_0$$
$$y' = y - y_0$$
$$z' = z - z_0$$

The new coordinates with the origin at the apex are

| Point | x | y | z |
|---|---|---|---|
| O | 0 | 0 | 0 |
| A | 2 | −2 | −6 |
| B | 0 | 3 | −10 |
| C | −3 | 3 | −4 |

The components of the applied force are

$$\mathbf{F}_x = \mathbf{F}\,(\cos\theta_x) = \mathbf{F}\,[\cos(0°)] = 1000$$
$$\mathbf{F}_y = 0$$
$$\mathbf{F}_z = 0$$

The direction cosines of the legs are found from the following table.

| Member | $x^2$ | $y^2$ | $z^2$ | $L^2$ | L | $\cos\theta_x$ | $\cos\theta_y$ | $\cos\theta_z$ |
|---|---|---|---|---|---|---|---|---|
| O–A | 4 | 4 | 36 | 44 | 6.63 | 0.3015 | −0.3015 | −0.9046 |
| O–B | 0 | 9 | 100 | 109 | 10.44 | 0 | 0.2874 | −0.9579 |
| O–C | 9 | 9 | 16 | 34 | 5.83 | −0.5146 | 0.5146 | −0.6861 |

From equations 11.65, 11.66, and 11.67, the equilibrium equations are

$$0.3015\mathbf{F}_A + \qquad -0.5146\mathbf{F}_C + 1000 = 0$$
$$-0.3015\mathbf{F}_A + 0.2874\mathbf{F}_B + 0.5146\mathbf{F}_C + \quad 0 = 0$$
$$-0.9046\mathbf{F}_A - 0.9579\mathbf{F}_B - 0.6861\mathbf{F}_C + \quad 0 = 0$$

The solution to this set of simultaneous equations is

$$\mathbf{F}_A = +1531 \text{ (tension)}$$
$$\mathbf{F}_B = -3480 \text{ (compression)}$$
$$\mathbf{F}_C = +2841 \text{ (tension)}$$

## 15 PROPERTIES OF AREAS

### A. CENTROIDS

The location of the *centroid* of a 2-dimensional area which is defined mathematically as $y = f(x)$ can be found from equations 11.68 and 11.69.[4] This is illustrated in example 11.13.

$$\overline{x} = \frac{\int x\,dA}{A} \qquad 11.68$$

$$\overline{y} = \frac{\int y\,dA}{A} \qquad 11.69$$

$$A = \int f(x)\,dx \qquad 11.70$$

$$dA = f(x)\,dx = f(y)\,dy \qquad 11.71$$

*Example 11.13*

Find the $x$ component of the centroid of the area bounded by the $x$ and $y$ axes, $x = 2$, and $y = e^{2x}$.

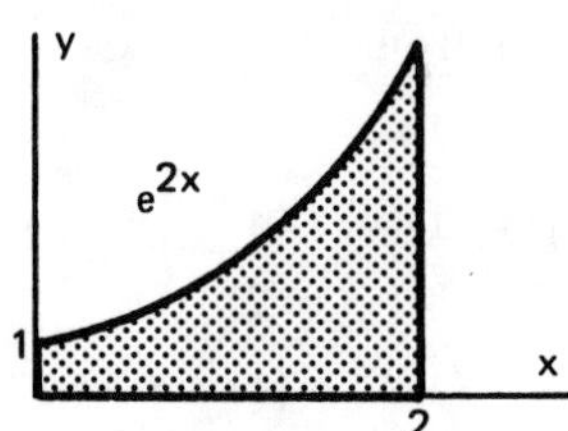


*step 1*: Find the area.

$$A = \int_0^2 e^{2x}dx = \left[\tfrac{1}{2}e^{2x}\right]_0^2$$
$$= 27.3 - 0.5 = 26.8$$

*step 2*: Put $dA$ in terms of $dx$.

$$dA = f(x)\,dx = e^{2x}dx$$

*step 3*: Use equation 11.68 to find $\overline{x}$.

$$\overline{x} = \frac{1}{26.8}\int_0^2 xe^{2x}dx = \frac{1}{26.8}\left[\tfrac{1}{2}xe^{2x} - \tfrac{1}{4}e^{2x}\right]_0^2$$
$$= 1.54$$

[4] The centroid also is known as the *first moment of the area*.

With few exceptions, most areas for which the centroidal location is needed will be either rectangular or triangular. The locations of the centroids for these and other common shapes are given as an appendix of this chapter.

The centroid of a complex 2-dimensional area which can be divided into the simple shapes in appendix A can be found from equations 11.72 and 11.73.

$$\overline{x}_{\text{composite}} = \frac{\sum(A_i\overline{x}_i)}{\sum A_i} \qquad 11.72$$

$$\overline{y}_{\text{composite}} = \frac{\sum(A_i\overline{y}_i)}{\sum A_i} \qquad 11.73$$

*Example 11.14*

Find the y-coordinate of the centroid for the object shown.

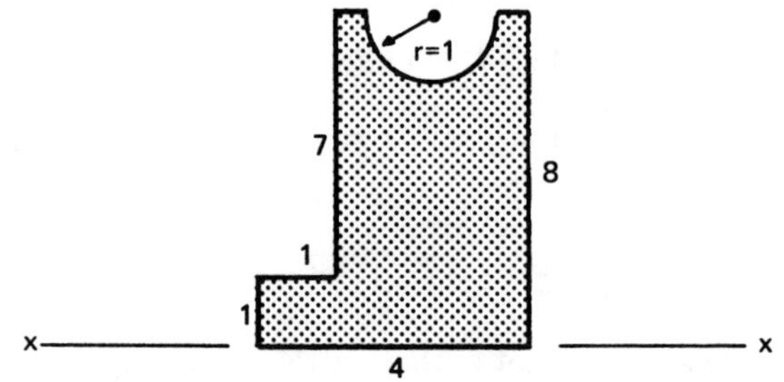


The object is divided into three parts: a 1 × 4 rectangle, a 3 × 7 rectangle, and a half-circle of radius 1. Then, the areas and distances from the $x$-$x$ axis to the individual centroids are found.

$$A_1 = (1)(4) = 4$$
$$A_2 = (3)(7) = 21$$
$$A_3 = \left(-\tfrac{1}{2}\right)\pi(1)^2 = -1.57$$
$$\overline{y}_1 = \tfrac{1}{2}$$
$$\overline{y}_2 = 4\tfrac{1}{2}$$
$$\overline{y}_3 = 8 - 0.424 = 7.576$$

Using equation 11.73,

$$\overline{y} = \frac{(4)\left(\frac{1}{2}\right) + (21)\left(4\frac{1}{2}\right) - (1.57)(7.576)}{4 + 21 - 1.57} = 3.61$$

## B. MOMENT OF INERTIA

The moment of inertia, $I$, of a 2-dimensional area is a parameter which often is needed in mechanics of materials problems. It has no simple geometric interpretation, and its units (length to the fourth power) add to the mystery of this quantity. However, it is convenient to think of the moment of inertia as a resistance to bending.

If the moment of inertia is a resistance to bending, it is apparent that this quantity always must be positive. Since bending of an object (e.g., a beam) can be in any direction, the resistance to bending must depend on the direction of bending. Therefore, a reference axis or direction must be included when specifying the moment of inertia.

In this chapter, $I_x$ is used to represent a moment of inertia with respect to the $x$ axis. Similarly, $I_y$ is with respect to the $y$ axis. $I_x$ and $I_y$ are not components of the "resultant" moment of inertia. The moment of inertia taken with respect to a line passing through the area's centroid is known as the *centroidal moment of inertia*, $I_c$. The centroidal moment of inertia is the smallest possible moment of inertia for the shape.

The moments of inertia of a function which can be expressed mathematically as $y = f(x)$ are given by equations 11.74 and 11.75.

$$I_x = \int y^2\, dA \qquad 11.74$$

$$I_y = \int x^2\, dA \qquad 11.75$$

In general, however, moments of inertia will be found from appendix A.

*Example 11.15*

Find $I_y$ for the area bounded by the y axis, $y = 8$, and $y^2 = 8x$.

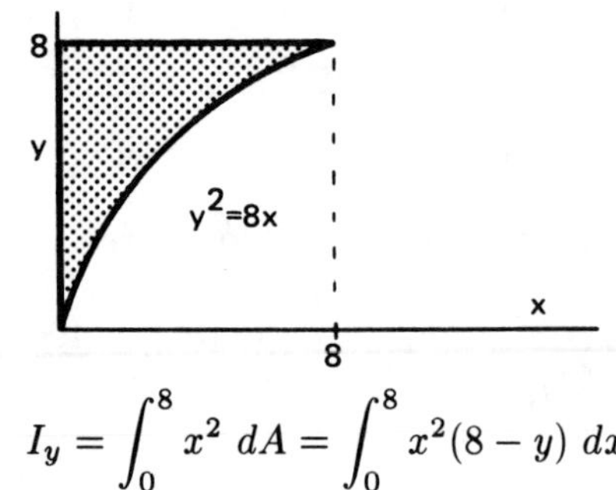


$$I_y = \int_0^8 x^2\, dA = \int_0^8 x^2(8 - y)\, dx$$

But $y = \sqrt{8x}$.

$$I_y = \int_0^8 \left(8x^2 - \sqrt{8}x^{\frac{5}{2}}\right) dx = 195.04 \text{ inches}^4$$

*Example 11.16*

What is the centroidal moment of inertia of the area shown?

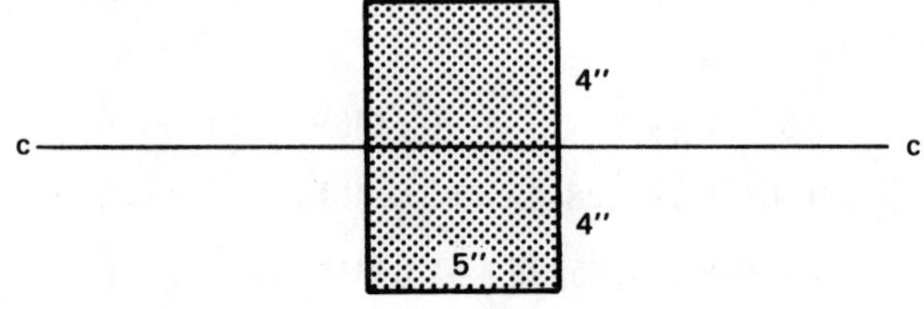

From appendix A,

$$I_c = \frac{(5)(8)^3}{12} = 213.3 \text{ inches}^4$$

The *polar moment of inertia* of a 2-dimensional area can be thought of as a measure of the area's resistance to torsion (twisting). Although the polar moment of inertia can be evaluated mathematically by equation 11.76, it is more expedient to use equation 11.77 if $I_x$ and $I_y$ are known.

$$J_z = \int (x^2 + y^2)\, dA \qquad 11.76$$

$$J_z = I_x + I_y \qquad 11.77$$

The *radius of gyration*, $r$, is a distance at which the entire area can be assumed to exist. The distance is measured from the axis about which the moment of inertia was taken.

$$I = r^2 A \qquad 11.78$$

$$r = \sqrt{\frac{I}{A}} \qquad 11.79$$

*Example 11.17*

What is the radius of gyration for the section shown in example 11.16? What is the significance of this value?

$$A = (5)(4+4) = 40$$

From equation 11.79,

$$r = \sqrt{\frac{213.3}{40}} = 2.31 \text{ inches}$$

2.31″ is the distance from the axis c-c that an infinitely long strip (with area of 40 square inches) would have to be located to have a moment of inertia of 213.3 inches$^4$.

The *parallel axis theorem* usually is needed to evaluate the moment of inertia of a composite object made up of several simple 2-dimensional shapes.[5] The parallel axis theorem relates the moment of inertia of an area taken with respect to any axis to the centroidal moment of inertia. In equation 11.80, $A$ is the 2-dimensional object's area, and $d$ is the distance between the centroidal and new axes.

$$I_{\text{any parallel axis}} = I_c + Ad^2 \qquad 11.80$$

*Example 11.18*

Find the moment of inertia about the $x$ axis for the 2-dimensional object shown.

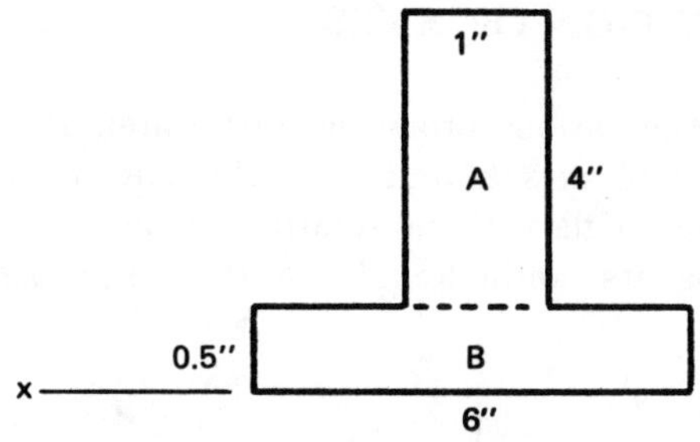


The T-section is divided into two parts: A and B. The moment of inertia of section B can be evaluated readily by using appendix A.

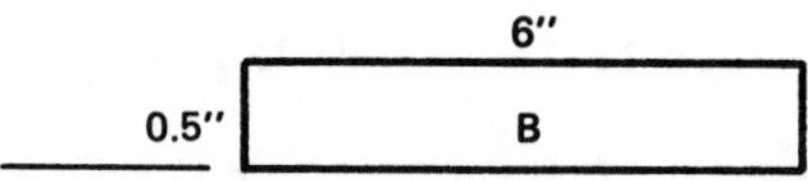


$$I_{x\text{-}x} = \frac{(6)(0.5)^3}{3} = 0.25$$

The moment of inertia of the stem about its own centroidal axis is

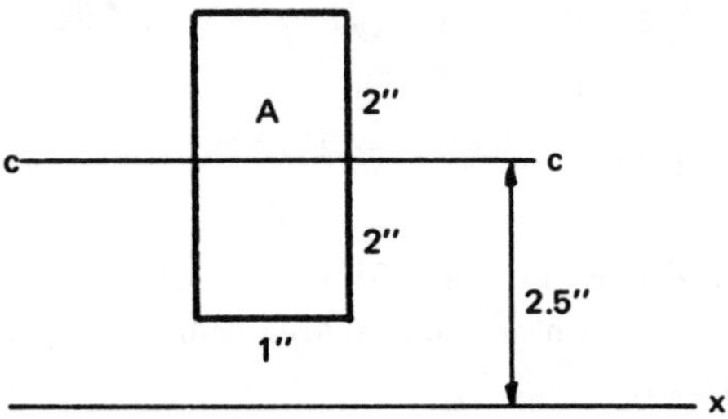


$$I_{c\text{-}c} = \frac{(1)(4)^3}{12} = 5.33$$

Using equation 11.80, the moment of inertia of the stem about the x-x axis is

$$I_{x\text{-}x} = 5.33 + (4)(2.5)^2 = 30.33$$

The total moment of inertia of the T-section is

$$0.25 + 30.33 = 30.58 \text{ inches}^4$$

## C. PRODUCT OF INERTIA

The *product of inertia* of a 2-dimensional object is found by multiplying each differential element of area times its $x$ and $y$ coordinates, and then integrating over the entire area.

$$P_{xy} = \int xy\, dA \qquad 11.81$$

The product of inertia is zero when either axis is an axis of symmetry. The product of inertia may be negative.

[5] This theorem also is known as the *transfer axis theorem*.

## 16 ROTATION OF AXES

Suppose the various properties of an area are known for one set of axes, $x$ and $y$. If the axes are rotated through an angle without rotating the area itself, the new properties can be found from the old properties.

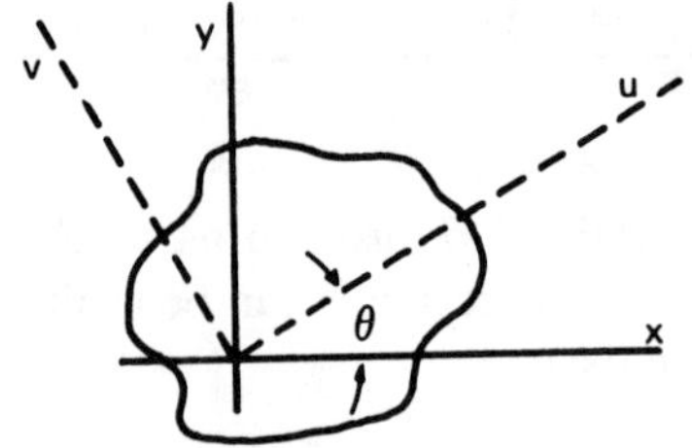


**Figure 11.15** Rotation of Axes

$$I_u = I_x \cos^2\theta - 2P_{xy}\sin\theta\ \cos\theta + I_y \sin^2\theta \quad 11.82$$

$$= \tfrac{1}{2}(I_x + I_y) + \tfrac{1}{2}(I_x - I_y)\cos 2\theta - P_{xy}\sin 2\theta \quad 11.83$$

$$I_v = I_x \sin^2\theta + 2P_{xy}\sin\theta\ \cos\theta + I_y \cos^2\theta \quad 11.84$$

$$= \tfrac{1}{2}(I_x + I_y) - \tfrac{1}{2}(I_x - I_y)\cos 2\theta + P_{xy}\sin 2\theta \quad 11.85$$

$$P_{uv} = I_x \sin\theta\ \cos\theta + P_{xy}(\cos^2\theta - \sin^2\theta) - I_y\sin\theta\ \cos\theta \quad 11.86$$

$$= \tfrac{1}{2}(I_x - I_y)\sin 2\theta + P_{xy}\cos 2\theta \quad 11.87$$

Since the polar moment of inertia about a fixed axis is constant, the sum of the two area moments of inertia is also constant.

$$I_x + I_y = I_u + I_v \quad 11.88$$

There is one angle that will maximize the moment of inertia, $I_u$. This angle can be found from calculus by setting $dI_u/d\theta = 0$. The resulting equation defines two angles, one of which maximizes $I_u$, the other of which minimizes $I_u$.

$$\tan 2\theta = -\frac{2P_{xy}}{I_x - I_y} \quad 11.89$$

The two angles which satisfy equation 11.89 are 90° apart. These are known as the *principal axes.* The moments of inertia about these two axes are known as the *principal moments of inertia.* These principal moments are given by equation 11.90.

$$I_{\text{max, min}} = \tfrac{1}{2}(I_x + I_y) \pm \sqrt{\tfrac{1}{4}(I_x - I_y)^2 + P_{xy}^2} \quad 11.90$$

## 17 PROPERTIES OF MASSES

### A. CENTER OF GRAVITY

The *center of gravity* in 3-dimensional objects is analogous to centroids in 2-dimensional areas. The center of gravity can be located mathematically if the object can be described by a mathematical function.

$$\overline{x} = \frac{\int x\ dm}{m} \quad 11.91$$

$$\overline{y} = \frac{\int y\ dm}{m} \quad 11.92$$

$$\overline{z} = \frac{\int z\ dm}{m} \quad 11.93$$

The location of the center of gravity often is obvious for simple objects. It always is located on an axis of symmetry. If the object is complex or composite, the overall center of gravity can be found from the individual centers of gravity of the constituent objects.

$$\overline{x}_{\text{composite}} = \frac{\sum(m_i\overline{x}_i)}{\sum m_i} \quad 11.94$$

$$\overline{y}_{\text{composite}} = \frac{\sum(m_i\overline{y}_i)}{\sum m_i} \quad 11.95$$

$$\overline{z}_{\text{composite}} = \frac{\sum(m_i\overline{z}_i)}{\sum m_i} \quad 11.96$$

### B. MASS MOMENT OF INERTIA

The mass moment of inertia can be thought of as a measure of resistance to rotational motion. Although it can be found mathematically from equations 11.97, 11.98, and 11.99, it is more expedient to use appendix B to evaluate simple objects.

$$I_x = \int (y^2 + z^2)\ dm \quad 11.97$$

$$I_y = \int (x^2 + z^2)\ dm \quad 11.98$$

$$I_z = \int (x^2 + y^2)\ dm \quad 11.99$$

The *centroidal mass moment of inertia* is found by evaluating the moment of inertia about an axis passing through the object's center of gravity. Once this centroidal mass moment of inertia is known, the parallel axis theorem can be used to find the moment of inertia about any parallel axis.

$$I_{\text{any parallel axis}} = I_c + md^2 \quad 11.100$$

The radius of gyration of a 3-dimensional object is defined by equation 11.101.

$$r = \sqrt{\frac{I}{m}} \quad 11.101$$

$$I = r^2 m \quad 11.102$$

## 18 FRICTION

Friction is a force which resists motion or attempted motion. It always acts parallel to the contacting surfaces. The frictional force exerted on a stationary object is known as *static friction* or *coulomb friction.* If the object is moving, the friction is known as *dynamic friction.* Dynamic friction is less than static friction in most situations.

The actual magnitude of the frictional force depends on the *normal force* and the *coefficient of friction*, $f$, between the object and the surface. For an object resting on a horizontal surface, the normal force is the weight, $w$.

$$\mathbf{F}_f = f\mathbf{N} = fw \tag{11.103}$$

If the object is resting on an inclined surface, the normal force will be

$$\mathbf{N} = w\cos\theta \tag{11.104}$$

The frictional force again is equal to the product of the normal force and the coefficient of friction.

$$\mathbf{F}_f = f\mathbf{N} = fw\cos\theta \tag{11.105}$$

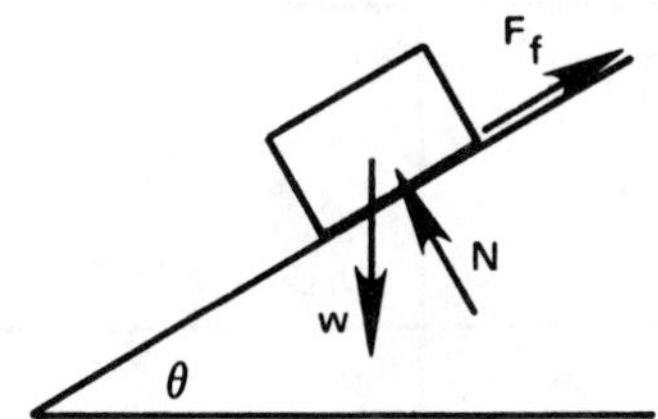


**Figure 11.16** Frictional Force

The object shown in figure 11.16 will not slip down the plane until the angle reaches a critical angle known as the *angle of repose.* This angle is given by equation 11.106.

$$\tan\theta = f \tag{11.106}$$

Typical values of the coefficient of friction are given in table 11.2.

Friction also exists between a belt, rope, or band wrapped around a drum, pulley, or sheave. If $\mathbf{T}_1$ is the tight side tension, if $\mathbf{T}_2$ is the slack side tension, and if $\phi$ is the contact angle in *radians*, the relationship governing *belt friction* is given by equation 11.107.

$$\frac{\mathbf{T}_1}{\mathbf{T}_2} = e^{f\phi} \tag{11.107}$$

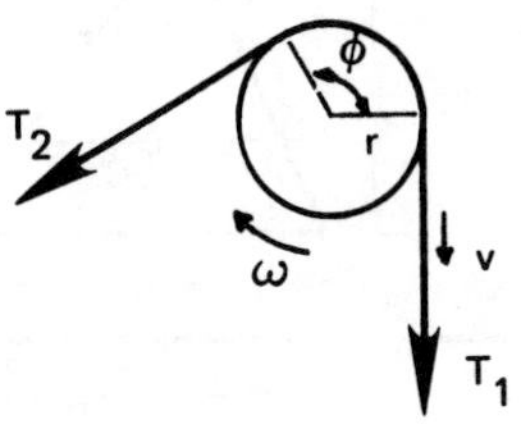


**Figure 11.17** Belt Friction

The transmitted torque in ft-lbf is

$$\text{torque} = (\mathbf{T}_1 - \mathbf{T}_2)\,r \tag{11.108}$$

The power in ft-lbf/sec transmitted by the belt running at speed $v$ in ft/sec is

$$\text{power} = (\mathbf{T}_1 - \mathbf{T}_2)\,v \tag{11.109}$$

**Table 11.2**
Approximate Coefficients of Friction

| material | condition | dynamic | static |
|---|---|---|---|
| cast iron on cast iron | dry | 0.15 | 1.10 |
| plastic on steel | dry | 0.35 | |
| grooved rubber on pavement | dry | 0.40 | 0.55 |
| bronze on steel | oiled | | 0.09 |
| steel on graphite | dry | | 0.21 |
| steel on steel | dry | 0.42 | 0.78 |
| steel on steel | oiled | 0.08 | 0.10 |
| steel on asbestos-faced steel | dry | | 0.15 |
| steel on asbestos-faced steel | oiled | | 0.12 |
| press fits (shaft in hole) | oiled | | 0.10–0.15 |

# Appendix A: Centroids and Area Moments of Inertia

| SHAPE | DIMENSIONS | CENTROID $(x_c, y_c)$ | AREA MOMENT OF INERTIA |
|---|---|---|---|
| Rectangle | y, y′, x′, x, h, C, O, b | (½b, ½h) | $I_{x'} = (1/12)bh^3$<br>$I_{y'} = (1/12)hb^3$<br>$I_x = (1/3)bh^3$<br>$I_y = (1/3)hb^3$<br>$J_C = (1/12)bh(b^2 + h^2)$ |
| Triangle | h, x′, C, x, b | $y_c = (h/3)$ | $I_{x'} = (1/36)bh^3$<br>$I_x = (1/12)bh^3$ |
| Trapezoid | y, b, $y'_c$, h, C, x′, x, O, B | $y'_c = \dfrac{h(2B + b)}{3(B + b)}$<br>Note that this is measured from the top surface. | $I_{x'} = \dfrac{h^3(B^2+4Bb+b^2)}{36(B+b)}$<br>$I_x = \dfrac{h^3(B+3b)}{12}$ |
| Quarter-Circle, of radius r | y, C, x, O | $((4r/3\pi), (4r/3\pi))$ | $I_x = I_y = (1/16)\pi r^4$<br>$J_O = (1/8)\pi r^4$ |
| Half Circle, of radius r | y, x′, C, x′, x, O | $(0, (4r/3\pi))$ | $I_x = I_y = (1/8)\pi r^4$<br>$J_O = ¼\pi r^4$  $I_{x'} = .11r^4$ |
| Circle, of radius r | y, O, x, C | (0,0) | $I_x = I_y = ¼\pi r^4$<br>$J_O = ½\pi r^4$ |
| Parabolic Area | 2a, C, h, x, O, y | $(0, (3h/5))$ | $I_x = 4h^3a/7$<br>$I_y = 4ha^3/15$ |
| Parabolic Spandrel | y, $y = kx^2$, C, h, O, x, a | $((3a/4),(3h/10))$ | $I_x = ah^3/21$<br>$I_y = 3ha^3/15$ |

# Appendix B: Mass Moments of Inertia

($m$ is in slugs; lengths are in feet)

| | | |
|---|---|---|
| Slender rod | 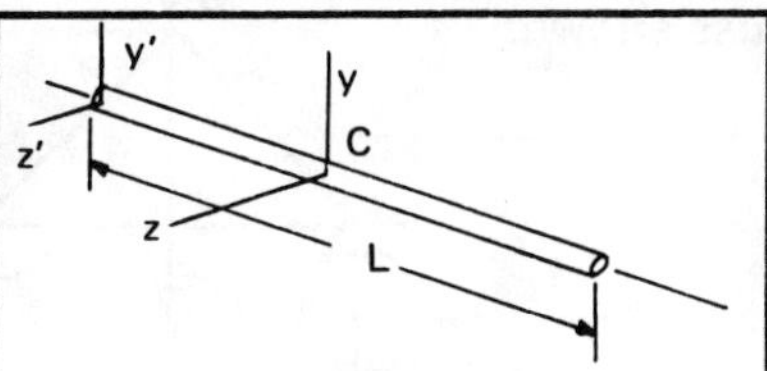 y', y, C, z', z, L | $I_y = I_z = (1/12)mL^2$<br>$I_{y'} = I_{z'} = (1/3)mL^2$ |
| Thin rectangular plate | 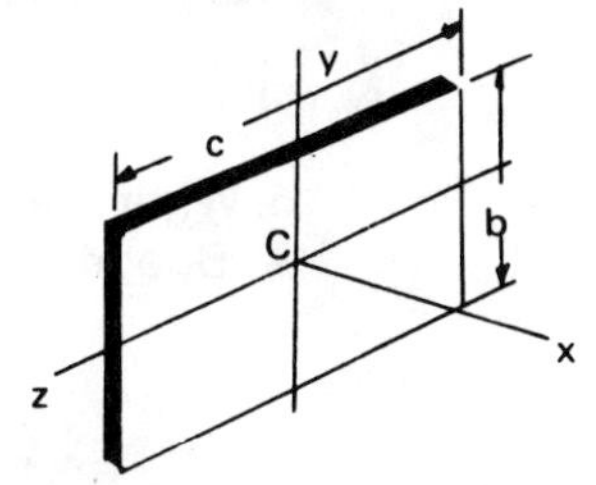 y, c, C, b, z, x | $I_x = (1/12)(b^2+c^2)\,m$<br>$I_y = (1/12)mc^2$<br>$I_z = (1/12)mb^2$ |
| Rectangular Parallelepiped | 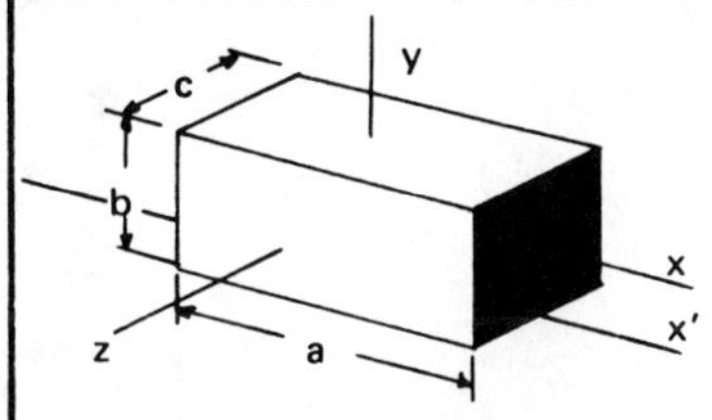 c, y, b, x, x', z, a | $I_x = (1/12)m(b^2+c^2)$<br>$I_y = (1/12)m(c^2+a^2)$<br>$I_z = (1/12)m(a^2+b^2)$<br>$I_{x'} = (1/12)m(4b^2+c^2)$ |
| Thin disk, radius r | 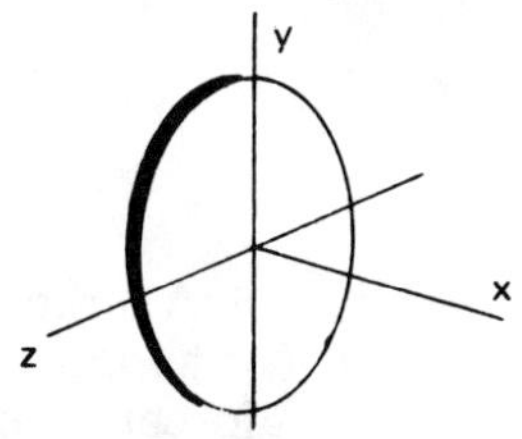 y, x, z | $I_x = \frac{1}{2}mr^2$<br>$I_y = I_z = \frac{1}{4}mr^2$ |
| Circular cylinder, radius r | 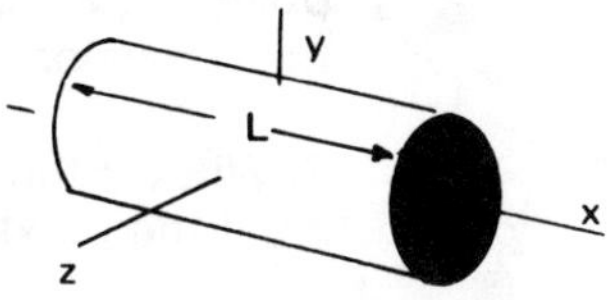 y, L, x, z | $I_x = \frac{1}{2}mr^2$<br>$I_y = I_z$<br>$= (1/12)m(3r^2+L^2)$ |
| Circular cone, base radius r | 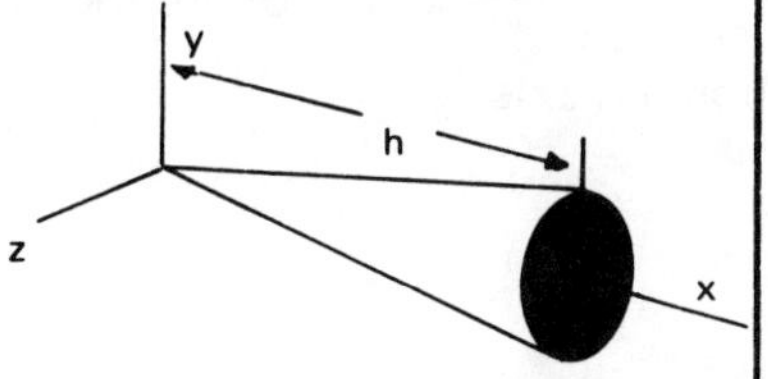 y, h, z, x | $I_x = (3/10)mr^2$<br>$I_y = I_z$<br>$= (3/5)m(\frac{1}{4}r^2+h^2)$ |
| Sphere, radius r | 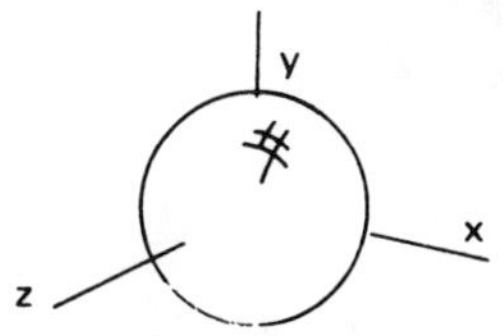 y, x, z | $I_x = I_y = I_z$<br>$= (2/5)mr^2$ |
| Hollow circular cylinder | 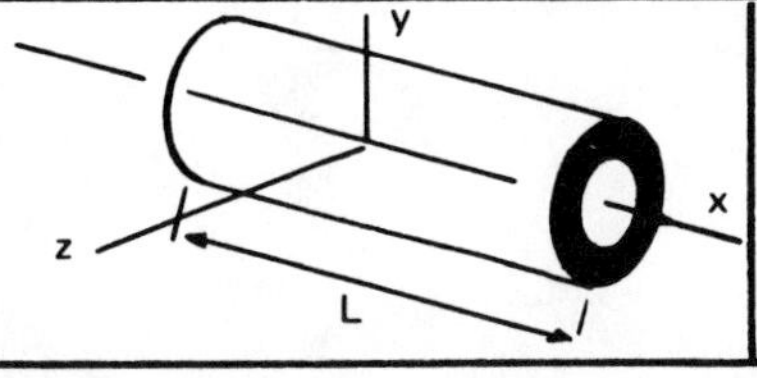 y, x, z, L | $I_x = \frac{1}{2}m(r_o^2 + r_i^2)$<br>$= \frac{\pi\rho L}{2}(r_o^4 - r_i^4)$ |

**Practice Problems: STATICS**

Untimed

1. Find the forces in all members of the truss shown.

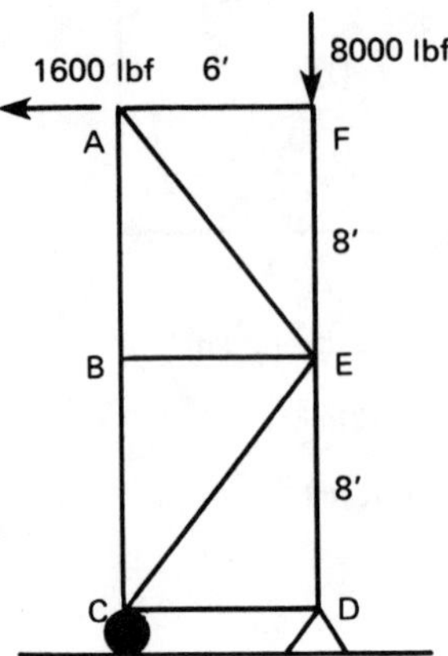


2. Find the forces in each of the legs.

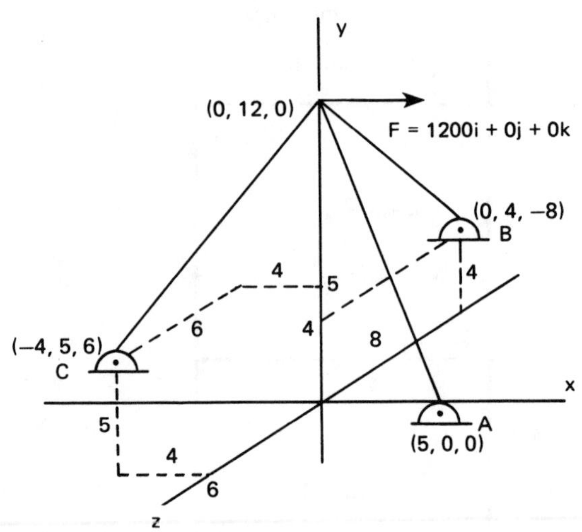


3. Find the centroidal moment of inertia about an axis parallel to the x-axis.

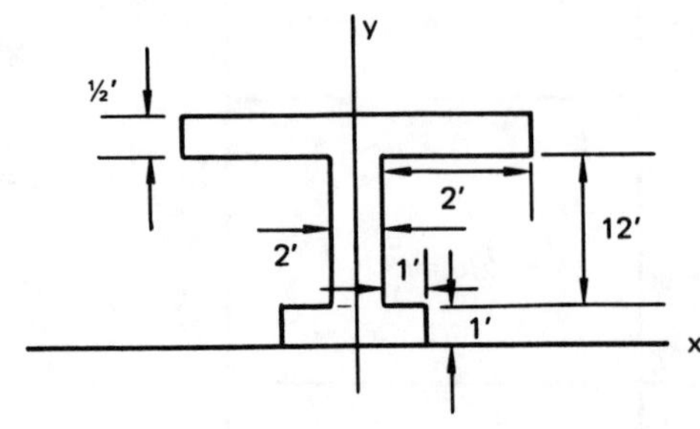


4. Find the forces in members DE and HJ.

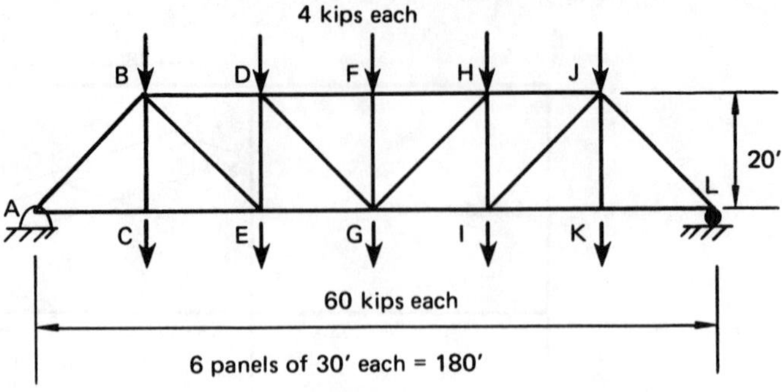


5. What are the x, y, and z components of the forces at A, B, and C?

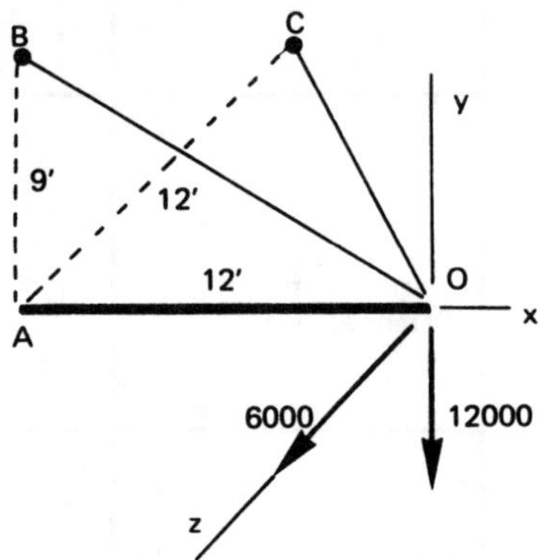


6. A power line weighs 2 pounds per foot of length. It is supported by two equal height towers over a level forest. The tower spacing is 100 feet and the mid-point sag is 10 feet. What are the maximum and minimum tensions?

7. What is the sag for the cable described in problem #6 if the maximum tension is 500 pounds?

8. Locate the centroid of the object shown.

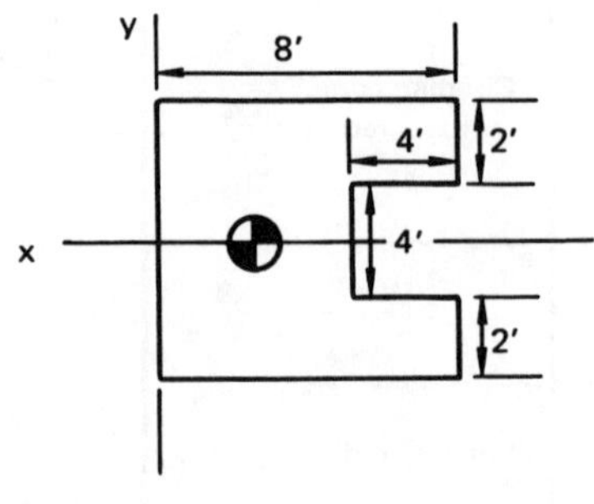

9. Replace the distributed load with three concentrated loads. Indicate the points of application.

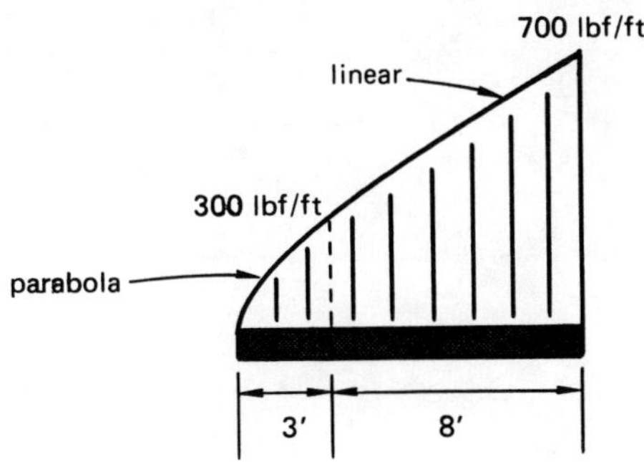


10. Find the forces in all members of the truss shown.

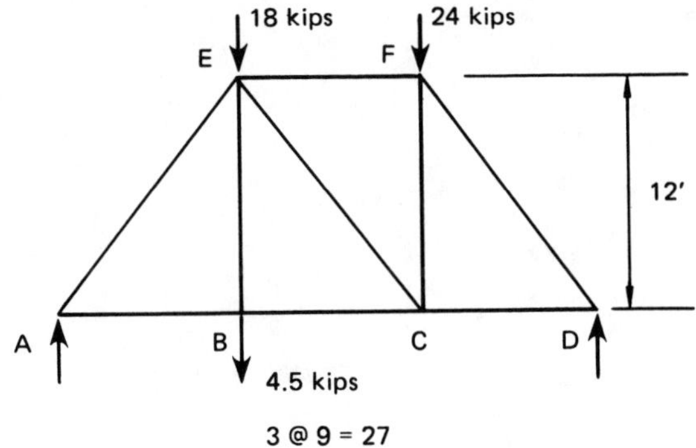

RESERVED FOR FUTURE USE

# 12 MECHANICS OF MATERIALS

## PART 1: Strength of Materials

*Nomenclature*

| | | |
|---|---|---|
| A | area | $in^2$ |
| b | width | in |
| c | distance from neutral axis to extreme fiber | in |
| C | correction | – |
| D | diameter | in |
| e | eccentricity | in |
| E | modulus of elasticity | psi |
| F | force, or load | lbf |
| F.S. | factor of safety | – |
| g | local gravitational acceleration | $ft/sec^2$ |
| $g_c$ | gravitational constant (32.2) | $\frac{lbm\text{-}ft}{lbf\text{-}sec^2}$ |
| G | shear modulus | psi |
| I | moment of inertia | $in^4$ |
| J | polar moment of inertia | $in^4$ |
| k | spring constant | lbf/in |
| K | stress concentration factor, or end restraint coefficient | |
| L | length | in |
| m | mass | lbm |
| M | moment | in-lbf |
| n | ratio, rotational speed, or number | –, rpm, – |
| N | number of cycles | – |
| p | pressure | psi |
| Q | statical moment | $in^3$ |
| r | radius, or radius of gyration | in |
| S | strength, or axial load | psi, lbf |
| t | thickness | in |
| T | temperature, or torque | °F, in-lbf |
| u | virtual truss load | lbf |
| U | energy | in-lbf |
| V | shear, or volume | lbf, $in^3$ |
| w | load per unit length, or width | lbf/in, in |
| W | work | in-lbf |
| x | distance, or displacement | in |
| y | deflection, or distance | in |
| Z | section modulus | $in^3$ |

*Symbols*

| | | |
|---|---|---|
| $\delta$ | elongation, or displacement | in |
| $\theta$ | angle | degrees |
| $\phi$ | angle | radians |
| $\sigma$ | normal stress | psi |
| $\alpha$ | coefficient of linear thermal expansion | 1/°F |
| $\beta$ | coefficient of volumetric thermal expansion | 1/°F |
| $\gamma$ | coefficient of area thermal expansion | 1/°F |
| $\tau$ | shear stress | psi |
| $\epsilon$ | strain | – |
| $\mu$ | Poisson's ratio | – |

*Subscripts*

| | |
|---|---|
| a | allowable |
| b | bending |
| br | bearing |
| c | centroidal, or compressive |
| e | endurance, Euler, or equivalent |
| ext | external |
| h | hoop |
| i | inside |
| L | long |
| o | original, or outside |
| p | pull |
| s | shear |
| t | transformed, tension, or temperature |
| th | thermal |
| T | torsion |
| u | ultimate |
| y | yield |

## 1 PROPERTIES OF STRUCTURAL MATERIALS

### A. THE TENSILE TEST

Many material properties can be derived from the standard tensile test. In a tensile test, a material sample is loaded axially in tension, and the elongation is measured as the load is increased. A graphical representation of typical test data for steel is shown in figure 12.1, in which the elongation, $\delta$, is plotted against the applied load, $F$.

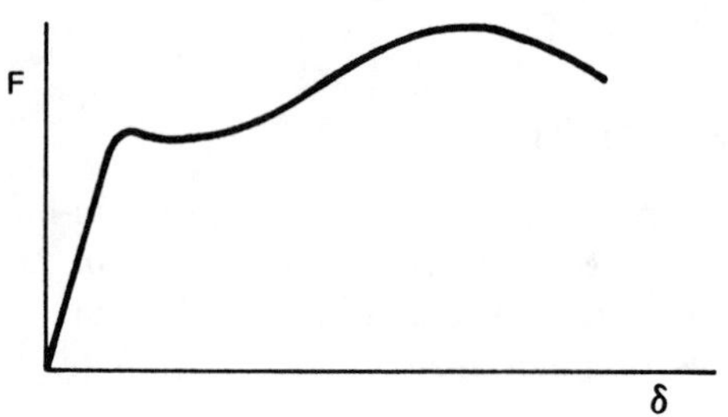


**Figure 12.1** Typical Tensile Test Results for Steel

Since this graph is applicable only to an object with the same length and area as the test sample, the data are converted to *stresses* and *strains* by use of equations 12.1 and 12.2. $\sigma$ is known as the *normal stress*, and $\epsilon$ is known as the *strain*. Strain is the percentage elongation of the sample.

$$\sigma = \frac{F}{A} \qquad 12.1$$

$$\epsilon = \frac{\delta}{L} \qquad 12.2$$

The stress-strain data also can be graphed, and the shape of the resulting curve will be the same as figure 12.1 with the scales changed.

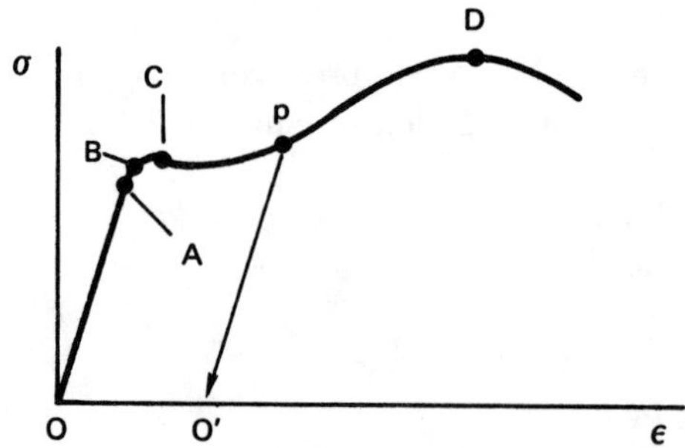


**Figure 12.2** A Typical Stress-Strain Curve for Steel

The line O-A in figure 12.2 is a straight line. The relationship between the stress and the strain is given by *Hooke's law*, equation 12.3. $E$ is the *modulus of elasticity* (*Young's modulus*) and is the slope of the line segment O-A. The stress at point $A$ is known as the *proportionality limit.* The modulus of elasticity for steel is approximately 3 EE7 psi.

$$\sigma = E\epsilon \qquad 12.3$$

Slightly above the proportionality limit is the *elastic limit* (point $B$). As long as the stress is kept below the elastic limit, there will be no permanent strain when the applied stress is removed. The strain is said to be *elastic*, and the stress is said to be in the *elastic region.*

If the elastic limit stress is exceeded before the load is removed, recovery will be along a line parallel to the straight line portion of the curve, as shown in the line segment p-O′. The strain that results (line O-O′) is permanent and is known as *plastic strain* or *permanent set.*

The *yield point* (point $C$) is very close to the elastic limit. For all practical purposes, the *yield stress*, $S_y$, can be taken as the stress which accompanies the beginning of plastic strain. Since permanent deformation is to be avoided, the yield stress is used in calculating safe stresses in ductile materials such as steel. A36 structural steel has a minimum yield strength of 36,000 psi.

$$\sigma_a = \frac{S_y}{F.S.} \qquad 12.4$$

Some materials, such as aluminum, do not have a well-defined yield point. This is illustrated in figure 12.3. In such cases, the yield point is taken as the stress which will cause a 0.2% *parallel offset.*

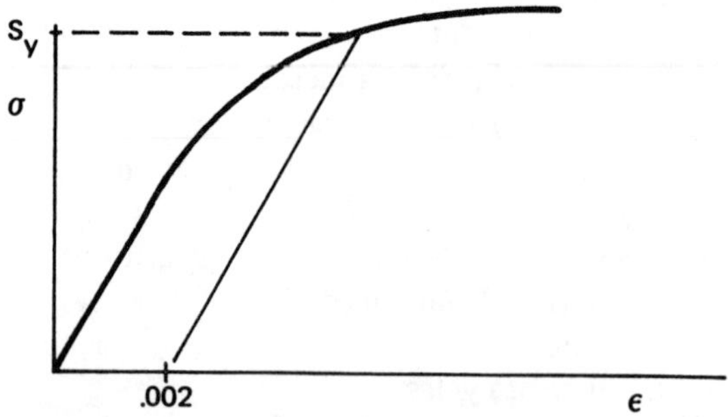


**Figure 12.3** A Typical Stress-Strain Curve for Aluminum

The *ultimate tensile strength*, point $D$ in figure 12.2, is the maximum load-carrying ability of the material. However, since stresses near the ultimate strength are accompanied by large plastic strains, this parameter should not be used for the design of ductile materials such as steel and aluminum.

As the sample is elongated during a tensile test, it also will decrease in thickness (width or diameter). The ratio of the lateral strain to the axial strain is known as *Poisson's ratio*, $\mu$. $\mu$ typically is taken as 0.3 for steel and as 0.33 for aluminum.

$$\mu = \frac{\epsilon_{\text{lateral}}}{\epsilon_{\text{axial}}} = \frac{\frac{\Delta D}{D_o}}{\frac{\Delta L}{L_o}} \qquad 12.5$$

B. FATIGUE TESTS

A part may fail after repeated stress loading even if the stress never exceeds the ultimate fracture strength of the material. This type of failure is known as *fatigue failure.*

The behavior of a material under repeated loadings can be evaluated in a fatigue test. A sample is loaded repeatedly to a known stress, and the number of applications of that stress is counted until the sample fails. This procedure is repeated for different stress levels. The results of many of these tests can be graphed, as is done in figure 12.4.

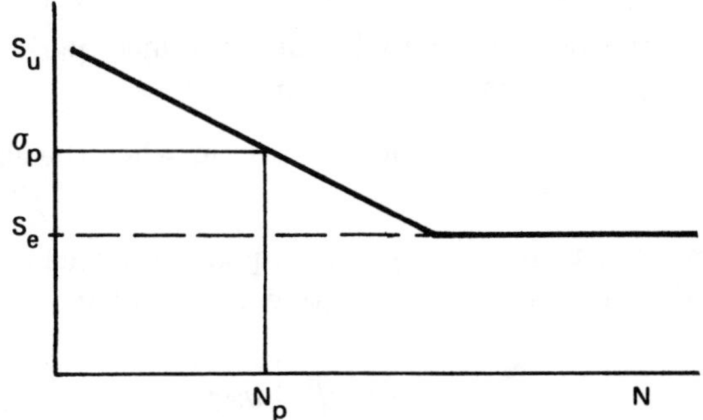


**Figure 12.4** Results of Many Fatigue Tests for Steel

For any given stress level, say $\sigma_p$ in figure 12.4, the corresponding number of applications of the stress which will cause failure is known as the *fatigue life.* That is, the fatigue life is just the number of cycles of stress required to cause failure. If the material is to fail after only one application of stress, the required stress must equal or exceed the ultimate strength of the material.

Below a certain stress level, called the *endurance limit* or the *endurance strength*, the part will be able to withstand an infinite number of stress applications without experiencing failure. Therefore, if a dynamically loaded part is to have an infinite life, the applied stress must be kept below the endurance limit.

Some materials, such as aluminum, do not have a well-defined endurance limit. In such cases, the endurance limit is taken as the stress that will cause failure at EE8 or 5 EE8 applications of the stress.

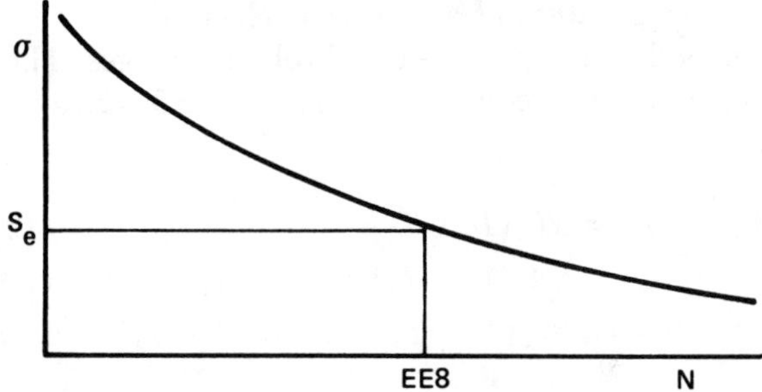


**Figure 12.5** Fatigue Test Results for Aluminum

C. ESTIMATES OF MATERIAL PROPERTIES

Although the properties of a material will depend on its classification (ASTM, AISC, etc.), average values are given in table 12.1.

## 2 DEFORMATION UNDER LOADING

Equation 12.2 can be rearranged to give the elongation of an axially loaded member in compression or tension.

$$\delta = L\epsilon = \frac{L\sigma}{E} = \frac{LF}{AE} \qquad 12.6$$

A tension load is taken as positive, and a compressive load is taken as negative. The actual length of a member under loading is given by equation 12.7 where the

**Table 12.1**
Typical Material Properties

| material | E (psi) | G (psi) | $\mu$ | $\rho$ (pcf) | $\alpha$ (1/°F) |
|---|---|---|---|---|---|
| steel (hard) | 30 EE6 | 11.5 EE6 | 0.30 | 489 | 6.5 EE−6 |
| steel (soft) | 29 EE6 | 11.5 EE6 | 0.30 | 489 | 6.5 EE−6 |
| aluminum alloy | 10 EE6 | 3.9 EE6 | 0.33 | 173 | 12.8 EE−6 |
| magnesium alloy | 6.5 EE6 | 2.4 EE6 | 0.35 | 112 | 14.5 EE−6 |
| titanium alloy | 15.4 EE6 | 6.0 EE6 | 0.34 | 282 | 4.9 EE−6 |
| cast iron (class 20) | 20 EE6 | 8 EE6 | 0.27 | 442 | 5.6 EE−6 |

algebraic sign of the deformation must be observed.

$$L_{\text{actual}} = L_o + \delta \qquad 12.7$$

The energy stored in a loaded member is equal to the work required to deform it. Below the proportionality limit, this energy is given by equation 12.8.

$$U = \frac{1}{2}F\delta = \frac{1}{2}\left(\frac{F^2L}{AE}\right) \qquad 12.8$$

## 3 THERMAL DEFORMATION

If the temperature of an object is changed, the object will experience length, area, and volume changes. These changes can be predicted by equations 12.9, 12.10, and 12.11.

$$\Delta L = \alpha L_o(T_2 - T_1) \qquad 12.9$$

$$\Delta A = \gamma A_o(T_2 - T_1) \approx 2\alpha A_o(T_2 - T_1) \qquad 12.10$$

$$\Delta V = \beta V_o(T_2 - T_1) \approx 3\alpha V_o(T_2 - T_1) \qquad 12.11$$

If equation 12.9 is rearranged, an expression for the *thermal strain* is obtained. Thermal strain is handled in the same manner as strain due to an applied load.

$$\epsilon_{th} = \frac{\Delta L}{L_o} = \alpha\,(T_2 - T_1) \qquad 12.12$$

For example, if a bar is heated but is not allowed to expand, the stress will be given by equation 12.13.

$$\sigma_{th} = E\,\epsilon_{th} \qquad 12.13$$

## 4 SHEAR AND MOMENT DIAGRAMS

It was illustrated in chapter 9 that, for an object in equilibrium, the sums of forces and moments are equal to zero everywhere. For example, the sum of moments about point $A$ for the beam shown in figure 12.6 is zero.

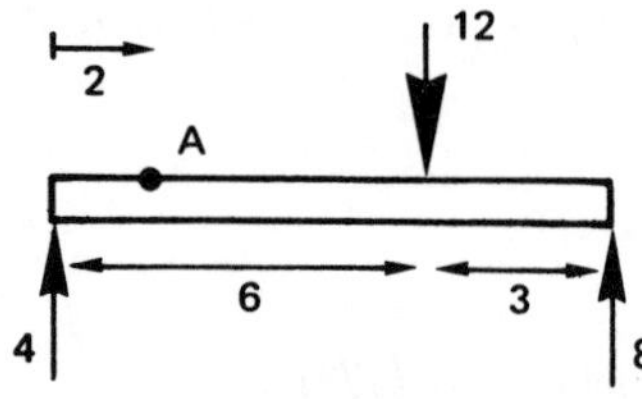


**Figure 12.6** A Beam in Equilibrium

Nevertheless, the beam shown in figure 12.6 will bend under the influence of the forces. This bending is evidence of the stress experienced by the beam. Since the sum of moments about any point is zero, the moment used to find stresses and deflection is taken from the point in question to one end of the beam only. This is called the *one-way moment.* The absolute value of the moment will not depend on the end used. This can be illustrated by the beam shown in figure 12.6.

$$\sum M_{A\,(\text{to right end})} = -8(7) + 4(12) = -8$$

$$\sum M_{A\,(\text{to left end})} = 4(2) = +8$$

The moment obtained will depend on the location chosen. A graphical representation of the one-way moment at every point along a beam is known as a *moment diagram.* The following guidelines should be observed in constructing moment diagrams.

- Moments should be taken from the left end to the point in question. If the beam is cantilever, place the built-in end at the right.
- Clockwise moments are positive. The *left-hand rule* should be used to determine positive moments.
- Concentrated loads produce linearly increasing lines on the moment diagram.
- Uniformly distributed loads produce parabolic lines on the moment diagram.
- The maximum moment will occur when the shear ($V$) is zero.
- The moment at any point is equal to the area under the shear diagram up to that point. That is,

$$M = \int V\,dx \qquad 12.14$$

- The moment is zero at a free end or hinge.

Similarly, the sum of forces in the $y$ direction on a beam in equilibrium is zero. However, the shearing stress at a point along the beam will depend on the sum of forces and reactions from the point in question to one end only.

A *shear diagram* is drawn to represent graphically the shear at any point along a beam. The following guidelines should be observed in constructing a shear diagram.

- Loads and reactions acting up are positive.
- The shear at any point is equal to the sum of the loads and reactions from the left end to the point in question.
- Concentrated loads produce straight (horizontal) lines on the shear diagram.

- Uniformly distributed loads produce straight sloping lines on the shear diagram.
- The magnitude of the shear at any point is equal to the slope of the moment diagram at that point.

$$V = \frac{dM}{dx} \qquad 12.15$$

*Example 12.1*

Draw the shear and moment diagrams for the following beam.

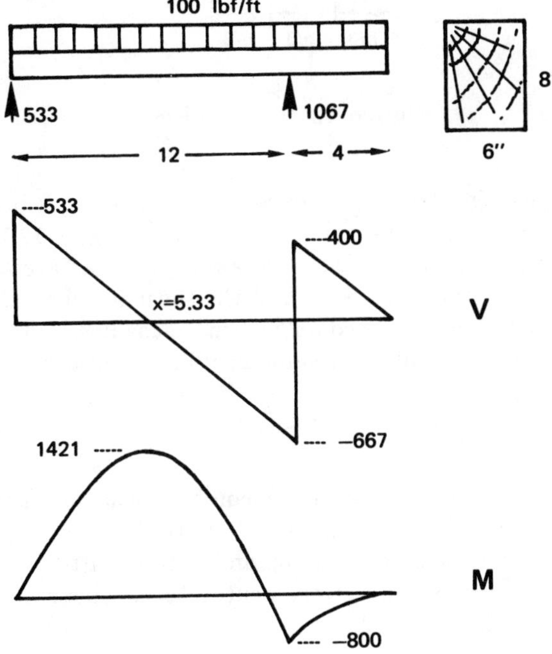


## 5 STRESSES IN BEAMS

### A. NORMAL STRESS

*Normal stress* is the type of stress experienced by a member which is axially loaded. The normal stress is the load divided by the area.

$$\sigma = \frac{F}{A} \qquad 12.16$$

Normal stress also occurs when a beam bends, as shown in figure 12.7. The lower part of the beam experiences normal tensile stress (which causes lengthening). The upper part of the beam experiences a normal compressive stress (which causes shortening). There is no stress along a horizontal plane passing through the centroid of the cross section. This plane is known as the *neutral plane* or the *neutral axis.*

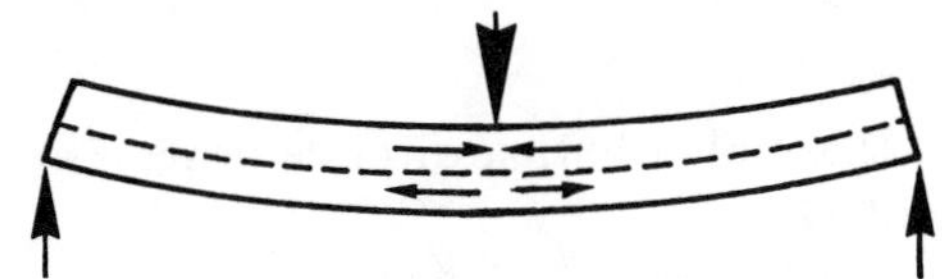

**Figure 12.7** Normal Stress Due to Bending

Although it is a normal stress, the stress produced by the bending usually is called *bending stress* or *flexure stress.* Bending stress varies with position within the beam. It is zero at the neutral axis, but it increases linearly with distance from the neutral axis.

$$\sigma_b = \frac{-My}{I_c} \qquad 12.17$$

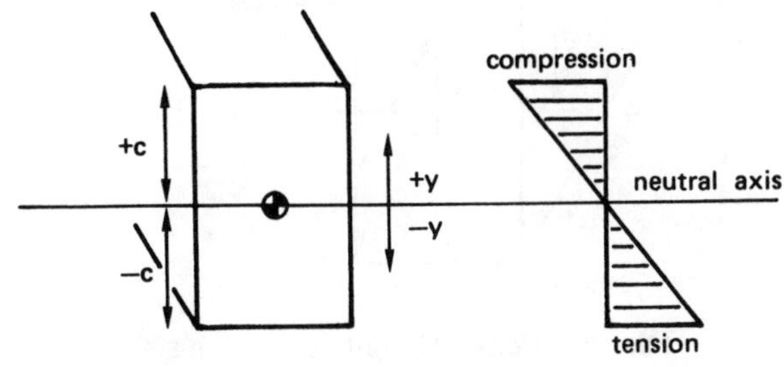


**Figure 12.8** Bending Stress Distribution in a Beam

The moment, $M$, used in equation 12.17 is the *one-way moment* previously discussed. $I_c$ is the centroidal moment of inertia of the beam's cross sectional area. The negative sign in equation 12.17 typically is omitted. However, it is required to be consistent with the convention that compression is negative.

Since the maximum stress will govern the design, $y$ can be set equal to $c$ to obtain the maximum stress. $c$ is the distance from the neutral axis to the *extreme fiber.*

$$\sigma_{b,\max} = \frac{Mc}{I_c} \qquad 12.18$$

For any given structural shape, $c$ and $I_c$ are fixed. Therefore, these two terms can be combined into the *section modulus,* $Z$.

$$\sigma_{b,\max} = \frac{M}{Z} \qquad 12.19$$

$$Z = \frac{I_c}{c} \qquad 12.20$$

For most beams, the section modulus, $Z$, is constant along the length of the beam. Equation 12.19 shows that the maximum stress along the length of a beam is proportional to the moment at that point. The location of the maximum bending moment is called the *dangerous section.* The dangerous section can be found directly from a moment or shear diagram of the beam.

If an axial member is loaded eccentrically, it will experience axial stress (equation 12.16) as well as bending stress (equation 12.17). This is illustrated by figure 12.9, in which a load is not applied to the centroid of a column's cross sectional area.

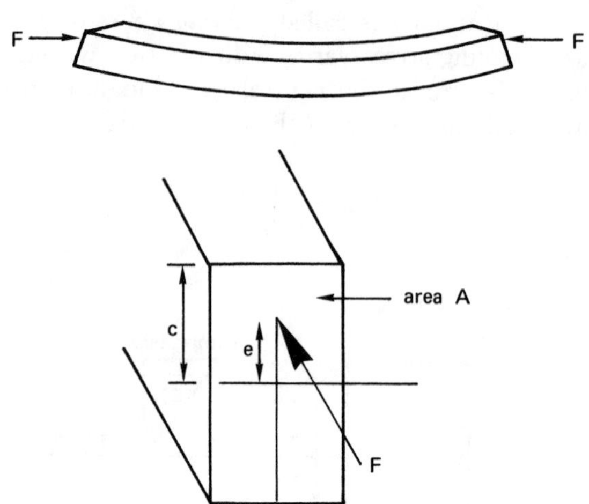


**Figure 12.9** Eccentric Loading of an Axial Member

Because the beam bends and supports a compressive load, the stress produced is a sum of bending and normal stress.

$$\sigma_{\text{max, min}} = \frac{F}{A} \pm \frac{Mc}{I_c} = \frac{F}{A} \pm \frac{Fec}{I_c} \qquad 12.21$$

If a cross section is loaded with an eccentric compressive load, part of the section can be in tension. This is illustrated in example 12.3. There will be no stress sign reversal, however, as long as the load is applied within a diamond-shaped area formed from the middle-thirds of the centroidal axes. This area is known as the *kern* or the *kernel.* It is particularly important to keep eccentric compressive loads within the kern on concrete and masonry piers since these materials do not tolerate tension. (The kern of a circular shaft or a round beam is outlined by a circle whose radius is one-quarter of the shaft radius.)

The *elastic strain energy* stored in a beam experiencing a moment (bending) is

$$U = \frac{1}{2} \int \frac{M^2}{EI} dx \qquad 12.22$$

B. SHEAR STRESS

Normal stress is produced when a load is absorbed by an area normal to it. *Shear stress* is produced by a load being carried by an area parallel to the load. This is illustrated in figure 12.11.

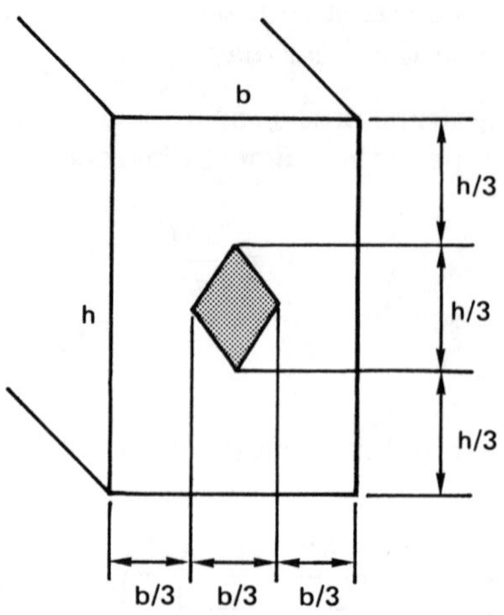


**Figure 12.10** The Kern

The average shear stress experienced by a pin, a bolt, or a rivet in single shear (as illustrated in figure 12.11) is given by equation 12.23. Because it gives an average value over the cross section of the shear member, this equation should be used only when the loading is low or when there is multiple redundancy in the shear group.

$$\tau = \frac{F}{A} \qquad 12.23$$

The actual shear stress in a beam is dependent on the location within the beam, just as was the bending stress. Shear stress is zero at the top and bottom surfaces of a beam and maximum at the neutral axis. This is illustrated in figure 12.12.

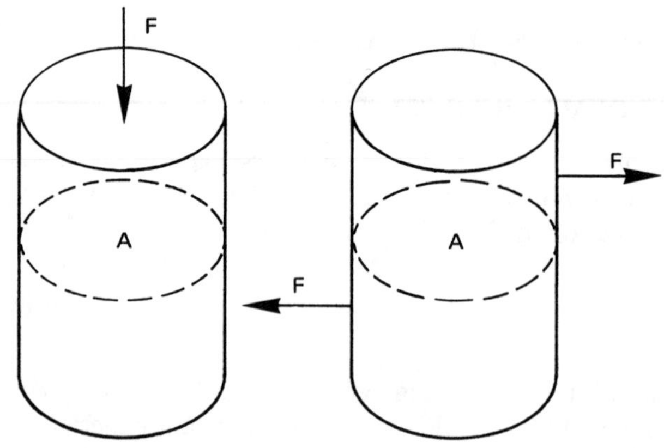


**Figure 12.11** Normal and Shear Stresses

The shear stress distribution within a beam is given by equation 12.24.

$$\tau = \frac{QV}{Ib} \qquad 12.24$$

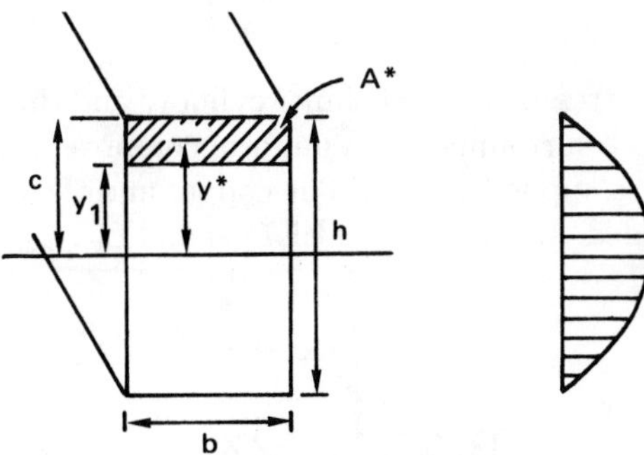


**Figure 12.12** Shear Stress Distribution in a Rectangular Beam

$V$ is the shear (in pounds) at the section where the shear stress is wanted. $V$ can be found from a shear diagram. $I$ is the beam's centroidal moment of inertia. $b$ is the width of the beam at the depth $y_1$ within the beam where the shear stress is wanted. $Q$ is the *statical moment*[1], as defined by equation 12.25.

$$Q = \int_{y_1}^{c} y \, dA \qquad 12.25$$

For rectangular beams, $dA = bdy$. Equation 12.25 can be simplified to equation 12.26 for rectangular beams.

$$Q = y^* A^* \qquad 12.26$$

Equation 12.26 says that the statical moment at a location $y_1$ within a rectangular beam is equal to the product of the area above $y_1$ and the distance from the centroidal axis to the centroid of $A^*$.

The maximum shear stress in a rectangular beam is

$$\tau_{\text{max}} = \frac{3V}{2A} = \frac{3V}{2bh} \qquad 12.27$$

For a round beam of radius $r$ and area $A$, the maximum shear stress is

$$\tau_{\text{max}} = \frac{4V}{3A} = \frac{4V}{3\pi r^2} \qquad 12.28$$

The shear, $V$, used in equations 12.27 and 12.28 is the *one-way shear.*

*Example 12.2*

What are the maximum shear and bending stresses for the beam shown in example 12.1?

From the shear diagram, the maximum shear is $-667$ pounds. From equation 12.27, the maximum shear stress is

$$\tau_{\text{max}} = \frac{(3)(-667)}{(2)(6)(8)} = -20.8 \text{ psi}$$

[1] The statical moment also is known as the *first moment of the area.*

From the moment diagram, the maximum moment is +1421 ft-lbf. The centroidal moment of inertia is

$$I_c = \frac{(6)(8)^3}{12} = 256 \text{ in}^4$$

From equation 12.18, the maximum bending stress is

$$\sigma_{b,\text{max}} = \frac{(1421)(12)(4)}{256} = 266.4 \text{ psi}$$

*Example 12.3*

The chain hook shown carries a load of 500 pounds. What are the minimum and maximum stresses in the vertical portion of the hook?

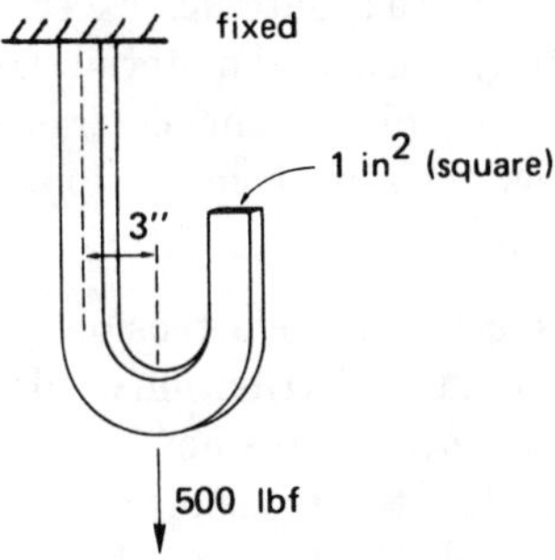


The hook is loaded eccentrically because the load and the supporting force are not in line. The centroidal moment of inertia of the $1'' \times 1''$ section is

$$I_c = \frac{bh^3}{12} = \frac{(1)(1)^3}{12} = 0.0833 \text{ in}^4$$

From equation 12.21,

$$\begin{aligned}\sigma_{\text{max, min}} &= \frac{500}{1} \pm \frac{(500)(3)(0.5)}{0.0833} \\ &= 500 \pm 9000 \\ &= +9500 \text{ and } -8500\end{aligned}$$

The 500 psi direct stress is tensile. However, the flexural compressive stress of 9000 psi counteracts this tensile stress, resulting in 8500 psi compressive stress at the outer face of the hook. The stress is 9500 psi tension at the inner face.

## 6 STRESSES IN COMPOSITE STRUCTURES

A *composite structure* is one in which two or more different materials are used, each carrying a part of the load. Unless all the various materials used have the same modulus of elasticity, the stress analysis will be dependent on the assumptions made.

Some simple composite structures can be analyzed using the assumption of *consistent deformations*. This is illustrated in examples 12.4 and 12.5. The technique used to analyze structures for which the strains are consistent is known as the *transformation method.*

*step 1*: Determine the modulus of elasticity for each of the materials used in the structure.

*step 2*: For each of the materials used, calculate the ratio

$$n = \frac{E}{E_{\text{weakest}}} \qquad 12.29$$

$E_{\text{weakest}}$ is the smallest modulus of elasticity of any of the materials used in the composite structure.

*step 3*: For all of the materials except the weakest, multiply the actual material stress area by $n$. Consider this expanded (*transformed*) area to have the same composition as the weakest material.

*step 4*: If the structure is a tension or compression member, the distribution or placement of the transformed area is not important. Just assume that the transformed areas carry the axial load. For beams in bending, the transformed area can add to the width of the beam, but it cannot change the depth of the beam or the thickness of the reinforcing.

*step 5*: For compression or tension numbers, calculate the stresses in the weakest and stronger materials.

$$\sigma_{\text{weakest}} = \frac{F}{A_t} \qquad 12.30$$

$$\sigma_{\text{stronger}} = \frac{nF}{A_t} \qquad 12.31$$

*step 6*: For beams in bending, proceed through step 9. Find the centroid of the transformed beam.

*step 7*: Find the centroidal moment of inertia of the transformed beam, $I_{ct}$.

*step 8*: Find $V_{\text{max}}$ and $M_{\text{max}}$ by inspection or from the shear and moment diagrams.

*step 9*: Calculate the stresses in the weakest and stronger materials.

$$\sigma_{\text{weakest}} = \frac{Mc_{\text{weakest}}}{I_{ct}} \qquad 12.32$$

$$\sigma_{\text{stronger}} = \frac{nMc_{\text{stronger}}}{I_{ct}} \qquad 12.33$$

*Example 12.4*

Find the stress in the steel inner cylinder and the copper tube which surrounds it if a uniform compressive load of 100 kips is applied axially. The copper and the steel are well bonded. Use $E_{\text{steel}} = 3$ EE7 psi and $E_{\text{copper}} = 1.75$ EE7 psi.

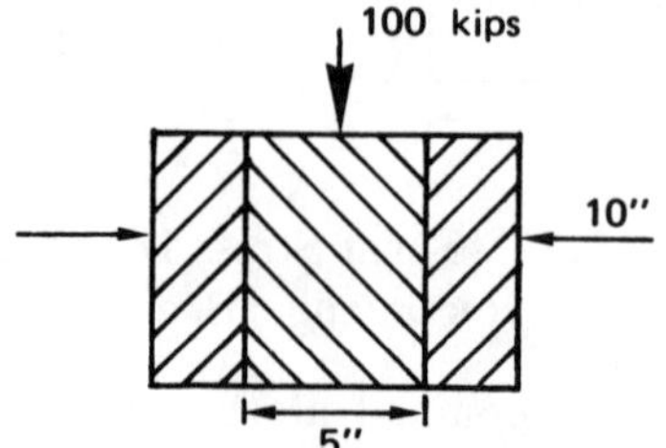


$$n = \frac{3\text{ EE7}}{1.75\text{ EE7}} = 1.714$$

The actual steel area is $\frac{1}{4}\pi(5)^2 = 19.63\text{ in}^2$.

The actual copper area is $\frac{1}{4}\pi[(10)^2 - (5)^2] = 58.9\text{ in}^2$.

The transformed area is $A_t = 58.9 + 1.714\,(19.63) = 92.55\text{ in}^2$.

$$\sigma_{\text{copper}} = \frac{100{,}000}{92.55} = 1080.5\text{ psi}$$

$$\sigma_{\text{steel}} = (1.714)(1080.5) = 1852.0\text{ psi}$$

*Example 12.5*

Find the maximum bending stress in the steel-reinforced wood beam shown at a point where the moment is 40,000 ft-lbf. Use $E_{\text{steel}} = 3$ EE7 psi and $E_{\text{wood}} = 1.5$ EE6 psi.

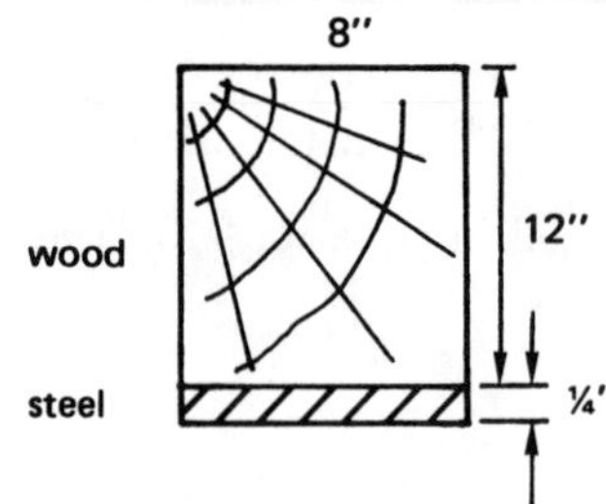


$$n = \frac{3\text{ EE7}}{1.5\text{ EE6}} = 20$$

The actual steel area is $(0.25)(8) = 2$.

The area of the steel is expanded to $20(2) = 40$. Since the depth of beam and reinforcement cannot be increased, the width must increase. The $160''$ dimension is arrived at by dividing the area of 40 square inches by the thickness of $\frac{1}{4}''$.

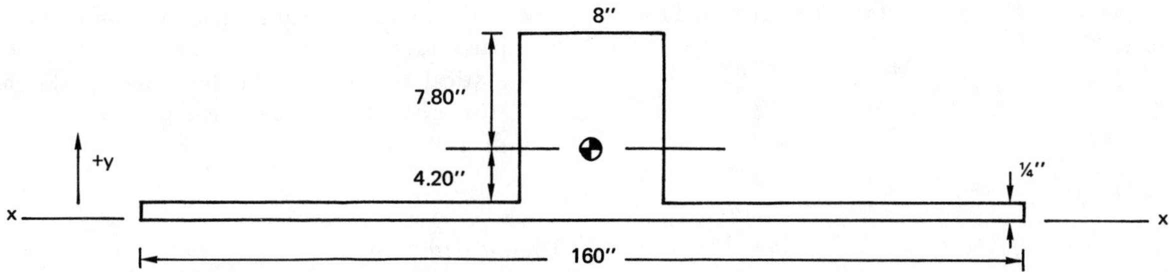


The centroid is located at $\bar{y} = 4.45$ inches from the x-x axis. The centroidal moment of inertia of the transformed section is $I_c = 2211.5$ in$^4$. Then, from equations 12.32 and 12.33,

$$\sigma_{\text{max, wood}} = \frac{(40,000)(12)(7.8)}{(2211.5)} = 1692 \text{ psi}$$

$$\sigma_{\text{max, steel}} = \frac{(20)(40,000)(12)(4.45)}{2211.5} = 19,320 \text{ psi}$$

## 7 ALLOWABLE STRESSES

Once the actual stresses are known, they must be compared to allowable stresses. If the allowable stress is calculated, it should be based on the yield stress and a reasonable factor of safety. This is known as the *allowable stress design method* or the *working stress design method.*

$$\sigma_a = \frac{S_y}{F.S.} \qquad 12.34$$

For steel, the factor of safety ranges from 1.5 to 2.5, depending on the type of steel and application.

The allowable stress method is being replaced in structural work by the *load factor design method,* also known as the *ultimate strength method* and the *plastic design method.* In this method, the applied loads are multiplied by a load factor. The product must be less than the structural member's ultimate strength, usually determined from a table.

## 8 BEAM DEFLECTIONS

### A. DOUBLE INTEGRATION METHOD

The deflection and the slope of a loaded beam are related to the applied moment and shear by equations 12.35 through 12.38.

$$y = \text{deflection} \qquad 12.35$$

$$y' = \frac{dy}{dx} = \text{slope} \qquad 12.36$$

$$y'' = \frac{d^2y}{dx^2} = \frac{M}{EI} \qquad 12.37$$

$$y''' = \frac{d^3y}{dx^3} = \frac{V}{EI} \qquad 12.38$$

If the moment function, $M(x)$, is known for a section of the beam, the deflection at any point can be found from equation 12.39.

$$y = \frac{1}{EI}\int\left[\int M(x)\,dx\right]dx \qquad 12.39$$

In order to find the deflection, constants must be introduced during the integration process. These constants can be found from table 12.2.

**Table 12.2**
Beam Boundary Conditions

| end condition | y | y′ | y″ | V | M |
|---|---|---|---|---|---|
| simple support | 0 | | | | |
| built-in support | 0 | 0 | | | |
| free end | | | 0 | 0 | 0 |
| hinge | | | | | 0 |

*Example 12.6*

Find the tip deflection of the beam shown. $EI$ is 5 EE10 lbf-in$^2$ everywhere on the beam.

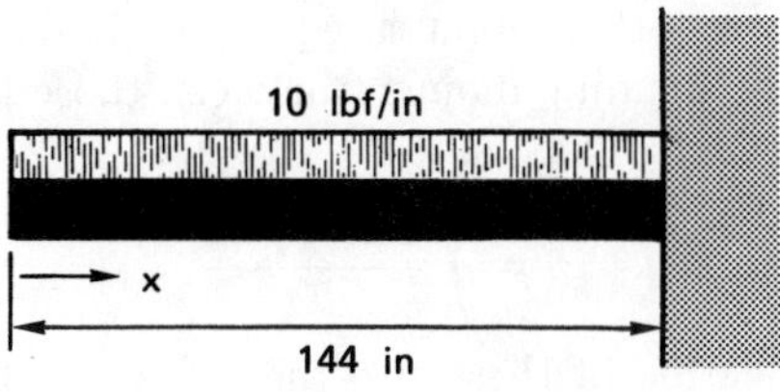


The moment at any point $x$ from the left end of the beam is

$$M(x) = (-10)(x)\left(\frac{1}{2}x\right) = -5x^2$$

This is negative by the left-hand rule convention. From equation 12.37,

$$y'' = \frac{M}{EI}$$

So,

$$EIy'' = -5x^2$$

$$EIy' = \int -5x^2\, dx = -\frac{5}{3}x^3 + C_1$$

Since $y' = 0$ at a built-in support (table 12.2) and $x = 144$ inches at the built-in support,

$$0 = -\frac{5}{3}(144)^3 + C_1$$

$$C_1 = 4.98 \text{ EE6}$$

$$EIy = \int \left(-\frac{5}{3}x^3 + 4.98 \text{ EE6}\right) dx$$

$$= -\frac{5}{12}x^4 + (4.98 \text{ EE6})x + C_2$$

Again, $y = 0$ at $x = 144$, so $C_2 = -5.38$ EE8.

Therefore, the deflection as a function of $x$ is

$$y = \left(\frac{1}{EI}\right)\left[\left(-\frac{5}{12}\right)x^4 + (4.98 \text{ EE6})x - 5.38 \text{ EE8}\right]$$

At the tip $x = 0$, so the deflection is

$$y_{\text{tip}} = \frac{-5.38 \text{ EE8}}{5 \text{ EE10}} = -0.0108 \text{ inches}$$

## B. MOMENT AREA METHOD

The moment area method is a semi-graphical technique which is applicable whenever slopes of deflection beams are not too great. This method is based on the following two theorems.

**Theorem I**: The angle between tangents at any two points on the *elastic line* of a beam is equal to the area of the moment diagram between the two points divided by $EI$. That is,

$$\theta = \int \frac{M(x)\, dx}{EI} \qquad 12.40$$

**Theorem II**: One point's deflection away from the tangent of another point is equal to the *statical moment* of the bending moment between those two points divided by $EI$. That is,

$$y = \int \frac{xM(x)\, dx}{EI} \qquad 12.41$$

The application of these two theorems is aided by the following two comments.

- If $EI$ is constant, the statical moment $\int xM(x)\, dx$ can be calculated as the product of the total moment diagram area times the distance from the point whose deflection is wanted to the centroid of the moment diagram.
- If the moment diagram has positive and negative parts (areas above and below the zero line), the statical moment should be taken as the sum of two products, one for each part of the moment diagram.

*Example 12.7*

Find the deflection, $y$, and the angle, $\theta$, at the free end of the cantilever beam shown.

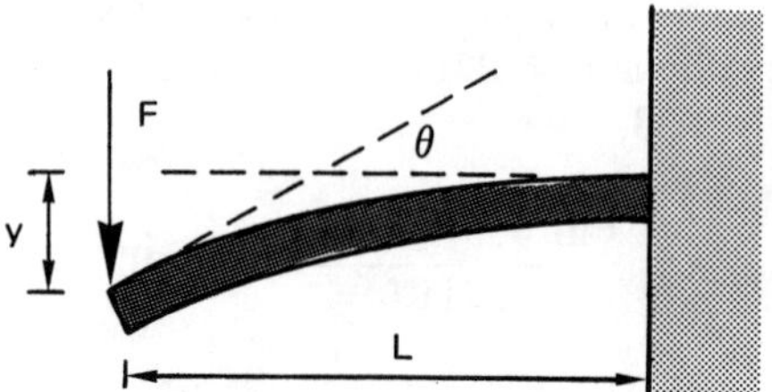


The deflection angle, $\theta$, is the angle between the tangents at the free and built-in ends (Theorem I). The moment diagram is

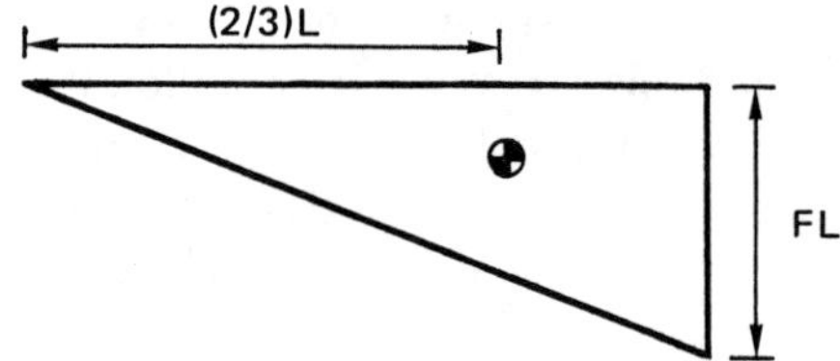


The area of the moment diagram is

$$\frac{1}{2}(FL)(L) = \frac{1}{2}FL^2$$

From Theorem I,

$$\theta = \frac{FL^2}{2EI}$$

From Theorem II,

$$y = \frac{FL^2}{2EI}\left(\frac{2}{3}L\right) = \frac{FL^3}{3EI}$$

*Example 12.8*

Find the deflection of the free end of the cantilever beam shown.

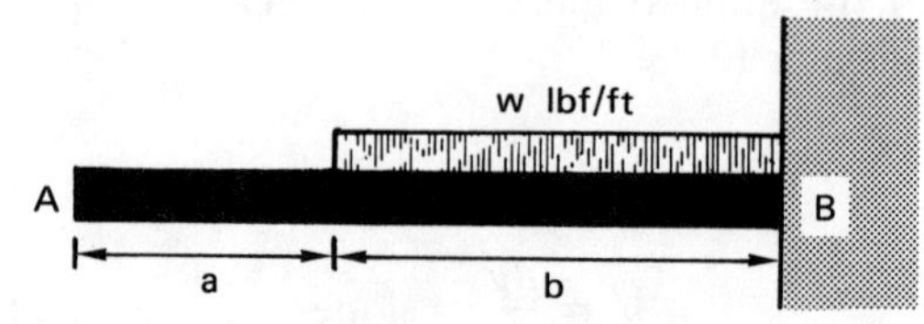

The distance from point $A$ (where the deflection is wanted) to the centroid is $(a + 0.75b)$. The area of the moment diagram is $(wb^3/6)$. From Theorem II,

$$y = \frac{wb^3}{6EI}(a + 0.75b)$$

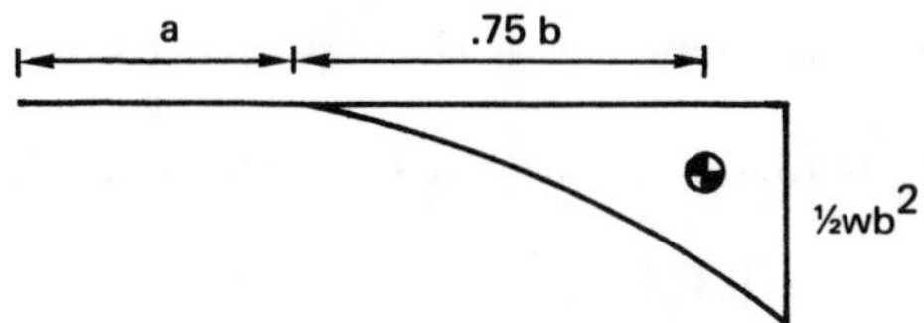


## C. STRAIN ENERGY METHOD

The deflection at a point of load application can be found by the strain energy method. This method equates the external work to the total internal strain energy as given by equations 12.8, 12.22, and 12.73. Since work is a force moving through a distance (which in this case is the deflection) we can write equation 12.42.

$$\frac{1}{2}Fy = \sum U \qquad 12.42$$

*Example 12.9*

Find the deflection at the tip of the stepped beam shown.

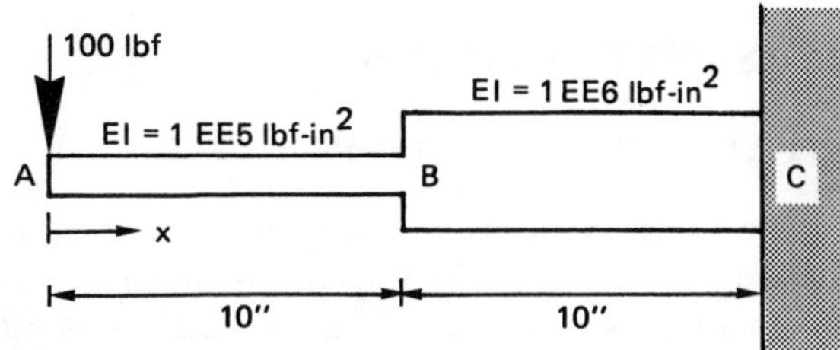


In section A–B: $M = 100x$ in-lbf

From equation 12.22,

$$U = \frac{1}{2}\int_0^{10} \frac{(100x)^2}{1\ \text{EE5}}dx = 16.67 \text{ in-lbf}$$

In section B-C: $M = 100x$

$$U = \frac{1}{2}\int_{10}^{20} \frac{(100x)^2}{1\ \text{EE6}}dx = 11.67 \text{ in-lbf}$$

Equating the internal work ($U$) and the external work,

$$16.67 + 11.67 = \frac{1}{2}(100)y$$
$$y = 0.567 \text{ in}$$

## D. CONJUGATE BEAM METHOD

The *conjugate beam method* changes a deflection problem into one of drawing moment diagrams. The method has the advantage of being able to handle beams of varying cross sections and materials. It has the disadvantage of not easily being able to handle beams with two built-in ends. The following steps constitute the conjugate beam method.

*step 1*: Draw the moment diagram for the beam as it is actually loaded.

*step 2*: Construct the $M/EI$ diagram by dividing the value of $M$ at every point along the beam by the product of $EI$ at that point. If the beam is of constant cross section, $EI$ will be constant, and the $M/EI$ diagram will have the same shape as the moment diagram. However, if the beam cross section varies with $x$, $I$ will change. In that case, the $M/EI$ diagram will not look the same as the moment diagram.

*step 3*: Draw a conjugate beam of the same length as the original beam. The material and the cross sectional area of this conjugate beam are not relevant.

(a) If the actual beam is simply supported at its ends, the conjugate beam will be simply supported at its ends.

(b) If the actual beam is simply supported away from its ends, the conjugate beam has hinges at the support points.

(c) If the actual beam has free ends, the conjugate beam has built-in ends.

(d) If the actual beam has built-in ends, the conjugate beam has free ends.

*step 4*: Load the conjugate beam with the $M/EI$ diagram. Find the conjugate reactions by methods of statics. Use the superscript, *, to indicate conjugate parameters.

*step 5*: Find the conjugate moment at the point where the deflection is wanted. The deflection is numerically equal to the moment as calculated from the conjugate beam forces.

*Example 12.10*

Find the deflections at the two load points. $EI$ has a constant value of 2.356 EE7 lbf-in$^2$.

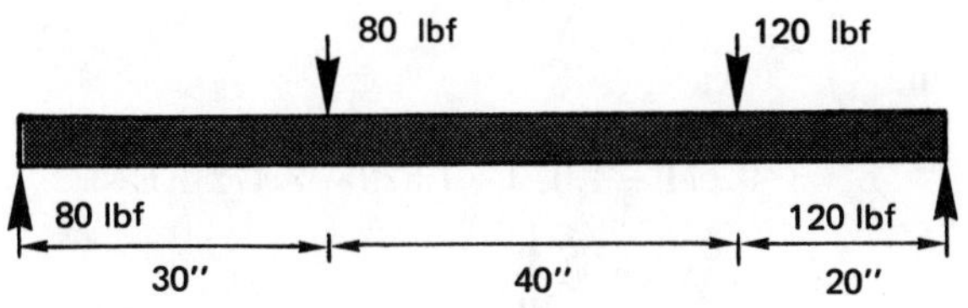

*step 1*: The moment diagram for the actual beam is

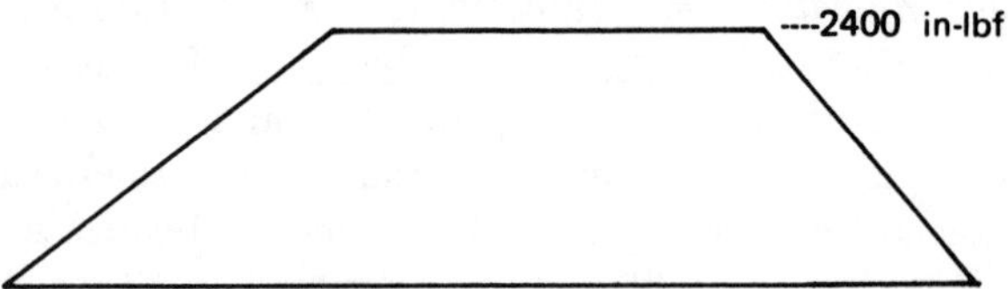


*steps 2, 3, and 4*: Since the cross section is constant, the conjugate load has the same shape as the original moment diagram. The peak load on the conjugate beam is

$$\frac{2400 \text{ in-lbf}}{2.356 \text{ EE7 lbf-in}^2} = 1.019 \text{ EE–4 (1/in)}$$

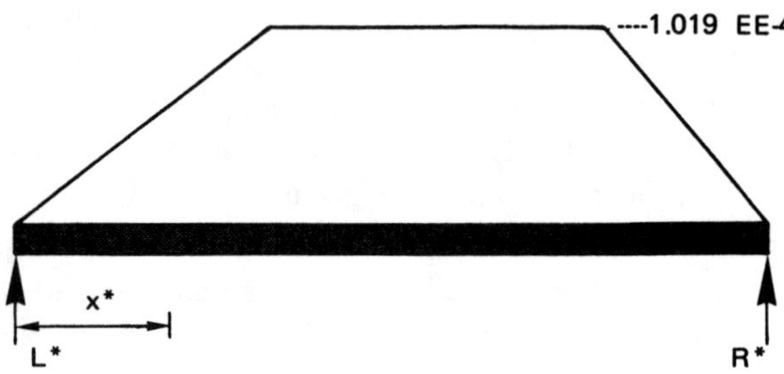


The conjugate reaction, $L^*$, is found by the following method. The loading diagram is assumed to be made up of a rectangular load and two "negative" triangular loads. The area of the rectangular load (which has a centroid at $x^* = 45$) is (90)(1.019 EE–4) = 9.171 EE–3.

Similarly, the area of the left triangle (which has a centroid at $x^* = 10$) is $\frac{1}{2}$(30)(1.019 EE–4) = 1.529 EE–3. The area of the right triangle (which has a centroid at $x^* = 83.33$) is $\frac{1}{2}$(20)(1.019 EE–4) = 1.019 EE–3.

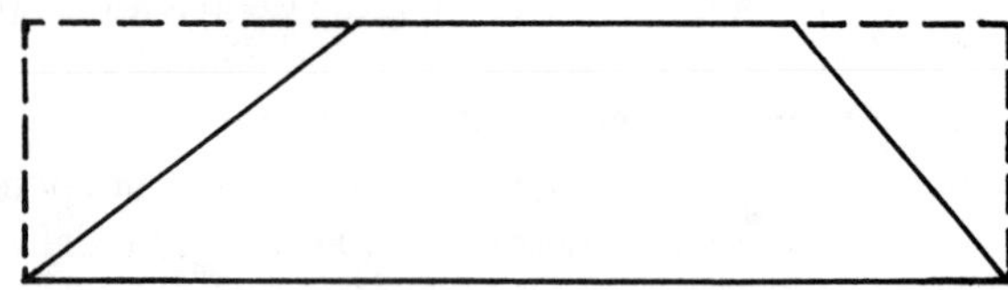

$$\begin{aligned}\sum M^*_{L^*} &= 90R^* + (1.019 \text{ EE–3})(83.3) \\ &\quad + (1.529 \text{ EE–3})(10) - (9.171 \text{ EE–3})(45) \\ &= 0\end{aligned}$$

$$R^* = 3.472 \text{ EE–3}\frac{1}{in}$$

Then,

$$\begin{aligned}L^* &= (9.171 - 1.019 - 1.529 - 3.472)\text{EE–3} \\ &= 3.151 \text{ EE–3}\frac{1}{in}\end{aligned}$$

*step 5*: The conjugate moment at $x^* = 30$ is

$$\begin{aligned}M^* &= (3.151 \text{ EE–3})(30) + (1.529 \text{ EE–3})(30 - 10) \\ &\quad - (9.171 \text{ EE–3})\left(\frac{30}{90}\right)(15) \\ &= 7.926 \text{ EE–2 in}\end{aligned}$$

The conjugate moment at the right-most load is

$$\begin{aligned}M^* &= (3.472 \text{ EE–3})(20) + (1.019 \text{ EE–3})(13.3) \\ &\quad - (9.171 \text{ EE–3})\left(\frac{20}{90}\right)(10) \\ &= 6.266 \text{ EE–2 in}\end{aligned}$$

### E. TABLE LOOK-UP METHOD

Appendix A is an extensive listing of the most commonly needed beam formulas. The use of these formulas is recommended whenever they can be applied singly or as part of a superposition solution.

### F. METHOD OF SUPERPOSITION

If the deflection at a point is due to the combined action of two or more loads, the deflections at that point due to the individual loads can be added to find the total deflection.

## 9 TRUSS DEFLECTIONS

### A. STRAIN-ENERGY METHOD

The deflection of a truss at the point of a single load application can be found by the *strain-energy method* if all member forces are known. This method is illustrated by example 12.11.

*Example 12.11*

Find the vertical deflection of point $A$ under the external load of 707 pounds. $AE$ = 10 EE5 pounds for all members. The internal forces have been determined.

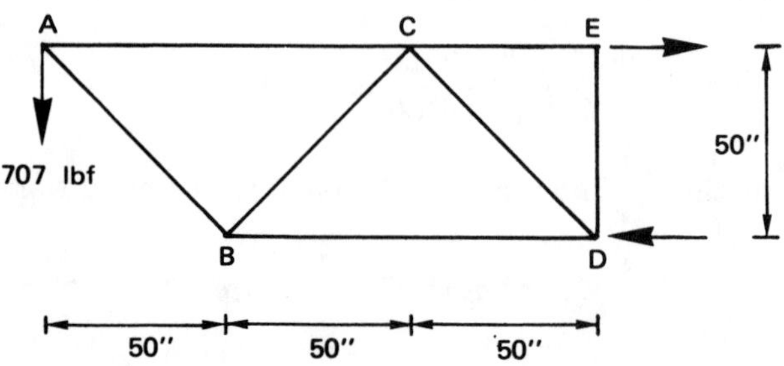

The length of member $AB$ is $\sqrt{(50)^2 + (50)^2} = 70.7$ inches. From equation 12.8, the internal strain energy in member $AB$ is

$$U = \frac{(-1000)^2(70.7)}{2\,(10\text{ EE5})} = 35.4\text{ in-lbf}$$

Similarly, the energy in all members can be determined.

| Member | L | F | U |
|---|---|---|---|
| AB | 70.7 | −1000 | +35.4 |
| BC | 70.7 | +1000 | +35.4 |
| AC | 100 | +707 | +25.0 |
| BD | 100 | −1414 | +100.0 |
| CD | 70.7 | −1000 | +35.4 |
| CE | 50 | +2121 | +112.5 |
| DE | 50 | +707 | +12.5 |
| | | | 356.2 |

The work done by a constant force $F$ moving through a distance $y$ is $Fy$. In this case, the force increases with $y$. The average force is $\frac{1}{2}F$. The external work is $W_{\text{ext}} = \frac{1}{2}(707)y$, so

$$\left(\frac{1}{2}\right)(707)y = 356.2$$

$$y = 1\,\text{inch}$$

## B. VIRTUAL WORK METHOD (HARDY CROSS METHOD)

An extension of the strain-energy method results in an easy procedure for computing the deflection of *any* point on a truss.

*step 1*: Draw the truss twice.

*step 2*: On the first truss, place all the actual loads.

*step 3*: Find the forces, $S$, due to the actual applied loads in all the members.

*step 4*: On the second truss, place a dummy one pound load in the direction of the desired displacement.

*step 5*: Find the forces, $u$, due to the one pound dummy load in all members.

*step 6*: Find the desired displacement from equation 12.43.

$$\delta = \sum \frac{SuL}{AE} \qquad 12.43$$

In equation 12.43, the summation is over all truss members which have non-zero forces in *both* trusses.

*Example 12.12*

What is the horizontal deflection of joint $F$ on the truss shown? Use $E = 3$ EE7 psi. Joint $A$ is restrained horizontally. (Member areas have been chosen for convenience.)

*steps 1 and 2*: Use the truss as drawn.

*step 3*: The forces in all the truss members are summarized in step 5.

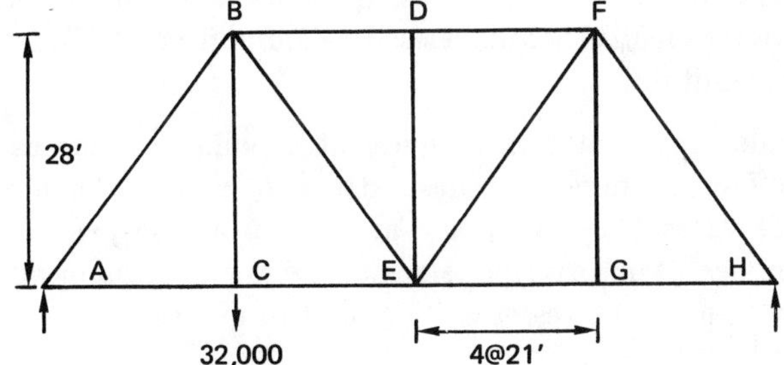


*step 4*: Draw the truss and load it with a unit horizontal force at point $F$.

*step 5*: Find the forces, $u$, in all members of the second truss. These are summarized in the following table. Notice the sign convention: + for tension and − for compression.

Table from example 12.12

| member | S(lbf) | u | L(ft) | A(in²) | $\frac{SuL}{AE}$(ft) |
|---|---|---|---|---|---|
| AB | −30,000 | 5/12 | 35 | 17.5 | −8.33 EE−4 |
| CB | 32,000 | 0 | 28 | 14 | 0 |
| EB | −10,000 | −5/12 | 35 | 17.5 | 2.78 EE−4 |
| ED | 0 | 0 | 28 | 14 | 0 |
| EF | 10,000 | 5/12 | 35 | 17.5 | 2.78 EE−4 |
| GF | 0 | 0 | 28 | 14 | 0 |
| HF | −10,000 | −5/12 | 35 | 17.5 | 2.78 EE−4 |
| BD | −12,000 | 1/2 | 21 | 10.5 | −4.00 EE−4 |
| DF | −12,000 | 1/2 | 21 | 10.5 | −4.00 EE−4 |
| AC | 18,000 | 3/4 | 21 | 10.5 | 9.00 EE−4 |
| CE | 18,000 | 3/4 | 21 | 10.5 | 9.00 EE−4 |
| EG | 6000 | 1/4 | 21 | 10.5 | 1.00 EE−4 |
| GH | 6000 | 1/4 | 21 | 10.5 | 1.00 EE−4 |
| | | | | | 12.01 EE−4 (ft) |

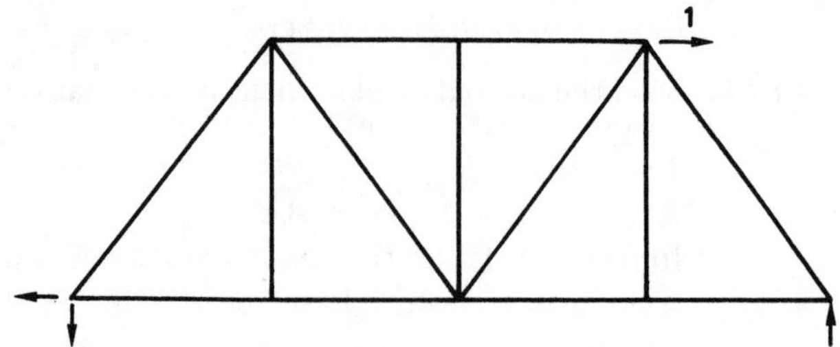


Since 12.01 EE–4 is positive, the deflection is in the direction of the dummy unit load. In this case, the deflection is to the right.

## 10 COMBINED STRESSES

Most practical cases of combined stresses have normal stresses on two perpendicular planes and a known shear stress acting parallel to these two planes. Based on knowledge of these stresses, the shear and the normal stresses on all other planes can be found from conditions of equilibrium.

Under any condition of stress at a point, a plane can be found where the shear stress is zero. The normal stresses on this plane are known as the *principal stresses*. The principal stresses are the maximum and the minimum stresses at the point in question.

The normal and shear stresses on a plane whose normal line is inclined an angle $\theta$ from the horizontal are given by equations 12.44 and 12.45.

$$\sigma_\theta = \frac{1}{2}(\sigma_x + \sigma_y) + \frac{1}{2}(\sigma_x - \sigma_y)\cos 2\theta + \tau \sin 2\theta \qquad 12.44$$

$$\tau_\theta = -\frac{1}{2}(\sigma_x - \sigma_y)\sin 2\theta + \tau \cos 2\theta \qquad 12.45$$

The maximum and minimum values of $\sigma_\theta$ and $\tau_\theta$ (as $\theta$ is varied) are the principal stresses. These are given by equations 12.46 and 12.47.

$$\sigma(\text{max, min}) = \tfrac{1}{2}(\sigma_x + \sigma_y) \pm \tau(\text{max}) \qquad 12.46$$

$$\tau(\text{max, min}) = \pm\tfrac{1}{2}\sqrt{(\sigma_x - \sigma_y)^2 + (2\tau)^2} \qquad 12.47$$

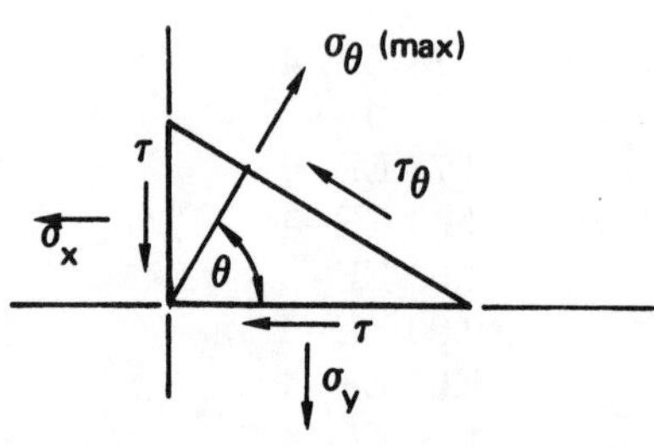


**Figure 12.13** Plane of Principal Stresses

The angles of the planes on which the normal stresses are minimum and maximum are given by equation 12.48. $\theta$ is measured from the $x$ axis, clockwise if negative and counterclockwise if positive. Equation 12.48 will yield two angles. These angles must be used in equation 12.44 to determine which angle corresponds to the minimum normal stress and which angle corresponds to the maximum normal stress.

$$\theta = \frac{1}{2}\arctan\left(\frac{2\tau}{\sigma_x - \sigma_y}\right) \qquad 12.48$$

The angles of the planes on which the shear stress is minimum and maximum are given by equation 12.49. The same angle sign convention used for equation 12.48 applies to equation 12.49.

$$\theta = \frac{1}{2}\arctan\left(\frac{\sigma_x - \sigma_y}{-2\tau}\right) \qquad 12.49$$

Proper sign convention must be adhered to when using equations 12.44 through 12.49. Normal tensile stresses are positive; normal compressive stresses are negative. Shear stresses are positive as shown in figure 12.14.

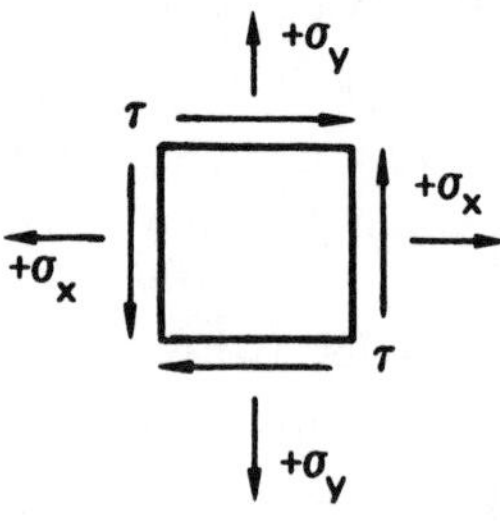


**Figure 12.14** Sign Convention

*Example 12.13*

Find the maximum shear stress and the maximum normal stress on the object shown.

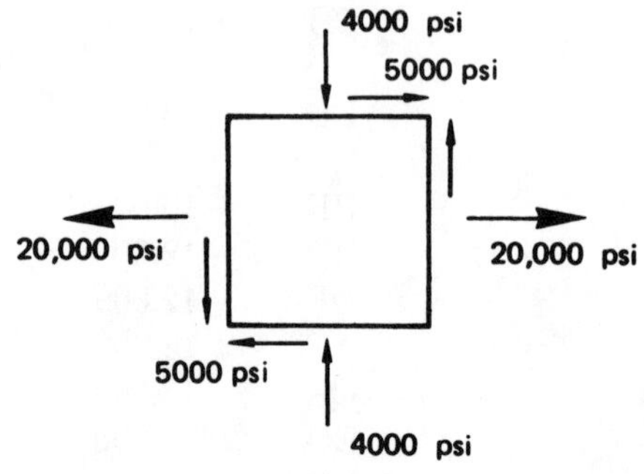

By the sign convention of figure 12.14, the 4000 psi is negative. From equation 12.47, the maximum shear stress is

$$\tau_{\text{max}} = \frac{1}{2}\sqrt{[20{,}000 - (-4000)]^2 + [(2)(5000)]^2}$$
$$= 13{,}000 \text{ psi}$$

From equation 12.46, the maximum normal stress is

$$\sigma_{\text{max}} = \frac{1}{2}[20{,}000 + (-4000)] + 13{,}000$$
$$= 21{,}000 \text{ psi (tension)}$$

## 11 DYNAMIC LOADING

If a load is applied suddenly to a structure, the transient response may create stresses greater than would normally be calculated from the concepts of statics and mechanics of materials alone. Although a *dynamic analysis* of the structure is appropriate, the procedure is extremely lengthy and complicated. Therefore, arbitrary dynamic factors are applied to the static stress. For example, if the load is applied quickly compared to the natural period of the structure, a dynamic factor of 2 can be used. This assumes that the load is applied as a ramp function.

## 12 INFLUENCE DIAGRAMS

Shear, moment, and reaction influence diagrams (influence lines) can be drawn for any point on a beam or truss. This is a necessary first step in the evaluation of stresses induced by moving loads. It is important to realize, however, that the influence diagram applies only to one point on the beam or truss.

### A. INFLUENCE DIAGRAMS FOR BEAM REACTIONS

In a typical problem, the load is fixed in position, and the reactions do not change. If a load is allowed to move across a beam, the reactions will vary. An influence diagram can be used to investigate the value of a chosen reaction as the load position varies.

To make the influence diagram as general in application as possible, the load is taken as one pound. As an example, consider a 20 foot, simply supported beam, and determine the effect on the left reaction of moving a one-pound load across the beam.

If the load is directly over the right reaction ($x = 0$), the left reaction will not carry any load. Therefore, the ordinate of the influence diagram is zero at that point. (Even though the right reaction supports one pound, this influence diagram is being drawn for one point only—the left reaction.) Similarly, if the load is directly over the left reaction ($x = L$), the ordinate of the influence diagram will be 1. Basic statics can be used to complete the rest of the diagram, as shown in figure 12.15.

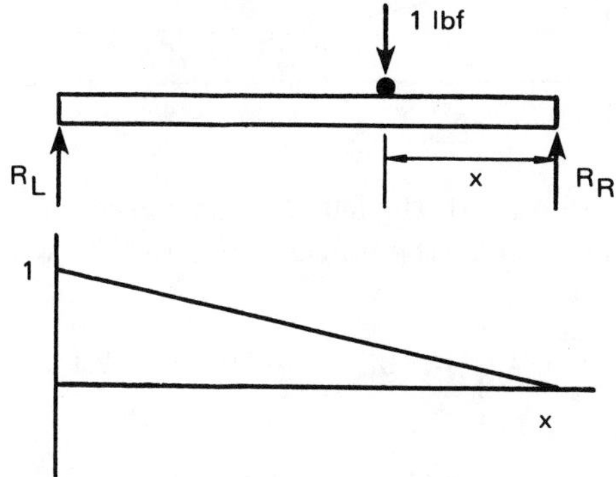


**Figure 12.15** Influence Diagram for Reaction of Simple Beam

We can use this rudimentary example of an influence diagram to calculate the left reaction for any placement of any load (not just one-pound loads) by multiplying the actual load by the ordinate of the influence diagram.

$$R_L = P \times \text{ordinate} \qquad 12.50$$

Even though the influence diagram was drawn for a point load, it can still be used when the beam carries a uniformly distributed load. In the case of a uniform load of $w$ lbf/ft distributed over the beam from $x_1$ to $x_2$, the left reaction can be calculated from equation 12.51.

$$R_L = \int_{x_1}^{x_2} (w \times \text{ordinate})dx$$
$$= w \times \text{area under curve} \qquad 12.51$$

*Example 12.14*

A 500 lbf load is placed 15 feet from the right end of a 20 foot, simply supported beam. Use the influence diagram to determine the left reaction.

Since the influence line increases linearly from 0 to 1, the ordinate is the ratio of position to length. That is, the ordinate is $15/20 = 0.75$. The left reaction is

$$R_L = (0.75)(500) = 375 \text{ lbf}$$

*Example 12.15*

A uniform load of 15 lbf/ft is distributed between $x = 4$ and $x = 10$ along a 20 foot, simply supported beam. What is the left reaction?

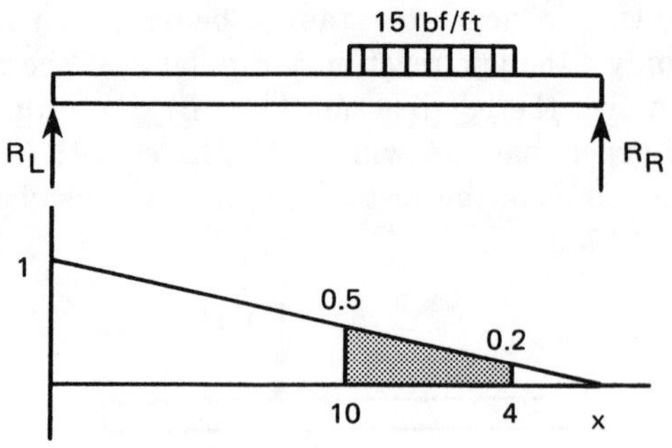


From equation 12.51, the left reaction can be calculated from the area under the influence diagram between the limits of loading.

$$\text{area} = \frac{1}{2}(10)(0.5) - \frac{1}{2}(4)(0.2) = 2.1$$

The left reaction is

$$R_L = (15)(2.1) = 31.5\,\text{lbf}$$

## B. FINDING REACTION INFLUENCE DIAGRAMS GRAPHICALLY

Since the reaction will always have a value of one when the unit load is directly over the reaction, and since the reaction is always directly proportional to the distance $x$, the reaction influence diagram can be easily determined from the following steps:

*step 1*: Remove the support being investigated

*step 2*: Displace (lift) the beam upward a distance of one unit at the support point. The resulting beam shape will be the shape of the reaction influence diagram.

*Example 12.16*

What is the approximate shape of the reaction influence diagram for reaction 2?

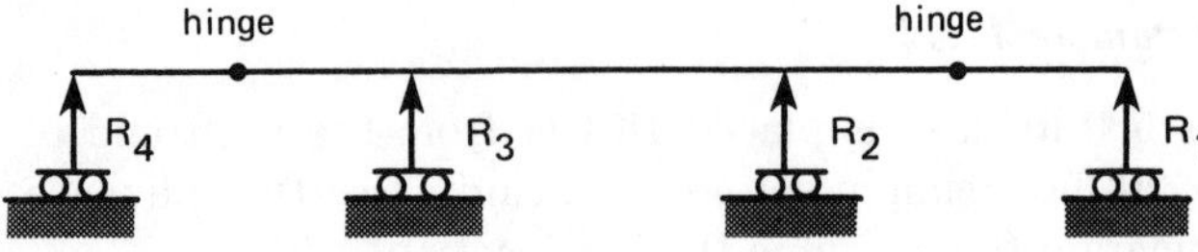


Pushing up at reaction 2 such that the deflection is one unit results in the shown shape.

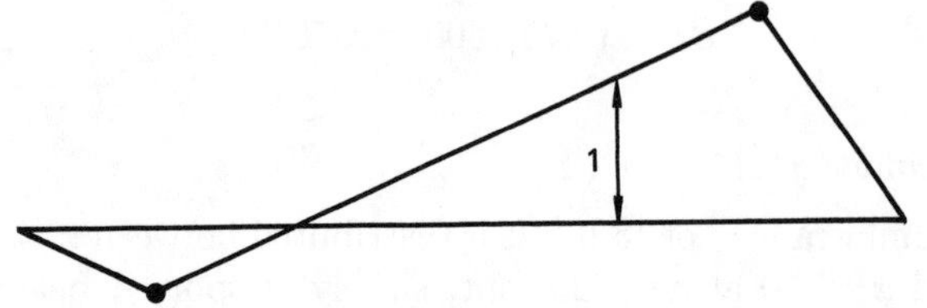


## C. INFLUENCE DIAGRAMS FOR BEAM SHEARS

A shear influence diagram (not the same as a shear diagram) illustrates the effect on the shear at a particular point in the beam of moving a load along the beam's length. As an illustration, consider point $A$ along the length of a 20 foot, simply supported beam.

In all cases, principles of statics can be used to calculate the shear at point $A$ as the sum of loads and reactions on the beam from point $A$ to the left end. (With the appropriate sign convention, summation to the right end could be used as well.) If the unit load is placed between the right end ($x = 0$) and point $A$, the shear at point $A$ will consist only of the left reaction, since there are no other loads between point $A$ and the left end. From the reaction influence diagram, we know that the left reaction varies linearly. At $x = 12$, the location of point $A$, the shear is $V = R_L = 12/20 = 0.6$.

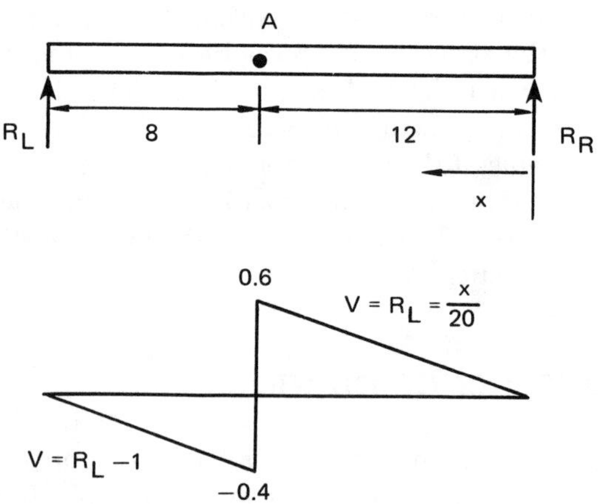


**Figure 12.16** Shear Influence Diagram for Simple Beam

When the unit load is between point $A$ and the left end, the shear at point $A$ is the sum of the left reaction (upward and positive) and the unit load itself (downward and negative). Therefore, $V = R_L - 1$. At $x = 12$, the shear is $V = 0.6 - 1 = -0.4$

Figure 12.16 is the shear influence diagram. In the diagram, notice that the shear goes through a reversal of 1. It is also helpful to note that the slopes of the two inclined sections are the same.

Shear influence diagrams are used in the same manner as reaction influence diagrams. The shear at point $A$ for any position of the load can be calculated by multiplying the ordinate of the diagram by the actual load. Distributed loads are found by multiplying the uniform load by the area under the diagram between the limits of loading. If the loading extends over positive and negative parts of the curve, the sign of the area is considered when performing the final summation.

If it is necessary to determine the distribution of loading which will produce the *maximum shear* at a point whose influence diagram is available, the load should be positioned in order to maximize the area under the diagram.[2] This can be done by "covering" either all of the positive area or all of the negative area.[3]

## D. SHEAR INFLUENCE DIAGRAMS BY VIRTUAL DISPLACEMENT

A difficulty in drawing shear influence diagrams for continuous beams on more than two supports is finding the reactions. The method of *virtual displacement* or *virtual work* can be used to find the influence diagram without going through that step.

*step 1*: Replace the point being investigated (i.e., point $A$) with an imaginary link with unit length. (It may be necessary to think of the link as having a length of 1 foot, but the link does not add to or subtract from any length of the beam.)

*step 2*: Push the two ends of the beam (with the link somewhere in between) a very small amount. The distance between supports does not change, but the linkage allows the beam sections to assume a slope. The sections to the left and right of the linkage displace $\delta_1$ and $\delta_2$ respectively from their equilibrium positions. The slope of both sections is the same. Points of support remain in contact with the beam.

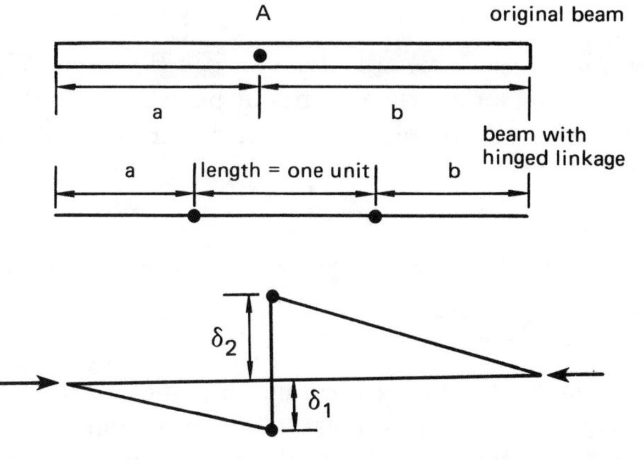


**Figure 12.17** Virtual Beam Displacements

*step 3*: Determine the ratio of $\delta_1$ and $\delta_2$. Since the slope on the two sections is the same, the longer section will have the larger deflection. If $L = a + b$ is the length of the beam, the relationships between the deflections can be determined from equations 12.52 through 12.54.

$$\delta_1 + \delta_2 = 1$$

$$\frac{\delta_1}{\delta_2} = \frac{a}{b} \qquad 12.52$$

$$\delta_1 = \left(\frac{a}{L}\right)\delta \qquad 12.53$$

$$\delta_2 = \left(\frac{b}{L}\right)\delta \qquad 12.54$$

Since $\delta = \delta_1 + \delta_2$ was chosen as one, equations 12.53 and 12.54 really give the relative proportions of the unit link which extend below and above the reference line in figure 12.17.

Knowing that the total shear reversal through point $A$ is one unit, and that the slopes are the same, the relative proportions of the reversal below and above the line will determine the shape of the displaced beam. The shape of the influence diagram is the shape taken on by the beam.

*step 4*: As required, use equations of straight lines to obtain the shear influence ordinate as a function of position along the beam.

*Example 12.17*

For the simply supported beam shown, draw the shear influence diagram for a point 10 feet from the right end.

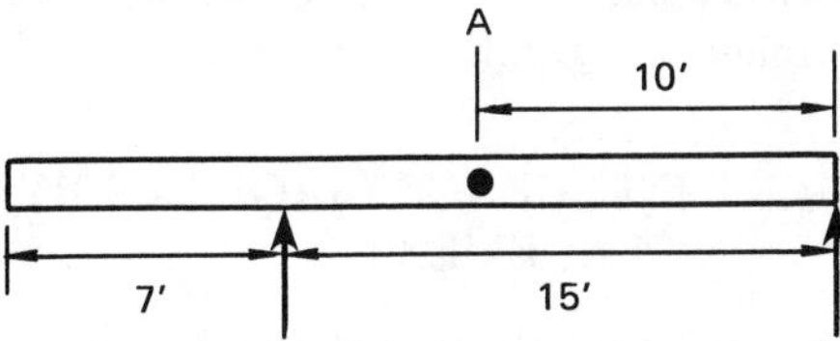


If a unit link is placed at point $A$ and the beam ends are pushed together, the following shape will result. Notice that the beam must remain in contact with the points of support, and that the two slopes are the same.

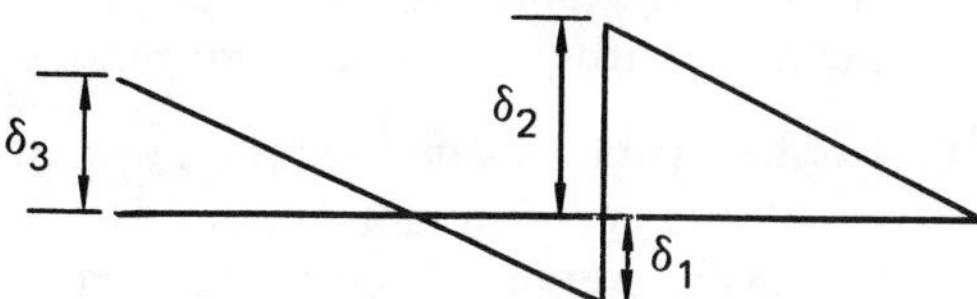


The overhanging seven feet of beam don't change the shape of the shear influence diagram between the sup-

---

[2] If the *minimum shear* is requested, the maximum negative shear is implied. The minimum shear is not zero in most cases.

[3] Usually, the dead load is assumed to extend over the entire length of the beam. The uniform live loads are distributed in any way which will case the maximum shear.

ports. The deflections can be evaluated assuming a 15′ long beam.

$$\delta_1 = \frac{5}{15} = 0.33$$

$$\delta_2 = \frac{10}{15} = 0.67$$

The slope in both sections of the beam is the same. This slope can be used to calculate $\delta_3$.

$$m = \frac{\delta_1}{a} = \frac{0.33}{5} = 0.066$$

$$\delta_3 = (7)(0.066) = 0.46$$

*Example 12.18*

Where should a uniformly distributed load be placed on the beam shown below to maximize the shear at section $A$?

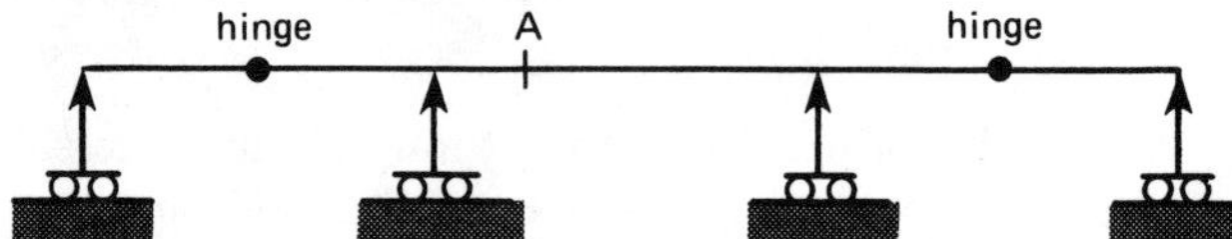


Using the principle of virtual displacement, the following shear influence diagram results by inspection. (It is not necessary to calculate the relative displacements to answer this question. It is only necessary to identify the positive and negative parts of the influence diagram.)

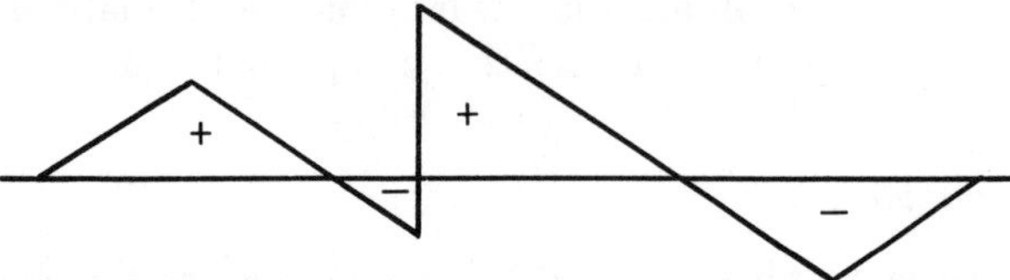


To maximize the shear, the uniform load should be distributed either over all positive or all negative sections of the influence diagram.

## E. MOMENT INFLUENCE DIAGRAMS BY VIRTUAL DISPLACEMENT

A moment influence diagram (not the same as a moment diagram) gives the moment at a particular point for any location of a unit load. The method of virtual displacement can be used in this situation to simplify finding the moment influence diagram.

*step 1*: Replace the point being investigated (i.e., point $A$) with an imaginary hinge.

*step 2*: Rotate the beam a unit rotation by applying equal but opposite moments to each of the two beam sections. Except where the point being investigated is at a support, this unit rotation can be achieved simply by "pushing up" on the beam at the hinge point.

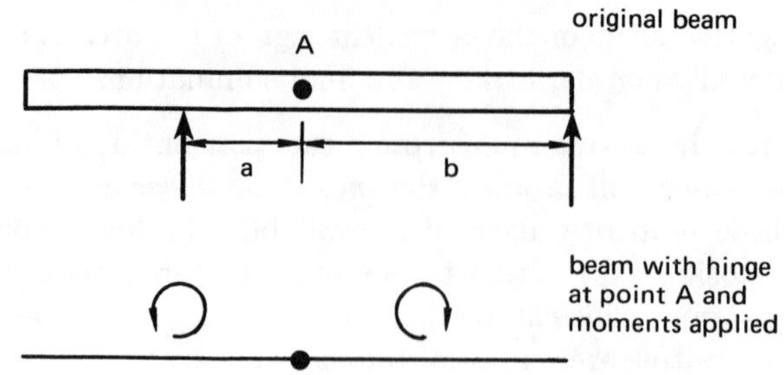


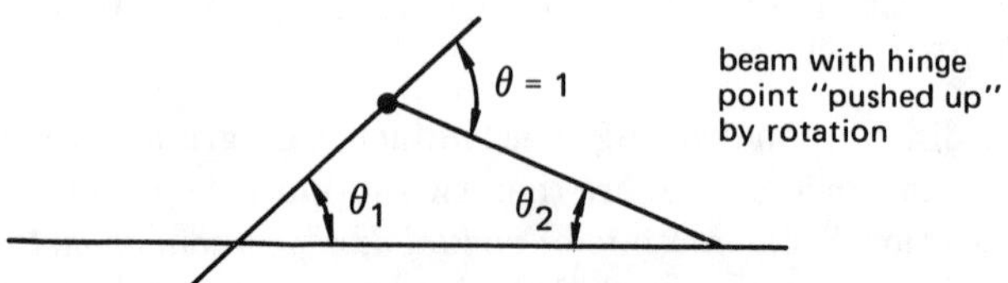


**Figure 12.18** Moment Influence Diagram by Virtual Displacement

*step 3*: The angles made by the sections on either side of the hinge will be proportional to the lengths of the opposite sections. (Since the angle is small for a virtual displacement, the angle and its tangent, or slope, are the same.)

$$\theta_1 = \frac{b}{L} \qquad 12.55$$

$$\theta_2 = \frac{a}{L} \qquad 12.56$$

$$L = a + b \qquad 12.57$$

*Example 12.19*

What are the approximate shapes of the moment influence diagrams for points $A$ and $B$ on the beam shown?

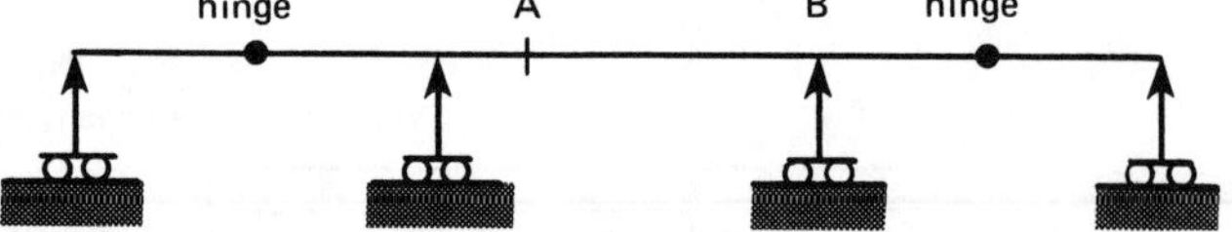


By placing an imaginary hinge at point $A$ and rotating the two adjacent sections of the beam, the following shape results.

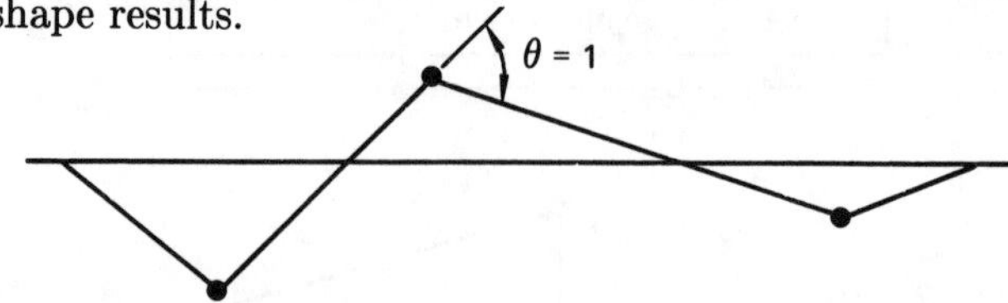


The moment influence diagram for point $B$ is found by placing an imaginary hinge at point $B$ and applying a rotating moment. Since the beam must remain in contact with all supports, and since there is no hinge between the two middle supports, the moment influence diagram must be horizontal in that region.

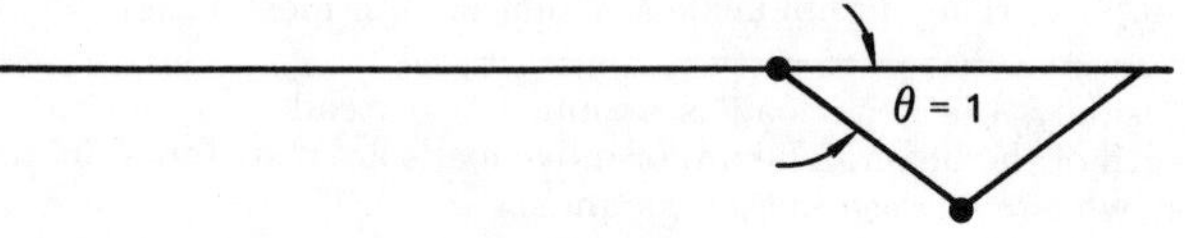

## F. SHEAR INFLUENCE DIAGRAMS ON CROSS-BEAM DECKS

When girder type construction is used to construct a road or bridge deck, the loads probably will not be applied directly to the girder. Rather, the loads will be transmitted to the girder at panel points from cross beams. Figure 12.19 shows a typical construction detail involving girders and cross beams.

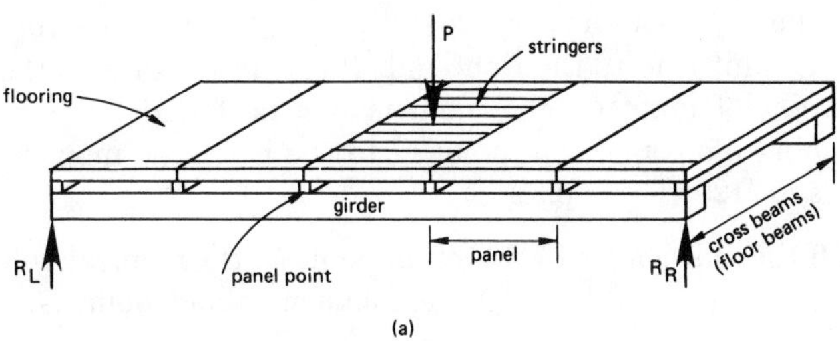


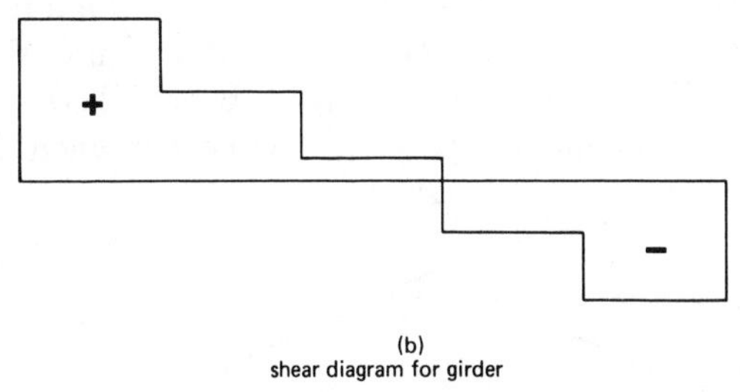


**Figure 12.19** Cross Beam Decking

A load applied to the deck stringers will be transmitted to the girder only at the panel points. Because the girder experiences a series of concentrated loads, the shear between panel points is horizontal. Since the shear is always constant between panel points, we speak of *panel shear* rather than shear at a point. Accordingly, shear influence diagrams are drawn for a panel, not for a point. Moment influence diagrams are similarly drawn for a panel.

## G. INFLUENCE DIAGRAMS ON CROSS-BEAM DECKS

Shear and moment influence diagrams for girders with cross beams are identical to simple beams, except for the panel being investigated. Once the influence diagram has been drawn for the simple beam, the influence diagram ordinates at the ends of the panel being investigated are connected to obtain the influence diagram for the girder. This is illustrated in figure 12.20.

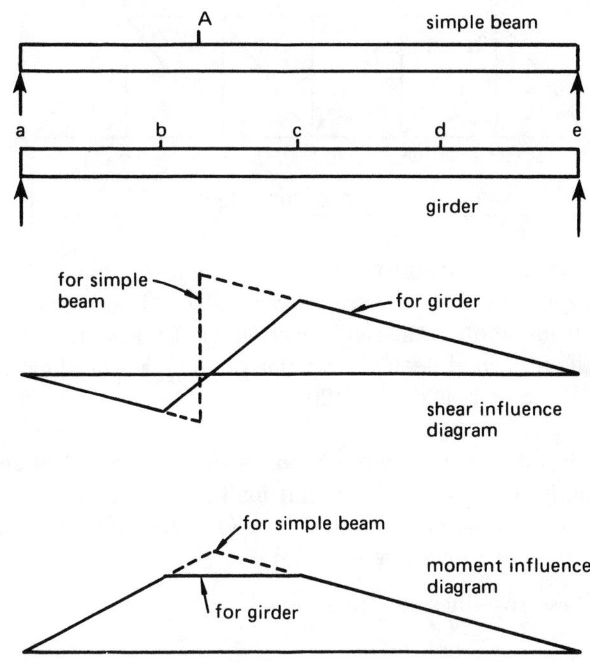


**Figure 12.20** Comparison of Influence Diagrams for Simple Beams and Girders

## H. INFLUENCE DIAGRAMS FOR TRUSS MEMBERS

Since members in trusses are initially assumed to be axial members, they cannot carry shears or moments. Therefore, shear and moment influence diagrams don't exist for truss members. However, it is possible to obtain an influence diagram showing the variation in axial force in a given truss member as the load varies in position.

There are two general cases for finding forces in truss members. The force in a horizontal truss member is proportional to the moment across the member's panel. The force in an inclined truss member is proportional to the shear across that member's panel.

So, even though we may only want the axial load in a truss member, it is still necessary to construct the shear and moment influence diagrams for the entire truss in order to determine the applications of loading on the truss which produce the maximum shear and moment across the member's panel.

*Example 12.20*

Draw the influence diagram for vertical shear in panel $DF$ of the through truss shown. What is the maximum force in member $DG$ if a 1000 pound load moves across the truss?

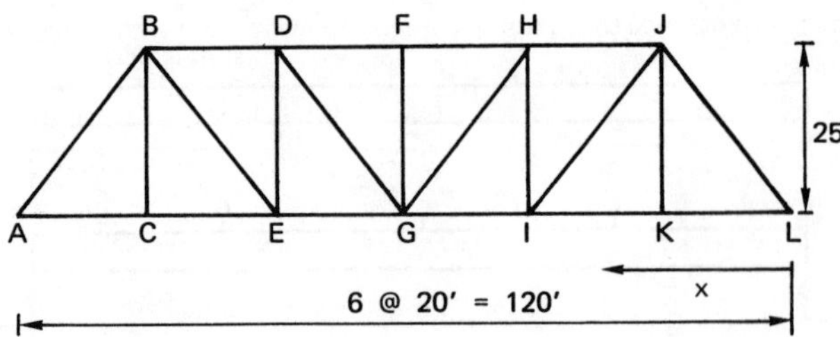


Allow a unit load to move from joint $L$ to joint $G$ along the lower chords. If the unit vertical load is at a distance $x$ from point $L$, the right reaction will be $+[1-(x/120)]$. The unit load itself has a value of $(-1)$, so the shear at distance $x$ is just $(-x/120)$.

Allow a unit load to move from joint $A$ to joint $E$ along the lower chords. If the unit load is a distance $x$ from point $L$, the left reaction will be $(x/120)$, and the shear at distance $x$ will be $[(x/120) - 1]$.

These two lines can be graphed.

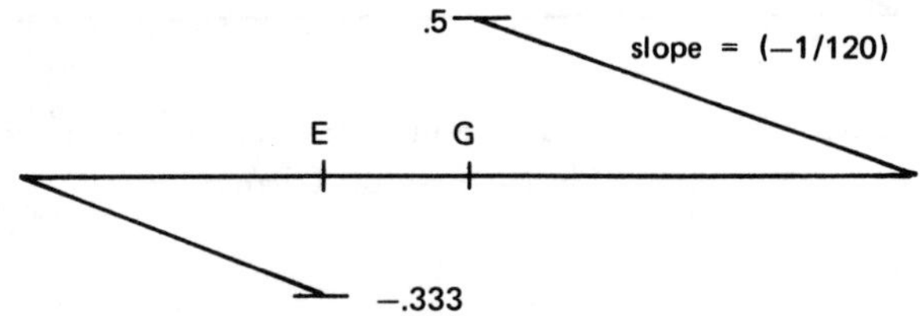


The influence line is completed by connecting the two lines as shown. Therefore, the maximum shear in panel $DF$ will occur when a load is at point $G$ on the truss.

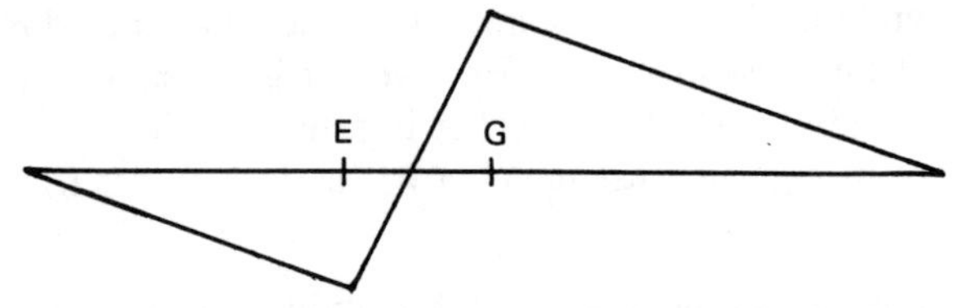


If the 1000 pound load is at point $G$, the two reactions at points $A$ and $L$ will each be 500 pounds. The cut-and-sum method can be used to calculate the force in member $DG$ simply by evaluating the vertical forces on the freebody to the left of point $G$.

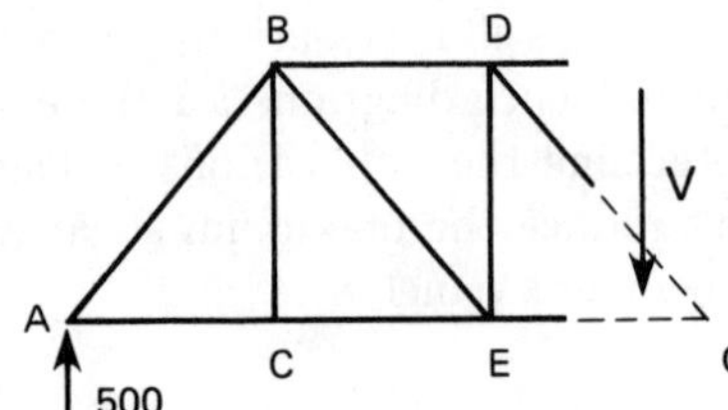


For equilibrium to occur, $V$ must be 500. This vertical shear is entirely carried by member $DG$. The length of member $DG$ is

$$\sqrt{(20)^2 + (25)^2} = 32$$

The force in member $DG$ is

$$DG = \left(\frac{32}{25}\right) 500 = 640$$

*Example 12.21*

Draw the moment influence diagram for panel $DF$ on the truss shown in example 12.20. What is the maximum force in member $DF$ if a 1000 pound load moves across the truss?

The left reaction is $(x/120)$ where $x$ is the distance from the unit load to the right end. If the unit load is to the right of point $G$, the moment can be found by summing moments from point $G$ to the left. The moment is $(x/120)(60) = 0.5x$.

If the unit load is to the left of point $E$, the moment will again be found by summing moments about point $G$.

$$\left(\frac{x}{120}\right)(60) - (1)(x - 60) = 60 - 0.5x$$

These two lines can be graphed. The moment for a unit load between points $E$ and $G$ is obtained by connecting the two end points of the lines derived above. Therefore, the maximum moment in panel $DF$ will occur when the load is at point $G$ on the truss.

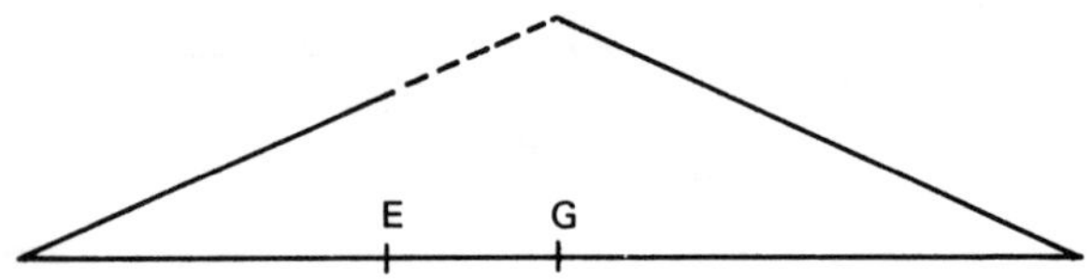


If the 1000 pound load is at point $G$, the two reactions at points $A$ and $L$ will each be 500 pounds. The method of sections can be used to calculate the force in member $DF$ by taking moments about joint $G$.

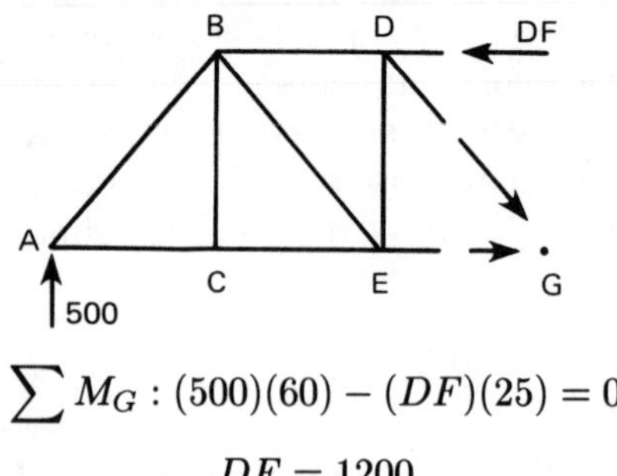


$$\sum M_G : (500)(60) - (DF)(25) = 0$$

$$DF = 1200$$

## 13 MOVING LOADS ON BEAMS

### A. GLOBAL MAXIMUM MOMENT ANYWHERE ON BEAM

If a beam supports a single moving load, the maximum bending and shearing stresses at any point can be found by drawing the moment and shear influence diagrams for that point. Once the positions of maximum moment and maximum shear are known, the stresses at the point in question can be found from equations 12.18 and 12.24.

If a simply-supported beam carries a set of moving loads (which remain equidistant as they travel across the beam), the following procedure can be used to find the *dominant load.* The dominant load is the one which occurs directly over the point of maximum moment.

*step 1*: Calculate and locate the resultant of the load group.

*step 2*: Assume that one of the loads is dominant. Place the group on the beam such that the distance from one support to the assumed dominant load is equal to the distance from the other support to the resultant of the load group.

*step 3*: Check to see that all loads are on the span and that the shear changes sign under the assumed dominant load. If the shear does not change sign under the assumed dominant load, the maximum moment may occur when only some of the load group is on the beam. If it does change sign, calculate the bending moment under the assumed dominant load.

*step 4*: Repeat steps 2 and 3, assuming that the other loads are dominant.

*step 5*: Find the maximum shear by placing the load group such that the resultant is a minimum distance from a support.

### B. PLACEMENT OF LOAD GROUP TO MAXIMIZE LOCAL MOMENT

In the design of specific members or connections, it is necessary to place the load group in a position which will maximize the load on those members or connections. The procedure for finding these positions of local maximum loadings is different from the global maximum procedures.

The solution to the problem of local maximization is somewhat trial and error oriented. It is aided by use of the influence diagram. In general, the variable being evaluated (reaction, shear, or moment) is maximum when one of the wheels is at the location or section of interest.

When there are only two or three wheels in the load group, the various alternatives can be simply evaluated by using the influence diagram for the variable being evaluated. When there are many loads in the load group (e.g., a train loading), it may be advantageous to use heuristic rules for predicting the dominant wheel.

## 14 COLUMNS

The *Euler load* is the theoretical maximum load that an initially straight column can support without buckling. For columns with pinned ends, this load is given by equation 12.58.

$$F_e = \frac{\pi^2 EI}{L^2} = \frac{(r\pi)^2 EA}{L^2} \qquad 12.58$$

The corresponding column stress is

$$\sigma_e = \frac{F_e}{A} = \frac{\pi^2 E}{\left(\frac{L}{r}\right)^2} \qquad 12.59$$

Equations 12.58 and 12.59 assume that the column is long so that the Euler stress is reached before the yield stress is reached. If the column is short, the yield stress of the material may be less than the Euler stress. In that case, short-column curves based on test data are used to predict the allowable column stress.

The value of $L/r$ at the point of intersection of the short column and the Euler curves is known as the critical *slenderness ratio.* The critical slenderness ratio becomes smaller as the compressive yield stress increases. The region in which the short column formulas apply is determined by tests for each particular type of column and material. Typical critical slenderness ratios range from 80 to 120.

In general, the Euler allowable stress formulas can be used if the stress obtained from equation 12.59 does not exceed the compressive yield stress.

*Example 12.22*

An S-type, 4 × 9.5 A36 steel I-beam 8.5′ long is used as a column. What is the working stress for a safety factor of 3? Use $E = 2.9$ EE7 psi. The yield stress for A36 steel is 36,000 psi. The required properties of the I beam are $A = 2.79\ \text{in}^2, I = 0.903\ \text{in}^4$, and $r = 0.569$ in.

From equation 12.59, the Euler stress is

$$\sigma_e = \frac{\pi^2(2.9\ \text{EE7})}{\left[\frac{(8.5)(12)}{0.569}\right]^2} = 8907\ \text{psi}$$

Since 8907 is less than 36,000, the Euler formula is valid. The allowable working stress is

$$\sigma_a = \frac{8907}{3} = 2969\ \text{psi}$$

An ultimate load for any column can be found by using the *secant formula*. The secant formula is particularly suited for use when the column is intermediate in length.

$$\sigma = \frac{F}{A} = \frac{S_y}{1 + \frac{ec}{r^2}\sec\phi} \qquad 12.60$$

$$\phi = \frac{1}{2}\left(\frac{L}{r}\right)\sqrt{\frac{F}{AE}} \qquad 12.61$$

The formula is solved by trial and error for $F$ with the given eccentricity, $e$. If the value of $e$ is not known, the eccentricity ratio $(ec/r^2)$ is taken as 0.25. Substituting this value and $E$ = 2.9 EE7 for steel and 1.00 EE7 for aluminum, respectively, the following formulas result which converge quickly to the known $L/r$ ratio when assumed values of $F$ are substituted.

$$\phi = \arccos\left(\frac{0.25F}{S_yA - F}\right) \qquad 12.62$$

$$\frac{L}{r} = 2\phi\sqrt{\frac{EA}{F}} \qquad 12.63$$

$$\left(\frac{L}{r}\right)_{\text{steel}} = \frac{10,770(\phi)}{\sqrt{\frac{F}{A}}} \qquad 12.64$$

$$\left(\frac{L}{r}\right)_{\text{aluminum}} = \frac{6325(\phi)}{\sqrt{\frac{F}{A}}} \qquad 12.65$$

*Example 12.23*

A steel member ($S_y$ = 36,000 psi, $A$ = 17.9 in$^2$, least $r$ = 2.45 in) is used as a 20-foot column. Use the secant formula and a factor of safety of 2.5 to determine the maximum concentric load.

Even though the loading is intended to be concentric, use $ec/r^2 = 0.25$ to account for uncertainties in construction and loading.

The slenderness ratio is

$$\frac{L}{r} = \frac{(20)(12)}{2.45} = 98$$

Assume a critical load of $F$ = 300,000 lbf. From equation 12.62,

$$\phi = \arccos\left[\frac{(0.25)(300)}{(36)(17.9) - 300}\right] = 1.35 \text{ radians}$$

From equation 12.64,

$$\frac{L}{r} = \frac{(10,770)(1.35)}{\sqrt{\frac{300,000}{17.9}}} = 112.3$$

Since $L/r$ will be smaller when $F$ is larger, try $F$ = 350,000 lbf.

$$\phi = \arccos\left[\frac{(0.25)(350)}{(36)(17.9) - 350}\right] = 1.27$$

$$\frac{L}{r} = \frac{(10,770)(1.27)}{\sqrt{\frac{350,000}{17.9}}} = 97.8 \quad \text{(close enough)}$$

$$F_{\text{allowable}} = \frac{350,000}{2.5} = 140,000 \text{ lbf}$$

All the preceding column formulas are for columns with frictionless round or pinned ends. For other end conditions, the *effective length* $L'$ should be used in place of $L$.

$$L' = KL \qquad 12.66$$

$K$ is the *end restraint coefficient* which varies from 0.5 to 2. For practical columns, $K$ smaller than 0.7 should not be used because infinite stiffness of the support structure is not normally achievable.

**Table 12.3**
Theoretical End-Restraint Coefficients
(Also see table 15.5)

| illus. | end conditions | $K$ |
|---|---|---|
| (a) | both ends pinned | 1 |
| (b) | both ends built in | 0.5 |
| (c) | one end pinned, one end built in | 0.707 |
| (d) | one end built in, one end free | 2 |
| (e) | one end built in, one end fixed against rotation but free | 1 |
| (f) | one end pinned, one end fixed against rotation but free | 2 |

(a) (b) (c) (d) (e) (f)

# PART 2: Application to Design

## 1 SPRINGS

Springs are assumed to be perfectly elastic within their working range. *Hooke's law* can be used to predict the amount of compression experienced when a load is placed on a spring.

$$\mathbf{F} = kx \qquad 12.67$$

$k$ is the *spring constant.* It has units of pounds per unit length.

When a spring is compressed, it stores energy. This energy can be recovered by restoring the spring's original length. It is assumed that no energy is lost through friction or hysteresis when a spring returns to its original length. The energy storage in a spring is given by equation 12.68. This energy is the same as the work required to compress the spring.

$$W = \Delta U = \tfrac{1}{2}kx^2 \qquad 12.68$$

If a weight is dropped from height $h$ onto a spring, the compression can be found by equating the change in potential energy to the energy storage.

$$m\left(\frac{g}{g_c}\right)(h+x) = \tfrac{1}{2}kx^2 \qquad 12.69$$

## 2 THIN-WALLED CYLINDERS

A cylinder can be considered *thin-walled* if its wall thickness-to-diameter ratio is less than 0.1. The circumferential *hoop stress* for internal pressure can be derived easily from the free-body diagram of a cylinder half.[4] This hoop stress is

$$\sigma_h = \frac{pr}{t} \qquad 12.70$$

Since the cylinder is assumed to be thin-walled, the radius used in equation 12.70 is taken as the inside radius.

If the cylinder is part of a tank, the axial force on the end plates produces an axial stress. The axial force is equal to the tank pressure times the end plate area. The stress produced is at right angles to the hoop stress. Accordingly, it is called *longitudinal stress* or *long stress.*

$$\sigma_L = \frac{pr}{2t} \qquad 12.71$$

Equation 12.71 also gives the stress in a spherical tank. In a spherical tank, the hoop and long stresses are the same.

[4] There is no easy method of evaluating stresses in thin-walled cylinders under external pressure, since failure is by collapse, not elongation. However, empirical equations exist for predicting the collapsing pressure.

The hoop and long stresses are principal stresses. They do not combine into a larger stress.

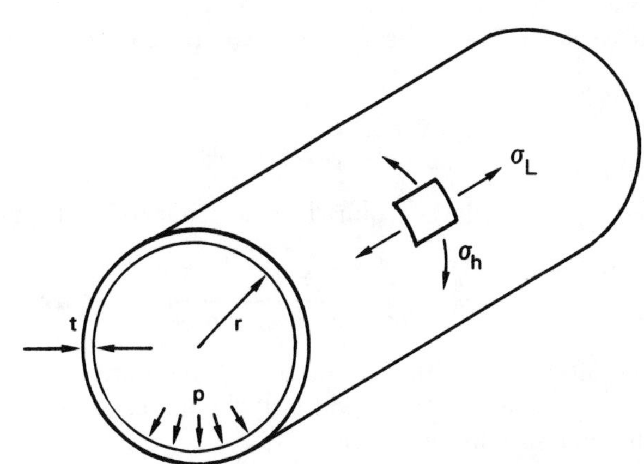


**Figure 12.21** Hoop and Long Stresses

## 3 RIVET AND BOLT CONNECTIONS

A *tension splice* using rivets or bolts can fail in one of three ways: bearing failure, shear failure, or tension failure. All three failure mechanisms must be checked to determine the maximum load the splice can carry.

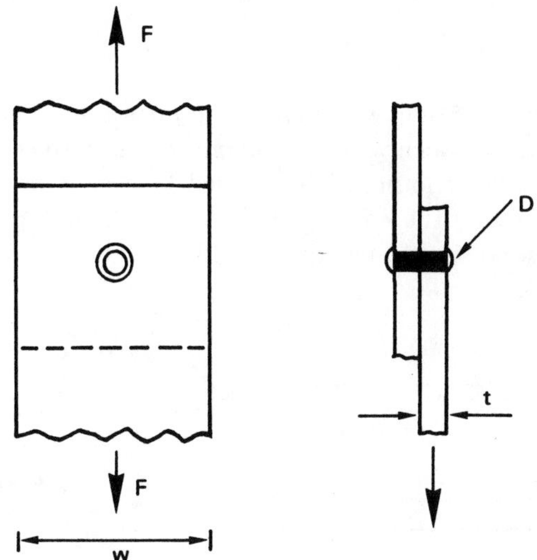


**Figure 12.22** A Tension Splice

The plate can fail in bearing. For one connector, the *bearing stress* in the plate is

$$\sigma_{br} = \frac{F}{Dt} \qquad 12.72$$

The number of rivets required to keep the actual bearing stress below the allowable bearing stress is

$$n_{br} = \frac{\sigma_{br}}{\text{allowable bearing stress}} \qquad 12.73$$

The rivet can fail in shear. The shear stress in the rivet is

$$\tau = \frac{F}{\frac{1}{4}\pi D^2} \qquad 12.74$$

The number of rivets required, as determined by shear, is

$$n_s = \frac{\tau}{\text{allowable shear stress}} \qquad 12.75$$

The plate also can fail in tension. If there are $n$ rivet holes in a line across the width of the plate, the minimum area in tension will be

$$A_t = t(w - nD) \qquad 12.76$$

The tensile stress in the plate at the minimum section is

$$\sigma_t = \frac{F}{A_t} \qquad 12.77$$

The maximum number of rivets across the plate width must be chosen to keep the tensile stress less than the allowable stress.

## 4 FILLET WELDS

The most common weld type is the *fillet weld*, shown in figure 12.23. Such welds commonly are used to connect one plate to another. The applied load is assumed to be carried by the *effective weld throat* which is related to the weld size, $y$, by equation 12.78.

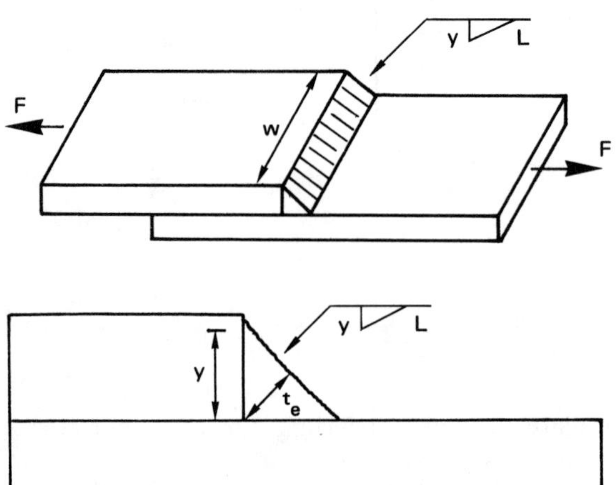


**Figure 12.23** Fillet Lap Weld and Symbol

The effective weld throat size is

$$t_e = (0.707)y \qquad 12.78$$

Weld sizes ($y$) of $\frac{3}{16}''$, $\frac{1}{4}''$, and $\frac{5}{16}''$ are desirable because they can be made in a single pass. However, fillet welds from $\frac{3}{16}''$ to $\frac{1}{2}''$ in $\frac{1}{16}''$ increments are available. The increment is $\frac{1}{8}''$ for larger welds.

Neglecting any effects due to eccentricity, the shear stress in the fillet lap weld shown in figure 12.23 is

$$\tau = \frac{F}{wt_e} \qquad 12.79$$

## 5 SHAFT DESIGN

Shear stress occurs when a shaft is placed in torsion. The shear stress at the outer surface of a bar of radius, $r$, which is torsionally loaded by a torque, $T$, is[5]

$$\tau = \frac{Tr}{J} \qquad 12.80$$

The total strain energy due to torsion is

$$U = \frac{T^2 L}{2GJ} \qquad 12.81$$

$J$ is the shaft's polar moment of inertia, as defined in chapter 11. For a solid round shaft, $J$ is

$$J = \frac{\pi r^4}{2} = \frac{\pi D^4}{32} \qquad 12.82$$

For a hollow round shaft, the polar moment of inertia is

$$J = \frac{\pi}{2}[r_o^4 - r_i^4] \qquad 12.83$$

If a shaft of length $L$ carries a torque $T$, the angle of twist (in radians) will be

$$\phi = \frac{TL}{GJ} \qquad 12.84$$

$G$ is the *shear modulus*, approximately equal to 11.5 EE6 psi for steel. The shear modulus can be calculated from the modulus of elasticity by using equation 12.85.

$$G = \frac{E}{2(1+\mu)} \qquad 12.85$$

[5] Shear stress in steel shafts commonly is limited to approximately 6000 psi. This represents a factor of safety of approximately 3 based on the torsional yield strength.

The torque, $T$, carried by a shaft spinning at $n$ revolutions per minute is related to the transmitted horsepower.[6]

$$T = \frac{(63{,}025)(\text{horsepower})}{n} \qquad 12.86$$

## 6 ECCENTRIC CONNECTOR ANALYSIS

An eccentric torsion connection is illustrated in figure 12.24. This type of connection gets its name from the load's tendency to rotate the bracket. This rotation must be resisted by the shear stress in the connectors.

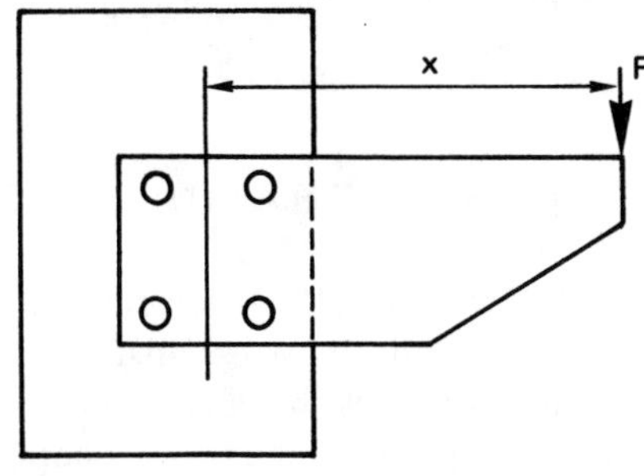


**Figure 12.24** Torsion Resistance

An extension of equation 12.80 can be used to evaluate the maximum stresses in the connector group. To use equation 12.80, the following changes in definition must be made.

- The torque, $T$, is replaced by the moment on the bracket. This moment is the product of the eccentric load, $F$, and the distance from the load to the centroid of the fastener group, $x$.
- $r$ is taken as the distance from the centroid of the fastener group to the critical fastener. The critical fastener is the one for which the vector sum of the vertical and torsional shear stresses is the greatest.
- $J$ is based on the parallel-axis theorem. As bolts and rivets have little resistance to twisting in their holes, their polar moments of inertia, $J_i$, are omitted.

  $$J = \sum r_i^2 A_i \qquad 12.87$$

  $r_i$ is the distance from the fastener group centroid to the $i$th fastener, which has an area of $A_i$.
- The vertical shear stress in the critical fastener must be added in a vector sum to the torsional shear stress. This vertical shear stress is

  $$\text{vertical shear stress} = \frac{F/\#\text{ of fasteners}}{A_{\text{critical}}} \qquad 12.88$$

*Example 12.24*

For the bracket shown, find the load on the most critical fastener. All fasteners have a nominal $\frac{1}{2}''$ diameter.

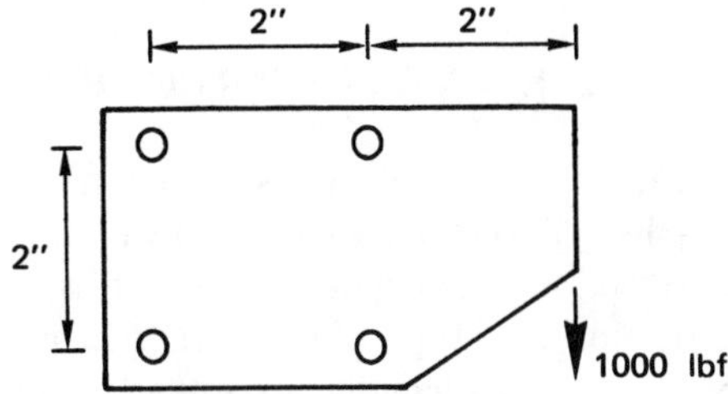


Since the fastener group is symmetrical, the group centroid is centered within the 4 fasteners. This makes the eccentricity of the load equal to 3 inches. Each fastener is located $r$ from the centroid, where

$$r = \sqrt{(1)^2 + (1)^2} = 1.414$$

The area of each fastener is

$$A_i = \tfrac{1}{4}\pi(0.5)^2 = 0.1963$$

Using the parallel axis theorem for polar moments of inertia,

$$J = 4[0.1963(1.414)^2] = 1.570 \text{ in}^4$$

The torsional stress on each fastener is

$$\tau_T = \frac{(1000)(3)(1.414)}{(1.570)} = 2702 \text{ psi}$$

This torsional shear stress is directed perpendicularly to a line connecting each fastener with the centroid.

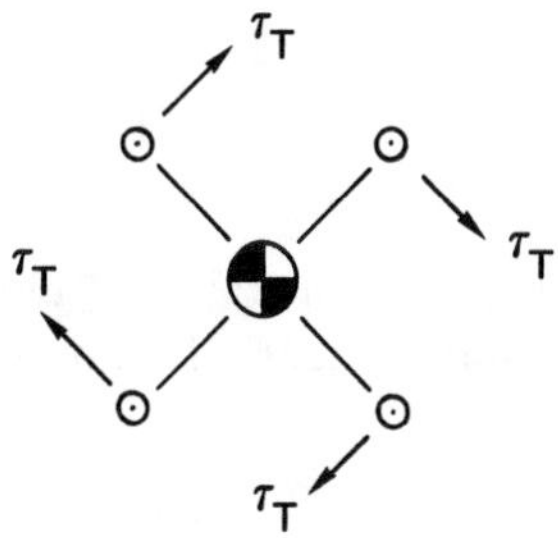


$\tau_T$ can be divided into horizontal stresses of $\tau_{T,x}$ and vertical stresses of $\tau_{T,y}$. Both of these components

[6] The torque is assumed to be steady, as would be supplied by a belt or a pulley. If the load varies, or if the shaft also carries a bending moment, a more complex method is required.

are equal to 1911 psi. In addition, each fastener carries a vertical shear load equal to (1000/4) = 250 pounds. The vertical shear stress due to this load is (250/.1963) = 1274.

The two right fasteners have vertical downward components of $\tau_T$ which add to the vertical downward stress of 1274. Thus, both of the two right fasteners are critical. The total stress in each of these fasteners is

$$\tau = \sqrt{(1911)^2 + (1911 + 1274)^2} = 3714 \text{ psi}$$

## 7 SURVEYOR'S TAPE CORRECTIONS

The standard surveyor's tape consists of a flat steel ribbon with a length very close to 100 feet. Such tapes are standardized at a particular temperature and with a specific tension and type of support. Since the tape cannot be used in the conditions under which it was standardized, corrections are needed.

### A. TEMPERATURE CORRECTION

If the tape is not used at the standardized temperature, the change in length will be

$$C_t = \alpha (T - T_{std}) L \qquad 12.89$$

$\alpha$ for steel has an approximate value of 6.5 EE–6 1/°F, although low-coefficient tapes containing nickel can reduce this expansion 75%. The correction given by equation 12.89 can be positive or negative, depending on the values of $T$ and $T_{std}$. The correction is applied to the distance according to the algebraic operations listed in table 12.4.

**Table 12.4**
Corrections for Surveyors' Tapes

| measuring a distance | setting out points |
|---|---|
| add $C_t$ | subtract $C_t$ |
| add $C_p$ | subtract $C_p$ |

### B. TENSION CORRECTION

The correction due to non-standard pull (tension) can be found from equation 12.90. It can be either positive or negative.

$$C_p = \frac{(F - F_{std})L}{AE} \qquad 12.90$$

The correction is applied to the distance according to the algebraic conventions listed in table 12.4.

*Example 12.25*

A steel surveyor's tape is standardized at 68°F. It is used at 50°F to place two monuments exactly 79 feet apart. What should be the tape reading used to place the monuments?

From equation 12.89,

$$C_t = (6.5 \text{ EE–6})(50 - 68)(79) = -9.2 \text{ EE–3}$$

From table 12.4,

$$\begin{aligned} \text{tape reading} &= 79.0000 - (-9.2 \text{ EE–3}) \\ &= 79.0092 \end{aligned}$$

(The tape cannot be read to the degree of precision indicated by this answer.)

## 8 STRESS CONCENTRATION FACTORS

Stress concentration factors are correction factors used to account for nonuniform stress distributions within objects.[7] Nonuniform distributions result from nonuniform shapes. Examples of nonuniform shapes requiring stress concentration factors are stepped shafts, plates with holes, shafts with keyways, etc.

The actual stress experienced is the product of the stress concentration factor and the ideal stress. Values of $K$ always are greater than 1.0, and they typically range from 1.2 to 2.5 for most designs. The exact values must be determined graphically from published results of extensive experimentation.

$$\sigma' = K\sigma \qquad 12.91$$

## 9 CABLES

Cables (*wire ropes*) can be obtained in a wide variety of materials and cross sections to suit the application. Strength and weight properties of steel *standard hoisting rope* (6 strands of 19 wires each) are given in table 12.5.

In addition to the primary tension load, the design of cables should include the significant effects of bending, friction, and the weight of the cable. Appropriate dynamic factors should be applied to allow for acceleration, deceleration, stops, and starts. In general, the working stress should not exceed 20% of the breaking strength (i.e., a factor of safety of 5).

The stress due to bending a cable, such as bending around a drum, is included as an equivalent added tension load. (For good design, the diameter of the drum on which a cable is wound should be 45 to 90 times the

[7] Stress concentration factors also are known as *stress risers.*

cable diameter.) If $d$ is the cable diameter in inches, $R$ is the bending radius in inches, and $N$ is the number of wires in the cable (114 for a 6×19 cable), the equivalent tensile load from bending is approximately

$$F = \frac{2.8\ \text{EE9}d^3}{N^2R} \quad 12.92$$

**Table 12.5**
6 × 19 (Standard Hoisting) Wire Ropes

| diam. inches | approx. weight per ft., pounds | breaking strength tons of 2000 pounds: impr. plow steel | plow steel | mild plow steel |
|---|---|---|---|---|
| 1/4 | 0.10 | 2.74 | 2.39 | 2.07 |
| 5/16 | 0.16 | 4.26 | 3.71 | 3.22 |
| 3/8 | 0.23 | 6.10 | 5.31 | 4.62 |
| 7/16 | 0.31 | 8.27 | 7.19 | 6.25 |
| 1/2 | 0.40 | 10.7 | 9.35 | 8.13 |
| 9/16 | 0.51 | 13.5 | 11.8 | 10.2 |
| 5/8 | 0.63 | 16.7 | 14.5 | 12.6 |
| 3/4 | 0.90 | 23.8 | 20.7 | 18.0 |
| 7/8 | 1.23 | 32.2 | 28.0 | 24.3 |
| 1 | 1.60 | 41.8 | 36.4 | 31.6 |
| 1 1/8 | 2.03 | 52.6 | 45.7 | 39.8 |
| 1 1/4 | 2.50 | 64.6 | 56.2 | 48.8 |
| 1 3/8 | 3.03 | 77.7 | 67.5 | 58.8 |
| 1 1/2 | 3.60 | 92.0 | 80.0 | 69.6 |
| 1 5/8 | 4.23 | 107 | 93.4 | 81.2 |
| 1 3/4 | 4.90 | 124 | 108 | 93.6 |
| 1 7/8 | 5.63 | 141 | 123 | 107 |
| 2 | 6.40 | 160 | 139 | 121 |
| 2 1/8 | 7.23 | 179 | 156 | |
| 2 1/4 | 8.10 | 200 | 174 | |
| 2 1/2 | 10.0 | 244 | 212 | |
| 2 3/4 | 12.1 | 292 | 254 | |

For ropes with steel cores, add 7 1/2% to the above strengths.

For galvanized ropes, deduct 10% from the above strengths.

*Example 12.26*

What is the factor of safety when a $\frac{1}{2}$ inch, mild plow steel, 6 × 19 standard hoisting cable carrying 10,000 pounds is bent around a 24 inch sheave? Is the factor of safety adequate?

$$F_{\text{bending}} = \frac{(2.8\ \text{EE9})(0.5)^3}{(114)^2(12)} = 2240\ \text{lbf}$$

$$F_{\text{total}} = 10{,}000 + 2240 = 12{,}240\ \text{lbf}$$

$$\text{breaking strength} = 8.13\,(2000) = 16{,}260\ \text{lbf}$$

$$\text{factor of safety} = \frac{16{,}260}{12{,}240} = 1.33$$

Even without including the cable weight, this is not adequate, since a factor of safety of at least 5 is recommended.

## 10 THICK-WALLED CYLINDERS UNDER EXTERNAL AND INTERNAL PRESSURE

### A. STRESSES

The theory of thick-walled cylinders, *Lamé's solution*, is a continuation of the theory of thin-walled cylinders. The thick-walled cylinder is assumed to be made up of thin laminar rings. The strain variation through the wall is determined such that all the rings are in equilibrium, and the stresses and the deformations are consistent at the boundaries between the rings.

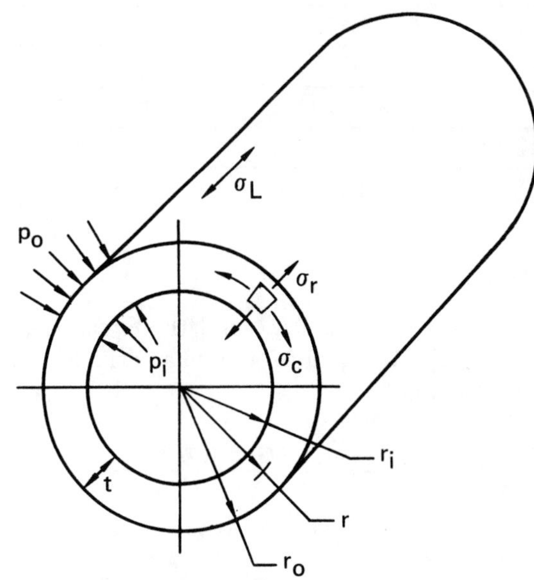


**Figure 12.25** Thick-Walled Cylinder

The general equations for stress are given here.[8]

$$\sigma_c = \frac{r_i^2 p_i - r_o^2 p_o + \frac{(p_i - p_o) r_i^2 r_o^2}{r^2}}{r_o^2 - r_i^2} \quad 12.93$$

$$\sigma_r = \frac{r_i^2 p_i - r_o^2 p_o - \frac{(p_i - p_o) r_i^2 r_o^2}{r^2}}{r_o^2 - r_i^2} \quad 12.94$$

The cases of main interest are those of internal or external pressure only. The stress formulas for these cases are summarized in table 12.6.

The maximum shear and normal stresses occur at the inner surface for both external and internal pressure.

[8] It is essential that compressive stresses be given a negative sign in all thick-walled cylinder equations, including those for deflection.

When a longitudinal stress due to internal pressure acting against end plates exists, the longitudinal stress is

$$\sigma_L = \frac{pr_i^2}{(r_o + r_i)t} \qquad 12.95$$

The longitudinal stress is assumed to be uniform across the wall. The magnitude and the location of the maximum shear and normal stress is not changed due to the addition of the longitudinal stress.

At every point in the cylinder, $\sigma_c$, $\sigma_r$, and $\sigma_L$ are the principal stresses.

**Table 12.6**
Stresses in Thick-Walled Cylinders

| stress | external pressure, $p$ | internal pressure, $p$ |
|---|---|---|
| $\sigma_{co}$ | $\frac{-(r_o^2 + r_i^2)p}{(r_o + r_i)t}$ | $\frac{2r_i^2 p}{(r_o + r_i)t}$ |
| $\sigma_{ro}$ | $-p$ | 0 |
| $\sigma_{ci}$ | $\frac{-2r_o^2 p}{(r_o + r_i)t}$ | $\frac{(r_o^2 + r_i^2)p}{(r_o + r_i)t}$ |
| $\sigma_{ri}$ | 0 | $-p$ |
| $\tau_{max}$ | $\left(\frac{1}{2}\right)\sigma_{ci}$ | $\left(\frac{1}{2}\right)(\sigma_{ci} + p)$ |

B. STRAINS

The *diametral strain*, $\Delta D/D$, and the *circumferential strain*, $\Delta C/C$, are equal in a circular cylinder under pressure loading.

$$\frac{\Delta D}{D} = \frac{\Delta C}{C} = \frac{\Delta r}{r} = \frac{\sigma_c - \mu(\sigma_r + \sigma_L)}{E} \qquad 12.96$$

*Example 12.27*

A steel cylinder of 1″ I.D. and 2″ O.D. is pressurized internally to 10,000 psi. (a) There are no end caps. What is the change in the inside diameter? (b) What would be the effect on the inside diameter of adding end caps? Use $E = 2.9$ EE7 and $\mu = 0.3$.

From table 12.6,

$$\sigma_{ci} = \frac{(1^2 + 0.5^2)(10{,}000)}{(1 + 0.5)0.5} = 16{,}667 \text{ psi}$$

$$\sigma_{ri} = -10{,}000 \text{ psi}$$

$$\sigma_L = 0$$

From equation 12.96,

$$\frac{\Delta D}{D} = \frac{[16{,}667 - 0.3(-10{,}000 + 0)]}{2.9 \text{ EE7}} = 0.000678$$

$$\Delta D = 0.000678\,(1) = 0.000678 \text{ inch}$$

(b)

$$\sigma_L = \frac{(10{,}000)(0.5)^2}{(1 + 0.5)(0.5)} = 3333 \text{ psi}$$

$$\frac{\Delta D}{D} = \frac{[16{,}667 - 0.3(-10{,}000 + 3333)]}{2.9 \text{ EE7}} = 0.000506$$

$$\Delta D = 0.000506\,(1) = 0.000506 \text{ in}$$

C. PRESS FITS

If two cylinders are pressed together with an initial interference, $I$, the pressure, $p$, acting between them expands the outer cylinder and compresses the inner one. *Interference* usually means diametral interference. Although the total interference can be allocated to inner and outer cylinders in almost any combination, the total interference usually is given to the outer disk in the case of shafts with disks.

$$I = |\Delta D_i|_{\text{outer cylinder}} + |\Delta D_o|_{\text{inner cylinder}} \qquad 12.97$$

If the cylinders are the same length, this method can be used to find the stress conditions after assembly. A stress concentration factor as high as 4 may be needed if the lengths are different.

In the special case where the shaft and the hub have the same modulus of elasticity and the shaft is solid, the total interference can be found from equation 12.98. The interference pressure, $p$, can be found if $I$ is known.

$$I = \frac{4r_{\text{shaft}}p}{E}\left[\frac{1}{1 - \left(\frac{r_{\text{shaft}}}{r_{\text{hub}}}\right)^2}\right] \qquad 12.98$$

When pieces are pressed together, the assembly force can be calculated as a sliding frictional force based on the normal force.

$$F_{\text{max, assembly}} = fN = 2\pi r_{\text{shaft}}Lpf \qquad 12.99$$

The coefficient of friction for press fits is highly variable, having been reported in the range of 0.03 to 0.33.

If the hub is acted upon by a torque or a torque-causing force, the maximum resisting torque can be calculated.

$$T_{\text{max}} = 2\pi r_{\text{shaft}}^2 Lpf \qquad 12.100$$

*Example 12.28*

A brass inner cylinder and an aluminum alloy outer cylinder (both 2.0″ long) have been pressed together with an interference of 0.004 in. What is the maximum shear stress in the brass? Assuming a coefficient of friction of 0.25, what is the force required to press them apart?

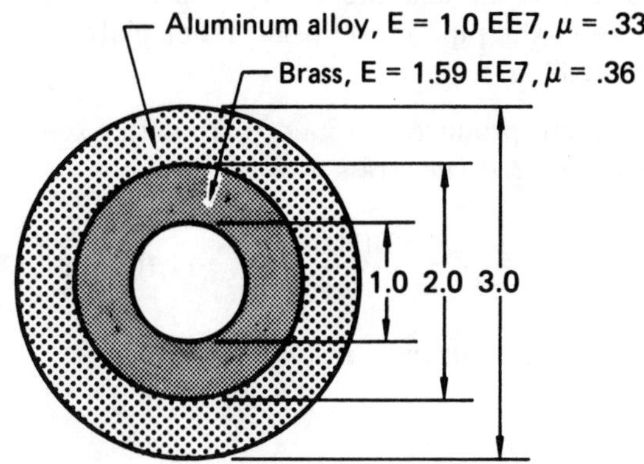


Aluminum alloy, internal pressure:

$$\sigma_{ci} = \frac{(1.5^2 + 1^2)p}{(1.5+1)(0.5)} = 2.6p$$

$$\sigma_{ri} = -p$$

$$\left(\frac{\Delta D}{D}\right)_i = \frac{[2.6p - 0.33(-p)]}{1.0 \text{ EE7}} = (2.93 \text{ EE–7})p$$

$$\Delta D_i = (2.93 \text{ EE–7})p(2) = (5.86 \text{ EE–7})p$$

Brass, external pressure:

$$\sigma_{co} = \frac{-(1^2 + 0.5^2)p}{(1+0.5)(0.5)} = -1.667p$$

$$\sigma_{ro} = -p$$

$$\left(\frac{\Delta D}{D}\right)_o = \frac{[-1.667p - 0.36(-p)]}{1.59 \text{ EE7}} = (-0.822 \text{ EE–7})p$$

$$\Delta D_o = (-0.822 \text{ EE–7})p(2) = (-1.644 \text{ EE–7})p$$

From equation 12.97,

$$|(5.86 \text{ EE–7})|p + |(-1.644 \text{ EE–7})|p = 0.004$$

$$p = 5330 \text{ psi}$$

From table 12.6,

$$\sigma_{\text{ci, brass}} = \frac{-2(1^2)5330}{(1+0.5)(0.5)} = 14{,}213 \text{ psi}$$

$$\tau_{\max} = \frac{1}{2}(14{,}213) = 7106.5 \text{ psi}$$

The force to separate the cylinders is

$$F = (5330)[2\pi(1)(2)](0.25) = 16{,}745 \text{ lbf}$$

**Table 12.7**
Flat Plates Under Uniform Pressure of p psi

| shape | edge condition | maximum stress | deflection at center |
|---|---|---|---|
| circular | simply supported | $(3/8)pr^2(3+\mu)/t^2$ at center | $(3/16)pr^4(1-\mu)(5+\mu)/(Et^3)$ |
| | built-in | $(3/4)pr^2/t^2$ at edge | $(3/16)pr^4(1-\mu^2)/(Et^3)$ |
| rectangular | simply supported | $C_1pb^2/t^2$ at center | $C_2pb^4/(Et^3)$ |
| | built-in | $C_3pb^2/t^2$ at centers of long edges | $C_4pb^4/(Et^3)$ |

| a/b | 1.0 | 1.2 | 1.4 | 1.6 | 1.8 | 2 | 3 | 4 | 5 | ∞ |
|---|---|---|---|---|---|---|---|---|---|---|
| $C_1$ | 0.287 | 0.376 | 0.453 | 0.517 | 0.569 | 0.610 | 0.713 | 0.741 | 0.748 | 0.750 |
| $C_2$ | 0.044 | 0.062 | 0.077 | 0.091 | 0.102 | 0.111 | 0.134 | 0.140 | 0.142 | 0.142 |
| $C_3$ | 0.308 | 0.383 | 0.436 | 0.487 | 0.497 | 0.500 | 0.500 | 0.500 | 0.500 | 0.500 |
| $C_4$ | 0.0138 | 0.0188 | 0.023 | 0.025 | 0.027 | 0.028 | 0.028 | 0.028 | 0.028 | 0.028 |

## 11 FLAT PLATES UNDER UNIFORM PRESSURE LOADING

Formulas for stresses and deflections of flat plates are given in table 12.7. Their application is subject to the following constraints.

- The plates are of medium thickness, meaning that the thickness is equal to or less than 1/4 the minimum width dimension, and the maximum deflection is equal to or less than 1/2 the thickness.
- The plates are constructed of isentropic, elastic material, and the elastic limit is not exceeded under the applied loading.

*Example 12.29*

The end of a 10″ inside diameter pipe is capped by welding on an end plate made from mild steel. The safe stress in the end cap is 11,100 psi. The internal pressure in the pipe is 500 psia. What plate thickness is required?

The welding produces a round plate with fixed edges. From table 12.7, the stress is

$$\sigma = \frac{3pr^2}{4t^2} = \frac{(3)(500)(5)^2}{4t^2} = 11{,}100 \text{ psi}$$

Solving for the thickness results in $t = 0.92''$.

# Appendix A: Beam Formulas

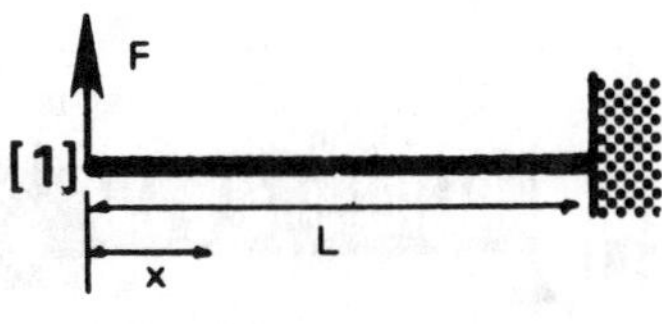


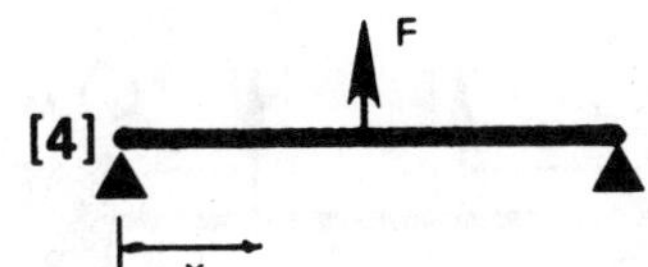


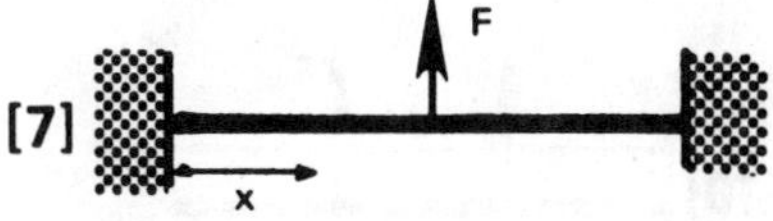


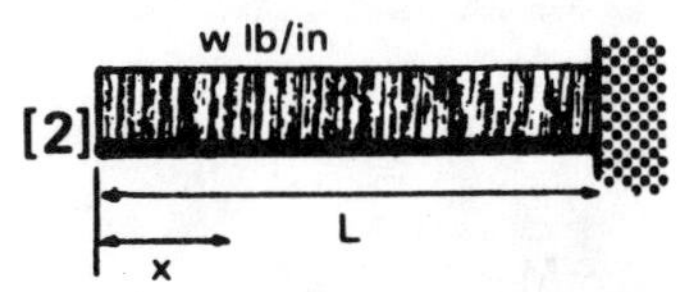

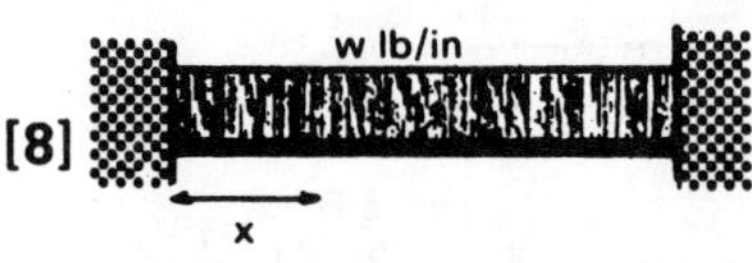


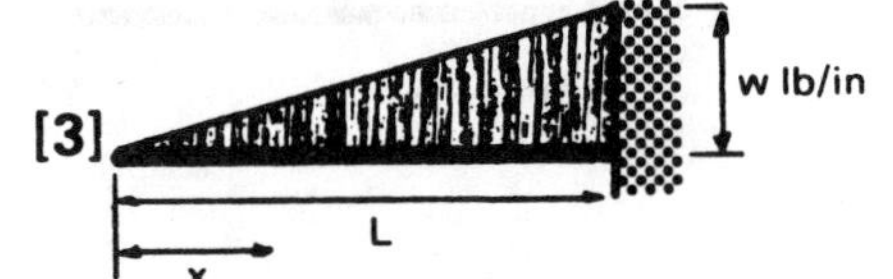

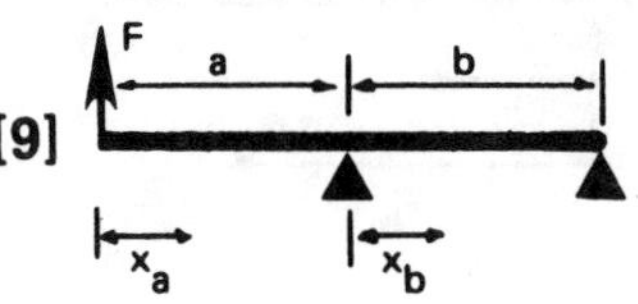


| case | moment | deflection |
|---|---|---|
| 1 | $M = Fx$<br>$M_{\text{max}} = FL$ | $y = (F/6EI)(2L^3 - 3L^2x + x^3)$<br>$y_{\text{max}} = FL^3/3EI$ |
| 2 | $M = \frac{1}{2}wx^2$<br>$M_{\text{max}} = \frac{1}{2}wL^2$ | $y = (w/24EI)(3L^4 - 4L^3x + x^4)$<br>$y_{\text{max}} = wL^4/8EI$ |
| 3 | $M = wx^3/6L$<br>$M_{\text{max}} = wL^2/6$ | $y = (w/120EIL)(4L^5 - 5L^4x + x^5)$<br>$y_{\text{max}} = wL^4/30EI$ |
| 4 | $M = -\frac{1}{2}Fx$<br>$M_{\text{max}} = -\frac{1}{4}FL$ | $y = (Fx/48EI)(3L^2 - 4x^2)$<br>$y_{\text{max}} = FL^3/48EI$ |
| 5 | $M = \left(\frac{1}{2}wx\right)(x - L)$<br>$M_{\text{max}} = -wL^2/8$ | $y = (wx/24EI)(L^3 - 2Lx^2 + x^3)$<br>$y_{\text{max}} = 5wL^4/384EI$ |
| 6 | $M = (-wx/6L)(L^2 - x^2)$<br>$M_{\text{max}} = -0.064wL^2$ at $x = 0.5774L$ | $y = (wx/360EIL)(7L^4 - 10L^2x^2 + 3x^4)$<br>$y_{\text{max}} = 0.00652wL^4/EI$ at $x = 0.5193L$ |
| 7 | $M = \frac{1}{2}F\left[\left(\frac{1}{4}L\right) - x\right]$<br>$M_{\text{max}} = FL/8$ at $x = 0$<br>$M_{\text{max}} = -FL/8$ at $x = \frac{1}{2}L$ | $y = (Fx^2/48EI)(3L - 4x)$<br>$y_{\text{max}} = FL^3/192EI$ |
| 8 | $M = \left(\frac{1}{2}wL^2\right)[(1/6) - (x/L) + (x/L)^2]$<br>$M_{\text{max}} = wL^2/12$ at $x = 0$ and $x = L$<br>$M = -wL^2/24$ at $x = \frac{1}{2}L$ | $y = (wx^2/24EI)(L - x)^2$<br>$y_{\text{max}} = wL^4/384EI$ |
| 9 | $M_a = Fx_a$<br>$M_b = (Fa/b)(b - x_b)$<br>$M_{\text{max}} = Fa$ at $x_a = a$ | $y_a = (F/3EI)[(a^2 + ab)(a - x_a) + (x_a/2)(x_a^2 - a^2)]$<br>$y_b = (Fax_b/6EI)[3x_b - (x_b^2/b) - 2b]$<br>$y_{\text{tip}} = (Fa^2/3EI)(a + b)$ (max up)<br>$y_{\text{max}} = (0.06415)Fab^2/EI$ at $x_b = 0.4226b$ (max down) |

## Appendix A, continued

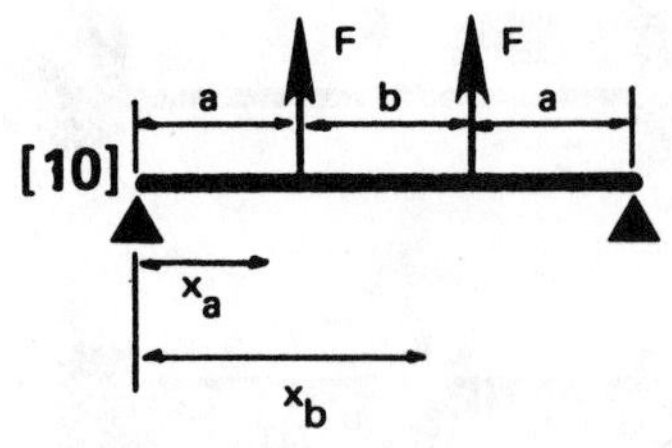


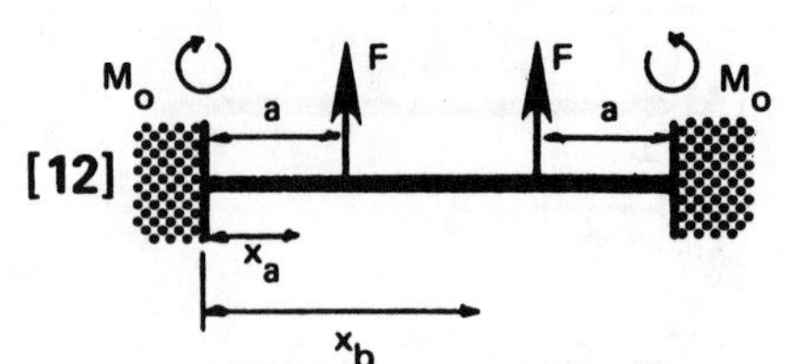


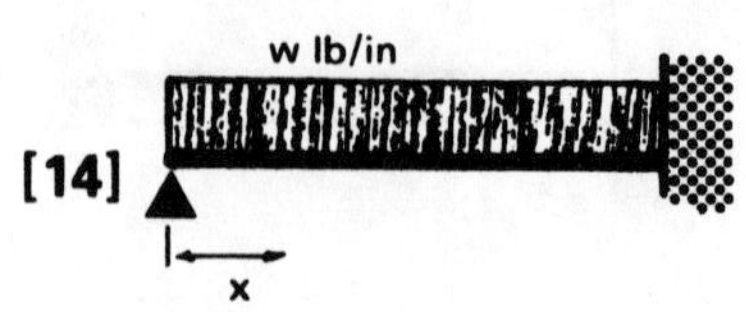


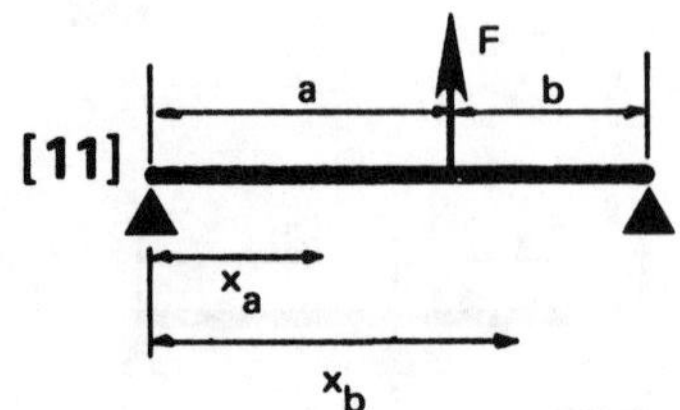


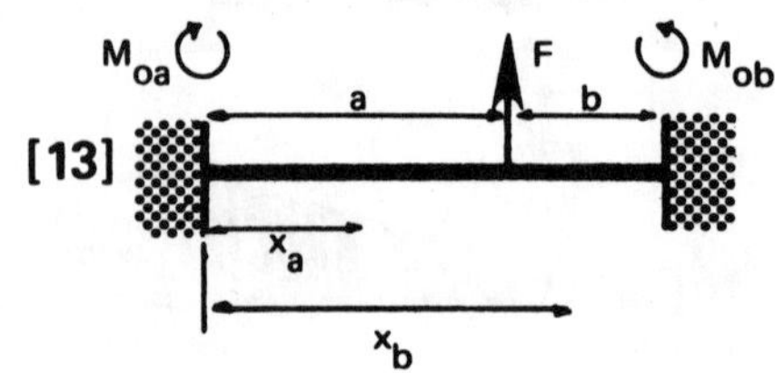


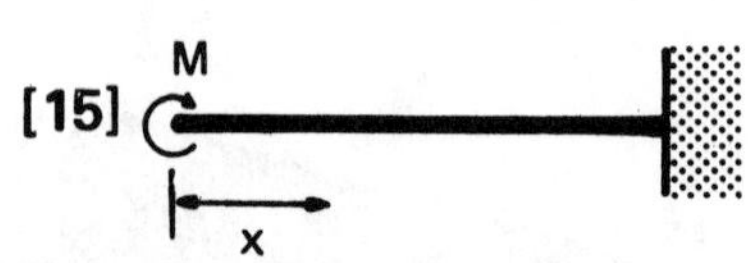


| case | moment | deflection |
|---|---|---|
| 10 | $M_a = -Fx_a$<br>$M_b = -Fa$<br>$M_{\text{max}} = -Fa$ (everywhere between loads) | $y_a = (Fx_a/6EI)[(3a)(L-a) - x_a^2]$<br>$y_b = (Fa/6EI)[3x_b(L-x_b) - a^2]$<br>$y_{\text{max}} = (Fa/24EI)[3L^2 - 4a^2]$ |
| 11 | $M_a = -Fbx_a/L$<br>$M_b = -Fa(L-x_b)/L$<br>$M_{\text{max}} = -Fab/L$ at $x_a = a$ | $y_a = (Fbx_a/6EIL)(L^2 - b^2 - x_a^2)$<br>$y_b = (Fb/6EIL)[(L/b)(x_b - a)^3 + (L^2 - b^2)x_b - x_b^3]$<br>$y = Fa^2b^2/3EIL$ at $x_a = a$<br>$y_{\text{max}} = (0.06415Fb/EIL)(L^2 - b^2)^{3/2}$ at $x = \sqrt{a(L+b)/3}$ |
| 12 | $M_a = (Fa/L)(L-a) - Fx_a$<br>$M_b = Fa^2/L$<br>$M_o = (Fa/L)(L-a)$ | $y_a = (Fx_a^2/2EI)[a(1-(a/L)) - (x_a/3)]$<br>$y_b = (Fa^2/2EI)[x_b - (x_b^2/L) - (a/3)]$<br>$y_{\text{max}} = (Fa^2/24EI)(3L - 4a)$ at $x = \frac{1}{2}L$ |
| 13 | $M_a = (Fb^2/L^3)[aL - x_a(L+2a)]$<br>$M_b = (Fa^2/L^3)[bL - (L-x_b)(L+2b)]$<br>$M_{oa} = Fab^2/L^2$ (max when $a < b$)<br>$M_{ob} = Fa^2b/L^2$ (max when $a > b$)<br>$M = -2Fa^2b^2/L^3$ at $x_a = a$ | $y_a = (Fx_a^2b^2/6EIL^3)[3aL - x_a(3a+b)]$<br>$y_b = (F(L-x_b)^2a^2/6EIL^3)[3bL - (L-x_b)(3b+a)]$<br>$y = Fa^3b^3/3EIL^3$ at $x_a = a$<br>$y_{\text{max}} = 2Fa^3b^2/[3EI(L+2a)^2]$ at $x = 2aL/(L+2a)$ |
| 14 | $M = (3wLx/8) - \frac{1}{2}wx^2$<br>$M_{\text{max}} = wL^2/8$ at $x = L$ | $y = (wx/48EI)[L^3 - 3Lx^2 + 2x^3]$<br>$y_{\text{max}} = wL^4/185EI$ at $x = 0.4215L$ |
| 15 | $M = M$ everywhere | $y = (M/2EI)(L^2 - 2xL + x^2)$<br>$y_{\text{max}} = ML^2/2EI$ at free end |

# Appendix B: Centroids and Area Moments of Inertia

| SHAPE | DIMENSIONS | CENTROID $(x_c, y_c)$ | AREA MOMENT OF INERTIA |
|---|---|---|---|
| Rectangle | 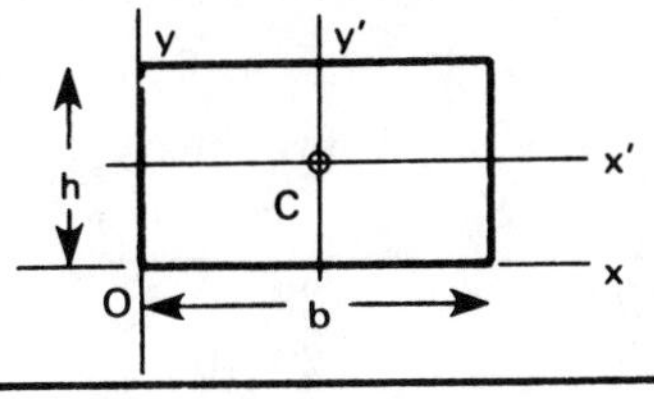 | $(\frac{1}{2}b, \frac{1}{2}h)$ | $I_{x'} = (1/12)bh^3$<br>$I_{y'} = (1/12)hb^3$<br>$I_x = (1/3)bh^3$<br>$I_y = (1/3)hb^3$<br>$J_C = (1/12)bh(b^2 + h^2)$ |
| Triangle | 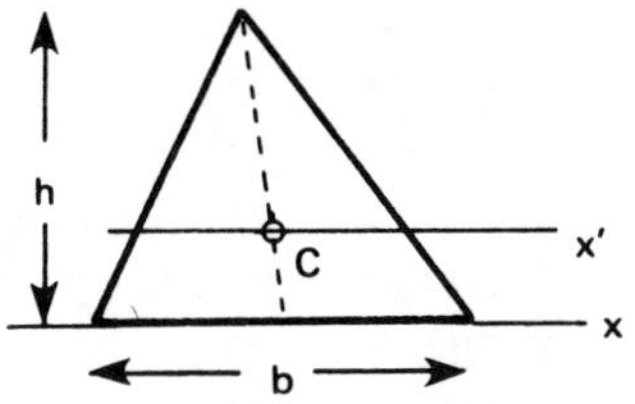 | $y_c = (h/3)$ | $I_{x'} = (1/36)bh^3$<br>$I_x = (1/12)bh^3$ |
| Trapezoid | 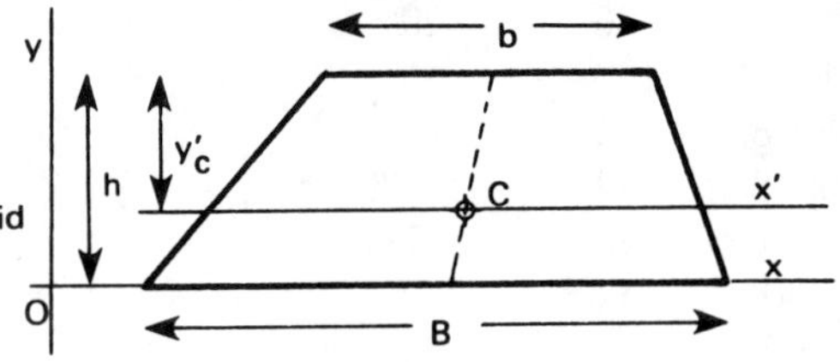 | $y'_c = \dfrac{h(2B + b)}{3(B + b)}$<br>Note that this is measured from the top surface. | $I_{x'} = \dfrac{h^3(B^2+4Bb+b^2)}{36(B+b)}$<br>$I_x = \dfrac{h^3(B+3b)}{12}$ |
| Quarter-Circle, of radius r | 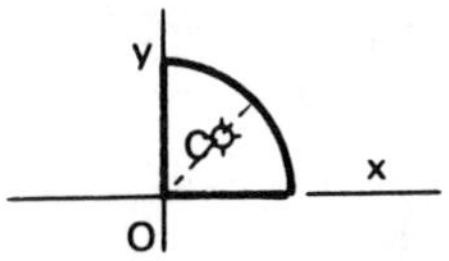 | $((4r/3\pi), (4r/3\pi))$ | $I_x = I_y = (1/16)\pi r^4$<br>$J_O = (1/8)\pi r^4$ |
| Half Circle, of radius r | 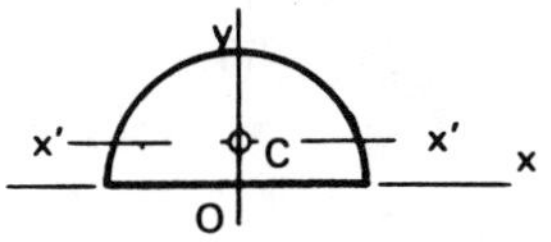 | $(0, (4r/3\pi))$ | $I_x = I_y = (1/8)\pi r^4$<br>$J_O = \frac{1}{4}\pi r^4 \quad I_{x'} = .11r^4$ |
| Circle, of radius r | 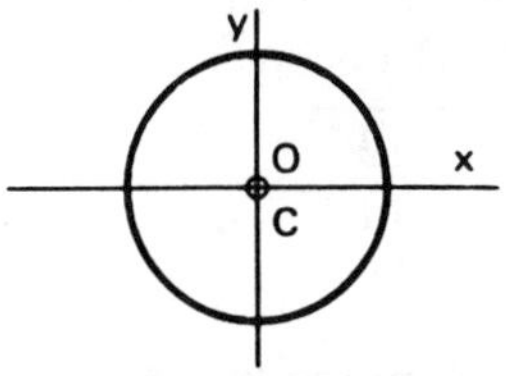 | $(0,0)$ | $I_x = I_y = \frac{1}{4}\pi r^4$<br>$J_O = \frac{1}{2}\pi r^4$ |
| Parabolic Area | 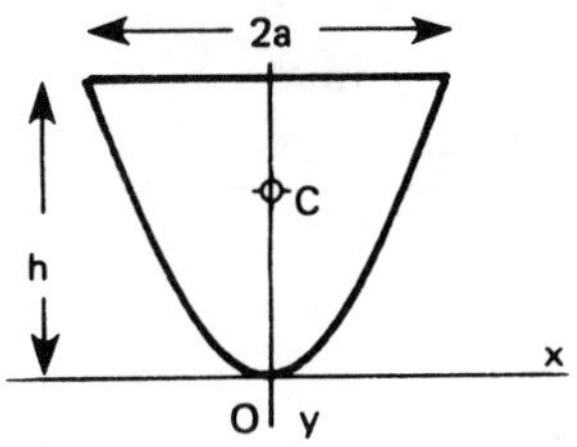 | $(0, (3h/5))$ | $I_x = 4h^3a/7$<br>$I_y = 4ha^3/15$ |
| Parabolic Spandrel | 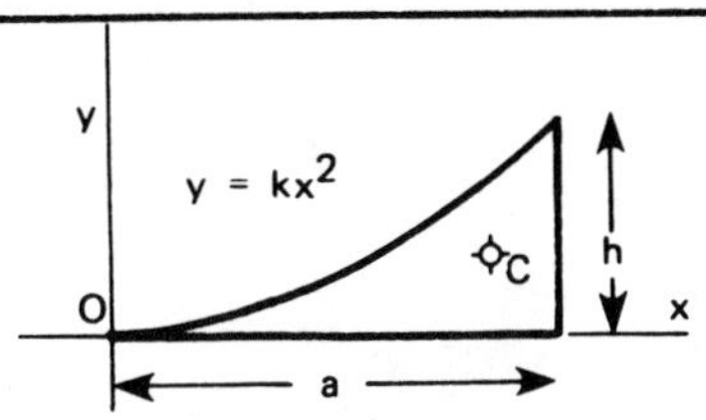 $y = kx^2$ | $((3a/4),(3h/10))$ | $I_x = ah^3/21$<br>$I_y = 3ha^3/15$ |

# Appendix C: Typical Properties of Structural Steel, Aluminum, and Magnesium (ksi)

| Designation | Application | $S_u$ | $S_y$ | Approximate $S_e$ |
|---|---|---|---|---|
| A36-70a | shapes | 58–80 | 36 | 29–40 |
| | plates | 58–80 | 36 | 29–40 |
| A53-72a | pipe | 60 | 35 | 30 |
| A242-70a | shapes | 70 | 50 | 35 |
| | plates to 3/4″ | 70 | 50 | 35 |
| A440-70a | shapes | 70 | 50 | 35 |
| | plates to 3/4″ | 70 | 50 | 35 |
| A441-70a | shapes | 70 | 50 | 35 |
| | plates to 3/4″ | 70 | 50 | 35 |
| A500-72 | tubes | 45 | 33 | 22 |
| A501-71a | tubes | 58 | 36 | 29 |
| A514-70 | plates to 3/4″ | 115–135 | 100 | 55 |
| A529-72 | shapes | 60–85 | 42 | 30–42 |
| | plates to 1/2″ | 60–85 | 42 | 30–42 |
| A570-72 | sheet/strip | 55 | 40 | 27 |
| A572-72 | shapes | 60 | 42 | 30 |
| | plates | 60 | 42 | 30 |
| A588-71 | shapes | 70 | 50 | 35 |
| | plates to 4″ | 70 | 50 | 35 |
| A606-71 | hot rolled sheet | 70 | 50 | 35 |
| | cold rolled sheet | 65 | 45 | 32 |
| A607-70 | sheet | 60 | 45 | 30 |
| A618-71 | shapes | 70 | 50 | 35 |
| | tubes | 70 | 50 | 35 |

Typical Properties of Structural Aluminum (ksi)

| Designation | Application | $S_u$ | $S_y$ | Approximate $S_e$ (EE8 cyc.) |
|---|---|---|---|---|
| 2014-T6 | shapes/bars | 63 | 55 | 19 |
| 6061-T6 | all | 42 | 35 | 14.5 |

Typical Properties of Structural Magnesium (ksi)

| Designation | Application | $S_u$ | $S_y$ | Approximate $S_e$ (EE7 cyc.) |
|---|---|---|---|---|
| AZ31 | shapes | 38 | 29 | 19 |
| AZ61 | shapes | 45 | 33 | 19 |
| AZ80 | shapes | 55 | 40 | |

**Practice Problems: MECHANICS OF MATERIALS**

Untimed

1. A 14 foot simple beam is uniformly loaded with 200 pounds per foot over its entire length. If the beam is 3.625" wide and 7.625" deep, what is the maximum bending stress? What is the maximum shear stress?

2. A reinforced concrete beam is illustrated. It is subjected to a maximum moment of 8125 ft-lbf. The total cross-sectional area of steel is 1 square inch. Take the modulus of elasticity for concrete to be 2 EE6 psi. Disregard the area of concrete in tension. Use the transformation method to calculate the maximum stress in the steel and concrete.

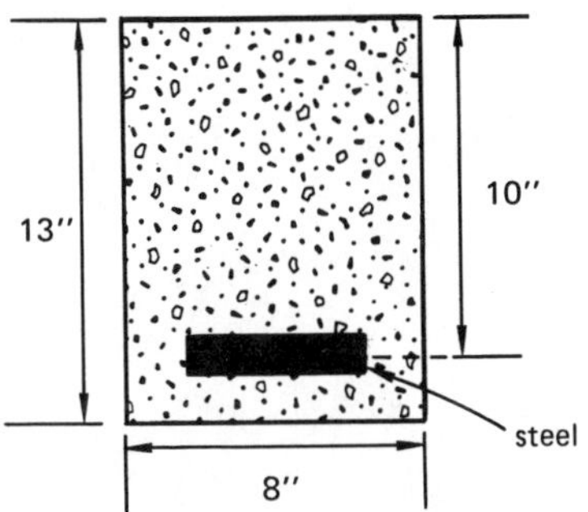


3. A steel truss with pinned joints is constructed as shown. The length/area ratio for each member is 50 $in^{-1}$. Support $R_1$ is a pin. Support $R_2$ is a roller. What is the vertical deflection at the point where the 10 kip load is applied? Use E = 2.9 EE7 psi.

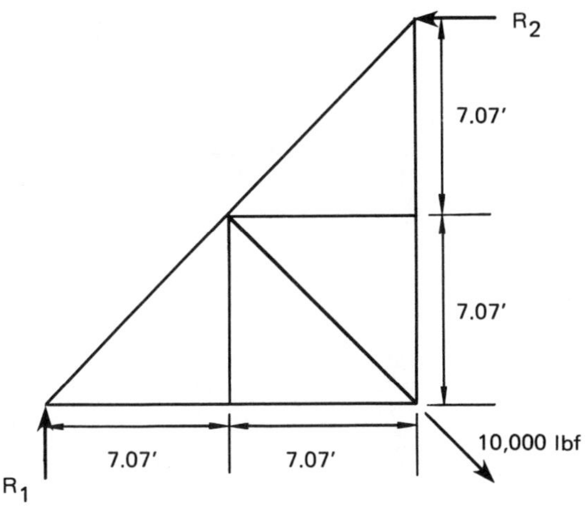


4. A beam 25 feet long is simply supported at the left end and 5 feet from the right end. A uniform load of 2 kips/ft extends over a 10 foot length starting from the left end. There is also a concentrated 10 kip load at the right end. Draw the shear and moment diagrams.

5. A 40 foot long steel beam with moment of inertia I is reinforced along the middle 20 feet, leaving the two 10 foot ends unreinforced. The moment of inertia of the reinforced section is 2I. A 20,000 pound load is applied mid-span along the beam, 20 feet from each end. What is the deflection in terms of EI?

6. The truss shown carries a moving uniform live load of 2 kips per foot and a moving concentrated live load of 15 kips. What are the maximum forces in members Bb and BC?

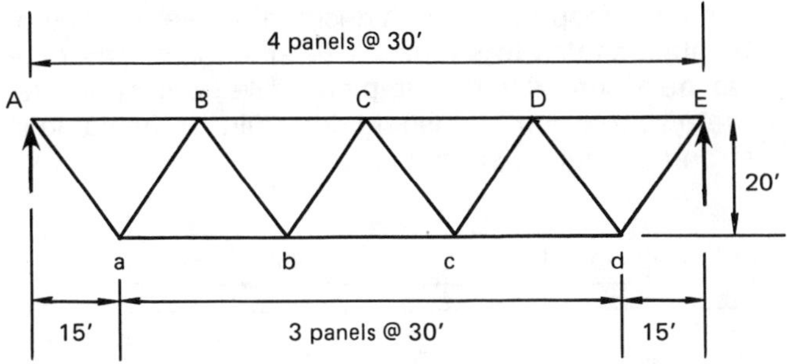


7. A 24 foot, 4 inch × 4 inch white oak timber (E = 1.5 EE6 psi) is used to support a sign. One end is fixed in a deep concrete base, the other end supports a sign 9 feet above the ground. Neglect wind effects and find the critical buckling sign weight.

8. What is the mid-span deflection for the steel beam shown? The cross sectional moment of inertia is 200 $inches^4$.

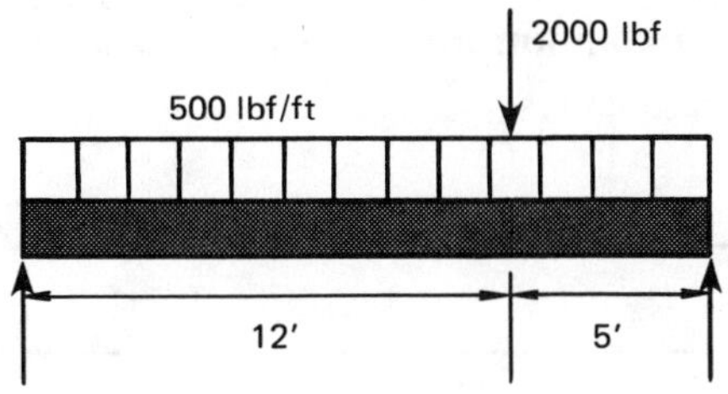


9. The truss shown carries a group of live loads along its bottom chord. What is the maximum force in member CD?

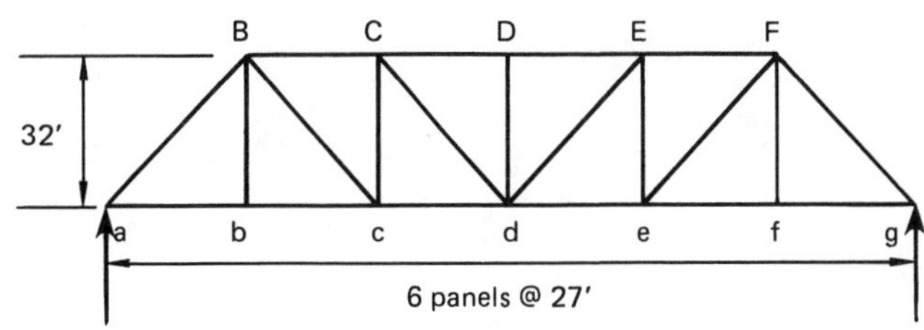


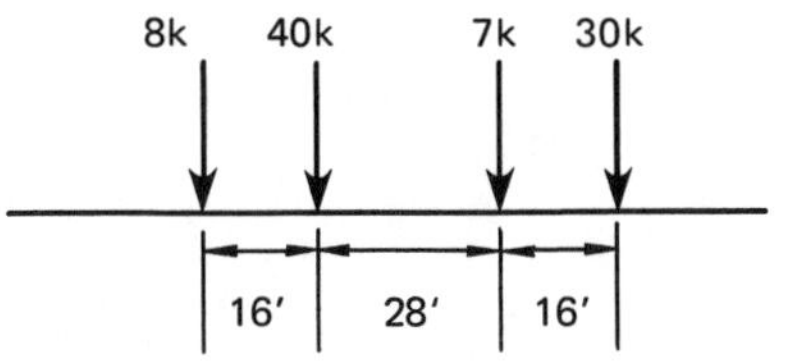

10. The allowable stress in a steel beam is 20,000 psi. If the maximum applied moment is 1.5 EE5 ft-lb, what is the required section modulus?

## Timed

1. Wet concrete is needed 200 feet from the nearest pump location, which is separated from the pour location by a deep ravine. It is decided to use a series of ½″ steel cables to support a 6″ steel pipe. The pipe can be assumed to be completely filled with concrete. State your assumptions and determine if the cable is strong enough to support the pipe.

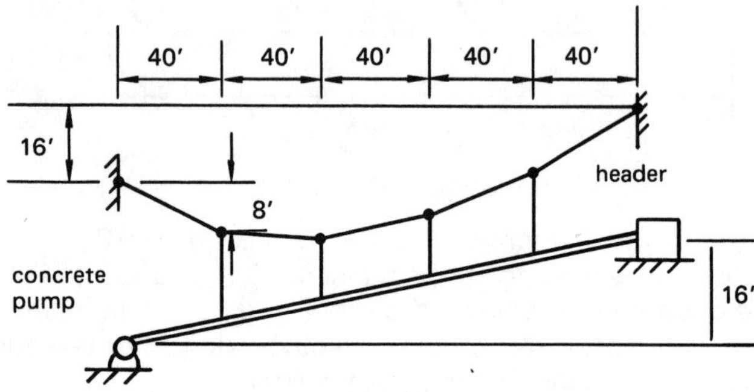


2. A moving load, consisting of two 30 kip loads separated by a constant 6 feet, travels over a two-span bridge. The bridge has an interior expansion joint that can be considered to be a hinge. Find the maximum moment and shear at point B.

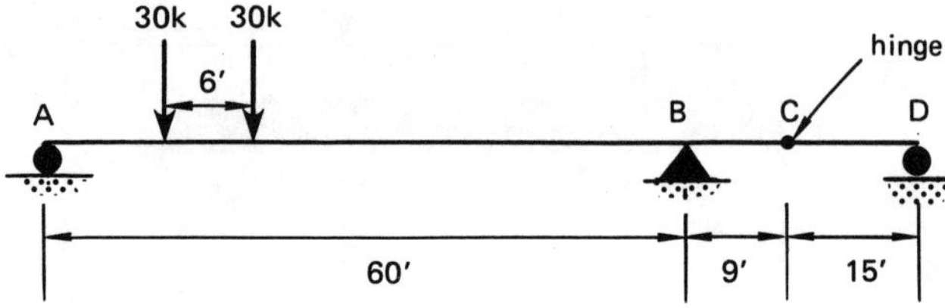

# 13 INDETERMINATE STRUCTURES

*Nomenclature*

| | | |
|---|---|---|
| A | area | $ft^2$ |
| C | carry-over factor | – |
| COM | carry-over moment | ft–lbf |
| D | distribution factor | – |
| E | modulus of elasticity | $lbf/ft^2$ |
| F | fixity factor | – |
| I | moment of inertia | $in^4$ |
| k | spring constant | ft–lbf/radian |
| K | stiffness factor | ft–lbf |
| L | length | ft |
| M | moment | ft-lbf |
| P | load | lbf |
| R | relative stiffness, or reaction | ft–lbf, or lbf |
| S | force in truss member | lbf |
| T | temperature | °F |
| u | dummy force in truss member | lbf |
| w | continuous load | lbf/ft |
| x* | distance to centroid of moment diagram | ft |

*Symbols*

| | | |
|---|---|---|
| $\delta$ | deflection | ft |
| $\sigma$ | stress | $lbf/ft^2$ |
| $\alpha$ | coefficient of thermal expansion | 1/°F |
| $\theta$ | angle | radians |
| $\Delta$ | deflection | ft |

*Subscripts*

| | |
|---|---|
| BM | bending moment |
| c | concrete |
| M | due to moment M |
| P | due to force P |
| s | steel |

## 1 INTRODUCTION

A structure that is statically indeterminate is one for which the equations of statics are not sufficient to determine all reactions, moments, and internal stress distributions. Additional formulas involving deflection are required to completely determine these unknowns. Although there are a large number of problem types which are statically indeterminate, this chapter is primarily concerned with the following cases:

- Beams on more than two supports
- Trusses with more than 2(# joints)–3 members
- Rigid frames

The *degree of redundancy* is equal to the number of reactions or members that would have to be removed in order to make the structure statically determinate. For example, a two-span beam on three simple supports is redundant to the first degree.

## 2 CONSISTENT DEFORMATION

The method of *consistent deformation* can be used to evaluate simple structures consisting of two or three members in tension or compression. This method is simple to learn and apply. The method makes use of geometry to develop relationships between the deflections (deformations) between different members or locations on the structure.

*Example 13.1*

A pile is constructed of concrete with a steel jacket. What are the stresses in the steel and concrete if a load $P$ is applied? Assume the end caps are rigid and the steel-concrete bond is perfect.

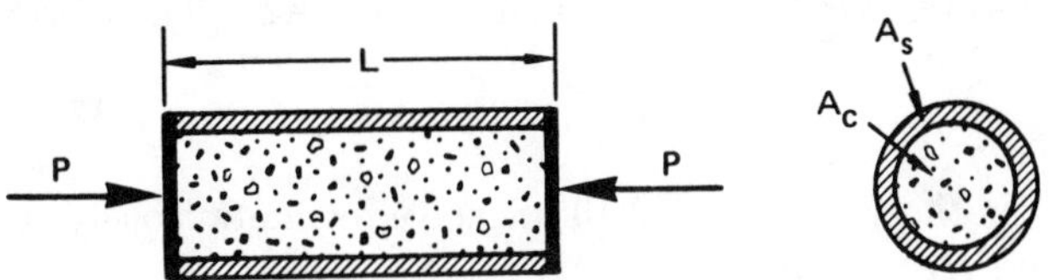

Let $P_c$ and $P_s$ be the loads carried by the concrete and steel respectively. Then,

$$P_c + P_s = P \qquad 13.1$$

The deformation of the steel is

$$\delta_s = \frac{P_s L}{A_s E_s} \qquad 13.2$$

Similarly, the deflection of the concrete is

$$\delta_c = \frac{P_c L}{A_c E_c} \qquad 13.3$$

But, $\delta_c = \delta_s$ since the bonding is perfect. Therefore,

$$\frac{P_c L}{A_c E_c} - \frac{P_s L}{A_s E_s} = 0 \qquad 13.4$$

Equations 13.1 and 13.4 are solved simultaneously to determine $P_c$ and $P_s$. The respective stresses are:

$$\sigma_s = \frac{P_s}{A_s} \qquad 13.5$$

$$\sigma_c = \frac{P_c}{A_c} \qquad 13.6$$

*Example 13.2*

A uniform bar is clamped at both ends and the axial load applied near one of the supports. What are the reactions?

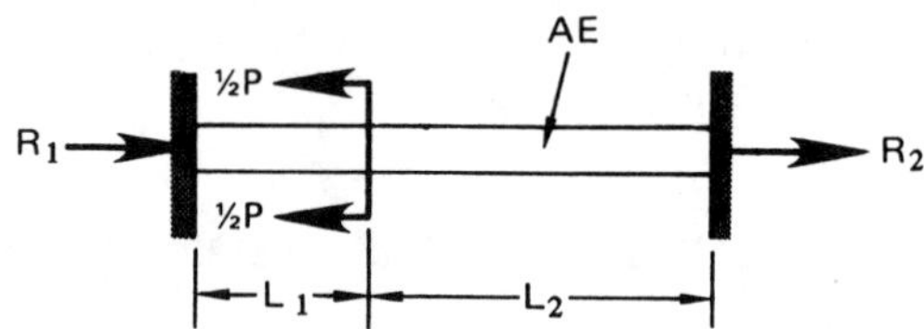


The first required equation is

$$R_1 + R_2 = P \qquad 13.7$$

The shortening of section 1 due to the reaction $R_1$ is

$$\delta_1 = \frac{-R_1 L_1}{AE} \qquad 13.8$$

The elongation of section 2 due to the reaction $R_2$ is

$$\delta_2 = \frac{R_2 L_2}{AE} \qquad 13.9$$

However, the bar is continuous, so $\delta_1 = -\delta_2$. Therefore,

$$R_1 L_1 = R_2 L_2 \qquad 13.10$$

Equations 13.7 and 13.10 are solved simultaneously to find $R_1$ and $R_2$.

*Example 13.3*

The non-uniform bar shown is clamped at both ends. What are the reactions if a temperature change of $\Delta T$ is experienced?

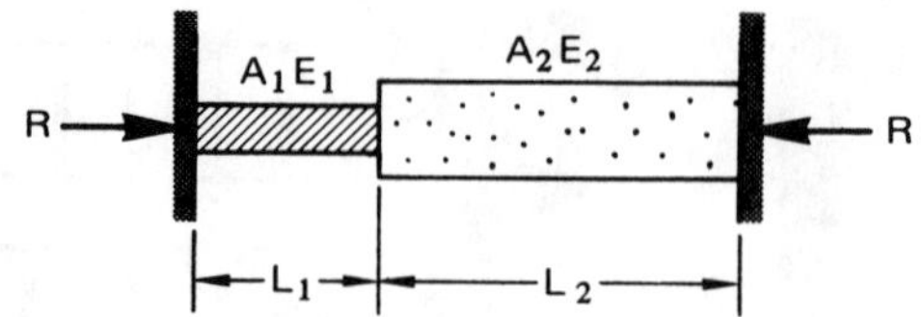


The thermal deformations of sections 1 and 2 can be calculated directly.

$$\delta_1 = \alpha_1 L_1 \Delta T \qquad 13.11$$

$$\delta_2 = \alpha_2 L_2 \Delta T \qquad 13.12$$

The total deformation is $\delta = \delta_1 + \delta_2$. However, the deformation can also be calculated from the principles of mechanics.

$$\delta = \frac{RL_1}{A_1 E_1} + \frac{RL_2}{A_2 E_2} \qquad 13.13$$

Combining equations 13.11 through 13.13 produces an equation that can be solved directly for $R$.

$$(\alpha_1 L_1 + \alpha_2 L_2)\Delta T = \left(\frac{L_1}{A_1 E_1} + \frac{L_2}{A_2 E_2}\right) R \qquad 13.14$$

*Example 13.4*

The beam shown is supported by dissimilar members. What are the forces in the members? Assume the bar is rigid and remains horizontal.[1]

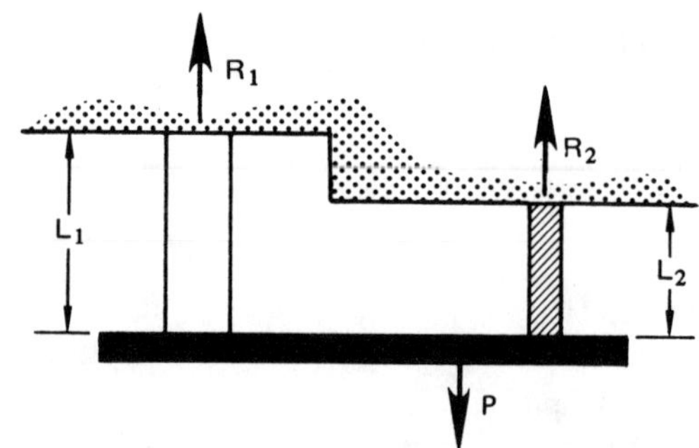


The required equilibrium condition is

$$R_1 + R_2 = P \qquad 13.15$$

The elongations of the two tension members are

$$\delta_1 = \frac{R_1 L_1}{A_1 E_1} \qquad 13.16$$

$$\delta_2 = \frac{R_2 L_2}{A_2 E_2} \qquad 13.17$$

[1] This example is easily solved by summing moments about a point on the horizontal beam.

If the horizontal bar remains horizontal, then $\delta_1 = \delta_2$.

$$\frac{R_1 L_1}{A_1 E_1} = \frac{R_2 L_2}{A_2 E_2} \qquad 13.18$$

Equations 13.15 and 13.18 are solved simultaneously to find $R_1$ and $R_2$.

*Example 13.5*

The beam shown is supported by dissimilar members. The bar is rigid, but is not constrained to remain horizontal. What are the reactions in the vertical members?

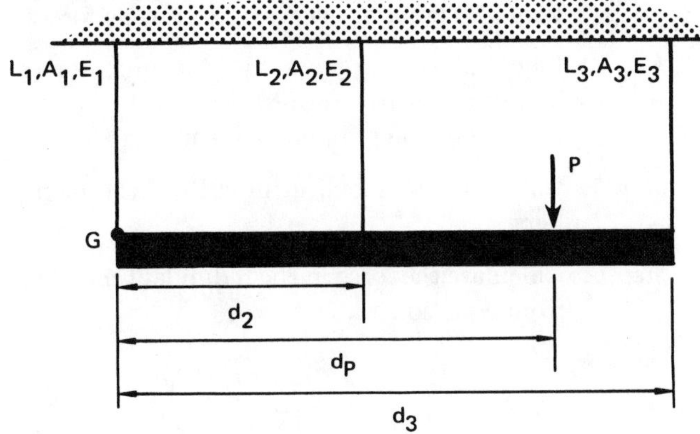


The forces in the supports are $R_1$, $R_2$, and $R_3$. Any of these may be tensile (positive) or compressive (negative).

$$R_1 + R_2 + R_3 = P \qquad 13.19$$

The changes in lengths are

$$\delta_1 = \frac{R_1 L_1}{A_1 E_1} \qquad 13.20$$

$$\delta_2 = \frac{R_2 L_2}{A_2 E_2} \qquad 13.21$$

$$\delta_3 = \frac{R_3 L_3}{A_3 E_3} \qquad 13.22$$

Since the bar is rigid, the deflections will be proportional to the distance from point $G$.

$$\delta_2 = \delta_1 + \frac{d_2}{d_3}(\delta_3 - \delta_1) \qquad 13.23$$

Moments can be summed about point $G$ to give a third equation.

$$M_G = R_3 d_3 + R_2 d_2 - P d_P = 0 \qquad 13.24$$

*Example 13.6*

Find the forces in the three tension members.

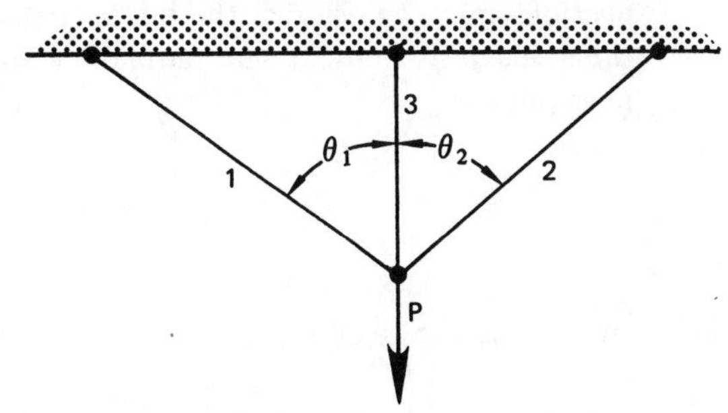


The equilibrium requirement is

$$P_{1y} + P_3 + P_{2y} = P \qquad 13.25$$

$$P_1 \cos\theta_1 + P_3 + P_2 \cos\theta_2 = P \qquad 13.26$$

The vertical elongations of all three tension members are the same at the junction.

$$\frac{P_1 L_1}{A_1 E_1}(\cos\theta_1) = \frac{P_3 L_3}{A_3 E_3} = \frac{P_2 L_2}{A_2 E_2}(\cos\theta_2) \qquad 13.27$$

Equations 13.26 and 13.27 can be solved simultaneously to find $P_1$, $P_2$, and $P_3$. (It may be necessary to work with the $x$-components of the deflections in order to find a third equation.)

## 3 USING SUPERPOSITION WITH STATICALLY INDETERMINATE BEAMS

Continuous beams and propped cantilevers that are indeterminate to the first degree can often be solved by superposition.[2] This method requires finding the deflections with one or more of the supports removed, and then satisfying the given boundary conditions.

*step 1*: Remove the redundant supports to reduce the structure to a statically determinate condition.

*step 2*: Calculate the deflections at the previous locations of redundant supports. Use consistent sign conventions.

*step 3*: Apply each redundant support as a load, and find the deflections at the redundant support points as functions of the redundant support forces.

[2] Actually, this method can also be used with higher order indeterminate problems. However, the simultaneous equations that must be solved may make this method unattractive for manual solutions.

*step 4*: Use superposition to combine (i.e., add) the deflections due to the actual loads and the redundant support "loads." The total deflections must agree with the known deflections (usually zero) at the redundant support points.

*Example 13.7*

Determine the reaction, $S$, at the prop.

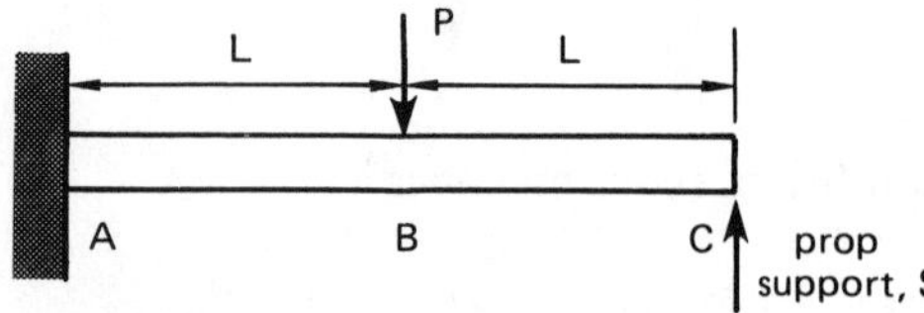


Start by removing the unknown prop reaction at point $C$. The cantilever that results is statically determinate. The deflection and slope at point $B$ can be found or derived from the beam equations.

$$\delta_B = \frac{-PL^3}{3EI} \qquad 13.28$$

$$y'_B = \frac{-PL^2}{2EI} \qquad 13.29$$

The slope remains constant to the right of point $B$. Therefore, the deflection at point $C$ due to the load at point $B$ is

$$\delta_{C,P} = \delta_B + y'_B L$$

$$= \frac{-5PL^3}{6EI} \qquad 13.30$$

The upward deflection at the cantilever tip due to the prop support, $S$, alone is

$$\delta_{C,S} = \frac{S(2L)^3}{3EI} = \frac{8SL^3}{3EI} \qquad 13.31$$

Now, it is known that the actual deflection at point $C$ is zero (the boundary condition). Therefore, the prop support, $S$, can be determined as a function of the applied load.

$$\delta_{C,S} + \delta_{C,P} = 0 \qquad 13.32$$

$$\frac{8SL^3}{3EI} - \frac{5PL^3}{6EI} = 0 \qquad 13.33$$

$$S = \frac{5P}{16} \qquad 13.34$$

## 4 INDETERMINATE TRUSSES: DUMMY UNIT LOAD METHOD

Due to the time required, it is unlikely that you would manually work with an indeterminate truss with more than one redundant member. Therefore, the following method is written specifically for trusses that are indeterminate to the first degree.

*step 1*: Draw the truss twice. Omit the redundant member on both trusses. (There may be a choice of redundant members.)

*step 2*: Load the first truss (which is now determinate) with the actual loads.

*step 3*: Calculate the forces, $S$, in all of the members. Assign a positive sign to tensile forces.

*step 4*: Load the second truss with two unit forces acting colinearly towards each other along the line of the redundant member.

*step 5*: Calculate the force, $u$, in each of the members.

*step 6*: Calculate the force in the redundant member from equation 13.35.

$$S_{\text{redundant}} = \frac{-\sum\left(\frac{SuL}{AE}\right)}{\sum\left(\frac{u^2L}{AE}\right)} \qquad 13.35$$

If $AE$ is the same for all members,

$$S_{\text{redundant}} = \frac{-\sum SuL}{\sum u^2L} \qquad 13.36$$

The true force in member $j$ of the truss is

$$P_{j,\text{true}} = S_j + (S_{\text{redundant}})u_j \qquad 13.37$$

*Example 13.8*

Find the force in members $BC$ and $BD$. $AE = 1$ for all members except for $CB$, which is 2, and $AD$, which is 1.5.

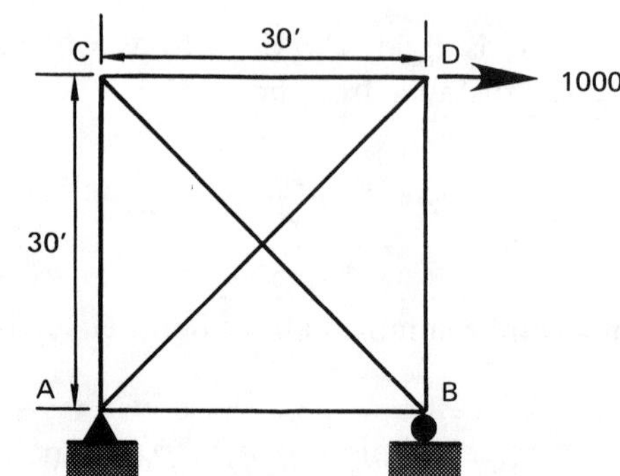

The two trusses are shown appropriately loaded.

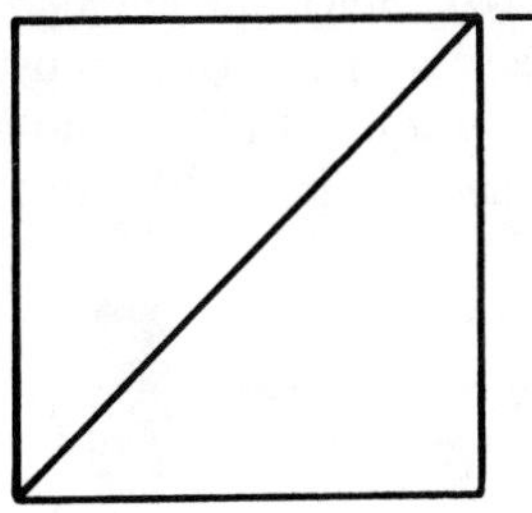


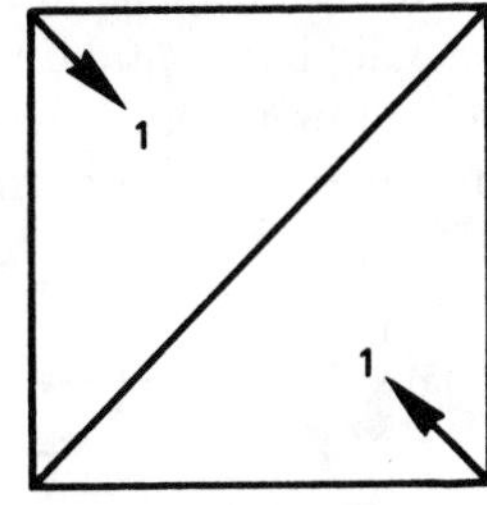


| member | $L$ | $AE$ | $S$ | $u$ | $\frac{SuL}{AE}$ | $\frac{u^2L}{AE}$ |
|---|---|---|---|---|---|---|
| $AB$ | 30 | 1 | 0 | −0.707(C) | 0 | 15 |
| $BD$ | 30 | 1 | −1000(C) | −0.707 | 21,210 | 15 |
| $DC$ | 30 | 1 | 0 | −0.707 | 0 | 15 |
| $CA$ | 30 | 1 | 0 | −0.707 | 0 | 15 |
| $CB$ | 42.43 | 2 | 0 | 1.0 | 0 | 21.22 |
| $AD$ | 42.43 | 1.5 | 1414(T) | 1.0 | 39,997 | 28.29 |
| | | | | | 61,207 | 109.51 |

From equation 13.35,

$$S_{BC} = \frac{-61{,}207}{109.51} = -558.9\,(\mathrm{C})$$

From equation 13.37,

$$P_{BD,\mathrm{true}} = -1000 + (-558.9)(-0.707) = -604.9\,(\mathrm{C})$$

## 5 INTRODUCTION TO THE MOMENT DISTRIBUTION METHOD[3]

The moment distribution method is extremely powerful. It can be used on beams of almost any complexity. It converges rapidly to a solution, despite the fact that the method is essentially an iterative process. Furthermore, the moment distribution method is easily learned.

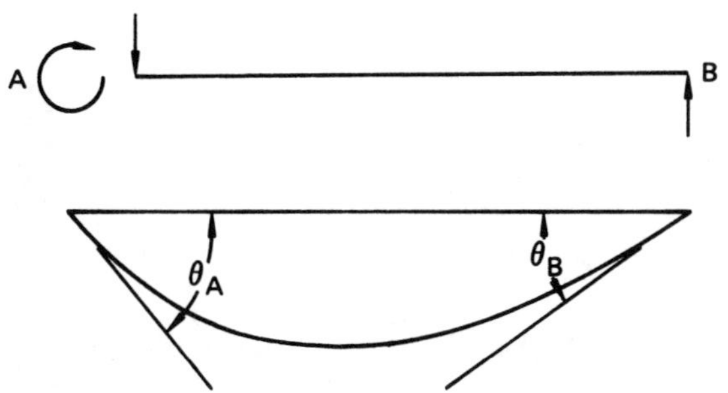


**Figure 13.1** Simply-Supported Beam with a Couple

Consider the simply-supported beam shown in figure 13.1. (The type of loading is not important, as only general relationships are being developed.) The deflection angles $\theta_A$ and $\theta_B$ can be found from the moment-area method.

If a simply-supported beam is acted upon by a clockwise couple (a moment) at one end as shown in figure 13.1, the angles of rotation will be given by equations 13.38 and 13.39.

$$\theta_A = \frac{M_A L}{3EI} \qquad 13.38$$

$$\theta_B = \frac{-M_A L}{6EI} \qquad 13.39$$

Figure 13.1 is particularly important because the equilibrium of the beam is not dependent on whether the moment is applied and the forces react, or whether the forces are applied and the moment reacts. If the forces are applied to the beam in figure 13.1, the moment develops to keep the beam end horizontal. A moment that is required to keep a beam end horizontal is known as a *fixed-end moment.*[4]

Any moment that would rotate an end of a beam clockwise is positive. Similarly, any moment that would rotate an interior joint clockwise is positive.

Being able to find the fixed-end moments at both ends of a loaded beam is absolutely necessary to the success of the moment distribution method. Extensive tables are available that contain the fixed-end moments for almost every conceivable loading. Appendix B is one such collection.

*Example 13.9*

The beam shown is acted upon by a moment at the right end. What is the relationship between the two fixed-end moments?

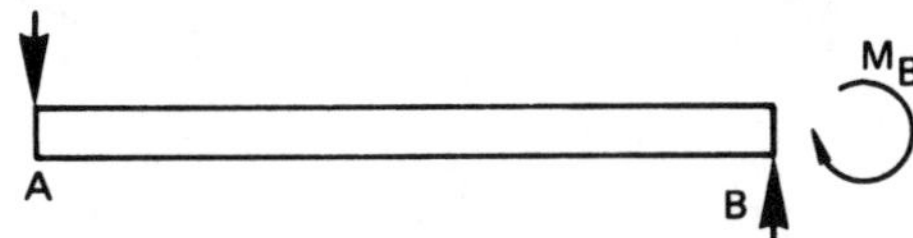


From equation 13.39, the angle of rotation at the left end due to a moment at the right end is

$$\theta_{A,M_B} = \frac{-M_B L}{6EI} \qquad 13.40$$

The angle at the left end due to the fixed-end moment at the left end is given by equation 13.38.

$$\theta_{A,M_A} = \frac{M_A L}{3EI} \qquad 13.41$$

[3] The moment distribution method was, at one time, also known as the *Cross method*, having been named after Professor Hardy Cross.

[4] Fixed-end moments are usually given double subscripts to indicate the location of the moment as well as the loaded span. For example, $M_{AB}$ would be a fixed-end moment at end $A$ caused by forces along the span $AB$.

However, $\theta_A = 0$, so $\theta_{A,M_B} + \theta_A M_A = 0$. Or,

$$\frac{M_B L}{6EI} = \frac{M_A L}{3EI} \quad 13.42$$

$$M_A = \tfrac{1}{2} M_B \quad 13.43$$

## 6 SIGNS OF FIXED-END MOMENTS

Determining the proper sign for fixed-end moments can sometimes be confusing. The phrase "positive moments make the beam smile" simply does not work with end moments.

End moments are the moments that the support must apply to the beam to keep the beam (or joint) from rotating. The common convention is that clockwise moments are positive. What is confusing is that a beam acted upon by a single force can have both positive and negative end moments.

Consider the beam in figure 13.2(a). If the ends were simply supported, the deflection curve would appear as figure 13.2(b). Since the actual deflection curve is as shown in figure 13.2(c), there must be moments acting on the ends (i.e., end moments). The moment at the left end is counterclockwise, and therefore, negative. The moment at the right end is clockwise, and therefore, positive.

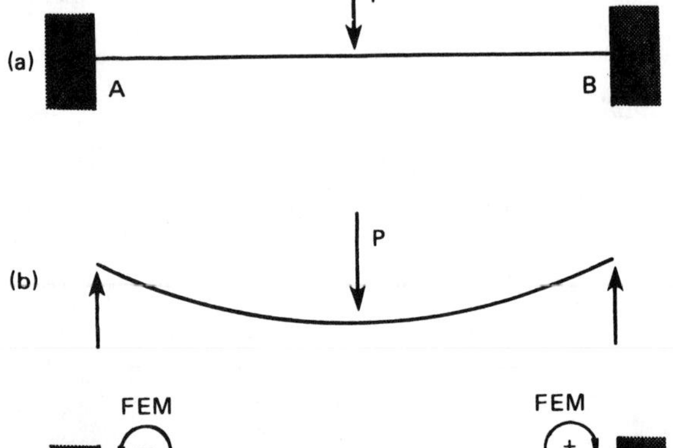


**Figure 13.2** A Simple Case of End Moments

Even though the beam in figure 13.2 "smiles," and we usually say it is acted upon by a positive moment, the moments in the fixed supports are both positive and negative.

Another case is that of a moment applied directly to a beam, as in figure 13.3(a). The same rules and sign conventions can be used to determine the signs of the end moments.

If the beam had simply supported ends, the deflection curve would be as in figure 13.3(b). Since the ends are actually horizontal due to the fixed support, there must be end moments preventing the rotation. Both of these end moments are positive, as shown in figure 13.3(c).

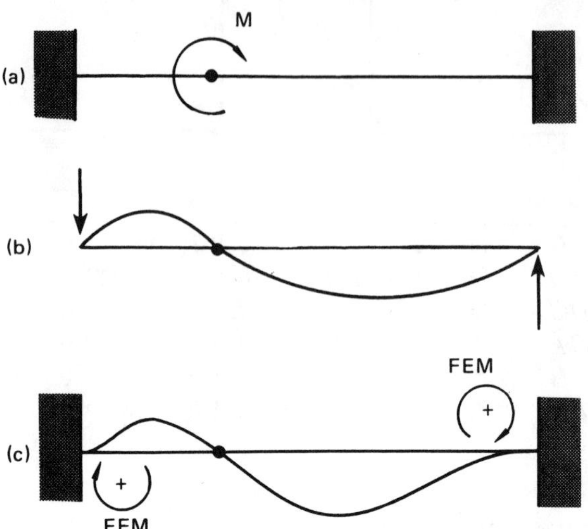


**Figure 13.3** End Moments Due to Direct Moment

Finally, consider the case of a force on a cantilever span. There is no moment at the free end, of course. However, the built-in end is fixed against rotation. Therefore, there is a fixed-end moment acting. As is the convention, clockwise moments are positive.

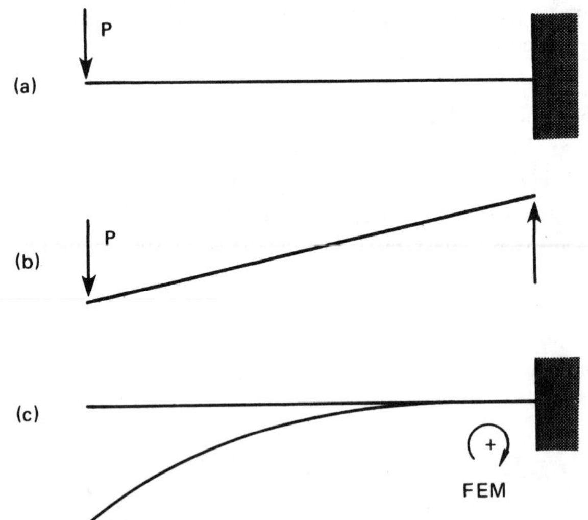


**Figure 13.4** End Moment on a Cantilever Support

## 7 BEAM STIFFNESS

From example 13.9, it is seen that a moment at one end of a simply-supported beam is partially transmitted to a fixed end. This transmitted moment is known as a *carry-over moment*, and the ratio of the carry-over moment to the original moment is known as the *carry-over factor*, $C$. In example 13.9, $C$ was $\frac{1}{2}$.

The carry-over factor can be determined for various degrees of *fixity*. The fixity is specified by a *fixity factor*, $F$, which varies from 0 (for a pin end) to 1 (for a built-in end).[5]

$$C = \frac{2F}{3+F} \qquad 13.44$$

The product $EI$ is known as the *stiffness* of the beam. The *relative stiffness* is

$$R = \frac{EI}{L} \qquad 13.45$$

The *angular spring constant* is known as the *stiffness factor*, $K$.

$$K = \frac{M}{\theta} = (3+F)R \qquad 13.46$$

It is possible to generalize about the moment required to produce a given angle of rotation. Equation 13.47 gives the moment for a simply supported beam.

$$M_{AB} = \frac{3EI\theta}{L} \qquad 13.47$$

Equation 13.47 could have been derived by using equation 13.46. Since $F = 0$ for a pinned end or simply-supported end,

$$M_{AB} = K\theta = 3R\theta = \frac{3EI\theta}{L} \qquad 13.48$$

If the beam is built-in at an opposite end, the moment required to produce an angle $\theta$ can be derived using equation 13.46 with $F = 1$.

$$M_{AB} = \frac{4EI\theta}{L} \qquad 13.49$$

Therefore, beams with built-in ends are 4/3 as stiff as beams with pinned ends.

## 8 DISTRIBUTION FACTORS

Consider the joint $B$ illustrated in figure 13.5(a). The joint rigidly connects a complex beam that is simply supported at ends $A$ and $D$, but is fixed at end $C$. The relative stiffnesses are $R_1, R_2$, and $R_3$ respectively for beams $BA, BC$, and $BD$. A moment $U$ is applied to the joint $B$. The joint will rotate as shown in figure 13.5(b).

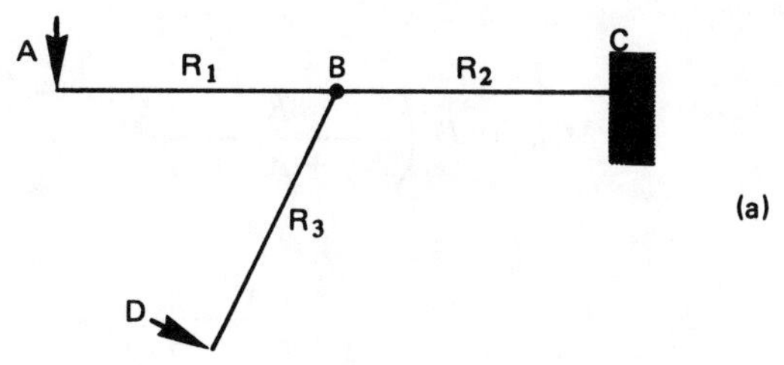


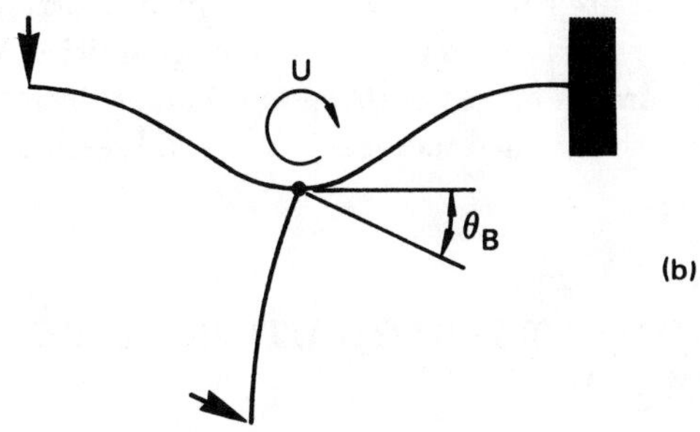


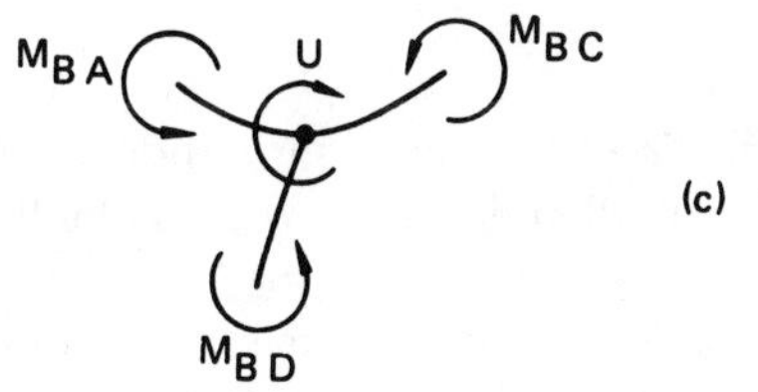


**Figure 13.5** A General Joint

A free body diagram of joint $B$ is shown in figure 13.5(c). It is clear that the rotation will continue until the moment $U$ is resisted by the stiffness of the beams.

$$U = M_{BA} + M_{BC} + M_{BD} \qquad 13.50$$

However, knowing $\theta_B$ gives the values of $M_{BA}, M_{BC}$, and $M_{BD}$ from equations 13.48 and 13.49.

$$U = 3R_1\theta_B + 4R_2\theta_B + 3R_3\theta_B \qquad 13.51$$

The individual resisting moments, $M_{BA}, M_{BC}$, and $M_{BD}$, can be found. $M_{BA}$, for example, is

$$\frac{M_{BA}}{U} = \frac{3R_1\theta_B}{3R_1\theta_B + 4R_2\theta_B + 3R_3\theta_B} \qquad 13.52$$

Since $\theta_B$ cancels,

$$M_{BA} = U\left(\frac{K_1}{K_1 + K_2 + K_3}\right) \qquad 13.53$$

[5] This is fixity against rotation, not fixity against translation. A simply-supported beam is completely fixed against translation, but it is still free to rotate.

Similarly,

$$M_{BC} = U\left(\frac{K_2}{K_1 + K_2 + K_3}\right) \quad 13.54$$

$$M_{BD} = U\left(\frac{K_3}{K_1 + K_2 + K_3}\right) \quad 13.55$$

From equations 13.53, 13.54, and 13.55, the moment $U$ is distributed to the three spans in proportion to the ratios of the stiffness factors. The quantities in parentheses are known as *distribution factors* since they distribute the applied moment, $U$, over all resisting beams.

## 9 MOMENT DISTRIBUTION METHOD PROCEDURE

The following steps constitute the moment distribution method.[6]

*step 1*: Divide the beam into independent spans, with all ends assumed to be built-in.

*step 2*: Calculate the relative stiffness of each member. Overhanging spans have $R = 0$ by definition. Both $E$ and $I$ can be given a value of 1.0 if they are constant along the entire span.

$$R = \frac{EI}{L} \quad 13.56$$

*step 3*: Determine the fixity factors, $F$, for the beam joints. $F = 0$ when the joint is simply supported or is otherwise free to rotate (i.e., the joint cannot impart a resisting moment). $F = 1$ when the joint is built-in or is a continuous support.

*step 4*: Calculate the stiffness factors, $K$. $F$ in equation 13.57 is the value associated with the opposite or facing joint, not the joint's own value of $F$. Therefore, $F = 0$ is used when the joint faces a simply-supported free end.

$$K = (3 + F_{\text{opposite}})R \quad 13.57$$

*step 5*: Calculate the distribution factor based on the stiffness factors for all spans tributary to the joint. By definition, $D = 0$ at built-in ends, since the support absorbs all the load. Also, $D = 1$ at any simply-supported end.

$$D = \frac{K}{\sum K} \quad 13.58$$

*step 6*: Determine the carry-over factors for ends. As in step 4, the values of $F$ used are from the beam's opposite ends.

$$C = \frac{2F_{\text{opposite}}}{3 + F_{\text{opposite}}} \quad 13.59$$

*step 7*: Calculate the fixed-end moments using appendix B and the actual transverse loading on each span.

*step 8*: At each joint, balance any unbalanced moments by distributing a counter moment among the connecting members according to their distribution factors. Add the following counter moments for balancing:

| | |
|---|---|
| at a built-in end: | zero |
| at an interior joint: | –(unbalance) |
| at a pinned end: | –(unbalance) |
| at a partially restrained end: | –(1–$F$)(unbalance) |

*step 9*: These distributed balancing moments produce carry-over moments at the opposite ends of members equal to the distributed moment times the carry-over factors, with the same signs as the distributed balancing moment. The carry-over moment appears at the opposite end from the location where the balancing moment was applied.

*step 10*: Repeat steps 6 and 7 until sufficient accuracy is obtained. Unless there is just one distribution (as in a 2-span beam), end the distribution process with a distribution, not a carry-over. The final fixed-end moment at the end of a member is equal to the algebraic sum of the original fixed-end moments and all distributed balancing and carry-over moments. The final moment is the moment on the beam end, not the moment on the support due to the loads.

[6] There are several different versions of the moment distribution method. The standard method is to set $C = \frac{1}{2}$ for all joints, and to always calculate $K = 4R$. The procedure presented here uses modified stiffness factors (i.e., modified by $F$) and converges more quickly than the standard method. However, the distribution factors calculated by the two methods will not be the same, even though the final fixed-end moments will be.

*Example 13.10*

Find the moments at points $A, B$, and $C$.

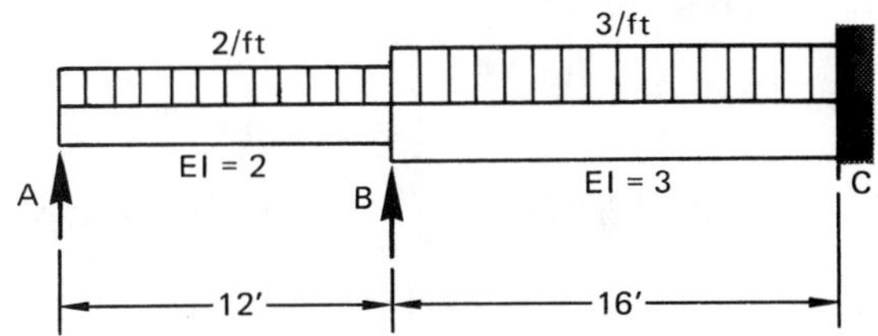


*step 1*: Divide the beam into spans $A$–$B$ and $B$–$C$.

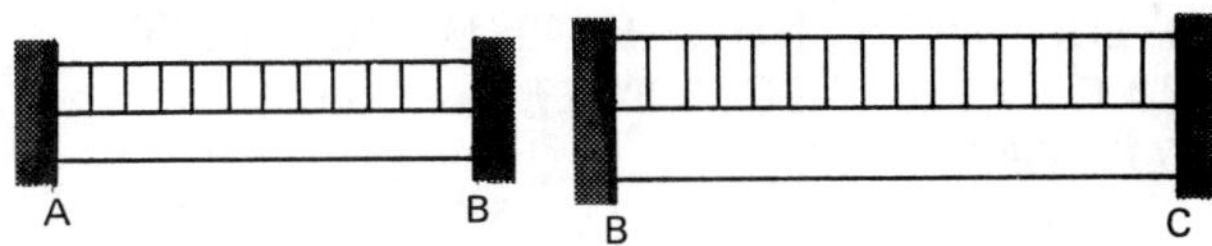


*step 2*:

$$R_{BA} = 2/12 = 1/6$$
$$R_{BC} = 3/16$$

*step 3*:

$F_{AB} = 0$ since the end is free to rotate.

$$F_{BA} = 1$$
$$F_{BC} = 1$$
$$F_{CB} = 1$$

*step 4*: Calculate $K_{AB}$ based on $F_{BA}$ (the fixity factor for the opposite end of span $AB$).

$$K_{AB} = (3+1)\left(\tfrac{1}{6}\right) = \tfrac{2}{3}$$

Calculate $K_{BA}$ based on $F_{AB}$.

$$K_{BA} = (3+0)\left(\tfrac{1}{6}\right) = \tfrac{1}{2}$$

$K_{BC}$ and $K_{CB}$ are both the same.

$$K_{BC} = K_{CB} = (3+1)\left(\tfrac{3}{16}\right) = \tfrac{12}{16} = \tfrac{3}{4}$$

*step 5*: $D = 1$ for joint $AB$, since any moment there must be transmitted to some other joint. $D = 0$ for joint $CB$, since the built-in support absorbs all moments. The only other joint in the structure is $B$.

$$D_{BC} = \frac{\frac{3}{4}}{\frac{3}{4}+\frac{1}{2}} = \tfrac{3}{5} = 0.6$$

$$D_{BA} = \frac{\frac{1}{2}}{\frac{3}{4}+\frac{1}{2}} = \tfrac{2}{5} = 0.4$$

*step 6*:

$$C_{AB} = \tfrac{1}{2}$$
$$C_{BA} = 0$$
$$C_{BC} = \tfrac{1}{2}$$
$$C_{CB} = \tfrac{1}{2}$$

*step 7*: Work with span $BC$, assuming both ends are fixed. The moment at $B$ is caused by the applied load. The applied moment is clockwise, and therefore it is positive. However, fixed-end moments are the resisting moments, and the resisting moment at $B$ is counterclockwise. From appendix B, the fixed-end moment at joint B is

$$M_{BC} = \frac{-wL^2}{12} = \frac{-(3)(16)^2}{12} = -64$$
$$M_{CB} = +64$$

Now, work with span $BA$, still assuming both ends are fixed.

$$M_{AB} = \frac{-(2)(12)^2}{12} = -24$$
$$M_{BA} = +24$$

At this time, it is necessary to start a table to keep track of the results of the calculations.

| | $M_{AB}$ | $M_{BA}$ | $M_{BC}$ | $M_{CB}$ |
|---|---|---|---|---|
| *FEM* | −24 | 24 | −64 | 64 |

*step 8*: At joint $A$, there is a −24 moment that is not balanced. Since joint $A$ is a pinned end that cannot carry any moment, the moment there must be removed. This is done by adding +24 to that joint. Since joint $A$ has only one 'side,' the entire +24 is added to $M_{AB}$.

| | $M_{AB}$ | $M_{BA}$ | $M_{BC}$ | $M_{CB}$ |
|---|---|---|---|---|
| *FEM* | −24 | 24 | −64 | 64 |
| Balance | +24 | | | |

At joint $B$, there is an unbalance of $(24-64) = -40$. So, this must be balanced by adding +40 to joint $B$. However, joint $B$ goes in two directions (towards $A$ and $C$), so the +40 must be distributed to spans $BA$ and $BC$ in proportion to their stiffnesses. The distribution factors were previously found to be 0.4 and 0.6 for $M_{BA}$ and $M_{BC}$ respectively.

$$\text{Balance}\, M_{BA} = (0.4)(40) = 16$$

$$\text{Balance}\, M_{BC} = (0.6)(40) = 24$$

| | $M_{AB}$ | $M_{BA}$ | $M_{BC}$ | $M_{CB}$ |
|---|---|---|---|---|
| *FEM* | −24 | 24 | −64 | 64 |
| Balance | 24 | 16 | 24 | |

The support at joint $C$ was assumed fixed in step 1. Since it actually is fixed, no correction is required.

| | $M_{AB}$ | $M_{BA}$ | $M_{BC}$ | $M_{CB}$ |
|---|---|---|---|---|
| *FEM* | −24 | 24 | −64 | 64 |
| Balance | 24 | 16 | 24 | 0 |

*step 9*: The balancing moments produce carry-over moments. If we add +24 to $M_{AB}$, we must also add $\frac{1}{2}(24) = 12$ to $M_{BA}$ since the carry-over factor $C_{AB} = \frac{1}{2}$. Similarly, the rest of the following table is prepared.

| | $M_{AB}$ | $M_{BA}$ | $M_{BC}$ | $M_{CB}$ |
|---|---|---|---|---|
| *FEM* | −24 | 24 | −64 | 64 |
| Balance | 24 | 16 | 24 | 0 |
| COM | 0 | 12 | 0 | 12 |

*step 10*: However, now the addition of the carry-over moments has unbalanced the beam again. So, the process must be repeated from step 8.

*step 8, repeated*: *At joint A*: No carry-over moment was applied, so no balancing is needed.

*At joint B*: The total carry-over moment applied was $(12 + 0) = 12$. So, a balancing moment of −12 is needed. The −12 is distributed to $M_{BA}$ and $M_{BC}$ according to

$$M_{BA} = (0.4)(-12) = -4.8$$
$$M_{BC} = (0.6)(-12) = -7.2$$

*At joint C*: Joint $C$ is actually fixed, so the carry-over moment of 12 is balanced by the beam's support. No balancing moment is needed.

*step 9, repeated*: Using the carry-over factors, we add 0 to $M_A$ and −3.6 to $M_{CB}$.

| | $M_{AB}$ | $M_{BA}$ | $M_{BC}$ | $M_{CB}$ |
|---|---|---|---|---|
| *FEM* | −24 | 24 | −64 | 64 |
| 1st Balance | 24 | 16 | 24 | 0 |
| 1st COM | 0 | 12 | 0 | 12 |
| 2nd Balance | 0 | −4.8 | −7.2 | 0 |
| 2nd COM | 0 | 0 | 0 | −3.6 |

*step 10, repeated*: Since the −3.6 is balanced by the external support at joint $C$, and since all other terms are zero, no further work is required.

$$M_{AB} = -24 + 24 = 0$$

$$M_{BA} = 24 + 16 + 12 - 4.8 = 47.2$$

$$M_{BC} = -64 + 24 - 7.2 = -47.2$$

$$M_{CB} = 64 + 12 - 3.6 = 72.4$$

## 10 DRAWING SHEAR AND MOMENT DIAGRAMS FROM FIXED-END MOMENTS

It is generally easy to obtain the shear and moment diagrams from the distributed fixed-end moments. Calculate the reactions at the ends of each span by taking moments about each end in turn. Be sure to consider the signs of the span end-moments as determined by the distribution.

*Example 13.11*

Determine the reactions, moment diagram, and shear-diagram for the beam evaluated in example 13.10.

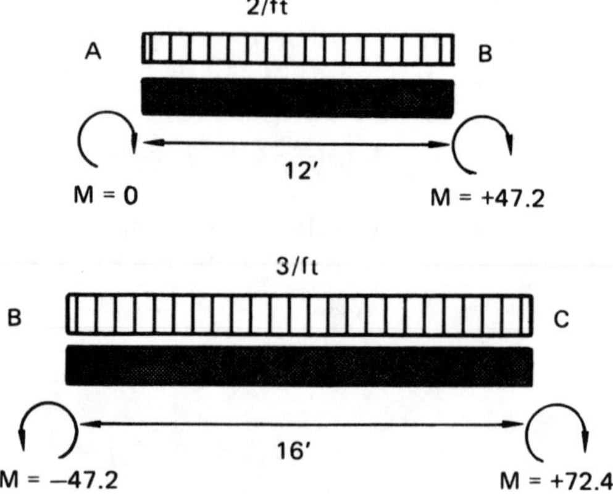


First, work with the left span. Sum moments about point $A$.

$$\sum M_A : 0 + 47.2 + \left(\tfrac{1}{2}\right)(2)(12)^2 - 12R_B = 0$$
$$R_B = 15.93$$

To find the reaction at point $A$, sum moments about point $B$.

$$\sum M_B : 0 + 47.2 - \left(\tfrac{1}{2}\right)(2)(12)^2 + 12R_A = 0$$
$$R_A = 8.07$$

Now, work with the right span. Sum moments about point $B$ to find the reaction at point $C$.

$$\sum M_B : -47.2 + 72.4 + \left(\tfrac{1}{2}\right)(3)(16)^2 - 16R_C = 0$$
$$R_C = 25.58$$

Sum moments about point $C$ to find the reaction at point $B$.

$$\sum M_C : -47.2 + 72.4 - \left(\tfrac{1}{2}\right)(3)(16)^2 + 16R_B = 0$$
$$R_B = 22.43$$

22.43
8.07
shear
– 15.93
– 25.58
$M_A = 0$
moment
$M_B = -47.2$
$M_C = -72.4$

*Example 13.12*

Draw the moment diagram for the beam shown.

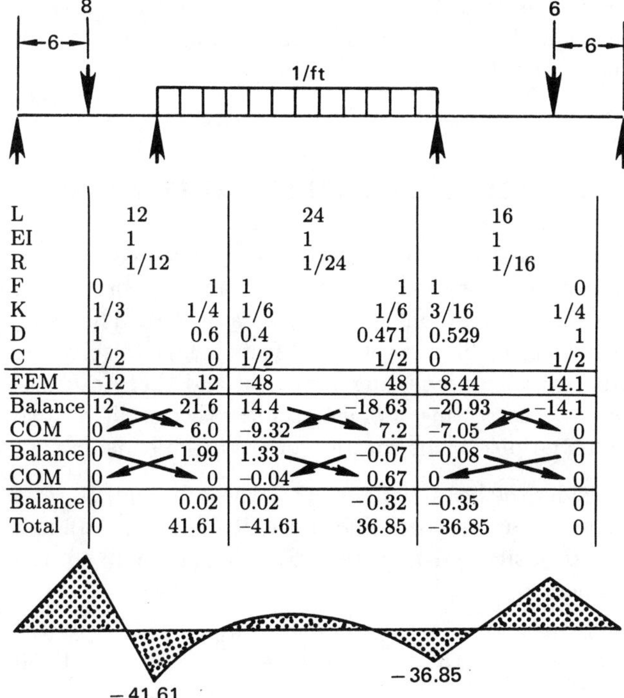


| | | | | | | |
|---|---|---|---|---|---|---|
| L | 12 | | 24 | | 16 | |
| EI | 1 | | 1 | | 1 | |
| R | 1/12 | | 1/24 | | 1/16 | |
| F | 0 | 1 | 1 | 1 | 1 | 0 |
| K | 1/3 | 1/4 | 1/6 | 1/6 | 3/16 | 1/4 |
| D | 1 | 0.6 | 0.4 | 0.471 | 0.529 | 1 |
| C | 1/2 | 0 | 1/2 | 1/2 | 0 | 1/2 |
| FEM | -12 | 12 | -48 | 48 | -8.44 | 14.1 |
| Balance | 12 | 21.6 | 14.4 | -18.63 | -20.93 | -14.1 |
| COM | 0 | 6.0 | -9.32 | 7.2 | -7.05 | 0 |
| Balance | 0 | 1.99 | 1.33 | -0.07 | -0.08 | 0 |
| COM | 0 | 0 | -0.04 | 0.67 | 0 | 0 |
| Balance | 0 | 0.02 | 0.02 | −0.32 | -0.35 | 0 |
| Total | 0 | 41.61 | -41.61 | 36.85 | -36.85 | 0 |

## 11 DIRECT MOMENTS AND FREE ENDS

When a moment is applied directly to a beam or joint, or when the beam has a cantilever end-span, the moment distribution procedure is changed slightly. If a concentrated moment is applied mid-span, the fixed-end moments at the span ends needed to keep the beam from rotating are added to fixed-end moments from other loads. The distances to the span ends on either side of the point of application determine the fraction of the direct moment distributed to each span end.

If a moment is applied directly to a joint, it may be unclear which direction the fixed-end moment acts to counter the rotation. In this case, it may be easier to consider the applied moment as an initial unbalance (i.e., place it in the BALANCE row) to be distributed immediately to both sides of the joint without change in sign.[7]

Spans with loaded cantilever sections can be replaced with an equivalent load and moment. All of the vertical load is carried by the adjacent support. The moment can be considered to be applied directly to the BALANCE row, without change of sign. Since the cantilever section is not included in the moment distribution process, the adjacent support becomes the end of the beam. The entire applied moment, therefore, is given to the inward-facing side of the first support.

After distribution, the moment at the first support will be zero, since it was assumed to be a free end. The moment at this support will have to be corrected by the amount of the original applied moment, since that moment is known to exist at that point.

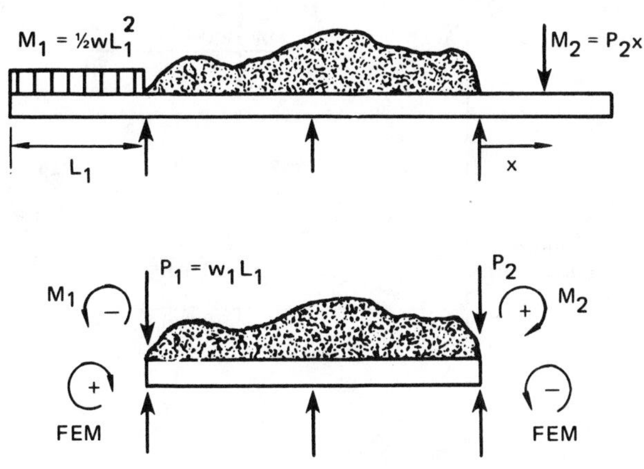


**Figure 13.6** Beam with Cantilever Ends and Equivalent Beam

[7] Obviously, putting the fixed-end moment in the FEM row or reversing the sign and putting it in the BAL column have the same effect. However, the rule to "... put the applied moment in the BAL row ..." is easy to remember.

*Example 13.13*

Find the bending moments at the ends of members $OA$ and $OB$. Joint $O$ is rigid. The vertical load of 6000 lbf is applied on a lever arm of length 3 feet.

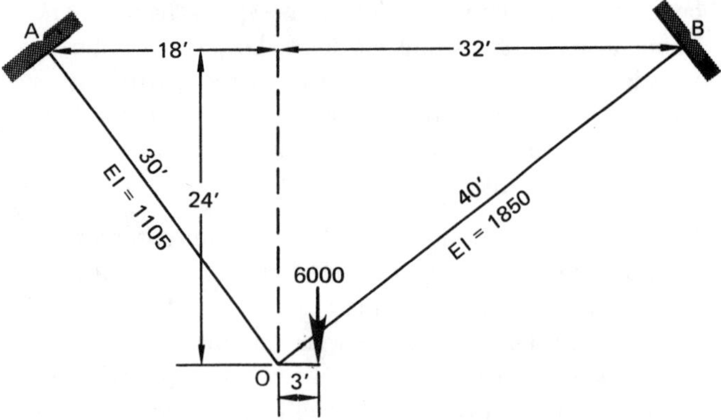


The applied moment is positive. This will require positive fixed-end moments about connections $A$ and $B$, as shown, to counteract the beam's rotation.

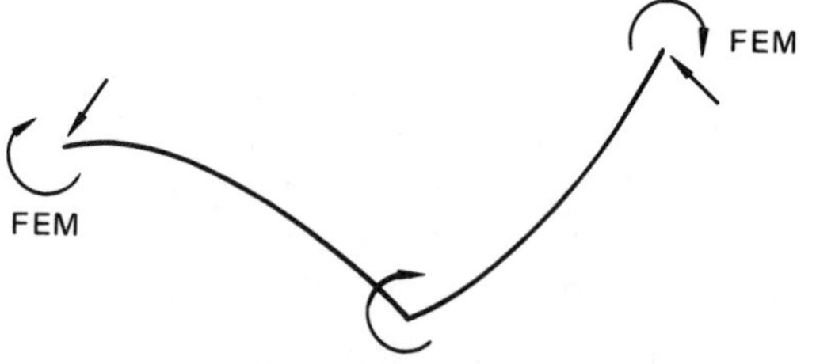


The amount of the applied moment distributed to legs $OA$ and $OB$ depends on the distribution factors, $D$. Since the fixed-end moments are known to be positive, a positive moment is distributed to each leg.

| | AO | OA | OB | BO |
|---|---|---|---|---|
| L | 30 | | 40 | |
| EI | 1105 | | 1850 | |
| R | 36.83 | | 46.25 | |
| F | 1 | 1 | 1 | 1 |
| K | 147.3 | 147.3 | 185.0 | 185.0 |
| D | 0 | 0.443 | 0.557 | 0 |
| C | 1/2 | 1/2 | 1/2 | 1/2 |
| Applied Moment | +18,000 | | | |
| Distributed BAL | | 7974 | 10,026 | |
| COM | 3987 | | | 5013 |
| Total | 3987 | 7974 | 10,026 | 5013 |

*Example 13.14*

Distribute the moments on the continuous beam shown. $EI$ is a constant 1 EE6 lbf-ft$^2$ along all spans.

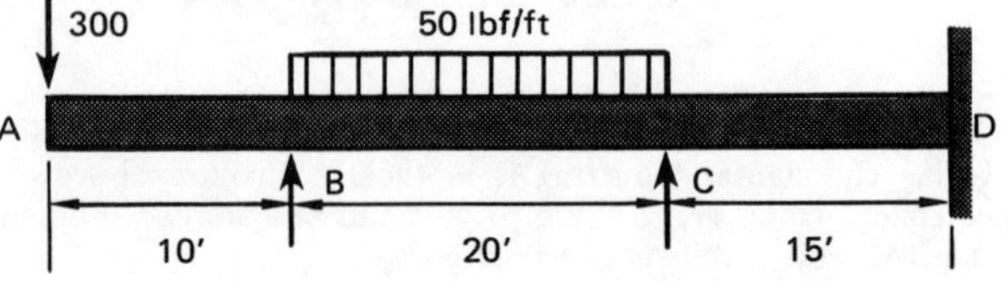


The magnitude of the moment at point $B$ due to the cantilever load is $(300)(10) = 3000$. The fixed-end moment at $B$ required to keep the continuous joint $B$ horizontal is clockwise, and therefore positive.

The fixed-end moments for the central span are

$$M_{BC} = -\frac{(50)(20)^2}{12} = -1667$$

$$M_{CB} = +1667$$

Span $AB$ can be omitted from the analysis. Its fixed-end moment is added to $M_{BC}$.

$$M'_{BC} = -1667 + 3000 = 1333$$

| | BC | CB | CD | DC |
|---|---|---|---|---|
| L | 20 | | 15 | |
| EI | 1 EE6 | | 1 EE6 | |
| R | 5 | | 6.67 | |
| F | 0 | 1 | 1 | 1 |
| K | 20 | 15 | 26.67 | 26.67 |
| D | 1 | 0.36 | 0.64 | 0 |
| C | 1/2 | 0 | 1/2 | 1/2 |
| FEM | 1333 | 1667 | 0 | 0 |
| Balance | −1333 | −600 | −1067 | 0 |
| COM | 0 | −667 | 0 | −533.5 |
| Balance | 0 | 240 | 427 | 0 |
| COM | 0 | 0 | 0 | 213.5 |
| Total | 0 | 640 | −640 | −320 |

$M_{BC}$ must be corrected by the amount of the original moment, since joint $B$ is not actually the end of the beam. On span $AB$, a clockwise moment is required to keep end $B$ horizontal. Therefore, $M_{BA} = +3000$. To maintain balance, $M_{BC} = -3000$.

## 12 FLEXIBLE AND YIELDING SUPPORTS

The moment distribution procedure presented assumes that the supports are unyielding. If any support moves, the distributed moments must be adjusted. The usual method of including yielding effects is first to solve the problem with unyielding supports, and then to calculate the correcting moments, since there will be two contributions to the rotation of the beam ends.

Consider the beam in figure 13.7 whose far end, $B$, rests on a yielding support. End $A$ is built-in (i.e., fixed), and end $B$ is simply supported. From equation 13.46, the stiffness is

$$K = \frac{M_A}{\theta} = 3R = \frac{3EI}{L} \qquad 13.60$$

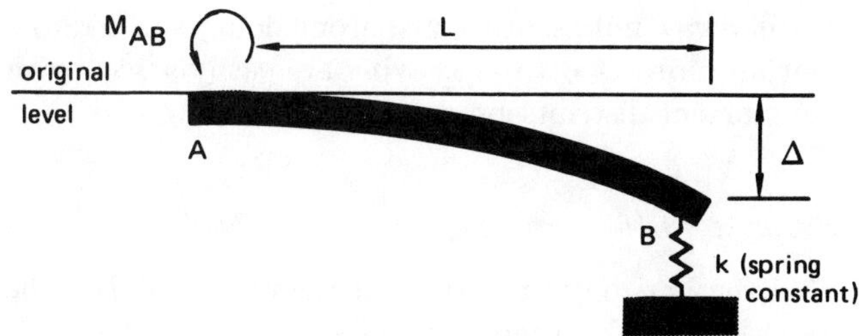


**Figure 13.7** Yielding Support Free to Rotate

However, $\theta \approx \tan\theta = \Delta/L$, so the moment causing deflection is

$$M_{AB} = \frac{3EI\Delta}{L^2} \qquad 13.61$$

Other relationships can be derived using the basic spring equations. Notice that the spring constant, $k$, is not the same as the beam stiffness, $K$.

$$M = PL \qquad 13.62$$

$$P = k\Delta \qquad 13.63$$

If ends $A$ and $B$ are both fixed, but end $B$ is free to translate (i.e., deflect due to the yielding support), the moment causing deflection will be

$$M_{AB} = M_{BA} = \frac{6EI\Delta}{L^2} \qquad 13.64$$

This moment appears at ends $A$ and $B$ with the same sign at both.

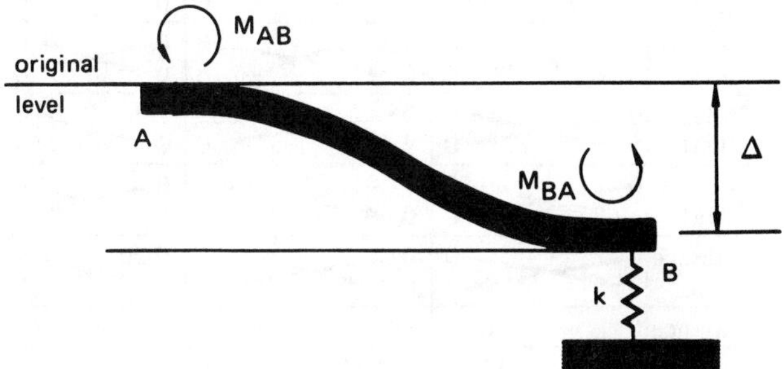


**Figure 13.8** Yielding Support Constrained to Horizontal

The fixed-end moments required to keep settled beam ends horizontal are no different than any other fixed-end moments, and the same sign conventions apply.

*Example 13.15*

Repeat example 13.14 assuming that support $C$ settles 0.5 inch.

Example 13.14 has already distributed the moments due to the applied loads. The settlement will not change these distributed moments, and it is not necessary to repeat those steps.

The settlement does, however, allow the beam to rotate. Therefore, additional moments are produced in the joints. Section $CD$ is fixed at both ends, and equation 13.64 is used to calculate the fixed-end moments. The deflection curve shows that positive moments are required to keep the beam ends horizontal.

$$M_{DC} = M_{CD} = \frac{(6)(1\,\text{EE}6)(0.5/12)}{(15)^2} = 1111$$

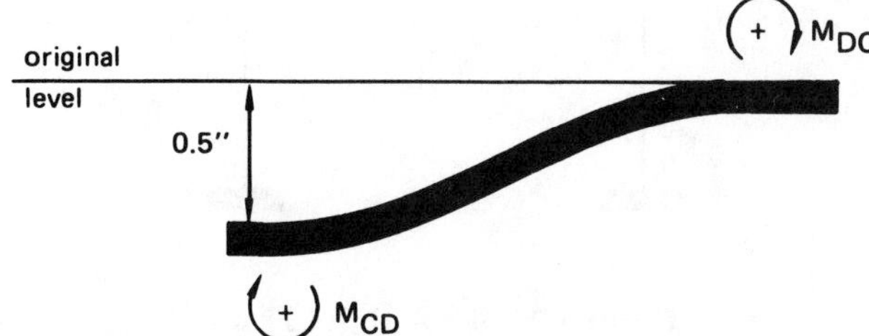


The support at joint $B$ is essentially a simple support, and span $AB$ is ignored. Therefore, equation 13.61 is used. $M_{CB}$ is negative because a counterclockwise moment is required to keep joint $C$ horizontal.

$$M_{CB} = \frac{-(3)(1\,\text{EE}6)(0.5/12)}{(20)^2} = -313$$

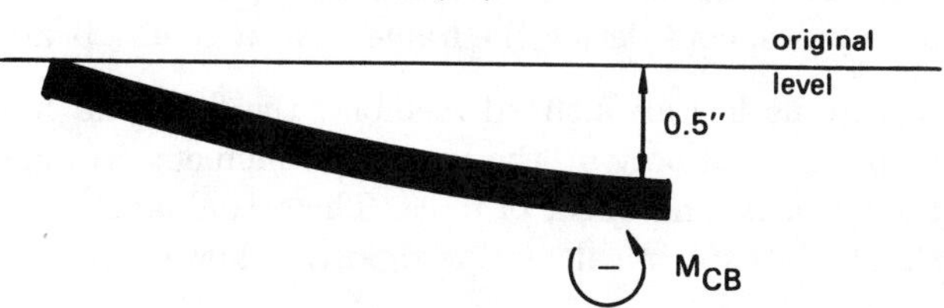


These moments are distributed in the usual manner.

| | BC | CB | CD | DC |
|---|---|---|---|---|
| L | 20 | | 15 | |
| EI | 1 EE6 | | 1 EE6 | |
| R | 5 | | 6.67 | |
| F | 0 | 1 | 1 | 1 |
| K | 20 | 15 | 26.67 | 26.67 |
| D | 1 | 0.36 | 0.64 | 0 |
| C | 1/2 | 0 | 1/2 | 1/2 |
| FEM | 0 | −313 | 1111 | 1111 |
| Balance | 0 | −287 | −511 | 0 |
| COM | 0 | 0 | 0 | −255.5 |
| Total | 0 | −600 | 600 | 855.5 |

The moments from this distribution are added to the moments from the distribution of example 13.14.

$$M_{AB} = 0$$
$$M_{BA} = +3000$$
$$M_{BC} = -3000 + 0 = -3000$$
$$M_{CB} = 640 - 600 = 40$$
$$M_{CD} = -640 + 600 = -40$$
$$M_{DC} = -320 + 855.5 = 535.5$$

## 13 FRAMES WITHOUT SIDESWAY

The rigid *portal frame* or *bent* is a fundamental structural unit. Therefore, an ability to analyze portal frames is essential.

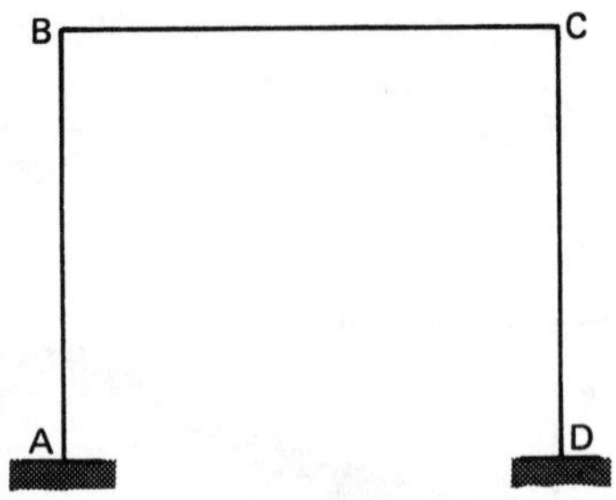


**Figure 13.9** A Rigid Frame

Because of the rigid joints at $B$ and $C$ in figure 13.9, it follows that all members meeting at a joint rotate through the same angle at that joint. Furthermore, any moment generated on a member will be transmitted through the joint to a connecting member. The moment distribution method can be applied directly in most cases, considering the frame to be a "bent" beam.

If a frame has an inclined member, the fixed-end moments depend only on the horizontal moment arm and the vertical component of force. The actual lengths, $L$, should be used in calculating rigidities, however.

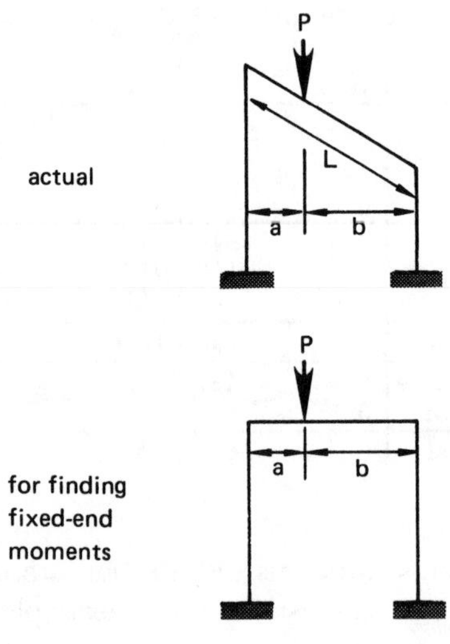


**Figure 13.10** Frame with Inclined Member

If a portal frame is symmetrical with respect to its leg lengths and loading, it will not sway or tend to move horizontally. If the frame or loading is unsymmetrical, it will sway unless prevented from doing so. Frames that are prevented from swaying are easily solved with the moment distribution method.

*Example 13.16*

Find the horizontal reactions at points $A$ and $D$. The frame is kept from moving laterally by a passive force at $C$. That is, the reaction at $C$ resists lateral movement, but it does not add any load or moment to the structure. Determine the moments and reactions at all joints, including the wall reaction at $C$.

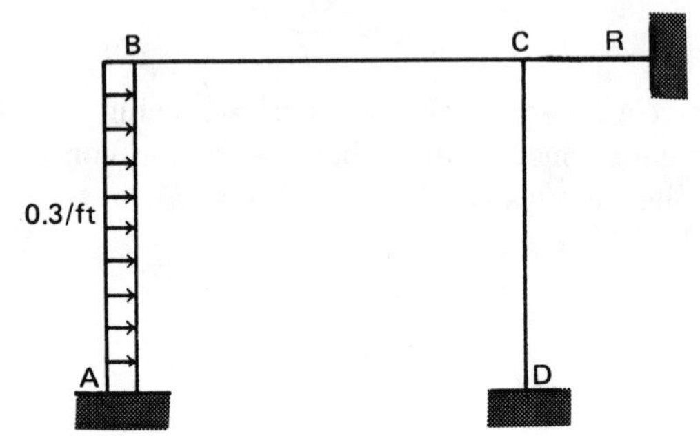


| | AB | BA | BC | CB | CD | DC |
|---|---|---|---|---|---|---|
| L | 25 | | 70 | | 25 | |
| EI | 1000 | | 10,500 | | 1000 | |
| R | 40 | | 150 | | 40 | |
| F | 1 | 1 | 1 | 1 | 1 | 1 |
| K | 160 | 160 | 600 | 600 | 160 | 160 |
| D | 0 | 0.21 | 0.79 | 0.79 | 0.21 | 0 |
| C | 1/2 | 1/2 | 1/2 | 1/2 | 1/2 | 1/2 |
| FEM | -15.63 | 15.63 | 0 | 0 | 0 | 0 |
| Balance | 0 | -3.28 | -12.35 | 0 | 0 | 0 |
| COM | -1.64 | 0 | 0 | -6.18 | 0 | 0 |
| Balance | 0 | 0 | 0 | 4.88 | 1.30 | 0 |
| COM | 0 | 0 | 2.44 | 0 | 0 | 0.65 |
| Balance | 0 | -0.51 | -1.93 | 0 | 0 | 0 |
| COM | -0.25 | 0 | 0 | -0.97 | 0 | 0 |
| Balance | 0 | 0 | 0 | 0.77 | 0.20 | 0 |
| COM | 0 | 0 | 0.38 | 0 | 0 | 0.10 |
| Balance | 0 | -0.08 | -0.30 | 0 | 0 | 0 |
| Total | -17.52 | 11.76 | -11.76 | -1.50 | 1.50 | 0.75 |

The free-bodies of the sections are used to determine the reactions.

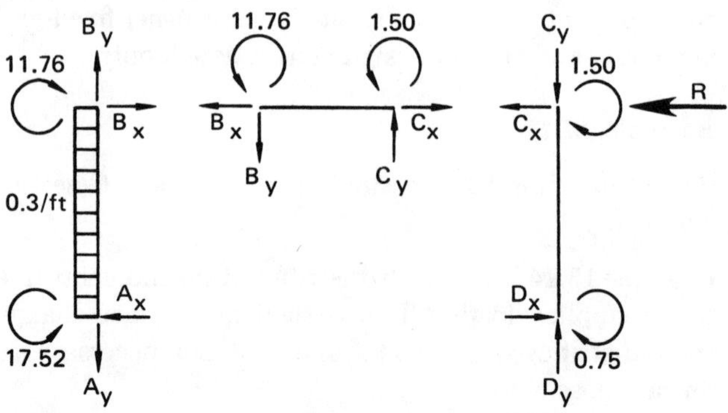

*Member AB*. To find $B_x$, sum moments about point $A$.

$$\sum M_A = \tfrac{1}{2}(25)^2(0.3) + 11.76 - 17.52 + 25(B_x) = 0$$
$$B_x = -3.52 \text{ (to the left)}$$

To find $A_x$, sum moments about point $B$.

$$\sum M_B = -\tfrac{1}{2}(25)^2(0.3) + 11.76 - 17.52 + 25(A_x) = 0$$
$$A_x = 3.98$$

*Member DC*. $C_x = -B_x = 3.52$ (to the right). To find $D_x$, sum moments about point $C$.

$$\sum M_C = 1.50 + 0.75 - 25(D_x) = 0$$
$$D_x = 0.09 \text{ (to the right)}$$

The wall reaction, $R$, is found from the horizontal equilibrium of freebody $CD$. (Alternatively, moments could be taken about point $D$.)

$$\sum F_x = 3.52 + 0.09 - R = 0$$
$$R = -3.61 \text{ (to the left)}$$

## 14 FRAMES WITH SIDESWAY

If an unsymmetrically-loaded frame is not braced against sidesway, lateral movement will occur until the generated shear force in the vertical columns just equals the horizontal force needed to prevent sidesway.

Superposition is used to evaluate frames with sidesway. First, the frame is analyzed based on the assumption that sidesway is prevented. This results in moments at all the joints. These moments and the artificial restraining force can be evaluated using moment distribution.

Then, all loads are removed from the frame and the restraining force is assumed to act alone opposite to its original direction. The moments from this distribution are added to the moments from the restrained distribution to obtain the true joint moments.

The following alternate procedure can also be used to calculate the fixed-end moments for a frame with sidesway.

*step 1*: Analyze the frame assuming sidesway is prohibited.

*step 2*: Calculate the reaction, $R$, needed to prevent sidesway.

*step 3*: Remove the actual loads, and deflect the frame an arbitrary amount, $\Delta$. There will be fixed-end moments $M_1$ and $M_2$ set up at both ends of the vertical members to resist this deflection.[8]

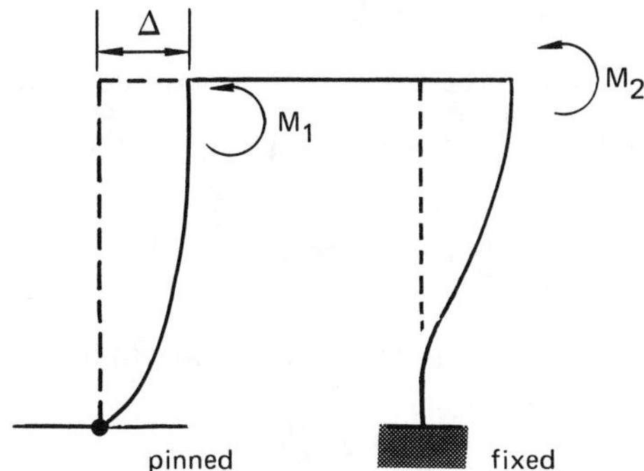


**Figure 13.11** Frame with Arbitrary Deflection

Equations 13.61 and 13.64 relate the fixed-end moments to the deflection. For the pinned member,

$$M_1 = \frac{3EI\Delta}{L^2} \qquad 13.65$$

For the built-in member,

$$M_2 = \frac{6EI\Delta}{L^2} \qquad 13.66$$

The total resisting moment, $M_1 + M_2$ is unknown, but it can be assumed to be some value (e.g., 100 or 1000). This total resisting moment is divided between the vertical members in proportion to the values of $M_1$ and $M_2$. The signs of the divided moments at the ends of both vertical members will be the same.

*step 4*: Analyze the frame using the moment distribution method.

*step 5*: Find the force, $R'$, causing the sidesway. If the directions of $R$ and $R'$ are the same, then the sign of the moments $M_1$ and $M_2$ used in step 3 was chosen incorrectly. In that case, merely reverse the sign of $M_{\text{from step 4}}$ in equation 13.67.

*step 6*: Calculate the true moment at each joint.

$$M_{\text{true}} = M_{\text{restrained}} + \left(\frac{R}{R'}\right) M_{\text{from step 4}} \qquad 13.67$$

[8] In effect, we have applied an equal but opposite force, $-R$ (to the right), which causes the deflection, $\Delta$. This equal but opposite force cancels the imaginary force used to prohibit sidesway initially.

*Example 13.17*

Find the moment at end $D$ for the frame in example 13.16. Sidesway is permitted.

*step 1*: See example 13.16.

*step 2*: See example 13.16.

*step 3*: Load the frame with a total moment of 100.

For member $AB$, the fixed-end moments are proportional to

$$M_1 = \frac{(6)(1000)}{(25)^2} = 9.6$$

For member $DC$, the fixed-end moments are the same.

$$M_2 = 9.6$$

If the total resisting moment is assumed to be $-100$, the fraction taken by member $AB$ is

$$M_{AB} = (-100)\left(\frac{9.6}{9.6+9.6}\right)$$

Member $DC$ takes the other 50 percent of the total resisting moment.

The sign of the FEM's applied to each span end are determined by observing the unconstrained deflection shape. In order to keep the ends of members $AB$ and $CD$ vertical, counterclockwise (negative) moments need to be applied. Therefore, $-50$ is the moment distributed to each member.

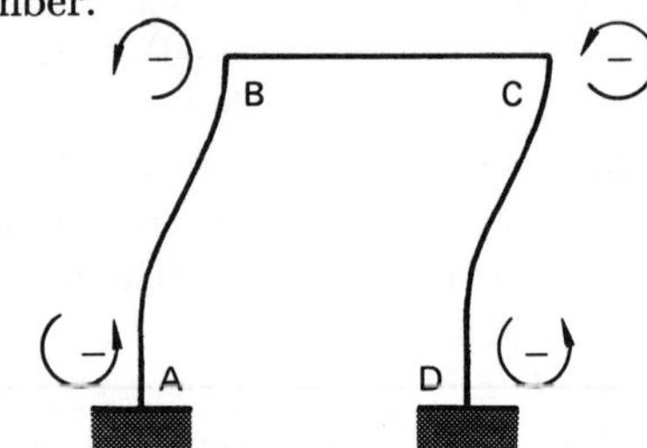


*step 4*:

| | AB | BA | BC | CB | CD | DC |
|---|---|---|---|---|---|---|
| L | 25 | | 70 | | 25 | |
| EI | 1000 | | 10,500 | | 1000 | |
| R | 40 | | 150 | | 40 | |
| F | 1 | 1 | 1 | 1 | 1 | 1 |
| K | 160 | 160 | 600 | 600 | 160 | 160 |
| D | 0 | 0.21 | 0.79 | 0.79 | 0.21 | 0 |
| C | 1/2 | 1/2 | 1/2 | 1/2 | 1/2 | 1/2 |
| $EI/L^2$ | 1.6 | | not loaded | | 1.6 | |
| FEM | −50 | −50 | 0 | 0 | −50 | −50 |
| Balance | 0 | 10.5 | 39.5 | 39.5 | 10.5 | 0 |
| COM | 5.25 | 0 | 19.75 | 19.75 | 0 | 5.25 |
| Balance | 0 | −4.15 | −15.60 | −15.60 | −4.15 | 0 |
| COM | −2.07 | 0 | −7.8 | −7.8 | 0 | −2.07 |
| Balance | 0 | 1.64 | 6.16 | 6.16 | 1.64 | 0 |
| COM | 0.82 | 0 | 3.08 | 3.08 | 0 | 0.82 |
| Balance | 0 | −0.65 | −2.43 | −2.43 | −0.65 | 0 |
| Total | −46.00 | −42.66 | 42.66 | 42.66 | −42.66 | −46.00 |

*step 5*: Take member $AB$ as a freebody.

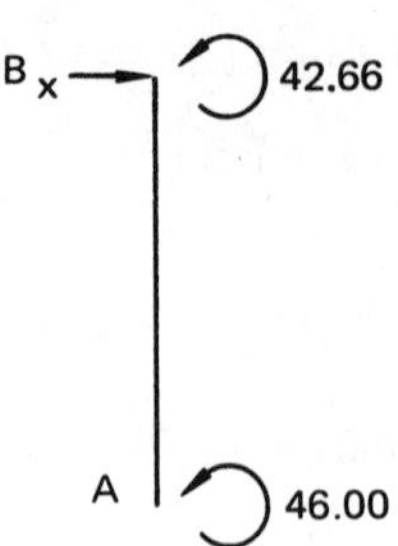


$$\sum M_A = -42.66 - 46.00 + B_x(25) = 0$$
$$B_x = 3.55 \text{ (to the right)}$$

Take member $DC$ as a freebody. $C_x = -B_x = -3.55$ (to the left).

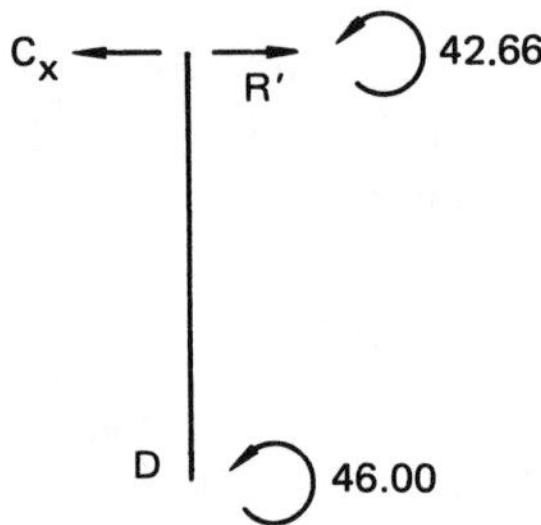


$$\sum M_D = -42.66 - 46.00 - 3.55(25) + R'(25)$$
$$R' = 7.10 \text{ (to the right)}$$

*step 6*:

$$M_{DC} = 0.75 + \left(\frac{3.61}{7.10}\right)(-46.0) = -22.6$$

## 15 FRAMES WITH CONCENTRATED JOINT LOADS

A frame that carries a concentrated load at a joint, as in figure 13.12, does not develop any fixed-end moments due to transverse loading since there are no transverse loads. However, the load does cause the frame to sway, and hence, moments are generated at the joints.

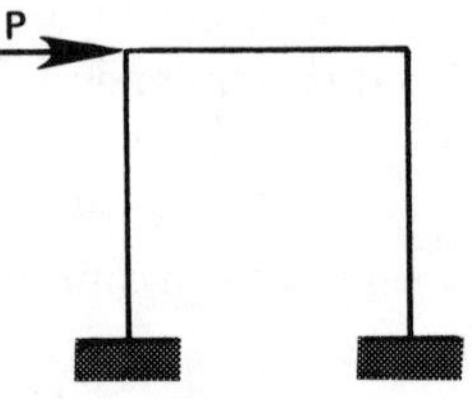


**Figure 13.12** Joint Loaded Frame

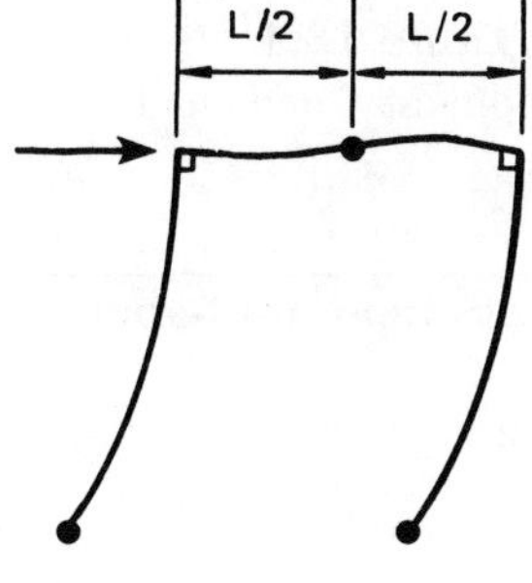


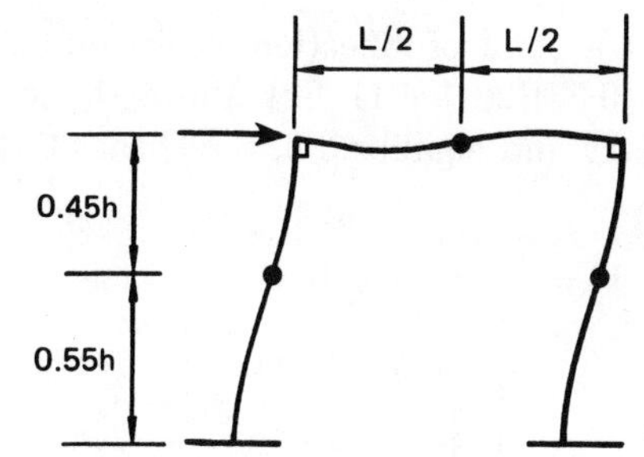


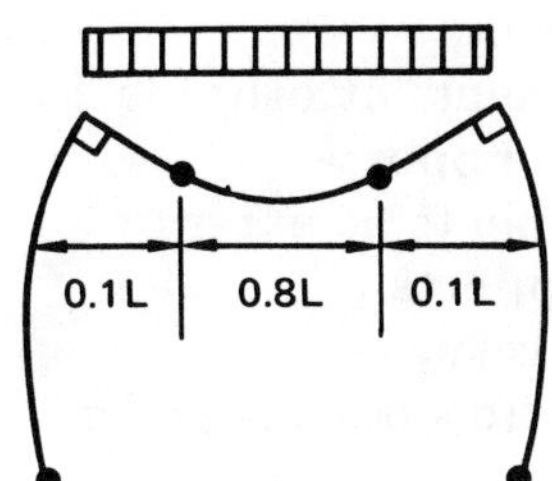


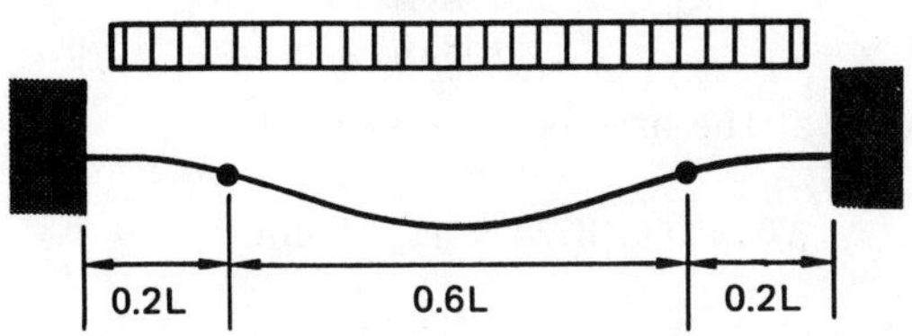


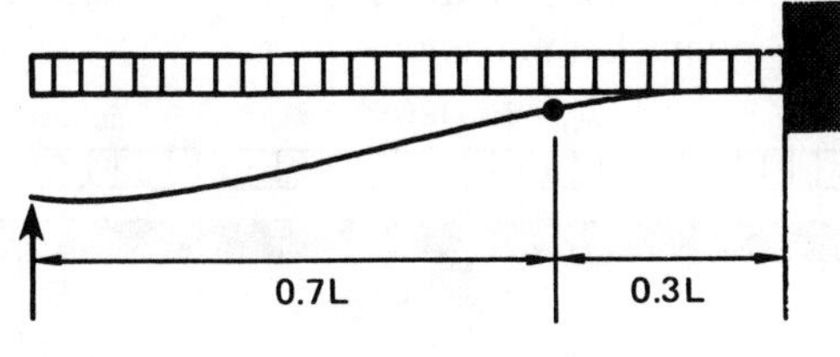


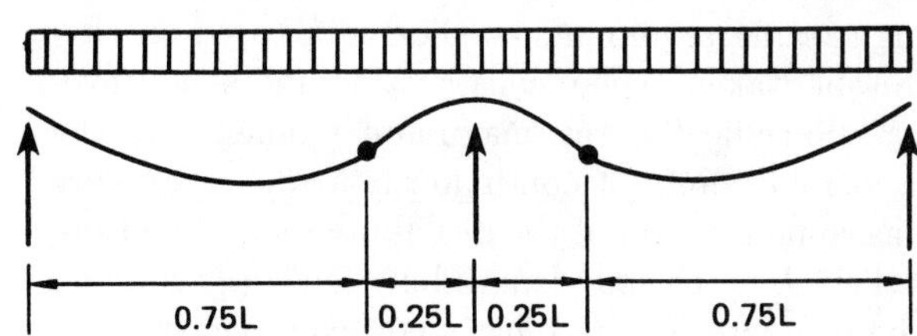


**Figure 13.13** Approximate Inflection Points

The solution procedure is similar to that used for frames with sidesway, except that the initial step is skipped. It is unnecessary to distribute moments from transverse loading, because there is no transverse loading. In effect, the applied load is equal to the force needed to prevent sidesway.

## 16 APPROXIMATE METHODS

When an exact solution is unnecessary, when time is short, or when it is desired to quickly check an exact solution, approximate methods may be used.

### A. ASSUMED INFLECTION POINTS

At points of inflection, the curvature will be changing from positive to negative. Accordingly, the moment at such a point will be zero. It is possible to assume a hinge exists at an inflection point, and that no moment is transferred across the hinge.

The accuracy of this method depends on being able to accurately predict the location of the inflection point. Figure 13.13 provides reasonable predictions of these locations.

*Example 13.18*

Determine the fixed-end moment for joint $A$.

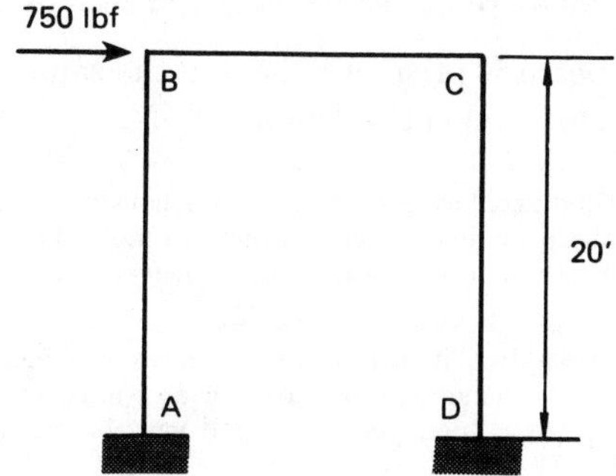

It is assumed that a point of inflection occurs on each vertical member $(0.55)(20) = 11$ feet above the supports. By symmetry and equilibrium, the shear at the inflection point is

$$V = \frac{750}{2} = 375 \text{ lbf}$$

The moment at the base is

$$M = (11)(375) = 4125 \text{ ft-lbf}$$

The fixed-end moment is counterclockwise; therefore, $M_{AB} = -4125$.

B. USE OF DESIGN MOMENTS (MOMENT COEFFICIENTS)[9]

For continuous beams and other similar structures meeting specific requirements, shortcuts based on theory can be taken when constructing moment envelopes.[10] Specifically, the maximum moments at the ends and mid-points of continuously-loaded spans are taken as some fraction of the distributed load function, $wL^2$, where $L$ is the span length between supports, and $w$ is the ultimate factored load. For example, the moment at some point along the beam would be taken as

$$M = C_1 w L^2 \qquad 13.68$$

The coefficient, $C_1$, in equation 13.68 is known as the *moment coefficient.* The method of moment coefficients can be used in concrete design when the following conditions are met:

- The load is continuously distributed.
- Construction is not prestressed.
- There are two or more spans.
- All spans are the same length, ±20% of $L$.
- The beam is prismatic, having the same cross section along its entire length.

[9] This method is allowed in ACI-318 for continuous beams and one-way slabs provided the conditions are met. This method is sometimes referred to as the *direct design method.*

[10] Strictly speaking, it is improper to draw the moment envelope using both the maximum positive and maximum negative moments, since the load placement will vary between these two extremes.

**Table 13.1**
ACI Moment Coefficients

| condition | $C_1$ |
|---|---|
| **positive moments near mid-span** | |
| end spans | |
| simple support | 1/11 |
| built-in | 1/14 |
| interior spans | 1/16 |
| **negative moments at exterior face of first interior support** | |
| 2 spans | 1/9 |
| 3 or more spans | 1/10 |
| **negative moments at other faces or interior supports** | 1/11 |
| **negative moments at exterior built-in supports** | |
| support is a column | 1/16 |
| support is a cross beam or girder | 1/24 |

*Example 13.19*

Draw the critical moment diagram for the uniformly-loaded, continuous beam shown.

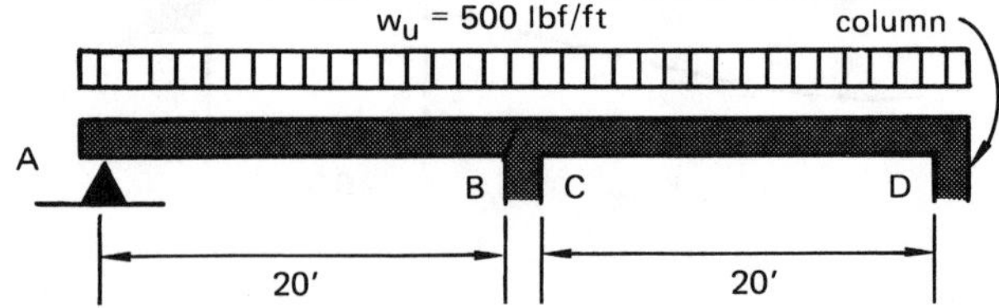


Point $A$ is a simple support. The moment there must be zero.

Point $B$ is an exterior face of the first interior support over a 2-span beam. The moment there is

$$M_B = -\frac{1}{9}(500)(20)^2 = -22,222 \text{ ft-lbf}$$

Point $B$ is also an exterior face of the first interior support (counting from the opposite end).

$$M_C = M_B = -22,222 \text{ ft-lbf}$$

Point $D$ is an exterior column support.

$$M_D = -\frac{1}{16}(500)(20)^2 = -12,500 \text{ ft-lbf}$$

The left span is a simply-supported end span. The maximum positive moment is

$$M_L = \frac{1}{11}(500)(20)^2 = 18,182 \text{ ft-lbf}$$

The right span is an end span with a built-in support. The maximum positive moment for it is

$$M_R = \frac{1}{14}(500)(20)^2 = 14,286 \text{ ft-lbf}$$

The critical moment diagram is

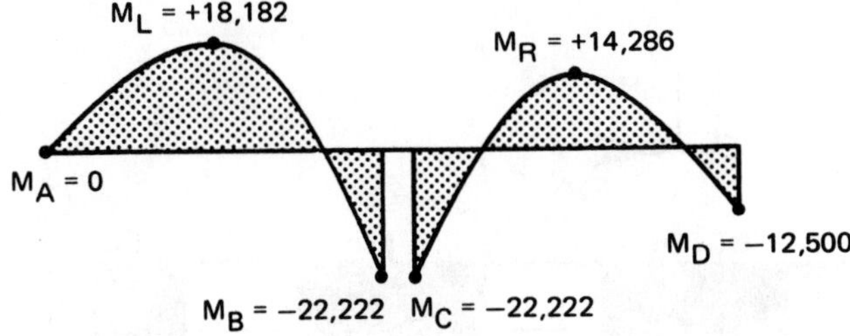


C. USE OF DESIGN SHEARS (SHEAR COEFFICIENTS)

When the conditions of the preceding section are met, it is possible to predict the critical shear in continuous beams and slabs on the basis of design coefficients. For shear, the design coefficient is used with the average span loading, $\frac{1}{2}wL$. That is,

$$V = C_2\left(\frac{wL}{2}\right) \qquad 13.69$$

$C_2$ for shear in end members at the first interior support is 1.15. For shear at the face of all other supports, $C_2 = 1.0$.

# Appendix A: Moment Distribution Worksheet

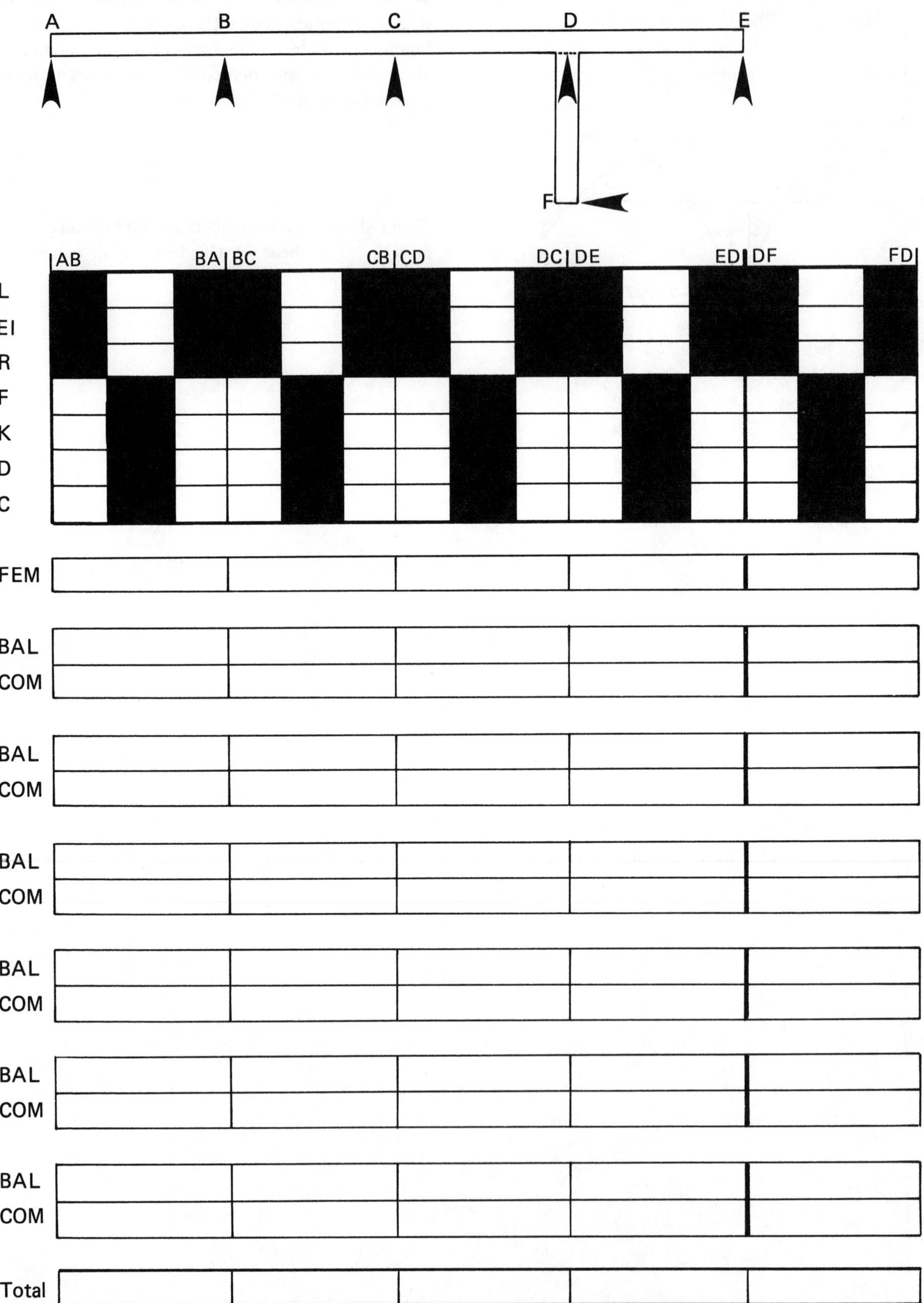

# Appendix B: Fixed-End Moments

(clockwise is positive)

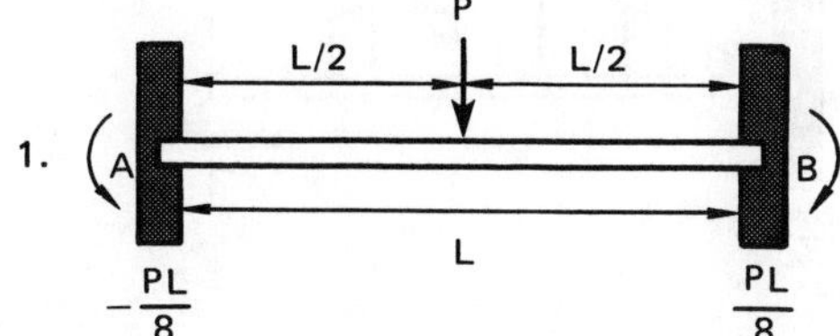


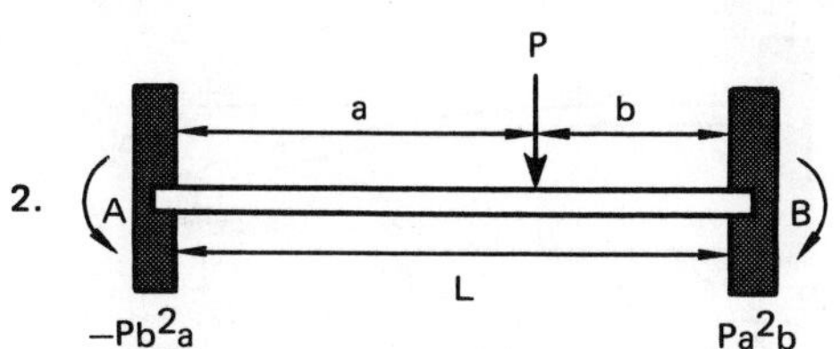


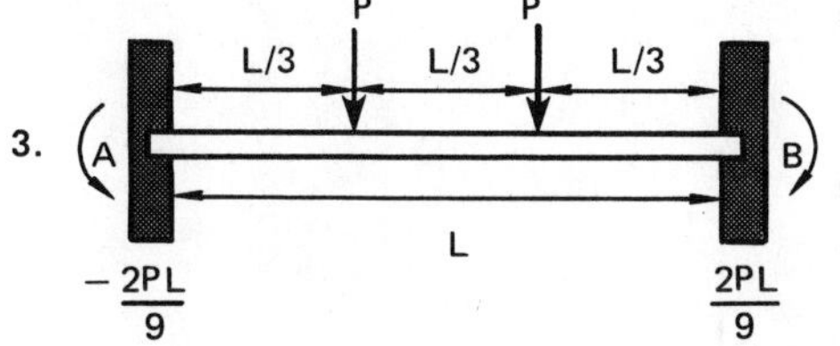


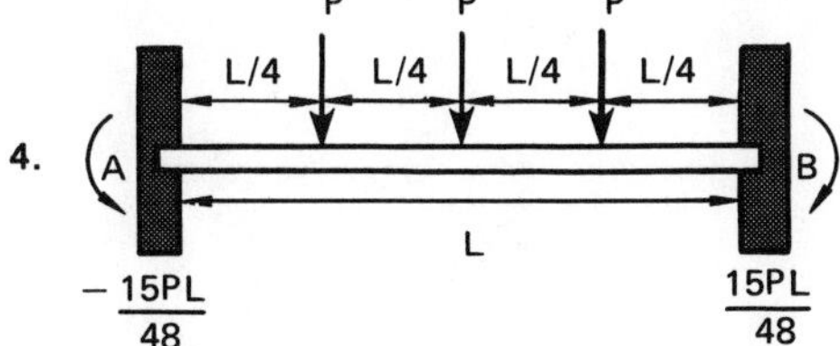


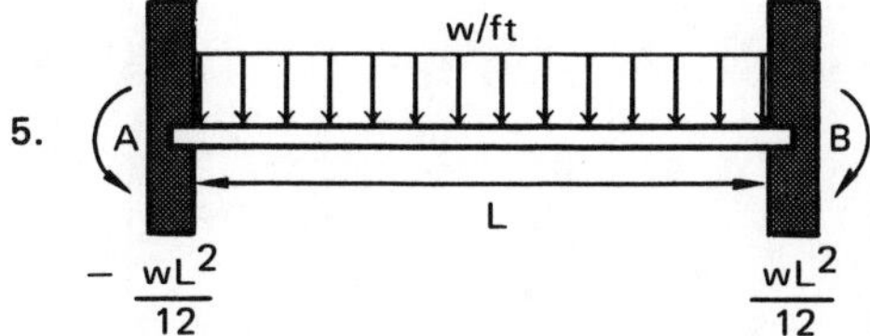


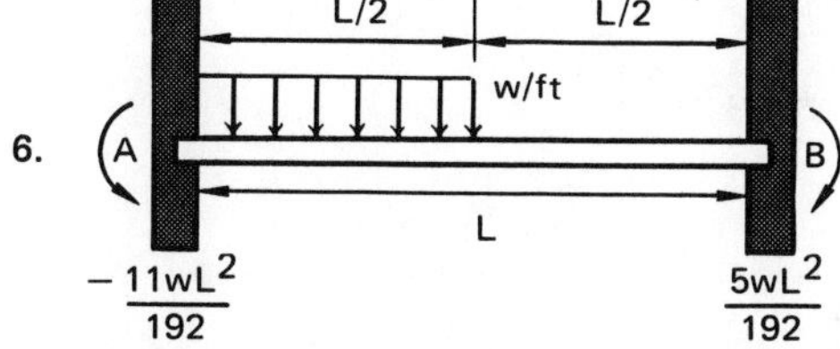


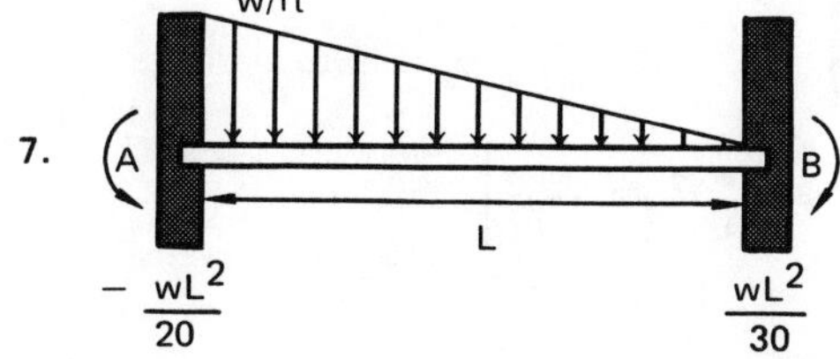


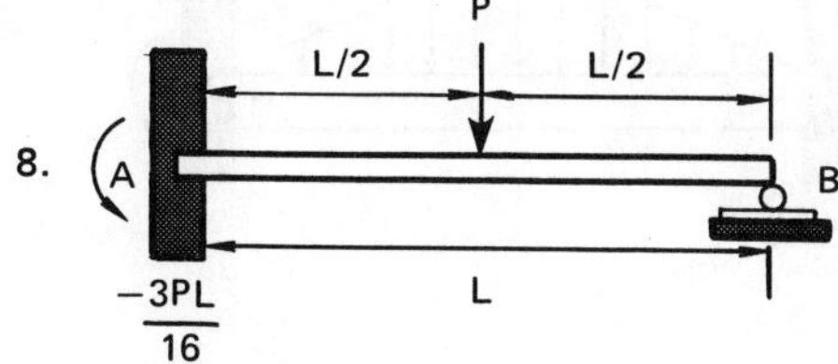


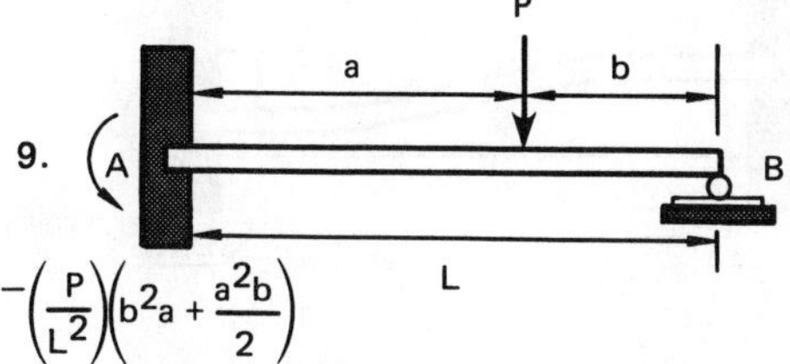


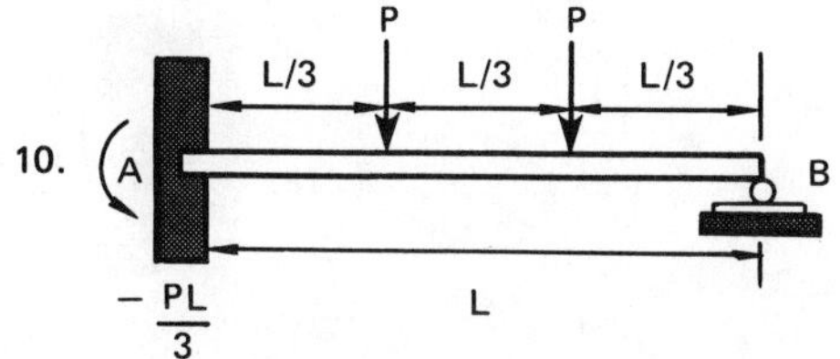


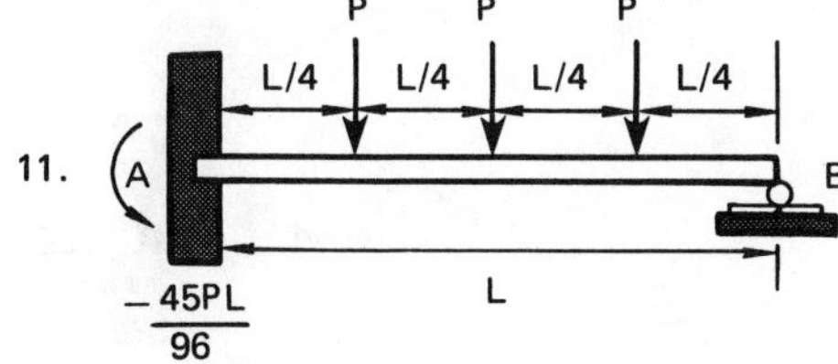


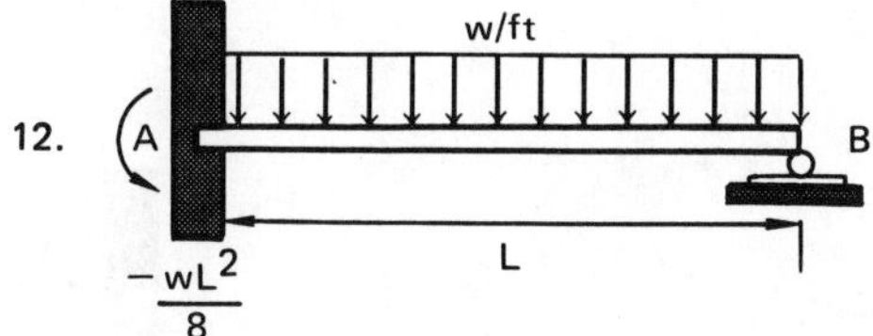


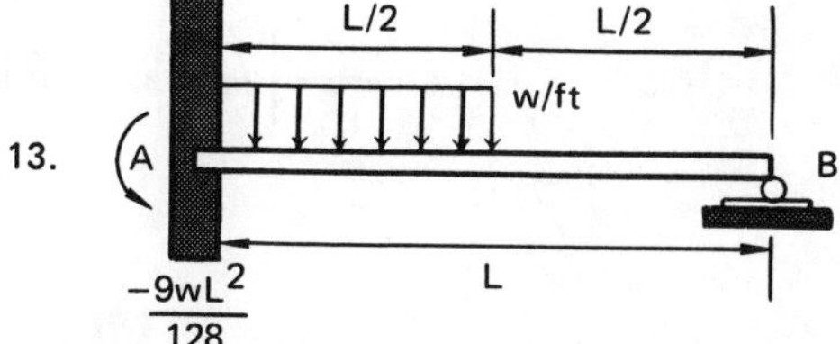


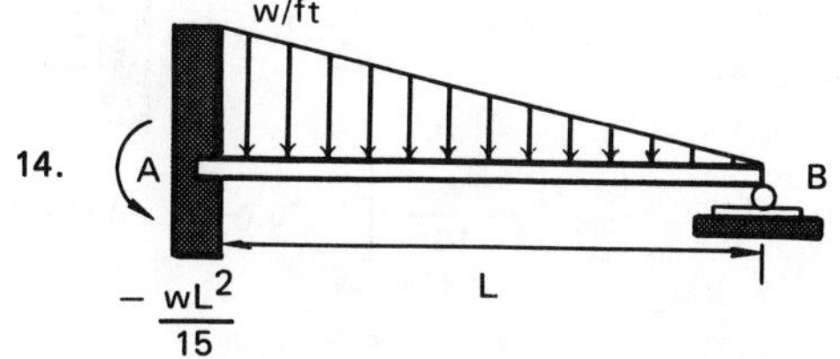

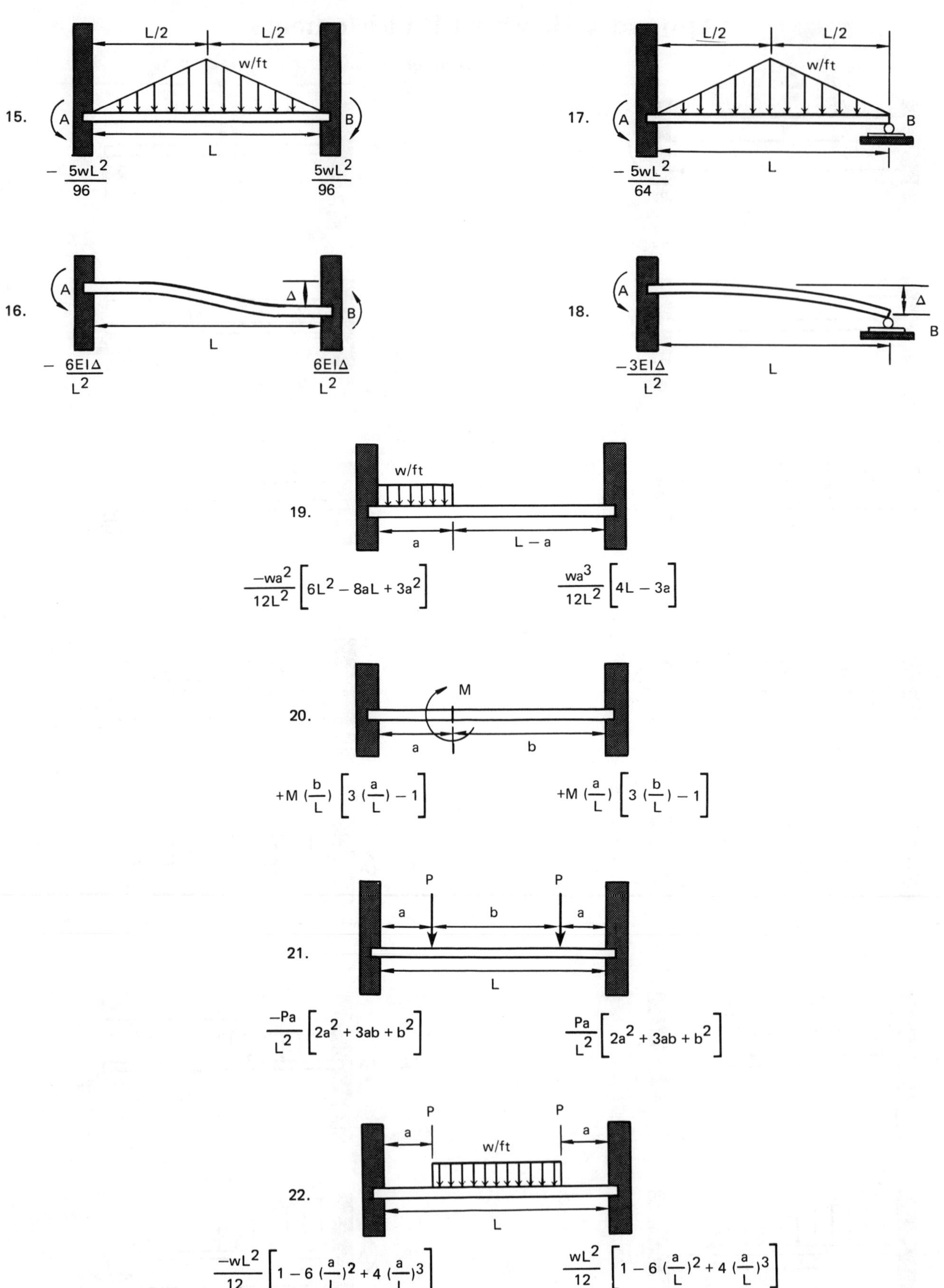

15.
L/2
L/2
w/ft
A
B
L
$-\frac{5wL^2}{96}$
$\frac{5wL^2}{96}$
17.
L/2
L/2
w/ft
A
B
L
$-\frac{5wL^2}{64}$
16.
A
Δ
B
L
$-\frac{6EI\Delta}{L^2}$
$\frac{6EI\Delta}{L^2}$
18.
A
Δ
B
L
$\frac{-3EI\Delta}{L^2}$
19.
w/ft
a
L − a
$\frac{-wa^2}{12L^2}\left[6L^2 - 8aL + 3a^2\right]$
$\frac{wa^3}{12L^2}\left[4L - 3a\right]$
20.
M
a
b
$+M\left(\frac{b}{L}\right)\left[3\left(\frac{a}{L}\right) - 1\right]$
$+M\left(\frac{a}{L}\right)\left[3\left(\frac{b}{L}\right) - 1\right]$
21.
P
P
a
b
a
L
$\frac{-Pa}{L^2}\left[2a^2 + 3ab + b^2\right]$
$\frac{Pa}{L^2}\left[2a^2 + 3ab + b^2\right]$
22.
P
P
a
a
w/ft
L
$\frac{-wL^2}{12}\left[1 - 6\left(\frac{a}{L}\right)^2 + 4\left(\frac{a}{L}\right)^3\right]$
$\frac{wL^2}{12}\left[1 - 6\left(\frac{a}{L}\right)^2 + 4\left(\frac{a}{L}\right)^3\right]$

# Appendix C: Beam Formulas

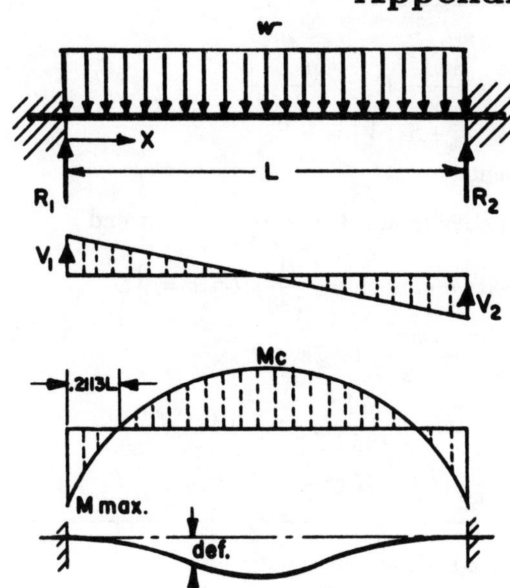


Uniformly distributed load, $w$ lb/ft
Total load $W = wL$

Reactions: $R_1 = R_2 = \dfrac{W}{2}$

Shear forces: $V_1 = +\dfrac{W}{2}$

$V_2 = -\dfrac{W}{2}$

Maximum (negative) bending moment

$$M_{\text{max}} = -\frac{wL^2}{12} = -\frac{WL}{12}, \text{ at end}$$

Maximum (positive) bending moment

$$M_c = \frac{wL^2}{24} = \frac{WL}{24}, \text{ at center}$$

$$\text{Maximum deflection} = \frac{wL^4}{384\,EI} = \frac{WL^3}{384\,EI}, \text{ at center}$$

$$\text{def.} = \frac{wx^2}{24\,EI}(L-x)^2,\ 0 \le x \le L$$

---

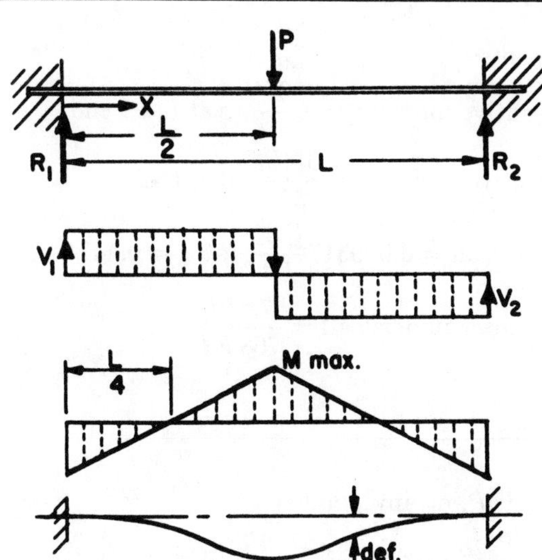


Concentrated load, $P$, at center

Reactions: $R_1 = R_2 = \dfrac{P}{2}$

Shear forces: $V_1 = +\dfrac{P}{2}$; $V_2 = -\dfrac{P}{2}$

Maximum bending moment

$$M_{\text{max}} = \frac{PL}{8}, \text{ at center}$$

$$M_{\text{max}} = -\frac{PL}{8}, \text{ at ends}$$

$$\text{Maximum deflection} = \frac{PL^3}{192\,EI}, \text{ at center}$$

$$\text{def.} = \frac{Px^2}{48\,EI}(3L - 4x),\ 0 \le x \le \frac{L}{2}$$

---

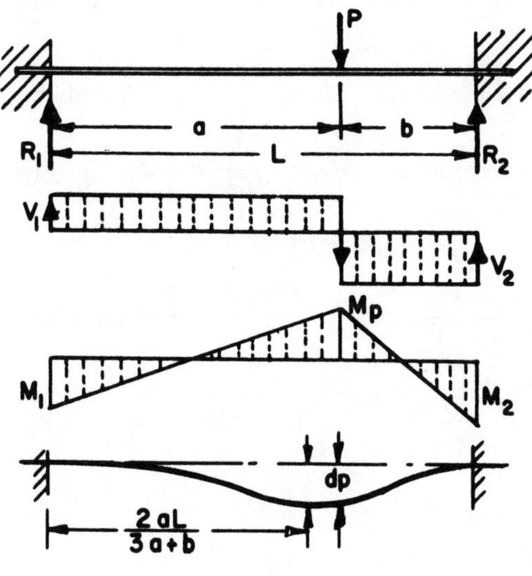


Concentrated load, $P$, at any point

Reactions: $R_1 = \dfrac{Pb^2}{L^3}(3a + b)$

$R_2 = \dfrac{Pa^2}{L^3}(3b + a)$

Shear forces: $V_1 = R_1$; $V_2 = -R_2$

Bending moments:

$$M_1 = -\frac{Pab^2}{L^2}, \text{ max. when } a < b$$

$$M_2 = -\frac{Pa^2b}{L^2}, \text{ max. when } a > b$$

$$M_p = +\frac{2Pa^2b^2}{L^3}, \text{ at point of load}$$

$$\text{Deflection} = \frac{Pa^3b^3}{3\,EIL^3}, \text{ at point of load}$$

$$\text{Max. def.} = \frac{2Pa^3b^2}{3\,EI(3a+b)^2}, \text{ at } x = \frac{2aL}{3a+b}, \text{ for } a > b$$

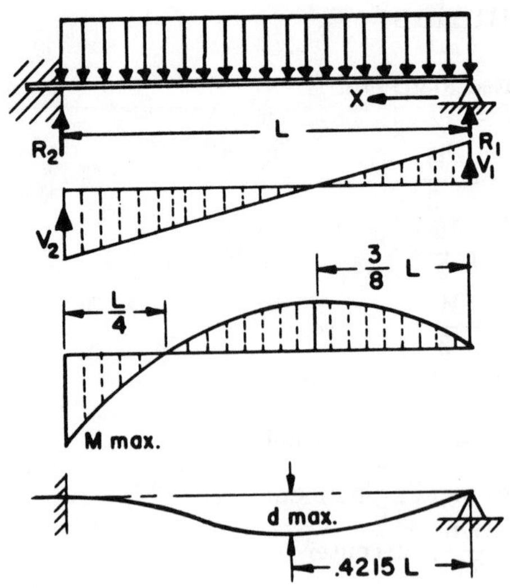


Uniformly distributed load, $w$ lb/ft

Total load $W = wL$

Reactions: $R_1 = \frac{3wL}{8}$, $R_2 = \frac{5wL}{8}$

Shear forces: $V_1 = +R_1$; $V_2 = -R_2$

Bending moments:

Max. negative moment $= -\frac{wL^2}{8}$, at left end

Max. positive moment $= \frac{9}{128}wL^2, x = \frac{3}{8}L$

$$M = \frac{3wLx}{8} - \frac{wx^2}{2},\ 0 \le x \le L$$

Maximum deflection $= \frac{wL^4}{185\,EI}$, $x = 0.4215L$

$$\text{def.} = \frac{wx}{48\,EI}(L^3 - 3Lx^2 + 2x^3),\ 0 \le x \le L$$

---

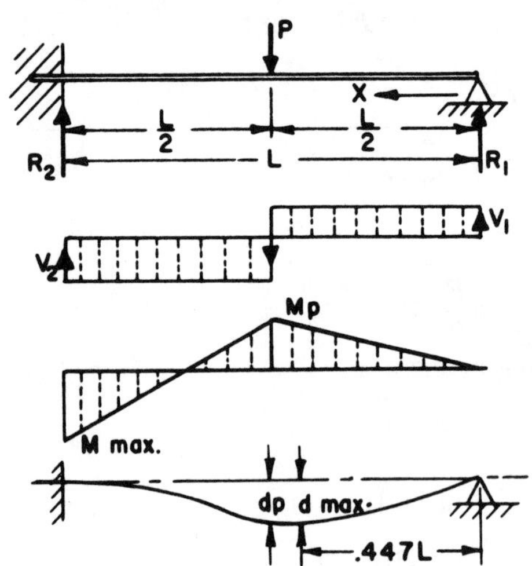


Concentrated load, $P$, at center

Reactions: $R_1 = \frac{5}{16}P$; $R_2 = \frac{11}{16}P$

Shear forces: $V_1 = R_1$; $V_2 = -R_2$

Bending moments:

Max. negative moment $= -\frac{3PL}{16}$, at fixed end

Max. positive moment $= \frac{5PL}{32}$, at center

Maximum deflection $= 0.009317\frac{PL^3}{EI}$, at $x = 0.447L$

Deflection at center under load $= \frac{7PL^3}{768\,EI}$

---

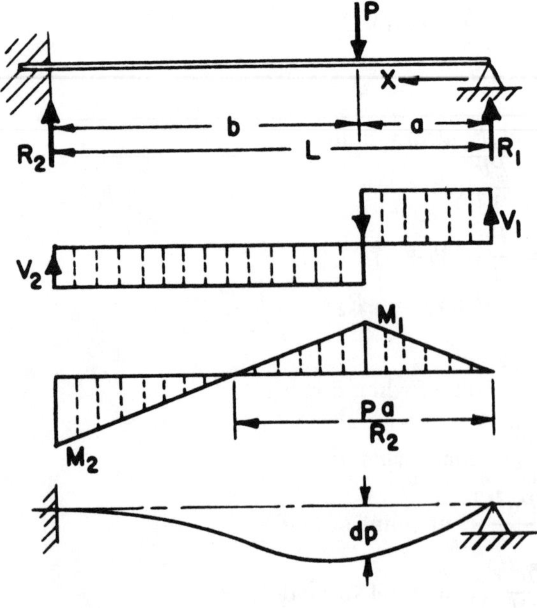


Concentrated load, $P$, at any point

Reactions: $R_1 = \frac{Pb^2}{2L^3}(a + 2L)$, $R_2 = \frac{Pa}{2L^3}(3L^2 - a^2)$

Shear forces: $V_1 = R_1$; $V_2 = -R_2$

Bending moments:

Max. negative moment, $M_2 = -\frac{Pab}{2L^2}(a + L)$, at fixed end

Max. positive moment, $M_1 = \frac{Pab^2}{2L^3}(a + 2L)$, at load

Deflections: $d_p = \frac{Pa^2b^3}{12\,EIL^3}(3L + a)$, at load

$$d_{\text{max}} = \frac{Pa(L^2 - a^2)^3}{3\,EI(3L^2 - a^2)^2},\ \text{at } x = \frac{L^2 + a^2}{3L^2 - a^2}L,\ \text{when } a < 0.414L$$

$$d_{\text{max}} = \frac{Pab^2}{6\,EI}\sqrt{\frac{a}{2L + a}},\ \text{at } x = L\sqrt{\frac{a}{2L + a}},\ \text{when } a > 0.414L$$

Continuous beam of two equal spans—equal concentrated loads, $P$, at center of each span

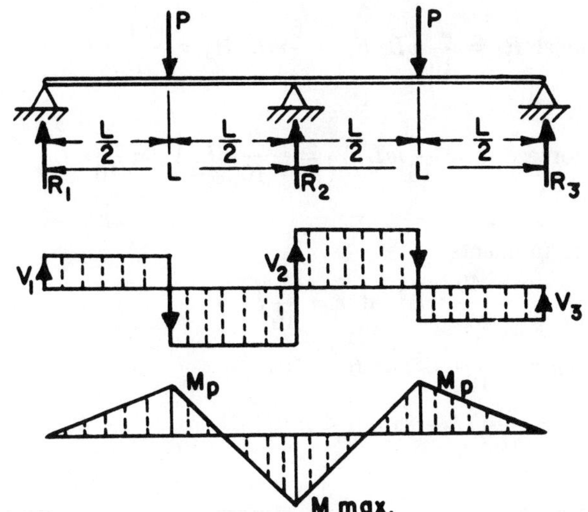


Reactions: $R_1 = R_3 = \frac{5}{16}P$

$R_2 = 1.375P$

Shear forces: $V_1 = -V_3 = \frac{5}{16}P$

$V_2 = \pm\frac{11}{16}P$

Bending moments:

$M_{\text{max}} = -\frac{6}{32}PL$, at $R_2$

$M_P = \frac{5}{32}PL$, at point of load

---

Continuous beam of two equal spans—concentrated loads, $P$, at third points of each span

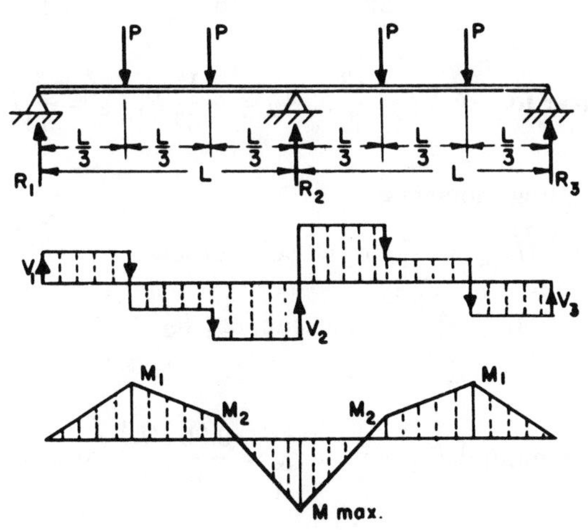


Reactions: $R_1 = R_3 = \frac{2}{3}P$

$R_2 = \frac{8}{3}P$

Shear forces: $V_1 = -V_3 = \frac{2}{3}P$

$V_2 = \pm\frac{4}{3}P$

Bending moments:

$M_{\text{max}} = -\frac{1}{3}PL$, at $R_2$

$M_1 = \frac{2}{9}PL$

$M_2 = \frac{1}{9}PL$

---

Continuous beam of two equal spans—uniformly distributed load of $w$ lb/ft

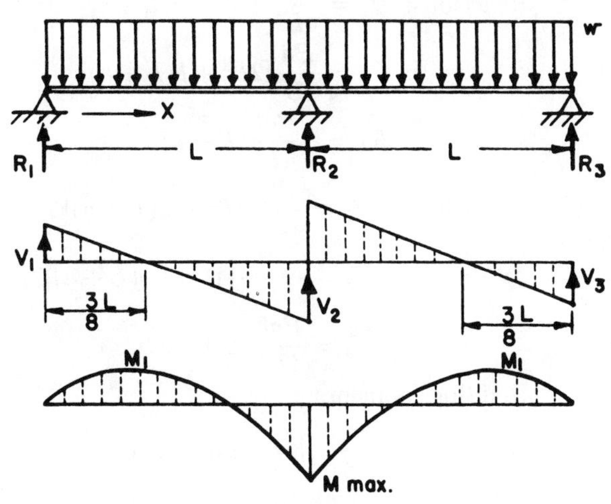


Reactions: $R_1 = R_3 = \frac{3}{8}wL$

$R_2 = 1.25wL$

Shear forces: $V_1 = -V_3 = \frac{3}{8}wL$

$V_2 = \pm\frac{5}{8}wL$

Bending moments:

$M_{\text{max}} = -\frac{1}{8}wL^2$

$M_1 = \frac{9}{128}wL^2$

Maximum deflection $= 0.00541\frac{wL^4}{EI}$

at $x = 0.4215L$

Def. $= \frac{w}{48\,EI}(L^3x - 3Lx^3 + 2x^4)$, $0 \le x \le L$

Continuous beam of two equal spans—uniformly distributed load of $w$ lb/ft on one span

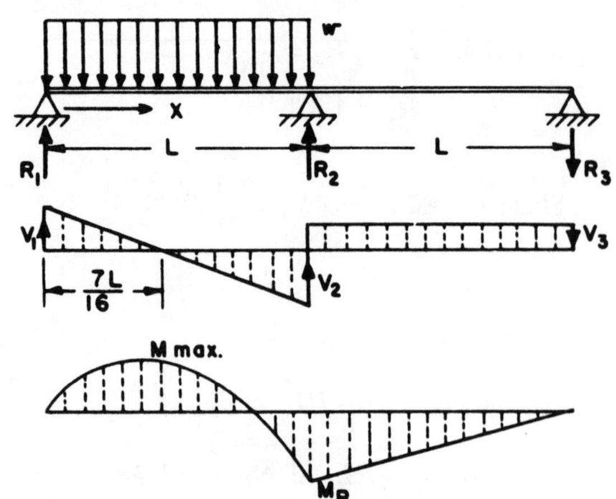


Reactions: $R_1 = \frac{7}{16}wL$, $R_2 = \frac{5}{8}wL$, $R_3 = -\frac{1}{16}wL$

Shear forces: $V_1 = \frac{7}{16}wL$, $V_2 = -\frac{9}{16}wL$, $V_3 = \frac{1}{16}wL$

Bending moments:

$$M_{\text{max}} = \frac{49}{512}wL^2, \text{ at } x = \frac{7}{16}L$$

$$M_R = -\frac{1}{16}wL^2, \text{ at } R_2$$

$$M = \frac{wx}{16}(7L - 8x),\ 0 \le x \le L$$

---

Continuous beam of two equal spans—concentrated load, $P$, at center of one span.

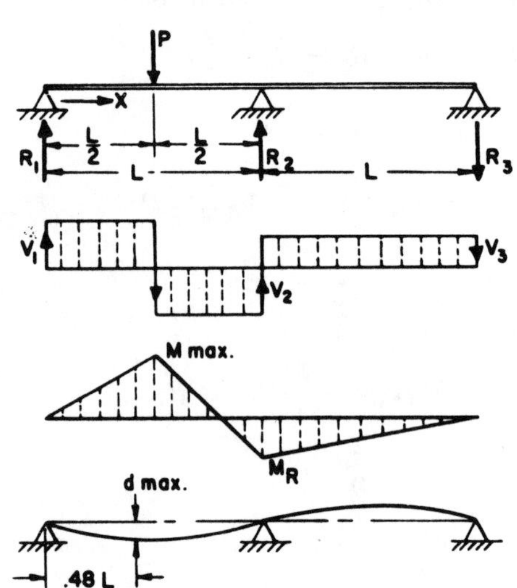


Reactions: $R_1 = \frac{13}{32}P$, $R_2 = \frac{11}{16}P$, $R_3 = -\frac{3}{32}P$

Shear forces: $V_1 = \frac{13}{32}P$, $V_2 = -\frac{19}{32}P$, $V_3 = \frac{3}{32}P$

Bending moments:

$$M_{\text{max}} = \frac{13}{64}PL, \text{ at point of load}$$

$$M_R = -\frac{3}{32}PL, \text{ at support } R_2$$

Maximum deflection: $d_{\text{max}} = \dfrac{0.96\,PL^3}{64\,EI}$, at $x = 0.48L$

---

Continuous beam of two equal spans—concentrated load, $P$, at any point on one span.

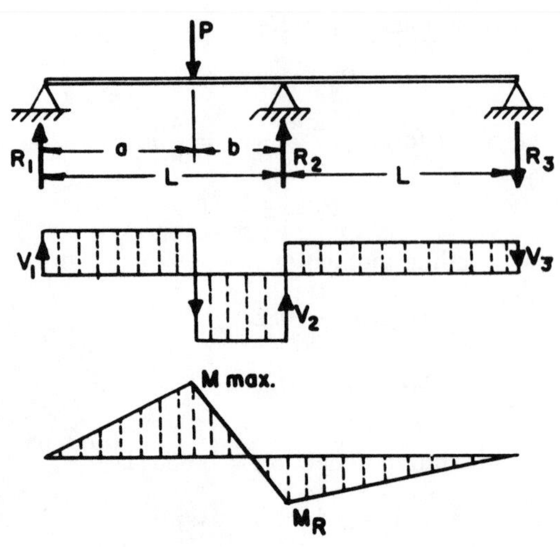


Reactions:

$$R_1 = \frac{Pb}{4L^3}[4L^2 - a(L + a)]$$

$$R_2 = \frac{Pa}{2L^3}[2L^2 + b(L + a)]$$

$$R_3 = -\frac{Pab}{4L^3}(L + a)$$

Shear forces:

$$V_1 = \frac{Pb}{4L^3}[4L^2 - a(L + a)]$$

$$V_2 = -\frac{Pa}{4L^3}[4L^2 + b(L + a)]$$

$$V_3 = \frac{Pab}{4L^3}(L + a)$$

Bending moments:

$$M_{\text{max}} = \frac{Pab}{4L^3}[4L^2 - a(L + a)]$$

$$M_R = -\frac{Pab}{4L^2}(L + a)$$

Continuous beam of three equal spans—concentrated load, $P$, at center of each span

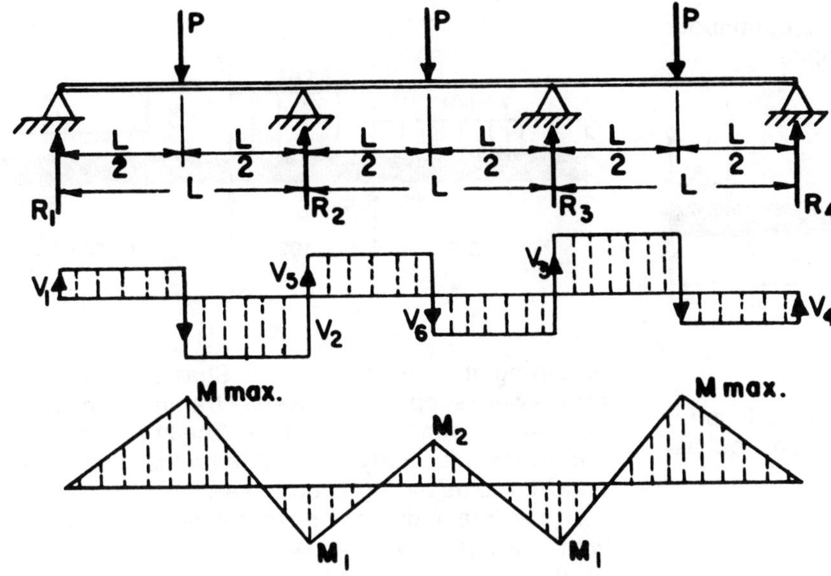


Reactions:

$$R_1 = R_4 = \frac{7}{20}P$$

$$R_2 = R_3 = \frac{23}{20}P$$

Shear forces:

$$V_1 = -V_4 = \frac{7}{20}P$$

$$V_3 = -V_2 = \frac{13}{20}P$$

$$V_5 = -V_6 = \frac{P}{2}$$

Bending moments:

$$M_{\text{max}} = \frac{7}{40}PL$$

$$M_1 = -\frac{3}{20}PL$$

$$M_2 = \frac{1}{10}PL$$

---

Continuous beam of three equal spans—concentrated loads, $P$, at third points of each span

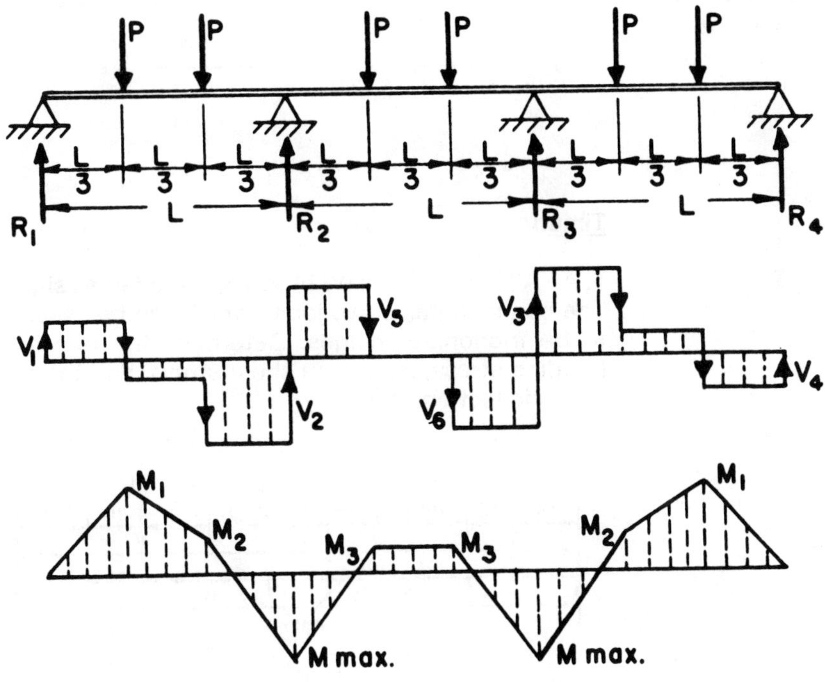


Reactions:

$$R_1 = R_4 = \frac{11}{15}P$$

$$R_2 = R_3 = \frac{34}{15}P$$

Shear forces:

$$V_1 = -V_4 = \frac{11}{15}P$$

$$V_3 = -V_2 = \frac{19}{15}P$$

$$V_5 = -V_6 = P$$

Bending moments:

$$M_{\text{max}} = -\frac{12}{45}PL$$

$$M_1 = \frac{11}{45}PL$$

$$M_2 = \frac{7}{45}PL$$

$$M_3 = \frac{3}{45}PL$$

## Practice Problems: INDETERMINATE STRUCTURES

Untimed

1. What are the reactions supporting the continuous beam shown? The reactions are all simple.

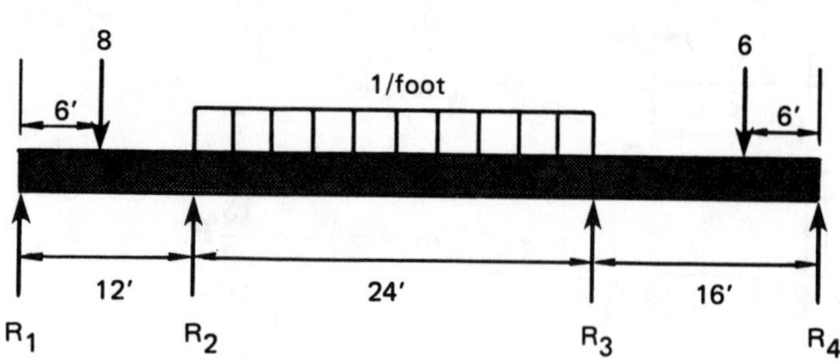


2. What are the joint moments and the reactions for the rigid frame shown? The supports may be assumed to be pinned.

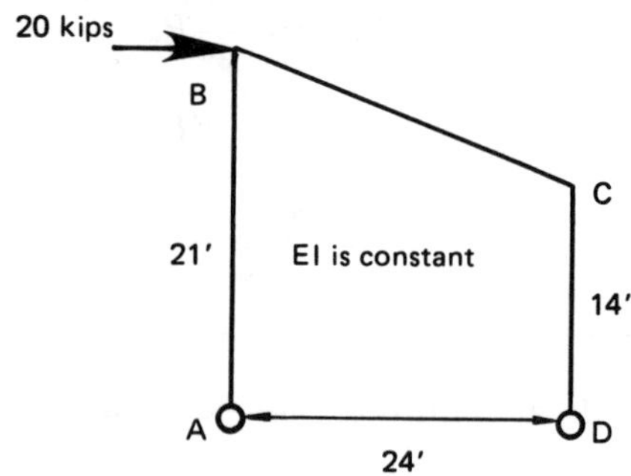


3. A frame with rigid joints is shown. Draw the moment diagrams for members AB, BC, CE, and CD.

$$I_{AB} = 240 \text{ in}^4$$
$$I_{BE} = 300 \text{ in}^4$$
$$I_{CD} = 300 \text{ in}^4$$

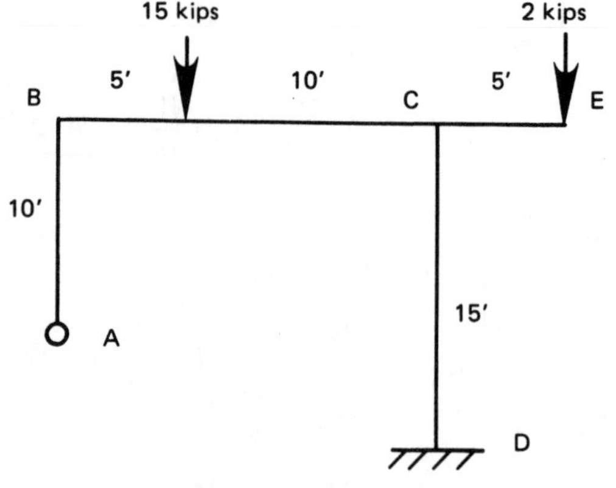


4. What are the maximum moments on the spans AB, BC, and CD?

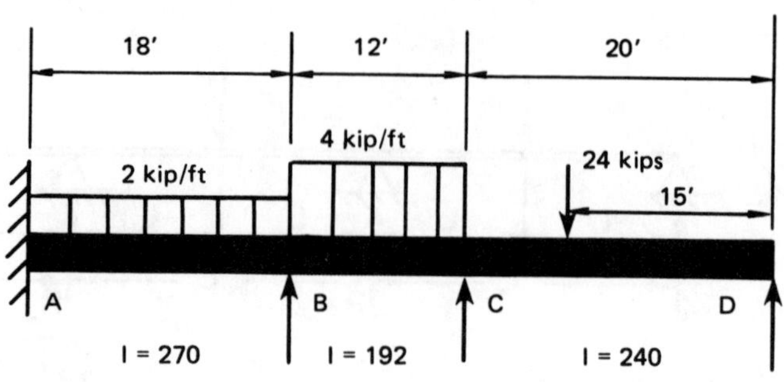


5. During the construction of Shasta dam, access ramps were supported by a steel frame which projected from the concrete dam face. The horizontal members were sufficiently embedded so that they may be considered as having fixed bases.

At the free ends, the horizontal members were tied with a vertical strut as shown. Calculate the deflection of the frame assuming the tied connections are (a) pinned and (b) rigid.

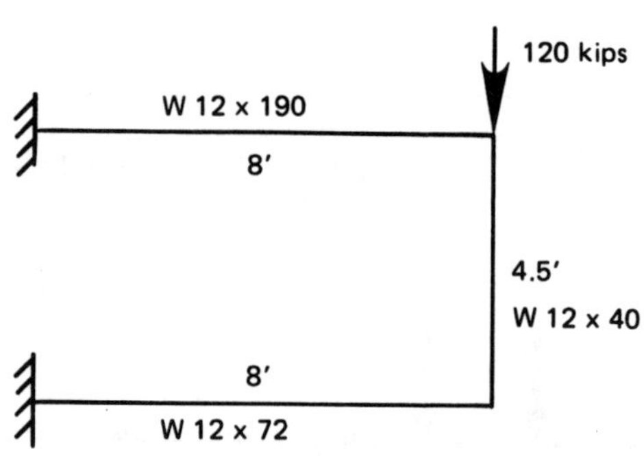


Timed

1. An A36 steel beam (W24 × 76) is loaded as shown. The beam contains two joints which can be assumed to be frictionless hinges. Determine the maximum bending stress, maximum shear stress, and the midpoint deflection.

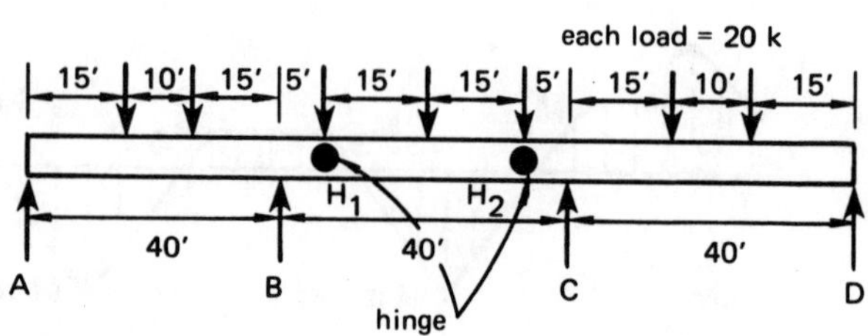

2. A steel beam is loaded as shown and is laterally supported only at the three reaction points. The beam carries 1400 lbm/ft over its total span. (a) Determine the reactions. (b) Draw the shear and moment diagrams. (c) Choose the lightest W14 beam that is capable of developing a minimum of $0.60F_y$ bending stress for this application. (d) Specify the maximum unbraced length for the beam you choose.

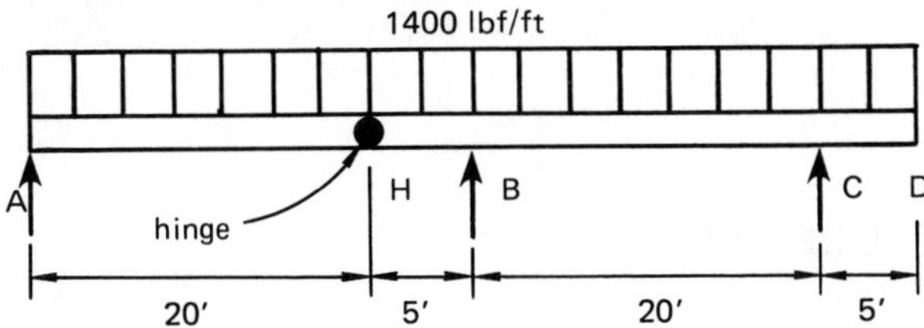


3. For the rigid-joint structure shown, (a) determine the reactions at A, C, and D, and (b) draw the moment diagram on the tension side of the frame. (Neglect the beam weights. Reaction at C is passive and does not load the frame.)

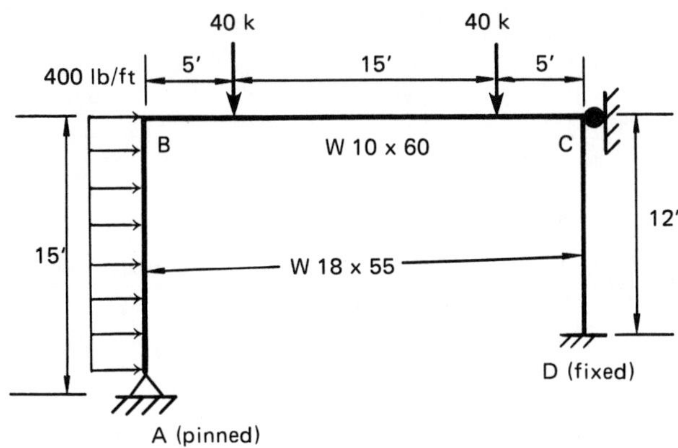

RESERVED FOR FUTURE USE

# 14 REINFORCED CONCRETE DESIGN[1]

*Nomenclature*

| | | |
|---|---|---|
| a | short side length, or height of stress block | in |
| A | area | $in^2$ |
| b | long side length, or width | in |
| c | distance from neutral axis to extreme compression fiber | in |
| C | compressive force | lbf |
| d | depth or diameter | in |
| e | eccentricity | in |
| E | modulus of elasticity | psi |
| f | stress or strength | psi |
| $f'_c$ | compressive strength of concrete | psi |
| $f_r$ | modulus of rupture | psi |
| $f_y$ | yield strength of steel | psi |
| h | height | in |
| I | moment of inertia | $in^4$ |
| j | a fraction | – |
| k | a fraction, or end condition factor | – |
| l | length | in |
| $l_d$ | development length | in |
| m | a factor | – |
| M | moment | in-lbf |
| n | modular ratio $(E_{st}/E_c)$ | – |
| p | factored soil pressure | psi |
| r | radius of gyration | in |
| $R_u$ | coefficient of resistance | psi |
| s | spacing | in |
| SR | slenderness ratio | – |
| t | thickness | in |
| T | tensile force | lbf |
| v | shear stress | psi |
| V | shear, or shear strength | lbf |
| w | density, or load per unit length | pcf or lbf/in |
| z | crack parameter | kips/in |

*Symbols*

| | | |
|---|---|---|
| $\alpha$ | ratio of column to footing area | – |
| $\beta$ | ratio of long side to short side | – |
| $\beta_1$ | a factor | – |
| $\delta$ | magnification factor | – |
| $\epsilon$ | fraction eccentricity | – |
| $\xi$ | timespan factor | – |
| $\rho$ | reinforcement ratio | – |
| $\lambda$ | long term deflection factor | – |
| $\theta$ | angle from the horizontal | – |
| $\phi$ | capacity reduction factor | – |

*Subscripts*

| | |
|---|---|
| A | active |
| b | base, bar, bending, or bearing |
| c | concrete, column, or cover |
| cr | cracked |
| D | dead |
| e | effective |
| f | footing or flexure |
| g | gross |
| h | horizontal |
| i | initial |
| L | live |
| n | nominal or clear |
| o | zero eccentricity |
| p | prestress |
| r | rupture |
| s | stem or spiral |
| st | steel |
| t | tension |
| u | ultimate |
| v | shear |
| w | service load |
| y | yield |

[1] This chapter is no substitute for the ACI code, and covers only a fraction of the relevant material.

### SIONS

| ply | by | to obtain |
|---|---|---|
| s | 1.356 | kN-m |
| ft-lbf | 1.356 | N-m |
| in-lbf | 0.113 | N-m |
| kips | 4.448 | kN |
| kips | 1000 | lbf |
| kips/ft | 14.59 | kN/m |
| kips/ft$^2$ | 47.88 | kN/m$^2$ |
| kN | 0.2248 | kips |
| kN/m | 0.06852 | kips/ft |
| kPa | 1000 | Pa |
| ksi | 6.895 EE6 | Pa |
| lbf | 0.001 | kips |
| lbf | 4.448 | N |
| N-m | 0.7376 | ft-lbf |
| N-m | 8.851 | in-lbf |
| N-m$^2$ | 1.0 | Pa |
| Pa | 1.0 | N/m$^2$ |
| Pa | EE–6 | MPa |
| Pa | 0.001 | kPa |
| Pa | 1.45 EE–7 | ksi |
| Pa | 1.45 EE–4 | psi |
| Pa | 0.02089 | psf |
| psf | 47.88 | Pa |
| psi | 6.895 EE3 | Pa |

## 2 DEFINITIONS

Absorption: The process by which a liquid is drawn into and tends to fill permeable pores in a porous body. Also, the increase in weight of a porous solid body resulting from the penetration of liquid into its permeable pores.

Admixture: A material other than water, aggregates, and portland cement that is used as an ingredient in concrete and is added to the batch immediately before or during its mixing.

Aggregate: Inert material which is mixed with portland cement and water to produce concrete.

Aggregate, coarse: Aggregate which is retained on a #4 sieve.

Aggregate, fine: Aggregate passing the #4 sieve and retained on the #200 sieve.

Aggregate, lightweight: Aggregate having a dry density of 70 pounds per cubic foot or less.

Bleeding: The autogenous flow of mixing water within, or its emergence from, recently placed concrete.

Cement factor: The number of bags or cubic feet of cement per cubic yard of concrete.

Column: An upright compression member, the length of which exceeds three times its least lateral dimension.

Combination column: A column in which a structural steel member, designed to carry the principal part of the load, is encased in concrete which carries the remainder of the load.

Composite column: A column in which a steel structural member is completely encased in concrete containing spiral and longitudinal reinforcement.

Composite concrete flexural construction: A precast concrete member and cast-in-place reinforced concrete so interconnected that the component elements act together as a flexural member.

Concrete: A mixture of portland cement, fine aggregate, coarse aggregate, and water.

Concrete, normal weight: Concrete having a hardened density of approximately 150 pcf which is made from aggregate of approximately the same density.

Concrete, plain: Concrete that is not reinforced with steel.

Concrete, precast: A plain or reinforced concrete element cast in other than its final position in the structure.

Concrete, prestressed: Reinforced concrete in which there have been introduced internal stresses of such magnitude and distribution that the stresses resulting from service loads are counteracted to a desired degree.

Concrete, reinforced: Concrete containing steel reinforcement.

Concrete, structural lightweight: A concrete containing lightweight aggregate.

Crushed gravel: The product resulting from artificial crushing of gravel with substantially all fragments having at least one fracture face.

Crushed stone: The product resulting from the artificial crushing of rocks where substantially all faces have resulted from the crushing operation.

Deformed bar: A reinforcing bar with ridges to increase bonding with the concrete.

Double reinforcement: A concrete beam with steel on both sides of the neutral axis to resist tension and compression.

Fineness modulus: An empirical factor obtained by adding the total percentages of a sample of the aggregate retained on each of a specified series of sieves, and dividing the sum by 100.

Formwork: The wood molds used to hold concrete during the curing and pouring processes.

Gravel: Granular material retained on a #4 sieve which is the result of natural disintegration of rock.

Heat of hydration: The exothermic heat given off by concrete as it cures.

Hydration: The chemical reaction which occurs when the cement ions attach themselves to the water molecules of crystallization.

Monolithic construction: Constructed as one piece.

Pedestal: An upright compression member whose height does not exceed three times its average least lateral dimension.

Plain bar: Reinforcement that does not conform to the definition of a deformed bar. That is, reinforcing bars without raised ridges.

Rebar: Steel reinforcing bar.

Sand: Granular material passing through a #4 sieve but predominantly retained on a #200 sieve.

Saturated Surface Dry: A condition of an aggregate which holds as much water as it can without having any free surface water between the aggregate particles.

Slab: A cast concrete member of uniform thickness.

Slump: The decrease in height of wet concrete when a supporting mold is removed.

Spiral column: A column with a continuous spiral of wire around the longitudinal steel.

SSD: See 'Saturated Surface Dry.'

Surface water: Water carried by an aggregate in addition to that held by absorption within the aggregate particles themselves. Water in addition to SSD water.

Tied column: A column which has individual loops of wire around the longitudinal steel.

Two-way: Construction with steel reinforcing running in two perpendicular directions.

Water-cement ratio: The ratio of water weight to cement weight. Alternatively, the number of gallons of water per 94 pound sack of cement.

## 3 CONCRETE MIXING

### A. TYPES OF CONCRETE

Concrete is a mixture of mineral aggregates locked into a solid structure by a binding material. The concrete is produced by adding water to the aggregate and binder, and then casting the mixture in place. The semi-fluid mixture hardens by chemical action.

The usual binding material is portland cement, which is manufactured from lime, silica, and alumina (with a small amount of plaster of paris) in the appropriate amounts. There are five types of portland cement, and the choice is dependent on the application.

- **Type I: Normal portland cement:** This is a general-purpose cement used whenever sulfate hazards are absent and when the heat of hydration will not produce objectionable rises in temperature. Typical uses are sidewalks, pavement, beams, columns, and culverts.
- **Type II: Modified portland cement:** This cement has a moderate sulfate resistance, but is generally used in hot weather in the construction of large concrete structures. Its heat rate and total heat generation are lower than for normal portland cement.
- **Type III: High-early strength portland cement:** This type develops its strength quickly. It is suitable for use when the structure must be put into early use or when long-term protection against cold temperatures is not feasible. Its shrinkage rate, however, is higher than for types I and II; and extensive cracking may result.
- **Type IV: Low-heat portland cement:** For extensive concrete structures, such as gravity dams, low-heat cement is required to minimize the curing heat. The ultimate strength also develops more slowly than for the other types.
- **Type V: Sulfate-resistant portland cement:** This type of cement is applicable when exposure to severe sulfate concentration is expected. This typically occurs in states having highly alkaline soils.

**Table 14.1**
Relative Strengths of Concrete Types

| | compressive strength, % of normal strength concrete | | |
|---|---|---|---|
| | 3 days | 28 days | 3 months |
| type I | 100 | 100 | 100 |
| type II | 80 | 85 | 100 |
| type III | 190 | 130 | 115 |
| type IV | 50 | 65 | 90 |
| type V | 65 | 65 | 85 |

## B. AGGREGATES

The bulk of concrete consists of sand and rock particles that have been added to the cement-water mixture to increase the weight and volume. These sand and rock particles are known as *aggregate.* Sand and other particles that will pass through a #4 sieve (less than 0.25″) are known as *fine aggregate.* Any particles that are larger than this are known as *coarse aggregate.*

Aggregate having a density of 70 pcf or less is known as *lightweight aggregate.* This type is used in the production of *lightweight concrete.* Lightweight concrete in which only the coarse aggregate is lightweight is known as *sand-lightweight concrete.* If both the coarse and fine aggregates are lightweight, it is known as *all-lightweight concrete.* Unless noted otherwise, concrete in this chapter is assumed to be *normal weight concrete.*

## C. ADMIXTURES

Anything added to the concrete to improve its workability, hardening, or strength characteristics is known as an *admixture.* Hydrated lime, diatomaceous silica, fly ash, and bentonite are added to concrete which has too little fine aggregate. These admixtures separate the coarse aggregate and reduce the friction of the mixture. Calcium chloride can be added as a curing accelerator. It is also an anti-freeze, but its use as such is not recommended.[2]

Sulfonated soaps and oils, as well as natural resins, increase *air entrainment.* This increases the durability of the concrete while decreasing the strength and weight only slightly.

Styrofoam can be added to increase the insulating values of the concrete.

## D. PROPORTIONING CONCRETE

The proportions of a concrete mixture are usually designated as a ratio of cement, fine aggregate, and coarse aggregate, in that order. For example, 1:2:3 means that one part of cement, two parts of fine aggregate, and three parts of coarse aggregate are to be combined. The ratio can be either in terms of weight or volume. Weight ratios are more common.

[2] The ACI code does not allow chloride to be added to concrete used in prestressed construction or in concrete with aluminum embedments. Natural chloride ions are also limited.

The amount of water used is called out in terms of gallons of water per 94 pound sack of cement. This is known as the *water-cement ratio.*[3]

## E. IN-PLACE VOLUME

The amount of concrete that can be made from mixing known quantities of ingredients can be found from the *absolute* or *solid volume method.* This assumes that there will be no voids in the placed concrete. Therefore, the amount of concrete is the sum of the solid volumes of the cement, sand, coarse aggregate, and water.

To use the absolute volume method, it is necessary to know the solid densities of the constituents. In the absence of other information, the following data can be used for solid densities: cement 195 pcf; fine aggregate, 165 pcf; coarse aggregate, 165 pcf; water, 62.4 pcf. A sack of cement weighs 94 pounds, and 7.48 gallons of water make a cubic foot. Alternatively, there are 239.7 gallons per ton of water.

If the mix proportions are volumetric, it will be necessary to multiply the ratio values by the bulk densities to get the weights of the constituents. Then, weight ratios may be calculated and the absolute volume method applied directly.

The problem may be complicated by *air entrainment* and/or the water content of the aggregate. The following guidelines should be observed,

- The yield is increased by the addition of air. This can be accounted for by dividing the solid yield by (1 − air percentage).
- Any water content in the aggregate above the *saturated surface dry* (SSD) water content must be subtracted from the water requirements.
- Any porosity (water affinity) below the SSD water content must be added to the water requirement.
- The densities used in the calculation of yield should be the SSD densities.

*Example 14.1*

A mix is designed as 1:1.9:2.8 by weight. The water-cement ratio is 7 gallons of water per sack of cement. (a) What is the concrete yield in cubic feet per sack of cement? (b) How much of each constituent is needed to make 45 cubic yards of concrete?

[3] Alternatively, the water-cement ratio may be given in pounds of water to pounds of cement.

(a) The solution can be tabulated as follows:

| material | ratio | weight per sack cement | solid density | absolute volume (ft$^3$/sack) |
|---|---|---|---|---|
| cement | 1.0 | $1 \times 94 = 94$ | 195 | $94/195 = 0.48$ |
| sand | 1.9 | $1.9 \times 94 = 179$ | 165 | $179/165 = 1.08$ |
| coarse | 2.8 | $2.8 \times 94 = 263$ | 165 | $263/165 = 1.60$ |
| water | | | | $7/7.48 = 0.94$ |
| | | | | 4.10 |

The yield is 4.1 cubic feet of concrete per sack cement.

(b) The number of one-sack batches is

$$\frac{(45)\,\text{yd}^3(27)\,\text{ft}^3/\text{yd}^3}{(4.1)\,\text{ft}^3/\text{sack}} = 296.3\,\text{sacks} \quad (\text{say } 297)^4$$

Order 297 sacks of cement.

$$\frac{(297)(1.9)(94)}{2000} = 26.5 \text{ tons of sand}$$

$$\frac{(297)(2.8)(94)}{2000} = 39.1 \text{ tons of coarse aggregate}$$

$$(297)(7) = 2079 \text{ gallons of water}$$

*Example 14.2*

50 cubic feet of 1:2$\frac{1}{2}$:4 (by weight) concrete are to be produced. The constituents have the following properties:

| constituent | SSD density (pcf) | moisture (dry basis from SSD) |
|---|---|---|
| cement | 197 | – |
| sand | 164 | 5% excess |
| coarse aggregate | 168 | 2% deficit |

What are the required order quantities if 5.5 gallons of water are to be used per sack and the mixture is to have 6% entrained air?

| constituent | ratio | weight per sack cement | SSD density | absolute volume |
|---|---|---|---|---|
| cement | 1.0 | 94 | 197 | 0.477 |
| sand | 2.5 | 235 | 164 | 1.433 |
| coarse | 4.0 | 376 | 168 | 2.238 |
| water | | 5.5/7.48 | | = 0.735 |
| | | | | 4.883 ft$^3$/sack |

The yield with 6% air is

$$\frac{4.883}{1 - 0.06} = 5.19\,\text{ft}^3/\text{sack}$$

[4] Don't round down, or the volume will be short.

The number of one sack batches is

$$\frac{50}{5.19} = 9.63$$

(In practice, this would be rounded up.)

The required sand weight (ordered as is, not SSD) is

$$\frac{(9.63)(1.05)(94)(2.5)}{2000} = 1.19\,\text{tons}$$

The required coarse aggregate weight (ordered as is, not SSD) is

$$\frac{(9.63)(0.98)(94)(4)}{2000} = 1.77\,\text{tons}$$

The excess water contained in the sand is

$$(1.19)\left(\frac{0.05}{1.05}\right)(239.7)\,\text{gal/ton} = 13.58\,\text{gallons}$$

The water needed to bring the coarse aggregate to SSD conditions is

$$(1.77)\left(\frac{0.02}{0.98}\right)(239.7)\,\text{gal/ton} = 8.66\,\text{gallons}$$

The total water needed is

$$(5.5)(9.63) + 8.66 - 13.58 = 48.0\,\text{gallons}$$

## 4 PROPERTIES OF CONCRETE

### A. SLUMP

The *slump test* is a measure of consistency of the plastic concrete mass prior to hardening. This test consists of filling a slump cone in three layers of about one-third of the mold volume. Each layer is rodded with a round rod. The mold is then removed by raising it carefully in the vertical direction. The slump is the difference in the mold height and the resulting concrete pile height.

Concrete mixtures that do not slump appreciably are known as *stiff mixtures*. Stiff mixtures are inexpensive because of the large amount of coarse aggregate. However, placing time and workability are impaired. Mixtures with large slumps are known as *wet* or *watery mixtures*. Such mixtures are needed for thin castings and structures with extensive reinforcing.

Recommended slumps for concrete that is hand vibrated are given in table 14.2. If high-frequency vibrators are used, the values given should be reduced by about one-third.

**Table 14.2**
Recommended Slumps

| application | slumps, inch maximum | minimum |
|---|---|---|
| reinforced foundations and footings | 3 | 1 |
| plain footings and substructure walls | 3 | 1 |
| slabs, beams, and reinforced walls | 4 | 1 |
| columns, reinforced | 4 | 1 |
| pavement and slabs | 3 | 1 |
| heavy mass construction | 2 | 1 |

B. COMPRESSIVE STRENGTH

Cylinders of the concrete are cast for each mixture ratio to be tested. After curing, the cylinders are placed in a compressive tester. A load is applied at a constant rate. The compressive strength is found by dividing the maximum load carried (at failure) by the area. Typical compressive strengths, $f'_c$, vary from 2000 psi to 8000 psi at 28 days, although 6000 psi is a common upper limit.

Since the strength of concrete increases with time, all values of $f'_c$ should be referenced to the age. If no age is given, a standard 28 day age is assumed.

The ultimate compressive strength is highly dependent on the water/cement ratio, and it is fairly independent of the proportion of the mixes. The compressive strength of concrete varies directly with the cement/water ratio, provided that the mix is of a workable consistency.[5]

C. TENSILE STRENGTH

Because concrete is not used to resist tension, tensile tests are seldom performed on normal weight concrete. However, the ACI code does specify a splitting tensile test for structural lightweight concrete. The ultimate tensile strength usually varies between 7% and 10% of the ultimate compressive strength.

D. SHEAR STRENGTH

The shear strength of concrete can be determined by torsion tests. Such tests vary widely in method and results. The shear strength will be between one-sixth and one-quarter of the ultimate compressive strength.

[5] This is *Abram's strength law*, named after Dr. Duff Abrams who formulated it in 1918.

E. DENSITY

The weight density (also known as *unit weight*) of concrete can vary from 100 pcf to 160 pcf, depending on the mixture ratios and the specific gravities of the constituents. Generally, the range will be 140 pcf to 160 pcf. Steel reinforced concrete will be between 3% and 5% higher than similar plain concrete. An average of 150 pcf can be used in most calculations.

F. MODULUS OF ELASTICITY

Typical results of a standard compressive test are shown in figure 14.1. The slope of the stress-strain line varies with the applied stress, and there are several ways of evaluating the modulus of elasticity. These methods give the *initial tangent modulus*, *actual tangent modulus*, and *secant modulus*. The secant modulus is most frequently used in design work.

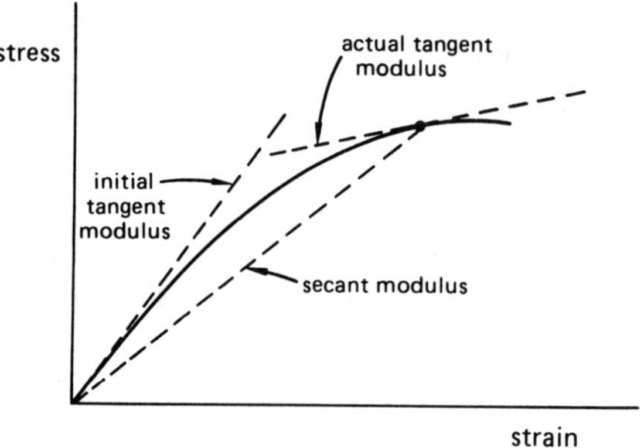


**Figure 14.1** Compressive Test Results

The ACI code calculates the secant modulus of elasticity based on the compressive strength. Equation 14.1 is for normal weight concrete only, with a density of $w = 90$ to 155 pcf.

$$E_c = w^{1.5}(33)\sqrt{f'_c} \qquad 14.1$$

The modulus of elasticity is greatly dependent on age, quality, and proportions. It can vary from 1 EE6 to 5 EE6 psi at 28 days.

## 5 STEEL REINFORCING

Since concrete is essentially incapable of resisting tension, steel reinforcing is used. Figure 14.2 shows typical reinforcing in concrete beams.

Straight bars resist flexural tension in the central part of the beam. Since the bending moment is smaller near the ends of the beam, less reinforcing is necessary. Some of the bar may be bent up to resist diagonal shear near

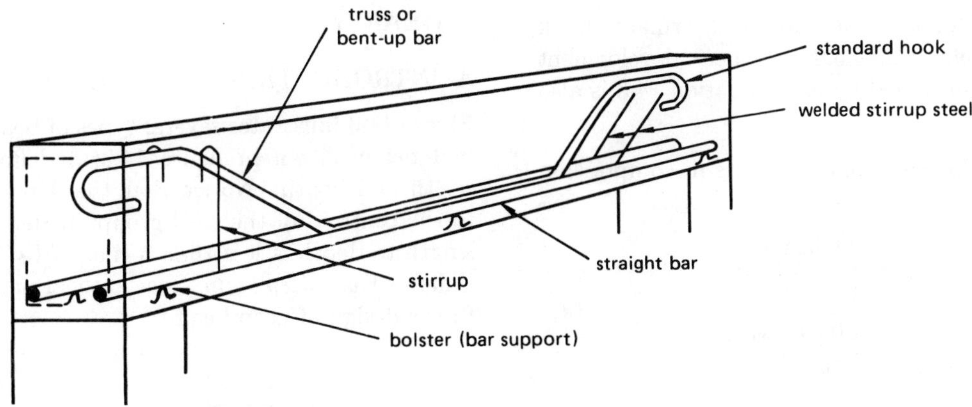


**Figure 14.2** Beam Reinforcement

the beam ends. (In continuous beams, the horizontal upper parts of the bent-up bars are continued on to the next span.)

*Stirrups* are used to resist the diagonal tension. These pass underneath the bottom steel for anchoring or are welded to the bottom steel.

The horizontal steel is supported on *bolsters* (*chairs*), of which there are a variety of designs.

Reinforcing steel, known as *rebar*, is available in a number of sizes, as well as in the form of wire for spiral wrapping, and wire mesh for shrinkage and thermal expansion control.

Steel grade is specified by its yield strength. Grade 60 is most common, although grade 40 was widely used in the past, and grades 50 and 80 are also available. However, not all sizes may be available in every grade.

**Table 14.3**
ASTM Standard Reinforcing Bars[6]

| size | weight (lb/ft) | diameter (in) | area (in²) | perimeter (in) |
|---|---|---|---|---|
| #2 | 0.167 | 0.250 | 0.05 | 0.786 |
| #3 | 0.376 | 0.375 | 0.11 | 1.178 |
| #4 | 0.668 | 0.500 | 0.20 | 1.571 |
| #5 | 1.043 | 0.625 | 0.31 | 1.963 |
| #6 | 1.502 | 0.750 | 0.44 | 2.356 |
| #7 | 2.044 | 0.875 | 0.60 | 2.749 |
| #8 | 2.670 | 1.000 | 0.79 | 3.142 |
| #9 | 3.400 | 1.128 | 1.00 | 3.544 |
| #10 | 4.303 | 1.270 | 1.27 | 3.990 |
| #11 | 5.313 | 1.410 | 1.56 | 4.430 |
| #14 | 7.65 | 1.693 | 2.25 | 5.32 |
| #18 | 13.60 | 2.257 | 4.00 | 7.09 |

[6] There is no difference between ACI and ASTM sizes and grades.

## 6 DEVELOPMENT LENGTH

The bond between the concrete and the reinforcement must be sufficient to keep the reinforcement from being pulled or pushed through the concrete. This is accomplished, in the ACI code, by specifying the minimum length of bar required to develop the full strength. This minimum length is known as the *development length.*

The *basic development length* for unhooked bars in tension with sizes up to and including #11 is[7]

$$l_d = \max\begin{cases} \dfrac{0.04A_b f_y}{\sqrt{f'_c}} \\ 0.0004 f_y d_b \\ 12'' \end{cases} \qquad 14.2$$

If at least 3″ of cover is provided, and if the spacing is 6″ or more, the basic development length may be multiplied by 0.8. If more steel reinforcing is used than is required, the development length may also be reduced.

$$l'_d = l_d\left(\frac{A_{st,\text{required}}}{A_{st,\text{actual}}}\right) \qquad 14.3$$

If standard hooks are used, the basic development length is

$$l_d = \max\begin{cases} \dfrac{1200 d_b f_y}{60{,}000\sqrt{f'_c}} \\ 8d_b \\ 6'' \end{cases} \qquad 14.4$$

[7] Different rules apply for larger bars. The basic development length given in equation 14.2 must be modified for use with top steel, lightweight concrete, and steel yield strengths in excess of 60,000 psi.

If the cover on the hook is at least $2\frac{1}{2}''$ (normal to hook plane) and $2''$ (behind the hook), the basic development length may be multiplied by 0.7. Equation 14.3 is also applicable.[8]

The basic development length for bars in compression is

$$l_d = \max \begin{cases} \dfrac{0.02 d_b f_y}{\sqrt{f'_c}} \\ 0.0003 d_b f_y \\ 8'' \end{cases} \qquad 14.5$$

Equation 14.3 applies. However, hooks are not considered effective in reducing development length for bars in compression.

## 7 ULTIMATE STRENGTH DESIGN

The ultimate strength design method is known as the *strength design method* in the ACI code. The actual (i.e., service) loads are increased by multiplicative safety factors. These *factored loads* are compared to the loads that would cause failure.[9]

The ultimate strength is also multiplied by a *capacity reduction factor*, $\phi$, to account for workmanship and understrength in the materials. Values of $\phi$ are given in table 14.4.

**Table 14.4**
Strength Reduction Factors

| type of stress | $\phi$ |
|---|---|
| flexure | 0.90 |
| axial tension | 0.90 |
| shear | 0.85 |
| torsion | 0.85 |
| axial compression with spiral reinforcement | 0.75 |
| axial compression with tied reinforcement | 0.70 |
| bearing on concrete | 0.70 |

## 8 BEAMS

### A. INTRODUCTION

Figure 14.3 illustrates several types of beams. The normal beam shown in figure 14.3(a) defines the beam width and depth. Notice that the depth is measured to the centroid of the steel group. If the ratio of beam length to depth is less than 5 (i.e., $l/d < 5$), the beam is defined as a *deep beam*. Special shear provisions apply to the design of deep beams.

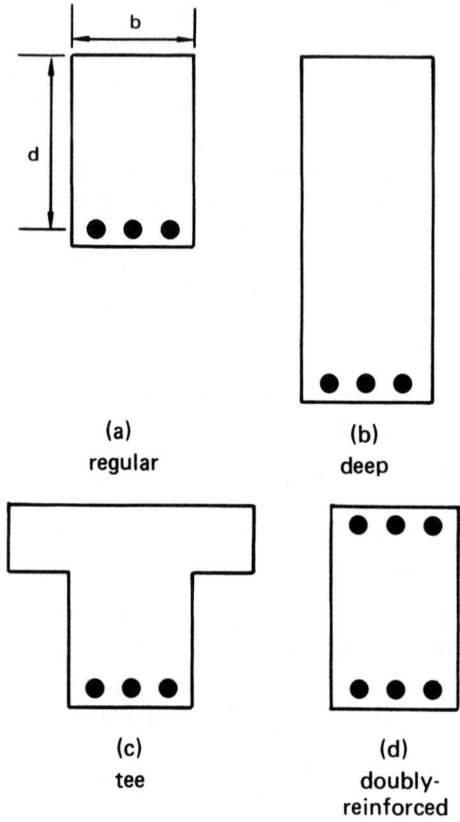


**Figure 14.3** Types of Beams

The *t-beam* shown in figure 14.3(c) is usually a portion of a larger monolithic slab-and-girder system. The *doubly-reinforced beam* shown in figure 14.3(d) is needed in continuous construction where the moment near supports is negative. Not shown are beams of prestressed concrete construction.

### B. FACTORED LOADS

The ultimate strength multipliers (i.e., *overload factors*) of 1.4 and 1.7 are used for dead and live loads, respectively. Thus, the factored moment and factored shear at a particular point are

$$M_u = 1.4M_D + 1.7M_L \qquad 14.6$$

$$V_u = 1.4V_D + 1.7V_L \qquad 14.7$$

[8] Other rules apply for hooks enclosed by ties or stirrups and lightweight concrete.

[9] Actually, the ACI code determines beam size and dimensions on the basis of ultimate strength, but calculates the moments on beams based on elastic theory. Redistribution of moments, which is taken into account in steel design, is not considered in concrete design.

## C. GENERAL PROVISIONS OF BEAM DESIGN

- The *reinforcement ratio*, $\rho$, is defined by equation 14.8.

$$\rho = \frac{A_{st}}{A_c} = \frac{A_{st}}{bd} \qquad 14.8$$

The ACI code places limits on the reinforcement ratio, as defined by equation 14.9 (ACI sections [10.3.3] and [10.5.1]).[10]

$$\frac{200}{f_y} \leq \rho \leq 0.75\rho_{\text{balanced}} \qquad 14.9$$

The *balanced reinforcement ratio* is given by equation 14.10.

$$\rho_{\text{balanced}} = \frac{(0.85)\beta_1 f'_c}{f_y}\left[\frac{87{,}000}{87{,}000 + f_y}\right] \qquad 14.10$$

The factor $\beta_1$ is defined by equations 14.11 and 14.12. $\beta_1$ may not be less than 0.65.

$$\beta_1 = 0.85 - 0.05\left(\frac{f'_c - 4000}{1000}\right) \quad 4000 < f'_c \leq 8000 \qquad 14.11$$

$$\beta_1 = 0.85 \quad f'_c \leq 4000 \qquad 14.12$$

- $b$ and $d$ can take on any reasonable values. However, small values of $d$ will produce small moments of inertia, leading to large deflections. Since large deflections produce cracks, most beams are designed to satisfy equation 14.13.[11]

$$1.75 < \frac{d}{b} < 2 \qquad 14.13$$

- All steel, including shear reinforcement, must be adequately covered by concrete. Appendix A lists the cover on reinforcing steel in detail. However, a minimum of $1\frac{1}{2}''$ of cover is generally required.

- The clear distance between bars is the maximum of one bar diameter or 1″.

- The minimum clear distance between layers is 1″.

## D. SPACING OF REINFORCEMENT IN BEAMS

Table 14.5 is easily derived from the clear spacing and depth of cover requirements of the ACI code.[12] The table assumes that #3 stirrups will be used, and cover is provided for them. If no stirrups are used, deduct 3/4″ from the table values. For additional bars horizontally, increase the beam width by adding the value in the last column.

## E. LOCATING THE NEUTRAL AXIS

Figure 14.4 shows a simple concrete beam. The neutral axis is located a distance $c$ down from the top of the beam.[13] To determine $c$, it is necessary to balance the areas above and below the neutral axis.

[10] Actually, it is possible to have flexural reinforcement ratios less than $200/f_y$ as long as the reinforcement provided is one-third more than is required for strength.

[11] However, this is not an ACI code requirement.

[12] See appendix A for spacing and cover requirements.

[13] With the older working stress method, it was common to refer to the distance $c$ as $kd$.

**Table 14.5**
Minimum Beam Widths
(with #3 stirrups)

| size of bars | number of bars in single layer of reinforcement: 2 | 3 | 4 | 5 | 6 | 7 | 8 | add for each added bar |
|---|---|---|---|---|---|---|---|---|
| #4 | 6.1 | 7.6 | 9.1 | 10.6 | 12.1 | 13.6 | 15.1 | 1.50 |
| #5 | 6.3 | 7.9 | 9.6 | 11.2 | 12.8 | 14.4 | 16.1 | 1.63 |
| #6 | 6.5 | 8.3 | 10.0 | 11.8 | 13.5 | 15.3 | 17.0 | 1.75 |
| #7 | 6.7 | 8.6 | 10.5 | 12.4 | 14.2 | 16.1 | 18.0 | 1.88 |
| #8 | 6.9 | 8.9 | 10.9 | 12.9 | 14.9 | 16.9 | 18.9 | 2.00 |
| #9 | 7.3 | 9.5 | 11.8 | 14.0 | 16.3 | 18.6 | 20.8 | 2.26 |
| #10 | 7.7 | 10.2 | 12.8 | 15.3 | 17.8 | 20.4 | 22.9 | 2.54 |
| #11 | 8.0 | 10.8 | 13.7 | 16.5 | 19.3 | 22.1 | 24.9 | 2.82 |
| #14 | 8.9 | 12.3 | 15.6 | 19.0 | 22.4 | 25.8 | 29.2 | 3.39 |
| #18 | 10.5 | 15.0 | 19.5 | 24.0 | 28.6 | 33.1 | 37.6 | 4.51 |

The "strength" of the concrete compressive block is $E_c cb$. The centroid of the compressive block is $c/2$ from the neutral axis.

Similarly, the "strength" of the steel tensile area is $E_{st}A_{st}$, and the steel's centroid is located $d-c$ from the neutral axis.

It is necessary to equate these two strengths and solve for $c$ to find the neutral axis.

$$E_c cb\left(\frac{c}{2}\right) = E_{st}A_{st}(d-c) \qquad 14.14$$

If the *modular ratio* is known, it can be incorporated into equation 14.14. $n$ is typically 8 for non-prestressed construction.

$$\frac{1}{2}bc^2 = nA_{st}(d-c) \qquad 14.15$$

$$n = \frac{E_{st}}{E_c} \qquad 14.16$$

As an alternative, equation 14.17 can be used to solve for $c$ directly.

$$c = \frac{nA_{st}}{b}\left[\sqrt{1+\frac{2bd}{nA_{st}}}-1\right] \qquad 14.17$$

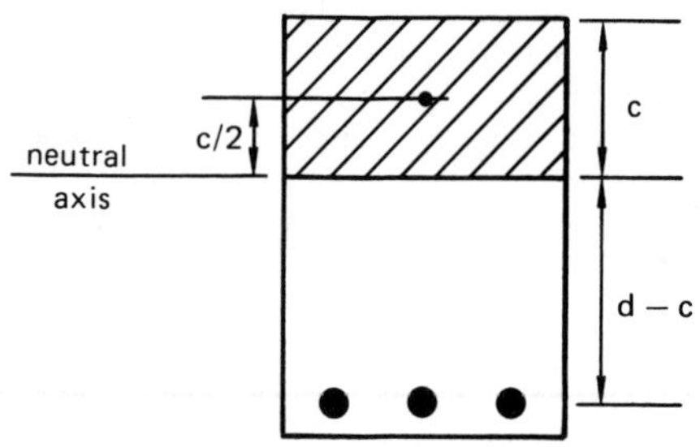


**Figure 14.4** The Neutral Axis

## F. INTRODUCTION TO STRENGTH THEORY[14]

Failure of concrete beams does not occur when steel reaches its yield point. Therefore, inelastic effects must be considered to predict the ultimate strength of beams. Since the location of the neutral axis shifts as the beam is stressed inelastically, much of the strength design theory is empirically derived.

The ACI code permits the assumption that the compressive stress distribution in the concrete is any shape that produces results in agreement with tests. It is commonly assumed that the distribution is rectangular with a value of $0.85f'_c$ at failure.[15]

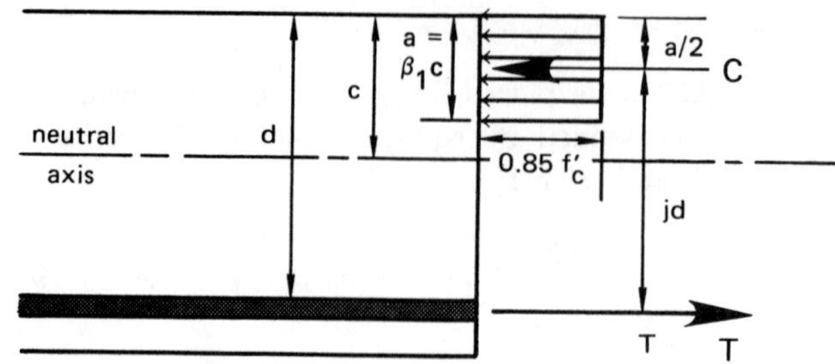


**Figure 14.5** Ultimate Strength Stress Distribution

In figure 14.5, the tensile and compressive forces are equal. It follows from $C = T$ that the height of the compressive block is

$$a = \frac{A_{st}f_y}{0.85f'_c b} = \frac{\rho f_y d}{0.85f'_c} \qquad 14.18$$

$$\rho = \frac{A_{st}}{A_c} = \frac{A_{st}}{bd} \qquad 14.19$$

The *nominal moment carrying ability* of the beam is

$$M_n = Tjd = Cjd \qquad 14.20$$

In equation 14.20, $jd$ is easily calculated from $d$ and $a$ (see figure 14.5). The steel is assumed to be stressed to yielding at failure. That is,

$$T = A_{st}f_y \qquad 14.21$$

$$C = 0.85f'_c ab \qquad 14.22$$

$$jd = d - \tfrac{1}{2}a \qquad 14.23$$

The *nominal strength* of the beam can be rewritten as

$$M_n = A_{st}f_y(d-\tfrac{1}{2}a) = 0.85f'_c ab(d-\tfrac{1}{2}a) \qquad 14.24$$

Substituting equation 14.18 into 14.24,

$$M_n = A_{st}f_y\left[d - 0.59\frac{A_{st}f_y}{f'_c b}\right]$$
$$= \rho bd^2 f_y\left[1-\frac{0.59\rho f_y}{f'_c}\right] \qquad 14.25$$

[14] T-beams and doubly-reinforced beams are not covered in this section.

[15] This is known as the *Whitney assumption*. The rectangular block is an approximation to a parabolic block that extends over the entire distance, $c$. Equations 14.11 and 14.12 give the factor $\beta_1$ used in figure 14.5.

The ACI code builds on this theory. It adds the *capacity reduction factor*, $\phi$, and restricts $\rho$.[16] (See equation 14.9.) The ultimate moment is calculated according to equation 14.26.

$$M_u = 1.4M_D + 1.7M_L \qquad 14.26$$

The strength requirement for flexure is

$$M_n \geq \frac{M_u}{\phi} = \frac{1.4M_D + 1.7M_L}{\phi} \qquad 14.27$$

*Example 14.3*

What is the maximum uniform live load that the beam shown can carry? The span is 120 inches long, and the ends are built-in. Use $f'_c = 3500$ psi and $f_y = 40{,}000$ psi. (The cross section of the beam shown is near the ends. Near the mid-span, steel will be required near the bottom.)

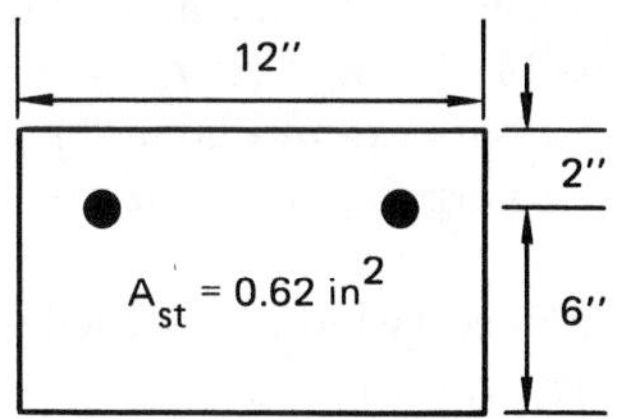


Near the ends, the beam is bent in reverse curvature, and $d = 6''$. From equation 14.19, the reinforcement ratio is

$$\rho = \frac{0.62}{(12)(6)} = 0.00861$$

From equation 14.18,

$$a = \frac{(0.00861)(40{,}000)(6)}{(0.85)(3500)} = 0.695 \text{ inch}$$

From equation 14.24, the nominal strength of the beam is

$$M_n = (0.62)(40{,}000)\left(6 - \frac{0.695}{2}\right)$$
$$= 140{,}200 \text{ in-lbf}$$

For a beam fixed at both ends, the maximum moment occurs at the ends.

$$M_{\text{max}} = \frac{wL^2}{12}$$

16 $\phi$ =0.90 for flexure.

Solving for the total factored unit loading,

$$w_{\text{total}} = \frac{12M_{\text{max}}}{L^2} = \frac{(12)(140{,}200)}{(120)^2} = 116.8 \text{ lbf/in}$$
$$= 1402 \text{ lbf/ft}$$

Assuming 150 pcf concrete and ignoring the steel weight, the dead load per foot of beam length is

$$w_D = \frac{(12)(8)(150)}{(144)(1)} = 100 \text{ lbf/ft}$$

From equation 14.6,

$$w_L = \frac{w_{\text{total}}\phi - 1.4w_D}{1.7}$$
$$= \frac{(1402)(0.90) - (1.4)(100)}{1.7} = 660 \text{ lbf/ft}$$

## G. STRENGTH DESIGN OF BEAMS

The following procedure can be used to design a rectangular beam with tension reinforcement (only).

*step 1*: Select a tension reinforcement ratio, $\rho$, according to equation 14.9. Cracking will be minimized by keeping $\rho$ to less than half the maximum allowable (i.e., 0.375 of $\rho_{\text{balanced}}$), particularly when grade 40 or higher steel is used. A general rule used to design economical beams is to try to satisfy equation 14.28.

$$\frac{\rho f_y}{f'_c} \approx 0.18 \text{ (beams only)} \qquad 14.28$$

*step 2*: Calculate the *coefficient of resistance*, $R_u$.

$$R_u = \rho f_y(1 - \tfrac{1}{2}\rho m) \qquad 14.29$$

$$m = \frac{f_y}{0.85f'_c} \qquad 14.30$$

*step 3*: Calculate the required value of $bd^2$ from equation 14.31. $M_u$ is the factored moment from equation 14.6. $\phi = 0.90$ for flexure. Be sure to use consistent units.

$$bd^2 = \frac{M_u}{\phi R_u} = \frac{M_n}{R_u} \qquad 14.31$$

*step 4*: Size the member so that $bd^2$ is approximately equal to the value calculated from equation 14.31. A good choice is to keep $d/b$ between 1.75 and 2.0. Size the beam to the nearest $\frac{1}{4}''$.[17]

*step 5*: If the actual $bd^2$ quantity is greatly different from the $bd^2$ value calculated in step 3, recalculate $\rho$.[18]

$$R_{u,\text{revised}} = \frac{M_n}{bd^2} = \frac{M_u}{\phi bd^2} \qquad 14.32$$

$$\rho_{\text{revised}} \approx \rho_{\text{old}} \left( \frac{R_{u,\text{revised}}}{R_{u,\text{old}}} \right) \qquad 14.33$$

*step 6*: Calculate the required steel area.

$$A_{st} = \rho_{\text{revised}} bd \qquad 14.34$$

*step 7*: Select the reinforcement steel bars to satisfy the distribution and placement requirements of the ACI code. Refer to tables 14.5 and 14.6.

[17] Actually, final beam dimensions ($b$ and $h$) are usually rounded to the nearest whole inch (slabs to the nearest half-inch). Regardless of the choice of $b$ and $d$, the section must ultimately be checked for excessive cracking and deflection.

[18] The relationship between $R_u$ and $\rho$ is linear only over small variations in either variable. However, it is not difficult to extract $\rho$ from equation 14.29 and obtain an exact value of the revised reinforcement ratio. See equation 14.142.

*step 8*: Specify the depth of concrete protection to satisfy the ACI code. (See Appendix A in this chapter.) Cover should be measured from the lowest stirrup surface. Note that concrete protection and cover depth, $d_c$, are different.

*step 9*: Design shear reinforcement (stirrups).

*step 10*: Verify the capacity of the beam. (See example 14.3.)

*step 11*: Check for cracking.

*step 12*: Check deflection.

*Example 14.4*

Design a rectangular beam with tension reinforcement (only) to carry service moments of 34,300 ft-lbf (dead) and 30,000 ft-lbf (live). Use $f'_c = 3500$ psi and $f_y = 40{,}000$ psi. #3 bars are available for shear reinforcing. (Do not design stirrup placement.)

*step 1*: From equation 14.10, the balanced reinforcement ratio is

$$\rho_{\text{balanced}} = \frac{(0.85)(0.85)(3500)}{40{,}000} \left( \frac{87{,}000}{87{,}000 + 40{,}000} \right)$$

$$= 0.0433$$

**Table 14.6**
Total Steel Areas for Various Numbers of Bars[19]

| bar size | nominal diameter (in.) | weight (lbm/ft) | number of bars 1 | 2 | 3 | 4 | 5 | 6 | 7 | 8 | 9 | 10 |
|---|---|---|---|---|---|---|---|---|---|---|---|---|
| #3 | 0.375 | 0.376 | 0.11 | 0.22 | 0.33 | 0.44 | 0.55 | 0.66 | 0.77 | 0.88 | 0.99 | 1.10 |
| #4 | 0.500 | 0.668 | 0.20 | 0.40 | 0.60 | 0.80 | 1.00 | 1.20 | 1.40 | 1.60 | 1.80 | 2.00 |
| #5 | 0.625 | 1.043 | 0.31 | 0.62 | 0.93 | 1.24 | 1.55 | 1.86 | 2.17 | 2.48 | 2.79 | 3.10 |
| #6 | 0.750 | 1.502 | 0.44 | 0.88 | 1.32 | 1.76 | 2.20 | 2.64 | 3.08 | 3.52 | 3.96 | 4.40 |
| #7 | 0.875 | 2.044 | 0.60 | 1.20 | 1.80 | 2.40 | 3.00 | 3.60 | 4.20 | 4.80 | 5.40 | 6.00 |
| #8 | 1.000 | 2.670 | 0.79 | 1.58 | 2.37 | 3.16 | 3.95 | 4.74 | 5.53 | 6.32 | 7.11 | 7.90 |
| #9 | 1.128 | 3.400 | 1.00 | 2.00 | 3.00 | 4.00 | 5.00 | 6.00 | 7.00 | 8.00 | 9.00 | 10.00 |
| #10 | 1.270 | 4.303 | 1.27 | 2.54 | 3.81 | 5.08 | 6.35 | 7.62 | 8.89 | 10.16 | 11.43 | 12.70 |
| #11 | 1.410 | 5.313 | 1.56 | 3.12 | 4.68 | 6.24 | 7.80 | 9.36 | 10.92 | 12.48 | 14.04 | 15.60 |
| #14 | 1.693 | 7.65 | 2.25 | 4.50 | 6.75 | 9.00 | 11.25 | 13.50 | 15.75 | 18.00 | 20.25 | 22.50 |
| #18 | 2.257 | 13.60 | 4.00 | 8.00 | 12.00 | 16.00 | 20.00 | 24.00 | 28.00 | 32.00 | 36.00 | 40.00 |

[19] #14 and #18 bars are typically used in columns only.

The maximum and minimum reinforcement ratios (from equation 14.9) are

$$\rho_{\text{max}} = 0.75 \times 0.0433 = 0.0325$$

$$\rho_{\text{min}} = \frac{200}{40{,}000} = 0.005$$

From equation 14.28, choose an approximate reinforcement ratio.

$$\rho = \frac{(0.18)(3500)}{40{,}000} = 0.0158 \quad \text{(ok)}$$

*step 2*: The coefficient of resistance is

$$m = \frac{40{,}000}{(0.85)(3500)} = 13.45$$

$$R_u = (0.0158)(40{,}000)\left(1 - \frac{(0.0158)(13.45)}{2}\right) = 564.8$$

*step 3*: The factored moment is

$$M_u = [(1.4)(34{,}300) + (1.7)(30{,}000)] \times 12 = 1.19\,\text{EE6 in lbf}$$

$$bd^2 = \frac{(1.19\,\text{EE6})}{(0.90)(564.8)} = 2341\ \text{in}^3$$

*step 4*: Choose a ratio of $d/b = 1.8$.

$$d = \sqrt[3]{\frac{d}{b}bd^2} = \sqrt[3]{(1.8)(2341)} = 16.15 \quad (\text{say } 16.25'')$$

$$b = \frac{16.15}{1.8} = 8.97 \quad (\text{say } 9'')$$

*step 5*: The revised reinforcement ratio will not be significantly different.

$$bd^2 = (9)(16.25)^2 = 2377$$

$$R_{u,\text{revised}} = \frac{1.19\,\text{EE6}}{(0.90)(2377)} = 556.3$$

$$\rho_{\text{revised}} = \frac{(0.0158)(556.3)}{564.8} = 0.0156$$

*step 6*: The required steel area is

$$A_{st} = (0.0156)(9)(16.25) = 2.28\ \text{in}^2$$

*step 7*: Select 3 #8 bars (2.37 in$^2$).

*step 8*: Use 1.5″ of concrete protection. The cover depth (using #3 stirrups and measuring from the center of the steel layer) is

$$d_c = \frac{1.00}{2} + 0.375 + 1.5 = 2.375$$

The total beam depth will be

$$h = d + d_c = 16.25 + 2.375 = 18.625 \quad (\text{say } 19'')$$

## H. CRACK CHECKING

When the ultimate strength design method is used with steels having a yield strength in excess of 40,000 psi, cracking can be a problem. The ACI code provides a simplified method of determining if cracking will be a significant factor in a beam's performance.[20]

The check is performed by calculating the limiting parameter, $z$, from equation 14.35. If $z$ exceeds 145 kips/in (exterior service) or 175 kips/in (interior service), cracking will be significant and the beam should be redesigned.

$$z = f_{st}\sqrt[3]{d_cA_e} = f_{st}\sqrt[3]{\frac{d_c(2d_{st}b)}{\#\text{ of bars}}} \qquad 14.35$$

$d_c$ is the *cover depth*—the distance from the center of the bottom steel layer to the bottom of the beam. For a beam with $1\frac{1}{2}''$ of cover over a stirrup, the cover depth would be

$$d_c = 1.5'' + d_{\text{stirrup}} + \tfrac{1}{2}d_b \qquad 14.36$$

$A_e$ in equation 14.35 is the effective concrete area in the tension block, divided by the total number of steel bars.

$f_{st}$ in equation 14.35 is the steel stress at the service level. It may be found in one of two ways. The first method is an obvious simplification.

$$f_{st} = 0.60f_y \qquad 14.37$$

If $z$ is acceptable using $f_{st}$ as defined in equation 14.37, no further checking will be required. However, if $z$ exceeds the limits specified, it may still be possible that the beam is adequate, and $f_{st}$ should be properly calculated.

[20] Actually, this method checks to see if the steel is sufficiently distributed throughout the tension region of the beam, rather than being clustered in one spot.

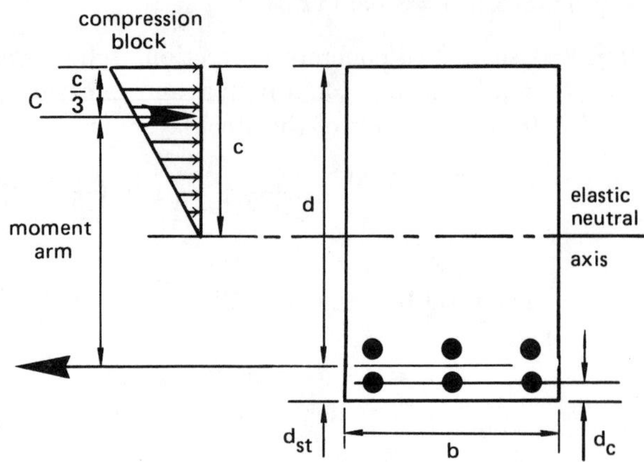


**Figure 14.6** Parameters in Crack Checking

In order to calculate the steel service stress accurately, it is necessary to know the location of the neutral axis. $M_w$ in equation 14.38 is the *service moment* (i.e., the actual moment on the beam at the location being investigated, not the factored moment).

$$f_{st} = \frac{M_w}{A_{st} \times \text{moment arm}} \quad 14.38$$

The moment arm in equation 14.38 is the distance between the tension and compression resultants. The moment arm can be derived from the location of the neutral axis.

$$\text{moment arm} = d - \frac{c}{3} \quad 14.39$$

If it is concluded that unacceptable cracking will occur, then smaller reinforcement bars must be used to redistribute the steel around a greater area.

## I. BEAM DEFLECTIONS

The ACI code controls beam deflections only in general terms. If deflections are not calculated at all, the limits on beam thickness in table 14.7 apply. If the deflections are calculated by appropriate methods, the limits in appendix A apply. These limits are presented as a fraction of the beam length between supports, and are designed to avoid the appearance of excessive deflection.

Beam deflections are calculated for concrete beams in the same way as for any beam—with the standard beam equations. Beam deflection equations contain two important parameters: $E$ and $I$. The modulus of elasticity, $E$, is the *secant modulus* specified by equation 14.1. The moment of inertia is calculated by considering only the portion of the beam which is uncracked and contributes to bending strength.

**Table 14.7**
Minimum Beam Thicknesses[21]

(deflections not computed)

| construction | minimum $h$ (fraction of span length) |
|---|---|
| simply supported | 1/16 |
| one end continuous | 1/18.5 |
| both ends continuous | 1/21 |
| cantilever | 1/8 |

The uncracked portion of the beam varies along the beam's length. Near supports in continuous beams, the cracks are above the neutral axis. Near beam centers, the cracks are below the neutral axis. At inflection points where there is no moment, there are no cracks. To circumvent the problem of deciding what the uncracked moment of inertia should be, the ACI codes specifies an *effective moment of inertia*, as defined by equation 14.40, to be used with the maximum moment and the standard beam deflection equations.[22]

$$I_e = \left(\frac{M_{cr}}{M_{\max}}\right)^3 I_g + \left[1 - \left(\frac{M_{cr}}{M_{\max}}\right)^3\right] I_{cr} \leq I_g \quad 14.40$$

$$M_{cr} = \frac{f_r I_g}{y_t} = 1.25bh^2\sqrt{f'_c} \quad 14.41$$

$$f_r = 7.5\sqrt{f'_c} \quad 14.42$$

The *cracked transformed moment of inertia*, $I_{cr}$, is the moment of inertia neglecting the concrete below the neutral axis of the service section, but including the transformed steel. As equation 14.40 indicates, $I_e$ must be less than the *gross moment of inertia*, $I_g$, ignoring reinforcing.

$$I_g = \frac{bh^3}{12} \quad 14.43$$

$M_{cr}$ is the *cracking moment.* $M_{\max}$ is the maximum unfactored service moment on the beam. $f_r$ is the *modulus of rupture.* $y_t$ is the distance from the neutral axis of the gross section to the extreme concrete fiber in tension. That is,

$$y_t = \frac{1}{2}h \quad 14.44$$

[21] The same limits apply to ribbed one-way slabs. However, different limits apply to solid one-way slabs.

[22] Refer to the ACI code if lightweight concrete is used.

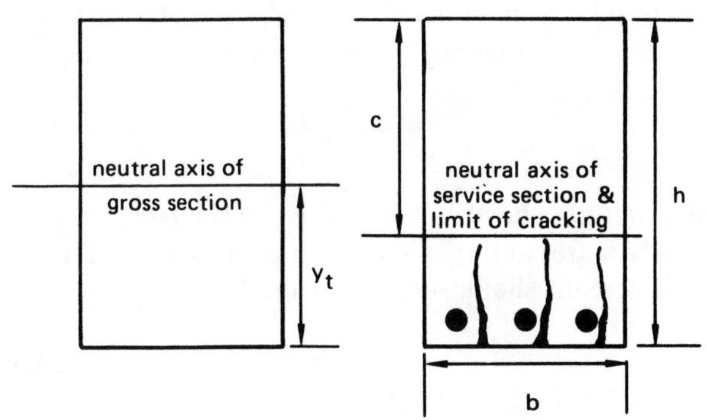


**Figure 14.7** Parameters for Calculating the Effective Moment of Inertia

The deflection calculated from $I_e$ and the secant modulus gives the *instantaneous deflection* (due to sustained loads). The additional *long-term deflection* can be calculated from equation 14.45.

$$\delta_{\text{long term}} = \lambda\delta_{\text{instantaneous}} \qquad 14.45$$

$$\lambda = \frac{\xi}{1+50\rho'} \qquad 14.46$$

$$\rho' = \frac{A_{st,\text{compression}}}{bd} \qquad 14.47$$

In equation 14.46, $\xi$ depends on the timespan being considered, and has values of 1.0 (3 months), 1.2 (6 months), 1.4 (12 months), and 2.0 (5 years or more). $\rho'$ is the reinforcement ratio for compression steel only.

The total deflection is the sum of the instantaneous and long term deflections.

*Example 14.5*

Determine the gross, cracked, and effective moments of inertia for the beam shown. The service moment is 100 ft-kips. Use $f'_c = 4000$ psi and $n = 8$.

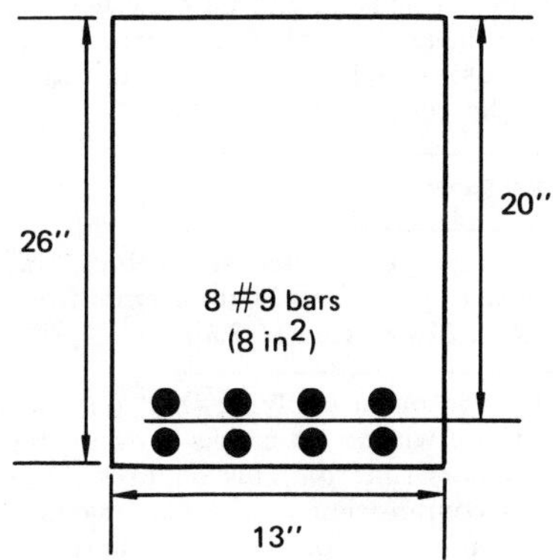


The gross moment of inertia is

$$I_g = \frac{(13)(26)^3}{12} = 19{,}041\ \text{in}^4$$

To find the cracked moment of inertia, it is necessary to locate the neutral axis. Tension cracks are expected below the neutral axis, and the concrete in the tension region should be disregarded when calculating the cracked moment of inertia. Equation 14.17 can be used to quickly locate the neutral axis.

$$\begin{aligned} c &= \frac{(8)(8)}{13}\left[\sqrt{1+\frac{(2)(13)(20)}{(8)(8)}}-1\right] \\ &= 9.95\ \text{in} \end{aligned}$$

From the parallel axis theorem with the 8 in$^2$ steel expanded by $n$ (8), the moment of inertia considering the compression concrete above the neutral axis and the transformed steel below the neutral axis is

$$\begin{aligned} I_{cr} &= \frac{(13)(9.95)^3}{3} + (8)(8)(20-9.95)^2 \\ &= 10{,}733\ \text{in}^4 \end{aligned}$$

The modulus of rupture is

$$f_r = 7.5\sqrt{4000} = 474\ \text{psi}$$

$y_t$ is half of the beam thickness, 13″. The cracking moment is

$$M_{cr} = \frac{(474)(19{,}041)}{(13)(12)(1000)} = 57.9\ \text{ft-kips}$$

$$\left(\frac{M_{cr}}{M_{\max}}\right)^3 = \left(\frac{57.9}{100}\right)^3 = 0.194$$

From equation 14.40, the effective moment of inertia to be used in calculating deflections is

$$\begin{aligned} I_e &= (0.194)(19{,}041) + [1-0.194]\,(10{,}733) \\ &= 12{,}345\ \text{in}^4 \end{aligned}$$

## J. SHEAR REINFORCEMENT IN BEAMS

Consider a simply-supported, single-span beam. The moment will be close to zero near the supports, while the shear will be maximum. Where the moment is zero, the flexural stress is also zero. Near the beam ends, a diagonal tension stress equal to the shear stress and inclined at 45° exists.[23] The stress on the beam at those points is in a state of nearly pure shear (tension).

[23] This is easily shown with the combined stress equation using $\sigma_x = 0$ and $\sigma_y = 0$.

Figure 14.8 shows a concrete beam with cracks that have formed below the neutral axis due to the diagonal shear. To reduce crack propagation, *shear reinforcement* in the form of stirrups attached to the longitudinal reinforcement and extending into the compression region are used.[24] This reinforcement is also known as *web reinforcement* and *diagonal reinforcement.*

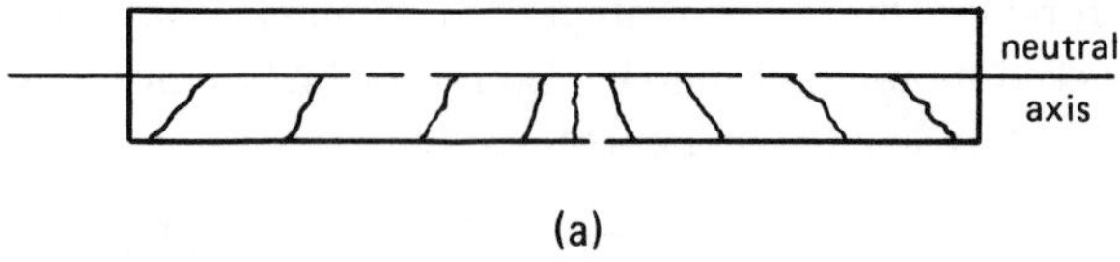


(a)

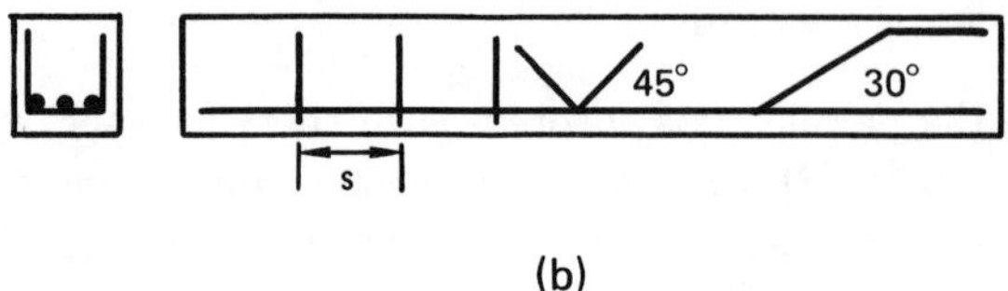


(b)

**Figure 14.8** Shear Cracking and Reinforcement

Shear reinforcement can take on several forms. Vertical stirrups are used most often, but stirrups inclined as little as 45° from the horizontal are occasionally seen, as is the use of welded wire fabric. Longitudinal steel that is bent up at 30° or more and enters the compression zone also contributes to shear reinforcement.[25]

If stirrups in the shape of a U are used, each bent stirrup contributes twice the stirrup bar area.

For the purpose of designing shear reinforcement, the live load should be considered variable in location, and should be moved to positions which will maximize the shear on the beam. An approximate *shear envelope* can be drawn. This envelope is a straight line drawn between the maximum possible shear at the midpoint of the support (determined by placing the live load to maximize the support reaction) and the maximum possible shear at midspan (determined by placing the live load to maximize the midspan shear).[26] Influence diagrams are valuable in determining the proper placement. Of course, the dead load shear must be added to the live load shear envelope.

[24] Shear reinforcement does not prevent crack formation. It is used to limit the length of cracks.

[25] When inclined stirrups and bent longitudinal bars are used, every 45° line extending downward from depth $\frac{1}{2}$d to the tension reinforcement must be crossed by at least one line of that reinforcement. However, only the center three-fourths of the inclined sections of longitudinal steel bars are considered effective in resisting shear.

[26] With a uniform live load, the maximum midspan shear typically occurs when only one-half of the beam is loaded.

Concrete has a nominal amount of shear strength of its own, even when no shear reinforcement is present.[27]

$$V_c = 2\sqrt{f'_c}\, bd \qquad 14.48$$

Shear reinforcement is required when the factored shear exceeds a fraction of this nominal strength. That is, reinforcement should be provided when[28]

$$V_u \geq \phi V_c \qquad 14.49$$

Reinforcement is used to make up the difference between $V_u$ and $\phi V_c$. The nominal shear strength to be provided by the steel is

$$V_{st} = V_n - V_c = \frac{V_u}{\phi} - V_c \qquad 14.50$$

If the required shear reinforcement is too great, the beam should be redesigned. The maximum shear reinforcement is

$$V_{st,\mathrm{max}} = 8\sqrt{f'_c}\, bd \qquad 14.51$$

A minimum amount of shear reinforcement is also required when[29]

$$\phi V_c > V_u > \tfrac{1}{2}\phi V_c \qquad 14.52$$

This minimum shear reinforcement is [ACI 318 formula 11-14]

$$A_v = \frac{50bs}{f_y} \qquad 14.53$$

Equation 14.53 implies that reinforcement must supply a minimum of 50 psi of shear strength. The corresponding shear contribution is

$$V_{st} = 50bd \qquad 14.54$$

To satisfy equation 14.49, checking for critical shear can begin a distance $d$ from the support face.[30] Whatever

[27] The ACI code contains an alternate method of calculating the nominal concrete shear strength. However, this method is usually only used when justifying larger spacings at critical sections than would normally be permitted otherwise.

[28] $\phi = 0.85$ for shear.

[29] **Thin slab-like members, such as footings, slabs, and beams with a total depth of less than 10" are exempt from this minimum requirement. [ACI 318 section 11.5.5]**

[30] Care must be taken in applying this concession. The implication is that there will be no cracks in the compression region closer to the support than *d*. This requires the reaction on the beam to induce compression in the end regions of the beams. Not all beam supports do so. Beams supported from above are particularly suspect.

shear exists at that location can be used to design the shear reinforcement and spacing between the support and that point.[31]

For nonprestressed members, the maximum spacing between stirrups is[32]

$$s = \min\begin{Bmatrix} 24'' \\ d/2 \end{Bmatrix} \qquad 14.55$$

The maximum spacing given by equation 14.55 is to be reduced by one-half when the shear reinforcement requirement is greater than twice the nominal concrete shear strength. Specifically, when

$$V_{st} > 4\sqrt{f'_c}bd \qquad 14.56$$

The actual spacing required decreases towards midspan. However, spacing is not varied continuously. Groups of stirrups should be evenly spaced based on the maximum shear within the region covered by that spacing. Judgment is used to determine when the spacing should be decreased.

If $A_v$ is known, equation 14.53 can be used to calculate the spacing.

For shear reinforcement perpendicular to the beam's axis and with area $A_v$ within the spacing distance $s$, the shear strength contribution is given by equation 14.57.[33]

$$V_{st} = \frac{A_v f_y d}{s} \qquad 14.57$$

For straight stirrups inclined at an angle $\theta$ from the horizontal, the shear strength contribution is

$$V_{st} = \frac{A_v f_y (\sin\theta + \cos\theta)d}{s} \qquad 14.58$$

If shear reinforcement is provided by a single bar or a single group of parallel bars, all bent up at the same distance from the support, the shear strength contribution is

$$V_{st} = A_v f_y \sin\theta \leq 3\sqrt{f'_c}bd \qquad 14.59$$

## K. ANCHORAGE OF SHEAR REINFORCEMENT

In order to develop its full tensile capacity, shear reinforcement must be anchored at both ends.[34] In a U-shaped stirrup, one end of the shear reinforcement is bent around the longitudinal steel, anchoring that end. The other end may be considered securely anchored by mere bonding with the concrete if the stirrup is sufficiently long and extends past the middepth of the member. In equation 14.60, $l_d$ is the standard *development length* for deformed bars.

$$\begin{matrix}\text{extension past} \\ \text{middepth}\end{matrix} = \min\begin{Bmatrix} l_d \\ 24d_b \\ 12'' \end{Bmatrix} \qquad 14.60$$

If the extension past middepth is at least $l_d/2$ but not the full $l_d$, a standard hook at the upper stirrup end can be used to make up the difference.[35] For #5 and smaller stirrups with $f_y$ greater than 40,000 psi, an extension of as little as $0.33l_d$ is allowed if the stirrup is bent at least 135° around longitudinal reinforcement in the compression region. [ACI 12.13.2.3]

Figure 14.9 illustrates these three cases.

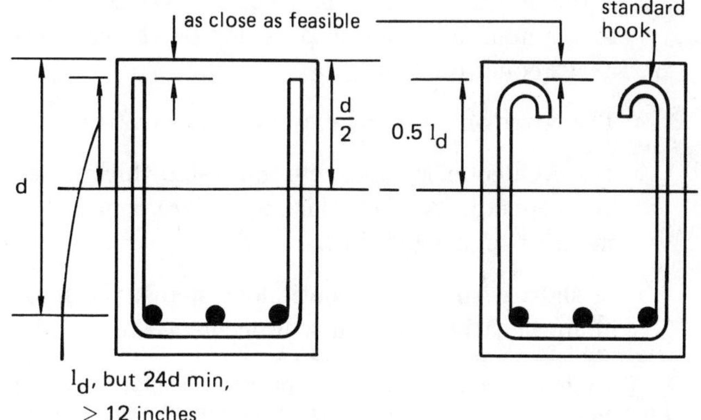


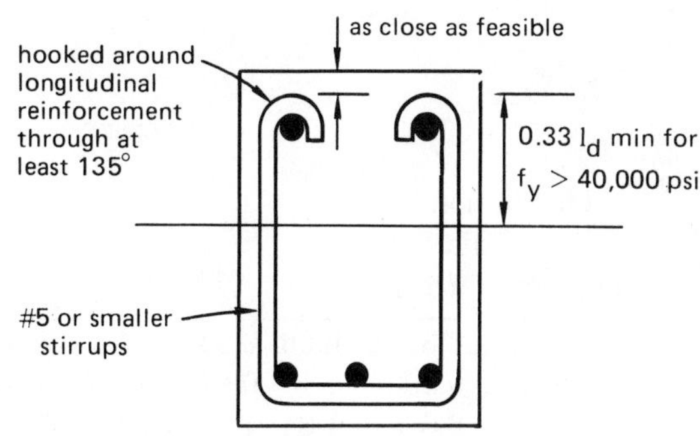


**Figure 14.9** Anchorage Methods for Shear Reinforcement

[31] It is common to place the first stirrup a distance $d/2$ from the face of the support. However, the first stirrup can be placed a full space from the face.

[32] Though not specified by the code, the minimum practical spacing is about 3″. If stirrups need to be closer than 3″, a larger bar size should be used.

[33] The yield strength of shear reinforcement should not exceed 60,000 psi.

[34] The ACI code requires that the shear reinforcement extend into the compression and tension regions as far as cover requirements and proximity to other bars permit.

[35] Wrapping around a longitudinal bar in the compression region is optional.

## 9 ONE-WAY FLOOR SLABS

A floor slab is typically supported on all four sides. If the slab length is more than twice the slab width, it can be assumed that the primary support will be from the beams along the length of the slab, and the slab is known as a *one-way slab*.[36]

One-way slabs can be designed and evaluated as beams. The beam width, $b$, is taken as 12″, which fixes one of the dimensions. The ultimate moment is computed per foot of slab width. Table 14.8 is used to determine the steel per foot of slab width.

If table 14.8 doesn't cover a specific application, equation 14.61 can be used to calculate the average steel per foot of width.

$$A_{st} = \frac{12A_b}{s} \qquad 14.61$$

Although slabs are designed as beams (the procedure given for beams can be used), there are some important differences. Important ACI code requirements and general design guidelines are presented here.

- Due to thickness limitations, shear reinforcement is seldom used. Therefore, shear stress is limited to the nominal strength provided by the concrete itself (equation 14.48).
- The cover on the steel may be as low as 3/4″.
- The ACI code specifies minimum slab thicknesses. See appendix A. Slab thicknesses are rounded to the next higher half-inch.
- In slabs with several continuous spans, the maximum negative moment will occur at the exterior face of the first interior support. This moment should be used to calculate the slab thickness.
- Span length is measured from the centers of supports for the purpose of determining fixed-end moments and stiffness factors. However, moment and shear coefficients may be used to design slabs, in which case the clear span (distance between faces of integral supporting beams) is used to calculate the moment on the slab.
- To limit deflection, the reinforcement ratio should be approximately one-half of the maximum allowed. That is,

$$\rho \approx 0.375\rho_{\text{balanced}} \qquad 14.62$$

The effective moment of inertia (equation 14.40) is used to calculate deflection.

- Development length may be a factor in parts of the reinforcement design.
- Where the slab is supported by a beam, some of the reinforcement is bent up to carry tension in the upper half of the slab. (For thin slabs, under 5″, straight bars are used.)
- The minimum bar size used is commonly #4.
- Shrinkage and temperature reinforcement perpendicular to the main reinforcement is required. For slabs reinforced in one direction only, the minimum reinforcement ratio for shrinkage and temperature control is 0.0014 (up to but excluding grade 40 steel), 0.0020 (grade 40), and 0.0018 (grade 60).[37] The maximum steel spacing is

$$s = \min\left\{\begin{matrix} 18'' \\ 5 \times t_{\text{slab}} \end{matrix}\right\} \qquad 14.63$$

[36] The primary reinforcement also runs perpendicular to the long direction, in one direction. However, this is not why the slabs are called *one-way*.

[37] The code has other provisions for higher grade steel as well.

**Table 14.8**
Average Steel Area per Foot of Width

| bar size number | nominal diameter (in.) | spacing of bars in inches: 2 | 2½ | 3 | 3½ | 4 | 4½ | 5 | 5½ | 6 | 7 | 8 | 9 | 10 | 12 |
|---|---|---|---|---|---|---|---|---|---|---|---|---|---|---|---|
| 3 | 0.375 | 0.66 | 0.53 | 0.44 | 0.38 | 0.33 | 0.29 | 0.26 | 0.24 | 0.22 | 0.19 | 0.17 | 0.15 | 0.13 | 0.11 |
| 4 | 0.500 | 1.18 | 0.94 | 0.78 | 0.67 | 0.59 | 0.52 | 0.47 | 0.43 | 0.39 | 0.34 | 0.29 | 0.26 | 0.24 | 0.20 |
| 5 | 0.625 | 1.84 | 1.47 | 1.23 | 1.05 | 0.92 | 0.82 | 0.74 | 0.67 | 0.61 | 0.53 | 0.46 | 0.41 | 0.37 | 0.31 |
| 6 | 0.750 | 2.65 | 2.12 | 1.77 | 1.51 | 1.32 | 1.18 | 1.06 | 0.96 | 0.88 | 0.76 | 0.66 | 0.59 | 0.53 | 0.44 |
| 7 | 0.875 | 3.61 | 2.88 | 2.40 | 2.06 | 1.80 | 1.60 | 1.44 | 1.31 | 1.20 | 1.03 | 0.90 | 0.80 | 0.72 | 0.60 |
| 8 | 1.000 | | 3.77 | 3.14 | 2.69 | 2.36 | 2.09 | 1.88 | 1.71 | 1.57 | 1.35 | 1.18 | 1.05 | 0.94 | 0.78 |
| 9 | 1.128 | | 4.80 | 4.00 | 3.43 | 3.00 | 2.67 | 2.40 | 2.18 | 2.00 | 1.71 | 1.50 | 1.33 | 1.20 | 1.00 |
| 10 | 1.270 | | | 5.06 | 4.34 | 3.80 | 3.37 | 3.04 | 2.76 | 2.53 | 2.17 | 1.89 | 1.69 | 1.52 | 1.27 |
| 11 | 1.410 | | | 6.25 | 5.36 | 4.69 | 4.17 | 3.75 | 3.41 | 3.12 | 2.68 | 2.34 | 2.08 | 1.87 | 1.56 |

## 10 T-BEAMS

### A. INTRODUCTION

Although a t-shaped beam can be constructed individually, it usually occurs as part of a monolithic beam-slab system. Since the flexural stresses in the beam "flange" are due to slab action, the slab flexural stresses are at right angles to the beam flexural stresses. Any interaction between the two stress distributions is ignored, and the slab (thickness) design and t-beam design are independent.

Equation 14.64 can be used to determine if a beam is rectangular. This equation contains several approximations (i.e., $\beta_1 = 0.85$ and $d - \frac{1}{2}a = 0.90d$), but is sufficiently accurate for most purposes.[38] The beam is rectangular if

$$\frac{t}{d} \geq \frac{1.71M_u}{f'_c b_e d^2} \qquad 14.64$$

$M_u$ must be expressed in consistent (inch) units in equation 14.64.

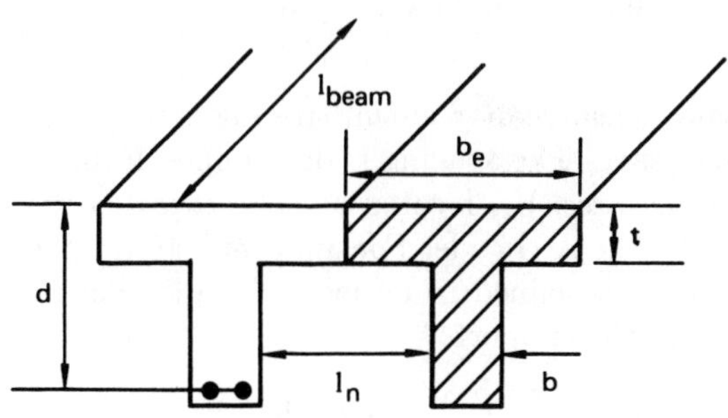


**Figure 14.10** T-Beam Dimensions

What appears to be a t-beam may not be a t-beam. To be a t-beam, the neutral axis must fall within the web, placing part of the web in compression. If the neutral axis falls within the flange (i.e., the slab), none of the web is in compression. Since concrete in tension is ignored in the calculation of moment of inertia, and since only the steel-to-neutral axis distance is relevant, the beam is not t-shaped.[39]

On the other hand, beams that appear to be odd-shaped may, indeed, be t-shaped. Figure 14.11 shows an I-beam shape that is actually a t-beam. Since all of the concrete below the neutral axis is in tension, only the steel-to-neutral axis distance is relevant in the design.

[38] The worst case occurs when the neutral axis is near the bottom of the flange. However, in that case, rectangular beam and t-team equations produce the same design.

[39] The beam should be designed as a rectangular beam in this instance.

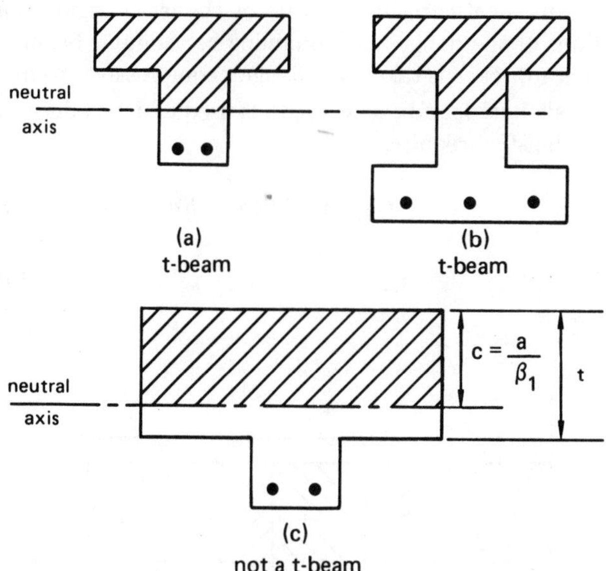


**Figure 14.11** T-Beams and Non-T-Beams

### B. FLANGE WIDTH LIMITATIONS

Although it is assumed that the stress is uniform across the entire flange width, the stress actually decreases at great distances from the web. Therefore, the ACI code limits the width of flanges for the purpose of calculating properties of the shape. The effective width is[40]

$$b_e \leq \min \begin{Bmatrix} \frac{1}{4}l_{\text{beam}} \\ b + 16t \\ b + l_n \end{Bmatrix} \qquad 14.65$$

### C. T-BEAM DESIGN AND ANALYSIS

Design of T-beams is similar, in many respects, to the design of rectangular beams. The moment at ultimate strength (see figure 14.10) is

$$M_u = \phi f_y A_{st} \left(d - \tfrac{1}{2}a\right) \qquad 14.66$$

If $c = \dfrac{a}{\beta_1}$ is less than $t$, the neutral axis falls within the flange, and the beam is rectangular. If $c$ is greater than $t$, the neutral axis falls within the web, and the beam is T-shaped. Ignoring the small compressive strength contributed by the web, the strength of the beam is

$$M_u = \phi f_y A_{st}(d - \tfrac{1}{2}t) \qquad 14.67$$

[40] L-shaped beams, such as would occur at the edges of a beam-slab system, and isolated t-beams are governed by different limitations.

In designing a t-beam, it may be desirable to separate the moment-carrying abilities of the overhanging portions of the flange and remaining rectangular beam. If this approach is taken, the flange's compressive strength is calculated, and this strength is used to determine part of the steel required.

$$C_f = 0.85 f'_c (b_e - b) t \qquad 14.68$$

$$A_{st,f} = \frac{C_f}{f_y} \qquad 14.69$$

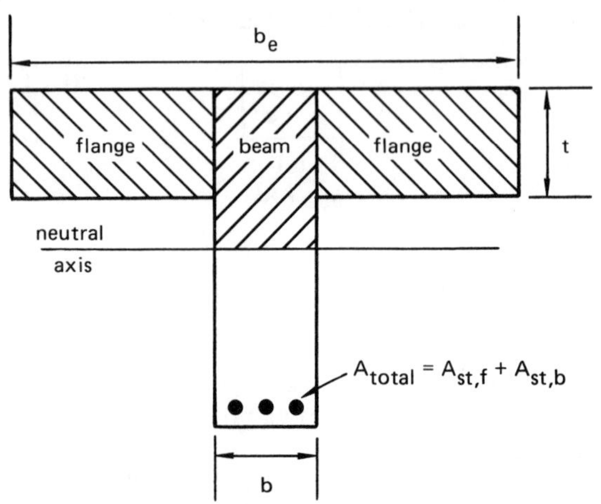


**Figure 14.12** Dual Beam Approach to T-Beam Design

Next, the design moment which the flange is capable of carrying alone is determined.

$$\phi M_{u,f} = \phi [A_{st,f} f_y (d - \tfrac{1}{2} t)] \qquad 14.70$$

The "beam" must carry the remaining design load.

$$M_{u,b} = M_u - \phi M_{u,f} \qquad 14.71$$

The remainder of the design procedure is the same as for any other rectangular beam, including that of determining the location of the neutral axis, height of stress block, and steel requirement, $A_{st,b}$.

The total steel required is

$$A_{\text{total}} = A_{st,f} + A_{st,b} \qquad 14.72$$

Additional steps required are: checking the reinforcement quantity against the maximum permitted, checking for excessive cracking, and checking required web width to meet cover requirements.

## 11 DEEP BEAMS

Deep beams are defined by the ACI code [section 10.7.1] to have depth-to-clear span $(d/l_n)$ ratios greater than 0.4 for continuous spans and 0.8 for simple spans.[41] Because of the expected orientation of cracks in deep beams, shear reinforcement is required horizontally as well as vertically.

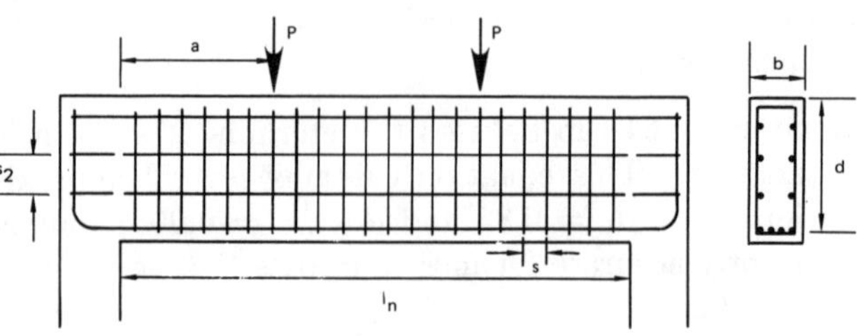


**Figure 14.13** A Deep Beam

The concrete's shear strength is the same as for normal-depth beams.[42]

$$V_c = 2\sqrt{f'_c}\, bd \qquad 14.73$$

The same shear reinforcement spacing is used throughout the beam. The spacing is determined from the *design shear force*, $V_u$, located at the *critical section*, as defined by the ACI code. For uniform loading, the critical section is defined to be located a distance $l_c$ from the face of the support.

$$l_c = \min \begin{Bmatrix} 0.15 l_n \\ d \end{Bmatrix} \qquad 14.74$$

For concentrated loading, the location of the critical section is given by equation 14.75. In equation 14.75, $a$ is the *shear span*, the distance from the face of the support to the closest concentrated load.

$$l_c = \min \begin{Bmatrix} 0.50 a \\ d \end{Bmatrix} \qquad 14.75$$

If $V_n$ is greater than $\frac{1}{2} V_c$, the ACI code requires a minimum amount of shear reinforcement. (Below $\frac{1}{2} V_c$, no shear reinforcement is needed.) Two types of reinforcement are used: vertical and horizontal. The following

[41] Actually, special shear reinforcement is required by the code if $l_n/d < 5$ $\left(d/l_n > 0.2\right)$ and the beam is loaded on its compression face. This limit effectively determines when a beam is *deep*. [ACI 318 section 11.8.1]

[42] The ACI code contains a more detailed method of calculating $V_c$ for deep beams, as it does for regular depth beams. However, these two detailed methods are not the same. A special *multiplier on the concrete shear strength* of normal depth beams is allowed. Use of the more detailed procedure is optional. However, its use will more than likely reduce the amount of shear reinforcement required.

equations give the minimum reinforcement areas and the maximum spacing. In these equations, $b$ is the width of the beam at the compression face.

Tension steel (i.e., flexural steel) for deep beams is designed the same way as for normal depth beams.

A deep beam's nominal shear strength is computed the same as for normal depth beams.

$$V_n = V_c + V_{st} = \frac{V_u}{\phi} \qquad 14.76$$

However, the maximum shear strength attributable to a deep beam is

$$V_n \leq 8\sqrt{f'_c}bd \quad (l_n/d \leq 2) \qquad 14.77$$

$$V_n \leq \frac{2}{3}\left(10 + \frac{l_n}{d}\right)\sqrt{f'_c}\,bd \quad (5 > l_n/d > 2) \qquad 14.78$$

$$A_v \geq 0.0015bs \qquad 14.79$$

$$s \leq \min\left\{\begin{array}{c} d/5 \\ 18'' \end{array}\right\} \qquad 14.80$$

$$A_{vh} \geq 0.0025bs_2 \qquad 14.81$$

$$s_2 \leq \min\left\{\begin{array}{c} d/3 \\ 18'' \end{array}\right\} \qquad 14.82$$

If $V_n$ is greater than $V_c$ (i.e., $V_u > \phi V_c$), equation 14.83 is used to calculate the strength contribution of the shear reinforcement. Since there are two areas, $A_v$ and $A_{vh}$, different reinforcement arrangements are possible.

$$V_{st} = \left[\frac{A_v}{s}\left(\frac{1 + \frac{l_n}{d}}{12}\right) + \frac{A_{vh}}{s_2}\left(\frac{11 - \frac{l_n}{d}}{12}\right)\right] f_y d \qquad 14.83$$

As in equations 14.79 and 14.81, $A_v$ and $A_{vh}$ in equation 14.83 are the areas of reinforcement within spans of $s$ and $s_2$ respectively.

## 12 DOUBLY-REINFORCED BEAMS

### A. INTRODUCTION

Figure 14.14 illustrates a beam which has steel in the compression region of the beam. Such steel can be used to increase the design strength of the beam or to reduce deflection. However, it is frequently used near the ends of beams where the expected moment is negative. Without compression reinforcement, the upper portion of the beam would be in tension, which is not acceptable.[43]

In addition, top steel is advisable in construction designed to resist earthquakes, due to the stress reversals expected with oscillatory motion.

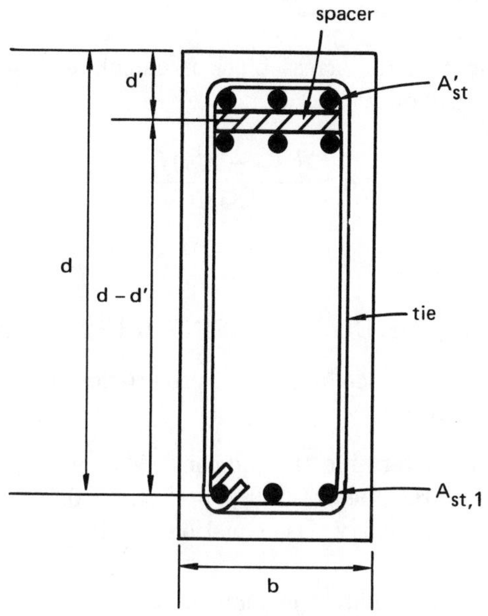


**Figure 14.14** Doubly-Reinforced Beam

In a doubly-reinforced beam, the total steel is made up of two components: the *compression steel* (or *top steel*) and the *tension steel.*

$$A_{st} = A_{st,1} + A'_{st} \qquad 14.84$$

The need for top steel is easily determined. The maximum reinforcement ratio for tension steel is [ACI 10.3.3]

$$\rho_{\text{max}} = 0.75\rho_{\text{balanced}} \qquad 14.85$$

The moment that can be resisted with the quantity of tension (top) steel is

$$M_n = \rho bdf_y\left[d - \frac{0.59\rho df_y}{f'_c}\right] \qquad 14.86$$

If $M_u > \phi M_n$, compression steel is needed to carry the difference.

[43] In monolithic slabs and beams, the beam web (not the slab "flange") would be in compression where the moment is negative. Depending on the reinforcement ratio, this might result in compression failure in the web before tensile failure in the slab occurs. In any case, top steel is the remedy.

B. CAPACITY OF DOUBLY-REINFORCED BEAMS

The ultimate moment which a doubly reinforced beam can carry is

$$\begin{aligned} M_u &= \phi M_n \\ &= \phi[(A_{st} - A'_{st})(f_y)\left(d - \tfrac{1}{2}a\right) \\ &\quad + A'_{st} f_y (d - d')] \end{aligned} \quad 14.87$$

The height of the stress block is calculated using the tension steel only.

$$a = \frac{(A_{st} - A'_{st}) f_y}{0.85 f'_c b} \quad 14.88$$

C. DESIGN OF DOUBLY-REINFORCED BEAMS

The following procedure can be used to design a doubly-reinforced beam.

*step 1:* Determine the moment, $M_{nc}$, which a concrete beam with $\rho = \rho_{\text{max}} = 0.75\rho_{\text{balanced}}$ can carry. Use equation 14.31.

*step 2:* Calculate the moment which the compression reinforcement must carry.

$$M'_u = M_u - \phi M_{nc} \quad 14.89$$

*step 3:* Assume compression reinforcement yields before the concrete.[44] The compression reinforcement ratio is

$$\rho' = \frac{M'_u}{\phi f_y (d - d') bd} \quad 14.90$$

*step 4:* The total reinforcement ratio is[45]

$$\rho = 0.75\rho_{\text{balanced}} + \rho' \quad 14.91$$

*step 5:* Calculate the steel areas.

$$A'_{st} = \rho' bd \quad 14.92$$

$$A_{st,1} = (\rho - \rho')bd = 0.75\rho_{\text{balanced}} bd \quad 14.93$$

*step 6:* Check the assumption that the compression reinforcement yields. Equation 14.94 must be satisfied.[46]

$$\frac{A_{st} - A'_{st}}{bd} \geq \frac{0.85\beta_1 f'_c d'}{f_y d}\left[\frac{87{,}000}{87{,}000 - f_y}\right] \quad 14.94$$

*step 7:* Select the reinforcement to meet the $A'_{st}$ and $A_{st,1}$ requirements.

*step 8:* Complete the design by checking for flexural cracking (the tension steel only), specifying depth of cover, checking beam width, and designing shear reinforcement.

## 13 PRESTRESSED AND POST-TENSIONED BEAM DESIGN

A. INTRODUCTION

Having a compressive stress in the tension region of a beam prior to normal loading produces a stiffer beam (i.e., there is less deflection). Alternatively, the decrease in expected deflection can be used to design smaller sections, thereby reducing both cost and dead weight. Most or all of the cracks at service load levels are eliminated, which is particularly valuable in corrosive environments. The finished beam also has greater impact and live loading capacities.

The disadvantages of prestressed and post-tensioned construction are higher cost and increased inspection. Materials, accessories, and labor costs are higher than for traditional construction. The close inspection and stricter quality control also add to the cost.

The basic approach is to use steel tendons which pass through the beam. Two methods are used to apply tension to the steel: pretensioning and post-tensioning.

In *pretensioned construction*, a form is built between two anchors, and steel is tensioned and secured to the two anchors. When the concrete poured into the form cures, it bonds to the steel. When the ends of the steel are cut from the anchors, the load transfers to the concrete.

In *post-tensioned construction*, the forms are built with hollow metal or plastic tubes running the length of the beam. (Alternatively, the steel tendons can be coated with grease or another release agent to prevent bonding.) The concrete is poured and allowed to cure. If

[44] This is known as *condition 1.* In condition 1, both the compression and tension steels yield before the concrete fails (i.e., the concrete strain exceeds 0.003). In *condition 2*, the tension steel yields, but the compression steel does not yield before the concrete fails.

[45] The ACI code specifies that, for members with compression reinforcement, the portion of $\rho_{\text{balanced}}$ contributed by compression reinforcement does not need to be reduced by the 0.75 factor.

[46] If equation 14.94 is not satisfied, there will be a condition 2 failure. The stress in the steel at failure will be $\epsilon'_{st} E_{st}$, where $\epsilon'_{st}$ can be found by elastic assumptions. The completion of condition 2 calculations is not covered in this chapter.

a tendon/tube assembly wasn't used, a steel tendon is placed in the tube. The tendon is tensioned to apply the prestress. Finally, the tubes are either filled with pumped grout, or the ends of the tendon are anchored in some manner.

In both pretensioned and post-tensioned construction, the tendons are draped to reduce their eccentricity near the beam ends. At the beam ends, the dead load moment is insufficient to counteract the tension prestress. Reducing the eccentricity eliminates tension in the concrete.

Several effects tend to reduce the effective prestress, including creep in steel and concrete, concrete shrinkage, elastic compression of the concrete, anchorage seating movement, and friction between the tendon and tube. These losses are significant, averaging 15% to 20%. Since this loss amounts to 25,000 or 35,000 psi, regular grade steel with a yield strength of 40,000 psi (approximately) cannot be used for prestressing. The tendons which are used for prestressing are manufactured from steel with an approximate ultimate strength ($f_{pu}$) of 250,000 psi.

It is common to use a *high early-strength concrete*, such as type III, with a compressive strength, $f'_c$, of 4000 to 6000 psi. This concrete sets up quickly and suffers smaller elastic compression losses. Due to the higher quality concrete, the *modular ratio*, $n = E_s/E_c$, is somewhat lower than in conventional construction—on the order of 6 or 7.

## B. ACI CODE PROVISIONS FOR STRESS

The maximum tensile stress in the prestressing tendons prior to transfer of stress to the concrete is

$$f_t = \min\begin{Bmatrix} 0.94f_{py} \\ 0.85f_{pu} \end{Bmatrix} \qquad 14.95$$

Immediately after the prestress is transferred to the concrete, the maximum tensile stress in the tendons is

$$f_t = \min\begin{Bmatrix} 0.82f_{py} \\ 0.74f_{pu} \end{Bmatrix} \qquad 14.96$$

In post-tensioned construction, the maximum tendon tensile stress immediately after tendon anchorage is

$$f_t = 0.70f_{pu} \qquad 14.97$$

The maximum extreme fiber compressive stress in the concrete immediately after prestress transfer (but before any time-dependent losses) is given by equation 14.98. $f'_{ci}$ is the compressive strength of the concrete at the time of initial prestress.

$$f_c = 0.60f'_{ci} \qquad 14.98$$

The maximum extreme fiber tensile stress in the concrete immediately after prestress transfer is given by equation 14.99. For tension at the ends of simply supported members, the limit is twice that of equation 14.99.

$$f_t = 3\sqrt{f'_{ci}} \qquad 14.99$$

If the tensile stresses exceed the values given by equations 14.97 and 14.99, the ACI code contains provisions for providing additional reinforcement in the tensile zone.

At service loads and after all prestress losses, the maximum extreme fiber stresses in the concrete are[47]

$$f_c = 0.45f'_c \qquad 14.100$$

$$f_t = 6\sqrt{f'_c} \qquad 14.101$$

## C. ANALYSIS OF PRESTRESSED CONSTRUCTION

The procedure which follows is based on a standard analysis of a simply-supported, single-span beam. There are many ACI code-related provisions which could be applied if desired.

*step 1*: Calculate or estimate the dead weight per foot of the beam, $w_D$.

*step 2*: Determine the maximum moment on the beam due to the dead load. If it is assumed that the dead load is a uniform load (as it would be for a beam of uniform construction), the moment is

$$M_D = \frac{w_D L^2}{8} \quad \text{(simple supports)} \qquad 14.102$$

*step 3*: Assume that all of the beam is in compression. Use the transformed area method to locate the neutral axis.[48]

*step 4*: Determine the centroidal moment of inertia about the neutral axis. Since the entire beam is in compression, all of it contributes to $I$.

*step 5*: Calculate the eccentricity, $e$, of the strands.

[47] The ACI code provides for doubling the limit given by equation 14.100 if additional deflection analysis is performed.

[48] The steel area is transformed only for pretensioned construction and post-tensioned construction with tendons grouted in place. If the tendons are merely anchored at their ends, only the plain concrete section is used to calculate the neutral axis and moment of inertia.

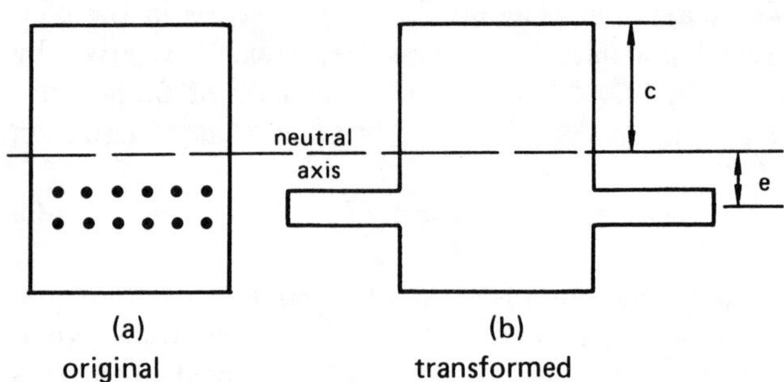


**Figure 14.15** Transformed Prestressed Beam

*step 6*: Calculate the stress (distribution) due to prestressing. This will consist of the basic compressive stress superimposed on the bending stress. $P$ is the *net prestress* acting on the beam, after losses.[49] In equation 14.103, compressive stresses are negative, and tensile stresses are positive.

$$f_{pc} = \frac{-P}{A_c} \pm \frac{Pec}{I} \qquad 14.103$$

*step 7*: Calculate the stress distribution due to the dead load.

$$f_b = \pm \frac{M_D c}{I} \qquad 14.104$$

*step 8*: The stress distribution immediately after transfer is

$$f = f_{pc} + f_b \qquad 14.105$$

*step 9*: Check to see that the tensile stress (if any) is less than the limit set by equation 14.96.[50]

*step 10*: Check to see that the compressive stress is less than the limit set by equation 14.98.

*step 11*: Calculate the stress distribution due to the live loading.

$$f_L = \frac{M_L c}{I} \qquad 14.106$$

*step 12*: The service load stress distribution is

$$f_{\text{total}} = f_{pc} + f_b + f_L \qquad 14.107$$

*step 13*: Check the long-term stress against the limits set by equation 14.100 and 14.101.

[49] The loss is frequently assumed to be 35,000 psi for prestressed construction and 25,000 psi for post-tensioned construction.

[50] Ordinarily, there is no need to calculate the steel stress since the stresses due to live loads never even make up the original loss of prestress.

**Table 14.9**
ASTM Standard Prestressing Tendons

(strand grade is $f_{pu}$ in ksi)

| type | nominal diameter, in. | nominal area, sq in. | nominal weight, lb per ft |
|---|---|---|---|
| grade 250 | 1/4 (0.250) | 0.036 | 0.12 |
| seven-wire | 5/16 (0.313) | 0.058 | 0.20 |
| strand | 3/8 (0.375) | 0.080 | 0.27 |
| | 7/16 (0.438) | 0.108 | 0.37 |
| | 1/2 (0.500) | 0.144 | 0.49 |
| | (0.600) | 0.216 | 0.74 |
| grade 270 | 3/8 (0.375) | 0.085 | 0.29 |
| seven-wire | 7/16 (0.438) | 0.115 | 0.40 |
| strand | 1/2 (0.500) | 0.153 | 0.53 |
| | (0.600) | 0.215 | 0.74 |
| prestressing | 0.192 | 0.029 | 0.098 |
| wire | 0.196 | 0.030 | 0.10 |
| | 0.250 | 0.049 | 0.17 |
| | 0.276 | 0.060 | 0.20 |
| smooth | 3/4 | 0.44 | 1.50 |
| prestressing | 7/8 | 0.60 | 2.04 |
| bars | 1 | 0.78 | 2.67 |
| | 1-1/8 | 0.99 | 3.38 |
| | 1-1/4 | 1.23 | 4.17 |
| | 1-3/8 | 1.48 | 5.05 |
| deformed | 5/8 | 0.28 | 0.98 |
| prestressing | 3/4 | 0.42 | 1.49 |
| bars | 1 | 0.85 | 3.01 |
| | 1-1/4 | 1.25 | 4.39 |
| | 1-3/8 | 1.56 | 5.56 |

## 14 COLUMNS

### A. INTRODUCTION

Tied and spiral columns are the main vertical load carrying members in buildings.[51] Figure 14.16 illustrates the reinforcement in *tied columns* and *spiral columns*.

The procedure used to design or evaluate a concrete column depends on whether the column is long or short (as defined by the ACI code) and the amount of eccentricity associated with the load.

[51] Pedestals and composite columns are not covered in this chapter.

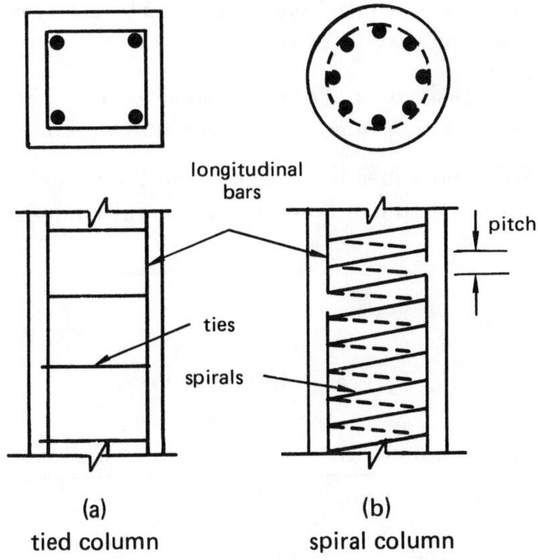


**Figure 14.16** Tied and Spiral Columns

Most column design makes use of ultimate strength concepts. The established procedures are based on experiments and empirical evidence. Thus, stress relationships are only valid when the column is loaded to near its ultimate strength. Since columns should not be loaded past their elastic limit, no attempt is made to calculate the stresses. The amount of steel used for ties or spirals is based on empirical data.

The *ultimate axial load*, $P_u$, which columns are designed to support is the factored load calculated from equation 14.108.[52]

$$P_u = 1.4 \times \text{dead load} + 1.7 \times \text{live load} \qquad 14.108$$

In addition to $P_u$, other load variables are used in column calculations. $P_o$ is the *nominal axial load strength* of a column with zero eccentricity. $P_n$ is the nominal axial load strength with an eccentricity of $n$. $\phi P_n$ is the *design axial load strength* which must exceed $P_u$. That is, equation 14.109 is the design criterion for columns.

$$\phi P_n \geq P_u \qquad 14.109$$

In simple column design, the major assumption made about columns at failure is that the steel is stressed to $f_y$ and the concrete is stressed to $0.85f'_c$. Then, the maximum load on the column would be

$$P_o = 0.85f'_c A_{\text{concrete}} + f_y A_{st} \qquad 14.110$$

[52] The effects of wind and earthquake loading are not covered in this chapter.

In equation 14.110, $A_{\text{concrete}}$ includes all core and cover concrete in the column.

$$A_{\text{concrete}} = A_g - A_{st} \qquad 14.111$$

The *reinforcement ratio* is

$$\rho_g = \frac{A_{st}}{A_g} \qquad 14.112$$

Using the definition of concrete area and reinforcement ratio, equation 14.110 can be rewritten.

$$P_o = A_g\left[0.85f'_c(1-\rho_g) + f_y\rho_g\right] \qquad 14.113$$

Equation 14.113 is theoretically correct. However, several additional factors are added to account for a minimum amount of eccentricity and the load transfer to core and steel when the outside cover spalls.

B. TIED COLUMNS

The following construction details are specified by the ACI code for tied columns.

- Longitudinal bars must have a clear distance of at least $1\frac{1}{2}$ times the bar diameter, but not less than $1\frac{1}{2}''$.
- The minimum tie wire size is #3 if the longitudinal bars are #10 or smaller. The minimum tie wire size is #4 for larger longitudinal bars.[53]
- The concrete covering should be at least $1\frac{1}{2}''$ thick over the outermost surface of the tie steel.
- At least 4 longitudinal bars are needed for columns with square or circular ties. [ACI 318 section 10.9.2]
- The ratio of longitudinal steel area to the gross column area must be between 0.01 and 0.08.[54]

$$0.01 \leq \rho_g \leq 0.08 \qquad 14.114$$

- Center-to-center spacing of the ties must not exceed the smallest of 16 longitudinal bar diameters, 48 tie bar diameters, or the least gross (outside) column dimension.
- Every corner and alternating longitudinal bar should be supported by a tie corner.[55]

[53] The maximum practical tie size is #5 bar.

[54] The lower limit keeps the column from being designed as all-concrete. The upper limit keeps the column from being too slender. 0.02 or 0.025 is typically chosen as a starting value of $\rho_g$.

[55] Tie corners are not relevant to tied columns with longitudinal bars spaced in a circular pattern.

- In tied columns, no longitudinal bar can be more than 6″ away from a tie corner supported longitudinal bar. Figure 14.17 illustrates extra ties.
- Tie corners should not make a bend of more than 135°.

Figure 14.17 illustrates several typical tied column reinforcement and tie corners.

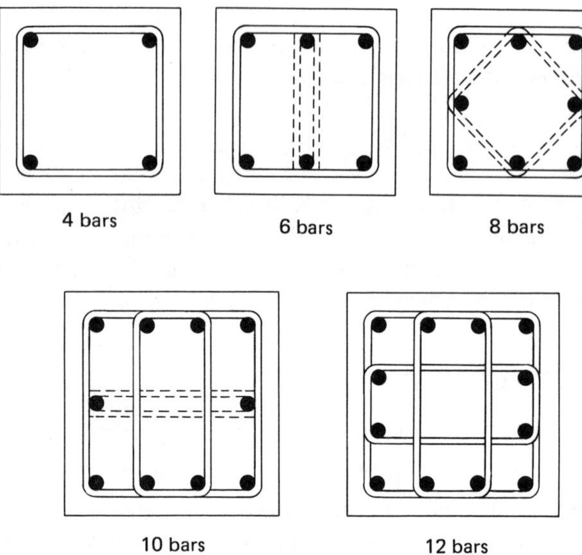


**Figure 14.17** Ties and Tied Corners

The design axial load strength for tied columns is given by equation 14.115. $\phi = 0.70$ for tied columns.[56]

$$P_u = \phi P_n = 0.80\phi P_o \qquad 14.115$$

C. SPIRAL COLUMNS

The following construction details are specified for spiral columns.

- Longitudinal bars must have a clear distance of at least $1\frac{1}{2}$ times the bar diameter, but not less than $1\frac{1}{2}''$.
- The minimum spiral wire diameter is 3/8″.[57]
- The clear distance between spirals should not exceed 3″, and it should not be less than 1″.
- The concrete covering should be at least $1\frac{1}{2}''$ thick over the outermost surface of the spiral steel.

[56] Some people find the form of equation 14.115 confusing, since $\phi$ appears on both sides of the equation. Since the design criterion is $P_u \leq \phi P_n$, it is only necessary to compare the ultimate factored load ($P_u$, as defined in equation 14.108) against the right-hand side of equation 14.115.

[57] The maximum practical spiral wire size is 5/8″.

- At least 6 longitudinal bars are to be used for spiral columns. [ACI 318 section 10.9.2]
- The ratio of longitudinal steel area to the gross column area must be between 0.01 and 0.08.

The ACI code does not specify spiral wire size. However, table 14.10 can be used for general guidelines.

**Table 14.10**
Typical Spiral Wire Sizes

| column diameter | spiral wire size |
|---|---|
| up to 15″, using #10 bars or smaller | 3/8″ |
| up to 15″, using #11 bars or larger | 1/2″ |
| 16″–22″ | 1/2″ |
| 23″ and up | 5/8″ |

The design axial load strength for spiral columns is given by equation 14.116. $\phi = 0.75$ for spiral columns.

$$P_u = \phi P_n = 0.85\phi P_o \qquad 14.116$$

The ACI code also requires the ratio of spiral reinforcement volume to column core volume to be greater than the value of equation 14.117.

$$\rho_s \geq 0.45\left(\frac{A_g}{A_c} - 1\right)\frac{f'_c}{f_y} \qquad 14.117$$

$A_c$ and $D_c$ are measured to the outside diameter of the spiral wire. $f_y$ in equation 14.117 is the *spiral steel* yield point, and may not exceed 60,000 psi.

If the spiral wire diameter is known, the *spiral pitch* can be found. $A_s$ in equation 14.118 is the cross sectional area of the spiral wire.

$$s \approx \frac{4A_s}{\rho_s D_c} \qquad 14.118$$

The *clear distance between spirals* will be the difference between the spiral pitch and the spiral wire diameter.

$$\text{clear distance} = s - D_s \qquad 14.119$$

D. COLUMN ECCENTRICITY

If the loads are designed to be applied axially to a column, and if there are no external moments acting on the column, the column is said to be a *low-eccentricity column*. Actually, a small amount of eccentricity is permissible. By specifying the strength reduction factors of 0.80 and 0.85 in the equations for design strength (equations 14.115 and 14.116), the code is compensating for up to a 5% eccentricity for spiral columns and a 10% eccentricity for tied columns. *Percent eccentricity* is defined as

$$\epsilon = \frac{e}{\text{column width}} \times 100 \qquad 14.120$$

Eccentricity can be introduced by external moments, as well as by off-center loading. The *eccentricity* due to a moment on the column is

$$e = \frac{M_u}{P_u} \qquad 14.121$$

Equations 14.115 and 14.116 should be used to design and evaluate short columns with low eccentricities.

*Example 14.6*

Calculate the design strength of the short spiral column shown. Assume low eccentricity loading, $f_y = 40{,}000$ psi, and $f'_c = 3500$ psi.

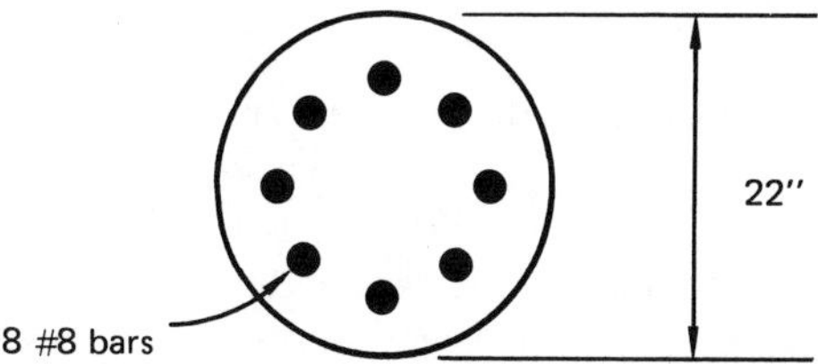


The gross area of the column is

$$A_g = \frac{1}{4}\pi(22)^2 = 380.1\,\text{in}^2$$

Each #8 longitudinal bar contributes 0.79 in$^2$ to the steel area, so the total steel area is

$$A_{st} = 8 \times 0.79 = 6.32\,\text{in}^2$$

The reinforcement ratio is

$$\rho_g = \frac{6.32}{380.1} = 0.0166$$

The nominal axial load strength is

$$\begin{aligned} P_o &= 380.1[(0.85)(1 - 0.0166)(3500) \\ &\quad + (40{,}000)(0.0166)] \\ &= 1.364\,\text{EE6 lbf} \end{aligned}$$

From equation 14.116, the design strength is

$$(0.85)(0.75)(1.364\,\text{EE6}) = 8.7\,\text{EE5 lbf}$$

*Example 14.7*

Design a spiral column to carry a factored load of 375,000 pounds. Use 3000 psi concrete and 40,000 psi steel.

Assume $\rho_g = 0.02$, which is in the allowable range. Larger values will decrease the column gross area. However, slenderness effects are being neglected, and it is desirable to keep the column as heavy as possible.

From equations 14.113 and 14.116,

$$\begin{aligned} 375{,}000 &= A_g(0.85)(0.75)[(0.85)(1 - 0.02)(3000) \\ &\quad + (40{,}000)(0.02)] \\ A_g &= 178.3\,\text{in}^2 \\ D_g &= \sqrt{\frac{4A_g}{\pi}} = 15.07'' \quad (\text{say } 15.25'') \end{aligned}$$

With a $1\frac{1}{2}''$ cover, the core diameter will be

$$D_c = 15.25 - 3.0 = 12.25''$$

The required steel area is

$$A_{st} = (0.02)(178.3) = 3.57\,\text{in}^2$$

The ACI code requires at least 6 bars to be used. Some possibilities which meet the required steel areas are 6 #7 bars, 9 #6 bars, or 12 #5 bars. Choose #6 bars after checking the clear spacing (not shown here).

From equation 14.117, substituting $D^2$ for $A$, the ratio of spiral reinforcement must be greater than

$$0.45\left[\left(\frac{15.25}{12.25}\right)^2 - 1\right]\frac{3000}{40{,}000} = 0.0186$$

Assume 3/8" spiral wire ($A_s = 0.11\,\text{in}^2$) with a $1\frac{1}{2}''$ pitch. From equation 14.118, the actual spiral reinforcement ratio is

$$\rho_s = \frac{(4)(0.11)}{(1.5)(12.25)} = 0.024''$$

Since 0.024 > 0.0186, the spiral spacing is adequate.

The clear space between spirals is

$$1.5 - 0.375 = 1.125''$$

This is between 1" and 3".

## E. LARGE ECCENTRICITIES

As a moment carried by a column increases, the axial load that the column can support decreases. A graph of allowable axial load versus the applied moment is known as an *interaction diagram.* Figure 14.18 illustrates a general interaction diagram.

Construction of an interaction diagram for a given column design is tedious, and requires calculating loads and moments for multiple values of eccentricity.[58] The diagram boundary represents the envelope of acceptable loadings. Inside the boundary, the design is acceptable but overdesigned. Outside the boundary, the design is unacceptable.

Radial lines outward from the origin represent different values for the percent eccentricity. Point $A$ defines the maximum eccentricity that the column can support and still be considered a low-eccentricity column. Point $C$ is the bending strength of the member.

Point $B$ defines the eccentricity separating compression and tensile failures. Above the line of *balanced eccentricity*, the column will fail by concrete crushing. Below the line, the column will fail by tension steel yielding.

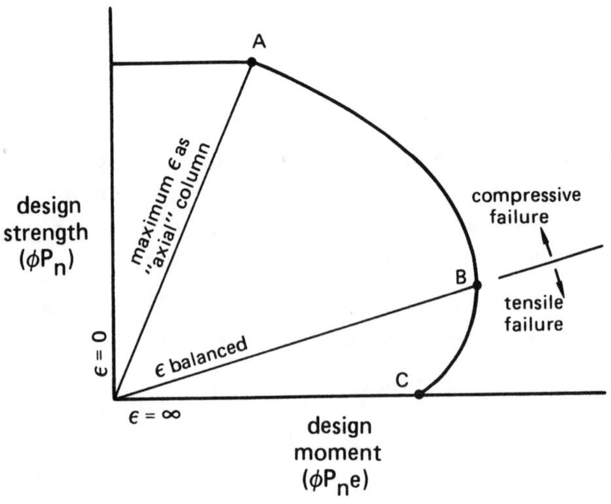


**Figure 14.18** Interaction Diagram for Large Eccentricity Column

To further complicate the design and analysis of large eccentricity columns, the strength reduction factor varies with the design axial load strength.

To assist in the design and analysis of large eccentricity columns, various sources (including ACI) have produced sets of interaction diagrams for common column configurations. These charts are primarily analysis aids, and design still requires a trial and error approach, since the percent eccentricity isn't known until the column is sized.

[58] Construction of interaction diagrams is not covered in this book.

F. SLENDERNESS EFFECTS

The effects of slenderness can be disregarded for short columns. The ACI code provides for determining when a column is short or slender, and for increasing the required strength of slender columns.[59]

The *slenderness ratio* is defined as

$$SR = \frac{kL_u}{r} \qquad 14.122$$

In equation 14.122, $k = 1$ for columns braced against sidesway, unless analysis shows that a lower value can be used. $k > 1$ for columns not braced against sidesway. $L_u$ is the unbraced length, typically taken as the clear distance between floor slabs, beams, or other members capable of providing lateral support.

The *radius of gyration*, when a rigorous analysis is not performed, is defined by the ACI code. The column dimension in equation 14.124 is the overall dimension in the direction stability is being evaluated.

$$r = 0.25d_g \quad \text{(round)} \qquad 14.123$$

$$r = 0.3 \times \text{column dimension} \quad \text{(rectangular)} \qquad 14.124$$

The effects of slenderness can be disregarded in columns not braced against sidesway if the slenderness ratio is less than 22. The determination is somewhat more difficult for columns braced against sidesway, being dependent on the bending moments at both ends of the column. In equation 14.125, $M_{1b}$ is the value of the smaller factored moment at the end of the compression member.[60] It is positive if the column is bent in single curvature, and negative if the column is bent in double curvature. $M_{2b}$ is the value of the larger factored end moment, and is always positive.

$$SR_{\text{max}} = 34 - 12\left(\frac{M_{1b}}{M_{2b}}\right) \qquad 14.125$$

If it is determined that a column is slender, as defined by the ACI code, then the design must consider the factored axial load, $P_u$, and a magnified factored moment, $M_c = \delta M$. The ACI code describes the procedure for calculating the *magnification factor*, $\delta$.

[59] Columns between floors of a building are almost always short columns.

[60] If the moments at both ends of the column are zero or the same, $M_{1b}/M_{2b} = 1$, and the maximum slenderness ratio is 22.

## 15 FOOTING DESIGN

### A. FAILURE MECHANISMS

There are two primary failure modes for footings: shear and flexure. For square footings, the total shear is used to find the unit stress on the critical section, and shear is known as *two-way shear*. The *critical section* for *two-way shear* is assumed to be located $\frac{1}{2}d$ from the column face, where $d$ is the depth of the reinforcement.

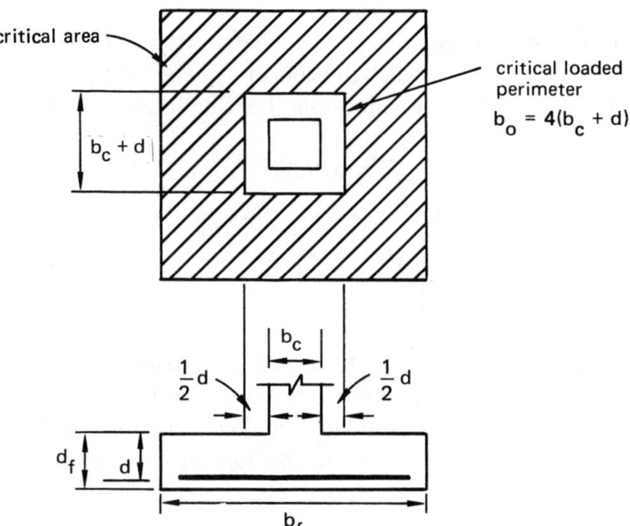


**Figure 14.19** Critical Area for Two-Way Shear

The critical section for *one-way shear*, which is applicable to rectangular footings, is assumed to be located at distance $d$ from the column face.

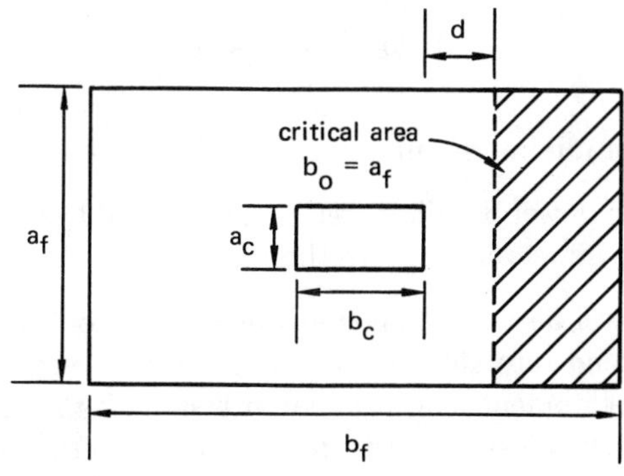


**Figure 14.20** Critical Area for One-Way Shear

Footings can also fail in flexure. The critical section for bending is at the face of the column.[61]

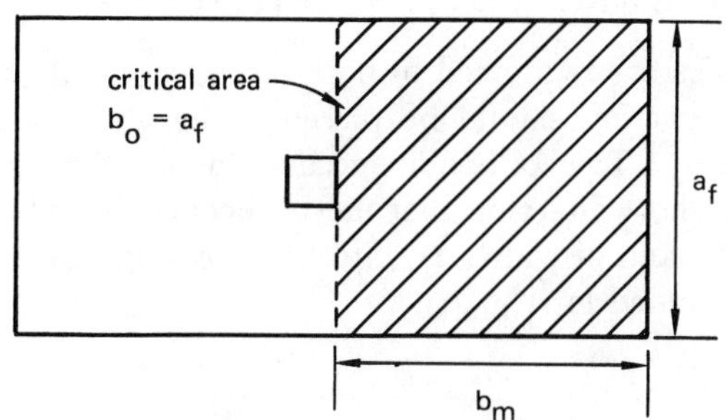


**Figure 14.21** Critical Area for Flexure

### B. FACTORED LOAD

The factored load used to design footings is

$$P_u = 1.4P_D + 1.7P_L \tag{14.126}$$

This factored load is used to calculate the net soil pressure beneath the footing.[70]

$$p_{\text{net}} = \frac{P_u}{A_f} = \frac{P_u}{a_f b_f} \tag{14.127}$$

For the purpose of checking shear, the ultimate shear is

$$\begin{aligned} V_u &= p_{\text{net}} \times \text{ critical area} \\ &= p_{\text{net}} \times [a_f^2 - (b_c + d)^2] \quad \text{(square)} \\ &= p_{\text{net}} \times \left[a_f\left(\frac{b_f - b_c}{2} - d\right)\right] \quad \text{(rectangular)} \end{aligned} \tag{14.128}$$

### C. NOMINAL SHEAR STRENGTH

The nominal shear strength in concrete without reinforcement (i.e., *plain footings*) for *single action* (i.e., one-way shear) is[62]

$$V_c = 2\sqrt{f'_c}b_o d \tag{14.129}$$

For *double action* (i.e., two-way shear), the nominal shear strength is

$$V_c = \left(2 + \frac{4}{\beta_c}\right)\sqrt{f'_c}b_o d \tag{14.130}$$

$$\beta_c = \min\begin{cases} \frac{b_c}{a_c} \\ 2 \end{cases} \tag{14.131}$$

In equations 14.129 and 14.130, $b_o$ is the length of the critical section.

[61] Footings supporting masonry walls and footings loaded through a steel base plate have different critical sections.

[62] The more detailed method of calculating $V_c$ may be used with footings also.

D. REINFORCEMENT IN FOOTINGS

Footings are reinforced in both directions. Bars in the long direction should be placed uniformly across the width $a_f$. The ACI code specifies the fraction of steel in the short direction that must be concentrated in the center band of width $a_f$. Steel in the long direction is not included in $A_{st,\text{total}}$.

$$\frac{A_{st,\text{band}}}{A_{st,\text{total}}} = \frac{2}{\beta_f + 1} \qquad 14.132$$

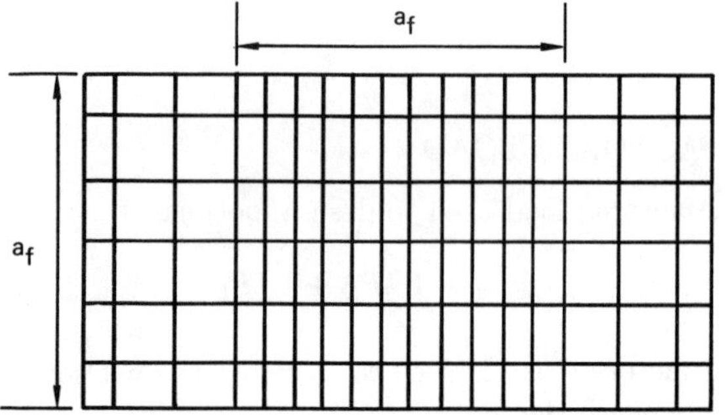


**Figure 14.22** Reinforcement in Center Band

Vertical dowels (i.e., *dowel bars*) can be used to provide horizontal shear resistance and to transfer the column load to the footing by bond instead of bearing. Stirrups are not usually used, since the allowable shear stress is kept low. Hooks are not usually used at the ends of bars in footings due to the danger of top cover failure.

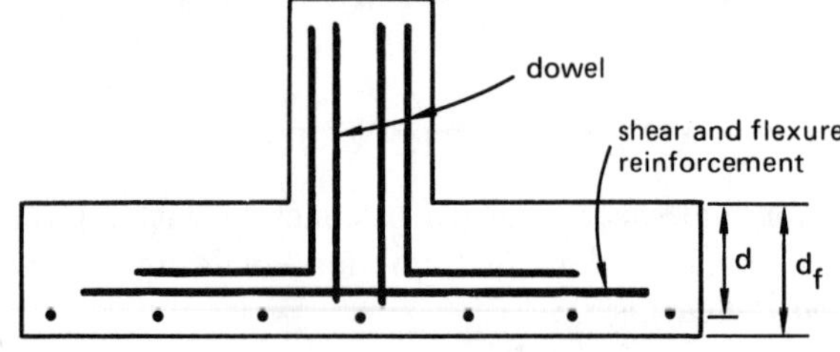


**Figure 14.23** Types of Footing Reinforcement

E. ACI CODE PROVISIONS FOR FOOTINGS

The following provisions are applicable to footings designed in accordance with the ACI code.

- The minimum depth, $d$, for reinforced concrete on soil is 6″. For footings on piles, the minimum depth is 12″.
- The minimum cover below bars is 3″.
- Bars in two layers can be in contact, and no clear distance between the layers is required.[63]

[63] With two layers, the average depth between the two layers is typically used for both layer orientations. However, for footings less than approximately 15 inches deep, the more conservative value of $d$ is appropriate.

F. BEARING PRESSURE

The portion of column load that is not transferred by bond (i.e., through dowels or extensions of the column steel) must be transferred by bearing. The nominal ultimate bearing stress, $f_b$, that the column concrete can withstand is[64]

$$f_b = 0.85 f'_c \qquad 14.133$$

The nominal compressive strength based on the column area is

$$P_n = f_b A_c = 0.85 f'_c a_c b_c = \frac{P_u}{\phi} \qquad 14.134$$

Since the footing area is much larger than the gross column area, the redistribution of stresses will permit an increase in the allowable bearing stress. [ACI 10.15.1.1]

$$f_b = \alpha_b 0.85 f'_c \qquad 14.135$$

$$\alpha_b = \min\begin{cases} \sqrt{\dfrac{A_f}{A_c}} = \sqrt{\dfrac{a_f b_f}{a_c b_c}} \\ 2 \end{cases} \qquad 14.136$$

In equation 14.136, the $A_f$ term is not strictly correct. The actual area to be used is the portion of the footing that is geometrically similar to and concentric with the loaded area. For square footings loaded by concentric square columns, this is $A_f$.

G. DOWEL BARS

The ACI code requires a minimum amount of dowel bar area, even when the bearing pressures are not excessive. The following provisions are applicable.

- The minimum dowel bar area is

$$A_{st} = 0.005 A_c = 0.005 a_c b_c \qquad 14.137$$

- The minimum number of dowel bars is 4.
- The dowel bar diameter cannot exceed the column bar diameter by more than 0.15″.
- Dowel bars must be embedded in the footing (and extend into the column) a distance exceeding the development length in compression. If the footing depth is less than the development length, smaller bars must be used. If a bent dowel bar (see figure 14.23) is used, the development length must be achieved over the vertical portion, since bends and hooks are not considered effective in compression.

[64] This analysis does not assume that the column and footing are constructed from the same concrete. It is possible to construct columns with higher-strength concrete than is used in the footing. In that case, the bearing stress on the footing may control. Otherwise, only the column check is needed.

If dowel bars are required to transfer load to the footing, the area of dowel steel is easily calculated.[65]

$$\text{excess } P_u = P_u - \phi f_y A_c \qquad 14.138$$

$$A_{st} = \frac{\text{excess } P_u}{\phi f_y} \qquad 14.139$$

In equation 14.139, $\phi = 0.70$ for bearing on concrete.

## H. FOOTING ANALYSIS[66]

Shear

*step 1*: Determine the effective depth, $d$, of the reinforcement.

*step 2*: Use equation 14.127 to calculate the net earth pressure.

*step 3*: Use equation 14.128 to calculate the ultimate shear. Square footings should always be checked for both one-way shear and two-way shear. Rectangular footings are usually controlled by one-way shear, but almost-square footings may be exceptions. For assurance, always check both.

*step 4*: Use equations 14.130 and 14.131 to calculate the nominal strength of the concrete without reinforcement.

*step 5*: Compare the ultimate shear with the nominal shear. If $V_u > \phi V_c$, shear reinforcement is required.[67] Estimate or specify the footing thickness as $d + d_b$ + cover.

Flexure

*step 6*: The bending moment based on the loaded area shown in figure 14.21 is

$$M_u = \frac{p_{\text{net}} a_f b_m^2}{2} \qquad 14.140$$

*step 7*: Calculate the required *coefficient of resistance.*[68]

$$R_u = \frac{M_u}{\phi a_f d^2} \qquad 14.141$$

*step 8*: Calculate the required reinforcement ratio.[69]

$$\rho = \frac{0.85 f'_c}{f_y}\left[1 - \sqrt{1 - \frac{2R_u}{0.85 f'_c}}\right] \qquad 14.142$$

*step 9*: Calculate the required steel area.

$$A_{st} = \rho d a_f \qquad 14.143$$

*step 10*: Choose reinforcement to satisfy the required area and spacing requirements. Since concrete cast next to earth must have a 3″ cover on steel, all steel must fit within the space of $a_f - 6''$.

*step 11*: Check development length for reinforcement. The distance from the column face to the end of the reinforcement must exceed the development length in *tension.*

## I. FOOTING DESIGN

Footing design is similar to footing analysis, although the order of the steps is different. The major difference is in the need to determine the footing area and depth of reinforcement.

The footing area should be determined in the normal manner—from the allowable soil pressure and the load to be carried.[70] Then, the net soil pressure is used to determine the required shear strength. The footing thickness is determined such that additional shear reinforcement is unnecessary. In equation 14.144, 150 is the estimated density of concrete.

$$\begin{aligned} p_{\text{net}} &= \frac{p_D + p_L}{A_f} \\ &= p_a - 150 \times \text{ footing thickness} \\ &\quad - \text{ overburden pressure} \end{aligned} \qquad 14.144$$

## J. ECCENTRICALLY-LOADED FOOTINGS

A footing which is loaded eccentrically will carry a moment in addition to an axial load. The key to analyzing such a footing is determining the shape of the soil pressure distribution. Once this is done, the shear on the critical area, as well as the bending moment on the footing, can be determined. Analysis is the same as for a concentrically-loaded footing.

---

[65] The minimum area may be sufficient to carry the excess.

[66] This assumes that the footing dimensions, including depth $d$, are known in advance. For continuous or wall footings, use a 12″ portion in the analysis.

[67] In a design environment, it may be easier to increase $d$ than to add shear reinforcement. $\phi$ = 0.85 in shear.

[68] $\phi =$ 0.90 for bending.

[69] The minimum reinforcement ratio of $200/f_y$ may not apply to footings. However, the temperature and shrinkage requirement is applicable for the short direction of rectangular footings. This requirement may govern.

[70] The ACI code specifies that the load to be carried is not to be factored when sizing the footing. [ACI 318 15.2.2]

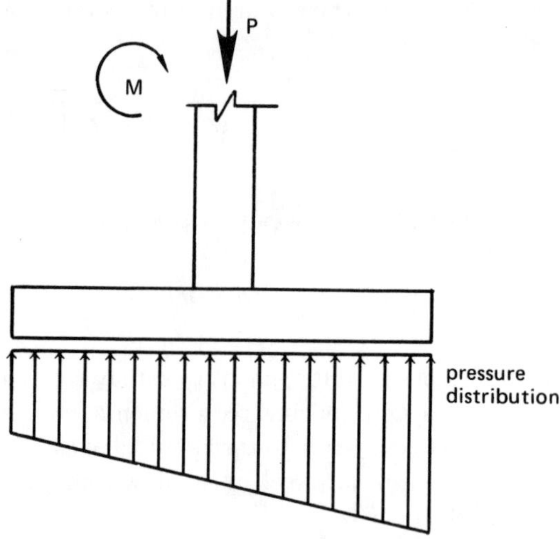


**Figure 14.24** Footing with Moment

*Example 14.8*

A 111″ × 111″ square footing is to support 82,560 pounds dead load and 75,400 pounds live load. The bearing block (column) is 12″ × 12″, concentric with the center of the footing. $f'_c = 4000$ psi and $f_y = 60{,}000$ psi. Design the footing.

*step 1*: See step 3.

*step 2*: The factored load is

$$P_u = (1.4)(82{,}560) + (1.7)(75{,}400) = 243{,}770 \text{ lbf}$$

The net soil pressure is

$$p_{\text{net}} = \frac{243{,}770}{(111)^2} = 19.78 \text{ psi}$$

*step 3*: For two-way shear, the critical shear line is located $\frac{1}{2}d$ from the column face. Since $d$ is unknown, estimate $d = 12''$. Then, the length of the critical perimeter will be (refer to the diagram)

$$b_o = (4)(24) = 96''$$

The critical area contributing to shear is

$$A_v = (111)^2 - (24)^2 = 11{,}745$$

The ultimate shear is

$$V_u = (19.78)(11{,}745) = 232{,}320 \text{ lbf}$$

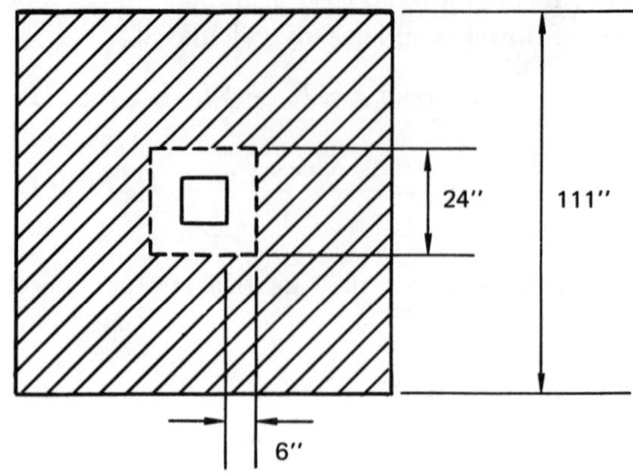


*step 4*: The nominal concrete strength in shear is

$$\beta_c = 12/12 = 1 \text{ (must be 2 or more)}$$

$$V_c = \left(2 + \frac{4}{2}\right)(0.85\sqrt{4000})(96)(12) = 291{,}436$$

$$\phi V_c = (0.85)(291{,}436) = 247{,}720$$

*step 5*: Since 232,320 < 247,720, no shear reinforcement is required. The footing thickness could be reduced somewhat.

The shear strength of the footing depends on the footing thickness. Assuming $d = 12''$, $d_b = 1''$, and cover $= 3''$, the thickness is

$$\text{thickness} = 12 + \tfrac{1}{2} + 3 = 15.5''$$

*step 6*: The bending moment is

$$M_u = \frac{(19.78)(111)(49.5)^2}{2} = 2.69 \text{ EE6 in-lbf}$$

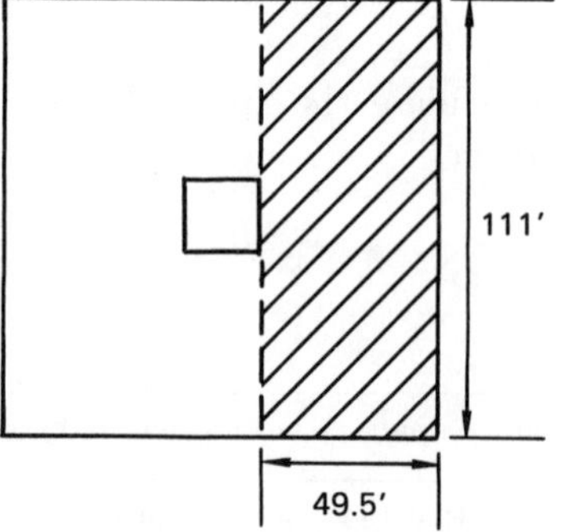


*step 7*: The coefficient of resistance is

$$R_u = \frac{2.69 \text{ EE6}}{(0.9)(111)(12)^2} = 187 \text{ psi}$$

*step 8*: From equation 14.142, the required reinforcement ratio is

$$\rho = \frac{(0.85)(4000)}{60{,}000}\left[1 - \sqrt{1 - \frac{(2)(187)}{(0.85)(4000)}}\right]$$
$$= 0.00321$$

*step 9*: The required steel area is

$$A_{st} = (0.00321)(12)(111) = 4.28\,\text{in}^2$$

*step 10*: The reinforcement must fit in 111−6 = 105″. Arbitrarily use 10″ bar spacing. The number of bars is

$$\frac{105}{10} = 10.5\,(\text{say } 11)$$

The bar area should be approximately

$$A_b \approx \frac{4.28}{11} = 0.39\,\text{in}^2$$

Choose #6 bars (0.44 in² each) and space evenly on 10″ centers in both directions (since the footing is square).

The minimum reinforcement ratio from equation 14.9 is

$$\rho_{\text{min}} = \frac{200}{60{,}000} = 0.00333$$

The actual reinforcement ratio is

$$\rho = \frac{(11)(0.44)}{(111)(12)} = 0.00363 \text{ (ok)}$$

(See footnote 69.)

*step 11*: Check development length. (Not done here.)

## 16 RETAINING WALLS

### A. GENERAL DESIGN CHARACTERISTICS

The following characteristics are typical of retaining walls. They are not code requirements, but can be used as starting points for subsequent design.

- The base should be proportioned so that

$$0.40 < \frac{b}{h} < 0.65 \qquad 14.145$$

- The stem thickness at the base should be approximately

$$\frac{1}{12} < \frac{t_s}{h} < \frac{1}{8} \qquad 14.146$$

- The toe should project approximately $b/3$ beyond the stem face.
- The stem thickness should decrease $\frac{1}{4}''$ to $\frac{1}{2}''$ per vertical foot.[71]
- The minimum stem thickness at the top is approximately 12″.

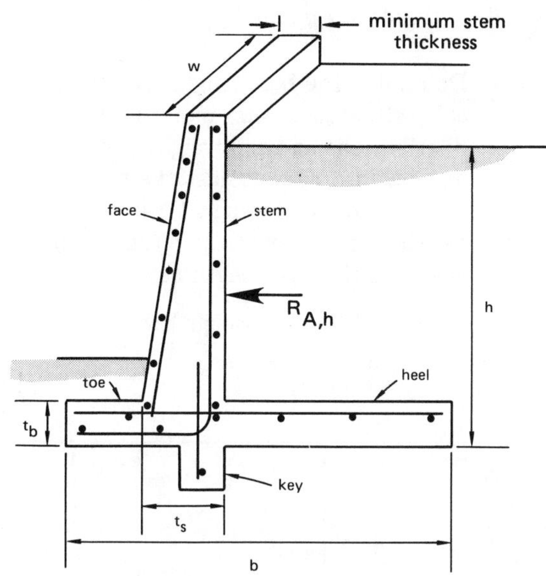


**Figure 14.25** Retaining Wall Nomenclature

- The bottom of the base must be below the frost level.
- To minimize the vertical earth pressure reaction, the resultant of the vertical loads should fall in the middle third of the base.
- The base thickness should be approximately equal to the thickness of the stem at the base, with a minimum of 12″. Alternatively, proportion the base such that

$$0.07 < \frac{t_b}{h} < 0.10 \qquad 14.147$$

### B. GENERAL DESIGN PROCEDURE

A retaining wall is most likely to fail structurally at its base due to the applied moment.[72] In this regard, it is similar to a cantilever beam with a non-uniform load. The following procedure presents one possible approach

[71] This decrease in stem thickness is known as *batter*, and is used primarily to disguise bending (deflection) that would otherwise make it appear as if the wall were failing. The batter serves no other purpose. Usually, the *batter decrement* is $\frac{1}{4}''$ per foot.

[72] The wall can also fail non-structurally. The subjects of resistance to sliding and overturning are covered in another chapter.

to design. It is assumed that sufficient soil data exists to calculate the active and other soil pressure distributions.

*step 1:* Choose the height of the wall, $h$, on the basis of frost penetration or other data. Estimate a base thickness, $t_b$.

*step 2:* Estimate $R_{A,h}$ and $R_{A,v}$.

*step 3*: Determine the base length, $b$. One analytical method is to sum moments from active distributions and soil weight about point $A$ and balance that against the moment from the unknown soil weight (in figure 14.26) to get the distance $x$.[73] Once that is obtained, calculate the base length as

$$\frac{R_A h}{3} \approx \frac{(\text{soil weight})x}{2}$$
$$b \approx 1.5 \times x \quad 14.148$$

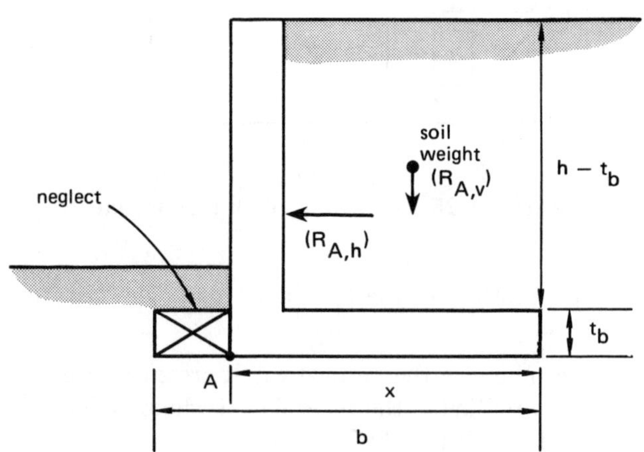


**Figure 14.26** Finding the Base Length

*step 4*: Select a reinforcement ratio to control deflections.

$$\rho \approx \tfrac{1}{2}\rho_{\text{max}} = 0.375\rho_{\text{balanced}} \quad 14.149$$

*step 5*: Calculate the moment at the base of the stem due to the active pressure distributions. Be sure to include the 1.7 overload factor for live loads when calculating $M_u$.

*step 6*: From the value of $\rho$ chosen, calculate the coefficient of resistance.

$$R_u = \rho f_y \left(1 - \frac{\rho f_y}{(2)(0.85)f'_c}\right) \quad 14.150$$

*step 7*: Calculate the required stem thickness at the base.[74] In equation 14.151, $w$ is the wall length, which may be taken as 12″ if flexure calculations are done on a per foot basis.

$$d = \sqrt{\frac{M_u}{\phi R_u w}} \quad 14.151$$

The total thickness of the stem at the base is

$$t_s = d + 2'' \text{ cover} + \tfrac{1}{2}d_b \quad 14.152$$

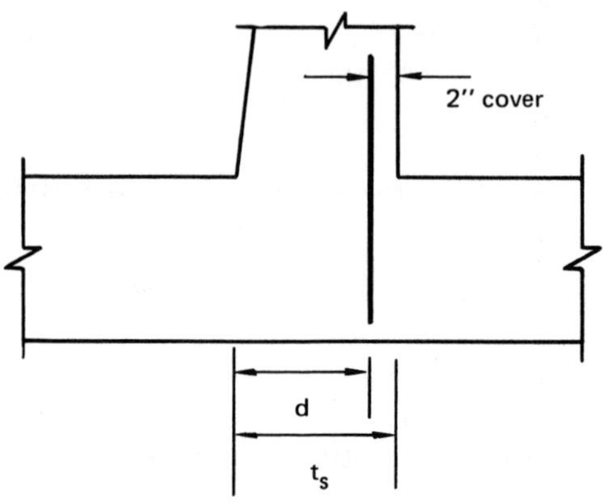


**Figure 14.27** Base of Stem Details

*step 8*: Choose the batter decrement. $\frac{1}{4}''$ per foot is typical.

*step 9*: Calculate the nominal shear strength of one foot of wall.

$$V_n = \phi V_c = 2\phi\sqrt{f'_c}(12)d \quad 14.153$$

In equation 14.153, the factor 12 is the width, in inches, of a one-foot section of wall. $t'_s$ is the stem thickness a distance $d$ (for one-way shear) from the bottom of the stem. If the stem does not decrease in thickness, $t'_s = t_s$. Otherwise, $t'_s$ will depend on the batter decrement.

*step 10*: Calculate the shear on the stem on the critical section. As in step 8, the critical section is assumed to be located a distance $d$ up from the bottom of the stem. It will be necessary to integrate the active pressure distributions

[73] Include surcharge distributions, if present, but disregard passive distributions.

[74] Since $\rho$ was chosen somewhat arbitrarily, $d$ is not unique. Use these calculations as guidelines. $\phi =$ 0.90 for flexure.

from the top of the retaining wall down to this critical level to obtain the shear.

$$V_u = 1.7\,V_{\text{critical section}} \qquad 14.154$$

*step 11*: If $V_u < \phi V_c$, shear reinforcement will not be needed, and the stem can be considered a slab.[75]

*step 12*[76]: Design the heel cantilever as a beam, and check for shear. The critical section is located approximately at the face of the stem. For ease of calculations, consider the pressure distributions of soil, surcharge (if any), and footing concrete for a 12″ strip of retaining wall. Use an overcapacity factor of 1.4 for the soil and footing concrete, and use 1.7 for the surcharge to calculate the ultimate moment on the heel. Disregard the upward soil pressure distribution as well as any effect a key might have on the distributions.

For shear checking, the critical section is taken at the face without considering the upward soil pressure distribution.[77] Therefore, the entire heel contributes to the ultimate shear, $V_u$. The nominal shear strength $V_n = \phi V_c$ is calculated and compared to $V_u$. If necessary, changes to $t_b$ and $d$ can be made to decrease $V_n$.

The new value of $t_b$ can be used to recalculate the footing weight, and a second iteration of this procedure will refine the values still further.

Once $t_b$ is stable, the new $M_u$ can be used to calculate $R_u$, from which $\rho$ and $A_{st}$ can be obtained.[78]

Complete the heel reinforcement design by calculating the development length. The distance from face of stem to end of heel reinforcement must equal or exceed the development length. The heel reinforcement must also extend a distance equal to the development length past the stem reinforcement into the toe.

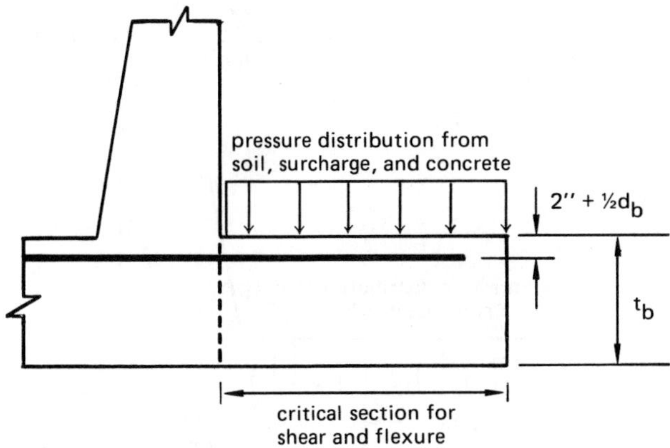


**Figure 14.28** Heel Construction Details

*step 13*: The design of the toe cantilever is similar to the heel design. The loading on the toe is assumed to be from the toe weight and the upward soil pressure distribution only, unless soil above the toe is permanent.[79] For flexure, the critical section is at the outer face of the wall. For shear (one-way shear), the critical section is located a distance $d$ from the wall face.

When calculating the ultimate moment, $M_u$, and ultimate shear, $V_u$, an overcapacity factor of 1.4 should be used for the concrete weight, and 1.7 should be used for the upward soil pressure distribution based on the service loads.[80]

It is likely that shear will determine the toe thickness. The toe thickness can be different from the heel thickness, although most designs maintain a constant base thickness.

*step 14*: Calculate the required steel area for flexure reinforcement in the stem. (If the base thickness has been changed from step 7, then $h$ will also have changed. In that case, recalculate $R_u$ from equation 14.155. (The shear

[75] From a practical design standpoint, the thickness should be increased rather than supplying shear reinforcement.

[76] Designing the heel and toe cantilevers may change some of the assumptions made about the base (e.g., the base thickness, etc.). If the base is not of interest, skip steps 11 and 12.

[77] Since tension occurs in the heel, the critical section cannot be taken as a distance $d$ from the face of the stem. (See ACI code 11.1.3.)

[78] The ACI code exempts slabs from the requirement of $200/f_y$ minimum reinforcement. However, it is arguable that a retaining wall is a beam, not a slab, in this situation. Therefore, the minimum reinforcement ratio or alternatively, one-third more than is required for strength, may apply.

[79] The soil on top of the toe may be removed for later work or repair on the retaining wall. Therefore, it cannot be counted on to reduce the toe stress.

[80] ACI 9.2.4 specifies that the factor is 1.4 for dead concrete and soil weights, and 1.7 for horizontal earth pressure. The toe soil pressure distribution is the result of the horizontal active soil pressure, so 1.7 is used.

was checked in step 10. Unless the stem thickness at the base has been changed, no additional shear checking is necessary.) In equation 14.155, $w$ may be taken as 12″ if all calculations are done per foot of wall.

$$R_u = \frac{M_u}{\phi w d^2} \qquad 14.155$$

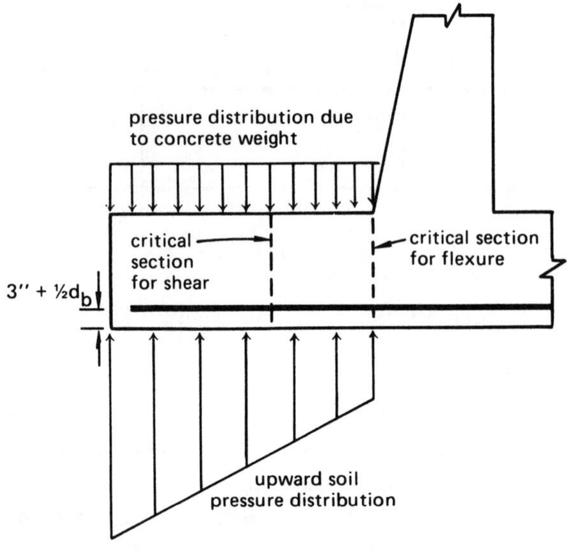


**Figure 14.29** Toe Construction Details

From $R_u$, obtain $\rho$ and $A_{st}$. Select reinforcement to meet distribution requirements.[81]

*step 15*: Check the development length of the stem reinforcement. The length into the stem itself (above the base) should be adequate without checking. However, it may be difficult to achieve full development in the base. The reinforcement may be extended into the key, if one is present. Otherwise, hooks, bends, or smaller bars should be used.

*step 16*: Calculate temperature and shrinkage reinforcement from $h$ and $w$. It is common to allocate 2/3 of this reinforcement to the outward face of the stem, since that surface is alternatively exposed to day and night. The remaining 1/3 is placed on the soil side of the stem, which is maintained at a more constant temperature by the soil insulation.

*step 17*: Provide for drainage if the design was based on a drained condition.[82] Weep holes should be placed regularly (i.e., approximately every 10 feet) along the length of the wall.

[81] This step only determines flexural reinforcement in the stem at the base. As the stem goes up, less reinforcement will be required. Similar calculations should be performed at one or two other locations (up the stem) to obtain a curve of $M_u$ versus location. Trial and error can be used to select reinforcement which meets the strength requirements.

[82] Unless there is a need to make the retaining wall act as a watertight bulkhead, drainage should always be part of the design.

# Appendix A: Miscellaneous ACI Detailing Requirements

### Minimum Cover on Non-prestressed Steel

- Cast against and permanently exposed to earth . . . 3″
- Exposed to earth or weather
  - #5 bars or smaller . . . $1\frac{1}{2}$″
  - #6 bars or larger . . . 2″
- Not exposed to weather or ground contact
  - Slabs, walls, or joists
  - #11 bars or smaller . . . 3/4″
  - #14 and #18 bars . . . $1\frac{1}{2}$″
  - Beams, girders, or columns . . . $1\frac{1}{2}$″

### Minimum Horizontal Clear Distance Between Bars in a Layer

- Beams: The maximum of one bar diameter or . . . 1″
- Walls: The maximum of one bar diameter or . . . 1″

### Minimum Vertical Clear Distance Between Bars in a Layer

- Beams: . . . 1″

### Maximum Rebar Spacing (clear distance)

- Walls and Slabs: The minimum of three times the wall or slab thickness, or . . . 18″

### Splices (Ratio of lap length to development length, $l_d$)[1,2]

- Tension splices [See ACI 318 section 12.15 for different classes.]
  - Class A . . . 1.0
  - Class B . . . 1.3
  - Class C . . . 1.7
  - The minimum tension splice lap length is . . . 12″
- Compression splices ($f'_c$ exceeds 3000 psi)
  - The minimum of 12″, $(0.0005)(f_y)$(bar diameter), and one development length

Many other provisions apply to both tension and compression lap splices. Welded splices are butt-welds developing an ultimate tensile strength of 125% of the yield strength of the bar.

### Minimum Bend Radii (In bar diameters)

- #3 to #8 bars . . . 6
- #9, #10, and #11 bars . . . 8
- #14 and #18 bars . . . 10

### Maximum Deflections

These limits are fractions of the distance between supports.

| element | when | is element attached to non-structural items likely to be damaged by large deflections? | limit |
|---|---|---|---|
| flat roof | immediate | no | 1/180 |
| floor | immediate | no | 1/360 |
| roof/floor | sustained | no | 1/240 |
| roof/floor | sustained | yes | 1/480 |

### Minimum Beam and Slab Thickness

(For use when deflections are not computed.) These thicknesses apply to normal weight concrete (density greater than 120 pcf) and grade 60 steel. For other grade steels, multiply the thickness by

$$0.4 + \frac{f_y}{100{,}000}$$

These fractions are the fraction of span length between supports.

| | simply supported | one end continuous | both ends continuous | cantilever |
|---|---|---|---|---|
| solid, one-way slabs | 1/20 | 1/24 | 1/28 | 1/10 |
| beams or ribbed one-way slabs | 1/16 | 1/18.5 | 1/21 | 1/8 |

### Maximum Aggregate Size

1/5 of the narrowest dimension between sides of the forms, or 1/3 of the slab depth, or 3/4 of the clear spacing between bars, whichever is greatest.

Notes:

[1] #14 and #18 bars may not be lap spliced. [Section 12.14.2]

[2] For 2-bar splices only. Bar bundles require additional lap length. [Section 12.14.2]

# Appendix B:
# The Alternate (Working Stress) Method of Design

ACI-318 permits the following stresses when the alternate design method is used.

CONCRETE (normal weight)

**flexure**

| | | |
|---|---|---|
| compressive stress on extreme fiber | | $0.45f'_c$ |

**shear**

| | | |
|---|---|---|
| beams (concrete carries all shear) | | $1.1\sqrt{f'_c}$ |
| one-way footings and one-way slabs | | $1.1\sqrt{f'_c}$ |
| two-way footings and two-way slabs | minimum of | $\left\{ \begin{array}{l} 2.0\sqrt{f'_c} \\ \left(1+\frac{2}{\beta}\right)\sqrt{f'_c} \end{array} \right.$ |

| | | |
|---|---|---|
| **bearing on loaded area** | | $0.30\sqrt{f'_c}$ |

STEEL (tensile stress)

**beams**

| | | |
|---|---|---|
| grade 40 or 50 | | 20,000 psi |
| grade 60 and higher | | 24,000 psi |
| one-way slabs, less than 12 foot span, with #3 bars or smaller main reinforcement | minimum of | $\left\{ \begin{array}{l} 0.50 \text{ fy} \\ 30{,}000 \text{ psi} \end{array} \right.$ |

The alternate method can be used to design beams if the following assumptions are made:

- Plane sections remain plane during bending.
- Steel and concrete both remain in the elastic range.
- Concrete carries all compressive loads in an area above the neutral axis.
- Steel carries all tension.
- Stress in the concrete is proportional to the distance from the neutral axis.

For maximum stresses in the concrete and steel, define the location of the neutral axis.

$$k = \frac{1}{1 + (f_s/nf_c)} = \sqrt{2n\rho + (n\rho)^2} - n\rho$$

$$\rho = \frac{A_s}{bd}$$

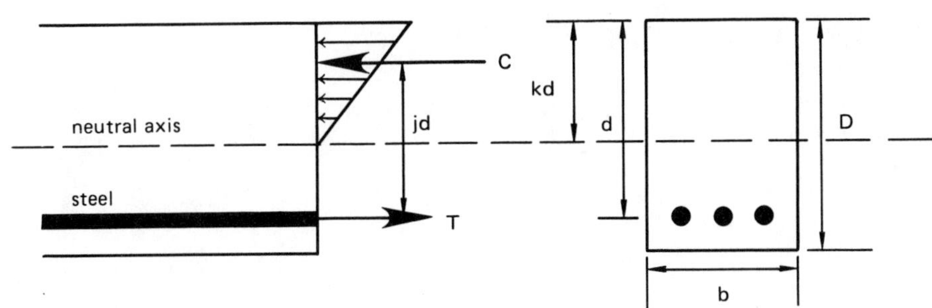


The total compressive force is

$$C = \frac{1}{2}f_c bkd$$

The total tensile force is

$$T = A_s f_s = C$$

Since the compressive force, $C$, acts one-third down from the top of the beam,

$$j = 1 - \frac{k}{3}$$

The maximum moments that the concrete and steel can carry are

$$M_c = Cjd = \frac{1}{2} f_c jkbd^2$$
$$M_s = Tjd = A_s f_s jd = f_s j \rho bd^2$$

The required steel area is

$$A_s = \frac{M_s}{f_s jd}$$

If the design is *balanced*, then $M_c = M_s$. Therefore,

$$\frac{1}{2} f_c jkbd^2 = f_s j \rho bd^2$$

**Practice Problems: CONCRETE DESIGN**

Untimed

1. 1.5 cubic yards of portland cement concrete are needed. Using the specifications below, determine the number of pounds of water, cement, fine aggregate, and course aggregate needed.

| | |
|---|---|
| cement content: | 6.5 bags per cubic yard |
| water cement ratio: | 5.75 gallons per bag |
| cement specific gravity: | 3.10 |
| fine aggregate specific gravity: | 2.65 |
| coarse aggregate specific gravity: | 2.00 |
| aggregate grading: | 30% fine, 70% coarse by volume |
| free moisture: | 1.5% in fine aggregate |
| coarse aggregate will absorb: | 3% moisture |
| entrained air: | 5% |

2. Design a square 2-way footing to carry a live load of 240,000 pounds which are transmitted through a 16" square column. The allowable soil pressure is 4000 psf and the top of the footing is level with the surrounding soil surface. Use 3000 psi concrete and 40,000 psi steel. Disregard dead loading.

3. Design a rectangular beam with tension reinforcement only using the ultimate strength method. The dead load moment is 50,000 ft-lb and the live load moment is 200,000 ft-lb. Use 3000 psi concrete and 50,000 psi steel. Do not design for shear, or check cracking or deflection.

4. Design a balanced beam meeting the following specifications. Use the working strength design method.

| | |
|---|---|
| maximum moment: | 50.4 ft-kips |
| maximum steel stress: | 20,000 psi |
| maximum concrete stress: | 1350 psi |
| $E_s/E_c$: | 10 |

5. Design a spiral column to carry a dead load of 175,000 pounds and a live load of 300,000 pounds. The loads are axial. Use 3000 psi concrete and 40,000 psi steel.

6. Design a short square tied column to carry a 100,000 pound dead load and a 125,000 pound live load. Include specifications for the ties. Use No. 5 bars or larger, 3500 psi concrete, and 40,000 psi steel.

7. An 18" square column supports 200,000 pounds and 145,000 pounds dead and live loads respectively. The allowable soil pressure is 4000 psf. Use the ultimate strength design method with 3000 psi concrete and 40,000 psi steel to design the footing. The column steel consists of 10 #9 bars.

8. A beam is being designed to withstand an ultimate factored moment of 400,000 ft-lb. 4000 psi concrete and 40,000 psi steel are available. Specify values of *b*, *d*, and the required steel area. How many layers of steel are needed? Do not check cracking or deflection.

9. 3000 psi concrete is used in the floor slab shown below. The slab must carry 20,000 ft-lb per foot of width. Use the transformed area method to evaluate the stresses in the concrete and steel.

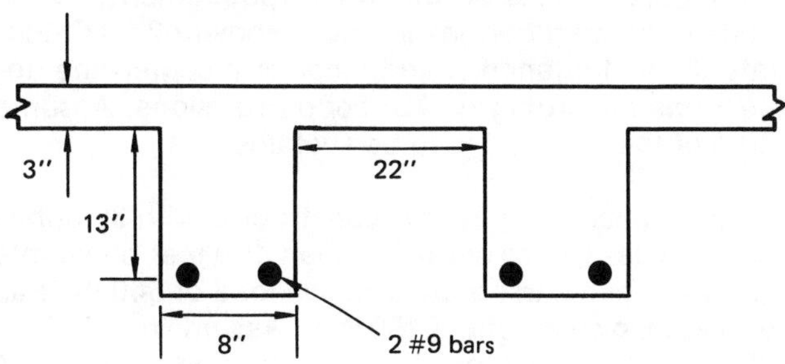


10. 100 pounds of aggregate were sieve graded. The weights retained on each sieve are shown below. What is the fineness modulus of the aggregate?

| sieve | weight retained |
|---|---|
| 4 | 4 |
| 8 | 11 |
| 16 | 21 |
| 30 | 22 |
| 50 | 24 |
| 100 | 17 |
| dust | 1 |

Timed

1. Use the latest ACI code to design the cross section and steel for the beam shown. Assume $f'_c = 4000$ psi and $f_y = 60,000$ psi. Only #11 steel is available. n = 8. All loads are dead loads. Neglect shear, but check for cracking under exterior conditions.

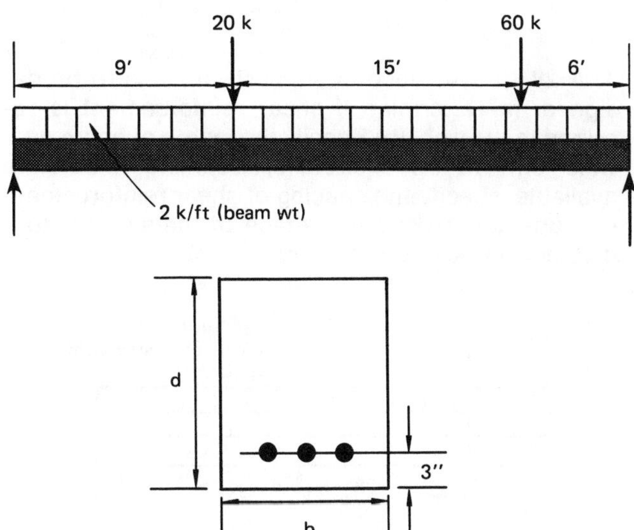

2. A reinforced concrete beam supporting a roof is needed to span 27 feet. It carries a dead load of 1 kip/ft (which includes the beam weight) and a live load of 2 kips/ft. Assume $f'_c$ = 4000 psi, $f_y$ = 60,000, and E = 3.6 EE6 psi. The beam is simply supported. Use the latest ACI code with the maximum steel percentage permitted. (a) Use #11 bars and design a rectangular beam with b = 14". Clear cover on the steel must be at least 1½". Do not design for shear. (b) Will the estimated crack sizes be within the requirements of the code if the member has interior exposure? (c) Calculate the instantaneous and long-term center-line deflections according to ACI code provisions. Assume 30% of the live load is to be sustained.

3. A pretensioned girder is constructed with 30 bonded strands. The girder must span 100 feet on simple supports and carry a uniform live load of 540 lb/ft as well as its own weight of 150 lb/ft. Assume n = 7, $f'_{ci}$ = 3800 psi, $f'_c$ = 5000 psi, $f_{ps}$ = 250 ksi, and $A_{ps}$ = 0.144 in²/strand. (a) Determine if the stresses in the concrete are acceptable immediately after transfer. Assume that only the dead weight acts on the beam and that prestress is 0.8 times $f_{pu}$, the tendon tensile strength. (b) Determine if the concrete stresses are acceptable at service loads. Assume the effective prestress is 0.6 times $f_{pu}$.

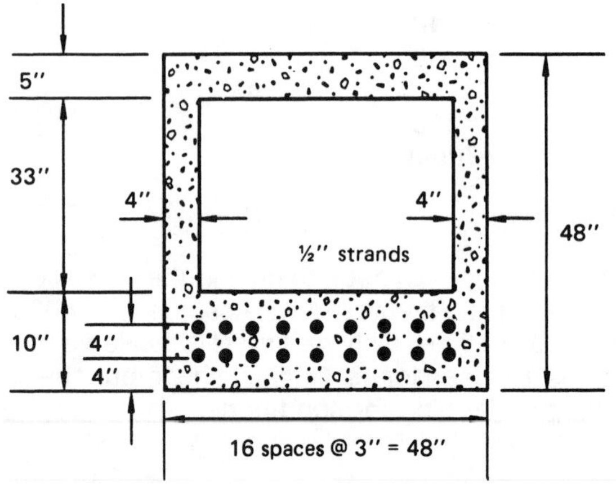


4. A 28′ (face-to-face of support) beam is to be designed. (a) Determine if shear reinforcement is required in the web. (b) Specify the length of beam over which stirrups are required. (c) Assuming #3 bars are available, specify the spacing of shear reinforcement (stirrups) located from the face of the support to a distance 21" (d) from the face.

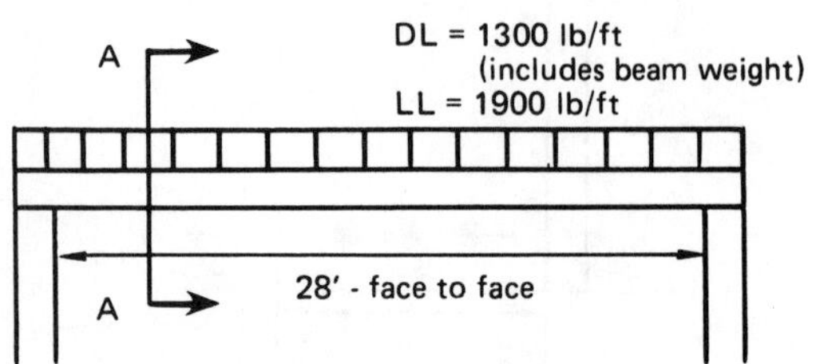


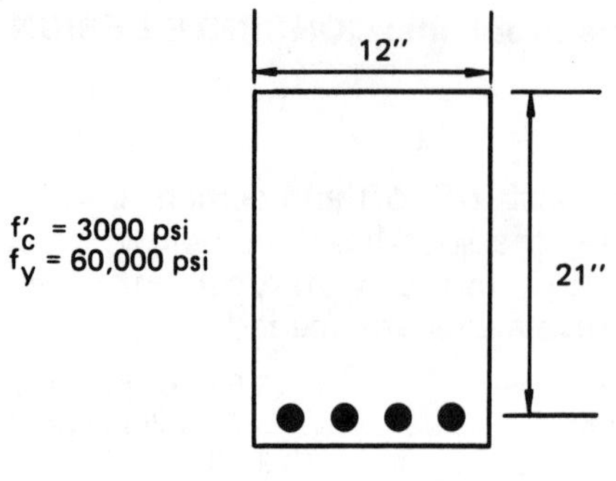


SECTION A-A

5. Given the steel-reinforced concrete combined footing shown, (a) find the dimension L which will result in a uniform soil pressure. (b) Draw complete shear and moment diagrams. (c) Should top steel in the pad be used?

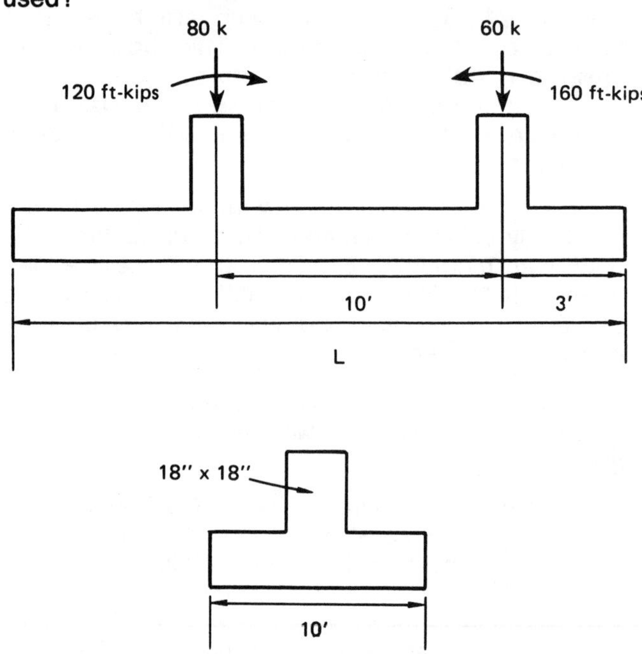


6. Near the end of the day, a long masonry pier is left unsupported at its upper end. (The lower end is firmly connected to the foundation.) During the night, a 30 psf wind starts blowing. Neglecting axial stresses due to dead load, determine the stresses in the steel and masonry.

(figure on next page)

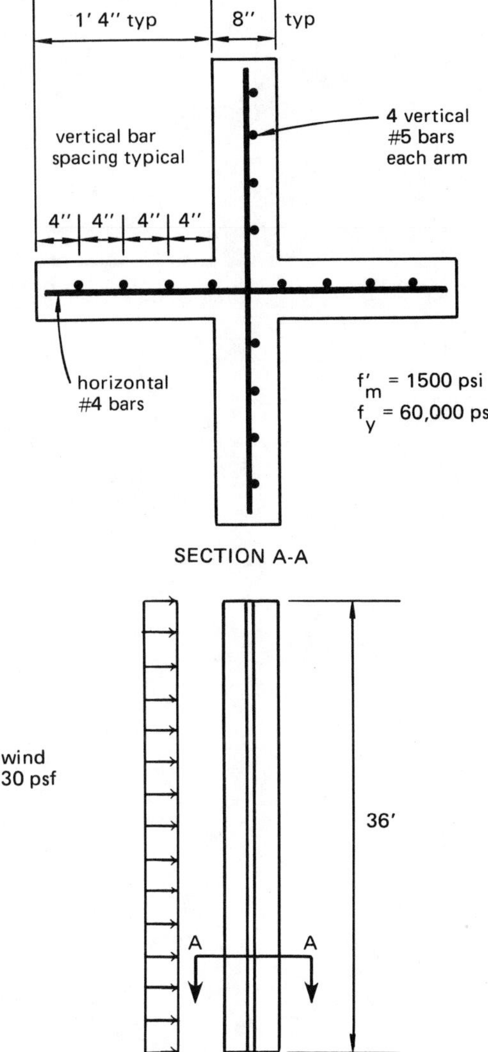


7. A 14″ square footing carries the live loads shown. The loads are not the result of wind or earthquake action. (Disregard dead loading.) The allowable soil load is 6000 psf. Use $f'_c = 3000$ psi for the concrete. (a) Size the footing. (b) What are the minimum and maximum pressures along the footing base? (c) What thickness of footing is required to avoid using shear reinforcement? (d) What is the maximum moment the footing must resist? (It is not necessary to design the steel reinforcement.)

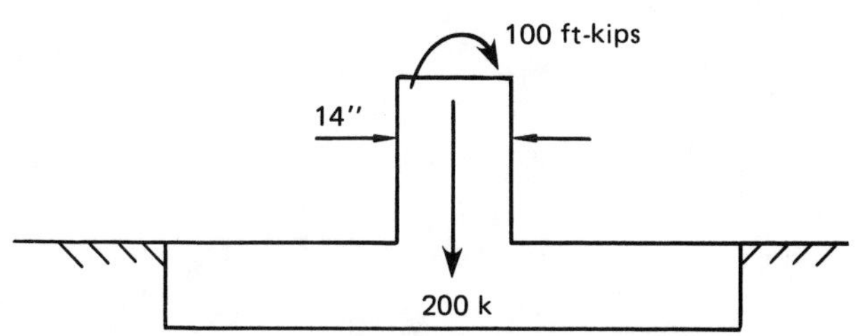


8. For the retaining wall shown, (a) what is the factor of safety against overturning? (b) What is the minimum theoretical heel depth? (c) Using the heel depth from part (b) and #8 steel, what bar spacing is required in the heel?

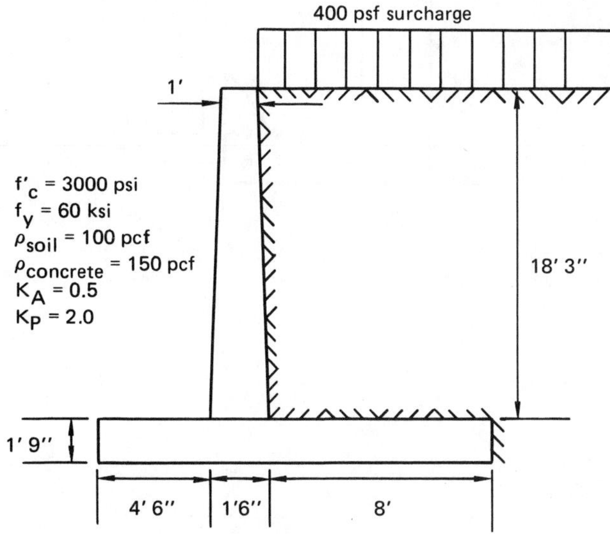


9. A 16″ wide beam is simply supported. It carries the loads shown. $f'_c = 3000$ psi. $f_y = 60$ ksi. (a) Design the beam, including the minimum beam depth. (b) Use USD and latest ACI code to detail the reinforcing. (c) Design any needed shear reinforcing using #3 bars.

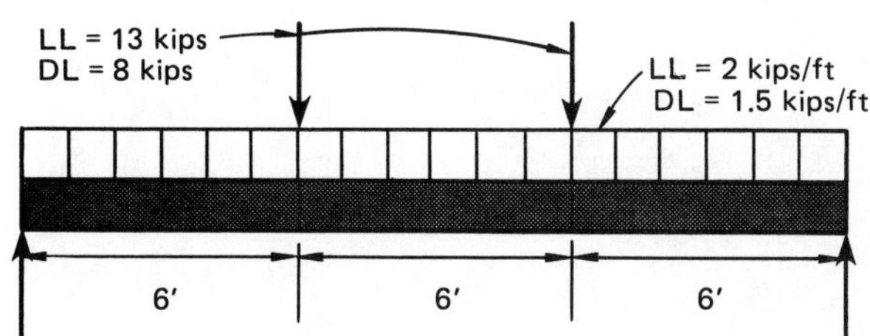


10. A deep concrete beam is to be designed as shown. Use $f'_c = 3000$ psi concrete and 60 ksi steel. Disregard self-weight, and consider the beam to be simply supported at its ends. (a) Find the nominal concrete shear strength, $v_c$, as given by the ACI codes for deep beams. (b) Find the amount of shear reinforcing, $A_v$, required for the loads as shown. (c) Determine the extent of this reinforcing.

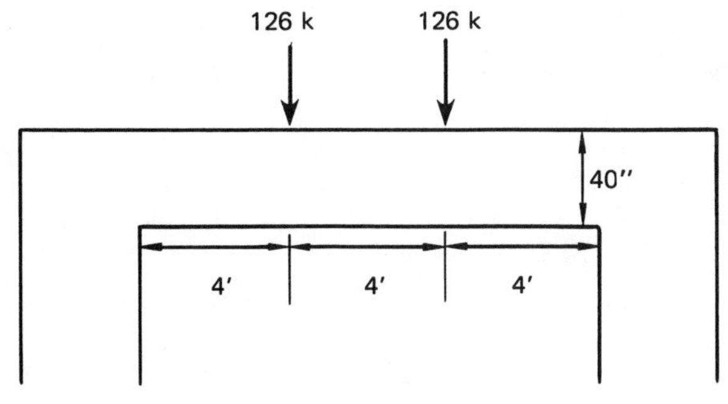


(additional figure on next page)

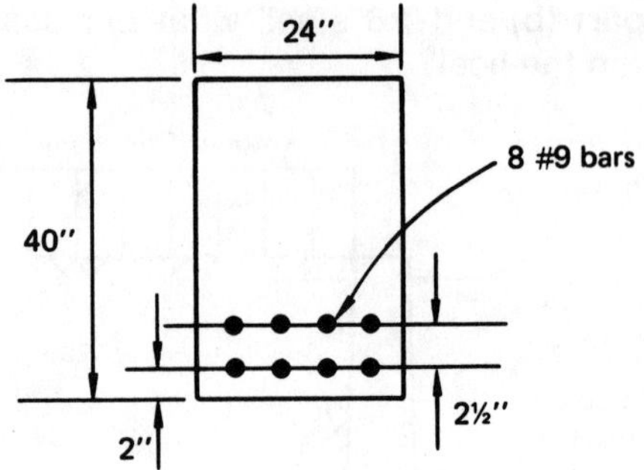
24"
40"
8 #9 bars
2½"
2"

# 15 STEEL DESIGN AND ANALYSIS[1]

*Nomenclature*

| | | |
|---|---|---|
| a | stiffener spacing | in |
| A | area | $in^2$ |
| b | width | in |
| c | distance to extreme fiber | in |
| C | coefficient | – |
| $C_c$ | critical slenderness ratio | – |
| d | depth or diameter | in |
| D | outside diameter | in |
| e | eccentricity | in |
| E | modulus of elasticity | psi |
| f | actual stress | ksi |
| F | strength or allowable stress | ksi |
| g | gage spacing | in |
| G | end condition coefficient | – |
| H | horizontal force | kips |
| I | area moment of inertia | $in^4$ |
| J | polar moment of inertia | $in^4$ |
| k | flange to web toe fillet distance, or spring constant | in, or lbf/in |
| K | end restraint coefficient | – |
| l | length between supports | in |
| L | length | in |
| M | moment | ft-kips |
| n | modular ratio | – |
| N | bearing length | in |
| P | force | lbf |
| r | radius of gyration, radius, or distance | in |
| $R_w$ | weld strength (resistance) | kips/in |
| s | pitch spacing | in |
| S | section modulus | $in^3$ |
| SR | slenderness ratio | – |
| t | thickness | in |
| T | tension | kips |
| V | shear | kips |
| w | weld size | in |
| Z | plastic modulus | $in^3$ |

1 This chapter cannot serve as a complete substitute for the American Institute of Steel Construction's *Manual of Steel Construction* or its accompanying *Specifications*. Throughout this chapter, references to the *AISC Specifications*, 8th edition, are listed in bold, square brackets. For example **[1.10-5]** is a reference to equation 1.10-5 in the *Specifications*, not to equations in this book. Also, tables, figures, and appendices listed in italic are part of the *AISC Manual* or *Specifications*. Thus, the *Allowable Stress Design Selection Table* would not be found in this volume.

*Subscripts*

| | |
|---|---|
| a | axial |
| b | bracing or bending |
| c | centroidal or concrete |
| cr | critical |
| e | effective |
| f | flange |
| g | gross |
| n | net |
| p | bearing or plastic |
| s | secondary or steel |
| st | stiffener |
| t | tension |
| u | ultimate or unbraced |
| v | shear |
| w | web |
| y | yield |

## 1 REVIEW OF STEEL NOMENCLATURE

It is traditional in steel design to use the upper case letter $F$ to indicate strength or allowable stress. Furthermore, such strengths or maximum stresses are specified in ksi (i.e., 1000's of psi). For example, $F_y = 36$ would imply a steel with a yield strength of 36 ksi. Similarly, $F_v$ is the allowable shear stress, and $F_b$ is the allowable bending stress, both in ksi.

Actual or computed stresses are given the symbol of lower case $f$. Computed stresses are also specified in ksi. For example, $f_t$ is a computed tensile stress in ksi. The symbol $\sigma$ is never used.

## 2 CONVERSIONS

| multiply | by | to obtain |
|---|---|---|
| ft-kips | 1.356 | kN-m |
| ft-lbf | 1.356 | N-m |
| in-lbf | 0.113 | N-m |
| kips | 4.448 | kN |
| kips | 1000 | lbf |
| kips/ft | 14.59 | kN/m |
| kips/ft$^2$ | 47.88 | kN/m$^2$ |
| kN | 0.2248 | kips |
| kN/m | 0.06852 | kips/ft |
| kPa | 1000 | Pa |
| ksi | 6.895 EE6 | Pa |
| lbf | 0.001 | kips |
| lbf | 4.448 | N |
| N-m | 0.7376 | ft-lbf |
| N-m | 8.851 | in-lbf |
| N-m$^2$ | 1 | Pa |
| Pa | 1.0 | N/m$^2$ |
| Pa | EE–6 | MPa |
| Pa | 0.001 | kPa |
| Pa | 1.45 EE–7 | ksi |
| Pa | 1.45 EE–4 | psi |
| Pa | 0.02089 | psf |
| psf | 47.88 | Pa |
| psi | 6.895 EE3 | Pa |

## 3 TYPES OF STEELS

ASTM A36 is the designation given to the all-purpose, carbon steel used for most projects. Most A36 shapes are hot rolled.

Other steels have higher strengths. Their use results in lower dead weights but higher material costs. The commonly available steels and their applications are listed in appendix A.

**Table 15.1**
Properties of A36 Steel

| | |
|---|---|
| $E$, modulus of elasticity: | 2.9 EE7 psi (up to 100°F) |
| $G$, shear modulus: | 11.5 EE6 psi |
| $\alpha$, coefficient of thermal expansion: | 6.5 EE–6 1/°F |
| $\mu$, Poisson's ratio: | 0.30 |
| $\rho$, density: | 490 lbm/ft$^3$ |
| $F_y$, yield strength: | 36 ksi (to 8″ thickness inclusive) |
| | 32 ksi (over 8″ thickness) |
| $F_u$, ultimate strength: | 58 ksi (minimum) |
| $F_e$, endurance limit: | approximately 30 ksi |

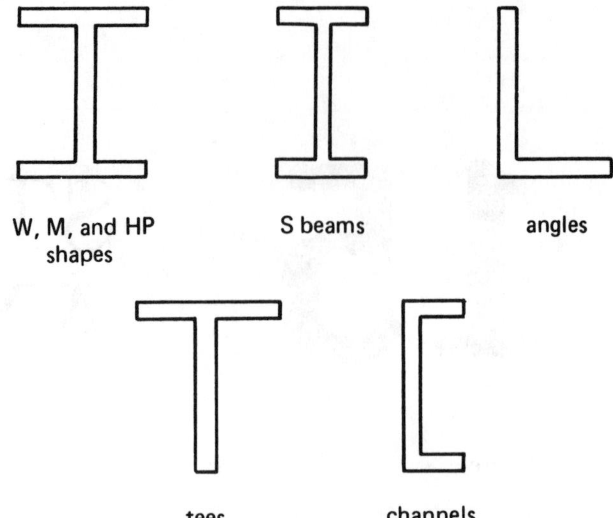


**Figure 15.1** Structural Shapes

## 4 STEEL PROPERTIES

Some properties of steel (such as the modulus of elasticity and density) are essentially independent of the type of steel. Other properties (such as the ultimate and yield strengths) depend not only on the type of steel but also on the size or thickness of the piece.

## 5 STRUCTURAL SHAPES

Many different structural shapes are available. The identifying dimension and weight must be appended to the designation to uniquely identify the shape. For example, W 30×132 means a W-shape with an overall depth of approximately 30 inches and which weighs 132 pounds per foot.

Table 15.2 lists structural shape designations.

**Table 15.2**
Structural Shape Designations

| shape | designation |
|---|---|
| wide flange beams | W |
| standard flanged beams | S |
| misc. flanged beams | M |
| American std. channels | C |
| bearing piles | HP |
| angles | L |
| tees | XT (cut from X) |
| plate | PL |
| bar | bar |
| pipe | pipe |
| structural tubing | TS |

## 6 STANDARD COMBINATIONS OF SHAPES

Figure 15.2 illustrates several combinations of shapes that are used in construction. The double angle combination is particularly useful for carrying axial loads. Combinations of W shapes and channels, channels with channels, or channels with angles are used for a variety of special applications, including struts and light crane rails. Properties for certain combinations have been tabulated in the *AISC Manual*.

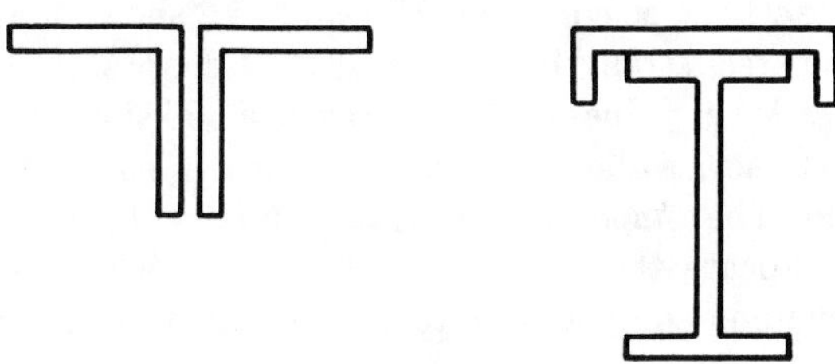

**Figure 15.2** Typical Combination Sections

## 7 REINFORCEMENT OF MILL SHAPES

Occasionally, it will be desirable to provide additional bending or compressive strength to a shape by adding plates. It is generally easy to calculate the properties of the built-up section from the properties of the shape and plate. No stress calculations are necessary. The following characteristics can be used when it is necessary to specify plate reinforcement.

- Plate widths should not be the same as $b_f$, due to difficulty in welding. Widths should be somewhat larger or smaller. It is better to keep plate width as close to $b_f$ as possible, as width-thickness ratios specified in the *AISC Manual* may govern.
- Width and length tolerances smaller than 1/8″ are not practical. Table 15.3 should be used when specifying the nominal plate width.

**Table 15.3**
Width Tolerance for Small Universal Mill Plates

| thickness (inches) | width (inches) 8–20, excl. | 20–36 excl. | 36 and above |
|---|---|---|---|
| 0–3/8, excl. | 1/8 | 3/16 | 5/16 |
| 3/8–5/8, excl. | 1/8 | 1/4 | 3/8 |
| 5/8–1, excl. | 3/16 | 5/16 | 7/16 |
| 1–2, incl. | 1/4 | 3/8 | 1/2 |
| 2–10, incl. | 3/8 | 7/16 | 9/16 |

- Not every plate exists in the larger thicknesses. Unless special plates are called for, the following thickness guidelines should be used:

  1/32″ increments up to 1/2″
  1/16″ increments from 9/16″ to 1″
  1/8″ increments from $1\frac{1}{8}$″ to 3″
  1/4″ increments for $3\frac{1}{4}$″ and above

*Example 15.1*

A W 30×124 shape must be reinforced to achieve the strong-axis bending strength of a W 30×173 shape by welding plates to both flanges. All steel is A36. The plates are welded continuously to the flanges. What size plate is required if all plate sizes are available?

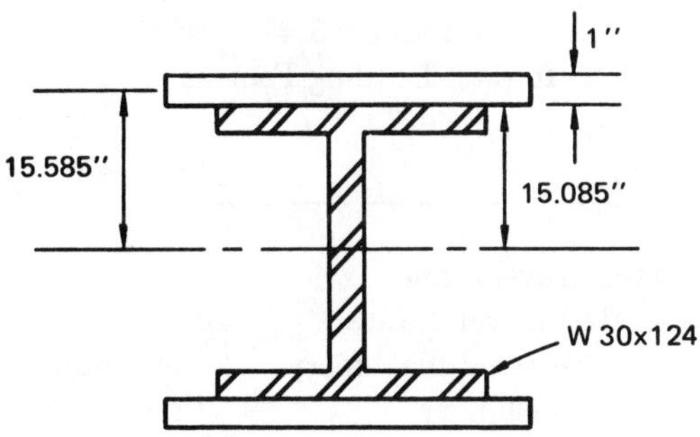


The moments of inertia are 8200 in$^4$ and 5360 in$^4$ for the two beams. The difference in bending resistance to be provided by the plates is

$$I_{\text{plates}} = 8200 - 5360 = 2840\,\text{in}^4$$

For ease of welding, assume the plate thickness will be approximately the same as the flange thickness. $t_f = 0.930''$, so choose a plate thickness of 1.0″.

The centroidal moment of inertia of the two plates acting together is

$$I_{c,\text{plates}} = 2\left[\frac{w(1)^3}{12}\right] = \frac{w}{6}$$

The depth of the W 30×124 beam is 30.17″. Therefore, the distance from the neutral axis to the plate centroid is

$$\frac{30.17}{2} + \frac{1}{2} = 15.585''$$

By the parallel axis theorem, the moment of inertia of the two plates about the neutral axis is

$$I_{\text{plates}} = \frac{w}{6} + (2)(w)(1)(15.585)^2 = 486.0w$$

The required moment of inertia is 2840. Therefore,

$$w = \frac{2840}{486} = 5.84'' \quad (\text{say } 6'')$$

## 8 FATIGUE LOADING

The effects of fatigue loading are generally not considered except for the case of connection design. If a load is to be applied and removed less than 20,000 times (as would be the case in a conventional building), no provision for repeated loading is necessary. (20,000 times is roughly equivalent to two times a day for 25 years.) However, some designs, such as for crane runway girders and supports, must consider the effects of fatigue. Design for fatigue loading is covered in Appendix B of the *AISC Specifications.*

## 9 ALLOWABLE STRESSES FOR IMPACT, WIND, AND EARTHQUAKE LOADS

The effects of impact are included by increasing the actual live load (but not the dead load) by the percentages contained in table 15.4.

**Table 15.4**
Impact Loading Factors

| supports for | % live load increase |
|---|---|
| elevators | 100 |
| cab operated travel cranes | 25 |
| travel operated travel cranes | 10 |
| shaft or motor-driven machinery | 20 minimum |
| reciprocating machinery | 50 minimum |
| floors and balconies | 33 |

Most allowable stresses, including those for connectors, columns, and beams, may be increased 1/3 for transitory wind and earthquake loading. This increase is applied to all stresses in the problem, not just the wind- or earthquake-induced stresses. The increase cannot be applied in fatigue loading problems. Also, the calculated section area cannot be less than the area required to carry the dead and live loads alone (without the 1/3 increase).

## 10 THE MOST ECONOMICAL SHAPE

Since a major part of the cost of using a rolled shape in construction is the cost of the raw materials, the lightest shape possible which will satisfy the structural requirements should always be used. Thus, the *most economical beam* is the structural shape which has the lightest weight per foot and which has the required strength. The *beam selection table* and *chart* are designed to make choosing economical shapes possible.

## 11 COMPACT SECTIONS

Compact sections have thicker webs and flanges than non-compact sections. Therefore, compact sections are afforded higher allowable stresses in many instances. To be compact, the flanges of a beam must be continuously connected to the web. Therefore, a built-up section or plate girder constructed with intermittent welds will not qualify. In addition, two conditions apply to standard rolled shapes without flange stiffeners. (Equation 15.2 assumes $f_a/F_y \leq 0.16$.)

$$\frac{b_f}{2t_f} \leq \frac{65.0}{\sqrt{F_y}} \qquad 15.1$$

$$\frac{d}{t_w} \leq \frac{640}{\sqrt{F_y}} \qquad 15.2$$

Compactness, as equations 15.1 and 15.2 show, depends on the steel strength. Most rolled W shapes are compact at lower values of $F_y$. However, a 36 ksi beam may be compact, while the same beam in 50 ksi steel may not be. The *shape tables* contain columns ($F_y'$ and $F_y'''$) that indicate the yield strengths at which the beams become non-compact due to the two conditions.

## 12 BEAM BENDING PLANES

The property tables for W shapes will show that each beam has two moments of inertia, $I_x$ and $I_y$. It is easy to use the wrong value, as there are several ways of referring to the plane of bending.

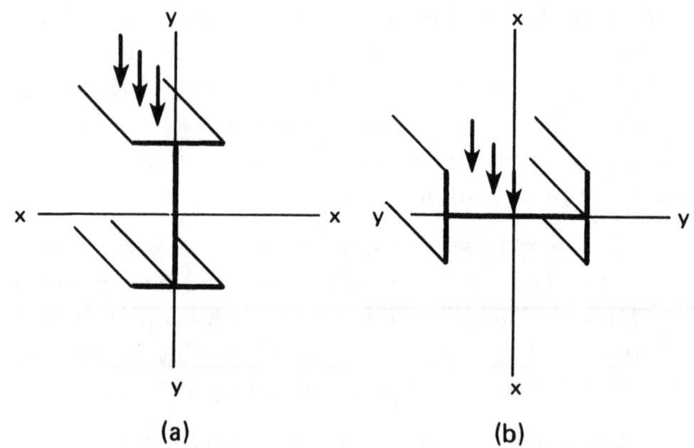


**Figure 15.3** Beam Bending Planes

Figure 15.3(a) shows a W shape beam used as it is typically. The value of $I_x$ should be used to calculate bending stress. This bending mode is referred to as "... loading in the plane of the web ...". However, it is also referred to as "... loading in the plane of the weak axis ...", which is confusing.

Figure 15.3(b) shows a W shape "... bending about the minor axis ...". This mode is also referred to as "... loading in the plane of the strong axis ...".

## 13 LATERAL BRACING

To prevent the lateral buckling illustrated in figure 15.4, a beam's compression flange must be supported at frequent intervals. Complete support is achieved when a beam is fully encased in concrete or has its flange welded or bolted along its full length. In many designs, however, the lateral support is at regularly spaced intervals.

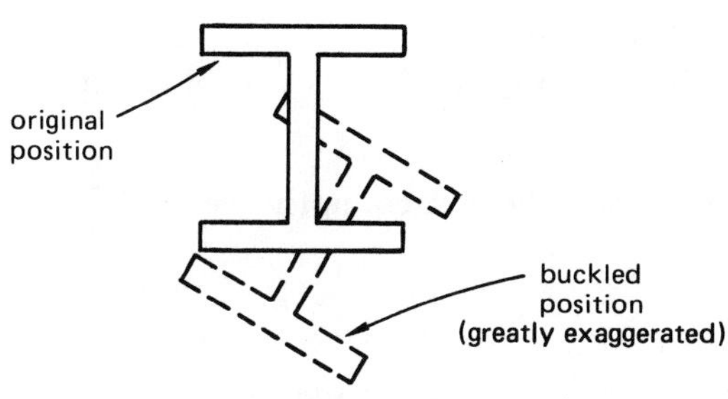


**Figure 15.4** Lateral Beam Buckling

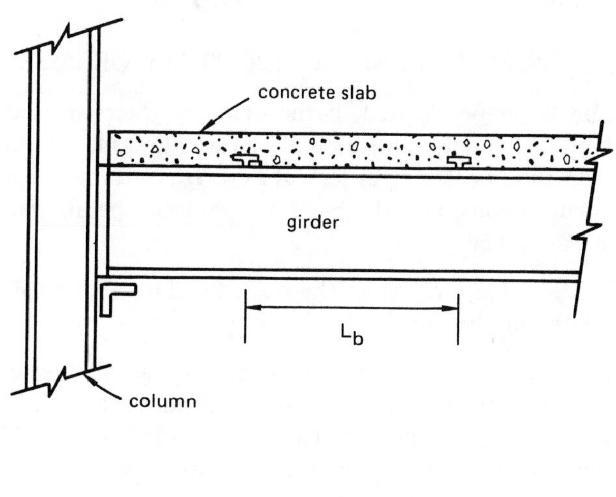


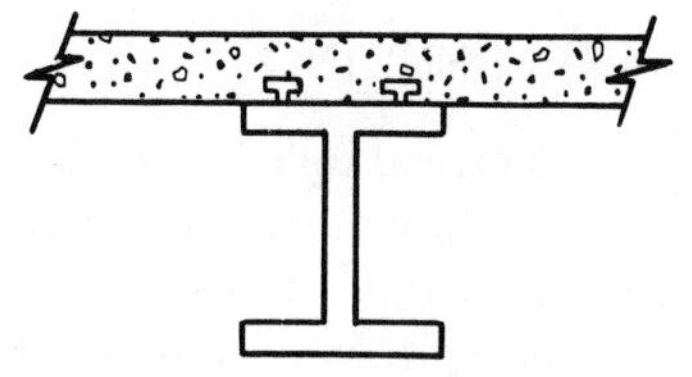

**Figure 15.5** Compression Flange Bracing Using Headed Studs

Lateral bracing is assumed to be provided if the compression flange is adequately supported. Such support is assumed to exist at all reaction points and at the point where the compression flange is bolted or welded to another member. However, simple clips between steel girders and precast concrete floors are not considered to provide sufficient lateral support.

The actual spacing between lateral bracing is designated as $L_b$. For the purpose of determining allowable stresses, two limits are placed on the spacing: $L_c$ and $L_u$. A higher allowable bending stress in beams is allowed if the actual spacing is less than $L_c$. [AISC sections 1.5.1.4.1(5) and 1.5.1.4.5(2)]

$$L_c = \frac{76b_f}{\sqrt{F_y}} \quad \text{(inches)} \tag{15.3}$$

$$L_u = \frac{20{,}000C_b}{\left(\dfrac{d}{A_f}\right)F_y} \quad \text{(inches)} \tag{15.4}$$

$b_f$ and the ratio $d/A_f$ are both tabulated for each beam.[2] The *moment gradient multiplier*, $C_b$, is almost always assumed to be 1.0 (the conservative case). $C_b$ is never used with $L_c$. (See Appendix D of this chapter.)

## 14 BEAM DEFLECTIONS

Steel beam deflections are calculated using traditional beam equations. Deflection limitations are typically unique to each design situation. The *AISC Commentary* suggests, but does not require, some general guidelines to maintain appearance and occupant confidence in a structure. These guidelines are 1/290 of a uniformly-loaded span's length for floors and 1/232 of a uniformly-loaded span's length for roofs.[3] The ratio is 1/360 for live load deflections of beams supporting plastered ceilings.

## 15 BENDING STRESS IN STEEL BEAMS

Elastic design and analysis of one-span beams is carried out with simple bending theory equations. Unless the beam is very short, it should be sized or analyzed with the flexure stress equation and subsequently checked for shear. Since both $c$ and $I$ are constant for any specific beam, the *section modulus* $S$ can be used from the AISC tables.

$$f_b = \frac{Mc}{I} = \frac{Md}{2I} = \frac{M}{S} \tag{15.5}$$

[2] It is seldom necessary to actually calculate $L_c$ and $L_u$ for rolled shapes with standard values of $F_y$, since the *Allowable Stress Design Selection Table* lists these values.

[3] Special provisions are used to check for ponding on flat roofs.

## 16 ALLOWABLE BENDING STRESS

### A. W-SHAPES BENDING ABOUT MAJOR AXIS

For W-shapes loaded in the plane of their webs and bending about the major axis, the allowable bending stress will be $0.66F_y$ or less. If $L_b < L_c$, or if the bracing is continuous, and the beam is compact, the allowable stress is $0.66F_y$.

If $L_u \geq L_b \geq L_c$, then the basic bending allowance of $0.60F_y$ applies.

If $L_b > L_u$, or if there is no bracing at all between support points, then the allowable bending stress is less than $0.60F_y$. Equations **[1.5-6a]**, **[1.5-6b]**, and **[1.5-7]** must be used to determine the actual value of $F_b$.[4]

### B. WEAK-AXIS BENDING

If a doubly-symmetrical rolled shape is placed such that bending will occur about its weak axis, the allowable bending stress is $0.75F_y$.[5] The shape must be compact, and other conditions may also apply. However, this is not an efficient use of the beam, so this configuration is seldom used.[6]

$0.75F_y$ is also the allowable bending stress for solid square, solid round, and solid rectangular shapes bending about the weak axis.

### C. INTERMEDIATE CASES

For both strong-axis and weak-axis bending, special cases exist for non-compact beams with $L_b < L_c$. In such instances, the so-called *blending formulas* (**[1.5-5a]** and **[1.5-5b]**) are used to produce intermediate values of $F_b$.[7]

## 17 SHEAR STRESS IN STEEL BEAMS

Only the web is assumed to carry shear in W shapes. The statical moment concept is almost never used in steel design, and the average shear stress is compared against the shear stress limitations.

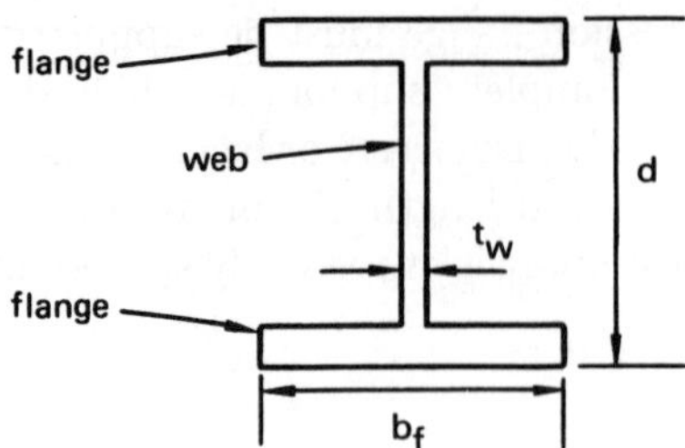


**Figure 15.6** A Steel Beam

The average shear stress in the web is

$$f_v = \frac{V}{A_w} = \frac{V}{dt_w} \qquad 15.6$$

The maximum allowable shear stress in the web of a beam is $F_v = 0.40F_y$.[8]

## 18 LOCAL BUCKLING[9]

Local buckling is a factor in the vicinity of large concentrated loads. Such loads may occur where a column frames into a supporting girder, or at reaction points. *Vertical buckling* and *web crippling*, two types of local buckling, can both be reduced or eliminated by use of *stiffeners*.

**Figure 15.7** Local Buckling

If the load is applied uniformly over a large enough area (say, along $N$ inches of beam flange or more), no stiffeners will be required. If the maximum stress at the toe of the beam fillet is to be kept below $0.75F_y$, the minimum length, $N$, is specified by equation 15.7 (**[1.10-8]**) for interior loads, and equation 15.8 (**[1.10-9]**) for reactions at beam ends. $k$ is the flange-to-web toe fillet distance tabulated in the shape tables.

$$N_{\text{min}} = \frac{\text{load}}{0.75F_y t_w} - 2k \quad \text{(interior)} \qquad 15.7$$

$$N_{\text{min}} = \frac{\text{reaction}}{0.75F_y t_w} - k \quad \text{(ends)} \qquad 15.8$$

[4] Actually, the allowable tensile bending stress remains at $0.60F_b$. Only the allowable compressive bending stress is reduced. For traditional design using symmetrical shapes, however, the tensile stress and the compressive stress are the same.

[5] It may seem curious that $F_b$ is larger for bending about the weak axis. The reason is to account for different failure modes. About the strong axis, the beam will fail by buckling of the compression flange and twisting about the minor axis. When bent about the weak axis, the beam fails by yielding.

[6] Nevertheless, weak-axis bending may occur, particularly in the case of beam-columns. Allowable stress for weak-axis bending is a factor in interaction equations such as equation 15.48.

[7] The name *blending formula* is used because the allowable stress for strong-axis bending will be between $0.60F_y$ and $0.66F_y$.

[8] Different limitations apply to shear stress in plate girders, bolts, and rivets.

[9] Lateral buckling has been covered previously.

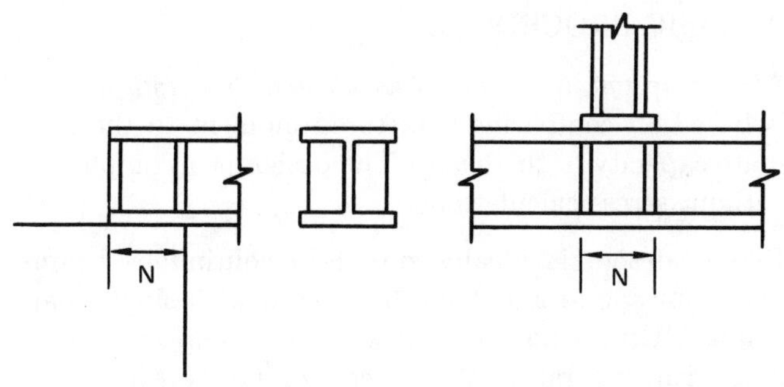


**Figure 15.8** End and Interior Bearing Stiffeners

*Intermediate stiffeners* (i.e., web stiffeners spaced throughout a stock rolled shape) are never needed with rolled shapes, but are typically present in plate girders. (For built-up beams, diagonal buckling requirements should also be checked.)

*Bearing stiffeners* are typically *web stiffeners* constructed as plates welded to the webs and flanges of rolled sections. *Flange stiffeners* are typically angles placed at the web-flange corner used to keep the flange perpendicular to the web. Flange stiffeners cannot be used in place of bearing stiffeners.

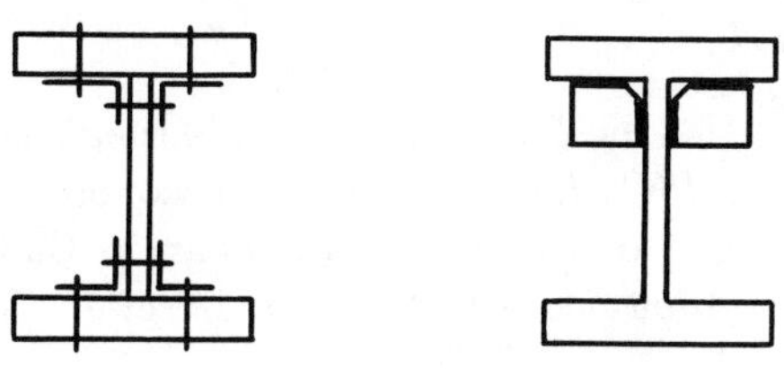

**Figure 15.9** Flange Stiffening

## 19 BEAM DESIGN AND ANALYSIS

Steel beams can be designed and analyzed with *allowable stress* or *plastic design procedures.*[10] Simple one-span beams are usually designed by the allowable stress method since there is no advantage to using load factor design procedures for single-span beams.

Beams are almost always constructed from $F_y = 36$ or $F_y = 50$ ksi steel.

Both beam design and analysis by the allowable stress method have the same standard procedures:

- Check or design for bending stress[11]
- Check shear stress
- Check deflections
- Check compactness (unnecessary if beam table or beam chart was used)
- Design for local buckling

## 20 BEAM DESIGN BY TABLE AND CHART

Since a trial and error solution to a design problem would be time consuming, the *AISC Manual* provides two simple methods of choosing beams: the *Allowable Stress Design Selection Table* and the *Allowable Moments in Beams Chart.*

### A. BEAM TABLE USE

The *Allowable Stress Design Selection Table* is easy to use, and it provides a method of quickly selecting economical beams. Its use assumes that either the required section modulus, $S$, or required resisting moment, $M_R$, is known, and uses one or the other of these criteria to select the beam. It does not check shear stress or deflection, and it is up to the designer to make sure that $L_b \leq L_c$.

Beams are arranged in groups in the beam selection table. Within a group, the most economical shape will be listed at the top in bold print. This is the beam that should generally be used, even if the moment resisting capacity and section modulus are greater than necessary. The weight of the most economical beam will be less than the beam in the group which most closely meets the structural requirements.

Care must be taken not to confuse the 50 ksi columns with the 36 ksi columns. Both steel strengths are listed in the table, and it is easy to use one when the other is needed.

### B. ALLOWABLE MOMENTS CHART

When the unbraced length, $L_b$, is greater than $L_c$, the *Allowable Moments in Beams* chart should be used. This chart can be used with unbraced lengths up to 42 feet. Each beam that is plotted on the chart is shown with a profile similar to that in figure 15.10.

[10] The allowable stress method is also known as *elastic design* and *working stress design.* The *load factor method* is also known as *plastic design* and *ultimate strength design.* Load factor design is also appropriate when design must be "... in accordance with part 2 of the *AISC Specifications* ..."

[11] If the live load is known but the self-load of the beam is not, it will be necessary to assume a beam weight prior to the selection. A second iteration can be used to obtain a different beam if necessary.

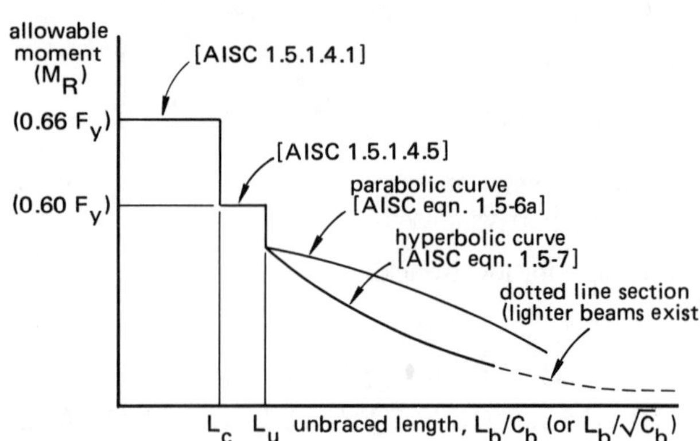


**Figure 15.10** Allowable Moments for a Single Beam

The chart is entered knowing $L_b$ and $M_R$.[12] (An allowance for self-weight should have been included when calculating $M_R$.) The nearest solid line above the intersection of the $L_b$ and $M_R$ values is the most economical beam. Dotted line sections mean that a lighter beam exists which has the same capacity.

As with the beam selection table, shear stress and deflection must still be checked. Care must be taken, also, not to mix up the 36 ksi and 50 ksi portions of the chart.

*Example 15.2*

Select a W shape ($F_y$ = 36 ksi) to carry a maximum moment of 140,000 ft-lbf. The compression flange is braced every 6 feet. Disregard deflection and shear stress criteria.

This problem is perfect for using the beam chart, since both the unbraced length and required moment are known.

Entering the chart with $L_b = 6'$ and $M_R$ = 140 ft-kip, a W 21×44 beam is selected. This beam has a moment carrying capacity of approximately 163 ft-kips, and several beams were skipped which had smaller capacities, but which still met the 140 ft-kip requirement. However, this is the lightest beam.

If the beam table had been used, the same beam would have been selected. However, it would have been necessary to verify that $L_b < L_c$.

[12] If unbraced length varies along the beam span, use the longest unbraced length with this chart.

## 21 PLASTIC BEAM DESIGN

A. BASIC PROCEDURE[13]

*Plastic design*, also known as *ultimate strength design*, is based on comparing a factored moment to the ultimate capacity of the beam. The design is accomplished without stress calculations.

Plastic design is ideally suited to continuous beams and frames, and is not normally used with single-span beams. It must not be used with non-compact shapes, crane runway rails, A514 steel, or steels with yield strengths in excess of 65 ksi.

*step 1*: Multiply the dead and live loads by 1.7 to obtain the *factored loading* on the beam.[14]

$$\text{factored load} = 1.7 \times \text{dead load} + 1.7 \times \text{live load} \qquad 15.9$$

*step 2*: Based on the factored loading, calculate the maximum (plastic) moment, $M_p$. (This moment cannot be calculated from elastic moment diagrams.)

*step 3*: If beam selection is to be made according to the required *plastic section modulus*, calculate $Z_x$.

$$Z_x = \frac{M_p}{F_y} \qquad 15.10$$

*step 4*: Use the *Plastic Design Selection Table* in the *AISC Manual* to select an economical beam. Use of this table is similar to that of the *Allowable Stress Design Selection Table.*

*step 5*: Check that the maximum factored shear based on plastic failure does not exceed the allowable shear.

$$V_{\text{factored, max}} = \leq 0.55 F_y t_w d \qquad 15.11$$

*step 6*: Specify web (intermediate) stiffeners at loading points where plastic hinges are expected.

*step 7*: Determine lateral bracing requirements. Compression flange support (lateral bracing) is required at points where plastic hinges will form. In addition, the distance between points of lateral support must be less than

[13] The procedure presented here for beam selection assumes that the *Plastic Design Selection Table* will be used, and therefore, compactness is assumed. The procedure also omits thickness ratio checks that must be performed when the beam carries axial loads in addition to its transverse loading.

[14] Other factors are used when wind and earthquake loads are present.

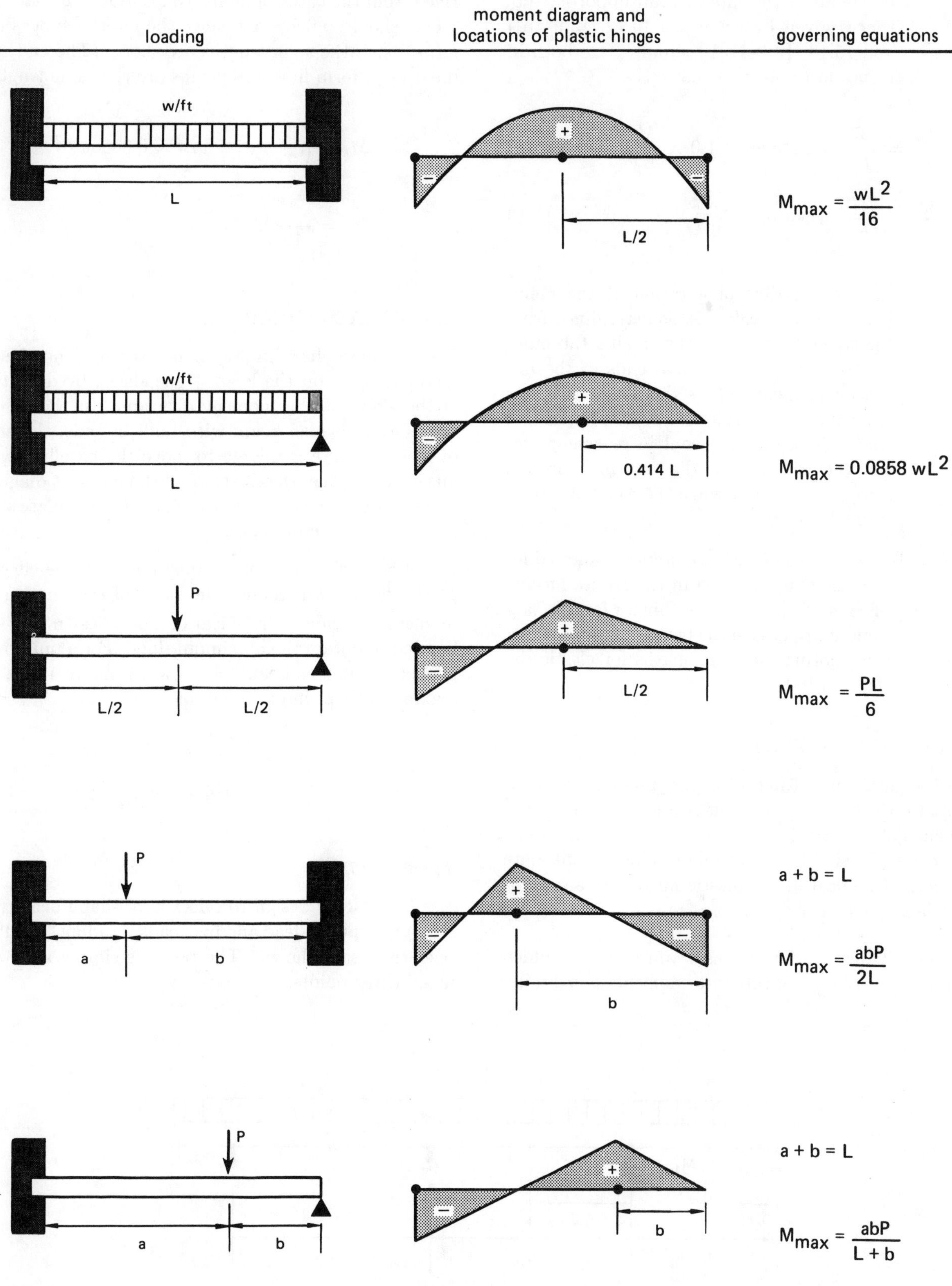


**Figure 15.11** Maximum Plastic Moments

(• = plastic hinge)

or equal to the laterally unsupported distances given by equations 15.12 ([**2.9-1a**]) and 15.13 ([**2.9-1b**]). $l_{cr}$ and $r_y$ are both in inches in these two equations.

$$\frac{l_{cr}}{r_y} = \frac{1375}{F_y} + 25 \text{ when } +1.0 > \frac{M}{M_p} > -0.5 \quad 15.12$$

$$\frac{l_{cr}}{r_y} = \frac{1375}{F_y} \text{ when } -0.5 \geq \frac{M}{M_p} > -1.0 \quad 15.13$$

$r_y$ is the radius of gyration of the member about its weak axis, as determined from the shape table. $M$ is the smaller (absolute value) moment at the two ends of the unbraced segment. $M_p$ is the maximum plastic moment on the span. $M/M_p$ is positive when the segment is bent in reverse curvature (i.e., the moment diagram goes through zero), and negative when the segment is bent in single curvature.

Some easing of the $l_{cr}$ distance is allowed for the last hinge to form in the failure mechanism, since having other hinges form is tantamount to having the beam in failure. However, normal bracing lengths for elastic design must still be met.

## B. ULTIMATE MOMENTS

Step 2 requires calculating the maximum moment based on plastic theory. This moment is not the same as the moment determined from elastic theory. Figure 15.11 can be used in simple cases to determine the ultimate moment, $M_p$. Locations of plastic hinges are also indicated in the figure.

A uniformly-loaded, continuous beam is a case which occurs frequently. The ultimate moments can be derived from the cases in figure 15.11. Both ultimate moments should be checked, since the interior hinges may form first if the span lengths are short. (The end span hinges will form first if all spans are the same length.)

$$M_1 = \left(\frac{3}{2} - \sqrt{2}\right) wL_1^2 \approx 0.0858wL_1^2$$

$$M_2 = \frac{wL^2}{16}$$

## C. ULTIMATE SHEARS

The ultimate shear in step 5 may or may not be the factored shear on the beam. The shear, by definition, is the slope of the moment diagram. In the case of a uniformly-loaded beam with built-in ends, the effect of plastic failure is merely to move the baseline, without changing the overall shape of the moment diagram. Since there is no change in the slope, the ultimate shear is merely the factored shear.

In the case of continuous beams, however, the effect of plastic failure will change the slope of the lines in the moment diagram. Graphical or analytical means can be used to obtain the maximum slope. For a uniformly-loaded continuous beam, as shown in figure 15.12, the maximum shear induced will be

$$\begin{aligned} V_{\text{max}} &= \frac{wL}{2} + \frac{M_{\text{max}}}{L} \\ &= 0.5858wL \end{aligned} \quad 15.14$$

*Example 15.3*

Use load factor design to select a W shape beam (A36 steel) to support dead and live loads totalling 4000 lbf/ft over the beam shown. The beam is simply supported at all three points.

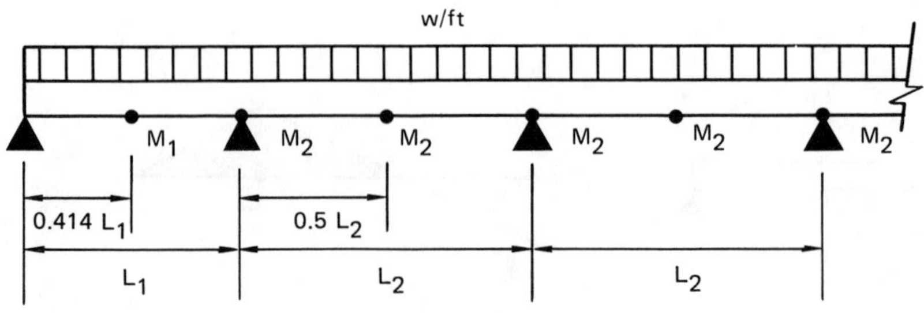


**Figure 15.12** Ultimate Moments on a Uniformly-Loaded Continuous Beam

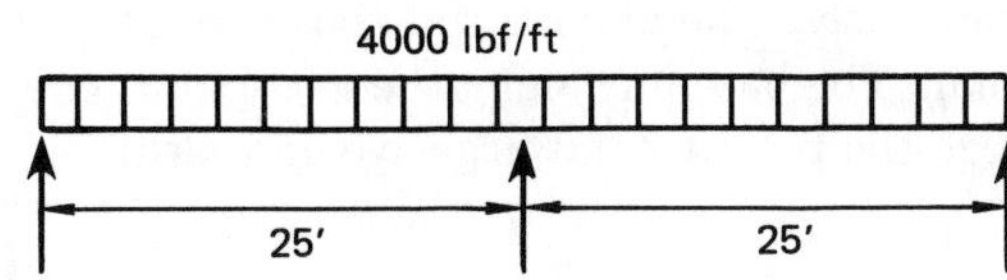


*step 1*: The factored load is

$$w = (1.7)(4000) = 6800\,\text{lbf/ft}$$

*step 2*: The maximum moment on an end-span of a uniformly-loaded continuous beam is

$$\begin{aligned} M_{\text{max}} &= 0.0858wL^2 = (0.0858)(6800)(25)^2 \\ &= 3.65\,\text{EE5 ft-lbf} \end{aligned}$$

*step 3*: The selection can be made on the basis of required plastic moment. This step is skipped.

*step 4*: Entering the *Plastic Design Selection Table*, a W 24×55 beam is chosen. The capacity is 402 ft-kips, more than required. However, this is the economical beam with a capacity exceeding the requirements.

*step 5*: The maximum shear on an end span of a uniformly-loaded, continuous beam is given by equation 15.14.

$$\begin{aligned} V_{\text{max}} &= 0.5858wL = (0.5858)(6800)(25) \\ &= 99,600\ \text{lbf} \end{aligned}$$

From the shape tables, $d$ = 23.57 inches, $t_w$ = 0.395 inches. From equation 15.11,

$$\begin{aligned} V_{\text{allowable}} &= (0.55)(36,000)(0.395)(23.57) \\ &= 184,340\ \text{lbf} \end{aligned}$$

Since $V_{\text{max}} < V_{\text{allowable}}$, shear is not a problem.

## D. CONSTRUCTING PLASTIC MOMENT DIAGRAMS

There are several methods of determining the shape of the plastic moment diagram. Most of these methods are based on theory. The method presented here is a graphical approach which can be used to quickly draw moment diagrams and locate points where plastic hinges will form.

*step 1*: Consider the beam as a series of simply-supported spans.

*step 2*: Draw the elastic bending moment on each span.

*step 3*: Construct the modified base line. This is a jointed set of straight lines that meets the following conditions:

- The base line meets the horizontal axis at simply-supported exterior ends.[15] This is consistent with the requirement that the moment be zero at free and simply-supported ends.
- The slope of the base line changes only at points of support.
- The base lines for all spans connect at points of support.

The base line is located to minimize the maximum ordinate along the entire length of the beam (along all spans).[16] Therefore, one span will control.

*step 4*: The maximum ordinate determines $M_p$.

*step 5*: Hinges form at points of maximum ordinates, and wherever else required to support full rotation. The moments at hinges which form simultaneously are identical.

*Example 15.4*

Draw the plastic moment diagram for the uniformly-loaded, two-span beam shown.

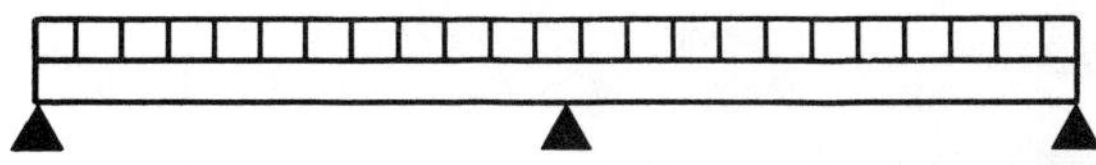

*steps 1 & 2* The moment diagram of a simply-supported, uniformly-loaded single span is drawn twice.

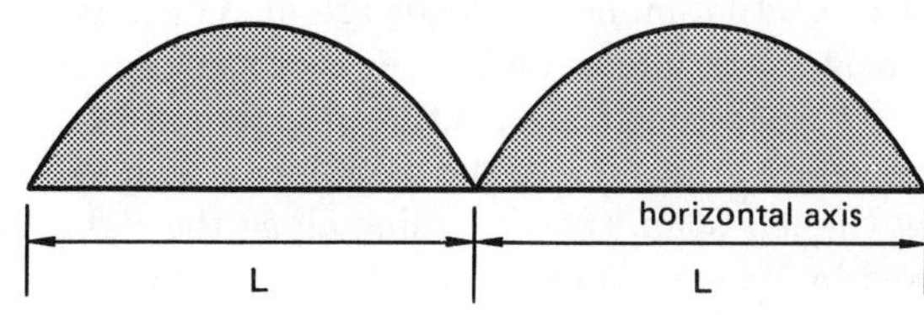


*step 3*: The modified base line is chosen to minimize the distance between the curved line and the base line.

[15] It isn't necessary for the base line to reach the horizontal axis at built-in ends. In fact, the moment is usually non-zero at those points.

[16] Don't consider the distance between the base line and the horizontal axis when minimizing the maximum ordinate.

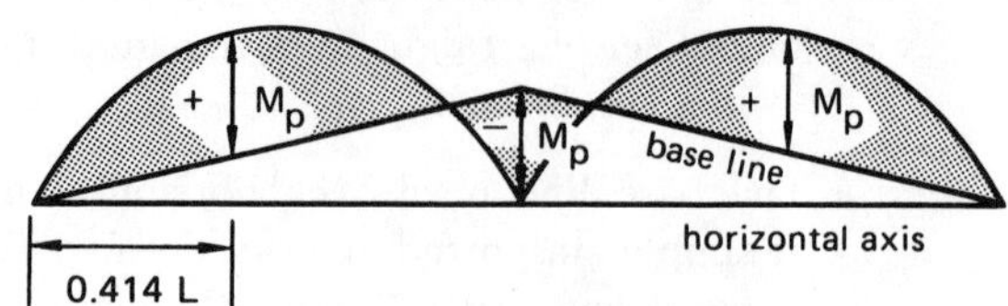


*step 4*: The maximum ordinate is determined visually to be near the middle of the end spans.

*step 5*: If hinges form near the middle of the end spans, a hinge must also form over the center support. Otherwise, the beam could not rotate in failure.

The final moment diagram can be drawn by "straightening out" the modified base line.

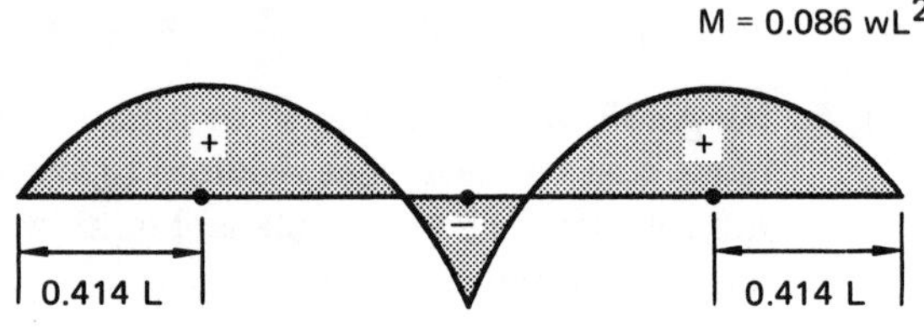


*Example 15.5*

Draw the plastic moment diagram for the beam shown.

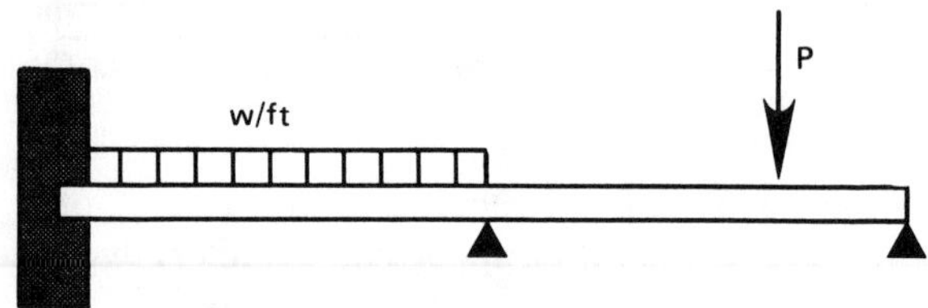


Since the actual numerical loads are unknown, it is not possible to determine which of the two spans controls.[17] If the left span controls, then three hinges must form simultaneously. This forces the base line to be horizontal along the left span. The base line along the right span is fixed by its endpoints.

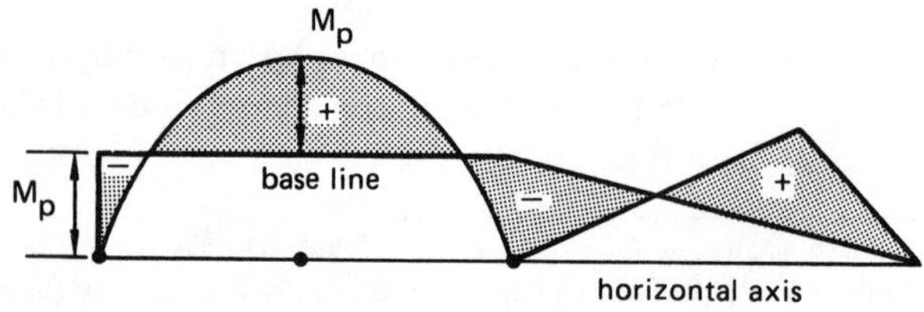


[17] The two alternate methods by which the beam can fail are known (in plastic theory) as *mechanisms.*

If the right span controls, then two hinges must form simultaneously. The moments at these two hinge points are equal. The base line along the left span is chosen to minimize the positive and negative ordinates.

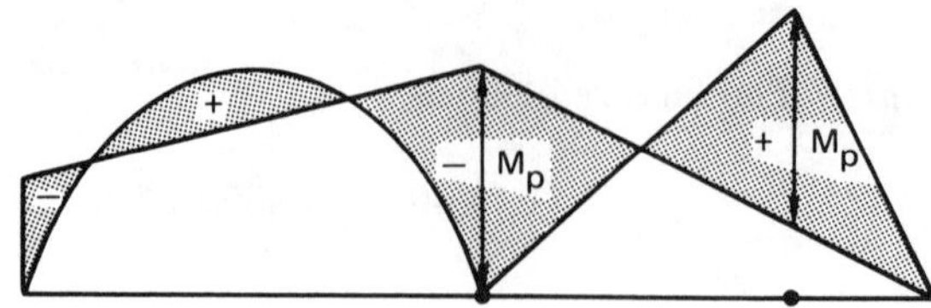


## 22 COLUMNS WITH AXIAL LOADS

### A. INTRODUCTION

Column design and analysis is greatly dependent on the Euler buckling load theory. Specific factors of safety and slenderness ratio limitations separate the design and analysis procedures from purely theoretical concepts, however.

### B. GEOMETRIC TERMINOLOGY

Figure 15.13 shows a $W$ shape used as a column. The $I_y$ moment of inertia is smaller than $I_x$. Therefore, the beam would be said to "... buckle about the minor axis ..." if the failure (buckling) mode is as shown. Since this is the expected buckling mode, bracing for the minor axis is usually provided, even if major axis bracing is not.

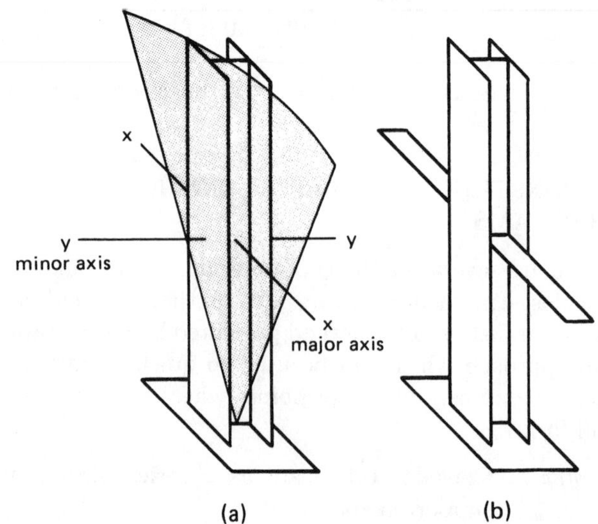


**Figure 15.13** Minor Axis Buckling and Bracing

Associated with each column are two unbraced lengths, $L_x$ and $L_y$.[18] In figure 15.13(b), $L_x$ is the full column height. However, $L_y$ is half the column height, assuming that the brace is placed at mid-height. It is not necessary for $L_x$ and $L_y$ to be identical.

Another important geometric feature is the *radius of gyration.* Since there are two moments of inertia, there are also two radii of gyration. Since $I_y$ is smaller than $I_x$, $r_y$ will be smaller than $r_x$. $r_y$ is known as the *least radius of gyration.*

## C. EFFECTIVE LENGTH

Since the restraints placed on column ends greatly affect a column's stability, an *end-restraint coefficient* is used to modify the unbraced length.[19] Thus, $KL$ is the product of the end-restraint coefficient and the unbraced length, and is known as the *effective length* of the column.

Values of $K$ depend on the conditions at both ends of the column. Either end can experience complete fixity (which is, in practice, impossible to achieve) to zero fixity (as in a free-standing sign post or flagpole). Table 15.5 lists recommended values of $K$ for use with steel columns.[20]

**Table 15.5**
End-Restraint Coefficients, $K$
(Also see table 12.3)

| end #1 | end #2 | $K$ |
|---|---|---|
| built-in | built-in | 0.65 |
| built-in | pinned | 0.80 |
| built-in | rotation fixed, translation free | 1.2 |
| built-in | free | 2.1 |
| pinned | pinned | 1.0 |
| pinned | rotation fixed, translation free | 2.0 |

The values of $K$ specified in table 15.5 do not require a prior knowledge of the column size or shape designation. However, if an existing design for which all column and framing members are known is being evaluated, the alignment charts in figure 15.14 and 15.15 can be used to obtain a more accurate end-restraint coefficient.

[18] Even if the column is braced in one or both directions, the *unbraced length* is the distance between braces.

[19] This coefficient is also known as the *effective length coefficient.*

[20] These are not the theoretical values often quoted for use with Euler's equation. They are slightly different, as recommended by the American Institute of Steel Construction in its *Commentary.*

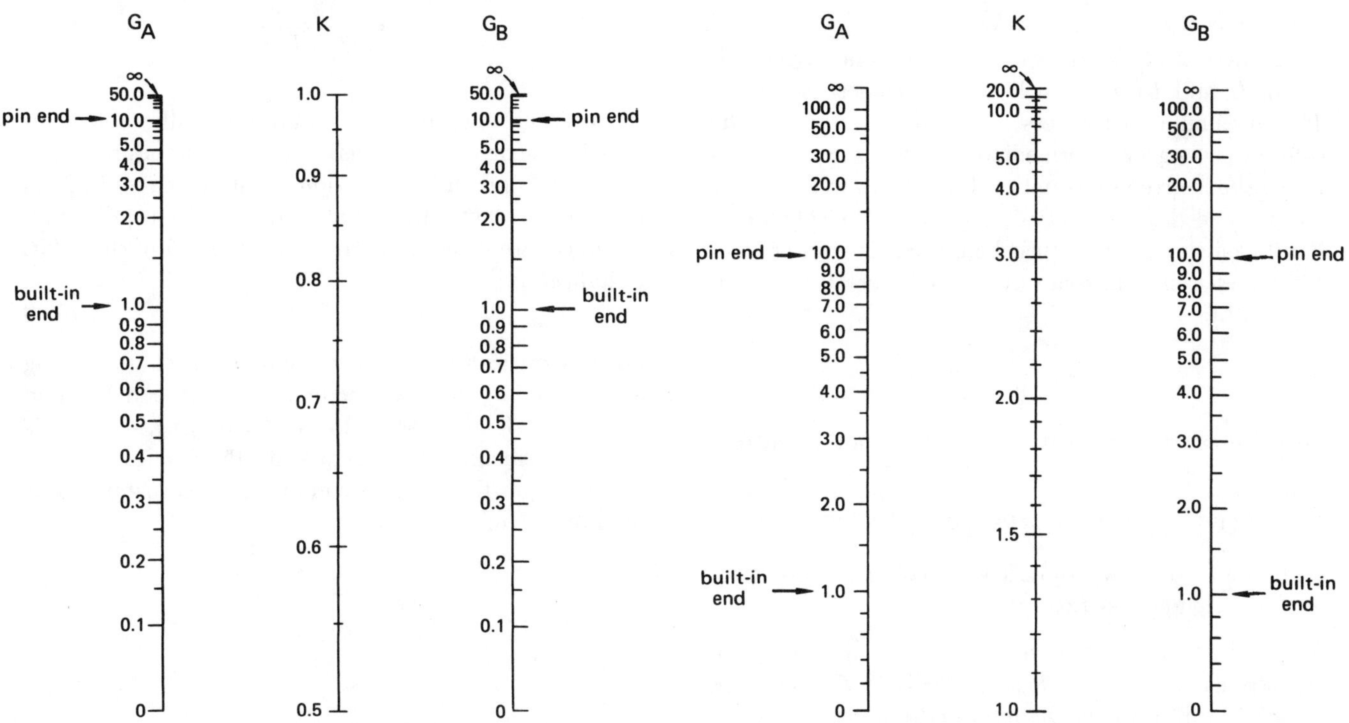


**Figure 15.14** Alignment Chart When Sidesway is Inhibited

**Figure 15.15** Alignment Chart When Sidesway is Uninhibited

To use the alignment charts, the *end condition coefficients*, $G_A$ and $G_B$, need to be calculated for the two column ends, $A$ and $B$. (The alignment charts are symmetrical. It is not important which end is labeled $A$ or $B$.)

$$G = \frac{\sum\limits_{\text{columns}} \left(\frac{I}{L}\right)}{\sum\limits_{\text{beams}} \left(\frac{I}{L}\right)} \qquad 15.15$$

In calculating $G$, only beams and columns which are in the expected plane of bending (i.e., which resist the tendency to buckle or bend) are included in the summation. Also, only beams and columns rigidly attached are included, since pinned connections do not resist moments.

For ground level columns, one of the column ends will not be framed to beams or other columns. In that case, $G = 10$ (theoretically $G = \infty$) is used for pinned ends, and $G = 1$ (theoretically, $G = 0$) is used for rigid footing connections.

D. SLENDERNESS RATIO

Steel columns are divided into *long columns* and *intermediate columns*, depending on their slenderness ratios. This slenderness ratio is calculated from equation 15.16.

$$SR = \frac{KL}{r} \qquad 15.16$$

Since there are two values of $r$ (and accompanying values of $K$ and $L$), there will be two slenderness ratios. The maximum slenderness ratio will determine if the column is long or intermediate. The *critical slenderness ratio* is given by equation 15.17. However, there is really no need to calculate $C_c$ for standard steel grades. For 36 ksi steel, a column is long if $SR$ is greater than 126.1 and is intermediate otherwise. For 50 ksi steel, $C_c = 107.0$.

$$C_c = \sqrt{\frac{2\pi^2 E}{F_y}} \qquad 15.17$$

Slenderness ratios of greater than 200 are not permitted.

E. ALLOWABLE COMPRESSIVE STRESS

As figure 15.16 shows, the allowable column stress varies with the slenderness ratio.[21]

The easiest method of determining allowable compressive stress is to use the *Allowable Stress for Compression Members* tables in the *AISC Specifications.* However, the allowable stress can also be calculated for steels with non-standard yield strengths.

[21] Very short compression members, those less than about 2 feet in effective length, are governed by different requirements.

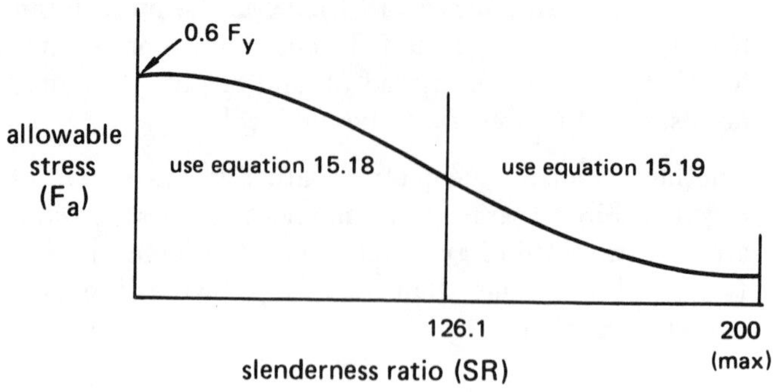


**Figure 15.16** Allowable Column Compressive Stress ($F_y = 36$ ksi)

Up to a slenderness ratio of $C_c$, equation 15.18 ([**1.5-1**]) gives the allowable stress.

$$F_a = \frac{\left[1 - \frac{(KL/r)^2}{2C_c^2}\right] F_y}{\frac{5}{3} + \frac{3(KL/r)}{8C_c} - \frac{(KL/r)^3}{8C_c^3}} \qquad 15.18$$

When the slenderness ratio exceeds $C_c$, equation 15.19 ([**1.5-2**]) must be used.

$$F_a = \frac{12\pi^2 E}{23(KL/r)^2} \qquad 15.19$$

An alternate method of calculating the allowable compressive stress for steels with non-standard yield strengths is possible. Equation 15.20 calculates $F_a$ from the yield strength and a reduction coefficient, $C_a$. Values of $C_a$ are listed in *Table 4, Appendix A* of the *AISC Specifications.*

$$F_a = C_a F_y \qquad 15.20$$

For less critical compressive members, such as bracing and *secondary members*, equation 15.21 can be used when the slenderness ratio is 120 or greater. $F_a$ is the value calculated from equation 15.18 or 15.19 using $K = 1$ regardless of end conditions. Consistent units for $L$ and $r$ must be used.

$$F_{a,s} = \frac{F_a}{1.6 - \frac{L}{200r}} \qquad 15.21$$

F. COLUMN ANALYSIS

The procedure for analyzing the adequacy of a column is essentially one of verifying that the actual compressive stress does not exceed the allowable stress.

*step 1*: Obtain the shape properties $A$, $r_x$, and $r_y$, as well as the unbraced lengths $L_x$ and $L_y$.

*step 2*: Obtain $K_x$ and $K_y$ from table 15.5 or from the alignment charts (figures 15.14 and 15.15).

*step 3*: Use equation 15.22 to calculate the maximum slenderness ratio.

$$SR = \max\left\{\begin{array}{c}\frac{K_x L_x}{r_x} \\ \frac{K_y L_y}{r_y}\end{array}\right\} \qquad 15.22$$

*step 4*: Use equation 15.17 to calculate the critical slenderness ratio, $C_c$. (For 36 ksi steel, $C_c = 126.1$; for 50 ksi steel, $C_c = 107.0$.)

*step 5*: Use the *Allowable Stress for Compression Members* table to obtain the allowable compressive stress. Alternatively, use either equation 15.18 or 15.19 (depending on $SR$).

*step 6*: Compare the actual load to the maximum allowable load on the column.[22]

$$P_{\text{max}} = F_a A \qquad 15.23$$

## G. COLUMN DESIGN

Trial and error column selection is next to impossible. Accordingly, the *AISC Manual* provides column selection tables which make it fairly easy to select a simple column based on the required column capacity.

*step 1*: Determine the load to be carried. Include an allowance for the column weight.

*step 2*: Based on preliminary choices for the column design, determine the end-restraint coefficients, $K_y$ and $K_x$, for the column. Calculate the effective length assuming that buckling will be about the minor axis.[23]

$$\text{effective length} = K_y L_y \qquad 15.24$$

*step 3*: Enter the table and locate a column which will support the required load with an effective length of $K_y L_y$.

*step 4*: Check for buckling in the strong direction. Calculate $L'_x$ from equation 15.25 and the ratio $r_x/r_y$ tabulated in the column tables.

$$L'_x = \frac{K_x L_x}{r_x/r_y} \qquad 15.25$$

*step 5*: If $L'_x < K_y L_y$, the column is adequate and the procedure is complete. Go to step 8.

*step 6*: If $L'_x > K_y L_y$ but the beam chosen can support the load at a length of $L'_x$, the column is adequate and the procedure is complete. Go to step 8.

*step 7*: Choose a larger member which will support the load at a length of $L'_x$. (The ratio $r_x/r_y$ is essentially constant.)

*step 8*: If sufficient information on other members framing into the column is available, use the alignment charts (figures 15.14 and 15.15) to check the values of $K$ used.

*Example 15.6*

Choose a W14 shape to support a 2000 kip concentric load. The unbraced column length is 11 feet in both directions. Use $K_y = 1.2$ and $K_x = 0.80$. $F_y = 36$ ksi.

*step 1*: Assume a column weight of approximately 500 lbf/ft. The load to be carried is

$$P = 2000 + \frac{(500)(11)}{1000} \approx 2005 \text{ kips}$$

*step 2*: From equation 15.24, the effective length for minor axis bending is $(1.2)(11) = 13.2'$ (say $13'$).

*step 3*: From the column table for 36 ksi steel, select a W 14×370 shape with a capacity of 2121 kips.

*step 4*: From the table, $r_x/r_y = 1.66$. Using equation 15.25,

$$L'_x = \frac{(0.80)(11)}{1.66} = 5.30$$

*step 5*: Since $5.30 < 13$, the column selected is adequate.

[22] The gross area of the column is used to calculate the allowable stress. *Lacing bars* do not contribute to gross area. However, *cover plates* may be used to carry column load.

[23] If buckling will occur about the major axis, the table cannot be used directly. See steps 4 through 7.

*Example 15.7*

Design a 25 foot long A36 W shape main member column to support a 375,000 pound live load. The base is rigidly framed in both directions. The top is rigidly framed in the weak direction and fixed against rotation in the strong direction, but translation in the strong direction is possible.

Assume the column dead weight will be about 2000 pounds. Then, the actual load will be 375,000 + 2000 = 377,000 pounds.

From the information about framing, the end restraint coefficients are $K_y = 0.65$ and $K_x = 1.2$.

The effective lengths are

$$L_y = (0.65)(25) = 16.25 \text{ ft}$$
$$L_x = (1.2)(25) = 30 \text{ ft}$$

Use the column selection table to find a column capable of supporting 377 kips with an effective length of 16 feet. Try a W 12×79 beam. This beam has $r_x/r_y = 1.75$. Then, from equation 15.25,

$$L'_x = \frac{30}{1.75} = 17.1$$

Since 17.1 > 16.25, the strong axis controls. Enter the table looking for a 17 foot (effective length) column capable of supporting 377 kips. The same column has sufficient capacity, and the column selection is complete.

## 23 STABILITY OF PLATES IN COMPRESSION

It is possible that *local buckling* in one of the plate elements of a rolled shape or built-up compression member may occur before buckling based on the slenderness ratio becomes the governing factor. The load-carrying ability of the column will be reduced if such local buckling occurs.

The ability of plate sections to carry compressive loads without buckling is determined by the *width-thickness ratio*, b/t. For the purpose of specifying limiting width-thickness ratios, compression elements are divided into stiffened elements and unstiffened elements. *Stiffened elements* are supported along two edges. Examples are webs of W shapes and sides of box beams. *Unstiffened elements* are supported along one edge only. Flanges of *W* shapes and sides of angles are unstiffened elements.[24]

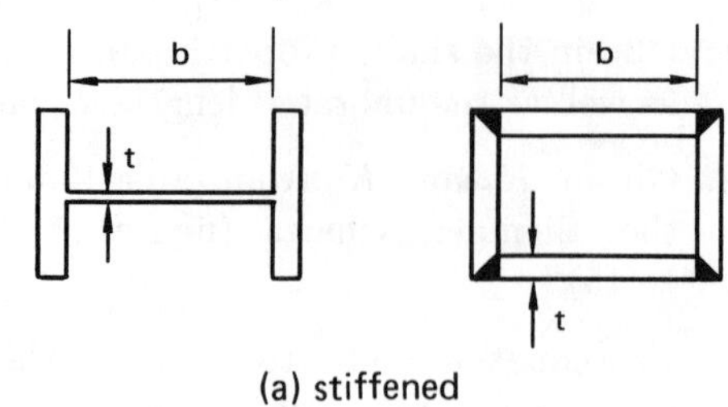


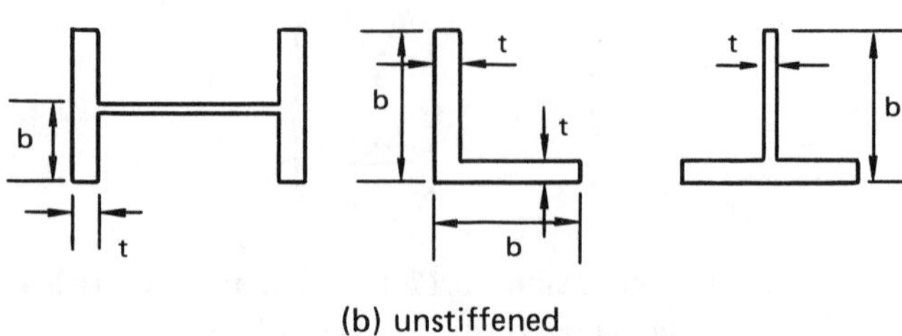


**Figure 15.17** Stiffened and Unstiffened Elements

To prevent local buckling, the *AISC Specifications* (*section 1.9*) require equation 15.26 to be met if the plate element is to be considered fully effective.[25]

$$\frac{b}{t} \le \frac{H}{\sqrt{F_y}} \qquad 15.26$$

Values of $H$ are listed in table 15.6.

**Table 15.6**
*H* Values for Width-Thickness Ratios

| element | $H$ |
|---|---|
| unstiffened elements | |
| stems of tees | 127 |
| double angles in contact | 95 |
| compression flanges of beams | 95 |
| angles or plates projecting from girders, columns, or other compression members | 95 |
| stiffeners on plate girders | 95 |
| flanges of tees and I-beams (use $\frac{1}{2}b_f$) | 95 |
| single angle struts or separated double angle struts | 76 |
| stiffened elements | |
| square and rectangular box sections | 238 |
| cover plates with multiple access holes | 317 |
| other uniformly compressed elements | 253 |

[24] If a *W* shape is selected from the column selection tables in the *AISC Manual*, the width-thickness ratios do not generally need to be evaluated. However, compression members constructed from double tees, structural tubing, and plate girders must be checked.

[25] The width-thickness ratios are applicable for elastic (working stress) design, and should not be used for inelastic design. Other provisions govern the width-thickness ratio of plate girder flanges, as well.

Circular tubular sections are considered to be fully effective according to the ratio of outside diameter to wall thickness.

$$\frac{D}{t} \leq \frac{3300}{F_y} \qquad 15.27$$

If the width-thickness ratios are exceeded by unstiffened compression members, the allowable compressive stress is reduced by a strength reduction factor, $Q_s$.[26]

$$F'_a = Q_s F_a \qquad 15.28$$

Stiffened compression elements exceeding the width-thickness ratios are handled differently. An *effective width*, $b_e$, is used in place of the actual width when calculating the flexural design properties and the permissible axial stress.

The specific provisions for determining $Q_s$ and $b_e$ are contained in Appendix C of the *AISC Specifications.*

*Example 15.8*

Two A36 L 9×4×$\frac{1}{2}$ angles are used with a 3/8″ gusset plate to produce a truss compression member. The short legs are back-to-back, making the long legs unstiffened elements. Can the combination fully develop compressive stresses?

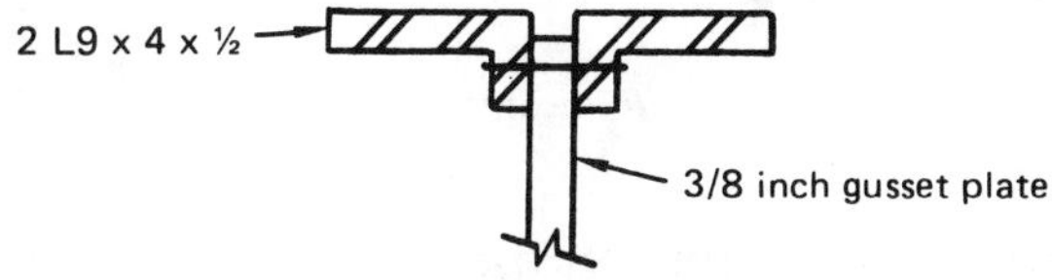


The limitation on the unstiffened separated double angles is

$$\frac{76}{\sqrt{F_y}} = \frac{76}{\sqrt{36}} = 12.67$$

The actual width-thickness ratio is

$$\frac{b}{t} = \frac{9}{0.5} = 18$$

Since 18 > 12.67, local buckling will control, and a reduced stress factor, $Q_s$, must be used when calculating the allowable compressive load on the truss member.

[26] Equation 15.28 implies that $F_a$ is the same as for members that meet the width-thickness ratios. Actually, $Q_s$ is also incorporated into the calculation of the critical slenderness ratio, $C_c$, and the allowable compressive stress, $F_a$.

## 24 MISCELLANEOUS COMBINATIONS IN COMPRESSION

Miscellaneous shapes, including round and rectangular tubing, single and double angles, and built-up sections, can be used as compression members. Generally, sections with distinct strength advantages in one plane (e.g., double angles or tees) should be used when bending is confined to that plane. For example, a double angle member could be used as a compression strut in a truss or as a spreader bar used for hoisting large loads.

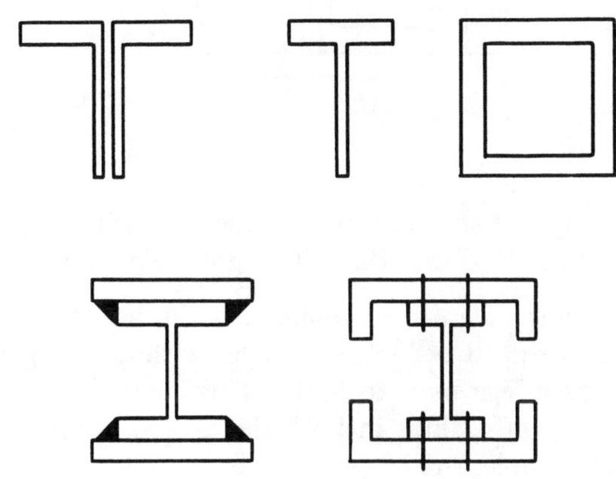

**Figure 15.18** Miscellaneous Compression Members

Design and analysis of these compression members is similar to the design and analysis of *W* shape columns. The same equations are used for calculating the allowable stress, $F_a$. It is, however, essential to check the width-thickness ratios for all elements in the compression members, including both flanges and stems of tees.

Where spot welding or stitch riveting is used to combine two shapes into one (as is done with double angles), the spacing of the connections must be sufficient to prevent premature buckling of one of the shapes. Once the maximum slenderness ratio has been determined (based on $x$- and $y$-directions), the spacing between connections can be calculated from equation 15.29, in which $r_z$ is the least radius of gyration for a single angle (as read from the shape table).

$$L_b = \frac{(SR_{\text{max}})r_z}{K} \qquad 15.29$$

Column load tables have been prepared for many combinations of double angles, tees, round pipe, and structural tubing. Single angle compression members are difficult to load concentrically, and special methods must be used for them.

*Example 15.9*

Design a double-angle strut, 10 feet long, to support an axial load of 40 kips. Use A36 steel and assume $K = 1$ for all cases.

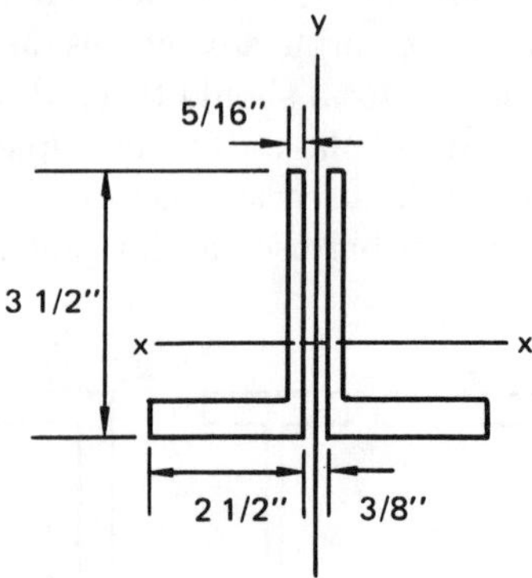


The weight of the strut will be insignificant compared to the axial load. So, the self-weight is ignored.

From the *Double Angle column tables* in the *AISC Manual*, try two L $3\frac{1}{2}\times2\frac{1}{2}\times5/16$ angles with a 3/8" gusset plate, long legs back to back. This configuration has a capacity of 42 kips in both the x- and y-directions. From the table, the properties are

$$A = 3.55$$

$$r_x = 1.11$$

$$r_y = 1.10$$

The controlling slenderness ratio is the larger value of KL/r.

$$\frac{K_xL}{r_x} = \frac{(1)(10)(12)}{1.11} = 108.1$$

$$\frac{K_yL}{r_y} = \frac{(1)(10)(12)}{1.10} = 109.1 \quad \text{(controls)}$$

The allowable compressive stress can be calculated from equation 15.18 or read directly from *AISC Specification Table 3-36, Appendix A.* $F_a = 11.81$ kips. The actual compressive stress is

$$f_a = \frac{40}{3.55} = 11.3\,\text{ksi} \quad \text{(ok)}$$

The width thickness ratio is easily calculated from the angle designation.

$$\frac{b}{t} = \frac{3.5}{5/16} = 11.2$$

The maximum width-thickness ratio for separated double angle struts is

$$\frac{76}{\sqrt{F_y}} = \frac{76}{\sqrt{36}} = 12.67 \quad \text{(ok)}$$

From the shape table, the least radius of gyration is $r_z = 0.54$. Since the maximum slenderness ratio is 109.1, the distance between connections between the two angles should not exceed

$$L = \frac{(109.1)(0.54)}{1} = 58.9''$$

## 25 MEMBERS IN BEARING

### A. PROJECTED AREAS IN CONNECTIONS

The maximum *bearing stress* on projected areas of bolts and rivets in shear connections is

$$F_p = 1.50F_u \qquad 15.30$$

If the connectors, plates, or shapes have different strengths, then the lower value of $F_u$ will govern.

### B. PROJECTED AREAS IN PINNED CONNECTIONS

When a connection is made with a pin in a reamed, drilled, or bored hole, the maximum bearing stress on projected areas is

$$F_p = 0.90F_y \qquad 15.31$$

### C. BEARING STIFFENERS

The bearing stress on the ends (tops and bottoms) of *bearing stiffeners*, and the stress on the contact area between mill sections and those bearing stiffeners, must not exceed

$$F_p = 0.90F_y \qquad 15.32$$

Only the part of the stiffeners outside the flange angle fillet or flange-to-web welds is considered effective in bearing.

### D. BEARING ON MASONRY SUPPORTS AT BEAM ENDS

Beams terminating at bearing connections on masonry, brick, or concrete supports are limited to bearing stresses based on the support material.

- sandstone and limestone:

$$F_p = 0.40\,\text{ksi} \qquad 15.33$$

- brick in cement mortar:

$$F_p = 0.25\,\text{ksi} \qquad 15.34$$

- on the full area of a concrete support:

$$F_p = 0.35 f'_c \qquad 15.35$$

For beams resting on only a portion of a concrete support, the allowable bearing pressure is given by equation 15.36. $F_p$ is not to exceed $0.7f'_c$, however.

$$F_p = 0.35 f'_c \sqrt{\frac{A_{\text{support}}}{A_{\text{bearing}}}} \qquad 15.36$$

Bearing plate area is determined by the load and $F_p$. Plate length, $N$, is found from equation 15.7 or 15.8. Width, $B$, is easily calculated as area/$N$. Bearing base plates are assumed to distribute the load to the masonry support as long as the base plate stress does not exceed $F_b = 0.75F_y$. Assuming a uniformly loaded cantilever span of length $n = \frac{1}{2}B - k$ supporting an actual bearing pressure of $F_b$, the required plate thickness is

$$t = n\sqrt{\frac{3f_p}{F_b}} = n\sqrt{\frac{4f_p}{F_y}} \qquad 15.37$$

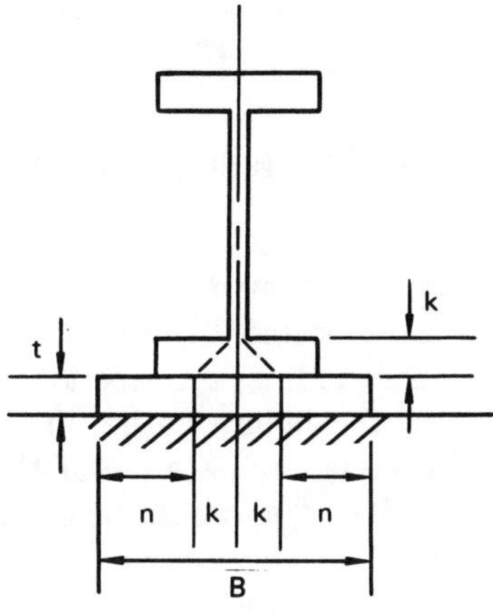


**Figure 15.19** Bearing Plate Thickness

E. COLUMN BASE PLATES

Column loads transmitted to foundations must meet the same masonry bearing pressure limitations as beams.[27] The load is transmitted from the column to the concrete foundation through a ***base plate***, as shown in figure 15.20.

[27] More likely than not, other codes or specifications will control the allowable concrete bearing pressure.

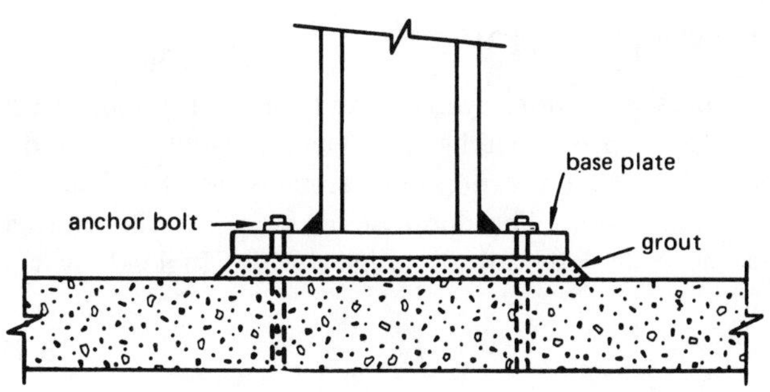


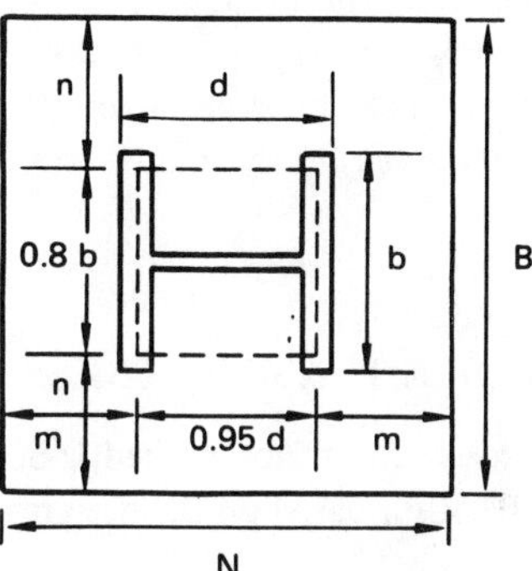


**Figure 15.20** Column Base Plate Design

The required base plate area can be found from the total column load and allowable bearing pressure.

$$A_{\text{plate}} = \frac{\text{column load}}{F_p} \qquad 15.38$$

It is common to specify base plate dimensions in whole inches. Therefore, the actual plate area will be somewhat larger than the required plate area. The actual bearing pressure is

$$f_p = \frac{\text{column load}}{\text{actual } A_{\text{plate}}} \qquad 15.39$$

It is assumed that part of the base plate outboard from a $0.95d \times 0.8b$ rectangle acts as a uniformly loaded cantilever. In order to limit the bending stress, $F_b$, to less than $0.75F_y$, the thickness of the base plate must be made sufficiently large. Once the plate size has been determined, the distances $m$ and $n$ can be determined. Then, equation 15.40 (similar to equation 15.37) can be used to calculate the required plate thickness. The larger plate thickness is required, since either $m$ or $n$ will be larger.

$$t = \{\text{larger of } m \text{ or } n\} \times \sqrt{\frac{4f_p}{F_y}} \qquad 15.40$$

## 26. TENSION MEMBERS

### A. INTRODUCTION

Wire cables, rods, eyebars, and structural shapes are typical tension members. Tension members are designed so that the nominal stress is less than the allowable stress. The nominal stress is just the average stress, calculated by dividing the design load by the area.

$$f = \frac{P}{A} \qquad 15.41$$

The area to be used is the actual area in tension. This is generally known as the *gross area.* For riveted or bolted connections, however, the area is taken as the gross area for checking against the yield strength, and it is taken as the *effective area* for checking against the ultimate strength.

### B. ALLOWABLE TENSILE STRESS

The tensile stress must not exceed $0.60F_y$ based on the gross area, or $0.50F_u$ based on the net area.[28]

### C. SLENDERNESS RATIOS FOR TENSION MEMBERS

Where structural shapes are used in tension, the *AISC Specifications* lists preferred (but does not require) maximum slenderness ratios of 240 for main members and 300 for bracing and other secondary members. Rods and wires are excluded from these limitations.

### D. THREADED MEMBERS IN TENSION

Tension on threaded parts made from approved steels must not exceed $0.33F_u$ when subjected to static loading, regardless of whether or not threads are present in the shear plane.[29] The area to be used with this allowable stress when calculating the maximum tensile load is the gross or *nominal area,* as determined from the outer extremity of the threaded section. That is, the nominal area is calculated from the nominal bolt diameter.

When threads are included in the shear plane, the lower of the loads based on $0.33F_u$ and $0.60F_y$ controls the design.

[28] Eyebars (pin-connected plates) are designed to a smaller allowable stress. There are other important factors affecting eyebar design. The subject is not covered in this book.

[29] Bolts in tension are governed by other limitations.

### E. PLATES AND MEMBERS WITH HOLES

Plates and members are connected to other plates and members by welding, riveting, and bolting. If rivets and bolts are used, the *effective area* of the tension member is less than the gross area. In some instances where there are several rows of fasteners and multiple redundancy, it will be difficult to determine the effective area.

There are actually three different areas used in tensile member calculations. The *gross area* (for the plate shown in figure 15.21) is

$$A_g = bt \qquad 15.42$$

The *net area* is calculated as the net width times the thickness.

$$A_n = b_n t \qquad 15.43$$

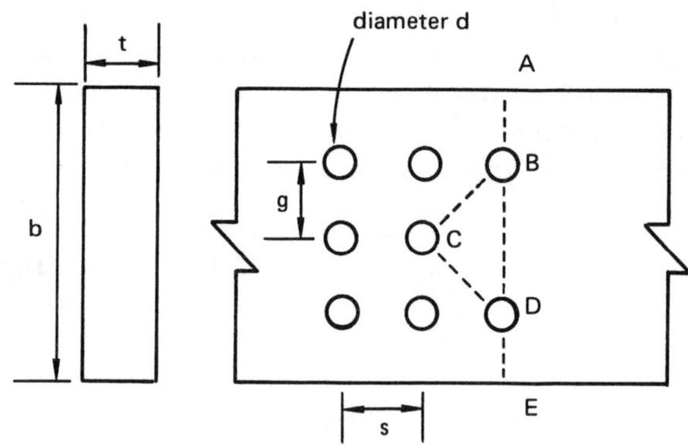


**Figure 15.21** Illustration of Net Area Calculation

The *net width* is calculated by subtracting the hole diameters in the expected failure path from the gross width, and then adding a correction factor for each diagonal leg in the failure path. (There are two diagonal legs in the failure path ABCDE in figure 15.21. Thus, the quantity $s^2/4g$ would be added twice.)

$$b_n = b - \sum_{\text{holes}} d + \sum_{\text{diag. paths}} \frac{s^2}{4g} \qquad 15.44$$

Bolt clearance holes should be 1/16″ larger than nominal fastener dimensions.[30]

$s$ in equation 15.44 is the fastener pitch or longitudinal spacing. $g$ is the gage or transverse spacing.

The chain of holes to be used as the expected failure path is the one which gives the minimum net length,

[30] The so-called *standard hole* is 1/16″ larger than the nominal fastener dimension. However, due to difficulties in producing uniform holes in field-produced assemblies, another 1/16″ could be added.

$b_n$. This minimum length chain is usually found by checking several possible paths. In figure 15.21, paths ABCDE and ABDE must both be checked.

When a tension member frames into a supporting member, some of the load carrying ability will be lost unless all connectors are in the same plane, and all elements of the tension member are connected to the support. (An angle connected to its support only by one of its legs is an example of lost load carrying ability.) Therefore, a further reduction coefficient, $C_t$, is used to calculate the effective net area.

$$A_e = C_t A_n \quad 15.45$$

$C_t$ can be taken as 1.0 if all cross-sectional elements are connected to the support to transmit the tensile force. For $W$, $M$, or $S$ structural shapes and structural tees cut from these shapes with connections to flange or flanges only, with 3 or more fasteners per line, and for which flange width/section depth ratio is 2/3 or greater, $C_t = 0.90$. For connections to built-up sections with 3 or more fasteners per line, $C_t = 0.85$. For all shapes not covered, and with at least 2 fasteners per line, $C_t = 0.75$. (*AISC Specifications* section 1.14.2.2.)

In addition to the reduction in the net area by $C_t$, the effective net area for *splice* and *gusset plates* must not exceed 85% of the gross area, regardless of the number of holes. Tests have shown that as few as one hole in a plate will reduce the strength of a plate by at least 15%.[31] It is a good idea to limit the effective net area to 85% of the gross area for all (not just splice and gusset plates) connections with holes in plates.

*Example 15.10*

Choose a 25 foot long $W$ shape (A36 steel) to carry dead and live tensile loads of 468 kips. The shape will be used as a main member with loads transmitted to framing members through the flanges only. Use $K = 1$.

Without additional information on the hole pattern, the gross and net areas are taken as the same. The allowable tensile stress in the member is the minimum of

$$0.6F_y = (0.6)(36) = 21.6 \quad \text{(controls)}$$

$$0.5F_u = (0.5)(58) = 29.0$$

[31] If an end row of connectors has fewer fasteners than interior rows, the tensile strength of the second row effective net area should be checked against a reduced load. The total load should be reduced in proportion to the number of fasteners in the end row. In figure 15.21, there are 2 fasteners in the end row, and there are 8 total fasteners. The load which the second row must carry is 6/8 = 3/4 of the total load.

Since the web does not transmit the tensile stress, $C_t = 0.90$ is used. The required area is

$$A_n = \frac{468}{(21.6)(0.90)} = 24.07 \text{ in}^2$$

A W 21×83 member has an area of 24.3. The minimum radius of gyration is $r_y = 1.83$. So, the maximum slenderness ratio is

$$SR = \frac{(1)(25)(12)}{1.83} = 163.9$$

This slenderness ratio is less than the suggested limit of 240.

*Example 15.11*

A long tensile member is constructed from two shorter plates as shown. Each plate is $1/2'' \times 9''$. The fasteners are $1/2''$ nominal bolts. The steel is A36. Determine the maximum tensile load the connection can support. Disregard the shear strength of the bolts.

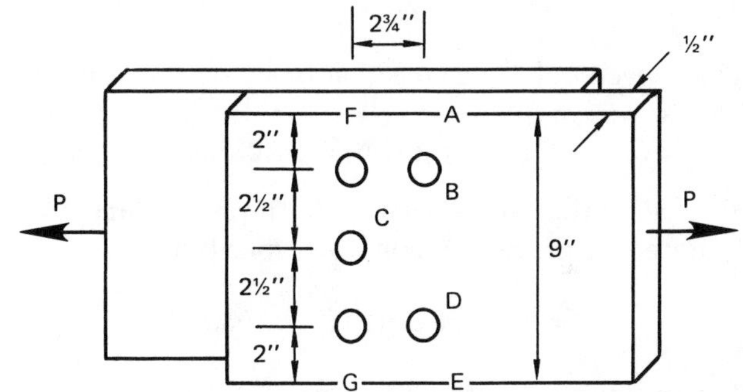


The allowable stress on the gross section is

$$F_t = 0.60F_y = (0.60)(36) = 21.6 \text{ ksi}$$

The allowable stress on the effective net section is

$$F_t = 0.50F_u = (0.50)(58) = 29.0 \text{ ksi}$$

The gross area of the plate is

$$A_g = (0.5)(9) = 4.5 \text{ in}^2$$

The effective hole diameter includes $1/8''$ allowance for clearance and manufacturing tolerances.

$$d = 0.5 + 0.125 = 0.625$$

The net area of the connection must be evaluated in three ways: paths ABDE, ABCDE, and FCG. Path ABDE doesn't have any diagonal runs. The net area is

$$A_{n,ABDE} = (0.5)(9 - 2 \times 0.625) = 3.875\,\text{in}^2$$

To determine the net area of path ABCDE, the quantity $s^2/4g$ must be calculated. $s$ is the longitudinal pitch, shown as 2.75″ in the figure. $g$ is the transverse gage, shown as 2.5″ in the figure.

$$\frac{s^2}{4g} = \frac{(2.75)^2}{(4)(2.5)} = 0.756$$

The net area of path ABCDE is

$$\begin{aligned} A_{n,ABCDE} &= (0.5)(9 - 3 \times 0.625 + 2 \times 0.756) \\ &= 4.319\,\text{in}^2 \end{aligned}$$

The net area of path FCG is

$$A_{n,FCG} = (0.5)(9 - 3 \times 0.625) = 3.5625\,\text{in}^2$$

The smallest area is $A_{n,FCG}$, which is less than 85% of $A_g$.

The capacity of the connection based on the gross section is

$$P_g = (4.5)(21.6) = 97.2\,\text{kips}$$

Since all of the connections are in the same plane, $C_t = 1$. Based on the net section, the capacity is

$$P_g = (1)(3.5625)(29.0) = 103.3\,\text{kips}$$

The capacity is the smaller value, 97.2 kips.

Since the end fastener row has fewer connectors than the second row (2 compared to 3), a further capacity check is required. If holes $B$ and $D$ carry 2/5 of the load, the stress in section FCG will be

$$f_t = \frac{\left(\frac{3}{5}\right)(97.2)}{3.5625} = 16.4\,\text{ksi} \quad \text{(ok)}$$

## 27 BEAM-COLUMN ANALYSIS

### A. INTRODUCTION

A compression member, such as a column, which is also acted upon by a bending moment is known as a *beam-column*. The bending moment can be due to an eccentric load or a true lateral load. Design and analysis of members with combined bending and axial loads generally attempt to transform moments into equivalent axial loads (i.e., the *equivalent axial compression method*) or make use of *interaction equations*. Interaction type equations are best suited for beam-column analysis and validation, since so much (e.g., area, moment of inertia, etc.) needs to be known about a shape. Equivalent axial compression methods are better suited for design.

### B. SMALL AXIAL COMPRESSIONS

When the axial load is small, the member is essentially a beam. When $f_a/F_a$ does not exceed 0.15, a simplified interaction criterion can be used.

*step 1*: Calculate the axial stress due to the axial load acting alone.

$$f_a = \frac{P}{A} \qquad 15.46$$

*step 2*: Calculate the slenderness ratios for both bending modes.

*step 3*: Based on the largest slenderness ratio, determine the allowable column stress. Use equation 15.18 or 15.19. Alternatively, $F_a$ can be obtained directly from the *Allowable Stress for Compression Members* table.

*step 4*: Calculate the ratio of $f_a/F_a$, which must be 0.15 or less to use this method.

*step 5*: Determine the adequacy of design by use of the simplified interaction criterion, equation 15.47 (**[1.6-2]**).

$$\frac{f_a}{F_a} + \frac{f_{bx}}{F_{bx}} + \frac{f_{by}}{F_{by}} \leq 1.0 \qquad 15.47$$

In equation 15.47, $f_b$ is the actual maximum compressive bending stress, as calculated from the standard bending stress equation. $F_b$ is the compressive bending stress that would be permitted if the bending moment acted alone.

### C. LARGE AXIAL COMPRESSIONS

When $f_a/F_a$ exceeds 0.15, the *stability interaction criterion* must be used. Two equations must be satisfied. The first is basically an interaction equation using $0.60F_y$ as the allowable compressive stress. Equation 15.48 (**[1.6-1b]** is the first equation.

$$\frac{f_a}{0.60F_y} + \frac{f_{bx}}{F_{bx}} + \frac{f_{by}}{F_{by}} \leq 1.0 \qquad 15.48$$

Equation 15.49 ([**1.6-1a**]) is the *stability criterion.*

$$\frac{f_a}{F_a}+\frac{C_{mx}f_{bx}}{\left(1-\frac{f_a}{F'_{ex}}\right)F_{bx}}+\frac{C_{my}f_{by}}{\left(1-\frac{f_a}{F'_{ey}}\right)F_{by}}\leq 1.0 \quad 15.49$$

$$F'_e=\frac{12\pi^2E}{23\left(\frac{KL_b}{r_b}\right)^2} \quad 15.50$$

In equation 15.49, $F_a$ is the axial compressive stress that would be allowed if there was no other bending stress. It depends on the maximum slenderness ratio. $C_m$ is an *equivalent moment factor.* If the compression member is subject to joint translation (i.e., sidesway), then $C_m = 0.85$. If the compression member is braced against joint translation in the plane of loading and also experiences transverse loading between the supports, $C_m$ may be taken as 0.85 if the member's ends are restrained, and 1.0 otherwise.

If a compression member is braced against joint translation in the plane of loading and does not experience transverse loads between the supports, then $C_m$ is calculated from equation 15.51.

$$C_m = 0.6-0.4\frac{M_1}{M_2} \quad 15.51$$

$M_1/M_2$ is the ratio of the smaller to larger moments at the ends of the compression member (in the plane of bending). The ratio is positive when the member is bent in reverse curvature, and negative when bent in single curvature.

In equation 15.50, $F'_e$ is the Euler stress divided by the basic factor of safety. $L_b$ and $r_b$ are the unbraced length and radius of gyration in consistent units, both in the plane of bending.

*Example 15.12*

A W 14×120 (A36) shape has been chosen to carry an axial compressive load of 200,000 pounds and a 250,000 ft-lbf moment about its strong axis. The member's unsupported length is 20 feet. Sidesway is permitted in the direction of bending. Use $K = 1$. Determine if the column is adequate.

Check to see if the axial load can be considered small. The column properties are:

$$A = 35.3\,\text{in}^2$$
$$r_y = 3.74$$
$$L_c = 15.5$$
$$L_u = 44.1$$
$$r_x = 6.24$$
$$S_x = 190$$

The axial stress is

$$f_a=\frac{P}{A}=\frac{200}{35.3}=5.67\,\text{ksi}$$

The maximum slenderness ratio is

$$SR=\frac{KL}{r}=\frac{(1)(20)(12)}{3.74}=64.2 \quad \text{(say 64)}$$

The allowable compressive stress from equation 15.18 (or from the *Allowable Stress for Compression Members* table) is 17.04 ksi.

The stress ratio is

$$\frac{f_a}{F_a}=\frac{5.67}{17.04}=0.33>0.15$$

Therefore the large axial compression criteria must be used. Since sidesway is permitted, $C_{\text{mx}} = 0.85$. Since the unbraced length of 20 feet is greater than $L_c$ but less than $L_u$,

$$F_b = 0.60\times F_y=(0.60)(36)=21.6\,\text{ksi}$$

From equation 15.50,

$$F'_e=\frac{12\pi^2(2.9\,\text{EE}7)/1000}{23\left[\frac{(1)(20)(12)}{6.24}\right]^2}=100\,\text{ksi}$$

The bending stress due to the applied moment is

$$f_{bx}=\frac{M}{S_x}=\frac{\frac{(250,000)(12)}{1000}}{190}=15.79$$

From equation 15.48 (the first criterion),

$$\frac{5.67}{(0.60)(36)}+\frac{15.79}{21.6}=0.99 \quad \text{(ok)}$$

From equation 15.49 (the second criterion),

$$\frac{5.67}{17.04}+\frac{(0.85)(15.79)}{\left(1-\frac{5.67}{100}\right)(21.6)}=0.99 \quad \text{(ok)}$$

D. WEB STIFFENERS

In most instances, beams and girders will transmit moments to the flanges of column members. If these moments are very large, it will be necessary to reinforce the web with web stiffeners. The column design tables and *AISC Specifications* should be consulted to determine when such stiffening is required.

## 28 BEAM-COLUMN DESIGN

In order that the shape selection process for combined axial and bending loads not be so much of a trial and error procedure, equations 15.47, 15.48, and 15.49 can be modified to allow the use of the column tables in the *AISC Manual.* Nevertheless, such modifications still require a considerable amount of calculation.

The *AISC Manual* suggests an alternate procedure which determines an *equivalent axial load.*[32]

*step 1*: Determine the effective length, $KL$, based on the weak axis bending and bracing.

*step 2*: Use *table B* in *Part 3, "Column Design"*, of the *AISC Manual* to obtain a value of the *equivalency factor*, $m$, for the first iteration.

*step 3*: Assume a value of $U$ to be 3. (Values of $U$ for subsequent iterations can be read directly from the column table.)

*step 4*: Calculate the equivalent axial load. $P$ is the actual axial load.

$$P_{\text{eff}} = P + M_x m + M_y m U \qquad 15.52$$

*step 5*: Select a column from the column table to support $P_{\text{eff}}$ with an effective length of $KL$.

*step 6*: Return to *table B* in the *AISC Manual* to obtain a revised value of $m$.

*step 7*: Read the $U$ value for the column from the column table.

*step 8*: Repeat steps 4 through 7 to reduce member weight until $m$ and $U$ remain constant.

*Example 15.13*

Select a W shape using A36 steel to carry 200,000 pounds axially and a moment of 250,000 ft-lbf about its strong axis. The unsupported length is 20 feet. Assume $K = 1$.

*step 1*: $KL = (1)(20) = 20$ ft

*step 2*: $m = 2.0$ (all shapes, 20 feet)

*step 3*: $U = 3$ (not needed in this example)

*step 4*: $P_{\text{eff}} = 200 + \dfrac{(250,000)(2)}{1000} = 700$ kips

*step 5*: Try W 14×132, with a 20 foot capacity of 662 kips (close to 700 kips).

*step 6*: $m = 1.7$

*step 7*: $P_{\text{eff}} = 200 + (1.7)(250) = 625 \text{ kips} < 662 \text{ kips}$ (O.K.)

## 29 PLATE GIRDERS

When beams with moments of inertia larger than standard mill shapes are required, plate girders can be used. The discussion on plate girders which follows is presented in an order which allows the design of a member. However, the individual sections can also be used if specific aspects of a given plate girder are to be evaluated.

A. DIMENSIONS AND NOMENCLATURE

Figure 15.22 illustrates two plate girders. One is built up and bolted or riveted together. The other is welded. The depth, $d$, is generally chosen to be around 1/10 of the span, although the ratio can vary between 1/5 to 1/15.

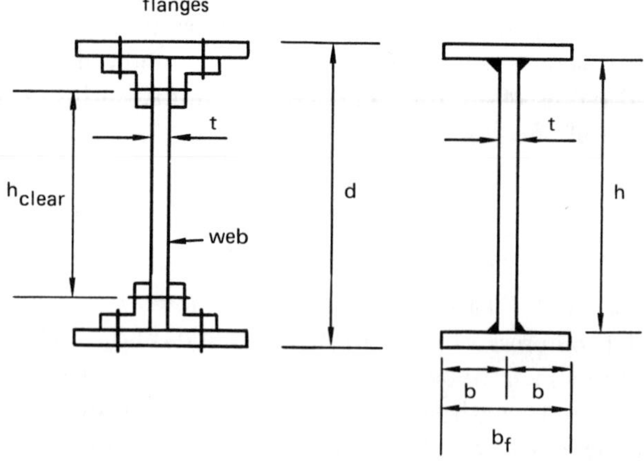


**Figure 15.22** Elements of a Plate Girder

Depending on the purpose, two different definitions of the *web depth*, $h$, are used.[33] For shear stress calculations, the full $d$ is used. For depth-thickness ($h/t$)

[32] This procedure has a tendency to oversize beam-columns. Therefore, it may be possible to use somewhat lighter members if the member initially selected is used as a starting point for a subsequent trial-and-error reduction study.

[33] In fact, there are two different definitions of depth—the *web depth*, *h*, and the *girder depth*, *d*.

ratios, the *AISC Specifications* require $h$ to be the clear distance between flanges. For bolted or riveted flanges, this distance is the separation between the last row of fasteners. For welded flanges, the distinction is not made.

B. DEPTH-THICKNESS RATIOS

The web thickness to be used depends on how closely *intermediate stiffeners* are spaced. If intermediate stiffeners are spaced no more than $1.5d$, where $d$ is the girder depth, then the maximum permissible depth-thickness ratio is given by equation 15.53. If $h$ is known, the required thickness, $t$ can be obtained.

$$\frac{h}{t} \le \frac{2000}{\sqrt{F_y}} \qquad 15.53$$

If intermediate stiffeners are not provided as often as $1.5d$, then equation 15.54 governs the depth-thickness ratio.

$$\frac{h}{t} \le \frac{14,000}{\sqrt{F_y(F_y + 16.5)}} \qquad 15.54$$

If $h/t$ satisfies equation 15.54 and is also less than 260, and if the shear stress is less than the allowable value, no intermediate stiffeners will be required.[34] In other cases, actual stiffener spacing will depend on the shear stress.

C. SHEAR STRESS

The maximum shear stress on the section can be calculated from $h$ and $t$.

$$f_v = \frac{V_{\max}}{ht} \qquad 15.55$$

The allowable web shear, $F_v$, may not exceed $0.40F_y$ or the value calculated from equation 15.56 (**[1.10-1]**).[35]

In equation 15.59, $a$ is the clear distance between stiffeners.

$$F_v = \frac{C_v F_y}{2.89} \qquad 15.56$$

$$C_v = \frac{45,000k}{F_y(h/t)^2} \quad (C_v < 0.8) \qquad 15.57$$

$$C_v = \frac{190}{h/t}\sqrt{\frac{k}{F_y}} \quad (C_v > 0.8) \qquad 15.58$$

$$k = 4.00 + \frac{5.34}{(a/h)^2} \quad (a/h < 1.0) \qquad 15.59$$

$$k = 5.34 + \frac{4.00}{(a/h)^2} \quad (a/h > 1.0) \qquad 15.60$$

When $a/h$ is very large, $k = 5.34$. This situation corresponds to the case of no intermediate stiffeners. If that is the case, equations 15.61 and 15.62 give the allowable shear stress. ($F_v$ is still limited to $0.4F_y$, however.)

$$F_v = \frac{152\sqrt{F_y}}{h/t} \quad \left(h/t < \frac{548}{\sqrt{F_y}}\right) \qquad 15.61$$

$$F_v = \frac{83,150}{(h/t)^2} \quad \left(h/t > \frac{548}{\sqrt{F_y}}\right) \qquad 15.62$$

An alternate method of calculating the web thickness based on achieving full flange compression (equation 15.63) will produce much thicker webs, and result in larger weight and cost. However, this equation can be used to determine if the allowable flange stress will be reduced. If the thickness is less than that determined from equation 15.63, the allowable flange stress will be reduced from the basic bending allowance.

D. DESIGN OF GIRDER FLANGES

To design the girder flanges, some initial estimate of the allowable bending stress in the flanges must be made. Theoretically, $F_b = 0.66F_y$ could be chosen. However, the allowable bending stress is a function of the depth-thickness ratio when

$$\frac{h}{t} > \frac{760}{\sqrt{F_b}} \qquad 15.63$$

Therefore, if $h/t$ is greater than $760/\sqrt{F_b}$, a lower value of $F_b$ should be used.[36]

Equation 15.64 is easily derived from basic mechanics principles, and gives an initial estimate of the required flange area.

$$A_f = \frac{M_x}{F_b h} - \frac{th}{6} \qquad 15.64$$

Once the flange area is known, trial flange widths and thicknesses can be evaluated.[37]

$$A_f = b_f t_f \qquad 15.65$$

[34] In any case, bearing stiffeners are still required at reaction and loading points.

[35] The *AISC Specifications* contains an alternative equation for calculating the allowable shear stress if intermediate stiffeners are used and other conditions are met.

[36] That is, $F_b$ can be reduced 10% or so. Very large reductions in bending stress underutilize the flange area.

[37] Girder flanges do not have to be the same thickness along the entire plate girder length. It is possible to substitute thinner flanges near the ends of beams, or to add cover plates near the mid-points of beams.

The limitations in available plate thicknesses should be considered when choosing $t_f$. Alternatively, $b_f$ can be chosen based on $b_f/d$ ratios, which typically vary between 0.2 and 0.3. Flange plate widths are typically rounded to the nearest 2″.

## E. WIDTH-THICKNESS RATIOS

The width-thickness ratios specified by equation 15.26 and table 15.6 apply to plate girders as well. Specifically, for unstiffened plates such as plate girder flanges, equation 15.66 should be used.

$$\frac{b}{t_f} < \frac{95}{\sqrt{F_y}} \qquad 15.66$$

$$b = \frac{1}{2}b_f \qquad 15.67$$

For stiffened plates, such as plate girder webs, equation 15.68 is used. Also, see equation 15.53.

$$\frac{h}{t_w} \leq \frac{14{,}000}{\sqrt{F_y(F_y + 16.5)}} \qquad 15.68$$

## F. REDUCTION IN FLANGE STRESS

As already mentioned, the allowable bending stress must be reduced if $h/t > 760/\sqrt{F_b}$. Equation 15.69 gives the reduced stress. $A_w$ is the area of the web, and $A_f$ is the area of the compression flange.

$$F_b' \leq F_b \left[1.0 - 0.0005\,\frac{A_w}{A_f}\left(\frac{h}{t} - \frac{760}{\sqrt{F_b}}\right)\right] \qquad 15.69$$

$$A_w = ht \qquad 15.70$$

$$A_f = b_f t_f \qquad 15.71$$

## G. FINAL CHECK

Once a trial design of the plate girder has been made, its moment of inertia can be determined by standard means. The expected stress based on assumed elastic behavior is calculated and compared to the allowable stress, $F_b'$.

## H. LOCATION OF FIRST (OUTBOARD) STIFFENERS

The first intermediate stiffeners can be located where the shear stress exceeds equation 15.56 ([**1.10-1**]). In practice, a trial distance $a$ ($a/h < 1.0$) is selected as the separation between the end panel and the first intermediate stiffener. Then, $F_v$ is calculated from equation 15.56 and compared to $f_v$ from equation 15.55.[38] If $F_v$ is greater than $f_y$, the location is adequate. Otherwise, a smaller $a$ should be tried.

## I. LOCATION OF INTERIOR STIFFENERS

The spacing, $a$, of interior intermediate stiffeners should not exceed the value determined from equation 15.72.

$$\frac{a}{h} \leq \left[\frac{260}{h/t}\right]^2 \qquad 15.72$$

The spacing, $a$, determined from equation 15.72 is valid as long as the shear stress does not exceed the value calculated from equation 15.56. In beams where the maximum shear occurs near the ends, the spacing chosen for the first interior stiffener will be adequate for the entire beam length. However, this illustrates the need to work with the largest shear when establishing an initial trial spacing.

If equation 15.72 is met, and if $C_v \leq 1.0$, then equation 15.73 ([**1.10-2**]) can be used in place of equation 15.56 ([**1.10-1**]). This is the so-called *tension field action equation.* Use of equation 15.73 places an additional restraint on the allowable bending stress in the girder web.

$$F_v = \frac{F_y}{2.89}\left[C_v + \frac{1 - C_v}{1.15\sqrt{1 + (a/h)^2}}\right] \leq 0.40F_y \qquad 15.73$$

## J. MAXIMUM BENDING STRESS

Plate girders which have been designed according to tension field concepts (i.e., have used equation 15.73 to determine stiffener spacing), are limited to the web bending stress in equation 15.74 ([**1.10-7**]). $F_v$ in equation 15.74 is calculated from the *tension field action equation*, equation 15.73. If the bending stress is excessive, or if the quantity in parentheses is greater than 0.6, the stiffener spacing must be reduced.

$$f_b \leq \left(0.825 - 0.375\frac{f_v}{F_v}\right) F_y < 0.6F_y \qquad 15.74$$

## K. DESIGN OF INTERMEDIATE STIFFENERS

*Intermediate stiffeners* are used to support the flange and prevent buckling. They may be constructed from plates or angles, either singly or in pairs. Intermediate stiffeners do not need to extend completely from the top to bottom flanges, but they must be fastened to the compression flange to resist uplift.[39] The *AISC Specifications* contains limitations on weld and rivet spacing.

Intermediate stiffeners are sized by their gross steel area, as calculated from the steel area in contact with the compression flange. Thus, the width and thickness

[38] This procedure is greatly aided by a shear diagram for the beam.

[39] Stiffeners which transmit loads and reactions must extend from flange to flange.

are used to calculate the stiffener area, not the width and depth. Equation 15.75 ([**1.10-3**]) gives the steel area at a particular location. This area can be divided between two stiffeners or given to a single stiffener. In equation 15.75, $D$ is 1.0 for stiffeners furnished in pairs, and 2.4 for single plate stiffeners. $D$ is 1.8 for single angle stiffeners.

$$A_{st} = \frac{1-C_v}{2}\left[\frac{a}{h} - \frac{(a/h)^2}{\sqrt{1+(a/h)^2}}\right]\left(\frac{F_{y,\text{web}}}{F_{y,\text{stiffener}}}\right) Dht_w \quad 15.75$$

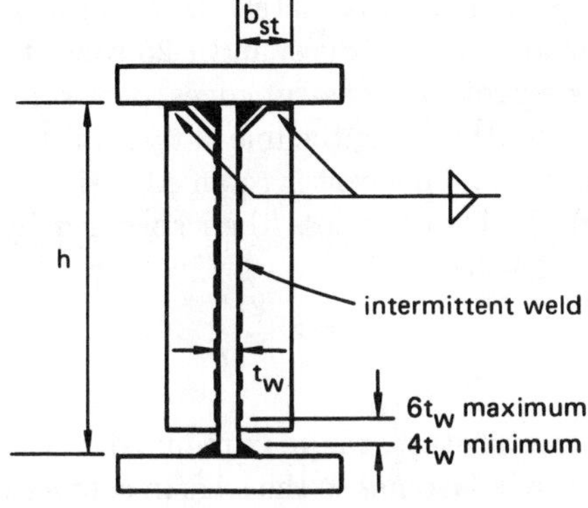


**Figure 15.23** Intermediate Stiffeners

If the actual shear stress, $f_v$, at the point where the bearing stiffener is located is less than the allowable shear stress, $F_v$, as calculated from equation 15.73, then the stiffener area may be reduced proportionally. That is,

$$A'_{st} = \left(\frac{f_v}{F_v}\right) A_{st} \quad 15.76$$

The moment of inertia, $I_{st}$, is taken with respect to an axis in the plane of the girder web. If two stiffeners are used, $b_{st}$ is the total of both their widths.

$$I_{st} = \frac{t_{st} b_{st}^3}{12} \quad 15.77$$

To be significant, the stiffener must have sufficient stiffness itself. Equation 15.78 is a lower limit on the moment of inertia.

$$I_{st} \geq \left(\frac{h}{50}\right)^4 \quad 15.78$$

## L. DESIGN OF BEARING STIFFENERS

The bearing pressure on stiffeners is limited to $0.90F_y$, and such stiffeners must essentially extend from the web to the edge of the flanges. Therefore, this criterion establishes one method of determining the bearing stiffener thickness. However, there are other criteria that must also be met. (The width is essentially fixed by the flange dimension. So, only the thickness needs to be determined.)

Since the stiffener is loaded as a column, it must satisfy the width-thickness ratio for an unstiffened element.

$$\frac{b_{st}}{t_{st}} = \frac{95}{\sqrt{F_y}} \quad 15.79$$

The stiffener should be designed as a column. For the purpose of determining the slenderness ratio, the effective length is taken as $0.75h$. The radius of gyration, $r$, can be determined exactly, or it can be approximated as 0.25 times the stiffener edge-to-edge distance. That is,

$$\frac{L}{r} = \frac{0.75h}{0.25(2b_{st} + t_w)} \quad 15.80$$

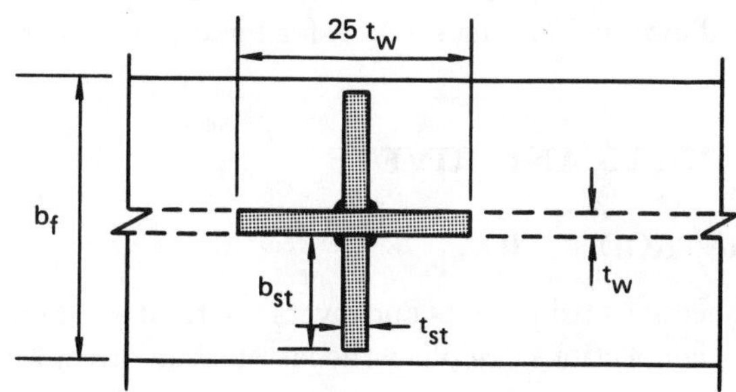


**Figure 15.24** Bearing Stiffener Design

Once the $L/r$ ratio is known, it can be used (as $KL/r$) to determine the allowable compressive stress, $F_a$. However, some of the web also supports the load. Specifically, 25 times the web thickness is the contributing area. Therefore, the required stiffener thickness based on column stress is

$$t_{st} = \frac{\frac{\text{load}}{F_a} - 25t_w^2}{2b_{st}} \quad 15.81$$

Another factor determining the thickness is possible compression yielding. Compressive stress is limited to $0.60F_y$. Therefore, the thickness is

$$t_{st} = \frac{\frac{\text{load}}{0.60F_y}}{2b_{st}} \quad 15.82$$

Stiffener thickness is the maximum thickness determined from the bearing stress, width-thickness ratio, column stress, and compression yield criteria.[40]

---

[40] If column stability is not the factor controlling stiffener thickness, the larger thickness could conceivably increase the slenderness ratio and reduce the allowable compressive stress even further. This should be checked, but is not likely to be a factor.

M. WEB CRIPPLING AT POINTS OF LOADING

The *AISC Specifiations* contains provisions (*section 1.10.10.2*) to determine if the web is capable of supporting the loads (concentrated or distributed) without experiencing web crippling. The equations ([**1.10-10**] and [**1.10-11**]) determine the maximum load that can be applied between stiffeners of a given spacing. The spacing should be reduced to meet these additional requirements, if necessary.

## 30 BENDING WITH AXIAL TENSION

Equation 15.48 should be used to size sections that are subject to axial tension and bending combined. $f_b$ is the bending stress that would exist if axial tension were not present. The effect of the axial tension is not allowed to produce bending stresses which, if acting alone, exceed the allowable bending stresses for flexural members.

## 31 BOLTS AND RIVETS

A. INTRODUCTION

Connections using bolts and rivets are treated similarly. Such connections can place the fasteners in direct shear, torsional shear, tension, or any combination of shear and torsion. Theoretical methods based on elastic design can be used for design, and in some instances, procedures based on ultimate strength concepts are available.

A *concentric connection* is one for which the applied load passes through the centroid of the fastener group. If the load is not directed through the fastener group centroid, the connection is said to be an *eccentric connection.*

At low loading, the distribution of forces among the fasteners is very non-uniform, since friction carries some of the load. However, at higher stresses (near yielding), the load is carried equally by all fasteners in the group.

Allowable stresses in fasteners can be increased by 1/3 for temporary exposure to wind and seismic loading.

*Stress concentration factors* are not normally applied to connections with multiple redundancy.

In many connection designs, materials with different strengths will be used. The material with the minimum strength is known as the *critical part*, and the critical part controls the design.

The analysis presented here for connection design and analysis assumes static loading. Other provisions for fatigue loading are contained in the *AISC Specifications, Appendix B.*

B. HOLE SPACING AND EDGE DISTANCES

The minimum distance between centers of fastener holes is 8/3 times the nominal fastener diameter. In addition, along the longitudinal connection direction (along the line of applied force), the distance between the centers of standard holes is given by equation 15.83 ([**1.16-1**]). $P$ is the force carried by one fastener in the connector group.

$$s \geq \frac{2P}{F_u t} + \frac{d}{2} \qquad 15.83$$

The minimum distance from the hole center to the edge of a member is approximately 1.75 times the nominal diameter for sheared edges, and 1.25 times the nominal diameter for rolled or gas-cut edges, both rounded to the nearest 1/8″.[41] Along the line of transmitted force, the distance from a hole center to an edge (in the direction of force) shall not be less than specified by equation 15.84 ([**1.16-2**]).

$$s \geq \frac{2P}{F_u t} \qquad 15.84$$

For parts in contact, the maximum edge distance from the center of a fastener to the edge in contact is 12 times the plate thickness or 6″, whichever is less.

C. STRESS AREA OF FASTENERS

The allowable load on a fastener is determined by multiplying its area by an allowable stress.[42] Except for rods with upset ends, the area is calculated simply from the fastener's nominal (unthreaded and undriven) dimension.

D. ALLOWABLE CONNECTOR STRESSES

Table 15.7 lists allowable stresses for tension and shear for common connector types. (Allowable bearing stress is covered elsewhere.) Reductions are required for use with oversized holes and connector patterns longer than 50 inches.

For fasteners made from approved steels, including A449, A572, and A588 alloys, the allowable tensile stress is $F_t = 0.33F_u$, whether or not threads are in the shear plane. For bearing type connections using the same approved steels, the allowable shear stress is $F_v = 0.17F_u$ if threads are present in the shear plane, and $F_v = 0.22F_u$ if threads are excluded from the shear plane.

41 For exact values, refer to *Table 1.16.5.1* in the *AISC Specifications.*

42 The *AISC Manual* also contains tables of allowable loads for common connector types and materials.

**Table 15.7**
Allowable Connector Stresses for Static Loading[43]
(All stresses are in ksi)

| type of connector | $F_t$ | $F_v$ friction connection | $F_v$ bearing connection |
|---|---|---|---|
| A 502, hot driven rivets, | | | |
| grade 1 | 23.0 | don't use | 17.5 |
| grade 2 | 29.0 | don't use | 22.0 |
| A 307, ordinary bolts | 20.0 | don't use | 10.0 |
| A 325, structural bolts | | | |
| no threads in the shear plane | 44.0 | 17.5 | 30.0 |
| threads in the shear plane | 44.0 | 17.5 | 21.0 |
| A 490, structural bolts | | | |
| no threads in the shear plane | 54.0 | 22.0 | 40.0 |
| threads in the shear plane | 54.0 | 22.0 | 28.0 |

If fasteners are exposed to both tension and shear, *AISC Specifications Table 1.6.3* should be used to determine the maximum tensile stress.

For A325 and A490 bolts used in friction-type connections with tension, the allowable shear stresses given in table 15.7 must be multiplied by a reduction factor. In equation 15.85, $f_t$ is the average tensile stress in the bolt group, $A_b$ is the nominal body area of the bolt, and $T_b$ is the bolt *pretension.*[44]

$$\text{reduction factor} = \left(1 - \frac{f_t A_b}{T_b}\right) \qquad 15.85$$

A distinction is made between bearing and friction connections. A *bearing connection* relies on the shearing resistance of the fasteners to resist loading. In effect, it is assumed that the fasteners are loose enough to allow the plates to slide slightly, bringing the fastener shank into contact with the hole. The area surrounding the hole goes into bearing, hence the name. Connections using rivets, welded studs, and A307 bolts are always bearing type connections.

If the fasteners are constructed from high-strength steel, a high *preload* can be placed on the bolts. This preload will clamp the plates together, and friction alone will keep the plates from sliding. The fastener shanks never come into contact with the plate holes. Bolts constructed from A325 and A490 steels are suitable for *friction connections.* However, high-strength bolts can also be used in bearing connections.

## E. CONCENTRIC TENSION CONNECTIONS

The number of fasteners required in the connection is determined by considering the fasteners in shear, the plate in bearing, and the effective net area of the plate in tension.

*Example 15.14*

Two $\frac{1}{4}$×8 A36 plates are joined with a lap joint using 3/4″ grade 1 A502 rivets. Design the connection to carry a concentric load of 25 kips. Disregard effects of eccentricity.

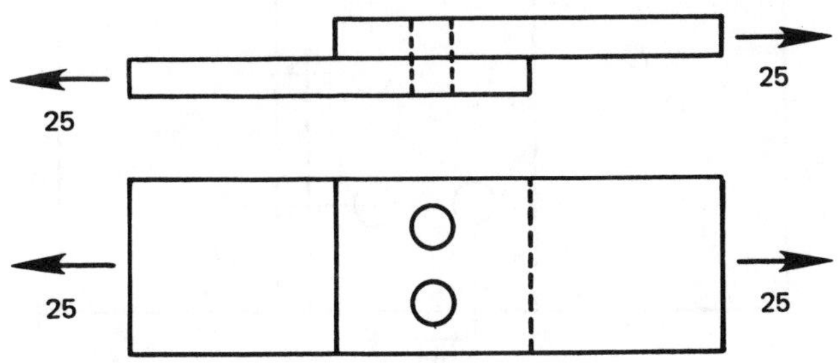


The area of a 3/4″ rivet is

$$A_v = \frac{1}{4}\pi(0.75)^2 = 0.442\,\text{in}^2$$

From table 15.7, the allowable shear stress is 17.5 ksi. So, the number of rivets determined by the shear stress criterion is

$$n = \frac{25}{(17.5)(0.442)} = 3.23 \quad \text{(shear criterion)}$$

The bearing area is

$$A_p = (0.75)(0.25) = 0.1875\,\text{in}^2$$

The allowable bearing stress is

$$f_p = 1.5F_u = (1.5)(58) = 87\,\text{ksi}$$

The number of rivets determined by bearing is

$$n = \frac{25}{(87)(0.1875)} = 1.53 \quad \text{(bearing criterion)}$$

Shear governs. 4 rivets are used.

[43] This is based on *AISC Specifications Table 1.5.2.1*, which contains greater detail.

[44] Minimum bolt tensions are given in *AISC Specification Table 1.23.5.*

The minimum distance between holes is

$$s_{\text{min}} = \frac{8}{3} \times \frac{3}{4} = 2''$$

The longitudinal spacing is further limited by equation 15.83.

$$s_{\text{min}} = \frac{(2)(25/4)}{(58)(3/4)} + \frac{3/4}{2} = 0.66''$$

Assuming sheared edges, the minimum edge spacing is $1.75 \times 3/4 \approx 1\frac{1}{4}''$. The first row cannot be closer to the short edge than

$$s_{\text{min}} = \frac{(2)(25/4)}{(58)(3/4)} = 0.29''$$

Based on these requirements, a trial layout is made.

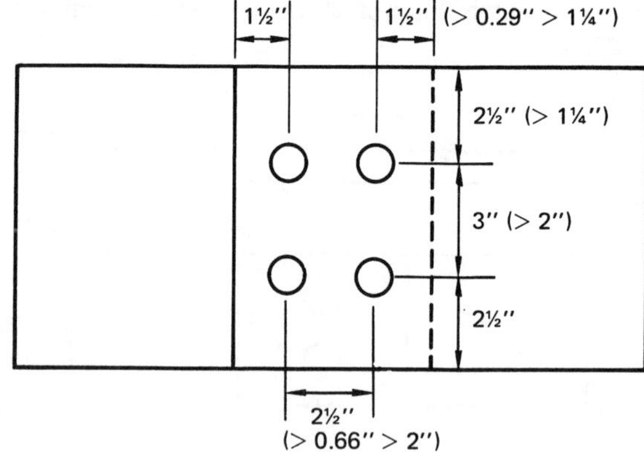


Assuming a hole which is 1/16″ larger than the nominal fastener diameter, each hole subtracts 13/16″ from the net effective area in tension. The tension area across the first row of holes is

$$A_t = (1/4)\left(8 - 2 \times \frac{13}{16}\right) = 1.39 \text{ in}^2$$

The maximum tensile stress on the net effective section is

$$F_t = 0.50F_u = (0.50)(58) = 29 \text{ ksi}$$

The maximum allowable load based on the net effective section is

$$P = (1.39)(29) = 40.3 \text{ ksi} \quad \text{(ok)}$$

Similarly, the maximum tensile stress in the gross section is

$$F_t = 0.60F_y = (0.60)(36) = 21.6 \text{ ksi}$$

The maximum allowable load based on the gross section is

$$P = \left(8 \times \frac{1}{4}\right)(21.6) = 43.2 \text{ ksi} \quad \text{(ok)}$$

## F. TENSION EFFECTS DUE TO ECCENTRICITY

Consider the simple framing connection shown in figure 15.25. If the vertical shear is assumed to act along line $A$, the bolts in the column connection will be put into tension. The most highly-stressed (in tension) will be the top-most fasteners.

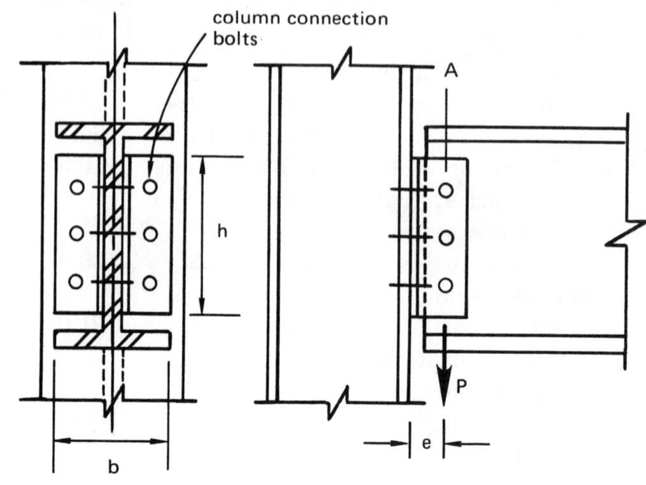


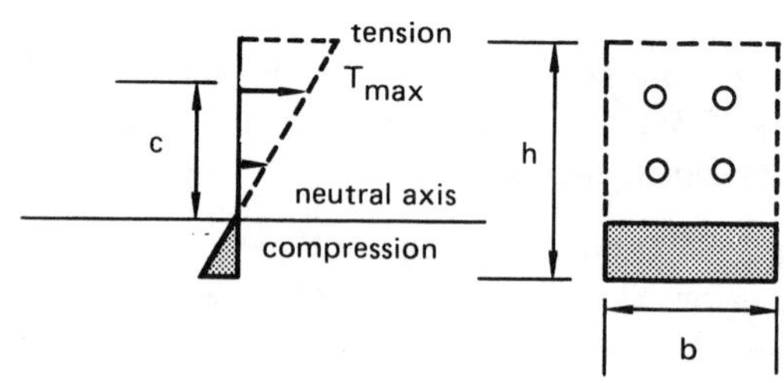


**Figure 15.25** Tension in Simple Connections

By summing moments about the neutral axis (for tension) of the fastener group, it is possible to determine the maximum tensile stress. The neutral axis is located approximately 1/6 or 1/7 up from the bottom of the connector group area, such that the area of bolts in tension equals the area of the support in compression. A few trials may be necessary to locate the neutral axis.

Once the neutral axis is located, the moment of inertia of the fastener group is found by use of the parallel axis theorem. The applied moment is known: $M = Pe$. The stress in the fasteners farthest from the neutral axis is

$$f_{t,\text{max}} = \frac{Mc}{I} = \frac{Pec}{I} \qquad 15.86$$

If necessary, the tensile force in the farthest fastener can be found.

$$T_{\text{max}} = f_{t,\text{max}} A_{\text{bolt}} \qquad 15.87$$

## G. BOLT PRELOADING

Preloading is an effective method of reducing the alternating stress in bolted connections. The initial tension produces a larger mean stress, but the overall result may be to produce a satisfactory design.

Consider the bolted connection shown in figure 15.26. The load varies from $P_{min}$ to $P_{max}$. If the bolt is initially snug but without initial tension, the force in the bolt also will vary from $P_{min}$ to $P_{max}$.

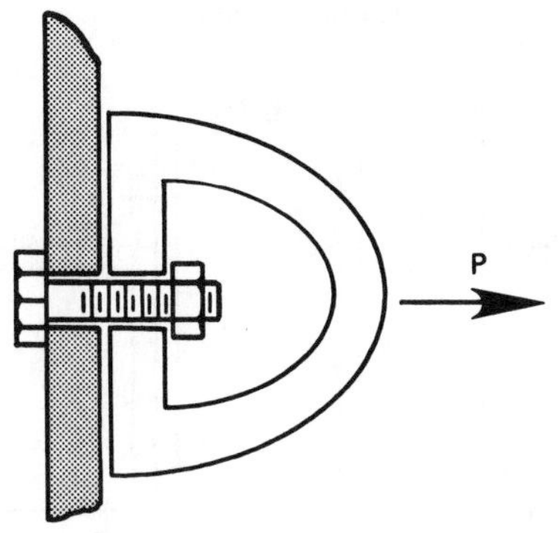


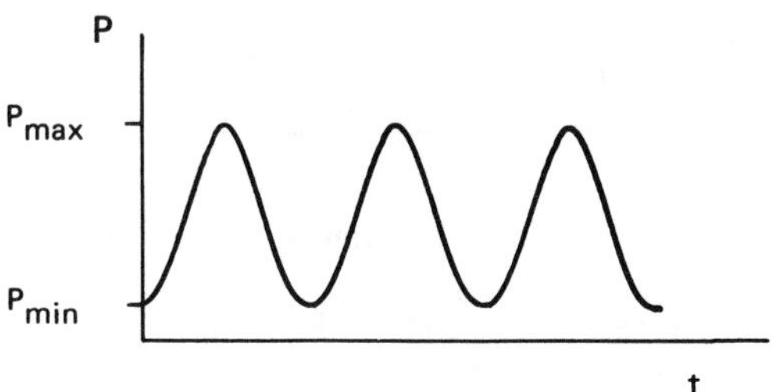


**Figure 15.26** A Bolted Joint

The stress in the bolt depends on the load carrying area of the bolt. It is convenient to define the *spring constant of the bolt.* The effects of the threads usually are ignored, so the area is based on the major (nominal) diameter. The *grip, L,* is the thickness of the parts being connected by the bolt. It is not the bolt length.

$$f_{bolt} = \frac{P}{A} \qquad 15.88$$

$$k_{bolt} = \frac{P}{\Delta L} = \frac{A_{bolt}E_{bolt}}{L_{bolt}} \qquad 15.89$$

If the bolt is tightened so that there is an initial force, $F_i$, in addition to the applied load, the members being held together will be in compression. The amount of compression will vary since the applied load varies. The clamped members will carry some of the applied load, since this varying load has to "uncompress" the members as well as lengthen the bolt. The net result is to reduce the variation of the bolt force.[45]

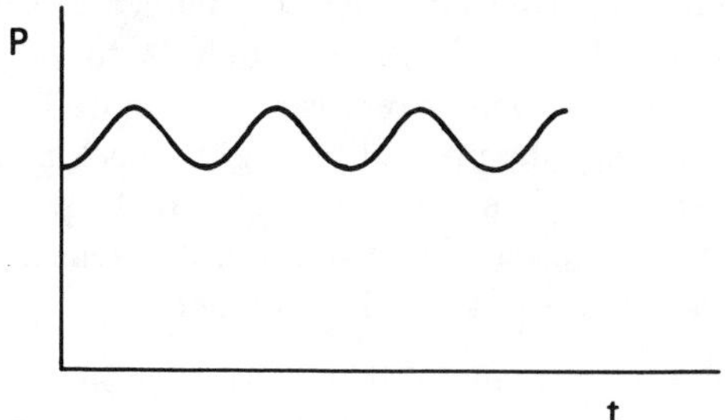


**Figure 15.27** A Bolted Joint with Preloading

The spring constant for each of the bolted parts is somewhat difficult to determine if the clamped area is not well defined. If the clamped parts are simply plates, it can be assumed that the bolt force spreads out to three times the bolt diameter. Of course, the hole diameter needs to be considered in calculating the effective force area. If $E_{part} = E_{bolt}$, this larger area results in the parts being 8 times more stiff than the bolts.

$$k_{parts} = \frac{A_{parts}E_{parts}}{L_{parts}} \qquad 15.90$$

If the clamped parts have different elasticities ($E$ values), the composite spring constant can be found from equation 15.91. (If a "soft" washer or gasket is used, its spring constant may control equation 15.91.)

$$\frac{1}{k_{composite}} = \frac{1}{k_1} + \frac{1}{k_2} + \frac{1}{k_3} + \cdots \qquad 15.91$$

Both the bolt and the clamped parts share the applied load.

$$P_{bolt} = P_i + \frac{k_{bolt}P_{applied}}{k_{bolt} + k_{parts}} \qquad 15.92$$

$$P_{parts} = \frac{k_{parts}P_{applied}}{k_{bolt} + k_{parts}} - P_i \qquad 15.93$$

Of course, if the applied load varies, the forces in the bolt and the parts also will vary. In that case, analysis by a Goodman diagram is called for.

For static loading, recommended amounts of preloading often are specified in terms of a percentage (e.g., 90%) of the tensile yield strength. The term *proof strength* (i.e.,

[45] It is assumed that the initial tension, $P_i$, is greater than $P_{max}$. If the clamped members separate, the bolt once again carries the entire load.

*proof load* divided by bolt area) can be used in place of tensile yield strength. For fatigue loading, the preload must be determined from an analysis of a Goodman diagram.

Tightening of a tension bolt will induce a torsional stress in the bolt. Where the bolt is to be locked in place, the torsional stress can be removed without greatly affecting the preload by slightly backing off the bolt. If the bolt is subject to cyclic loading, the bolt probably will slip back by itself, and it is reasonable to neglect the effects of torsion in the bolt.

More important than the effects of torsion are stress concentrations at the root. Although stress concentrations frequently are neglected for static loading of ductile connectors, there will be a significant reduction in the endurance limit for cyclic loading. Therefore, the alternating stress should be multiplied by an appropriate factor (e.g., 2.0 to 4.0).

## 32 FRAMING CONNECTIONS

### A. INTRODUCTION

Three types of framing (beam-to-column, beam-to-beam, etc.) connections are defined.

- Type 1, *rigid frame connections*: These connections are intended to transmit moments from one member to another. Type 1 connections can be designed by both working stress and plastic design methods.
- Type 2, *simple framing connections*: These connections transmit vertical loads, but essentially no (i.e., less than 20%) moment transfer occurs. Generally, only working stress methods are used to design type 2 connections.
- Type 3, *semi-rigid connections*: When the moment transfer across a connection is significant but not total (i.e., between 20% and 90%), the connection is semi-rigid. Plastic design is not used for type 3 connections. Since it is difficult to determine the amount of moment transfer, type 3 connections are rarest.

### B. SIMPLE (TYPE 2) FRAMING CONNECTIONS

Simple framing connections, also known as *flexible connections*, are designed to be as flexible as possible. Design is fairly predetermined by use of the standard tables of framed beam connections in the *AISC Manual.* Construction methods include beam seats and clips to beam webs. Figure 15.28 illustrates several type 2 connections.

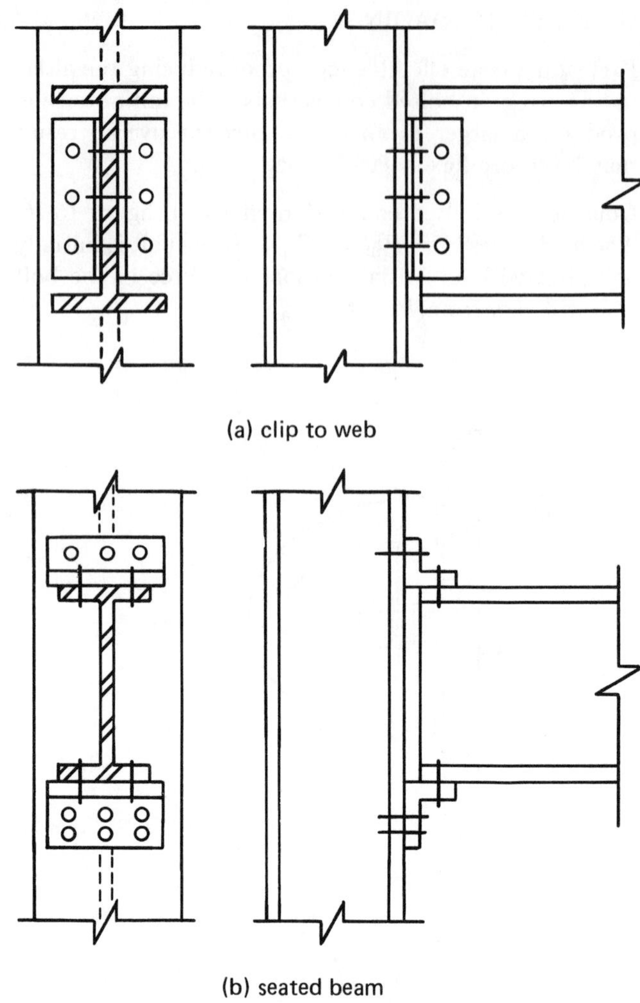


**Figure 15.28** Type 2 (Simple) Framing Connections

Connections to beam and column can be either by bolting or welding, and such fastening methods can be used either on the beam or column.[46] Welded connections, "stiffened" connections, and use of top seats do not necessarily imply a moment-resisting connection. Coping, where used moderately, does not reduce the shear capacity of members.

Direct shear determines the number of bolts required in the column connection. It is common to neglect the effect of eccentricity in determining shear stresses in riveted and bolted beam connections to webs. It is also common to neglect the effect of eccentricity in determining shear stresses in riveted and bolted column connections. However, this eccentricity can be considered, which will add tension stress to the shear stress in column fasteners.

Angle thickness must be checked for allowable bearing pressure. Angles should be checked for direct shear, as

[46] Usually, the shop connection will be welded. The field connection, however, can be either welded or bolted.

well. Since the connection is designed to rotate, the angle (for seated beams) should be checked for bending stress as its free lip bends.

### C. MOMENT-RESISTING (TYPE 1) FRAMING CONNECTIONS

Moment resisting connections transmit their vertical (shearing) load through the same types of connections as type 2 connections, typically through connections at the beam web. However, the flanges of the beam are also rigidly connected to the column. These top and bottom flange connections are in tension and compression respectively, and serve to transmit the moment.

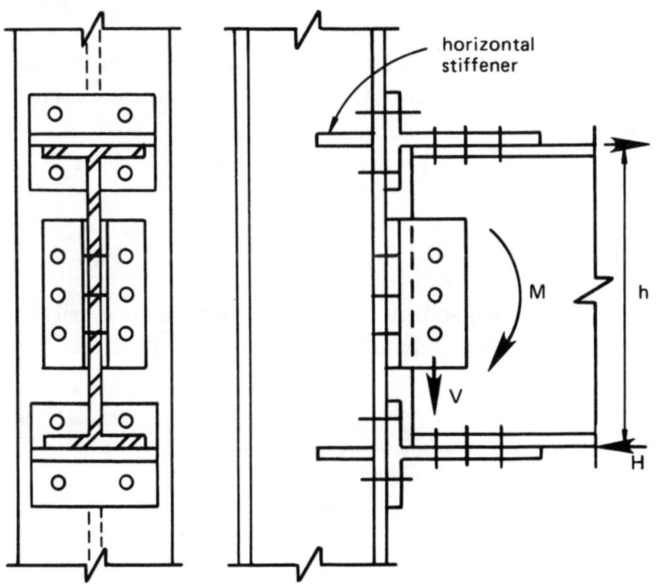


**Figure 15.29** Type 1 (Moment Resisting) Framing Connections

Since moment transfer is through the flanges by tension and compression connections, design of such connections involves ensuring adequate strength in tension and compression. Design of the shear transfer mechanism (i.e., the web connections) is essentially the same as for type 2 connections. Also, in order to prevent localized buckling of the column flanges, *horizontal stiffeners* (between the column flanges) may be needed.

The moment-resisting ability of a type 2 connection can be increased to essentially any desired value by increasing the distance between the tension and compression connections. Figure 15.30 illustrates how this could be accomplished by using an intermediate plate between the beam flanges and column. The horizontal tensile and compressive forces, $H$, can be calculated from equation 15.94.

$$H = \pm \frac{M}{h} \qquad 15.94$$

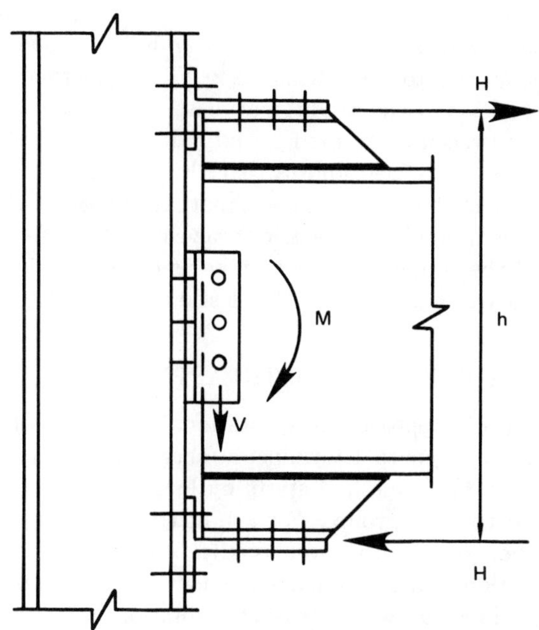


**Figure 15.30** Increasing Moment-Resisting Ability

## 33 BOLTED AND RIVETED ECCENTRIC SHEAR CONNECTIONS

### A. INTRODUCTION

An *eccentric shear connection* is illustrated in figure 15.31. This type of connection gets its name from the tendency of the bracket to rotate around the centroid of the fastener group, shearing the fasteners. The tendency to rotate is resisted by the shear stress in the connectors. Friction is not assumed to contribute to the rotational resistance of the connection.

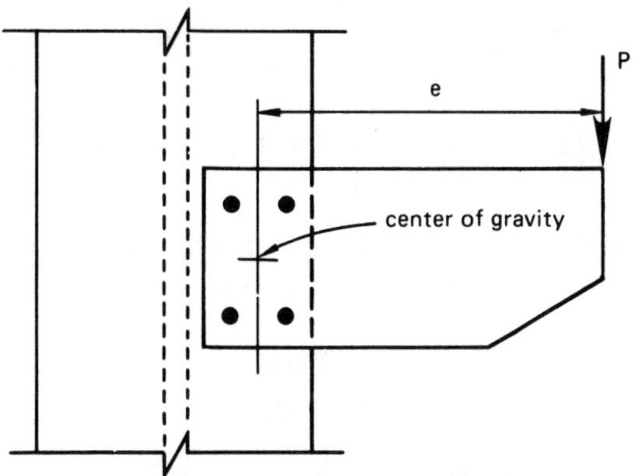


**Figure 15.31** Eccentric Shear Connection

The moment tending to cause rotation is

$$M = Pe \qquad 15.95$$

It is clear that the fasteners must resist this moment by shear. However, it is not clear how the shearing resistance is to be calculated.[47] Three methods are available: (1) the traditional elastic approach, (2) a reduced-eccentricity model, and (3) ultimate strength analysis. It is well known that the traditional elastic approach will greatly underestimate the capacity of (or, will overdesign) an eccentric shear connection. In some cases, the actual capacity may be as much as twice the capacity calculated from the elastic model.

## B. TRADITIONAL ELASTIC APPROACH

The elastic approach uses traditional mechanics of materials concepts to determine the shearing stress in each fastener. This method starts by locating the centroid of the fastener group. For symmetrical (rectangular) fastener groups, this can usually be accomplished by inspection. Once the centroid is located, the *polar moment* of inertia (which resists the rotation) is calculated from the nominal fastener area and distance from the centroid to fastener. In equation 15.96, the summation is over all fasteners in the group.

$$J = \sum_i r_i^2 A_i \qquad 15.96$$

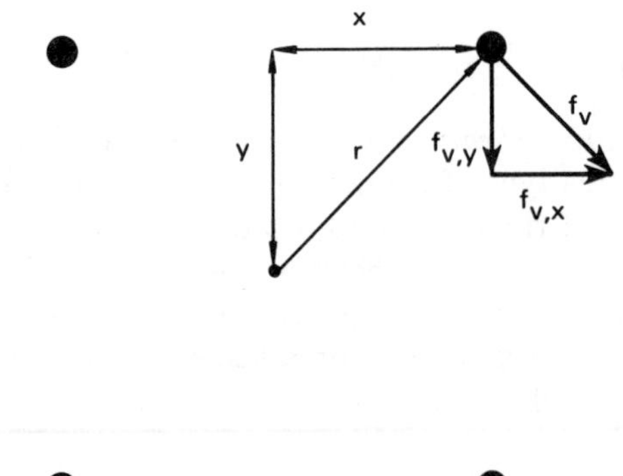


**Figure 15.32** Fastener Group Analysis

The shear stress in each member is calculated from the standard torsional stress equation:

$$f_v = \frac{Mr_{\text{critical}}}{J} \qquad 15.97$$

The *critical fastener*, typically the highest, right-most fastener, is the one whose vector sum of direct and eccentric shear stresses will be the largest. It is usually found by inspection.

[47] And, the *AISC Manual* does not require a particular method, either.

The shear stress in each member is directed perpendicular to the line connecting the center of rotation and the fastener. This shear stress must be converted to $x$- and $y$-components to be combined with the direct shear stress. If the distance, $r$, has been broken into $x$- and $y$-components, it will be easy to calculate the components of shear.

$$f_{v,y} = \frac{f_v x}{r} \qquad 15.98$$

$$f_{v,x} = \frac{f_v y}{r} \qquad 15.99$$

The *direct shear stress* $f_{v,d}$, is merely the load divided by the total area of all fasteners.

$$f_{v,d} = \frac{P}{\sum A_i} \qquad 15.100$$

Since the direct shear stress acts downward, and there is no $x$-component, the total stress in the critical fastener is

$$f_{v,\text{total}} = \sqrt{(f_{v,d} + f_{v,y})^2 + f_{v,x}^2} \qquad 15.101$$

*Example 15.15*

For the bracket shown, find the load on the most critical fastener. All fasteners have a nominal $\frac{1}{2}''$ diameter.

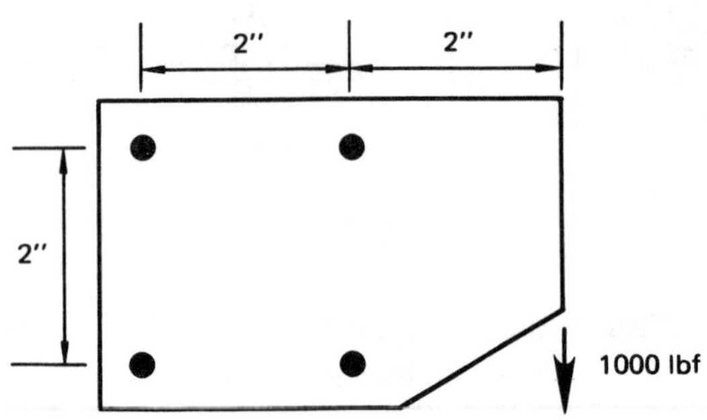


Since the fastener group is symmetrical, the group centroid is centered within the 4 fasteners. This makes the eccentricity of the load 3''. Each fastener is located $r$ from the centroid, where

$$r = \sqrt{(1)^2 + (1)^2} = 1.414$$

The area of each fastener is

$$A_i = \frac{1}{4}\pi(0.5)^2 = 0.1963$$

The polar moment of inertia is

$$J = 4[0.1963(1.414)^2] = 1.570\,\text{in}^4$$

The eccentric shear stress on each fastener is

$$f_v = \frac{(1000)(3)(1.414)}{(1.570)} = 2702\,\text{psi}$$

The $x$- and $y$-components of the shear stresses are

$$f_{v,y} = \frac{\sqrt{2}}{2} \times 2702 = 1911\,\text{psi}$$

$$f_{v,x} = 1911\,\text{psi}$$

The fasteners carry a direct vertical shear load of 1000/4 = 250 pounds each. The vertical shear stress due to this load is

$$f_{v,d} = \frac{250}{0.1963} = 1274$$

The two right fasteners have vertical downward components of $f_{v,y}$ which add to the vertical downward stress of 1274. Thus, both of the two right fasteners are critical. The total stress in each of these fasteners is

$$f_v = \sqrt{(1911)^2 + (1911 + 1274)^2} = 3714\,\text{psi}$$

C. REDUCED ECCENTRICITY MODELS

Since the elastic approach so greatly underestimates the capacity of eccentric shear connections, although not an AISC code requirement, it is logical to reduce the eccentricity when calculating the moment to be resisted.[48] Equations 15.102 and 15.103 give reduced eccentricities known as *effective lengths.* Equation 15.102 is for use when there is only one vertical line of fasteners. Equation 15.103 is for the more common situation of two or more vertical lines of fasteners. $n$ is the number of fasteners in one vertical line.

$$e_{\text{eff}} = e - \frac{1+2n}{4}\bigg|_{\text{1 fastener line}} \qquad 15.102$$

$$e_{\text{eff}} = e - \frac{1+n}{2}\bigg|_{\text{2 or more fastener lines}} \qquad 15.103$$

D. ULTIMATE STRENGTH ANALYSIS

Design and analysis using ultimate strength is preferred over elastic methods. Several ultimate strength theories have been proposed. In practice, however, the application of such theories involves more table look-up than theory. Most methods end up calculating the allowable load on a fastener group from the product of tabulated coefficients and allowable fastener loads. For example, in equation 15.104, $F_v$ is the allowable shear load on the fasteners, and $C$ is a tabulated coefficient.

$$P = CA_bF_v \qquad 15.104$$

The ability to use equation 15.104 depends on having tables of $C$ values for the fastener configuration and eccentricity needed. The *AISC Manual* contains such tables.

[48]The allowable AISC loads on high-strength bolts is already high, and using reduced eccentricity models decreases the factor of safety below 2.5. Therefore, use of reduced eccentricity should be restricted to connections with rivets and low-strength bolts which have conservative allowable loads.

## 34 WELDS

The most widely-used weld type is the *fillet weld* shown in figure 15.33. The applied load is assumed to be carried by the *weld throat*, which has an effective dimension of $t_e$.

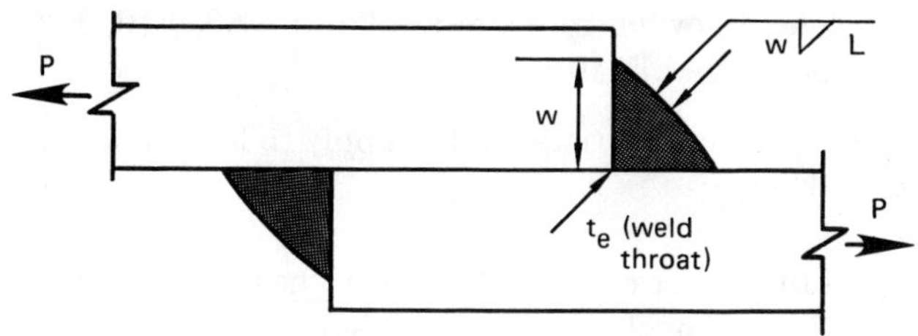


**Figure 15.33** Fillet Weld

The *effective weld throat size*, $t_e$, depends on the type of welding used. For hand-held *manually shielded arc welding (SMAW)* processes,

$$t_e = 0.707w \qquad 15.105$$

If *submerged arc welding (SAW)* is used,

$$t_e = 0.707w + 0.11 \quad (w \geq 7/16'') \qquad 15.106$$

$$t_e = w \quad (w \leq 3/8'') \qquad 15.107$$

Weld sizes, $w$, of 3/16″, 1/4″, and 5/16″ are desirable because they can be made in a single pass.[49] However, fillet welds from 3/16″ to 1/2″ can be made in 1/16″ increments. For welds larger than 1/2″, every 1/8″ weld size can be made.

Allowable shear stress on the weld fillet throat is[50]

$$F_v = 0.30F_{u,\text{rod}} \qquad 15.108$$

Shear stress in the base material may not be greater than $0.40F_y$. For tensile loads, the allowable stress parallel to the weld axis is the same as for the base metal:

$$F_t = 0.60F_y \qquad 15.109$$

To simplify analysis and design of certain types of welded connections, it is convenient to define the *resistance per unit length of weld.* Tensile resistances can

[49] The 5/16″ limitation is appropriate for shielded metal arc weld (typically using hand-held rods). If a submerged arc process is used, up to 1/2″ welds can be made in one pass.

[50] In almost all instances, loads are transmitted through welds by shear stresses, regardless of the weld group orientation.

be found similarly.

$$R_w = \min \begin{cases} t_e F_{v,\text{rod}} & = t_e(0.30)F_{u,\text{rod}} \\ w F_{v,\text{member}} & = w(0.40)F_{y,\text{member}} \end{cases} \qquad 15.110$$

The ultimate strength of a welding rod is part of the rod designation. Thus, $F_u$ for an E70 welding rod is 70 ksi. The following rods are available: E60, E70, E80, E90, E100, and E110.

Several special restrictions that apply to fillet welds are given here.

- Minimum weld sizes depend on the thickness of the thickest of the two parts joined.

**Table 15.8**
Minimum Fillet Weld Size

(all dimensions in inches)

| larger part thickness | minimum $w$ |
|---|---|
| to $\frac{1}{4}$ inclusive | $\frac{1}{8}$ |
| over $\frac{1}{4}$ to $\frac{1}{2}$ | $\frac{3}{16}$ |
| over $\frac{1}{2}$ to $\frac{3}{4}$ | $\frac{1}{4}$ |
| over $\frac{3}{4}$ | $\frac{5}{16}$ |

- The maximum weld size along edges of connecting parts is equal to the edge thickness for materials less than $\frac{1}{4}''$ thick. For materials thicker than $\frac{1}{4}''$, the maximum size must be $\frac{1}{16}''$ less than the material thickness.
- The minimum length weld for full strength analysis is 4 times the weld size.[51]
- If the required strength of a welded connection is less than would be obtained from a full-length weld of the smallest size, then an *intermittent weld* can be used. The minimum length of an intermittent weld is 4 times the weld size or $1\frac{1}{2}$ inches, whichever is greater.
- The minimum weld length for lap joints (as illustrated in figure 15.33) is 5 times the thinner plate's thickness, but not less than 1 inch.

[51] If this criterion is not met, the weld size is downgraded to $\frac{1}{4}$ of the weld length.

## 35 WELDED CONNECTIONS

### A. CONCENTRIC TENSION CONNECTIONS

If the weld group centroid is in line with the applied load, the loading is concentric. Equation 15.111 can be used to design or evaluate the connection.

$$f_v = \frac{P}{A_{\text{weld}}} = \frac{P}{L_{\text{weld}} t_e} \qquad 15.111$$

*Example 15.16*

Two $\frac{1}{2}$×8 plates (A36 steel) are lap welded using E70 electrodes and a shielded arc process. Size the weld to carry a concentric tensile loading of 50 kips.

Good design will weld both joining ends to the base plate. The total weld length will be

$$L_{\text{weld}} = (2)(8) = 16''$$

This length meets the 5″ minimum length specification.

From table 15.8, the minimum weld thickness for a $\frac{1}{2}''$ plate is 3/16″.

The required strength per inch of weld is

$$\frac{50}{16} = 3.125 \text{ kips/in}$$

From equation 15.110,

$$R_w = \min \begin{cases} (0.707)(3/16)(0.30)(70) = 2.78 \\ (3/16)(0.40)(36) = 2.7 \end{cases}$$

Since $R_w < 3.125$, a larger weld is needed. Try $w = 1/4''$.

$$R_w = \min \begin{cases} (0.707)(1/4)(0.30)(70) = 3.71 & \text{(ok)} \\ (0.25)(0.40)(36) = 3.6 & \text{(ok)} \end{cases}$$

The maximum weld size allowed is $\frac{1}{2}'' - 1/16'' = 7/16''$. The 5/16″ weld can be used.

### B. MOMENT RESISTING CONNECTIONS

The *AISC Manual* contains several procedures for designing moment connections (both type 1 and type 3).[52] These procedures assume top plate or end plate construction to transmit moments. Split beam tee construction, and the accompanying problem of prying action stress, is not covered.

[52] See "*Moment Connections*" in the *AISC Manual*.

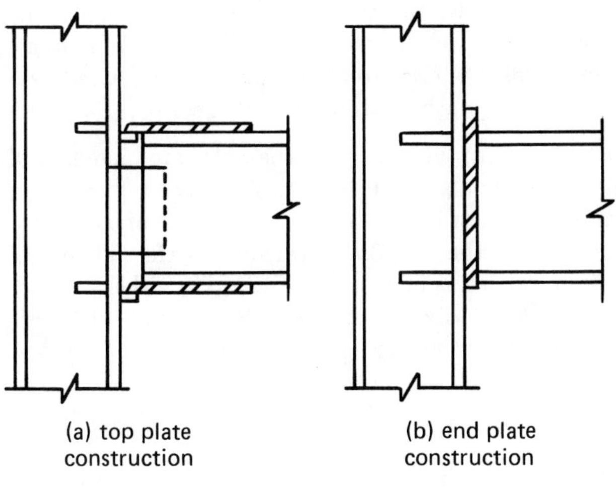


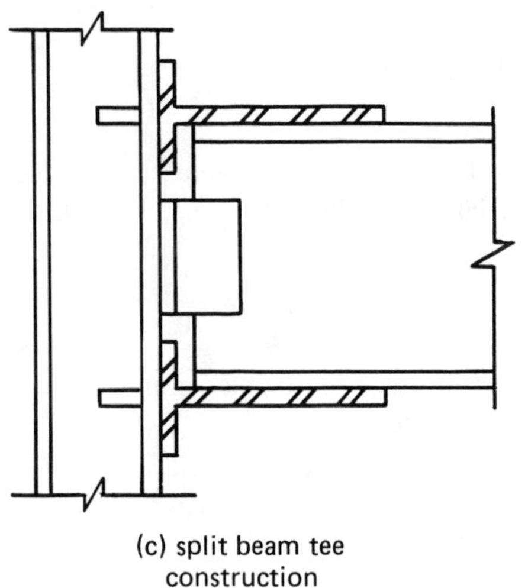


**Figure 15.34** Welded Moment Resisting Connections

## C. BALANCING WELD GROUPS

When tension is applied to an unsymmetrical member, the tensile force will act along a line passing through the centroid of the member. In that instance, it may be desirable to design a *balanced weld*, and have the force pass through the centroid of the weld group.[53] Figure 15.35 shows a tensile force being transmitted through a single angle.

The procedure for designing the unequal weld lengths assumes that the end weld (if present) acts at full shear stress with a resultant passing through its mid-height ($d/2$ in figure 15.35). The forces in the other welds are assumed to act along the edges of the angle. Moments can be taken about a point on the line of action of either longitudinal weld, with the following results.

$$P_3 = T\left(1 - \frac{y}{d}\right) - \frac{P_2}{2} \qquad 15.112$$

$$P_2 = R_w L_2 = R_w d \qquad 15.113$$

$$P_1 = T - P_2 - P_3 \qquad 15.114$$

$$L_1 = \frac{P_1}{R_w} \qquad 15.115$$

$$L_3 = \frac{P_3}{R_w} \qquad 15.116$$

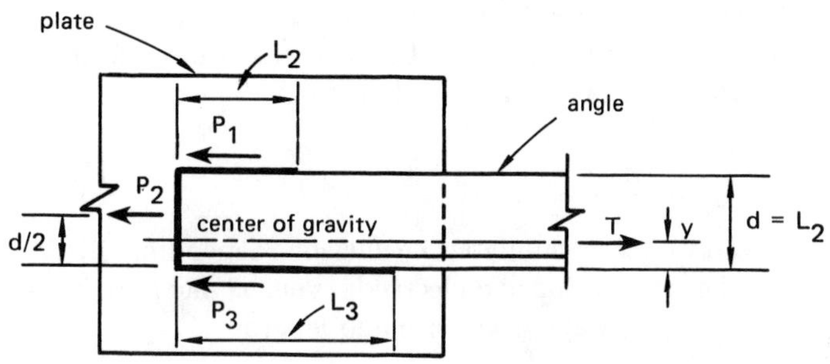


**Figure 15.35** A Balanced Weld Group

## D. COMBINING SHEAR AND BENDING STRESSES

Figure 15.36 illustrates a case where a welded connection must support both direct shear and a bending moment. Even though the maximum shear and maximum bending moment do not actually occur at the same place, simplifying assumptions are made to combine the nominal stresses as vectors.

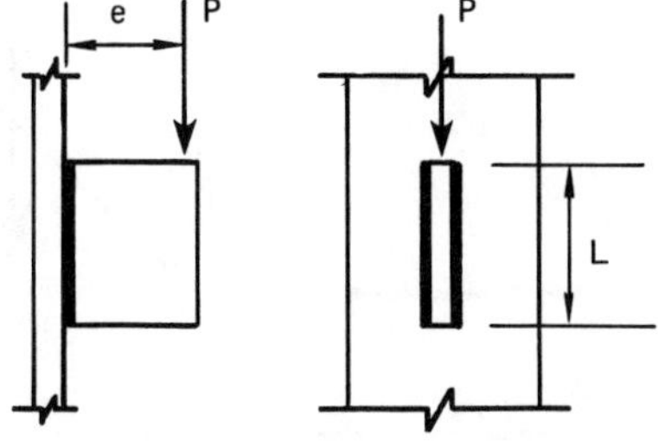


**Figure 15.36** Welds in Combined Bending and Shear

The nominal shear stress is

$$f_v = \frac{P}{A_{\text{weld}}} = \frac{P}{2Lt_e} \qquad 15.117$$

[53] *AISC Specification section 1.15.3* exempts single angles and double angles from the need to balance welds when loading is static. All other static loading configurations, and angles subjected to fatigue, must have balanced welds.

The nominal bending stress is easily calculated if the section modulus of the weld group can be obtained from appendix C.

$$f_b = \frac{Mc}{I} = \frac{M}{S} = \frac{Pe}{S} \qquad 15.118$$

The resultant stress is

$$f = \sqrt{f_v^2 + f_b^2} \qquad 15.119$$

The resultant stress must be less than the allowable stress, as calculated from equation 15.108.

The *AISC Manual* contains tables enabling the capacity of such eccentric loads to be calculated from a formula of the form in equation 15.120.[54]

$$P = CC_1Dl \qquad 15.120$$

$C$ and $C_1$ are tabulated coefficients depending on the load configuration and electrode type. $D$ and $l$ are functions of the weld group size and length.

## E. TORSION CONNECTIONS

The complexity of the stress distribution in torsion connections (such as are shown in figure 15.37) makes accurate analysis and design based on pure mechanics principles impossible. Therefore, simplifications are made and the assumed shear stress in the most critical location is calculated.

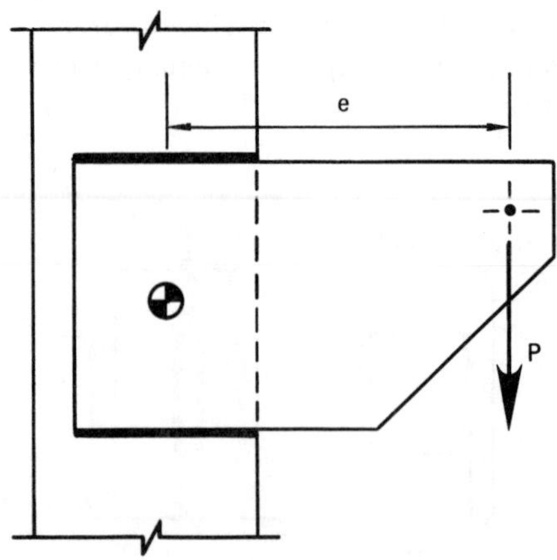


**Figure 15.37** A Welded Torsion Connection

The solution procedure (for both analysis and design) requires finding the centroid of the weld group. This centroid can be located in the normal manner, by weighting the weld's areas by the distances from an assumed axis. Alternatively, a weld can be treated as a line and its length used in place of the area.

Once the centroid is located, the polar area moment of inertia is calculated. Calculating the *polar moment of inertia* is easiest if the coordinate moments of inertia are known. For the purpose of calculating the moment of inertia, the welds may also be treated as either areas or lines[55] (Appendix C lists the polar moment of inertia for many common weld group configurations.)

$$J = I_x + I_y \qquad 15.121$$

The distance from the weld group centroid to the critical location in the weld group is then determined. This distance is used to calculate the torsional shear.

$$f_v = \frac{Mr}{J} = \frac{Per}{J} \qquad 15.122$$

The direct shear is easily calculated from the total weld area (or length, if the throat size is not known).

$$f_{v,d} = \frac{P}{A} \qquad 15.123$$

Vector addition is used to combine the torsional and direct shear stresses.

$$f = \sqrt{(f_{v,y} + f_{v,d})^2 + f_{v,x}^2} \qquad 15.124$$

*Example 15.17*

A 50 ksi plate bracket is welded with E70 electrodes to the face of a 50 ksi column as shown. What size fillet weld is required? Neglect buckling and bending effects of the plate and column.

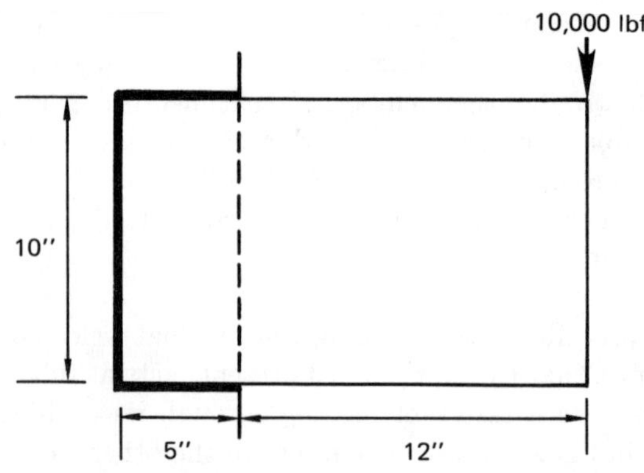


[54] See "*Eccentric Loads of Weld Groups*." This method is based on ultimate strength design concepts. As with bolted connections based on ultimate design, capacities much greater than would be expected from elastic analysis are obtained.

[55] In an analysis of a weld group, the throat size will be known, and welds can be treated as areas. In design, the throat size will be unknown, and it is common to treat the welds as lines. Alternatively, a variable weld size can be used and carried along in a design problem. However, the results will be essentially identical.

*step 1*: Assume the weld has thickness $t$.

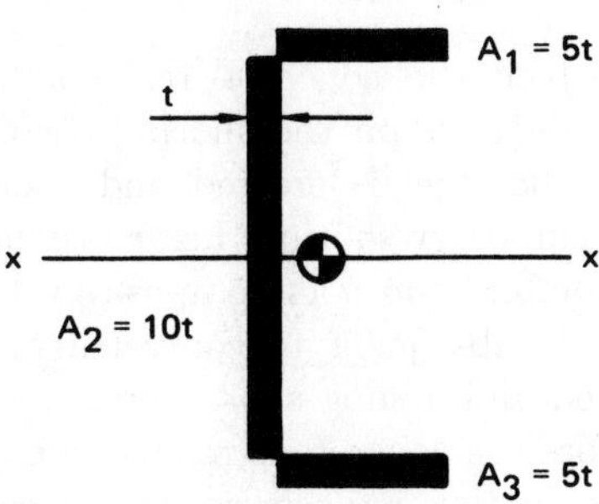


*step 2*: Find the centroidal location of the weld group. By inspection, $\overline{y}_c = 0$. For the three welds,

$$A_1 = 5t$$
$$\overline{x}_1 = 2.5$$
$$A_2 = 10t$$
$$\overline{x}_2 = 0$$
$$A_3 = 5t$$
$$\overline{x}_3 = 2.5$$

$$\text{So, } \overline{x}_c = \frac{5t(2.5) + 10t(0) + 5t(2.5)}{5t + 10t + 5t} = 1.25$$

*step 3*: Determine the centroidal moment of inertia of the weld group about the $x$-axis. Use the parallel axis theorem for areas 1 and 3.

$$I_x = \frac{t(10)^3}{12} + 2\left[\frac{5(t)^3}{12} + 5t(5)^2\right]$$
$$= 333.33t + 0.833(t)^3$$

Since $t$ will be small (probably less than 0.5″), the $t^3$ term can be neglected. So, $I_x = 333.33t$.

*step 4*: Determine the centroidal moment of inertia of the weld group about the $y$-axis.

$$I_y = \frac{10(t)^3}{12} + (10t)(1.25)^2 + 2\left[\frac{t(5)^3}{12} + (5t)(1.25)^2\right]$$
$$= 0.833t^3 + 52.08t \approx 52.08t$$

*step 5*: The polar moment of inertia is

$$J = I_x + I_y = 333.33t + 52.08t = 385.4t$$

*step 6*: By inspection, the maximum shear stress will occur at point **a**.

$$r = \sqrt{(3.75)^2 + (5)^2} = 6.25$$

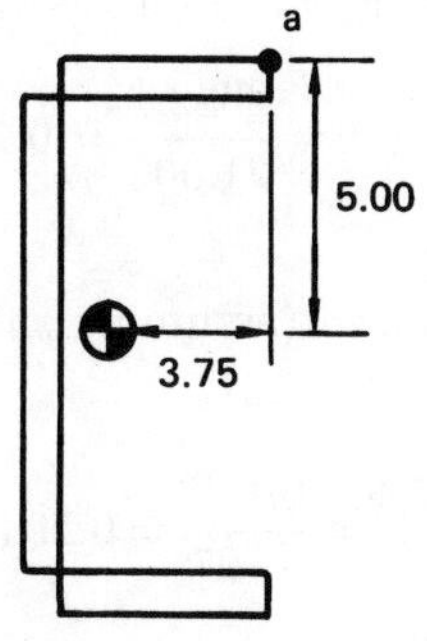


*step 7*: The applied moment is

$$M = (10{,}000)(12 + 3.75) = 157{,}500 \text{ in-lbf}$$

*step 8*: The torsional shear stress is

$$f_v = \frac{Mr}{J} = \frac{(157{,}500)(6.25)}{385.4t} = \frac{2554.2}{t} \text{ psi}$$

This shear stress is directed at right angles to the line $r$. The $x$- and $y$-components of the stress can be determined from geometry.

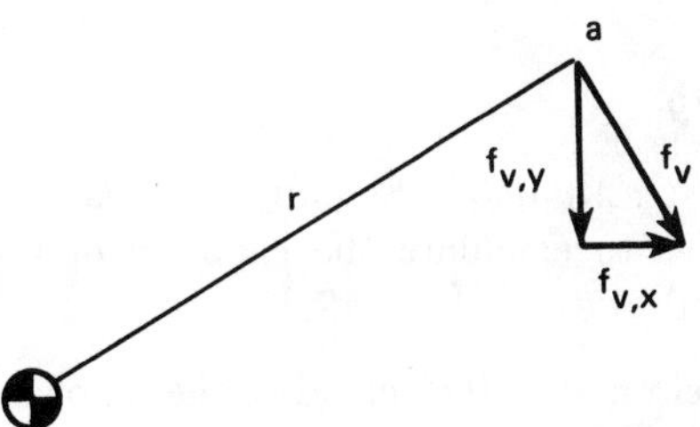


$$f_{v,y} = \left(\frac{3.75}{6.25}\right)\left(\frac{2554.2}{t}\right) = \frac{1532.5}{t}$$
$$f_{v,x} = \left(\frac{5.00}{6.25}\right)\left(\frac{2554.2}{t}\right) = \frac{2043.4}{t}$$

*step 9*: The direct shear is

$$f_{v,d} = \frac{10{,}000}{5t + 10t + 5t} = \frac{500}{t}$$

*step 10*: The resultant shear stress at point **a** is

$$f_v = \sqrt{\left(\frac{2043.4}{t}\right)^2 + \left(\frac{1532.5 + 500}{t}\right)^2} = \frac{2882.1}{t}$$

*step 11*: For 50 ksi base metal, the weld strength controls the allowable stress. (See equation 15.110.) The allowable stress is

$$F_v = 0.30F_u = (0.30)(70) = 21 \text{ ksi}$$

The weld throat size is

$$t = \frac{2882.1}{21{,}000} = 0.137''$$

*step 12*: The weld size (assuming shielded arc welding) is

$$w = \frac{t}{0.707} = \frac{0.137}{0.707} = 0.194'' \quad (\text{say } 1/4'')$$

*Example 15.18*

Use appendix C to calculate the polar moment of inertia of the weld group in example 15.17.

$$b = 5$$

$$d = 10$$

$$J = \left[\frac{(8)(5)^3 + (6)(5)(10)^2 + (10)^3}{12} - \frac{(5)^4}{(2)(5) + 10}\right] t$$

$$= 385.42t$$

(Example 15.17 calculated the moment of inertia to be 385.4$t$.)

*Example 15.19*

Use the *AISC Manual* "*Eccentric Loads on Weld Groups*" tables to calculate the capacity of the weld group designed in example 15.17.

For a 1/4" weld and E70 electrodes, the following coefficients are needed.

$$l = 10''$$

$$k = \frac{5}{10} = 0.5$$

$$a = \frac{12 + 3.75}{10} = 1.575$$

$$C = \mathbf{0.419} \quad (\text{interpolated})$$

$$C_1 = 1 \quad (\text{for E70 electrodes})$$

$$D = \frac{\frac{1}{4}}{\frac{1}{16}} = 4$$

From equation 15.120, the capacity of the connection is

$$P = CC_1Dl = (0.419)(1)(4)(10) = 16.76 \text{ kips}$$

## 36 COMPOSITE CONSTRUCTION

*Composite construction* usually means that a concrete slab is bonded to steel girders below. Although there are varying degrees of composite action, the concrete usually becomes the compression "flange" of the composite beam, and the steel carries tension.

The degree to which the concrete-steel combination acts compositely depends on the shoring used during construction. If the steel is erected and concrete is poured without temporary shoring (i.e., construction is *unshored*), the combination acts compositely to carry only loads applied subsequent to concrete curing. If the steel is erected and a snug shore is put into place at that time, before the concrete is poured (i.e., *partial shoring*), the combination acts compositely to carry the concrete weight as well as live loads.

In fully *shored construction*, a temporary support carries the steel and concrete weights both until the concrete has cured. When the shore is removed, the beam acts compositely to carry the steel and concrete weights in addition to live loads applied later.

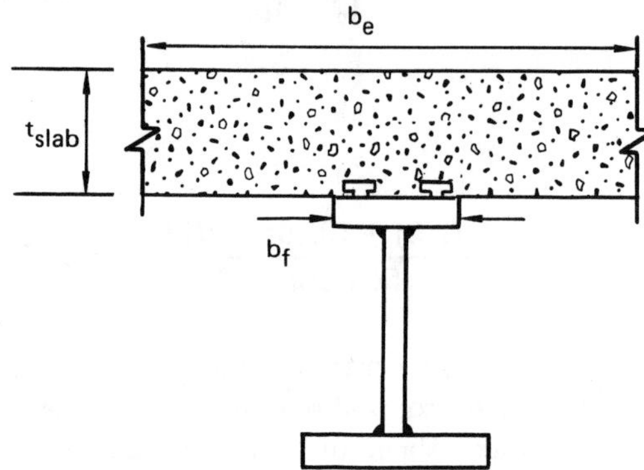


**Figure 15.38** Typical Composite Construction with Plate Girder

Analysis of composite steel-concrete beams is governed by AISC section 1.11. It is necessary to know the modulus of elasticity for the concrete and steel in order to calculate the *modular ratio, n*.[56]

$$n = \frac{E_s}{E_c} \qquad 15.125$$

For the purpose of calculating the moment of inertia, the effective width of the compression flange must be known. For interior girders with slabs extending on both sides, the effective width is[57]

$$b_e = \min \begin{Bmatrix} \frac{1}{4} \times \text{girder length} \\ \text{girder spacing} \\ b_f + 16t_{\text{slab}} \end{Bmatrix} \qquad 15.126$$

[56] The ACI formula for *secant modulus* is used to calculate $E_c$.

[57] Other rules apply to exterior girders with slabs extending to one side only.

# Appendix A: Steel Used for Buildings and Bridges

| designation | grade (if any) | $F_y$ minimum yield stress (ksi) | $F_u$ tensile strength (ksi) | maximum thickness for plates (in.) | use |
|---|---|---|---|---|---|
| A36 | | 32<br>36 | 58–80<br>58–80 | over 8<br>to 8 | general structural purposes; bolted and welded, mainly for buildings |
| A53 | A<br>B | 30<br>35 | 48<br>60 | | welded and seamless pipe |
| A242 | | 42<br>46<br>50 | 63<br>67<br>70 | over $1\frac{1}{2}$ to 4<br>over $\frac{3}{4}$ to $1\frac{1}{2}$<br>to $\frac{3}{4}$ | welded and bolted bridge construction where corrosion resistance is desired; essentially superseded by A709, grade 50W |
| A440 | | 42<br>46<br>50 | 63<br>67<br>70 | over $1\frac{1}{2}$ to 4<br>over $\frac{3}{4}$ to $1\frac{1}{2}$<br>to $\frac{3}{4}$ | bolted construction; essentially superseded by A572 for buildings and A709 for bridges |
| A441 | | 40<br>42<br>46<br>50 | 60<br>63<br>67<br>70 | over 4 to 8<br>over $1\frac{1}{2}$ to 4<br>over $\frac{3}{4}$ to $1\frac{1}{2}$<br>to $\frac{3}{4}$ | welded construction; largely superseded by A572 for buildings and A709 for bridges |
| A500 | A<br>B<br>C<br><br>A<br>B<br>C | 33<br>42<br>46<br><br>39<br>46<br>50 | 45<br>58<br>62<br><br>45<br>58<br>62 | round<br><br>shaped | cold-formed welded and seamless round and shaped tubing for general structural purposes |
| A501 | | 36 | 58 | | hot-formed welded and seamless round and shaped tubing for general structural purposes |
| A514 | | 90<br>100 | 100–130<br>110–130 | over $2\frac{1}{2}$ to 6<br>to $2\frac{1}{2}$ | alloy steel plates for welded construction; superseded by A709 for bridges |
| A529 | | 42 | 60–85 | to $\frac{1}{2}$ | pre-engineered rigid frames |
| A570 | A<br>B<br>C<br>D<br>E | 25<br>30<br>33<br>40<br>42 | 45<br>49<br>52<br>55<br>58 | | cold-formed sections |
| A572 | 42<br>50<br>60<br>65 | 42<br>50<br>60<br>65 | 60<br>65<br>75<br>80 | to 6<br>to 2<br>to $1\frac{1}{4}$<br>to $1\frac{1}{4}$ | welded and bolted construction for buildings; welded bridges in grades 42, and 50 only; essentially superseded by A709, grade 50 for bridges |

| designation | grade (if any) | $F_y$ minimum yield stress (ksi) | $F_u$ tensile strength (ksi) | maximum thickness for plates (in.) | use |
|---|---|---|---|---|---|
| A588 | | 42 | 63 | over 5 to 8 | weathering steel for welded |
| | | 46 | 67 | over 4 to 5 | and bolted construction; essentially |
| | | 50 | 70 | to 4 | superseded by A709, grade 50W for bridges |
| A606 | | 45 | 65 | | hot- and cold-rolled sheet |
| | | 50 | 70 | (hot-rolled cut lengths only) | and strip steel available in coils or cut lengths, used for cold-formed sections |
| A607 | 45 | 45 | 60 | | hot-rolled and cold-rolled |
| | 50 | 50 | 65 | | sheet and strip steel in coils |
| | 55 | 55 | 70 | | or cut lengths, used in cold- |
| | 60 | 60 | 75 | | formed sections |
| | 65 | 65 | 80 | | |
| | 70 | 70 | 85 | | |
| A611 | A | 25 | 42 | | cold-rolled sheet steel |
| | B | 30 | 45 | | for cold-formed sections |
| | C | 33 | 48 | | |
| | D | 40 | 52 | | |
| | E | 80 | 82 | | |
| A618 | I | 50 | 70 | | hot-formed welded and seamless |
| | II | 50 | 70 | | tubing for general structural |
| | III | 50 | 65 | | purposes |
| A709 | 36 | 32 | 58 | over 8 | bridge construction: grade 36 |
| | | 36 | 58–80 | to 8 | is approximately the same as A36; |
| | 50 | 50 | 65 | to 2 | grade 50 as A441; grade 50W as |
| | 50W | 50 | 70 | to 4 | A588; and grade 100 as A514 |
| | 100 & 100W | 90 | 100–130 | over $2\frac{1}{2}$ to 4 | |
| | 100 & 100W | 100 | 110–130 | to $2\frac{1}{2}$ | |

# Appendix B: Properties of Structural Steel at High Temperatures

| type of steel | temperature (degF) | yield strength 0.2% offset (ksi) | tensile strength (ksi) |
|---|---|---|---|
| ASTM A36 | 80 | 36.0 | 64.0 |
| | 300 | 30.2 | 64.0 |
| | 500 | 27.8 | 63.8 |
| | 700 | 25.4 | 57.0 |
| | 300 | 21.5 | 44.0 |
| | 1100 | 16.3 | 25.2 |
| | 1300 | 7.7 | 9.0 |
| ASTM A242 | 80 | 54.1 | 81.3 |
| | 200 | 50.8 | 76.2 |
| | 400 | 47.6 | 76.4 |
| | 600 | 41.1 | 81.3 |
| | 800 | 39.9 | 76.4 |
| | 1000 | 35.2 | 52.8 |
| | 1200 | 20.6 | 27.6 |
| ASTM A588 | 80 | 58.6 | 78.5 |
| | 200 | 57.3 | 79.5 |
| | 400 | 50.4 | 74.8 |
| | 600 | 42.5 | 77.7 |
| | 800 | 37.6 | 70.7 |
| | 1000 | 32.6 | 46.4 |
| | 1200 | 17.9 | 23.3 |

# Appendix C: Properties of Welds Treated as Lines

| weld configuration | centroid location | section modulus $S = I_x/\overline{y}$ | polar moment of inertia $J = I_x + I_y$ |
|---|---|---|---|
|  | | $\frac{d^2}{6}$ | $\frac{d^3}{12}$ |
| 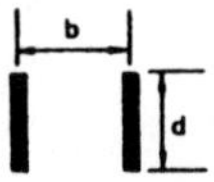 | | $\frac{d^2}{3}$ | $\frac{d(3b^2+d^2)}{6}$ |
| 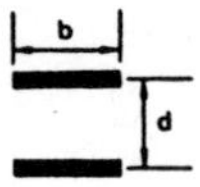 | | $bd$ | $\frac{b(3d^2+b^2)}{6}$ |
| 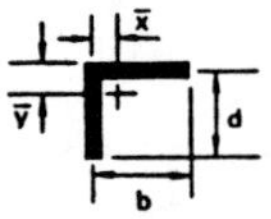 | $\overline{y} = \frac{d^2}{2(b+d)}$<br>$\overline{x} = \frac{b^2}{2(b+d)}$ | $\frac{4bd+d^2}{6}$ | $\frac{(b+d)^4 - 6b^2d^2}{12(b+d)}$ |
| 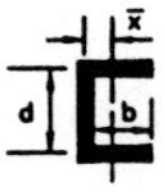 | $\overline{x} = \frac{b^2}{2b+d}$ | $bd + \frac{d^2}{6}$ | $\frac{8b^3+6bd^2+d^3}{12} - \frac{b^4}{2b+d}$ |
| 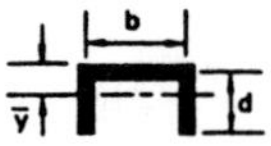 | $\overline{y} = \frac{d^2}{b+2d}$ | $\frac{2bd+d^2}{3}$ | $\frac{b^3+6b^2d+8d^3}{12} - \frac{d^4}{2d+b}$ |
| 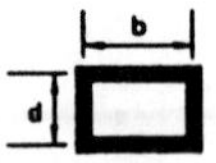 | | $bd + \frac{d^2}{3}$ | $\frac{(b+d)^3}{6}$ |
| 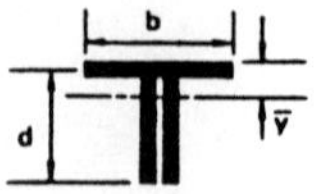 | $\overline{y} = \frac{d^2}{b+2d}$ | $\frac{2bd+d^2}{3}$ | $\frac{b^3+8d^3}{12} - \frac{d^4}{b+2d}$ |
| 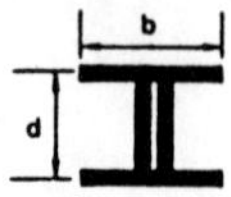 | | $bd + \frac{d^2}{3}$ | $\frac{b^3+3bd^2+d^3}{6}$ |
|  | | $\pi r^2$ | $2\pi r^3$ |

# Appendix D: The Moment Gradient Multiplier

If a concentrated load is applied to a beam at a point of lateral support, there will be less tendency for the beam to buckle than if the load is applied between points of lateral support. The moment gradient multiplier, $C_b$, is an optional refinement that accounts for the less-critical conditions. The moment gradient multiplier increases the maximum allowable spacing between supports, $L_u$. $C_b$ varies between 1.0 and 2.3 (its maximum). Conservative designs use $C_b = 1.0$.

$$C_b = 1.75 + 1.05\left(\frac{M_1}{M_2}\right) + 0.3\left(\frac{M_1}{M_2}\right)^2 \quad \text{[AISC 1.5.1.4.5]}$$

$M_1$ is the absolute value of the smaller moment at the ends of an unbraced beam section. $M_2$ is the absolute value of the larger moment at the opposite end of the unbraced section. The quantity $(M_1/M_2)$ is negative for simple bending (i.e., single curvature) and positive if the moment line goes through zero within the span (i.e., reverse curvature).

If the moment on a section of beam is maximum between the points of lateral support, $C_b = 1$.

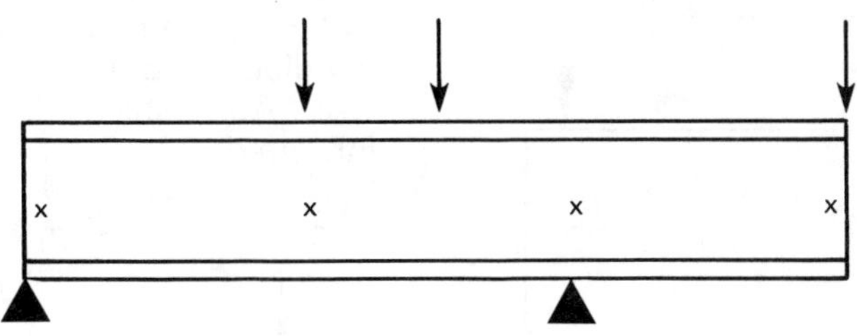


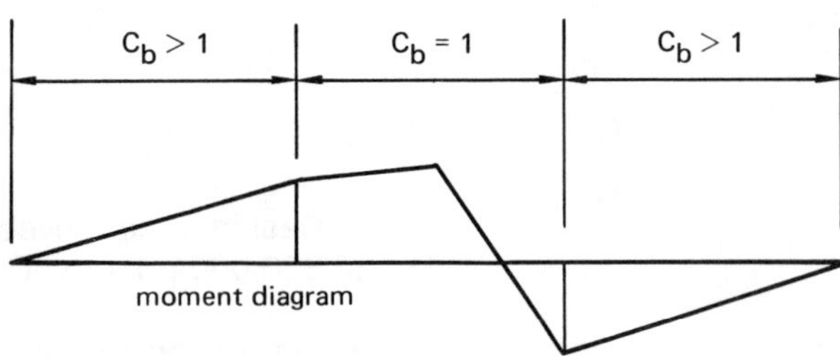


**Figure 15.39** Moment Gradient Multiplier

In evaluating $F_b$ for use with beam columns (equation 15.49 or [**1.6-1a**]), $C_b = 1$ is used when joint translation is prohibited. When joint translation is possible, calculate $C_b$ according to the equation given.

## Practice Problems: STEEL DESIGN

Untimed

1. A 25 foot long beam is simply supported at its left end and 7 feet from its right end. A load of 3000 pounds per foot is uniformly distributed over its entire length. Lateral support is provided only at the reactions. Choose an economical W shape for this application. Use A36 steel.

2. Select the lightest W section of A36 steel to serve as a main member 30 feet long and to carry an axial load of 160,000 pounds. The member is pinned at its top and bottom. It is supported at mid-height in its weak direction. The member is vertical.

3. Determine the stresses in bolts A, B, C, and D. All connectors are ¾″ bolts.

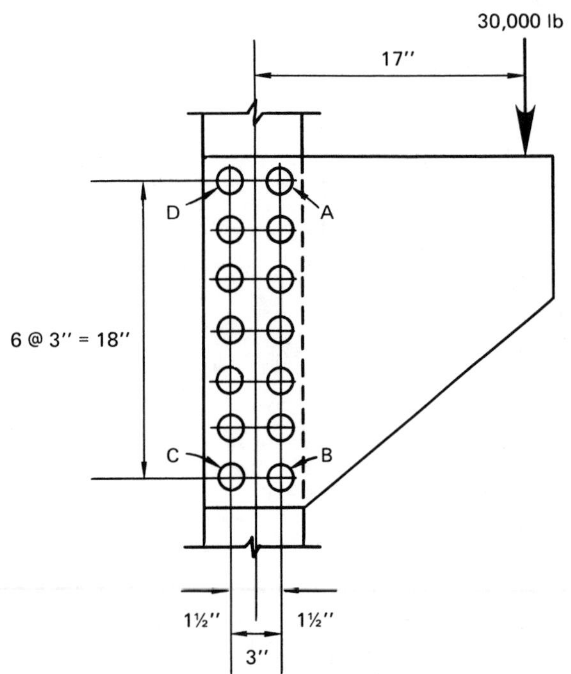


4. Determine the size of the fillet weld required to connect the plate bracket to the column face. The steel is A36. The electrodes used are E70XX.

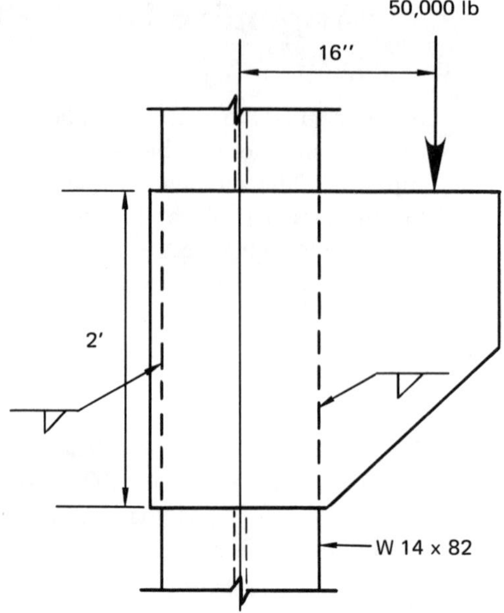


5. Design the interior columns of the frame shown. Columns are braced continuously in the plane perpendicular to the frame. Girders are W12 x 96. Use A36 steel. Left-to-right sidesway is uninhibited.

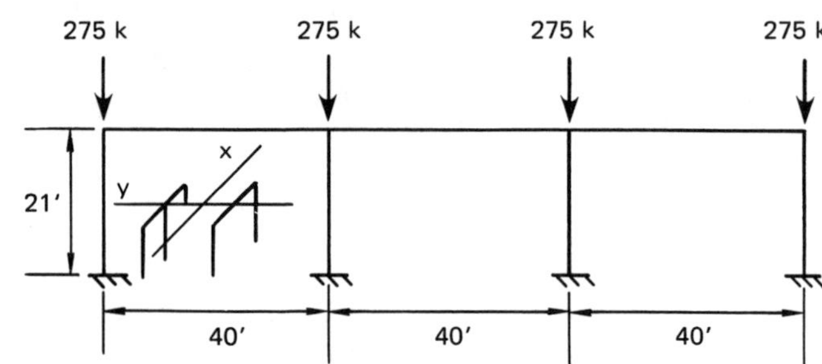


6. Design a plate eyebar to carry a static tensile load of 300,000 pounds. Use A36 steel.

7. Determine the tensile capacity of the connection shown. A325, ⅞″ diameter bolts are used. The steel is grade 50. The connection is a friction type with no threads in the shear plane.

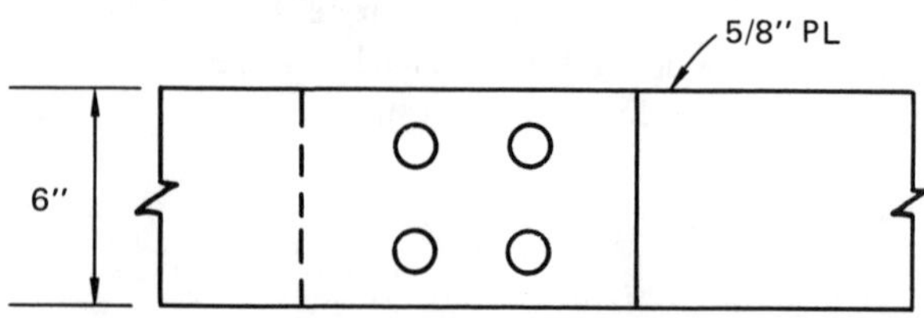


8. Select an A36 W shape with lateral support at 5½ foot intervals to span 20 feet and carry a uniformly distributed load of 1000 lb/ft. The load includes a uniform dead load allowance of 40 lb/ft. Limit the maximum deflection to (L/240).

9. Select an economical A36 W shape (completely laterally braced) to span 24 feet such that the maximum deflection is (L/300). The span carries a uniformly distributed load of 800 lb/ft over its entire length. This load does not include the dead weight of the beam.

10. A 16 foot column is acted upon by an axial gravity load of 17,000 lb and a uniform wind load of *w* lb/ft. The W12 x 58 column is made of A 441 steel. The lower end is built in. The upper end is not supported in the weak direction. (a) Find the required spacing of lateral bracing for the maximum lateral (uniform) loading. (b) Find the maximum lateral (uniform) loading allowed with the spacing calculated in part (a).

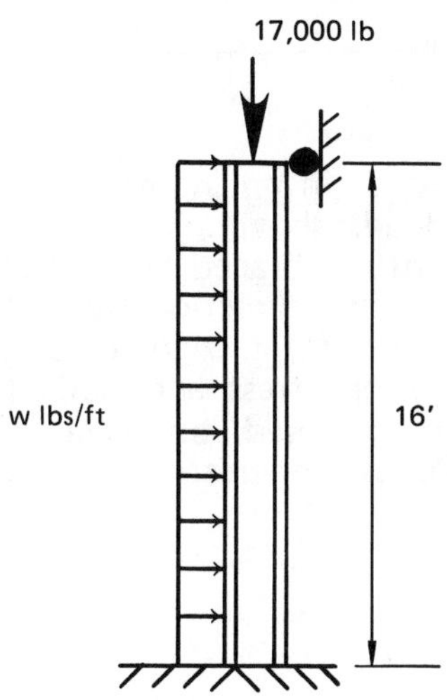


Timed

1. The steel beam shown carries 1.4 kips/ft over its entire span. Assume lateral bracing only at the reaction points. (a) Draw shear and moment diagrams. (b) Choose the lightest W18 beam assuming A36 steel is used. (c) If the beam you chose in part (b) is constructed of 42 ksi steel, what is the maximum unbraced length?

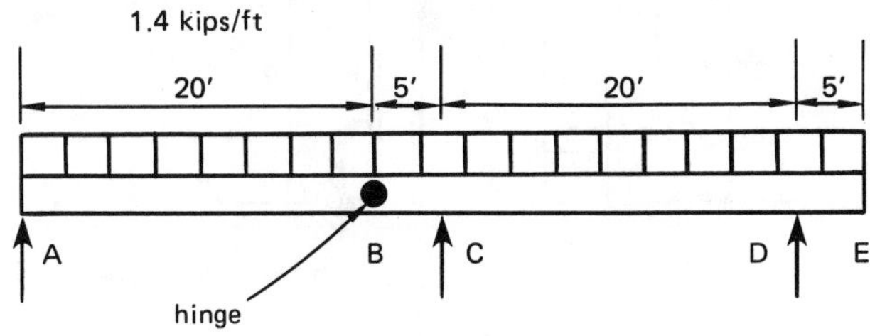


2. A column carries 525 kips. The beams framing into it apply a moment of 225 ft-kips. (a) If the column is W14 x 82 (50 ksi steel) and the beams are W14 x 53 (50 ksi), determine if the column is adequate. (b) If the design is inadequate, specify an acceptable W14 (50 ksi) column. Bracing in the weak axis is provided and bending is about the strong axis only.

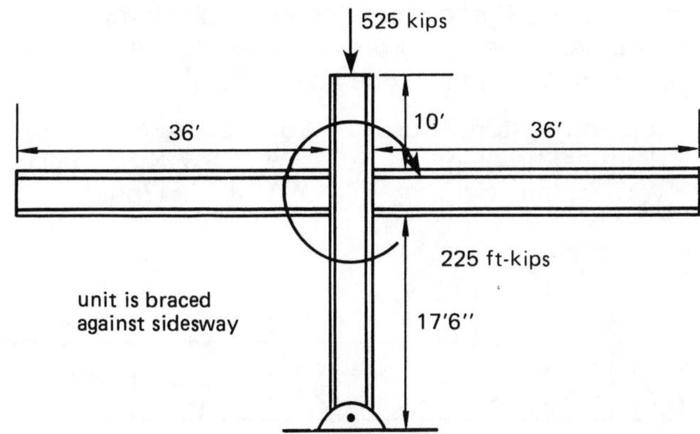


3. Two angles are welded to an intermediate wide-flange section which is bolted to a column. (a) Using the maximum weld size permitted, design the weld between the angles and the intermediate section. Use E70 electrodes. The design should also specify the weld locations. (b) Determine if the bolts are adequate in quantity and location. No threads are in the shear plane.

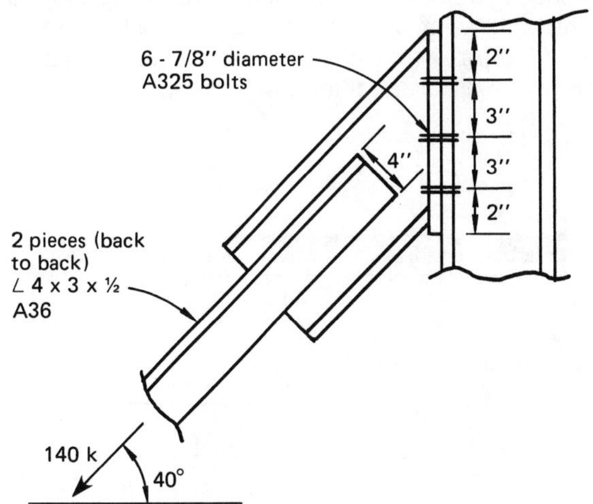


4. A column of 20 feet high is constructed from a channel and a wide flange beam as shown. The column is constrained against bending in the x-plane by bracing at the 10 foot point on the column. Both ends are pinned and free to rotate. Determine the buckling load.

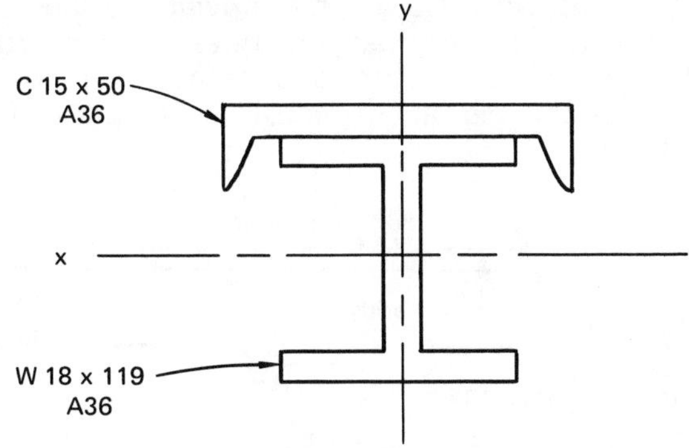


5. The truss shown is constructed of a variety of members, all of which are assumed to be pin-ended. Use

the latest AISC code to determine if all members are adequate. If members are not adequate, specify replacements with the same nominal depth.

| | |
|---|---|
| top and bottom chords: | W8 × 28 (A36) |
| diagonal bracing: | L3½ × 3½ × 5/16 (pair) |
| vertical members: | L4 × 4 × 3/8 (pair) |

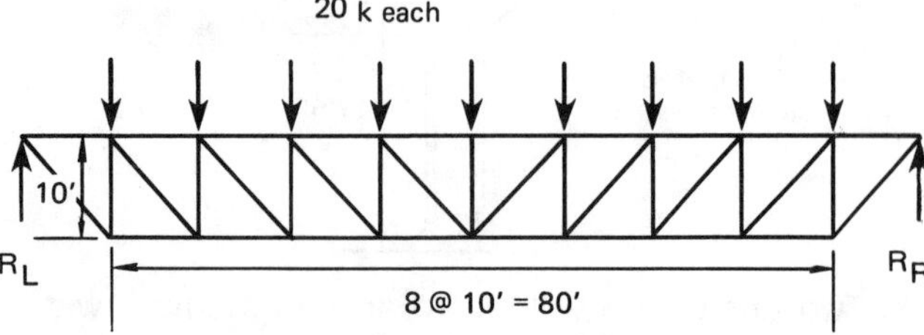


6. Two equal loads remain 3' apart as they move across a 40' span. The beam is constructed from a W shape and a C shape as shown. Sufficient lateral support is provided. (a) Find the maximum allowable load, *P*. Consider only beam bending and shear stresses. Do not consider fatigue. (b) If the weld is intermittent, what spacing is required using the minimum fillet size? (c) Do the loads computed in part (a) satisfy the AISC specification for web crippling?

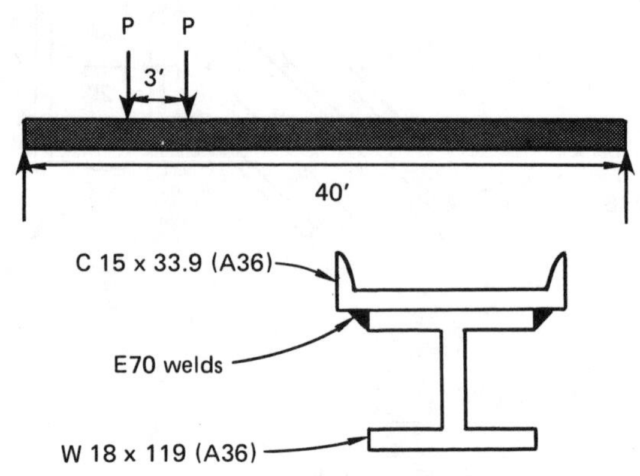


7. A bridge is to be supported by a plate girder. Loads are resisted by composite action. The bridge is 80 feet long. It will not be braced during construction. Flange support will be provided at 20 foot intervals. Consider a moving 40 kip load and a 1 kip/ft dead load (includes the concrete). Determine if the plate girder (assumed to be simply supported at its ends) is adequate. Use n = 10.

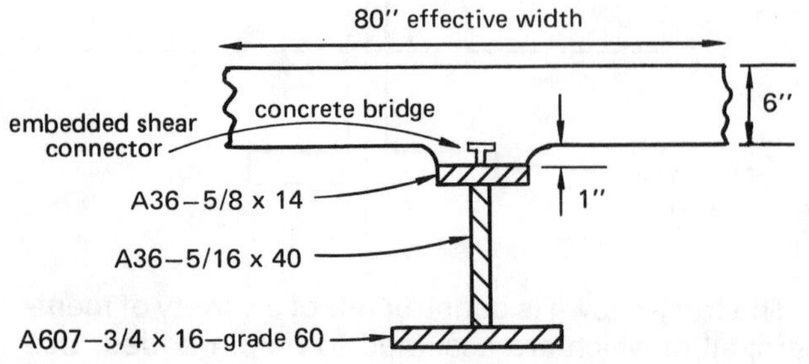


8. Design column CD as a minimum-size wide flange member per Part 2 of the AISC. Assume continuous lateral support of the column. All loads shown are ultimate.

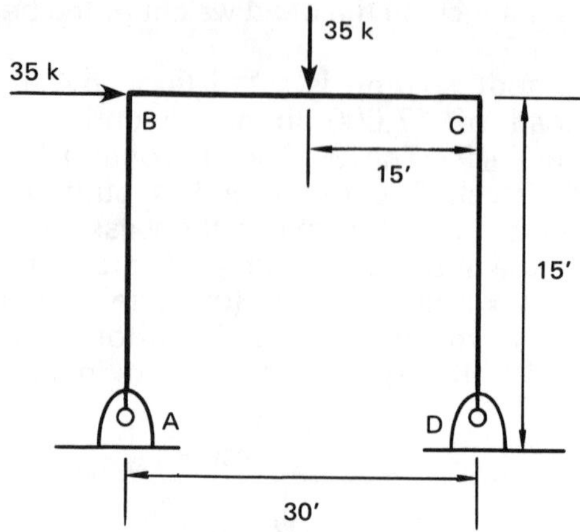


9. A tall column in an industrial storage building carries the loads shown. The lateral wind loads are carried into a braced frame. The column is braced in its weak direction by a strut 24' from the hinged end. (a) Design an economical W24 section based on stress and deflection. (b) Size the base plate based on axial loading only. Assume 3000 psi concrete is used for the resting support.

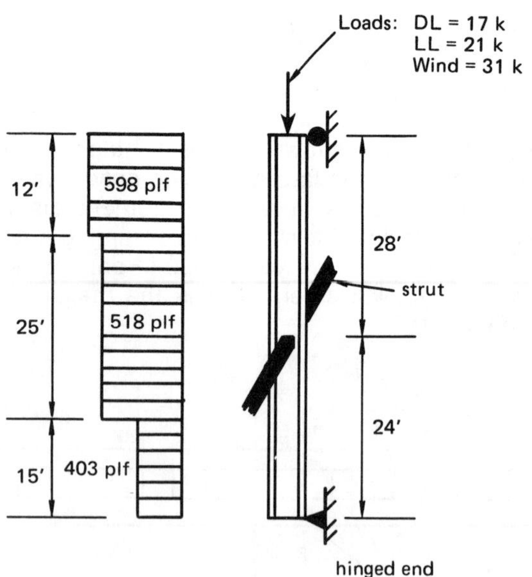


10. A plate girder spans 40 feet and is constructed with 3/8" A36 steel. The flange is 16" wide, and the web is 48" deep. A uniform but unknown load is carried across the girder's entire length, which is simply supported at both ends. (a) Based on shear consideration only, determine the maximum uniform load (in kips/ft) that the beam can carry. (b) Design the web stiffeners located 4 feet from the supports.

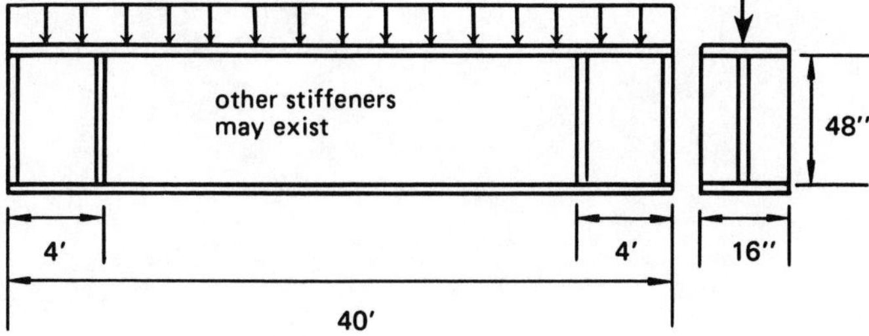
other stiffeners
may exist
48″
4′
4′
16″
40′

RESERVED FOR FUTURE USE

# 16 TRAFFIC ANALYSIS, TRANSPORTATION, AND HIGHWAY DESIGN

*Nomenclature*

| | | |
|---|---|---|
| a | acceleration, or layer coefficient | $ft/sec^2$, – |
| A | area | sq ft |
| ADT | average daily traffic | vpd |
| c | capacity, or distance from neutral axis to extreme fiber | vphpl, in |
| d | separation distance | ft |
| D | directional factor, or density | –, vpmpl |
| DDHV | directional design hourly volume | vph |
| DHV | design hourly volume | vph |
| E | passenger car equivalent | – |
| EAL | equivalent axle load | lbf |
| EWL | equivalent wheel load | lbf |
| f | coefficient of friction, factor, or fraction | –, –, – |
| $f_t$ | allowable working stress | psi |
| F | force | lbf |
| FF | fatigue fraction | – |
| FS | factor of safety | – |
| g | acceleration due to gravity (32.2) | $ft/sec^2$ |
| G | grade | decimal |
| $G_f$ | gravel equivalent factor | – |
| GE | gravel equivalent | ft |
| I | moment of inertia | $in^4$ |
| k | modulus of subgrade reaction, or ratio of DHV to ADT | psi/in, – |
| L | slab length | ft |
| LC | length of curve | ft |
| LSF | load safety factor | – |
| m | mass | slugs |
| M | middle ordinate | ft |
| MR | modulus of rupture | psi |
| MSF | maximum service flow rate | vphpl |
| n | number of cycles | cycles |
| N | number of lanes, normal force, or fatigue life | –, lbf, cycles |
| $p_t$ | terminal serviceability | – |
| r | curve radius | ft |
| R | soil resistance value | – |
| s | distance | ft |
| S | sight distance, or speed | ft, mph |
| $S_c$ | compressive strength | psi |
| SF | service flow rate | vph |
| SN | structural number | – |
| t | time, or thickness | seconds, inches, feet |
| TI | traffic index | – |
| v | velocity, or volume | fps, vphpl |
| w | vehicle mass, or slab width | lbm, ft |

*Symbols*

| | | |
|---|---|---|
| $\theta$ | angle | degrees |
| $\mu$ | coefficient of friction | – |

*Subscripts*

| | |
|---|---|
| b | base |
| B | bus |
| c | centrifugal |
| f | frictional or free-flow |
| HV | heavy vehicle |
| j | jam |
| n | net |
| o | initial |
| p | perception or population |
| R | recreational vehicle |
| s | steel |
| sb | subbase |
| t | tangential |
| T | truck |
| w | at the wheels or width |

## 1 DEFINITIONS

AASHTO: American Association of State and Highway Transportation Officials. Previously known as AASHO.

Abandonment: The reversion of title to the owner of the underlying fee where an easement for highway purposes is no longer needed.

Access control: See 'Control of access.'

Acquisition: The process of obtaining right of way.

Arterial highway: A general term denoting a highway primarily for through traffic usually on a continuous route.

Auxiliary lane: The portion of a roadway adjoining the traveled way for truck climbing, speed change, or for other purposes supplementary to through traffic movement.

Base course: The bottom portion of a pavement where the top and bottom portions are not of the same composition.

Base: A layer of selected, processed, or treated aggregate material of planned thickness and quality placed immediately below the pavement and above the subbase or basement soil.

Belt highway: An arterial highway carrying traffic partially or entirely around an urban area.

CBD: Abbreviation for *central business district.*

Cement treated base: A base layer constructed with good quality, well-graded aggregate mixed with up to 6% cement.

Channelization: The separation or regulation of conflicting traffic movements into definite paths of travel by use of pavement markings, raised islands, or other means.

Condemnation: The process by which property is acquired for public purposes through legal proceedings under power of eminent domain.

Control of access: The condition where the right of owners or occupants of abutting land or other persons to access in connection with a highway is fully or partially controlled by public authority.

Divided highway: A highway with separated roadbeds for traffic in opposing directions.

Easement: A right to use or control the property of another for designated purposes.

Embankment: A raised structure constructed of natural soil from excavation or borrow sources.

Eminent domain: The power to take private property for public use without the owner's consent upon payment of just compensation.

Emulsion: A mixture with water. Asphalt emulsions are produced by adding a small amount of emulsifying soap to asphalt and water. When the water evaporates, the asphalt sets.

Encroachment: Use of the highway right-of-way for non-highway structures or other purposes.

Flexible pavement: A pavement having sufficiently low bending resistance to maintain intimate contact with the underlying structure, yet having the required stability furnished by aggregate interlock, internal friction, and cohesion to support traffic.

Freeway: A divided arterial highway with full control of access.

Frontage road: A local street or road auxiliary to, and located on, the side of an arterial highway for service to abutting property and adjacent areas, and for control of access.

Gore: The area immediately beyond the divergence of two roadways bounded by the edges of those roadways.

Inverse condemnation: The legal process that may be initiated by a property owner to compel the payment of just compensation when his property has been taken or damaged for a public purpose.

Median: The portion of a divided highway separating the traveled ways for traffic in opposite directions.

Median lane: A lane within the median to accommodate left-turning vehicles.

Parkway: An arterial highway for non-commercial traffic, with full or partial control of access, usually located within a park or a ribbon of parklike development.

Penetration treatment: Application of light liquid asphalt to the road-bed material. It is used primarily as a dust reducer on detours, medians, and parking areas.

Plant mix: An asphalt concrete mixture that is not prepared at the paving site.

Prime coat: The initial application of a low viscosity liquid asphalt to an absorbent surface, preparatory to any subsequent treatment, for the purpose of hardening or toughening the surface and promoting adhesion between it and the superimposed constructed layer.

Resurfacing: A supplemental surface or replacement placed on an existing pavement to restore its riding qualities or increase its strength.

Right of access: The right of an abutting land owner for entrance to or exit from a public road.

Rigid pavement: A pavement having sufficiently high bending resistance to distribute loads over a comparatively large area.

Road-mixed asphalt surfacing: A lower-quality surfacing used when plant mixes are not available or not economically feasible. Liquid asphalts are normally used. Road-mixed asphalt surfacing is used on low-traffic volume roads where higher quality surfacing is not required for traffic volume.

Roadbed: That portion of the roadway extending from curb line to curb line or shoulder line to shoulder line. Divided highways are considered to have two roadbeds.

Seal coat: A bituminous coating, with or without aggregate, applied to the surface of a pavement for the purpose of water-proofing and preserving the surface, rejuvenating a previous bituminous surface, altering the surface texture of the pavement, providing delineation, or providing resistance to traffic abrasion.

Structural section: The planned layers of specific materials, normally consisting of subbase, base, and pavement, placed over the basement soil.

Subbase: A layer of aggregate of planned thickness and quality placed on the basement soil as a foundation for the base.

Subgrade: The portion of a roadbed surface, which has been prepared as specified, upon which a subbase, base, base course, or pavement is to be placed.

Tack coat: The initial application of bituminous material to an existing surface to provide bond between the existing surface and the new material.

Traveled way: The portion of the roadway for the movement of vehicles exclusive of shoulders and auxiliary lanes.

## 2 TRANSLATIONAL DYNAMICS

Newton's second law can be used to relate the net tractive force on a vehicle to its acceleration.

$$F_n = ma \qquad 16.1$$

$$m = \frac{w}{g} \qquad 16.2$$

The *net tractive force* is the difference between the applied force at the wheels and the frictional force.

$$F_n = F_w - F_f \qquad 16.3$$

The *net force* can be found directly from the velocity of the vehicle and the horsepower expenditure at that velocity.

$$F_n = \frac{(550)(HP)}{v} \qquad 16.4$$

The frictional force is a combination of dynamic, rolling, turning, and aerodynamic forces which act to oppose motion.

The relationships between position, velocity, and acceleration as functions of time for linear motion are given here.

$$a = \frac{dv}{dt} = \frac{d^2s}{dt^2} \qquad 16.5$$

$$v = \frac{ds}{dt} = \int a\,dt \qquad 16.6$$

$$s = \int v\,dt = \iint a\,dt^2 \qquad 16.7$$

If the acceleration is uniform, table 16.1 can be used to determine values of unknown variables. Acceleration is negative for vehicles with decreasing velocities.

*Example 16.1*

A 4000 pound car traveling at 80 mph locks up its wheels and slides 580 feet before stopping. (a) How much time does it take to stop? (b) What is the deceleration? (c) What is the retarding force? (d) What is the coefficient of friction between the tires and the road?

(a) The initial velocity is

$$v_o = (80)\left(\frac{5280}{3600}\right) = 117.3\text{ ft/sec}$$

The stopping distance is $s = 580$ feet. The final velocity is $v = 0$. From table 16.1,

$$t = \frac{2s}{v_o + v} = \frac{(2)(580)}{117.3 + 0} = 9.89\text{ sec}$$

(b) From table 16.1,

$$a = \frac{v - v_o}{t} = \frac{0 - 117.3}{9.89} = -11.9\text{ ft/sec}^2$$

(c) $F_f = ma = \left(\frac{4000}{32.2}\right)(11.9) = 1480 \text{ lbf}$

(d) $\mu = \frac{F_f}{N} = \frac{1480}{4000} = 0.37$

**Table 16.1**
Uniform Acceleration Formulas

| to find | given these | use this equation |
|---|---|---|
| $t$ | $a\,v_o\,v$ | $t = \frac{v - v_o}{a}$ |
| $t$ | $a\,v_o\,s$ | $t = \frac{\sqrt{2as + v_o^2} - v_o}{a}$ |
| $t$ | $v_o\,v\,s$ | $t = \frac{2s}{v_o + v}$ |
| $a$ | $t\,v_o\,v$ | $a = \frac{v - v_o}{t}$ |
| $a$ | $t\,v_o\,s$ | $a = \frac{2s - 2v_o t}{t^2}$ |
| $a$ | $v_o\,v\,s$ | $a = \frac{v^2 - v_o^2}{2s}$ |
| $v_o$ | $t\,a\,v$ | $v_o = v - at$ |
| $v_o$ | $t\,a\,s$ | $v_o = \frac{s}{t} - \frac{1}{2}at$ |
| $v_o$ | $a\,v\,s$ | $v_o = \sqrt{v^2 - 2as}$ |
| $v$ | $t\,a\,v_o$ | $v = v_o + at$ |
| $v$ | $a\,v_o\,s$ | $v = \sqrt{v_o^2 + 2as}$ |
| $s$ | $t\,a\,v_o$ | $s = v_o t + \frac{1}{2}at^2$ |
| $s$ | $a\,v_o\,v$ | $s = \frac{v^2 - v_o^2}{2a}$ |
| $s$ | $t\,v_o\,v$ | $s = \frac{1}{2}t(v_o + v)$ |

## 3 SIMPLE ROADWAY BANKING

If a vehicle travels in a circular path with instantaneous radius $r$ and tangential velocity $v_t$, it will experience an apparent centrifugal force given by equation 16.8.

$$F_c = \frac{mv_t^2}{r} \qquad 16.8$$

This centrifugal force must be resisted by a combination of roadway banking (superelevation) and sideways friction. If it is desirable to bank the roadway so that little or no friction is required to resist the centrifugal force, the angle of superelevation is given by equation 16.9. (Generally, it is not desirable to rely on roadway banking alone, since such banking will be applicable at only one speed.)

$$\tan\theta = \frac{v^2}{g\,r} \qquad 16.9$$

In general, a lower banking angle is used in urban areas than in rural areas. For arterial streets in downtown areas, the maximum superelevation should be 0.04 to 0.06. For arterial streets in suburban areas and freeways where there is no snow or ice, the maximum is 0.10 to 0.12. For arterial streets and freeways that experience snow and ice, the maximum is 0.06 to 0.08.

Equation 16.10 is the basic formula used for determining superelevation when friction is relied upon to counteract some of the centrifugal force.

$$\tan\theta = \frac{v^2}{g\,r} - f \qquad 16.10$$

If $v$ is expressed in mph, equation 16.10 becomes

$$\tan\theta = \frac{(\text{MPH})^2}{15r} - f \qquad 16.11$$

$f$ is the *side friction factor.* It is usually assumed to be 0.16 for speeds of 30 mph and under. For higher speeds,

$$f = 0.16 - 0.01\left(\frac{\text{MPH} - 30}{10}\right) \quad \text{(up to 50 mph)} \qquad 16.12$$

$$f = 0.14 - 0.02\left(\frac{\text{MPH} - 50}{10}\right) \quad \text{(50 to 70 mph)}$$

Since the maximum $\tan\theta$ is usually 0.08 or 0.10, equation 16.10 can be used to calculate the minimum curve radius if the speed is known.

Transitions from flat to superelevated sections should be gradual.

*Example 16.2*

A 4000 pound car travels at 40 mph around a banked curve with a radius of 500 feet. What should be the banking angle such that tire friction is not needed to prevent the car from sliding?

$$v = (40)(1.467)\frac{\text{fps}}{\text{mph}} = 58.68 \text{ fps}$$

From equation 16.9,

$$\theta = \arctan\left[\frac{(58.68)^2}{(32.2)(500)}\right] = 12.07°$$

## 4 SIGHT AND STOPPING DISTANCES

*Sight distance* is the length of roadway that the driver can see. It is assumed that the driver's eyes are 3.75 feet above the surface of the roadway. The sight distance should be long enough to allow a driver traveling at the maximum speed to stop before coming upon an observed object. This required distance is known as the *stopping sight distance.* It is assumed that the object being observed has a height of 0.5 feet.

Since distance is covered during the driver's reaction period as well as during the deceleration period, the stopping sight distance includes both of these distances. The coefficient of friction is usually evaluated for wet pavement. For straight-line travel on a constant grade, $G$, equation 16.13 can be used. $G$ is a decimal, and it is negative if the roadway is downhill.

$$S = (1.47)(t_p)(\text{MPH}) + \frac{(\text{MPH})^2}{(30)(f+G)} \qquad 16.13$$

**Table 16.2**
Coefficients of Skidding Friction

*BC*: bituminous concrete, dry
*SA*: sand asphalt, dry
*RA*: rock asphalt, dry
*CC*: portland cement concrete, dry
wet: all wet pavements

| condition | *BC* | *SA* | *RA* | *CC* | wet |
|---|---|---|---|---|---|
| new tires | | | | | |
| 11 mph | 0.74 | 0.75 | 0.78 | 0.76 | |
| 20 | 0.76 | 0.75 | 0.76 | 0.73 | 0.40 |
| 30 | 0.79 | 0.79 | 0.74 | 0.78 | 0.36 |
| 40 | 0.75 | 0.75 | 0.74 | 0.76 | 0.33 |
| 50 | | | | | 0.31 |
| 60 | | | | | 0.30 |
| 70 | | | | | 0.29 |
| badly worn tires | | | | | |
| 11 mph | 0.61 | 0.66 | 0.73 | 0.68 | |
| 20 | 0.60 | 0.57 | 0.65 | 0.50 | 0.40 |
| 30 | 0.57 | 0.48 | 0.59 | 0.47 | 0.36 |
| 40 | 0.48 | 0.39 | 0.50 | 0.33 | 0.33 |
| 50 | | | | | 0.31 |
| 60 | | | | | 0.30 |
| 70 | | | | | 0.29 |

If the design speed is used in equation 16.13, $S$ is known as a *desirable value.* If the speed is less than the design value, $S$ is known as a minimum value. The minimum speed to be used is

$$\text{MPH}_{\text{min}} = \text{MPH}_{\text{design}} - 0.2(\text{MPH}_{\text{design}} - 20) \qquad 16.14$$

The desirable value should be used in most cases. These are listed in table 16.3 for various design speeds.

The *braking reaction-perception time,* $t_p$ in equation 16.13, has a median value of approximately 0.90 seconds for unexpected (not anticipated) events.[1] However, this time varies widely from subject to subject. Individuals with slow reactions may require up to 2.5 seconds.[2]

The *passing sight distance* is applicable only to 2-lane, 2-way highways. It is the length of roadway ahead necessary to pass without meeting an oncoming vehicle.[3] Minimum passing sight distances are given in table 16.3. The values should be increased 18% for downgrades steeper than 3% and longer than one mile.

If a vehicle locks its brakes and skids to a stop, the deceleration will be $(f)(g) = (f)(32.2)$ ft/sec². The skidding distance will be

$$\text{skidding distance} = \frac{(\text{MPH})^2}{(30)(f+G)} \qquad 16.15$$

If a vehicle does not lock its brakes, its deceleration will be dependent on its brakes. The distance traveled during deceleration is given by equation 16.16.

$$\text{stopping distance} = \frac{v^2}{2a} = \frac{(1.08)(\text{MPH})^2}{a} \qquad 16.16$$

**Table 16.3**
**AASHTO Minimum Sight Distances**

| | stopping sight distance wet pavements | | passing sight distance | |
|---|---|---|---|---|
| design speed, mph | initial speed, mph | minimum desirable distance, ft | assumed passing speed, mph | distance 2-lane highway, ft |
| 30 | 28–30 | 200 | 36 | 1100 |
| 40 | 36–40 | 275–325 | 44 | 1500 |
| 50 | 44–50 | 400–475 | 51 | 1800 |
| 60 | 52–60 | 525–650 | 57 | 2100 |
| 70 | 58–70 | 625–850 | 64 | 2500 |

[1] $t_p$ is also known as the *PIEV time.* This name is an acronym for the various elements of reaction time, including perception, identification, emotion, and volition.

[2] For the purpose of determining minimum stopping sight distances, $t_p$ is taken as 2.5 seconds. For determining passing sight distances, $t_p$ is taken as 3.5 to 4.5 seconds.

[3] It is assumed that the driver's eyes are 3.50 feet above the surface of the roadway. The object being viewed (e.g., an oncoming car) is assumed to be at a height of 4.25 feet.

## 5 LENGTH OF CIRCULAR HORIZONTAL CURVE FOR STOPPING DISTANCE

A horizontal curve on level ground is shown in figure 16.1. A typical design problem is to design a curve (i.e., specify a radius) that will simultaneously provide the required sight stopping distance while maintaining a clearance from a roadside obstruction.

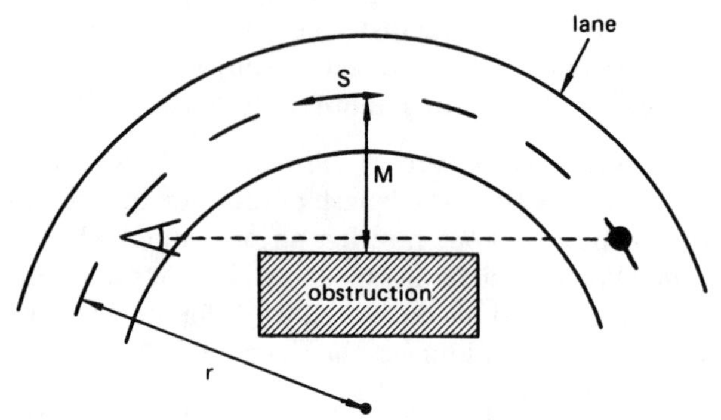


**Figure 16.1** Stopping Distance

The governing equations are given assuming $S \leq LC$. In this analysis, the stopping sight distance and length along the curve are the same. The angles are in degrees.

$$S = \frac{r}{28.65}\left[\arccos\left(\frac{r-M}{r}\right)\right] \qquad 16.17$$

$$M = r\left[1 - \cos\left(\frac{28.65S}{r}\right)\right] \qquad 16.18$$

## 6 LENGTH OF VERTICAL CURVES FOR SIGHT DISTANCES

The curve length may be shorter or longer than the safe passing or stopping sight distances. (Passing sight distance is not relevant on multi-lane highways.) Table 16.4 can be used to calculate lengths of curves. Table 16.4 is used by calculating the curve length for both assumptions that $S < LC$ and $S > LC$.

*Example 16.3*

A car is traveling at 40 mph up a hill with a +1.25% grade. The descending grade is −2.75%. What is the required length of curve for proper stopping sight distance?

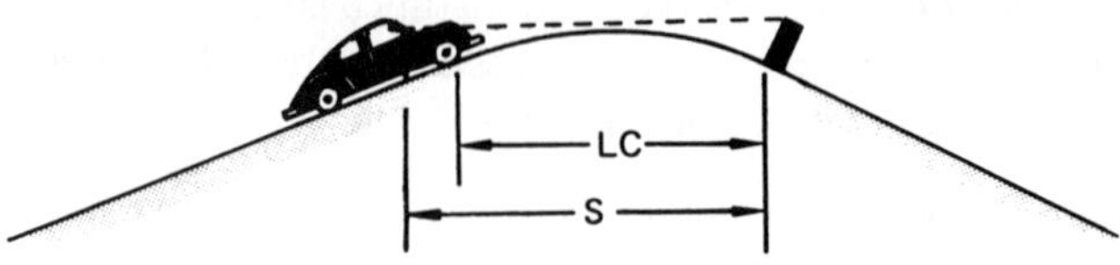


From table 16.3, the minimum stopping sight distance is 275 feet at 40 mph.

Using table 16.4 and assuming that $275 > LC$:

$$LC = 2(275) - \frac{1329}{1.25 - (-2.75)} = 217.8$$

Using table 16.4 and assuming $275 < LC$:

$$LC = \frac{[1.25 - (-2.75)](275)^2}{1329} = 227.6$$

Since 227.6 is less than 275, the second assumption is not valid. The required curve length is 217.8 feet.

## 7 SPEED PARAMETERS

Many different measures of vehicle speed are used.

- Running Speed: The distance traveled divided by the running time without delays. This parameter can be averaged over all traffic.
- Average Spot Speed: This is the average instantaneous speed of all vehicles at a particular location.
- Overall Travel Speed: If delays and stops at intersections are included in running time, the overall travel speed can be calculated from the distance traveled.

**Table 16.4**
**Required Lengths of Curves on Grades**

| assuming | stopping sight distance (crest curves) | passing sight distance (crest curves) | stopping sight distance (sag curves) |
|---|---|---|---|
| $S < LC$ | $\frac{(G_1 - G_2)S^2}{1329}$ | $\frac{(G_1 - G_2)S^2}{3093}$ | $\frac{(G_2 - G_1)S^2}{400 + 3.5S}$ |
| $S > LC$ | $2S - \frac{1329}{G_1 - G_2}$ | $2S - \frac{3093}{G_1 - G_2}$ | $2S - \frac{400 + 3.5S}{G_2 - G_1}$ |

- Operating Speed: The highest overall speed at which a driver can travel under favorable weather conditions while driving in a safe manner.
- Design Speed: The maximum safe speed when conditions are so favorable that the design features of the highway govern.
- Average Highway Speed: The weighted average of the design speeds over a section of highway.

## 8 DESIGN SPEEDS

Most elements of roadway design depend on the design speed. The design speed is the maximum safe maintainable speed on a roadway under the design conditions. Typical design speeds are given in table 16.5.

**Table 16.5**
Recommended Design Speeds (mph)

| type of facility | level | rolling | mountainous |
|---|---|---|---|
| freeways | | | |
| rural | 70 | 60 | 50 |
| urban | 50 | 50 | 50 |
| rural arterial highways | | | |
| 50 <ADT≤750, | | | |
| DHV< 200 | 50 | 40 | 30 |
| DHV> 200 | 70 | 60 | 40 |
| urban arterial | | | |
| highways | 30–40 | 30–40 | 30–40 |
| suburban arterial highways | 40–50 | 40–50 | 40–50 |
| rural roads and streets | | | |
| ADT< 250 | 40 | 30 | 20 |
| 250 <ADT< 400 | 50 | 40 | 30 |
| ADT> 400, | | | |
| DHV> 100 | 50 | 40 | 30 |
| urban roads and streets | | | |
| collectors | 30–40 | 30–40 | 30–40 |
| local | 20–30 | 20–30 | 20–30 |

## 9 VOLUME PARAMETERS

There are many volume parameters in use. Not all parameters will be needed in every capacity or strength calculation. It is particularly important to note if volumes are for both directions combined, or for one or more lanes combined.

ADT: The current *average daily traffic.* ADT may be one- or two-way traffic.

DHV: The peak *design hourly volume* in the design year. DHV is usually the 30th highest hourly expected volume in the design year. It is not an average or a maximum. DHV is two-way unless noted otherwise.

k: Ratio of DHV to ADT.

D: A *directional factor.* D is the percentage in the dominant flow direction. D may range up to 80% for rural roadways at peak hours down to 50% for central business district traffic.

DDHV: The *directional design hourly volume.* It is calculated as the product of the directional factor, D, and DHV.

$$\text{DDHV} = (\text{D})(\text{DHV}) \qquad 16.19$$

Design Capacity: The maximum volume of traffic that the roadway can handle.

MSF: The *maximum service flow rate* per lane under ideal conditions.

Logical methods, including straight-line extrapolation, should be used in the estimation of future traffic counts. Expansion factors should be determined for each of the axle classifications. Considerable judgment is needed to develop realistic expansion factors.

It also is necessary to estimate the distribution of truck traffic on the various lanes of a multi-lane facility. Traffic is usually lightest in the *inside lane* (*lane 1*, or the *fast lane*). The following lane distribution factors can be used.

**Table 16.6**
Lane Distribution Factors for Multi-Lane Freeways

| number of lanes in one direction | lane 1 | lane 2 | lane 3 | lane 4 |
|---|---|---|---|---|
| 1 | 1.0 | | | |
| 2 | 1.0 | 1.0 | | |
| 3 | 0.2 | 0.8 | 0.8 | |
| 4 | 0.2 | 0.2 | 0.8 | 0.8 |

## 10 TRUCK, BUS, AND RV EQUIVALENTS

Since buses and trucks take up more space on a road than cars do, and since buses and trucks tend to travel more slowly up grades, bus and truck volumes are converted to equivalent passenger car volumes. Table 16.7 lists *passenger car equivalents* for various conditions.

**Table 16.7**
Passenger Car Equivalents
for Freeways and Multi-Lane Highways

| terrain | $E_B$ (buses) | $E_T$ (trucks) | $E_R$ (RV's) |
|---|---|---|---|
| level | 1.5 | 1.7 | 1.6 |
| rolling | 3 | 4 | 3 |
| mountainous | 5 | 8 | 4 |
| grade | | | |
| 0–3% | 1.6 | | |
| 4% | 1.6 | | |
| 5% | 3.0 | | |
| 6% | 5.5 | | |

Passenger car equivalents for trucks and RV's on grades depend on the grade, the length of the grade, the number of lanes, and the percentage of trucks and buses.[4]

## 11 LEVEL OF SERVICE

Conditions on a highway are classified into levels $A$ through $F$. Level $A$ represents a condition where there are no physical restrictions on operating speed. Since there are few vehicles on the freeway, operation at highest speeds is possible. However, the traffic volume is small. Level $F$ represents a stop-and-go, low-speed condition with poor safety and maneuverability.

**Table 16.8**
Levels of Service for Freeways
and Multi-Lane Highways

| level | density (pc/mi-ln) | description |
|---|---|---|
| $A$ | ≤12 | free flow, with low volumes and high speeds |
| $B$ | 13–20 | stable flow, but speeds are beginning to be restricted by traffic conditions |
| $C$ | 21–30 | stable flow, but most drivers cannot select their own speed |
| $D$ | 31–42 | approaching unstable flow, and drivers have little room in which to maneuver |
| $E$ | 43–67[5] | unstable flow with short stoppages |
| $F$ | ≥68 | forced flow at slow speeds; lines of vehicles at certain locations |

[4] Tables are provided in the *Highway Capacity Manual.*

[5] 67 passenger cars per mile per lane is generally considered to be the *critical density.* Maximum flow (i.e., *capacity flow*) occurs at the *critical density* within level of service $E$.

The desired design condition is between levels $A$ and $F$. Economic considerations favor lower levels and their higher traffic volume per lane. However, political considerations favor higher levels. Typically, levels $B$ and $C$ are chosen for initial designs.

The actual level of service experienced on a freeway is determined by comparing the actual density with the density limits in table 16.8.

## 12 CALCULATION OF FREEWAY CAPACITY

The maximum capacity under ideal conditions, $c$, is taken as the maximum service flow rate per lane for level of service. Table 16.9 contains MSF values. In most cases, $c_{max} = 2000$ pc/hr-ln.

$$c_{\text{max}} = \text{MSF}_E \qquad 16.20$$

**Table 16.9**
Maximum Service Flow Rates
for Freeways[6]

(vehicles per hour)

| service level | maximum service volume at average freeway speed per lane: 70 mph | 60 mph | 50 mph |
|---|---|---|---|
| $A$ | 700 | — | — |
| $B$ | 1100 | 400 | — |
| $C$ | 1550 | 1000 | 1300 |
| $D$ | 1850 | 1700 | 1600 |
| $E$ | 2000 | 2000 | 1900 |
| $F$ | — | — | — |

For any other level of service, the maximum *volume-to-capacity ratio* for level of service $i$ is

$$(v/c)_i = \frac{\text{MSF}_i}{c_{\text{max}}} \qquad 16.21$$

The *service flow rate* can be calculated from the number of lanes ($N$), the lane width adjustment factor ($f_w$), the factor to adjust for the presence of heavy vehicles such as buses, trucks, and recreational vehicles ($f_{HV}$), and a

[6] This table is not valid for rural, multi-lane highways. Refer to the *Highway Capacity Manual.*

**Table 16.10**
Adjustment Factor for Restricted Lane Width and Lateral Clearance for Freeways[7]

| distance of obstruction from traveled pavement (ft) | obstructions on one side of the roadway | | | | obstructions on both sides of the roadway | | | |
|---|---|---|---|---|---|---|---|---|
| | lane width (ft) | | | | | | | |
| | 12 | 11 | 10 | 9 | 12 | 11 | 10 | 9 |
| | 4-lane freeway (2 lanes each direction) | | | | | | | |
| ≥6 | 1.00 | 0.97 | 0.91 | 0.81 | 1.00 | 0.97 | 0.91 | 0.81 |
| 5 | 0.99 | 0.96 | 0.90 | 0.80 | 0.99 | 0.96 | 0.90 | 0.80 |
| 4 | 0.99 | 0.96 | 0.90 | 0.80 | 0.98 | 0.95 | 0.89 | 0.79 |
| 3 | 0.98 | 0.95 | 0.89 | 0.79 | 0.96 | 0.93 | 0.87 | 0.77 |
| 2 | 0.97 | 0.94 | 0.88 | 0.79 | 0.94 | 0.91 | 0.86 | 0.76 |
| 1 | 0.93 | 0.90 | 0.85 | 0.76 | 0.87 | 0.85 | 0.80 | 0.71 |
| 0 | 0.90 | 0.87 | 0.82 | 0.73 | 0.81 | 0.79 | 0.74 | 0.66 |
| | 6- or 8-lane freeway (3 or 4 lanes each direction) | | | | | | | |
| ≥6 | 1.00 | 0.96 | 0.89 | 0.78 | 1.00 | 0.96 | 0.89 | 0.78 |
| 5 | 0.99 | 0.95 | 0.88 | 0.77 | 0.99 | 0.95 | 0.88 | 0.77 |
| 4 | 0.99 | 0.95 | 0.88 | 0.77 | 0.98 | 0.94 | 0.87 | 0.77 |
| 3 | 0.98 | 0.94 | 0.87 | 0.76 | 0.97 | 0.93 | 0.86 | 0.76 |
| 2 | 0.97 | 0.93 | 0.87 | 0.76 | 0.96 | 0.92 | 0.85 | 0.75 |
| 1 | 0.95 | 0.92 | 0.86 | 0.75 | 0.93 | 0.89 | 0.83 | 0.72 |
| 0 | 0.94 | 0.91 | 0.85 | 0.74 | 0.91 | 0.87 | 0.81 | 0.70 |

factor to adjust for the effect of the driver population ($f_p$).[8]

$$\begin{aligned} SF_i &= \text{MSF}_i \times N \times f_w \times f_{HV} \times f_p \\ &= c_{\max} \times \left(\frac{v}{c}\right)_i \times N \times f_w \times f_{HV} \times f_p \end{aligned} \quad 16.22$$

The width adjustment factor, $f_w$, is taken from table 16.10.

The *heavy vehicle factor* is a function of the truck, bus, and recreational vehicle fractions.

$$f_{HV} = \frac{1}{1+\begin{pmatrix}\text{fraction}\\\text{trucks}\end{pmatrix}(E_T-1)+\begin{pmatrix}\text{fraction}\\\text{buses}\end{pmatrix}(E_B-1)+\begin{pmatrix}\text{fraction}\\\text{RV's}\end{pmatrix}(E_R-1)} \quad 16.23$$

$E_B, E_T$, and $E_R$ are the *passenger car equivalents* of a bus, truck, or recreational vehicle, respectively. Table 16.7 provides these values.

The *population adjustment factor*, $f_p$, is 1.0 for weekday or commuter traffic. It is 0.75 to 0.90 for weekend, recreational, and other types of traffic that do not use the available space as efficiently. Engineering judgment is required in selecting $f_p$ in those instances.

## 13 SPEED, FLOW, AND DENSITY RELATIONSHIPS FOR UNINTERRUPTED FLOW

The speed of travel will decrease as the number of cars occupying the freeway increases. The *density*, $D$, is defined as the number of vehicles per mile per lane (vpmpl). The *jam density* is the density when the vehicles are all at a standstill. Then, the speed for any given density can be related to the *free-flow speed* by equation 16.24.

$$S = S_f\left(1 - \frac{D}{D_j}\right) \quad 16.24$$

[7] Certain types of obstructions, such as high-type median barriers in particular, do not decrease the flow. Exercise judgment in applying these factors.

[8] From a practical standpoint, the *service flow rate*, $SF$, is set equal to the actual peak flow rate. The peak flow rate can be calculated from the actual flow rate and the *peak hour factor*: $SF = v/PHF$.

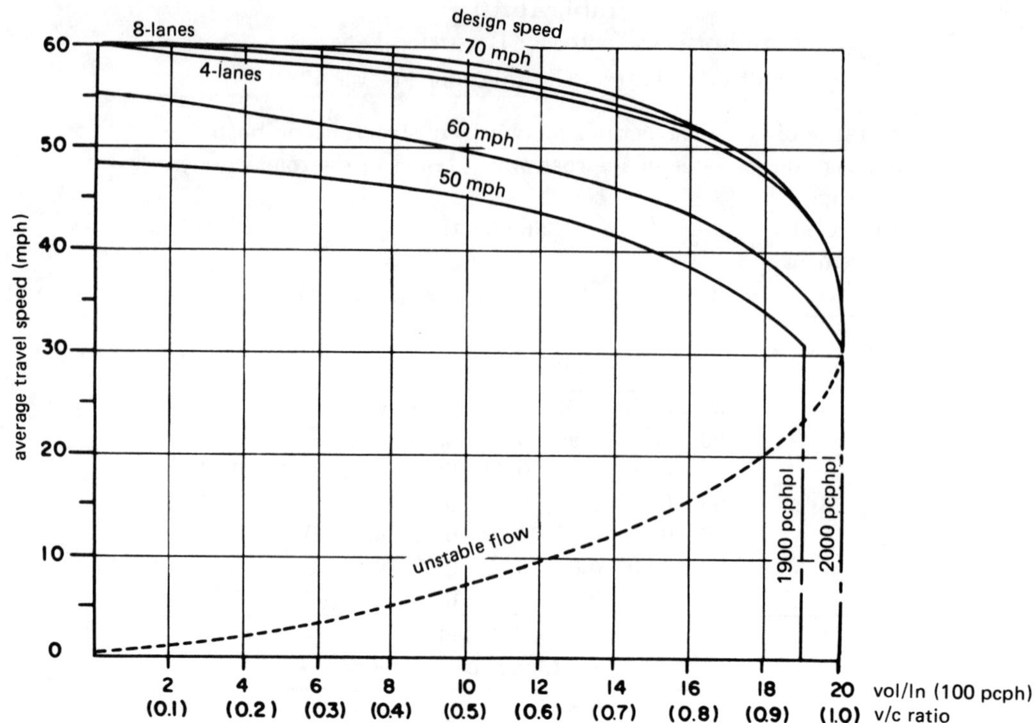


**Figure 16.5** Freeway Speed-Flow Relationships Under Ideal Conditions

(v/c ratio is based on 2000 pcphpl and is valid only for 60- and 70-mph speeds.)

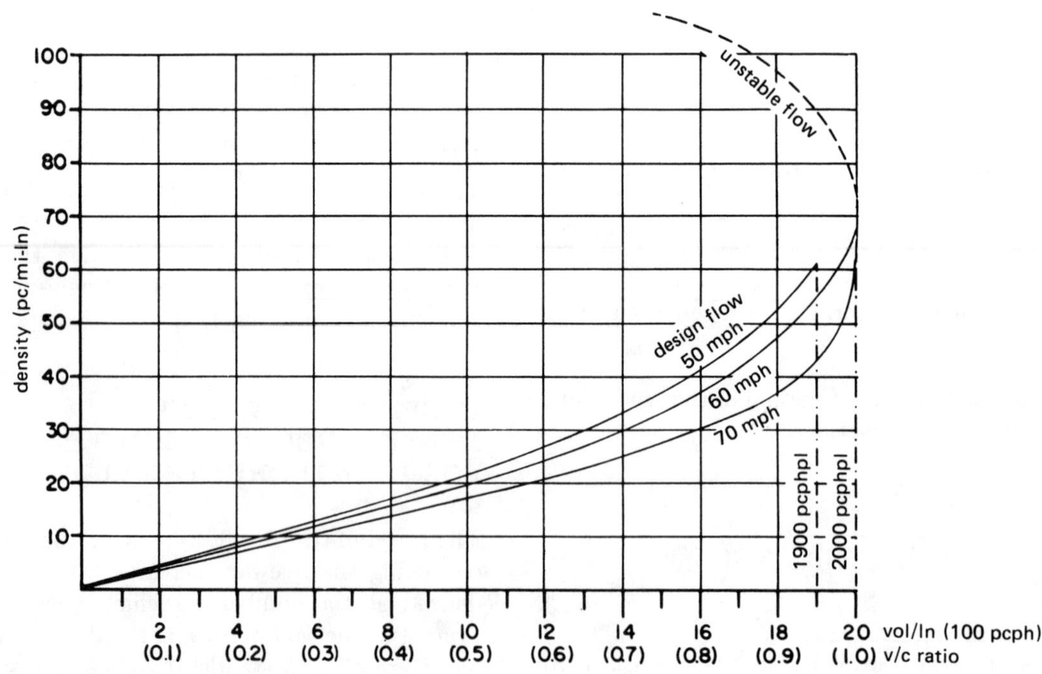


**Figure 16.6** Freeway Density-Flow Relationships Under Ideal Conditions

(v/c ratio is based on 2000 pcphpl and is valid only for 60- and 70-mph speeds.)

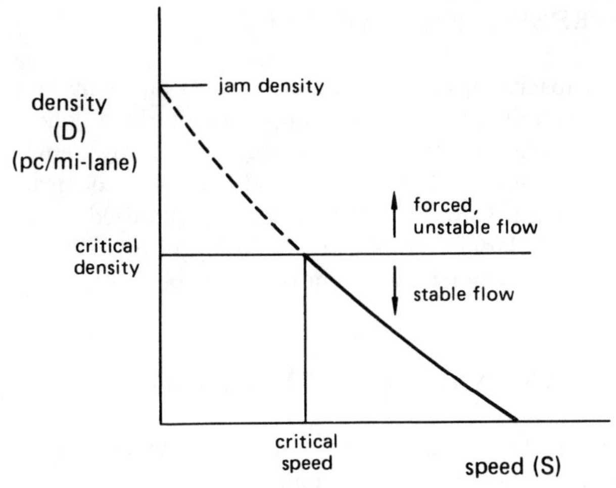


**Figure 16.2** Speed Versus Density

The number of vehicles (volume) crossing a point per hour per lane (vphpl) is

$$v = SD \tag{16.25}$$

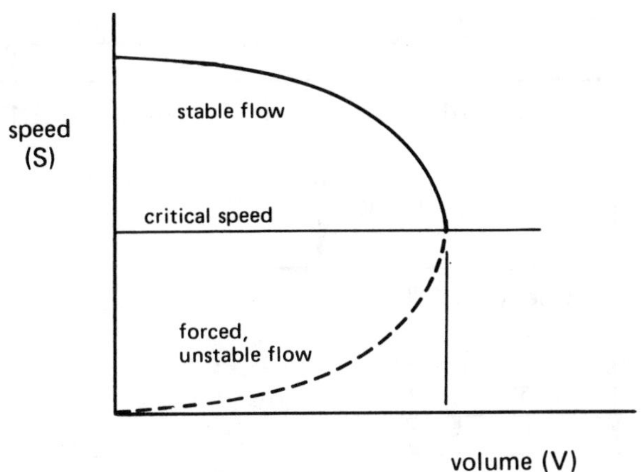


**Figure 16.3** Speed Versus Volume

The *headway* is the time between successive vehicles. *Spacing* is the distance between common points (e.g., the front bumper) on successive vehicles.

$$\text{spacing (ft/veh)} = \frac{5280 \text{ (ft/mi)}}{D \text{ (vpmpl)}} \tag{16.26}$$

$$\text{headway (sec/veh)} = \frac{\text{spacing (ft/veh)}}{\text{speed (ft/sec)}} \tag{16.27}$$

$$\text{volume or flow rate (vph)} = \frac{3600 \text{ sec/hr}}{\text{headway (sec/veh)}} \tag{16.28}$$

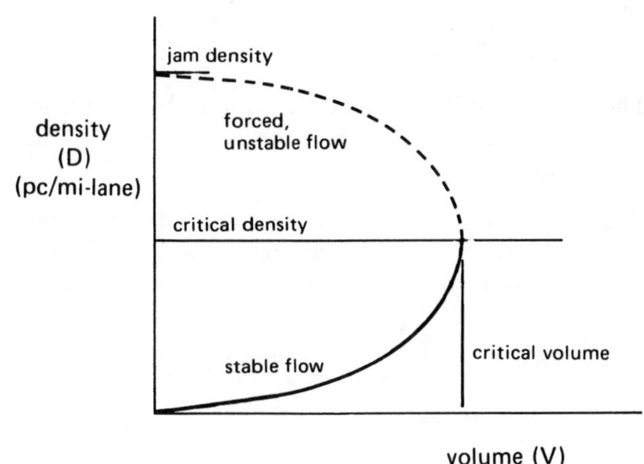


**Figure 16.4** Density Versus Volume

## 14 DETERMINING THE LEVEL OF SERVICE

The level of service is determined primarily from the density criteria in table 16.8. However, it is also possible to correlate the level of service with the v/c ratio. In this case, it is necessary to know the design speed, $v_{\text{design}}$, for which the freeway was constructed. Table 16.11 correlates v/c ratios with different levels of service.

**Table 16.11**
Level of Service Versus Maximum v/c Ratio for Freeways

| level of | design speed (mph) | | |
|---|---|---|---|
| service | 70 | 60 | 50 |
| *A* | 0.35 | — | — |
| *B* | 0.54 | 0.49 | — |
| *C* | 0.77 | 0.69 | 0.67 |
| *D* | 0.93 | 0.84 | 0.83 |
| *E* | 1.00 | 1.00 | 1.00 |

*Example 16.4*

A four-lane freeway is constructed with 11-foot lanes, no shoulders, and retaining walls at the pavement edge. The average vehicle speed is 60 mph. The terrain is rolling. The actual service volume is 1500 vehicles per hour, with 3% buses and 5% trucks. What is the level of service? What is the capacity volume?

The base volume is 2000 vphpl (table 16.9, level of service $E$).

The number of lanes is 2.

The width adjustment factor is 0.87 (table 16.10, assuming retaining walls are at the outer edges of both slow lanes).

The passenger car equivalents of the trucks and buses in rolling terrain are obtained from table 16.7. $E_t = 4$. $E_B = 3$. From equation 16.23, the heavy vehicle factor is

$$\begin{aligned} f_{HV} &= \frac{1}{1 + (0.05)(4-1) + (0.03)(3-1)} \\ &= 0.826 \end{aligned}$$

Assuming weekday traffic ($f_p = 1$), the capacity volume is given by equation 16.22.

$$\begin{aligned} SF &= 2000 \times 2 \times 0.87 \times 0.826 \times 1 \\ &= 2874 \text{ vph} \end{aligned}$$

The v/c ratio is

$$\frac{v}{c} = \frac{1500}{2874} = 0.52$$

From table 16.11, this v/c ratio corresponds to level of service $C$ for a design speed of 60 mph.

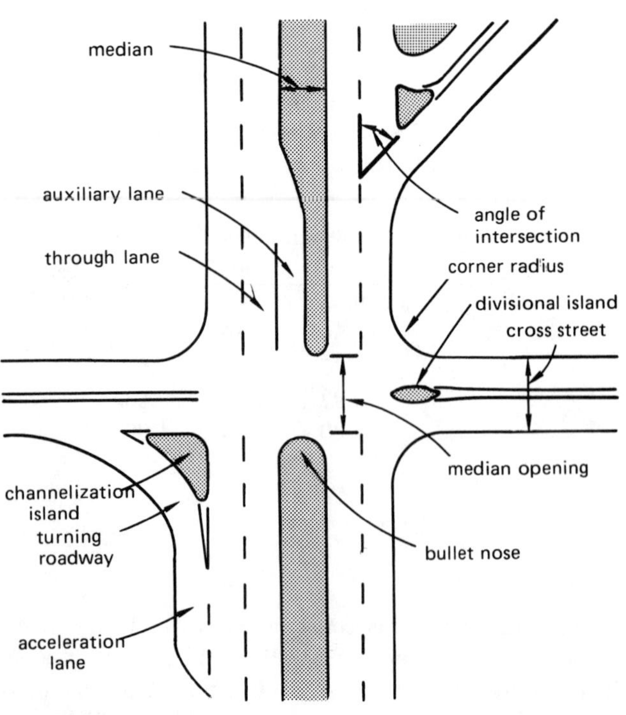


**Figure 16.7** Elements of an Intersection

## 15 AT-GRADE SIGNALIZED INTERSECTION CAPACITY

The capacity of an intersection depends on many factors, including the width of approach, parking conditions, traffic direction (i.e., one-way or two-way), environment, bus and truck traffic, and percentage of turning vehicles. Capacity calculations for signalized intersections is largely graphical. Inclusion of all relevant graphs is not practical within the scope of this book.[9]

## 16 STANDARD TRUCK LOADINGS

Table 16.12 and figure 16.8 illustrate standard truck loads commonly used for design. In cases where the separation between axles varies, the distance that produces the maximum stress in the section should be used.

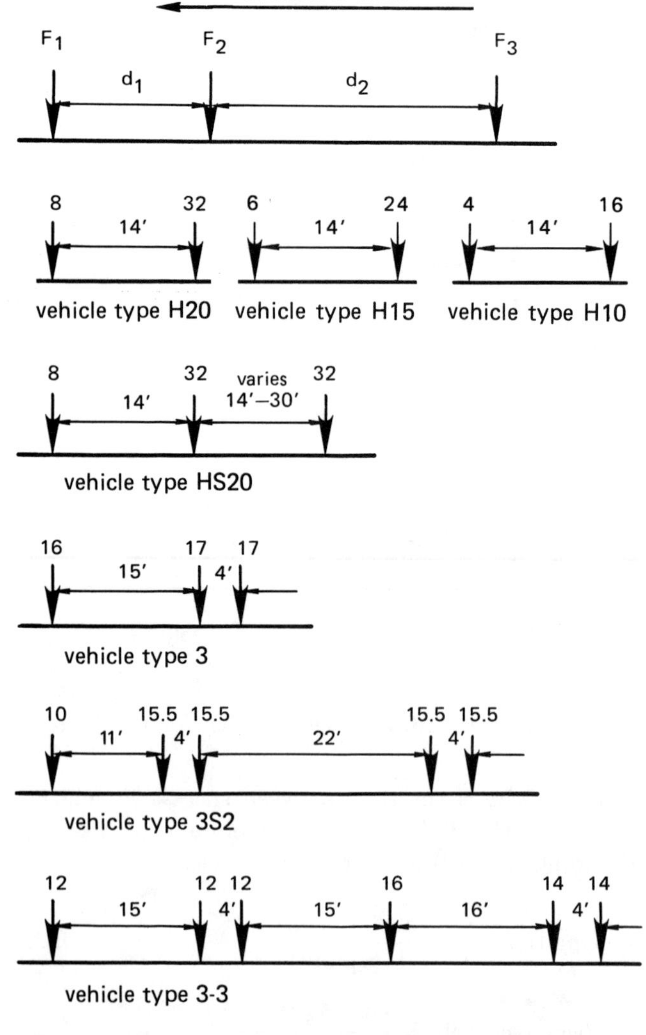


**Figure 16.8** Standard Truck Loadings

[9] The authoritative reference on this subject is the *Highway Capacity Manual.*

**Table 16.12**
Standard Truck Loadings

(All loads are axle loads)

| load designation | $F_1$ | $F_2$ | $F_3$ | $d_1$ | $d_2$ |
|---|---|---|---|---|---|
| H20–44 | 8000 | 32,000 | 0 | 14′ | |
| H15–44 | 6000 | 24,000 | 0 | 14′ | |
| H10–44 | 4000 | 16,000 | 0 | 14′ | |
| HS20–44 | 8000 | 32,000 | 32,000 | 14′ | 14′ to 30′ |
| HS15–44 | 6000 | 24,000 | 24,000 | 14′ | 14′ to 30′ |
| 3 | 16,000 | 17,000 | 17,000 | 15′ | 4′ |
| 3S2 | See figure 16.8 | | | | |
| 3–3 | See figure 16.8 | | | | |

Some traffic analysis studies define design vehicles according to the following categories: passenger car (P), single unit truck (SU), single unit bus (BUS), intermediate length semitrailer (WB-40), large semitrailer (WB-50), and semitrailer/full trailer combination (WB-60). In the case of WB vehicles, the number represents the wheelbase distance between the front (cab) axle and the last trailer axle.

Classification into axle types also is common. An axle may be *single* or *tandem*, and each axle may have single or dual tires.[10] *Spread tandem axles*, where two axles are separated by more than 96 inches, generally are classified as two single axles.

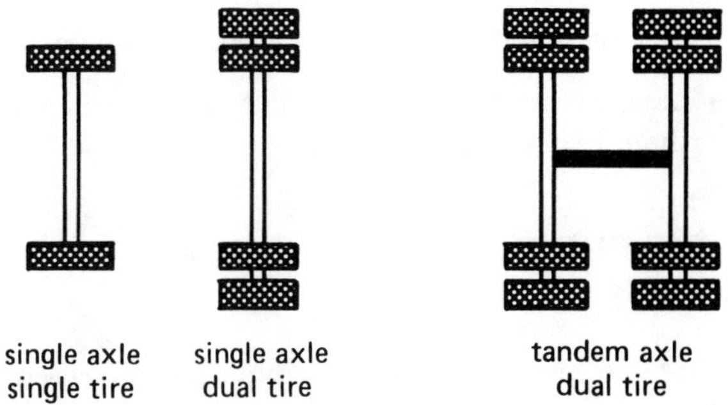


**Figure 16.9** Types of Axles and Axle Sets

Except in theoretical stress studies, no attempt is made to account for the number of tires per axle. Although it is true that stresses at shallow depths are caused principally by individual wheels acting singly, stresses at greater depths are maximum midway between wheels. Deep stresses due to dual-wheeled axles are approximately the same as for single-wheeled axles. Therefore, required pavement thickness is determined by the total axle load.[11]

## 17 TYPES OF PAVEMENT

### A. RIGID PAVEMENT

Portland cement concrete is used almost exclusively where rigid pavement is called for. Typical applications are high volume traffic lanes, freeway-to-freeway connections, and exit ramps that experience heavy traffic.

Cement concrete pavement has excellent durability and long service life. It provides good contrast with asphalt surfaces. It will also withstand repeated flooding and subsurface water without deterioration.

It has three primary disadvantages: (1) It may lose its original nonskid surface during use. (2) It must be used with an even subgrade and only where uniform settling is expected. (3) It may rise (fault) at transverse joints.

Portland cement concrete is placed with slip-form construction methods. A stiff concrete ("no-slump" with slumps less than 1") is placed in front of the paving train. The paving train then distributes, vibrates, screeds, and finishes the layer while traveling at a slow, continuous speed.

Rigid pavement can be reinforced or unreinforced.[12] With unreinforced construction, *contraction joints* should be placed at regular intervals. A maximum spacing of 15 feet or 30 times the thickness of the slab is a typical rule for unreinforced pavement.[13]

If steel is used, it is assumed not to contribute to the structural section. It is used only to control crack growth. Steel is usually deformed bar but an equivalent amount of steel mesh can be used. Construction joints are required in reinforced pavement, as well. However, the spacing is greater—every 50 to 100 feet, depending on reinforcement.

AASHTO has specified a formula for calculating the quantity of longitudinal steel required in a concrete highway slab.[14] The area is specified in square inches per foot of slab width. In equation 16.29, $F$ is the *coefficient of resistance* between the slab and subgrade

[10] Tandem axles have two axles separated by 40 to 96 inches. However, up to five axle sets are used for heavy loads.

[11] However, pavement stress calculations should be based on the individual wheel loads.

[12] California has many miles with unreinforced concrete.

[13] California uses a random spacing following the sequence of 12, 15, 13, 14 feet. Sawn joints are placed with a diagonal skew of 2 feet in 12 feet of width.

[14] The formula can also be used to calculate transverse reinforcement requirements if $L$ is taken as the distance between the slab edges.

(typically taken as 1.5), $L$ is the distance in feet between transverse joints, $w$ is the weight of the pavement slab (typically based on 150 lbf/ft$^3$), and $f_s$ is the allowable working stress in the steel (approximately two-thirds of the yield strength). Based on experience, the fraction of steel required will vary between 0.5 and 0.8 percent of the cross-sectional area of the pavement.

$$A_s = \frac{FLw}{2f_s} \qquad 16.29$$

AASHTO also has specified spacing and sizes for round steel dowels used as load transfer devices between slabs. 18″ dowels placed 12″ apart should be used. For 6″ thick pavement, a 3/4″ diameter bar should be used. For 7″ and 8″ pavements, 1″ bars are called for. $1\frac{1}{4}$″ bars are used for all pavements thicker than 8″.

If possible corrosion of the steel bars is a concern, the steel can be epoxy coated. Other additives, such as *latex* or *silica fume* (*microsilica*), can be used to protect the concrete against penetration of deicing salts.

B. PRESTRESSED CONCRETE PAVEMENT

Prestressed concrete pavements have been used in Europe since the 1950's. However, they have seen little use in the United States. One of the reasons is that the primary advantage of prestressed pavements—that of requiring thinner sections—results in overall savings only with thicker sections such as used in airports.

In addition to requiring sections that are only 40% to 60% as thick as conventional pavements, prestressing also reduces transverse joint requirements to one every 300 to 600 feet. Fewer joints increase the ride quality and decrease maintenance. Also, prestressing requires as little as 20% of the steel used in continuously reinforced pavement. A service life of 30 years is expected with prestressed pavements.

Demonstration projects in the United States have shown that prestressed pavements for highways are competitive on a first-cost basis. Performance to date indicates lower maintenance costs. Normal paving procedures can be used.

C. FLEXIBLE ASPHALT CONCRETE

Typical uses of asphalt concrete are traffic lanes, auxiliary lanes, ramps, parking areas, frontage roads, and shoulders.

Asphalt concrete pavement has the advantage of adjusting to limited amounts of differential settlement. It is easily repaired, and additional thicknesses can be placed at any time to withstand increased usage and loading. Its non-skid properties do not deteriorate to a great extent.

However, asphalt concrete loses some of its flexibility and cohesion with time, and it will have to be resurfaced sooner than would cement concrete. It would not normally be chosen where water is expected.

D. FULL-DEPTH ASPHALT PAVEMENT

If asphalt mixtures are used for all courses above the subgrade, the pavement is said to be a *full-depth asphalt pavement.* Since asphalt bases are stronger than untreated bases, the pavement thickness is less. Other advantages of full-depth pavements include a potential decrease in trapped water within the pavement, a decrease in the moisture content of the subgrade, and little or no reduction in subgrade strength.[15]

E. DEEP-LIFT ASPHALT PAVEMENT

If the asphalt layer is thicker than 4″ and is placed all in one lift, or if lift layers are thicker than 4″, the construction is said to be *deep lift.* Using deep lifts to place hot-mix asphalt concrete is advantageous for several reasons.

- Thicker layers hold heat longer, and it is easier to roll the layer to the required density.
- Lifts can be placed in cooler weather.
- One lift of a given thickness is more economical to place than multiple lifts equalling the same thickness.
- Placing one lift is faster than placing several.
- Less distortion of the asphalt course will result than if thin lifts are rolled.

## 18 SULFUR EXTENDED ASPHALT BINDER

To become less dependent on imported oil (used to make asphaltic binders), and to use the byproduct of desulfurized oil, sulfur has been added to asphalt concrete in a number of instances.[16] The sulfur percentage varies from 10% to 50% of the total binder content (by weight), although 30% to 40% is common.

At standard temperatures and pressures, ordinary sulfur is an odorless and tasteless yellow solid. Its specific gravity is 2.07 and the melting point is 238°F. Sulfur

15 Subgrade drains are still required, however.

16 This is not a new idea. *Sulfur concrete* was being used at least as far back as the 1920's.

is not a hazardous material. The practices for hauling, heating, and storing molten sulfur are well established.

The working range for molten sulfur corresponds well to the working range for paving grade asphalt: 255°F to 300°F. Above 315°F, sulfur becomes viscous and cannot be easily worked. Liquid sulfur is hot, and in that respect, poses the same dangers as hot asphalt (or any other hot liquid). When heated, the concentration of eye-irritating sulfur fumes is low (below 300°F), but increases rapidly above 300°F. Molten sulfur at 300°F will ignite, as will asphalt. Therefore, all sources of ignition near liquid sulfur should be removed.

Sulfur-asphalt pavement binder consists primarily of a very fine dispersion of sulfur in asphalt, with asphalt forming the continuous phase. The mixing is performed in a colloid or pug mill using molten sulfur and hot asphalt.

When mixed well, sulfur measuring approximately 20% by weight of the asphalt will dissolve in the asphalt.

The remainder of the sulfur forms a dispersion in the asphalt. Both the dissolved and dispersed sulfur modify the properties of the binder.

1 to 2 ppm of silcone can be added to stabilize the sulfur-asphalt emulsion and make the mixture easier to work. (Normally, 2 ppm added to the hot asphalt will improve moisture release from the hot mix and reduce pulling and tearing at the screed of the paving machine.)

The paving operation does not need any special equipment. Standard pavers, rollers, haul trucks, emission control equipment, and testing equipment can be used. However, special equipment is required to store and handle sulfur, and a special sulfur-asphalt emulsion mixing plan is necessary.

Vapor given off during the preparation and placement of sulfur-asphalt mixtures contains a certain amount of elemental sulfur.[17] There is no practical way to eliminate this pollutant. Sulfur is virtually nontoxic, and there is no evidence of systemic poisoning resulting from inhalation. However, sulfur does irritate open cuts and the inner surface of the eyelids. Goggles and gloves can reduce such irritations.

The properties of the resulting *sulfur-asphalt concrete* are superior to those of pure asphalt concrete. A high-strength concrete results. Since there is no alkaline cement binder (e.g., portland cement) in the mix, sulfur concrete is an acid-resistant structural material. It achieves 80% of its ultimate compressive strength in a few hours, and 100% within 24 hours.

[17] Above 300°F, sulfur dioxide and hydrogen sulfide become the dominant pollutants. Therefore, the molten sulfur temperature must be kept below 300°F. 280°F to 290°F is the preferred operating range.

## 19 MINIMUM LAYER THICKNESSES

It generally is impractical and uneconomical to place surface, base, and subbase courses with thicknesses less than minimum values. These minimum thicknesses are 1.5″–2″ for asphalt surface courses, and 4″ for cement-, lime-, and asphalt-treated bases and subbases.

## 20 PAVEMENT DESIGN PARAMETERS

### A. LAYER STRENGTHS

Once the materials for the surface, base, and subbase layers have been selected, their strengths can be determined by testing. It may be necessary to convert one strength parameter into another for use with a particular design procedure. Appendices A, B, and C can be used for this purpose.

The quality of the basement soil is a required factor. Since there may be a considerable range in these values, a design value must be chosen. If the range is small, the lowest value should be selected. If there are a few exceptionally low values that come from one area, it may be possible to specify replacing that area's soil with borrow soil. If there are changing geological formations along the route which modify the value, it may be necessary to design different pavement sections.

### B. EQUIVALENT AXLE LOADINGS

The pavement thickness will be dependent on the axle loadings that are estimated for the truck traffic predicted for the design period. (The effects of passenger cars, pickups, and two-axle trucks with single rear tires are not considered.) Both the number of trucks and the axle loading of those trucks must be known.

The usual sources of truck volume data are traffic counts. These may be actual counts on existing roadways needing resurfacing, or they may be redistributed counts from other highways. The trucks are classified according to the number of axles.

Pavement thickness should be chosen to serve the estimated one-way truck traffic for a period of 20 years. A shorter period, not to be less than 10 years, may be used for temporary construction or for other justifiable reasons.

An 18-kip axle load (two 9-kip wheel loads) is used as the standard loading in highway section designs. This reduces all traffic data into the number of 18-kip axle passes that would cause the same structural damage. The analysis is made difficult by the number of ways that traffic can be reported. Traffic counts can be accurate (as when actual highway loadometer data is used) or approximate (as when only the number of trucks is available).

*Method A*: Use actual highway loadometer data and get the *equivalency factor* assuming a *structural number* (SN) of 3 from appendix D or E. Each number of axle passes is multiplied by an equivalency factor to convert it to a number of 18-kip axle passes. The choice of SN = 3 is arbitrary. When the design is completed, a more accurate SN will be determined.[18]

*Method B*: Convert other equivalent axle loads (e.g., 12-kip axle loads) to 18-kip axle loads.[19,20]

$$\text{EAL}_{\text{18-kips}} = \text{EAL}_{\text{n-kips}} \left( \frac{n \text{ in kips}}{18} \right)^4 \qquad 16.30$$

*Method C*: Convert equivalent axle load data from other durations.

$$\text{EAL}_{\text{20 years}} = \text{EAL}_{\text{n years}} \left( \frac{20}{n \text{ in years}} \right) \qquad 16.31$$

*Method D*: Approximate the equivalent 18-kip axle loading from the number of trucks passing *per day*. The coefficients are known as *truck constants*.[21]

$$\begin{aligned} \text{EAL}_{\text{18-kip, 20 years}} &= (1380) \left( \begin{matrix} \text{\#2-axle} \\ \text{trucks} \end{matrix} \right) \\ &+ (3680) \left( \begin{matrix} \text{\#3-axle} \\ \text{trucks} \end{matrix} \right) + (5880) \left( \begin{matrix} \text{\#4-axle} \\ \text{trucks} \end{matrix} \right) \\ &+ (13,780) \left( \begin{matrix} \text{\#5-axle} \\ \text{trucks} \end{matrix} \right) \end{aligned} \qquad 16.32$$

*Method E*: Convert 5000-pound *equivalent wheel loads* (EWL) to 18-kip EAL values.

$$\text{EAL} = \frac{\text{EWL}}{11.8} \qquad 16.33$$

*Method F*: Convert *design index* (DI) to 20-year EAL using table 16.13.[22]

**Table 16.13**
Design Index versus EAL

| DI | 20-year EAL |
|---|---|
| 1 | 7300–36,500 |
| 2 | 36,501–146,000 |
| 3 | 146,001–547,500 |
| 4 | 547,501–1,825,000 |
| 5 | 1,825,001–6,570,000 |
| 6 | 6,570,001–21,900,000 |

*Method G*: Convert *traffic index* (TI) to 20-year EAL using equation 16.36.

C. PREDICTING TRAFFIC GROWTH

If the 20-year EAL is to be predicted from the current (first) year EAL, and if a constant growth rate of $i\%$ per year is assumed, economic analysis tables can be used to simplify the calculation.

$$\text{EAL}_{20} = \text{EAL}_{\text{first year}} \times (F/A, i\%, 20) \qquad 16.34$$

## 21 AASHTO METHOD OF FLEXIBLE PAVEMENT DESIGN

This method is based on tests performed in Illinois between 1958 and 1960 by the American Association of State Highway and Transportation Officials (AASHTO).

*step 1*: Estimate the desired *terminal serviceability*, $p_t$. This parameter is a numerical rating of road quality after 20 years. (It is assumed that the first major resurfacing will occur at that time.) $p_t$ actually can be calculated from the condition of an existing road surface (i.e., from rut depths, percentage of cracked area, and percentage of patched areas, among other factors). However, for design work, it is common to specify $p_t = 2.5$ for highways and $p_t = 2$ for low traffic roads.

[18] All axles in the truck must be included. For example, a tractor/trailer would contribute three quantities to EAL—one for the tractor front axle, one for the tractor tandem axle, and one for the trailer tandem axle.

[19] This is essentially the same as using the load equivalency tables for single-axle loads as described in method A.

[20] The exponent 4 in equation 16.30 is also reported as 4.2 in literature.

[21] The truck constants in equation 16.32 are not unique. They represent typical factors from California statewide truck weighings. Other states or authorities may specify different truck constants.

[22] Actually, design index is usually correlated in military studies with *one-day EAL*. However, this chapter uses only 20-year EAL, and table 16.13 reflects this.

**Table 16.14**
Terminal Serviceabilities

| $p_t$ | condition |
|---|---|
| 0–1 | very poor |
| 1–2 | poor |
| 2–3 | fair |
| 3–4 | good |
| 4–5 | very good |

*step 2*: Convert the traffic volume to the number of equivalent 18-kip single axle loads.

*step 3*: Determine the *regional factor*, *R*. (This is not the same as the soil R-value.) This factor recognizes that a load does more damage when the ground is saturated, as during a spring thaw, compared to solid dry or solid frozen support. It is difficult to specify the regional factor, but a range of 0.5 to 4.0 should not be exceeded for the continental U.S.[23]

**Table 16.15**
Regional Factors

| *R* | condition |
|---|---|
| 0.2–1.0 | Roadbed frozen to depth of 5″ or more |
| 0.3–1.5 | Roadbed dry, summer and fall; no winter freezing |
| 0.5 | Sandy desert |
| 1.5 | Roadbed subject to frost, but fairly dry |
| 4.0–5.0 | Roadbed wet, spring break-up thaw, high water table, soil saturated |

*step 4*: Get the *soil support value*, *S*, based on the quality of the subgrade (basement) soil. *S* varies from 1 (theoretically, 0) through 10 for crushed rock. Silty clay (soil type A-6) has a soil support value of approximately 3. For most studies, it will be necessary to correlate S with some other known soil property, such as a California Bearing Ratio (CBR), modulus of subgrade reaction (k), or Hveem's Resistance value (R). These correlations are not exact, and they may vary from authority to authority. Appendix A, B, or C can be used.

[23] If $R = 1$, the regional factor will not have an effect on the structural number. $R = 1$ is appropriate for all of California.

**Table 16.16**
Approximate Modulus of Subgrade Reaction for Paving Materials

| AASHTO soil group | category | $k$, (psi/in) |
|---|---|---|
| A-1-a | CTB, BTB | 400 and up |
| A-1-b | | 250 and up |
| A-2-4, A-2-5 | gravels | 300 and up |
| A-2-6, A-2-7 | | 175–325 |
| A-3 | sand, clay gravel | 200–325 |
| A-4 | silt, silty clay | 100–300 |
| A-5 | plastic clay | 50–175 |
| A-6 | | 50–225 |
| A-7-5, A-7-6 | | 50–225 |

*step 5*: Determine the *structural number* from the terminal serviceability ($p_t$), 20-year, 18-kip equivalent axle loading (EAL), soil support value (S), and regional factor (R) from figure 16.10 or figure 16.11. If the structural number differs from what was used in step 2, repeat from step 2.

*step 6*: Determine the *layer coefficients* (also known as *strength coefficients*) for the subbase, base, and surface course materials. These coefficients vary from state to state. However, table 16.17 can be used for general calculations.

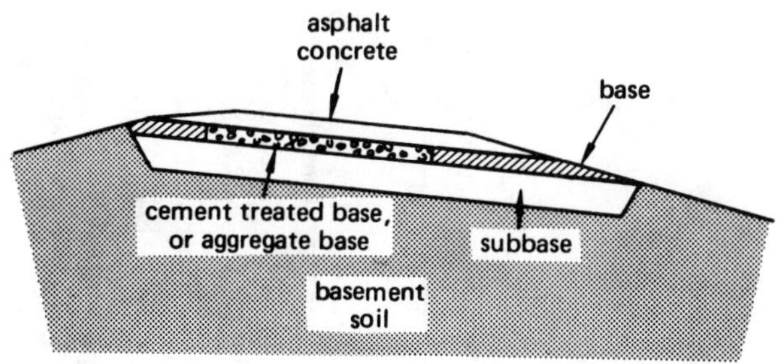


**Figure 16.12** Asphalt Concrete Pavement Cross Section

*step 7*: Write the *layer-thickness equation.* Values of *t* in equation 16.35 are in inches. If a subbase is not to be used, omit the third term.

$$t_1a_1 + t_2a_2 + t_3a_3 = SN \qquad 16.35$$

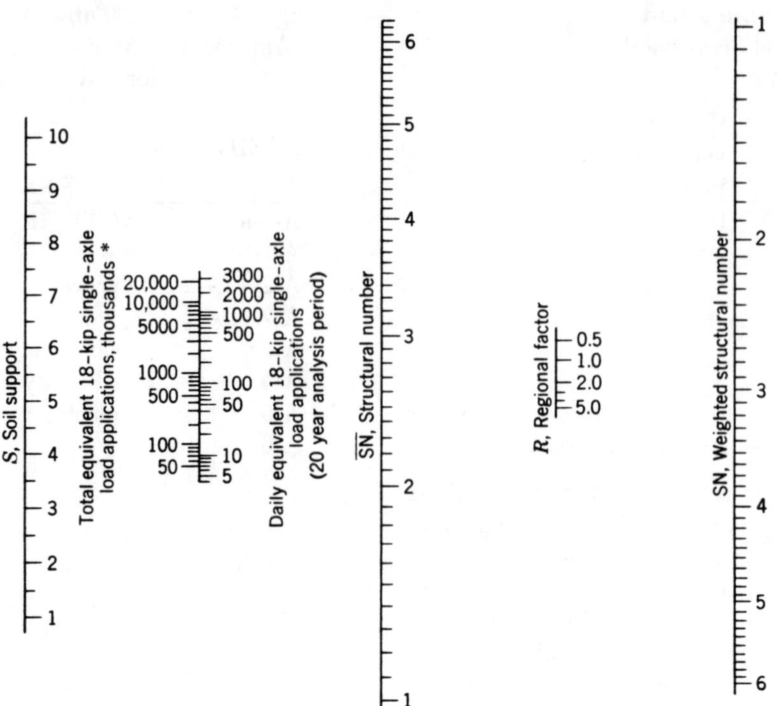


**Figure 16.10** AASHTO Flexible Pavement Design Nomograph *

(Terminal Serviceability, $p_t$ = 2.0)

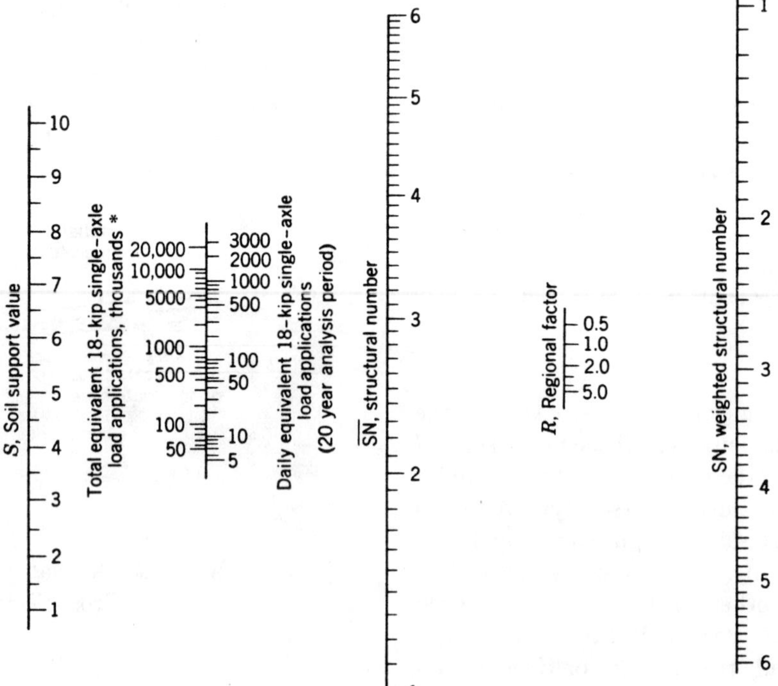


**Figure 16.11** AASHTO Flexible Pavement Design Nomograph

(Terminal Serviceability, $p_t$ = 2.5)

* Total equivalent 20 year load = daily equivalent load x 365 x 20 years

Theoretically, any combinations of thicknesses that satisfy equation 16.35 will work. However, minimum layer thicknesses result from construction techniques and strength requirements.

**Table 16.17**
Layer Coefficients from Various Sources

| | |
|---|---|
| **subbase coefficient, $a_3$** | |
| sandy gravel | 0.11 |
| sand, sandy clay | 0.05–0.10 |
| lime-treated soil | 0.11 |
| lime-treated clay, gravel | 0.14–0.18 |
| **base coefficient, $a_2$** | |
| sandy gravel | 0.07 |
| crushed stone | 0.14 |
| cement treated base (CTB) | |
| $f'_{c,\text{7-day}} > 650$ psi | 0.23 |
| 400–650 | 0.20 |
| $< 400$ | 0.15 |
| bituminous treated base (BTB) | |
| coarse | 0.34 |
| sand | 0.30 |
| lime treated base | 0.15–0.30 |
| soil cement | 0.20 |
| lime/fly ash base | 0.25–0.30 |
| **surface course coefficient, $a_1$** | |
| plant mix | 0.44 |
| road mix | 0.20 |
| sand asphalt | 0.40 |

*Example 16.5*

It is desired to verify the adequacy of a flexible pavement design for a well-traveled highway in California.

R value of basement soil: 10

subbase: aggregate, thickness 17″

base: aggregate, thickness 12″

asphalt concrete: thickness 8″

The surface is expected to carry the following traffic.

2-axle trucks: ADT = 935

3-axle trucks: ADT = 550

4-axle trucks: ADT = 225

5-axle trucks: ADT = 1025

*step 1*: Assume a terminal serviceability of $p_t = 2.5$.

*step 2*: Use equation 16.32.

$$\begin{aligned} EAL &= (1380)(935) + (3680)(550) + (5880)(225) \\ &\quad + (13{,}780)(1025) \\ &= 18.8\,\text{EE6} \end{aligned}$$

*step 3*: Choose $R = 1$ for all of California.

*step 4*: From appendix B, the soil support value, S, is approximately 3 for an R value of 10.

*step 5*: From figure 16.11, the required structural number is approximately 6.

*step 6*: Use the following layer coefficients: $a_3 = 0.11, a_2 = 0.14, a_1 = 0.44$.

*step 7*: Calculate the actual structural number from equation 16.35.

$$(0.44)(8) + (0.14)(12) + (0.11)(17) = 7.07$$

Since 7.07 is greater than 6, the pavement has adequate strength.

## 22 CALTRANS METHOD OF FLEXIBLE PAVEMENT DESIGN

The California Department of Transporation (CALTRANS) has adopted a method of designing flexible pavements which works downward from the asphalt concrete layer. Starting with the top layer, and dealing with each lower layer in turn, the equivalent thickness of a gravel layer is calculated from the traffic volume and strength of the layer below. This *gravel equivalent* is then converted into an actual layer thickness according to the strength of the layer material.

*step 1*: Determine the 18-kip equivalent axle loading for the surface.

*step 2*: Calculate the *traffic index*, TI, from equation 16.36. Round the value up or down to the nearest 0.5. Greater accuracy is not justified.

$$TI = (9.0)(\text{EAL}/10^6)^{0.119} \qquad 16.36$$

*step 3*: Choose the base material. Aggregate base is most commonly used with an R-value of 78. Cement treated bases (CTB) are also used. They are constructed with a good quality, well-graded aggregate mixed with up to 6% portland cement. Other bases can also be used to treat poor soils. These include lime treated (stabilized) bases (LTB), bituminous treated bases (BTB), and soil cement bases (CS).

Class B CTB is often used to increase the R-value of the structural section. If the aggregate R-value was 60 or greater before mixing with cement, the ultimate R-value will exceed 80. Therefore, 80 is used as the minimum value for class B CTB.

Class A CTB is used under asphalt to provide added strength to heavily-traveled surfaces. $f'_c$ will exceed 750 psi after curing for 7 days. Class A CTB is rated by $f'_c$, and R-values are not assigned.

*step 4*: Calculate the total *gravel equivalent* (GE) thickness required for the layer being designed and all layers above it.[24] (The first time through this step, the gravel equivalent of the asphalt concrete surface layer will be determined.) $R$ is for the material below the layer being designed.

$$GE = 0.0032(TI)(100 - R) \qquad 16.37$$

*step 5*: Calculate the thickness of the layer by dividing the gravel equivalent, GE, by the *gravel equivalent factor*, $G_f$. Table 16.18 lists gravel equivalent factors.

$$t_{\text{layer}} = \frac{GE}{G_f} \text{ (in feet)} \qquad 16.38$$

*step 6*: Select the material for the subbase. Subbase aggregate is classified according to the distribution of sizes. Minimum R-values are listed in table 16.19. (Class 2 aggregate is most common.)

**Table 16.19**
Aggregate Subbase Classes

| class | % passing # 4 sieve | minimum R-value |
|---|---|---|
| 1 | 30–75 | 60 |
| 2 | 35–95 | 50 |
| 3 | 45–100 | 40 |

*step 7*: Determine the thickness of the base. Calculate the total gravel equivalent from the R-value of the subbase and equation 16.37. Subtract the surface layer's gravel equivalent from the total gravel equivalent. Calculate the base's thickness from equation 16.38.

*step 8*: Determine the thickness of the subbase. Calculate the total gravel equivalent from the R-value of the basement soil and equation 16.37. Subtract the base's and surface layer's gravel equivalents (corresponding to their actual thicknesses) from the total gravel equivalent. Calculate the subbase's thickness from equation 16.38.

**Table 16.18**
Gravel Equivalent Factors[25]

| layer material | $G_f$ |
|---|---|
| **surface layer (asphalt concrete)** | |
| TI: 5.0 and below | 2.50 |
| 5.5 and 6.0 | 2.32 |
| 6.5 and 7.0 | 2.14 |
| 7.5 and 8.0 | 2.01 |
| 8.5 and 9.0 | 1.89 |
| 9.5 and 10.0 | 1.79 |
| 10.5 and 11.0 | 1.71 |
| 11.5 and 12.0 | 1.64 |
| 12.5 and 13.0 | 1.57 |
| 13.5 and 14.0 | 1.52 |
| 14.5 and above | 1.50 |
| **bases** | |
| aggregate | 1.10 |
| Class A CTB | 1.70 |
| Class B CTB, BTB, LTB, CS | 1.20 |
| lean concrete | 1.90 |
| **subbase** | |
| aggregate | 1.00 |

*Example 16.6*

Use the CALTRANS method to design a flexible pavement lane over a basement soil with an R-value of 10 to carry the following average daily traffic.

2 axle trucks: ADT = 935

3 axle trucks: ADT = 550

4 axle trucks: ADT = 225

5 axle trucks: ADT = 1025

[24] A peculiarity of the CALTRANS method is that a safety factor of 2″ to 3″ is added to the gravel equivalent to account for deficient layer thickness. The sophistication is omitted here.

[25] The gravel equivalent factors for asphalt concrete surface courses are calculated from the following equation:

$$G_f = 2.5\sqrt{\frac{5.14}{TI}}$$

*step 1*:

$$EAL = (1380)(935) + (3680)(550) + (5880)(225) + (13{,}780)(1025) = 18.8\,\text{EE6}$$

*step 2*: $$TI = (9.0)\left(\frac{18.8\,\text{EE6}}{10^6}\right)^{0.119} = 12.76 \quad \text{(say 12.5)}$$

*step 3*: Choose an aggregate base with an R-value of 78.

*step 4*: The gravel equivalent of the asphalt concrete surface layer is

$$GE = 0.0032(12.5)(100 - 78) = 0.88\,\text{ft of gravel}$$

*step 5*: The actual thickness of the asphalt concrete layer is calculated from equation 16.38.

$$t_1 = \frac{0.88}{1.57} = 0.56\,\text{ft}$$

*step 6*: Select a class 2 aggregate subbase with an R-value of 50.

*step 7*: The gravel equivalent of the base and asphalt layers combined is

$$GE = 0.0032(12.5)(100 - 50) = 2.0\,\text{ft of gravel}$$

Since the surface course has already provided 0.88 ft of equivalent gravel layer, the gravel thickness of the base is

$$2.0 - 0.88 = 1.12\,\text{ft}$$

From equation 16.38, the base thickness is

$$t_2 = \frac{1.12}{1.10} = 1.02\,\text{ft}$$

*step 8*: The gravel equivalent of the subbase, base, and asphalt layers combined is

$$GE = 0.0032(12.5)(100 - 10) = 3.6\,\text{ft of gravel}$$

The surface and base layers have already provided $0.88 + 1.12 = 2.0$ ft of gravel equiv-

**Table 16.20**
Asphalt Institute Traffic Classifications

| traffic class | EAL | type of street or highway | approximate range—number of heavy trucks expected during design period |
|---|---|---|---|
| I | $5 \times 10^3$ | • parking lots, driveways<br>• light traffic residential streets<br>• light traffic farm roads | ≤ 7000 |
| II | $10^4$ | • residential streets<br>• rural farm and residential roads | 7000–15,000 |
| III | $10^5$ | • urban minor collector streets<br>• rural minor collector roads | 70,000–150,000 |
| IV[26] | $10^6$ | • urban minor arterial and light industrial streets<br>• rural major collector and minor arterial highways | 700,000–1,500,000 |
| V[26] | $3 \times 10^6$ | • urban freeways, expressways and other principal arterial highways<br>• rural interstate and other principal arterial highways | 2,000,000–4,500,000 |
| VI[26] | $10^7$ | • urban interstate highways<br>• some industrial roads | 7,000,000–15,000,000 |

[26] Whenever possible, the traffic analysis and design procedures given in the Asphalt Institute manual, *Thickness Design—Asphalt Pavements for Highways and Streets* (MS-1), should be used for roads and streets in traffic category IV or higher.

alent. Therefore, the gravel equivalent thickness of the subbase is

$$3.6 - 2.0 = 1.6\,\text{ft}$$

The subbase thickness is

$$t_3 = \frac{1.6}{1.0} = 1.6\,\text{ft}$$

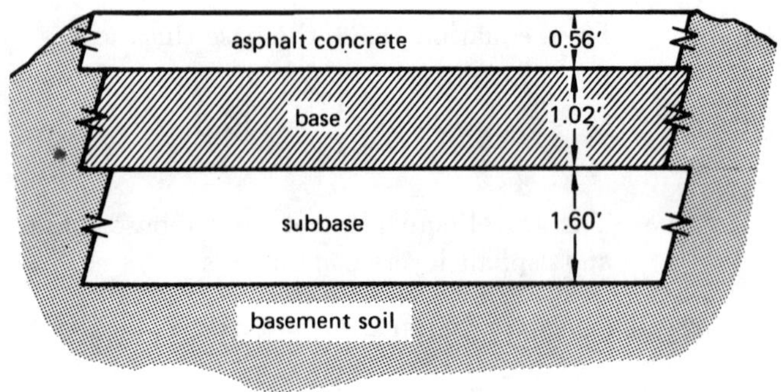


## 23 FULL-DEPTH ASPHALT PAVEMENTS—SIMPLIFIED ASPHALT INSTITUTE METHOD[27]

*step 1*: Determine the 20-year, 18-kip equivalent axle loading (EAL) for the pavement.

*step 2*: Use table 16.20 to convert EAL to a traffic class.

*step 3*: Use table 16.21 to classify the subgrade soil into poor, medium, or good-to-excellent categories.

**Table 16.21**
Subgrade Soil Categories

| | typical values | | |
|---|---|---|---|
| category | resilient modulus[28] | CBR | R-value |
| poor | 4500 psi | 3 | 6 |
| medium | 12,000 | 8 | 20 |
| good-to-excellent | 25,000 | 17 | 43 |

[27] This method cannot be used to design pavement sections with asphalt concrete directly over untreated aggregate bases.

[28] The *resilient modulus* is the same as the *modulus of elasticity* of the soil. It is not the same as the modulus of subgrade reaction, although the two are related. For positive values of the resilient modulus, MR $\approx 485 \times k - 5010$. The resilient modulus is also approximately 1500 times the California Bearing Ratio (CBR).

*step 4*: Choose the base and subbase materials. Emulsified asphalt base mixtures are divided into three types. Type I is plant mixed base made with processed dense-graded aggregates. Type II bases are made with semi-processed, crusher-run, pit-run, or bank-run aggregates. Type III bases are made with sands or silty sands. Both type II and type III can be either plant or road mixed.

Untreated aggregate subbase materials can also be used. Special quality and minimum strength requirements apply in those instances.[29]

*step 5*: Use table 16.22 or table 16.23 to design the pavement. If the EAL or approximate number of trucks places the design between traffic classes, interpolation between thicknesses in the tables is allowed.

## 24 DESIGNING PORTLAND CEMENT CONCRETE PAVEMENT

On projects with three or more lanes in one direction, separate designs usually are made for the inside and outside lanes. This results in steps at the bottoms of pavement and base. It is cheaper to construct stepped sections than uniform or tapered sections, which result in increased soil removal. However, in order to provide a uniform grading plane, it is permissible to increase the thickness of the subbase under the inside lanes.[30] This total thickness of subbase should be used to design the pavement for the inside lanes.

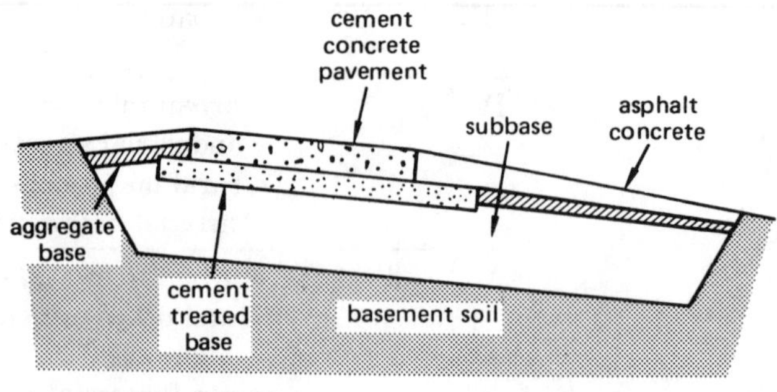


**Figure 16.13** Typical Portland Cement Concrete Section

[29] Specifically, untreated aggregate must have a minimum CBR of 20 or a minimum R-value of 55 to be used as a subbase. It must have a minimum CBR of 80 of a minimum R-value of 78 to be used as a base. Other requirements exist also.

[30] If the subbase is omitted, as it is in many rigid pavement designs, increase the thickness of the base.

**Table 16.22**
Full-Depth Asphalt Pavements Using Asphalt Concrete or Emulsified Asphalt Base Mixes

| subgrade class | pavement section | I | II | III | IV | V | VI |
|---|---|---|---|---|---|---|---|
| | | traffic classification thickness in inches | | | | | |
| | | full-depth asphalt concrete | | | | | |
| poor | asphalt concrete surface | 1.0 | 1.0 | 1.5 | 2.0 | 2.0 | 2.0 |
| | asphalt concrete base | 3.5 | 4.0 | 5.5 | 8.0 | 10.5 | 13.0 |
| | total: | 4.5 | 5.0 | 7.0 | 10.0 | 12.5 | 15.0 |
| medium | asphalt concrete surface | 1.0 | 1.0 | 1.5 | 2.0 | 2.0 | 2.0 |
| | asphalt concrete base | 3.0 | 3.0 | 3.5 | 6.0 | 8.0 | 11.0 |
| | total: | 4.0 | 4.0 | 5.0 | 8.0 | 10.0 | 13.0 |
| good to excellent | asphalt concrete surface | 1.0 | 1.0 | 1.5 | 2.0 | 2.0 | 2.0 |
| | asphalt concrete base | 3.0 | 3.0 | 2.5 | 4.0 | 6.5 | 9.0 |
| | total: | 4.0 | 4.0 | 4.0 | 6.0 | 8.5 | 11.0 |
| | | emulsified asphalt mix type I | | | | | |
| poor | asphalt concrete surface | 1.0 | 1.0 | 1.5 | 2.0 | 2.0 | 2.0 |
| | type I base | 3.5 | 4.0 | 5.5 | 8.0 | 10.5 | 13.0 |
| | total: | 4.5 | 5.0 | 7.0 | 10.0 | 12.5 | 15.0 |
| medium | asphalt concrete surface | 1.0 | 1.0 | 1.5 | 2.0 | 2.0 | 2.0 |
| | type I base | 3.0 | 3.0 | 3.5 | 7.0 | 9.0 | 11.5 |
| | total: | 4.0 | 4.0 | 5.0 | 9.0 | 11.0 | 13.5 |
| good to excellent | asphalt concrete surface | 1.0 | 1.0 | 1.5 | 2.0 | 2.0 | 2.0 |
| | type I base | 3.0 | 3.0 | 2.5 | 4.5 | 7.5 | 10.0 |
| | total: | 4.0 | 4.0 | 4.0 | 6.5 | 9.5 | 12.0 |
| | | emulsified asphalt mix type II | | | | | |
| poor | asphalt concrete surface | 2.0 | 2.0 | 2.0 | 3.0 | 3.5 | 4.0 |
| | type II base | 2.5 | 3.0 | 5.5 | 9.0 | 11.0 | 13.5 |
| | total: | 4.5 | 5.0 | 7.5 | 12.0 | 14.5 | 17.5 |
| medium | asphalt concrete surface | 2.0 | 2.0 | 2.0 | 3.0 | 3.5 | 4.0 |
| | type II base | 2.0 | 2.0 | 3.0 | 7.0 | 9.0 | 11.5 |
| | total: | 4.0 | 4.0 | 5.0 | 10.0 | 12.5 | 15.5 |
| good to excellent | asphalt concrete surface | 2.0 | 2.0 | 2.0 | 3.0 | 3.5 | 4.0 |
| | type II base | 2.0 | 2.0 | 2.0 | 5.0 | 7.0 | 9.5 |
| | total: | 4.0 | 4.0 | 4.0 | 8.0 | 10.5 | 13.5 |
| | | emulsified asphalt mix type III | | | | | |
| poor | asphalt concrete surface | 2.0 | 2.0 | 2.0 | 3.0 | 3.5 | 4.0 |
| | type III base | 4.5 | 5.0 | 8.5 | 12.5 | 15.0 | 19.0 |
| | total: | 6.5 | 7.0 | 10.5 | 15.5 | 18.5 | 23.0 |
| medium | asphalt concrete surface | 2.0 | 2.0 | 2.0 | 3.0 | 3.5 | 4.0 |
| | type III base | 2.0 | 2.5 | 5.5 | 9.5 | 12.0 | 15.5 |
| | total: | 4.0 | 4.5 | 7.5 | 12.5 | 15.5 | 19.5 |
| good to excellent | asphalt concrete surface | 2.0 | 2.0 | 2.0 | 3.0 | 3.5 | 4.0 |
| | type III base | 2.0 | 2.0 | 3.0 | 6.5 | 9.0 | 12.0 |
| | total: | 4.0 | 4.0 | 5.0 | 9.5 | 12.5 | 16.0 |

**Table 16.23**
Asphalt Pavements with Untreated Aggregate Base and Subbase

| subgrade class | pavement section | I | II | III | IV | V | VI |
|---|---|---|---|---|---|---|---|
| | | traffic classification (thickness in inches) | | | | | |
| poor | asphalt concrete surface | 1.0 | 1.0 | 1.5 | 2.0 | 2.0 | 2.0 |
| | asphalt concrete base | 2.5 | 3.0 | 5.0 | 8.0 | 10.0 | 12.5 |
| | untreated aggregate base | 4.0 | 4.0 | 4.0 | 4.0 | 4.0 | 4.0 |
| | total: | 7.5 | 8.0 | 10.5 | 14.0 | 16.0 | 18.5 |
| medium | asphalt concrete surface | 1.0 | 1.0 | 1.5 | 2.0 | 2.0 | 2.0 |
| | asphalt concrete base | 2.0 | 3.0 | 2.5 | 5.0 | 7.5 | 10.0 |
| | untreated aggregate base | 4.0 | 4.0 | 4.0 | 4.0 | 4.0 | 4.0 |
| | total: | 7.0 | 8.0 | 8.0 | 11.0 | 13.5 | 16.0 |
| good to excellent | asphalt concrete surface | 1.0 | 1.0 | 1.5 | 2.0 | 2.0 | 2.0 |
| | asphalt concrete base | 2.0 | 3.0 | 2.5 | 3.0 | 5.0 | 8.0 |
| | untreated aggregate base | 4.0 | 4.0 | 4.0 | 4.0 | 4.0 | 4.0 |
| | total: | 7.0 | 8.0 | 8.0 | 9.0 | 11.0 | 14.0 |

## 25 AASHTO METHOD OF RIGID PAVEMENT DESIGN

The AASHTO method of designing structural sections is based on modifications to the *Westergaard theory of stress distribution* in rigid slabs. There is no provision or adjustment for environment or weather (i.e., there is no regional factor), nor does the method specifically design the base or lower layers. It is assumed that the strengths assumed for the lower layers will be achieved by properly constructing those layers.[31]

*step 1*: Select the terminal serviceability, $p_t$. As with the AASHTO method for designing flexible pavements, $p_t = 2.5$ is appropriate for highways.

*step 2*: Determine the 20-year, 18-kip equivalent axle loading (EAL).

*step 3*: Select or determine the subbase material. AASHTO has specified six types of subbases, as listed in table 16.24, although others could be used. All types except type F can be used satisfactorily for the top 4 inches of subbase. (Type F can be used below the top 4 inches.)

*step 4*: Determine the *modulus of subgrade reaction* (also known as *gross k*) for the roadbed soil.

[31] Actually, only one support layer is assumed with the AASHTO method. AASHTO refers to this layer as the *subbase*, and omits the base layer. Others may refer to the layer as the *base* and omit the subbase.

**Table 16.24**
AASHTO Rigid Pavement Subbase Materials

| type | description |
|---|---|
| A | open graded |
| B | dense graded |
| C | cement treated (CTB) |
| D | lime treated (LTB) |
| E | bituminous treated (BTB) |
| F | granular |

*step 5*: Determine the *modulus of elasticity* of the concrete used in the rigid pavement. Typically, static compression tests on concrete cylinders are used to determine this parameter. (The original AASHTO studies used $E_c = 4.2$ EE6 psi.)

*step 6*: Determine the allowable working stress in the concrete. This is determined by dividing the 28-day flexural strength (i.e., *modulus of rupture*) from *third point loading tests* by a factor of safety.

$$f_t = \frac{MR}{FS} \tag{16.39}$$

FS is generally taken as 1.33, particularly where the 18-kip, 20-year EAL is less than 1,000,000. Where surface replacement would be inconvenient, or where local conditions require a thicker pavement, use FS = 2.0.

An alternate method of calculating the concrete working stress is to take 75% of the modulus of rupture.

$$f_t = 0.75 \times MR \qquad 16.40$$

*step 7*: Read the slab thickness from figure 16.14. Round up to the next whole number of inches.

*step 8*: If the slab thickness is less than 8 inches, specify steel reinforcement.

## 26 FATIGUE STRENGTH METHOD OF DESIGN

As long as the stress induced in the pavement is less than the *fatigue strength*, an infinite number of repetitions of that stress can be applied without damage to the pavement. For any given stress above the fatigue strength, there is an allowable *fatigue life*, or number of allowable repetitions. If the actual number of applications of a stress taken over 20 years is divided by the fatigue life for that stress, the *fatigue fraction* is obtained.

$$FF_i = \frac{n_i}{N_i} \qquad 16.41$$

If the sum of all fatigue fractions corresponding to different stresses is less than approximately 1.0, the pavement is adequate.

This method has been adopted by CALTRANS, and is based on fatigue strength ratios proposed by the Portland Cement Association. It is better suited for analysis of proposed pavement sections than for design, since layer thicknesses are required in the procedure.

*step 1*: Decide on a proposed pavement design. Determine the materials to be used and their thicknesses.

*step 2*: Obtain the values of all axle loads from the loadometer survey. The actual distribution of axle loads is needed.

*step 3*: If desired, multiply all of the axle loads by a *load safety factor* to provide a margin of safety for impact.

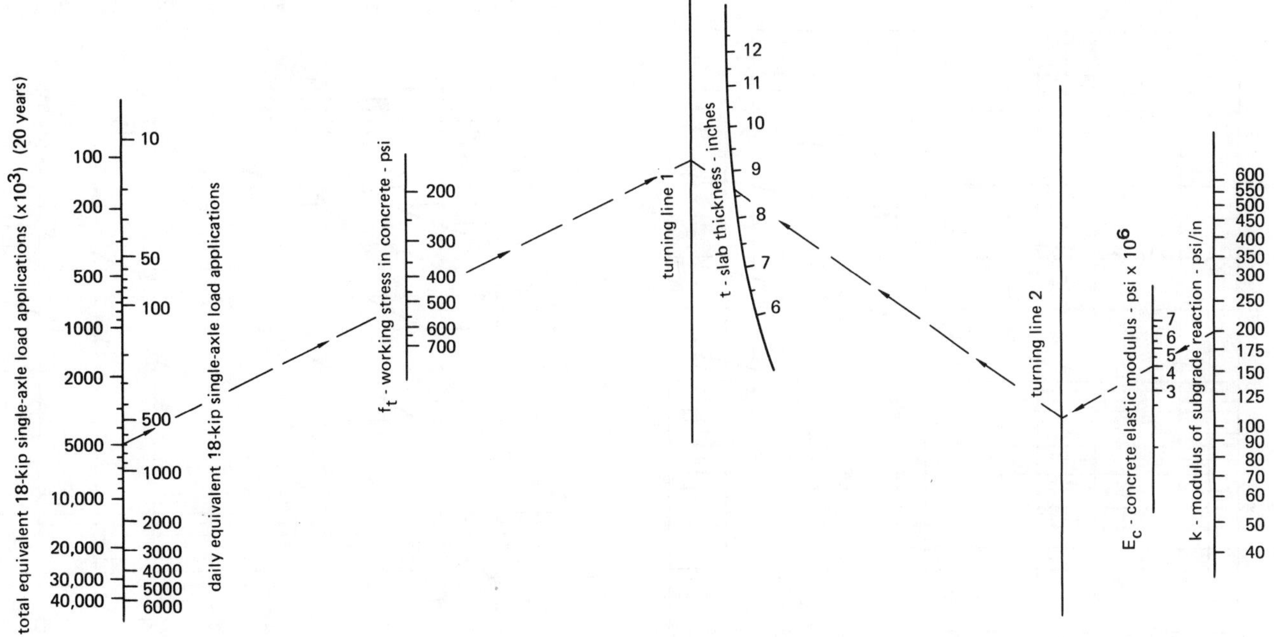


**Figure 16.14** AASHTO Rigid Pavement Design Chart ($p_t = 2.5$)

**Table 16.25**
Load Safety Factors

| type of facility | LSF |
|---|---|
| airport aprons, taxiways, runway ends, hangar floors | 1.7 to 2.0 |
| central portions of runways, and high-speed taxiways | 1.4 to 1.7 |
| outside lanes of multi-lane facilities with high truck traffic | 1.3 |
| inside lanes of multi-lane facilities with high truck traffic, and all lanes with moderate truck traffic | 1.2 |
| minor highways, frontage roads, and all streets with low truck traffic | 1.1 |
| residential streets or roads with occasional truck traffic | 1.0 |

*step 4*: Obtain the modulus of subgrade reaction, $k_s$, for the soil.

*step 5*: Based on the modulus of subgrade reaction of the soil, $k_s$, and the thickness of the subbase, determine a total modulus of reaction, $k_{sb}$, for the combined strength of subgrade soil and subbase. (If there is no subbase, this step is skipped and $k_s$ is used for $k_{sb}$.) Figure 16.15 can be used for this purpose.

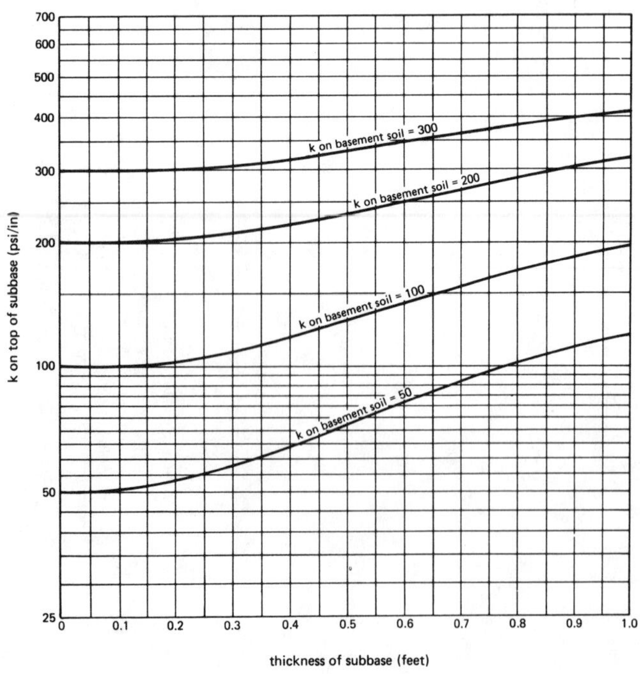


**Figure 16.15** $k$-Values for Base and Basement Soil Combination

(Do not use if subbase is omitted.)

*step 6*: Using the modulus of reaction for the subbase, $k_{sb}$, and the thickness of the base, use figure 16.16 or figure 16.17 to determine the total modulus of reaction on the base (under the rigid pavement).

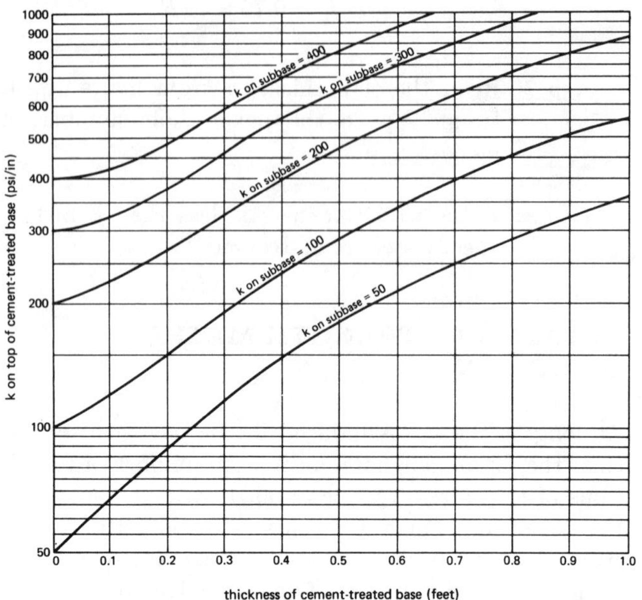


**Figure 16.16** Total $k$-Values Using Cement Treated Base

($k$ values are for basement soil if subbase is omitted.)

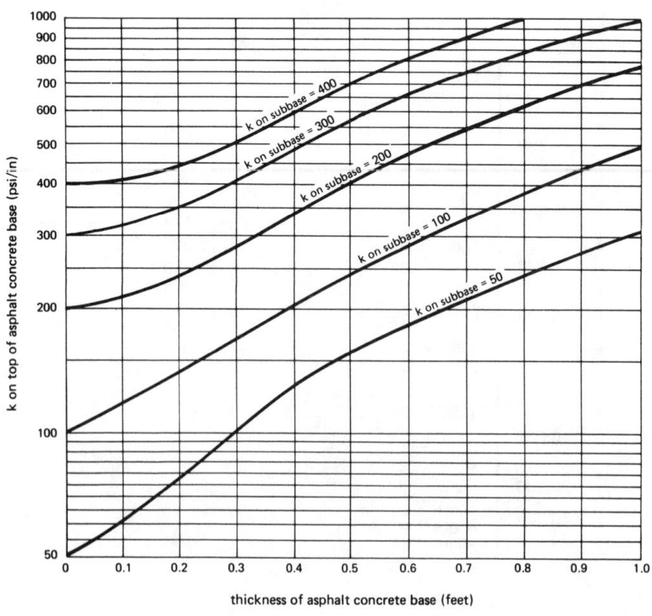


**Figure 16.17** Total $k$-Values Using Asphalt Treated Base

($k$ values are for basement soil if subbase is omitted.)

*step 7*: Determine the stress induced in the pavement from each category of axle loads. Figures 16.18 and 16.19 can be used for this purpose.

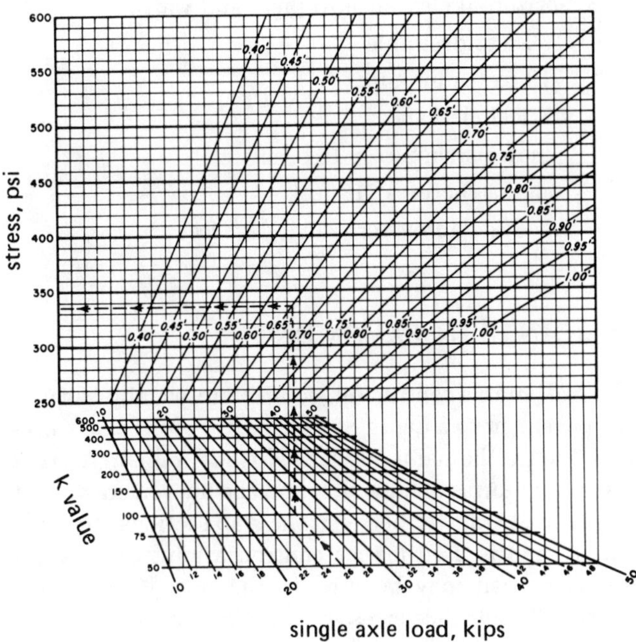


**Figure 16.18** Stress Chart for Single Axle Loads

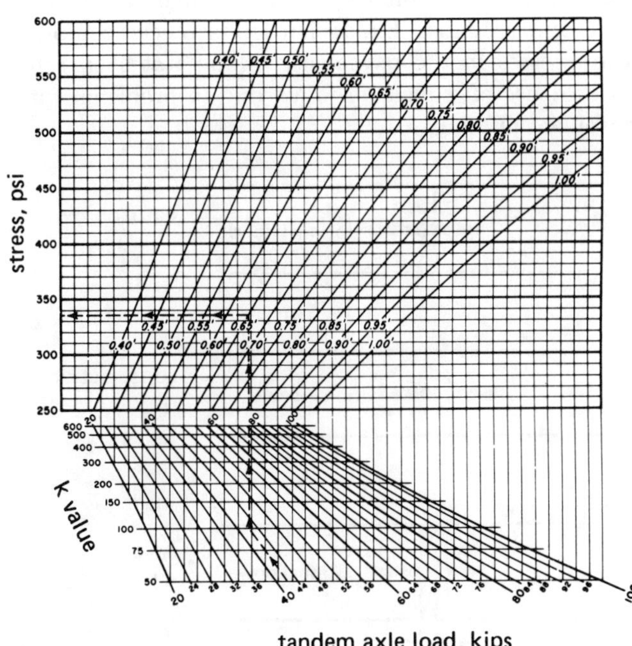


**Figure 16.19** Stress Chart for Tandem Axle Loads

*step 8*: Determine the 28-day *modulus of rupture* for the concrete to be used in the surface layer. (A minimum of 550 psi for a 28-day strength should be specified.)[32] The modulus of rupture is taken as the extreme fiber stress under the breaking load in a beam-breaking test.[33]

$$MR = \frac{Mc}{I} \quad 16.42$$

*step 9*: Divide each category of stress value by the modulus of rupture. Record these numbers to the nearest 0.01. Values of 0.50 or less do not need to be calculated or recorded since this corresponds to the endurance strength of concrete, and unlimited repetitions are allowed.

*step 10*: For each category of stress, determine the fatigue life (allowable number of repetitions). Table 16.26 can be used for this purpose.

**Table 16.26**
Allowable Load Repetitions for Concrete

| stress ratio | allowable repetitions | stress ratio | allowable repetitions |
|---|---|---|---|
| 0.51 | 400,000 | 0.71 | 1500 |
| 0.52 | 300,000 | 0.72 | 1100 |
| 0.53 | 240,000 | 0.73 | 850 |
| 0.54 | 180,000 | 0.74 | 650 |
| 0.55 | 130,000 | 0.75 | 490 |
| 0.56 | 100,000 | 0.76 | 360 |
| 0.57 | 75,000 | 0.77 | 270 |
| 0.58 | 57,000 | 0.78 | 210 |
| 0.59 | 42,000 | 0.79 | 160 |
| 0.60 | 32,000 | 0.80 | 120 |
| 0.61 | 24,000 | 0.81 | 90 |
| 0.62 | 18,000 | 0.82 | 70 |
| 0.63 | 14,000 | 0.83 | 50 |
| 0.64 | 11,000 | 0.84 | 40 |
| 0.65 | 8000 | 0.85 | 30 |
| 0.66 | 6000 | 0.86 | 23 |
| 0.67 | 4500 | 0.87 | 17 |
| 0.68 | 3500 | 0.88 | 13 |
| 0.69 | 2500 | 0.89 | 10 |
| 0.70 | 2000 | 0.90 | 8 |

32 The 90-day modulus of rupture is 110% of the 28-day modulus of rupture.

33 Strictly speaking, the stress is not within the elastic range when the concrete fails. However, equation 16.42 is used anyway.

*step 11*: Divide the estimated number of repetitions over 20 years of each load category by the allowable number of repetitions from step 10 to obtain the *fatigue fraction.*[34]

*step 12*: Add all of the fatigue fractions to determine the fraction of fatigue strength used. The total fraction used should be between 1.0 and 1.1.[35,36]

$$1.0 < \frac{n_1}{N_1} + \frac{n_2}{N_2} + \frac{n_3}{N_3} + \cdots < 1.1 \qquad 16.43$$

## 27 ROADWAY DETAILING

The following geometric details are recommended.

- lane width: 12 feet, all freeways
  11 feet for restricted areas
- crown slope:
  portland cement concrete: 2%
  bituminous mix pavement: 2%
  penetration treated earth and gravel: $2\frac{1}{2}\% - 3\%$
  unsurfaced, graded: $2\frac{1}{2}\% - 3\%$
- shoulders: to the right of traffic: 10 feet (minimum of 6 feet)
  to the left of traffic:
  4 and 6 lanes: 5 feet
  8 lanes: 8 feet
- shoulder slope: 5% away from median
- maximum grade:
  3% freeways
  6% other roads
  2% steeper allowed in rugged terrain
- side slopes on adjacent cuts:
  freeways: 2:1 max (*h*:*v*)
  other roads: $1\frac{1}{2}$:1 max (*h*:*v*)
- cut-to-right-of-way clearance:
  10 feet minimum
  50 feet maximum
  20 feet for cuts 30 to 50 ft high
  25 feet for cuts 50 to 75 ft high
  (1/3) cut height above 75 feet
- divided median width:
  urban area freeways: 30 feet
  rural area freeways: 46 feet
  costly right of way and bridges: 4 feet
- median valley slopes: 10:1 to 20:1 (*h*:*v*)
- horizontal clearance to piers and walls:
  30 feet desirable
  10 feet minimum
- vertical clearance:
  major structures: $16\frac{1}{2}$ feet
  sign structures: 18 feet
  pedestrian overcrossing: $18\frac{1}{2}$ feet

## 28 JOINTS IN PAVEMENT

*Control joints* (*contraction joints*) are usually sawn in the pavement to a depth of one quarter of the slab thickness (hence the name, *weakened plane joints*). A crack eventually forms at that site, and the uneven crack joint allows load transfer between the slab sections. In areas where sand is used for ice control, a joint sealer compound or pad may be used in the slot. Contraction joints relieve tensile stresses in the pavement.

*Construction joints* (*contact joints*) are used at the end of a pour (as at the end of a day) or between pavement lanes.

*Isolation joints* (*expansion joints*) are used with premolded compressible fillers. Isolation joints separate slabs from wall, gutters, columns, and other non-pavement sections. Expansion joints relieve compressive stresses in the pavement. Many states have restricted use of expansion joints due to problems with pavement pumping.

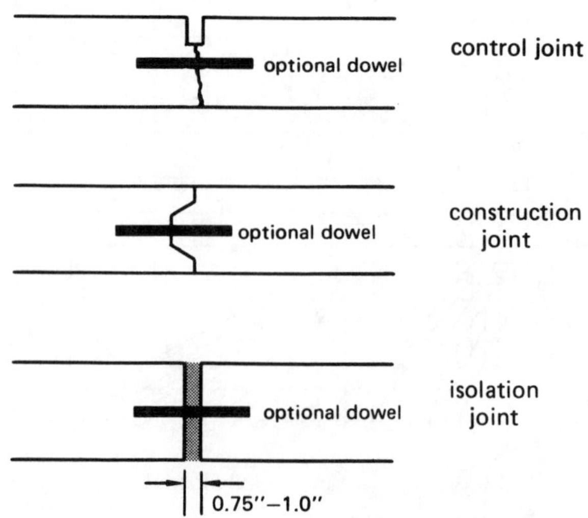


**Figure 16.20** Types of Joints

[34] If loadometer data provide current-day counts, include some provision for traffic growth. Economic analysis tables can be used if the annual growth rate is known.

[35] If the fraction of fatigue strength used is significantly less than 1.0, the pavement is overdesigned.

[36] California allows the maximum to be 1.25.

If *dowel bars* are used as load transfer devices in expansion joints, one side of the dowel should be lubricated. This will allow the slab to slip over the dowel as the slab expands and contracts.

*Hinge joints* (also known as *warping joints*) are similar to control joints (and, they may also be constructed as any of the other types of joints). However, hinge joints are longitudinal joints, generally placed along the centerline of the highway.

## 29 GROOVING PAVEMENTS

A proven method of increasing skid resistance and reducing hydroplaning is by grooving. Grooves drain water laterally, permit water to escape under tires, prevent build-up of surface water, and increase the pavement texture. Grooving should be used only with structurally adequate pavements free from defects. If the pavement is in poor condition, rehabilitation is required prior to grooving.

Grooves should be continuous. In the case of special surfaces, such as airfields, transverse rather than longitudinal grooves may be used. The recommended groove is a square cut with $\frac{1}{4}'' \times \frac{1}{4}''$ dimensions, and a center-to-center spacing of $1\frac{1}{4}''$. The minimum spacing is $1\frac{1}{8}''$; the maximum is $2''$.

## 30 GEOTEXTILES

Geotextiles are support and filter fabrics that are placed in contact with the soil to stabilize and retain the soil. Geotextiles are also known as *filter cloth, reinforcing fabric,* and *support membrane.* Modern geotextiles are not subject to biological and chemical degradation. They can be made from wood pulp (rayon and acetate), silica (fiberglass), and petroleum (polyamides, polyester, and polypropylene). There are woven and non-woven varieties.

Geotextiles are used to prevent dissimilar materials, such as an aggregate base and subbase or soil, from mixing, thereby reducing the support strength. They also reinforce layers, since geotextiles may contribute a substantial tensile strength. Geotextiles provide a filtering function, allowing free passage of water while restraining soil movement.

A significant fraction of all geotextiles is used in highway repair. A layer of geotextile spread at the base of a new road saves on materials such as sand and gravel, while helping to control drainage. Geotextiles can also be used to prevent the infiltration of fine clays into underdrains. In another application, geotextiles appear to strengthen flexible pavements when placed directly under the surface layer.

## 31 SUBGRADE DRAINAGE

Subgrade drains should be considered whenever the following conditions exist:

- high ground-water levels which may reduce subgrade stability and provide a source of water for frost action
- subgrade soils of silts and very fine sands which may become quick or spongy when saturated
- water seeping from underlying water-bearing strata
- cuts in terrain that intercept the natural drainage path of higher elevations
- sag curves with low-permeability subgrade soil

Figure 16.21 illustrates subgrade drains. In general, drains should not be located too close to the pavement (to prevent damage to one when working on the other), and some provision should be made to prevent the infiltration of silt and fines into the drain. (Roofing felt or geotextiles can be used for this purpose.)

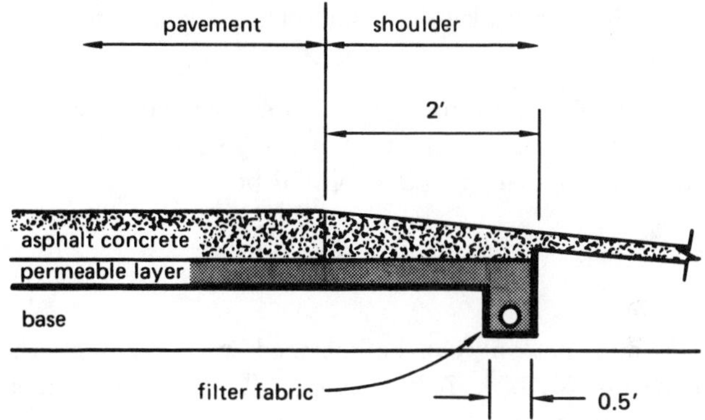


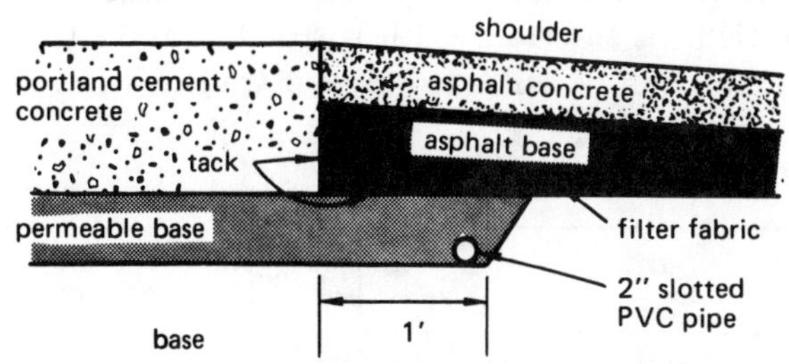


**Figure 16.21** Typical Subgrade Drain Details

## 32 CONTROLLING FROST DAMAGE

Frost heaving and reduced subgrade strength (and accompanying pumping) during spring thaw can quickly destroy a pavement. The following techniques are used to reduce damage done in frost susceptible areas:

- constructing stronger (thicker) pavement sections
- lowering the water table by use of subdrains and drainage ditches
- using layers of coarse sands or waterproof sheets to reduce capillary action
- removing and replacing frost susceptible materials down through the zone of frost penetration
- using rigid foam sheets to insulate and reduce the depth of frost penetration

## 33 PARKING DESIGN

Parking stall width is commonly taken as 9.0 feet, although widths from 8.0 to 9.5 feet are also used. (8.0 foot stalls should be limited to attendant-parked lots. 8.5 feet generally is the accepted minimum. 9.5 foot stalls are appropriate in shopping market areas where package loading is expected.) The minimum length of a stall should be 20 feet to accommodate large luxury cars. Shorter stalls (e.g., 18.5 feet) can be used for compact parking.

In designing parking lots, an area per car of 320 to 380 $ft^2$ will allow for access through lots. Thus, the maximum capacity of a lot would be

$$\text{capacity} \doteq \frac{\text{lot area}}{320} \qquad 16.44$$

Diagonal parking can be specified with angles to the curb of 45°, 60°, 75°, and 90°. The effects of diagonal parking on lane width can be determined from trigonometry.

Figure 16.22 illustrates parallel parking near an intersection. If the stall width is specified as 12 feet, the parking can be converted to an extra lane when necessary.

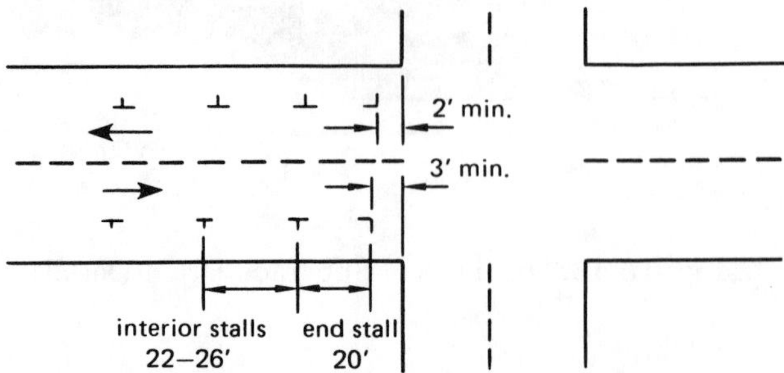


**Figure 16.22** Parallel Parking Design[37]

[37] If the space is available, leave an 8 to 10 foot opening every two spaces. This will provide sight through from the driver to the sidewalk and cross streets.

## 34 INTERSECTION SIGNALING

### A. CONDITIONS REQUIRING SIGNALING

Signaling should be considered only when one of the following conditions is present.[38]

- High traffic volume. For major streets, this is 500 to 600 vehicles per hour, counted over both directions and all lanes. For minor streets, this is 150 to 200 vehicles per hour counted over one direction, all lanes.
- Interruption of traffic. Cross street traffic cannot enter or cross the main traffic flow.
- High crosswalk usage
- Nearby school crossing
- Need to regulate speed of cars and prevent "platooning" of flow
- Excessive accident activity
- Need to combine two roads into one, or allow orderly entrance into a higher-speed road

### B. SIGNAL CONTROLLERS

*Fixed-time controllers* are the least expensive and simplest to use. They are most efficient if the traffic can be accurately predicted. (With multi-dial controllers, cycle lengths can change at different times of the day.) Fixed-time controllers are necessary if sequential intersections or intersections less than one-half mile apart are to be coordinated.

*On-demand controllers* (traffic-activated controllers) are more expensive. However, activated controllers do a better job of controlling flow, and they are better accepted by drivers.

### C. DETERMINING FIXED-TIME CYCLE LENGTHS

In general, the maximum cycle length is 120 seconds. No cycle should be less than 35 seconds.

Green cycle lengths with fixed-time controllers should be chosen to clear all waiting traffic in 95% of the cycles. (Usually, the 85th percentile speed is used in preliminary studies.) Since the green cycle must handle peak loads, the efficiency of the installation is sacrificed during the rest of the day, unless the cycle length is changed.

Although queuing models and simulation can be used to determine cycle lengths, the following simplified procedure can be worked manually.

[38] These reasons are known as *warrants*.

**Table 16.27**
Cycle Length Chart
(Time in Seconds)

| vehicles per hour y-direction | vehicles per hour, x-direction 100 | 200 | 300 | 400 | 500 | 600 | 700 | 800 | 900 | 1000 |
|---|---|---|---|---|---|---|---|---|---|---|
| 100 | 35 | 35 | 35 | 35 | 35 | 40 | 45 | 50 | 55 | 60 |
| 200 | 35 | 35 | 35 | 35 | 40 | 45 | 55 | 65 | 75 | 85 |
| 300 | 35 | 35 | 35 | 40 | 45 | 60 | 75 | 90 | 110 | — |
| 400 | 35 | 35 | 40 | 45 | 60 | 80 | 100 | 120 | — | — |
| 500 | 35 | 40 | 45 | 60 | 80 | 110 | — | — | — | — |
| 600 | 40 | 45 | 60 | 80 | 110 | — | — | — | — | — |
| 700 | 45 | 55 | 75 | 100 | — | — | — | — | — | — |
| 800 | 50 | 65 | 90 | 120 | — | — | — | — | — | — |
| 900 | 55 | 75 | 110 | — | — | — | — | — | — | — |
| 1000 | 60 | 85 | — | — | — | — | — | — | — | — |

*step 1*: Determine the lanes carrying the greatest number of vehicles for both $x$- and $y$-directions. All other lanes will carry fewer vehicles, and these lanes will have enough time.

*step 2*: Determine the number of *car equivalents* for the two maximum lanes. Different methods exist for converting buses and trucks to car equivalents. Equation 16.45 is a simplified approach.[39]

$$E = \#\text{cars} + (1.5)(\#\text{buses}) + (1.5)(\#\text{trucks}) + (1.6)\begin{pmatrix}\#\text{ vehicles}\\ \text{turning left}\end{pmatrix} \qquad 16.45$$

*step 3*: Use table 16.27 to obtain the cycle length.

*step 4*: Determine the time split between the $x$- and $y$-directions. The split is proportional to the traffic flow. For example, the $x$-direction combined green and amber time is proportional to $E_x/(E_x + E_y)$.

*step 5*: Specify the amber time. Usually, the yellow time is 3 to 6 seconds. Higher speeds should be given the higher yellow times.

*step 6*: Provide for a short all-red clearance interval after the yellow light to clear the intersection.

[39] The passenger car equivalents used in left-turn analyses are not the same as were used in freeway capacity analyses.

*step 7*: Check for pedestrian needs. People walk approximately 4 ft/sec and require a starting time of approximately 5 seconds.[40] Determine if pedestrians can cross the street within the green cycle time. (A short all-way red clearance interval is suggested after the green walk signal terminates.)

## D. DETERMINING ON-DEMAND TIMING

Since cycle changes with on-demand (activated) controllers depend on the arriving traffic volume, no traffic counts are needed. Four parameters must be specified: the initial allowance, vehicle time allowance, maximum time allowance, and clearance allowance.

- **Initial Period:** The initial period must allow enough time for traffic stopped between the stop line and detector to begin moving. If a 20 foot car length is used, the number of cars between the stop line and detector is

$$\begin{array}{c}\#\text{cars in}\\ \text{initial period}\end{array} = \frac{\text{distance between line and detector}}{20} \qquad 16.46$$

The first car can be assumed to cross the stop line 5 seconds after the green light appears. The next car requires 3 seconds more. Subsequent cars between

[40] Actually, some people can walk as fast as 7 ft/sec. However, only 30 to 40% of people walk slower than 4 ft/sec. If a significant percentage of walkway users are elderly, the design speed for walking should be reduced to 3 ft/sec.

the detector and stop line require $2\frac{1}{4}$ seconds.[41] The initial period includes time for cars between the two lines.

- **Vehicle Period:** The vehicle period must be long enough to allow a car crossing the detector (moving at the slowest reasonable speed) to get to the intersection before the yellow light appears. (It is not necessary to have the vehicle get through the intersection. That is the purpose of the yellow period.) This period can be calculated from distance and velocity. In a 30 mph zone, a speed of approximately 20 mph into the intersection is reasonable.
- **Maximum Period:** The maximum period is the maximum delay that the opposing traffic can tolerate. 60 seconds is typical for a main street. 30 to 40 seconds is appropriate for a side street. It should never be greater than 120 seconds.
- **Yellow Period:** The yellow clearance period can be determined from the time required to perceive, brake, and stop the vehicle, and the assumed average speed into the intersection.
- **Green Period:** The green period would be the smaller of the sum of initial and vehicle periods and the maximum period.

## E. TIME-SPACE DIAGRAMS

Time-space diagrams are used to coordinate successive fixed-length controllers. This requires setting the controller's *offset*. Offset is the time from the end of one controller's green cycle to the end of the next controller's green cycle. To minimize travel delay, the offset should be minimized.

*step 1*: Choose a scale. 1″ = 100′ and 1″ = 200′ are typical scales.

*step 2*: Draw the main and intersecting streets to scale. For signals separated by short distances, draw the cross-street widths accurately.

*step 3*: Assume or obtain the average travel speed along the main street.

*step 4*: Make an initial assumption for the cycle length. For a two-way street, the cycle length should be either two times (*alternate mode*) or four times (*double alternate mode*) the travel time at the average speed between intersections of average separation. The general timing guidelines (35 second minimum, 120 second maximum) apply here. Two times the travel time is preferred over four times.

*step 5*: Check to see if the assumed cycle can handle the heaviest traveled intersection. Use table 16.27 by entering with the main street flow and cycle time. Find the cross street flow. If the cycle time does not allow sufficient cross street flow, a different cycle time will be required.

*step 6*: For each intersection, determine the *split*. The procedure for determining the split is the same as for the isolated fixed signal—on the basis of the equivalent car traffic.

*step 7*: Try to minimize the conflict between green and red (with yellow) periods at adjacent intersections. This can be done graphically by cutting strips of paper for each intersection, and marking off green, yellow, and red periods according to the time scale. The strips are placed on the time-space diagram with offsets determined by the average speed and average intersection separation.

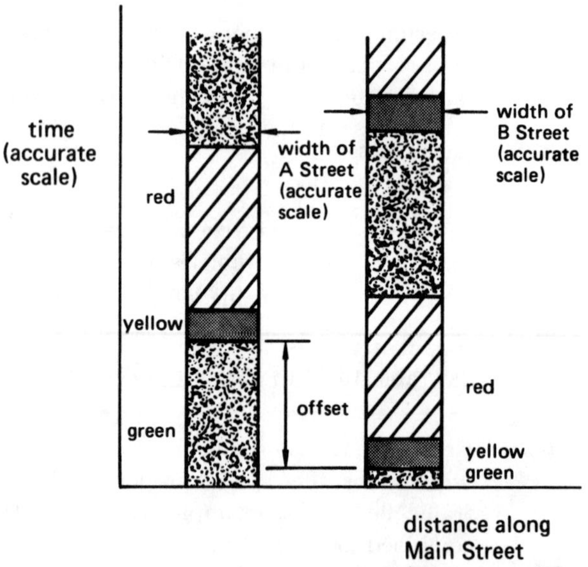


**Figure 16.23** A Time-Space Diagram Showing Offset for Two Streets

For alternate mode operation, every other signal will be green at the same time. For double alternate mode operation, two adjacent pairs of signals will have the same color (i.e., either red or green) while the following

[41] Actual studies have shown the average *start-up lost time* and average *arrival headway* to be approximately half of these values. These values allow for the slowest of drivers.

two adjacent signals will have the opposite color (i.e., either green or red).

Figure 16.24 shows a completed time-space diagram with diagonal lines drawn between green cycle limits. The diagonal lines indicate the *green window* of travel.

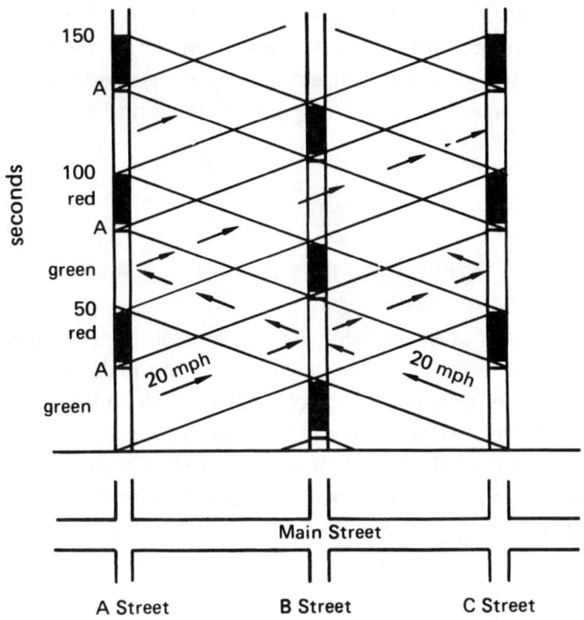


**Figure 16.24** Time-Space Diagram

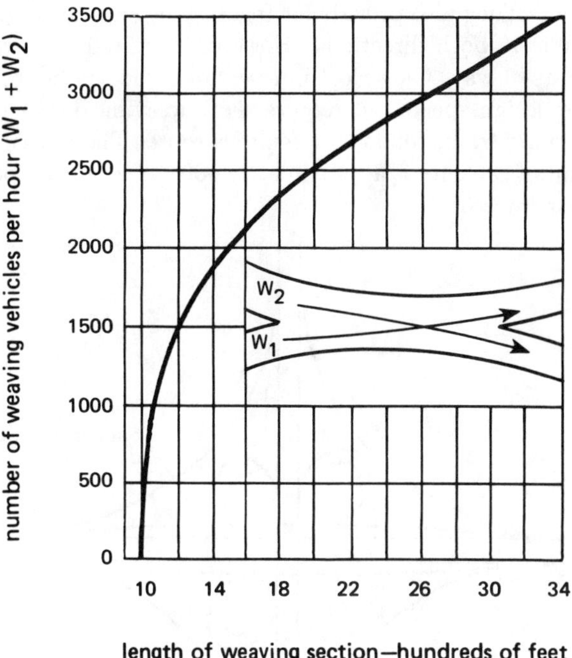


**Figure 16.25** Recommended Weaving Section Lengths[42]

## 35 HIGHWAY INTERCHANGE DESIGN

An interchange allows traffic to enter or leave a highway. Locations of interchanges are affected by the volume of traffic expected on the interchange, convenience, and required land use area. The frequency of interchanges along a route should be sufficient to allow weaving between the interchanging traffic. Figure 16.25 can be used to determine the minimum distance required to merge traffic entering the highway from one on-ramp with traffic headed for the off-ramp of the next interchange.

*Diamond interchanges* are suitable for major road-minor road intersections. They lend themselves to stage construction. The frontage roads and/or ramps can be constructed, while the freeway lanes can be built at a later date. The right-of-way costs are low, since little additional area around the freeway is required. If designed correctly, diamond interchanges will handle fairly large volumes.

Diamond interchanges force traffic using the ramp to substantially reduce its speed. They cannot be used for highway-to-highway intersections, because the capacity per ramp is limited.[43]

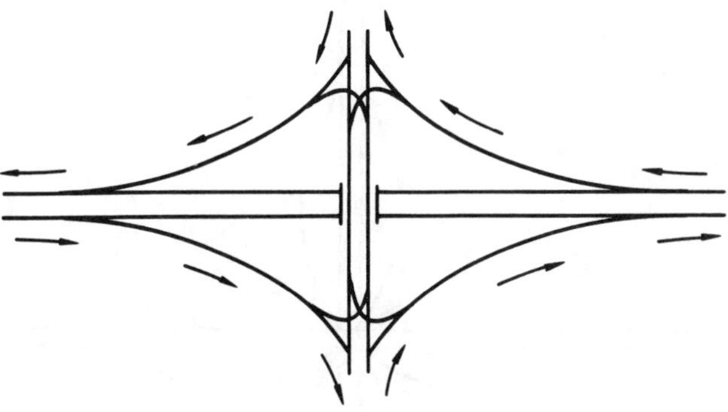

**Figure 16.26** Diamond Interchange

[42] The *Highway Capacity Manual* presents a much more detailed method of analyzing weaving sections.

[43] Up to 1000 vehicles per hour may be able to use a ramp. However, the limiting factor may be the intersection beyond the ramp.

*Cloverleaf interchanges* allow non-stop left-turn movement. They also provide for free flow by separating the traffic in both directions. However, they require large rights-of-way. Cloverleaf intersections slow traffic from the design speed and require short weaving distances. Turning traffic follows a circuitous route. The practical limit of capacity is approximately 800–1000 vehicles per hour.

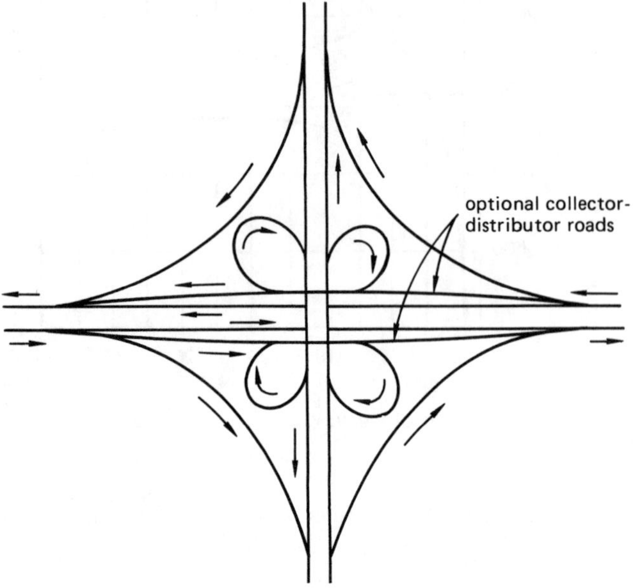


**Figure 16.27** Cloverleaf Interchange

The cloverleaf interchange should not be used to connect two freeways, since it cannot handle the large volumes of turning traffic.

**Figure 16.28** A Four-Layer Directional Interchange

Both diamond and cloverleaf intersections can be improved by various means, such as the use of a third level and *collector-distributor roads* to increase the speed and volume of weaving sections.

*Directional interchanges* allow direct or semi-direct connections for left-turn movements. The design speeds of connections normally are near the design speed of through lanes, and large traffic volumes can be handled. However, large rights-of-way are required. Directional interchanges are expensive because of the structures and capacities provided. Directional interchanges may be of three-layer, four-layer, or rotary bridge design.

## 36 PEDESTRIAN LEVELS OF SERVICE

The level of service for pedestrians in walkways and queuing areas can be categorized in much the same way as is done for freeway and highway vehicles. Table 16.28 relates important parameters to the level of service (LOS). The primary criterion for determining pedestrian level of service is space (the inverse of density).

**Table 16.28**
Pedestrian Levels of Service in Walkways and Queuing Areas

| | walkways | | queuing areas | |
|---|---|---|---|---|
| LOS | space ($ft^2$/ped) | flow rate (ped/min-ft)* | area (sq ft/person) | spacing (ft) |
| A | $\geq 130$ | $\leq 2$ | $\geq 13$ | $\geq 4$ |
| B | $\geq 40$ | $\leq 7$ | 10–13 | 3.5–4.0 |
| C | $\geq 24$ | $\leq 10$ | 7–10 | 3.0–3.5 |
| D | $\geq 15$ | $\leq 15$ | 3–7 | 2.0–3.0 |
| E | $\geq 6$ | $\leq 25$ | 2–3 | $\leq 2$ |
| F | $\leq 6$ | – | $\leq 2$ | – |

* pedestrians per minute per foot width of walkway

## 37 ECONOMIC JUSTIFICATION OF HIGHWAY SAFETY FEATURES

Much can be done to improve the safety of some sections of highway. Features such as breakaway poles, cushioned barriers, barriers separating two directions of traffic, and direction channeling away from abutments are common. These features must be economically jus-

**Table 16.29**
Economic Analysis of Accidents
(average dollar values)

| accident element | U.S. National Safety Council (1972)* | U.S. Department of Defense (1975) | Office of Management and Budget (1984) | OSHA (1984) | OSHA/OMB (1985) |
|---|---|---|---|---|---|
| fatality | 330,000 | 287,000 | 1,000,000 | 3,500,000 | 2–5,000,000 |
| non-fatal injury | 3400 | 8100 | | | |
| property damage | 480 | 520 | | | |

*Average over rural and urban accidents

tified, particularly where they are to be retrofitted to existing highways. The justification is usually the value of personal and property damage avoided by the installation of such features.[44]

There are three general classifications of accidents—those involving death with property damage, injury with property damage, and property damage only. The cost of each element (i.e., death, injury, and property damage) can be evaluated from insurance records, court awards, state disability records, and police records. Also, federal agencies such as the U.S. National Safety Council and OSHA are active in monitoring such costs.

Table 16.29 lists representative values which have been associated with the three elements of accidents.

There are also significant variations in the accident values, depending on the age, sex, and location of the accident. Fatalities of women and children may be valued at rates as low as 60% of their male counterparts. Individuals over 55 have been valued as low as 15% of their prime-aged counterparts. Accidents in developed and urban areas bring awards as much as twice the average award, and three times the equivalent accident in a rural setting.

## 38 QUEUING MODELS

*Special Nomenclature*

- L expected system length (includes service)
- $L_q$ expected queue length
- p{n} probability of $n$ customers in the system
- s number of parallel servers
- W expected time in the system (includes service)
- $W_q$ expected time in the queue
- $\lambda$ mean arrival rate
- $\rho$ traffic intensity $= (\lambda/\mu)$ and must be less than $s$
- $\mu$ mean service rate per server

[44] Nobody likes the concept of saying a human life "... is worth such and such an amount ..." This subject is included, however, to illustrate how economic justifications are made.

*Queue* is a technical word for a waiting line. Queueing theory can be used to predict the length of waiting time, the average time a customer can expect to spend in the queue, and the probability that a given number of customers will be in the queue.

Many queueing models have been developed. Most of these models are fairly specialized and complex. However, two models are important. The relationships given below are for steady state operation, which means that the service facility has been open and in operation for some time.

### A. GENERAL RELATIONSHIPS

The following simple relationships are valid for all queueing models.

$$L = \lambda W \qquad 16.47$$

$$L_q = \lambda W_q \qquad 16.48$$

$$W = W_q + \frac{1}{\mu} \qquad 16.49$$

$$\lambda < \mu s \qquad 16.50$$

$$\text{average service time} = \frac{1}{\mu} \qquad 16.51$$

$$\text{average time between arrivals} = \frac{1}{\lambda} \qquad 16.52$$

### B. THE M/M/1 SYSTEM

It is assumed that the following are true for the M/M/1 system.

- There is only one *server* ($s = 1$).

- The *calling population* is infinite.
- The *service times* are *exponentially distributed* with mean $\mu$. That is, the probability of a customer's remaining service time exceeding $h$ (after already spending time with the server) is given by equation 16.53.

$$p\{t > h\} = e^{-\mu h} \qquad 16.53$$

Notice that equation 16.53 is independent of the time already spent with the server. This result holds true regardless of the elapsed service time. The specific *service time distribution* is

$$f(t) = \mu e^{-\mu t} \qquad 16.54$$

- The *arrival rate* is distributed as *Poisson* with mean $\lambda$. The probability of $x$ customers arriving in the next period is

$$p\{x\} = \frac{e^{-\lambda}\lambda^x}{x!} \qquad 16.55$$

The following relationships describe the M/M/1 system.

$$p\{0\} = 1 - \rho \qquad 16.56$$

$$p\{n\} = p\{0\}(\rho)^n \qquad 16.57$$

$$W = \frac{1}{\mu - \lambda} = W_q + \frac{1}{\mu} = \frac{L}{\lambda} \qquad 16.58$$

$$W_q = \frac{\rho}{\mu - \lambda} = \frac{L_q}{\lambda} \qquad 16.59$$

$$L = \frac{\lambda}{\mu - \lambda} = L_q + \rho \qquad 16.60$$

$$L_q = \frac{\rho\lambda}{\mu - \lambda} \qquad 16.61$$

*Example 16.7*

Given an M/M/1 system with $\mu = 20$ customers per hour and $\lambda = 12$ per hour, find the steady state value of $W$, $W_q$, $L$, and $L_q$. What is the probability that there will be 5 customers in the system?

$$\rho = \frac{12}{20} = 0.6$$

$$W = \frac{1}{20 - 12} = 0.125 \text{ hours}$$

$$W_q = \frac{0.6}{20 - 12} = 0.075 \text{ hours}$$

$$L = \frac{12}{20 - 12} = 1.5 \text{ customers}$$

$$L_q = \frac{(0.6)(12)}{20 - 12} = 0.9 \text{ customers}$$

$$p\{0\} = 1 - 0.6 = 0.4$$

$$p\{5\} = 0.4(0.6)^5 = 0.031$$

C. THE M/M/s SYSTEM

The same assumptions are used for the M/M/s system as were used for the M/M/1 system except that there are $s$ servers instead of only 1. Each server has a mean service rate $\mu$. Each server draws from a single line so that the first person in line goes to the first (any) server that is available. Each server does not have its own line.

However, if customers are allowed to change the lines they are in so that they go to any available server, this model also can predict the performance of a multiple server system where each server has its own line.

$$W = W_q + \frac{1}{\mu} \qquad 16.62$$

$$W_q = \frac{L_q}{\lambda} \qquad 16.63$$

$$L_q = \frac{p\{0\}\,\rho^{s+1}}{s!(1-\rho)^2} \qquad 16.64$$

$$L = L_q + \rho \qquad 16.65$$

$$p\{0\} = \frac{1}{\dfrac{(\rho)^s}{s!(1-(\rho/s))} + \sum\limits_{j=0}^{s-1}\dfrac{(\rho)^j}{j!}} \qquad 16.66$$

$$p\{n\} = \frac{p\{0\}(\rho)^n}{n!} \quad (n \le s) \qquad 16.67$$

$$p\{n\} = \frac{p\{0\}(\rho)^n}{s!s^{n-s}} \quad (n > s) \qquad 16.68$$

# Appendix A: Approximate Correlation between California Bearing Ratio and Subgrade Modulus

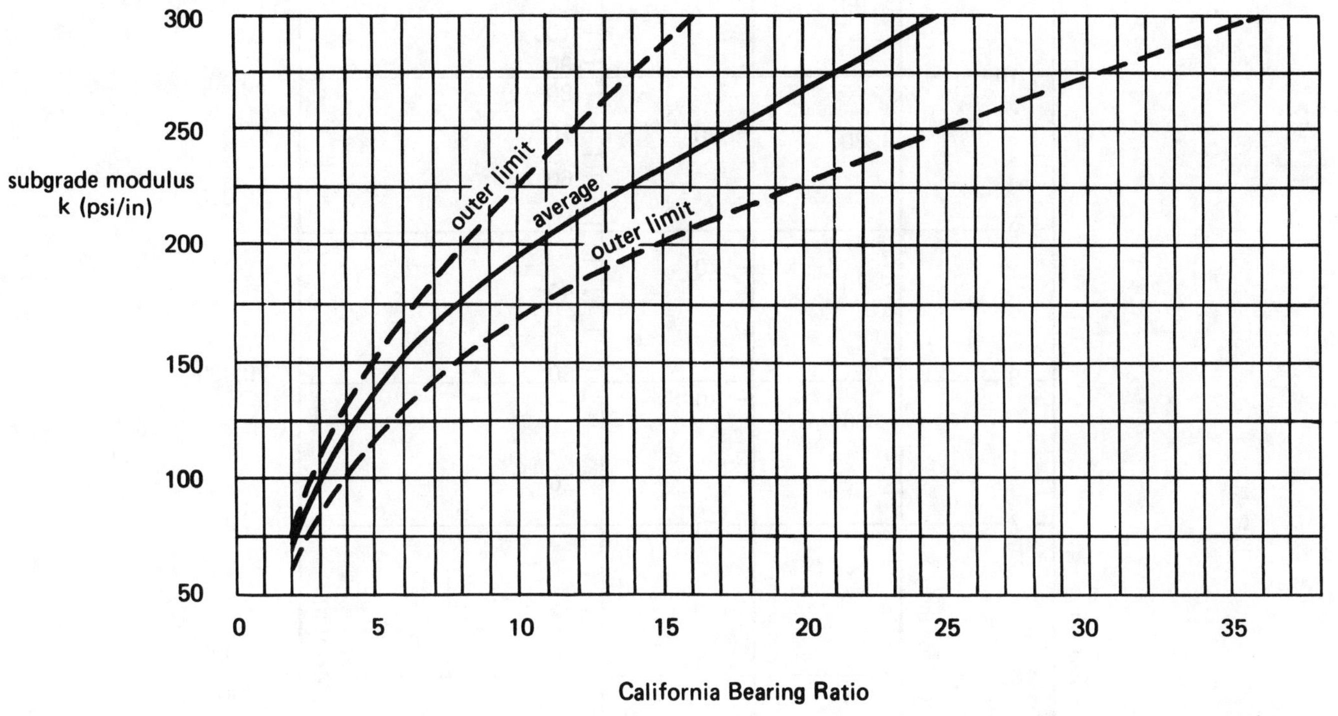

# Appendix B: Revised Soil Support Correlations

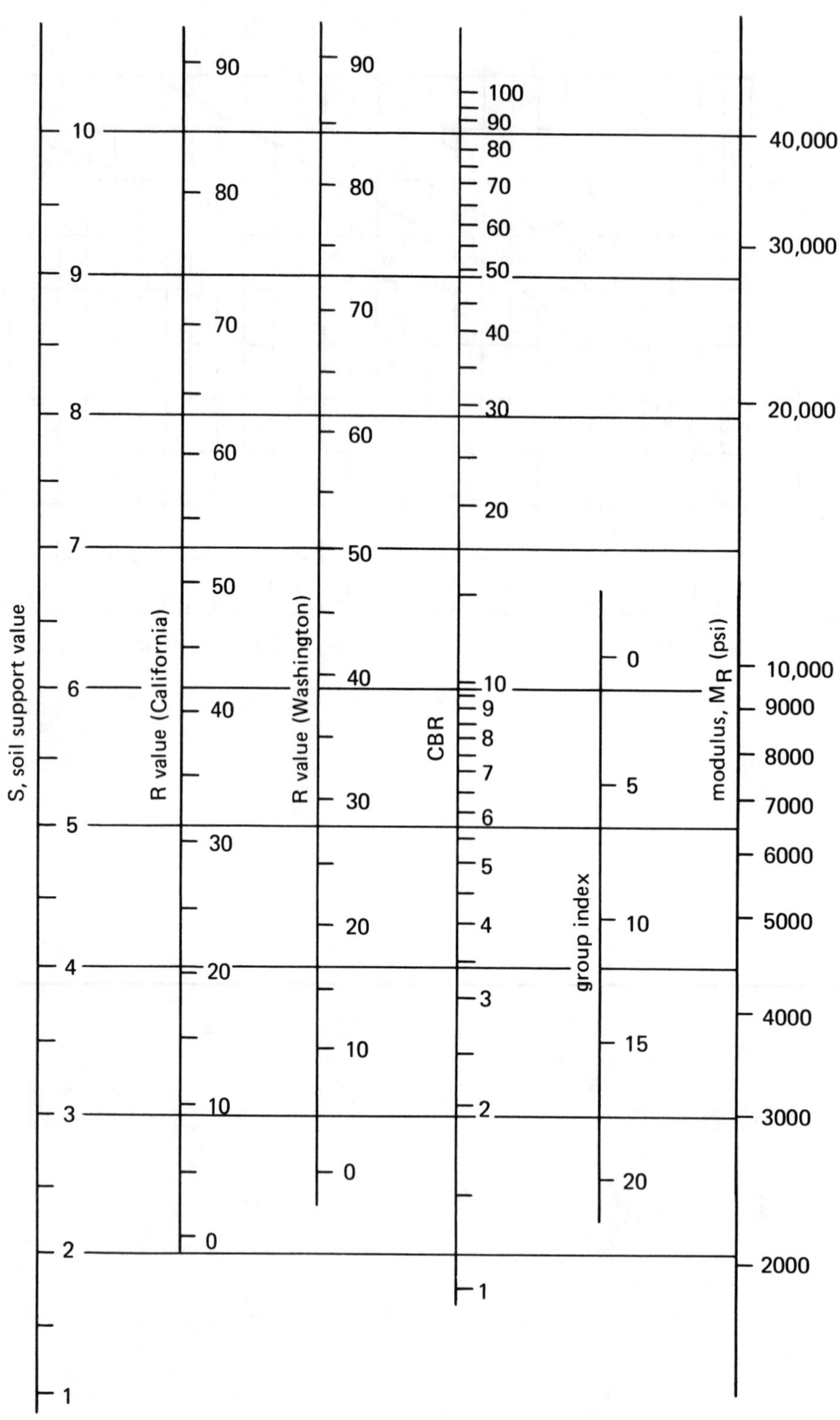

## Appendix C: Approximate Correlation between Subgrade Modulus and Soil R-value

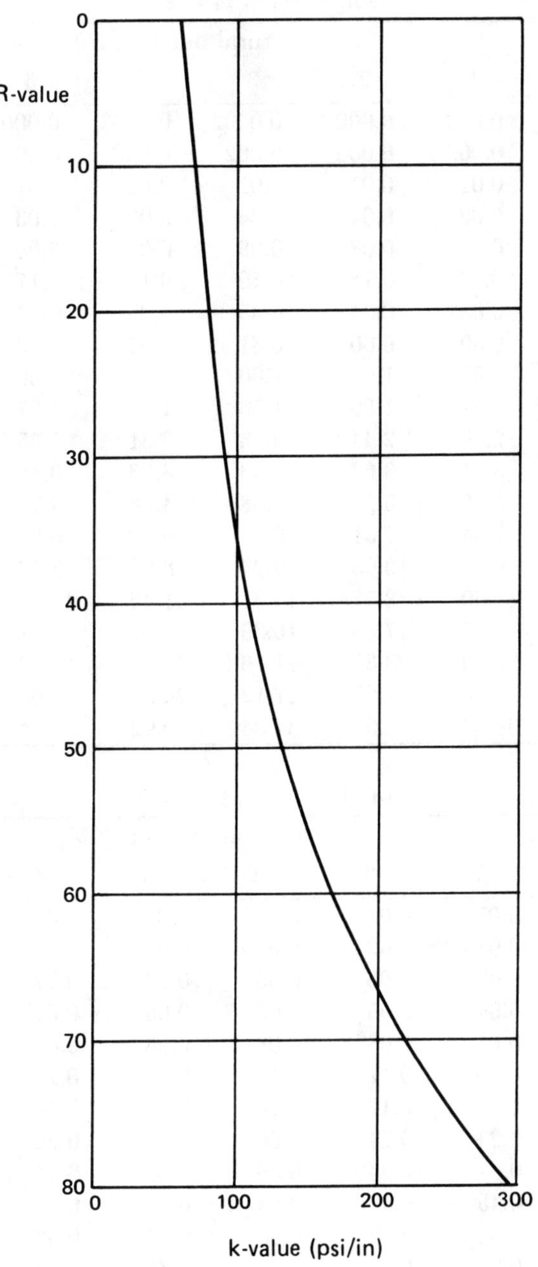

# Appendix D: AASHTO Equivalence Factors—Flexible Pavement

single axles, $p_t = 2.0$

| axle load | structural number, $SN$ | | | | | |
|---|---|---|---|---|---|---|
| (kips) | 1 | 2 | 3 | 4 | 5 | 6 |
| 2 | 0.0002 | 0.0002 | 0.0002 | 0.0002 | 0.0002 | 0.0002 |
| 4 | 0.002 | 0.003 | 0.002 | 0.002 | 0.002 | 0.002 |
| 6 | 0.01 | 0.01 | 0.01 | 0.01 | 0.01 | 0.01 |
| 8 | 0.03 | 0.04 | 0.04 | 0.03 | 0.03 | 0.03 |
| 10 | 0.08 | 0.08 | 0.09 | 0.08 | 0.08 | 0.08 |
| 12 | 0.16 | 0.18 | 0.19 | 0.18 | 0.17 | 0.17 |
| 14 | 0.032 | 0.34 | 0.35 | 0.35 | 0.34 | 0.33 |
| 16 | 0.59 | 0.60 | 0.61 | 0.61 | 0.60 | 0.60 |
| 18 | 1.00 | 1.00 | 1.00 | 1.00 | 1.00 | 1.00 |
| 20 | 1.61 | 1.59 | 1.56 | 1.55 | 1.57 | 1.60 |
| 22 | 2.49 | 2.44 | 2.35 | 2.31 | 2.35 | 2.41 |
| 24 | 3.71 | 3.62 | 3.43 | 3.33 | 3.40 | 3.51 |
| 26 | 5.36 | 5.21 | 4.88 | 4.68 | 4.77 | 4.96 |
| 28 | 7.54 | 7.31 | 6.78 | 6.42 | 6.52 | 6.83 |
| 30 | 10.38 | 10.03 | 9.24 | 8.65 | 8.73 | 9.17 |
| 32 | 14.00 | 13.51 | 12.37 | 11.46 | 11.48 | 12.07 |
| 34 | 18.55 | 17.87 | 16.30 | 14.97 | 14.87 | 15.63 |
| 36 | 24.20 | 23.30 | 21.16 | 19.28 | 19.02 | 19.93 |
| 38 | 31.14 | 29.95 | 27.12 | 24.55 | 24.03 | 25.10 |
| 40 | 39.57 | 38.02 | 34.34 | 30.92 | 30.04 | 31.25 |

tandem axles, $p_t = 2.0$

| axle load | structural number, $SN$ | | | | | |
|---|---|---|---|---|---|---|
| (kips) | 1 | 2 | 3 | 4 | 5 | 6 |
| 10 | 0.01 | 0.01 | 0.01 | 0.01 | 0.01 | 0.01 |
| 12 | 0.01 | 0.02 | 0.02 | 0.01 | 0.01 | 0.01 |
| 14 | 0.02 | 0.03 | 0.03 | 0.03 | 0.02 | 0.02 |
| 16 | 0.04 | 0.05 | 0.05 | 0.05 | 0.04 | 0.04 |
| 18 | 0.07 | 0.08 | 0.08 | 0.08 | 0.07 | 0.07 |
| 20 | 0.10 | 0.12 | 0.12 | 0.12 | 0.11 | 0.10 |
| 22 | 0.16 | 0.17 | 0.18 | 0.17 | 0.16 | 0.16 |
| 24 | 0.23 | 0.24 | 0.26 | 0.25 | 0.24 | 0.23 |
| 26 | 0.32 | 0.34 | 0.36 | 0.35 | 0.34 | 0.33 |
| 28 | 0.45 | 0.46 | 0.49 | 0.48 | 0.47 | 0.46 |
| 30 | 0.61 | 0.62 | 0.65 | 0.64 | 0.63 | 0.62 |
| 32 | 0.81 | 0.82 | 0.84 | 0.84 | 0.83 | 0.82 |
| 34 | 1.06 | 1.07 | 1.08 | 1.08 | 1.08 | 1.07 |
| 36 | 1.38 | 1.38 | 1.38 | 1.38 | 1.38 | 1.38 |
| 38 | 1.76 | 1.75 | 1.73 | 1.72 | 1.73 | 1.74 |
| 40 | 2.22 | 2.19 | 2.15 | 2.13 | 2.16 | 2.18 |
| 42 | 2.77 | 2.73 | 2.64 | 2.62 | 2.66 | 2.70 |
| 44 | 3.42 | 3.36 | 3.23 | 3.18 | 3.24 | 3.31 |
| 46 | 4.20 | 4.11 | 3.92 | 3.83 | 3.91 | 4.02 |
| 48 | 5.10 | 4.98 | 4.72 | 4.58 | 4.68 | 4.83 |

# Appendix E: AASHTO Equivalence Factors—Flexible Pavement

single axles, $p_t = 2.5$

| axle load | structural number, $SN$ | | | | | |
|---|---|---|---|---|---|---|
| (kips) | 1 | 2 | 3 | 4 | 5 | 6 |
| 2 | 0.0004 | 0.0004 | 0.0003 | 0.0002 | 0.0002 | 0.0002 |
| 4 | 0.003 | 0.004 | 0.004 | 0.003 | 0.003 | 0.002 |
| 6 | 0.01 | 0.02 | 0.02 | 0.01 | 0.01 | 0.01 |
| 8 | 0.03 | 0.05 | 0.05 | 0.04 | 0.03 | 0.03 |
| 10 | 0.08 | 0.10 | 0.12 | 0.10 | 0.09 | 0.08 |
| 12 | 0.17 | 0.20 | 0.23 | 0.21 | 0.19 | 0.18 |
| 14 | 0.33 | 0.36 | 0.40 | 0.39 | 0.36 | 0.34 |
| 16 | 0.59 | 0.61 | 0.65 | 0.65 | 0.62 | 0.61 |
| 18 | 1.00 | 1.00 | 1.00 | 1.00 | 1.00 | 1.00 |
| 20 | 1.61 | 1.57 | 1.49 | 1.47 | 1.51 | 1.55 |
| 22 | 2.48 | 2.38 | 2.17 | 2.09 | 2.18 | 2.30 |
| 24 | 3.69 | 3.49 | 3.09 | 2.89 | 3.03 | 3.27 |
| 26 | 5.33 | 4.99 | 4.31 | 3.91 | 4.09 | 4.48 |
| 28 | 7.49 | 6.98 | 5.90 | 5.21 | 5.39 | 5.98 |
| 30 | 10.31 | 9.55 | 7.94 | 6.83 | 6.97 | 7.79 |
| 32 | 13.90 | 12.82 | 10.52 | 8.85 | 8.88 | 9.95 |
| 34 | 18.41 | 16.94 | 13.74 | 11.34 | 11.18 | 12.51 |
| 36 | 24.02 | 22.04 | 17.73 | 14.38 | 13.93 | 15.50 |
| 38 | 30.90 | 28.30 | 22.61 | 18.06 | 17.20 | 18.98 |
| 40 | 39.26 | 35.89 | 28.51 | 22.50 | 21.08 | 23.04 |

tandem axles, $p_t = 2.5$

| axle load | structural number, $SN$ | | | | | |
|---|---|---|---|---|---|---|
| (kips) | 1 | 2 | 3 | 4 | 5 | 6 |
| 10 | 0.01 | 0.01 | 0.01 | 0.01 | 0.01 | 0.01 |
| 12 | 0.02 | 0.02 | 0.02 | 0.02 | 0.01 | 0.01 |
| 14 | 0.03 | 0.04 | 0.04 | 0.03 | 0.03 | 0.02 |
| 16 | 0.04 | 0.07 | 0.07 | 0.06 | 0.05 | 0.04 |
| 18 | 0.07 | 0.10 | 0.11 | 0.09 | 0.08 | 0.07 |
| 20 | 0.11 | 0.14 | 0.16 | 0.14 | 0.12 | 0.11 |
| 22 | 0.16 | 0.20 | 0.23 | 0.21 | 0.18 | 0.17 |
| 24 | 0.23 | 0.27 | 0.31 | 0.29 | 0.26 | 0.24 |
| 26 | 0.33 | 0.37 | 0.42 | 0.40 | 0.36 | 0.34 |
| 28 | 0.45 | 0.49 | 0.55 | 0.53 | 0.50 | 0.47 |
| 30 | 0.61 | 0.65 | 0.70 | 0.70 | 0.66 | 0.63 |
| 32 | 0.81 | 0.84 | 0.89 | 0.89 | 0.86 | 0.83 |
| 34 | 1.06 | 1.08 | 1.11 | 1.11 | 1.09 | 1.08 |
| 36 | 1.38 | 1.38 | 1.38 | 1.38 | 1.38 | 1.38 |
| 38 | 1.75 | 1.73 | 1.69 | 1.68 | 1.70 | 1.73 |
| 40 | 2.21 | 2.16 | 2.06 | 2.03 | 2.08 | 2.14 |
| 42 | 2.76 | 2.67 | 2.49 | 2.43 | 2.51 | 2.61 |
| 44 | 3.41 | 3.27 | 2.99 | 2.88 | 3.00 | 3.16 |
| 46 | 4.18 | 3.98 | 3.58 | 3.40 | 3.55 | 3.79 |
| 48 | 5.08 | 4.80 | 4.25 | 3.98 | 4.17 | 4.49 |

## Practice Problems: TRAFFIC ANALYSIS AND HIGHWAY DESIGN

Untimed

1. A flexible pavement with a 3″ thick asphalt concrete surface layer is to be designed for a state primary road. The subgrade is well drained and is not considered susceptible to frost action. Give the recommended thicknesses of the base and subbase. Use the CALTRANS method with the following information:

| | |
|---|---|
| maximum allowable load on a single axle: | 18,000 pounds |
| maximum aggregate size: | 2″ |
| base: | rolled stone with CBR = 90 |
| subbase: | soil-aggregate with CBR = 40 |
| subgrade: | CBR = 5 |
| traffic index: | 7.5 |

2. The outside lane of an interstate highway is being designed for a state where soil freezes to a depth of 5 inches. A flexible pavement is being considered. The projected life of the roadway is 20 years. The traffic which will use the roadway is as follows:

| | |
|---|---|
| mean ADT: | 20,000 passenger cars and 2200 trucks (over 4 lanes) |
| distribution: | 75% in outside lane |
| average gross truck load: | 22,000 pounds |
| average axle load: | 8800 pounds |
| axle load distribution: | |

| | % of avg. truck ADT | |
|---|---|---|
| axle load (lbs) | single axle | tandem axle |
| under 8000 | 36.3 | – |
| 8000 – 16000 | 28.4 | 4.5 |
| 16000 – 20000 | 12.9 | 10.5 |
| 20000 – 24000 | | 4.0 |
| 24000 – 30000 | | 3.1 |
| 30000 – 34000 | | .3 |

The roadway is to be constructed of the following materials:

| | |
|---|---|
| pavement: | high stability plant mix |
| base: | coarse graded crushed stone treated with asphalt, R = 50 |
| subbase: | crushed stone, R = 78 |
| subgrade: | CBR = 5, R = 28 |

3. Two cars are moving at 60 mph going in the same direction in the same lane. The cars are separated by 20 feet for each 10 mph of their speed. The coefficient of friction (skidding) between the tires and the roadway is .6. The reaction time is assumed to be .5 seconds. (a) If the lead car hits a parked truck, what is the speed of the second car when it hits the first (stationary) car? (b) At what speed does the rule of thumb of one car length per 10 mph become safe? (c) What should the rule actually be?

4. You have been hired by the owner of a demolished house to investigate the car crash that caused the damage. From the police report, you learn that the car was traveling down a 3% grade at an unknown speed. The skid marks are 185 feet long, and the pavement was dry at the time of the accident. The police report estimates from the visible damage to car and house that the initial speed of the car was 25 mph. The house owner doesn't believe the estimate of initial speed. You have the following test data from tests performed on level roadways:

| | | | | |
|---|---|---|---|---|
| initial speed (mph): | 30 | 40 | 50 | 60 |
| coef. of friction: | .59 | .51 | .45 | .35 |

(a) Find the minimum initial speed of the car. (b) If the police report was mistaken in assuming the road surface was dry, what was the minimum initial speed of the car?

5. One lane of a 2-lane road was observed for an hour during the day. The following data was gathered:

| | |
|---|---|
| average distance between front bumpers of successive cars: | 80 ft |
| average mean speed during the study: | 30 mph |
| space mean speed: | 29 mph |

(a) What is the average headway? (b) What is the density in vehicles per mile? (c) What is the traffic volume in vehicles per hour? (d) What is the maximum capacity of the lane? (e) Sketch the relationship between speed and volume. Label the axes and indicate the region of unstable flow. (f) Sketch a graph of the relationship of speed and density. (g) Sketch a graph of the relationship between volume and density. Label the axes and indicate the jam density. (h) Which is a more accurate parameter of traffic capacity: volume or density? Why? (i) What is the generally accepted capacity of one lane of multilane freeway?

6. The intersection of First Street and Main Street in the central business district is being investigated. The population of the metropolitan area is 250,000. The peak hour factor is .85. The signal cycle is 60 second, 2-phase. N/S lanes are 20 feet wide each. W lanes are 11 feet each, not including 11 feet of parking on each side.

| parameter | First (W) | Main (N) | Main (S) |
|---|---|---|---|
| green time | 27 sec | 27 sec | 27 sec |
| yellow time | 3 sec | 3 sec | 3 sec |
| parking | both sides | none | none |
| trucks | 7% | 5% | 5% |
| buses | 0% | 0% | 0% |
| left turns | 10% | 10% | 0% |
| right turns | 10% | 0% | 10% |

(a) Find the maximum service capacity on First Street for service level "E." (b) Find the maximum service capacity on Main Street (N) for service level "B." (c) Find the maximum service capacity on Main Street (S) for service level "E."

7. Two streets intersect at a stop sign: a 36 foot side street and a 44 foot arterial. There have been many complaints from users of the side street that traffic signals are needed at the intersection. Give a logical, step-by-step description of how you would investigate the problem. Discuss how you would arrive at your recommendations. If a signal is required, specify fixed timing or on-demand type decision criteria.

Timed

1. A series of commuter trains travels between stations spaced at regular one-mile locations. The headway is 5 minutes, and the trains can attain a maximum speed of 80 mph. Each train has five cars, with a maximum capacity of 220 people per car. When the train stops, it must wait one minute to allow for passenger movement. The uniform acceleration of the train is 5.5 ft/sec². Deceleration is more gentle at 4.4 ft/sec². (a) What is the top speed of the train? (b) What is the maximum capacity of the line in people per hour? (c) What is the average train speed? (d) What is the average running speed? (e) In regards to your answer in part (d), is the basis "time" or "spacing"? Explain your answer. (f) Suppose the maximum speed was increased to 100 mph. How much time would be saved between stations?

2. Four intersections and five segments of highway have been evaluated using prior years' accident data. (a) Calculate the number of accidents per million vehicles for each intersection. Rank the intersections in order of need for improvement. (b) For the highway segments, calculate the number of accidents per year per million vehicle miles. Rank the segments in order of need for improvement. (c) Explain why calculations in (a) and (b) should be done. (d) Assume that the intersections varied in terms of the fractions of accidents that were fatal, injury, and property damage only. How would you compare the safety needs at those intersections?

| intersection | ADT | # accidents/yr |
|---|---|---|
| A | 820 | 4 |
| B | 1200 | 5 |
| C | 1070 | 7 |
| D | 1400 | 6 |

| highway segment | ADT | # accidents/yr | length (miles) |
|---|---|---|---|
| 1 | 1900 | 1 | 1.50 |
| 2 | 2000 | 14 | 1.35 |
| 3 | 5500 | 18 | 4.50 |
| 4 | 3000 | 11 | 0.53 |
| 5 | 4000 | 30 | 2.48 |

3. Twenty miles of a rural 2-lane highway pass through mountainous (rolling) terrain. The traffic is 5% trucks, 15% buses, and 80% cars. The ADT is 36,000 for two lanes. The design speed is 50 mph. Over a particular segment of the highway, there are 2 miles of 4% grade. A passing sight distance of 1500 feet or greater is possible on 12 of the 20 miles. Lane width is 11 feet. There is a 4′ encroachment on one side. (a) What volume of traffic can be expected at service level "D"? (b) Is service level "B" possible? Explain. (c) What improvements would you recommend in the highway to raise the service level to "B"? (d) What is the maximum capacity of the highway section?

4. You are hired as a consultant to give expert evidence in a court action arising from a vehicle collision. The two vehicles collided head-on while traveling on a curve tangent with a 4% grade. Vehicle #1 (a 1976 Chevrolet Impala) skidded 195 feet downhill before colliding with vehicle #2 (a 1978 Honda). The Honda skidded 142 feet. The police report estimates that both vehicles were traveling at 25 mph prior to the collision. This estimate was based on vehicle deformation. (a) What were the respective speeds of the two vehicles prior to the application of brakes if the coefficient of friction is known to be .48? (b) Which of your assumptions produces the greatest possible error in your speed calculations?

5. A non-signalized blind intersection is shown. It is desired to prohibit parking near the intersection in order to provide adequate sight distance into the intersection. Vehicles travel on the major street at 35 mph or slower 85% of the time. PIEV time is 1.0 sec for the average driver. Vehicles accelerate according to the table given:

| Acceleration data | |
|---|---|
| distance | time |
| 10 feet | 2.3 seconds |
| 30 | 3.7 |
| 50 | 6.2 |
| 80 | 8.5 |
| 100 | 11.8 |

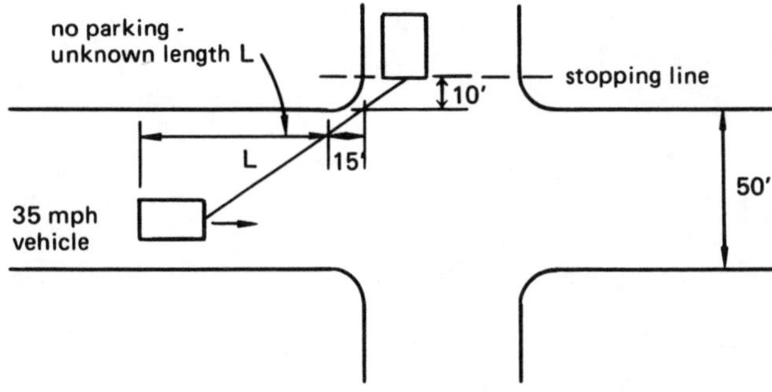


(a) Determine the length, L, needed to allow adequate sight distance. (b) How many parallel parking spaces will be lost?

6. A benefit/cost analysis is to be performed to justify the installation of safety-related road improvements (flexible barriers, break-away poles, etc.). Discuss how you would obtain a dollar value for a human life saved by the safety improvements.

7. Discuss the purpose of using sulfur in large proportions for road surfaces. What are the disadvantages? What special steps are taken in the installation? What special steps are taken to repair a sulfur-based road bed?

# 17 SURVEYING

*Nomenclature*

| | | |
|---|---|---|
| A | area | $ft^2$ |
| C | chord length, correction, or a constant | ft |
| d | cell width, or distance | ft, or stations |
| D | degree of curve | degrees/station |
| E | error, or external distance | ft |
| F | tension | lbf |
| g | grade, or acceleration due to gravity | %, $ft/sec^2$ |
| $g_c$ | gravitational constant | $ft\text{-}lbm/lbf\text{-}sec^2$ |
| h | height | ft |
| HI | height of the instrument | ft |
| I | intersection angle | degrees |
| IH | telescope height above ground | ft |
| k | number of observations | – |
| K | stadia interval factor | – |
| L | length | ft |
| LC | length of curve | ft or stations |
| LS | length of spiral | ft or stations |
| m | mass per length | lbm/ft |
| M | middle distance | ft |
| r | rate of grade change per station | %/station |
| R | rod reading, or radius | ft |
| s | sample standard deviation | various |
| T | tangent distance | ft |
| w | relative weight | – |
| x | distance or location | ft or stations |
| y | distance or elevation | ft |

*Symbols*

| | | |
|---|---|---|
| $\theta, \alpha, \beta$ | angles | degrees |
| $\mu$ | distribution mean | various |

*Subscripts*

| | |
|---|---|
| a | actual |
| c | curvature |
| p | probable |
| r | refraction |
| s | sag or spiral |

## 1 ERROR ANALYSIS

### A. MEASUREMENTS OF EQUAL WEIGHT

There are many opportunities for errors in surveying, although calculators and modern equipment have reduced the magnitude of most errors. The purpose of error analysis is not to eliminate errors, but rather to estimate their magnitudes and to assign them to the appropriate measurements.

The *expected value* (also known as the *most likely value* or the *probable value*) of a measurement is the value that has the highest probability of being correct. If a series of measurements is taken of a single quantity, the most probable value is the average (*mean*) of those measurements.

$$x_p = \frac{x_1 + x_2 + \cdots + x_k}{k} \qquad 17.1$$

For related measurements whose sum should equal some known quantity, the most probable values are the observed values corrected by an equal part of the total error.

Measurements of a given quantity are assumed to be normally distributed. If a quantity has a mean $\mu$ and a standard deviation $s$, the probability is 50% that a measurement of that quantity will fall within the range of $\mu \pm (0.6745)s$. The quantity $(0.6745)s$ is known as the *probable error*. The probable *ratio of precision* is $\mu/(0.6745)s$. The interval between the extremes is known as the 50% *confidence interval.*[1]

The standard deviation, $s$, is the *small sample standard deviation.*

[1] Other confidence limits are easily obtained from the normal tables. For example, for a 95% interval, replace 0.6745 with 1.96. For a 99% interval, use 2.57.

The *probable error of the mean* of $k$ observations of the same quantity is given by equation 17.2.

$$E_{\text{mean}} = \frac{0.6745s}{\sqrt{k}} = \frac{E_{\text{total, } k \text{ measurements}}}{\sqrt{k}} \qquad 17.2$$

*Example 17.1*

The interior angles of a traverse were measured as 63°, 77°, and 41°. Each measurement was made once, and all angles were measured with the same precision. What are the most probable interior angles?

The sum total of angles should equal 180. The error in the measurement is $63+77+41-180 = +1°$. Therefore, the correction required is $-1°$, which is proportioned equally among the three angles. The most probable values are 62.67°, 76.67°, and 40.67°.

*Example 17.2*

12 tapings were made of a critical distance. The mean value was 423.7 feet with a standard deviation ($s$) of 0.31 feet. What are the 50% confidence limits for the distance?

From equation 17.2, the standard error of the mean value is

$$E_{\text{mean}} = \frac{(0.6745)(0.31)}{\sqrt{12}} = 0.06 \text{ ft}$$

The probability is 50% that the true distance is within the limits of $423.7 \pm 0.06$ feet.

*Example 17.3*

The true length of a tape is 100 feet. The most probable error of a measurement with this tape is 0.01 feet. What is the expected error if the tape is used to measure out a distance of one mile?

The number of tapings will be $5280/100 = 52.8$, or 53 tapings. The most probable error will be $0.01 \times \sqrt{53} = 0.073$ feet.

## B. MEASUREMENTS OF UNEQUAL WEIGHT

Some measurements may be more reliable than others. It is not unreasonable to weight each measurement with its relative reliability. Such weights can be determined subjectively, but more frequently, they are determined from relative frequencies of occurrence or from the relative inverse squares of the probable errors.

The *probable error* and 50% confidence interval for weighted observations can be found from equation 17.3. $x_i$ represents the $i$th observation and $w_i$ represents its relative weight. The number of observations is $k$.

$$E_{p,\text{weighted}} = 0.6745\sqrt{\frac{\sum[w_i(\overline{x} - x_i)^2]}{(\sum w_i)(k-1)}} \qquad 17.3$$

For related weighted measurements whose sum should equal some known quantity, the most probable weighted values are corrected inversely to the relative frequency of observation.

Weights can also be calculated when the probable errors are known. These weights are the relative squares of the probable errors.

*Example 17.4*

An angle was measured five times by five equally competent crews on similar days. Two of the crews obtained a value of 39.77°, and the remaining three crews obtained a value of 39.74°. What is the probable value of the angle?

$$\theta = \frac{(2)(39.77) + (3)(39.74)}{5} = 39.75°$$

*Example 17.5*

A distance has been measured by three different crews. The measurements and probable errors are given. What is the most probable value?

$$\text{crew 1}: 1206.40 \pm 0.03 \text{ ft}$$
$$\text{crew 2}: 1206.42 \pm 0.05 \text{ ft}$$
$$\text{crew 3}: 1206.37 \pm 0.07 \text{ ft}$$

The sum of squared probable errors is

$$(0.03)^2 + (0.05)^2 + (0.07)^2 = 0.0083$$

The weights to be applied to the three measurements are

$$\frac{0.0083}{(0.03)^2} = 9.22$$

$$\frac{0.0083}{(0.05)^2} = 3.32$$

$$\frac{0.0083}{(0.07)^2} = 1.69$$

The most probable length is

$$\frac{(1206.40)(9.22) + (1206.42)(3.32) + (1206.37)(1.69)}{9.22 + 3.32 + 1.69}$$

$$= 1206.40 \text{ ft}$$

*Example 17.6*

What is the 50% confidence interval for the measured distance in example 17.5?

It is easier to work with the decimal part only.

$$\overline{x} = \frac{1}{3}(0.40 + 0.42 + 0.37) \approx 0.40$$

| $i$ | $x_i$ | $\overline{x} - x_i$ | $(\overline{x} - x_i)^2$ | $w$ | $w(\overline{x} - x_i)^2$ |
|---|---|---|---|---|---|
| 1 | 0.40 | 0 | 0 | 9.22 | 0 |
| 2 | 0.42 | −0.02 | 0.0004 | 3.32 | 0.0013 |
| 3 | 0.37 | 0.03 | 0.0009 | 1.69 | 0.0015 |
| | | | | 14.23 | 0.0028 |

From equation 17.3,

$$E_{p,\text{weighted}} = 0.6745\sqrt{\frac{0.0028}{(14.23)(3-1)}} = 0.0067$$

The 50% confidence interval is $1206.40 \pm 0.0067$ ft

*Example 17.7*

The interior angles of a triangular traverse were repeatedly measured. What is the most probable value for angle #1?

| angle | value | number of measurements |
|---|---|---|
| 1 | 63° | 2 |
| 2 | 77° | 6 |
| 3 | 41° | 5 |

The total of the angles is $63 + 77 + 41 = 181°$. So $-1°$ must be divided among the three angles. These corrections are inversely proportional to the number of measurements. The sum of the measurement inverses is

$$\frac{1}{2} + \frac{1}{6} + \frac{1}{5} = 0.867$$

The most probable value of angle #1 is

$$63° + \left(\frac{\frac{1}{2}}{0.867}\right)(-1) = 62.42°$$

*Example 17.8*

The interior angles of a triangular traverse were measured. What is the most probable value of angle #1?

| angle | value |
|---|---|
| 1 | $63° \pm 0.01°$ |
| 2 | $77° \pm 0.03°$ |
| 3 | $41° \pm 0.02°$ |

The total of the angles is $63 + 77 + 41 = 181°$. So $-1°$ must be divided among the three angles. The corrections are proportional to the square of the probable errors.

$$(0.01)^2 + (0.03)^2 + (0.02)^2 = 0.0014$$

The most probable value of angle #1 is

$$63 + \frac{(0.01)^2}{0.0014}(-1) = 62.93°$$

## C. ERRORS IN COMPUTED QUANTITIES

When independent quantities with known errors are added or subtracted, the error of the result is given by equation 17.4. The squared errors under the radical are added regardless of whether the calculation is addition or subtraction.

$$E_{\text{total}} = \sqrt{E_1^2 + E_2^2 + E_3^2 + \cdots} \qquad 17.4$$

The error in the product of two quantities $x_1$ and $x_2$ which have known errors $E_1$ and $E_2$ is given by equation 17.5.

$$E_{\text{product}} = \sqrt{x_1^2 E_2^2 + x_2^2 E_1^2} \qquad 17.5$$

*Example 17.9*

An EDM instrument manufacturer has indicated that the measurement error with a particular instrument is ±0.04 feet with another error of ±10 ppm. What is the expected error of measurement if the instrument is used to measure 3000 feet?

There are two independent errors here, since both parts can be positive or negative, and they may have opposite signs. The variable error is

$$E = \pm(3000)\left(\frac{10}{1{,}000{,}000}\right) = 0.03\ \text{ft}$$

From equation 17.4

$$E = \sqrt{(0.04)^2 + (0.03)^2} = 0.05\ \text{ft}$$

*Example 17.10*

The sides of a rectangular section were determined to be 1204.77±0.09 feet and 765.31±0.04 feet respectively. What is the probable error in the area?

From equation 17.5,

$$E_{\text{area}} = \sqrt{(1204.77)^2(0.04)^2 + (765.31)^2(0.09)^2} = 84.06\ \text{ft}^2$$

## 2 ORDERS OF ACCURACY

Traverse closure errors are ranked into *orders of accuracy.* The closure error is divided by the sum of all traverse leg lengths to determine the fractional error. Table 17.1 lists the maximum permissible errors for each order of accuracy.[2]

**Table 17.1**
Maximum Traverse Closure Errors

| order of accuracy | maximum error |
|---|---|
| first | 1/25,000 |
| second | 1/10,000 |
| third | 1/5000 |
| fourth | 1/3000 |
| fifth | 1/1000 |
| lowest | 1/500 |

## 3 DISTANCE MEASUREMENT

Lengths may be divided into 100 foot long sections called *stations.* Interval stakes along an established line are ordinarily laid down at 100 foot intervals called *full stations.* If a marker stake is placed anywhere else along the line, it is called a *plus station.* Thus, a stake placed 1500 feet from a reference point is labeled "15+00," and a stake placed 1325 feet from a reference point is labeled "13+25."

Tapes may be used for short distances. Steel tapes are relatively low in cost. Low-coefficient tapes made from nickel-steel alloys (e.g., *invar*) are not excessively sensitive to temperature changes.[3]

Tape readings are affected by temperature, tension, and sag. Corrections for temperature and tension have been presented in chapter 12. The correction for sag is given by equation 17.6. Since tapes can be standardized flat or suspended, the correction can be added or subtracted to the actual measurement.

$$C_s = \pm\frac{m^2L^3}{24F^2}\left(\frac{g}{g_c}\right)^2 \qquad 17.6$$

The correction $C_s$ is in units of feet when $m$ is in lbm/ft, $L$ is in feet, and the tension, $F$, is in lbf.

[2] The classification of maximum errors into first, second, third, etc., can vary from authority to authority. For example, some federal agencies specify a maximum error of 1/7500 for third order accuracy.

[3] The modulus of elasticity of invar is approximately 2.1 EE7 psi. The coefficient of thermal expansion is approximately 1 EE–7 1/°F.

**Table 17.2**
Corrections for Surveyor's Tapes

| | measured flat | measured suspended |
|---|---|---|
| standardized flat | none | subtract $C_s$ |
| standardized suspended | add $C_s$ | none |

Electronic distance measuring (EDM) equipment can be based on microwave or laser operation. Distance readings are generally direct.

Distance can also be measured tachymetrically. This method involves sighting through a small angle at a distant scale. The angle may be fixed and the length measured (*stadia method*), or the length may be fixed and the angle measured (*European method*).[4]

Stadia measurement consists of observing the apparent locations of the horizontal cross hairs on a distant stadia rod. The interval between the two rod readings is called the *stadia interval* or the *stadia reading.* This distance is directly related to the distance between the telescope and the rod. For rod readings $R_1$ and $R_2$ (both in feet), the separation distance is

$$x = K(R_2 - R_1) + C \qquad 17.7$$

$C$ is the sum of focal length and distance from plumb bob (center of instrument) to forward lens. It varies from 0.6 to 1.4 feet, but is typically set by the manufacturer at 1 foot. $C$ is zero for internal focusing telescopes. $K$ is the *stadia interval factor* which usually has a value of 100.

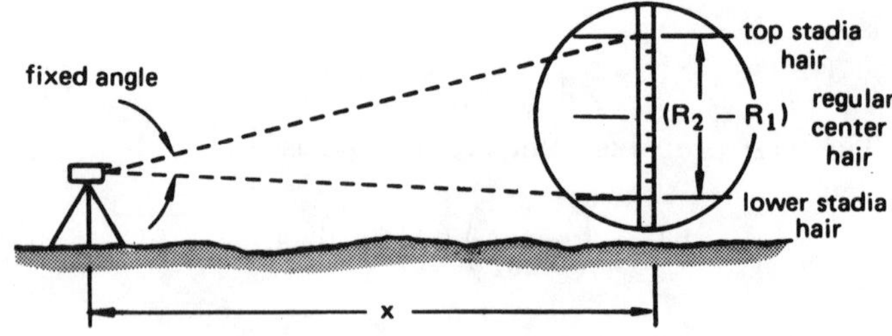


**Figure 17.1** Horizontal Stadia Measurement

If the sighting is inclined, as it is in figure 17.2, it will be necessary to find both the horizontal and vertical distances. These may be found from equations 17.8 and 17.9. $y$ is measured from the telescope to the sighting

[4] The word "stadia" means "temporary position."

rod center. The actual elevation difference will require knowledge of the instrument height.

$$x = K(R_2 - R_1)\cos^2\theta + C\cos\theta \qquad 17.8$$

$$y = \frac{1}{2}(K)(R_2 - R_1)(\sin 2\theta) + C\sin\theta \qquad 17.9$$

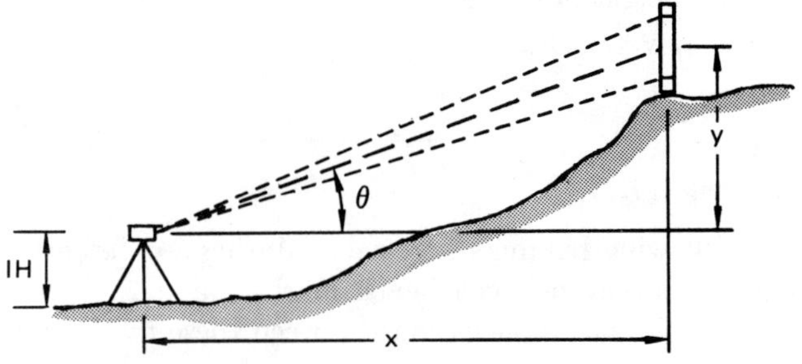


**Figure 17.2** Inclined Stadia Measurement

## 4 ELEVATION MEASUREMENT

*Leveling* is the act of using an engineer's level and rod to measure the vertical distance (the *elevation*) from an arbitrary level surface. Usually, the elevation is measured with respect to sea level.

### A. CURVATURE AND REFRACTION

If a level sighting is taken on an object with actual height $h_a$, the curvature of the earth will cause the object to appear taller by an amount $h_c$. In equation 17.10, $x$ is measured in feet along the curved surface of the earth.

$$h_c = 2.4\,\text{EE–8}(x^2) \qquad 17.10$$

Atmospheric refraction will make the object appear shorter by an amount $h_r$.

$$h_r = 3.0\,\text{EE–9}(x^2) \qquad 17.11$$

The corrected rod reading (actual height) is

$$h_a = R_{\text{observed}} + h_r - h_c$$
$$= R_{\text{observed}} - (2.1\ \text{EE} - 8)(x^2) \qquad 17.12$$

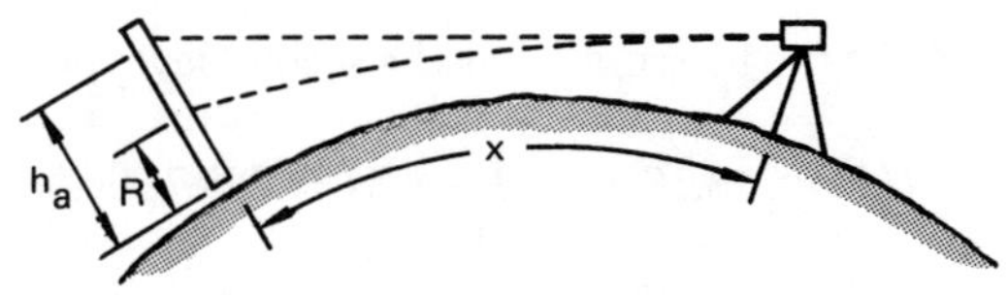


**Figure 17.3** Curvature and Refraction Effects

### B. DIRECT LEVELING

The most common method of determining the difference in elevations of two points is known as direct leveling. A level is set up at a point approximately midway between the two points whose difference in elevation is wanted. The vertical distances are observed by reading directly from the rod. Refer to figure 17.4 which uses the following nomenclature:

| | |
|---|---|
| $y_{A-B}$ | difference in elevations between points $A$ and $B$ |
| $y_{A-L}$ | difference in elevations between points $A$ and $L$ |
| $y_{L-B}$ | difference in elevations between points $L$ and $B$ |
| $R_A$ | rod reading at $A$ |
| $R_B$ | rod reading at $B$ |
| $h_{rc,A-L}$ | effects of curvature and refraction between points $A$ and $L$ |
| $h_{rc,B-L}$ | effects of curvature and refraction between points $B$ and $L$ |

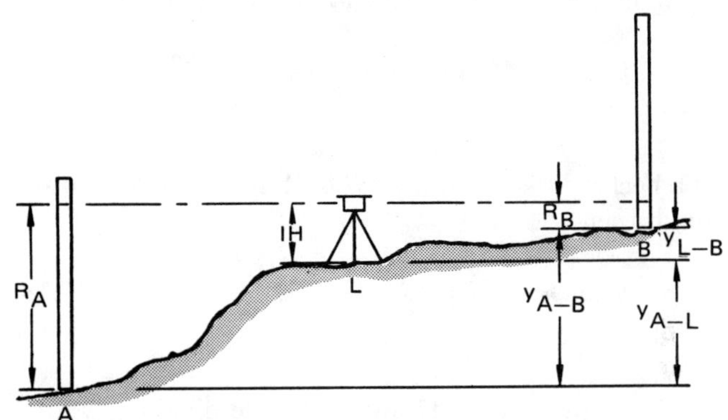


**Figure 17.4** Direct Leveling

By inspection,

$$y_{A-L} = R_A - h_{rc,A-L} - IH \qquad 17.13$$

$$y_{L-B} = R_B - h_{rc,L-B} - IH \qquad 17.14$$

The difference in elevations between points $A$ and $B$ is

$$y_{A-B} = y_{A-L} - y_{L-B}$$
$$= R_A - R_B + h_{rc,L-B} - h_{rc,A-L} \qquad 17.15$$

If the backsight and foresight distances are equal, then the effects of refraction and curvature cancel.

$$y_{A-B} = R_A - R_B \qquad 17.16$$

## C. INDIRECT LEVELING

*Indirect leveling* does not require a backsight (although one can be taken to eliminate the effects of curvature and refraction). A case of indirect leveling is illustrated in figure 17.5 where the difference in elevations between points $A$ and $B$ is sought. It is assumed that the distance $AC$ has been determined. Within the limits or ordinary practice, angle $ACB$ is 90°.

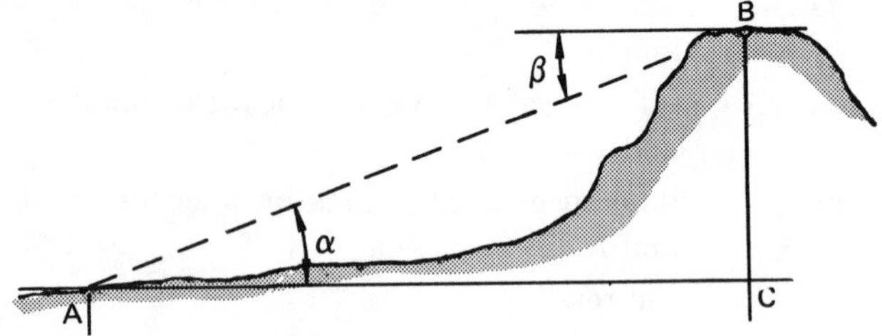


**Figure 17.5** Indirect Leveling

Including the effects of curvature and refraction,

$$y_{A-B} = AC(\tan\alpha) + 2.1\,\text{EE–8}(AC)^2 \qquad 17.17$$

If a backsight is taken from $B$ to $A$ and angle $\beta$ is measured, then

$$y_{A-B} = AC(\tan\beta) - 2.1\,\text{EE–8}(AC)^2 \qquad 17.18$$

Adding equations 17.17 and 17.18 and dividing by 2,

$$y_{A-B} = \frac{1}{2}(AC)(\tan\alpha + \tan\beta) \qquad 17.19$$

## D. DIFFERENTIAL LEVELING

*Differential leveling* is the consecutive application of direct leveling to the measurement of large differences in elevation. There is usually no attempt to exactly balance the foresights and backsights. Thus, there is no record made of the exact locations of the level positions. Furthermore, the path taken between points need not be along a straight line connecting them, as only the elevation differences are relevant.

If greater accuracy is desired without having to accurately balance the foresight and backsight distances, it is possible to eliminate most of the curvature and refraction error by balancing the sum of the foresights against the sum of the backsights.

The following abbreviations are used with differential leveling.

- $BM$ bench mark or monument
- $TP$ turning point
- $FS$ foresight (also known as a minus sight)
- $BS$ backsight (also known as a plus sight)
- $HI$ height of the instrument[5]
- $L$ level position

*Example 17.11*

The following readings were taken during a differential leveling survey between bench marks 1 and 2. What is the difference in elevations between these two bench marks?

| station | $BS$ | $HI$ | $FS$ | elevation |
|---|---|---|---|---|
| $BM1$ | 7.11 | | | 721.05 |
| $TP1$ | 8.83 | | 1.24 | |
| $TP2$ | 11.72 | | 1.11 | |
| $BM2$ | | | 10.21 | |

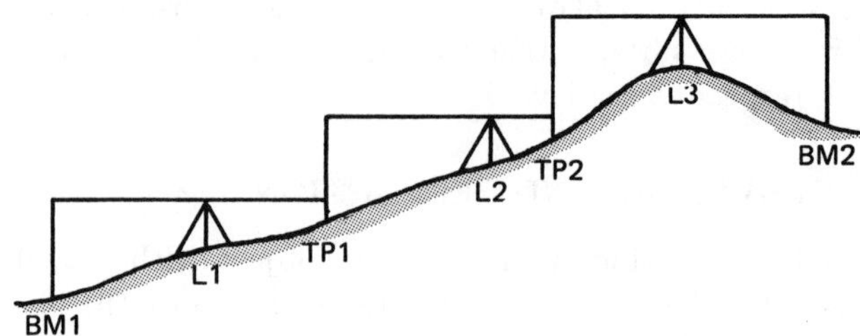


The first measurement is shown in larger scale. The height of the instrument is

$$HI = 721.05 + 7.11 = 728.16$$

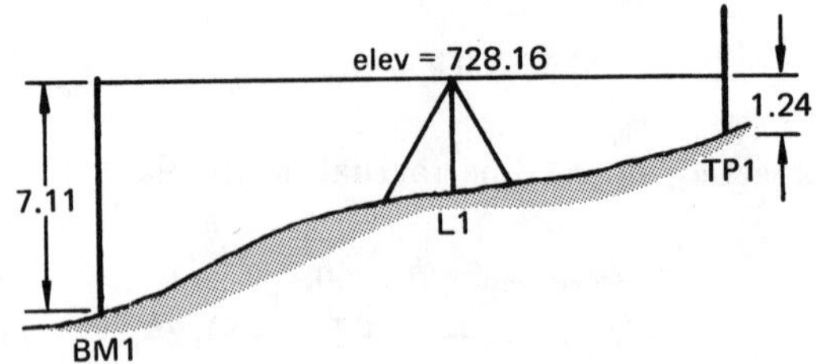


The height of the instrument at the second level position is

$$HI = 728.16 + 8.83 - 1.24 = 735.75$$

[5] Do not confuse the height of the instrument ($HI$) with the elevation at which the instrument is located. The $HI$ is the elevation of the line of sight of the telescope when the instrument is leveled.

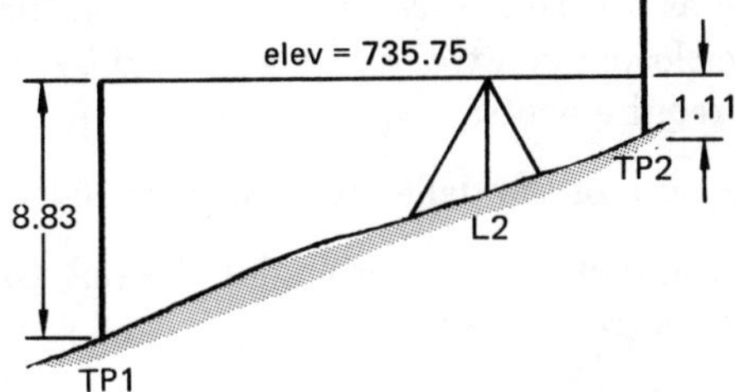


The height of the instrument at the third level position is

$$HI = 735.75 + 11.72 - 1.11 = 746.36$$

The elevation of $BM2$ is

$$746.36 - 10.21 = 736.15.$$

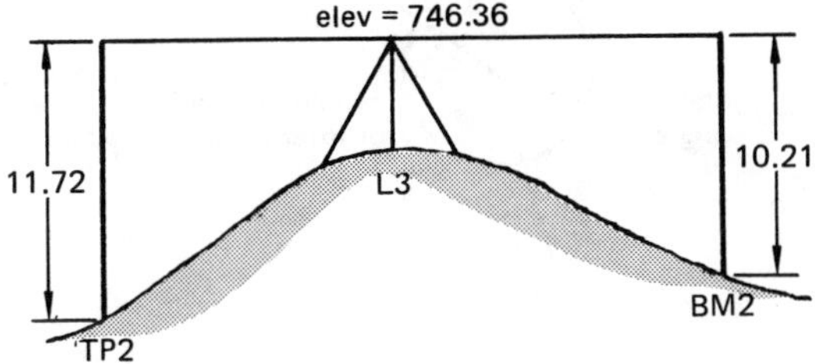


The difference in elevations is $736.15 - 721.05 = 15.1$.

The backsight sum is $711 + 8.83 + 11.72 = 27.66$.

The foresight sum is $1.24 + 1.11 + 10.21 = 12.56$.

The difference is $27.66 - 12.56 = 15.1$ (check)

## 5 EQUIPMENT AND METHODS USED TO MEASURE ANGLES

The *telescope* is actually part of the sighting equipment. A transit or theodolite is often referred to as a "telescope," even though the telescope is only a part of the total instrument.[6]

An *engineer's level* or *dumpy level* is a sighting device with a telescope rigidly attached to the level bar. A sensitive level vial is used to ensure level operation. The engineer's level can be rotated, but in its basic form, cannot be elevated.

In a *semi-precise level*, also known as a *prism level*, the level vial is visible from the eyepiece end. Other than that, the semi-precise and engineer's levels are similar.

A *precise level*, also known as a *geodetic level*, has even better control of horizontal angles. The bubble vial is magnified for greater accuracy.

[6] The *alidade* consists of the base and telescope part of the transit without including the tripod or leveling equipment.

An *engineer's transit* (or *surveyor's transit*) measures vertical angles as well as horizontal angles. It can, of course, be clamped vertically and used as a level. Angles are usually read by the naked eye by looking at a plate scale.

With a *theodolite*, angles are measured by looking into viewpieces. There may be up to four viewpieces, one each for the sighting-in telescope, the compass, the horizontal and vertical angles, and the optical plummet.

Transits and levels are used with rods. *Leveling rods* are used to measure the vertical distance between the line of sight and the point being observed. The *standard rod* is typically made of wood or fiberglass and is extendable.[7] *Precise rods*, made of wood-mounted invar, are typically constructed in one piece and are spring loaded in tension to avoid sagging.

## 6 ANGLE MEASUREMENT

The direction of any line can be specified by an angle between it and some reference line. The reference line is known as a *meridian.* If the meridian is arbitrarily chosen, it is called an *assumed meridian.* If the meridian is a true north-to-south line passing through the true north pole, it is called a *true meridian.* If the meridian is parallel to the earth's magnetic field, it is known as a *magnetic meridian.*[8]

A true meridian differs from a magnetic meridian by a *declination* (*magnetic declination* or *variation*). If the north end of a compass points to the west of the true meridian, the declination is said to be a *west declination* or *minus declination.* Otherwise, it is an *east declination* or *plus declination.*

Plus declinations are added to the magnetic compass azimuth to obtain the true azimuth. Minus declinations are subtracted from the magnetic compass azimuth.

The variation of a line from its meridian may be given in several ways:

- Azimuths: The azimuth of a line is the clockwise angle measured from the south branch of the meridian to the line. This is known as an *azimuth from the south.* (*Azimuths from the north* are also sometimes used.)
- Deflection Angles: The angle between the prolongation of a line and another line is a deflection angle. Such measurements must be labeled as 'right' (clockwise) or 'left' (counterclockwise).

[7] The standard rod may also be known as a *Philadelphia rod.*

[8] A rectangular grid can be drawn over a map with any arbitrary orientation. In such a case, the vertical lines are known as *grid meridians.* All angles are referenced to those grid meridians.

- Angles to the right: The angle to the right is a clockwise angle measured from the preceding to the following line.
- Azimuths from the back line: Same as angles to the right.
- Bearings: The bearing of a line is referenced to the quadrant in which the line falls and the angle that the line makes with the meridian in that quadrant. It is necessary to specify the two cardinal directions that define the quadrant in which the line is found. The north and south directions are always specified first.

These methods of angle measurement are illustrated in figure 17.6.

azimuth from the south: 157°

azimuth from the north: 337°

deflection angle: 23°L

angle to the right: 157°

bearing: N 23° W

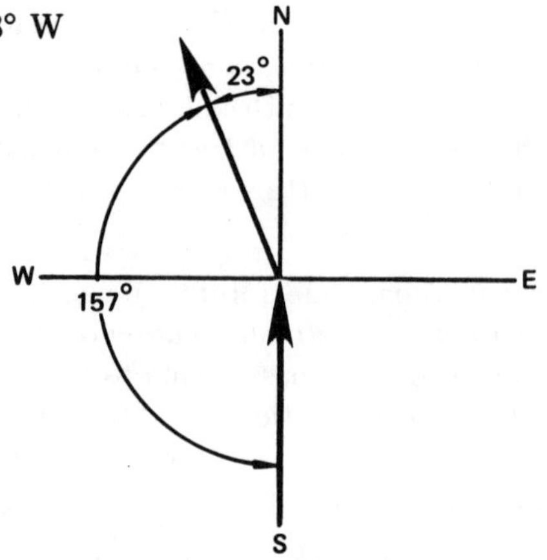

**Figure 17.6** Angle Measurements

*Example 17.12*

A band of hikers on the Pacific Crest Trail (magnetic declination of +17°) used a magnetic compass to site in on a distant geographic feature. What was the true azimuth if the observed angle was 42°?

A plus declination is added to the observed azimuth. Therefore, the true azimuth was $42° + 17° = 59°$.

## 7 CLOSED TRAVERSES

A *traverse* is a series of straight lines whose lengths and deflection angles (or other angle measurements) are known. A traverse that comes back to its starting point is known as a *closed traverse*. The polygon that results from the closing of a traverse is governed by the following two requirements:

- The sum of the deflection angles is 360°.
- The sum of the interior angles of a polygon with $n$ sides is $(n-2)180°$.

There are three ways to measure and record angles in traverses. Figure 17.7 illustrates the *station angle*, *deflection angle*, and *explement angle* (also known as the *interior angle*).

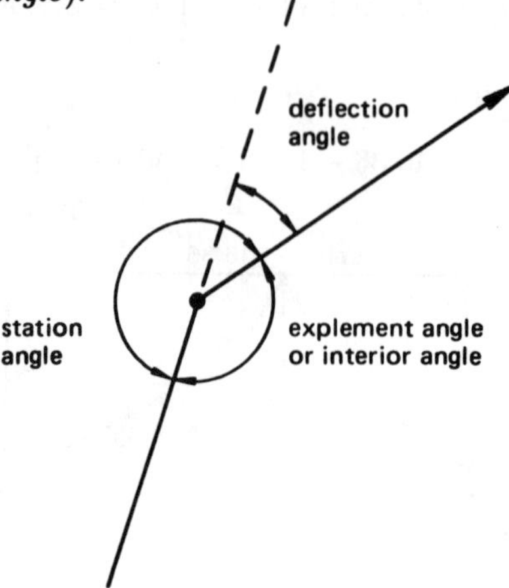


**Figure 17.7** Angles in Traverses

### A. ADJUSTING CLOSED TRAVERSE ANGLES

Due to errors, variations in magnetic declination, and local magnetic attractions, it is likely that the sum of angles making up the interior angles will not exactly equal $(n-2)180°$. The following procedure can be used to distribute the angle error of closure among the angles.

*step 1*: Calculate the interior angle of each station from the observed bearings.

*step 2*: Subtract $(n-2)180°$ from the sum of the interior angles.

*step 3*: Unless additional information in the form of numbers of observations or probable errors is available, assume the angle error of closure can be divided among all angles. Divide the error by the number of angles.

*step 4*: Find a line whose bearing is assumed correct. That is, find a line whose bearing appears unaffected by errors, variations in magnetic declination, and local attractions. Such a line may be chosen as one for which the forward and back bearings are the same. If there is no such line, take the line whose difference in forward and back bearings is the smallest.

*step 5*: Start with the assumed correct line and add (or subtract) the prorated error to each interior angle.

*step 6*: Correct all bearings except the one for the assumed correct line.

*Example 17.13*

Adjust the angles on the four-sided closed traverse whose magnetic foresights and backsights are listed.

| line | bearing |
|---|---|
| *AB* | N 25° E |
| *BA* | S 25° W |
| *BC* | S 84° E |
| *CB* | N 84.1° W |
| *CD* | S 13.1° E |
| *DC* | N 12.9° W |
| *DA* | S 83.7° W |
| *AD* | N 84° E |

*step 1*: The interior angles are calculated from the bearings.

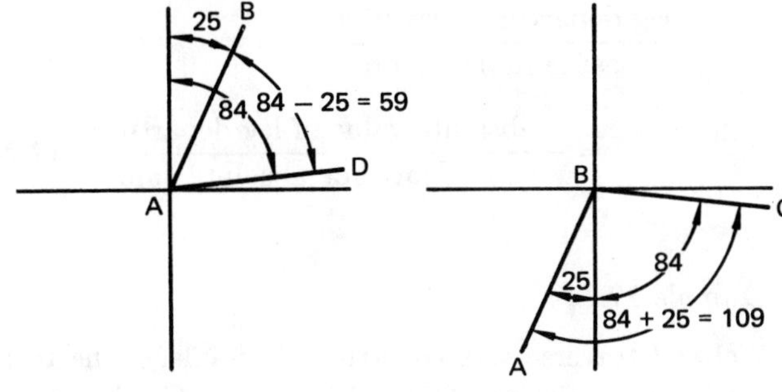


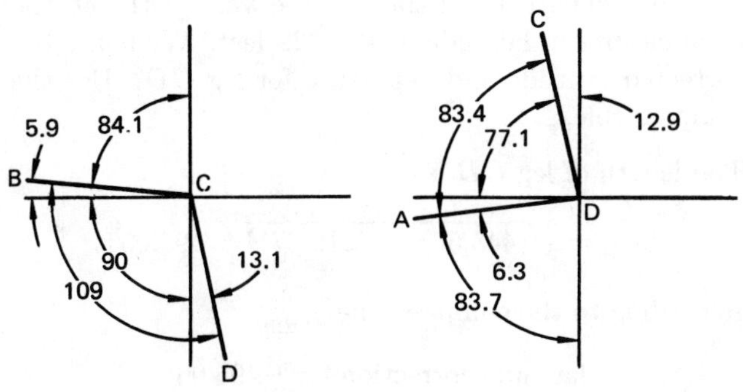


*step 2*: The sum of the angles is: $59 + 109 + 109 + 83.4 = 360.4°$. For a four-sided traverse, the sum of interior angles should be 360°, so a correction of (−0.4°) must be divided up evenly among the four angles. The original and corrected traverses are shown.

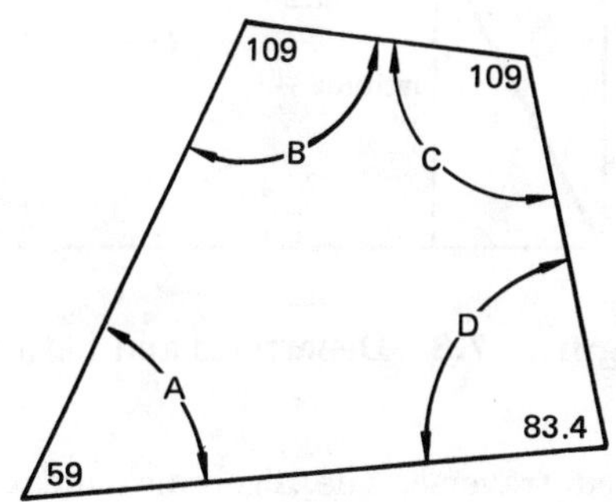


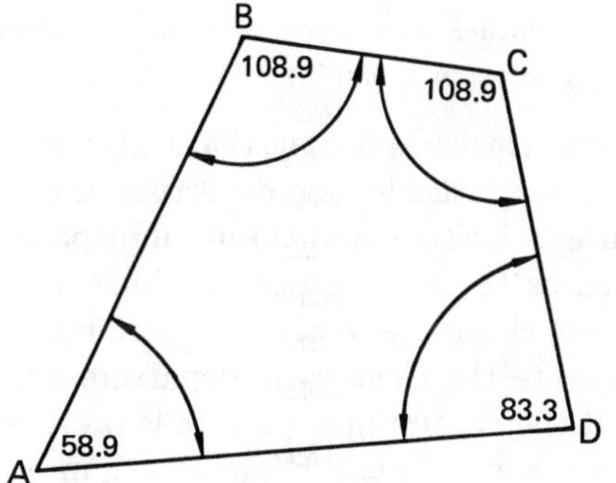


Since the backsight and foresight bearings of line *AB* are the same, it is assumed that the 25° bearings are the most accurate. The remaining bearings are then adjusted to use the corrected angles.

| line | bearing |
|---|---|
| *AB* | N 25° E |
| *BA* | S 25° W |
| *BC* | S 83.9° E |
| *CB* | N 83.9° W |
| *CD* | S 12.8° E |
| *DC* | N 12.8° W |
| *DA* | S 83.9° W |
| *AD* | N 83.9° E |

## B. LATITUDES AND DEPARTURES

The *latitude* of a line is the distance that the line extends in a north or south direction. A line that runs towards the north has a positive latitude; a line that runs towards the south has a negative latitude.

The *departure* of a line is the distance that the line extends in an east or west direction. A line that runs towards the east has a positive departure; a line that runs towards the west has a negative departure.

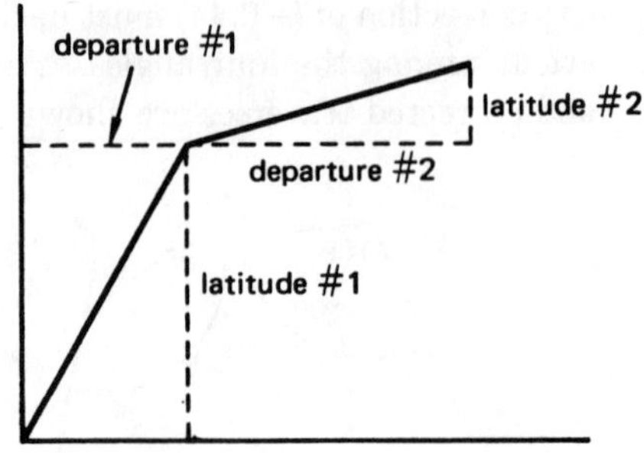


**Figure 17.8** Departures and Latitudes

In a closed traverse, the algebraic sum of latitudes should be zero. The algebraic sum of departures should also be zero. These sums, which are distances in feet with actual values near zero, are called *closure in latitude* and *closure in departure* respectively.

The *traverse closure* is the line that will exactly close the traverse. Since latitudes and departures are orthogonal, the closure in latitude and closure in departure can be considered as the rectangular coordinates to calculate the traverse closure length. The coordinates will have signs opposite the closures in departure and latitude. That is, if the closure in departure is positive, point $A$ will lie to the left of point $A'$, as shown in figure 17.9.

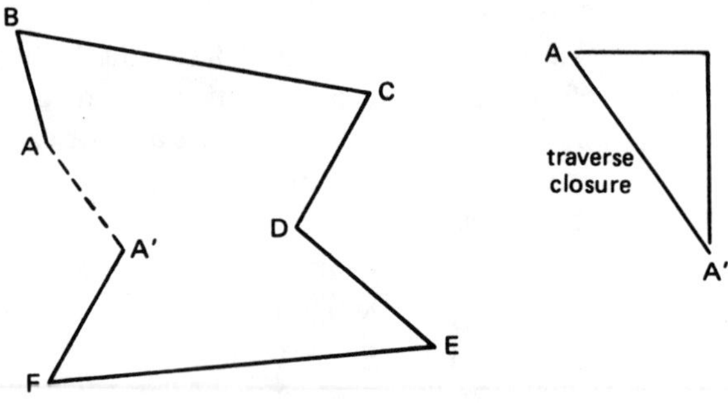


**Figure 17.9** Traverse Closure

The length of a traverse closure is

$$L = \sqrt{\begin{pmatrix}\text{closure in}\\ \text{departure}\end{pmatrix}^2 + \begin{pmatrix}\text{closure in}\\ \text{latitude}\end{pmatrix}^2} \qquad 17.20$$

## C. ADJUSTING CLOSED TRAVERSE LATITUDES AND DEPARTURES

To balance a closed traverse, the traverse closure must be divided among the various legs of the traverse. This correction requires that the latitudes and departures be known for each leg of the traverse.

The most common method used to balance the traverse legs is known as the *compass rule*.[9] This rule states that the correction to a leg of the traverse is to the total traverse correction as the leg length is to the total traverse length, with the signs reversed.

$$\frac{\text{leg departure correction}}{\text{closure in departure}} = -\frac{\text{leg length}}{\text{total traverse length}} \qquad 17.21$$

$$\frac{\text{leg latitude correction}}{\text{closure in latitude}} = -\frac{\text{leg length}}{\text{total traverse length}} \qquad 17.22$$

It is appropriate to use the compass rule when the angles and distances in the traverse are considered equally precise. If the angles are precise, but the distances are less precise (e.g., as in taping through rugged terrain), the *transit rule* is a preferred alternative. The transit rule distributes the closure error in proportion to the absolute values of the latitudes and departures.

$$\frac{\text{leg latitude correction}}{\text{closure in latitude}} = -\frac{\text{absolute value of leg latitude}}{\text{sum of latitude absolute values}} \qquad 17.23$$

$$\frac{\text{leg departure correction}}{\text{closure in departure}} = -\frac{\text{absolute value of leg departure}}{\text{sum of latitude absolute values}} \qquad 17.24$$

*Example 17.14*

A closed traverse was constructed of 7 legs, the total of whose lengths was 2705.13 feet. Leg $CD$ has a departure of 443.56 and a latitude of 219.87. The total closure in departure for the traverse was +0.41 feet; the total closure in latitude was −0.29 feet. What are the corrected latitude and departure for leg $CD$? Use the compass rule.

The length of leg $CD$ is

$$L_{CD} = \sqrt{(443.56)^2 + (219.87)^2} = 495.06 \text{ ft}$$

According to the compass rule,

$$\frac{\text{latitude correction}}{-0.29} = -\frac{495.06}{2705.13}$$

[9] Use of one method or the other to allocate error to legs is arbitrary. If you know that one leg in particular was poorly measured due to difficult terrain, you may decide to give all of the error to that leg.

The latitude correction is 0.05 feet.

$$\frac{\text{departure correction}}{+0.41} = -\frac{495.06}{2705.13}$$

The departure correction is $-0.08$ ft.

The corrected latitude is $219.87 + 0.05 = 219.92$ ft.

The corrected departure is $443.56 - 0.08 = 443.48$ ft.

## D. RECONSTRUCTING MISSING SIDES AND ANGLES

If one or more sides or angles of a traverse are missing or cannot be determined by measurement, they will have to be reconstructed. The procedures for three common cases are listed below. These procedures draw primarily upon the subjects of geometry and trigonometry.

**One leg missing**: One leg is missing in figure 17.10. However, the line $EA$ can be reconstructed easily from its components $E$–$E'$ and $E'$–$A$. These components are equal to the sum of the departures and sum of the latitude respectively, with the signs changed. The angle can be determined from the ratio of the sides, and the length $E$–$A$ can be found from equation 17.20.

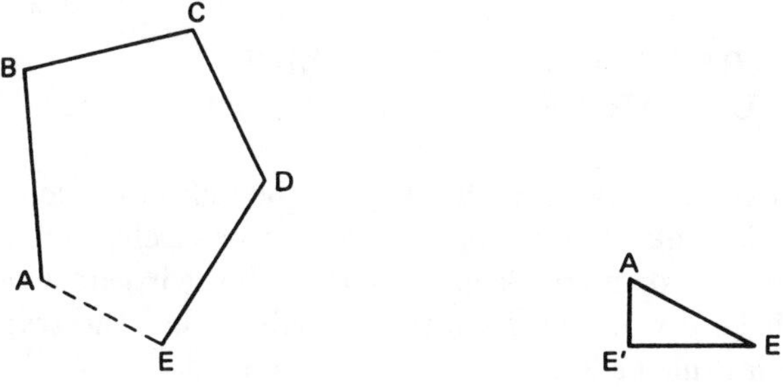


**Figure 17.10** One Missing Leg

**Adjacent legs missing**: Figure 17.11 shows a traverse that has two adjacent legs missing. The traverse can be closed as long as some length/angle information is available. The technique is to close the traverse by using the method presented in case 1 above. This will give the line $D$–$A$. Then, the triangle $E$–$A$–$D$ can be completed by using whatever information is available.

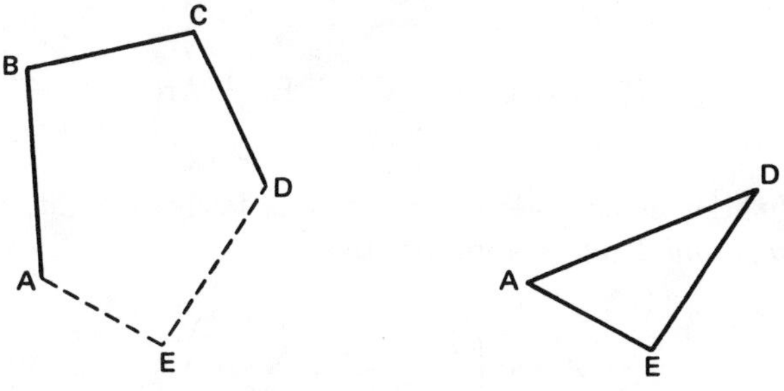


**Figure 17.11** Two Adjacent Legs Missing

**Two non-adjacent legs missing**: Figure 17.12 shows a traverse with two non-adjacent legs missing. Since the latitudes and departures of two parallel lines are equal, this can be solved by closing up the traverse and shifting one missing leg to be adjacent to the other missing leg. This reduces the problem to the previous case.

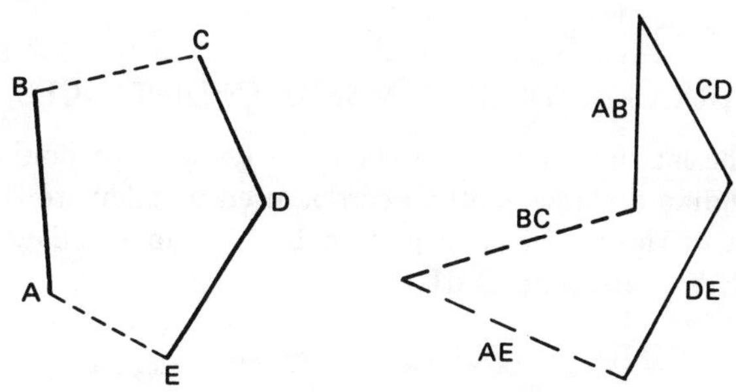


**Figure 17.12** Two Non-Adjacent Legs Missing

## E. AREA OF A TRAVERSE

An area of a traverse can always be found by dividing the traverse into a number of geometric shapes (rectangles, triangles, etc.) and summing the area of each subdivision. If the coordinates of the traverse leg end points are known, the *method of coordinates* can be used. The coordinates can be $x$–$y$ coordinates referenced to some arbitrary set of axes, or they can be departure-latitude coordinates.

$$A = \left| \frac{1}{2} \left( \sum_{i=1}^{n} y_i (x_{i-1} - x_{i+1}) \right) \right| \qquad 17.25$$

In equation 17.25, $x_0$ is replaced with $x_n$. Similarly, $x_{n+1}$ is replaced with $x_1$.

The area calculation is simplified if the coordinates are written in the following form:

$$\frac{x_1}{y_1} \times \frac{x_2}{y_2} \times \frac{x_3}{y_3} \times \frac{x_4}{y_4} \times \frac{x_1}{y_1} \quad \text{etc.}$$

Then, the area is

$$A = \frac{1}{2} \left| \left( \sum \text{full line products} - \sum \text{dotted line products} \right) \right| \qquad 17.26$$

*Example 17.15*

Calculate the area of a triangle with coordinates of its corners given: (3,1), (5,1), and (5,7).

$$\frac{3}{1} \times \frac{5}{1} \times \frac{5}{7} \times \frac{3}{1}$$

$$A = \frac{1}{2}\Big[(3)(1) + (5)(7) + (5)(1) - (1)(5)$$
$$-(1)(5) - (7)(3)\Big]$$
$$= \frac{1}{2}[43 - 31] = 6$$

## F. AREAS BY DOUBLE MERIDIAN DISTANCES

If the latitudes and departures are known, the double meridian distance method can be used to calculate the area of the traverse. Equation 17.27 defines a *double meridian distance* (DMD).

$$\text{DMD}_{\text{leg i}} = \text{DMD}_{\text{leg i-1}} + \text{departure}_{\text{leg i-1}} + \text{departure}_{\text{leg i}} \qquad 17.27$$

Special rules are required to handle the first and last courses. The DMD of the first course is defined as the departure of that course. The DMD of the last course is the negative of its own departure.

The traverse area is calculated from equation 17.28.

$$A = \frac{1}{2}\left|\sum_i \text{latitude}_{\text{leg i}} \times \text{DMD}_{\text{leg i}}\right| \qquad 17.28$$

The next example illustrates the tabular approach often preferred with the double meridian distance method.

*Example 17.16*

The latitudes and departures of a six-leg traverse have been calculated. Use the double meridian distance method to calculate the traverse area. (The latitudes and departures are given in the table.) All values are in feet.

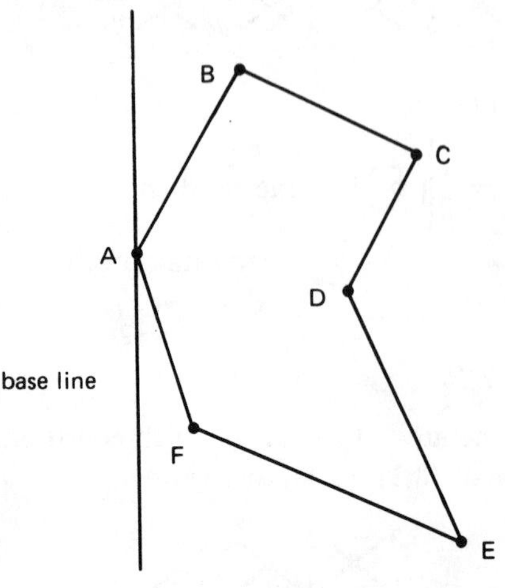


By the special rule, the DMD of the first leg, $AB$, is the departure of leg $AB$. The DMD of leg $BC$ is

$$\begin{aligned}\text{DMD}_{BC} &= \text{DMD}_{AB} + \text{departure}_{AB} + \text{departure}_{BC}\\ &= 200 + 200 + 200\\ &= 600\end{aligned}$$

The other DMD's are calculated similarly.

| leg | latitude | departure | DMD | lat ×DMD |
|---|---|---|---|---|
| $AB$ | 200 | 200 | 200 | 40,000 |
| $BC$ | −100 | 200 | 600 | −60,000 |
| $CD$ | −200 | −100 | 700 | −140,000 |
| $DE$ | −300 | 200 | 800 | −240,000 |
| $EF$ | 200 | −400 | 600 | 120,000 |
| $FA$ | 200 | −100 | 100 | 20,000 |
| | | | TOTAL | −260,000 |

The area is

$$A = \frac{1}{2}(260{,}000) = 130{,}000\ \text{ft}^2$$

## 8 AREAS WITH IRREGULAR BOUNDARIES

Areas of sections with irregular boundaries, such as creek banks, cannot be determined precisely, and approximation methods must be used. If the irregular side can be divided into a series of cells, either the trapezoidal rule or Simpson's rule can be used.

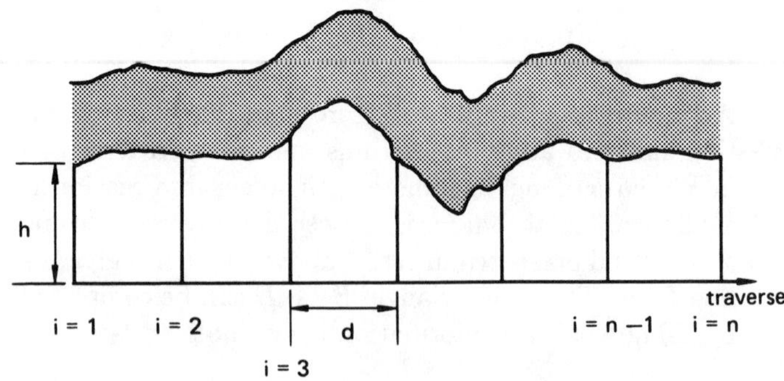


**Figure 17.13** Creek Bank Areas

If the irregular side of each side is fairly straight, the *trapezoidal rule* is appropriate.

$$A = d\left(\frac{h_1 + h_n}{2} + \sum_{i=2}^{n-1} h_i\right) \qquad 17.29$$

If the irregular side of each side is curved (parabolic), *Simpson's rule* should be used.[10]

$$A = \frac{d}{3}\left[h_1 + h_n + 2\sum_{\substack{i\,\text{odd}\\ i=3}} h_i + 4\sum_{\substack{i\,\text{even}\\ i=2}} h_i\right] \qquad 17.30$$

## 9 CURVES

Roads, rail lines, and water courses are usually designed to be straight lines. Where a direction change is needed, a curve is used. The straight lines connected by a curve are known as *tangents* or *tangent lines.* A curve on level ground changing the direction of two tangents is known as a *horizontal curve.* Horizontal curves are usually arcs of circles.

Curves must also be used to connect roads and rail lines that change grade (slope). Such curves are called *vertical curves.* A curve that connects an upgrade tangent to a downgrade tangent is known as a *crest curve*, whereas a curve that connects a downgrade tangent to an upgrade tangent is known as a *sag curve.* Vertical curves are usually parabolic in shape.

**Figure 17.14** Sag and Crest Vertical Curves

## 10 HORIZONTAL CURVES

### A. INTRODUCTION TO SYMBOLS

The elements of circular curves and their standard abbreviations are given here.

- $R$ radius of the curve
- $V$ vertex of the tangent intersection point
- $PI$ point of intersection
- $I$ interior angle
- $PC$ point of curvature—the place where the first tangent ends and the curve begins
- $PT$ point of tangency—the place where the curve ends and the second tangent begins
- $POC$ any point on the curve
- $LC$ length of the arc—the length of the curve from $PC$ to $PT$
- $T$ tangent distance from $V$ to $PC$ or from $V$ to $PT$
- $C$ the long chord—the straight distance from $PC$ to $PT$
- $E$ the external distance—the distance from $V$ to the midpoint of the curve
- $M$ the middle ordinate—the distance from the curve midpoint to the midpoint of the long chord
- $D$ the degree of the curve

The following alternative designations are also sometimes used.

- $PVI$ point of vertical intersection (same as $PI$)
- $TC$ a change from a tangent to a curve (same as $PC$)
- $CT$ a change from a curve to a tangent (same as $PT$)
- $BC$ the beginning of a curve (same as $PC$)
- $EC$ the end of a curve (same as $PT$)
- $\Delta$ the intersection angle (same as $I$)

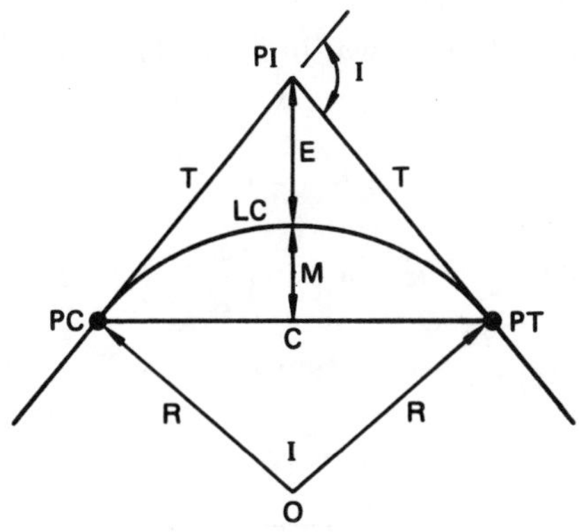


**Figure 17.15** Elements of a Circular Curve

Equations 17.31 through 17.35 can be used to solve problems involving circular curves.

$$T = R\tan\left(\tfrac{1}{2}I\right) \qquad 17.31$$

$$E = R\left(\tan\tfrac{1}{2}I\right)\left(\tan\tfrac{1}{4}I\right) = R\left(\sec\left(\tfrac{1}{2}I\right) - 1\right) \qquad 17.32$$

$$M = R\left(1 - \cos\tfrac{1}{2}I\right) = \tfrac{1}{2}C\left(\tan\tfrac{1}{4}I\right) \qquad 17.33$$

$$C = 2R\left(\sin\tfrac{1}{2}I\right) = 2T\left(\cos\tfrac{1}{2}I\right) \qquad 17.34$$

$$LC = R(I\text{ in radians}) = R(I\text{ in degrees})\left(\frac{2\pi}{360}\right) = 100\left(\frac{I}{D}\right) \qquad 17.35$$

[10] Also known as *Simpson's 1/3 rule.*

The curvature of city streets, property boundaries, and some highways is usually specified by the radius, $R$. The curvature may also be specified (in degrees) by the *degree of curve*, $D$.

$$D = \frac{360 \times 100}{2\pi R} = \frac{5729.6}{R} \qquad 17.36$$

In most highway work, the length of the curve is understood to be the actual arc, and the degree of the curve is the angle subtended by an arc of 100 feet.[11] Therefore, the degree of curve can be expressed in degrees per station.

Stationing is continuous every 100 feet along a highway and around the curve. However, when the initial route is laid out between $PI$'s, the curve is undefined. The route distance is measured from $PI$ to $PI$. Therefore, each $PI$ will have two stations associated with it. The *forward station* is equal to the $PC$ station plus the tangent length. The *back station* is equal to the $PT$ minus the tangent length.

## B. TANGENT OFFSETS FOR CIRCULAR CURVES

The tangent offset shown in figure 17.16 can be calculated from equation 17.38.

$$y = R(1 - \cos\alpha) \qquad 17.38$$

$$\alpha = \arcsin\left(\frac{x}{R}\right) \qquad 17.39$$

$$x = R\sin\alpha \qquad 17.40$$

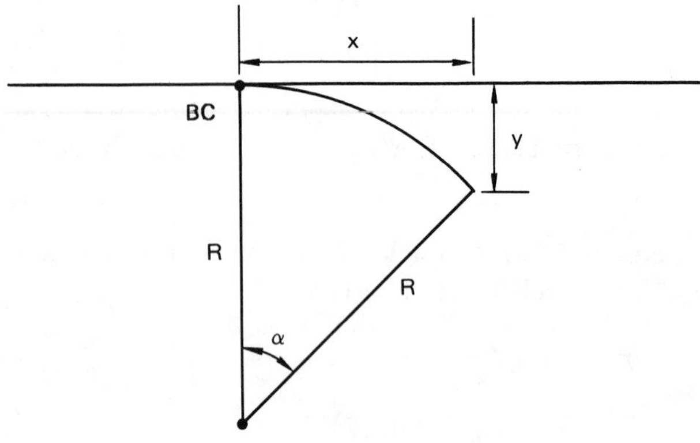


**Figure 17.16** Tangent Offset

[11] When the degree of curve is related to an arc of 100 feet, it is said to be calculated on an *arc basis*. In railroad surveys, the *chord basis* is used, and the degree of curve is the angle subtended by a chord of 100 feet. In that case, the degree of curve and radius are related by

$$\sin\left(\frac{D}{2}\right) = \frac{50}{R} \qquad 17.37$$

Where the radius is large (4° curves or smaller), the difference between the arc and chord methods is insignificant.

## C. DEFLECTION ANGLES

The surveyor must stake out the curve so that the road crew knows where to put the road. Stakes should be put at the $PC$, $PT$, and at all full stations. If the curve is sharp, stakes may also be required at +25, +50, and +75 stations. The *deflection angle method* is the most common method used for staking out the curve. In this method, the curve distance is usually assumed to start from 00 + 00 at the $PC$.

The *deflection angle* is defined as the angle between the tangent and a chord. This is illustrated in figure 17.17. The deflection angles are calculated using the following theorems.

- The deflection angle between a tangent and a chord is half the subtended arc.
- The angle between two chords is half the subtended arc.

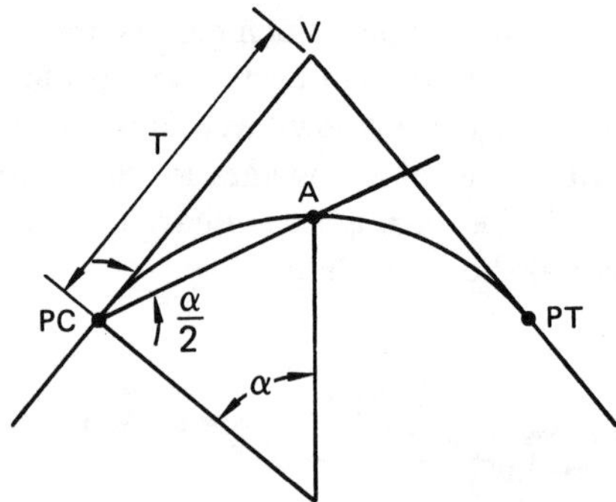


**Figure 17.17** Circular Curve Deflection Angle

In figure 17.17, angle $V$–$PC$–$A$ is a deflection angle between a tangent and a chord. The angle is

$$V\text{–}PC\text{–}A = \tfrac{1}{2}\alpha \qquad 17.41$$

Angle $\alpha$ can be found from the following relationships.

$$\frac{\alpha}{360} = \frac{\text{arc length}\,(PC\text{–}A)}{2\pi R} \qquad 17.42$$

$$\frac{\alpha}{I} = \frac{\text{length}\,(PC\text{–}A)}{LC} \qquad 17.43$$

The chord length $PC$–$A$ is given by equation 17.44.

$$PC\text{–}A = 2R\sin\left(\tfrac{1}{2}\alpha\right) \qquad 17.44$$

The entire curve can be laid out from the $PC$ by sighting the deflection angle $V$–$PC$–$A$ and taping distance $PC$–$A$. The $PC$ and $PT$ can be found by solving for $T$ and starting at $V$.

*Example 17.17*

A circular curve is to be constructed with a 225 foot radius and an interior angle of 55°. Determine where the stakes should be placed if the separation between stakes along the arc is 50 feet. Specify the first and last deflection angles and chord lengths.

The length of the curve is given by equation 17.35.

$$LC = (225)(55)\left(\frac{2\pi}{360}\right) = 215.98\text{ ft}$$

The last stake will be 215.98 − 200 = 15.98 feet from the next to the last stake. The central angle for an arc of 50 feet is given by equation 17.42.

$$\alpha_{\text{degrees}} = \left(\frac{360}{2\pi}\right)\left(\frac{50}{225}\right) = 12.732°$$

12.732° goes into 55° four times with a remainder of 4.072°. From equation 17.44, the required chord lengths are

$$(2)(225)\sin\left(\frac{12.732}{2}\right) = 49.90\text{ ft}$$

$$(2)(225)\sin\left(\frac{4.072}{2}\right) = 15.98\text{ ft}$$

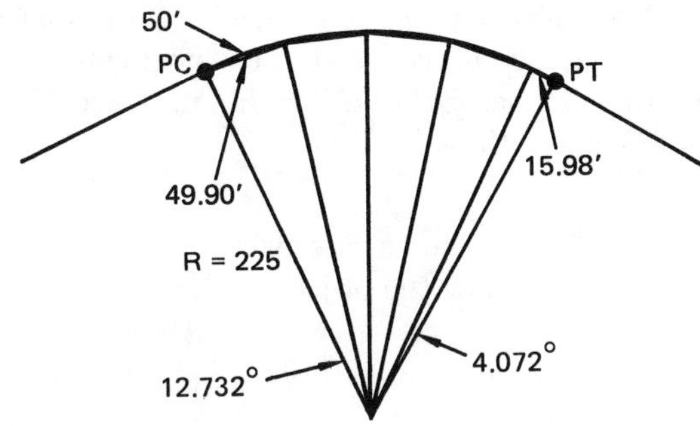


*Example 17.18*

An interior angle of 8.4° is specified for a horizontal curve. The *PI* station is 64+27.46. Use a 2° curve and locate the *PC* and *PT* stations.

From equation 17.36,

$$R = \left(\frac{360}{2}\right)\left(\frac{100}{2\pi}\right) = 2864.79\text{ ft}$$

From equations 17.31 and 17.35

$$T = (2864.79)\tan\left(\frac{8.4}{2}\right) = 210.38$$

$$LC = (2864.79)(8.4)\left(\frac{2\pi}{360}\right) = 420.00$$

Then, the *PC* and *PT* points are located.

$$PC = (64{+}27.46) - (2{+}10.38) = 62{+}17.08$$

$$PT = (62{+}17.08) + (4{+}20.00) = 66{+}37.08$$

## 11 VERTICAL CURVES

Vertical curves are used to change the grade of a highway. *Equal-tangent parabolic curves* are usually used for this purpose. A vertical sag curve connecting two grades is shown in figure 17.18. Since the grades are very small, the actual arc length of the curve is approximately equal to the chord length *BVC–EVC.*

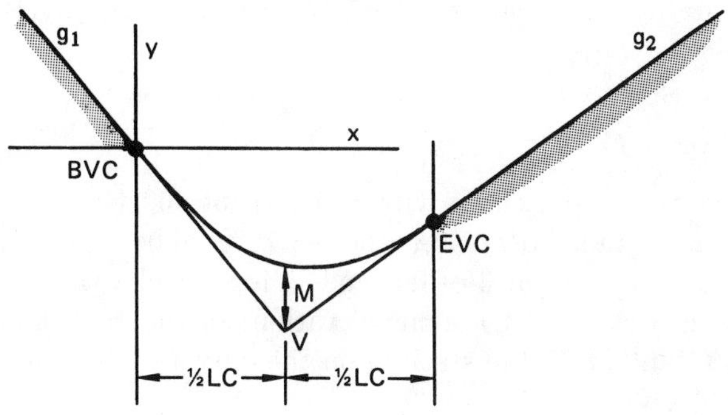


**Figure 17.18** A Vertical Curve

The following standard and optional abbreviations are used.

- $LC$ the horizontal length of the curve, in stations
- $g_1$ the grade from which the stationing starts, in percent
- $g_2$ the grade towards which the stationing heads, in percent
- $V$ the vertex—the intersection of the two tangents
- $PVI$ same as $V$
- $BVC$ beginning of the vertical curve
- $EVC$ end of the vertical curve
- $PVC$ same as $BVC$
- $PTT$ same as $EVC$
- $M$ the middle ordinate

A vertical parabolic curve is completely specified by the two grades and the curve length. Alternately, the *rate of grade change* per station can be used in place of the curve length.

$$r = \frac{g_2 - g_1}{LC} \qquad 17.45$$

Equation 17.46 defines an equal-tangent parabolic curve. $x$ is measured in stations beyond $BVC$. $y$ is measured in feet, with the same reference point used to measure all elevations.

$$\text{elev}_x = \left(\frac{r}{2}\right)x^2 + g_1 x + \text{elevation}_{BVC} \qquad 17.46$$

The maximum or minimum elevation will occur at the *turning point*. The turning point is not located directly above or below $V$, but is found at

$$x = \frac{-g_1}{r} \text{ (in stations)} \qquad 17.47$$

The *middle ordinate* is

$$M = \frac{(g_1 - g_2)(LC)}{8} \qquad 17.48$$

*Example 17.19*

A vertical crest curve with a length of 400 feet is to connect grades of +1.0% and −1.75%. The vertex is located at station 35 + 00, and it has an elevation of 549.20 feet. What are the elevations of the $BVC$ and $EVC$, and at all full stations on the curve?

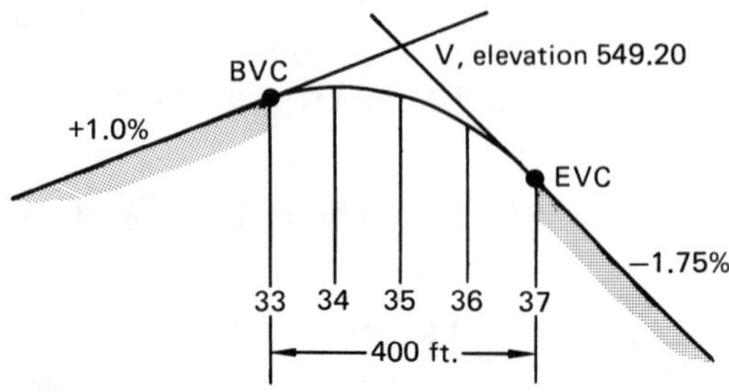


The elevation at $BVC$ is 549.20 − 1(2) = 547.20.

The elevation at $EVC$ is 549.20 − 1.75(2) = 545.70.

$$r = \frac{-1.75 - 1}{4} = -0.6875 \text{ percent}$$

$$\frac{1}{2}r = -0.3438$$

The equation of the curve is

$$y = -0.3438x^2 + x + 547.20$$

At station 34, $x = (34 - 33) = 1$. So,

$$y_{34} = -0.3438(1)^2 + 1 + 547.20 = 547.86$$

Similarly,

$$y_{35} = -0.3438(2)^2 + 2 + 547.20 = 547.82$$
$$y_{36} = -0.3438(3)^2 + 3 + 547.20 = 547.11$$

## 12 VERTICAL CURVES WITH OBSTRUCTIONS

If a curve is to have some minimum clearance from an obstruction, as in figure 17.19, the length of the curve, $BVC$, and $EVC$ will generally not be known in advance. The problem of finding the curve length can be solved by using the following procedure. The procedure can also be used when the curve is placed over a feature and a minimum cover depth is required.

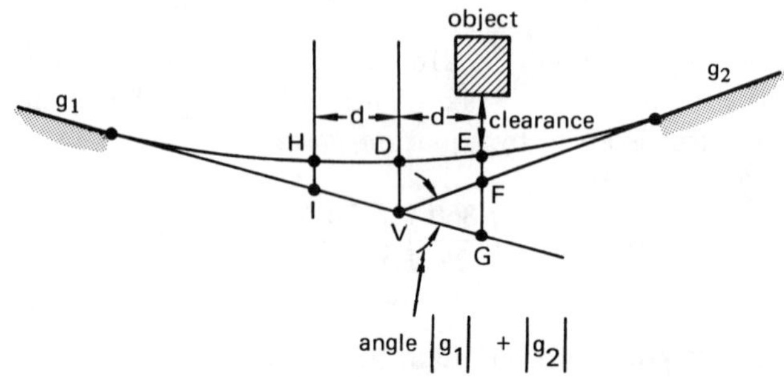


**Figure 17.19** A Curve with an Obstruction

This procedure is based on the fact that a vertical distance from a tangent line to a point on a curve (i.e., the *tangent offset*) is proportional to the square of the horizontal distance along the tangent. (See equation 17.46.) This fact, combined with the curve symmetry of an equal-tangent parabola ($EF = HI$ in figure 17.19), permits direct solutions.

*step 1*: Calculate the curve elevation of point $E$ directly below the object.

$$\text{elevation}_E = \text{elevation}_{\text{object}} - \text{clearance} \qquad 17.49$$

*step 2*: Calculate distance $EG$. ($g_1$ is negative as shown in figure 17.19. $d$ is in stations.)

$$EG = \text{elevation}_E - \text{elevation}_V - (d)(g_1) \qquad 17.50$$

*step 3*: Calculate the distance $EF$.

$$EF = \text{elevation}_E - \text{elevation}_V - (d)(g_2) \qquad 17.51$$

*step 4*: Solve simultaneously for $LC$. Both $d$ and $LC$ are in units of stations.

$$\frac{EG}{\left(\frac{LC}{2} + d\right)^2} = \frac{EF}{\left(\frac{LC}{2} - d\right)^2} \qquad 17.52$$

## 13 SPIRAL CURVES

*Spiral curves* are used to produce a gradual transition from tangents to curves. A spiral curve is a curve of gradually changing radius and gradually increasing degree of curvature. Figure 17.20 illustrates a spiral curve, showing the *tangent to spiral* ($TS$) point, *length of spiral* ($LS$), and *spiral to curve* ($SC$) point.

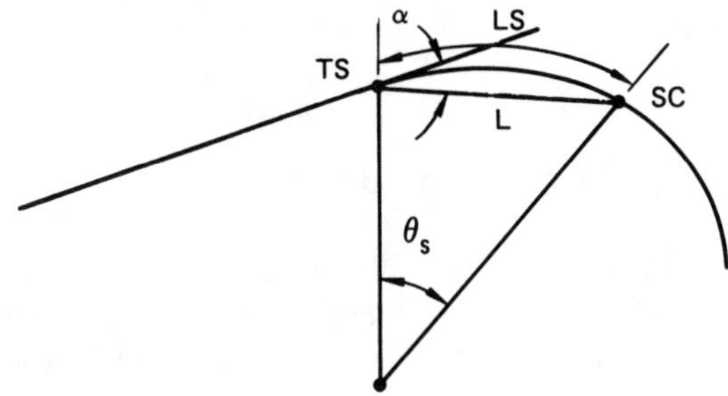


**Figure 17.20** A Simple Spiral Curve

The *degree of spiral* changes continuously, so the average degree of spiral is used to calculate the length. ($\theta_s$ is in degrees in equation 17.53.)

$$LS = \frac{100 \times \theta_s}{\frac{D}{2}} \qquad 17.53$$

Deflection angles to be used when setting out points on the spiral curve can be calculated from equation 17.54.

$$\alpha = \frac{\theta_s}{3}\left(\frac{L}{LS}\right)^2 \qquad 17.54$$

Experiments suggest that lengths of spiral curves should be based on the speed of traffic entering the curve and the radius of the curve being approached.

$$LS = \frac{(1.6)(\text{mph})^3}{R \text{ in feet}} \qquad 17.55$$

## 14 PHOTOGRAMMETRY

Photogrammetry uses photography to obtain distance measurements. Both vertical and oblique photographs can be used, as shown in figure 17.21.

The *scale* of a vertical photograph is the ratio of dimensions on the photograph to the dimensions on the ground.

$$\text{scale} = \frac{\text{flight altitude (feet)}}{\text{focal length (inches)}} \qquad 17.56$$

The scale is constantly changing since the elevation above ground level depends on the surface terrain. Also, the distance from the camera to the ground directly below it will be less than the distance from the camera to points on the outer fringes of the photograph. Therefore, an average distance from camera to ground level is used as the flight altitude.

The number of photographs and the number of flight paths required depend on the film size and lap percentages, as illustrated in example 17.20.

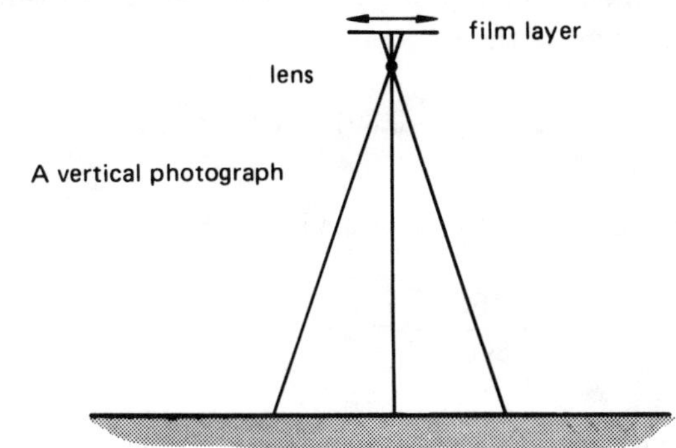


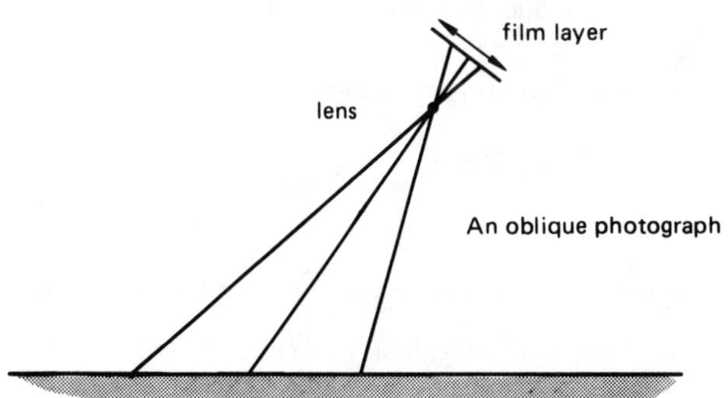


**Figure 17.21** Vertical and Oblique Photographs

*Example 17.20*

Aerial mapping is being used with a scale of 1″:1000′. The film holder uses 8″ × 8″ film. The focal length of the camera is 6″. The area to be mapped is 6 miles on each side (square). A side lap of 30% is desired, as well as an end lap of 60%. The plane travels at 150 mph.

(a) How many square feet does each photograph cover?

(b) At what altitude should the plane fly while taking photographs?

(c) How far apart will the flight paths be?

(d) How many flight paths are required?

(e) How many photographs are required per flight path?

(f) How many photographs will be taken altogether?

(g) How frequently should the photographs be taken?

(a) The photograph area is $8'' \times 8'' = 64\ \text{in}^2$. The area covered by each photograph is $64 \times 1000^2 = 64{,}000{,}000\ \text{ft}^2$.

(b) The flight altitude is calculated from equation 17.56.

$$\text{altitude} = 6'' \times 1000' = 6000\ \text{ft}$$

(c) Each photograph covers $64{,}000{,}000\ \text{ft}^2$, or an area 8000 ft on each side. With a 30% overlap, the distance between flight paths is

$$8000(1 - 0.30) = 5600\ \text{ft}$$

(d) The number of flight paths is

$$\frac{(6)(5280)}{5600} = 5.7 \quad (\text{say } 6)$$

(e) The distance between photographs along the flight path is

$$8000(1 - 0.60) = 3200\ \text{ft}$$

The number of photographs per flight path is

$$\frac{(6)(5280)}{3200} = 9.9 \quad (\text{say } 10)$$

(f) The total number of photographs will be $6 \times 10 = 60$.

(g) At 150 mph, the frequency of camera shots will be

$$\frac{3200}{\frac{150 \times 5280}{3600}} = 14.5\ \text{seconds/photograph}$$

## 15 TRIANGULATION

Triangulation is a method of surveying in which the positions of survey points are determined by measuring the angles of triangles. In triangulation, the survey lines form a network of triangles. Each survey point (monument) is at a corner of one or more triangles. The three angles of each triangle are measured. Lengths of triangle sides are trigonometrically calculated. The positions of the points are established from the measured angles and the computed sides.

Triangulation is used primarily for geodetic surveys, such as those performed by the National Geodetic Survey. Most first- and second-order control points in the national control network have been established by triangulation procedures. The use of triangulation for transportation surveys is minimal. Generally, its use is limited to strengthening traverses for control surveys.

## 16 TRILATERATION

Trilateration is similar to triangulation in that the survey lines form triangles. In trilateration, however, the lengths of the triangles' sides are measured, and the angles are computed. Orientation of the survey is established by selected sides whose directions are known or measured. The positions of trilaterated points are determined from the measured distances and the computed angles.

## 17 SOLAR OBSERVATIONS

The direction of a survey line can be determined by making horizontal and vertical angular measurements to the sun at a known time and from a point of known latitude. Such measurements are called ***solar observations.***

Generally, solar observations are required only in areas where horizontal control points are sparse or control data are unavailable. In such areas, solar observations provide azimuth control (starting, closing, and check azimuths). In some cases, the entire orientation of a survey might be based on solar observations. In other surveys, they might simply provide orientation checks. For example, long traverses can require intermediate azimuth checks where record points do not exist. Another common use is orientation for retracement surveys when searching for corners in rugged terrain or in heavy vegetation.

Azimuths determined by solar observations are adequate for third-order surveys. Greater accuracy is impractical because exact pointings can not be made on the relatively fast-moving sun. If more accurate azimuths must be determined by astronomical observations, Polaris sightings should be used.

Azimuths should be reliable within 12 seconds. However, many factors can affect the accuracy. Two of the more important factors are the skill of the observer and the care used in making the observations. (Care in leveling the instrument and in pointing on the sun is especially important.) Other factors are the hour of the day and the time of the year the observations are made, the precision of the theodolite, the accuracy of the time measurements, and the latitude of the point of observation and the accuracy to which it is determined.

## 18 CELESTIAL BASIS

An azimuth determined by solar observations is based on the solution of a spherical triangle, the ***astronomical*** *triangle.* Figure 17.22 shows the points on the celestial sphere that form this triangle. These three points are

- **P**-north pole of the celestial sphere
- **Z**-observer's zenith point. This is the point on the celestial sphere directly above the point of observation.
- **S**-sun

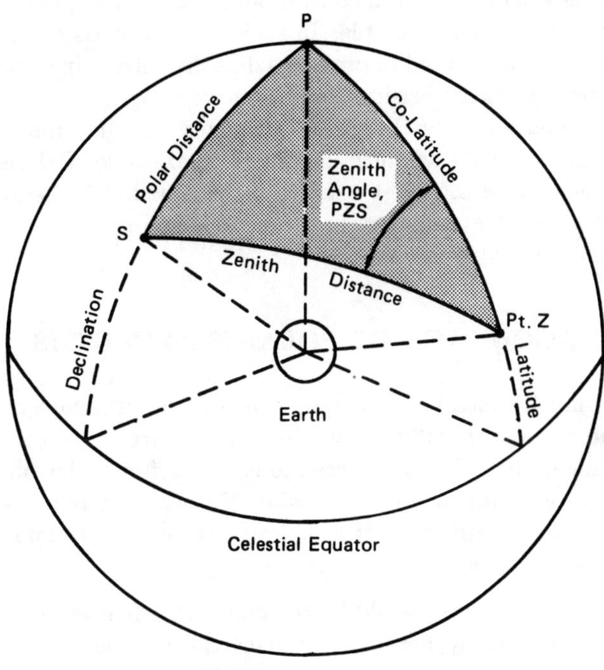


**Figure 17.22** Astronomical Triangle for Solar Observation

In figure 17.22, triangle $PZS$ is the *astronomical triangle.* Angle $PZS$ is known as the *zenith angle.*

## 19 SIDES (ANGLES) OF THE ASTRONOMICAL TRIANGLE

Side $S$-$P$ is known as the *polar distance.* It can be found as 90° minus the sun's declination. *Declination* is the angular distance the sun is above or below the celestial equator. The value is typically determined from an *ephemeris.*

Side $P$-$Z$ is known as the *co-latitude.* It is found as 90° minus the latitude of the point from which the observation is made.

Side $Z$-$S$ is known as the *zenith distance.* It is 90° minus the true altitude of the sun. The *true altitude* is the measured altitude corrected for refraction and parallax. Tables of corrections are also available in ephemerides. Figure 17.23 illustrates how the refraction correction ($r$) is added to the field-measured zenith angle, and the parallax correction ($p$) is subtracted from it.

$$\text{true altitude} = \text{measured altitude} - \text{refraction} + \text{parallax} \qquad 17.57$$

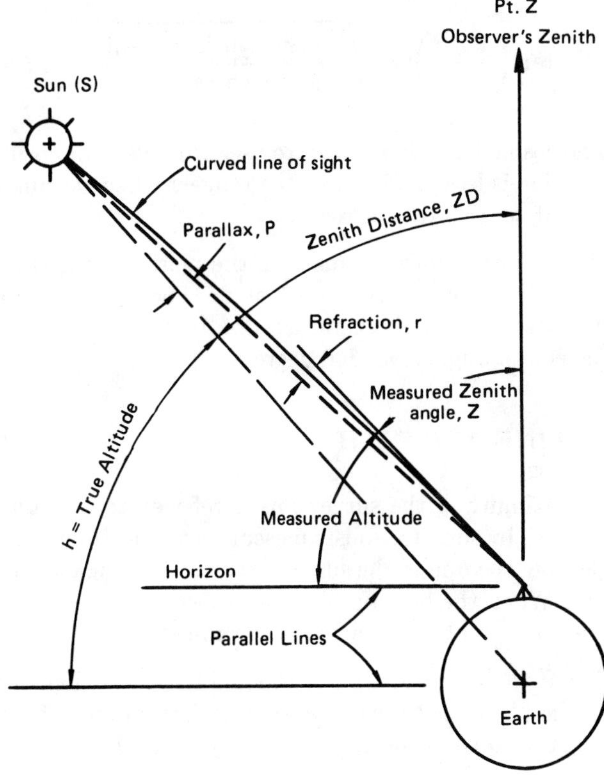


**Figure 17.23** Zenith Distance and True Altitude

## 20 ZENITH ANGLE

With three sides of the spherical triangle known, the angle between the north pole and the sun at the observer's zenith can be calculated. This angle is known as the *zenith angle*, $PZS$.[12] It establishes the azimuth of the sun (line $Z$–$S$), and therefore is also known as the *azimuth angle.*

Two equations can be used to calculate the zenith angle $PZS$ from its trigonometric functions. In equations 17.58 and 17.59, the following nomenclature is used:

$PZS$ = zenith angle.

$d$ = sun's declination.

$h$ = sun's true altitude.

[12] Note that the term "zenith angle" also refers to the angle $Z$ in a vertical plane that is measured with a theodolite that has a vertical circle oriented at zero degrees when the telescope is pointed on the zenith. See figure 17.23.

$l$ = latitude of the point of observation.

$p$ = polar distance = $90° - d$.

$s = 1/2(l + h + p)$

$$\cos PZS = \frac{(\sin d) - [(\sin h)(\sin l)]}{(\cos h)(\cos l)} \quad 17.58$$

If $\cos PZS$ is negative, $PZS$ is greater than 90 degrees.

$$\tan\left(\frac{PZS}{2}\right) = \sqrt{\frac{[\sin(s-l)][\sin(s-h)]}{[\cos s][\cos(s-p)]}} \quad 17.59$$

The cosine formula is easier to use. However, the cosine function is less precise than the tangent when the angle is near zero or 180 degrees.

If the observations are made in the morning, the sun's azimuth is equal to the zenith angle. For afternoon observations, the sun's azimuth is obtained by subtracting the zenith angle from 360 degrees.

## 21 LINE AZIMUTH

The azimuth of the survey line is referred to the sun's azimuth by simultaneously measuring the horizontal angle from the line to the sun and the vertical angle to the sun. When the horizontal angle is measured clockwise, the azimuth of the survey is determined as follows:

- If the horizontal angle is equal to or less than the sun's azimuth, subtract the horizontal angle from the sun's azimuth.
- If the horizontal angle is greater than the sun's azimuth, subtract the horizontal angle from 360 degrees and add the remainder to the sun's azimuth.

*Example 17.21*

A horizontal angle was measured during a survey as 336°. The zenith angle $PZS$ was measured as 100°. What was the azimuth of the survey line $A$–$B$?

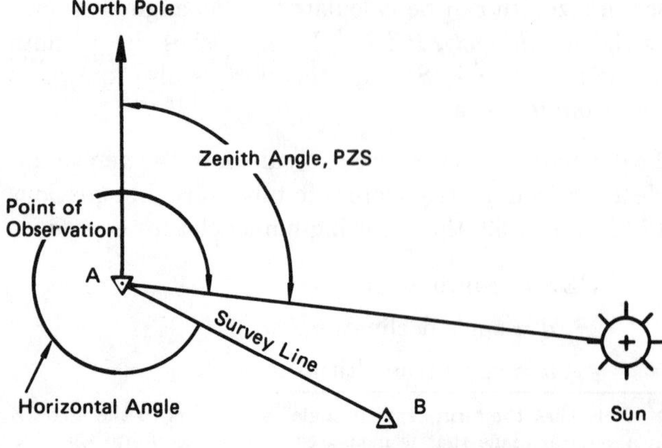


azimuth $A$–$B$ = 100 + 360 − 336 = 124°

## 22 POLARIS OBSERVATIONS

Azimuths of survey lines can be determined by making horizontal angular measurements to *Polaris*, the *North Star*, at known, exact times and from points of known latitude. Such measurements also can be used when the latitude is unknown. In this case, horizontal and zenith observations are made simultaneously.

Generally, Polaris observations are required only in areas where horizontal control points are sparse or control data are unavailable. In such areas, Polaris observations can be used to provide azimuth control. In some cases, the entire orientation of a survey might be based on Polaris observations. In other surveys, they might simply provide azimuth checks. For example, Polaris observations might be used to provide azimuth checks on a long traverse that has numerous courses between existing control points.

## 23 ACCURACY OF POLARIS AZIMUTHS

Azimuth control can be determined by observations on the sun. Generally, solar observations are more convenient than Polaris observations. Therefore, solar observations are usually made when the accuracy requirements are third-order or less. Polaris observations must be used for second-order accuracy.

Polaris azimuths should be reliable within five seconds. However, many factors can affect the accuracy. The more important factors (in approximate order of importance) are:

- The skill of the observer and the care used in making the observations. Care in leveling the instrument and in noting when the star is bisected by the cross hair is especially important.
- The accuracy of the time measurement.
- The hour of the observations.
- The latitude, and its accuracy, of the occupied point. Or, if the star's altitude is measured, the accuracy of the zenith angle measurements.

## 24 CELESTIAL BASIS

Polaris is a relatively bright star in the northern sky. Its apparent position is always very close to the vertical projection of the North Pole. The actual angular distance from the pole (the *polar distance*) is approximately 51 minutes and is almost a constant. It varies less than one minute during the year.

As viewed from the earth, Polaris appears to revolve counterclockwise around the pole. One "revolution" is completed each sidereal day (time measured relative to the stars). Unlike the sun and some other stars, Polaris does not "rise" and "set." It is always visible. Because of its location and relatively slow apparent motion, Polaris is the star commonly used for determining azimuths and latitudes by stellar observations.

An azimuth determined by Polaris observations is based on the solution of a spherical triangle which is called the *astronomical triangle.* Figure 17.24 shows the points on the celestial sphere that form this triangle. These three points are:

- **P**-north pole of the celestial sphere
- **Z**-observer's zenith point. This is the point on the celestial sphere directly above the point of observation.
- **S**-pole star Polaris

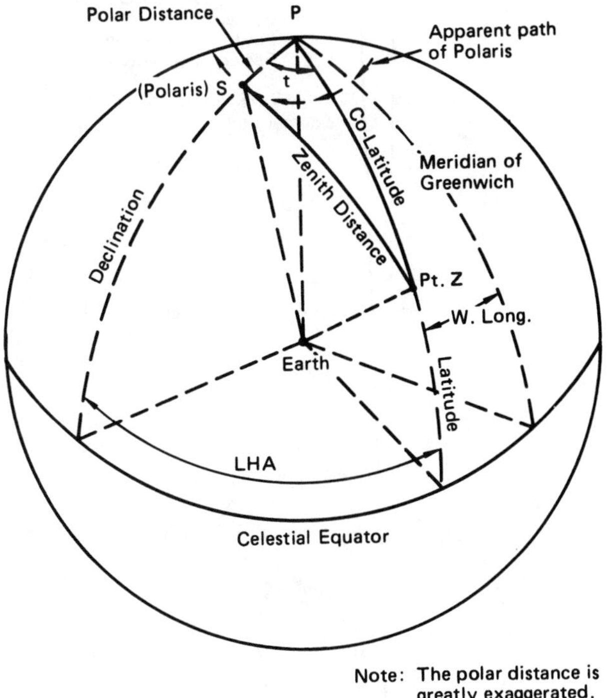


**Figure 17.24** Astronomical Triangle for Polaris Observations

## 25 SIDES (ANGLES) OF THE ASTRONOMICAL TRIANGLE

In figure 17.24, side $S$-$P$ is known as the *polar distance.* It is calculated as 90° minus Polaris's declination. This *declination* is the angular distance above or below the celestial equator that Polaris is. It is found from an ephemeris.

Side $P$-$Z$ is the *co-latitude.* It is calculated as 90° minus the latitude of the point of observation.

Side $S$-$Z$ is the *zenith distance.* It is 90° minus the true altitude of Polaris. The *true altitude* is the measured altitude corrected for refraction. The *measured altitude* is 90° minus the zenith angle of Polaris. Tables for refraction are contained in ephemerides.

In addition to the three sides, the angle at the pole between the zenith point and Polaris can be determined. This angle ($SPZ$) is called the *meridian angle.* The meridian angle is commonly designated by the letter "$t$" as shown in figure 17.24. It is computed by comparing the actual time of the observation with the time at which upper culmination of Polaris occurs at the observer's meridian.[13] (*Upper culmination* is the upper crossing of a meridian by a celestial body.) The meridian angle is computed as $t = LHA$ or $t = LHA - 360$, whichever is smaller in magnitude.

$LHA$ is the *local hour angle,* equal to the time of observation minus the time of upper culmination ($LCT$) plus the correction for sidereal time. (*Sidereal time* is based on the rotation of the earth relative to the stars.)

$LCT$ is the *local civil time,* which is the mean time based on the observer's meridian. Mean time is based on the average rotation of the earth relative to the sun.

## 26 ZENITH ANGLE FOR POLARIS OBSERVATIONS

With three elements of the spherical triangle known, the angle between the North Pole and Polaris at the observer's zenith can be calculated. This angle is known as the *zenith angle, PZS.*[14] It establishes the azimuth of Polaris (line $Z$-$S$), and is, therefore, sometimes called the *azimuth angle.*

The zenith angle $PZS$ can be calculated from equations 17.60 and 17.61. In both equations, the algebraic sign of $t$ is ignored. (It should be remembered that the sign of the cosine function is negative when the angle is between 90 and 270 degrees.) The following nomenclature is used in these two equations.

[13] The times at which upper culmination of Polaris occur are tabulated in ephemerides. Some ephemerides use a different procedure to compute the meridian angle. Thus, they will tabulate some value other than the time of upper culmination.

[14] The term "zenith angle" can also be used for the angle in a vertical plane that is measured with a theodolite that has a vertical circle oriented at zero degrees when the telescope is pointed on the zenith.

$PZS$ = zenith angle
$d$ = Polaris's declination
$l$ = latitude of the point of observation
$t$ = meridian angle
$h$ = Polaris's true altitude

$$\tan(PZS) = \frac{\sin t}{(\cos l)(\tan d) - (\sin l)(\cos t)} \qquad 17.60$$

$$\sin(PZS) = \frac{(\sin t)(\cos d)}{\cos h} \qquad 17.61$$

Because of the size and limited variance in the angles involved, the sine equation can be very closely approximated by equation 17.62. In equation 17.62, $p$ is the polar distance, $90° - d$.

$$\angle PZS = \frac{\sin t}{\cos h}(90 - d) = \frac{\sin t}{\cos h}p \qquad 17.62$$

When $t$ is negative (0 to 12 hours prior to upper culmination), the azimuth of Polaris is equal to the zenith angle. When $t$ is positive (0 to 12 hours after upper culmination), the azimuth is obtained by subtracting the zenith angle from 360 degrees.

## 27 LINE AZIMUTH FROM POLARIS OBSERVATIONS

The azimuth of the survey line is referred to the azimuth of Polaris by the horizontal angle measurements. When the horizontal angle is measured clockwise, the azimuth of the survey line is determined as follows:

1. If the horizontal angle ($H$) is equal to or less than the azimuth of Polaris ($Az$), subtract the horizontal angle from the azimuth of Polaris.

2. If the horizontal angle is greater than the azimuth of Polaris, then subtract the horizontal angle from 360 degrees and add the result to the azimuth of Polaris.

*Example 17.22*

**Referring to figure 17.25, suppose the zenith angle was 0°44′0″ and the horizontal angle was 304°30′0″. What would be the azimuth of the survey line?**

$$360° - H = 55°30'0''$$

$$\text{azimuth of line } A\text{–}B = 55°30'0'' + 0°44'0'' = 56°14'0''$$

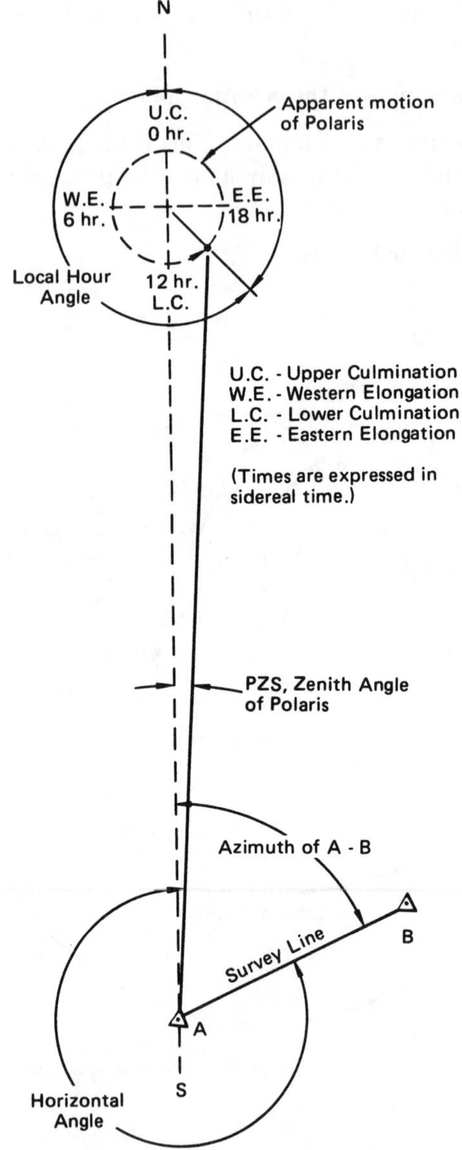


**Figure 17.25** Polaris observation

# Appendix A: Oblique Triangle Equations

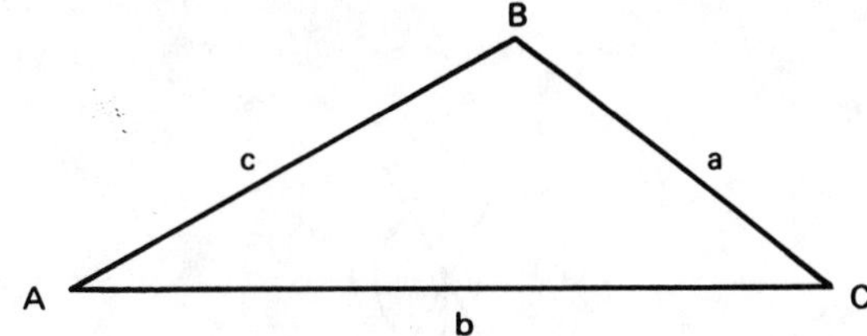


| given | equation |
|---|---|
| $A, B, a$ | $C = 180° - (A + B)$ |
| | $b = \dfrac{a}{\sin A} \times \sin B$ |
| | $c = \dfrac{a}{\sin A} \times \sin(A + B) = \dfrac{a}{\sin A} \times \sin C$ |
| | $\text{area} = \frac{1}{2} ab \sin C = \dfrac{a^2 \sin B \sin C}{2 \sin A}$ |
| $A, a, b$ | $\sin B = \dfrac{\sin A}{a} \times b$ |
| | $C = 180° - (A + B)$ |
| | $c = \dfrac{a}{\sin A} \times \sin C$ |
| | $\text{area} = \frac{1}{2} ab \sin C$ |
| $C, a, b$ | $c = \sqrt{a^2 + b^2 - 2ab \cos C}$ |
| | $\frac{1}{2}(A + B) = 90° - \frac{1}{2}C$ |
| | $\tan \frac{1}{2}(A - B) = \dfrac{a - b}{a + b} \times \tan \frac{1}{2}(A + B)$ |
| | $A = \frac{1}{2}(A + B) + \frac{1}{2}(A - B)$ |
| | $B = \frac{1}{2}(A + B) - \frac{1}{2}(A - B)$ |
| | $c = (a + b) \times \dfrac{\cos \frac{1}{2}(A + B)}{\cos \frac{1}{2}(A - B)} = (a - b) \times \dfrac{\sin \frac{1}{2}(A + B)}{\sin \frac{1}{2}(A - B)}$ |
| | $\text{area} = \frac{1}{2} ab \sin C$ |
| $a, b, c$ | Let $s = \dfrac{a + b + c}{2}$ |
| | $\sin \frac{1}{2}A = \sqrt{\dfrac{(s - b)(s - c)}{bc}}$ |
| | $\cos \frac{1}{2}A = \sqrt{\dfrac{s(s - a)}{bc}}$ |
| | $\tan \frac{1}{2}A = \sqrt{\dfrac{(s - b)(s - c)}{s(s - a)}}$ |
| | $\sin A = 2\sqrt{\dfrac{s(s - a)(s - b)(s - c)}{bc}}$ |
| | $\cos A = \dfrac{b^2 + c^2 - a^2}{2bc}$ |
| | $\text{area} = \sqrt{s(s - a)(s - b)(s - c)}$ |

# Appendix B: Circle and Circular Curve Geometry

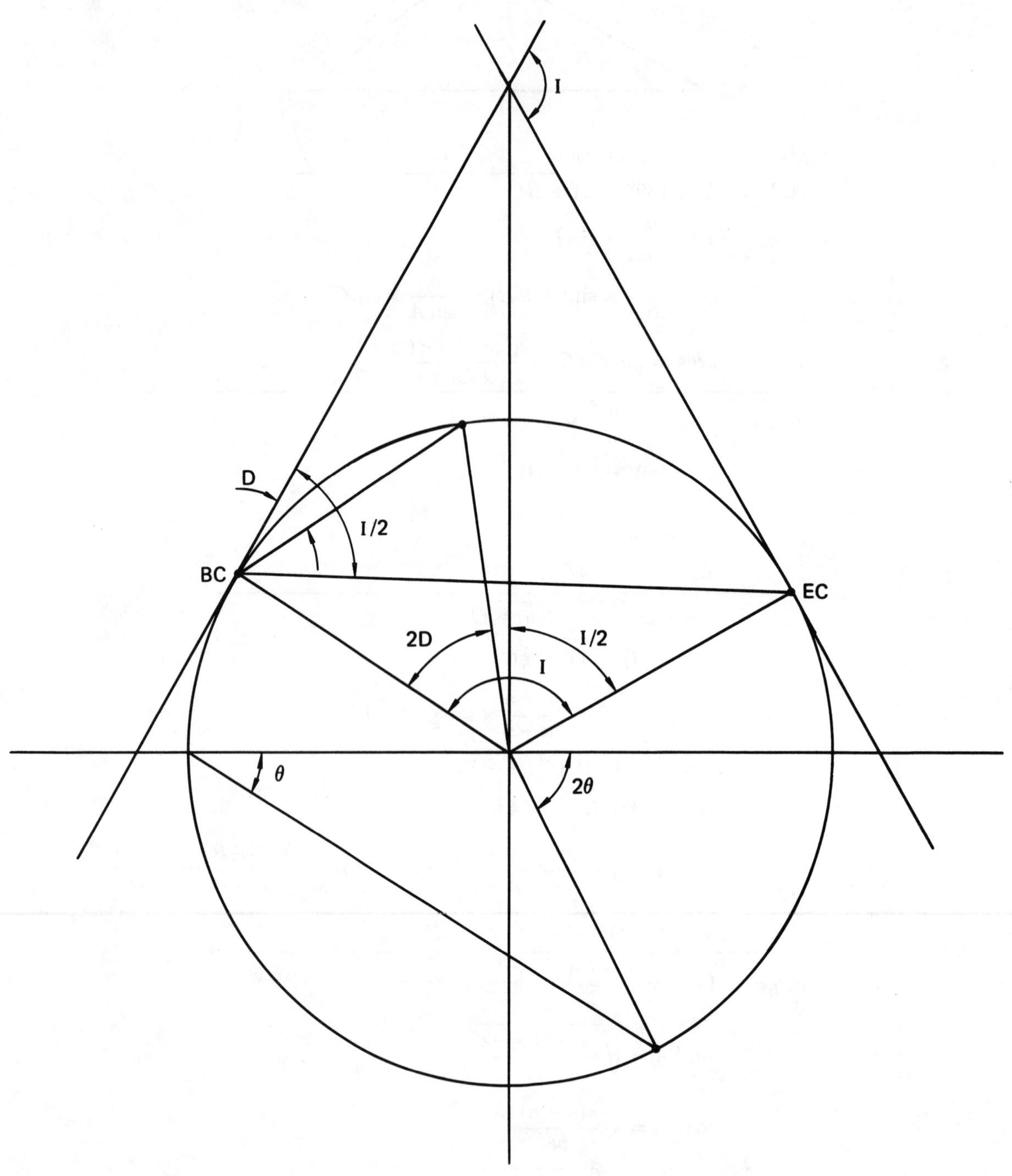

# Appendix C: Surveying Conversion Factors

| multiply | by | to obtain |
|---|---|---|
| acres | 43,560 | square feet |
| | 10 | square chains |
| | 4046.87 | square meters |
| acre-feet | 43,560 | cubic feet |
| | 1233.49 | cubic meters |
| chain | 66 | feet |
| | 22 | yards |
| | 4 | rods |
| day (mean solar) | 86,400 | seconds |
| day (sidereal) | 86,164.09 | seconds |
| degrees (angle) | 0.0174533 | radians |
| | 17.77778 | mils |
| engineer's link | 1 | feet |
| feet (U.S. Survey) | 0.3048006 | meters |
| grads | 0.9 | degrees (angle) |
| | 0.01570797 | radians |
| hectare | 2.47104 | acres |
| | 10,000 | square meters |
| inches | 25.4 | millimeters |
| labors | 177.14 | acres |
| leagues | 4428.40 | meters |
| link—see 'engineer's link' and 'surveyor's link' | | |
| mils | 0.05625 | degrees (angle) |
| | 3,037,500 | minutes |
| miles (statute) | 5280 | feet |
| | 80 | chains (surveyor's) |
| | 320 | rods |
| | 0.86839 | miles (nautical) |
| square miles | 640 | acres |
| | 27,878,400 | square feet |
| minutes (angle) | 0.29630 | mils |
| | 0.000290888 | radians |
| minutes (mean solar) | 60 | seconds |
| minutes (sidereal) | 59.83617 | seconds |
| outs | 330 | feet |
| | 10 | 33-foot chains |
| radians | 57.2957795 | degrees (angle) |
| | 57°17′44.806″ | degrees (angle) |
| rods | 16.5 | feet |
| | 1 | perches |
| | 1 | poles |
| seconds (angle) | 4.848137 EE–6 | radians |
| seconds (sidereal) | 0.9972696 | seconds (mean solar) |
| surveyor's link | 0.66 | feet |
| | 7.92 | inches |
| VARA (California) | 33 | inches |
| VARA (Texas) | 33.333 | inches |
| yard (U.S.) | 0.914402 | meters |

# Appendix D: Critical Constants

| | | |
|---|---|---|
| 0.0000001 per °F | = | coefficient of expansion invar tape |
| 0.00000645 per °F | = | coefficient of expansion steel tape |
| 0.6745 | = | coefficient for 50% standard deviation |
| 1.15 miles | = | 1 minute of latitude |
| 6 miles | = | length and width of township |
| 10 square chains | = | 1 acre |
| 15 degrees longitude | = | width of one time zone |
| 23 degrees 26.5 minutes | = | maximum declination of the sun at solstice |
| 24 hours | = | 360 degrees of longitude |
| 36 | = | number of sections in a township |
| 69.1 miles | = | 1 degree latitude |
| 100 | = | usual stadia ratio |
| 101 feet | = | 1 second of latitude |
| 400 grads | = | 360 degrees |
| 480 chains | = | width and length of township |
| 640 acres | = | 1 normal section |
| 4046.9 square meters | = | 1 acre |
| 6400 mils | = | 360 degrees |
| 43,560 square feet | = | 1 acre |
| 20,906,000 feet | = | mean radius of earth |

## Practice Problems: SURVEYING

Untimed

1. A downgrade of 4% meets a rising grade of 5% in a sag curve. At the start of the curve the level is 123.06 at location 4034+20. At 4040+20 there is an overpass with an underside level of 134.06. If the curve is designed to afford a clearance of 15 feet under the overpass at this point, calculate the required length.

2. An existing length of road consists of a rising gradient of 1 in 20 followed by a vertical parabolic summit curve 300 feet long, and then a falling gradient of 1 in 40. The curve joins both gradients tangentially and the elevation of the highest point on the curve is 173.07 feet. Visibility is to be improved over this stretch of road by replacing this curve with another parabolic curve 600 feet long. (a) Find the depth of excavation required at the mid-point of the curve. (b) Tabulate the elevations of points at 100 foot intervals on the new curve.

3. Two straights intersecting at a point B have azimuths BA 270° and BC 110°. They are to be joined by a circular curve which must pass through a point D, 350 feet from B. The azimuth of BD is 260°. Find the required radius, tangent lengths, length of curve, and setting out angle for a 50 foot chord.

4. Three points (A, B, and C) were selected on the centerline of an existing road curve as a first step in determining the curve radius. The telescope was set horizontally at point B. Readings were taken on a vertical staff at points A and C. The readings are summarized below. The instrument has constants of 100 and 0. (a) Calculate the radius of the circular curve. (b) If the trunnion axis was 4.7 feet above the road at B, find the gradients AB and BC.

| staff at | horizontal bearing | stadia readings | | |
|---|---|---|---|---|
| A | 0.00° | 4.851 | 3.627 | 2.403 |
| C | 195.57° | 7.236 | 5.778 | 4.320 |

5. Four level circuits were run over four different routes to determine the elevation of a benchmark. The observed elevations and probable errors for each circuit are given. What is the most probable value for the elevation of the benchmark?

Route 1: 745.08 ± 0.03
Route 2: 745.22 ± 0.01
Route 3: 745.45 ± 0.09
Route 4: 745.17 ± 0.05

6. A transit with an interval factor of 100 and an instrument factor of 1.0 foot was used to take stadia sights. The instrument height was 4.8 feet. The location where the instrument was set up had an elevation of 297.8 feet. Using the data below, find the distance AB and the elevation of point B.

| object | azimuth | rod interval ($R_2-R_1$) | middle hair reading | vertical angle |
|---|---|---|---|---|
| A | 42.17° | 3.22 | 5.7 | −6.3° |
| B | 222.17° | 2.60 | 10.9 | +4.17° |

7. The balanced latitudes and departures of the legs of a closed traverse are given below. What is the traverse area?

| leg | latitude | departure |
|---|---|---|
| AB | N 350 | E 0 |
| BC | N 550 | E 600 |
| CD | S 250 | E 1200 |
| DE | S 750 | E 200 |
| EF | S 550 | W 1100 |
| FA | N 650 | W 900 |

8. Balance and adjust (to the nearest 0.1 ft) the traverse given below.

| leg | bearing | length |
|---|---|---|
| AB | N | 500.0 |
| BC | N45.00°E | 848.6 |
| CD | S69.45°E | 854.4 |
| DE | S11.32°E | 1019.8 |
| EF | S79.70°W | 1118.0 |
| FA | N54.10°W | 656.8 |

9. Determine the elevation of BM 11 and BM 12.

| station | backsight | foresight | elevation |
|---|---|---|---|
| BM 10 | 4.64 | | 179.65 |
| TP 1 | 5.80 | 5.06 | |
| TP 2 | 2.25 | 5.02 | |
| BM 11 | 6.02 | 5.85 | |
| TP 3 | 8.96 | 4.34 | |
| TP 4 | 8.06 | 3.22 | |
| TP 5 | 9.45 | 3.71 | |
| TP 6 | 12.32 | 2.02 | |
| BM 12 | | 1.98 | |

10. A five-leg closed traverse is taped and scoped in the field, but obstructions make it impossible to collect all readings. It is known that the general direction of EA is easterly. Complete the table of information below.

| leg | north azimuth | distance |
|---|---|---|
| AB | 106.22° | 1081.3 |
| BC | 195.23° | 1589.5 |
| CD | 247.12° | 1293.7 |
| DE | 332.37° | — |
| EA | — | 1737.9 |

Timed

1. An existing 6° curve connects PC and $PT_1$ as shown. It is desired to avoid having vehicles pass close to a historical monument, so a proposal has been made to relocate the curve 120 feet forward. The PC will remain the same.

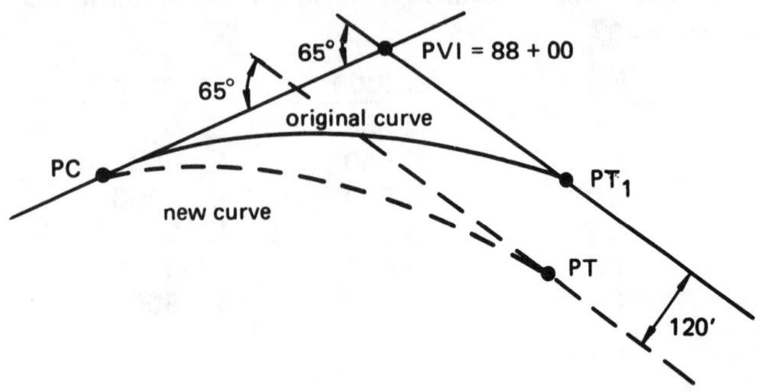


(a) Find the PC station. (b) Find the radius of curvature for the new curve. (c) Find PT of the new curve.

2. Two highways are planned, each running perpendicular to the other. The details of the lower highway are already fixed. The design process is continuing for the upper highway, which must maintain a 25 foot clearance over the highway below (as measured from the highway centerlines).

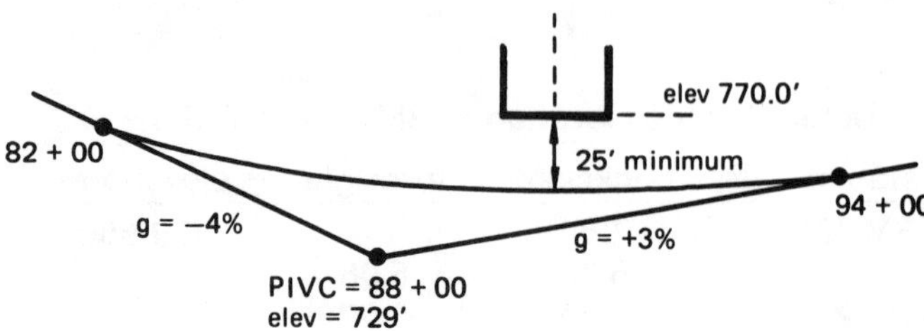


(a) Locate the minimum station where the new highway can be constructed. (b) Locate the maximum station where the new highway can be constructed. (c) What is the station of the lowest point on the existing curve? (d) What is the elevation of the lowest point on the existing curve?

3. A historical monument must be avoided near a proposed highway. A compound (reverse) curve is being considered. Both curves will have the same curvature. Refer to the diagram for details of the proposed curves. (a) What are the stations of the PC, PRC, and PT? (b) What is the biggest hazard associated with reverse curves? (c) What is the deflection angle from PC to the center of the first curve?

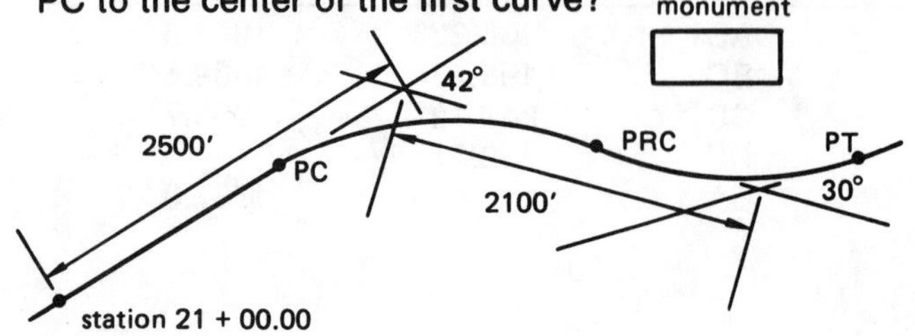


4. An equal tangent vertical curve is described as follows:

BVC: station 110+00
PVI: station 116+00; elevation 1262
EVC: station 122+00
$g_1 = +3\%$; $g_2 = -2\%$

(a) Find the centerline roadway elevations at 50 foot intervals. (b) The road continues on to a horizontal curve 5% to the right. Superelevation of 8% is used. The road is 60 feet wide at all points. The horizontal curve is supported by three pile bents symmetrical about the PVI of the curve. The pile bents are located at stations 115+50, 116+00, and 116+50. Each pile bent consists of 7 piles symmetrical about the roadway centerline and spaced at 10 foot intervals. The tops of the piles are 3.5 feet below the roadway. What are the elevations of each pile?

5. The tangent of a horizontal curve is relocated 10 feet west of its original position. The PC remains the same. Using the information given in the diagram, find R and R'.

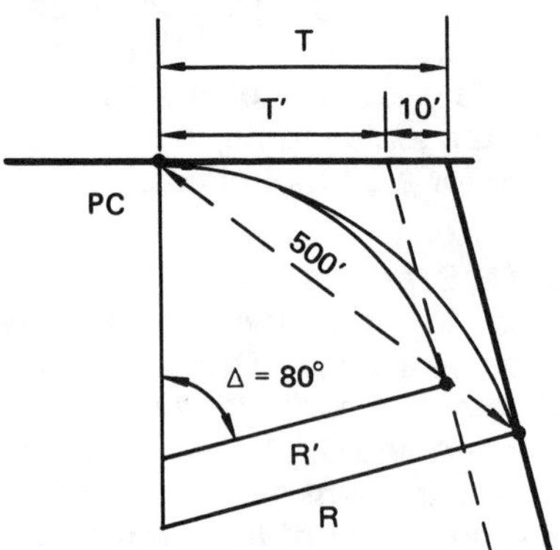


6. A new highway can go either over or under (perpendicular to) an existing highway at station 75+00. The elevation of the existing highway at station 75+00 is 510.00 ft. 22 feet of clearance is required if the new highway goes over the existing highway. 20 feet of clearance is required if the new highway goes under the old highway. (a) What is the shortest vertical curve that will pass over the existing highway? (b) What is the longest vertical curve that will pass under the existing highway?

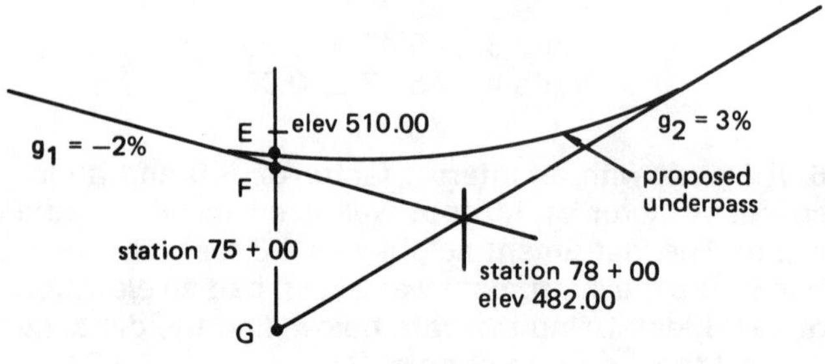

7. The tangent of a horizontal curve is parallel to a railroad line 150 feet away. The two curve tangents intersect at station 182+27.52. Using the other information on the diagram, find (a) the station of the railroad line's intersection with the highway, (b) the stations of the PC and PT, (c) the middle ordinate (M), and (d) the external ordinate (E).

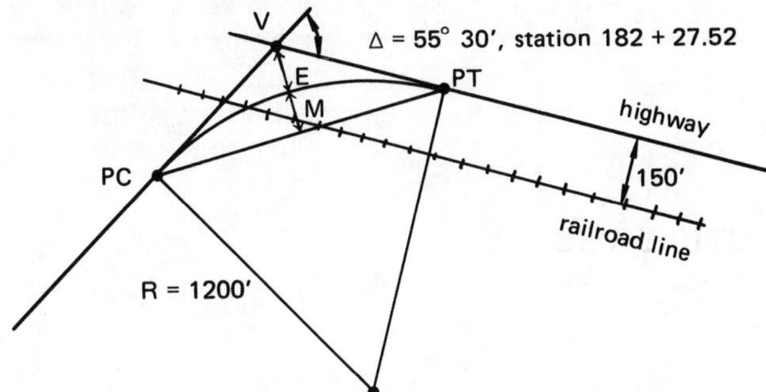


8. The grades on a vertical curve intersect at station 105+00, elevation 350 ft. A drain grating exists at station 105+00 and elevation 358.30 ft. (a) What is the minimum continuous curve length such that no point on the curve will be lower than the drain grating? (b) If a length of 1900 feet is used, what are the stations of BVC and EVC? (c) If a length of 1900 feet is used, what is the elevation of the lowest point on the curve? (d) If a length of 1900 feet is used, what is the elevation at station 106+40.00?

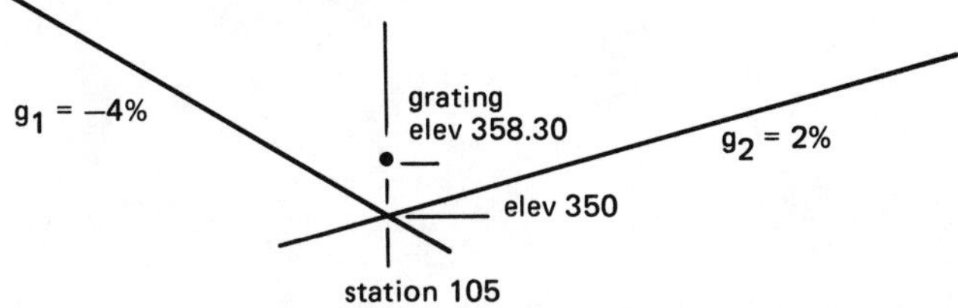


9. A highway is being constructed without disturbing an existing monument. The tangent alignments are shown. The centerline of the road can be on either side of the monument. Find the range of radii of circular arcs which will miss the monument by 1 foot. (Consider the monument to be a point.) The road has a 150 foot right-of-way.

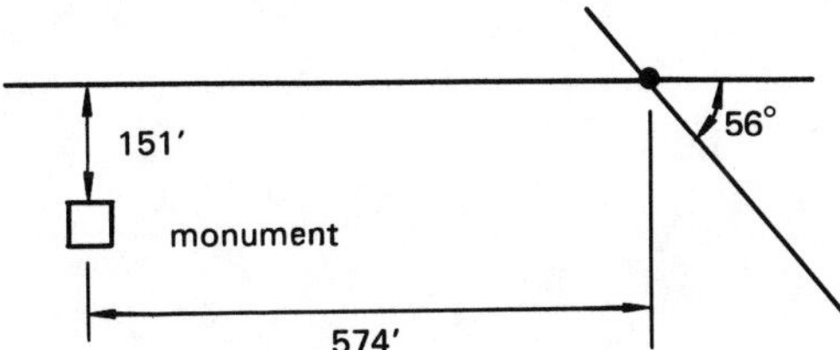

RESERVED FOR FUTURE USE

# 18 MANAGEMENT THEORIES

## 1 INTRODUCTION

Effective management techniques are based on behavioral science studies. Behavioral science is an outgrowth of the *human relations theories* of the 1930's, in which the happiness of employees was the goal (i.e., a happy employee is a productive employee ...). Current behavioral science theories emphasize minimizing tensions that inhibit productivity.

There is no evidence yet that employees want a social aspect to their jobs. Nor is there evidence that employees desire job enlargement or autonomy. Behavioral science makes these assumptions anyway.[1]

## 2 BEHAVIORAL SCIENCE KEY WORDS

Cognitive system: method we use to interpret our environment.

Collaboration: influence through a mutual agreement, relationship, respect, or understanding, without a formal or contractual authority relationship.

Equilibrium: maintaining the status quo of a group or an individual.

Job enrichment: letting employees have more control over their activities and working conditions.

Manipulation: influencing others by recognizing and building upon their needs.

MBO: management by objectives—setting job responsibilities and standards for each group and employee.

Normative judgment: judging others according to our own values.

Paternalism: corporate subsidy—showering employees with benefits and expecting submission in return.

Personal map: a person's expectations of his environment.

Selective perception: seeing what we want to see—a form of defense mechanism, since things first must be perceived to be ignored.

Superordinate goals: goals which are outside of the individual, such as corporate goals.

## 3 HISTORY OF BEHAVIORAL SCIENCE STUDIES AND THEORIES

### A. HAWTHORNE EXPERIMENTS

During 1927–1932, the Hawthorne (Chicago) Works of the Western Electric Company experimented with working conditions in an attempt to determine what factors affected output.[2] Six average employees were chosen to assemble and inspect phone relays.

Many factors were investigated in this exhaustive test. After weeks of observation without making any changes (to establish a baseline), Western Electric varied the number and length of breaks, the length of the work day, the length of the work week, and the illumination level of the lighting. Group incentive plans were tried, and in several tests, the company even provided food during breaks and lunch periods.

The employees reacted in the ways they thought they should. Output (as measured in relay production) increased after every change was implemented. In effect, the employees reacted to the attention they received, regardless of the working conditions.

Western Electric concluded that there was no relationship between illumination and other conditions to productivity. The increase in productivity during the test-

[1] In an exhaustive literature survey up through 1955, researchers found no conclusive relationships between satisfaction and productivity. There was, however, a relationship between lack of satisfaction and absenteeism and turnover.

[2] Experiments were conducted by Elton Mayo from the Harvard Business School.

ing procedure was attributed to the sense of value each employee felt in being part of an important test. The employees also became a social group, even after hours. Leadership and common purpose developed, and even though the employees were watched more than ever, they felt no supervision anxiety since they were, in effect, free to react in any way they wanted.

One employee summed up the test when she said, "It was fun."

## B. BANK WIRING OBSERVATION ROOM EXPERIMENTS

In an attempt to devise an experiment which would not suffer from the problems associated with the Hawthorne studies, Western Electric conducted experiments in 1931 and 1932 on the effects of wage incentives.

The group of nine wiremen, three soldermen, and two inspectors was interdependent. This was supposed to prevent any individual from slacking off. However, wage incentives failed to improve productivity. In fact, fast employees slowed down to protect their slower friends. Illicit activities, such as job trading and helping, also occurred.

The group was reacting to the notion of a ***proper day's work.*** When the day's work (or what the group considered to be a day's work) was assured, the whole group slacked off. The group also varied what it reported as having been accomplished and claimed more unavoidable delays than actually occurred. The output was essentially constant.

Western Electric concluded that social groups form as protection against arbitrary management decisions, even when such decisions have never been experienced. The effort to form the social groups, to protect slow workers, and to develop the notion of a proper day's work is not conscious. It develops automatically when the company fails to communicate to the contrary.

## C. NEED HIERARCHY THEORY

During World War II, Dr. Abraham Maslow's ***need hierarchy theory*** was implemented into leadership training for the U.S. Air Force. This theory claims that certain needs become dominant when lesser needs are satisfied. Although some needs can be sublimated and others overlap, the need hierarchy theory generally requires the lower-level needs to be satisfied before the higher-level needs are realized. (The ego and self-fulfillment needs rarely are satisfied.)

The need hierarchy theory explains why money is a poor motivator of an affluent individual. The theory does not explain how management should apply the need hierarchy to improve productivity.

**Table 18.1**
The Need Hierarchy

(In order of lower to higher needs)

1. Physiological needs: air, food, water.
2. Safety needs: protection against danger, threat, deprivation, arbitrary management decisions. Need for security in a dependent relationship.
3. Social needs: belonging, association, acceptance, giving and receiving of love and friendship.
4. Ego needs: self-respect and confidence, achievement, self-image. Group image and reputation, status, recognition, appreciation.
5. Self-fulfillment needs: realizing self potential, self development, creativity.

## D. THEORY OF INFLUENCE

In 1948, the Human Relations Program (under the direction of Donald C. Pelz) at Detroit Edison studied the effectiveness of its supervisors. The most effective supervisors were those who helped their employees benefit. Supervisors who were close to their employees (and sided with them in disputes) were effective only if they were influential enough to help the employees. The study results were formulated into the ***theory of influence.***

- Employees think well of supervisors who help them reach their goals and meet their needs.
- An influential supervisor will be able to help employees.
- An influential supervisor who is also a disciplinarian will breed dissatisfaction.
- A supervisor with no influence will not be able to affect worker satisfaction in any way.

The implication of the theory of influence is that whether or not a supervisor is effective depends on his influence. Training of supervisors is useless unless they have the power to implement what they have learned. Also, increases in supervisor influence are necessary to increase employee satisfaction.

## E. HERZBERG MOTIVATION STUDIES

Frederick W. Herzberg interviewed 200 technical personnel in 11 firms during the late 1950's. Herzberg was especially interested in exceptional occurrences resulting in increases in job satisfaction and performance. From those interviews, Herzberg formulated his *motivation-maintenance theory.*

According to this theory, there are satisfiers and dissatisfiers which influence employee behavior. The *dissatisfiers* (also called *maintenance/motivation factors*) do not motivate employees; they can only dissatisfy them. However, the dissatisfiers must be eliminated before the satisfiers work. Dissatisfiers include company policy, administration, supervision, salary, working conditions (environment), and interpersonal relations.

*Satisfiers* (also known as *motivators*) determine job satisfaction. Common satisfiers are achievement, recognition, the type of work itself, responsibility, and advancement.

An interesting conclusion based on the motivation/maintenance theory is that fringe benefits and company paternalism do not motivate employees since they are related to dissatisfiers only.

### F. THEORY X AND THEORY Y

During the 1950's Douglas McGregor (Sloan School of Industrial Management at MIT) introduced the concept that management had two ways of thinking about its employees. One way of thinking, which was largely pessimistic, was theory X. The other theory, theory Y, was largely optimistic.

Theory X is based on the assumption that the average employee inherently dislikes and avoids work. Therefore, employees must be coerced into working by threats of punishment. Rewards are not sufficient. The average employee wants to be directed, avoids responsibility, and seeks the security of an employer-employee relationship.

This assumption is supported by much evidence. Employees exist in a continuum of wants, needs, and desires. Many of the need satisfiers (salary, fringe benefits, etc.) are effective only off the job. Therefore, work is considered a punishment or a price paid for off-the-job satisfaction.

Theory X is pessimistic about the effectiveness of employers to satisfy or motivate their employees. By satisfying the physiological and safety (lower level) needs, employers have shifted the emphasis to higher level needs which they cannot satisfy. Employees, unable to derive satisfaction from their work, behave according to theory X.

Theory Y, on the other hand, assumes that the expenditure of effort is natural and is not inherently disliked. It assumes that the average employee can learn to accept and enjoy responsibility. Creativity is widely distributed among employees, and the potentials of average employees are only partially realized.

Theory Y places the blame for worker laziness, indifference, and lack of cooperation in the lap of management, since the integration of individual and organization needs is required. This theory is not fully validated, nor is its full use ever likely to be implemented.[3]

## 4 JOB ENRICHMENT

In an effort to make their employees happier, companies have tried to enrich the jobs performed by employees. Enrichment is a subjective result felt by employees when their jobs are made more flexible or are enlarged. Adding flexibility to a job allows an employee to move from one task to another, rather than doing the same thing continually. Horizontal job enlargement adds new production activities to a job. Vertical job enlargement adds planning, inspection, and other non-production tasks to the job.

There are advantages to keeping a job small in scope. Learning time is low, employee mental effort is reduced, and the pay rate can be lower for untrained labor. Supervision is reduced. Such simple jobs, however, also result in high turnover, absenteeism, and lower pride in job (and subsequent low quality rates).

Job enlargement generally results in better quality products, reduces inspection and material handling, and counteracts the disadvantages previously mentioned. However, training time is greater, tooling costs are higher, and inventory records are more complex.

## 5 QUALITY IMPROVEMENT PROGRAMS

### A. ZERO DEFECTS PROGRAM

Employees have been conditioned to believe that they are not perfect and that errors are natural. However, we demand zero defects from some professions (e.g., doctors, lawyers, engineers). The philosophy of a zero defect program is to expect zero defects from everybody.

Zero defects programs develop a constant, conscious desire to do the job right the first time. This is accomplished by giving employees constant awareness that their jobs are important, that the product is important, and that management thinks their efforts are important.

[3] Theory Y is not synonymous with soft management. Rather than emphasize tough management (as does theory X), theory Y depends on commitment of employees to achieve mutual goals.

Zero defects programs try to correct the faults of other types of employee programs.[4] Programs are based on what the employee has for his own: pride and desire. The programs present the challenge of perfection and explain the importance of that perfection. Management sets an example by expecting zero defects of itself. Standards of performance are set and are related to each employee. Employees are checked against these performance requirements periodically, and recognition is given when goals are met.

[4] Motivational programs are not honest, according to the zero-defects theory, since management tries to convince employees to do what management wants. Wage incentive programs encourage employee dishonesty and errors by emphasizing quantity, not quality. Theory X management, with its implied punitive action if goals are not achieved, never has been effective.

## B. QUALITY CIRCLES/TEAM PROGRAMS

Quality circle programs are voluntary or required programs in which employees within a department actively participate in measuring and improving quality and performance. It involves periodic meetings on a weekly or a monthly basis. Workers are encouraged to participate in volunteering ideas for improvement.

RESERVED FOR FUTURE USE

RESERVED FOR FUTURE USE

# 19 MISCELLANEOUS TOPICS

## PART 1: Accuracy and Precision

## Experiments

### 1 ACCURACY

An experiment is said to be *accurate* if it is unaffected by experimental error. In this case, *error* is not synonymous with *mistake*, but rather includes all variations not within the experimenter's control.

For example, suppose a gun is aimed at a point on a target and five shots are fired. The mean distance from the point of impact to the sight-in point is a measure of the alignment accuracy between the barrel and sights. The difference between the actual value and the experimental value is known as *bias*.

### 2 PRECISION

*Precision* is not synonymous with accuracy. Precision is concerned with the repeatability of the experimental results. If an experiment is repeated with identical results, the experiment is said to be precise.

In the previous example, the average distance of each impact from the centroid of the impact group is a measure of the precision of the experiment. Thus, it is possible to have a highly precise experiment with a large bias.

Most techniques applied to experiments to improve the accuracy of the experimental results (e.g., repeating the experiment, refining the experimental methods, or reducing variability) actually increase the precision.

Sometimes the word *reliability* is used with regards to the precision of an experiment. A reliable estimate is used in the same sense as a precise estimate.

### 3 STABILITY

*Stability* and *insensitivity* are synonymous terms. A stable experiment is insensitive to minor changes in the experiment parameters. Suppose the centroid of a bullet group is 2.1 inches away from the sight-in point at 65°F and 2.3 inches away at 80°F. The experiment's sensitivity to temperature changes would be $(2.3 - 2.1)/(80 - 65) = 0.0133$ inches/°F.

# PART 2: Dimensional Analysis

*Nomenclature*

| | | |
|---|---|---|
| $c_p$ | specific heat | BTU/lbm-°F |
| $C_i$ | a constant | – |
| D | diameter | ft |
| F | force | lbf |
| $g_c$ | gravitational constant (32.2) | lbm-ft/sec$^2$-lbf |
| $\bar{h}$ | average film coefficient | BTU/hr-ft$^2$-°F |
| J | Joule's constant (778) | ft-lbf/BTU |
| k | number of pi-groups $(m - n)$ | – |
| L | length | ft |
| m | number of relevant independent variables | – |
| M | mass | lbm |
| n | number of independent dimensional quantities | – |
| $N_{Nu}$ | Nusselt number | – |
| $N_{Pe}$ | Peclet number | – |
| $N_{Re}$ | Reynolds number | – |
| v | velocity | ft/sec |
| $x_i$ | the *i*th independent variable | various |
| y | dependent variable | various |

*Symbols*

| | | |
|---|---|---|
| $\rho$ | density | lbm/ft$^3$ |
| $\theta$ | time | sec |
| $\pi_i$ | *i*th dimensionless group | – |
| $\mu$ | viscosity | lbm/ft-sec |

Dimensional analysis is a means of obtaining an equation for some phenomenon without understanding the inner mechanism of the phenomenon. The most serious limitation to this method is the need to know beforehand which variables influence the phenomenon. Once these variables are known or are assumed, dimensional analysis can be applied by a routine procedure.

The first step is to select a system of primary dimensions. Usually the ML$\theta$T system (mass, length, time, and temperature) is used, although this choice may require the use of $g_c$ and $J$ in the final results. The dimensional formulas and symbols for variables most frequently encountered are given in table 19.1.

The second step is to write a functional relationship between the dependent variable and the independent variables, $x_i$.

$$y = \mathbf{f}(x_1, x_2, \ldots, x_m) \qquad 19.1$$

This function can be expressed as an exponentiated series.

$$y = C_1 x_1^{a_1} x_2^{b_1} x_3^{c_1} \cdots x_m^{z_1} + C_2 x_1^{a_2} x_2^{b_2} x_3^{c_2} \cdots x_m^{z_2} + \cdots \qquad 19.2$$

The $C_i$, $a_i$, $b_i$, $\cdots z_i$ in equation 19.2 are unknown constants.

The key to solving the above equation is that each term on the right-hand side must have the same dimensions as $y$. Simultaneous equations are used to determine some of the $a_i$, $b_i$, $c_i$, and $z_i$. Experimental data is required to determine the $C_i$ and the remaining exponents. In most analyses, it is assumed that the $C_i = 0$ for $i = 2$ and up.

*Example 19.1*

A sphere submerged in a fluid rolls down an incline. Find an equation for the velocity, $v$.

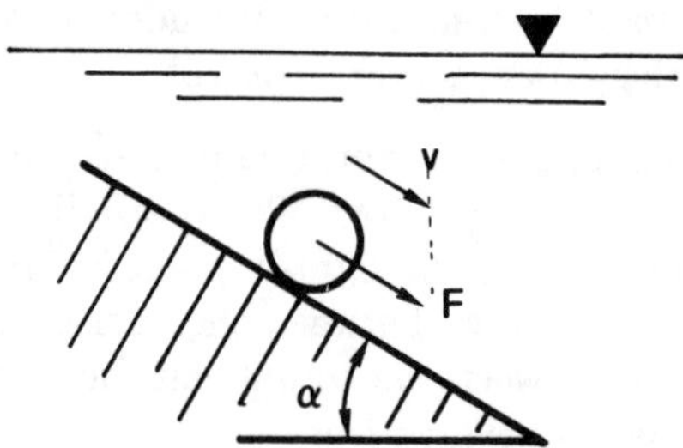


It is assumed that the velocity depends on the force, $F$, due to the inclination, the diameter of the sphere, $D$, the density of the fluid, $\rho$, and the viscosity of the fluid, $\mu$.

$$v = \mathbf{f}(F, D, \rho, \mu)$$

This equation can be written in terms of the dimensions of the variables.

$$\frac{L}{\theta} = C\left(\frac{ML}{\theta^2}\right)^a (L)^b \left(\frac{M}{L^3}\right)^c \left(\frac{M}{L\theta}\right)^d$$

Since $L$ on the left-hand side has an implied exponent of one, the necessary equation is

$$1 = a + b - 3c - d \qquad (L)$$

Similarly, the other necessary equations are

$$-1 = -2a - d \qquad (\theta)$$
$$0 = a + c + d \qquad (M)$$

**Table 19.1**
Units and Dimensions of Typical Variables

| Quantity | Symbol | MLθT System | MLθTFQ System | Units in Engineering System |
|---|---|---|---|---|
| length | L or x | L | L | ft |
| time | $\theta$ | $\theta$ | $\theta$ | sec or hour |
| mass | M | M | M | lbm |
| force | F | $ML/\theta^2$ | F | lbf |
| temperature | T | T | T | °F |
| heat | Q | $ML^2/\theta^2$ | Q | BTU |
| velocity | V | $L/\theta$ | $L/\theta$ | ft/sec |
| acceleration | a or g | $L/\theta^2$ | $L/\theta^2$ | ft/sec$^2$ |
| dimensional conversion factor | $g_c$ | none | $ML/\theta^2F$ | 32.2 lbm-ft/sec$^2$-lbf |
| energy conversion factor | J | none | FL/Q | 778 ft-lbf/BTU |
| work | W | $ML^2/\theta^2$ | FL | ft-lbf |
| pressure | p | $M/\theta^2L$ | $F/L^2$ | lbf/ft$^2$ |
| density | $\rho$ | $M/L^3$ | $M/L^3$ | lbm/ft$^3$ |
| internal energy and enthalpy | u, h | $L^2/\theta^2$ | Q/M | BTU/lbm |
| specific heat | c | $L^2/\theta^2T$ | Q/MT | BTU/lbm-°F |
| dynamic viscosity | $\mu_f$ | $M/L\theta$ | $F\theta/L^2$ | lbf-sec/ft$^2$ |
| absolute viscosity | $\mu$ | $M/L\theta$ | $M/L\theta$ | lbm/ft-sec |
| kinematic viscosity | $\nu = u/\rho$ | $L^2/\theta$ | $L^2/\theta$ | ft$^2$/sec |
| thermal conductivity | k | $ML/\theta^3T$ | $Q/LT\theta$ | BTU/hr-ft-°F |
| coefficient of expansion | $\beta$ | 1/T | 1/T | 1/°F |
| surface tension | $\sigma$ | $M/\theta^2$ | F/L | lbf/ft |
| stress | $\sigma$ or $\tau$ | $M/L\theta^2$ | $F/L^2$ | lbf/ft$^2$ |
| film coefficient | h | $M/\theta^3T$ | $Q/\theta L^2T$ | BTU/hr-ft$^2$-°F |
| mass flow rate | m | $M/\theta$ | $M/\theta$ | lbm/sec |

Solving simultaneously yields

$$b = -1$$
$$c = a - 1$$
$$d = 1 - 2a$$

or

$$v = C\left(\frac{\mu}{D\rho}\right)\left(\frac{F\rho}{\mu^2}\right)^a$$

$C$ and $a$ would have to be determined experimentally.

Since the above method requires working with $m$ different variables and $n$ different independent dimensional quantities (such as M, L, T, and $\theta$), an easier method is desirable. One simplification is to combine the $m$ variables into dimensionless groups, called *pi-groups.*

If these dimensionless groups are represented by $\pi_1$, $\pi_2$, $\pi_3, \cdots \pi_k$, the equation expressing the relationship between the variables is given by the *Buckingham $\pi$-theorem.*

$$\mathbf{f}(\pi_1, \pi_2, \pi_3, \ldots, \pi_k) = 0 \qquad 19.3$$

$$k = m - n \qquad 19.4$$

The dimensionless pi-groups usually are found from the $m$ variables according to an intuitive process. A formalized method is possible as long as the following conditions are met.

- The dependent variable and independent variables chosen contain all of the variables affecting the phenomenon. Extraneous variables can be included at the expense of obtaining extra pi-groups.
- The pi-groups must include all of the original $x_i$ at least once.
- The dimensions all must be independent.

The formal procedure is to select $n$ variables ($x_i$) out of the total $m$ as repeating variables to appear in all $k$ pi-groups. These variables are used in turn with the remaining variables in each successive pi-group. Each of the repeating variables must have different dimensions, and the repeating variables collectively must contain all of the dimensions. This procedure is illustrated in example 19.2.

*Example 19.2*

It is desired to determine a relationship giving the heat transfer to air flowing across a heated tube. The following variables affect the heat flow.

| Variable | Symbol | Dimensional Equation |
|---|---|---|
| tube diameter | D | L |
| fluid conductivity | k | $ML/\theta^3T$ |
| fluid velocity | v | $L/\theta$ |
| fluid density | $\rho$ | $M/L^3$ |
| fluid viscosity | $\mu$ | $M/L\theta$ |
| fluid specific heat | $c_p$ | $L^2/\theta^2T$ |
| film coefficient | $\overline{h}$ | $M/\theta^3T$ |

There are $m = 7$ variables and $n = 4$ primary dimensions (L, M, $\theta$, and T). Accordingly, there are $k = 7 - 4 = 3$ dimensionless groups that are required to correlate the data. The four repeating variables are chosen such that all dimensions are represented. Then the $\pi_i$ are written as functions of these repeating variables in turn with the remaining variables.

The repeating variables should not include any of the unknown quantities. For example, $\overline{h}$ should not be chosen as a repeating variable since it is directly related to the unknown heat flow. In addition, important material properties, such as $c_p$ and $k$, often are omitted. Trial and error is required to include all four primary dimensions.

Using trial and error, omitting $\overline{h}$ as a repeating variable, and representing all four primary dimensions, arbitrarily choose the variables as $D$, $k$, $v$, and $\rho$.

The pi-groups are

$$\pi_1 = D^{a_1} k^{a_2} v^{a_3} \rho^{a_4} \mu$$

$$\pi_2 = D^{a_5} k^{a_6} v^{a_7} \rho^{a_8} c_p$$

$$\pi_3 = D^{a_9} k^{a_{10}} v^{a_{11}} \rho^{a_{12}} \overline{h}$$

Since the $\pi_i$ are dimensionless, we write for $\pi_1$

$$0 = a_1 + a_2 + a_3 - 3a_4 - 1 \quad \text{(L)}$$

$$0 = a_2 + a_4 + 1 \quad \text{(M)}$$

$$0 = -3a_2 - a_3 - 1 \quad (\theta)$$

$$0 = -a_2 \quad \text{(T)}$$

Therefore,

$$a_2 = 0 \quad a_3 = -1 \quad a_4 = -1 \quad a_1 = -1$$

$$\pi_1 = \frac{\mu}{Dv\rho}$$

$\pi_1$ is the reciprocal of the Reynolds number. Proceeding similarly with $\pi_2$,

$$0 = a_5 + a_6 + a_7 = 3a_8 + 2 \quad \text{(L)}$$

$$0 = a_6 + a_8 \quad \text{(M)}$$

$$0 = -3a_6 - a_7 - 2 \quad (\theta)$$

$$0 = -a_6 - 1 \quad \text{(T)}$$

Therefore,

$$a_6 = -1 \quad a_7 = 1 \quad a_8 = 1 \quad a_5 = 1$$

$$\pi_2 = \frac{Dv\rho c_p}{k}$$

$\pi_2$ is the *Peclet number* (product of the Reynolds number and the Prandtl number).

$\pi_3$ is found to be $\frac{D\overline{h}}{k}$, which is the *Nusselt number.*

The seven original variables have been combined into three dimensionless groups, making data correlation much easier. The implicit equation for heat transfer is

$$f_1(\pi_1, \pi_2, \pi_3) = f_1(N_{Re}, N_{Nu}, N_{Pe}) = 0$$

Rearrangement of the pi-groups is needed to isolate the dependent variable (in this case, $\overline{h}$).

$$N_{Nu} = f_2(N_{Re}, N_{Pe}) = C(N_{Re})^{e_1}(N_{Pe})^{e_2}$$

$C$, $e_1$, and $e_2$ are found experimentally.

The selection of the repeating and non-repeating variables is the key step. The choice of repeating variables determines which dimensionless groups are obtained. The theoretical maximum number of valid dimensionless groups is

$$\frac{m!}{(n+1)!(m-n-1)!} \qquad 19.5$$

Not all dimensionless groups obtained are equally useful to researchers. For example, the Peclet number was obtained in the above example. However, researchers would have chosen $D$, $k$, $\rho$, and $\mu$ as repeating variables in order to obtain the Prandtl number as a dimensionless group. This choice of repeating variables is a matter of intuition.

# PART 3: Reliability

*Nomenclature*

| | | |
|---|---|---|
| f(t) | probability density function | – |
| F(t) | cumulative density function | – |
| k | minimum number for operation | – |
| MTBF | mean time before failure ($1/\lambda$) | time |
| n | number of items in the system | – |
| R* | system reliability | – |
| $\mathbf{R}_i$(t) | ith item reliability | – |
| t | time | time |
| x | number of failures | – |
| X | binary ith item performance variable | – |
| Y | arbitrary event | – |
| z(t) | hazard function | 1/time |

*Symbols*

| | | |
|---|---|---|
| $\lambda$ | constant failure or hazard rate (1/MTBF) | 1/time |
| $\phi$ | binary system performance variable | – |

## 1 ITEM RELIABILITY

*Reliability* as a function of time, $\mathbf{R}(t)$, is the probability that an item will continue to operate satisfactorily up to time $t$. Although other distributions are possible, reliability often is described by the *negative exponential distribution.* Specifically, it is assumed that an item's reliability is

$$\mathbf{R}(t) = 1 - \mathbf{F}(t) = e^{-\lambda t} = e^{-t/MTBF} \qquad 19.6$$

This infers that the probability of $x$ failures in a period of time is given by the Poisson distribution.

$$p\{x\} = \frac{e^{-\lambda}\lambda^x}{x!} \qquad 19.7$$

The negative exponential distribution is appropriate whenever an item fails only by random causes but never experiences deterioration during its life. This implies that the *expected future life* of an item is independent of the previous duration of operation.

The *hazard function* is defined as the conditional probability of failure in the next time interval given that no failure has occurred thus far. For the exponential distribution, the hazard function is

$$\mathbf{z}(t) = \lambda = \frac{1}{\text{MTBF}} \qquad 19.8$$

Since this is not a function of $t$, exponential failure rates are not dependent on the length of time previously in operation.

In general,

$$\mathbf{z}(t) = \frac{\mathbf{f}(t)}{\mathbf{R}(t)} = \frac{\frac{d\mathbf{F}(t)}{dt}}{1 - \mathbf{F}(t)} \qquad 19.9$$

The exponential distribution is summarized by equations 19.10 through 19.13.

$$\mathbf{f}(t) = \lambda e^{-\lambda t} \qquad 19.10$$

$$\mathbf{F}(t) = 1 - e^{-\lambda t} \qquad 19.11$$

$$\mathbf{R}(t) = 1 - \mathbf{F}(t) = e^{-\lambda t} \qquad 19.12$$

$$\mathbf{z}(t) = \frac{\lambda e^{-\lambda t}}{e^{-\lambda t}} = \lambda \qquad 19.13$$

*Example 19.3*

An item exhibits an exponential time to failure distribution with MTBF of 1000 hours. What is the maximum operating time such that the reliability does not drop below 0.99?

$$0.99 = e^{-t/1000}$$

$$t = 10.05 \text{ hours}$$

## 2 SYSTEM RELIABILITY

The binary variable, $X_i$, is defined as 1 if item $i$ operates satisfactorily and 0 otherwise. Similarly, the binary variable, $\phi$, is 1 only if the system operates satisfactorily. $\phi$ will be a function of the $X_i$.

### A. SERIAL SYSTEMS

The *performance function* for a system of $n$ serial items is

$$\phi = X_1X_2X_3\ldots X_n = \min\{X_i\} \qquad 19.14$$

Equation 19.14 implies that the system will fail if any of the individual items fail. The system reliability is

$$R^* = R_1R_2R_3\ldots R_n \qquad 19.15$$

*Example 19.4*

A block diagram of a system with item reliabilities is shown. What is the performance function and the system reliability?

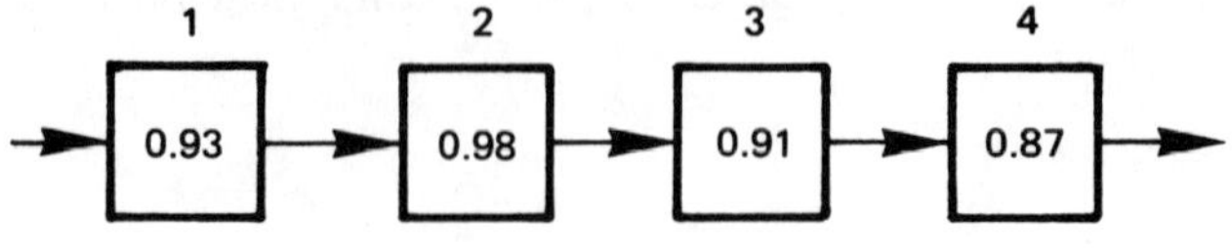


$$\phi = X_1X_2X_3X_4$$
$$R^* = (0.93)(0.98)(0.91)(0.87) = 0.72$$

B. PARALLEL SYSTEMS

A parallel system with $n$ items will fail only if all $n$ items fail. This property is called *redundancy*, and such a system is said to be redundant. Using redundancy, a highly reliable system can be produced from components with relatively low individual reliabilities.

The performance function of a redundant system is

$$\begin{aligned}\phi &= 1 - (1 - X_1)(1 - X_2)(1 - X_3)\cdots(1 - X_n)\\ &= \max\{X_i\}\end{aligned} \quad 19.16$$

The reliablity is

$$R^* = 1 - (1 - R_1)(1 - R_2)(1 - R_3)\cdots(1 - R_n) \quad 19.17$$

*Example 19.5*

What is the reliability of the system shown?

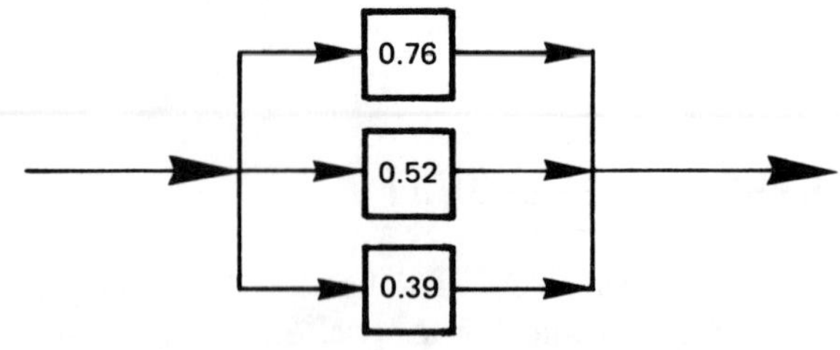


$$R^* = 1 - (1 - 0.76)(1 - 0.52)(1 - 0.39) = 0.93$$

C. k-out-of-n SYSTEMS

If the system operates with an $k$ of its elements operational, it is said to be a $k$-out-of-$n$ system. The performance function is

$$\phi = \begin{cases} 1 \text{ if } \Sigma X_i \geq k \\ 0 \text{ if } \Sigma X_i < k \end{cases} \quad 19.18$$

The evaluation of the system reliability is quite difficult unless all elements are identical and have identical reliabilities, **R**. In that case, the system reliability follows the binomial distribution.

$$R^* = \sum_{j=k}^{n} \binom{n}{j} R^j (1 - R)^{n-j} \quad 19.19$$

D. GENERAL SYSTEM RELIABILITY

A general system can be represented by a graphical network. Each path through the network from the starting node to the finishing node represents a possible operating path. For the 5-path network below, even if BD and AC are cut, the system will operate by way of path ABCD.

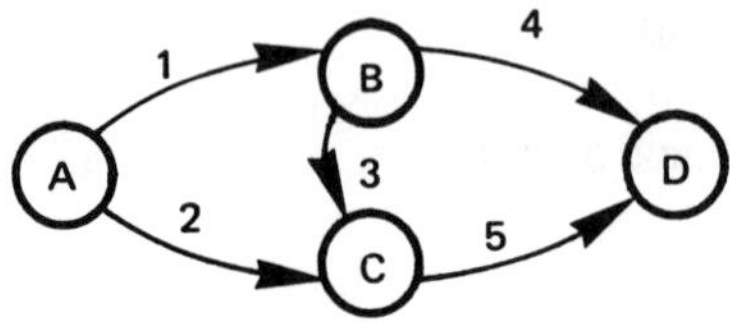


The reliability of the system will be the sum of the serial reliabilities, summed over all possible paths in the system. However, the concepts of minimal paths and minimal cuts are required to facilitate the evaluation of the system reliability.

A *minimal path* is a set of components that, if operational, will ensure the system's functioning. In the previous example, components [1 with 4] are a minimal path, as are [2 with 5] and [1 with 3 with 5]. A *minimal cut* is a set of components that, if non-functional, inhibits the system from functioning. Minimal cuts in the previous example are [1 with 2], [4 with 5], [1 with 5], and [2 with 3 with 4].

Since it usually is easier to determine all minimal paths, a method of finding the exact system reliability from the set of minimal paths is needed. In general, the probability of a union of $n$ events contains $(2^n - 1)$ terms and is given by

$$\begin{aligned}&p\{Y_1 \text{ or } Y_2 \text{ or } \cdots Y_n\} = p\{Y_1\} + p\{Y_2\} + p\{Y_3\}\\ &+ \cdots + p\{Y_n\} - p\{Y_1 \text{ and } Y_2\} - p\{Y_1 \text{ and } Y_3\}\\ &- \cdots - p\{Y_1 \text{ and } Y_n\} - p\{Y_1 \text{ and } Y_2 \text{ and } Y_3\}\\ &- p\{Y_1 \text{ and } Y_2 \text{ and } Y_4\} - \cdots - p\{Y_i \text{ and } Y_j \text{ and } Y_k\}\\ &\text{all } i \neq j \neq k\\ &+ \{-1\}^{n-1} p\{Y_1 \text{ and } Y_2 \text{ and } Y_3 \text{ and } \ldots \text{ and } Y_n\}\end{aligned} \quad 19.20$$

Returning to the 5-path example,

$$Y_1 = [1 \text{ with } 4]$$
$$Y_2 = [2 \text{ with } 5]$$
$$Y_3 = [1 \text{ with } 3 \text{ with } 5]$$

Then,

$$\begin{aligned} p\{\phi = 1\} &= p\{Y_1 \text{ or } Y_2 \text{ or } Y_3\} \\ &= p\{X_1X_4 = 1\} + p\{X_2X_5 = 1\} \\ &\quad + p\{X_1X_3X_5 = 1\} - p\{X_1X_2X_4X_5 = 1\} \\ &\quad - p\{X_1X_3X_4X_5 = 1\} \\ &\quad - p\{X_1X_2X_3X_5 = 1\} \\ &\quad + p\{X_1X_2X_3X_4X_5 = 1\} \end{aligned}$$

In terms of the individual item reliabilities, this is

$$\begin{aligned} R^* &= R_1R_4 + R_2R_5 + R_1R_3R_5 \\ &\quad - R_1R_2R_4R_5 - R_1R_3R_4R_5 - R_1R_2R_3R_5 \\ &\quad + R_1R_2R_3R_4R_5 \end{aligned}$$

In the 5-path example given,

$$R^* \leq p\{X_1X_4 = 1\} + p\{X_2X_5 = 1\} + p\{X_1X_3X_5 = 1\}$$

This method requires considerable computation, and an upper bound on $R^*$ would be sufficient. Such an upper bound is close to $R^*$ since the product of individual reliabilities is small. The upper bound is given by

$$p\{\phi = 1\} \leq p\{Y_1\} + p\{Y_2\} + \cdots + p\{Y_n\} \qquad 19.21$$

# PART 4: Replacement

*Nomenclature*

| | |
|---|---|
| $C_1$ | item replacement cost with group replacement |
| $C_2$ | item replacement cost after individual failure |
| **F**(t) | number of units failing in the interval ending at $t$ |
| **K**(t) | total cost of operating from $t = 0$ to $t = T$ |
| MTBF | mean time before failure |
| n | number of units in original system |
| p{$t$} | probability of failing in the interval ending at $t$ |
| **S**(t) | number of survivors at the end of time $t$ |
| t | time |
| v{$t$} | conditional probability of failure in the interval $(t-1)$ to $t$ given non-failure before $(t-1)$ |

## 1 INTRODUCTION

Replacement and renewal models determine the most economical time to replace existing equipment. Replacement processes fall into two categories, depending on the life pattern of the equipment, which either deteriorates gradually (becomes obsolete or less efficient) or fails suddenly.

In the case of gradual deterioration, the solution consists of balancing the cost of new equipment against the cost of maintenance or decreased efficiency of the old equipment. Several models are available for cases with specialized assumptions, but no general solution methods exist.

In the case of sudden failure, of which light bulbs are examples, the solution consists of finding a replacement frequency which minimizes the costs of the required new items, the labor for replacement, and the expected cost of failure. The solution is made difficult by the probabilistic nature of the life spans.

## 2 DETERIORATION MODELS

The replacement criterion with deterioration models is the present worth of all future costs associated with each policy. Solution is by trial and error, calculating the present worth of each policy and incrementing the replacement period by one time period for each iteration.

*Example 19.6*

Item A currently is in use. Its maintenance cost is $400 this year, increasing each year by $30. Item A can be replaced by item B at a current cost of $3500. However, the cost of B is increasing by $50 each year. Item B has no maintenance costs. Disregarding income taxes, find the optimum replacement year. Use 10% as the interest rate.

Calculate the present worth of the various policies.

*policy 1*: Replacement at $t = 5$ (starting the 6th year)

$$PW(A) = -400\left(\frac{P}{A}, 10\%, 5\right) - 30\left(\frac{P}{G}, 10\%, 5\right) = -1722$$

$$PW(B) = -[3500 + 5(50)]\left(\frac{P}{F}, 10\%, 5\right) = -2328$$

*policy 2*: Replacement at $t = 6$

$$PW(A) = -400\left(\frac{P}{A}, 10\%, 6\right) - 30\left(\frac{P}{G}, 10\%, 6\right) = -2033$$

$$PW(B) = -[3500 + 6(50)]\left(\frac{P}{F}, 10\%, 6\right) = -2145$$

*policy 3*: Replacement at $t = 7$

$$PW(A) = -400\left(\frac{P}{A}, 10\%, 7\right) - 30\left(\frac{P}{G}, 10\%, 7\right) = -2330$$

$$PW(B) = -[3500 + 7(50)]\left(\frac{P}{F}, 10\%, 7\right) = -1975$$

The present worth of $B$ drops below the present worth of $A$ at $t = 6$. Replacement should take place at that time.

## 3 FAILURE MODELS

The time between installation and failure is not constant for members in the general equipment population. Therefore, in order to solve a failure model, it is necessary to have the distribution of individual item lives (*mortality curve*). The *conditional probability of failure* in a small time interval, say from $t$ to $(t + \delta t)$, is calculated from the mortality curve. This probability is *conditional* since it is conditioned on non-failure up to time $t$.

The conditional probability can decrease with time (e.g., *infant mortality*), remain constant (as with an exponential reliability distribution and failure from random causes), or increase with time (as with items that deteriorate with use). If the conditional probability decreases or remains constant over time, operating items should never be replaced prior to failure.

It usually is assumed that all failures occur at the end of a period. The problem is to find the period which minimizes the total cost.

*Example 19.7*

100 items are tested to failure. Two failed at $t = 1$, five at $t = 2$, seven at $t = 3$, 20 at $t = 4$, 35 at $t = 5$, and 31 at $t = 6$. Find the probability of failure in any period, the conditional probability of failure, and the mean time before failure.

The MTBF is

$$\frac{(2)(1) + (5)(2) + (7)(3) + (20)(4) + (35)(5) + (31)(6)}{100}$$

$$= 4.74$$

| elapsed time t | failures F(t) | survivors S(t) | probability of failure $p\{t\} = 0.01\mathbf{F}(t)$ | conditional probability of failure $v\{t\} = \mathbf{F}(t)/\mathbf{S}(t-1)$ |
|---|---|---|---|---|
| 0 | 0 | 100 | — | — |
| 1 | 2 | 98 | 0.02 | 0.02 |
| 2 | 5 | 93 | 0.05 | 0.051 |
| 3 | 7 | 86 | 0.07 | 0.075 |
| 4 | 20 | 66 | 0.20 | 0.233 |
| 5 | 35 | 31 | 0.35 | 0.530 |
| 6 | 31 | 0 | 0.31 | 1.00 |

## 4 REPLACEMENT POLICY

The expression for the number of units failing in time $t$ is

$$\mathbf{F}(t) = n\left[p\{t\} + \sum_{i=1}^{t-1} p\{i\}p\{t-i\} + \sum_{j=2}^{t-1}\left[\sum_{i=1}^{j-1} p\{i\}p\{j-i\}\right]p\{t-j\} + \cdots\right] \quad 19.22$$

The term $np\{t\}$ gives the number of failures in time $t$ from the original group.

The term $n\sum p\{i\}p\{t-i\}$ gives the number of failures in time $t$ from the set of items which replaced the original items.

The third probability term times $n$ gives the number of failures in time $t$ from the set of items which replaced the first replacement set.

It can be shown that $\mathbf{F}$(t) with replacement will converge to a steady state limiting rate of

$$\overline{\mathbf{F}(t)} = \frac{n}{MTBF} \quad 19.23$$

The optimum policy is to replace all items in the group, including items just installed, when the total cost per period is minimized. That is, we want to find $T$ such that $\mathbf{K}$(T)/T is minimized.

$$\mathbf{K}(T) = nC_1 + C_2\sum_{t=0}^{T-1}\mathbf{F}(t) \quad 19.24$$

Discounting usually is not included in the total cost formula since the time periods are considered short. If the equipment has an unusually long life, discounting is required.

There are some cases where group replacement always is more expensive than replacing just the failures as they occur. Group replacement will be the most economical policy if equation 19.25 holds.

$$C_2[\overline{\mathbf{F}(t)}] > \left.\frac{\mathbf{K}(T)}{T}\right|_{\text{minimum}} \quad 19.25$$

If the opposite inequality holds, group replacement still may be the optimum policy. Further analysis is required.

# PART 5: FORTRAN Programming

The FORTRAN language currently exists in several versions. Although differences exist between compilers, these are relatively minor. However, some of the instructions listed in this chapter may not be compatible with all compilers.

This section is not intended as instruction in FORTRAN programming, but rather serves as a documentation of the language.

## 1 STRUCTURAL ELEMENTS

Symbols are limited to the upper-case alphabet, digits 0 through 9, the blank, and the following special characters.

+ = − * / ( ) , . $

Statements written with these characters generally are prepared in an 80-column format. Statements are executed sequentially regardless of the statement numbers.

| position | use |
|---|---|
| 1 | The letter *C* is used for a *comment.* Comments are not executed. |
| 2–5 | The statement number, if used, is placed in positions 2 through 5. Statement numbers can be any integers from 1 through 9999. |
| 6 | Any character except *zero* can be placed in position 6 to indicate a continuation from the previous statement. |
| 7–72 | The FORTRAN statement is placed here. |
| 73–80 | These positions are available for any use and are ignored by the compiler. Usually, the final debugged program is numbered sequentially in these positions. |

FORTRAN compilers pack the characters. Therefore, blanks can be inserted at any place in most statements. For example, the following statements are compiled the same way.

```
IF (AGE.LT.YEARS) GO TO 10
IF(AGE.LT.YEARS)GOTO10
```

## 2 DATA

Numerical data can be either real or integer. *Integers* usually are limited to nine digits. Unsigned integers and integers preceded by a plus sign are the same. Commas are not allowed in integer constants. For example, ninety thousand would be written as 90000, not 90,000.

*Real numbers* are distinguished from integers by a decimal point and may contain a fractional part. Scientific notation is indicated by the single letter *E*. Real numbers are limited to one decimal point and usually seven digits.

| value | FORTRAN Notation |
|---|---|
| 2 million | 2. E6 |
| 0.00074 | 7.4 E−4 |
| 2. | 2. |

## 3 VARIABLES

*Variable* names can be formed from up to six alphanumeric characters. The first character must be a letter. Variable names starting with the letters *I*, *J*, *K*, *L*, *M*, or *N* are assumed by the compiler to be integers unless defined otherwise by an *explicit typing statement.* All other variable names represent real variables, unless explicitly typed.

The type convention can be overridden in an explicit typing statement. This is done by defining the desired variable type in the first part of the program with an INTEGER or REAL statement. For example, the statements

```
INTEGER TIME,CLOCK
REAL INSTANT
```

would establish TIME and CLOCK as integer variables and INSTANT as a real variable. The order of such declarations is unimportant. Variables following the standard type convention (implicit typing) do not have to be declared.

*Subscripted variables* with up to seven dimensions are allowed. They always must be defined in size by the DIMENSION statement. For example, the statements

```
DIMENSION SAMPLE(5)
REAL DIMENSION INCOME(2,7)
```

would establish a $1 \times 5$ real *array* called SAMPLE and a $2 \times 7$ real array called INCOME. INCOME would have been an integer array without the REAL declaration.

Elements of arrays are addressed by placing the subscripts in parentheses.

```
SAMPLE(2)
INCOME(1,6)
```

The subscripts also can be variables. SAMPLE(K) would be permitted as long as $K$ was defined, was between 1 and 5, and was an integer.

Variables and arrays once defined and declared are not initialized automatically. If it is necessary to initialize a storage location prior to use, the DATA statement can be used. Consider the following statements.

```
REAL X,Y,Z
DIMENSION ONEDIM(5)
DIMENSION TWODIM(2,3)
DATA X,Y,Z/3*0.0/(ONEDIM(I),I = 1,5)
1/5*0.0/ ((TWODIM(I,J),J = 1,3),I = 1,2)
2/1.,2.,3.,4.,5.,6.
```

Variables X, Y, and Z will be initialized to 0.0. The entries in ONEDIM will have the values (0,0,0,0,0). The TWODIM array will be initialized to

$$\begin{pmatrix} 1.0 & 2.0 & 3.0 \\ 4.0 & 5.0 & 6.0 \end{pmatrix}$$

After being initialized with a DATA statement, variables can have their values changed by arithmetic operations.

## 4 ARITHMETIC OPERATIONS

FORTRAN provides for the usual arithmetic operations. These are listed in table 19.2.

**Table 19.2**
FORTRAN OPERATORS

| symbol | meaning |
|---|---|
| = | replacement |
| + | addition |
| − | subtraction |
| * | multiplication |
| / | division |
| ** | exponentiation |
| ( ) | preferred operation |

The *equals* symbol is used to replace one quantity with another. For example, the following statement is algebraically incorrect. However, it is a valid FORTRAN statement.

$$Z = Z + 1$$

Each statement is scanned from left-to-right (except that a right-to-left scan is made for exponentiation). Operations are performed in the following order.

exponentiation first
multiplication and division second
addition and subtraction last
Parentheses can be used to modify this hierarchy.

Each operation must be stated explicitly and unambiguously. Thus, $AB$ is not a substitute for $A^*B$. Two operations in a row, as in $(A + -B)$ also are unacceptable. Some FORTRAN compilers allow mixed-mode arithmetic. Others, ANSI FORTRAN among them, require all variables in an expression to be either integer or real. Where mixed-mode arithmetic is permitted, care must be taken in the conversion of real data to the integer mode.

Integer variables used to hold the results of a mixed-mode calculation will have their values truncated. This is illustrated in the following example.

*Example 19.8*

Evaluate $J$ in the following expression.

$$J = (6.0 + 3.0)^*3.0/6.0 + 5.0 - 6.0^{**}2.0$$

The expressions within parentheses are evaluated in the first pass.

$$J = 9.0^*3.0/6.0 + 5.0 - 6.0^{**}2.0$$

The exponentiation is performed in the second pass.

$$J = 9.0^*3.0/6.0 + 5.0 - 36.0$$

The multiplication and division are performed in the third pass.

$$J = 4.5 + 5.0 - 36.0$$

The addition and subtraction are performed in the fourth pass.

$$J = -26.5$$

However, $J$ is an integer variable, so the real number $-26.5$ is truncated and converted to integer. The final result is $-26$.

## 5 PROGRAM LOOPS

*Loops* can be constructed from IF and GO TO statements. However, the DO statement is a convenient method of creating loops. The general form of the DO statement is

$$\text{DO } s\ i\ = j, k, l$$

where $s$ is a statement number.

$i$ is the integer loop variable.

$j$ is the initial value assigned to $i$.

$k$, which must exceed $j$, is an inclusive upper bound on $i$.

$l$ is the increment for $i$, with a default value of 1 if omitted.

The DO statement causes the execution of the statements immediately following it through statement $s$ until $i$ equals $k$ or greater.[1] A loop can be *nested* by placing it within another loop. The loop variable may be used to index arrays.

When $i$ equals or exceeds $k$, the statement following $s$ is executed. However, the loop may be exited at any time before $i$ reaches $k$ if the logic of the loop provides for it.

## 6 INPUT/OUTPUT STATEMENTS

The READ, WRITE, and FORMAT statements are FORTRAN's main I/O statements. Forms of the READ and WRITE statements are

READ ($u_1$, $s$) **[list]**

WRITE ($u_2$, $s$) **[list]**

where

$u_1$ is the unit number designation for the desired input device, usually 5 for the card reader.

$u_2$ is the unit designation for the desired output device, usually 6 for the line printer.

$s$ is the statement number of an associated FORMAT statement.

**[list]** is a list of variables separated by commas whose values are being read or written.

The **[list]** also can include an implicit DO loop. The following example reads six values, the first five into the array PLACE and the last into SHOW.

READ (5,85) (PLACE(J),J = 1,5) SHOW

The purpose of the FORMAT statement is to define the location, size, and type of the data being read. The form of the FORMAT statement is

$s$ FORMAT **[field list]**

As before, $s$ is the statement number. **[field list]** consists of specifications, set apart by commas, defining the I/O fields. **[list]** can be shorter than **[field list]**.

The format code for integer values is $nIw$.

$n$ is an optional repeat counter which indicates the number of consecutive variables with the same format. $w$ is the number of character positions. The format codes for real values are

$$nFw.d \text{ or } nEw.d$$

Again, $w$ is the number of character positions allocated, including the space required for the decimal point. $d$ is the implied number of spaces to the right of the decimal point. In the case of input data, decimal points in any position take precedence over the value of $d$.

The $F$ format will print a total of $(w-1)$ digits or blanks representing the number. The $E$ format will print a total of $(w-2)$ digits or blanks and give the data in a standard scientific notation with an exponent.

Other formats which can be used in the FORMAT statement are

X horizontal blanks
/ skipping lines
H alphanumeric data
D double precision real
T position (column) indicator
Z hexadecimal
P decimal point modification
L logical data
' ' literal data

The usual output device is a line printer with 133 print positions. The first print position is used for *carriage control.* The data (control character) in the first output position will control the printer advance according to the rules in table 19.3.

**Table 19.3**
FORTRAN Printer Control Characters

| control character | meaning |
|---|---|
| blank | advance one line |
| 0 | advance two lines |
| 1 | skip to line one on next page |
| + | do not advance (overprint) |

Carriage control usually is accomplished by the use of literal data. Consider the following statements.

INTEGER K

K = 193

[1] A peculiarity of FORTRAN DO statements is that they are executed at least once, regardless of the values of $i$ and $j$.

```
WRITE (6,100) K
WRITE (6,101) K
100 FORMAT (' ',I3)
101 FORMAT (I3)
```

The above program would print the number 193 on the next line of the current page and the number 93 on the first line of the next page.

Data can be written to or read from an array by including the array subscripts in the I/O statement.

```
    DIMENSION CLASS (2,5)
    READ (5,15) ((CLASS(I,J),I = 1,2),J = 1,5)
15  FORMAT (10F3.0)
```

## 7 CONTROL STATEMENTS

The STOP statement is used to indicate the logical end of the program. The format is *s* STOP.

STOP should not be the last statement. When it is reached, program execution is terminated. The value of *s* is printed out or made available to the next program step. The use of STOP rarely is recommended.

The END statement is required as the last statement. It tells the compiler that there are no more lines in the program to be compiled. A program cannot be compiled or executed without an END statement.

The PAUSE statement will cause execution to stop temporarily. Its format is *s* PAUSE.

When the PAUSE statement is reached, the number *s* is transmitted to the computer operator. This gives the operator a chance to set various control switches on the console (the choice of switches being dependent on the value of *s* and the program logic), prior to pushing the START button. The PAUSE statement is used only if the programmer is operating the computer.

The CALL statement is used to transfer execution to a *subroutine*. CALL EXIT will terminate execution and turn control over to the operating system. The CALL EXIT and STOP statements have similar effects. The RETURN statement ends execution of a program called subroutine and passes execution to the main program. The CONTINUE statement does nothing. It can be used with a statement number as the last line of a DO loop.

The GO TO [**s**] statement transfers control to statement *s*.

The arithmetic IF statement is written

$$\text{IF}[\mathbf{e}]s_1, s_2, s_3$$

[**e**] is any numerical variable or arithmetic expression, and the $s_i$ are statement numbers. The transfer occurs according to the following table.

| [e] | statement executed |
|---|---|
| [e] $<$ 0 | $s_1$ |
| [e] $=$ 0 | $s_2$ |
| [e] $>$ 0 | $s_3$ |

The logical IF statement has the form

IF[**le**][**statement**]

[**le**] is a logical expression, and [**statement**] is any executable statement except DO and IF. Only if [**le**] is true will [**statement**] be executed. Otherwise, the next instruction will be executed.

The logical expression [**le**] is a relational expression using one of several operators.

Logical expressions also can incorporate the connectors .AND., .OR., and .NOT..

**Table 19.4**
FORTRAN Logical Operations

| operator | meaning |
|---|---|
| .LT. | less than |
| .LE. | less than or equal to |
| .EQ. | equal to |
| .NE. | not equal to |
| .GT. | greater than |
| .GE. | greater than or equal to |

*Example 19.9*

IF (A.GT.25.6) A = 27.0

Meaning: If A is greater than 25.6, set A equal to 27.0.

IF (Z.EQ.(T−4.0).OR.Z.EQ.0.) GO TO 17

Meaning: If Z is equal to (T−4.0) or if Z is equal to *zero*, go to statement 17.

## 8 LIBRARY FUNCTIONS

The following single-precision library functions are available. Most are accessed by placing the argument in parentheses after the function name. Placing the letter *D* before the function name will cause the calculation to be performed in double precision. Arguments for trigonometric functions are expressed in radians.

**Table 19.5**
Some FORTRAN Library Functions

| function | use |
|---|---|
| EXP | $e^x$ |
| ALOG | natural logarithm |
| ALOG10 | common logarithm |
| SIN | sine |
| COS | cosine |
| TAN | tangent |
| SINH | hyperbolic sine |
| SQRT | square root |
| ASIN | arcsine |
| MOD | remaindering modulus (integer) |
| AMOD | remaindering modulus (real) |
| ABS | absolute value (real) |
| IABS | absolute value (integer) |
| FLOAT | convert integer to real |
| FIX | convert real to integer |

## 9 USER FUNCTIONS

A user-defined function can be created with the FUNCTION statement. Such functions are governed by the following rules.

- The function is defined as a variable in the main program even though it is a function.
- When used in the main program, the function is followed by its arguments in parentheses.
- In the function itself, the function name is type-declared and defined by the word FUNCTION.
- The arguments (parameters) need not have the same names in the main program and function.
- Only the function has a RETURN statement.
- Both the main program and the function have END statements.
- The arguments (parameters) must agree in number, order, type, and length.

These construction rules are illustrated by example 19.10.

*Example 19.10*

```
REAL HEIGHT, WIDTH, AREA, MULT
HEIGHT = 2.5
WIDTH = 7.5
AREA = MULT(HEIGHT,WIDTH)
END

REAL FUNCTION MULT(HEIGHT,WIDTH)
REAL HEIGHT, WIDTH
MULT = HEIGHT*WIDTH
RETURN
END
```

## 10 SUBROUTINES

A subroutine is a user-defined subprogram. It is more versatile than a user-defined function as it is not limited to mathematical calculations. Subroutines are governed by the following rules.

- The subroutine is activated by the CALL statement.
- The subroutine has no type.
- The subroutine does not take on a value. It performs operations on the arguments (parameters) which are passed back to the main program.
- A subroutine has a RETURN statement.
- Both the main program and the subroutine have END statements.
- The arguments (parameters) need not have the same names in the main program and the subroutine.
- The arguments (parameters) must agree in number, order, type, and length.
- It is possible to return to any part of the main program. It is not necessary to return to the statement immediately below the CALL statement.

These rules are illustrated by example 19.11.

*Example 19.11*

```
    REAL HEIGHT, WIDTH, AREA
    CALL GET(HEIGHT, WIDTH)
    AREA = HEIGHT*WIDTH
    END

    SUBROUTINE GET(A,B)
    REAL A,B
    READ (5,100) A,B
100 FORMAT(2F3.1)
    RETURN
    END
```

Variables in functions and subroutines are completely independent of the main program. Subroutine and main program variables which have the same names will not have the same values. A link between the main program and the subroutine can be established, however, with the COMMON statement.

The COMMON statement assigns storage locations in memory to be shared by the main program and all of its subroutines. Even the COMMON statement, however, allows different names. It is the order of the common variables which fixes their position in upper memory.

*Example 19.12*

What are the values of $X$ and $Y$ in the subroutine?

```
COMMON X, Y        main program
  X = 2.0
  Y = 10.0

COMMON Y, X        subroutine
```

Since $Y$ is the first common subroutine variable which corresponds to $X$ in the main program, $Y = 2.0$. Similarly, $X = 10.0$.

If variables are to be shared with only some of the subroutines, the *named* COMMON statement is required. Whereas there can be only one regular COMMON statement, there can be multiple-named COMMON statements.

```
COMMON/PLACE/CAT, COW, DOG  main program
COMMON/PLACE/HORSE, PIG, EXPENSE
                             subroutine
```

# PART 6: Fire Safety Systems

## 1 INTRODUCTION

In many cases, the design of fire detection, fire alarm, and sprinkler systems is governed by state or local codes. It is necessary to review all applicable codes and to meet the most stringent of them. Generally, insurance carrier requirements are more stringent than code minimums. Although codes mandate minimum standards, common sense and professional prudence should be used in specifying the level of fire protection in a building.

In buildings that are partially or wholly sprinklered, provisions for alarm and evacuation as well as provisions for sprinkler supervision must be made.

The type of occupancy greatly affects the degree of protection. Nursing homes, schools, hospitals, and office buildings all have greatly different needs. Furthermore, multi-story buildings with limited escape routes affect the degree of protection.

## 2 DETECTION DEVICES

- **Manual Fire Alarm Stations**
  Mandatory in any system, large or small. Locate in the natural path of exit with the maximum traveling distance to any manual station of 200 feet. Identification of an activated manual station which when opened should be readily visible from the side, down a corridor for at least 200 feet.

- **Heat Detectors—Fixed Temperature**
  135°F in open spaces. 190°–200°F generally used in enclosed or confined spaces such as boiler rooms, closets, etc., where the heat build-up will be fast and confined.

- **Heat Detectors—Rate of Rise**
  Combination rate of rise and fixed temperature of 135° or 200°F. Rate of rise portion operates when the temperature rises in excess of 15° per minute. More sensitive than the fixed temperature detector.

- **Heat Detector—Rate Compensated**
  Considered to be the most responsive of all thermal detectors. Operates at 135° or 200°F. Detects both slow and fast developing fires by anticipating the temperature increase and moving towards the alarm point as the temperature gradually increases.

- **Photoelectric Smoke Detectors**
  Operates on a photo beam or light scattering principle. The photoelectric detector responds best to products of combustion or smoke with a particle size from approximately 10.0 microns down to 0.1 micron and of the proper concentration. Proper concentration is defined by Underwriters Laboratories as the ability to sense smoke in the 0.2 to 4.0 percent obscuration per foot range. Photoelectric detectors generally are considered to be the best for cold smoke fires.

- **Ionization Detectors**
  Ionization detectors detect products of combustion by sensing the disruption of conductivity in an ionized chamber due to the presence of smoke. The ionization detector responds best to fast burning fires where particle sizes range from approximately 1.0 micron down to 0.01 micron and of the proper concentration. Proper concentration is defined by Underwriters Laboratories as the ability to sense smoke from 0.2 to 4.0 percent obscuration per foot range. **Note**: Ionization and photoelectric detectors can be intermixed within a system to provide the best form of detection suitable to the environment.

- **Infrared Flame Detectors**
  Generally, infrared detectors respond to radiation in the 6500 to 8500 angstrom range. Good detectors will filter out solar interference and respond to radiation in the 4000 to 5500 angstrom range. It is preferrable to use a detector with a dual sensing circuit in order to discriminate unwanted or false alarms.

- **Ultraviolet Flame Detectors**
  Ultraviolet flame detectors respond to radiation in the spectral range of 1700 to 2900 angstroms. It is not sensitive to solar interference. Built-in time delays prevent false alarms.

- **Waterflow Detectors**
  Used in wet sprinkler systems to indicate a flow of water. Use on the main sprinkler risers and throughout the building to indicate sub-sections or floors of the building to locate sprinkler discharges quickly. These detectors employ a retard mechanism to prevent false alarms from water surges.

- **Pressure Switches**
  Used in dry or pre-action sprinkler systems to provide an alarm when water is discharged.

- **Valve Monitor Switches**
Closed water supply valves are a major weakness of sprinkler systems. This is the most frequently neglected and forgotten item in the sprinkler system. Closed valves also have accounted for countless millions of dollars worth of damage because the system would not operate.

- **Low Temperature Monitor Switches**
Low air temperature (under 40°F) or low water temperature is detrimental to a sprinkler system and creates a great nuisance by freezing and bursting the sprinkler system pipes.

## 3 ALARM DEVICES

- **Bells**
This is the most commonly accepted form of alarm. However, bells should not be used in schools where bells are used for other signaling purposes. Bells should not be used in any area where the same sound is used for any other function.

- **Chimes**
Single stroke devices in coded systems and used in certain types of applications such as quiet areas of hospitals or nursing homes.

- **Horns**
Horns generally are capable of producing a higher sound level than either a bell or a chime.

- **Alarm Lights**
Used individually as a fire alarm visual indicator or used in conjunction with a horn. The light can be either a flashing incandescent bulb or a flashing strobe. It sometimes is required by certain codes or types of occupancy in order to provide an alarm for handicapped persons.

- **Remote Annunciators**
Generally, these duplicate the main control panel and have a light for every fire alarm zone. They are used at the second entrance or at the main entrance to assist the fire department in locating the fire zone. Also used in nursing homes, at nursing stations in hospitals or in the engineering room of a factory or a building.

- **Speakers**
Used in emergency voice evacuation systems, generally in high-rise buildings. Specific quality, construction, and performance have been established by the NFPA code for speakers used in voice evacuation systems.

## 4 SIGNAL TRANSMISSION

- **Reverse Polarity**
Uses a dedicated leased telephone directly between the protected premises and the municipal fire department or a commercial central station.

- **Central Station**
These are central monitoring facilities which monitor fire and security alarms from protected premises for a monthly fee.

- **Telephone Dialers**
Tape dialers have had a very bad reputation due to high failure rate and susceptibility to false alarms. Solid state digital dialers with compatible solid state digital receivers have increased the reliability and dependability of this product very dramatically and are becoming more acceptable as an alternate form of signal transmission.

- **Radio Transmitters**
Radio transmitter boxes are used in certain applications to transmit alarms between the protected premises and some monitoring point. Before using, it is advisable to check that the line of transmission is clear from obstruction and interference.

## 5 AUXILIARY CONTROL

- **Smoke Doors**
Generally held open with floor or wall mounted electromagnets. Generally all doors close in all parts of the building on the first alarm.

- **Fire Doors**
Generally treated the same way as smoke doors.

- **Stairwell Exit Doors**
Sometimes for security reasons, these doors are held latched with a door strike. The door also must employ a panic exit bar to override and manually open the door.

- **Elevator Capture**
Generally accepted by all codes that the elevator immediately return to the first floor or to some designated alternate floor. The specification should designate that the elevator manufacturer is responsible for accepting low voltage signal or a dry contact from the fire alarm panel and programming the elevator to return to the designated floor.

- **Fire Dampers**
Either motorized or fusible link type are closed to prevent the spread of smoke to other areas. Care should be exercised in connecting these to a fire alarm system because a fire alarm system generally operates with small amounts of D.C. power. The responsibility should be stated for coordinating the

voltages and contact ratings necessary to do the job.

- **Fans—Supply and Exhaust**
  Sometimes all fans are shut down on the first alarm. However, some buildings and other considerations make it desirable to exhaust the smoke from the building and shut down the input supply fans.

- **Pressurization**
  Pressurization creates positive air pressure to inhibit the influx of smoke from adjacent areas or floors above or below. Exit stairwell escape routes out of the high-rise building should be pressurized to provide a smoke-free exit path.

## 6 SPECIAL CONSIDERATIONS

- **Handicapped Persons**
  There may be requirements from federal health officials (HEW) or OSHA to provide consideration for handicapped persons.

- **Weather Protection of Devices**
  Check to make sure that none of the devices will be exposed to adverse conditions such as excessive moisture.

- **Open Plenums**
  Drop ceilings that use the space above as an open air plenum present special problems and are treated as a potential hazard in most codes. Smoke detectors located in open air plenums need to be located properly, and it is advisable to locate them so that they are accessible for maintenance. A remote alarm lamp should be brought down and mounted below the ceiling level to indicate which detector is an alarm.

- **Duct Detectors**
  Duct heat detectors are considered to be of negligible value. Duct smoke detectors can provide warning of smoke in an air duct and can prevent costly damage to expensive HVAC equipment. Duct detectors are not a substitute for open area smoke detectors. Problem areas can develop with duct detectors from excessive humidity, poor or no maintenance, and improper location.

- **Sound Pressure Level of Alarm Devices**
  The sound pressure level required to provide an adequate alarm in the environment in which it is intended to operate generally is considered to have been accomplished if the alarm sound is 12 decibels above the normal ambient level.

- **Emergency Generators**
  If the fire alarm system does not use its own standby battery pack, it is advisable to coordinate the details of how the emergency generator feeds power back into the building distribution system and to insure that the fire alarm system will be provided with emergency power. In some cases, it is advisable to provide the fire alarm system with a small amount of standby battery power in order to keep the fire alarm system on line until the generator gets to full power.

- **Fire Pump Supervision**
  Insure that fire pumps, if used, are supervised adequately for all critical functions and that the building fire alarm system is provided with one or more zones to interface with fire pump signals.

## 7 SPECIAL SUPPRESSION SYSTEMS

- **Deluge Systems**
  The control panel for the water deluge system is specialized and generally is mounted in a hostile environment. The control equipment is mounted inside an enclosure which is watertight and dust tight. Deluge systems are actuated manually and also by thermal detectors or flame detectors. The entire system, including the electronic control panel, is provided by the sprinkler contractor.

- **Foam Systems**
  High expansion foam or water suppression systems are specialized systems requiring special handling. They generally are activated by infrared or ultraviolet flame detectors.

- **Halon Systems**
  Halon systems generally are used in clean environments such as computer rooms or any room that contains high value equipment. Examples of this are tapes, microfilms, computer records, laboratories for research and design, and medical laboratories. Generally, the criterion is the high value of the equipment or the data stored, and it is desirable that water not touch the equipment. Halon systems, because of their great expense, usually are engineered specially for the particular room in which they are to be located. The control panel uses cross-zoning techniques to prevent unnecessary discharges.

- **High/Low Pressure $CO_2$ Systems**
  $CO_2$ systems are used in industrial applications to suppress fires where the use of water, extinguishing powders, or Halon would be unsuitable because of expense, hazard, or the size of the equipment to be protected. Because $CO_2$ is hazardous to life, it has to be an engineered system designed for the particular application.

# PART 7: Nondestructive Testing

## 1 MAGNETIC PARTICLE TESTING

This procedure is based on the attraction of magnetic particles to leakage flux at surface flaws. The particles accumulate and become visible at the flaw. This method works for magnetic materials in locating cracks, laps, seams, and in some cases, subsurface flaws. The test is fast and simple and is easy to interpret. However, parts must be relatively clean and demagnetized. A high current (power) source must be available.

## 2 EDDY CURRENT TESTING

Alternating currents from a source coil induce eddy currents in metallic objects. Flaws and other material properties affect the current flow. The change in current flow is observed on a meter or a screen. This method can be used to locate defects of many types, including changes in composition, structure, and hardness, as well as locating cracks, voids, inclusions, weld defects, and changes in porosity.

Intimate contact between the material and the test coil is not required. Operation can be continuous, automatic, and monitored electronically. Therefore, this method is ideal for unattended continuous processing. Sensitivity is easily adjustable. Many variables, however, can affect the current flow.

## 3 LIQUID PENETRANT TESTING

Liquid penetrant (dye) is drawn into surface defects by capillary action. A developer substance then is used to develop the penetrant to aid in visual inspection. This method can be used with any nonporous material, including metals, plastics, and glazed ceramics. It locates cracks, porosities, pits, seams, and laps.

Liquid penetrant tests are simple to perform, can be used with complex shapes, and can be performed on site. Parts must be clean, and only small surface defects are detectable.

## 4 ULTRASONIC TESTING

Mechanical vibrations in the 0.1–25 MHz range are induced in an object. The transmitted energy is reflected and scattered by interior defects. The results are interpreted from a screen or a meter. The method can be used for metals, plastics, glass, rubber, graphite, and concrete. It detects inclusions, cracks, porosity, laminations, changes in structure, and other interior defects.

This test is extremely flexible. It can be automated and is very fast. Results can be recorded or interpreted electronically. Penetration of up to 60 feet of steel is possible. Only one surface needs to be accessed. However, rough surfaces or complex shapes may cause difficulties.

## 5 INFRARED TESTING

Infrared radiation emitted from objects can be detected and correlated with quality. The detection can be recorded electronically. Any discontinuity that interrupts heat flow, such as flaws, voids, and inclusions, can be detected.

Infrared testing requires access to only one side, and it is highly sensitive. It is applicable to complex shapes and assemblies of dissimilar components, but it is relatively slow. Results are affected by material variations, coatings, and colors, and hot spots can be hidden by cool surface layers.

## 6 RADIOGRAPHY

X-ray and gamma-ray sources can be used to penetrate objects. The intensity is reduced in passing through, and the intensity changes are recorded on film or screen. This method can be used with most materials to detect internal defects, material structure, and thickness. It also can be used to detect the absence of internal parts.

Up to 30 inches of steel can be penetrated by x-ray sources. Gamma sources, which are more portable and lower in cost than x-ray sources, are applicable to 10″ thickness of steel.

There are health and government standards associated with these tests. Electric power and cooling water may be required in large installations. Shielding and film processing also is required, making this the most expensive nondestructive test.

# PART 8: Environmental Impact Assessment

Most large scale construction projects must be assessed for potential environmental damage before being built. The assessment is usually developed in an *environmental impact report.* Such a report should evaluate all of the potential ways in which a project could affect the ecological balance and environment. The questions which follow can be used as an outline for such a report.

1. What is the nature of the proposed project?

2. What is the nature of the project area, including distinguishing natural and man-made characteristics?

3. How, and to what degree, will the earth be altered by any of the following means?

   - change in topology, including earth removal
   - off-site hauling, dust, smoke, or air pollutants generated during construction
   - chemical treatment
   - change in structural composition of the soil as a result of compacting, tilling, shoring, etc.
   - change in moisture content
   - change in slope

4. In what ways will the project affect natural drainage or flooding? Compare existing flooding and drainage conditions with those that would exist after the project is finished.

5. To what extent could the project result in the erosion of property both on or off the project site?

6. To what extent will the project affect the potential use, extraction, or conservation of the earth's resources, such as crops, minerals, and ground water?

7. How, and to what extent, will the project affect the quality and characteristics of soil, water, and air in the immediate project area and in the community? Include an estimate of the amounts of sewage, drainage, airborne particulate matter, and solid waste to be generated by the completed project.

8. Describe the plant and animal life presently on the site, and describe how, and to what extent, this life will be affected by the project.

9. Indicate the percentages of land use, both before and after the project is completed, for the following categories of land use:

   - residential
   - commercial
   - industrial
   - public

10. Indicate the percentages of land use, both before and after the project is completed, for the following categories of land use:

   - agricultural
   - street and highway
   - parking, parking driveways, loading, etc.
   - surface or aerial utilities
   - railroads
   - vacant and unimproved
   - landscaping

11. How will the project change the views of the site from points around the project?

12. How will the project change the views of the surrounding community from points within the project?

13. How, and to what extent, will the project affect wilderness, open space, landscaping, and other aesthetic and recreational considerations?

14. Are any of the natural or man-made features in the project unique (i.e., not found in any other parts of the city, county, state, or nation)?

15. What effect will the project have upon the health and safety of the people in the project area and in the community?

16. How much fresh water will be consumed as a result of the project? What will be the source of this water?

17. How, and to what extent, will the project add to existing noise levels on the site and in the community?

18. How many additional people will live in the project area? How many additional people will work there? How many people will be displaced from the project area?

19. How will the project add to, or subtract from, the availability of roads, highways, and other elements of the transportation system, or increase or decrease the burden upon such elements?

20. Could the project serve to encourage development of presently undeveloped areas? Could it intensify development of already developed areas?

21. Will the project involve the application, use, or disposal of hazardous materials?

22. Will the project involve the construction of facilities in areas of known earthquake faults or soil instability (e.g., subsidence, landslides, or severe erosion)?

23. Does the project affect a historically significant or archaeological site?

24. Are there any other significant environmental effects, either positive or negative, of the project?

# PART 9: Mathematical Programming

## 1 INTRODUCTION

Mathematical programming is a modeling procedure applicable to problems for which the goal and resource limitations can be described mathematically. If the *goal function* and all *resource constraints* are linear (polynomials of degree 1 only), the procedure is known as *linear programming.*

If the variables can take on only integer values, a procedure known as *integer programming* is required. If the polynomials are of any degree or contain other functions, a procedure known as *dynamic programming* is required.

## 2 FORMULATION OF A LINEAR PROGRAMMING PROBLEM

All linear programming problems have a similar format. Each has an *objective function* which is to be optimized. Usually the objective function is to be maximized, as in the case of a *profit function.* If the objective is to minimize some function, such as cost, the problem may be turned into a maximization problem by maximizing the negative of the original function.

*Example 19.13*

A cattle rancher buys three types of cattle food. The rancher wants to minimize the cost of feeding his cattle. Write the objective function for this problem on a per animal basis.

| food type | cost per pound |
|---|---|
| 1 | 1.5 |
| 2 | 2.5 |
| 3 | 3.5 |

Let $x_i$ be the number of pounds of food $i$ purchased per animal. Then, the objective function to be minimized is

$$Z = 1.5x_1 + 2.5x_2 + 3.5x_3$$

Each linear programming problem also has a set of limitation functions called *constraints.* Constraints are used to set the bounds for the objective function.

*Example 19.14*

The rancher is concerned with meeting published nutritional information on minimum daily requirements (MDR) given in milligrams per animal. The composition of each food type is known and the contributions for each vitamin in mg/pound are

| | | food type | | |
|---|---|---|---|---|
| vitamin | MDR (mg) | 1 | 2 | 3 |
| A | 100 | 1 | 7 | 13 |
| B | 200 | 3 | 9 | 15 |
| C | 300 | 5 | 11 | 17 |

It is also physically impossible for an animal to eat more than the following amounts per day.

| food type | maximum feeding |
|---|---|
| 1 | 50 lbs |
| 2 | 40 |
| 3 | 30 |

The constraints on this problem are

$$\begin{aligned} x_1 + 7x_2 + 13x_3 &\geq 100 \\ 3x_1 + 9x_2 + 15x_3 &\geq 200 \\ 5x_1 + 11x_2 + 17x_3 &\geq 300 \\ x_1 &\leq 50 \\ x_2 &\leq 40 \\ x_3 &\leq 30 \\ x_1 &\geq 0 \\ x_2 &\geq 0 \\ x_3 &\geq 0 \end{aligned}$$

Linear programming problems are generally solved by computer. Some simple problems can be solved by hand with a procedure known as the *simplex method.* Specialized methods allowing easy manual solutions are available for certain classes of problems, primarily the *transportation problem* and *assignment problem.*

Once a solution is found, it is possible to determine the effect on the objective function of changing one of the program parameters. This is known as *sensitivity analysis* and is very important in instances where the accuracy of collected data is unknown.

## 3 SOLUTION TO 2-DIMENSIONAL PROBLEMS

If a linear programming problem can be formulated in terms of only two variables, $x_1$ and $x_2$, it can be solved graphically by the following procedure:

*step 1:* Graph all of the constraints and determine the *feasible region.* Usually this will result in a *convex hull.*

*step 2*: Evaluate the objective function, $Z$, at each corner of the hull.

*step 3*: The values of $x_1$ and $x_2$ which optimize $Z$ are the coordinates of the corner at which $Z$ is optimized.

*Example 19.15*

Solve the following linear programming problem graphically.

$$
\begin{aligned}
\text{Max } Z = & \quad x_1 + 2x_2 \\
\text{such that } & \quad 4x_1 + x_2 \leq 24 \\
& \quad 2x_1 + x_2 \leq 14 \\
& \quad -x_1 + 2x_2 \leq 8 \\
& \quad -x_1 + x_2 \leq 3 \\
& \quad x_1 \geq 0 \\
& \quad x_2 \geq 0
\end{aligned}
$$

The region enclosed by the constraints is shown.

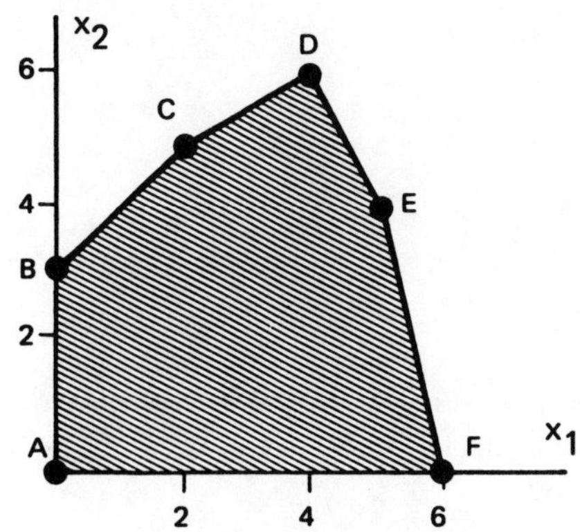


The coordinates and $Z$ value for each corner are

| corner | coordinates $(x_1, x_2)$ | $Z$ |
|---|---|---|
| A | (0,0) | 0 |
| B | (0,3) | 6 |
| C | (2,5) | 12 |
| D | (4,6) | 14 |
| E | (5,4) | 13 |
| F | (6,0) | 6 |

$Z$ is maximized when $x_1 = 4$ and $x_2 = 6$.

RESERVED FOR FUTURE USE

# 20 SYSTEMS OF UNITS

## 1 CONSISTENT SYSTEMS OF UNITS

A set of units used in a problem is said to be *consistent*[1] if no conversion factors are needed. For example, a moment with units of foot-pounds cannot be obtained directly from a moment arm with units of inches. In this illustration, a conversion factor of $\frac{1}{12}$ feet/inch is needed, and the set of units used is said to be *inconsistent.*

On a larger scale, a system of units is said to be consistent if Newton's second law of motion can be written without conversion factors. Newton's law states that the force required to accelerate an object is proportional to the amount of matter in the item.

$$F = ma \tag{20.1}$$

The definitions of the symbols, $F$, $m$, and $a$, are familiar to every engineer. However, the use of Newton's second law is complicated by the multiplicity of available unit systems. For example, $m$ may be in kilograms, pounds, or slugs. All three of these are units of mass. However, as figure 20.1 illustrates, these three units do not represent the same amount of mass.

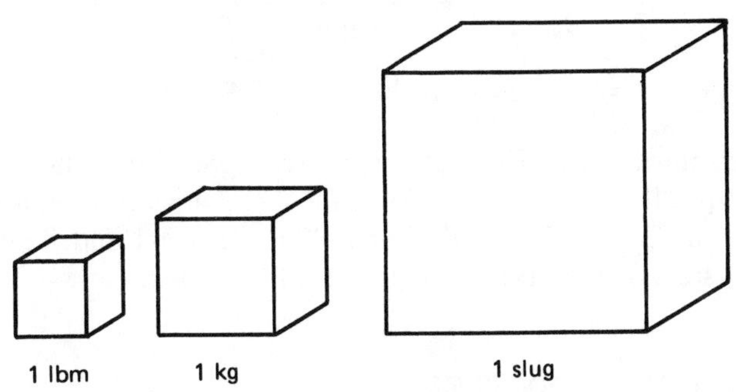


**Figure 20.1** Common Units of Mass

[1] The terms *homogeneous* and *coherent* also are used to describe a consistent set of units.

It should be mentioned that the decision to work with a consistent set of units is arbitrary and unnecessary. Problems in fluid flow and thermodynamics commonly are solved with inconsistent units. This causes no more of a problem than working with inches and feet in the calculation of a moment.

## 2 THE ABSOLUTE ENGLISH SYSTEM

Engineers are accustomed to using pounds as a unit of mass. For example, density typically is given in pounds per cubic foot. The abbreviation *pcf* tends to obscure the fact that the true units are pounds of *mass* per cubic foot.

If pounds are the units for mass, and feet per second squared are the units of acceleration, the units of force for a consistent system can be found from Newton's second law.

$$\begin{aligned} \text{units of } F &= (\text{units of } m)(\text{units of } a) \\ &= (\text{lbm})(\text{ft/sec}^2) = \frac{\text{lbm-ft}}{\text{sec}^2} \end{aligned} \tag{20.2}$$

The units for $F$ cannot be simplified any more than they are in equation 20.2. This particular combination of units is known as a *poundal.*[2]

The absolute English system, which requires the poundal as a unit of force, seldom is used, but it does exist. This existence is a direct outgrowth of the requirement to have a consistent system of units.

## 3 THE ENGLISH GRAVITATIONAL SYSTEM

Force frequently is measured in pounds. When the thrust on an accelerating rocket is given as so many pounds, it is understood that the pound is being used as a unit of force.

[2] A poundal is equal to 0.03108 pounds force.

If acceleration is given in feet per second squared, the units of mass for a consistent system of units can be determined from Newton's second law.

$$\text{units of } m = \frac{\text{units of } F}{\text{units of } a} = \frac{\text{lbf}}{\text{ft/sec}^2} = \frac{\text{lbf-sec}^2}{\text{ft}} \qquad 20.3$$

The combination of units in equation 20.3 is known as a *slug*.[3] Slugs and pounds-mass are not the same, as illustrated in figure 20.1. However, units of mass can be converted, using equation 20.4.

$$\#\ \text{slugs} = \frac{\#\ \text{lbm}}{g_c} \qquad 20.4$$

$g_c$[4] is a dimensional conversion factor having the following value.

$$g_c = 32.1740 \frac{\text{lbm-ft}}{\text{lbf-sec}^2} \qquad 20.5$$

32.1740 commonly is rounded to 32.2 when six significant digits are unjustified. That practice is followed in this book.

Notice that the number of slugs cannot be determined from the number of pounds-mass by dividing by the local gravity. $g_c$ is used regardless of the local gravity. However, the local gravity can be used to find the weight of an object. *Weight* is defined as the force exerted on a mass by the local gravitational field.

$$\text{weight in lbf} = (m \text{ in slugs})(g \text{ in ft/sec}^2) \qquad 20.6$$

If the effects of large land and water masses are neglected, the following formula can be used to estimate the local acceleration of gravity in ft/sec$^2$ at the earth's surface. $\phi$ is the latitude in degrees.

$$g_{\text{surface}} = 32.088[1 + (5.305 \text{ EE} - 3)\sin^2\phi - (5.9 \text{ EE} - 6)\sin^2 2\phi] \qquad 20.7$$

If the effects of the earth's rotation are neglected, the gravitational acceleration at an altitude, $h$, in miles is given by equation 20.8. $R$ is the earth's radius—approximately 3960 miles.

$$g_h = g_{\text{surface}}\left[\frac{R}{R+h}\right]^2 \qquad 20.8$$

[3] A slug is equal to 32.1740 lbm.

[4] Three different meanings of the symbol $g$ commonly are used. $g_c$ is the dimensional conversion factor given in equation 20.5. $g_o$ is the standard acceleration due to gravity with a value of 32.1740 ft/sec$^2$. $g$ is the local acceleration due to gravity in ft/sec$^2$.

## 4 THE ENGLISH ENGINEERING SYSTEM

Many thermodynamics and fluid flow problems freely combine variables containing pound-mass and pound-force terms. For example, the steady-flow energy equation (SFEE) used in chapter 6 mixes enthalpy terms in BTU/lbm with pressure terms in lbf/ft$^2$. This requires the use of $g_c$ as a mass conversion factor.

Newton's second law becomes

$$F \text{ in lbf} = \frac{(m \text{ in lbm})(a \text{ in ft/sec}^2)}{g_c \text{ in } \frac{\text{lbm-ft}}{\text{lbf-sec}^2}} \qquad 20.9$$

Since $g_c$ is required, the English Engineering System is inconsistent. However, that is not particularly troublesome, and the use of $g_c$ does not overly complicate the solution procedure.

*Example 20.1*

Calculate the weight of a 1.0 lbm object in a gravitational field of 27.5 ft/sec$^2$.

Since weight commonly is given in pounds-force, the mass of the object must be converted from pounds-mass to slugs.

$$F = \frac{ma}{g_c} = \frac{(1) \text{ lbm } (27.5) \text{ ft/sec}^2}{(32.2)\frac{\text{lbm-ft}}{\text{lbf-sec}^2}} = .854 \text{ lbf}$$

*Example 20.2*

A rocket with a mass of 4000 lbm is traveling at 27,000 ft/sec. What is its kinetic energy in ft-lbf?

The usual kinetic energy equation is $E_k = \frac{1}{2}mv^2$. However, this assumes consistent units. Since energy is wanted in foot-pounds-force, $g_c$ is needed to convert $m$ to units of slugs.

$$E_k = \frac{mv^2}{2g_c} = \frac{(4000) \text{ lbm } (27{,}000)^2 \text{ ft}^2/\text{sec}^2}{(2)(32.2)\frac{\text{lbm-ft}}{\text{lbf-sec}^2}} = 4.53 \text{ EE10 ft-lbf}$$

In the English Engineering System, work and energy typically are measured in ft-lbf (mechanical systems) or in British Thermal Units, BTU (thermal and fluid systems). One BTU equals 778.26 ft-lbf.

## 5 THE cgs SYSTEM

The cgs system has been used widely by chemists and physicists. It is named for the three primary units used to construct its derived variables. The centimeter, the gram, and the second form the basis of this system.

The cgs system avoids the lbm versus lbf type of ambiguity in two ways. First, the concept of weight is not used at all. All quantities of matter are specified in grams, a mass unit. Second, force and mass units do not share a common name.

When Newton's second law is written in the cgs system, the following combination of units results.

$$\text{units of force} = (m \text{ in } g)\left(a \text{ in } \frac{\text{cm}}{\text{sec}^2}\right) = \frac{g\text{–cm}}{\text{sec}^2} \qquad 20.10$$

This combination of units for force is known as a *dyne.*

Energy variables in the cgs system have units of dyne-cm or, equivalently, of

$$\frac{g\text{–cm}^2}{\text{sec}^2}$$

These combinations are known as an *erg.* There is no uniformly accepted unit of power in the cgs system, although calories per second frequently is used. Ergs can be converted to calories by multiplying by 2.389 EE−8.

The fundamental volume unit in the cgs system is the cubic centimeter (cc). Since this is the same volume as one thousandth of a liter, units of millimeters (ml) are used freely in this system.

## 6 THE mks SYSTEM

The mks system is appropriate when variables take on values larger than can be accomodated by the cgs system. This system uses the *m*eter, the *k*ilogram, and the *s*econd as its primary units. The mks system avoids the lbm versus lbf ambiguity in the same ways as does the cgs system.

The units of force can be derived from Newton's second law.

$$\text{units of force} = (m \text{ in kg})\left(a \text{ in } \frac{\text{m}}{\text{sec}^2}\right) = \frac{\text{kg–m}}{\text{sec}^2} \qquad 20.11$$

This combination of units for force is known as a *newton.*

Energy variables in the mks system have units of N-m or, equivalently, $\frac{\text{kg–m}^2}{\text{sec}^2}$. Both of these combinations are known as a *joule.* The units of power are joules per second, equivalent to a *watt.* The common volume unit is the liter, equivalent to one-thousandth of a cubic meter.

*Example 20.3*

A 10 kg block hangs from a cable. What is the tension in the cable?

$$F = ma = (10)\text{ kg } (9.8)\frac{\text{m}}{\text{sec}^2} = 98\frac{\text{kg–m}}{\text{sec}^2} = 98\ N$$

*Example 20.4*

A 10 kg block is raised vertically 3 meters. What is the change in potential energy?

$$\Delta E_p = mg\Delta h = (10)\text{ kg } (9.8)\ \frac{\text{m}}{\text{sec}^2}\ (3)\text{ m} = 294\frac{\text{kg–m}^2}{\text{sec}^2} = 294\ J$$

## 7 THE SI SYSTEM

Strictly speaking, both the cgs and the mks systems are *metric* systems. Although the metric units simplify solutions to problems, the multiplicity of possible units for each variable sometimes is confusing.

The SI system (International System of Units) was established in 1960 by the General Conference of Weights and Measures, an international treaty organization. The SI system is derived from the earlier metric systems, but it is intended to supersede them all.

The SI system has the following features.

(a) There is only one recognized unit for each variable.

(b) The system is fully consistent.

(c) Scaling of units is done in multiples of 1000.

(d) Prefixes, abbreviations, and symbol-syntax are defined rigidly.

Three types of units are used: base units, supplementary units, and derived units. The base units (table 20.2) are dependent on only accepted standards or reproducible phenomena. The supplementary units (table 20.3) have not yet been classified as being base units or derived units. The derived units (tables 20.4 and 20.5) are made up of combinations of base and supplementary units.

The expressions for the derived units in symbolic form are obtained by using the mathematical signs of multiplication and division. For example, units of velocity are $m/s$. Units of torque are $N \cdot m$ (not $N$-$m$ or $Nm$).

**Table 20.1**
SI Prefixes

| prefix | symbol | value |
|---|---|---|
| exa | E | EE18 |
| peta | P | EE15 |
| tera | T | EE12 |
| giga | G | EE9 |
| mega | M | EE6 |
| kilo | k | EE3 |
| hecto | h | EE2 |
| deca | da | EE1 |
| deci | d | EE−1 |
| centi | c | EE−2 |
| milli | m | EE−3 |
| micro | $\mu$ | EE−6 |
| nano | n | EE−9 |
| pico | p | EE−12 |
| femto | f | EE−15 |
| atto | a | EE−18 |

**Table 20.2**
SI Base Units

| quantity | name | symbol |
|---|---|---|
| length | meter | m |
| mass | kilogram | kg |
| time | second | s |
| electric current | ampere | A |
| temperature | kelvin | K |
| amount of substance | mole | mol |
| luminous intensity | candela | cd |

**Table 20.3**
SI Supplementary Units

| quantity | name | symbol |
|---|---|---|
| plane angle | radian | rad |
| solid angle | steradian | sr |

In addition, there is a set of non-SI units which can be used. This temporary concession is due primarily to the significance and widespread acceptance of these units. Use of the non-SI units listed in table 20.6 usually will create an inconsistent expression requiring conversion factors.

In addition to having standardized units, the SI system also specifies syntax rules for writing the units and combinations of units. Each unit is abbreviated with a specific *symbol*. The rules for writing these symbols should be followed.

(a) The symbols always are printed in roman type, irrespective of the type used in the rest of the text. The only exception to this is in the use of the symbol for *liter*, where the lower case *l* (ell) may be confused with the number 1 (one). In this case, *liter* should be written out in full or the script *l* used. There is no problem with such symbols as cl (centiliter) or ml (milliliter).

(b) Symbols are never pluralized: 1 kg, 45 kg (not 45 kgs).

(c) A period is not used after a symbol, except when the symbol occurs at the end of a sentence.

(d) When symbols consist of letters, there always is a full space between the quantity and the symbols; e.g. 45 kg (not 45kg). However, when the first character of a symbol is not a letter, no space is left, e.g. 32°C (not 32° C or 32 °C) or 42°12′45″ (not 42 ° 12 ′ 45 ″).

(e) All symbols are written in lower case, except when the unit is derived from a proper name. For example, m for meter, s for second, but A for ampere, Wb for weber, N for newton, W for watt. Prefixes are printed roman type without spacing between the prefix and the unit symbol, e.g., km for kilometer.

(f) In text, symbols should be used when associated with a number. When no number is involved, the unit should be spelled out. For example, the area of a carpet is 16 $m^2$, not 16 square meters, and carpet is sold by the square meter, not by the $m^2$.

(g) A practice in some countries is to use a comma as a decimal marker, while the practice in North America, the United Kingdom, and some other countries is to use a period (or a dot) as the decimal marker. Further, in some countries using the decimal comma, a dot frequently is used to divide long numbers into groups of three. Because of these differing practices, spaces must be used instead of commas to separate long lines of digits into easily-readable blocks of three digits with respect to the decimal marker, e.g. 32 453.246 072 5. A space (a half space is preferred) is optional with a four-digit number, e.g; 1 234, 1 234, or 1234.

(h) Where a decimal fraction of a unit is used, a zero should be placed before the decimal marker; e.g., 0.45 kg (not .45 kg). This practice draws attention to the decimal marker and helps avoid errors of scale.

(i) Some confusion may arise with the word *tonne* (1 000 kg). When this word occurs in French text of Canadian origin, the meaning may be a ton or 2 000 pounds.

**Table 20.4**
Some SI Derived Units with Special Names

| quantity | name | symbol | expressed in terms of other units |
|---|---|---|---|
| frequency | hertz | Hz | |
| force | newton | N | |
| pressure, stress | pascal | Pa | $N/m^2$ |
| energy, work, quantity of heat | joule | J | N·m |
| power, radiant flux | watt | W | J/s |
| quantity of electricity, electric charge | coulomb | C | |
| electric potential, potential difference, electromotive force | volt | V | W/A |
| electric capacitance | farad | F | C/V |
| electric resistance | ohm | Ω | V/A |
| electric conductance | siemen | S | A/V |
| magnetic flux | weber | Wb | V·s |
| magnetic flux density | tesla | T | $Wb/m^2$ |
| inductance | henry | H | Wb/A |
| luminous flux | lumen | lm | |
| illuminance | lux | lx | $lm/m^2$ |

**Table 20.5**
Some SI Derived Units

| quantity | description | expressed in terms of other units |
|---|---|---|
| area | square meter | $m^2$ |
| volume | cubic meter | $m^3$ |
| speed—linear | meter per second | m/s |
| angular | radian per second | rad/s |
| acceleration—linear | meter per second squared | $m/s^2$ |
| angular | radian per second squared | $rad/s^2$ |
| density, mass density | kilogram per cubic meter | $kg/m^3$ |
| concentration (of amount of substance) | mole per cubic meter | $mol/m^3$ |
| specific volume | cubic meter per kilogram | $m^3/kg$ |
| luminance | candela per square meter | $cd/m^2$ |
| dynamic viscosity | pascal second | Pa·s |
| moment of force | newton meter | N·m |
| surface tension | newton per meter | N/m |
| heat flux density, irradiance | watt per square meter | $W/m^2$ |
| heat capacity, entropy | joule per kelvin | J/K |
| specific heat capacity, specific entropy | joule per kilogram kelvin | J/(kg·K) |
| specific energy | joule per kilogram | J/kg |
| thermal conductivity | watts per meter kelvin | W/(m·K) |
| energy density | joule per cubic meter | $J/m^3$ |
| electric field strength | volt per meter | V/m |
| electric charge density | coulomb per cubic meter | $C/m^3$ |
| surface density of charge, flux density | coulomb per square meter | $C/m^2$ |
| permittivity | farad per meter | F/m |
| current density | ampere per square meter | $A/m^2$ |
| magnetic field strength | ampere per meter | A/m |
| permeability | henry per meter | H/m |
| molar energy | joule per mole | J/mol |
| molar entropy, molar heat capacity | joule per mole kelvin | J/(mol·K) |
| radiant intensity | watt per steradian | W/sr |

**Table 20.6**
Acceptable Non-SI Units

| quantity | unit name | symbol | relationship to SI unit |
|---|---|---|---|
| area | hectare | ha | 1 ha = 10 000 $m^2$ |
| energy | kilowatt-hour | kWh | 1 kWh = 3.6 MJ |
| mass | metric ton[5] | t | 1 t = 1000 kg |
| plane angle | degree (of arc) | ° | 1° = 0.017 453 rad |
| speed of rotation | revolution per minute | r/min | 1 r/min = $\frac{2\pi}{60}$ rad/s |
| temperature interval | degree Celsius | °C | 1°C = 1 K |
| time | minute | min | 1 min = 60 s |
| | hour | h | 1 h = 3600 s |
| | day (mean solar) | d | 1 d = 86 400 s |
| | year (calendar) | a | 1 a = 31 536 000 s |
| velocity | kilometer per hour | km/h | 1 km/h = 0.278 m/s |
| volume | liter[6] | $l$ | 1 $l$ = 0.001 $m^3$ |

Numbers in parentheses are the number of ESU or EMU units, per single SI unit, except for the permittivity and the permeability of free space, where each actual values of $\epsilon_o$ and $\mu_o$ are given.

[5] The international name for metric ton is *tonne*. The metric ton is equal to the *megagram*, Mg.

[6] The international symbol for liter is the lowercase "l," which can be confused easily with the numeral "1." Several English speaking countries have adopted the script $l$ as the symbol for liter in order to avoid any misinterpretation.

# Appendix A:
# Selected Conversion Factors to SI Units

| | SI Symbol | Multiplier to Convert From Existing Unit to SI Unit | Multiplier to Convert From SI Unit to Existing Unit |
|---|---|---|---|
| **Area** | | | |
| Circular Mil | μm² | 506.7 | 0.001 974 |
| Foot Squared | m² | 0.092 9 | 10.764 |
| Mile Squared | km² | 2.590 | 0.386 1 |
| Yard Squared | m² | 0.836 1 | 1.196 |
| **Energy** | | | |
| Btu (International) | kJ | 1.055 1 | 0.947 8 |
| Erg | μJ | 0.1 | 10.0 |
| Foot Pound-Force | J | 1.355 8 | 0.737 6 |
| Horsepower Hour | MJ | 2.684 5 | 0.372 5 |
| Kilowatt Hour | MJ | 3.6 | 0.277 8 |
| Meter Kilogram-Force | J | 9.806 7 | 0.101 97 |
| Therm | MJ | 105.506 | 0.009 478 |
| Kilogram Calorie (International) | kJ | 4.186 8 | 0.238 8 |
| **Force** | | | |
| Dyne | μN | 10. | 0.1 |
| Kilogram-Force | N | 9.806 7 | 0.101 97 |
| Ounce-Force | N | 0.278 0 | 3.597 |
| Pound-Force | N | 4.448 2 | 0.224 8 |
| KIP | N | 4 448.2 | 0.000 224 8 |
| **Heat** | | | |
| Btu Per Hour | W | 0.293 1 | 3.412 1 |
| Btu Per (Square Foot Hour) | W/m² | 3.154 6 | 0.317 0 |
| Btu Per (Square Foot Hour °F) | W/(m²·°C) | 5.678 3 | 0.176 1 |
| Btu Inch Per (Square Foot Hour °F) | W/(m·°C) | 0.144 2 | 6.933 |
| Btu Per (Cubic Foot °F) | MJ/(m³·°C) | 0.067 1 | 14.911 |
| Btu Per (Pound °F) | J/(kg·°C) | 4 186.8 | 0.000 238 8 |
| Btu Per Cubic Foot | MJ/m³ | 0.037 3 | 26.839 |
| Btu Per Pound | J/kg | 2 326. | 0.000 430 |
| **Length** | | | |
| Angstrom | nm | 0.1 | 10.0 |
| Foot | m | 0.304 8 | 3.280 8 |
| Inch | mm | 25.4 | 0.039 4 |
| Mil | mm | 0.025 4 | 39.370 |
| Mile | km | 1.609 3 | 0.621 4 |
| Mile (International Nautical) | km | 1.852 | 0.540 |
| Micron | μm | 1.0 | 1.0 |
| Yard | m | 0.914 4 | 1.093 6 |
| **Mass (weight)** | | | |
| Grain | mg | 64.799 | 0.015 4 |
| Ounce (Avoirdupois) | g | 28.350 | 0.035 3 |
| Ounce (Troy) | g | 31.103 5 | 0.032 15 |
| Ton (short 2000 lb.) | kg | 907.185 | 0.001 102 |
| Ton (long 2240 lb.) | kg | 1 016.047 | 0.000 984 2 |
| Slug | kg | 14.593 9 | 0.068 522 |
| **Pressure** | | | |
| Bar | kPa | 100.0 | 0.01 |
| Inch of Water Column (20°C) | kPa | 0.248 6 | 4.021 9 |
| Inch of Mercury (20°C) | kPa | 3.374 1 | 0.296 4 |
| Kilogram-force per Centimeter Squared | kPa | 98.067 | 0.010 2 |
| Millimeters of Mercury (mm·Hg) (20°C) | kPa | 0.132 84 | 7.528 |
| Pounds Per Square Inch (P.S.I.) | kPa | 6.894 8 | 0.145 0 |
| Standard Atmosphere (760 torr) | kPa | 101.325 | 0.009 869 |
| Torr | kPa | 0.133 32 | 7.500 6 |

# Appendix A (continued):

| | SI Symbol | Multiplier to Convert From Existing Unit to SI Unit | Multiplier to Convert From SI Unit to Existing Unit |
|---|---|---|---|
| **Power** | | | |
| Btu (International) Per Hour | W | 0.293 1 | 3.412 2 |
| Foot Pound-Force Per Second | W | 1.355 8 | 0.737 6 |
| Horsepower | kW | 0.745 7 | 1.341 |
| Meter Kilogram-Force Per Second | W | 9.806 7 | 0.101 97 |
| Tons of Refrigeration | kW | 3.517 | 0.284 3 |
| **Torque** | | | |
| Kilogram-Force Meter (kg·m) | N·m | 9.806 7 | 0.101 97 |
| Pound-Force Foot | N·m | 1.355 8 | 0.737 6 |
| Pound-Force Inch | N·m | 0.113 0 | 8.849 5 |
| Gram-Force Centimeter | mN·m | 0.098 067 | 10.197 |
| **Temperature** | | | |
| Fahrenheit | °C | 5/9 (°F − 32) | (9/5°C) + 32 |
| Rankine | K | (°F + 459.67)5/9 | (°C + 273.16)9/5 |
| **Velocity** | | | |
| Foot Per Second | m/s | 0.304 8 | 3.280 8 |
| Mile Per Hour | m/s | 0.447 04 | 2.236 9 |
| | *or* | *or* | *or* |
| | km/h | 1.609 34 | 0.621 4 |
| **Viscosity** | | | |
| Centipoise | mPa·s | 1.0 | 1.0 |
| Centistoke | μm²/s | 1.0 | 1.0 |
| **Volume (Capacity)** | | | |
| Cubic Foot | *l* (dm³) | 28.316 8 | 0.035 31 |
| Cubic Inch | cm³ | 16.387 1 | 0.061 02 |
| Cubic Yard | m³ | 0.764 6 | 1.308 |
| Gallon (U.S.) | *l* | 3.785 | 0.264 2 |
| Ounce (U.S. Fluid) | ml | 29.574 | 0.033 8 |
| Pint (U.S. Fluid) | *l* | 0.473 2 | 2.113 |
| Quart (U.S. Fluid) | *l* | 0.946 4 | 1.056 7 |
| **Volume Flow (Gas-Air)** | | | |
| Standard Cubic Foot Per Minute | m³/s | 0.000 471 9 | 2119. |
| | *or* | *or* | *or* |
| | l/s | 0.471 9 | 2.119 |
| | *or* | *or* | *or* |
| | ml/s | 471.947 | 0.002 119 |
| Standard Cubic Foot Per Hour | ml/s | 7.865 8 | 0.127 133 |
| | *or* | *or* | *or* |
| | μl/s | 7 866. | 0.000 127 |
| **Volume Liquid Flow** | | | |
| Gallons Per Hour (U.S.) | l/s | 0.001 052 | 951.02 |
| Gallons Per Minute (U.S.) | l/s | 0.063 09 | 15.850 |

# 21 ENGINEERING LICENSING

*Purpose of Registration*

As an engineer, you may have to obtain your professional engineering license through procedures which have been established by the state in which you reside. These procedures are designed to protect the public by preventing unqualified individuals from legally practicing as engineers.

There are many reasons for wanting to become a professional engineer. Among them are the following:

- You may wish to become an independent consultant. By law, consulting engineers must be registered.
- Your company may require a professional engineering license as a requirement for employment or advancement.
- Your state may require registration as a professional engineer if you use the title *engineer.*

*The Registration Procedure*

The registration procedure is similar in most states. You probably will take two 8-hour written examinations. The first examination is the *Engineer-in-Training* examination, also known as the *Intern Engineer* exam and the *Fundamentals of Engineering* exam. The initials E-I-T, I.E., and F.E. also are used. The second examination is the *Professional Engineering* (*P.E.*) exam, which differs from the E-I-T exam in format and content.

If you have significant experience in engineering, you may be allowed to skip the E-I-T examination. However, actual details of registration, experience requirements, minimum education levels, fees, and examination schedules vary from state to state. You should contact your state's Board of Registration for Professional Engineers.

*Reciprocity Among States*

All states use the NCEES P.E. examination.[1] If you take and pass the P.E. examination in one state, your certificate probably will be honored by other states which have used the same NCEES examination. It will not be necessary to retake the P.E. examination.

The simultaneous administration of identical examinations in multiple states has led to the term *Uniform Examination.* However, each state is free to choose its own minimum passing score or to add special questions to the NCEES examination. Therefore, this Uniform Examination does not automatically ensure reciprocity among states.

Of course, you may apply for and receive a professional engineering license from another state. However, a license from one state will not permit you to practice engineering in another state. You must have a professional engineering license from each state in which you work.

*Applying for the Examination*

Each state charges different fees, requires different qualifications, and uses different forms. Therefore, it will be necessary for you to request an application and an information packet from the state in which you reside or in which you plan to take the exam. It generally is sufficient to phone for this information. Telephone numbers for all of the U.S. state boards of registration are given in the accompanying table.

[1] The National Council of Examiners for Engineering and Surveying (NCEES) in Clemson, South Carolina, produces, distributes, and grades the national P.E. examinations. It does not distribute applications to take the P.E. examination.

**Phone Numbers of State Boards of Registration**

| | |
|---|---|
| Alabama | (205) 261-5568 |
| Alaska | (907) 465-2540 |
| Arizona | (602) 255-4053 |
| Arkansas | (501) 371-2517 |
| California | (916) 920-7466 |
| Colorado | (303) 866-2396 |
| Connecticut | (203) 566-3386 |
| Delaware | (302) 656-7311 |
| District of Columbia | (202) 727-7454 |
| Florida | (904) 488-9912 |
| Georgia | (404) 656-3926 |
| Guam | (671) 646-8643 |
| Hawaii | (808) 548-4100 |
| Idaho | (208) 334-3860 |
| Illinois | (217) 782-8556 |
| Indiana | (317) 232-1840 |
| Iowa | (515) 281-5602 |
| Kansas | (913) 296-3053 |
| Kentucky | (502) 564-2680 |
| Louisiana | (504) 568-8450 |
| Maine | (207) 289-3236 |
| Maryland | (301) 333-6322 |
| Massachusetts | (617) 727-3055 |
| Michigan | (517) 335-1669 |
| Minnesota | (612) 296-2388 |
| Mississippi | (601) 354-7241 |
| Missouri | (314) 751-2334 |
| Montana | (406) 444-4285 |
| Nebraska | (402) 471-2021 |
| Nevada | (702) 329-1955 |
| New Hampshire | (603) 271-2219 |
| New Jersey | (201) 648-2660 |
| New Mexico | (505) 827-9940 |
| New York | (518) 474-3846 |
| North Carolina | (919) 781-9499 |
| North Dakota | (701) 258-0786 |
| Ohio | (614) 466-8948 |
| Oklahoma | (405) 521-2874 |
| Oregon | (503) 378-4180 |
| Pennsylvania | (717) 783-7049 |
| Puerto Rico | (809) 722-2121 |
| Rhode Island | (401) 277-2565 |
| South Carolina | (803) 734-9166 |
| South Dakota | (605) 394-2510 |
| Tennessee | (615) 741-3221 |
| Texas | (512) 440-7723 |
| Utah | (801) 530-6632 |
| Vermont | (802) 828-2363 |
| Virginia | (804) 257-8512 |
| Virgin Islands | (809) 774-1301 |
| Washington | (206) 753-6966 |
| West Virginia | (304) 348-3554 |
| Wisconsin | (608) 266-1397 |
| Wyoming | (307) 777-6156 |

*Examination Format*

The NCEES Professional Engineering examination in Civil Engineering consists of two four-hour sessions separated by a one-hour lunch period. Both the morning and the afternoon sessions contain twelve problems. Most states do not have required problems.

Each examinee is given an exam booklet which contains problems for civil, mechanical, electrical, and chemical engineers. Some states, such as California, will allow you to work problems only from the civil part of the booklet. Other states, such as New York, will allow you to work problems from the entire booklet. Read the examination instructions on this point carefully.

In 1982, NCEES completed a task analysis of engineering activities. This study was intended to determine all the important activities in which engineers engage, and then to suggest changes in the NCEES examinations. The following subjects and numbers of problems resulted from that task analysis.

**Results of C.E. Task Analysis**

| examination subject | number of problems |
|---|---|
| transportation | 5 |
| structural | 7 |
| sanitary | 5 |
| hydraulics | 4 |
| soils | 2 |
| engineering economics | 1 |

These subjects are rather broad, and the NCEES examinations are not obligated to follow the task analysis guidelines.

Since the examination structure is not rigid, it is not possible to give the exact number of problems that will appear in each subject area. Only engineering economics can be considered a permanent part of the examination. (NCEES reduced the number of engineering economics problems from two to one in 1980.) There is no guarantee that any other single subject will appear.

The examination is open book. Usually, all forms of solutions aids are allowed in the examination, including nomographs, specialty slide rules, and pre-programmed and programmable calculators. Since their use says little about the depth of your knowledge, such aids should be used only to check your work. For example, very few points will be earned if a pre-programmed calculator is used to solve a surveying problem.

Most states do not limit the number and types of books you can bring into the exam.[2] Loose-leaf papers (including Post-It$^{tm}$ notes) and writing tablets are usually forbidden, although you may be able to bring in loose reference pages in a three-ring binder. References used in the afternoon session need not be the same as for the morning session.

Any battery-powered, silent calculator may be used. There are no restrictions on programmable or pre-programmed calculators. Printers cannot be used.

You will not be permitted to share books, calculators, or any other items with other examinees.

You will receive the results of your examination by mail. Allow 12-14 weeks for notification. Your score may or may not be revealed to you, depending on your state's procedure.

*Examination Dates*

The NCEES examinations are administered on the same weekend in all states. Each state decides independently whether to offer the examination on Thursday, Friday, or Saturday of the examination period. The upcoming examination dates are given in the accompanying table.

**National P.E. Examination Dates**

| year | Spring exam | Fall exam |
|---|---|---|
| 1990 | April 19–21 | October 25–27 |
| 1991 | April 11–13 | October 24–26 |
| 1992 | April 9–11 | October 29–31 |
| 1993 | April 15–17 | October 28–30 |
| 1994 | April 14–16 | October 27–29 |
| 1995 | April 6–8 | October 26–28 |
| 1996 | April 18–20 | October 24–26 |

*Objectively Scored Problems*

Objectively scored problems were added to the April 1988 professional engineering exam, and are expected to continue to appear in subsequent examinations. Such problems appear in multiple-choice, true/false, and data-selection formats.

The single remaining economics problem on the exam is one of the problems that has been converted to objective scoring. In addition, NCEES has specifically targeted the following subjects for objective scoring:

- sanitary
- hydraulics
- structural
- transportation
- highways
- traverses

Objectively scored problems appear in both the morning and afternoon parts of the examination. In all, approximately 25% (2 or 3 problems) of each session is objectively scored.

*Grading the Examination*

Full credit is achieved by correctly working four problems in the morning and four problems in the afternoon. You may not claim credit for more than eight worked problems or for more than four per session. All solutions are recorded in official solution booklets.

At its August 13, 1983, board meeting, NCEES adopted a new method of grading its professional engineering exams. The minutes of that meeting include the following paragraph.

> "The following is the method for determining the recommended passing standard for an applicant for the P.E. examination, effective April 1984. The Principles and Practice of Engineering Examination's recommended passing standard will be established as a minimum raw score of 48..."

NCEES has decided to name the new grading method the *Criterion-Referenced Method* to distinguish it from the old *Norm-Referenced Method.* The criteria are the specific elements you must include in your solution to receive credit for your solution. The criteria are determined in advance, prior to the administration of the examination.

As an example of a criterion-graded problem, consider the design of a long circular concrete culvert. In order to receive six points (the minimum required to pass the problem), the examinee might have to:

- consider the effects of friction on the hydraulic head
- calculate the flow rate based on the correct hydraulic head
- recognize that the flow was entrance controlled and the culvert did not flow full
- calculate the normal depth to within ±10% from the answer key
- verify that the normal depth was less than the critical depth

[2] Check with your state to see if review books can be brought into the examination. Most states do not have any restrictions. Some states ban only collections of solved problems, such as Schaum's Outline Series. A few prohibit all review books.

- set up all equations correctly, though not necessarily complete the problem or obtain the correct answer.

Getting ten points requires solving the problem correctly to completion, making no mathematical errors in the solution. For each mathematical error, a point or two would be lost from the total possible of ten. However, if all of the above six points were met, no less than six points would be received.

This grading method has been validated by correlating passing rates from previous examinations graded by both methods. (Previous examinations, however, used the old grading method.) It is not expected that passing percentages will change markedly. Only examinees with marginal scores will be affected.

Each state still is free to specify its own cut-off score and passing requirements. NCEES' recommended method or score does not have to be used (although most states do so).

*Preparing for the Exam*

You should develop an examination strategy early in the preparation process. This strategy will depend on your background. One of the following two general strategies is recommended:

- A broad approach has been successful for examinees who have recently completed academic studies. Their strategy has been to review the fundamentals of a broad range of undergraduate civil engineering subjects. The examination includes enough fundamental problems to give merit to this strategy.

- Working engineers who have been away from classroom work for a long time have found it better to concentrate on the subjects in which they have had extensive professional experience. By studying the list of examination subjects, they have been able to choose those which will give them a good probability of finding enough problems that they can solve.

Do not make the mistake of studying only a few subjects in hopes of finding enough problems to work. The more subjects you are familiar with, the better will be your chances of passing the examination. More important than strategy are fast recall and stamina. You must be able to recall quickly solution procedures, formulas, and important data; and this sharpness must be maintained for eight hours. You will not have time in the exam to derive solution methods; you must know them instinctively.

It is imperative that you develop and adhere to a review outline and schedule. If you are not taking a classroom review course where the order of preparation is determined by the lectures, you should develop an Outline of Subjects for Self-Study to schedule your preparation.

It is unnecessary to take a large quantity of books to the examination. The examination is very fast-paced. You will not have time to look up solution procedures, data, or equations with which you are not familiar. Although the examination is open-book, there is insufficient time to use books with which you are not thoroughly familiar.

To minimize time spent searching for often-used formulas and data, you should prepare a one-page summary of all important formulas and information in each subject area. You can then use these summaries during the examination instead of searching for the correct page in your book.

*Items to Get for the Examination*

- Obtain ten sheets of each of the following types of graph paper: 10 squares to the inch grid, semi-log (3 cycles × 10 squares to the inch grid), and full-log (3 cycles × 3 cycles).

- Obtain a flexible clear plastic ruler marked in tenths of an inch or centimeters.

- Obtain a package of tracing paper (approximately 20# weight).

*What to do Before the Exam*

The engineers who have taken the P.E. exam in previous years have made the suggestions listed below. These suggestions will make your examination experience as comfortable and successful as possible.

- Keep a copy of your examination application. Send the original application by certified mail and request a receipt of delivery.

- Visit the exam site the day before your examination. This is especially important if you are not familiar with the area. Find the examination room, the parking area, and the rest rooms.

- Plan on arriving at least 30 minutes before the examination starts. This will assure you a convenient parking place and adequate time for site, room, and seating changes.

- If you live a considerable distance from the examination site, consider getting a hotel room in which to spend the night before.

- Take off the day before the examination to relax. Don't cram the last night. Rather, get a good night's sleep.
- Be prepared to find that the examination room is not ready at the designated time. Take an interesting novel or magazine to read in the interim and at lunch.
- If you make arrangements for baby sitters or transportation, allow for a delayed completion.
- Prepare your examination kit the day before. Here is a checklist of items to take with you to the examination.

[ ] copy of your application

[ ] proof of delivery receipt

[ ] letter admitting you to the exam

[ ] photographic identification

[ ] other reference books

[ ] CIVIL ENGINEERING HANDBOOK (Merritt)

[ ] course notes in a binder

[ ] calculator and a spare

[ ] spare calculator batteries or battery pack

[ ] battery charger and 20' extension cord

[ ] chair cushions. A large, thick bath mat works well.

[ ] earplugs

[ ] desk expander. If you are taking the exam in theater chairs with tiny, fold-up writing surfaces, you should take a long, wide board to place across the arm rests.

[ ] a cardboard box cut to fit your references

[ ] twist-to-advance pencils

[ ] extra leads

[ ] large eraser

[ ] snacks such as raisins, nuts, or trail mix

[ ] thermos filled with hot chocolate

[ ] a light lunch

[ ] a collection of graph paper

[ ] scissors, stapler, and staple puller

[ ] construction paper for stopping drafts and sunlight

[ ] transparent and masking tapes

[ ] sunglasses

[ ] extra prescription glasses, if you wear them

[ ] aspirin

[ ] travel pack of Kleenex

[ ] Webster's dictionary

[ ] dictionary of scientific terms

[ ] $2 in change

[ ] a light comfortable sweater

[ ] comfortable shoes or slippers for the exam room

[ ] raincoat, boots, gloves, hat, and umbrella

[ ] local street maps

[ ] note to the parking patrol for your windshield

[ ] pad of scratch paper with holes for 3-ring binder

[ ] straightedge, ruler, compass, protractor, and French curves

[ ] battery-powered desk lamp

[ ] watch

[ ] extra car keys

*What to do During the Exam*

Previous examinees have reported that the following strategies and techniques have helped them considerably.

- Read through all of the problems before starting your first solution. In order to save you from rereading and reevaluating each problem later in the day, you should classify each problem at the beginning of the four hour session. The following categories are suggested:
  - problems you can do easily
  - problems you can do with effort
  - problems for which you can get partial credit
  - problems you cannot do
- Do all of the problems in order of increasing difficulty. All problems on the examination are worth

10 points. There is nothing to be gained by attempting the difficult or long problems if easier or shorter problems are available.

- Follow these guidelines when solving a problem:
  - Do not rewrite the problem statement.
  - Do not unnecessarily redraw any figures.
  - Use pencil only.
  - Be neat. (Print all text. Use a straightedge or template where possible.)
  - Draw a box around each answer.
  - Label each answer with a symbol.
  - Give the units.
  - List your sources whenever you use obscure solution methods or data.
  - Write on one side of the page only.
  - Use one page per problem, no matter how short the solution is.
  - Go through all calculations a second time and check for mathematical errors.
- Record the details of any problem which you think is impossible to solve with the information given. Your being able to point out an error may later give you the margin needed to pass.

# 22 POSTSCRIPTS

This chapter collects comments, revisions, and commentary which cannot be incorporated into the body of the text until the next edition. New postscript sections are added as needed when the *Civil Engineering Reference Manual* is reprinted. Subjects in this chapter are not necessarily represented by entries in the index. **It is suggested that you make a note in the appropriate text pages to refer to this chapter.**

*Update: October, 1986*

## VARIANCE

The statistical term *variance* is used in example 1.42 (page 1-27) but is not defined in the text. The variance is the square of the standard deviation. Since there are two standard deviations ($s$ and $\sigma$), there are two variances. The sample variance is $s^2$. The population variance is $\sigma^2$.

## WATER HAMMER IN DUCTILE PIPE

The speed of sound used in water hammer calculations (i.e., $c$ in equations 3.198 and 3.199) must account for the expansion of ductile pipe walls as the water pressure builds up. Equation 3.20 can be used to calculate $c$, but the modulus of elasticity used should include the elastic contributions of the water and pipe material both. In the equation below, $t_{\text{pipe}}$ is the pipe wall thickness, and $d_{\text{pipe}}$ is the inside diameter.

$$E = \frac{E_{\text{water}} t_{\text{pipe}} E_{\text{pipe}}}{t_{\text{pipe}} E_{\text{pipe}} + d_{\text{pipe}} E_{\text{water}}}$$

## HEATING VALUE OF DIGESTER GAS

The heating value of digester gas is listed (page 8-29) as 600 BTU/ft$^3$. This value is appropriate for digester gas with the composition given: 65% methane, 35% carbon dioxide. The actual heating value will depend on the fraction of combustible methane, as well as the temperature and pressure. At 60°F and 14.7 psia, pure methane has a lower heating value of 900 BTU/ft$^3$. Carbon dioxide does not contribute to the heating effect.

## NEUTRAL AXES IN CONCRETE BEAMS

Figures 14.4 and 14.5 (page 14-10) both contain the variable $c$, the distance from the neutral axis to the top of beam. However, these two distances are not the same, even though the symbol is the same. The location of the neutral axis changes as the ultimate strength is approached.

## MODULUS OF ELASTICITY FOR MASONRY

Masonry structures are generally designed using the alternate (working stress) design method, which requires knowing the modulus of elasticity. Equation 14.1 for concrete cannot be used. An approximate value can be found from the masonry's compressive strength.

$$E_m = 1000 f'_m \quad (< 3 \text{ EE6 psi})$$

## LENGTH OF VERTICAL CURVE

Equation 17.52 (page 17-16) can be solved without trial and error.

$$LC = 2d \left( \frac{\sqrt{\frac{EG}{EF}} + 1}{\sqrt{\frac{EG}{EF}} - 1} \right)$$

*Update: April, 1987*

## Chapter 5: CHANNEL SLOPE

In uniform flow, the slopes of the channel bottom, water surface, and energy grade line are identical. For uniform flow, then, equation 5.6 could be written using the geometric slope, $S_o$, instead of the hydraulic slope, $S$:

$$v = C\sqrt{r_H S_o}$$

Similarly, other equations in chapter 5 could also be written using $S_o$, but only under the condition of uniform flow. Using $S_o$, however, is clearly a special case of a general rule.

## Chapter 5: OPEN CHANNEL FLOW in SI

The factor 1.49 in equations 5.7–5.9 (and others) converts customary SI to English units. For problems in SI units (m/sec, etc.), replace the 1.49 with 1.00.

## Chapter 7: RAPID SAND FILTERS

Rapid sand filters usually operate with hydraulic heads (distance between water surfaces in filter and clearwell) of 9–12 feet.

Prior to backwashing with clear water, the filter material may be expanded by an *air prewash* of 1–8 (2–5 typical) cfm/ft$^2$ for 2–10 (3–5 typical) minutes.

## Chapter 8: SANITARY LANDFILLS

Many large-scale sanitary landfills do not apply daily cover to deposited solid waste. Time and cost are typically cited as the reasons that the landfill is not covered with soil. To account for the absence of such cover, the loading factor in equation 8.62 must have a value of 1.0.

## Chapter 9: PERCENT PASSING SIEVE

Equations 9.2 and 9.3, as well as various soils classification schemes, require knowing percentages of soil passing through specific sieve sizes. When sieve data is incomplete, the needed values can be interpolated by plotting the known data on a *particle size distribution chart.* (Also, see figure 9.2.)

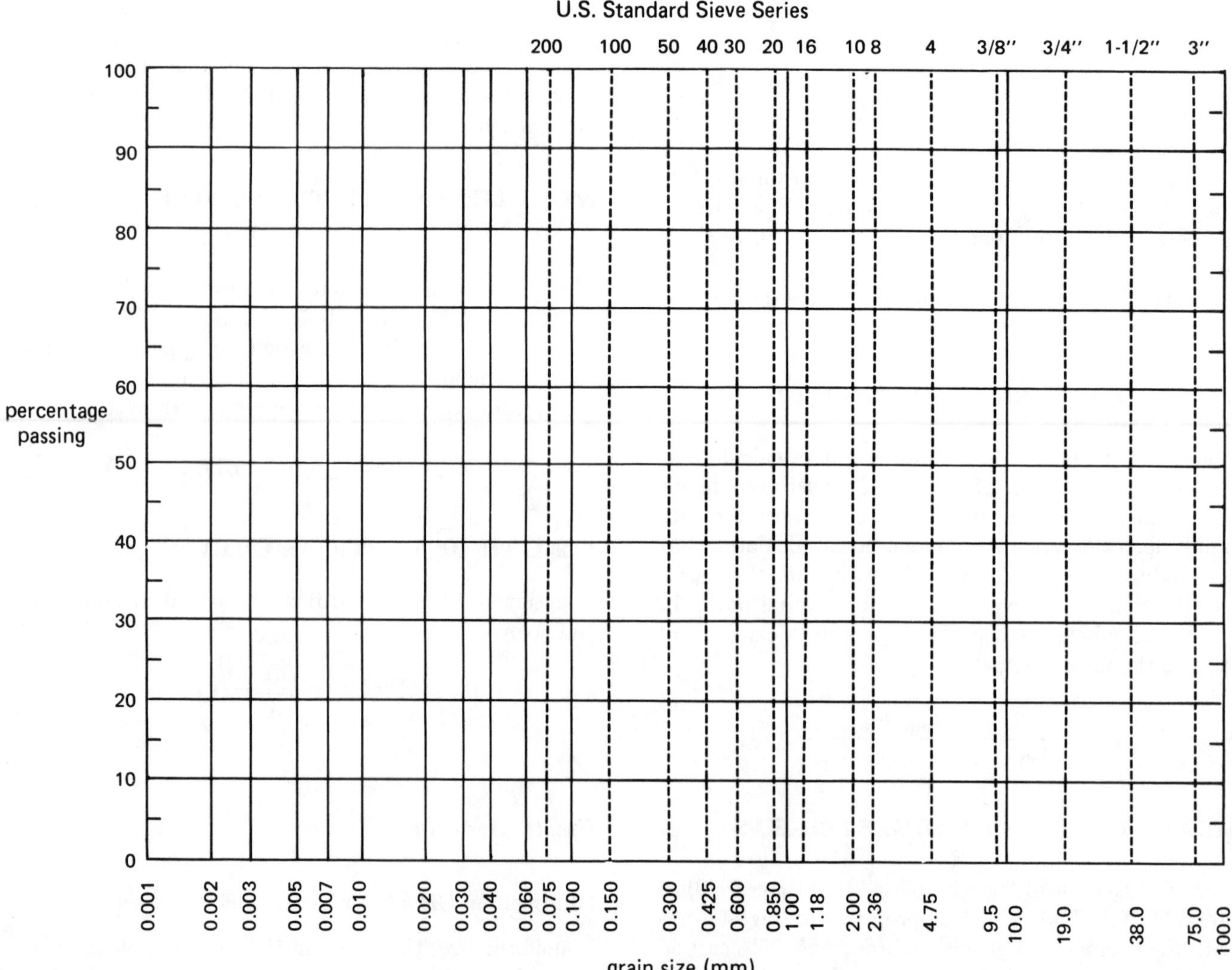

## THE UNIFIED SOIL CLASSIFICATION SYSTEM

| Major Division | | Group Symbol | Laboratory Classification Criteria: Finer than 200 Sieve (%) | Laboratory Classification Criteria: Supplementary Requirements | Soil Description |
|---|---|---|---|---|---|
| Coarse-grained (over 50% by weight coarser than No. 200 sieve) | Gravelly soils (over half of coarse fraction larger than No. 4) | GW | 0–5* | $D_{60}/D_{10}$ greater than 4 $D_{30}{}^2/(D_{60} \times D_{10})$ between 1 & 3 | Well-graded gravels, sandy gravels |
| | | GP | 0–5* | Not meeting above gradation for GW | Gap-graded or uniform gravels, sandy gravels |
| | | GM | 12 or more* | PI less than 4 or below A-line | Silty gravels, silty sandy gravels |
| | | GC | 12 or more* | PI over 7 and above A-line | Clayey gravels, clayey sandy gravels |
| | Sandy soils (over half of coarse fraction finer than No. 4) | SW | 0–5* | $D_{60}/D_{10}$ greater than 4, $D_{30}{}^2/(D_{60} \times D_{10})$ between 1 & 3 | Well-graded, gravelly sands |
| | | SP | 0–5* | Not meeting above gradation requirements | Gap-graded or uniform sands, gravelly sands |
| | | SM | 12 or more* | PI less than 4 or below A-line | Silty sands, silty gravelly sands |
| | | SC | 12 or more* | PI over 7 and above A-line | Clayey sands, clayey gravelly sands |
| Fine-grained (over 50% by weight finer than No. 200 sieve) | Low compressibility (liquid limit less than 50) | ML | Plasticity chart | | Silts, very fine sands, silty or clayey fine sands, micaceous silts |
| | | CL | Plasticity chart | | Low plasticity clays, sandy or silty clays |
| | | OL | Plasticity chart, organic odor or color | | Organic silts and clays of low plasticity |
| | High compressibility (liquid limit more than 50) | MH | Plasticity chart | | Micaceous silts, diatomaceous silts, volcanic ash |
| | | CH | Plasticity chart | | Highly plastic clays and sandy clays |
| | | OH | Plasticity chart, organic odor or color | | Organic silts and clays of high plasticity |
| Soils with fibrous organic matter | | Pt | Fibrous organic matter; will char, burn, or glow | | Peat, sandy peats, and clayey peat |

*For soils having 5 to 12% passing the No. 200 sieve, use a dual symbol such as GW–GC.

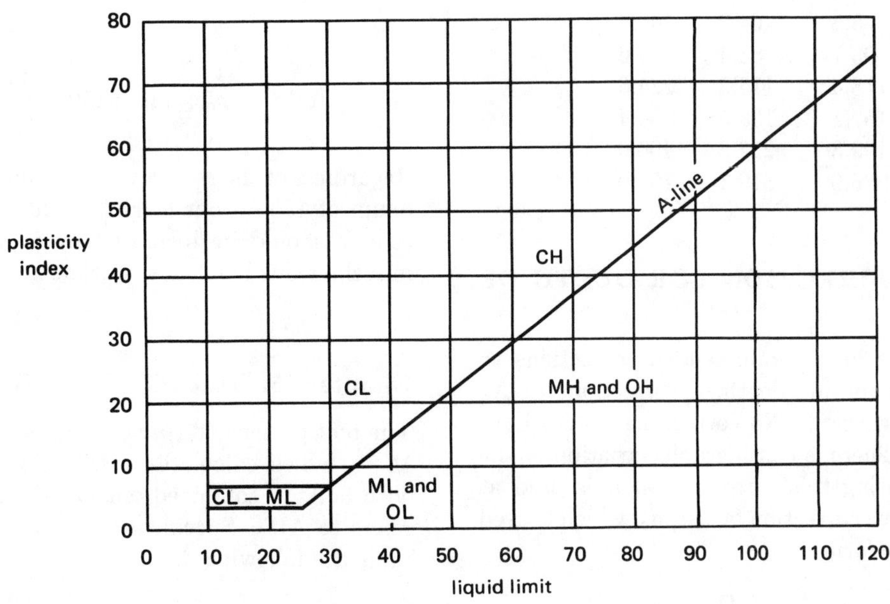


Plasticity chart for the classification of fine-grained soils. Tests made on fraction finer than No. 40 sieve, 0.425 mm.

### Chapter 9: CONSOLIDATION TESTS

It is common practice to plot *strain*, $\epsilon$, versus log $p$ for consolidation test data, as well as void ratio versus log $p$ as shown in figure 9.9. This eliminates having to know the initial void ratio, $e_o$. The slope of the $\epsilon$–log $p$ line is $C_{\epsilon c}$, which is related to the compression index, $C_c$ as follows:

$$C_{\epsilon c} = \frac{C_c}{1+e_o}$$

Analogous to equation 10.40, settlement can be calculated from

$$S = C_{\epsilon c} H \log_{10}\left(\frac{p_o+p_v}{p_o}\right)$$

### Chapter 10: BEARING CAPACITY FACTORS

The bearing capacity factors in table 10.3 are based on Terzaghi's 1943 studies. The following values are based on Meyerhof's (and others) 1955 studies, and have been widely used. Other values are also in use.

| $\phi$ | $N_c$ | $N_q$ | $N_\gamma$ |
|---|---|---|---|
| 0 | 5.14 | 1.0 | 0.0 |
| 5 | 6.5 | 1.6 | 0.5 |
| 10 | 8.3 | 2.5 | 1.2 |
| 15 | 11.0 | 3.9 | 2.6 |
| 20 | 14.8 | 6.4 | 5.4 |
| 25 | 20.7 | 10.7 | 10.8 |
| 30 | 30.1 | 18.4 | 22.4 |
| 32 | 35.5 | 23.2 | 30.2 |
| 34 | 42.2 | 29.4 | 41.1 |
| 36 | 50.6 | 37.7 | 56.3 |
| 38 | 61.4 | 48.9 | 78.0 |
| 40 | 75.3 | 64.2 | 109.4 |
| 42 | 93.7 | 85.4 | 155.6 |
| 44 | 118.4 | 115.3 | 224.6 |
| 46 | 152.1 | 158.5 | 330.4 |
| 48 | 199.3 | 222.3 | 496.0 |
| 50 | 266.9 | 319.1 | 762.9 |

### Chapter 10: CORRECTION FOR DEPTH OF FOOTING

Several researchers have recommended corrections to $N_c$ to account for footing depth. (Corrections to $N_q$ have also been suggested. No corrections to $N_\gamma$ have been suggested.) There is considerable variation in the method of calculating this correction, if it is used at all. A multiplicative correction factor, $d_c$, which is used most often has the form

$$d_c = 1 + \frac{KD_f}{B}$$

$K$ is a constant. Values of 0.2 and 0.4 have been proposed for $K$.

### Chapter 14: NOMINAL STRENGTHS

The term *nominal strength* has two applications in reinforced concrete design. Consider shear strength of a beam, for example. The unreinforced concrete has a resistance to shear given by equation 14.48. The quantity $V_c$ is known as the *nominal shear strength* of the concrete. However, the steel also contributes shear strength represented by $V_{st}$. The sum of these two quantities is the *nominal strength of the beam*:

$$V_n = V_c + V_{st}$$

Common usage attributes the term *nominal* to both $V_n$ and $V_c$.

### Chapter 14: CONCRETE PRESSURE ON FORMWORK

Formwork must be strong enough to withstand hydraulic loading from concrete during curing. The hydraulic load is greatest immediately after pouring. As the concrete sets up, it begins to support itself, and the lateral force is reduced. Publication ACI 347 predicts the maximum lateral pressure for regular (Type I) concrete with a 4″ slump (or less), ordinary work, and internal vibration. The maximum pressure depends on the temperature, $T_{°F}$, and the vertical rate of pour, $R_{\text{ft/hr}}$.

$$R \le 7 \text{ ft/hr}: \quad p_{\text{max,psf}} = 150 + 9000\left(\frac{R}{T}\right)$$

$$R > 7 \text{ ft/hr}: \quad p_{\text{max,psf}} = 150 + \frac{43{,}400}{T} + 2800\left(\frac{R}{T}\right)$$

Regardless of the rate of pour, $p$ cannot exceed the minimum of 150 × pour height, or 2000 psf. As a general rule, $p$ should be increased by 50 psf to account for miscellaneous live loads, workmen, and impact.

### Chapter 15: STRUCTURAL BOLT TENSIONS

The bolt pretension (preload) is needed to use equation 15.85, which calculates a reduction factor to be used with bolts in combined tension and shear. Footnote 44 refers to AISC Specifications Table 1.23.5, which contains the following data:

Minimum Bolt Tension

| bolt size | A325 bolts | A490 bolts |
|---|---|---|
| 1/2 (inches) | 12 kips | 15 kips |
| 5/8 | 19 | 24 |
| 3/4 | 28 | 35 |
| 7/8 | 39 | 49 |
| 1 | 51 | 64 |
| 1-1/8 | 56 | 80 |
| 1-1/4 | 71 | 102 |
| 1-3/8 | 85 | 121 |
| 1-1/2 | 103 | 148 |

For bolts of other sizes, or for bolts manufactured from other steels, the minimum pretension is

$$T_{\text{b,min}} = 0.70 A_b F_{ut} \quad \text{(rounded)}$$

## Chapter 16: LEVELS OF SERVICE

The 1985 edition of the *Highway Capacity Manual* prefers to categorize highway *level of service* by density, rather than the volume/capacity ratios listed in table 16.11. Although table 16.11 is correct, density should now be used wherever possible to determine level of service. In the following table, density is given in passenger cars per mile per lane.

| LOS | density |
|---|---|
| A | $\leq$ 12 pc/mi-ln |
| B | $\leq$ 20 |
| C | $\leq$ 30 |
| D | $\leq$ 42 |
| E | $\leq$ 67 |
| F | $>$ 67 |

## Chapter 17: SPIRAL CURVES

The length of a spiral curve can be adjusted to between 75% and 200% of the value calculated in equation 17.55. The distance for *superelevation runoff* frequently determines the spiral curve length. The runoff may be determined by a *time rule* (e.g., runoff shall be completed within 4 seconds at the design speed), or by a *speed rule* (e.g., 3 feet of runoff for every MPH in design speed, regardless of initial superelevation). Other time and speed rules are in use. The important point is that the spiral curve length can be adjusted to meet varying design criteria.

## Chapter 17: HORIZONTAL CURVES THROUGH POINTS

A special type of horizontal curve problem requires finding a curve radius to pass the curve through a given point. There are two variations of this problem, depending on how the point is located. However, both problems are solved in the same way. Angle $\Delta$ must be known.

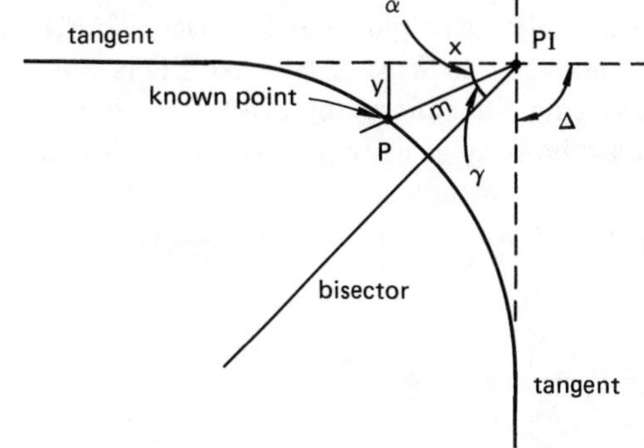


*step 1*: Get $\alpha$ and $m$ from $x$ and $y$. (If $\alpha$ and $m$ are known, skip this step.)

$$\alpha = \arctan \frac{y}{x}$$

$$m = \sqrt{x^2 + y^2}.$$

*step 2*: Calculate $\gamma = 90 - \frac{1}{2}\Delta - \alpha$.

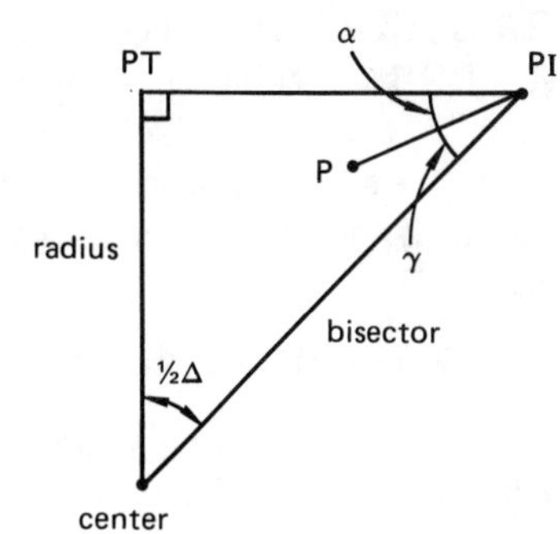


*step 3*: Calculate $\phi = 180 - \arcsin\left(\frac{\sin\gamma}{\cos\frac{1}{2}\Delta}\right)$.

*step 4*: Calculate $\theta = 180 - \gamma - \phi$.

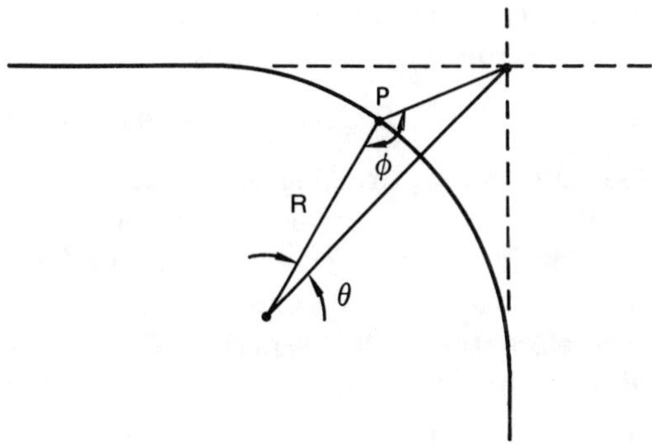


*step 5*: Calculate $R$ from the law of sines.

$$\frac{\sin\theta}{m} = \frac{\sin\phi}{R\sec\frac{1}{2}\Delta} = \frac{\sin\phi\cos\frac{1}{2}\Delta}{R}$$

### Chapter 21: CHANGE IN EXAM FORMAT

The National Council of Engineering Examiners (NCEE) has announced its intention to include objective (multiple-choice, true/false, and data-selection) questions in its Principles and Practices (P.E.) examinations as early as April 1988. NCEE is fine-tuning and validating the concept by pre-testing questions on engineers who have already passed their E-I-T and P.E. exams. If the concept is validated, one-quarter of the P.E. examination will initially be replaced by objective questions.

*Update: February, 1988*

### Chapter 2: PAYBACK PERIOD

It is tempting to apply discounting and compounding factors to calculations of payback period. Common practice, however, ignores interest, and calculates payback period simply as the number of years required for the net cash flow to equal the initial investment.

### Chapter 2: USING THE MARR

When a MARR is given, it is not necessary to calculate an alternative's ROR (to compare against) in order to qualify or disqualify the alternative. It is easier to calculate the alternative's present worth using the MARR as an interest rate. If the present worth is positive, then ROR > MARR.

### Chapter 5: THE FROUDE NUMBER

There is considerable confusion regarding the definition of the Froude number. Dimensional analysis determines it to be $v^2/gL$, a form which is also used in model similitude analysis. However, in open channel flow analysis, the Froude number is taken as the square root of the derived form (as defined in equation 5.47). Whether the derived form or its square root is used can be determined from the application. If the Froude number is squared (as in equation 5.48, or in $dE/dd = 1 - N_{Fr}^2$), then the square root form is necessary.

### Chapter 7: WATER QUALITY STANDARDS

The 1986 Safe Water Drinking Act established various schedules for regulating volatile organic chemicals (VOC's), fluoride, synthetic organic chemicals (SOC's), inorganic chemicals (IOC's), and microbial contaminants. Therefore, table 7.9 is expected to be in a state of flux during the 1987–1991 timetable established by the act.

### Chapter 8: TSS VENTILATION REQUIREMENTS FOR PUMP WELLS

Appendix A in chapter 8 should include ventilation for wet and dry wells. The Ten States' Standards (TSS) requirement is 30 complete air changes per hour for both wet and dry wells using intermittent ventilation. For continuous ventilation, the requirement is reduced to 12 (wet wells) or 6 (dry wells) air changes per hour. In general, dilution air should be forced in, as opposed to air extraction and replacement by infiltration.

### Chapter 12: DEFLECTION OF A PIER

Appendix A of chapter 12 omits the equation for calculating the performance of a fixed pier (i.e., a column with one end that can translate laterally, with both ends remaining vertical).

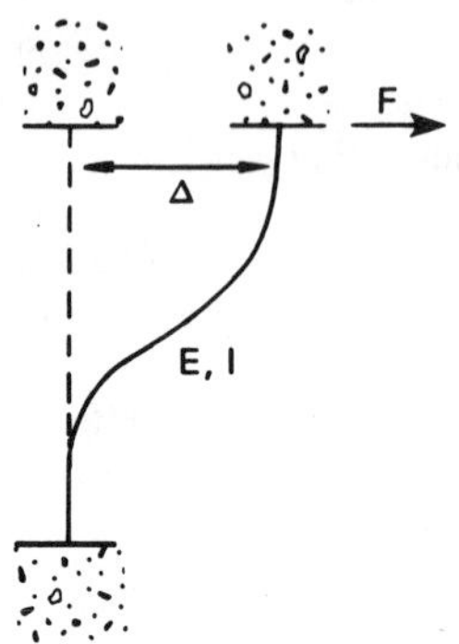


$$\Delta = \frac{FL^3}{12EI}$$

$$\text{stiffness} = \frac{F}{\Delta} = \frac{12EI}{L^3}$$

### Chapter 15: OPTIMIZED DESIGN IN COLUMN BASE PLATES

The required bearing plate area will be minimized when the allowable bearing pressure is maximized. This occurs when the area of the supporting concrete plane is more than 4 times the area of the base plate (or, the concrete is infinite in size). With $A_{\text{support}}/A_{\text{bearing}} = 4$, equation 15.36 gives $F_p = 0.70f'_c$, the maximum value of $F_p$ permitted. This value of $F_p$ should be used unless the support area is limited in some way, as it would be with a pedestal-mounted column.

Referring to figure 15.20, it is desired to have $m = n$. This approximately occurs when

$$A_1 = \frac{\text{column load}}{F_p}$$

$$N \approx \sqrt{A_1} + \tfrac{1}{2}(0.95d - 0.80b_f)$$

$$B = \frac{\sqrt{A_1}}{N}$$

Although equation 15.40 is theoretically correct, the AISC procedure imposes an additional limitation by specifying an additional parameter $n'$ for each specific column shape. The value of $n'$ must be obtained from the AISCM. Equation 15.40, then, becomes

$$t = \{\text{larger of } m,\ n,\ \text{or } n'\} \times \sqrt{\frac{4f_p}{F_y}}$$

*Update: October 1988*

## Chapter 14: MOISTURE EXCESS AND DEFICIT IN AGGREGATE

There is some confusion about what "5% moisture excess" means. (See example 14.2.) This confusion results from the use of both dry (i.e., SSD) and wet (i.e., with the excess) weights as the basis for the excess percentage. This problem has existed in civil engineering for a long time, and both methods are in field use. The important relationships are covered below, where $m$ is the variable for mass.

| | dry basis | wet basis |
|---|---|---|
| fraction moisture, $f$ | $\frac{m_{\text{excess water}}}{m_{\text{SSD sand}}}$ | $\frac{m_{\text{excess water}}}{m_{\text{SSD sand}} + m_{\text{excess water}}}$ |
| wet mass of sand, $m_{\text{wet sand}}$ | $m_{\text{SSD sand}} + m_{\text{excess water}}$<br>$(1+f)m_{\text{SSD sand}}$ | $m_{\text{SSD sand}} + m_{\text{excess water}}$<br>$\left(\frac{1}{1-f}\right) m_{\text{SSD sand}}$ |
| SSD mass of sand, $m_{\text{SSD sand}}$ | $\frac{m_{\text{wet sand}}}{1+f}$ | $(1-f) \times m_{\text{wet sand}}$ |
| mass of excess water, $m_{\text{excess water}}$ | $f \times m_{\text{SSD sand}}$<br>$= \frac{f \times m_{\text{wet sand}}}{1+f}$ | $\frac{f \times m_{\text{SSD sand}}}{1-f}$<br>$f \times m_{\text{wet sand}}$ |

## Chapter 14: STRENGTH DESIGN OF BEAMS

Equation 14.33 (used to calculate the reinforcement ratio from a trial beam size) is an approximation good only if $bd^2$ is not too different from the ideal value. The exact method is to use the following equation:

$$\rho_{\text{revised}} = \frac{1}{m}\left(1 - \sqrt{1 - \frac{2mR_{u,\text{revised}}}{f_y}}\right)$$

## Chapter 14: BEAM DEFLECTIONS

$I_{cr}$ in equation 14.40 is the *cracked moment of inertia.* It is calculated from the equation below. Its use is illustrated in example 14.5.

$$I_{cr} = \frac{bc^3}{3} + \frac{A_{st}E_{st}}{E_c}(d-c)^2$$

## Chapter 15: MULTIPLE FASTENER LAP CONNECTIONS

The following additional check is implied by the last part of example 15.11, but is not explicitly stated elsewhere in the text.

If the connectors (e.g., rivets or bolts) in a tension lap splice are arranged in two or more rows, and if the rows have unequal numbers of fasteners, each row should be checked for tension capacity assuming the previous rows have absorbed a proportionate share of the load. Consider the following diagram:

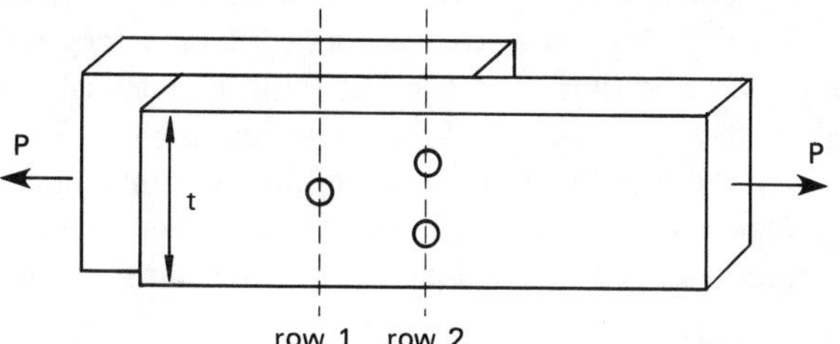


The net section (i.e., $t$ less two hole diameters) of row 2 should be checked for tension capacity when 2/3 of the load is applied. This assumes that row 1 carries 1/3 of the load. (For the second plate, the net section of row 1 could be checked assuming 1/3 of the load was carried. However, this is not the controlling case.)

## Chapter 15: CHECKING BEARING STRESS

It is obvious that bearing stress should be evaluated in bearing connections. Theoretically, bearing should not be a problem in friction-type connections because the bolts never bear on the pieces assembled. However, in the event there is slippage due to insufficient tension in the connectors, bearing should routinely be checked. The fact that the assembly may fail in some other manner if a friction connection becomes a bearing connection does not change the advisability of checking for bearing stress.

*Update: September, 1989*

## FIFTH EDITION

Hundreds of minor changes have been made to the fourth edition of the *Civil Engineering Reference Manual* during its six reprintings. These changes have brought the book into compliance with new codes and legislation, improved and clarified explanations, expanded tables of data, and corrected errata. All of the changes were made without changing the edition, something that has caused confusion among students using the book in a classroom setting.

With this latest printing, another hundred or so changes have been made in the book. Although there has been no substantial change in the organization of material, the edition has been changed to differentiate the current book from its earlier versions.

## Chapter 5: ALTERNATE AND CONJUGATE DEPTHS

The terms *alternate depth* and *conjugate depth* are not synonymous. Alternate depths (calculated in equation 5.41) are derived from the conservation of energy equation (i.e., a variation of the Bernoulli equation). Conjugate depths (calculated in equations 5.58 and 5.59) are derived from a conservation of momentum equation. Conjugate depths are calculated only when there has been an energy loss such as a hydraulic jump or drop.

## Chapter 5: TOTAL SPILLWAY ENERGY

Finding the depth of flow at the toe of a spillway (before a hydraulic jump) requires writing an energy balance. Neglecting friction, the total energy at the toe equals the total upstream energy before the spillway. The total upstream energy before the spillway is

$$E_1 = y + H + \frac{v^2}{2g}$$

The upstream velocity, $v$, is the velocity before the spillway (which is essentially zero), not the velocity over the brink. The velocity over the brink, if the brink depth is known, can be used with the continuity equation to calculate the upstream velocity, but it should not be used to determine total energy since $d_{\text{brink}} \neq H$.

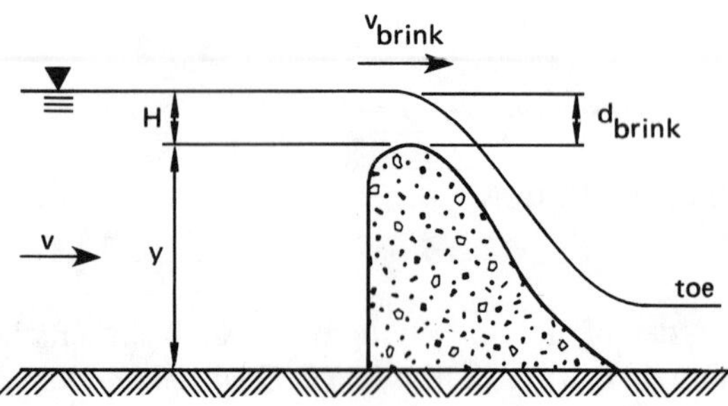


## Chapter 10: EFFECTS OF WATER TABLE LOCATION ON FOOTING DESIGN

*General Principle 1*: If the soil is cohesive ($\phi = 0°$), then the location of the water table does not affect the bearing capacity, and the effect of the water table is disregarded. (Strictly speaking, this is not true, but it is almost true. Here is why. If $\phi$ is zero, then $N_y = 0$. Also, $N_q = 1$, which is essentially zero. These two terms "zero out" the density terms in equation 10.1.)

*General Principle 2*: Use the submerged density ($\rho_{\text{dry}} - 62.4$) in the equation for bearing capacity (equation 10.1):

(a) When the water table is at the base of the footing, use the submerged density in the first term of equation 10.1 only.

(b) When the water table is at the surface, use the submerged density in both the first and third terms of equation 10.1.

(c) When the water table is between the base of the footing and the surface, use the submerged density in the first term as in (a) above. Calculate the third term in equation 10.1 as

$$\begin{aligned}&[p_q + \rho D_w + (\rho - 62.4)(D_f - D_w)]N_q \\ &\quad = [p_q + \rho D_f + 62.4(D_w - D_f)]N_q\end{aligned}$$

*General Principle 3*: If the water table depth, $D_w$, is greater than $D_f + B$ (i.e., more than a distance $B$ below the base of the footing), the bearing capacity is not affected. Calculate the bearing capacity from equation 10.1 as if there was no water table.

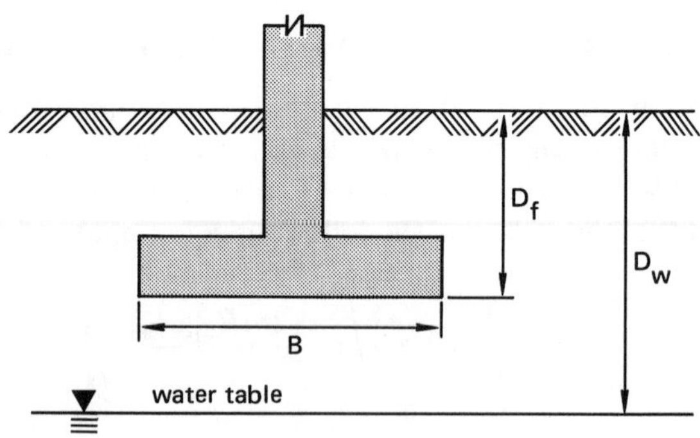


*Approximation to Exact Method*: Since the submerged density is approximately half of the dry density, it is commonly stated that the bearing capacity of a footing with the water table at the ground surface is half the dry bearing capacity and varies linearly to full strength a distance $B$ below the footing base. For a water table within a distance $B$ of the footing base, the dry bearing capacity can be multiplied by $C_w$ (equation 10.23):

$$C_w = 0.5 + 0.5\left(\frac{D_w}{D_f + B}\right) \qquad 0 \leq D_w \leq D_f + B$$

## Chapter 14: AIR ENTRAINMENT

Just as there is confusion with the meaning of "percentage moisture" in concrete batching (see page 22-7), there is also confusion with the meaning of "percentage air entrainment."

If "5%" (for example) air means that the concrete volume is 5% air, then the solid volume is only 95% of the final volume. The solid volume should be divided by 0.95 to obtain the final volume. If "5%" air means the concrete volume is increased by 5% when air is added, then the solid volume should be multiplied by 1.05 to obtain the final volume.

The meaning of "5%" air must be clarified in the problem. If it isn't clear, then either definition could apply. In any case, the final volume is not affected significantly by the interpretation.

## Chapter 14: DEPTH OF COVER AND PROTECTION COVER THICKNESS

[ACI 318 section 10] defines the variable $d_c$ as "thickness of *concrete cover* measured from extreme tension fiber to center of bar or wire located closest thereto." This definition is consistent with equation 14.36 (page 14-13) and figure 14.6 (page 14-14) in this book.

[ACI 318 section 7.7] also specifies the thickness of a concrete layer necessary to provide protection to the steel without giving the thickness a symbol. The term *concrete cover* is again used.

The term "concrete cover" is ambiguous. The symbol $d_c$ is not.

## Chapter 15: USING $C_b$ WITH THE ALLOWABLE MOMENT CHART

If used directly, the allowable moments chart implicitly assumes $C_b = 1.0$. However, the chart can be used with other values of $C_b$ if $L_b/C_b$ or $L_b/\sqrt{C_b}$ is used instead of $L_b$. The proper replacement for $L_b$ depends on the beam. Most beams are controlled by torsional strength and AISC equation 1.5-7 controls. When equation 1.5-7 controls, the curved portion of the allowable moments chart will be hyperbolic, as shown in figure 15.10. For these beams, replace $L_b$ with $L_b/C_b$.

In some cases, the curved line in the allowable moments chart will be parabolic (see figure 15.10) indicating the AISC equation 1.5-6a controls. In that case, replace $L_b$ with $L_b/\sqrt{C_b}$.

# DON'T GAMBLE!

## These books will extract every last point from the examination for you!

*Call (415) 593-9119 for current prices.*

### ENGINEERING LAW, DESIGN LIABILITY, AND PROFESSIONAL ETHICS

$8\frac{1}{2}'' \times 11''$ • soft cover • 120 pages

The most difficult problems are essay questions about management, ethics, professional responsibility, and law. Since these questions can ask for definitions of terms you're not likely to know, it is virtually impossible to fake it by rambling on. And yet, these problems are simple if you have the right resources. If you don't feel comfortable with such terms as *comparative negligence, discovery proceedings,* and *strict liability in tort,* you should bring **Engineering Law, Design Liability, and Professional Ethics** with you to the examination.

None of this material is in your review manual. And, nothing from your review manual has been duplicated here.

### EXPANDED INTEREST TABLES

$6'' \times 9''$ • soft cover • 112 pages

There's nothing worse than knowing how to solve a problem but not having the necessary data. Engineering economics problems are like that. You might know how to do a problem, but where do you get interest factors for non-integer interest rates? **Expanded Interest Tables** will prove indispensible for such problems. It has pages for interest rates starting at $\frac{1}{4}$% and going to 25% in $\frac{1}{4}$% increments. Factors are given for up to 100 years. There's no other book like it.

If you want to be prepared for an engineering economics problem with 11.75% interest, you need **Expanded Interest Tables**.

### ENGINEERING UNIT CONVERSIONS

$6'' \times 9''$ • hard cover • 152 pages

If you have ever struggled with converting grams to slugs, centistokes to square feet per second, or pounds per million gallons (lbm/MG) to milligrams per liter (mg/ℓ), you will immediately appreciate the time-saving value of this book. With more than 4500 conversions, this is the most complete reference of its kind. By covering traditional English, conventional metric, and SI units in the fields of civil, mechanical, electrical, and chemical engineering, this book puts virtually every engineering conversion at your fingertips.

This book belongs in your library, even if you are not taking the E-I-T/P.E. exam. But, if you are, you'll wonder how you ever got along without it.

# INDEX


## A

A36 structural steel 12-2, 15-2
AASHTO 9-3, 9-4, 16-2, 16-13, 16-16, 16-24
abandonment 16-2
Abram's strength law 14-6
ABS 8-10
absolute English system 20-1
absolute pressures 3-7
absolute viscosity 3-3
absolute volume method 14-4
absorption 14-2
abutment 10-2
accelerated cost recovery system 2-8
accelerated flow 5-2, 5-14
accelerated methods 2-8
acceleration 16-3
access control 16-2
accessibility 5-10
accidents 16-35
accuracy 19-1
ACI moment coefficients 13-18
acid 7-2, 7-7
acidity 7-10
acquisition 16-2
acrylonitrile-butadiene-styrene 8-10
activated carbon 7-24, 7-31
activated sludge 8-2, 8-22, 8-35
active earth pressure 10-2, 10-16
activity 1-37
activity-on-node model 1-37
addition and subtraction of matrices 1-9
adhesion 10-9
adjusting departures 17-10
adjusting latitudes 17-10
adjusting traverse angles 17-8
administrative expenses 2-3
admixture 9-2, 14-2, 14-4
adsorbed water 9-2
adsorbents 7-24
adsorption 9-2
ADT 16-7
aerated grit chambers 8-16
aerated lagoon 8-2, 8-21
aerating period 8-23
aeration 7-2, 7-21, 7-31
aeration methods 8-24
aeration tank 8-35
aerators 7-22
aerobic decomposition 8-31
aerobic digestion 8-29
aerobic process 7-2
affinity laws 4-11
aggregate 9-2, 14-2, 14-4, 14-38
aggregate, coarse 14-2
aggregate, fine 14-2
aggregate, lightweight 14-2
air break 7-2
air chamber 3-35
air changes per hour 22-6
air entrainment 14-4, 22-9
air prewash 22-2
air stripping 8-25
alarm devices 19-17
algae 7-2
algebra 1-4
alidade 17-7
alignment charts 15-14
alkalinity 7-10, 7-15
all-lightweight concrete 14-4
allowable bending stress 15-6
allowable compressive stress 15-14
allowable connector stresses 15-28
allowable moments chart 15-7, 22-9
allowable shear 14-29
allowable soil pressure 10-3, 10-4
allowable stress design method 12-9
allowable stresses 12-9, 15-7, 15-20
alternate depths 5-10, 22-8
alternate mode 16-32
altitude valve 7-2
alum 7-24, 7-31
aluminum 12-2
aluminum hydroxide floc 7-24
aluminum sulfate 7-24, 7-30, 7-31
amber period 16-32
American Insurance Association 7-17
American Petroleum Institute 3-2
ammonia alum 7-30
ammonia nitrogen 8-6
ammonia stripping 8-25
amortization 2-8
AMU 7-2
anabranch 6-2
anaerobic decomposition 8-12, 8-31
anaerobic digesters 8-35
anaerobic digestion 8-28, 8-29
anaerobic microbes 8-28
anaerobic process 7-2
analytic geometry 1-12
anchorage of shear reinforcement 14-17
anchored bulkheads 10-21
angle between two lines 1-13
angle measurement 17-7
angle of contact 3-5
angle of internal friction 9-16, 10-16
angle of repose 5-17, 10-16, 11-17
angles to the right 17-8
angular spring constant 13-7
anion 7-2
anionic polymers 7-24
anisotropic materials 1-17
annual amount 2-3
annual cost method 2-5
annual return method 2-5
annunciators 19-17
ANSI FORTRAN 19-11
antecedent moisture conditions 6-17
anti-freeze 14-4
anticlinal spring 6-2
API scale 3-2, 3-3
appurtenance 8-2
apron 5-2
aquiclude 6-2
aquifer 6-2, 6-6, 9-13
aquifuge 6-2
arc basis 17-14
Archimedes principle 3-11
area method 8-30
area of traverse 17-11
areal gross rain 6-19
areas 1-32
areas under the normal curve 1-25
arithmetic mean 1-26
array 19-10
arrival headway 16-32
arrival rate 16-36
arsenic 8-32
arterial highway 16-2
artesian formation 6-2, 6-7
asbestos-cement pipe 7-19, 8-10
asphalt concrete pavement 16-14
Asphalt Institute method 16-21
asphalt pavements 16-21
ASRC 2-8
assembly force 12-28
assignment problem 19-22
associative law for addition 1-4
associative law for multiplication 1-4
assumed meridian 17-7
ASTM 9-3, 9-5
astronomical triangle 17-18, 17-19, 17-21
atmospheric head 4-3
atmospheric refraction 17-5
atomic number 7-2
atomic weight 7-2
Atterberg limit 9-3, 9-12
auger-hole method 9-14
autotrophic bacteria 8-4
auxiliary lane 16-2
available combined residuals 8-6
average daily traffic 16-7
average highway speed 16-7
average spot speed 16-6
Avogadro's law 7-2
axial flow impellers 4-2
axial-flow turbines 4-18
axial members 11-5
azimuth angle 17-19, 17-21
azimuths 17-7
azimuths from the back line 17-8
azimuths from the north 17-7
azimuths from the south 17-7

## B

B. coli 7-2
back station 17-14
backfill 7-2, 10-16
backwashing 7-26, 22-2
backwater 5-2
balance eccentricity 14-28
balanced chemical equations 7-5
balanced reinforcement ratio 14-9
balanced weld 15-37
balloon payment 2-17
bank wiring observation room 18-2
banking 16-4
bar screens 8-35
base 7-2, 7-7, 16-2, 16-24
base circle failures 10-15
base course 16-2
base flow 6-2, 6-8
base plates 15-19, 22-6
base units 20-3

basic development length 14-7
basin efficiency 7-24
batter 14-33
batter decrement 14-33, 14-34
batter pile 10-2
Baume hydrometer 3-2
Bayes theorem 1-22
Baylis turbidimeter 7-14
BDH 7-11
beam bending planes 15-4
beam-column 15-22, 15-24
beam deflections 12-9, 14-14, 15-5, 22-7
beam design 15-7
beam stiffness 13-6
beam table 15-7
beams, concrete 14-8
beams, steel 15-7
beams, strength design of 22-7
bearing capacity factors 22-4
bearing connection 15-29
bearing pressure 14-30
bearing stiffeners 15-7, 15-18, 15-27
bearing stress 12-23, 15-18, 22-7
bearing-type connections 22-7
bearing value 9-18
bearings 17-8
bedding 7-21
behavioral science 18-1
bell 10-2
belt friction 11-17
belt highway 16-2
belt-line layout 7-2
bend radii 14-37
bending stress 12-5
bending stress in steel beams 15-5
benefit-cost ratio method 2-5
benefits 2-5
bent 13-14
bentonite slurry trenching fluid 8-33
bentonite 8-33, 9-2, 14-4
berm 10-2
Bernoulli's equation 3-14
beta ratio 3-31
bias 19-1
bifurcation ratio 6-2
Bilharziasis 7-14
binomial coefficient 1-23
binomial distribution 1-22
bioactivation process 8-2
biochemical oxygen demand 8-4
biological beds 8-17
biosorption 8-24
biosorption process 8-2
biota 8-2
bisection method 1-5
bituminous treated bases 16-20
blade 3-34
bleeding 14-2
blending formulas 15-6
blind drainage 6-2
boards of registration 21-1
BOD 8-4
BOD loading 8-17
BOD removal rate constant 8-21
bolsters 14-7
bolt preloading 15-31
bolts 12-23, 15-28
bonding 14-7
borrow soil 9-8
Boussinesq's equation 7-20, 10-10, 10-12
braced cuts 10-21
braided stream 6-2
brake horsepower 4-8
braking reaction-perception time 16-5
branch sewer 8-2
brass piping 3-39
break-even analysis 2-14
break-even point 2-14
breakpoint 8-6
breakpoint chlorination 7-2, 8-6
brick in cement mortar 15-19
bridge deck 12-19
brink depth 5-13, 22-8
brink velocity 22-8
broad-crested weir 5-8, 5-13
broad fill 7-20
broken slope 10-16
bromine 7-14, 8-8
bromine chloride 8-8
bubble 3-4
Buckingham pi theorem 19-3
bulk density 9-6
bulk modulus 3-5
bulkhead 10-21
bulking (sludge) 8-2, 8-23
buoyancy 3-11
burden 2-13
burden budget variance 2-14
burden capacity variance 2-14
burden variance 2-14
Bureau of Soils 9-3
buried pipes 7-18
BTB 16-20
BUS 16-13

## C

cables 11-9, 12-26
cables under concentrated loads 11-9
cables under distributed loads 11-9
$CaCO_3$ equivalent 7-6
caisson 9-2
calcium chloride 14-4
calculators 21-3
calculus 1-28, 1-31
California Bearing Ratio 9-17, 16-22
California Department of Transportation 16-19, 16-25
calling population 16-36
CALTRANS 16-19, 16-25
Camp formula 8-16
cantilever footing 10-2
capacity factors, bearing 22-4
capacity flow 16-8
capacity of an intersection 16-12
capacity reduction factor 14-8, 14-11
capillarity 3-5
capillary constants 3-5
capillary water 6-2
capita 7-2
capital gain 2-10
capital losses 2-10
capital recovery method 2-5
capitalized cost 2-5
car equivalents 16-7, 16-9, 16-31
carbonaceous demand 8-2
carbonate alkalinity 7-11
carbonate hardness 7-2, 7-11
carbonic acid 7-10
carriage control 19-12
carry-over factor 13-7
carry-over moment 13-6
Casagrande method 9-15
cased hole 10-2
cash flow diagrams 2-1
cast iron pipe 8-10
catalyst 7-8
catena 9-2
catenary 11-10
cation 7-2, 9-2
cationic polymers 7-24
Cauchy-Schwartz theorem 1-15
cavitation 4-5, 5-17
cavitation coefficient 4-12
cavitation number 4-12
CBD 16-2
CBR 9-17, 16-22
celestial basis 17-18, 17-20
cell 8-30
cell height 8-31
cell yield coefficient 8-27
cement bentonite trench 8-33
cement factor 14-2
cement treated base 16-2
center of buoyancy 3-12
center of gravity 11-16
center-radius form 1-19
central business district 16-2
central limit theorem 1-27
central station 19-17
central tendency 1-26
centrifugal pump 4-2, 4-10
centrifuge thickening 8-28
centroid 11-13
centroidal mass moment of inertia 11-16
centroidal moment of inertia 11-14
cesspools 8-14
cgs system 20-2
chairs 14-7
challenger 2-6
channel slope 22-1
channel time 6-13
channelization 16-2
characteristic 1-7
check 5-2
chemical flocculation 8-17
chemical oxygen demand 8-6
chemical precipitation 8-2
chemical sedimentation 8-17
Chezy equation 5-3
Chezy-Manning equation 5-4, 5-5
chloramine 7-2, 7-13, 7-31, 8-6
chloramine destruction 8-6
chloride content 7-13
chloride ions 14-4
chlorinated copperas 7-24, 7-30
chlorination 8-35
chlorine 7-26
chlorine-ammonia treatment 7-31
chlorine demand 7-2, 8-6
chlorine dioxide 7-15, 7-27, 7-31, 8-8
chlorine doses 8-6
chlorine gas 8-8
chlorine residuals 7-27
chlorophenol compounds 7-13
choke 5-14
choked flow 5-14
cholera 7-14
chord basis 17-14
chromium 8-32
chromium removal 8-15
Cipoletti weir 5-8
circle 1-2, 1-19
circular arc method 10-14
circular channel 5-5
circular channel ratios 5-4
circular sector 1-2
circular segment 1-2
circumferential strain 12-28
clarification 7-22, 8-17
clarifiers 8-25
class A CTB 16-20

class B CTB 16-20
classical adjoint 1-10
clay 9-2, 10-2, 10-4
clay liners 8-33
clay pipe 8-10
clean-out 8-2
clear distance 14-37
clear distance between spirals 14-26
clear well 7-2, 7-26
closed traverses 17-8
closure in departure 17-10
closure in latitude 17-10
cloverleaf interchanges 16-34
co-latitude 17-19, 17-21
coagulants 7-24, 7-30
coagulation 7-2, 8-17
coarse aggregate 14-4
$CO_2$ systems 19-18
COD 8-6
coefficient of active earth pressure 10-16
coefficient of compressibility 10-13
coefficient of consolidation 10-12, 10-13
coefficient of curvature 9-5
coefficient of discharge 3-23
coefficient of drag 3-36
coefficient of earth pressure at rest 10-16
coefficient of friction 11-17
coefficient of friction for press fits 12-28
coefficient of lateral earth pressure at failure 10-9
coefficient of passive earth pressure 10-16
coefficient of permeability 6-6
coefficient of resistance 14-11, 16-13
coefficient of secondary compression 10-14
coefficient of thermal expansion 15-2
coefficient of variation 1-27
coefficient of velocity 3-23
coefficients of lift 3-36
cofactor 1-10
cognitive system 18-1
coherent system 20-1
cohesion 9-16, 10-16
cold clear water 4-6
coliform 7-2
collaboration 18-1
collector-distributor roads 16-34
colloid 5-2, 7-2
colon bacilli 7-2
color units 7-13
colorimeter 7-12
column 12-21, 14-2, 14-24, 15-14
column base plates 15-19, 22-6
column design 15-15
column eccentricity 14-26
column footing 10-2
columns with axial loads 15-12
combination 1-21
combination column 14-2
combined available chlorine 7-26
combined footing 10-2
combined residuals 7-2
combined residuals 7-13, 7-26, 8-6
combined stresses 12-14
combined system 8-2
comment 19-10
comminutor 8-2, 8-16
common ion effect law 7-9
common logarithm 1-7
COMMON statement 19-15
commutative law for addition 1-4
commutative law for multiplication 1-4
compact sections 15-4
compacted density 9-6
compass rule 17-10
complete mixing 7-24, 8-2, 8-24
complexione 7-11
components 11-3
composite concrete construction 14-2, 14-24, 15-40
composite structure 12-7
compound 7-2
compound amount factor 2-2
compressibility 3-5
compressible fluids 3-10, 3-39
compression index 9-15, 10-12
compression steel 14-21
compressive strength 9-12, 14-6
concentric connection 15-28
concrete 14-2
concrete batching 22-7
concrete cover 22-9
concrete (liquid) pressure 22-4
concrete, normal weight 14-2
concrete, plain 14-2
concrete, precast 14-2
concrete, prestressed 14-2
concrete, reinforced 14-2
concrete, structural lightweight 14-2
concrete mixing 14-3
concrete mixture 14-4
concrete pipe 7-19, 8-10
concrete support 15-19
condemnation 16-2
condition "1" 14-22
condition "2" 14-22
conditional probability of failure 19-8
conditions of equilibrium 11-3
conductivities of clay liners 8-33
conductivity coefficient 9-14
cone of depression 6-2, 6-6
cone penetrometer 9-9
confidence interval 17-1
confidence level 1-28
confined compression tests 9-14
confined water 6-2
confirmed test 7-2
conic sections 1-19
conjugate beam method 12-11
conjugate depth 5-2, 5-16, 22-8
connate water 6-2
connection 15-33, 15-36
connector 15-28
connector analysis 12-25
consistent deformation 12-8, 13-1
consistent systems of units 20-1
consolidated clay 9-14
consolidation test 9-14, 22-4
constant coefficient 1-33
constant-head tests 9-13
constant value dollars 2-15
constraints 19-22
construction joints 16-28
consumer loans 2-16
contact joints 16-28
contact stabilization 8-24
contact tank 8-24
continuity equation 3-19
continuous footing 10-2
continuous probability density 1-23
contracted weirs 5-7
contraction of flow 3-23, 5-2, 5-7
contraction joints 16-13, 16-28
control joints 16-28
control of access 16-2
control statements 19-13
controller offset 16-32
controls on flow 5-12, 5-14
conventional aeration 8-24
conversion factors 1-41
convex hull 19-22
conveyance 5-4
Cook equation 6-16
copper drainage tube 3-39
copper pipe 3-39
copper sulfate treatment 7-31
copper water tubing 3-39
copperas 7-24, 7-31
correlation coefficient 1-13
cost accounting 2-13
cost per unit 2-15
costs 2-5
Coulomb friction 11-17
Coulomb's earth-pressure theory 10-16
Coulomb's equation 9-15
counterflexure 10-21
couple 11-2
cover depth 14-13, 22-9
cover on steel 14-37
cover plates 15-15
CPM 1-37
crack checking 14-13
cracked transformed moment of inertia 14-14
cracking moment 14-14
Cramer's rule 1-6
cranes 15-4
crash durations 1-37
crest curve 17-13
criterion-referenced method 21-3
critical density 16-8
critical depth 5-2, 5-11, 10-9
critical fastener 15-34
critical flow 5-2, 5-13
critical oxygen deficit 8-13
critical part 15-28
critical path method 1-37
critical path techniques 1-37
critical point 8-13
critical section 14-20, 14-29
critical slenderness ratio 15-14
critical slope 5-2
critical velocity 5-2, 5-12, 7-23
cross beams 12-19
cross connection 7-2
Cross method 13-5
cross product 1-16
crown 5-18
crown slope 16-28
crushed gravel 14-2
crushed stone 14-2
crushing strength 7-20
CS (soil cement) 16-20
cubic equations 1-5
culvert 3-24
cunette 8-2
curb inlets 8-9
curvature 17-5
curve 17-13
curve number 6-17
curvilinear interpolation 1-17
cut 10-21
cut-and-sum method 11-7
cut-to-right-of-way clearance 16-28
cyanide removal 8-15
cyclic loading 15-32

# D

damaging floods 6-5
dangerous section 12-5

Darcy formula 3-19
Darcy's law 6-6, 9-13
dates 21-4
DBF (design basis flood) 6-5
DDHV 16-7
dead load 10-2
deadman 10-21
decagon 1-3
decay problems 1-35
dechlorination 7-15, 7-26, 7-31
deciles 1-26
declination 17-7, 17-19, 17-21
declining balance depreciation 2-8
decoloration time 8-5
decomposition 7-4
deep beam 14-8, 14-20
deep-lift asphalt pavement 16-14
deep well injection 8-30
defender 2-6
deflection 12-9, 14-37, 15-5
deflection angle 17-7, 17-8, 17-14
deflection angle method 17-14
deflection of a pier 22-6
deflections, beam 12-9, 14-14, 15-5, 22-7
defluoridation 7-27
deformation under loading 12-3
deformed bar 14-2
degree of compaction 9-10
degree of consolidation 10-14
degree of curve 17-13, 17-14
degree of saturation 9-6
degree of spiral 17-17
deluge systems 19-18
demand (water) 7-16
demand multipliers 7-17
demineralization 7-32
density 3-2, 15-2
density (of traffic) 16-9
density at 100% compaction 9-9
density index 9-8
density of a gas 3-2
density of concrete 14-6
density of solid constituents 9-6
deoxygenation 8-2, 8-12
deoxygenation rate constant 8-4
departure 17-9
depletion 2-8
depletion allowance 2-8
depreciation 2-7
depreciation recovery 2-8
depression storage 6-2
depth, alternate 5-10, 22-8
depth-area-duration analysis 6-2
depth, conjugate 5-2, 5-16, 22-8
depth correction factors 22-4
depth factor 10-14
depth of cover 14-13, 22-9
depth-thickness ratios 15-25
derivative 1-28
derivative operator 1-28, 1-33
derived units 20-3
desalination 7-32
design axial load strength 14-25
design basis flood 6-5
design capacity 16-7
design flow quantity 8-8
design hourly volume 16-7
design index 16-16
design moments 13-18
design shear force 14-20
design shears 13-19
design speed 16-5, 16-7
design-storm frequency 6-4
desirable value 16-5
detention period 7-23
detention time 7-2, 7-23
deterioration models 19-8
determinant 1-6, 1-10
determinate 11-6
detritus 7-2
detritus tank 8-15
Detroit Edison 18-2
devastating flood 6-5
development length 14-7, 14-17
dewatering 8-2
DHV 16-7
diagonal matrix 1-9
diagonal parking 16-30
diagonal reinforcement 14-16
diametral interference 12-28
diametral strain 12-28
diamond interchange 16-33
diatomaceous earth filters 7-26
diatomaceous silica 14-4
dichloramines 8-6
differential cost 2-14
differential equations 1-32
differential leveling 17-6
digester gas heating value 22-1
digestion 8-2, 8-28
dilatancy 9-2
dilatancy test 9-3
dilution disposal 8-2
dilution purification 8-12
dimensional analysis 19-2
dimple spring 6-2
dioxin 8-32
direct combination 7-4
direct design method 13-18
direct leveling 17-5
direct material costs 2-13
direct moments 13-11
direct reduction loans 2-17
direct shear stress 15-34
direction angles 1-16
direction cosines 1-16, 11-1
direction numbers 1-16
directional derivative 1-30
directional design hourly volume 16-7
directional factor 16-7
directional interchanges 16-34
directrix 1-19
disbenefits 2-5
disbursements 2-2
discharge coefficient 5-19
discharge from an orifice 3-23
discharge from tanks 3-23
discount factors 2-3
disinfection 7-26, 8-8
dispersion test 9-3
dissatisfiers 18-3
dissolved air floatation 8-28
dissolved oxygen 8-3
dissolved solids 7-14, 8-7, 8-26
distance between two points 1-13, 1-16
distillation 7-2, 7-32
distributed load 11-2
distribution factors 13-7, 13-8
distributive law 1-4
divided highway 16-2
division of matrices 1-9
DMD 17-12
DO statement 19-12
domestic use 7-2
domestic waste 8-2
dominant load 12-21
dot product 1-15
double action 14-29
double alternate mode 16-32
double angles 15-17
double declining balance 2-8
double decomposition 7-5
double extra-strong pipe 3-39
double integration method 12-9
double main system 7-2
double meridian distance 17-12
double reinforcement 14-2
double suction impellers 4-2
double suction pump 4-3
doubly-reinforced beam 14-8, 14-21
dowel bars 14-30, 16-14, 16-29
downpull 5-2
downstream control 5-14
draft, reservoir 6-22
draft tube 4-18
drag 3-36, 7-24
drainage density 6-2
drains 16-29
drawdown 6-2, 6-6
dredge level 10-2
Druit equation 6-7
dry density 9-6
dry strength test 9-3
dry weather flow 6-2, 6-8
dry well air 22-6
dual-media filters 7-26
duct detectors 19-18
ductile pipe 7-19
dummy nodes 1-37
dummy unit load method 13-4
dumpy level 17-7
duration 1-37
dynamic analysis 12-15
dynamic discharge head 4-4
dynamic friction 11-17
dynamic head 3-14, 4-5
dynamic loading 12-15
dynamic programming 19-22
dynamic similarity 3-38
dynamic suction head 4-4
dynamic suction lift 4-4
dynamic viscosity 3-3
dynamically similar pump 4-12
dyne 20-3
dysentery 7-14

## E

E. coli 7-2
EAL 16-15
earliest finish 1-38
earliest start 1-38
earth pressure 10-16
earthquake loading 15-4
easement 16-2
east declination 17-7
eccentric connection 15-28
eccentric loading 12-6
eccentric torsion connection 12-25, 15-33
eccentrically-loaded footings 14-31
eccentricity 14-27
eccentricity of an ellipse 1-20
echelon matrix 1-9
economic indicator 2-15
economic life 2-13
economic order quantity 2-16
Economic Recovery Act of 1981 2-8
economical shape 15-4
eddy current testing 19-19
edge distances 15-28
EDM 17-4
EDTA 7-11
effective area 15-20
effective grain size 9-5

effective hydrostatic density 10-18
effective hydrostatic loading 10-18
effective interest rate 2-11
effective length 12-22, 15-13, 15-35
effective moment of inertia 14-14
effective soil pressure 9-16, 10-12
effective stress parameters 9-16
effective weld throat 12-24, 15-35
effective width 5-8, 15-17
effective width of the compression flange 15-40
efficiency, pump 4-8
efficiency, turbine 4-18
efficient cross section 5-6
effluent 8-2
effluent disposal 8-30
effluent stream 6-2
effluent strength 8-6
ego needs 18-2
elastic design 15-7
elastic limit 12-2
elastic line 12-10
elastic region 12-2
elastic strain energy 12-6
electrical horsepower 4-8
electrodialysis 7-2, 7-32
electrolysis 7-3
electronic distance measuring 17-4
element 7-2
elevation 17-5
elevation measurement 17-5
elevator capture 19-17
elevators 15-4
eliminating the constant term 1-7
ellipse 1-2, 1-20
e-log p curve 9-14
elutriation 8-2
Elton Mayo 18-1
embankment 16-2
embankment fill 7-20
embedment 10-21
eminent domain 16-2
emulsified oil removal 8-15
emulsion 16-2
encapsulation 8-32
encroachment 16-2
end condition coefficient 12-22, 15-13
end-point deflection 9-18
end-restraint coefficient 12-22, 15-13
endothermic reaction 7-7
endurance limit 12-3, 15-2, 15-32
endurance strength 12-3
energy gradient 5-2
energy storage 12-23
energy stored in a member 12-4
Engineer-In-Training test 21-1
engineer's level 17-7
engineer's transit 17-7
English engineering system 20-2
English gravitational system 20-1
English system 20-1
enteric 7-3
entrained air 22-9
envelope of rupture 9-15
environmental impact report 19-20
Environmental Protection Agency 8-8
EOQ 2-16
ephemeral stream 6-2
ephemeris 17-19
epoxy coating 16-14
equal-tangent parabolic curves 17-15
equilibrium 18-1
equilibrium constant 7-8
equipotential lines 6-7
equivalence 2-1
equivalency factor 15-24, 16-16
equivalent axial compression method 15-22
equivalent axial load 15-24
equivalent axle loadings 16-15
equivalent diameter 3-16
equivalent lengths 3-22
equivalent moment factor 15-23
equivalent uniform annual cost 2-5
equivalent weight 7-3, 7-5
equivalent wheel loads 16-16
equivalents for trucks 16-8
erg 20-3
eriochrome black T 7-12
erodible canals 5-17
error 19-1
error analysis 17-1
errors in computed quantities 17-3
Escherichia coli 7-3
estimating economic life 2-13
Euler load 12-21
European method 17-4
eutrophication 7-3, 7-13
evaporation pans 6-21
evapotranspiration 6-2
even symmetry 1-37
event 1-37
event time 1-37
EWL 16-16
examination dates 21-4
examination strategy 21-4
exceedance problem 1-27
excess moisture 22-7
exothermic reaction 7-7
exotics 8-8
expansion joints 16-28
expected future life 19-5
expected value 2-11, 17-1
explement angle 17-8
explicit probability problems 2-12
explicit typing statement 19-10
exponential 1-23
exponential distribution 1-23, 16-36, 19-5
exponential failure 19-5
exponentiation 1-7
extended aeration 8-24
external distance 17-13
extra-strong 3-39
extraneous roots 1-6
extrema and optimization 1-30
extreme fiber 12-5
eyebars 15-20

## F

factor of safety against overturning 10-19
factor of safety against sliding 10-20
factored load 14-8, 14-29, 15-8
factorial 1-1
facultative 7-3
failure models 19-8
falling limb 6-8
falling-head tests 9-13
fans, supply and exhaust 19-18
fast lane 16-7
fasteners 15-28
fatalities 16-35
fatigue failure 12-3
fatigue fraction 16-25, 16-28
fatigue life 12-3, 16-25
fatigue loading 15-4
fatigue strength 16-25
fatigue tests 12-3
feasible region 19-22
ferric chloride 7-24, 7-30
ferric sulfate 7-30, 7-31
ferrous sulfate 7-24, 7-30
field density test 9-12
fiftieth percentile 1-26
fillet weld 12-24, 15-35, 15-36
filter 7-26
filter cloth 16-29
filterable solids 7-14
final clarifiers 8-25, 8-35
fine 9-2
fine aggregate 14-4
fineness modulus 14-2
fire alarm stations 19-16
fire dampers 19-17
fire doors 19-17
fire fighting 7-17
fire hydrants 7-17
fire pump supervision 19-18
fire safety systems 19-16
first derivative 1-28
first moment of the area 11-13, 12-7
first order differential equations 1-32
first order linear 1-32
first stage demand 8-2
first stage recarbonation 7-28
first stage treatment 7-27
fish populations 8-3
fixed-end moment 13-5, 13-6
fixed percentage method 2-8
fixed solids 7-14
fixed-time controllers 16-30
fixed-time cycle 16-30
fixity 13-7
fixity factor 13-7
flange stiffeners 15-7
flange-to-web toe fillet 15-6
flange width limitations, concrete 14-19
flash mixer 7-24
flat plates 12-29
flexible anchored bulkheads 10-21
flexible asphalt concrete 16-14
flexible connections 15-32
flexible membrane liners 8-32
flexible pavement 16-2, 16-16, 16-19, 16-22
flexible pipes 7-19
flexible supports 13-12
flexure stress 12-5
float 1-37
floatation 8-2
floatation thickening 8-28
floating-cover digester 8-29
floc 7-3
flocculation additives 7-24
flocculation clarifier 7-25
flocculator 7-24
flood plains 5-6
flooding 6-5
floor slab 14-18
flow choking 5-14
flow coefficient 3-32
flow energy 3-14
flow measurement 3-31
flow measuring devices 3-29
flow nets 6-7
flow-through period 7-24
flow work 3-14
flowing well 6-2
fluid density 3-2
fluid energy 3-14
fluid masses under acceleration 3-12
fluid mixture problems 1-34
fluid shear stress 3-3
flume 5-2
fluoridation 7-12, 7-27

fluoride content 7-12
fluorosis 7-12
flush hydrant 7-3
fly ash 14-4
FML's 8-32
foam systems 19-18
focus 1-20
food-to-microorganism ratio 8-23, 8-27
footing 10-2
footing analysis 14-31
footing depth factors 22-4
footing design 14-29, 14-31
footings on clay 10-4
footings on rock 10-6
footings on sand 10-5
footings with water tables 22-8
force main 8-2
forced vortex 4-10
forces 11-1
forebay 4-18, 5-2, 6-2
formality 7-6
FORMAT statement 19-12
formwork 14-3
formwork, pressure on 22-4
FORTRAN operators 19-11
FORTRAN programming 19-10
forward station 17-14
Fourier analysis 1-36
frame 13-14
frames with concentrated joint loads 13-16
frames with sidesway 13-15
frames without sidesway 13-14
framing connections 15-32
free available chlorine 7-26
free-body diagram 11-4
free chlorine residuals 7-13
free-flow speed 16-9
free residuals 7-3, 7-26, 7-31
freeboard 5-2
freeway 16-2
freeway capacity 16-8
freeze (of piles) 10-2
frequency distribution 1-24
frequency histogram 1-24
frequency polygon 1-24
friable 9-2
friction 3-19, 11-17
friction angles 10-9
friction connections 15-29
friction head 4-3
friction horsepower 4-8
friction loss 5-5
friction piles 10-8
friction-type connections 15-29, 22-7
frontage road 16-2
frost damage 16-29
frost heaving 16-29
frost susceptibility 9-2
Froude number 3-39, 5-12, 22-6
full-depth asphalt pavement 16-14
full stations 17-4
fully compensated foundation 10-7
fulvic acid 7-14
functions of the related angles 1-11
fundamental equation of fluid statics 3-10
fundamental frequency 1-36
fundamental theorem of calculus 1-31
Fundamentals of Engineering test 21-1
fungus 7-3
future worth 2-2

## G

gage 15-20
gage pressures 3-7
gain 2-10
gain on the sale of a depreciated asset 2-10
gamma radiation 8-8
gamma-ray testing 19-19
gap graded 9-2
garnet 7-26
gas heat content 8-29
gear motors 4-9
gel polymers 7-29
general form of equation 1-12
general shear 10-4
general triangles 1-12
geodetic level 17-7
geometric mean 1-26
geometric similarity 3-38
geometric slope 22-2
geotextiles 16-29
girder 12-19, 15-24
girder depth 15-24
girder flanges 15-25
glacial till 9-2
glauconite 7-29
global maximum moment 12-20
goal function 19-22
Goodman diagram 15-32
gore 16-2
gradient 5-2
gradient vector 1-29
grading method 21-3
grain properties 9-5
granular activated carbon 7-14
granular quicklime 7-27
grate inlets 8-10
gravel 14-3
gravel equivalent 16-19, 16-20
gravitational fields 3-12
gravitational head 3-14
gravitational water 6-2
gravity 3-39, 20-2
gravity distribution 7-3, 7-17
gravity filter 7-25
gravity thickening 8-27
gray cast iron pipe 7-19
grease 8-7
Greek alphabet 1-1
green period 16-32
green window 16-33
greensand 7-29
grid-iron layout 7-3
grid meridians 17-7
grillage 10-2
grinders 8-35
grip 15-31
grit 7-3
grit chambers 8-15, 8-35
grit clarifier 8-15
grooving pavements 16-29
gross area 15-20
gross bearing capacity 10-3
gross "k" 16-24
gross moment of inertia 14-14
gross pressure 10-3
gross rain 6-8, 6-19
groundwater 6-2, 6-8
group index 9-4
growth of traffic 16-16
GS&A 2-3
gumbo 9-2
gusset plates 15-21
gutter 8-10

## H

half-life 1-35
half-thickness 10-13
half-wave symmetry 1-37
halon systems 19-18
handicapped persons 19-18
hardness (of water) 7-3, 7-11, 7-15
Hardy Cross method 3-28, 12-13
harmonic mean 1-26
Harvard Business School 18-1
Hawthorne experiments 18-1
Hawthorne Works 18-1
hazard function 19-5
hazardous waste disposal 8-31
hazardous waste landfills 8-32
hazardous waste spill 8-34
Hazen uniformity coefficient 9-5
Hazen-Williams formulas 3-19
head 3-14
head terms 4-3
headwall 5-2
headway 16-10
heat detectors 19-16
heat horsepower 4-8
heat of hydration 14-3
heating value, digester gas 22-1
heavy metal removal 8-15
heavy vehicle factor 16-9
Henry's law 7-7
hepatitis 7-14
heptagon 1-3
Herzberg motivation studies 18-2
heterotrophic bacteria 8-4
hexagon 1-3
high early-strength concrete 14-3, 14-23
high purity oxygen aeration 8-25
high rate aeration 8-24
high rate filters 8-17
highly significant 1-28
Highway Capacity Manual 16-8, 16-12, 16-33, 22-5
highway interchange design 16-33
highway safety features 16-34
hinge joints 16-29
histogram is 1-24
hole spacing 15-28
homogeneous equations 1-32
homogeneous system 20-1
homologous pump 4-10, 4-12
hook, concrete 14-17
Hooke's law 12-2, 12-23
hoop stress 12-23
horizon 9-2
horizon of the project 2-1
horizontal curve 16-6, 17-13
horizontal curves through points 22-5
horsepower expenditure 16-3
human relations theories 18-1
humic acid 7-14
humus 8-2
hundred year flood 6-5
Hveem's resistance value test 9-19
hydrated lime 7-27, 14-4
hydration 14-3
hydraulic conductivity 6-6
hydraulic depth 5-3
hydraulic drop 5-13
hydraulic grade line 3-16
hydraulic horsepower 4-8
hydraulic jump 5-2, 5-16
hydraulic loading 8-17
hydraulic mean depth 5-2, 5-3
hydraulic radius 1-4, 3-17, 5-3

hydraulic slope 22-2
hydraulically long 5-17
hydraulically short 5-17
hydrocarbons 4-6
hydrogen ion 7-3
hydrogen sulfide removal 8-15
hydrograph 6-8
hydrograph analysis 6-8
hydrograph separation 6-8
hydrograph synthesis 6-12
hydrological cycle 6-2
hydrolyzing metal ions 7-24
hydrometer 3-2
hydrometeor 6-2
hydronium ion 7-3
hydrophilic 7-3
hydrostatic density 10-18
hydrostatic paradox 3-10
hydrostatic pressure 3-8, 11-3
hygroscopic water 6-2
hyperbola 1-20
hyperbolic functions 1-12
hypochlorites 8-8
hypothesis test 1-27

## I

ideal fluids 3-2
identity matrix 1-9
IME 2-13
Imhoff cone 7-14
Imhoff tank 8-14
impact energy 3-15
impact factors (soils) 7-20
impact loading factors 15-4
impact pressure 3-30
impeller 4-2
impervious layer 6-2
implicit differentiation 1-29
implicit probability problems 2-12
impossible event 1-22
impulse 3-33
impulse-momentum principle 3-33
impulse turbine 4-16
in-place density test 9-12
in-place volume 14-4
incineration 8-30
income tax 2-7, 2-10
incremental analysis 2-6
indefinite integrals 1-31
indeterminate trusses 11-6, 13-4
indicator solutions 7-11
indirect leveling 17-6
indirect manufacturing expenses 2-13
indirect material costs 2-13
induction motors 4-8, 4-9
industrial effluents 8-5
industrial wastes 8-15
infant mortality 19-8
infiltration 6-2, 6-8, 8-2, 8-9
inflation 2-15
inflection points 13-17
influence chart (soils) 10-11
influence diagrams 11-5, 12-15
influence diagrams for beam moments 12-18
influence diagrams for beam reactions 11-5, 12-15
influence diagrams for beam shears 12-16
influence diagrams for truss members 12-19
influence diagrams on cross-beam decks 12-19
influence value (soils) 10-11
influent 6-2, 8-2
infrared flame detectors 19-16
infrared testing 19-19
initial abstraction 6-20
initial loss 6-3
initial period 16-31
initial tangent modulus 14-6
injection wells 8-33
inlet 4-2
inlet control 5-18
inorganic chemicals 22-6
inorganic salt removal 8-26
input/output statements 19-12
insensitivity 19-1
inside lane 16-7
instantaneous deflection 14-15
Insurance Services Office 7-17
intangible property 2-7
integer programming 19-22
integers 19-10
integrating factor 1-32
integration by parts 1-31
intensity-duration-frequency curves 6-14
interaction diagram 14-27
interaction equations 15-22
intercept form 1-12
interception 6-3
interceptors 8-10
interchange 16-33
interest paid on unpaid balance 2-17
interference 12-28
interference pressure 12-28
interflow 6-3
interior angle 17-13
intermediate clarifiers 8-25
intermediate columns 15-14
intermediate stiffeners 15-7, 15-25 15-26
intermittent sand filters 8-21
intermittent weld 15-36
Intern Engineer exam 21-1
International System of Units 20-3
interpolation 1-17
intersection signaling 16-30
intrinsic water 7-3
inverse 1-10
inverse condemnation 16-2
inverse transform 1-34
invert 5-18
inverted siphon 8-2
investment tax credit 2-10
IOC 22-6
iodine 8-8
ion exchange method 7-20, 7-29
ion exchange 7-32
ionic concentration 7-7
ionization constant 7-8
ionization detectors 19-16
ions 7-3
iron content 7-12, 7-27
irregular boundaries 17-12
isohyetal method 6-4
isolated footing 10-2
isolation joints 16-28
isotopes 7-3
Izzard formula 6-13

## J

Jackson candle apparatus 7-14
Jackson turbidity units 7-14
jam density 16-9
jet on blade 3-34
jet on flat plate 3-34
jet propulsion 3-34
job cost accounting 2-13
job enrichment 18-1, 18-3
joint efficiency 3-39
joints in pavement 16-28
juvenile water 6-3

## K

k-out-of-n systems 19-6
kern 12-6
kernel 12-6
kilo-volt-amp ratings 4-9
kilograin 7-30
kinematic viscosity 3-3
kinetic pumps 4-1
Kraus process 8-2
KVA ratings 4-9

## L

L-shaped beams 14-19
labor costs 2-13
labor variance 2-13
lacing bars 15-15
lagging 10-2
lagging storm method 6-12
lagoons 8-35
Lagrangian interpolating polynomial 1-17
Lamé's solution 12-27
laminar flow 3-16
lamp holes 8-3
landfill(s) 8-30, 22-2
lane "1" 16-7
lane width 16-28
lap connections 22-7
Laplace transforms 1-33, 1-40
lateral (pipes) 8-3
lateral bracing 15-5
latest finish 1-38
latest start 1-38
latex 16-14
latitude 17-9
latus recta 1-20
law of cosines 1-12
law of mass action 7-8
law of sines 1-12
layer coefficients 16-17
layer strengths 16-15
layer-thickness equation 16-17
Le Chatelier's principle 7-7, 7-9
leachate collection systems 8-32
leaching cesspools 8-14
learning curves 2-15
least radius of gyration 15-13
least squares 1-13
length of a curve 1-32, 22-1
length of spiral 17-17
length ratio 3-38
level of service 16-8, 16-10, 16-34, 22-5
leveling 17-5
leveling rods 17-7
library function 19-13
lift 3-36, 8-30
lightweight aggregate 14-4
lightweight concrete 14-4
lime 7-27
lime and soda ash softening 7-27
lime treated bases 16-19, 16-20
limestone 15-18
limit slope 5-2

line azimuth 17-20
line load 10-17
line of seepage 6-7
linear algebra 1-9
linear programming 19-22
linearity theorem 1-34
liners for disposal sites 8-32
liquid limit 9-12
liquid penetrant testing 19-19
liquidity index 9-12
lithium compounds 8-33
live load 10-2
load factor design method 12-9
load safety factor 16-25
loans 2-16
local buckling 15-6, 15-16
local civil time 17-21
local gravity 20-2
local hour angle 17-21
local moment 12-21
local shear 10-4
loess 9-2
logarithm identities 1-8
logarithms 1-7
long columns 15-14
long stress 12-23
long-term deflection 14-15
longitudinal stress 12-23
loops 19-11
LOS 16-34, 22-5
loss coefficient 3-22
low-eccentricity column 14-26
low-heat portland cement 14-3
low-temperature monitor switches 19-17
LTB 16-20
lysimeter 6-3

## M

M/M/1 system 16-35
M/M/s system 16-36
Mach number 5-12
macroporous polymers 7-29
macroporous structures 7-29
magnetic declination 17-7
magnetic meridian 17-7
magnetic particle testing 19-19
magnification factor 14-28
main (pipes) 8-3, 8-10
maintenance/motivation factors 18-3
malodorous 8-3
management techniques 18-1
manganese content 7-12, 7-27
manganese zeolite process 7-27
manholes 8-10
manipulation 18-1
Manning equation 5-4, 5-19
Manning roughness constant 5-4
manometer 3-7, 3-30
mantissa 1-7
marginal cost 2-14
MARR 2-4, 2-6, 22-6
Marston's formula 7-18
masonry, modulus of elasticity 22-1
masonry supports 15-18
mass action equation 7-9
mass diagram 6-22
mass moment of inertia 11-16
mat 10-7
material variance 2-13
mathematical programming 19-22
matrix 1-9
matrix algebra 1-9
maxima 1-30
maximum contaminant level 7-14, 7-16
maximum dry density 9-9
maximum error 1-5
maximum grade 16-28
maximum lives 2-12
maximum normal shear 12-14
maximum period 16-32
maximum service flow rate 16-7
Mayo, Elton 18-1
MBO 18-1
MCL 7-14, 7-16
mean (average) 1-26, 17-1
mean arrival rate 16-35
mean service rate 16-35
mean time before failure 1-23, 19-5
mean velocity 5-2
mean velocity gradient 7-25
meandering stream 6-3
measured altitude 17-21
mechanically similar 3-38
mechanisms 15-12
median 1-26, 16-2
median lane 16-2
median valley slopes 16-28
median width 16-28
medium strength wastewater 8-5
Meinzer units 6-6
meniscus 3-5
mensuration 1-2
meridian 17-7
meridian angle 17-21
mers 7-24
mesophilic bacteria 8-3
metacenter 3-12
metacentric height 3-12
meteoric water 6-3
methane 8-29, 8-31
methane heating value 8-29
method of coordinates 17-11
method of joints 11-6
method of sections 11-8
method of slices 10-14
methyl orange alkalinity 7-11
metric systems 20-3
Meyerhof capacity factors 22-4
microsilica 16-14
microstrainers 7-26, 7-31
middle distance 17-13
middle ordinate 17-16
milligram equivalent weights 7-6
milligrams per liter 7-6
minima 1-30
minimal cut 19-6
minimal path 19-6
minimum attractive rate of return 2-4, 2-6, 22-6
minimum layer thickness 16-15
minimum normal stress 12-14
minimum shear 12-17
minor losses 3-20
minus declination 17-7
missing sides 17-11
mistake 19-1
MIT 9-3
mixed flow impellers 4-2
mixed liquor 8-22
mixed liquor suspended solids 8-22
mixing paddles 7-24
mixing velocity 7-24
mixture 7-3
mks system 20-3
ML 8-22
MLSS 8-22
mode 1-26
model scale 3-38
modified portland cement 14-3
modified Proctor test 9-12
modular ratio 14-10, 14-23, 15-40
modulus of elasticity 12-2, 15-2, 16-22, 16-24
modulus of elasticity, masonry 22-1
modulus of elasticity of concrete 14-6
modulus of rupture 14-14, 16-24, 16-27
modulus of subgrade reaction 9-18, 16-24
Mohlman index 8-3, 8-23
Mohr-Coulomb equation 9-15
Mohr's circle 9-15
moisture deficit 22-7
moisture-density relationships 9-9
molarity 7-6
mole 7-3
mole fraction 7-6
molecular weight 7-3
molecule 7-3
moment 11-1
moment area method 12-10
moment arm 11-1
moment coefficients 13-18
moment diagram 12-4, 13-10
moment distribution method 13-5, 13-8
moment gradient multiplier 15-5, 22-9
moment of inertia 11-14
moment-resisting connections 15-33, 15-36
moment vectors 11-2
moments on footings 10-6
momentum 3-33
monochloramines 8-6
monolithic construction 14-3
Monte Carlo simulation 6-21
Moody friction factor chart 3-19
morality curve 19-8
most efficient cross section 5-6
most likely value 17-1
motivation-maintenance theory 18-3
motivators 18-3
motor sizes 4-9
motor-reducer drives 4-9
moving load on beams 12-20
MSF 16-7
MTBF 1-23, 19-5
mud line 10-2
multiple stage pump 4-3
multiplication of matrices 1-9

## N

N-value 9-9, 10-6
naperien logs 1-7
nappe 5-7
National Council of Engineering Examiners 21-1
National Research Council 8-18
natural logs 1-7
natural watercourse 5-6
NCEE 21-1, 21-2
need hierarchy theory 18-2
negative boundaries 6-3
negative exponential distribution 19-5
nephelometer 7-14
nested 19-12
net area 15-20
net force 16-3
net inlet pressure required 4-5
net positive suction head 4-5
net positive suction head available 4-5
net positive suction head required 4-5
net prestress 14-24
net rain 6-3, 6-8, 6-19
net soil pressure 10-4

net thickness 15-20
neutral axis 14-9
neutral axis, concrete beams 22-1
newton (unit) 20-3
Newton's Law of Cooling 1-35
Newton's second law 16-3, 20-1
Newtonian fluids 3-2, 4-7
NIPR 4-5
nitrification and denitrification process 8-25
nitrification demand 8-4
nitrogen content 7-13
nitrogen conversion 8-25
nitrogenous demand 8-3, 8-4
nominal area 15-20
nominal axial load strength 14-25
nominal beam strength 22-4
nominal concrete strength 22-4
nominal interest rate 2-11
nominal moment carrying ability 14-10
nominal shear strength 22-4
nominal strength 14-10, 22-4
non-carbonate hardness 7-11
non-clog pumps 4-16
non-filterable solids 7-14
non-linear data 1-14
non-Newtonian fluids 3-2
non-pathogenic weight 7-3
non-plastic soils 9-12
non-sequential drought method 6-21
non-uniform flow 5-1
nonagon 1-3
nondestructive testing 19-19
nonionic polymers 7-24
Norm-Referenced Method 21-3
normal depth 5-2, 5-4
normal distribution 1-23
normal durations 1-37
normal force 11-17
normal form 1-12
normal line function 1-30
normal portland cement 14-3
normal stress 12-2, 12-5
normal vector 1-18
normal weight concrete 14-4
normality 7-6
normally distributed 17-1
normally loaded (soil) 9-2, 9-14, 10-12
normative judgment 18-1
North Star 17-20
nozzle loss 4-16
NPSHA 4-5
NPSHR 4-5
nuisance flooding 6-5
numerical event 1-22
Nusselt number 19-4

## O

O&M 2-3
objective function 19-22
O'Conner and Dobbins formula 8-12
octagon 1-3
odd symmetry 1-37
odor control 7-31, 8-16
oedometer tests 9-14
offset 16-32
ogee 5-16
ogee spillways 5-8
oil separation 8-15
on-demand controllers 16-30
on-demand timing 16-31
one-day EAL 16-16
one percent flood 6-5
one-tail test 1-28
one-way drainage 10-13
one-way floor slabs 14-18
one-way moment 12-4, 12-5
one-way shear 12-7, 14-29
open jet 3-34
open jet on inclined plate 3-36
open plenums 19-18
operating and maintenance costs 2-3
operating point 4-14
operating pressure 3-39
operating speed 16-7
opportunity interest cost 2-8
optimum water content 9-9
orders of accuracy 17-4
organic clay 9-3
organic loading 8-17
organic matter 9-3
organic polymers 7-24
organic precursors 7-14
organic silt 9-3
orifice coefficients 3-24
orifice plate 3-31
orthogonal 1-16
orthophosphates 7-13
OSHA 16-35
osmosis 7-3
outfall 8-3
outlet control 5-18
overall efficiency 4-8
overall travel speed 16-6
overburden 10-4, 10-6, 14-31
overchute 5-2
overconsolidated (soil) 9-15
overconsolidation pressure 9-15
overconsolidation ratio 9-15
overflow rate 7-23
overflow spillways 5-16
overhead variance 2-14
overhead 2-13
overland flow 6-3, 6-8
overload factors 14-8
oxidants 7-24
oxidation 7-3
oxidation number 7-3, 7-5
oxidation pond 8-21
oxidation-reduction 7-5
oxygen deficit 8-3, 8-12
oxygen sag curve 8-12
oxygen saturation coefficient 8-23
ozone 7-15, 7-27, 8-8

## P

paddles 7-24
pan 6-3
anel shear 12-19
parabola 1-2, 1-19
parabolic cables 11-10
paraboloid of revolution 1-3
parallel axis theorem 11-5, 11-16
parallel lines 1-13
parallel pipe systems 3-25
parallel systems (reliability) 19-6
parallelogram 1-3
parametric equations 1-18
parking design 16-30
parkway 16-2
Parshall flume 5-9
"partial compensation" 10-7
partial differentiation 1-29
partial emission pump 4-10
partial fractions 1-8
partial shoring 15-40
"partial similarity" 3-39
partial treatment 8-3
particle size 9-2
particle size distribution 9-5, 22-2
parts per million 7-6
Pascal-second 3-3
passenger car equivalents 16-7, 16-9, 16-31
passing sight distance 16-5
passive earth pressure 10-2, 10-16
paternalism 18-1
pathogenic organisms 7-3
pavement 16-13
payback period 22-6
peak hour factor 16-9
peak runoff 6-10, 6-13, 6-15
Peclet number 19-4
pedestal 14-3, 14-24
pedestrian levels of service 16-34
pedestrians 16-31
pedology 9-2
penetration resistance test 9-9
penetration test 10-6
penetration treatment 16-2
penstock 4-18
pentagon 1-3
percent eccentricity 14-27
percent pore space 9-6
percentage of compaction 9-10
percentile ranks 1-26
perched spring 6-3
percolation 6-3
performance curve 4-12
performance function 19-5
permanent hardness 7-3, 7-11
permanent set 12-2
permeability 9-13
permeators 9-13
permutation 1-21
perpendicular lines 1-13
personal map 18-1
personal property 2-7
PERT 1-39
pH 7-3
pH adjustment 8-15
phenol 7-13
phenol removal 8-15
phenolphthalein alkalinity 7-11
Philadelphia rod 17-7
phosphorous content 7-13
phosphorus removal 8-25
photoelectric detector 19-16
photogrammetry 17-17
phreatic zone 6-3, 6-6
phreatophytes 6-3
physiological needs 18-2
pi-groups 19-3
pier deflection 22-6
pier shaft 10-2
piers 10-7, 10-8
PIEV time 16-5
piezometer tap 3-29, 3-30
piezometric level 6-3
pile 10-8, 10-21
pile group 10-10
pile group efficiency 10-10
pipe 3-39, 8-10
pipe bends 3-35
pipe entrance 3-22
pipe exit 3-22
pipe materials 7-18
pipe networks 3-28
pipe, water hammer 22-1
piping materials 3-39
pitot tube 3-30
plain bar 14-3

plain footings 14-29
plain sedimentation 7-22, 8-17
plane 1-18
plane of failure 9-16
plane of maximum principle stress 9-16
plant mix 16-2
plastic design 12-9, 15-7, 15-8
plastic limit 9-12
plastic moment diagram 15-11
plastic pipe 7-19, 8-10
plastic section modulus 15-8
plastic strain 12-2
plasticity index 9-12
plasticity test 9-3
plat 6-3
plate bearing value test 9-18
plate girders 15-24
plates 15-3
plates in compression 15-16
plug flow 7-24, 8-24
plus declination 17-7
plus stations 17-4
PMF 6-5
pOH 7-3
point-bearing piles 10-8
point load 10-17
point of curvature 17-13
point of intersection of two lines 1-13
point of tangency 17-13
point-slope form 1-12
points of counterflexure 10-21
points of inflection 1-30
poise 3-3
Poisson distribution 1-23, 16-36
Poisson's ratio 12-3, 15-2
polar distance 17-19, 17-20, 17-21
polar form 1-12
polar moment of inertia 11-15, 15-34
15-38
Polaris 17-20
Polaris observations 17-20
polio 7-14
polished water 7-3
polishing filter 8-21
polyelectrolytes 7-24
polyhedrons 1-4
polymers 7-24, 7-29
polynomial equations 1-4
polyphosphates 7-13
polystyrene resins 7-29
polyvinyl chloride 8-10
ponding 15-5
population adjustment factor 16-9
population equivalent 8-5
population standard deviation 1-27
population variance 22-1
pore pressure 9-16, 10-9
porosity 6-3, 9-6
portal frame 13-14
portland cement 14-3
Portland Cement Association 16-25
portland cement concrete 16-13
portland cement concrete pavement
16-22
positive displacement pumps 4-1
post-chlorination 8-3
post hydrant 7-3
post-tensioned construction 14-22
potable 7-3
potash alum 7-30
potassium permanganate 7-15
potential head 3-14
poundal 20-1
powerhouse 4-18
pre-chlorination 8-3
precedence relationships 1-37
precedence table 1-37
precipitation 6-3
precipitation excess 6-19
precipitation softening 7-29
precise level 17-7
precise rods 17-7
precision 19-1
preconsolidated clay 9-15
preconsolidation pressure 9-15
preliminary treatment 8-15
preload (bolts) 15-29, 15-31
preloaded clay 9-15
present worth 2-2
present worth factor 2-2
present worth method 2-4
press fits 12-28
pressure 3-7, 3-14
pressure energy 3-14
pressure filters 7-26
pressure head 3-14, 4-3
pressure loads 11-3
pressure switches 19-16
pressurization 19-18
prestressed concrete pavement 16-14
prestressed construction 14-22
presumptive test 7-3
pretension (bolts) 15-29
pretensioned construction 14-22
prewash, air 22-2
price index 2-15
primary consolidation 10-14
primary settlement 10-12
primary treatment 8-17
prime coat 16-2
prime cost 2-13
principal axes 11-16
principal moments of inertia 11-16
principal stresses 12-14
printer control characters 19-12
probabilistic problems 2-11
probability density functions 1-22
probability rules 1-22
probable error 17-1, 17-2
probable error of the mean 17-2
probable maximum flood 6-5
probable maximum rainfall 6-3
probable value 17-1
process cost accounting 2-13
Proctor test 9-9, 9-12
product of inertia 11-15
professional engineer 21-1
Professional Engineering test 21-1
profit function 19-22
program evaluation and review
technique 1-39
progressive method 8-30
project completion time 1-39
proof load 15-32
proof strength 15-31
proper day's work 18-2
property 2-7
proportion problem 7-5
proportionality limit 12-2
proportioning concrete 14-4
protozoa 7-3
pump 3-23, 4-1, 8-35
pump performance curves 4-12
pump similarity 4-12
pumping hydrocarbons 4-6
pumps in parallel 4-15
pumps in series 4-15
pumps used in wastewater plants 8-11
putrefaction 8-3, 8-12
PVC 8-10
pycnometer 9-2
Pythagorean theorem 1-11

## Q

Q-curve 5-2
Q-test 9-16
quadrant 1-11
quadratic equations 1-5
quality circle programs 18-3
quality improvement programs 18-3
quarter-wave symmetry 1-37
quartiles 1-26
queue 16-35
queuing models 16-35
quick mixer 7-24
quick test 9-16

## R

R-value 9-19
racks 8-35
radial flow impellers 4-2
radial flow turbines 4-18
radians 1-11
radical 7-3, 7-4
radio transmitters 19-17
radiography 19-19
radius of gyration 11-15, 14-28, 15-13
raft 10-7
rafts on clay 10-7
rafts on sand 10-7
rain gage 6-3
rainfall intensity 6-4
random numbers 6-21
range 1-26
ranger 10-2
Rankine theory 10-16
rapid flow 5-2, 5-10
rapid sand filter 7-25, 22-2
rapid-mix flash units 7-25
rate constant 7-8
rate of grade change 17-15
rate of oxygen transfer 8-22
rate of reaction 7-8
rate of return 2-4, 2-6
rate of return (sludge) 8-23
rate of return on added investment 2-6
rate of rise 19-16
rating curve 5-2
ratio of precision 17-1
rational formula 6-13
RCRA 8-31
reach 5-2
reaction 11-4
reaction period 16-5
reaction turbines 4-17
reaction velocity 7-8
real fluids 3-2
real numbers 19-10
real property 2-7
rebar 14-3, 14-7
rebar spacing 14-37
rebound curve 9-15
recarbonation 7-3, 7-28
receipts 2-2
recession 6-8
reciprocating action pumps 4-1
recirculation 4-7, 8-17, 8-19
recirculation ratio 8-17
Recommended Standards for Sewage
Works 8-35
reconsolidation index 9-15

rectangle 1-3
rectangular channels 5-6
rectangular distribution 2-12
rectangular hyperbola 1-20
recurrence interval 6-4, 6-22
redox reaction 7-3, 7-5
reduced eccentricity models 15-35
reduction 7-3
redundancy 13-1, 19-6
refractory pollutants 8-3, 8-8
regenerants 7-29
regional factor 16-17
regression 1-13
regular gain 2-10
regular losses 2-10
regular polygon 1-3
regular polyhedron 1-3
regulator 8-3
Rehbock equation 5-8
reinforcement of mill shapes 15-3
reinforcement ratio 14-9, 14-11, 14-25
reinforcing fabric 16-29
reinforcing steel 14-7
rejuvenating solution 7-29
relative density 9-8, 10-9
relative dispersion 1-27
relative paddle velocity 7-24
relative stability 8-5
relative stiffness 13-7
reliability 19-1, 19-5
reloading curve 9-15
renewal 19-8
reoxygenation 8-3, 8-12
reoxygenation coefficient 8-3
replacement 2-6, 19-8, 19-9
reservoir branching systems 3-27
reservoir seepage 6-21
reservoir sizing 6-20
reservoir yield 6-20
residential property 2-7
residual 7-3
resilient modulus 16-22
resin exchange process 7-29
resistance against sliding 10-19
resistance per unit length of weld 15-35
Resource Conservation Recovery Act 8-31
resource constraints 19-22
resurfacing 16-3
retaining walls 10-18, 14-33
retardance coefficient 6-13
retarded flow 5-2, 5-14
retention period 7-3, 7-23, 7-24
retention time 7-3, 7-23, 7-24
return on investment 2-4
reverse osmosis 7-3, 7-32
reverse polarity 19-17
reversible reactions 7-7
Reynold's number 3-16, 3-38, 19-4
right circular cone 1-3
right circular cylinder 1-3
right-hand rule 11-2
right of access 16-3
right triangles 1-11
rigid frame connections 15-32
rigid pavement 16-3, 16-13, 16-24
rigid pipes 7-18
ring permutation 1-21
rip rap 10-2
Rippl diagram 6-22
rising limb 6-8
rivets 12-23, 15-28
road-mixed asphalt 16-3
roadbed 16-3
roadway banking 16-4
roadway detailing 16-28
rock flour 9-2
rods (welding) 15-20, 15-36
ROI 2-4
root(s) 1-6, 15-32
root-mean-squared (rms) value 1-26
ROR 2-4, 2-6
ROT 8-22
rotary action pumps 4-1
rotary bridge 16-34
rotary drum filter 8-28
rotating biological contactors 8-20
rotating biological reactors 8-20
rotation of a function 1-32
rotation of axes 11-16
rotational symmetry 1-37
running speed 16-6
runoff potenial 6-17
rupture line 9-15

## S

S-curve 6-12, 6-13
S-tests 9-16
sack of cement 14-3
safe yield 6-3
safety needs 18-2
Safe Water Drinking Act 22-6
sag pipe 8-3
sag vertical curve 8-9, 8-13, 17-13
salt 7-4
sample standard deviation 1-27
sample variance 22-1
sand 10-2, 14-3
sand drying beds 8-28
sand filters 7-26
sand filters, rapid 22-2
sand-lightweight concrete 14-4
sand trap 5-3
sandstone 15-18
sanitary landfills 8-30, 22-2
sanitary sewer 8-8
satisfiers 18-3
saturated sand 10-18
saturated surface dry 14-3, 14-4, 22-7
SAW 15-35
scalar matrix 1-9
scale buildup 3-19
scale of photograph 17-17
schedule 3-39
Schistosomiasis 7-14
scour 5-3, 5-17
scouring velocity 8-10, 8-16
screens 8-15, 8-35
SCS methods 6-15, 6-17
SCS unit hydrograph 6-11
seal coat 16-3
secant formula 12-22
secant modulus 14-6, 14-14, 15-40
second derivative 1-28
second order differential equations 1-32, 1-33
second stage demand 8-3
second stage recarbonation 7-28
second stage treatment 7-28
secondary compression index 10-14
secondary consolidation 10-14
secondary members 15-14
secondary treatment 8-17
Seconds Saybolt Furol 3-3
Seconds Saybolt Universal 3-3
section modulus 12-5, 15-5
sedimentation tanks 8-17, 8-25, 8-35
seed 8-3
seed organisms 8-5
seep 6-3
seepage 6-7
selective perception 18-1
self-cleaning pipe 6-13
self-cleansing velocity 8-10, 8-16
self-fulfillment needs 18-2
self purification 8-12, 8-13
semi-circular cross section 5-6
semi-precise level 17-7
semi-rigid connections 15-32
sensitivity 19-22
sensitivity analysis 2-18
sensitivity test 9-12
separate system 8-3
septic (definition) 8-3
septic tank 8-14
serial systems 19-5
series pipe systems 3-25
service connections 8-10
service flow rate 16-8, 16-9
service levels 22-5
service moment 14-14
service times 16-36
setting (turbines) 4-18
settleable solids 7-14
settlement 10-12
settling basin 5-3
settling velocity 7-22
sewer pipes 8-10
shaft design 12-24
shape factor 12-24
sharp-crested weirs 5-7
shear coefficients 13-19
shear connections 15-33
shear diagram 12-4, 13-10
shear envelope 14-16
shear influence diagrams on cross-beam decks 12-19
shear modulus 12-24, 15-2
shear reinforcement 14-15, 14-16
shear span 14-20
shear strength 14-6, 14-16
shear stress 12-6, 15-6
sheeted pit 10-2
shielded arc welding 15-35
shooting flow 5-3
shored construction 15-40
shoulders 16-28
shredders 8-16, 8-35
shrinkage limit 9-6
SI base units 20-4
SI prefixes 20-4
SI supplementary units 20-4
SI system 20-3
SI units 22-2
side friction factor 16-4
side slopes on adjacent cuts 16-28
sidereal time 17-21
sidesway 13-14, 13-15
sieve, soils 22-2
sight distance 16-5
signal controllers 16-30
signal transmission 19-17
signaling 16-30
significant digits 1-4
silica fume 16-14
siliceous-gel zeolite 7-29
silicone 16-15
sill 5-3
silts 9-2
silver oxide 8-8
similarity 3-38, 4-12
simple framing connections 15-32
simple interest 2-16
simplex method 19-22

Simpson's rule 17-13
simultaneous linear equations 1-6
simultaneous quadratic equations 1-7
single action 14-29
single axle 16-13
single displacement 7-5
single main system 7-4
single suction impellers 4-2
single suction pump 4-3
sinking fund method 2-8
sinking fund plus interest on first cost 2-8
sinuosity 6-3
siphons 3-24
skewness 1-27
skidding distance 16-5
skimming tanks 8-16
skin friction 10-8, 10-9
slab 14-3
slab thickness 14-38
slack time 1-37
slenderness effects 14-28
slenderness ratio 12-21, 14-28, 15-14
slenderness ratios for tension members 15-20
slickenside 10-2
slip 4-8
slope 1-13, 5-3
slope circle failures 10-15
slope form (straight lines) 1-12
slope height 10-14
slope of channel 22-1
slope stability 10-14
sloped backfill 10-16
slow sand filter 7-26, 8-21
sludge age 8-23
sludge bulking 8-3, 8-23
sludge dewatering 8-28
sludge disposal 8-26
sludge drying beds 8-35
sludge quantities 8-26
sludge thickening 8-27
sludge volume index 8-23
sludge 7-24
slug (unit of mass) 20-2
sluiceways 5-17
slump 14-3
slump test 14-5
slurry trench containment 8-33
small sample standard deviation 17-1
SMAW 15-35
smoke detectors 19-16
smoke doors 19-17
Snyder synthetic hydrograph 6-10
SOC 22-6
social needs 18-2
soda ash 7-27
sodium aluminate 7-30
sodium bisulfate 7-15, 7-26
sodium polyphosphate 7-28
soffitt 5-18
soil bearing pressures 10-3, 10-10
soil bentonite trench 8-33
soil cement bases 16-20
soil classification 9-3
soil covers 8-31
soil density 9-6
soil indexing 9-5
soil support value 16-17
solar observations 17-18
soldier pile 10-2
solid contact units 7-25
solid densities 14-4
solid volume 22-9
solid volume method 14-4
solubility product 7-9
solution 7-4
solutions of gases in liquids 7-7
solutions of solids in liquids 7-6
solvents 8-32
sound pressure level 19-18
spacing of reinforcement 14-9
specific energy 3-14, 5-10
specific energy diagram 5-10
specific gravity 3-2
specific roughness 3-20
specific speed 4-10, 4-18
specific volume 3-2
specific weight 3-12
specific yield 6-3
speed of sound 3-5
speed rule 22-5
SPF 6-5
sphere 1-3, 1-21
spillway(s) 5-16, 22-8
spiral column 14-3, 14-24, 14-26
spiral curve(s) 17-17, 22-5
spiral pitch 14-26
spiral steel 14-26
spiral to curve transition 17-17
splice 14-37, 15-21
split casing 4-16
split chlorination 8-3
split process 7-28
split spoon sampler 9-9
splitting tensile test 14-6
spread footing 10-2
spread tandem axles 16-13
spring (water) 6-3
spring (wire) 12-23
spring constant 12-23
spring constant of a bolt 15-31
spring thaw 16-29
square 1-9
SSD 14-3, 14-4, 22-7
stability 19-1
stability criterion 15-22, 15-23
stability number 10-14
stability of floating objects 3-12
stabilization 8-28
stabilization ponds 8-21
stabilization time 8-5
stabilometer test 9-19
stadia interval 17-4
stadia interval factor 17-4
stadia method 17-4
stadia reading 17-4
stage 4-3
stagnation 3-15
stagnation pressure 3-30
stairwell exit doors 19-17
standard acceleration due to gravity 20-2
standard BOD 8-4
standard deviation 1-26, 17-1, 22-1
standard deviation of the sample 1-27
standard error of the mean 1-27
standard gravitational field 3-13
standard hoisting rope 12-26
standard hole 15-20
standard hook 14-7
standard normal variable 1-23
standard penetration test 9-9
standard project flood 6-5
standard rod 17-7
standard truck loadings 16-12
standing wave 5-3, 5-12
start-up lost time 16-32
State Boards of Registration 21-1
static discharge head 4-4
static energy 3-14
static friction 11-17
static head 3-14
static probe 3-29
static suction head 4-3
static suction lift 4-3
statical moment 12-7, 12-10
statically indeterminate 13-1
station angle 17-8
statistical analysis 1-23
steady flow 5-1, 5-3, 5-9
Steel formula 6-4
steel grade 14-7
steel pipe 3-39, 7-19
steel properties 15-2
steel reinforcing 14-6
steel tendons 14-22
stem thickness 14-33
step flow 8-24
step interval 1-24
stiff mixtures 14-5
stiffened elements 15-16
stiffeners 15-6
stiffness 13-7
stiffness factor 13-7
stilling basin 5-3
stirrups 14-7, 14-16
stock speeds 4-9
stoke (unit of viscosity) 3-3
Stoke's law 3-37, 7-22
stoichiometry 7-4, 7-5
stopping distance 16-5
stopping sight distance 16-5
storage capacity 6-20
storm drains 8-9
storm duration 6-12
straight line 1-12
straight-line interpolation 1-17
straight-line method 2-8
straight-line plus average interest method 2-9
straight-line plus interest on first cost 2-9
strain 12-2
strain energy method 12-11, 12-12
stratum 9-2
stream coordinates 3-23
stream gaging 5-2, 5-3
stream order 6-3
streamlines 6-7
Streeter-Phelps equations 8-12
strength coefficients 16-17
strength design method 14-8, 14-11
strength design of beams 22-7
strength of concrete 14-3
strength reduction factors 14-8
strength theory 14-10
stress 12-2
stress area of fasteners 15-28
stress concentration factors 12-26, 15-28, 15-32
stress contour charts 10-12
stress risers 12-26
Strickler's equation 5-4
stringer 10-2
strong wastewater 8-5
structural bolts 22-4
structural number 16-16, 16-17
structural section 16-3
structural shapes 15-2
styrofoam 14-4
SU 16-13
subbase 16-3, 16-24
subcritical flow 5-3, 5-10
subgrade 16-3

subgrade drainage 16-29
subgrade modulus 9-18
submain 8-3
submerged condition 10-5
submerged density 22-8
submerged metal arc welding 15-35
subroutine 19-13, 19-14
subscripted variables 19-10
substitution 1-6
subsurface runoff 6-3
subsurface water 6-6
suction side 4-2
suction specific speed 4-10
sudden contractions 3-22
sudden enlargements 3-22
sulfate-resistant portland cement 14-3
sulfonated soaps 14-4
sulfur-asphalt concrete 16-14, 16-15
sulfur concrete 16-14
sulfur dioxide 7-26
sulfur extended asphalt 16-14, 16-15
sum-of-the-years' digits 2-8
sunk costs 2-2
superchlorination 7-4, 7-31
supercritical flow 5-10
superelevation 16-4
superelevation runoff 22-5
supernatant 8-3, 8-29
superordinate goals 18-1
superposition 11-8, 12-12, 13-3
superposition theorem 1-34
supplementary units 20-3
support membrane 16-29
suppressed weir 5-7
surcharge loading 10-2, 10-17
surface detention 6-3
surface evaporation 1-36
surface flow 6-8
surface loading 7-23
surface retention 6-3
surface runoff 6-3, 6-8
surface temperature 1-35
surface tension 3-4
surface water 14-3
surface wave 5-12
surfaces of revolution 1-32
surge 3-35
surge chamber 3-35, 4-18
surge relief valve 3-35
surge supressor 3-35
surge tank 3-35, 4-18
surge wave 5-12
surveyor's tape corrections 12-26
surveyor's transit 17-7
suspended solids 7-14, 8-7, 8-25
SVI 8-23
swelling index 9-15
symmetry 1-37
synchronous speed 4-8
syndets 7-4, 7-13
synthesis 7-4
synthetic detergents 7-4, 7-13
synthetic drought 6-22
synthetic hydrograph 6-10
synthetic organic chemicals 22-6
synthetic polymers 7-24
synthetic storm 6-17
system curves 4-14
system reliability 19-5, 19-6

## T

T-beam 14-8, 14-19
tack coat 16-3
tail race 4-18, 5-3
tail water 4-18, 5-3
tandem axle 16-13
tangent lines 17-13
tangent modulus 14-6
tangent offset 17-16
tangent plane function 1-30
tangent to spiral transition 17-17
tangents 17-13
tangible property 2-7
tapered aeration 8-24
tapes (surveying) 17-4
task analysis 21-2
taste control 7-31
Taylor chart 10-14
team programs 18-3
telephone dialers 19-17
telescope 17-7
temporary hardness 7-4, 7-11
ten-states' standards 8-35, 22-6
tendon/tube assembly 14-23
tensile strength 14-6
tensile test 12-2
tension connections 15-29
tension field action equation 15-26
tension members 15-20
tension splice 12-23
tension steel 14-21
tensors 1-17
terminal serviceability 16-16, 16-24
tertiary treatment 8-25
Terzaghi-Meyerhof equation 10-3
theodolite 17-7
theory of influence 18-2
theory X 18-3
theory Y 18-3
thermal conditioning 8-29
thermal deformation 12-4
thermal strain 12-4
thermocline 7-4
thermophilic bacteria 8-3
thick-walled cylinders 12-27
thickening of sludge 8-27
Thiessen method 6-4
thin-walled cylinders 12-23
third point loading tests 16-24
thixotropic 9-2
THM 7-14
threaded members in tension 15-20
three-dimensional structures 11-11
throttling valve 4-6, 4-15
tied column 14-3, 14-24, 14-25
till 9-2
time base 6-8, 6-12
time factor 10-14
time of concentration 6-3, 6-13
time rate of primary consolidation 10-13
time rule 22-5
time-space diagrams 16-32
tip speed 7-24
titration 7-10
toe 14-33
toe circle failures 10-15
top steel 14-21
torsion connections (welded) 15-38
torsion resistance 12-25
total discharge head 4-4
total energy 3-15
total energy line 3-14
total head 3-14, 3-15, 4-5
total head available 4-16
total pressure 3-30, 9-16
total solids 7-14, 8-7
total specific energy 3-14
total static head 4-5
total suction head 4-4
total suction lift 4-4
tracer compounds 8-33
tractive force 16-3
traditional elastic approach 15-34
traffic-activated controllers 16-30
traffic counts 16-15
traffic growth 16-16
traffic index 16-16, 16-19
traffic intensity 16-35
tranquil flow 5-3, 5-10
transcendental functions 1-28
transfer axis theorem 11-15
transfer efficiency 7-22
transformation method 12-8
transformed area 12-8
transit 17-7
transit rule 17-10
transition width 7-20
transmissivity 6-6
transpiration 6-3
transportation problem 19-22
transpose of a matrix 1-9
transverse loading 11-3
transverse spacing 15-20
trapezoid 1-3
trapezoidal channels 5-6
trapezoidal rule 17-12
trapezoidal weir 5-8
trash racks 8-15
traveled way 16-3
traverse 17-8
traverse closure 17-10
trench method 8-30
trench width 7-19
trial wedge method 10-16
triangle 1-2, 1-3, 1-11
triangular matrix 1-9
triangular weirs 5-8
triangulation 17-18
triaxial stress tests 9-15
trichloramines 8-6
trickling filters 8-17, 8-35
trigonometric functions 1-11
trigonometric identities 1-11
trigonometry 1-11
trihalomethanes 7-14
trilateration 17-18
triple-media filters 7-26
tripod solution 11-12
tritium 8-33
truck constants 16-16
truck loads 16-12
true altitude 17-19, 17-21
true meridian 17-7
trunks 8-10
truss deflections 12-12
truss pipe 8-10
trusses 11-6
Tschebotarioff trapezoidal pressure distribution 10-22
TSS 8-35
TTHM 7-14
turbidity 7-4, 7-13, 7-16, 7-30
turbine 4-17
turbine specific speed 4-18
turbulent flow 3-16, 3-19
turning point 17-16
turnout 5-3
two-angle formulas 1-12
two percent parallel offset 12-2
two-point form 1-12
two-tail test 1-28
two-way construction 14-3
two-way drainage 10-13

two-way shear 14-29
type 1 connections 15-32, 15-33
type 2 connections 15-32
type 3 connections 15-32
typhoid fever 7-14

U

U.S. National Safety Council 16-35
U.S. Soil Conservation Service 6-11, 6-15
ultimate axial load 14-25
ultimate compressive strength 14-6
ultimate moments 15-10
ultimate shears 15-10
ultimate strength 15-2
ultimate strength connections 15-35
ultimate strength design 12-9, 14-8, 15-7, 15-8
ultimate strength multipliers 14-8, 15-9
ultimate tensile strength 12-2
ultrasonic testing 19-19
ultraviolet flame detectors 19-16
ultraviolet radiation 8-8
unavailable combined residuals 8-6
unbraced length 14-28, 15-13
unconfined compressive strength test 9-12
unconsolidated-undrained test 9-16
undrained case 10-4
Unified Soil Classification system 22-3
uniform acceleration 16-3
Uniform Examination 21-1
uniform flow 5-1, 5-3, 5-4
uniform gradient factor 2-3
uniform series factors 2-3
uniformity coefficient 9-5
unit circle 1-11
unit hydrograph 6-10
unit stream power 6-3
unit vectors 1-15
unit weight 9-6
units of mass 20-1
unshored construction 15-40
unstiffened elements 15-16
upflow tanks 7-25
upper culmination 17-21
upstream control 5-14
USC (soil classification) 22-3
USDA 9-3
USDA triangular chart 9-4
user functions 19-14

V

V-belts 4-9
vaclose water 6-3
vadose zone 6-3, 6-6
valence 7-4
valve monitor switches 19-17
valves and fittings 3-22
vapor pressure 3-4
vapor pressure head 4-3
variable names 19-10
variable transformation 1-14
variance 2-13, 22-1
variation 17-7
varied flow 5-3, 5-14
vector 1-15
vector magnitude 1-15
vector multiplication 1-15
vector operations 1-15
vehicle period 16-32
vehicle type 3 16-12
vehicle type 3-3 16-12
vehicle type 3S2 16-12
vehicle type H20 16-12
vehicle type HS20 16-12
velocity head 3-14, 4-3
velocity head coefficient 5-19
velocity of approach factor 3-31
ventilation requirements 22-6
venturi meter 3-32
versene 7-11
vertex 1-19
vertical buckling 15-6
vertical clearance 16-28
vertical curve length 22-1
vertical curves 16-6, 17-13, 17-15
vertical dowels 14-30
virgin branch 9-14
virgin consolidation line 9-14
virtual displacement 12-17, 12-18
virtual work 12-13, 12-17
viruses 7-26
viscometer 3-3
viscosity 3-3
viscosity conversions 3-4
vitrified clay pipe 8-10
VOC 22-6
void ratio 9-6
volatile acids 8-7
volatile organic chemicals 22-6
volatile solid 7-14, 8-3, 8-7, 8-28
volume parameters 16-7
volume-to-capacity ratio 16-8

W

W-shape 15-2
wale 10-2
walking speed 16-31
wall footing 10-2
warping joints 16-29
warrants 16-30
wastewater quality standards 8-8
wasteway 5-3
water affinity 14-4
water-cement ratio 14-3, 14-4
water content 9-6
water demand 7-16
water distribution 7-17
water hammer 3-35, 22-1
water hardness 7-11
water horsepower 3-23, 4-8
Water Pollution Control Act 8-8
water quality standards 7-16
water softening 7-27
water table 6-3, 10-8, 22-8
waterflow detectors 19-16
watery mixtures (of concrete) 14-5
watt 20-3
WB-40 vehicle 16-13
WB-50 vehicle 16-13
WB-60 vehicle 16-13
weak-axis bending 15-6
weak wastewater 8-5
weakened plane joints 16-28
weaving 16-33
web crippling 15-6, 15-28
web depth 15-24
web reinforcement 14-16
web shear 15-25
web stiffeners 15-7, 15-24
Weber numbers 3-39
wedge theories 10-16
weight 20-2
weighted measurements 17-2
weighting agents 7-24
weighting factor 8-19
weir 5-7
weir loading 7-24
weld 12-24, 15-35
weld intermittent 15-36
weld throat 15-35
welded connections 15-36
welding rod 15-36
well drawdown 6-6
west declination 17-7
Westergaard theory of stress distribution 10-12, 16-24
Western Electric Company 18-1, 18-2
wet mixtures (of concrete) 14-5
wet oxidation 8-28
wet well 8-3
wet well air 22-6
wetted perimeter 5-3
Whitney assumption 14-10
width-thickness ratio 15-16, 15-26
wire ropes 12-26, 15-20
working stress design method 12-9, 15-7

X

X-ray 19-19
xerophytes 6-3

Y

year-end convention 2-2
yield coefficient 8-27
yield point 12-2
yield strength 15-2
yield stress 12-2
yielding support 13-12
Young's modulus 12-2

Z

zenith angle 17-19, 17-21
zenith distance 17-19, 17-21
zenith point 17-19
zeolite 7-4, 7-29
zeolite process 7-29
zero air voids curve 9-10
zero air voids density 9-6
zero defects program 18-3
zone of aeration 6-3
zone of saturation 6-3
zooglea 8-3

# INDEX OF FIGURES AND TABLES


**A**
A36 steel properties 15-2
AASHTO equivalent factors 16-40, 16-41
AASHTO flexible pavement nomograph 16-18
AASHTO rigid pavement nomograph 16-25
AASHTO soil classification 9-4
ACI detailing requirements 14-37
ACI moment coefficients 13-18
acceleration formulas 16-4
accident values 16-35
active component charts 10-23, 10-24
aeration characteristics 8-25
air properties 3-41
alkalinity versus hardness 7-15
alternate design method, concrete 14-39
altitude versus atmospheric pressure 4-19
angle of friction, soil 9-17, 10-9
areas under normal curve 1-25
Asphalt Institute pavements 16-21, 16-23
atmospheric pressure versus altitude 4-19
atomic weights 7-36

**B**
beam formulas 12-31, 12-32, 13-23
beam widths (concrete) 14-9
bearing capacity factors 10-3, 10-4
bearing pressure, soil 10-3
bedding classes 7-21
BOD, typical 8-5
bolt stresses 15-9
Boussinesq stress contours 10-25, 10-26

**C**
$CaCO_3$ conversions 7-35
California bearing ratio 9-18
cast iron pipe 3-51
CBR 9-18
CBR-to-subgrade modulus chart 16-37
cell yield coefficients 8-27
centroids 11-18, 12-33
chemicals in water treatment 7-37
chlorine doses 8-6
chlorine residuals 7-27
circular channel ratios 5-4, 5-25
circular curve geometry 17-24
circular pipe ratios 3-43
closure errors, traverse 17-4
coefficients, discharge 3-33
coefficients, drag 3-37
coefficients, flow 3-32
coefficients, orifice 3-24
coefficients, rational runoff 6-24
coefficients, Steel formula 6-5
coefficients of friction 11-17
coefficients of permeability 6-6
coefficients of skidding 16-5
cohesion, soil 9-17
column alignment charts 15-13
compaction, typical values 9-11
concrete detailing 14-37
concrete sewer pipe 3-50
concrete types 14-3
connector stresses 15-29
constants, fundamental 1-43
constants, Hazen-Williams 3-20
constants, Manning's *n* 5-23
constants, minor loss 3-22, 3-53, 5-20
constants, reoxygenation 8-12
constants, Snyder hydrograph 6-11
constants, surveying 17-26
conversion, general 1-41
conversions, concrete 14-2
conversions, fluid 3-42
conversions, foundations 10-1
conversions, hydrology 6-1
conversions, mg/l as $CaCO_3$ 7-35
conversions, SI 20-7, 20-8
conversions, soils 9-1
conversions, steel 15-2
conversions, surveying 17-25
conversions, viscosity 3-4
conversions, volumetric 4-22
conversions, waste-water 8-1
conversions, water supply 7-1
copper tubing 3-48, 3-49
correction factors, pump 4-23
critical depths (circular channels) 5-26
culvert flow 5-19
cumulative storm distribution 6-20
curve numbers 6-18, 6-19
cycle length chart 16-31
cylinder (thick) formulas 12-28

**D**
design speeds 16-7
detailing (concrete) 14-37
diameters, equivalent 3-17
discharge coefficients 3-33
discount factor equations 2-3, 2-11
distribution of storm 6-20
drag coefficients 3-37
drainage coefficients 6-17

**E**
efficiency, pumps 4-8
end-restraint coefficients 12-22, 15-13
equivalent diameters 3-17
equivalent lengths of pipe 3-22, 3-53
exchange materials, synthetic 7-30

**F**
fillet weld sizes 15-36
fire fighting durations 7-18
fire hydrant placement 7-18
fitting losses 3-22, 3-53
fixed-end moments 13-21, 13-22

flat plate formulas 12-29
flow coefficients 3-32
fluid conversions 3-42
fluoride, fluorine 7-12, 7-27
friction angles 9-17, 10-9
friction coefficients 11-17
friction factor chart 3-21
fundamental constants 1-43

G
gravel equivalent factors 16-20
Greek alphabet 1-1

H
hardness classifications 7-12
hardness versus alkalinity 7-15
Hazen-Williams constants 3-20
Hazen-Williams nomograph 3-54
high temperature properties, steel 15-43
hoisting rope 12-27
horsepower, hydraulic 4-8
hydraulic horsepower 4-8
hydraulic jump lengths 5-16
hydraulic radius, circular pipe 3-18, 3-43

I
impact factors 7-20
impact loading factors (steel) 15-4
impeller types 4-10
indicator solutions 7-11
influence chart, soil 10-11
integrals 1-31
ionic valences 7-4
iron removal processes 7-28

L
lane distribution factors 16-7
Laplace transforms 1-40
levels of service 16-8, 16-11, 16-34
limits of specific speeds 4-20, 4-21
load repetitions, concrete 16-27

M
Manning nomograph 3-55, 5-24
Manning's $n$-constant 5-23
mass moments of inertia 11-19
material properties 12-3, 12-34
maximum velocity, open channel flow 5-18
mensuration 1-2
metric system (see SI)
minor loss constants 3-22, 3-53, 5-20
modulus of subgrade reaction 16-17
moment coefficients (ACI) 13-18
moment distribution worksheet 13-20
moments, maximum plastic 15-9
moments of inertia 11-18, 11-19, 12-33
Moody friction factor chart 3-21
motor sizes 4-9

N
normal curve 1-25
NPSHR corrections, hydrocarbons 4-7

O
oblique triangle equations 17-23
optimum moisture, typical values 9-11
orders of accuracy 17-4
orifice coefficients 3-24
oxygen, saturated in water 8-36

P
Parshall flume $K$-values 5-9
passenger car equivalents 16-8
periodic chart 7-36
permeabilities, typical 9-13
permeability coefficients 6-6
pipe, cast iron 3-51
pipe, concrete sewer 3-50
pipe, steel 3-44
pipe bedding classes 7-21
pipe load coefficients 7-19
pipe materials 3-40, 7-19
piping symbols 3-52
plastic moments 15-9
plate (flat) formulas 12-29
polygons 1-3
polyhedrons 1-4
prestressing tendons 14-24
properties of air 3-41
properties of materials 12-3, 12-34
properties of steel 12-34, 15-41
properties of water 3-41
properties of welds as lines 15-44
pump correction factors 4-23
pump efficiencies 4-8
pumps, types 4-2
pumps, wastewater 8-11

R
$R$-value-to-subgrade modulus chart 16-39
$R$-values, soil 9-19
radical valences 7-4
rainfall map 6-26
random numbers 6-25
rational runoff coefficients 6-24
rebar sizes 14-7
relative stability 8-5
reoxygenation constants 8-12
retaining wall charts 10-23, 10-24
rivet stresses 15-29
runoff curve numbers 6-18

S
saturated oxygen in water 8-36
serviceabilities, terminal 16-17
settling velocities 7-23
sewage strengths 8-8
SI conversions 20-7, 20-8
SI units 20-4, 20-5
side slopes, open channel flow 5-17
sieve sizes 9-5
sight distances 16-5
signal cycle length chart 16-31
skidding coefficients 16-5
slumps, concrete 14-6

soil formulas 9-7
soil pressure, allowable 10-3
soil strengths, typical 9-17
soil support correlations 16-38
specific roughness 3-20
specific speed limits 4-20, 4-21
speeds, design 16-7
standard normal curve 1-25
standard truck loadings 16-12, 16-13
steel areas for number of bars 14-12, 14-18
Steel formula coefficients 6-5
steel pipe dimensions 3-44
steel prestressing tendons 14-24
steel properties 12-34, 15-41, 15-43
steel properties (A36) 15-2
steel reinforcing bar 14-7
steel types 15-41
strength reduction factors 14-8
stress contour charts 10-25, 10-26
subgrade modulus, soil 9-19
subgrade modulus-to-CBR chart 16-37
subgrade modulus-to-*R*-value chart 16-39
surveying constants 17-26
symbols, piping 3-52
synchronous speeds 4-8

**T**

Taylor chart 10-15
Ten-States' standards 8-35
tendon sizes 14-24
terminal serviceabilities 16-17
thick-walled cylinder formulas 12-28
transforms, Laplace 1-40
traverse closure errors 17-4
treatments, wastewater 8-37
treatments, water supply 7-32, 7-33
triangle chart 9-3
triangles, oblique 17-23
truck loadings 16-12, 16-13
tubing dimensions, copper 3-48, 3-49
types of flow, culvert 5-19

**U**

uniform acceleration formulas 16-4
USDA triangle chart 9-3

**V**

valences 7-4
value of accidents 16-35
vapor pressure 3-4
varying *n* (circular channel ratios) 5-25
velocity, maximum, open channel flow 5-18
velocity of flow 8-11
viscosity conversions 3-4
viscosity of water 3-42
viscosity units 3-3
volumetric conversion factors 4-22

**W**

water-borne organisms 7-14
water demand 7-16, 7-17
water horsepower 4-8
water pipe materials 7-19
water properties 3-41
water quality standards 7-16, 8-9
water viscosity 3-42
welds as lines, properties 15-44
width of beams (concrete) 14-9
width-thickness ratios 15-16
wire-rope 12-27
working stress method (concrete) 14-39

# Quick – I need additional study materials!

Please rush me the review materials I have checked. I understand any item may be returned for a full refund within 30 days. I have provided my bank card number as method of payment, and I authorize you to charge your current prices against my account.

| | Solutions Manuals: |
|---|---|
| **For the E-I-T Exam:** | |
| [ ] Engineer-In-Training Review Manual | [ ] |
| [ ] Engineering Fundamentals Quick Reference Cards | |
| [ ] E-I-T Mini-Exams | |
| **For the P.E. Exams:** | |
| [ ] Civil Engineering Reference Manual | [ ] |
| [ ] Civil Engineering Sample Examination | |
| [ ] Civil Engineering Quick Reference Cards | |
| [ ] Seismic Design | |
| [ ] Timber Design | |
| [ ] Structural Engineering Practice Problem Manual | |
| [ ] Mechanical Engineering Review Manual | [ ] |
| [ ] Mechanical Engineering Quick Reference Cards | |
| [ ] Mechanical Engineering Sample Examination | |
| [ ] Electrical Engineering Review Manual | [ ] |
| [ ] Chemical Engineering Reference Manual | [ ] |
| [ ] Chemical Engineering Practice Exam Set | |
| [ ] Land Surveyor Reference Manual | [ ] |
| **Recommended for all Exams:** | |
| [ ] Expanded Interest Tables | |
| [ ] Engineering Law, Design Liability, and Professional Ethics | |

**SHIP TO:**

Name ______________________

Company ______________________

Street ______________________ Apt. No. ________

City ______________ State _____ Zip ________

Daytime phone number ______________________

**CHARGE TO (required for immediate processing):**

______________________ ________

VISA/MC/AMEX account number — expiration date

______________________

name on card

______________________

signature

# Send more information

Please send me descriptions and prices of all available E-I-T and P.E. review books. I understand there will be no obligation.

______________________

______________________

______________________

______________________

A friend of mine is taking the exam too. Send additional literature to:

______________________

______________________

______________________

______________________

# I disagree...

I think there is an error on page ________. Here is the way I think it should be.

Title of this book: ______________________

[ ] Please tell me if I am correct.

Contributed by (optional):

______________________

______________________

______________________

Affix
Postage

PROFESSIONAL PUBLICATIONS, INC.
1250 Fifth Avenue
Belmont, CA 94002

Affix
Postage

PROFESSIONAL PUBLICATIONS, INC.
1250 Fifth Avenue
Belmont, CA 94002

Affix
Postage

PROFESSIONAL PUBLICATIONS, INC.
1250 Fifth Avenue
Belmont, CA 94002